NUCLEATION AND ATMOSPHERIC AEROSOLS

Nucleation and Atmospheric Aerosols

17th International Conference, Galway, Ireland, 2007

Edited by

Colin D. O'Dowd
National University of Ireland
Galway
Ireland

Paul E. Wagner
University of Vienna
Austria

Springer

Colin D. O'Dowd
National University of Ireland
Galway
Ireland

Paul E. Wagner
University of Vienna
Austria

ISBN 978-1-4020-6474-6 e-ISBN 978-1-4020-6475-3

Library of Congress Control Number: 2007934296

Printed on acid-free paper.

9 8 7 6 5 4 3 2 1

springer.com

ICNAA 2007
Galway, Ireland

The **Galway Hooker** is a traditional sailing boat used in Galway Bay off the west coast of Ireland. The hooker was developed for the strong seas there. It is identified by the sail formation, which is extremely distinctive and quite beautiful. It consisted of a single mast with a main sail and two foresails. Traditionally, the boat is black (being coated in pitch) and the sails are a dark red-brown.
Recently there has been a major revival, and renewed interest in the Galway hooker, and the boats are still being painstakingly constructed.
The hooker refers to four classes of boats. All are named in Irish. The *Bád Mór* (big boat) ranges in length from 35 to 44 feet (10.5–13.5 metres). The smaller *Leath Bhád* (half boat) is about 28 feet (10 metres) in length. These boats were used to carry turf to be used as fuel across Galway Bay from Connemara and County Mayo to the Aran Islands and the Burren. The boats often brought limestone on the return journeys, to neutralise the acid soils of Connemara and Mayo. The *Gleoiteog* ranges in length from 24 to 28 feet and has the same sails and rigging as the larger boats. They were used for fishing and carrying cargo. Another boat, the *Púcán*, is similar in size to the *Gleoiteog* but has one mainsail and one foresail.

Preface

The 17th International Conference on Nucleation and Atmospheric Aerosols (ICNAA) is held at the National University of Ireland, Galway, from 13 to 17 August 2007. This is the third time the conference is taking place in Ireland. The first ICNAA was held in Dublin in 1955 and the ninth was held in Galway 30 years ago. A short history of ICNAA is found on the following pages.

Since 1988, this series of conferences has been held jointly with the Nucleation Symposia to further strengthen interaction between scientists working in related fields.

At the 17th ICNAA, 245 papers were accepted in the following eleven categories:

- Nucleation: Experiment and Theory (54 papers)
- Nucleation: Binary, homogeneous and heterogeneous (14)
- Nucleation: Ion and cluster properties (13)
- Ice Nucleation (12)
- Air Quality – Climate (12)
- Aerosol–Cloud Interactions (15)
- Aerosol Characterisation and Properties Aerosols (51)
- Aerosol Formation, Dynamics and Growth (35)
- Marine Aerosol Production (16)
- Remote Sensing of Aerosols and Clouds (10)
- Chemical Composition and Cloud Drop Activation (13)

Together with the International Advisory Committee, we have identified current topics of high interests and the following leading scientists have accepted Plenary Lecture invitations to present the state of the art in their respective fields of nucleation and atmospheric aerosols: Paul DeMott, Terry Deshler, Maria Cristina Facchini, Markku Kulmala, Surabi Menon, Kikuo Okuyama, Spyros Pandis, V. Ramanathan, Yinon Rudich, Gerald Wilemski, Barbara E. Wyslouzil. As conference chairs, we would like to thank our sponsors and the help of the local organising committee. Welcome to the City of the Tribes – Galway.

National University of Ireland, Galway
May 2007

Colin D. O'Dowd
Paul E. Wagner

International Advisory Committee

D. Boulaud, France
K. S. Carslaw, UK
M. C. Facchini, Italy
B. N. Hale, USA
K. Hämeri, Finland
R. Jaenicke, Germany
Y. S. Mayya, India
J. Gras, Australia
S. G. Jennings, Ireland
G. Raga, Mexico
J. H. Seinfeld, USA
H. Tammet, Estonia
Paul DeMott, USA
B. E. Wyslouzil, USA
P. Artaxo, Brazil

Local Organising Committee

Colin D. O'Dowd
Aodhagan Roddy
Saji Varghese
Merry Zacharias
Gerard Jennings
Harald Berresheim
Clarie Scannell
Yvonne Barrett
Tom O'Connor
Robert Flanagan
Darius Ceburnis
Philip McVeigh

Sponsored by

International Union on Geophysics & Geodesy (IUGG)
International Association on Meteorology & Atmospheric Science (IAMAS)
International Commission on Clouds & Precipitation (ICCP)
Committee on Nucleation & Atmospheric Aerosols (CNAA)
Atmospheric Composition Change – A European Network of Excellence (ACCENT)
National Science Foundation
Faitle Ireland
Environmental Protection Agency, Ireland
National University of Ireland, Galway: Registrars Office and Department of Physics.
Environmental Change Institute.

CNAA
Committee on Nucleation & Atmospheric Aerosols

ACCENT
ATMOSPHERIC COMPOSITION CHANGE
THE EUROPEAN NETWORK OF EXCELLENCE

NSF
National Science Foundation
WHERE DISCOVERIES BEGIN

Previous proceedings in the Series of International Conferences on Nucleation and Atmospheric Aerosols (ICNAA)

	Year	Location	Publisher	ISBN
16th	2004	Kyoto, Japan	Kyoto University Press	4-87698-635-5
15th	2000	Rolla, Missouri, USA	American Institute of Physics	1-56396-958-0
14th	1996	Helsinki, Finland,	Pergamon Press	008-0420303
13th	1992	Salt Lake City, USA,	A. Deepak Publishing	0-937194-26-3
12th	1988	Vienna, Austria,	Springer-Verlag	3-540-50108-8
11th	1984	Budapest, Hungary	Hungarian Met. Service	84-392

Plenary Lectures

Paul DeMott	Progress and Issues in Quantifying Ice Nucleation Involving Atmospheric Aerosols	USA
Terry Deshler	Stratospheric Aerosol: Measurements, Importance, Life Cycle, Anomalous Aerosol	USA
Maria Cristina Facchini	Organic Marine Aerosol: State of the Art and New Findings	Italy
Markku Kulmala	Atmospheric Nucleation	Finland
Surabi Menon	Current understanding of Aerosol-Climate Effects	USA
Kikuo Okuyama	Size Analysis of Ion and Aerosol Nanoparticles in Nucleation and Growth	Japan
Spyros Pandis	Regional Air Quality-Atmospheric Nucleation Interactions	Greece/USA
V. Ramanathan	Global Dimming by Air Pollution and Global Warming by Greenhouse Gases: Global and Regional Perspectives	USA
Yinon Rudich	Laboratory studies on the properties and processes of complex organic aerosols	Israel
Gerald Wilemski	Homogeneous Binary Nucleation Theory And The Structure of Binary Nanodroplets	USA
Barbara E. Wyslouzil	Homogeneous nucleation rate measurements and the properties of critical clusters	USA

International Nucleation and Atmospheric Aerosol Conference Series: A Brief History 1954–2007

The return of the International Conference on Nucleation and Atmospheric Aerosols to Galway in Ireland for the 17th conference in the series provides a good opportunity to consider the history of the development of these conferences over their first 50 years. They may be said to have begun with what R. Siskna (1966) described as the "Zeroth Conference" at the General Assembly of the IUGG in Rome in 1954. A group of four met in a cafe to hear an informal lecture by L.W. Pollak on his work in Dublin developing condensation nucleus counters. They prevailed on him to organize the Symposium on Condensation Nuclei which may be called the first international conference. It was held during 26–28 April 1955 and was attended by 34 persons from eight countries. With 26 papers presented over five half-day sessions there was plenty of time for discussions and viewing the exhibition and demonstrations of various photoelectric and photographic condensation nucleus counters in Pollak's laboratories. The main organizers were L.W. Pollak and P.J. Nolan.

The Second Symposium on Condensation Nuclei was held in Switzerland at Basle and Locarno during 1–4 October 1956. This second conference brought together about 50 persons from 12 countries and 28 papers were presented. It introduced the happy idea, repeated on four occasions, of holding the conference in two locations so that participants learned more about various centers of research. The main organizers were F. Verzár, M. Bider and J.C. Thams.

The third International Symposium on Condensation Nuclei was held in the UK at the Cavendish Laboratory of the University of Cambridge during 16–18 July 1958. It was attended by 50 participants from nine countries and 24 papers were presented. It was organized by T.W. Wormell.

The fourth International Conference on Atmospheric Condensation Nuclei was held at two locations in Germany during 24–28 May 1961. The first part was held in Frankfurt/Mainz and was concerned with fundamental aspect of atmospheric aerosols. The second part was in Heidelberg and concentrated mainly on radioactivity in the atmosphere. The main organizers were H.W. Georgii and G. Schumann. Some 74 persons from 14 countries presented 37 papers. The proceedings of the first four conferences were published as special volumes of the journal *Geofisica Pura e Applicata* (PAGEOPH), Volumes 31, 36, 42, and 50. They give only the abstract of some presentations as they were published in full in other journals.

France was the host country for the fifth International Conference on Condensation Nuclei and Ice Nuclei during 13–17 May 1963. For the first time ice nuclei featured in the title of the conference. There were 54 registered participants from 14 countries. The principal organizer was H. Dessens. The first day was held in the University of Clermont-Ferrand and was devoted to ice nuclei. On 14 May the participants made the 400 km journey to Toulouse, with a memorable break for a 16-course lunch at Aubrac en route. The second part in the University of Toulouse covered condensation nuclei, their origin and measurement, and radioactivity. The conference concluded on 17 May with a visit to l'Institut de Physique de l'Atmosphere at Lannemezan and a spectacular demonstration of the Meteortron, producing convection in the atmosphere by burning oil at the rate of 1000 litres per minute! There were 34 presentations to the conference. All of the abstracts and 16 of the papers, in either English or French, were published in the *Journal Recherches Atmosphériques*, Volume 1, 1963, as well as a note giving references to where another 15 papers appeared in other journals.

The sixth International Symposium on Condensation Nuclei was held at two venues in the USA during 9–13 May 1966. The first part was held in the State University of New York at Albany, NY, and the second part was in Pennsylvania State University at University Park, PA. The main organizers were V.J. Schaefer and C.L. Hosler. There were 86 delegates from 14 countries and the 55 presentations were published in a special number of the *Journal Recherches Atmosphérique*, Volume 2, pp. 50–436 (1966). The proceedings were dedicated to the memory of L.W. Pollak, who died in November 1964. The organization of these conferences over the first ten years had been undertaken on a voluntary basis by interested individuals, without statutes or rules or officers or offices. It was then decided to entrust the general organization of future conferences to a sub-committee on nucleation of the International Commission on Cloud Physics within the International Association of Meteorology and Atmospheric Physics. This official status facilitated wider publicity for the conferences and easier participation for some in them through official channels, without disrupting the close links of personal friendship, which had become a feature of these meetings.

The seventh International Conference on Condensation and Ice Nuclei was held at two locations in Central Europe – Prague and Vienna – during18–24 September 1969. The host institutions were the Institute for Physics of the Atmosphere of the Czechoslovak Academy of Science and the Institute for Physics of the University of Vienna. The main organizers were J. Podzimek, K. Spurny, and O. Preining. Participants from 26 countries presented 80 papers which were published as a special book of the proceedings and a supplement by the Czechoslovak Academy of Science.

The eighth conference in the series was held in Russia at Leningrad (St. Petersburg) during 24–29 September 1973. It was called the Eighth International Conference on Nucleation and was the first conference to be devoted almost entirely to the problems of ice nucleation. The main organizer was I.I. Gaivoronsky. There were 74 presentations from scientists from 18 countries. These were published, mainly in English and some in Russian, in a book by Gidrometeoizdat in Moscow in 1975.

The ninth International Conference on Atmospheric Aerosols, Condensation and Ice Nuclei was held in Galway in the west of Ireland during 21–27 September 1977. The principal local organizers were T.C.O'Connor and A.F. Roddy. The widened scope of the conference, as seen in the title, was reflected in the attendance. Some 142 scientists from 22 countries presented 101 papers. The proceedings were published as a book entitled *Atmospheric Aerosols and Nuclei* by Galway University Press.

The tenth International Conference on Condensation and Ice Nuclei was held in Hamburg during 26–28 August 1981 as part of the third Scientific Assembly of the International Association of Meteorology and Atmospheric Physics (IAMAP). It was organized by the Nucleation Committee of the International Commission on Cloud Physics. The convener was H.W. Georgii. There was not a registration for the conference separate from that for IAMAP so the number and origin of the participants are not known exactly. The programme and abstracts of 58 presentations were included in the IAMAP proceedings and 26 papers were published as a special issue of *Időjárás*, the journal of the Hungarian Meteorological Service, Volume 86(2–4), 1982.

The eleventh International Conference on Atmospheric Aerosols, Condensation, and Ice Nuclei was held in Budapest during 3–8 September 1984. It was sponsored by IAMAP through its International Commission on Cloud Physics and the Commission on Atmospheric Chemistry and Global Pollution. The principal organizers were E. Meszaros and G. Vali. There were sessions in memory of N.A. Fuchs and J.E. Jiusto. There were 110 presentations many of which were included in the proceeding in a special issue of *Időjárás* in 1985.

The twelfth International Conference on Atmospheric Aerosols and Nucleation was held in Vienna during 22–27 August 1988. It was held for the first time jointly with the Nucleation Symposium to stimulate contact and an exchange of information between scientists in these closely related disciplines. The principal organizers were P.E. Wagner and G. Vali. Scientists from 28 countries presented 167 papers, which were published by Springer-Verlag as number 309 of their *Lecture Notes in Physics* series under the title *Atmospheric Aerosols and Nucleation*.

The thirteenth International Conference on Nucleation and Atmospheric Aerosols was again held in the USA in conjunction with the Nucleation Symposium. It was organized in Salt Lake City by N. Fukuta of the University of Utah. Ninety papers were presented by scientists from 22 countries. The proceedings, edited by N. Fukuta and P.E. Wagner, were published in 523 pages by A. Deepak Publishing under the title *Nucleation and Atmospheric Aerosols*.

The fourteenth International Conference on Nucleation and Atmospheric Aerosols was held in Helsinki during 26–30 August 1996. The principal organizers were M. Kulmala and P.E. Wagner. There were 264 participants from 24 countries, including 52 with the title student. The Nobel Laureate P.J. Crutzen gave the opening lecture on "The Role of Particulate Matter in Ozone Photochemistry". There were ten plenary lectures, 87 oral presentations in 17 platform sessions and 101 posters plus some late breaking posters included in the proceedings. Of these many presentations, 16 papers were published in a special issue of the *Journal Atmospheric Research*, Volume 46(3–4), 1998.

The fifteenth International Conference on Nucleation and Atmospheric Aerosols returned to the USA to the University of Missouri at Rolla, MO, during 6–11 August 2000. The principal organizers were B.N. Hale and M. Kulmala.. Authors from 27 countries presented over 200 papers which are contained in the first electronically generated Proceedings. These were published by the American Institute of Physics in the AIP Conference Proceedings series and include a special symposium in honour of Howard Reiss.

For the sixteenth International Conference on Nucleation and Atmospheric Aerosols the venue was Kyoto University in Japan. It was held during 26–30 July 2004. The Proceedings contain 174 papers on 848 pages. They were edited by M. Kasahara and M. Kulmala and are published by the Kyoto University Press in 2004.

The seventeenth conference is held again in Ireland at the National University of Ireland at Galway during 13–17 August 2007. Be part of this history

Tom O'Connor

Contents

Part IV Ice Nucleation

Part V Air Quality – Climate

Part VI Aerosol – Cloud Interactions

Part I
Nucleation: Experiment and Theory

Homogeneous Nucleation Rate Measurements and the Properties of Critical Clusters

Barbara E. Wyslouzil[1], Reinhard Strey[2], Judith Wölk[2], Gerald Wilemski[3], and Yoojeong Kim[4]

Abstract By combining a range of experimental techniques, quantitative nucleation rate measurements can now be made over ~20 orders of magnitude. These rates can be used to directly test the predictions of nucleation theories or scaling laws. They can also provide direct information regarding the properties of the critical clusters – the first fragments of the new phase that are in unstable equilibrium with the supersaturated mother phase. This paper reviews recent progress in the field of vapor phase nucleation with a special focus on integrating the results from supersonic nozzle and nucleation pulse chamber studies.

Keywords Homogeneous nucleation, nucleation theorem, critical cluster properties

Introduction

Phase transitions play a critical role in many natural and anthropogenic processes. They occur, even in the absence of impurities or surfaces, when the original phase is supersaturated with respect to a more stable one. If the supersaturation is moderate, the first step in the phase transition is nucleation, the formation of microscopic regions, clusters, or nuclei of a thermodynamically more stable phase from the metastable phase. As the growing nuclei reduce the supersaturation, the nucleation rate approaches zero. The phase transition is completed as particles grow and age, via coagulation and Ostwald ripening, until equilibrium is reestablished. Of these three processes – nucleation, growth, and aging – nucleation is still the least well understood, the hardest to predict, and the hardest to measure.

[1] *Department of Chemical Engineering, Ohio State University, Columbus, OH 43210, USA*

[2] *Institut für Physikalische Chemie, Universität zu Köln, D-50939 Köln, Germany*

[3] *Department of Physics, University of Missouri – Rolla, Rolla, MO 65409, USA*

[4] *Triton Systems Inc., 200 Turnpike Road, Chelmsford, MA 01824, USA*

C.D. O'Dowd and P.E. Wagner (eds.), *Nucleation and Atmospheric Aerosols*, 3–13.

Our interest is in phase transitions that occur in the absence of any preexisting nucleation centers, i.e., homogeneous nucleation. In particular, we are interested in vapor-to-liquid phase transitions that take place in excess carrier gas. This well-defined system has been extensively investigated[1,2] using a wide range of experimental devices that now cover ~20 orders of magnitude in nucleation rate J. They include thermal diffusion cloud chambers ($10^{-3} < J/\text{cm}^{-3}\text{s}^{-1} < 10^{2}$),[3–6] laminar flow tube reactors ($10^{3} < J/\text{cm}^{-3}\text{s}^{-1} < 10^{8}$),[7–11] single piston expansion cloud chambers ($10^{2} < J/\text{cm}^{-3}\text{s}^{-1} < 10^{5}$),[12,13] piston-expansion tubes ($10^{4} < J/\text{cm}^{-3}\text{s}^{-1} < 10^{9}$),[14,15] nucleation pulse chambers ($10^{5} < J/\text{cm}^{-3}\text{s}^{-1} < 10^{9}$),[16–20] shock tubes($10^{7} < J/\text{cm}^{-3}\text{s}^{-1} < 10^{9}$),[21,22] expansion wave tubes ($10^{8} < J/\text{cm}^{-3}\text{s}^{-1} < 10^{11}$),[23,24] and supersonic nozzles ($10^{16} < J/\text{cm}^{-3}\text{s}^{-1} < 10^{18}$).[25–33] Although quantitative rate measurements are now routine, it is often difficult for all devices to measure rates for a common compound at the same temperature. Combining data from one class of experiments, expansion devices for example, is somewhat easier because the physical properties that make a compound suitable for study in one apparatus make it suitable for most apparatuses in that class.

The emphasis of our recent work has been to combine data from the nucleation pulse chamber (NPC) and supersonic nozzles (SN) in order to generate consistent data sets covering ~10 orders of magnitude in rate in the same temperature range. We then use these data to test the predictions of nucleation theories and nucleation scaling laws, and to determine the properties of the critical clusters down to clusters containing fewer than 10 molecules.

Nucleation Theory

It is difficult to discuss nucleation experiments in detail in the absence of nucleation theory, because designing and analyzing the former requires a basic understanding of the latter. Classical nucleation theory (CNT), formulated in 1935 by Becker and Döring,[34] is probably still the theory most widely used to predict nucleation rates. A key assumption in this theory is that the critical nucleus has the same properties as a macroscopic liquid drop and, therefore, calculating a rate only requires values for the bulk physical properties of the species undergoing the phase transition. In its standard form the classical nucleation rate can be written as

$$J_{\text{CNT}} = \sqrt{\frac{2\sigma}{\pi m}} v_m \left(\frac{p_v}{kT} \right) \exp\left(-\frac{16\pi}{3} \frac{v_m^2 \sigma^3}{(kT)^3 (\ln S)^2} \right) \tag{1}$$

where σ is the surface tension, m and v_m the molecular mass and volume, k the Boltzmann constant, T the temperature, and p_v the actual vapor pressure. The supersaturation S is the ratio of the actual vapor pressure p_v to the equilibrium vapor pressure $p_{ve}(T)$

$$S = \frac{p_v}{p_{ve}(T)}. \tag{2}$$

Despite its simplicity, for most materials there is a range of temperatures where CNT does a remarkably good job. On the other hand, a major weakness of the theory is the observation that the temperature dependence of J_{CNT} is higher than that of the experiments. This observation led to the development of simple, but quite robust correction functions that modify CNT to bring it into much better agreement with the experimental observations. In particular, Wölk and Strey[19] suggested the following form of a correction function to the classical nucleation rate based on their NPC rate measurements for the isotopes of water,

$$J_{Corrected} = J_{CNT} \exp\left(\frac{A}{T} - B\right). \tag{3}$$

Another, reasonably successful approach to predicting nucleation rates over a wide range of temperature and supersaturation is the scaled nucleation model of Hale.[37–39] Here the nucleation rate expression is given by

$$J_{\mathrm{scaled}} = J_{0\mathrm{c}} \exp\left(\frac{16\pi}{3}\Omega^3\left[\frac{T_{\mathrm{c}}}{T} - 1\right]^3 \frac{1}{(\ln S)^2}\right), \tag{4}$$

where J_{0c}, the kinetic prefactor, is set equal to $J_{0c} = 10^{26}$ cm^{-3}s^{-1} and T_c is the critical temperature. The parameter Ω can be thought of as the surface entropy per molecule (divided by k), and is calculated from the constants found by fitting the bulk physical properties of the condensible material. In particular, $\Omega \times T_C$ is the slope of a straight line fit to a plot of $\sigma/(\tilde{n}^{2/3}kT)$ versus 1/T, where ñ is the liquid number density.[40] Another convenient feature of this equation is that it can be used to directly compare experimental measurements, made under very different temperature and supersaturation regimes, to test the consistency of these data sets and determine an effective value of Ω.

Nucleation Experiments

To conduct a homogeneous nucleation experiment, we must first generate a well-defined supersaturated state. Isentropic expansions are a convenient way to rapidly cool vapor – carrier gas mixtures, and as the expansion proceeds, the supersaturation increases because the equilibrium vapor pressure of the condensible decreases more quickly than its partial pressure. The expansion rate, or cooling rate, determines the maximum supersaturation that can be achieved and, thus, the characteristic range of nucleation rates. Cooling rates in the NPC are on the order of 10^4 Ks^{-1}, and nucleation rates can reach as high as 10^9 cm^{-3}s^{-1}, while in the SN cooling rates are ~5×10^5 Ks^{-1} and characteristic rates are close to 10^{17} cm^{-3}s^{-1}.

In the NPC, expansions are deliberately terminated before the spontaneous onset of condensation, and the length of the nucleation pulse is defined by recompressing the gas mechanically. The temperature corresponding to the nucleation event is determined from the measured pressure change and the initial temperature using the isentropic relationships. A significant advantage of controlling the supersaturation and the nucleation pulse length is that one can access up to four orders of magnitude in nucleation rate at a fixed temperature.

A similar approach was used to make the first nucleation rate measurements in a SN, where the recompression was introduced by shaping the nozzle blocks.[30] Although the approach worked, data analysis was often confounded by the occurrence of a second, spontaneous, nucleation burst downstream of the recompression that produced a bimodal aerosol distribution. Since this complication can be avoided by working with conventional SNs, i.e., nozzles that expand monotonically in the supersonic flow region, our focus has shifted to the latter.[31–33]

In the NPC, the length of the nucleation pulse is determined by analyzing the shape of the imposed pressure pulse.[18] In a conventional SN, the nucleation pulse occurs because the rapid increase in supersaturation is terminated when the spontaneous onset of condensation rapidly depletes the vapor. To determine the characteristic time associated with the peak nucleation rate in a SN, we measure the axial pressure profiles and integrate the diabetic flow equations to determine the temperature T and supersaturation S as a function of time, t, where $t = 0$ at the throat. The characteristic time associated with the peak nucleation rate Δt_{Jmax} is then determined by evaluating

$$\Delta t_{J\,max} = \frac{J(S,T)\,dt}{J_{max}} \tag{5}$$

where $J(S,T)$ is the nucleation rate predicted by any reasonable nucleation rate expression. The values of Δt_{Jmax} determined this way, and the values of S and T associated with J_{max}, have been shown to be rather insensitive to the choice of nucleation rate expression.[7, 33, 41]

Figure 1 illustrates typical nucleation pulses for the NPC and in one of our nozzles calculated using CNT, Eq. (1) and CNT incorporating the temperature correction function of Wölk and Strey, Eq. (3). The difference in characteristic time, calculated by these two models with rather different temperature dependencies, is less than ~30%. In the NPC, the characteristic times are on the order of a few milliseconds while in the nozzles the characteristic times for water nucleation range from 8–15 µs.[31–33]

The last step is to determine the number density N of the aerosol formed during the nucleation event. In the NPC, the droplets formed grow to sizes greater than ~1 µm, and N can be determined by constant angle Mie scattering using light from a He–Ne laser, (Figure 2). In contrast, the size of droplets formed in the SN are only 1–20 nm,

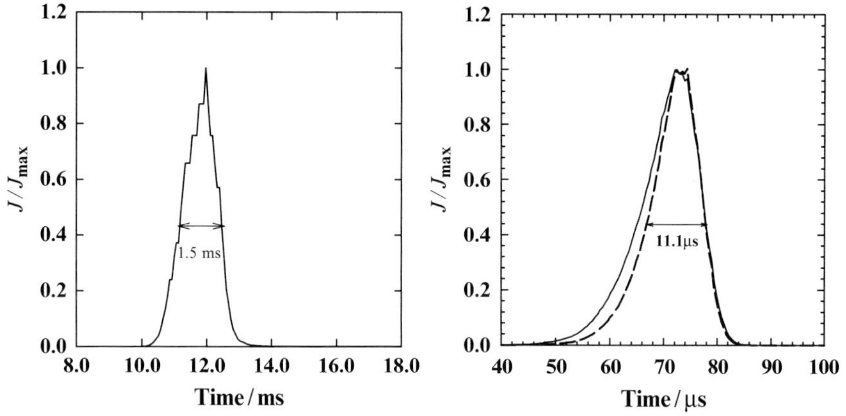

Figure 1 Typical nucleation pulses in a NPC (left) and a conventional SN (right). The solid line is calculated using Eq. (3) while the dashed line corresponds to Eq. (1). We used the physical property correlations given in Wölk and Strey[19] to evaluate the nucleation rates

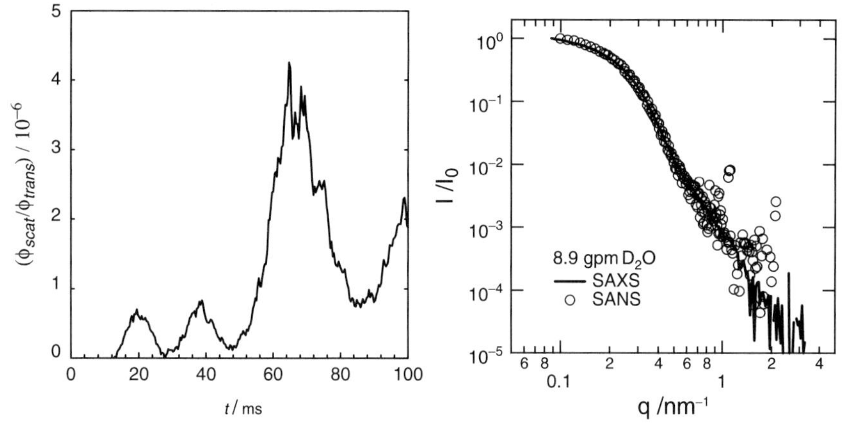

Figure 2 **Left**: A typical light scattering signal from the NPC. The height of the first peak is used to determine *N*. **Right**: Under identical operating conditions at a fixed position in the nozzle, the scaled small angle neutron scattering (SANS) and small angle x-ray scattering (SAXS) spectra agree quantitatively. A fit to the scattering curves is used to determine *N*

and, thus, shorter wavelength radiation, e.g., neutrons or x-ray scattering, is required to adequately characterize the aerosol. As illustrated in Figure 2, under identical operating conditions, the normalized scattering spectra for neutrons and x-rays are essentially identical. Because most compounds scatter x-rays more intensely than neutrons, and the flux of x-rays at a synchrotron sources is much higher than the flux of neutrons at a neutron source, data acquisition times have been reduced by more than three orders of magnitude and we are now in a position to investigate particle formation in SNs far more systematically than was possible using neutron scattering.

Finally, if we assume that there is a one-to-one correspondence between the droplet number density and the critical cluster number density, i.e., that coagulation and Ostwald ripening can be ignored on the timescales of the experiments, the nucleation rate in either experimental system is calculated as

$$J = \frac{N}{\Delta t_{J\max}} f \tag{6}$$

where f is a factor that corrects for any difference in the gas density between the time the particles were formed and the time they are detected. In the NPC f is essentially 1, while in the SN f is typically between 1.1 and 1.3.

Nucleation Theorems

Once accurate nucleation rates have been measured they can be used to determine the properties of the critical clusters, in particular the number of molecules in the critical cluster, n^*, and the excess internal energy of the critical cluster, $E_x(n^*)$ with respect to the bulk liquid. For nucleation of a single species, these quantities are derived from the measured nucleation rates by applying the first[42–44]

$$\left(\frac{d(\ln J)}{d(\ln S)}\right)_T \cong n^*, \tag{7}$$

and second[45,46]

$$\left(\frac{d(\ln J)}{dT}\right)_{\ln S} \cong \frac{1}{kT^2}\left[L - kT + E_x(n^*)\right], \tag{8}$$

nucleation theorems respectively. In Eq. (8) L is the latent heat of vaporization for the bulk liquid. Alternatively, McGraw[47] recently showed that the temperature derivative of the nucleation rate at constant vapor number density directly yields the excess internal energy of the cluster with respect to the vapor.

Nucleation Rate Measurements

Nucleation rates have now been measured in the NPC and the SN for a number of materials including both isotopes of water, a series of n-alcohols, and a series of n-alkanes. Our most extensive joint studies have been conducted for the isotopes of water, in particular for D_2O. Figure 3 illustrates the D_2O rates measurements of Wölk and Strey,[19] those of Kim et al. based on small angle neutron scattering (SANS),[33] and our recent measurements based on small angle x-ray scattering (SAXS). The SN data demonstrate good repeatability, independent of the aerosol characterization method, and they also illustrate one of the limitations of these devices: the range of accessible nucleation rates is quite limited even when nozzles with different cooling rates are used.

Although the SN and NPC can measure nucleation rates in the same temperature range, the rates in the SN are typically 10 orders of magnitude higher than in the NPC and, thus, to determine whether the data sets are consistent we must rely on theory. In Figure 3, the two theoretical curves correspond to the temperature corrected CNT, Eq. (3) (solid line) and Hale theory, Eq. (4) with $\Omega = 1.476$ (dashed line). Both models do a remarkably good job of quantitatively predicting the NPC and SN data, except perhaps at the lowest temperatures and highest supersaturations where, as discussed later in this paper, the critical clusters are rather small.

One difficulty with the water nucleation rate measurements is that the experimental temperature range corresponds to the supercooled liquid state. Thus, not all of the physical properties are measured over the entire range leading to uncertainty in comparisons with models. This is not the case for the *n*-alcohols, and was one reason why the nucleation community selected *n*-pentanol in 1995 for a series of

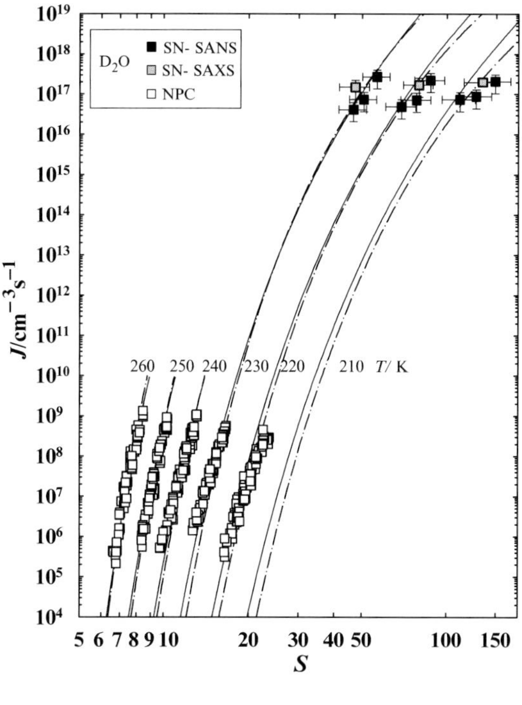

Figure 3 D_2O nucleation rates measured in the NPC (Ref 19) and the SN (Ref 33 and current work)

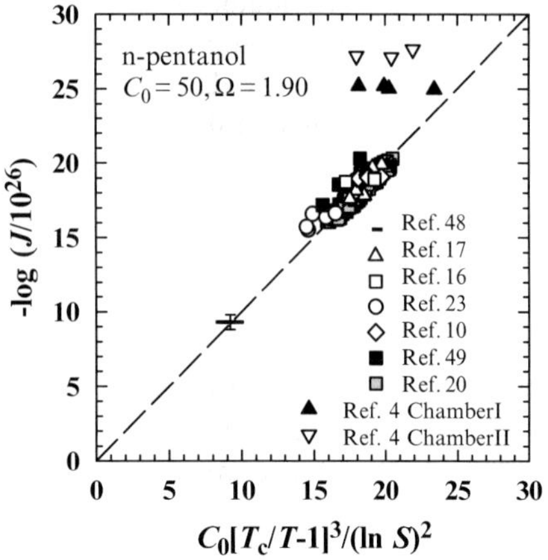

Figure 4 A Hale plot illustrates there is good agreement between many of the *n*-pentanol measurements made using a wide range of nucleation measurement devices

experiments to measure homogenous nucleation over as wide a range of nucleation rates as possible (~20 orders of magnitude).[10] Although the target temperature was 260 K, most experiments covered a range of temperatures and, thus, to compare all of the data simultaneously we use a Hale plot, Figure 4. All of the measured nucleation rates lie close to the 45° line corresponding to Ω❑= 1.90, except those made using a thermal diffusion cloud chamber (open downward facing triangles and filled upward pointing triangles). Here the data point closest to the 45° line corresponds to 260 K, the remaining points are at higher temperatures, and the deviations increase as the temperature is raised. Hale's model was developed for $T/T_c < 0.5$, and given that T_c for *n*-pentanol is 588 K, this may explain part of the deviation observed for the high-temperature data.

Critical Cluster Properties

A major goal of our effort is to determine the properties of the critical clusters by applying the nucleation theorems to our data. Figure 5 summarizes the values obtained for the number of molecules in the critical cluster n^* for both isotopes of water and combines the data from the SN, the NPC, and the laminar flow tube reactor (LFTR). The vertical error bars associated with the measured values of n^* are calculated by propagation of error. The horizontal error bars reflect the range in n^* predicted by the Gibbs–Thomson equation when we incorporate the uncertainty in *S*. Almost all of the experimental values for n^* lie within ±20% of the predictions of the Gibbs–Thomson equation. Only when the critical cluster contains fewer than ~8 molecules do the data begin to deviate systematically from this criterion.

We then applied the 2nd nucleation theorem to the data generated in the SN and the NPC and determined $E_x(n^*)$, the excess internal energies of the clusters with respect to the bulk liquid. In their respective regions of overlap, the NPC and SN data sets for the two isotopes of water are in remarkably good agreement. To bridge the gap between the two data sets we rely on the expression developed by Ford,[45,46] who showed that within the framework of the capillarity approximation, $E_x(n^*)$ is given by

$$E_{xc}(n^*) = \left(\sigma - T\frac{d\sigma}{dT}\right) A_1 (n^*)^{2/3} \tag{9}$$

where $A_1(n^*)^{2/3}$ is the surface area of a spherical droplet containing n^* molecules. As illustrated in Figure 5, most of the NPC data agree quantitatively with the predictions of Eq. (9), and even the SN data are consistent down to $n^* \approx 8$ molecules.

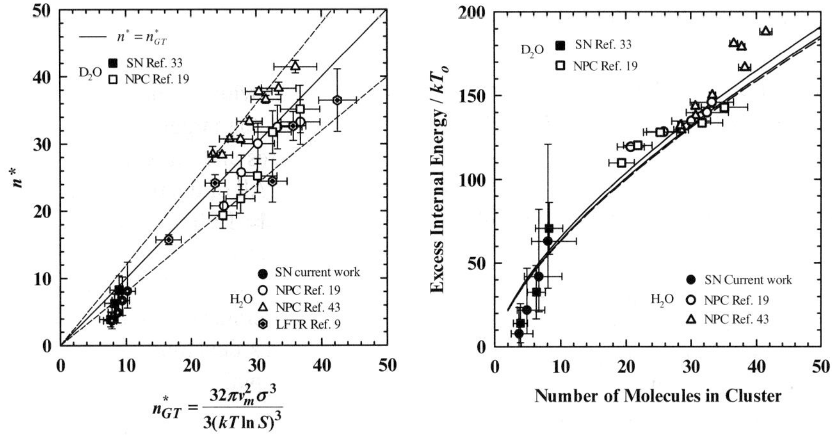

Figure 5 **Left**: A critical test of the Gibbs–Thomson equation. **Right**: The experimental values of $E_x(n^*)$ are compared to the predictions of Eq. (9). The upper and lower solid lines correspond to Eq. (9) evaluated for D_2O at 260 and 210 K respectively – the extreme temperature values for all of the experiments – and for the 50° change in temperature the curves differ by only 3%. The value of T_0 used to normalize $E_x(n^*)$ is $T_0 = 273.15$ K

Table 1 Empirical temperature correction coefficients based on NPC experiments[19,35,36]

Substance	*A*/K	*B*
H_2O	6,550	28
D_2O	8,600	36
Ar	3,630	27
N_2	4,270	55
Ethanol	16,932	66
1-propanol	2,464	6
1-butanol	12,725	28
1-pentanol	15,643	39

Our ability to develop consistent nucleation rate data sets that span large ranges of rate and temperature, and to use the information to generate information about the critical clusters, should help spur the development of more robust nucleation rate theories. Table 1 summarizes the values of *A* and *B* for a several substances including the isotopes of water,[19] several straight chain alcohols,[35] and cryogenic gases.[36]

Acknowledgments This work was supported by the National Science Foundation under Grant numbers CHE-0410045 & CHE-0518042 and by the Donors of the Petroleum Research Fund administered by the American Chemical Society. Use of the Advanced Photon Source was supported by the US Department of Energy, Office of Science, Office of Basic Energy Sciences, under Contract No. W-31–109-ENG-38. We thank R. Winans and S. Seifert for their help conducting the SAXS experiments.

References

1. Pound, G.M., *J. Phys. Chem. Ref. Data* **1**, 119 (1972).
2. Heist, R.H. and He, H., *J. Phys. Chem. Ref. Data*, **23**, 781 (1994).
3. Katz, J.L. and Ostermier, B.J., *J. Chem. Phys.*, **47**, 478 (1967).
4. Rudek, M.M. and Katz, J.L., Vidensky, I.V., Ždímal, V., Smolik, J., *J. Chem. Phys.*, **111**, 3623 (1999).
5. Heist, R.H., Ahmed, J. and M. Janjua, *J. Phys. Chem.*, **98**, 4443 (1994).
6. Bertelsmann, A. and Heist, R.H. *Aerosol Sci. Technol.*, **28**, 259 (1998).
7. Hämeri, K., Kulmala, M., Krissinel, E. and Kodenyov, G., *J. Chem. Phys.*, **105**, 7683 (1996).
8. Mikheev, V.B., Laulainen, N.S., Barlow, S.E., Knott, M., and Ford, I., *J. Chem. Phys.*, **113**, 3704 (2000).
9. Mikheev, V.B., Irving, P.M., Laulainen, N.S., Barlow, S.E., and Pervukhin, V.V., *J. Chem. Phys.*, **116**, 10772 (2002).
10. Lihavainen, H., Viisanen, Y., and Kulmala, M., *J. Chem. Phys.*, **114**, 10031 (2001).
11. Anisimov, M.P., Hopke, P.K., Shandakov, S. D., and Schvets, I.I., *J. Chem. Phys.*, **113**, 1971 (2000).
12. Adams, G.W., Schmitt, J. L., and Zalabsky, R.A., *J. Chem. Phys.*, **81**, 5074 (1984).
13. Schmitt, J.L., and Doster, G.J., *J. Chem. Phys.*, **116**, 1976 (2002).
14. Rodemann, T., and Peters, F., *J. Chem. Phys.*, **105**, 5168 (1996).
15. Graßmann, A., and Peters, F., *J. Chem. Phys.*, **116**, 7617 (2002).
16. Strey, R., Wagner, P.E., and Schmeling, T., *J. Chem. Phys.*, **84**, 2325 (1986).
17. Hruby, J., Viisanen, Y., and Strey, R., *J. Chem. Phys.*, **104**, 5181 (1996).
18. Strey, R., Wagner, P.E., and Viisanen, Y., *J. Phys. Chem.*, **98**, 7748 (1994).
19. Wölk, J. and Strey, R., *J. Phys. Chem. B*, **105**, 11683 (2001).
20. Iland, K., Wedekind, J., Wölk, J., Wagner, P.E., and Strey, R., *J. Chem. Phys.*, **121**, 12259 (2004).
21. Lee, C.F., Condensation of H_2O and D_2O in Argon in the Centered Expansion Wave in a Shock Tube, In Condensation in High Speed Flows, edited by A.A. Pouring, ASME, pp.83–96. *Am. Soc. Mech. Eng.*, NY (1977).
22. Peters, F. and Paikert, B., *Exp. Fluids*, **7**, 521 (1989).
23. Luijten, C.C.M., Baas, O.D.E., and van Dongen, M.E.H., *J. Chem. Phys.*, **106**, 4152 (1997).
24. Holten, V., Labetski, D.G. and van Dongen, M.E.H., *J. Chem. Phys.*, **123**, 104505 (2005).
25. Oswatitsch, K., *ZAMM*, **22**, 1 (1942).
26. Hill, P.G. *J. Fluid Mech.*, **25**, 593 (1966).

27. Stein, G.D. and Wegener, P.P., *J. Chem. Phys.*, **46**, 3685 (1967).
28. Stein, G.D. and Moses, C.A., *J. Coll. Int. Sci.*, **39**, 504 (1972).
29. Moses, C.A. and Stein, G.D., *J. Fluids Eng.*, **100**, 311 (1978).
30. Streletzky, K.A., Zvinevich, Y., Wyslouzil, B.E., and Strey, R., *J. Chem. Phys.*, **116**, 4058 (2002).
31. Khan, A., Heath, C.H., Dieregsweiler, U.M., Wyslouzil, B.E., and Strey, R., *J. Chem. Phys.*, **119**, 3138 (2003).
32. Heath, C.H., Streletzky, K.A., Wyslouzil, B.E. Wölk, J., and Strey, R., *J. Chem. Phys.*, **118**, 5465 (2003).
33. Kim, Y.J., Wyslouzil, B.E., Wilemski, G., Wölk, J., and Strey, R., *J. Phys. Chem. A*, **108**, 4365 (2004).
34. Becker, R. and Döring, W., *Ann. Phys.*, **24**, 719 (1935).
35. Wedekind, J., Diploma thesis, Universität zu Köln (2003).
36. Iland, K., Ph.D. thesis, Universität zu Köln (2004).
37. Hale, B.N., *Phys. Rev. A*, **33**, 4156 (1986).
38. Hale, B.N.,*Metall. Trans. A*, **23**, 1863 (1992).
39. Hale, B.N., *J. Chem. Phys.*, **122**, 204509 (2005).
40. Hale, B.N., personal communication (2004).
41. Wagner, P. and Anisimov, M., *J. Aerosol Sci.*, **24 S1**, 103 (1993).
42. Kashchiev, D., *J. Chem. Phys.*, **76**, 5098 (1982).
43. Viisanen, Y., Strey, R. and Reiss, H., *J. Chem. Phys.*, **99**, 4680 (1993).
44. Oxtoby, D.W. and Kashchiev, D., *J. Chem. Phys.*, **100**, 7665 (1994).
45. Ford, I., *J. Chem. Phys.*, **105**, 8324 (1996).
46. Ford, I., *Phys. Rev. E.*, **56**, 5615 (1997).
47. McGraw, R., *Report Series in Aerosol Science No. 80* (edited by M. Kulmala and T. Petaja) Finnish Association for Aerosol Science, 69 (2006).
48. Gharibeh, M., Kim, Y., Dieregsweiler, U., Wyslouzil, B.E., Ghosh, D., and Strey, R., *J. Chem. Phys.*, **122**, 94512–94521 (2005).

Size Analysis of Ion and Aerosol Nanoparticles in Nucleation and Growth

Kikuo Okuyama[1], Takafumi Seto[2], and Chan Soo Kim[3]

Abstract The size distribution of nanometer-sized aerosol and ion clusters is one of the most important parameters to measure when evaluating the "onset" of nucleation events in atmospheric environments. Nanoparticles have properties distinct from those of bulk materials and are attracting considerable attention. In this lecture we introduce various instruments for the size analysis of nanoparticles and ion clusters during nucleation and growth. These instruments are applied to measure dynamic changes that take place in the size distribution as the process of nucleation and growth advances. Several topics on the application of ion clusters will also be introduced.

Keywords Nanoparticle, ion cluster, plasma, ion-induced nucleation, differential mobility analyzer

Introduction

Drastic improvements in aerosol measurement instruments, most notably the differential mobility analyzer (DMA), the condensation nucleus counter (CNC), the particle size magnifier (PSM), and the Faraday cup electrometer (FCE), now permit the simultaneous measurement of the size distribution of ion clusters and nanoparticles. With these advances, aerosol instruments can be applied for the analysis of not only the nucleation and growth of nanometer-size aerosols, but also their "size-dependent" properties.

[1] *Graduate School of Engineering, Hiroshima University 1-4-1 Kagamiyama, Higashi-Hiroshima 739-8527, Japan*

[2] *National Institute of Advanced Industrial Science and Technology (AIST) 1-2-1 Namiki, Tsukuba 305-8564, Japan*

[3] *Seoul National University, Korea*

C.D. O'Dowd and P.E. Wagner (eds.), *Nucleation and Atmospheric Aerosols*, 14–25.

In this lecture we introduce various instruments for the size analysis of nanoparticles and ion clusters in nucleation and growth. In the first part we introduce recent topics to do with various types of aerosol chargers and review the progress of aerosol instruments such as the high-resolution DMA and mixing-type CNC. In the second part we report the experimental application of these instruments for the measurement of dynamic changes in size distribution during nucleation and growth during processes such as chemical and physical vapor condensation. We also introduce the ion clusters and nanoparticles produced by the energy beam irradiation of gas and electro-atomization processes.

Instruments for the Size Analysis of Nanoparticles

Aerosol measurement systems generally fall into two categories: atmospheric-pressure systems and low-pressure systems. The most important requirement for the atmospheric system is to obtain high resolution and high sensitivity by suppressing the effect of "Brownian" diffusion. In this section we introduce a system for size distribution analysis using a high-resolution DMA and mixing-type CNC (PSM). A key requirement for the low-pressure system is to design a pumping system appropriate for the pressure balance. To address this, we review the application of DMA and CNC to a low-pressure system. The electrical charging of aerosol is important as a pretreatment for electrostatic manipulation and the detection of aerosol in both the low- and atmospheric-pressure systems. We also report the recent progress of several new types of aerosol chargers and neutralizers.

Measurement of Ion Cluster Under Atmospheric Pressure

A DMA is a powerful tool for sizing nanometer- to submicron-sized particles in the gas phase. The application of DMA has recently been expanded to sub- to single-nanometer particles through design improvements enabling reductions in the Brownian diffusion within DMAs. De la Mora and coworkers developed a DMA capable of running under a very high Reynolds number flow (high-resolution DMA: HR-DMA).[1] The HR-DMA has a molecular resolution suitable for analyzing atmospheric ion clusters generated by the electrospray ionization of liquid solutions (Figure 1(a)). The resolution ($=\Delta Z_p/Z_p$) of the HR-DMA is improved by the optimized sheath air flow rate and DMA design. Figure 1(b) shows the ion mobility analyzer system using an HR-DMA (cluster DMA) and mass spectrometer.[2]

The combination of a DMA and CNC is often used for the analysis and detection of nanoparticle sizes. The measurement is restricted to particles larger than actual primary particles, however, as the current instrumentation is incapable of measuring nanoparticles of 3 nm diameter or lower.

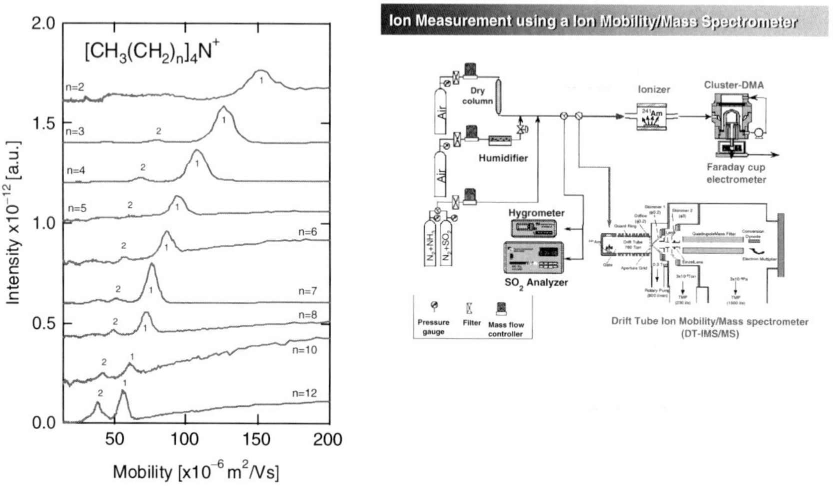

Figure 1 Mobility spectrum of electrosprayed tetra alkyl ammonium cations with alkyl chains of variable lengths (**left**), and a schematic illustration of a high-resolution (cluster) DMA coupled with mass spectrometer (**right**)

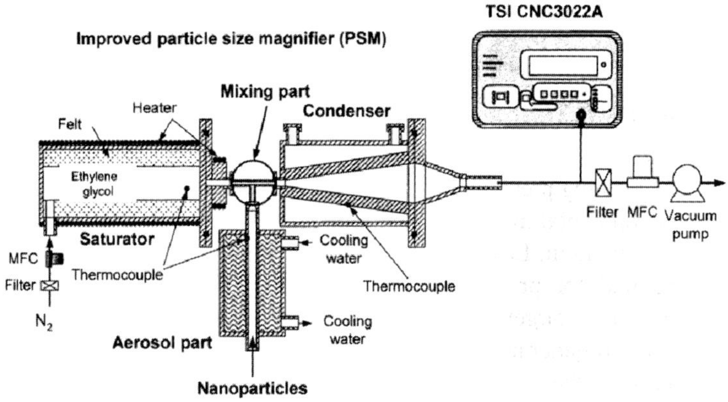

Figure 2 Schematic diagram of the PSM/CNC system

Figure 2 shows a schematic illustration of an improved PSM.[3] This device helps to increase the counting efficiency of CNC, especially when particle diameters are below 10 nm. We measured the concentration of the growth particles at the exit of the PSM using a commercial CNC (TSI model 3022A) capable of detecting particle diameters as low as 4 nm. As shown in Figure 3, our research on the performance of an improved PSM showed a measurement sensitivity of 100% for particles with mobility diameters of 1.6 nm (=1.2 nm in mass diameter). This new technique promises to be particularly useful for experimental investigations of nucleation and particle growth. Later in this report we describe the use of this PSM/CNC system

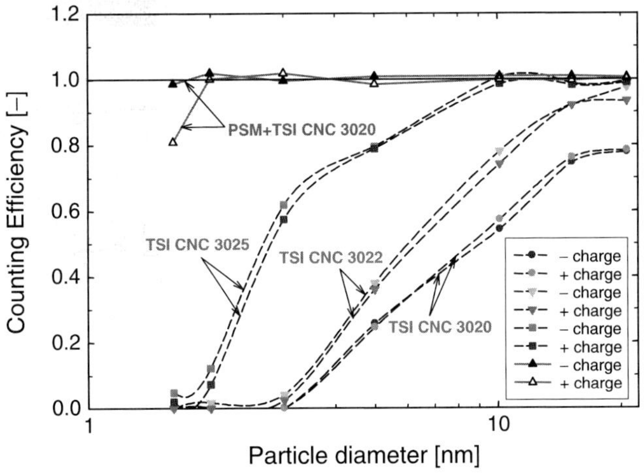

Figure 3 Schematic diagram of the PSM/CNC system

to detect primary particles formed by nucleation during chemical vapor condensation.

Measurement of Nanoparticles Under Low Pressure

The reduced operating pressure of the DMA makes it possible to apply DMA technology to environmental monitoring and various material processes such as plasma CVD and laser ablation. Laser ablation at low pressure is a novel and particularly promising method for producing very high-purity nanoparticles using systems comprised of only a target material and inert gas species. Previous experiments were performed to generate high-purity silicon nanoparticles laser ablation while actively controlling the size of the generated nanoparticles using an Low Pressure DMA (LPDMA) technique.[4]

In the present study we used an LPDMA and electrometer to measure the effect of the background gas pressure on the size distribution of naturally charged nanoparticles (Figure 4). The mobility equivalent diameter, d_m, was calculated from the Stokes–Einstein equation. The number concentration was converted from the electrical current of the charged nanoparticles on the assumption that the number of the charge, p, was unity. The particles were charged during the particle formation process by the collision of ions and electrons in the laser-induced plasma plume. As shown in Figure 4, a slight change in the pressure resulted in a drastic change in the distribution. The particles generated under a pressure of 2.1 Torr were broadly distributed, with a small concentration peak at about 5 nm. The laser plume was large, exceeding the distance between the target and sampling tube (about 50 mm). Presumably this explained why most of the ablated atoms were directly deposited

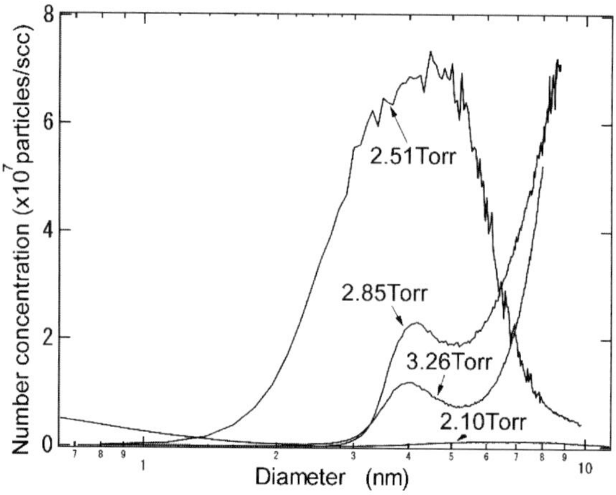

Figure 4 Onset of a nucleation event in laser ablation with variable background gas pressures

on the surface of the sampling tube and chamber, rather than being formed into nanoparticles by the gas phase nucleation. The number concentration of the particles peaked at 4 nm at 2.5 Torr, a pressure that might mark the onset of the gas phase nucleation. At a pressure higher than 2.9 Torr, the number concentration at diameter of 4 nm decreased, then rose to a second peak at a diameter larger than 6 nm. Remarkably, the first peak at 4 nm was independent of background pressures between 2.5 and 3.3 Torr. The second peak might have been the result of the agglomeration of primary particles or droplet-like particles.

Low-Pressure CNC

A mixing-type condensation nucleus counter (MTCNC) was used to detect nanometer-sized particles suspended under low-pressure conditions.[5] The experimental results show that MTCNC is capable of measuring nanoparticles smaller than 10 nm at low pressures above 65 Torr (8.644 kPa) by adjusting the temperature, the rate of aerosol flow, and the amount of vapor in accordance with the desired operation pressure. At pressures above 160 Torr (21.33 kPa), i.e., under the optimal operation condition for an MTCNC, variables such as the saturator temperature, condenser temperature, and mixing ratio are similar to those used at atmospheric pressure. The residence time is a strong determinant of the condensational growth of particles. As a consequence, the current MTCNC will require improvements if the growth time is to be increased even under lower-pressure conditions. At optimal operation conditions, the counting efficiencies for nanoparticle size 3 nm particles are 85% at atmospheric pressure and 40% at an operating pressure of 115 Torr (15.33 kPa).

Neutralizers and Chargers for Aerosol Nanoparticles

With the advance of nanoparticle technology in recent years, increasing attention has been focused on the charge distribution of aerosols in research applications involving size measurement, sampling, transport, and material processing. The charge neutralization of aerosol particles is an important parameter in distribution analyses in which the well-known function of charge distribution is applied as a means of reducing the CPC counts into the size distribution function. In most cases, the equilibrium charge distribution of aerosol particles can be achieved by colliding particles with bipolar ions generated by radioactive sources such as ^{241}Am, ^{85}Kr, and ^{210}Po. The ease and simplicity of use make radioactive sources a popular favorite, but the approach also has several disadvantages. In addition to requiring expensive safety precautions, radioactive sources are subjected to severe legal restrictions in some localities. Moreover, their limited ion concentrations severely limit their charging efficiency for highly concentrated particles. As an alternative to radioactive sources, some researchers are attempting to apply different physical principles to attain an equilibrium charge distribution.

Soft x-ray Aerosol Charger

Our group attempted to replace the radioactive source by applying soft x-ray for the generation of bipolar ions for aerosol charging (Figure 5).[6] The soft x-ray bipolar charger produced bipolar ions with a concentration 3.5 times higher than those of a ^{241}Am α-ray source charger, hence the resulting particles attained an equilibrium bipolar charging state within a shorter residence time. The system was incapable, however, of clearly describe the charging probability for particles larger than 30 nm. This was a serious drawback, as it precluded the determination of the equilibrium charging state for the particle size range. Later we tried using an x-ray charger to measure the size distribution of polydispersed NaCl particles with a peak diameter of 20 nm and a size range of 1–100 nm.[7] We observed almost equal ion mobilities of positive and negative ions generated by soft x-ray discharging, and thus obtained almost identical size distributions of positively and negatively charged particles.

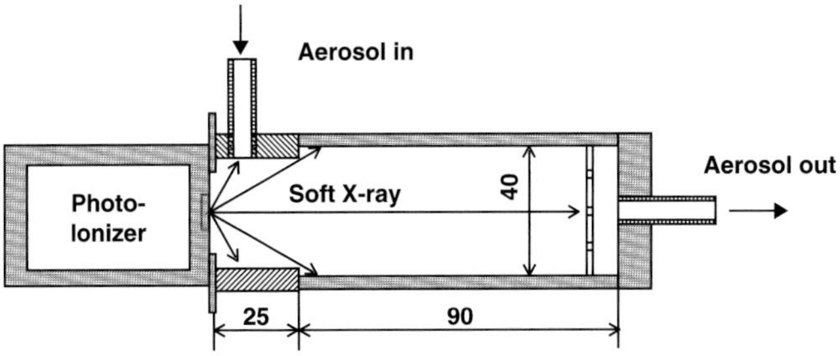

Figure 5 Soft x-ray aerosol charger

Surface-Discharge Microplasma Aerosol Charger (SMAC)

Intrigued by the advantages of the dielectric barrier discharge (DBD) approach, most notably the high concentration of bipolar ions, absence of an electric field, and simple controllability and applicability, our group attempted to apply a prototype DBD charger as an aerosol charger on a microscale (hereafter we refer to this device as the "SMAC" (Surface-discharge Microplasma Aerosol Charger); Figure 6).[8, 9] Positive and negative DC pulse voltages were applied on the positive and negative electrodes (dual electrode) for discharging, respectively. Surface discharge was induced through the dielectric barrier, and thus the aerosol particles were charged by positive or negative ions originating from the surface discharge. The positive or negative ion concentration was adjusted by the amplitude of the offset voltage. The charging probability of the SMAC for nanometer-sized particles was determined for each particle size and compared with the charging probability of the radioactive source and the theoretically predicted value. The equilibrium charge distributions obtained by the equal concentration of bipolar ions adjusted by the SMAC were in general agreement with the diffusion charging theory.

Unipolar Chargers

The unipolar charging of aerosol is especially important for the measurement of particles smaller than 10 nm. Figure 7 compares the extrinsic charging efficiencies of the SMAC (used as an unipolar ion source) with those of previous studies.[10] When applied for the positive unipolar charging of particles smaller than 6 nm, the SMAC achieved a good charging efficiency in comparison with charging efficiencies reported by Hernandez-Sierra et al.,[11] Buscher et al.,[12] and Kruis and Fissan.[13] With particles of over 6 nm, however, the unipolar positive charging of the system became relatively inefficient. The system from Chen and Pui[14] was the most efficient for charging particles of less than 6 nm, but their system lost efficiency as

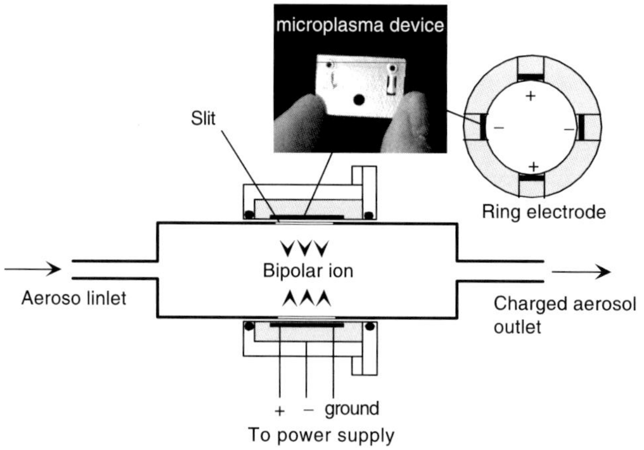

Figure 6 Surface-discharge microplasma aerosol charger (SMAC)

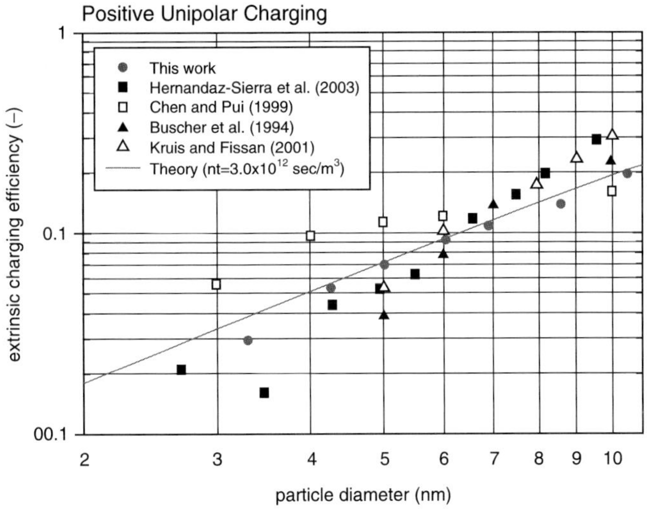

Figure 7 Unipolar charging efficiency of various chargers as a function of particle diameter

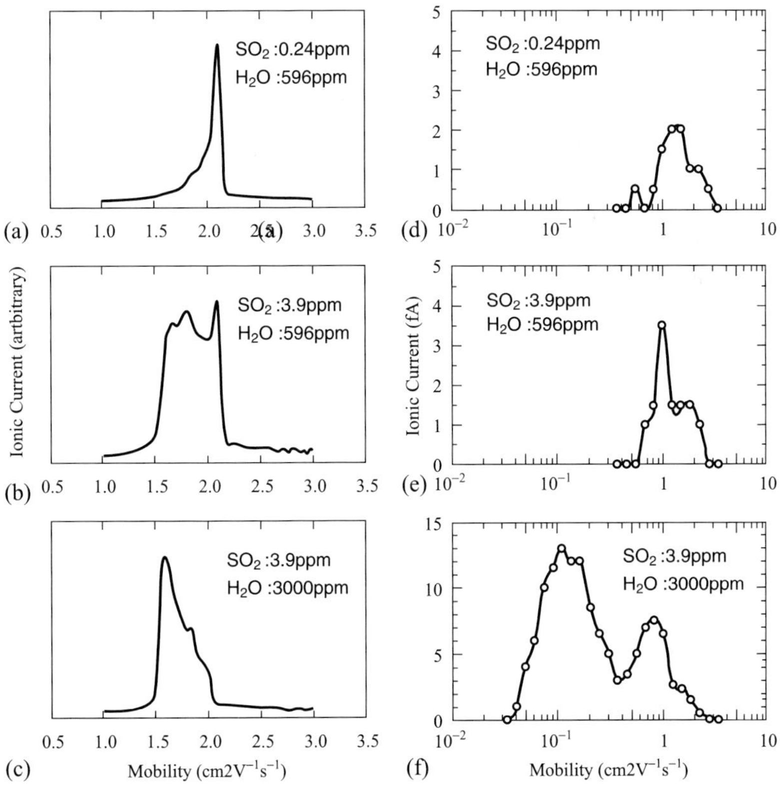

Figure 8 Negative ion mobility spectra for SO_2/H_2O/air mixtures obtained by DT-IMS/MS (a–c) and by C-DMA (d–f)

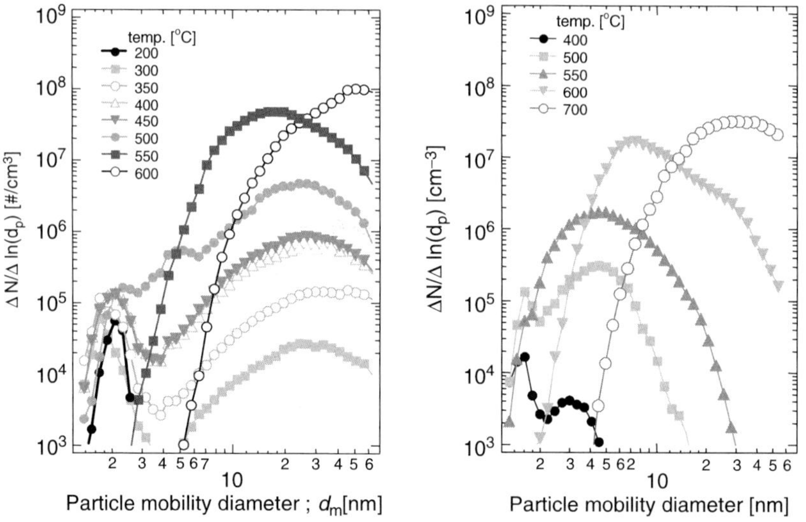

Figure 9 Nucleation mode in the titanium nanoparticle generated by chemical vapor condensation

the particle size increased to 10 nm. On the basis of our experiments, we find that the SMAC can achieve size-dependent charging efficiencies predictable by Fuchs charging theory.

Size Analysis of Ions and Nanopartilces During Nucleation Process

Mobility Distribution of Ion Clusters Generated by Irradiation of α-Ray

Nagato et al. developed an experimental system using a drift tube ion mobility spectrometer/mass spectrometer (DT-IMS/MS, shown in Figure 1 (right)) and a cluster-differential mobility analyzer with a Faraday cup electrometer (C-DMA/FCE) for the study of ion-induced nucleation.[2] Both the timescales and size scales of the ions to be measured can be extended significantly through this approach. The system offers potential for tracing the entire process of ion-induced nucleation initiated by the radioactive ionization of air. The main processes are the formation of stable cluster ions and the growth of the cluster ions to nanometer-sized ultrafine particles. Experiments were performed to examine the negative ions and negatively charged nanoparticles generated by α-ray radiolysis in SO_2/H_2O/air mixtures.

High-resolution mobility spectra in which ions were distributed in a relatively high mobility range of 2.1–1.5 cm^2 V^{-1} s^{-1} were obtained by DT-IMS/MS. Alternatively, more aged ions with mobilities as low as 0.05 cm^2 V^{-1} s^{-1} were observed in the C-DMA spectra. The differences between the two spectra reflect the mechanism by which cluster ions grow into nanoparticles. In addition to mobility measurements, the DT-IMS/MS is also capable of obtaining the mass spectra of ions and thereby identifying the nucleating ions and relevant ion chemistry to be investigated. SO_2^-, SO_3^-, SO_4^-, and SO_5^- were observed as major ions in the mass spectra at low SO_2 and H_2O concentrations, thus indicating that SO_2 is preferentially ionized by ion–molecule reactions with primary negative ions. HSO_4^- and HSO_4^-. H_2SO_4 ions became prominent when the concentrations of SO_2 and H_2O were increased, thus suggesting an enhancement of the oxidization of SO_2 into H_2SO_4, probably due to the enhanced production of OH radicals by ion–molecule reactions involving H_2O. On the basis of these findings, we speculate that the role of SO_2 in particle formation differs as the concentrations of SO_2 and H_2O change. At low SO_2 and H_2O concentrations, SO_2 tends to form core ions that lead to ion-induced particle formation. At higher SO_2 and H_2O concentrations, more of the SO_2 molecules are transformed into gaseous H_2SO_4 and a homogenous nucleation involving H_2SO_4 and H_2O is enhanced.

Nucleation Mode in Chemical Vapor Condensation

The nucleation in chemical vapor condensation is triggered by supersaturation generated by a chemical reaction of gaseous precursor. In 2004, two of the authors used the DMA/PSM/CNC system to measure the nucleation mode size distribution of TiO_2 nanoparticles prepared using two different precursors (titanium tetraisopropoxide and $TiCl_4$).[15] The system revealed three stages or regimes of particle generation: first, the nucleation control regime, when the concentration of nuclei remained low; second, the nucleation/surface reaction/coagulation control regime, when the concentration of nuclei was sufficiently large; and third, the coagulation sintering control regime, when the chemical reactions proceeded to completion.

Ion/Nanoparticle Generation by Electron Beam Irradiation

We also studied the nucleation of aerosols in a field of high-density free radicals, ions, and secondary thermalized electrons by irradiating electron beams into air, N_2, O_2, and Ar gases intermixed with variable amounts of water. In 2003, two of the authors used a C-DMA equipped with a FCE to analyze charged nanoparticles in irradiated gases.[16] All experiments with humidified gases formed both positively and negatively charged particles with mean mobility equivalent diameters (D_m) in a range of 7–10 nm, with an unchanging number concentration (N), and with a water ion cluster D_m in the range of 1.0–1.1 nm. The *N* of these large particles increased in step with the water content and absorbed dose. According to the experimental result, these large charged particles contained hydrogen peroxide. In the presence of ppbv-level benzene, the D_m and *N* of the large charged particles increased in step with the benzene concentration, although their D_m values were constant at different doses.

Charge Distribution of Electrosprayed Droplets – Nanoe

We evaluated the size distribution of fine water droplets produced by electrostatic atomization using a DMA and a CNC. A broad size distribution with a peak of mobility diameter of around 15–20 nm was observed for the neutralized polydispersed water droplets. We analyzed the number of the electrical charge using a tandem DMA system. The water droplets were highly charged overall, though the number of the charge was dependent on the particle size.

Conclusions

We reviewed the recent progress in aerosol instrumentation for the size analysis of nanoparticles and ion clusters in nucleation and growth. New types of DMA and CNC devices enable high-resolution and high-sensitivity measurement of ion and nanoparticles under both atmospheric- and reduced-pressure conditions. To explore the interactions between ion clusters and nanoparticles, we introduced various types of aerosol chargers and compared their performances. These instruments were applied to measure the dynamic change of the size distribution in the nuclei mode during nucleation and growth.

Acknowledgments This research was supported by a grant from the New Energy and Industrial Technology Development Organization of Japan (NEDO) for "Evaluating risks associated with manufactured nanomaterials (P06041)." The authors are grateful to Professor M. Adachi, Professor M. Shimada, Professor K. Nagato, and Dr. S.B. Kwon for their valuable discussions and experimental support.

References

1. de Juan, L. and de la Mora, J.F., *J. Aerosol Sci.*, **29**, 589–599 (1998).
2. Nagato, K., Kim, C.S., Adachi, M., and Okuyama K., *J. Aerosol Sci.*, **36**, 1036–1049 (2005).
3. Kim, C.S., Okuyama, K., and de la Mora, J.F., *Aerosol Sci. Tech.*, **37**, 791–803 (2003).
4. Seto, T., Orii, T., Hirasawa, M., and Aya, N., *Thin Solid Films*, **437**, 230–234 (2003).
5. Kim, C.S., Okuyama, K., and Shimada, M., *J. Aerosol Sci.*, **33**, 1389–1404 (2002).
6. Shimada, M., Han, B.W., Okuyama K., and Otani, Y., *J. Chem. Eng. Jpn*, **35**, 786–793 (2002).

7. Lee, H.M., Kim, C.S., Shimada, M., and Okuyama, K., *J. Aerosol Sci.*, **37**, 813–829 (2005).
8. Kwon, S.B., Sakurai, H., Seto, T., and Kim, Y.J., *J. Aerosol Sci.*, **37**, 483–499 (2006).
9. Kwon, S.B., Fujimoto, T., Kuga, Y., Sakurai, H., and Seto, T., *Aerosol Sci. Tech.*, **39**, 750–759 (2005).
10. Kwon, S.B., Sakurai, H., and Seto, T., *J. Nanoparticle Res.*, in press (2007).
11. Hernandez-Sierra, A., Alguacil, F.J., and Alonso, M., *J. Aerosol Sci.*, **34**, 733–745 (2003).
12. Büscher, P., Schmidt-Ott, A., and Wiedensohler, A., *J. Aerosol Sci.*, **25**, 651–663 (1994).
13. Kruis, F.E. and Fissan, H., *J. Nanopart. Res.*, **3**, 39–50 (2001).
14. Chen, D.-R. and Pui, D.Y.H., *J. Nanopart. Res.*, **1**, 115–126 (1999).
15. Kim, C.S., Okuyama, K., Nakaso, K., and Shimada, M., *J. Chem. Eng. Jpn.*, **37**, 1379–1389 (2004).
16. Hakoda, T., Kim, H. H., Okuyama, K., and Kojima, T., *J. Aerosol Sci.*, **34**, 977–991 (2003).

Classical and Generalized Gibbs' Approaches and the Work of Critical Cluster Formation in Nucleation Theory

Jürn W.P. Schmelzer

Abstract In the theoretical interpretation of the kinetics of first-order phase transitions, thermodynamic concepts developed long ago by Gibbs are widely employed giving some basic qualitative insights into these processes. However, from a quantitative point of view, the results of such analysis, based on the classical Gibbs' approach and involving in addition the capillarity approximation, are often not satisfactory. Some progress can be reached here by the van der Waals' and more advanced density functional methods of description of thermodynamically heterogeneous systems having, however, its limitations in application to the interpretation of experimental data as well. Moreover, both mentioned theories – Gibbs' and density functional approaches – lead to partly contradicting each other results. As shown in preceding papers, by generalizing Gibbs' approach, existing deficiencies and internal contradictions of these two well-established theories can be removed and a new generally applicable tool for the interpretation of phase formation processes can be developed. In the present analysis, a comparative analysis of the basic assumptions and predictions of the classical and the generalized Gibbs' approaches is given. It is shown, in particular, that – interpreted in terms of the generalized Gibbs approach – the critical cluster as determined via the classical Gibbs approach corresponds not to a saddle but to a ridge point of the appropriate thermodynamic potential hypersurface. By this reason, the classical Gibbs approach (involving the classical capillarity aproximation) overestimates as a rule the work of critical cluster formation in nucleation theory and, in general, considerably.

Keywords Homogeneous nucleation, work of critical cluster formation, Gibbs' theory

Institut für Physik, Universität Rostock, 18051 Rostock, Germany

C.D. O'Dowd and P.E. Wagner (eds.), *Nucleation and Atmospheric Aerosols*, 26–30.

Introduction

Nucleation-growth and spinodal decomposition processes are two basic classical model mechanisms by which first-order phase transitions may proceed. They determine the kinetics of self-structuring of matter from nanoscale up to galactic dimensions with a wide spectrum of possible applications both in fundamental and applied research and technology.

In the interpretation of experimental results on the dynamics of first-order phase transitions starting from metastable initial states, up to now predominantly the classical nucleation theory is used treating the respective processes in terms of cluster formation and growth. In the specification of the cluster properties, thermodynamic methods are intensively employed based in the majority of cases on the thermodynamic description of heterogeneous systems as developed by Gibbs. As one additional assumption it is assumed hereby frequently that the bulk properties of the clusters are widely similar to the properties of the newly evolving macroscopic phases.

This or similar assumptions are supported by the results of Gibbs' classical thermodynamic theory of heterogeneous systems. Indeed, following Gibbs' thermodynamic treatment one comes to the conclusion that the critical clusters have properties widely similar to the properties of the newly evolving macroscopic phases. As a second additional assumption, the interfacial specific energy of the critical clusters is supposed to be equal in a first approximation to the respective values for an equilibrium coexistence of both phases at planar interfaces. In order to come to an agreement between experimental and theoretical data on the value of the nucleation rates, this second assumption is often released by introducing a curvature dependence of the surface tension or specific interfacial energy. However, such assumption leads to another internal contradictions of the theory which cannot be resolved if one remains inside Gibbs' thermodynamic treatment of cluster properties. This way, Gibbs' classical treatment of surface phenomena is confronted with serious principal difficulties in application to nucleation.

Gibbs employed in his approach a simplified but theoretically fully consistent model considering the cluster as a homogeneous body divided from the otherwise homogeneous ambient phase by a sharp interface of zero thickness. The alternative continuum's concept of a thermodynamic description of heterogeneous systems was developed by van der Waals. It has been reinvented and applied for the first time to an analysis of nucleation by Cahn and Hilliard. In application to nucleation-growth processes, Cahn and Hilliard came to the conclusion that the bulk state parameters of the critical clusters may deviate considerably from the respective values of the evolving macroscopic phases and from the predictions of Gibbs' theory. Such deviations occur, in particular, in the vicinity of the classical spinodal curve dividing thermodynamically metastable and thermodynamically unstable initial states of the systems under consideration. The results of the van der Waals approach were reconfirmed later on by more advanced density functional computations.

Morover, Cahn and Hilliard developed also the alternative to the nucleation-growth model approach, the description of spinodal decomposition. According to the common belief (having again its origin already in the classical papers by Gibbs) the nucleation-growth model works well for the description of phase formation starting from metastable initial states, while thermodynamically unstable states are believed to decay via spinodal decomposition. As one consequence, the problem arises, how one kinetic mode of transition (nucleation-growth) goes over into the alternative one (spinodal decomposition) if the state of the ambient phase is changed continuously from metastable to unstable initial states, i.e., how the transition proceeds in the vicinity of the classical spinodal curve. The classical Gibbs' theory predicts here some kind of singularity which is, however, not confirmed by the Cahn-Hilliard description, statistical-mechanical model analyses, and experiment. From a more general point of view, we are confronted here with an internal contradiction in the predictions of two well-established theories, which has to be, hopefully, resolved.

The resolution of the circle of problems sketched briefly above is possible in the framework of a generalization of Gibbs' classical method developed by us in recent years. The basic ideas underlying this generalization are summarized below. In detail, the consequences of this generalized Gibbs' treatment in application to the determination of the work of critical cluster formation are outlined.

Classical and Generalized Gibbs' Approaches

As it follows from Gibbs' classical treatment of interfacial phenomena, his treatment is restricted from the very beginning to systems in thermodynamic equilibrium states. By this reason, Gibbs never formulated, in terms of his thermodynamic theory, thermodynamic potentials for clusters or ensembles of clusters in thermodynamic non-equilibrium states. In addition (and as a consequence), in Gibbs' theory the surface tension or specific interfacial energy depends, under typical conditions, on $(k + 1)$ independent variables, only (k is here the number of components of the system under consideration), i.e., on the intensive state parameters of one of the coexisting phases. This assumption is correct for an equilibrium coexistence of the considered phases. However, in order to determine correctly singular points (including the parameters of the critical clusters) one has to formulate first the thermodynamic potentials for arbitrary states of a cluster or ensembles of clusters in the otherwise homogeneous ambient phase and only afterwards to search for the extrema or saddle points by known methods. In latter case, the specific interfacial energy has to depend, in general, on the intensive state parameters of both coexisting phases (i.e., on $2(k + 1)$ independent variables).

The generalized Gibbs' approach is a reformulation of Gibbs' classical theory allowing one to determine the thermodynamic potentials of clusters and ensembles of clusters in the otherwise homogeneous ambient phase for any thermodynamically defined nonequilibrium states. It generalizes, moreover, one of the basic assumptions of Gibbs' classical approach allowing one to incorporate into the description the required dependence of the specific interfacial energy on the state parameters of the two coexisting phases. This approach is described in detail in [1] and [2]. It leads to a variety of consequences with respect to the understanding of phase transition phenomena. One of them, the differences in the determination of the work of critical cluster formation is discussed in the subsequent section.

Classical and Generalized Gibbs' Approaches and the work of Critical Cluster Formation

The generalized Gibbs' approach leads to a set of equilibrium conditions, which deviate for small cluster sizes from Gibbs' classical results. As a consequence, the state parameters – determined via the generalized Gibbs' approach – deviate, as a rule, significantly from the values obtained via Gibbs' classical theory. As shown for model systems, the results are in agreement with van der Waals and more advanced density functional computations, resolving one of the principal problems discussed in the introduction.

Both in the classical and the generalized Gibbs' approaches, the work of critical cluster formation is given as $W = (1/3)\sigma A$, where σ is the interfacial tension (specific surface energy) and A the surface area of the critical clusters. However, both the values of the cluster size and the interfacial energy differ. **It can be shown that – as a rule – the classical Gibbs' approach overestimates the work of critical cluster formation And underestimates, consequently, the value of the steady-state nucleation rates**. The discussion of this result is the main aim of the contribution at the present conference.

Concluding Remarks

The generalized Gibbs' approach allows one to derive, in addition to above discussed result, a variety of conclusions concerning general features of the phase transformation kinetics. In particular it can be shown that both nucleation-growth and spinodal decomposition can be described in terms of a generalized thermodynamic cluster model taking into account both variations in the size parameters and the intensive state parameters of the clusters in the course of their evolution. Some of the results of this analysis are discussed in an accompanying contribution [3].

References

1. J. W. P. Schmelzer, G. Sh. Boltachev, and V. G. Baidakov, *J. Chem. Phys.* **124,** 194503 (2006).
2. J. W. P. Schmelzer (Ed.), *Nucleation Theory and Applications*, Wiley VCH, Berlin, Weinheim, (2005).
3. A. S. Abyzov and J. W. P. Schmelzer, Nucleation versus Spinodal Decomposition in Confined Binary Solutions, present volume.

Microstructure of Supersaturation Field and Nucleation Experiments

Sergey P. Fisenko

Abstract The microstructure of the supersaturation field, arising near growing droplets in nucleation zone, is discussed. Estimations show that this effect can reduce calculated nucleation rate on one–two orders of magnitude. For diffusion cloud chamber the microstructure of the supersaturation field helps to explain "pressure" effect.

Keywords Laminar flow diffusion chamber, diffusion cloud chamber, screening radius

Introduction

Well known that the interpretation of nucleation experiments is not straightforward problem because currently some important parameters of nonequilibrium mother phase cannot be measured. Experimentalists have to use mathematical model to calculate these parameters. Essential feature of all developed mathematical models is the assumption about spatial uniformity of supersaturation field near growing droplets. But this assumption is wrong one if the Knudsen number, Kn, is smaller than one. We define Knudsen number, $Kn = \lambda/2R$, where λ is the mean free path of vapor molecules, R is the droplet radius. In turn, for reliable optical measurements droplet radius has to be about $1\,\mu$m. For atmospheric conditions it means that the Knudsen number is about 0.1. During many nucleation experiments droplets reach much larger sizes than $1\,\mu$m. The supersaturation S is equal to one on the surface of the droplet. Thus we have microstructure of supersaturation field near droplet. The estimation of the influence of this microstructure of supersaturation on observable nucleation rates is the main aim of the report. I consider two examples of setups for

A.V. Luikov Heat& Mass Transfer Institute, National Academy of Sciences 15, P. Brovka St., Minsk, 220072, Belarus

C.D. O'Dowd and P.E. Wagner (eds.), *Nucleation and Atmospheric Aerosols*, 31–35.

nucleation studies: a laminar flow diffusion chamber and a diffusion cloud chamber. These setups have a finite size of the nucleation zone and the motion of growing droplets through this zone [1–3].

Motion of Growing Droplet in Supersaturated Vapor

Laminar Flow Diffusion Chamber (LFDC)

Let us consider the growing droplet in co-flow of a mixture of a carrier gas and supersaturated vapor [2]. For micron's size droplets the Knudsen number $Kn < 0.1$, therefore we can consider the diffusion regime of growth. The droplet temperature is practically equal to the local temperature of the gaseous mixture because concentration of the carrier gas is much higher than vapor one. The steady-state solution of the diffusion problem gives the formula for the spherically symmetrical supersaturation field $S(r)$:

$$S(r) = S_{\infty} + \frac{(1 - S_{\infty})R}{r}, \tag{1}$$

where S_{∞} is the supersaturation far from droplet. All mathematical models, developed for description of the performance of a laminar flow diffusion chamber, give only the value S_{∞}. It follows from (1) that we can neglect the local structure of supersaturation if the distance from the droplet $r \geq 10R$ [2]. To denote the effective screening radius near growing droplet, R_{ef}, which is equal to $10R$. Due to decreased supersaturation we can neglect nucleation process in the sphere with radius R_{ef} near this immobile droplet. In fact the effective screening radius R_{ef} is about 100λ. The expression (1) is correct if the following inequality is valid:

$$\left(DL / uR^2\right) \gg 1,$$

where D is vapor diffusion coefficient, L is length of nucleation zone in LFDC, u is the velocity of gaseous mixture. For micron size droplets in LFDC this inequality is valid. Characteristic time τ of the transition to steady-state regime is: $\tau \sim R^2_{ef}/\pi^2 D$. The numerical estimation gives that τ is about 1 μs. For droplet, which moves with gas flow, the screening volume Ω is:

$$\Omega = \neq R_{ef}^2 u \tag{2}$$

Accordingly to the naive picture of nucleation, we could expect formation $J\Omega$ droplets in this volume, where J is the calculated nucleation rate. Instead we will observe only one droplet. For relatively high nucleation rate we see a significant effect. Indeed, for typical values for LFDC: $u \sim 10^{-2}$m/s, $R_{ef} \sim 10^{-5}$m and for nucleation rate $J \sim 10^{14}$ particles/(m^{3}s), we have only one particle instead several hundreds.

Discrepancy between theoretical results and experimental ones is about two orders of magnitude. Thus the local effect of supersaturation field drastically affects on manifestation of nucleation [2,4].

We have another prediction related with described above effect. Let us denote as S_c the area of cross section of nucleation zone in LFDC [2]. Then the maximum of the total flow rate of droplets I is:

$$I = uS_c / \pi R_{ef}^2$$

It means [2] also that there is the maximum of the observable nucleation rate J_m in LFDC:

$$J_m = I / L = uS_c / L\pi R_{ef}^2$$

It is worth to note that effective screening radius depends on total pressure in the chamber; therefore the last expression gives a key for explanation some experimental effects. Effect of the maximum total nucleation rate was observed in [4], but did not receive the correct interpretation.

Diffusion Cloud Chamber (DCC)

The effect of the total pressure on results of nucleation experiments in DCC was discovered experimentally more than 10 years ago; the latest results are in [3,4, and references therein]. Physical explanation of this effect has not been obtained yet. For example the experimental data for the critical supersaturation of glycerol vapor are presented in Figure 1 (taken from [3]). The pressure effect is obvious one.

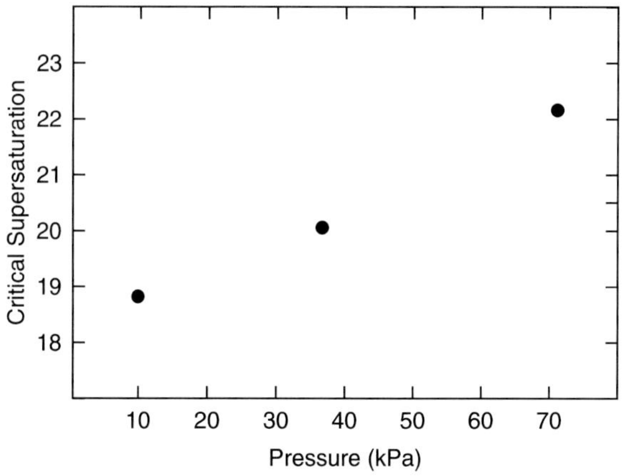

Figure 1 Critical supersaturation glycerol versus total pressure in DCC

For the same nucleation temperature the relationship between supersaturation S, total pressure P, and droplet radius R has been obtained in [1,3]:

$$R^4 \sim (S-1)/P$$

This formula helps to explain the data in Figure1, assuming the constant value of the laser radiation scattered by droplets. Nevertheless, there is still question: if we increase the supersaturation why additional droplets don't form? The main physical reason is the microstructure of the supersaturation field in the nucleation zone of DCC [1,3]. Let is estimate the effective screening radius. For calculation of the screening volume Ω we use the expression (1). The characteristic droplet radius in nucleation zone is about 10 µm, and velocity $v \sim 0.001$ m/s. Using the formulas (1–2), we can estimate that even huge increasing of nucleation rate about 10^{13} droplet/(sm^3) does not change the total number of droplets in the screening volume Ω (with logarithmic accuracy). Screening effect eliminates the increasing of the nucleation rate. The consideration above is valid only for steady-state regime of DCC performance. To remind that there is the counterflow of droplets and vapor diffusion flow in DCC, and increasing supersaturation without increasing total pressure leads to oscillatory nucleation [5]. The maximum of the total flow rate of droplets I is:

$$I \sim v / \pi R_{ef}^2 ,$$

and the maximum of the nucleation rate J_m:

$$J_m \simeq I_m / h ,$$

where h is the height of the nucleation zone; usually it is about 10% of chamber height.

Contributions of a diffusion interaction between growing droplets and vapor could be neglected if the average distance between falling droplets is about $10R_{ef}$. Therefore correct comparison between experimental results and theoretical predictions can be made only for such regime of experiments. Practically it means only for low nucleation rates.

Conclusions

It was shown that microstructure of the supersaturation field determines many important characteristics of two kinds of nucleation setups: the laminar flow diffusion chamber (LFDC) and the diffusion cloud chamber (DCC). The motion of growing droplets through nucleation zone drastically reduces local supersaturation, and therefore drastically reduces the number newly formed droplets. All mathematical model do not include this effect into consideration, therefore it may be one of key factors, which produce the deviation between theoretical predictions and experimental results.

For LDFC our estimations show that due to local microstructure of supersaturation field effective nucleation rate drops on one–two orders of magnitude in compare with standard calculation. For DCC our estimations show that the microstructure of supersaturation gives physical explanation of the "pressure effect" in DCC. The expansion of the conception of microstructure of a supersaturation field on other nucleation setups is under progress now. In particular, for expansion chambers the entrainment of growing droplets by acoustic waves could lead to the screening effect.

Acknowledgments Partial financial support was received from Belarus Foundation for Basic Researches (grant T06P-013).

References

1. Fisenko, S.P. and Heist, R.P., High pressure nucleation experiments in diffusion cloud chambers, in: *Nucleation Theory and Applications*, edited by J. Schmelzer, Dubna, Russia: Publishing Department of the Joint Institute for Nuclear Researches, pp. 146–164 (2002).
2. Fisenko, S.P. and Brin, A.A., *Intern. J. Heat & Mass Transfer*, **49**, 1004–1014 (2006).
3. Kane, D., Fisenko, S.P., Rusyniak, M., and El- Shall, M.S., *J. Chem. Phys.*, **111**, 8496–8502 (1999).
4. Brus, D., Hyvarinen, A.P., Zdimal, V., and Lihavainen, H. *J. Chem. Phys.*, **122**, 214506 (2005).
5. Fisenko, S.P., Rusyniak, M., Kane, D., and El- Shall M.S., *Intern. J. Heat & Mass Transfer*, **49**, 2044–2052 (2006).

Efficiency of Immersion Mode Ice Nucleation on Surrogates of Mineral Dust

C. Marcolli, S. Gedamke, and B. Zobrist

Abstract A differential scanning calorimeter (DSC) was used to explore heterogeneous ice nucleation of emulsified aqueous suspensions of two Arizona test dust (ATD) samples with nominal 0–3 and 0–7 μm particle diameters, respectively. Aqueous suspensions with ATD concentrations from 0.01–20 wt% have been investigated. The DSC-thermograms exhibit a homogeneous and a heterogeneous freezing peak whose intensity ratios vary with the ATD concentration in the aqueous suspensions. Depending on ATD concentration, heterogeneous ice nucleation occurred at temperatures up to 256 K or down to the onset of homogeneous ice nucleation (237 K). The measured heat release in the DSC was modeled with Classical Nucleation Theory (CNT), using experimentally determined ATD size and emulsion droplet volume distributions to quantify the heterogeneous surface present in the droplets. The observed dependence of the heterogeneous freezing temperatures on ATD concentrations could not be achieved with one constant contact angle but was reproduced by a model that describes heterogeneous ice nucleation occurring on active sites with variable ability to nucleate ice. This model implies that only a very minor fraction of the ATD surface is indeed effective as ice nuclei (IN).

Keywords Heterogeneous nucleation, ice nucleation, mineral dust, DSC

Introduction

Ice crystals in the atmosphere may form by homogeneous nucleation of cloud droplets and aqueous aerosol particles or by heterogeneous nucleation on foreign surfaces of so-called ice nuclei (IN). Cloud droplet freezing on heterogeneous IN has shown that often insoluble material such as mineral dust and fly ash act as IN.[1] Mineral surfaces are known to effectively initiate ice formation. Water droplets of

Institute for Atmospheric and Climate Science, ETH Zurich, Switzerland

C.D. O'Dowd and P.E. Wagner (eds.), *Nucleation and Atmospheric Aerosols*, 36–40.

kaolinite and montmorillonite suspensions were observed to freeze between −14°C and −30°C.[2] However, in these studies the concentration of the IN and consequently also the active surface was not quantified.

When water-in-oil emulsions are cooled in a differential scanning calorimeter (DSC), a high number of water droplets nucleate independently from each other and the freezing can be observed via the released latent heat. Usually the onset of the freezing peak is taken as the ice nucleation temperature. In the present study, we attempt a full analysis of the freezing curve in terms of Classical Nucleation Theory (CNT)[3,4] taking into account the volume distribution of the water droplets in the emulsions as well as the surface of IN within the droplets. With such an analysis the spread of IN activity in a sample can be assessed. We do this for two different size distributions of Arizona test dusts (ATD), which may serve as surrogate for natural mineral dust particles.

Experimental and Evaluation Procedure

CNT offers a framework to parameterize homogeneous and heterogeneous nucleation rates as a function of temperature. It describes the heterogeneous ice nucleation rate coefficient for supercooled water, $j_{het}(T)$,[3] by:

$$j_{het}(T) = \frac{kT}{h} \exp\left[-\frac{\Delta F_{diff}(T)}{kT}\right] \times n_s \exp\left[-\frac{\Delta G(T) \cdot f_{het}}{kT}\right] \quad (1)$$

where k and h are the Boltzmann and the Planck constant, respectively, T the absolute temperature, n_s ($\approx 10^{15}$ cm^{-2}) the number density of water molecules at the ice nucleus/water interface, $\Delta F_{diff}(T)$ the diffusion activation energy of a water molecule to cross the water/ice embryo interface, and $\Delta G(T)$ the Gibbs free energy for the formation of the critical ice embryo in the absence of a heterogeneous ice nucleus. The compatibility function f_{het} (≤ 1) describes the reduction of the Gibbs energy barrier due to the presence of an ice nucleus. The latter may be described as[5]:

$$f_{het} = \frac{1}{4}(2 + \cos\alpha)(1 - \cos\alpha)^2 \quad (2)$$

The parameter α formally represents the contact angle between the ice embryo and the ice nucleus in an aqueous medium.

The formulation of $j_{het}(T)$ is very similar to the one of the homogeneous ice nucleation rate coefficient, where f_{het} is removed and n_s is replaced by n_v, the number density of water molecules in the volume.[3]

The number of droplets freezing homogeneously and heterogeneously in an emulsion in a time interval t is given by

$$n_{hom}(t) = n_l(1 - exp(-Vj_{hom}t) \text{ and } n_{het}(t) = n_l(1 - exp(-A_d j_{het}t) \quad (3)$$

respectively, where n_l is the number of liquid droplets, V the droplet volume, and A_d the ATD surface present in a droplet.

The emulsions used in the DSC experiments consisted of 80 wt% of a mixture of lanolin and mineral oil and 20 wt% of water or an aqueous suspension of ATD. The volume distribution of water droplets in the emulsions was established by analyzing transmitted-light microscope images of emulsion samples prepared on an objective slide. Emulsion droplet freezing was studied with a DSC (TA Instruments Q10). A finer (0–3 µm particle diameters) and a coarser (0–7 µm) quality of commercially available ultra fine ATD (Powder Technology Inc.) were used in the experiments. Their size distributions were measured with a scanning mobility particle sizer (SMPS, TSI) and an aerodynamic particle spectrometer (APS, TSI).

Results and Discussion

The freezing of aqueous suspensions with ATD concentrations 0.01–20 wt% for the finer and 1–20 wt% for the coarser ATD have been investigated. Figure 1a shows the DSC-thermograms of emulsified suspensions with selected ATD contents of the finer ATD quality. Homogeneous freezing is observed in the temperature range from 232 to 237.4 K and heterogeneous freezing at temperatures up to 256 K and down to the onset of homogeneous ice nucleation. The intensity ratios of the homogeneous and heterogeneous freezing peaks vary with the concentration of ATD in the aqueous suspension, such that larger ATD concentrations lead to a larger heterogeneous peak. The heterogeneous freezing curves are quite narrow for high ATD contents and broaden when the number of ATD particles per water droplet decreases.

In Figure 1b the modeled freezing curves of the emulsified suspensions are shown, assuming that the ATD particles are randomly distributed in emulsion droplets and that all particles have an equal efficiency to nucleate ice. The modeled curves represent well the ratio of homogeneous-to-heterogeneous freezing, thus confirming the assumption of random distribution of ATD between the droplets. However, the modeled heterogeneous freezing curves for low ATD contents are much narrower than the measured ones. The decrease of heterogeneous surface with decreasing ATD content is not able to explain the strong shift to lower freezing temperatures and broadening of the freezing curves with decreasing ATD content.

Hence, in a second calculation, we assumed that the ability to nucleate ice differs between ATD particles. The contact angle was allowed to vary between particles, while it was kept constant for the whole surface of a particle. With this model, quite good agreement with the measured freezing curves was obtained. Since the assumption of contact angles varying between particles and being constant for the whole surface of a particle might not be realistic for a material like ATD, we performed a third calculation assuming that ice nucleation occurs on

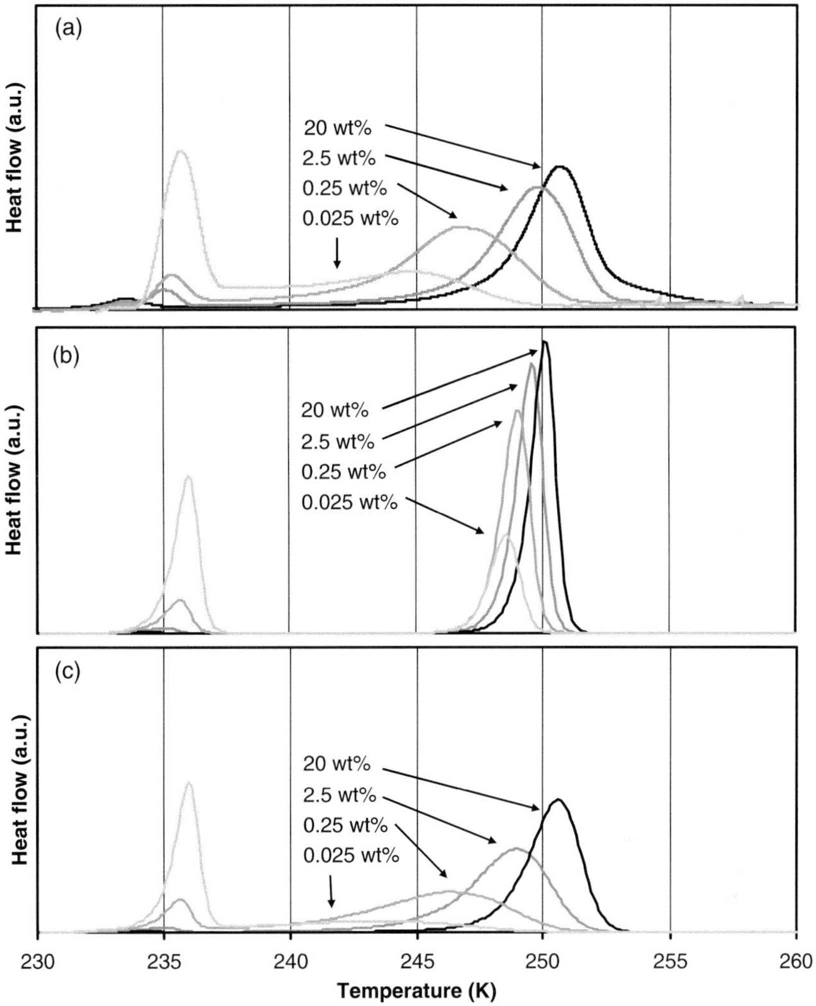

Figure 1 Ice nucleation of fine ATD (0–3 µm particle diameters). (a) DSC freezing experiment of emulsified aqueous suspensions with concentrations from 0.025–20 wt%; (b) Modeled freezing curve with a constant contact angle of 65°; (c) Modeled freezing curve with nucleation occurring on active sites

active sites[6] with varying IN quality (Figure 1c). It was found that the observed freezing curves are reproduced best when the occurrence probability of active sites is very low and decreases strongly with decreasing contact angle. The low number of active sites provides ATD particles with the wide range of IN activities that is needed to reproduce the large spread of freezing temperatures observed for ATD suspensions with low concentrations.

Conclusion

DSC freezing experiments with emulsified aqueous suspensions of ATD showed that this mineral dust acts as immersion mode IN. Heterogeneous ice nucleation strongly depends on ATD concentration with freezing temperatures up to 256 K and down to the onset of homogeneous ice nucleation. The intensity ratio of the homogeneous to the heterogeneous freezing peaks is consistent with the assumption of random distribution of ATD within the water droplets. The strong dependence of heterogeneous freezing temperatures on the ATD concentration can be reproduced in the model when ice nucle-ation occurs on few active sites of varying quality instead of the whole ATD surface. The active site model that yielded best agreement with the measured DSC curves implies that only a very minor fraction of the ATD surface is indeed effective as IN.

References

1. DeMott, P.J., Cziczo, D.J., Prenni, A.J., Murphy, D.M., Kreidenweis, S.M., Thomson, D.S., Borys, R., and Rogers, D.C., *Proc. Natl. Acad. Sci. U. S. A.*, **100**, 14655–14660 (2003).
2. Pitter R.L. and Pruppacher H.R., *Quart. J. R. Met. Soc.*, **99**, 540–550 (1973).
3. Pruppacher, H.R. and Klett, J.D., *Microphysics of clouds and precipitation*, Dordrecht: Kluwer, pp. 191–209 and 309–355 (1997).
4. Zobrist, B., Koop, T., Luo, B.P., Marcolli, C., and Peter, T., *J. Phys. Chem. C*, **111**, 2149–2155 (2007).
5. Seinfeld, J.H. and Pandis, S.N., *Atmospheric Chemistry and Physics*, New York: Wiley, pp. 545–595 (1998).
6. Fletcher, N.H., *J. Atmos. Sci.*, **26**, 1266–1271 (1969).

Do Highly Supercooled Liquids Freeze by Spinodal Decomposition?

Lawrence S. Bartell

Abstract Under what conditions might the free energy barrier to the freezing of pure liquids vanish? Whether pure liquids ever freeze by spinodal decomposition has been the subject of much conjecture. That spinodal decomposition does occur in the freezing of some solutions is well-established, however. Spinodals do not arise in capillary models of the freezing of pure liquids. Some mean-field treatments suggest that spinodals can arise but near-mean-field treatments and Ising models suggest they do not. Molecular dynamics (MD) simulations invoking realistic intermolecular interactions would seem to offer the most direct evidence. Some MD simulations, analyzed via certain order parameters, have indicated that the free energy barrier to freezing vanishes before the temperature of liquids is decreased to 60% of the bulk freezing point. Other simulations, to be described, show that freezing is a stochastic process leading to first-order kinetics of freezing down to temperatures considerably below 60% of the melting point. In these simulations, no recognizable spinodal was encountered. Whether spinodal decomposition can ever occur for pure liquids is still an open question but present evidence indicates that, for pure liquids, if spinodal decomposition does occur, the temperature must be much lower than 60% of the melting point.

Keywords Homogeneous nucleation, simulations, spinodal decomposition, theories of freezing

Background

First-order phase changes are ubiquitous and have been known since antiquity. Yet the understanding of the molecular mechanism underlying such transformations and the associated kinetics has been studied scientifically only comparatively recently. The work of many experimentalists and theorists has led to a rather satisfactory understanding of the condensation of vapor.[1] On the other hand, a

Department of Chemistry, University of Michigan, Ann Arbor, MI 48109, USA

C.D. O'Dowd and P.E. Wagner (eds.), *Nucleation and Atmospheric Aerosols*, 41–45.

satisfactory understanding of the freezing of pure liquids has proven to be more elusive. J. Willard Gibbs pointed out over a century ago that when one phase transforms into another, an interface possessing an excess interfacial free energy must form between the two phases, and this imposes a barrier to the phase change.[2] This observation led several theorists to formulate capillary models to treat condensation and later, freezing. These models enjoyed a qualitative success. Quantitative deficiencies became apparent when experiments became more precise.

The problem of condensation proved to be more tractable than freezing because the difference between the nuclei of condensation and the gaseous mother phase is considerably greater than is the difference between solid nuclei embedded in a supercooled liquid. In freezing, the two phases have comparable densities. It was Turnbull who devised a fruitful technique for studying the kinetics of freezing a half century ago.[3] He also extended the capillary theory to account for his findings.[4] Unfortunately, his experimental technique was so demanding to carry out with precision that it was seldom applied rigorously, and it was not easy to assess the validity of the theory of the day. The capillary treatment correctly accounted for qualitative aspects of the kinetics including the lowering of the free energy barrier to nucleation as critical nuclei became smaller as the supercooling deepened. The theory gave no indication that the barrier vanished at any degree of supercooling, however.

Alternative Theoretical Treatments

Statistical thermodynamicists have been creative in devising theories to treat the kinetics of freezing. These theories have included mean-field theories introduced originally by van der Waals[5] and extended to variants of density functional theory. Applied to alloys, results indicated that spinodal decomposition followed naturally from the postulates, and experimental verification soon followed.[6] Applied to pure liquids, however, results have not been so definitive. Some treatments led naturally to results in which the free energy barrier to the formation of nuclei vanished at a sufficiently low temperature. One near-mean-field theory was stated to give rise to "spinodal nucleation"[7] a term misinterpreted by some as predicting that a true spinodal arose, i.e., that the free energy barrier vanished.[8] The term had been intended, however, only to mean that the interfacial free energy decreases greatly as an unattainable spinodal is approached. Alternative treatments via the Ising model have also been applied. In these, no spinodal occurred in freezing.[9]

Do Liquids Freeze by Spinodal Decomposition?

In view of subtleties in theoretical treatments and difficulties in experimental studies, how might one decide what really happens? The most direct and physically intuitive possibility appears to be realistic molecular dynamics (MD) simulations of freezing. At temperatures high enough to make thermal energies large compared with the spacing between quantum energy levels, classical simulations can be expected to be satisfactory. Happily, this requirement still allows simulations of nucleation at very substantial degrees of supercooling. Various order parameters can be devised to recognize tiny solid-like nuclei in the presence of an overwhelming amount of liquid. Therefore, it is possible to follow the course of events in molecular detail. Unfortunately, not even MD simulations have settled the question to the satisfaction of everyone: does spinodal decomposition occur in freezing at fairly modest degrees of supercooling as forecast a few years ago?[10]

It has been asserted that at a supercooling to 60% of the freezing point, the free energy barrier to nucleation must surely vanish.[10] Apparently confirming this claim is a recent MD study of the Lennard–Jones system with simulations biased to generate trajectories passing close to crystallization events.[8] Analyzed by a plausible choice of order parameter, results suggested that a spinodal was reached even before the temperature fell to 60% of the freezing point. On the other hand, in massive unbiased MD simulations carried out in our laboratory it was found that the rate of formation of nuclei, $J(n, t)$ was stochastic for all nuclear sizes n, leading to first-order kinetics at supercoolings considerably deeper than 60% of the freezing point.[11–14] Different methods of analysis were adopted. The first was a conventional determination of $N(n, t)$, the number of nucleation events per unit volume at time t, in which nuclei of size n or larger had been formed In this case it was expected on theoretical grounds that the steady-state rate should be independent of nuclear size, and this result was confirmed. The second method was a new one of identifying the times of first appearance of nuclei of size n.[14, 15] In both cases the reduced flux,

$$\phi(n,t)=J(n,t)/J_{steadystate}$$

could be represented very well by two parameters charactering the transient regime: the time lag and the "reduced moment." The second method provided more information. While it yielded steady state nucleation rates for nuclei of all sizes, that rate depended upon the nuclear size, and this dependency revealed the critical size n^* and the Zeldovich factor Z. Results obtained in the MD simulations of thousands of spontaneous nucleation events, in runs down to 40% of the freezing point, never encountered a spinodal.

Some Computational Details

It should be noted that there are significant differences between our MD simulations and those indicating a spinodal at a temperature of 64% of the melting point.[8] First, the simulations of Trudo et al. were performed on the Lennard–Jones atomic system whereas ours were on a system of the polyatomic molecule selenium hexafluoride, a system we had also studied in experiments where minute droplets froze in supersonic flow. At temperatures of about 60% of the freezing point, our droplets froze to single crystals in our experiments, not to haphazard spinodal masses.

Although the Lennard–Jones and SeF_6 systems are conspicuously different, there are close thermodynamic similarities. Each has a rather ordinary ratio of boiling point to critical temperature but each has an uncommonly small difference – of only a few degrees – between the boiling point and the melting point. The two systems, being bonded by van der Waals forces, are much more similar to each other than to the system of salt which is bonded principally by electrostatic forces. And molten salt, which is very eager to crystallize, shows normal freezing in our MD simulations down to temperatures considerably below 60% of the melting point,[16,17] just as does liquid SeF_6. Another substantial difference between the simulations of the Lennard–Jones system and those of SeF_6, is that the former simulations were carried out imposing periodic boundary conditions, meaning that there was no free surface at which nucleation might take place, On the other hand, the simulations of SeF_6 were on clusters with free surfaces, with cluster sizes ranging, in different simulations, from 138 to 1,722 molecules. At temperatures close to 60% of the freezing point, nucleation occasionally occurred at the surface but mainly occurred in the interior. As supercooling deepened, the preference for interior nucleation decreased somewhat.[18] It is not clear that a free surface would inhibit spinodal decomposition if a spinodal would otherwise have been encountered. Moreover, only ten independent nucleation events were studied in the Lennard–Jones system whereas many thousands of independent nucleation events were characterized in our simulations. Our simulations were carried out in unbiased "brute force" runs instead of the biased simulations of Trudo et al., because our principal aim was to determine not only the steady state nucleation rate but also to characterize the transient regime, the regime before the steady state is reached, a regime to which the biased runs were blind. The reader must decide for himself whether the above differences in simulations were sufficient to account for the absence of a spinodal in our simulations and the apparent occurrence of a spinodal in the simulations of Ref. 8. In any event, it is our contention that spinodals do not occur in the freezing of pure liquids at temperatures as moderate as 60% of the freezing point. This does not answer the question of whether a spinodal can ever occur in the freezing of a pure liquid but it lowers the degree of supercooling necessary for spinodal decomposition to a much deeper value than had been previously postulated.

Acknowledgments I gratefully acknowledge helpful discussions with Professors James Gunton and William Klein whose views may not be reflected entirely accurately in the foregoing.

References

1. Nucleation and Atmospheric Aerosols 2000, edited by B. N. Hale, and M. Kulmala, AIP Conference Proceedings, New York: American Institute of Physics (2000).
2. Gibbs, J.W, *Transactions of the Connecticut Academy*, **III**, 343–524 (1877–1878).
3. Turnbull, D., *J. Chem. Phys.*, **18**, 198–203 (1949).
4. Turnbull, D. and Fisher, J.C., *J. Chem. Phys.*, **17**, 71–73 (1949).
5. van der Waals, J.D., *Verhand. Kon, Akad, v, Wetensch. Ie Sect.*, **1** (1893).

6. Gránásy, L., *Solid State Phenomena*, **56**, 67–106 (1997).
7. Klein, W. and Leyvraz, E., *Phys. Rev. Lett.*, **57**, 2845–2848 (1986).
8. Trudo, F., Donadio, D., and Parrinello, M., *Phys. Rev. Lett.*, **97**, 105701-1-105701–105704 (2006).
9. Bustillos, A.T., Heermann, D.W., and Cordeiro, C.E., *J. Chem. Phys.*, **121**, 4804–4809 (2004).
10. ten Wolde, P.R., Ruiz- Montero, M.J., and Frenkel, D., *J. Chem. Phys.*, **104**, 9932–9947 (1996).
11. Chushak, Y.G., Santikary, P., and Bartell, L.S., *J. Phys. Chem. A*, **103**, 5636–5644 (1999).
12. Chuskak, Y.G. and Bartell, L.S., *J. Phys. Chem. B*, **103**, 11196–11204 (1999).
13. Turner, G.W., Chuskak, Y.G., and Bartell, L. S., *J. Phys. Chem. A*, **103**, 1666–1670 (2004).
14. Turner, G.W. and Bartell, L.S., *J. Phys. Chem. B*, **108**, 19742–19747 (2004).
15. Bartell, L.S. and Wu, D.T., *J. Chem. Phys.*, **125**, 194503–1–194503–4 (2006).
16. Huang, J., Zhu, X., and Bartell, L. S., *J. Phys. Chem. A*, **102**, 2708–2715 (1998).
17. Bartell, L.S. and Huang J., *J. Phys. Chem. A*, **102**, 8722–8726 (1998).
18. Turner, G.W. and Bartell, L.S., *J. Phys. Chem. A*, **109**, 6877–6879 (2003).

The Balanced Nucleation and Growth (BNG) Model for Controlled Crystal Nucleation and Size Distribution

1. The Model

Ingo H. Leubner

Abstract The Balanced Nucleation and Growth (BNG) Model was first proposed and experimentally supported in 1980.[1,2,3,4] This model has correctly predicted previously unknown nucleation behaviors.

The derivation of the BNG model is based on the classical concepts of Becker and Doering, but replaces the kinetic nucleation rate with the introduction of crystal growth processes during the nucleation phase.

The BNG model predicts the experimental result that many crystallization processes lead to a limited number of crystals during a nucleation period which is followed by growth only. Further, the model predicts the rate of crystal formation during the nucleation phase. These and the crystal size distribution, maximum crystal size, number of crystals, and nucleation rate and time. Model variables are the molar reactant addition rate during the nucleation phase, nucleation efficiency, critical nucleus size, and crystal maximum crystal growth rate. These are related to experimental parameters. Without growth, the crystal number is proportional to addition rate and nucleation occurs continuously during the time of reactant addition.

The BNG model was originally derived for balanced double-jet batch precipitation, and later extended and experimentally confirmed for controlled double-jet continuous crystallizations.[5]

Keywords Heterogeneous nucleation, controlled batch crystallization, controlled double-jet precipitation, modeling, crystal nucleation, crystal growth, nucleation phase, nucleation rates, nucleation time, size distribution, crystal number, reactant variables

Crystallization Consulting
35 Hillcrest Dr., Penfield, NY 14526–2411/USA

C.D. O'Dowd and P.E. Wagner (eds.), *Nucleation and Atmospheric Aerosols*, 46–47.

Biographical Abstract

For the precision control of crystals in controlled crystallization, Dr. Leubner developed the BNG model. He has taught the concepts for the control of crystallization and precipitations at seminars, academic institutions, and industrial companies. These models have been successfully applied to the development of a wide variety of commercial products for his employers and clients. He applied the fundamental discoveries toward product development and lead teams to provide products with significantly improved features. As a consultant, he applies his knowledge in precipitation and chemical sciences to help solve problems in industrial research, product development, and manufacturing. His clients include, in addition to his work at Kodak, Dow, Xerox, J&J, Cabot, Southern Clay Products, TempTime, Sachem, William Blythe, and many other corporations.

Ingo H. Leubner received his Ph.D. in Science (Dr. rer. nat.) from the Technical University in Munich, Germany, with a major in physical chemistry and minors in inorganic and organic chemistry, and chemical engineering. After a Welch Research Fellowship at Texas Christian University in Fort Worth, Texas, he joined Eastman Kodak Company. He was awarded the Lieven–Gevaert Medal, the highest award in photographic science, by the Society of Imaging Science and Technology for his publications on the mechanisms of spectral and chemical sensitization, and dopants. His work included the controlled crystallization/precipitation of silver halides for the precision control crystal size, morphology, and photosensitivity. His work contributed to accelerated product development and resulted in important contributions to the manufacture of improved products. For a full list of his publications, see www.crystallizationcon.com.

References

1. Leubner, I.H., Jagannathan, R., and Wey, J.S., "Formation of Silver Bromide Crystals in Double-Jet Precipitations", *Photogr. Sci. Eng.*, **24**, 268 (1980).
2. Leubner, I.H., "Crystal Formation (Nucleation) under Kinetically and Diffusion Controlled Growth Conditions", *J. Phys. Chem.*, **91**, 6069 – 6073 (1987).
3. Leubner, I.H., "The Balanced Nucleation and Growth Model for Controlled Crystallizations", *J. Disp. Sci. Technol.*, **22**, 125 – 138 (2001).
4. Leubner, I.H., "The Balanced Nucleation and Growth Model for Controlled Crystal Size Distribution", *J. Disp. Sci. Technol.*, **23**, 577 – 590 (2002).
5. Leubner, I.H., "A New Crystal Nucleation Theory for Continuous Precipitation of Silver Halides", *J. Imaging Sci. Technol.*, **42**, 357 (1998).

The Balanced Nucleation and Growth (BNG) Model for Controlled Crystal Nucleation and Size in Controlled Precipitations

2. Experimental

Ingo H. Leubner

Abstract Based on the Balanced Nucleation and Growth (BNG) model, experiments are presented that quantitatively relate the number and size of crystals formed in controlled batch processes to reaction variables, which reaction addition rate, crystal solubility, and reaction temperature, and the effect of Ostwald ripening agents and crystal growth restrainers.

The models were initially developed and applied to the precision precipitations of silver halides in research, product development, and manufacturing. They have since been successfully applied to the precision precipitation of boron–fluoride salts, calcium carbonate, silicates, polyacetylenes, and others. Knowledge of the model is essential for the size control of crystallization of organic and inorganic materials from the laboratory to the manufacturing scale.

Keywords Heterogeneous nucleation, controlled batch crystallization, controlled double-jet precipitation, modeling, crystal nucleation, size distribution, crystal number, reactant variables, reactant addition rate, crystal solubility, reaction temperature, ripeners, restrainers

Biographical Abstract

For the precision control of crystals in controlled crystallization, Dr. Leubner developed the Balanced Nucleation and Growth (BNG) model. He has taught the concepts for the control of crystallization and precipitations at seminars, academic institutions, and industrial companies. These models have been successfully applied to the development of a wide variety of commercial products for his employers and clients. He applied the fundamental discoveries toward product development and

Crystallization Consulting
35 Hillcrest Dr., Penfield, NY 14526–2411/USA

C.D. O'Dowd and P.E. Wagner (eds.), *Nucleation and Atmospheric Aerosols*, 48–49.

lead teams to provide products with significantly improved features. As a consultant, he applies his knowledge in precipitation and chemical sciences to help solve problems in industrial research, product development, and manufacturing. His clients include, in addition to his work at Kodak, Dow, Xerox, J&J, Cabot, Southern Clay Products, TempTime, Sachem, William Blythe, and many other corporations.

Ingo H. Leubner received his Ph.D. in Science (Dr. rer. nat.) from the Technical University in Munich, Germany, with a major in physical chemistry and minors in inorganic and organic chemistry, and chemical engineering. After a Welch Research Fellowship at Texas Christian University in Fort Worth, Texas, he joined Eastman Kodak Company. He was awarded the Lieven–Gevaert Medal, the highest award in photographic science, by the Society of Imaging Science and Technology for his publications on the mechanisms of spectral and chemical sensitization, and dopants. His work included the controlled crystallization/precipitation of silver halides for the precision control crystal size, morphology, and photosensitivity. His work contributed to accelerated product development and resulted in important contributions to the manufacture of improved products. For a full list of his publications, see www.crystallizationcon.com.

Continuous Crystallization: Nucleation and Size Control as a Function of Crystal Solubility

Ingo H. Leubner

Continuous crystallization is an attractive method for large-scale preparation of crystalline materials. To control the crystal size of the product it is desirable to have a theory based on fundamental science that allows simple modeling and accurate predictions. The balanced nucleation and growth (BNG) theory combines nucleation and growth in crystallizations for batch and continuous reactors. It predicts the size dependence as a function of experimentally controlled parameters, like reactant addition rate, crystal solubility, temperature, concentration of ripening, and growth restraining agents. In the present work, the size dependence on crystal solubility is predicted and experimentally supported for continuous crystallization in CSTR (MSMPR) crystallizers.

For silver chloride as a model system, the solubility was varied from 0.81 to 8.3E-06 mole/l (60C, 3.0 min residence time). The correlation coefficient between model and crystal sizes (0.34–0.52 μm) was 0.9996. The model calculated the average maximum crystal growth rate (8.5 A/s) and the ratio of critical to average crystal size (0.39). It further calculated the sizes of nascent, newly formed, crystals (0.13–0.30 μm), critical crystal sizes (0.13–0.20 μm), and supersaturation ratios (1.0049–1.0075) for the experimental conditions. The model showed a reactant split ratio R_n/R_i (0.06–0.26) of incoming reactant addition rate, R_0, into growth (R_i) and nucleation streams (R_n). The average crystal size is predicted to be independent of reactant addition rate, suspension density, and reaction volume. This was experimentally confirmed for variation of suspension density from 0.05 to 0.4 mole/l (average crystal size 0.337 +/– 0.009 μm). The crystal number increased linearly with suspension density. The BNG-based continuous crystallization model thus correctly predicted and quantitatively correlated the crystal size-dependence on solubility and suspension density in the CSTR crystallizer without the need for arbitrary adjustable parameters. It allowed determining important reaction parameters that were previously not accessible.

Crystallization Consulting
35 Hillcrest Drive, Penfield, NY 14526-2411, USA
ileubner@crystallizationcon.com

C.D. O'Dowd and P.E. Wagner (eds.), *Nucleation and Atmospheric Aerosols*, 50–51.

Keywords Controlled continuous crystallization, controlled double-jet continuous precipitation, MSMPR reactor, CSTR, modeling, experiment, crystal nucleation, crystal size, crystal number, reactant variables, reactant addition rate, suspension density, crystal solubility

Biographical Abstract For the precision control of crystals in controlled crystallization, Dr. Leubner developed the balanced nucleation and growth (BNG) model. He has taught the concepts for the control of crystallization and precipitations at seminars, academic institutions, and industrial companies. These models have been successfully applied to the development of a wide variety of commercial products for his employers and clients. He applied the fundamental discoveries toward product development and lead teams to provide products with significantly improved features. As a consultant, he applies his knowledge in precipitation and chemical sciences to help solve problems in industrial research, product development, and manufacturing. His clients include, in addition to his work at Kodak, Dow, Xerox, J&J, Cabot, Southern Clay Products, TempTime, Sachem, William Blythe, and many other corporations.

Ingo H. Leubner received his Ph.D. in Science (Dr. rer. nat.) from the Technical University in Munich, Germany, with a major in physical chemistry and minors in inorganic and organic chemistry, and chemical engineering. After a Welch Research Fellowship at Texas Christian University in Fort Worth, Texas, he joined Eastman Kodak Company. He was awarded the Lieven–Gevaert Medal, the highest award in photographic science, by the Society of Imaging Science and Technology for his publications on the mechanisms of spectral and chemical sensitization, and dopants. His work included the controlled crystallization/precipitation of silver halides for the precision control crystal size, morphology, and photosensitivity. His work contributed to accelerated product development and resulted in important contributions to the manufacture of improved products. For a full list of his publications, see www.crystallizationcon.com.

Investigating the Role of Ammonia in Atmospheric Nucleation

Theo Kurtén[1], Leena Torpo[1], Hanna Vehkamäki[1], Markku R. Sundberg[2], Kari Laasonen[3], Veli-Matti Kerminen[4], Madis Noppel[5], Martta Salonen[1], and Markku Kulmala[1]

Abstract Recent quantum chemical studies indicate that ammonia significantly enhances the formation of sulfuric acid–water clusters containing multiple sulfuric acid molecules. Our calculations on clusters containing up to three sulfuric acid molecules indicate a lower limit of 1:3 for the NH_3:H_2SO_4 mole ratio of nucleating clusters in atmospheric conditions. However, computations on NH_3–H_2SO_4 cluster cores also indicate an upper limit of 1:1, which is unlikely to be exceeded in any atmospheric conditions. Ammonia is also predicted to be only weakly bound to the HSO_4^- ion, and should thus not play a significant role in ion-induced nucleation of the sulfuric acid–water system.

Keywords Nucleation, sulfuric acid, ammonia, quantum chemistry

The Role of Ammonia in Neutral Clusters

The role of ammonia in atmospheric sulfuric acid–water nucleation is controversial. Earlier quantum chemical studies[1] predicted the NH_3:H_2SO_4 mole ratio of nucleating clusters to be close to zero, while state of the art thermodynamics[2] predicted extensive ammonium bisulfate formation,[3] corresponding to a ratio of 1:1.

We have recently[4] calculated formation free energies for small sulfuric acid–water–ammonia clusters at the RI-MP2[5]/aug-cc-pV(T+d)Z//MPW1B95[6]/aug-cc-p V(D+d)Z level. The computed Gibbs free energies of formation for a set of clusters with one-to-three sulfuric acids are presented in Table 1. The first column shows to the standard ΔG° values, which correspond to a temperature of 298.15 K and reference pressures of 1 atm for all reactants, while the second column shows the ΔG values

[1] *Department of Physical Sciences, University of Helsinki, PL 64, 00014 Helsingin yliopisto, Finland*

[2] *Department of Chemistry, University of Helsinki, PL 55, 00014 Helsingin yliopisto, Finland*

[3] *Department of Chemistry, University of Oulu, PL 3000, 90014 Oulun yliopisto, Finland*

[4] *Finnish Meteorological Instititute, PL 503, 00101 Helsinki, Finland*

[5] *Institute of Environmental Physics, University of Tartu, 18 Ülikooli Str, 50090 Tartu, Estonia*

C.D. O'Dowd and P.E. Wagner (eds.), *Nucleation and Atmospheric Aerosols*, 52–56.

Table 1 Gibbs free energies of formation (from molecular precursors) of various clusters, computed at the RI-MP2/aug-cc-pV(T+d)Z//MPW1B95/aug-cc-pV(D+d)Z level. The first column "ΔG°" is the standard Gibbs free formation energy (at 298.15 K, with reference pressures of 1 atm for all monomers); the second column "ΔG(265 K)" is computed at atmospherically more representative conditions of T = 265 K, $p(H_2O)$ = 1.655 mbar, $p(H_2SO_4)$ = 0.7 ppt, and $p(NH_3)$ = 100 ppt

Cluster	ΔG°, kcal mol^{-1}	ΔG(265 K), kcal mol^{-1}
$H_2SO_4 \cdot H_2O$	−2.6	14.5
$H_2SO_4 \cdot (H_2O)_2$	−4.5	15.0
$H_2SO_4 \cdot NH_3$	−7.3	18.6
$H_2SO_4 \cdot H_2O \cdot NH_3$	−8.6	19.6
$(H_2SO_4)_2$	−8.3	20.1
$(H_2SO_4)_2 \cdot H_2O$	−11.2	19.4
$(H_2SO_4)_2 \cdot (H_2O)_2$	−13.8	19.2
$(H_2SO_4)_2 \cdot NH_3$	−22.4	16.8
$(H_2SO_4)_2 \cdot H_2O \cdot NH_3$	−26.2	15.4
$(H_2SO_4)_3$	−13.0	28.8
$(H_2SO_4)_3 \cdot H_2O$	−12.5	31.4
$(H_2SO_4)_3 \cdot (H_2O)_2$	−20.5	25.7
$(H_2SO_4)_3 \cdot NH_3$	−30.0	22.5
$(H_2SO_4)_3 \cdot H_2O \cdot NH_3$	−36.3	18.7

computed at atmospherically representative conditions of T=265 K, $p(H_2O)$= 1.655 mbar (corresponding to RH 50%), $p(H_2SO_4)$=0.7 ppt, and $p(NH_3)$=100 ppt.

It can be seen from Table 1 that the standard formation free energies for ammonia-containing clusters are always significantly lower than for ammonia-free clusters. However, due to the low concentration of ammonia in the atmosphere, this is not always the case for the formation free energies in atmospheric conditions. When the concentration differences of the monomer species are accounted for, the ammonia-containing one-acid clusters are predicted to be less stable than the ammonia-free clusters. The ammonia-containing two-acid clusters are predicted to be slightly more stable than the one-acid clusters. However, lower-level computations[7] on two-acid clusters containing more (up to seven) water molecules indicate that the fraction of ammonia-containing clusters in the fully hydrated two-acid cluster distribution may still be small. (In effect, water tends to out-compete ammonia due to mass-balance effects.) For the three-acid clusters, the difference in stabilities of ammonia-containing and ammonia-free clusters is even larger. For example, the difference in formation-free energies for the five-molecule clusters $(H_2SO_4)_3 \cdot H_2O \cdot NH_3$ and $(H_2SO_4)_3 \cdot (H_2O)_2$ in atmospheric conditions is 7 kcal/mol. This difference is so large that the binding of further water molecules is not expected to change the qualitative conclusion that the presence of ammonia significantly enhances the formation of three-acid clusters. This implies a lower limit of 1:3 for the NH_3:H_2SO_4 mole ratio of atmospheric H_2SO_4–H_2O–NH_3 clusters.

Estimating an Upper Limit to the NH_3:H_2SO_4 Ratio

Previous quantum chemical studies indicate that the presence of water affects the acid–ammonia binding relatively weakly.[1,7] The effects that do exist are mostly systematic, with the addition of multiple water molecules decreasing the binding of ammonia to the clusters. (The $(H_2SO_4)_3$ and $(H_2SO_4)_3{\cdot}H_2O$ clusters presented in Table 1 are exceptions to this rule. This is mainly due to the surprisingly low stability of the $(H_2SO_4)_3{\cdot}H_2O$ structure.) This observation can be used to set an approximate upper limit to the NH_3:H_2SO_4 ratio of atmospheric clusters by studying only the acid-ammonia "core" of the clusters, without explicitly including water molecules.

We have recently computed ammonia addition energies for clusters containing two sulfuric acid and up to four ammonia molecules at the RI-MP2/aug-cc-pV(T+d)Z//RI-MP2/aug-cc-pV(D+d)Z level.[8] The key results are shown in Table 2. The corresponding NH_3:H_2SO_4 mole ratios (calculated as a weighted average of the cluster distribution) are shown as a function of the ammonia concentration in Figure 1. To account for possible errors in the computational method (or in the assumptions about the role of hydration) we have also performed a sensitivity analysis, in which the electronic energy change of each ammonia addition reaction has been decreased by 2 kcal/mol, and the vibrational frequencies scaled by 0.75. The results of the sensitivity analysis are shown as dashed lines in Figure 1.

Figure 1 indicates that the formation of small ammonium sulfate clusters (NH_3:H_2SO_4 ratio 2:1) in any conditions encountered in the atmosphere can probably be ruled out. Even the formation of ammonium bisulfate clusters (1:1) requires either quite high (over 10 ppb) NH_3 concentrations or very low temperatures. The sensitivity analysis demonstrates that errors in the computational method might lead to an overestimation of the NH_3 concentration required for ammonium bisulfate to form by at most two orders of magnitude. However, the finding that ammonium sulfate molecular clusters do not form in the atmosphere is insensitive even to large systematic errors.

Based on our results, we can thus conclude that the NH_3:H_2SO_4 ratio of nucleating atmospheric clusters is probably between 1:3 and 1:1. This is lower than the typical experimental values measured for larger (>10 nm in diameter) particles.[9] This indicates that the chemical composition of nucleating clusters may differ significantly from that of larger particles.

Table 2 Gibbs free energies for ammonia addition reactions, computed at the RI-MP2/aug-cc-pV(T+d)Z//RI-MP2/aug-cc-pV(D+d)Z level, with electronic energies extrapolated to the RI-MP2 basis-set limit. All values correspond to 298.15 K and reference pressures of 1 atm

Reaction	$\Delta G°$, kcal mol^{-1}
$(H_2SO_4)_2 + NH_3 \leftrightarrow (H_2SO_4)_2{\cdot}NH_3$	−16.4
$(H_2SO_4)_2{\cdot}NH_3 + NH_3 \leftrightarrow (H_2SO_4)_2{\cdot}(NH_3)_2$	−8.2
$(H_2SO_4)_2{\cdot}(NH_3)_2 + NH_3 \leftrightarrow (H_2SO_4)_2{\cdot}(NH_3)_3$	−3.9
$(H_2SO_4)_2{\cdot}(NH_3)_3 + NH_3 \leftrightarrow (H_2SO_4)_2{\cdot}(NH_3)_4$	−4.3

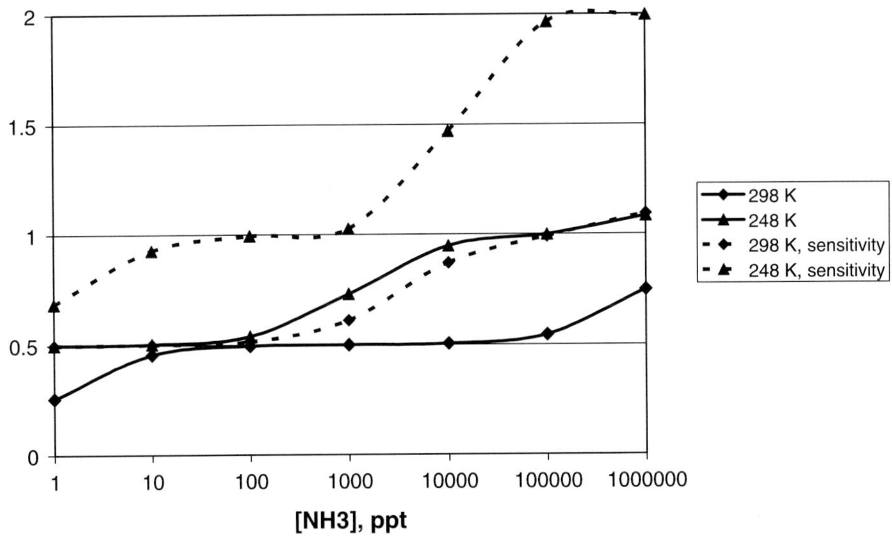

Figure 1 The estimated upper limit for the NH_3:H_2SO_4 mole ratio of two-acid cluster cores as function of the NH_3 concentration, at two different temperatures. Solid lines correspond to the data in Table 1, while dashed lines correspond to a sensitivity analysis (see text for details)

The Role of Ammonia in Ion-Induced Nucleation

High-level (MP2/aug-cc-pV(T+d)Z with MP4 corrections) calculations[10] on ion clusters show that NH_3 binds very weakly to the HSO_4^- ion. The standard free energy of formation of the $H_2SO_4 \cdot NH_3$ cluster is computed to be around −7 kcal/mol, while that of the $HSO_4^- \cdot NH_3$ cluster is only +0.7 kcal/mol. The binding of ammonia does not significantly change the water affinity of the HSO_4^- ion in either direction. This implies that ammonia will probably not play a major role in ion-induced nucleation of the sulfuric acid–water system.

All our computations have been done using the Gaussian 03[11] and Turbomole[12] v. 5.8. programs. The thermal contributions to the free energies have been computed using the standard rigid rotor and harmonic oscillator approximations.

Acknowledgments The authors thank the Scientific Computing Center (CSC) in Espoo, Finland for computing time and the Academy of Finland for funding.

References

1. Ianni, J.C. and Bandy, A.R., *J. Phys. Chem. A*, **103**, 2801–2811 (1999).
2. Clegg, S.L., Brimblecombe, P., and Wexler, A.S., *J. Phys. Chem. A*, **102**, 2137–2154 (1998).

3. Vehkamäki, H., Napari, I., Kulmala M., and Noppel., M., *Phys. Rev. Lett.*, **93**, 148501 (2004).
4. Torpo, L., Kurtén, T., Vehkamäki, H., Sundberg, M.R., Laasonen, K., and Kulmala, M., manuscript in preparation (2007).
5. Weigend, F. and Häser, M., *Theor. Chem. Acc.*, **97**, 331–340 (1997).
6. Zhao, Y. and Truhlar, D.G., *J. Phys. Chem. A*, **108**, 6908–6918 (2004).
7. Kurtén, T., Torpo, L., Ding, C.-G., Vehkamäki, H., Sundberg, M.R., Laasonen, K., and Kulmala, M., *J. Geophys. Res.*, in press (2007).
8. Kurtén, T., Torpo, L., Sundberg, M.R., Kerminen, V.-M., Vehkamäki, H., and Kulmala, M., *Atmos. Chem. Phys. Discuss.*, **7**, 2937–2960 (2007).
9. Feng, Y. and Penner, J.E., *J. Geophys. Res.*, **112**, D01304, doi:10.1029/2005JD006404 (2007).
10. Kurtén, T., Noppel, M., Vehkamäki, H., Salonen, M., and Kulmala, M, submitted to *Boreal Env. Res.* (2007).
11. Frisch, M.J., et al., Gaussian 03, Revision C.02, Gaussian, Inc., Wallingford CT, USA (2004).
12. Ahlrichs, R., Bär, M., Häser, M., Horn, H., and Kölmel, C., *Chem. Phys. Lett.*, **162**, 165–169 (1989).

Dependence of Nucleation Rates on Sulfuric Acid Vapor Concentration in Diverse Atmospheric Locations

Chongai Kuang[1], Peter H. McMurry[2], and Alon V. McCormick[1]

Abstract Correlations between concentrations of newly formed particles and sulfuric acid vapor have been analyzed for 13 nucleation events measured in various atmospheric environments. Models based on the assumption that nucleation rates ($J_{1\,nm}$) were equal to $K_{activation} \cdot [H_2SO_4]^1$ (the "activation" model) and $K_{kinetic} \cdot [H_2SO_4]^2$ (the "kinetic" model) were tested. Both models fit the measured data to some extent in all locations, but the quality of the fit and values of the pre-factors varied significantly from region to region.

Keywords New particle formation, sulfuric acid, kinetic, activation

Introduction

New particle formation by nucleation of gas phase species significantly influences the size distributions and number concentrations of atmospheric aerosol particles. These aerosol particles are believed to exert a considerable impact on global climate by affecting the earth's radiation balance directly through the scattering of solar radiation or indirectly through their role as cloud condensation nuclei.[1] The magnitudes of these effects are quite sensitive to the particle size distribution. Global climate models must accurately predict new particle formation in order to realistically capture aerosol climate effects.

New particle formation in the atmospheric boundary layer has been frequently observed in diverse locations including Boreal forests, European coastal environments, and rural and urban regions in North America. These observations indicate a close connection between H_2SO_4 vapor and new particle formation with boundary

[1]*Department of Chemical Engineering & Materials Science, University of Minnesota, 55455, Minneapolis, USA*

[2]*Department of Mechanical Engineering, University of Minnesota, 55455, Minneapolis, USA*

C.D. O'Dowd and P.E. Wagner (eds.), *Nucleation and Atmospheric Aerosols*, 57–61.

layer particle formation having a much weaker functional dependence on H_2SO_4 vapor concentration than predicted by binary nucleation involving H_2SO_4 and water.[2] Analysis of measurement campaigns from several European locations has led to the development of particle formation parameterizations that are able to quantitatively explain many features of the observed nucleation events and have subsequently been implemented in global climate models.[3]

The present study was motivated by similar correlations between the concentrations of H_2SO_4 vapor and newly formed particles observed during various measurement campaigns: MILAGRO (Megacity Initiative: Local and Global Research Observations), ACE 1 (First Aerosol Characterization Experiment), PEMT-A (Pacific Exploratory Mission Tropics – A),[4] ANARChE (Aerosol Nucleation and Real Time Characterization Experiment), and various field studies at Boulder, CO, Idaho Hill, CO, and Mauna Loa, HI. It is the goal of this study to analyze simultaneous measurements of vapor-phase H_2SO_4 and newly formed particles from a diverse set of atmospheric locations, develop geographically robust models for new particle formation, and analyze variability in modeled parameters as a function of location and other measured quantities.

Methods

Ultrafine particle concentrations were measured using the pulse height analysis (PHA) method for the Idaho Hill, Mauna Loa, PEMT-A, and ACE 1 measurement campaigns; the Nano Scanning Mobility Particle Sizer (Nano-SMPS) method was used for the MILAGRO, ANARChE, and Boulder measurement campaigns. H_2SO_4 vapor concentrations were measured with a chemical ionization mass spectrometer (CIMS).

Formation rates of 3 nm particles (J_{3nm}) were estimated from changes in the ultrafine particle concentrations between 3 and 6 nm (N_{3-6nm}) and particle growth rates from 1 to 3 nm (GR_{1-3nm}) according to the relation[2]:

$$J_{3\,nm} \cong \frac{N_{3-6\,nm}}{6\,nm - 3\,nm} \cdot GR_{1-3\,nm}. \tag{1}$$

N_{3-6nm} was obtained from size distribution measurements and GR_{1-3nm} was estimated from the observed time shift between increasing H_2SO_4 vapor and ultrafine particle concentrations, which is often interpreted as the time required for a hydrated H_2SO_4 vapor molecule of roughly 1 nm diameter to reach the lower detection limit of 3 nm.[5, 6] The appropriate time shift was selected by maximizing the correlation coefficient between N_{3-6nm} and either $[H_2SO_4]$ or $[H_2SO_4]^2$. The nucleation rate (J_{1nm}) was then estimated from J_{3nm} by accounting for the probability that a particle would grow from 1 to 3 nm before being scavenged by the preexisting aerosol according to the relation[5]:

$$J_{1\,nm} = J_{3\,nm} \cdot \exp\left[\frac{A_{Fuchs}}{GR_{1-3\,nm}} \cdot \Psi\right] \tag{2}$$

where A_{Fuchs} is the Fuchs surface area of the preexisting aerosol and ψ is a lumped parameter that is a function of temperature, aerosol density, initial (1 nm), and final (3 nm) particle diameter.

For each nucleation event, $J_{1\,nm}$ was compared with the corresponding temporal evolution of $[H_2SO_4]$ according to a simple power law expression of the form:

$$J_{1\,nm} = K_{region} \cdot [H_2SO_4]^n \tag{3}$$

where the pre-factor, K_{region}, is a parameter that is expected to vary with region and contains details about the nucleation process and the exponent, *n*, describes the molecularity of the nucleation mechanism. A least-squares analysis was then performed between $J_{1\,nm}$ and time-shifted $[H_2SO_4]$ according to Eq. (3) resulting in best fit values for K_{region} assuming either an activation ($n = 1$) or kinetic ($n = 2$) nucleation mechanism.

Results and Discussion

The correlation between $N_{3\text{–}6\,nm}$ and either time-shifted $[H_2SO_4]^1$ or $[H_2SO_4]^2$ was observed on a number of days during each of the measurement campaigns. A typical example of this correlation with time-shifted $[H_2SO_4]^1$ is shown in Figure 1.

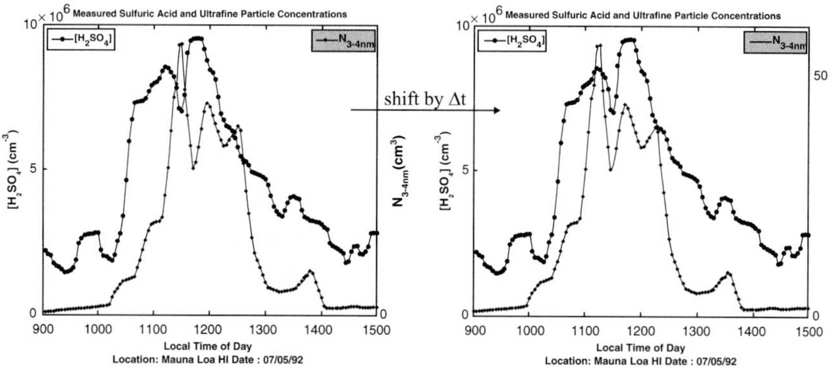

Figure 1 Number concentrations of 3–4 nm particles and sulfuric acid vapor during a nucleation event measured in Mauna Loa, Hawaii. The time shift between the two curves is indicated on the left plot

Table 1 Calculated quantities for correlation analysis

Day[a]	Δt (h)	$GR_{1-3\,nm}$ (nm/h)	R^b ($n = 1$)	R^c ($n = 2$)
Idaho Hill: 9/21/1993	1.15	2.17	0.929	0.946
Mauna Loa: 7/15/1992	0.30	8.33	0.847	0.875
PEMT-A: 9/23/1996	0.06	44.1	0.801	0.820
Boulder: 6/1/2004	0.45	5.56	0.846	0.888
Boulder: 6/2/2004	0.84	2.99	0.824	0.773
Boulder: 9/7/2004	0.51	4.86	0.859	0.846
Boulder: 9/8/2004	0.26	9.72	0.619	0.653
MIRAGE: 3/16/2006	0.13	19.4	0.931	0.880
MIRAGE: 3/22/2006	0.26	9.73	0.869	0.902
ANARChE: 7/31/2002	0.45	5.56	0.785	0.878
ANARChE: 8/01/2002	0.42	5.98	0.627	0.639
ANARChE: 8/10/2002	0.77	3.24	0.643	0.658

[a] ACE 1 data not included due to constant $[H_2SO_4]$ profile that accompanied measured particle burst

[b, c] R: correlation coefficient between $N_{3-6\,nm}$ and either $[H_2SO_4]^1$ or $[H_2SO_4]^2$

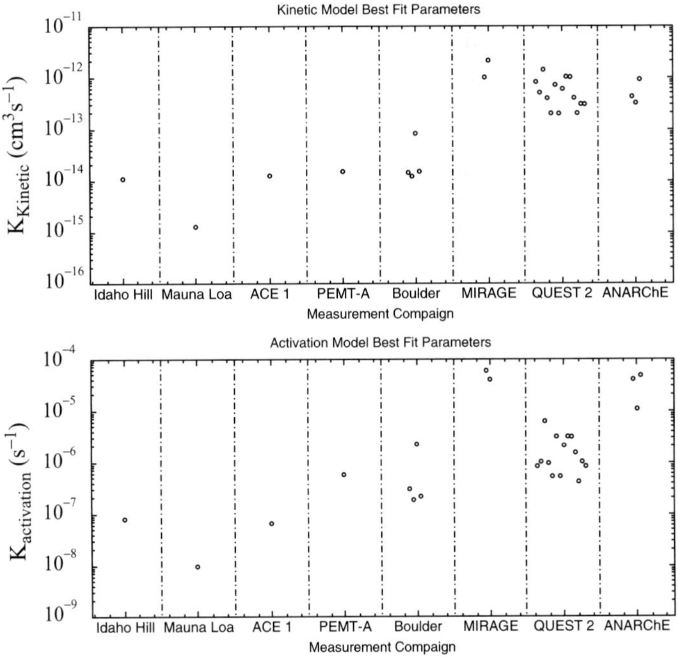

Figure 2 Values of activation and kinetic parameters determined from least-squares fits to experimental data. Analogous best-fit parameters from the QUEST 2 study are included for comparison[6]

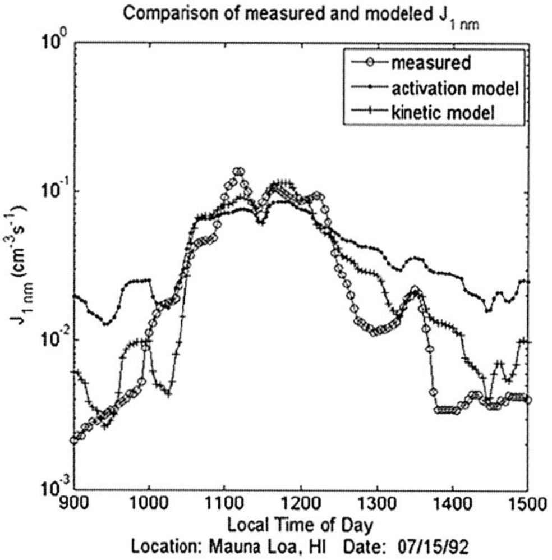

Figure 3 Nucleation rates estimated from measurements and modeled from $[H_2SO_4]$ assuming kinetic and activation models. Correlations between newly formed particles and sulfuric acid vapor have been analyzed for nucleation events in various atmospheric locations and have been shown to exhibit simple power law relationships, yielding best-fit parameter values that vary considerably with regional environment

The calculated time shifts, growth rates, and correlation coefficients for each analyzed nucleation event are detailed in Table 1.

The resulting best-fit values of the pre-factor ($K_{activation}$ for $n = 1$ and $K_{kinetic}$ for $n = 2$) are shown in Figure 2 where the best-fit parameters are seen to exhibit variability that spans three to four orders of magnitude over the entire data set and one order of magnitude over each region. A typical comparison of measured and modeled $J_{1\,nm}$ for the nucleation event in Mauna Loa is shown in Figure 3. For this particular event, assuming a kinetic mechanism for nucleation yielded modeled nucleation rates that were in better qualitative agreement with measured rates compared with nucleation rates calculated with an activation model.

Acknowledgments This work was supported in part by the IGERT Program of the National Science Foundation under Award Number DGE-0114372. Support was also provided by NSF Award Number ATM-0500674.

References

1. IPCC, *Climate Change 2001*, Cambridge: Cambridge University Press (2002).
2. Weber, R.J., et al., *Chem. Eng. Comm.*, **151**, 53–64 (1996).
3. Spracklen, D.V., et al., *Atmos. Chem. Phys.*, **5**, 5631–5648 (2006).
4. Clarke, A.D., et al., *Science*, **282**, 89–92 (1998).
5. Weber, R.J., et al., *J. Geophys. Res.*, **102(D4)**, 4375–4385 (1997).
6. Sihto, S.-L., et al., *Atmos. Chem. Phys.*, **6**, 4079–4091 (2006).

Evaluation of Surface Tension and Tolman Length as a Function of Droplet Radius from Experimental Nucleation Rate and Supersaturation Ratio

A.A. Onischuk[1], S.V. Vosel[1,2], P.A. Purtov[1], and A.M. Baklanov[1]

Abstract The critical embryo surface tension and Tolman length as a function of radius has determined from the experimental homogeneous nucleation rate as measured by this paper authors (Zn), and other researches (Hg, Mg, water, *n*-pentanol, *n*-nonane).

The thermodynamic description of small-scale systems exhibits peculiar features with respect to the macroscopic systems. The basis for thermodynamics of small particles has its origin in the Gibbs theory of capillarity. Instead of this real system, the theory describes an imaginary system composed of a small droplet of the homogeneous phase β and another homogeneous phase α separated by the so called surface of tension of radius R_S. The surface tension σ is attributed to this surface. The difference δ between the equimolar radius R_E and the radius R_S is called Tolman length. It is well known now that the surface tension is a strong function of radius for small droplets. Therefore, to describe the thermodynamics of nanoscale systems one should know both $\sigma(R_S)$ and the location of the surface of tension (δ). Both $\sigma(R_S)$ and δ can be evaluated from the experimentally measured homogeneous nucleation rate. To this aim two main shortcomings of the classical nucleation theory must be solved, that is, instead of the surface tension for the flat surface σ_∞ it should consider $\sigma(R_S)$; besides, replacement free energy correction factor is to be taken into account properly. The long-term discussion on the "Translation-Rotation Paradox in the Theory of Nucleation" resulted to the Reiss, Kegel, and Katz [1] (RKK) correction factor. Recently Nishioka and Kusaka [2] and Debenedetti and Reiss [3] have extended the Gibbs treatment to noncritical nucleus. This formalism results in a new expression for the reversible work W of noncritical embryo formation the extrema conditions for which give the Gibbs formula

$$W_{crit} = \frac{4\pi R_S^2 \sigma(R_S)}{3},$$

[1] *Institute of Chemical Kinetics and Combustion, Novosibirsk, 630090, Russia*

[2] *Institute of Mineralogy and Petrography, Novosibirsk, 630090, Russia*

C.D. O'Dowd and P.E. Wagner (eds.), *Nucleation and Atmospheric Aerosols*, 62–68.

where W_{crit} is the minimum work required to form a criticalembryo. Using the RKK correction factor and the theory of Nishioka and Kusaka and Debenedetti and Reiss the nucleation rate can be presented as [4]:

$$I \approx \frac{n_1^{Sat} S}{2\pi R_S} \sqrt{\frac{3lnS}{\rho\kappa}} \exp\left(-\frac{16\pi m^2}{3\rho^2 (lnS)^2} \left(\frac{\sigma(R_S)}{k_B T} \right)^3 \right), \tag{1}$$

where k_B is the Boltzmann constant, T is temperature, ρ (g/cm^3) denotes the density of the bulk phase β, m is mass of molecules, n_1^{sat} is saturated vapor concentration (cm^{-3}), S is supersaturation ratio, κ is isothermal compressibility of the liquid. Equation (1) is to be solved together with the Kelvin expression:

$$ln\, S = \frac{2\sigma(R_S) m}{k_B T \rho R_S}. \tag{2}$$

Thus, one may evaluate $\sigma(R_S)$ and R_S from the experimentally measured nucleation rate, supersaturation ratio, and temperature solving together Eqs. (1) and (2). Surface tension is connected with the Tolman length via the Gibbs-Tolman-Koenig (GTK) equation [5]:

$$\frac{d[\ln\sigma(R_S)]}{d[lnR_S]} = \frac{\frac{2\delta(R_S)}{R_S}\left[1 + \frac{\delta(R_S)}{R_S} + \frac{1}{3}\left(\frac{\delta(R_S)}{R_S}\right)^2\right]}{1 + \frac{2\delta(R_S)}{R_S}\left[1 + \frac{\delta(R_S)}{R_S} + \frac{1}{3}\left(\frac{\delta(R_S)}{R_S}\right)^2\right]}. \tag{3}$$

This equation was solved approximately in [4] in assumption that the Tolman length is governed by the formula [6]:

$$\delta(R_S) \approx \frac{\chi}{R_S} + \delta_\infty, \tag{4}$$

Thus, the solution of the GTK equation together with Eq. (4) gives [4]:

$$\frac{\delta(2R_S)}{R_S} \approx -\frac{ln\frac{\sigma(R_S)}{\sigma_\infty}}{2 + \beta ln\frac{\sigma(R_S)}{\sigma_\infty}}. \tag{5}$$

where β is constant close to unity, which can be determined from the experimental data on homogeneous nucleation.

In this paper we compare our experimental data on Zn vapor homogeneous nucleation with the literature nucleation data for bivalent metals as well as water, *n*-pentanol and *n*-nonane.

Figure 1 compares this paper authors data on Zn vapor nucleation in a flow diffusion chamber with the literature data for Hg [7] and Mg [8]. Solving Eqs. (1) and (2) for fixed nucleation rate and temperature we can determine the critical nucleus surface tension as well as radius R_S for each point in Figure 1. The evaluated ratios σ/σ_∞ are presented as functions of R_S in Figure 2. The bivalent metals Zn, Hg, Mg demonstrate $\sigma/\sigma_\infty < 1$. Moreover within the experimental accuracy the surface tension is independent on R_S for these elements in the studied range of critical nucleus radius. Using Eq. (5), one can determine $\delta(2R_S)$ directly from $\sigma(R_S)/\sigma_\infty$ presented in Figure 2. On the other hand, taking the derivative $\frac{d}{d(lnR_S)}\left(ln\frac{\sigma(R_S)}{\sigma_\infty}\right)$ from Figure 2 one can determine $\delta(R_S)$ using the GTK Eq. (3). The values of δ are presented in Figure 3. The determined quantities δ were fitted by Eq. (4) (solid lines in Figure 3). Then the evaluated functions $\delta(R_S)$ were used to reconstruct the σ/σ_∞ versus R_S dependence by the numerical integration of Eq. (3) (see solid lines in

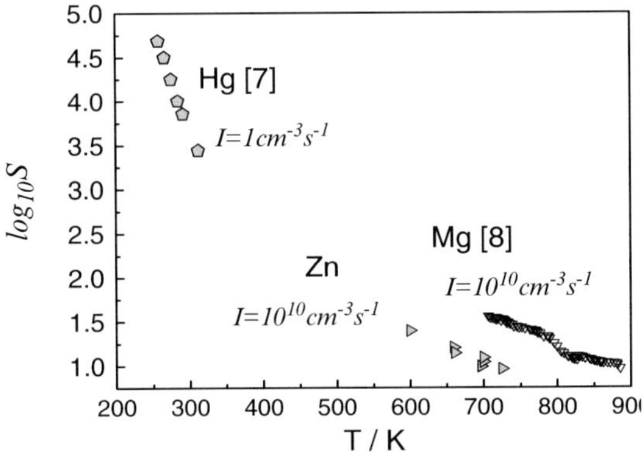

Figure 1 Supersaturation versus nucleation temperature

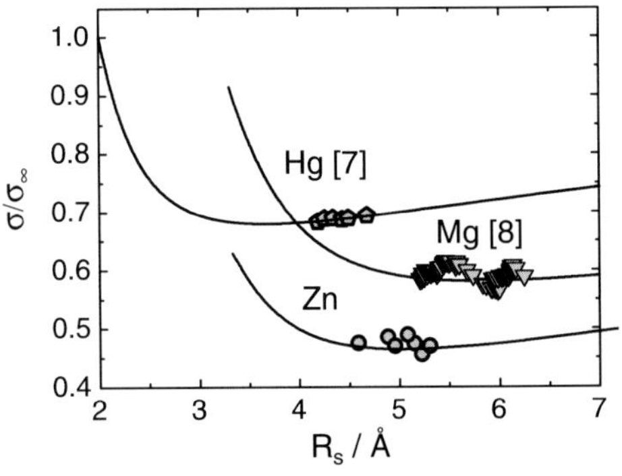

Figure 2 Droplet surface tension σ to flat interface surface tension σ_∞ ratio versus radius R_S of the surface of tension as determined from the nucleation data. Solid lines are determined by the numerical integration of Eq. (3) using $\delta(R_S)$ (**solid curves** from Figure 3)

Figure 3 Tolman length versus radius of the surface of tension as determined from σ/σ_∞ data (Figure 2) using Eq. (5): open symbols, and Eq. (3): semifilled symbols. Solid lines are governed by Eq. (4)

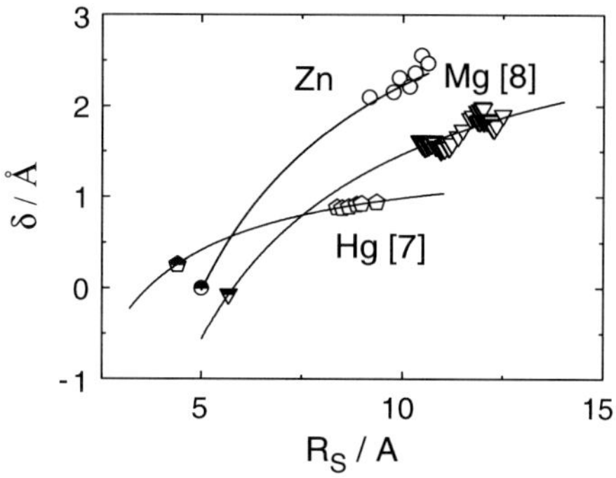

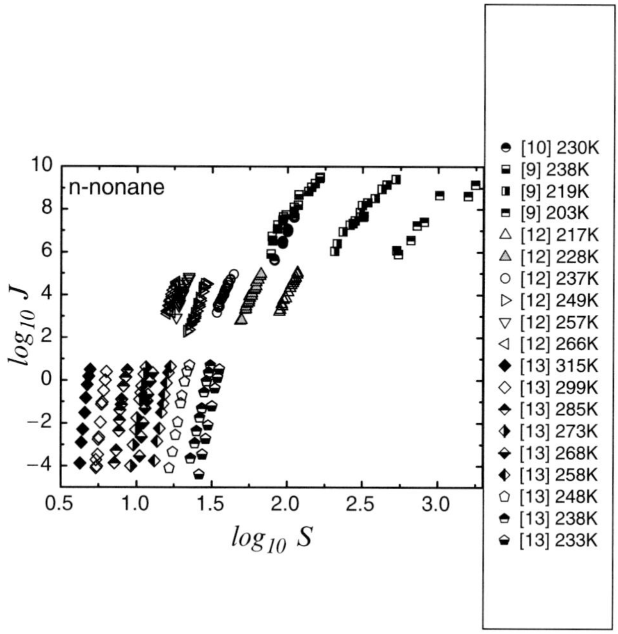

Figure 4 Nucleation rate (cm^{-3} s^{-1}) versus supersaturation ratio

Figure 2). The same approach was applied to *n*-nonane, water, and *n*-pentanol. These substances were studied widely by different research groups (see refs. for *n*-nonane [9–13], water [14–20], and *n*-pentanol [21–29]) which have measured the nucleation rate as a function of temperature and supersaturation ratio. A comparison of these groups data for *n*-nonane is given in Figure 4; the nucleus surface tension as a function of radius R_S for different temperatures is shown in Figure 5. For water σ/σ_∞ and δ versus R_S are illustrated in Figures 5 and 6 as examples. Figure 7 compares the temperature dependensies of σ for *n*-nonane, *n*-pentanol, and water.

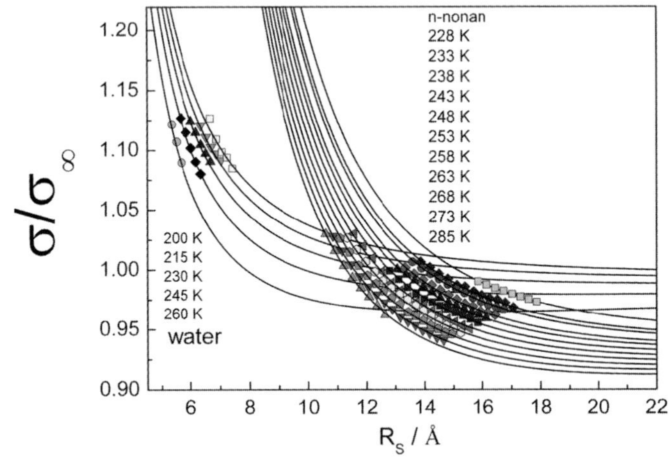

Figure 5 Droplet surface tension to flat interface surface tension ratio versus radius R_S of the surface of tension as determined from the nucleation data. Solid lines are determined by the numerical integration of Eq. (3) using evaluated $\delta(R_S)$ functions

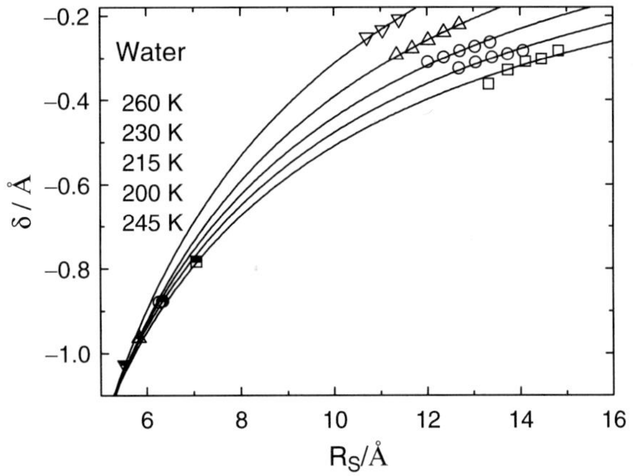

Figure 6 Tolman length versus radius of the surface of tension as determined from σ/σ_8 data (Figure 5) using Eq. (5): open symbols, and Eq. (3): semifilled symbols. Solid lines are governed by Eq. (4)

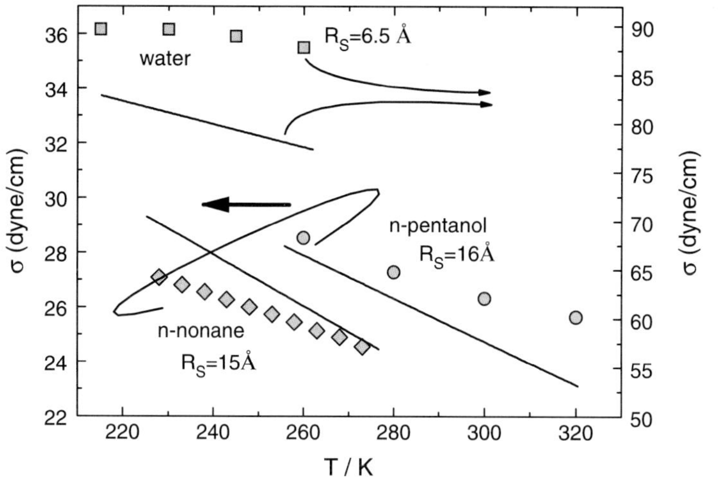

Figure 7 Droplet surface tension versus temperature. Flat interface surface tension is given for comparison (**solid lines**)

Acknowledgment Financial support for this work was provided by RFBR 05-03-90576-NSC_a, grant of SD RAS No 78, NSC_Taiwan-RFBR No. 94WFA0600016 _Contract No. RP05E15.

References

1. Reiss, H., Kegel, W.K., and Katz, J.L., *J. Phys. Chem., A* **102**, 8548–8555 (1998).
2. Nishioka, K. and Kusaka, I., *J. Chem. Phys.*, **96**, 5370–5376 (1992).
3. Debenedetti, P.G. and Reiss, H., *J. Chem. Phys.*, **108**, 5498–5505 (1998).
4. Onischuk, A.A.,_ Purtov, P.A., Baklanov, A.M., Karasev, V.V., Vosel, S.V., *J. Chem. Phys.*, **124**, 014506-1–014506-13 (2006).
5. Nishioka, K., Tomino, H., Kusaka, I., and Takai, T., *Phys. Rev., A*, **39**, 772–776 (1989).
6. Bartell, L.S., *J. Phys. Chem. B*, **105**, 11615–11618 (2001).
7. Uchtmann, H., Rademann, K., and Hensel, F., Annalen der Physic, **48**, S207–214 (1991).
8. Ferguson, F.T., Nuth III, J.A., and Lilleleht, L.U., *J. Chem. Phys.*, **104**, 3205–3210 (1996).
9. Wagner, P.E. and Strey R., *J. Chem. Phys.*, **80**, 5266–5275 (1984).
10. Viisanen, Y., Wagner, P.E., and Strey, R., *J. Chem. Phys.*, **108**, 4257 – 4266 (1998).
11. Garman, A.I., Napari, I., Winkler, P.M., Vehkamäki, H., Wagner, P. E., Strey, R., Viisanen, Y., and Kulmala, M., *J. Chem. Phys.*, **123**, 244502-1–244502-12 (2005).
12. Adams, G.W., Schmitt, J.L., and Zalabsky, R.A., *J. Chem. Phys.*, **81**, 5074 – 5078 (1984).
13. Hung, C.-H., Krosnopoler, M.J., Katz, J.L., *J. Chem. Phys.*, **90**, 1856 – 1865 (1989).
14. Miller, R.C., Anderson, R.J., Kassner, J.L., and Hagen, D.E., *J. Chem. Phys.*, **78**, 3204–3211 (1983).
15. Wölk, J. and Strey, R., *J. Phys. Chem.*, **105**, 11683–11701 (2001).
16. Heath, C.H., Streletzky, K.A., and Wyslouzil, B.E., *J. Chem. Phys.*, **118**, 5465–5473 (2003).
17. Peters, F. and Paikert, B., Experiments in Fluids, **7**, 521–530 (1989).

18. Schmitt, J.L., Van Brunt, K., and Doster, G.J., The homogeneous nucleation of water, in: *Nucleation and Atmospheric Aerosols 2000*, edited by B.N. Hale and M. Kulmala, Melville, NY: AIP , pp. 31–34 (2000).
19. Holten, V., Labetski, D.G., and van Dongen, M.E.H., *J. Chem. Phys.*, **123**, 104505-1–104505-9 (2005).
20. Mikheev, V.B., Irving, P.M., Laulainen, N.S., Barlow, S.E., and Pervukhin, V.V., *J. Chem. Phys.*, **116**, 10772–10786 (2002).
21. Iland, K., Wedekind, J., Wolk, J., Wagner, P.E., and Strey, R., *J. Chem. Phys.*, **121**, 12259–12264 (2004).
22. Luijten C.C.M., Baas O.D.E., and van Dongen M.E.H., *J. Chem. Phys.*, **106**, 4152–4156 (1997).
23. Gharibeh, M., Kim, M., Y., Dieregsweiler, U., Wyslouzil, B.E., Ghosh, D., and Strey, R., *J. Chem. Phys.*, **122**, 094512-1–094512-9 (2005).
24. Lihavinen, H., Viisanen, Y. and Kulmala, M., *J.Chem.Phys.*, **114**, 10031–10038 (2001).
25. Hruby, J., Viisanen, and Strey, Y.R., *J.Chem.Phys.*, **104**, 5181–5187 (1996).
26. Strey, R., Wagner, P.E., and Schmeling, T., *J.Chem.Phys.*, **84**, 2325–2335 (1986).
27. Schmitt, J.L., Doster, G.J., *J. Chem. Phys.*, **116**, 1976–1978 (2002).
28. Biet, T., and Strey, R., Homogeneous nucleation of *n*-pentanol and droplet growth: a quantitative comparison of experiment and theory, in: *Nucleation and Atmospheric Aerosols: 15-th Int.'l Conf.*, edited by B.N. Hale and M. Kulmala, Rolla, Missouri, **534**, pp. 249–252 (2000).
29. Anisimov, M.P. and Hopke, P.K., *J. Chem. Phys.*, **113**, 1971–1975 (2000).

Atmospheric H_2SO_4/H_2O Particle Formation: Mechanistic Investigations

T. Berndt, O. Böge, and F. Stratmann

Abstract The formation of H_2SO_4/H_2O particles has been investigated in a laboratory study using a flow tube operated at atmospheric pressure. The needed H_2SO_4 vapour was produced in three different ways: (i) in situ formation starting from OH + SO_2, (ii) reaction of SO_3 with water vapour, (iii) taking H_2SO_4 vapour from a liquid reservoir. The findings are discussed with respect to the mechanism of the formation of particle's precursors.

Keywords Atmospheric nucleation, H_2SO_4, mechanism

Introduction

Atmospheric particles have a strong impact on the Earth's radiation budget due to their radiative properties and the fact that they can act as cloud condensation nuclei. Field measurements at ground level show atmospheric nucleation events for H_2SO_4 concentrations of ~10^7 molecule cm^{-3} [1]. Despite intensive research activities in the last decade, the mechanism leading to new particles has not been unambiguously revealed yet.

In a previous investigation from our laboratory, experimental evidence for the formation of new particles in the system H_2SO_4/H_2O under near-atmospheric conditions with H_2SO_4 concentrations of ~10^7 molecule cm^{-3} was found [2]. Here, H_2SO_4 was produced in situ via the reaction OH + SO_2 in the presence of water vapour. Similar observations are reported from investigations in the 590 m^3 Calspan chamber [3] as well as from another flow-tube experiment [4].

In contrast, taking H_2SO_4 from a liquid sample reservoir ~10^{10} molecule cm^{-3} of H_2SO_4 are needed for significant new particle formation [5]. This observation is roughly in line with the prediction from binary nucleation theory H_2SO_4/H_2O [6].

Leibniz-Institut für Troposphärenforschung e.V., Permoserstr. 15, 04318 Leipzig, Germany

C.D. O'Dowd and P.E. Wagner (eds.), *Nucleation and Atmospheric Aerosols*, 69–72.

Subject of this study are mechanistic investigations on H_2SO_4/H_2O particle formation trying to explain the different threshold H_2SO_4 concentrations for nucleation, ~10^7 molecule cm^{-3} versus ~10^{10} molecule cm^{-3}.

Experimental

The experiments have been performed in the *IfT*-LFT (Institute for Tropospheric Research – Laminar Flow Tube; i.d. 8 cm; length 505 cm) at atmospheric pressure and 293 ± 0.5K using synthetic air as the carrier gas. Particle size distributions (dp > 2 nm) were determined using a differential mobility particle sizer (DMPS) consisting of a short Vienna-type differential mobility analyser (DMA) and an ultrafine particle counter (UCPC, TSI 3025). Total particle numbers were measured by means of different types of UCPC's to be directly attached at the outlet of the *IfT*-LFT. In the case of in situ H_2SO_4 formation via OH + SO_2, the needed OH radicals were formed either by O_3 photolysis in the presence of water vapour or by ozonolysis of *t*-butene (dark reaction), cf. [2]. Photolysis experiments were conducted using furan or CO for OH radical titration, i.e., conditions in the presence or absence of organic compound in the carrier gas. In experiments starting with SO_3, this species was produced outside the *IfT*-LFT in a pre-reactor oxidizing SO_2 in a catalytic reaction on a Pt surface at 525°C. The conversion of SO_2 to SO_3 was followed by online UV spectroscopy. For experiments using H_2SO_4 vapour from a liquid sample, H_2SO_4 concentrations were measured at the outlet of the saturator by means of a denuder system with subsequent analysis of SO_4^{2-} ions by ion chromatography.

Discussion

Gas-phase oxidation of SO_2 in the atmosphere is initiated by the attack of OH radicals. The following reaction sequence is currently accepted leading finally to H_2SO_4 vapour.

$$OH + SO_2 \rightarrow HOSO_2 \tag{1}$$

$$HOSO_2 + O_2 \rightarrow SO_3 + HO_2 \tag{2}$$

$$SO_3 + 2H_2O \rightarrow H_2SO_4 + H_2O \tag{3}$$

Produced $HOSO_2$ radicals from the primary reaction of OH radicals with SO_2 via pathway (1) react in a very fast consecutive reaction with O_2 leading to SO_3. For atmospheric reactant levels (OH: 2 • 10^6; SO_2: 10^{11}; O_2: 5 • 10^{18}; all in molecule cm^{-3}) a steady state $HOSO_2$ concentration of 0.1 molecule cm^{-3} follows assuming that no other reactions of $HOSO_2$ can compete with pathway (2). SO_3 reacts with two water molecules or a water dimer producing finally H_2SO_4. This fast pathway results in the very low steady state SO_3 concentrations of ·1 molecule cm^{-3} in the atmosphere. And also here, other pathways for SO_3 conversion have to be of less importance. That means that each attacked SO_2 molecule from pathway (1) is transformed to H_2SO_4 in the atmosphere according to the scheme given above.

It is to be noted that in competition to pathway (2) also the addition of O_2 on $HOSO_2$ can take place leading to $HOSO_2O_2$.

$$HOSO_2 + O_2 \rightarrow HOSO_2O_2 \tag{2a}$$

An assessment of the importance of pathway (2a) for the conversion of $HOSO_2$ seems to be difficult because experimental data in the literature are sparse. Stockwell and Calvert [7] derived from a chamber experiment that less than 20% of $HOSO_2$ reacted via pathway (2a) and more than 80% via pathway (2). Howard [8] measured $HOSO_2$ and HO_2 simultaneously in a flow-tube experiment and found a lower limit of 70% for pathway (2).

If pathway (2a) plays a significant role for the fate of $HOSO_2$ radicals, less than one molecule H_2SO_4 per attacked SO_2 is formed and $HOSO_2O_2$ can initiate other reaction pathways not considered so far.

Figure 1 shows the experimental results using the different approaches for H_2SO_4 formation.

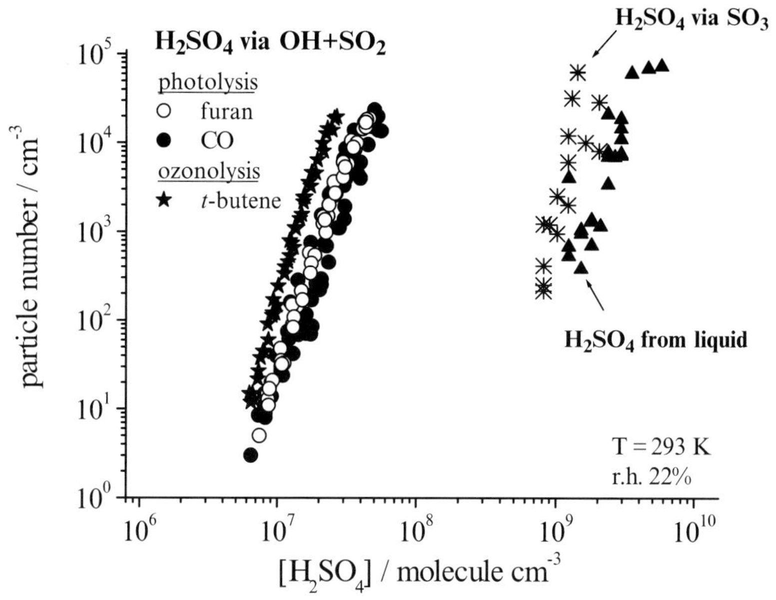

Figure 1 Measured particle numbers versus H_2SO_4 concentration for different formation pathways

Under conditions starting from OH + SO_2, all three series of measurements consistently demonstrate that new particle formation starts for H_2SO_4 concentrations of ~10^7 molecule cm^{-3}. Here, the small differences can be explained by uncertainties arising from different approaches for H_2SO_4 determination. On the other hand, using SO_3 or starting with H_2SO_4 from a liquid sample, H_2SO_4 concentrations of ~10^9 molecule cm^{-3} are needed for particle formation.

From this behaviour it can be speculated that in the course of SO_2 conversion to SO_3 via pathways (1) and (2) other, additional steps may be important, for instance pathway (2a). The products of this other steps could trigger the particle formation observed for conditions starting from OH + SO_2.

References

1. Weber, R.J., McMurry, P.H., Mauldin III, R.L., Tanner, D.J., Eisele, F.L., Clarke, A.D., and Kapustin, V.N., *Geophys. Res. Lett.*, **26**, 307–310 (1999).
2. Berndt, T., Böge, O., Stratmann, F., Heintzenberg, J., and Kulmala, M., *Science*, **307**, 698–700 (2005).
3. Verheggen, B., Determination of particle nucleation and growth rates from measured aerosol size distributions, *Thesis*, Toronto: York University (2004).
4. Friend, J.P., Barnes, R.A., and Vasta, R.M., *J. Phys. Chem.*, **84**, 2423–2436 (1980).
5. Ball, S. M., Hanson, D.R., Eisele, F.L., and McMurry, P.H., *J. Geophys. Res.* **104**(D19), 23709–23718, 10.1029/1999JD900411 (1999).
6. Kulmala, M., Laaksonen, A., and Pirjola, L., *J. Geophys. Res.*, **103**, 8301–8307 (1998).
7. Stockwell, W.R. and Calvert, J.G., *Atmos. Environ.* **17**, 2231–2235 (1983).
8. Howard, C.J. Communication at the 17th Symposium on Free Radicals, Granby, CO, 1985. Ref.(11) in: Martin, D., Jourdain, J.L., and Le Bras, G., *J. Phys. Chem.*, **90**, 4143–4147 (1986).

The Role of Ions in Condensation Particle Counting for Particle Diameters Below 2 nm

P.M. Winkler, A. Vrtala, and P.E. Wagner

Abstract Condensation particle counting for particle diameters below 2 nm may be influenced by the simultaneous presence of ions depending on the experimental procedure. Neutralization of electrostatically classified aerosols by means of a ^{241}Am bipolar diffusion charger causes comparatively high number concentrations of positive and negative ions which enter the measuring chamber simultaneously to the particle species of interest. In this paper we illustrate how the presence of ions influences the measured particle number concentration and how a removal of ions allows to measure number concentration of seed particles correctly.

Keywords Condensation particle counting, heterogeneous nucleation, bipolar diffusion charger

Introduction

Condensation particle counting is among others one of the most frequently used methods in aerosol physics. In condensation particle counters (CPCs) the number concentration of heterogeneously nucleated growing droplets in the micrometer-size range is optically determined. The measured droplet number concentration provides direct information on the number concentration of condensation nuclei of much smaller sizes which would otherwise not be optically detectable.[1] Depending on the size of the seed particles, corresponding onset vapor saturation ratios are required to heterogeneously nucleate the droplets in the measuring chamber. The Kelvin equation is often used to relate droplet curvature and onset saturation ratio. With decreasing seed particle diameter, increasing saturation ratios are needed in order to achieve heterogeneous nucleation. Thus, the ultimate limit for condensation particle counting is the onset of homogeneous nucleation. Accordingly, the lower limit for the cutoff diameter has been expected around 2 nm. However, in recent

Fakultät für Physik, Universität Wien, Boltzmanngasse 5, A-1090 Wien, Austria

C.D. O'Dowd and P.E. Wagner (eds.), *Nucleation and Atmospheric Aerosols*, 73–77.

experiments we have observed heterogeneous nucleation already well below the Kelvin prediction.[2,3] Therefore considerably smaller cutoff diameters are to be expected for condensation particle counting.

For studies of heterogeneous nucleation at saturation ratios close to the onset of homogeneous nucleation well-defined, clean experimental conditions are essential as any impurity may already act as condensation nucleus and thus influence the measuring signal. Besides, in the size range of only few nanometers, charge effects become increasingly significant thereby causing ion-induced nucleation. In this paper we will describe how ions obtained from a ^{241}Am bipolar diffusion charger influence the measured number concentration and how this influence can be avoided by proper removal of ions.

Experimental Method

The experimental system mainly consists of an aerosol generation unit, a vapor generation unit, and an expansion chamber for measuring droplet number concentration. A schematic diagram of the experimental arrangement can be seen in Figure 1. Vapor generation is performed by injecting a well-defined liquid beam from a syringe pump into a heating unit where the liquid is vaporized and mixed with a highly dried

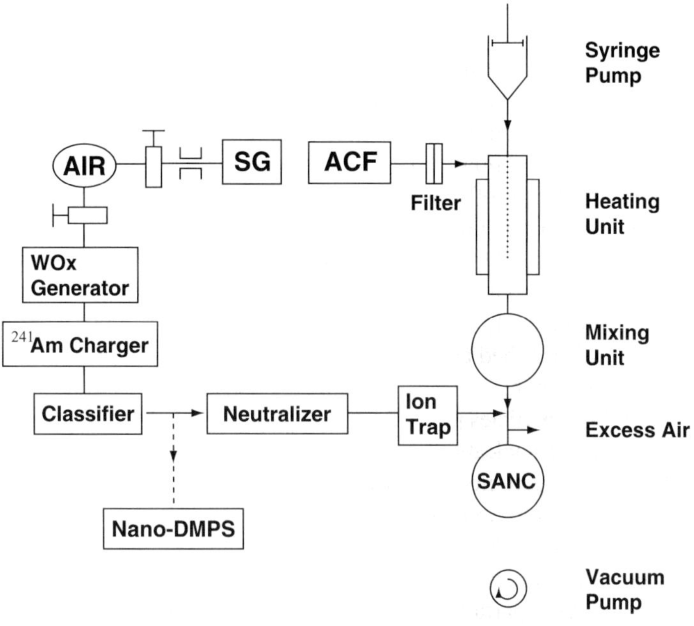

Figure 1 Experimental arrangement for measuring total particle number concentration of neutral particles with diameters below 2 nm. SG: silicagel gas dryer, ACF: active carbon filter

and purified well-defined air flow acting as carrier gas flow. In the present study *n*-propanol was used as the working fluid. Aerosol generation is done by means of a commercially available tungsten oxide generator (Grimm Aerosoltechnik, Modell 7860). WO_x particles are charged in a ^{241}Am bipolar diffusion charger and subsequently classified in a newly developed Differential Mobility Analyzer (DMA). The monodisperse aerosol fraction leaving the DMA is then neutralized by means of ^{241}Am bipolar diffusion charger in order to achieve charge equilibrium. Thereby, the radioactive material produces ions which equilibrate the preexisting charge on the WO_x particles. As the number of ions by far exceeds the number of WO_x particles they will partly remain in the system before being gradually removed by diffusion and recombination. In order to accelerate the removal an ion trap has been applied where ions can be removed by supplying sufficient voltage to the ion trap. Finally, the aerosol is mixed with the vapor and passed into the expansion chamber (SANC). Growth rates and number densities of the condensing droplets were measured using the CAMS detection method.[4] The size distribution of the seed particles was determined by means of a newly developed nano-DMPS system.

Results and Discussion

Using the SANC we have measured the number concentration of homogeneously and heterogeneously nucleated droplets at various vapor saturation ratios. For heterogeneous nucleation measurements, WO_x particles of well-defined sizes as well as bipolar ions have been used as condensation nuclei. Depending on the classifier and ion trap settings heterogeneous, ion-induced, or homogeneous nucleation has been detected. As can be seen from Figure 2, by setting the classifier voltage to zero and supplying 100 V to the ion trap, all WO_x particles and ions are removed from the aerosol flow and a steep increase in droplet number concentration beyond saturation ratios of 3.6 is observed, which is related to the onset of homogeneous nucleation. Good agreement with previously measured data by Wedekind et al.[5] is found. On the other hand, by turning off the ion trap, i.e., setting the voltage to 0 V, bipolar ions enter the measuring chamber and are clearly nucleated to growth already at saturation ratios well below the homogeneous limit. Clearly, a plateau is reached at saturation ratios beyond $S = 3.4$ where all ions have been nucleated to growth.

Letting now WO_x particles pass the classifier by setting a corresponding classifier voltage heterogeneous nucleation can be observed. As can be seen from Figure 3, for the case of 2 nm WO_x particles an increase in droplet number concentration is found starting already at saturation ratios of $S = 2.4$. If the ion trap is turned off the simultaneous presence of ions causes a further increase in particle number concentration at saturation ratios above $S = 3$. In this case the total number concentration of WO_x particles cannot be measured anymore and consequently determination of heterogeneous nucleation probabilities is impossible. However, if the ion trap is turned on, a plateau is reached with a mean particle number

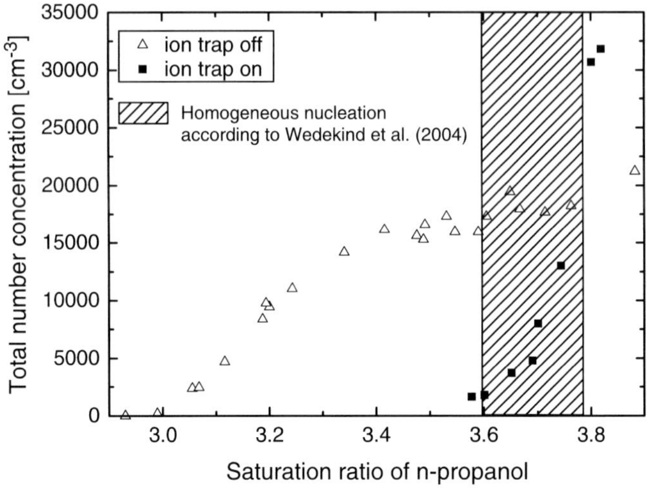

Figure 2 Total ion number concentration if no voltage is supplied to classifier and ion trap. Homogeneous nucleation is observed when ions are removed in the ion trap

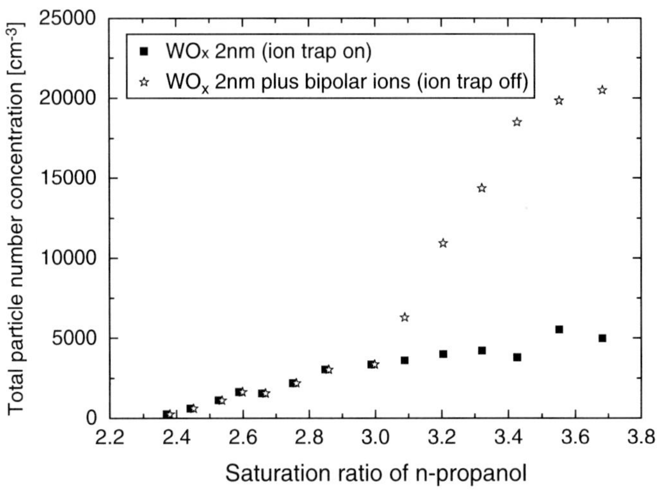

Figure 3 Total particle number concentration depending on experimental conditions. Ions are removed by supplying 100 V to the ion trap. See text for further details

concentration of approximately 4,000–5,000/cc. For that reason ions obtained from a bipolar diffusion charger need to be removed properly when measuring heterogeneous nucleation on neutral particles with diameters below 2–2.5 nm.

Acknowledgments This work was supported by the Austrian Science Foundation (FWF, Project No. P16958-N02). The authors want to thank G. W. Steiner and G. P. Reischl for valuable discussion.

References

1. Stolzenburg, M.R. and McMurry, P.H., *Aerosol Sci. Technol.* **14**, 48–65 (1991).
2. Winkler, P.M., Steiner, G.W., Reischl, G.P., Vrtala, A., Wagner, P.E., Gaman, A.I., Vehkamäki, H., and Kulmala, M., Observation of nucleation of organic vapours by nanoparticles already at particle sizes well below the Kelvin prediction, in: *Proceedings of 7th Int. Aerosol Conf.*, edited by P. Biswas et al., St. Paul, AAAR (2006), pp. 1615–1616.
3. Hienola, A.I., Winkler, P.M., Wagner, P.E., Vehkamäki, H., Lauri, A., Napari, I., and Kulmala, M., *J. Chem. Phys.*, **126**, 094705 (2007).
4. Wagner, P.E., *J. Colloid Interface Sci*, **105**, 456–467 (1985).
5. Wedekind, J., Iland, K., Wagner, P.E., and Strey, R., Homogeneous nucleation of 1-alcohol vapors, in: *Nucleation and Atmospheric Aerosols 2004*, edited by M. Kasahara and M. Kulmala, 16th ICNAA Conference proceedings, Kyoto: Kyoto University Press (2004), pp. 49–52.

n-Nonane Nucleation in the Presence of Carbon Dioxide

D.G. Labetski[1], J. Hrubý[2], and M.E.H. van Dongen[1]

Abstract An extensive study of *n*-nonane nucleation in methane/carbon dioxide has been performed. Nucleation rate data was obtained for methane/carbon dioxide carrier gas mixtures with a 0.01 molar fraction of carbon dioxide at three different conditions: 236 K and 10 bar, 240 K and 25 bar, 240 K and 40 bar. The obtained data is compared with the available data for the methane/*n*-nonane mixture. Furthermore, the nucleation rates of *n*-nonane in methane/carbon dioxide mixtures were determined at 235 K and 10 bar for initial molar fractions of carbon dioxide up to 0.3. Two series of such experiments have been performed: (1) the molar fraction of carbon dioxide is fixed, $y_{co2} = 0.25$, and the *n*-nonane fraction is varied; (2) the molar fraction of *n*-nonane is constant, $y_{c9} = 230$ ppm, and the carbon dioxide fraction is varied. A comparison with the methane/*n*-nonane data shows that carbon dioxide does influence nucleation when its molar concentration in the mixture exceeds 0.25.

Keywords Homogeneous ternary nucleation, natural gas, methane, carbon dioxide, *n*-nonane

Introduction

There is vast amount of natural gas reserves which have not yet been developed because they contain a high concentration of carbon dioxide: up to 0.2–0.3 molar fraction. The availability of a cheap and robust technology to purify natural gas with a high concentration of carbon dioxide will open these important energy resources for economical utilization. Controlled droplet generation and droplet

[1]*Department of Applied Physics, Eindhoven University of Technology, P.O.Box 513, 5600 MB, Eindhoven, The Netherlands, e-mail: D.G.Labetski@tue.nl, M.E.H.v.Dongen@tue.nl*

[2]*Department of Thermodynamics, Institute of Thermomechanics AS CR, Dolejskova 5,CZ-18200 Prague 8, Czech Republic, e-mail: Hruby@it.cas.cz*

C.D. O'Dowd and P.E. Wagner (eds.), *Nucleation and Atmospheric Aerosols*, 78–82.

growth is a very important technology for separation of methane from carbon dioxide or other gases. This technology is based on the fact that most gases have a higher critical temperature than methane. By forcing their nucleation and droplet growth with a consequent removal of droplets, the natural gas can be purified.

Homogeneous nucleation plays a key role in the formation of droplets and is determinate for mean droplet size and droplet number density. This report describes a continuation of research at TU/e, aimed at a quantitative and systematic study of nucleation rates in natural gas-related mixtures in a wide window of thermodynamic state variables.[1–4] The present study focuses on nucleation in mixtures which contain one heavy alkane vapor (*n*-nonane) and two noncondensable components, methane and carbon dioxide. It is expected that at relatively high pressures these noncondensable components will affect the nucleation process.

Experimental Setup

Nucleation has been studied with an expansion wave tube. The setup is described in detail by Looijmans and van Dongen,[5] Luijten et al.,[3] and Peeters et al.[4]

In this tube, the nucleation pulse principle is applied to split the nucleation and droplet growth processes in time. The gas–vapor mixture under study is adiabatically expanded, which brings the mixture into a supersaturated state, such that nucleation occurs. After a short period of time, the vapor supersaturation is slightly decreased by re-compression. Nucleation is effectively stopped, but droplet growth remains possible, because the vapor is still supersaturated. The droplets are detected by optical means using a combination of Mie scattering techniques. Droplet number densities, nucleation rates, and the thermodynamic state variables have been measured. The gas–vapor mixtures have been prepared in the mixture preparation part of the setup with a relatively high accuracy. The original setup has been modified to allow the preparation of ternary mixtures. The dry methane flow passes two saturators, which are half filled with liquid *n*-nonane, at controlled temperature and pressure. The output gas–vapor flow is diluted with the dry methane and methane/carbon dioxide gas flows to prepare a ternary methane/carbon dioxide/*n*-nonane mixture.

Results and Dicussions

We have restricted our study to carbon dioxide concentrations for which it remains subsaturated. First, experiments were performed with methane/carbon dioxide carrier gas mixtures with a 0.01 molar fraction of carbon dioxide at three different conditions: 236 K and 10 bar, 240 K and 25 bar, 240 K and 40 bar. Second, the nucleation rates of *n*-nonane in methane/carbon dioxide mixtures were determined at 235 K and 10 bar for initial molar fractions of carbon dioxide varying up to 0.3. Two series of such experiments have been performed: (1) the molar fraction of carbon dioxide is fixed, $y_{co2}=0.2$, and the n-nonane fraction is varied; (2) the molar fraction of n-nonane is constant, $y_{c9}=230$ ppm, and the carbon dioxide fraction is varied.

The new data on n-nonane nucleation in a methane/carbon dioxide (0.01) mixture are shown as nucleation isotherms in Figure 1 (filled circles). The actual temperature and pressure for a particular data point can differ from the average values depicted in the figure within 0.9 K and 0.3 bar intervals. The supersaturation of n-nonane S_{c9} is calculated as $S_{c9}=y_{c9}/y_{c9}^{eq}$ and the equilibrium molar fraction of n-nonane y_{c9}^{eq} is a result of a flash calculation with the NIST database program.[6] The first conclusion from Figure 1 is that with pressure increase the nucleation isotherms shift to lower supersaturations, the same trend as observed for methane/n-nonane. Also, the slopes of the nucleation isotherms are decreasing with pressure increase. According to the nucleation theorem,[7] this indicates that the number of n-nonane molecules in the critical cluster is decreasing. Because of the large scatter of data at 40 bar, it is difficult to define accurately the slope of the isotherm and to obtain the exact number of n-nonane molecules. This large scatter of data is caused by the scatter in the nucleation temperatures. In the experiments at lower pressures, 25 and 10 bar, the temperature scatter is lower.

The n-nonane nucleation data for methane/carbon dioxide(0.01) is compared with the methane/n-nonane data (open circles) obtained by Luijten[3] and by Peeters et al.[4] in Figure 1. The nucleation isotherms for both datasets at 10 and 25 bar almost coincide. At 40 bar the nucleation rates in methane are systematically higher than in methane/carbon dioxide. Also, it should be noted that the addition of a 0.01 molar fraction of carbon dioxide to the methane carrier gas almost does not change

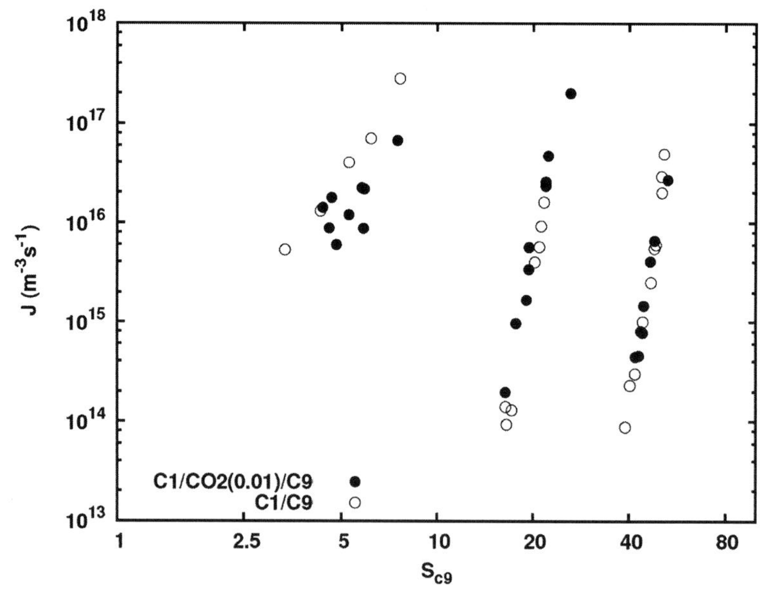

Figure 1 n-Nonane nucleation rate in methane/carbon dioxide (0.01) and in methane as functions of n-nonane supersaturation

the equilibrium *n*-nonane molar fraction y_{c9}^{eq} in the gas phase. For example, at 240 K and 40 bar the equilibrium molar fraction y_{c9}^{eq} is 18.0 ppm in methane/carbon dioxide(0.01) and $y_{c9}^{eq} = 17.9$ ppm in methane. As a result, the data shown in Figure 1 are obtained at the same initial molar fractions of *n*-nonane. Furthermore, phase equilibrium calculations show that a relatively small amount of carbon dioxide will dissolve in liquid *n*-nonane. For example, at 240 K and 40 bar the molar fraction of carbon dioxide in the liquid phase is approximately 0.011. The same trend can be expected for the critical cluster composition. Certainly, even a rather small amount of carbon dioxide in the cluster can influence the energy of cluster formation and so the nucleation rate. But, the observed nucleation rates suggest that this is not the case at least for low carbon dioxide concentration.

Next, the effect of an increase of carbon dioxide concentration on *n*-nonane nucleation has been studied. Two series of experiments have been performed for *n*-nonane nucleation in methane/carbon dioxide with relatively high, up to 0.3, molar fractions of carbon dioxide. The results of these experiments are shown in Figure 2. At the same supersaturation of *n*-nonane, the nucleation rate of *n*-nonane (filled circles) in methane/carbon dioxide(0.25) is approximately one order of magnitude higher than the *n*-nonane nucleation rate (open squares) in methane. Therefore, the presence of carbon dioxide stimulates nucleation of *n*-nonane. If the molar fraction of *n*-nonane is kept constant and we change the composition of the carrier gas by increasing the molar fraction of carbon dioxide, the nucleation rate of *n*-nonane will increase (filled triangles).

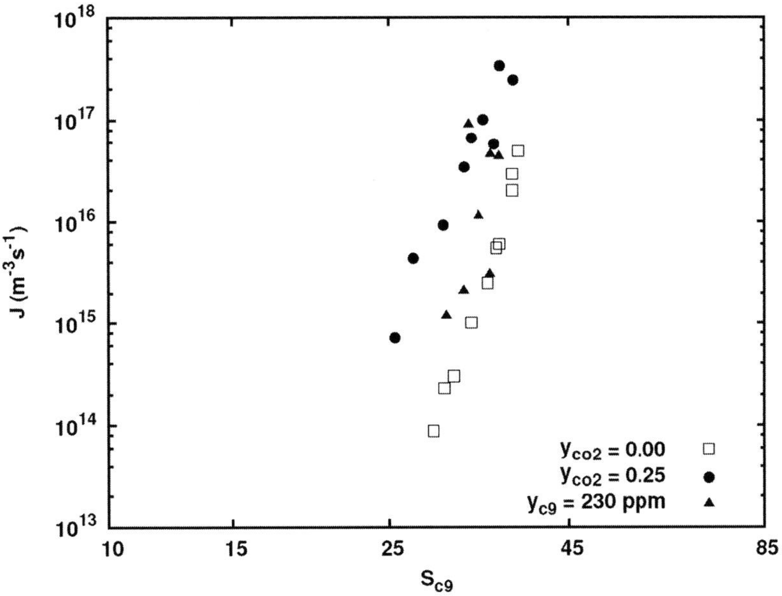

Figure 2 *n*-Nonane nucleation in methane/carbon dioxide mixtures and in methane at 235 K and 10 bar

A higher molar fraction of carbon dioxide in the carrier gas leads to a higher supersaturation of *n*-nonane S_{c9} This is because the addition of carbon dioxide decreases the *n*-nonane equilibrium molar fraction y_{c9}^{eq} and so increases supersaturation according to $S_{c9} = y_{c9}/y_{c9}^{eq}$. The increase of the nucleation rate with supersaturation is a trivial fact, but what is new and not trivial, is that the change of the *n*-nonane supersaturation by adding carbon dioxide affects the nucleation rate more strongly than the change of supersaturation by adding *n*-nonane. Evidently, an addition of carbon dioxide to the carrier gas changes not only the equilibrium state, but it also changes the energy of critical cluster formation.

References

1. Looijmans, K.N.H., Luijten, C.C.M., and van Dongen, M.E.H., *J. Chem. Phys.*, **103**(11), 1714–1719 1995.
2. Luijten, C.C.M., van Hooy, R.G.P., Janssen, J.W.F., and van Dongen, M.E.H., *J. Chem. Phys.*, **109**(9), 3553–3558 (1998).
3. Luijten, C.C.M., Peeters, P., and van Dongen, M.E.H., *J. Chem. Phys.*, **111**(18), 8535–8544 (1999).
4. Peeters, P., Hrubý, J., and van Dongen, M.E.H., *J. Phys. Chem. B*, **105**(47), 11763–11771 (2001).
5. Looijmans, K.N.H. and van Dongen, M.E.H., *Exp. Fluids*, **23**, 54–63 (1997).
6. Huber, M.L., *NIST Thermophysical Properties of Hydrocarbon Mixtures Database (SUPERTRAPP)* 2003, http://www.nist.gov/srd/webguide/4-3.1/2-04_3.1.htm.
7. Oxtoby, D.W. and Kashchiev, D., *J. Chem. Phys.*, **100**(10), 7665–7671 (1994).

Challenges in the Thermochemical Water-splitting Cycle Based on the ZnO/Zn Redox Pair: Rapid Quench and Nucleation of Zinc

Frederik Rütten, Ivo Alxneit, and Hans Rudolf Tschudi

Abstract For the optimization of the thermochemical water splitting cycle based on the redox pair Zn/ZnO a fundamental understanding of the nucleation and condensation process of zinc vapor is needed to optimize the separation of zinc and oxygen at the exit of a high temperature solar reactor. To obtain kinetic data for the nucleation and condensation of zinc vapor, an experiment based on a Laval nozzle was build. First experimental results of the nucleation and condensation of *n*-butanol as a test substance are shown.

Keywords Solar technology, homogeneous nucleation, condensation, zinc vapor, Laval nozzle

Introduction

The laboratory for Solar Technology at Paul Scherrer Institute is exploring the potential to produce chemical fuels using concentrated solar light with the aim to reduce the need of fossil fuels and the associated production of CO_2. An example of a process converting sunlight into fuel is the Zn/ZnO thermochemical water splitting cycle in which concentrated solar radiation is used for the direct decomposition of ZnO into its elements.[1] If elementary zinc formed can be separated from oxygen the former can be used in a water splitting cycle to obtain hydrogen and zinc oxide which can be recycled.[2]

At present, the separation of zinc and oxygen is implemented as a rapid quench. To optimize the efficiency of the quench process fundamental knowledge of the process of homogeneous nucleation and condensation is needed.[3]

Paul Scherrer Institute, 5232 Villigen PSI, Switzerland

C.D. O'Dowd and P.E. Wagner (eds.), *Nucleation and Atmospheric Aerosols*, 83–86.

Experimental Section

To provide experimental data for a fast nucleation process at high supersaturation, an experiment was built based on an adiabatic expansion in a Laval nozzle.[4] Nucleation and condensation are induced by an adiabatic, stationary expansion of vapor and nitrogen as an inert carrier gas through an optically transparent Laval nozzle. A laser beam propagates along the axis of the nozzle and the light scattered by the droplets is detected perpendicular to the nozzle axis. By using an intensified CCD (ICCD) camera, the scattered laser light can be detected spatially resolved. An analytical model is used to describe the central part of the flow along the nozzle's axis. This model incorporates the laws of fluid dynamics as well as the nucleation and condensation process. It predicts the particle size distribution at every position along the nozzle axis. Assuming Rayleigh scattering, the intensity of the scattered light can be calculated from the second momentum of the particle size distribution. Input parameters for the model such as the sticking probability (P_S) of a vapor molecule to the liquid surface and a scaling factor of the nucleation rate (q) are obtained by fitting calculated scattering curves to measured ones.

An overview of the setup is reported in Figure 1. The central part of the setup is the Laval nozzle of rectangular cross section (throat $2 \times 4\,mm^2$) made of stainless steel with inserted windows of fused silica (1). Two individually adjustable flows of hot nitrogen are fed to the nozzle. One of the nitrogen flow (4) is saturated with vapor in an oven (vaporizer); the second flow (5) is added to a labyrinth (6) before the entrance of the nozzle. The total nitrogen flow entering the nozzle is set to a defined value between 400 and $500\,l_N$/min. The partial pressure and thus the saturation

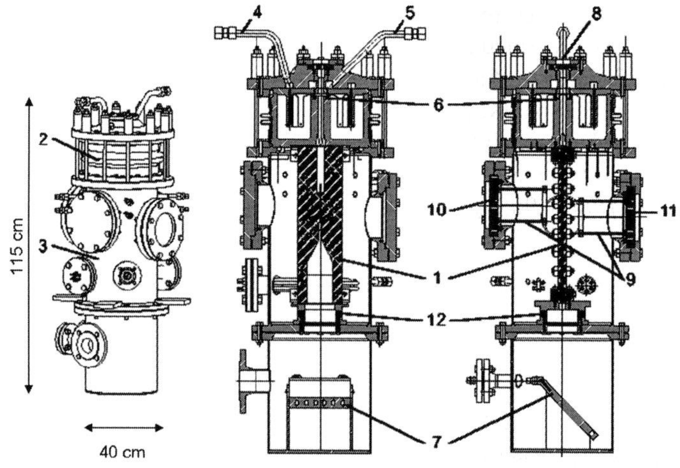

Figure 1 The setup of the nucleation experiment: nozzle (1), vaporizer (2), chamber (3), gas inlet (4, 5) labyrinth (6), condenser (7), fused silica windows (8, 10, 11), optional lens (9). (Reused with permission from [4]. Copyright 2007, American Institute of Physics.)

of the vapor inlet of the nozzle can be varied without changing the temperature of the vaporizer by varying the gas flow through the vaporizer. The laser beam of a He–Ne laser with a wavelength of 633 nm enters the setup through a fused silica window in the top flange and propagates along the axis of the nozzle. Inside the nozzle, light scattered on droplets formed is observed at an angle of 90° by an ICCD camera through one of the two fused silica windows (10, 11) installed in different observation positions.

For the experiments, the ICCD camera is programed to read out 100 × 512 pixels that image the nozzle with a spatial resolution of 0.07 mm/pixel. The image is flat-field corrected to compensate inhomogeneities in the sensitivity of the ICCD or the collection efficiency of the optics. Then the background recorded when a hot nitrogen flow of $500\,l_N$/min flushes the nozzle is subtracted. Finally the central 6 pixels covering the region of the nozzle's axis corresponding to a strip of 0.42 mm width where boundary effects due to the nozzle's walls can be neglected are averaged. This average defines the scattering curve, the basic output of the experiment.

To further improve the S/N ratio typically 60 scattering curves recorded under identical conditions are averaged. The Nelder–Mead simplex algorithm[6] is then used to determine numerical values for the model parameters from a large set of scattering curves corresponding to different stagnation conditions.

Results

First experiments to verify the measurement principle and data analyses were performed on *n*-butanol for which a solid data basis is available from the literature ([3] and references therein[5]). In Figure 2, we report the comparison of a measured scattering curve to the corresponding calculated curve, that is, the second moment of the particle size distribution delivered by our analytical model.

The scattering curves of a series of experiments that cover a range of stagnation conditions (T_0 = 380–400K, p_0 = 360–430 kPa, S_0 = 15–25%) were used to obtain data for the sticking probability (P_S), the effective opening angle of the nozzle (α), and the nucleation rate factor (q) of *n*-butanol. From a cursory evaluation we determined values of $\alpha = 0.96$, $P_S = 0.15$, $q = 2.02 \times 10^{12}$ comparable with values found in literature[3,5].

Outlook

A more detailed analysis of the data taking into account the evaporation kinetics of *n*-butanol in the evaporator is currently performed. In parallel, our setup is modified to allow experiments on the homogeneous nucleation of zinc. These studies will be complemented by small-angle neutron scattering measurements to determine the particle size distribution and the local velocity of the particles in the nozzle.

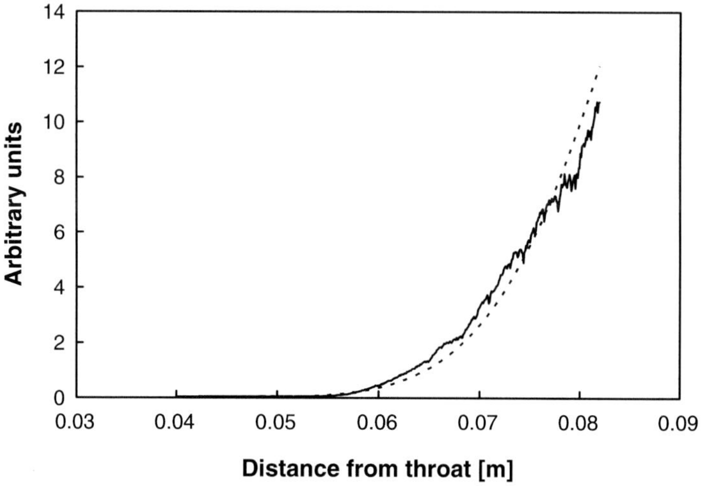

Figure 2 Comparison between an experimentally obtained scattering curve for *n*-butanol (dashed line; T_0 = 383K, p_0 = 412.5 kPa, S_0 = 20%) and a calculated scattering curve (solid curve) corresponding to the same conditions

Acknowledgment This work is supported by the Swiss Federal Office of Energy (SFOE).

References

1. Palumbo, R., Lédé, J., Boutin, O., Elorza Ricart, E., Steinfeld, A., Moeller, S., Weidenkaff, A., Fletcher, E.A., and Bielicki, J., *Chem. Eng. Sci.*, **53**, 2503–2518 (1998).
2. Steinfeld, A., Kuhn, P., Reller, A., Palumbo, R., Murray, J.P., and Tamaura, Y., *Int. J. Hydrogen Energy*, **23**, 43–53 (1998).
3. Karlsson, M., Nucleation and condensation in a stationary supersonic flow, *Dissertation, Eidgenössische Technische Hochschule Zürich*, http://e-collection.ethbib.ethz.ch/show?type=diss&nr=16542 (2006).
4. Karlsson, M., Alxneit, I., Rütten, F., Wuillemin, D., and Tschudi, H.R., *Rev. Sci. Instrum*, **78**, 31102 (2007).
5. Gharibeh, M., Kim, Y., Dieregsweiler, U., Wyslouzil, B.E., Ghosh, D., and Strey, R., *J. Chem. Phys.*, **122**, 094512 (2005).
6. Nelder, J.A. and Mead, R., *Computer Journal*, **7**, 308–315 (1965).

Relevance of Several Nucleation Theories in Different Environments

Michael Boy[1], Boris Bonn[1], Jan Kazil[2], Ned Lovejoy[2], Andrew Turnipseed[3], Jim Greenberg[3], Thomas Karl[3], Lee Mauldin[3], Edward Kusciuch[3], Jim Smith[3], Kelly Barsanti[3], Alex Guenther[3], Birgit Wehner[4], Olaf Hellmuth[4], Holger Siebert[4], Stefan Bauer[4], Alfred Wiedensohler[4], and Markku Kulmala[1]

Abstract The ability of different nucleation theories to explain new particle formation within the meteorological boundary layer was investigated with data from three field sites by a model to predict new aerosol formation in the lower troposphere (MALTE). In addition to the field studies, we used data obtained during intensive chamber experiments at NCAR in Boulder, Colorado to study the participation of organic vapours in the formation of clusters and their growth to larger particles in detail in order to draw conclusions for atmospheric processes.

Keywords Atmospheric nucleation, tropospheric aerosols, organic vapours, sulphuric acid

Introduction

The role of atmospheric aerosols is perhaps the biggest unknown concerning our climate and greenhouse warming (IPCC 2007). New particle formation has been observed at almost all sites, where both particle number concentrations and size distributions have been measured. A comprehensive summary of these studies is given by Kulmala et al. (2004). However, many questions remain regarding the extent to which these secondary aerosols influence radiative properties, climate, and human health. Most of these issues are linked to the chemical composition of the newly formed particles, which is currently not understood entirely.

Although many field campaigns, laboratory experiments, and new modelling approaches have led to increased understanding, detailed mechanisms responsible for the formation of new particles in the troposphere and their influence on health,

[1]*Department of Physical Sciences, University of Helsinki, P.O. Box 64, 00014 Helsinki, Finland*

[2]*NOAA Earth System Research Laboratory, 325 Broadway, Boulder, CO 80305, USA*

[3]*ACD, NCAR, 3450 Mitchell Lane, Boulder, CO 80301, USA*

[4]*Leibniz Institute for Tropospheric Research, Permoserstrasse 15, 04 318 Leipzig, Germany*

C.D. O'Dowd and P.E. Wagner (eds.), *Nucleation and Atmospheric Aerosols*, 87–91.

environment, and climate have still not been completely elucidated. In MALTE individually developed codes from different institutes around the globe merged into a one-dimensional model including aerosol dynamics, boundary layer meteorology, biology, and chemistry in order to investigate the formation and growth processes of Secondary Organic Aerosols (SOA) under realistic atmospheric conditions.

Model Description

MALTE is a one-dimensional model which includes several modules for the simulation of boundary layer dynamics and both chemical and aerosol dynamical processes. For the description of Planetary Boundary Layer (PBL) processes, a first-order closure technique is applied.The aerosol dynamics are solved by the size-segregated aerosol model, University of Helsinki Multicomponent Aerosol model (UHMA – Korhonen et al. 2004). For a detailed description of the model see Boy et al. 2006.

Our knowledge concerning the formation of very small particles or clusters (diameter < 2 nm) is still limited. The question of which molecules (sulphuric acid, ammonia, ions, and/or organic vapours) are involved in the atmospheric nucleation processes remains controversial within the aerosol community. MALTE takes several hypothetical pathways into account and currently includes six different nucleation theories for the formation of secondary aerosols (Figure 1).

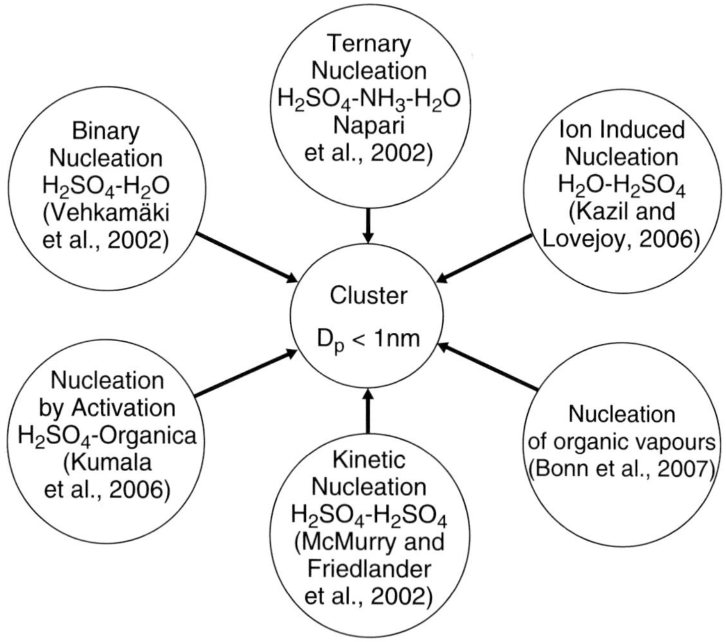

Figure 1 Nucleation mechanisms currently included in MALTE

Selected Environments

To test the different hypothesis concerning atmospheric nucleation, we used data from three field stations in different environments (SMEAR II in Hyytiälä, Finland; Melpitz field station of the Institute for Tropospheric Research in Leipzig, Germany and data from a field campaign at the Mountain Research Station in Niwot Ridge, Colorado, USA). Furthermore, we investigated the role of certain organic vapours (e.g., α-pinene, β-carophylene) in the particle formation processes during intensive laboratory chamber experiments at NCAR – National Centre for Atmospheric Research in Boulder, Colorado, USA.

Relevance of Different Nucleation Theories

Initial model simulations performed with the activation or kinetic nucleation mechanism predict nucleation rates that are strongly correlated with observed nucleation rates at background areas like boreal forest in Finland or somewhat more polluted areas east of Leipzig, Germany (Boy et al. 2006). In addition, many chamber experiments raise the possibility that organic vapours may form new particles by organic nucleation.

Regarding the possibility that organic vapours can nucleate, as apparently observed in several chamber experiments, we estimated with MALTE that even with low SO_2 concentrations, sufficient sulphuric acid is generated to explain observed particle number concentrations via the activation or kinetic nucleation mechanism alone. In our chamber experiments at NCAR we used a Chemical Ionization Mass Spectrometer (CIMS) instrument to measure sulphuric acid concentrations during the ozonolyses of different terpenes by using zero-air as carrier gas. The results showed that modelled and simulated H_2SO_4 concentrations agreed well by assuming SO_2 concentrations far below 0.1 ppb (detection limit of most SO_2 instruments).

To investigate the possibility that organic nucleation is still going on inside the chamber as well as in the real atmosphere, we included in MALTE a parameterization developed and published recently by Bonn et al. (2007a, b). Calculated nucleation rates with the organic nucleation code for data achieved during the Niwot Ridge campaign and the chamber experiments at NCAR showed high agreement with the measurements. It is currently not possible to draw a final conclusion if nucleation of organic vapours in the real atmosphere is present and if yes how important it is compared to other nucleation mechanisms.

Another open question is the importance of ion-induced nucleation at different heights in the troposphere, and whether it is necessary to include this mechanism when calculating aerosol number concentration. To explore this question, a newly developed parameterization for the production of stable clusters by ion-induced nucleation (Kazil and Lovejoy 2006) was implemented in MALTE. The simulations performed on four days for southern Finland show that ion-induced nucleation

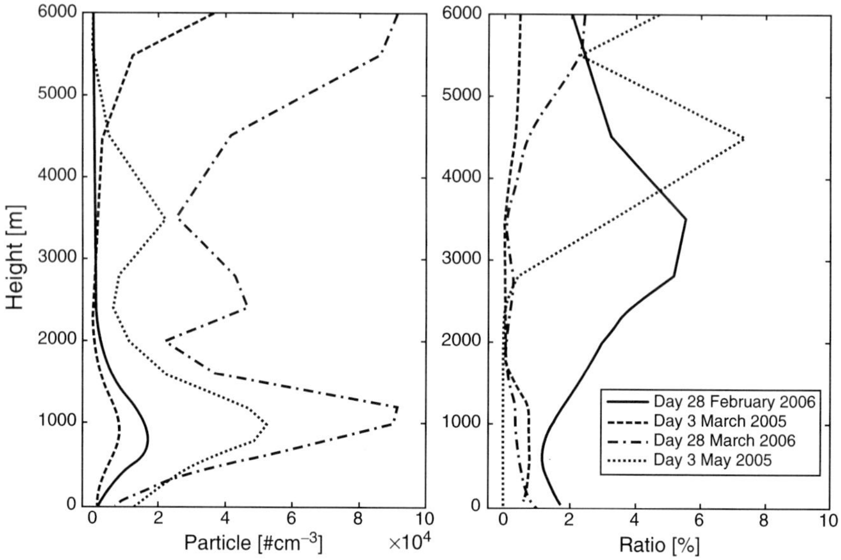

Figure 2 Vertical profiles at noon: a) Particle number concentration in the 3–20 nm size range; b) Ratio between particle number concentrations (3–20 nm) produced by ion-induced nucleation and kinetic nucleation

contributes maximal up to 8% of aerosol numbers (3–20 nm) above the boundary layer and to less than 4% within, depending on height, sulphuric acid concentrations, temperature, relative humidity, and background aerosol concentration (Figure 2, manuscript in preparation).

Ternary nucleation including sulphuric acid, ammonia, and water was tested for the SMEAR II site. The calculated nucleation rates exceed the observed rates by a factor of 10 when using measured ammonia concentrations. In the binary nucleation case with sulphuric acid and water, the simulations of the nucleation rates in the mixed layer are far below the measurements.

Conclusion

New available data sets from field campaigns and chamber experiments in combination with new hypothetical nucleation theories led to a more optimistic view concerning the understanding of atmospheric nucleation. Today we know that ion-induced nucleation is always participating, but our results and earlier publications showed that its contribution is limited to a few per cent in the lower troposphere. Ternary and binary nucleation theories with present understanding over- or underestimate measured atmospheric nucleation rates. The big open question today and for the near future is to what extend organic vapours are involved in the nucleation mechanism and is sulphuric acid the main steering parameter for the formation of new clusters or only activating the organic compounds.

References

1. Bonn, B., Petäjä, T., Boy, M., Korhonen, H., and Kulmala, M., Atmos. Chem. Phys. Diss., accepted (2007a).
2. Bonn, B., Kulmala, M., Riipinen, I., Sihto, S.-L., and Ruuskanen, T.M., *J. Aerosol Sci.*, submitted (2007b).
3. Boy, M., Hellmuth, O., Korhonen, H., Nillson, D., ReVelle, D., Turnipseed, A., Arnold, F., and Kulmala, M., *Atmos. Chem. Phys.*, **6**, 1–19 (2006).
4. IPCC, Intergovernmental Panel on Climate Change (IPCC). *Climate Change: The Scientific Basis*, UK: Cambridge University Press (2007).
5. Kazil, J. and Lovejoy, E.R., *Geophy. Res. Lett.*, accepted for publication (2006).
6. Kulmala, M., Vehkamaeki, H., Petaejae, T., Dal Maso, M., Lauri, A., Kerminen, V.-M., Birmili, W., and McMurry, P.H., *J. Aerosol Sci.*, **35**, 143–176 (2004).
7. Kulmala, M., Lehtinen, K.E.J., and Laaksonen, A., *Atmos. Chem. Phys.*, **6**, 787–793 (2006).
8. McMurry, P.H. and Friedlander, S.K., *Atmos. Environ.*, **13**, 1635–1651 (1979).
9. Napari, I., Noppel, M., Vehkamaeki, H., and Kulmala, M., *J. Chem. Phys.*, **116**, 4221–4227 (2002).
10. Sihto, S.-L., Kulmala, M., Kerminen, V.-M., Dal Maso, M., Petäjä, T., Riipinen, I., Korhonen, H., Arnold, F., Janson, R., Boy, M., Laaksonen, A., and Lehtinen, K.E.J., *Atmos. Chem. Phys.*, **6**, 4079–4091 (2006).
11. Vehkamäki, H., Kulmala, M., Napari, I., Lehtinen, K.E.J., Timmreck, C., Noppel, M., and Laaksonen, A., *J. of Geophys. Res.*, **107**, D22, 4622, doi: 10.1029/2002JD002184 (2002).

Kinetic Theory Applied to Nucleation and Droplet Growth

Vincent Holten and M.E.H. van Dongen

Abstract Condensation of liquid from the vapour phase has been studied by the numerical solution of the kinetic equations describing the time dependence of the cluster size distribution. The model has been applied to the nucleation pulse method; the resulting droplet population is investigated. The results are compared with results from a simplified theory that describes nucleation and droplet growth as separate processes. The simulations show that the simplified model is unable to replicate the results of the purely kinetic model.

Keywords Homogeneous nucleation, clusters, kinetics

Introduction

The condensation of liquid from the vapour phase plays an important role in numerous industrial processes. Condensation models are needed in fluid dynamic models of devices such as steam turbines, gas–vapour separators, and combustion engines.

Usually, condensation models are divided into a nucleation model and a macroscopic droplet growth model.[1,2] The nucleation model describes how many new droplets are formed per unit of time and volume, and the growth model describes how fast these droplets grow. However, from a physical point of view, a single process – the change in size of molecular clusters – is responsible for both formation and growth.

Therefore, condensation can be more accurately modelled by solving the kinetic master equations describing the time dependence of the cluster size distribution. This was done, for example, by Courtney[3] and Abraham[4] (for clusters smaller than 110 molecules) and more recently by Kožíšek et al.[5] (up to 100,000 molecules).

Eindhoven University of Technology, Department of Applied Physics, The Netherlands, P.O. Box 513, 5600 MB Eindhoven, e-mail: v.holten@tue.nl, m.e.h.v.dongen@tue.nl

C.D. O'Dowd and P.E. Wagner (eds.), *Nucleation and Atmospheric Aerosols*, 92–96.

In this study, we examine the differences between both methods, in particular the errors that are made when condensation is artificially split up into nucleation and droplet growth. We will examine a reference experiment, an application of the nucleation pulse method, and model the resulting droplet populations by the full kinetic method and by the simpler nucleation–growth model.

Kinetic Model

In the kinetic model, clusters can gain or lose single molecules, and cluster–cluster interactions are neglected. The rate of change of the n-cluster concentration is

$$\frac{df_n}{dt} = C_{n-1}f_{n-1} - \left(C_n + E_n\right)f_n + E_{n+1}f_{n+1} = J_{n-1} - J_n, \tag{1}$$

where f_n is the number density of clusters with n molecules. C_n is the rate at which molecules condense on an n-cluster and E_n is the rate at which molecules evaporate from such a cluster. The 'current' J_n, the number of clusters per unit time and volume which grow from size n to $n + 1$, is

$$J_n = C_n f_n - E_{n+1} f_{n+1}. \tag{2}$$

The condensation coefficient is the product of the sticking probability α (assumed to be unity), the collision frequency per unit area β and the cluster surface area a_n.

$$C_n = \alpha \cdot \beta \cdot a_n = \left(\frac{kT}{2\pi m}\right)\rho_1 \cdot a_1 n^{2/3}, \tag{3}$$

where $a_1 = (36\pi)^{1/3}\, v_1^{2/3}$ is the molecular surface area and m and v_1 are the mass and volume of a molecule, respectively. Further, k is the Boltzmann constant, T is the temperature and v_1 is the density of monomers.

The evaporation coefficient is found by the detailed balance equation: at saturation ($S = 1$, denoted by superscript 'sat') all J_n equal zero, so Eq. (2) becomes

$$E_{n+1} = C_n^{sat}\left(\rho_n^{sat} / \rho_{n+1}^{sat}\right). \tag{4}$$

The supersaturation S is defined as $S = p / p_{sat}$, with p the vapour pressure and p_{sat} the saturated vapour pressure. The size distribution ρ_n^{sat} is assumed to follow a Boltzmann form,

$$\rho_n^{sat} = \rho_1 \exp\left(-\Theta n^{2/3}\right), \tag{5}$$

where Θ is the dimensionless surface tension $\Theta = a_1\sigma/kT$ and σ is the surface tension.

The following boundary conditions are used. (1) The concentration of the smallest clusters considered, f_5, is taken constant and equal to[6] ρ_1 exp(51n $S - \Theta(5)^{2/3}$). This means that depletion of the vapour phase is not taken into account. (2) At $t = 0$, $f_n = 0$ for $n > 5$.

Numerical Solution

The present model should be able to simulate the cluster distribution for large values of n (of the order of 10^9 molecules). To reduce the size of the system (1), not each individual f_n is included in the model. Instead, only averages of ranges of f_n's are included. The J_n within these averaging intervals is not computed; only the fluxes at the edges of the intervals are required and are found by linear interpolation between two adjoining intervals. The size of the intervals is unity from the beginning of size space up to a certain size $n >> n^*$ (no averaging is done there) and then increases with increasing cluster size. For example, size space up to $n = 3 \times 10^9$ can be covered by 3×10^5 intervals; up to $n = 10{,}000$ the intervals have unit length, and the largest interval – located at the end – has a length of 3×10^4. In this way the number of equations is reduced by a factor 10,000.

The reduced system of differential equations is solved by numerical integration using the Crank–Nicolson method, using time steps in the range of 10^{-9}–10^{-8} s. After f_n has been obtained, it is converted to a distribution as a function of droplet radius $f(r,t)$:

$$f(r,t) = 3\left(\frac{4\pi}{3v_1}\right)^{1/3} n^{2/3} f_n(t). \tag{6}$$

Simplified Model

The simpler nucleation–growth model describes the evolution of the droplet distribution $f(r,t)$ macroscopically, as follows:

$$\frac{\partial f}{\partial t} = -\dot{r}\frac{\partial f}{\partial r}, \tag{7}$$

where $\dot{r}$ is the growth rate of clusters, which is assumed to follow the Hertz–Knudsen growth law:

$$\dot{r} = \frac{1}{3}\left(\frac{3v_1}{4\pi}\right)^{1/3} \frac{S-1}{S} C_1. \tag{8}$$

In this droplet growth expression, the Kelvin effect is not taken into account, that is, the expression for large droplets is used here.

Nucleation is included in the model as a boundary condition. Newly formed droplets enter size space at a nucleation rate J and with initial size r^*:

$$f\dot{r} = J, \text{ at } r = r^*. \tag{9}$$

Clusters that are smaller than the critical radius disappear immediately in this model, so

$$f = 0, \text{ for } r < r^*. \tag{10}$$

Results

The two models are evaluated using the condensation of water vapour at 228K in a nucleation pulse with a length of 0.05 ms. After the pulse, the temperature is increased to 253K and nucleation stops. The remaining conditions are listed in Table 1.

Droplet size distributions computed with the two models at the end of the pulse and after the pulse are plotted in Figure 1. At the end of the pulse (Figure 1a), the simple model approximates the kinetic model quite well; both the width and the height of f are reproduced. A notable difference is the lack of clusters with $r < r^* = 0.6$ nm in the simple model.

At 0.24 ms after the pulse (Figure 1b) there are greater differences between both models. At the end of the pulse, r^* increases from 0.6 to 1.6 nm. In the simple model, all clusters with $r < r^*$ vanish instantaneously. In the kinetic model, however, some of these clusters succeed in passing the r^* barrier, and grow further. This leads to a characteristic tail of the distribution.

Table 1 Conditions of the simulation

Quantity	During Pulse ($t < 0.05$ ms)	After Pulse ($0.05 < t$/ms < 0.29)
T (K)	228.0	253.2
p (Pa)	225.0	292.5
S	20.57	2.319
Θ	12.74	10.61
ρ_5 (m^{-3})	8.446×10^{11}	8.093×10^{10}
C_1^{sat} (s^{-1})	2.142×10^5	2.304×10^6
$\dot{r}$ (m/s)	2.721×10^{-4}	1.955×10^{-4}
J (m^{-3} s^{-1}), steady-state	2.849×10^{13}	2.642×10^{-80}

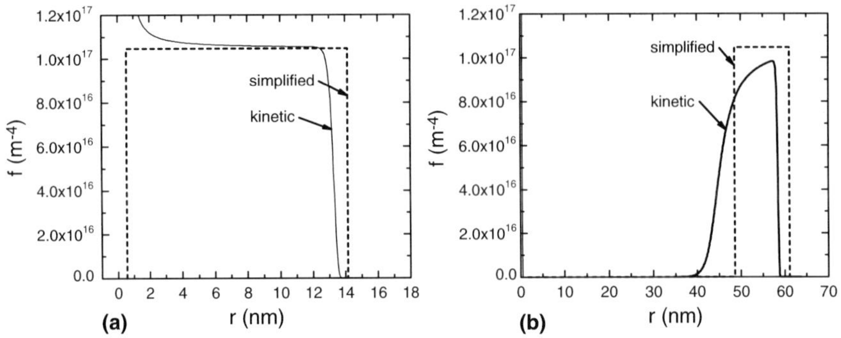

Figure 1 Droplet size distributions calculated with the kinetic and simplified models: **(a)** just prior to the end of the pulse at $t = 0.05$ ms; **(b)** 0.24 ms after the pulse; $t = 0.29$ ms

A further discrepancy between the two models is the position of the front edge of the distributions. The reason for this difference is that at low supersaturation, the Hertz–Knudsen growth law overestimates the growth rate for clusters close to r^*.

Discussion

The simulations show that a simplified nucleation–growth model is unable to replicate the results of a purely kinetic model. Additional computations will be done to investigate how the differences of the two models depend on the pulse conditions.

The simple model can be improved by including the Kelvin effect so that the growth rates of the two models agree better. However, because the growth rate is then zero at $r = r^*$, a value $r > r^*$ has to be taken as the initial size of the droplets.

References

1. Hagmeijer, R., IJzermans, R.H.A., and Put, F., *J. Fluid Mech.*, **17**, 056101 (2005).
2. Luo, X., Prast, B., van Dongen, M.E.H., Hoeijmakers, H.W.M., and Yang, J., *J. Fluid Mech.*, **548**, 403 (2006).
3. Courtney, W.G., *J. Chem. Phys.*, **36**, 2009 (1961).
4. Abraham, F.F., *J. Chem. Phys.*, **51**, 1632 (1969).
5. Kožíšek, Z., Sato, K., Demo, P., and Sveshnikov, A.M., *J. Chem. Phys.*, **120**, 6660 (2004).
6. Kashchiev, D., *Nucleation: Basic Theory with Applications*, Oxford: Butterworth-Heinemann (2004).

Computation of Nucleation Rates for *n*-Nonane Using the Gradient Theory

J. Hrubý[1], V. Vinš[1], D.G. Labetski[2], and M.E.H. Van Dongen[2]

Abstract The gradient theory is used to compute the work of formation of critical clusters of *n*-nonane, the dependence of the surface tension on the cluster size, and the rates of homogeneous vapor–liquid nucleation of *n*-nonane droplets. An adjustment of the Peng–Robinson (P–R) equation of state is suggested in order to achieve accurate gradient theory results.

Keywords Homogeneous nucleation, Gradient theory, surface tension, Tolman length, nonane

Introduction

Although a general agreement exists that the basic ideas behind the classical theory of homogeneous nucleation are physically correct, its use for quantitative predictions is limited. The dependency of the surface tension on the size of the droplet is neglected in the classical nucleation theory (CNT). Assuming a macroscopic surface tension for a cluster of 10–100 molecules is a crude approximation. Several researchers[1,2,3] developed modifications of the CNT including various estimates for the "microscopic surface tension." In this article we use the gradient theory[4,5,6] to compute the work of formation of clusters of *n*-nonane.

Gradient Theory Computations

In the gradient theory, the density profile $\rho(r)$, r being radial coordinate, for a droplet is obtained by solving the Euler–Lagrange differential equation[7]:

[1]*Institute of Thermomechanics AS CR, v.v.i., Dolejškova 1402/5, 182 00 Prague 8, Czech Republic*

[2]*Eindhoven University of Technology, Department of Applied Physics, P.O. Box. 513, 5600 MB Eindhoven, The Netherlands*

C.D. O'Dowd and P.E. Wagner (eds.), *Nucleation and Atmospheric Aerosols*, 97–101.

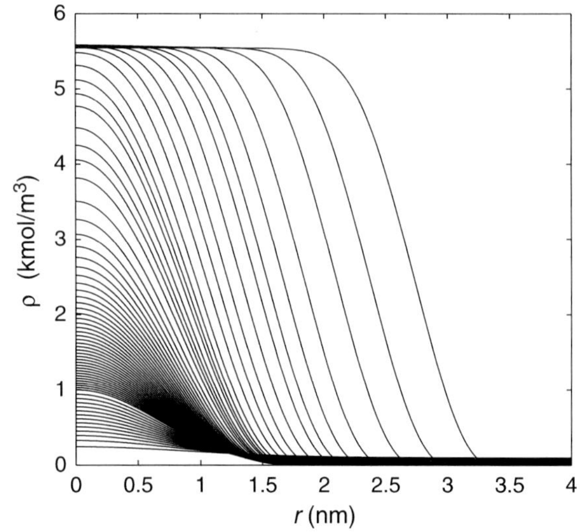

Figure 1 Computed profiles of *n*-nonane droplets at 230K

$$\frac{d^2\rho}{dr^2}+\frac{2}{r}\frac{d\rho}{dr}=\frac{1}{c}\Delta\mu(\rho),\quad \lim_{r\to\infty}\rho=\rho_V,\quad \left.\frac{d\rho}{dr}\right|_{r=0}=0. \tag{1}$$

In Eq. (1), c is the so-called influence parameter, obtained from experimental surface tension data[13]; $\Delta\mu \equiv \mu(\rho) - \mu_V$, where μ_V is the chemical potential of the vapor and $\mu(\rho)$ is chemical potential computed from an equation of state for given local density. The computed profiles are shown in Figure 1. The work of formation is computed as

$$\Delta\Omega=\int_0^{\infty}\left[\Delta\omega(\rho)+\tfrac{1}{2}c\left(d\rho/dr\right)^2\right]4\pi r^2 dr. \tag{2}$$

Here $\Delta\omega \equiv \phi - \rho\mu_V + p_V$; ϕ is local density of Helmholtz energy evaluated from the equation of state, and p_V is pressure of the vapor. The surface tension is evaluated as $\sigma = (3\Delta\Omega/16\pi\Delta p^2)^{1/3}$, where $\Delta p \equiv p_L - p_V$ and p_L is the pressure of a hypothetic bulk liquid phase in chemical equilibrium with the vapor phase. The pressure difference is related to the radius of the sphere of tension by the Laplace equation, $\Delta p = 2\sigma / r$. The computed surface tension is shown in Figure 2. For large clusters (small Δp), it is slightly higher than for the planar case ($\Delta p = 0$). For small clusters, the tendency reverses, the surface tension diminishes and it reaches zero at the vapor–liquid spinodal.

Figure 3 shows a comparison of the nucleation rates obtained using the gradient theory with experimental data[10,11,12] and with the CNT[14] and the internally consistent classical theory (ICCT)[15]. The gradient theory, CNT, and ICCT calculations are based on the P–R equation of state.[8] This equation is, however, significantly

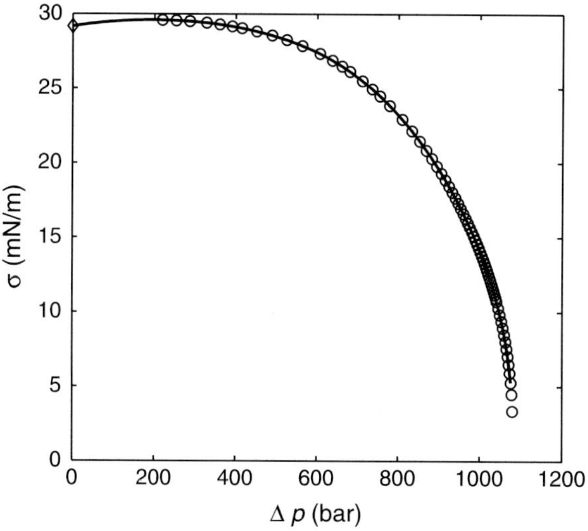

Figure 2 Surface tension of a *n*-nonane droplet as function of the pressure difference $\Delta p = 2\sigma / r_s$

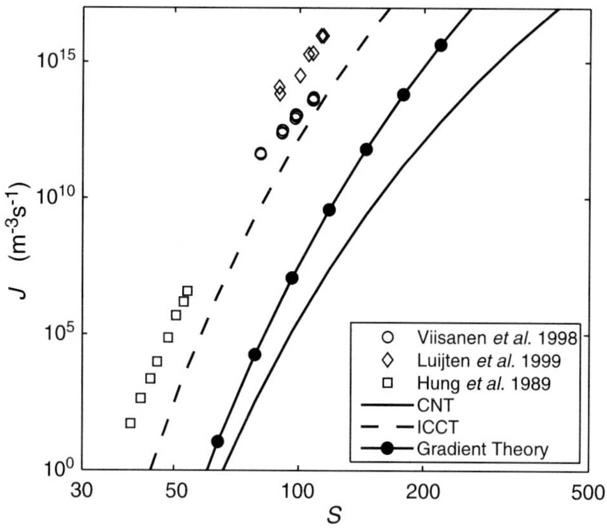

Figure 3 Nucleation rate *J* versus supersaturation *S*, comparison with experimental data

inaccurate. To enable accurate predictions, we introduce a volume-translated modification of the P–R equation:

$$p = \frac{RT}{v - C - B} - \frac{A}{(v-C)^2 + 2B(v-C) - B^2}. \tag{3}$$

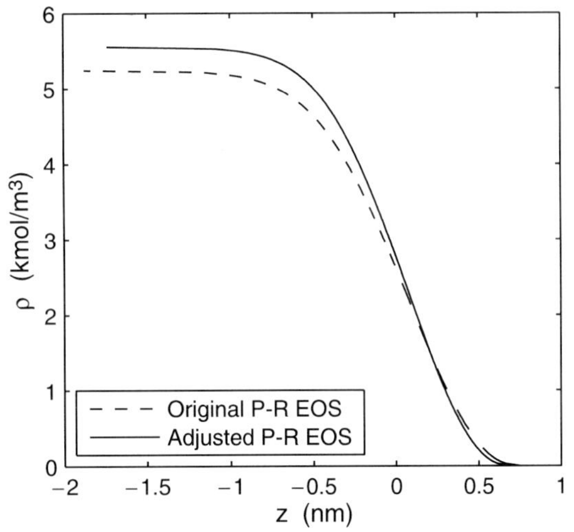

Figure 4 Surface tension of nonane dependent on temperature; comparison of modified EOS, Peng–Robinson EOS and Jasper's experimental data

Parameters A, B, and C are considered as temperature functions and they are computed by fitting saturated vapor pressure, density, and compressibility of saturated liquid to correlated experimental values.[9]

Figure 4 shows comparison of planar surface density profiles, compared with the original P–R equation and modified Eq. (3). The modified equation leads to a steeper density profile. This fact can have a significant influence on the size-dependency of the surface tension. In further work, work of formation, surface tension, and the nucleation rates will be computed for the modified equation.

Acknowledgments The authors gratefully acknowledge the support by grant No. ESF.6472 of the Technology Foundation STW, The Netherlands, and by grant No. 101/05/2214 of the Grant Agency of the Czech Republic.

References

1. Tolman, R.C., *J. Chem. Phys.*, **17**, 333–337 (1949).
2. Dillmann, A., and Meier, G.E.A., *J. Chem. Phys.*, **94**, 3872–3884 (1991).
3. Kalikmanov, V.I., *J. Chem. Phys.*, **124**, 124505 (2006).
4. Van der Waals, J.D., *Z. Phys. Chem. (Leipzig)*, **13**, 657–725 (1894).
5. Cahn, J.W. and Hilliard, J.E., *J. Chem. Phys.*, **28**, 258–267 (1958).
6. Cahn, J.W. and Hilliard, J.E., *J. Chem. Phys.*, **31**, 688–699 (1959).
7. Davis, H.T., *Statistical Mechanics of Phases, Interfaces, and Thin Films*, New York: VCH Publishers (1996).
8. Peng, D.-Y., and Robinson, D. B., *Ind. Eng. Chem., Fundam.* **15**, 59–64 (1976).
9. Lemmon, E.W. and Span, R., *J. Chem. Eng. Data*, **51**, 785–850 (2006).

10. Hung, C.-H., Krasnopoler, M.J., and Katz, J.L., *J. Chem. Phys.*, **90**, 1856–1865 (1989).
11. Viisanen, Y., Wagner, P.E., and Strey, R., *J. Chem. Phys.*, **108**, 4257–4266 (1998).
12. Luijten, C.C. M., Peeters, P., and Van Dongen, M.E.H., *J. Chem. Phys.*, **111**, 8335–8544 (1999).
13. Jasper, J.J., Kerr E.R., and Gregorich, F., *J. Am. Chem. Soc.*, **75**, 5252–5254 (1953).
14. Becker, R. and Döring, W., *Ann. Phys. 5. Folge*, **24**, 719–752 (1935).
15. Girshick, S.L. and Chiu, C.-P., *J. Chem. Phys.*, **93**, 1273–1277 (1990).

The Temperature of Nucleating Droplets

Jan Wedekind[1], David Reguera[1], and Reinhard Strey[2]

Abstract We investigate the temperature of nucleating and growing droplets using molecular dynamics simulations. We find that subcritically sized clusters tend to be colder than the bath temperature, while post-critically sized clusters are warmer. However, the average temperature of all clusters is found to be always higher than the bath temperature. Even though nucleation is highly sensitive to temperature, the results surprisingly show that the deviation of the cluster temperature from the bath temperature does not have a significant influence on the nucleation rate, the deviation being less than a factor of two.

Keywords Nucleation, temperature fluctuations, cluster temperature, carrier-gas, thermostats

Introduction

There have been many discussions on the role of the cluster temperatures and the influence of their inevitable fluctuations on the nucleation rate.[1–6] The underlying idea is that latent heat released during nucleation might heat up the system while the evaporation rate, the supersaturation S of the vapor and, thus, the nucleation rate are very sensitive to even the slightest change in temperature. Thus, both in experiment and in simulations, it is very important to keep the temperature of the system constant. In experiments, quasi-isothermal conditions are achieved by diluting the condensable in an inert carrier gas that serves as a heat bath. This heat bath absorbs the latent heat that is released during condensation and is able to keep a constant temperature, provided that the excess of carrier gas is large and the nucleation rate sufficiently low. Yet, in small nonequilibrium system such as clusters that are relevant in nucleation, the determination or even the definition of temperature is

[1]*Departament de Física Fonamental, Universitat de Barcelona, Barcelona, Spain*

[2]*Institut für Physikalische Chemie, Universität zu Köln, Cologne, Germany*

C.D. O'Dowd and P.E. Wagner (eds.), *Nucleation and Atmospheric Aerosols*, 102–106.

very subtle and not at all trivial. The main difficulty stems from the fact that during nucleation, a significant amount of latent heat is released and the system thus surely is not in equilibrium. Consequently, the old and newly forming phase do not necessarily and in fact will not generally have the same temperature.

We performed extensive MD simulations of Lennard–Jones atoms to study the temperature of nucleating and growing droplets. Details of the simulations are provided elsewhere.[7] The number of LJ-argon atoms was $N = 343$ in all simulations. Two different box volumes, $(16\,\text{nm})^3$ and $(18\,\text{nm})^3$, were chosen corresponding to a high supersaturation and a low supersaturation, respectively. The liquid clusters were identified using a Stillinger criterion with a threshold distance of $r_c = 1.8V$. First, the velocity scaling (VS) and the Nosé–Hoover (NH) thermostat were applied as direct thermostating methods.[8] Second, we repeated the simulations with three different ratios of vapor to carrier gas of roughly 1:1, 1:2, and 1:3. Here, LJ-helium atoms are added and only these atoms are subjected to a VS thermostat. We analyzed the simulation data using the new mean first-passage time method, which offers a very efficient, easy-to-implement, and rigorous procedure to evaluate simulations of activated processes.[9]

Results

We determined the kinetic temperature of the clusters from their mean kinetic energy. Figure 1a shows the temperature deviation ΔT of the *average* cluster temperature T_{avg} (filled symbols) from the imposed bath temperature of 50K as a function the cluster size n using the different thermostats for the case of a high supersaturation. As expected, growing clusters heat up due to the release of latent

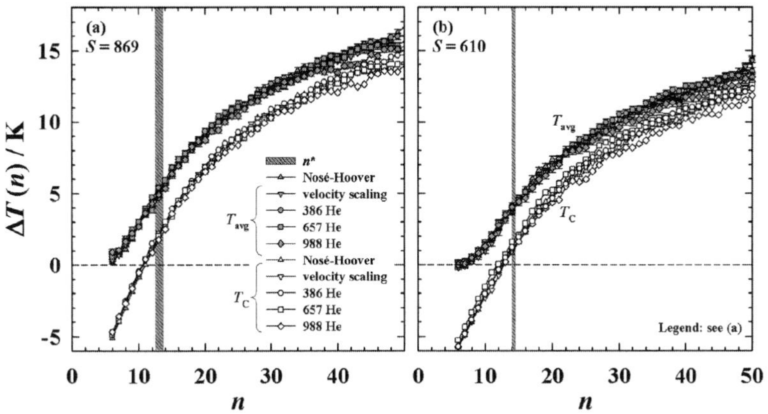

Figure 1 Average cluster temperatures (**filled symbols**) *Tavg* and peak cluster temperatures *TC* as a function of the cluster size n. (a) for $V = (16\,\text{nm})^3$; (b) for $V = (18\,\text{nm})^3$

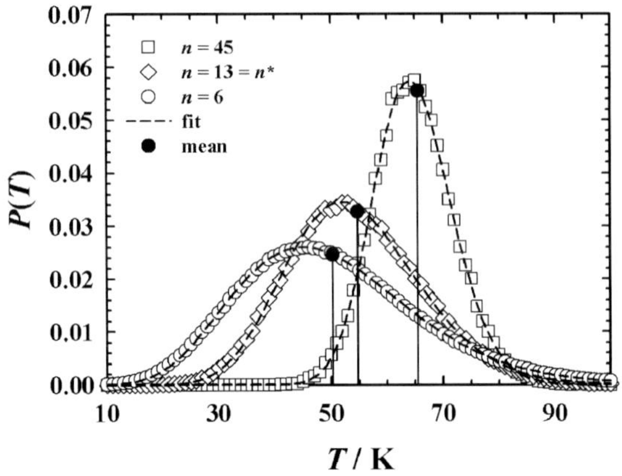

Figure 2 Cluster temperature distribution for subcritically, critically, and post-critically sized clusters taken from the simulations with velocity scaling (VS) and a high supersaturation. The distributions are clearly non-Gaussian for small cluster sizes and can be very well described by a fit to Eq. (2)

heat from the condensation. In addition, T_{avg} is always higher than the fixed bath temperature of 50K and, quite surprisingly, it turns out that that there is no noticeable difference in T_{avg} for all the different thermostating techniques that we have used. Figure 2b shows the results for the low supersaturation case, where we find the same qualitative picture, but slightly less-pronounced temperature deviations.

In addition to T_{avg}, a lot of information can be gained by looking at the full temperature distribution of clusters of a particular size n (Figure 2). The cluster temperatures fluctuate over a wide range of values. In addition, the temperature distribution is clearly not Gaussian for critically sized or smaller clusters. In this case, the average temperature of a cluster is not a good measure of the most likely temperature of clusters of that size. The distribution of equilibrium temperature fluctuations for a small cluster of a fixed size n around a given "bath temperature" T_0 was given by McGraw and Laviolette[2]:

$$P(T) = K_1 \exp\left(-W(T)/kT_0\right), \tag{1}$$

where K_1 is a normalization constant and $W(T) = C_V(T - T_0) + C_V T_0 \ln(T_0/T)$ is the reversible work required to bring the system from its equilibrium temperature T_0 to a temperature T; C_V is the heat capacity of the liquid clusters. These equations describe the equilibrium fluctuations of a small cluster of a fixed size around its equilibrium temperature T_0.

However, the nucleating clusters are *not* at equilibrium and their temperature does not coincide in general with the (global) bath temperature T_B, i.e., 50K in our case. The latent heat of growth or decay leads to a net shift in the temperature of the clusters. However, between successive condensation/evaporation events, the clusters have a comparatively long lifetime that allows them equilibrate their local temperature.[6] Hence, at steady-state conditions an ensemble of clusters of one size

n will be locally equilibrated around this shifted temperature. The distribution of fluctuations around this shifted (or local equilibrium) temperature will thus be described by an equation similar to Eq. (1), namely

$$P(T) = K_1 \exp\left(-(C_V(T - T_C) + C_V T_C \ln(T_C / T)) / kT_C\right), \tag{2}$$

where T_C now is the *local* equilibrium temperature of the cluster, which plays the same role as T_0 for an equilibrium system. We can use Eq. (2) to fit the temperature distributions in Figure 2, using T_C and C_V as free parameters, yielding excellent agreement with the simulation results. Figure 2a shows the resulting cluster temperatures T_C for all cluster sizes n. The important difference to T_{avg} is that the temperature of clusters smaller than the critical size is most likely *below* the enforced T_B of 50K. Therefore, the *average* temperature T_{avg} of the cluster is always higher than T_B, while the *local equilibrium* temperature T_C is lower than T_B for $n < n^*$, and larger for $n > n^*$. We also note that for larger clusters the distribution of temperatures is gradually becoming more and more Gaussian, also in agreement with theoretical predictions.

Thus, previous seemingly contradictory works on the role of nonisothermal effects and cluster temperature fluctuations[1,2,4] are found to be correct in their own right once the proper temperature, T_{avg} or T_C, is used. The cluster temperatures can be several degrees different from the bath temperature and these differences are more pronounced for highly supersaturated systems, where nucleation and growth occur so fast that the clusters do not have enough time to get rid of their latent heat. Yet, it turns out that the temperature of the clusters is shifted with respect to the bath temperatures in a way that compensates these temperature fluctuations so that, overall, the nucleation rate is mostly close to the isothermal one. The classical work of Feder et al.[10] offers simple analytical expressions to quantify this influence of nonisothermal effects on the nucleation rate and the cluster temperatures, and we will show that it provides a remarkable semiquantitative agreement with the data.

Table 1 Number of simulation runs, supersaturation, equilibrium vapor pressure p_{eq} at 50K, nucleation rate J, the ratio of J over J_{VS} of the velocity scaling system, and the critical cluster size n^* for each simulated system.

Property/Units Thermostat	Num. of sim.	S	p_{eq} Pa	J 10^{24}cm^{-3}s^{-1}	J/J_{VS}	n^*
Nosé–Hoover	300	869	66.5	9.9	0.8	13.7
Velocity Scaling	1,000			10.9	1.0	13.2
386 He	300			12.1	1.1	12.8
657 He	300			15.7	1.4	12.8
988 He	300			18.9	1.7	12.5
Nosé–Hoover	300	610	66.5	0.42	0.8	14.4
Velocity Scaling	300			0.55	1.0	14.1
386 He	300			0.58	1.1	14.1
657 He	300			0.76	1.4	14.0
988 He	200			0.77	1.4	14.0

Acknowledgements This work was supported through the "Acciones Integradas Hispano-Alemanas" program of the German Academic Exchange Service (DAAD) and the Spanish DGCyT.

References

1. Ford, I.J. and Clement, C.F., *J. Phys. A: Math. Gen.*, **22**, 4007–4018 (1989).
2. McGraw, R. and Laviolette, R.A., *J. Chem. Phys.*, **102**, 8983–8994 (1995).
3. Barrett, J.C., Clement, C.F., and Ford, I.J., *J. Phys. A: Math. Gen.*, **26**, 529–548 (1993).
4. Wyslouzil, B.E. and Seinfeld, J.H., *J. Chem. Phys.*, **97**, 2661–2670 (1992).
5. Kuni, F.M., Grinin, A.P., and Shchekin, A.K., *Physica A*, **252**, 67–84 (1998).
6. Napari, I. and Vehkamaki, H., *J. Chem. Phys.*, **124**, 024303 (2006).
7. Wedekind, J., Reguera, D., and Strey, R., *J. Chem. Phys.* (2007).
8. Frenkel, D. and Smit, B., *Understanding Molecular Simulation*, 2nd edn. San Diego, CA: Academic Press (2002).
9. Wedekind, J., Strey, R., and Reguera, D., *J. Chem. Phys.*, **126** (2007).
10. Feder, J., Russell, K.C., Lothe, J., and Pound, G.M., *Adv. Phys.*, **15**, 111–178 (1966).

Kinetic Reconstruction of the Nucleation Free Energy Landscape

David Reguera and Jan Wedekind

Abstract We present a new method to back-trace a lot of valuable information from molecular dynamics (MD) simulations of nucleation and activated processes in general. In particular, it is possible to reconstruct the free energy profile and the actual rate of attachment of molecules to a cluster of a given size from the pure kinetics of the process. We illustrate these ideas by their application to a real MD simulation of nucleation in a Lennard–Jones vapor. We then analyze the results of the simulations using this technique and compare them to the predictions of the classical nucleation theory.

Keywords Molecular dynamics simulations, homogenous nucleation, mean first-passage times

Introduction

Molecular Dynamics simulations offer an excellent way to investigate the dynamics of activated processes in general and nucleation in particular. Recently, it has also become possible to monitor the formation of clusters experimentally in real time using confocal microscopy, at least in the case of crystallization of colloids.[1] Yet, the activated nature of the nucleation process, characterized by the presence of a barrier that has to be overcome, limits the range of applicability of these techniques to situations in which the nucleation barrier is low or moderately high. Although more complicated *equilibrium* simulation techniques allow exploring higher barrier cases, the appeal of direct MD simulations is the possibility of monitoring the real dynamics without artifacts.

Typically, the information one seeks from these simulations or experiments is the nucleation rate. However, given the stochastic, nonequilibrium and rare nature

Departament de Física Fonamental, Universitat de Barcelona, Martí i Franquès 1, 08028- Barcelona, Spain

C.D. O'Dowd and P.E. Wagner (eds.), *Nucleation and Atmospheric Aerosols*, 107–111.

of the nucleation events, it is difficult to analyze the results of these experiments or simulations. Recently, we have shown how to extract the nucleation rate as well as the critical cluster size and the Zeldovich factor accurately and efficiently using the concept of mean first-passage time (MFPT).[2]

Here we present results that facilitate to trace back even more information of the process, again purely from its kinetics. In particular, we will show how to reconstruct the free energy landscape and even the rate of attachment of molecules directly from MD simulations. It is worth to emphasize that this method can also be applied to Brownian Dynamics simulations and experiments, and, in general, to any sort of activated process.

The technique requires two ingredients that can be easily calculated in the simulation. One is the steady-state probability distribution and the other one is the MFPT. The connection between these ingredients and the barrier is discussed theoretically below and then applied to the reconstruction of the free energy of cluster formation in MD simulations.

Theory

The dynamics of nucleation and, in general, of many activated processes,[3] can be described in terms of a Fokker–Planck equation

$$\frac{\partial P(n,t)}{\partial t}=\frac{\partial}{\partial n}\left(D(n)e^{-\beta\Delta G(n)}\frac{\partial}{\partial n}\left(P(n,t)e^{\beta\Delta G(n)}\right)\right), \tag{1}$$

where $P(n, t)$ is the probability of finding a cluster of size n at time t, $D(n)$ and $\Delta G(n)$ are the rate of attachment of molecules and the free energy of formation of a cluster of size n, and $\beta = 1/kT$, T being the temperature and k the Boltzmann's constant.

When the nucleation barrier is relatively high and after a short transient time, the system reaches a steady state characterized by a constant current J (the nucleation rate, which is independent of n), and a time-independent probability distribution $P^{st}(n)$. It is easy to show that this probability is connected to the free energy of cluster formation through the following relation

$$\beta\Delta G(n)=-\ln P^{st}(n)-J\int\frac{1}{D(n')P^{st}(n')}dn'+C, \tag{2}$$

where C is a constant. This formula establishes that one can reconstruct the free energy landscape of cluster formation from the knowledge of $P^{st}(n)$, $D(n)$ and the steady-state rate J. An even more useful result can be proven.

The steady-state rate is also related to another important quantity: MFPT. In our case, the MFPT $\tau(n)$ is defined as the average time that the system, starting out at size n_0, takes to reach size n for the first time, and it is given by the expression[2]

$$\tau(n) = \int_{n_0}^{n} \frac{\exp(\beta \Delta G(y))}{D(y)} dy \int_{a}^{y} \exp(-\beta \Delta G(z)) dz. \tag{3}$$

By combining the steady-state distribution with the MFPT, it is easy to obtain the following relation, which we give here without proof:

$$\beta \Delta G(n) = \ln f(n) - \int \frac{1}{f(n')} dn' + cte, \tag{4}$$

where

$$f(n) = \frac{1}{P^{st}(n)} \left[\int_0^n P^{st}(n') dn' - J\tau(n) \right]. \tag{5}$$

Therefore, from the knowledge of $P^{st}(n)$ and $\tau(n)$ it is possible to reconstruct the free energy landscape. This information can be obtained easily from MD simulations, as discussed in the next section.

Results

We have verified the feasibility of the kinetic reconstruction of the free energy profile for the case of CNT, by solving numerically the Fokker–Planck equation, Eq. (1), for the case in which both the free energy and the attachment rate $D(n)$ are given by the classical expressions, namely $\Delta G_{CNT}(n) = 2\Delta G^{*}(-x + 3/2x^{2/3})$, and $D_{CNT}(n) = A(n)p/\sqrt{2\pi mkT}$, where ΔG^{*} is the height of the nucleation barrier, $x \equiv n/n^{*}$, n^{*} is the critical cluster size, p is the pressure of the supersaturated vapor and $A(n)$ is the area of the cluster.

However, the importance of the method lies in its application to MD simulations. To illustrate its power, we have performed a set of simulations of $N = 343$ Lennard–Jones argon atoms in a volume of $V = (18\,\text{nm})^3$ at $T = 50\text{K}$ (details of the simulations are provided in Ref. 4). In these simulations, the MFPT and the steady-state probability can be easily calculated.

Figure 1 illustrates the resulting MFPT and steady-state probability distribution obtained from 200 realizations of our MD simulations. Using these two inputs, we can reconstruct the free energy landscape through Eq. (4) as plotted in Figure 2. For comparison, we have also plotted the free energy reconstructed using Eq. (2) with the CNT expression of $D(n)$. It is also possible to obtain an expression for $D(n)$ that can be evaluated directly from the MD simulations, although it is numerically a bit more challenging to calculate for all cluster sizes. We have evaluated $D(n)$ in the vicinities of the critical size, obtaining $D(n^{*}) = 4\,ns^{-1}$. This estimate coincides nicely with the value calculated by measuring the diffusivity around the critical size.[5]

We can see from the results that the nucleation barrier is just a few kT high. Such a value is indeed common in most direct MD simulation, where in order to facilitate the observation of nucleation events the supersaturation is so high that the actual

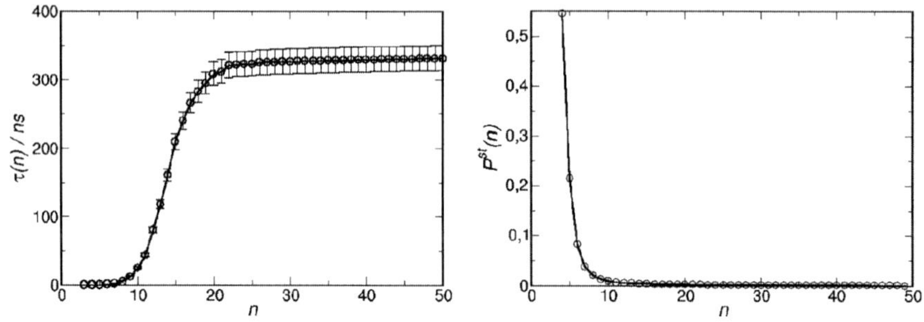

Figure 1 **(a)** Mean first-passage times as a function of the cluster size obtained from the MD simulations of LJ argon at $T = 50$K. **(b)** The steady-state probability distribution of cluster sizes obtained for the same set of simulations

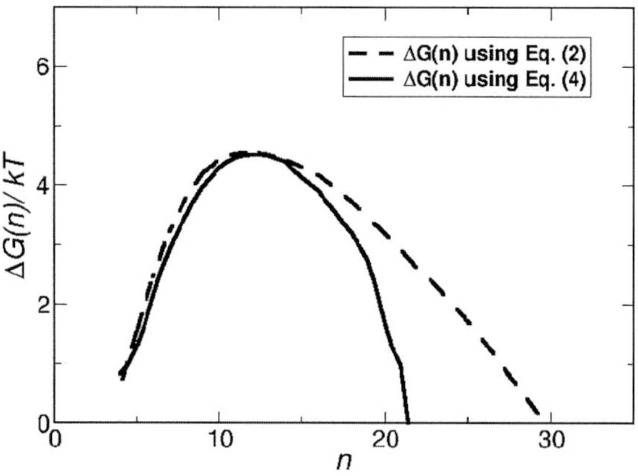

Figure 2 Kinetic reconstruction of the free energy of cluster formation obtained from the MD simulations using Eq. (4) (**solid line**), and using Eq. (2) and the $D(n)$ given by the CNT (**dashed lined**)

nucleation barrier can be of the order of 1 or $2\,kT$ or even vanishes. In that case we will not be strictly observing a nucleation process but rather something alike to a spinodal decomposition where, the kinetics will be controlled only by the rate of attachment of molecules to a cluster. The new method offers a way to reveal these situations and to realize which part of the kinetics of cluster formation actually corresponds to the nucleation barrier and which one to the prefactor.

In addition, the method opens up the possibility to compare the actual barriers appearing in the true kinetics of nucleation with the equilibrium barriers that can be obtained, for instance, from advanced Monte Carlo or umbrella sampling techniques. We will also discuss the comparison of the simulation results with the predictions of CNT.

Acknowledgments This work has been funded by the DGCyT of the Spanish government through grant No. FIS2005–01299. Financial support of the "Acciones Integradas Hispano-Alemanas" program of the German Academic Exchange Service (DAAD) and the Spanish DGCyT is also gratefully acknowledged.

References

1. Gasser, U. et al., *Science*, **292**, 258–262 (2001).
2. Wedekind, J., Strey, R., and Reguera, D., *J. Chem. Phys.*, **126**, accepted (2007).
3. Reguera, D., Rubí, J.M., and Vilar, J.M.G., *J. Phys. Chem. B*, **109**, 21502 (2005).
4. Wedekind, J., Reguera, D., and Strey, R., *J. Chem. Phys.*, submitted.
5. Auer, S. and Frenkel, D., *Nature*, **409**, 1020–1023 (2001).

Laboratory-measured Nucleation Rates of Sulfuric Acid and Water from the SO_2 + OH Reaction

David R. Benson and Shan-Hu Lee

Abstract Binary homogeneous nucleation of H_2SO_4/H_2O is the most basic nucleation system, yet its nucleation mechanisms are poorly understood. Here, we present results of the laboratory study of this binary nucleation system. H_2SO_4 was produced through the reaction of $SO_2 + OH \rightarrow HSO_3$ (R1) in the presence of SO_2, OH, O_2, and H_2O in a fast flow reactor at 288K and atmospheric pressure. OH was produced from the photolysis of H_2O. The power dependence of the nucleation rate (J) on sulfuric acid concentration ($[H_2SO_4]$) was 2–10 in the $[H_2SO_4]$ range from $3 \times 10^6 - 2 \times 10^9$ cm^{-3}. This power dependence increased with decreasing RH and increasing nucleation time. The measured aerosol sizes ranged from 4–20 nm. These aerosol sizes were larger at higher $[H_2SO_4]$ and higher RH, and with higher nucleation times. The effects of RH on aerosol growth were also more pronounced at higher $[H_2SO_4]$ and with higher nucleation times.

Keywords Binary nucleation, sulfuric Acid, CIMS

Introduction

Atmospheric observations have shown that new particle formation occurs in various atmospheric conditions, ranging from rural and urban areas to the upper troposphere and lower stratosphere [1]. Aerosol nucleation has been studied extensively by modeling simulations, but these numerical calculations contain large uncertainties because the current nucleation theories are not rigorously tested by experiments. Various nucleation theories, including binary, ternary, and ion-induced nucleation, have been used to validate field measurements, but most of these theories have also fallen short of accurate predictions. In particular, laboratory studies of nucleation are very limited because of various difficulties associated with experiments.

Department of Chemistry, Kent State University, Kent, OH 44240, USA

C.D. O'Dowd and P.E. Wagner (eds.), *Nucleation and Atmospheric Aerosols*, 112–116.

Laboratory studies of binary H_2SO_4/H_2O nucleation often investigated how the nucleation rate depends on sulfuric acid concentrations ($[H_2SO_4]$), relative humidity (RH), and temperature, as such information can be directly used for testing theoretical models. Most of the previous studies have produced sulfuric acid vapor from liquid H_2SO_4 samples [2–3]. Wyslouzil et al. [2] measured J values at a magnitude of $10^{-3} – 10^{2}$ cm^{-3} s^{-1} for calculated relative acidities between 0.04 and 0.46, RH between 0.6 and 65% and temperatures of 293K, 298K, and 303K. From these results, they estimated the number of sulfuric acid molecules (n_{H2SO4}) in the critical clusters to be approximately 4 – 30, and the number of water molecules (n_{H2O}) in the critical clusters to be about 9. Ball et al. [3] measured J from 10^{-2} to 10^{3} cm^{-3} s^{-1} at an $[H_2SO_4]$ range from 2.5×10^{9}–1.2×10^{10} cm^{-3}, RH from 2– 15% and 295K. They estimated an n_{H2SO4} of 7– 13 and n_{H2O} of 4 – 6. Berndt et al. (2005, 2006) have used the $SO_2 + OH \rightarrow HSO_3$ reaction (R1) to produce sulfuric acid in situ, and determined an n_{H2SO4} of 3–5.

Here, we present laboratory studies of the binary H_2SO_4/H_2O system at sulfuric acid concentrations relevant to atmospheric conditions (e.g., $[H_2SO_4]$ of 10^{6}– 10^{8} cm^{-3}). H_2SO_4 vapor was produced by in situ gas-phase reactions of SO_2, OH, O_2, and H_2O based on the above reaction (R1). A chemical ionization mass spectrometer (CIMS) was used to detect sulfuric acid concentrations. We will discuss the dependence of J and aerosol sizes on the $[H_2SO_4]$, RH, and nucleation time. Preliminary results of our nucleation studies are described by Young et al. [6].

Experiment Setup

Kent State University's nucleation experimental setup consists of four main components: (i) an OH generator where OH forms from water UV absorption, (ii) a fast-flow reactor in which nucleation occurs, (iii) a CIMS which measures concentrations of the equilibrium H_2SO_4 vapor, and (iv) a TSI scanning mobility particle sizer and an ultrafine water-based condensation particle counter (SMPS/UWCPC). This setup is described in detail elsewhere [6]. Table 1 summarizes experimental conditions, including flow reactor geometry, total flow rate, the initial $[SO_2]$, RH, and nucleation time.

Table 1 Experimental conditions used in the ksu laboratory nucleation studies. Temperature, 288K; atmospheric pressure

Experiment	Flow Reactor (ID; Length) (cm)	Total Flow (lpm)	Initial $[SO_2]$ (ppm)	RH (%)	Nucleation time (s)
J dependence on $[H_2SO_4]$	5.1; 82	1.52–5.0	1.5–13	11–42	20–66
J dependence on RH	2.5; 80	2–3	1–10	11–50	8–12,
	5.1; 82	2	0.5–5	15–45	50
Size Dependence on RH	5.1; 82	2–3	1.5–10	15–60	33–50

Results

J Dependence on [H_2SO_4]

Figure 1 shows the measured *J* as a function of [H_2SO_4] at different RH and nucleation times. Nucleation times were changed by varying the flow rate. Overall, *J* varied from 0.1 to 10,000 $cm^{-3}\ s^{-1}$, for [H_2SO_4] from 3×10^6 to $2 \times 10^9\ cm^{-3}$, RH from 11% to 38% and nucleation times from 20 to 66 s. This is a very promising result, because the [H_2SO_4] range used in our study (10^6–$10^9\ cm^{-3}$) is similar to that used in these previous studies [3–4]. In fact, some of our data fell between the 10^6 and $10^8\ cm^{-3}$ range [4–5] and the 10^9–$10^{10}\ cm^{-3}$ range [3], and yet the measured *J* values were similar to those from these studies. The slope of a plot of Log *J* versus Log [H_2SO_4] provides the value of n_{H2SO4} at the specific RH and temperature [3]. The derived n_{H2SO4} ranged from 2 to 10 at these conditions, again consistent with previous reports [2–6].

The n_{H2SO4} also increased with decreasing RH, similar to Ball et al. [3]. For example, with a nucleation time of 38 s, n_{H2SO4} was 5.6 at 30% RH, whereas n_{H2SO4} was 7.9 at 22% RH. Similarly, for a nucleation time of 66 s, n_{H2SO4} increased from 6.3 to 9.5 as RH decreased from 38% to 26%.

We found that nucleation times also affect the measured J. Comparing the two sets of experiments with similar RH values (22–30%) but different nucleation times (38 and 66 s), n_{H2SO4} was higher (8.5–9.5), at the longer nucleation time of 66 s,

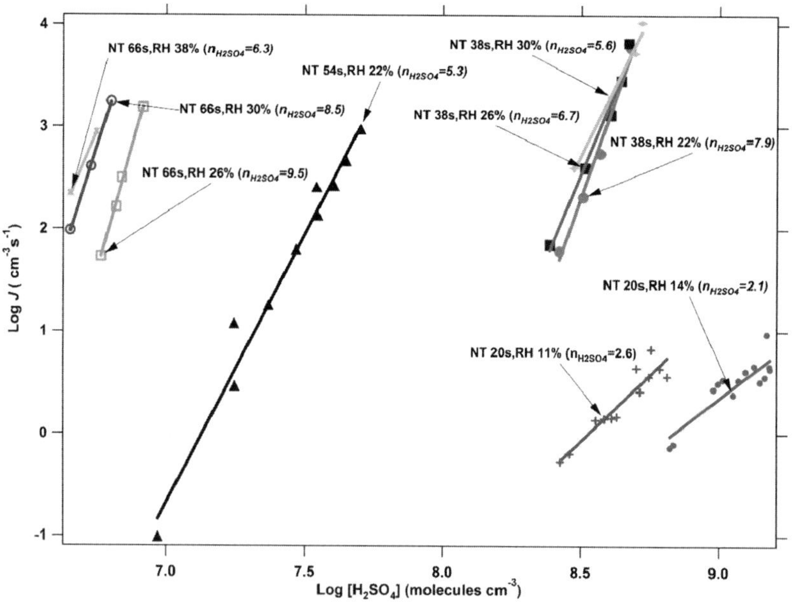

Figure 1 A summary of the laboratory measured *J* as a function of [H2SO4]. For each data set (**shown with the same colors**), the nucleation time and RH was constant while the [H2SO4] varied with different SO2 flow rates. NT indicates the nucleation time

compared to that (5.6–6.7) measured at the shorter nucleation time of 38 s. This is likely because of different wall losses at different nucleation times and hence different [H_2SO_4] (the measured [H_2SO_4] is rather the equilibrium concentration than the initial concentration). The measured [H_2SO_4], in fact, decreased with increasing nucleation times (within 20–66 s) (Figure 1). For example, [H_2SO_4] was at the 10^6 cm^{-3} range at the longer nucleation time (66 s), whereas [H_2SO_4] was at the 10^8 cm^{-3} range at the shorter nucleation time (38 s).

For the lower nucleation time (20 s) and lower RH (11–14%), the measured J was substantially lower (<1 cm^{-3} s^{-1}), compared to those at higher nucleation times and higher RH. This is because J is lower at lower RH (as shown above) and also because lower nucleation times (e.g., 20 s) may be insufficient for nucleation (and thus less particles produced).

Size Dependence on RH

Figure 2 shows the measured median aerosol size, maximum D_p, as a function of initial [SO_2], RH and nucleation time. The measured maximum D_p ranged from 4 to 10 nm, at [H_2SO_4] from 10^6 to 10^8 cm^{-3}, RH from 17% to 58%, and nucleation

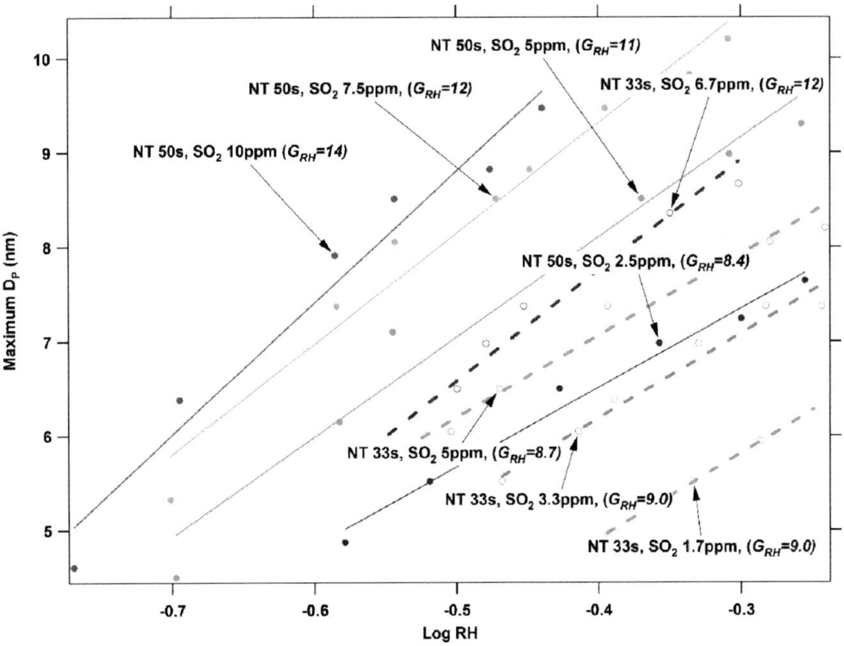

Figure 2 A summary of the laboratory measured maximum *Dp* as a function of RH. For each data set (**shown with the same colors**), the nucleation time and initial [SO2] was constant while the RH varied with different H2O flow rates. NT indicates the nucleation time

times from 33 to 50 s. In general, the maximum D_p was higher for higher initial $[SO_2]$ (hence higher initial $[H_2SO_4]$), higher RH, and higher nucleation times (Figure 2). The slope of a linear fitting curve (maximum D_p versus Log RH) shown in Figure 2 (G_{RH}) indicates the RH effects on the aerosol growth at the specific condition of the initial $[SO_2]$ and nucleation time. The G_{RH} was higher at the higher initial $[SO_2]$ and with higher nucleation time. For example, for the same initial $[SO_2]$ (5 ppmv), G_{RH} was 10.6 for a nucleation time of 50 s, whereas it was only 8.7 for a nucleation time of 33 s. These results imply that the RH effects on aerosol growth are more pronounced for higher $[H_2SO_4]$ and higher nucleation times.

Our results for the aerosol size dependence on $[H_2SO_4]$, RH, and nucleation time are unique because no other studies have presented such results, except for Wyslouzil et al. [2]. In Wyslouzil et al. [2] where liquid H_2SO_4 samples were used to produce sulfuric acid vapor, however, their results showed a decreasing effect of maximum D_p with increasing relative acidity, whereas our results showed an increasing maximum D_p with increasing initial $[SO_2]$.

Future Plans

We plan to do further experiments to determine the J dependence on RH. Also, further work will be done to determine the J dependence on nucleation time by keeping all other conditions constant. This is an important issue because the sulfuric acid wall loss is very sensitive to nucleation times as discussed here. We will also change the experimental setup to measure the initial $[H_2SO_4]$ (rather the equilibrium $[H_2SO_4]$) to see the correlation between J and the initial $[H_2SO_4]$, which includes the equilibrium $[H_2SO_4]$ and the sulfuric acid loss to the wall and particles. The *initial* $[H_2SO_4]$ range used here was also higher than the actual atmospheric conditions and we want to reduce it to the atmospheric concentrations. We also plan to quantitatively characterize the sulfuric acid wall loss.

Acknowledgments This study is supported by the NSF Career Award (ATM-0645567) and the KSU startup fund to SHL. We also thank Greg Huey and Dave Tanner at Georgia Tech for the CIMS construction.

References

1. Kulmala, M., et al., *J. Aerosol Sci.*, **35**, 143–176 (2004).
2. Wyslouzil, B.E., et al., *J. Chem. Phys.*, **94**, 6842–6850 (1991).
3. Ball, S.M., et al., *J. Geophys. Res.*, **104**, 23709–23718 (1999).
4. Berndt, T., et al., *Science*, **307**, 698–700 (2005).
5. Berndt, T., et al., *Geophys. Res. Lett.*, **33**, Doi:10.1029/2006GL026660 (2006).
6. Young, L.-H., et al., Laboratory studies of H2SO4/H2O binary homogeneous nucleation from SO2 + OH reactions: Preliminary results, a manuscript in preparation.

Simulation of the Growth and Decay of Isolated Lennard–Jones Clusters

Jonathan C. Barrett and Andrew P. Knight

Abstract Molecular dynamics simulations of isolated Lennard–Jones clusters were performed. Starting from two atoms with total energy E, atoms are injected at random intervals, consistent with a specified vapour saturation, and simulations are continued until the cluster re-evaporates back to a dimer. The maximum size achieved is recorded. Typically, between 10^7 and 10^9 simulations are performed and the fraction of these growing to sizes five and six is determined. This fraction decreases with increasing E, but is much larger than predicted by a treatment based on the capillary approximation. However, the mean energy of the resulting 5-mers and 6-mers is higher than that at equilibrium, indicating that the grown clusters are very unstable.

Keywords Homogeneous nucleation, Lennard–Jones clusters, molecular dynamics

Introduction

When discussing homogeneous nucleation of liquid droplets in a supersaturated vapour, it is usual to picture the vapour as consisting of a mixture of single molecules (monomers) and groups or "clusters" of molecules which grow or evaporate (decay) by gaining or losing monomers. Small clusters are more likely to decay than grow, but fluctuations can cause clusters to reach the critical size, above which growth is more likely than decay. The nucleation rate is proportional to the rate at which critical clusters are formed. Usually, growth rates are determined from decay rates using equilibrium thermodynamics or statistical mechanics, but this ignores the fact that nucleation is a non-equilibrium process. Direct computer simulation of the full nucleation process is now possible (for monomers interacting via simple potentials), but only at very high saturations where the critical size is

Nuclear Department, HMS Sultan, Military Road, Gosport, PO12 3BY, UK

C.D. O'Dowd and P.E. Wagner (eds.), *Nucleation and Atmospheric Aerosols*, 117–120.

small and nucleation rates are much higher than observed experimentally. Rather than attempting to simulate the full nucleation process, here we use computer simulation to investigate the growth and decay of small sub-critical clusters.

Simulation Procedure

We model the behaviour of very small clusters of atoms interacting via the Lennard–Jones potential, $u(r) = 4\varepsilon\{(\sigma/r)^{12} - (\sigma/r)^6\}$, where r is the atomic separation and ε and σ are the energy and length parameters. We start from a randomly generated arrangement of two atoms within a distance R_1 of their centre of mass and with total (kinetic plus potential) energy close to a specified value E. We determine a monomer-cluster collision time, t_1, randomly chosen from an exponential distribution with mean equal to the reciprocal of the average monomer-cluster collision frequency. We perform molecular dynamics simulations up to time t_1. If the cluster decays in this time, we start again, generating a new initial arrangement. If not, we inject a monomer from a distance R_{max}, reset the new cluster's centre of mass and centre of mass velocity to zero, and repeat the process of generating t_1 and performing molecular dynamics simulations until either the maximum simulation time is reached or the number of atoms in the simulation volume is two or fewer. We perform a large number of trials (between 10^7 and 10^9), recording the maximum cluster size achieved in each trial and so determine the fraction of trials that reach a given size as a function of initial dimer energy. A cluster is here defined as a group of atoms within a specified distance R_2 of their mutual centre of mass ($R_1 < R_2 < R_{max}$).

We can compare our results with the predictions of the capillary approximation, in which the monomer evaporation rate, α_i, is related to the gain rate β_i, by:

$$\alpha_i = \frac{\beta_{i-1}}{S}\exp\left(\frac{\gamma A_1}{kT}[i^{2/3} - (i-1)^{2/3}]\right), \tag{1}$$

where S is the vapour saturation, γ is the surface tension and A_1 is the monomer surface area in the bulk liquid. The fraction of dimers becoming clusters of size M can be written $\beta_{M-1}\, n^{ss}_{M-1}/(\tfrac{1}{2}\beta_1 n_1 + \alpha_3 n_3^{\,ss})$, where the steady-state cluster size distribution can be found iteratively from $n_i^{ss} = (J - \beta_{i-1} n^{ss}_{i-1}) \,/\, \alpha_i$, with the nucleation rate, J, given by the usual Becker–Döring expression, except truncated at $i = M{-}1$.

Results

We present some preliminary results from our simulations. We make all quantities dimensionless using appropriate combinations of the Lennard–Jones parameters ε and σ and the monomer mass, m. In dimensionless units, we have taken $R_1 = 1$, $R_2 = 2$, $R_{max} = 5$, and $T = 0.5$. Figure 1 shows the fraction of dimers that grow to clusters containing at least 5 or at least 6 atoms as a function of initial dimer energy for

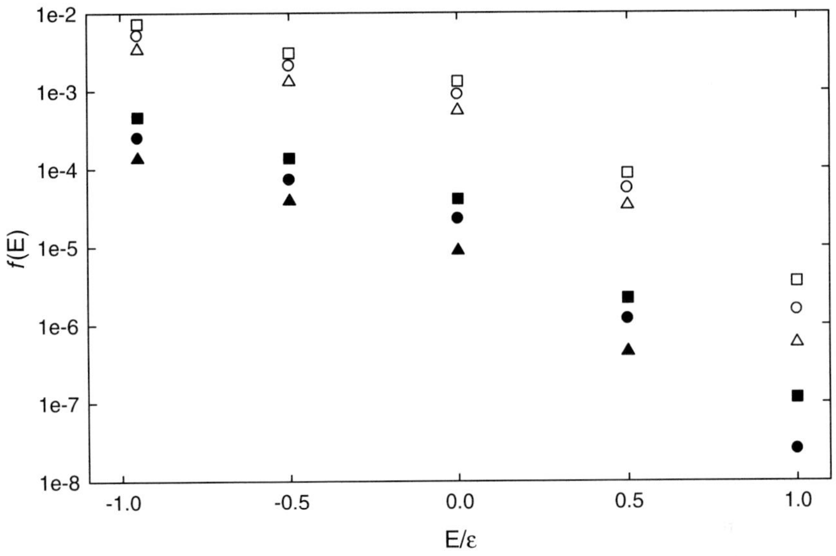

Figure 1 Fraction of dimers of energy E growing to size 5 (**open symbols**) and 6 (**filled symbols**) for monomer injection frequencies (in units of $(\varepsilon/m\sigma^2)^{1/2}$), $\nu R_{max}^{\ 2}$ = 0.1 (**triangles**), 0.125 (**circles**), and 0.15 (**squares**)

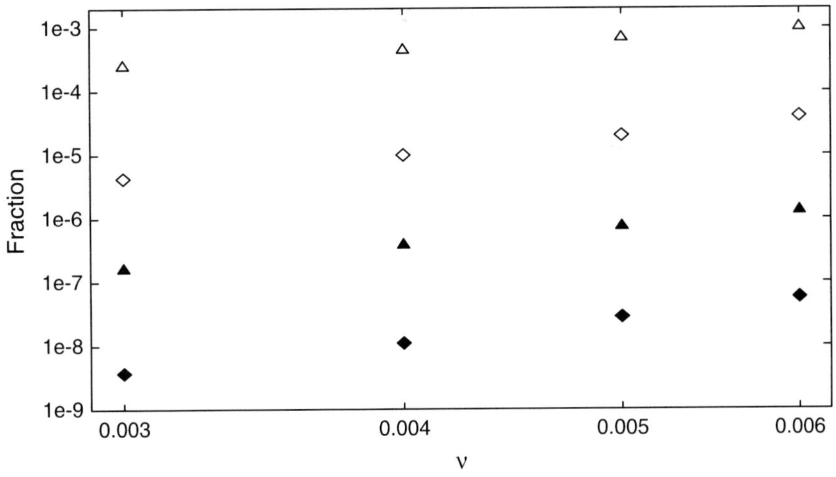

Figure 2 Fraction of all dimers growing to size 5 (**triangles**) and 6 (**diamonds**) for various monomer injection frequencies (divided by $R_{max}^{\ 2}$). Open symbols: simulation results. Filled symbols: capillary approximation predictions

various mean monomer injection frequencies. The fraction growing, $f(E)$, is seen to have a strong dependence on initial dimer energy and to increase with monomer-cluster collision frequency. Figure 2 shows the overall fraction achieving sizes $M = 5$ and 6. Also shown are the predictions of the capillary approximation, as

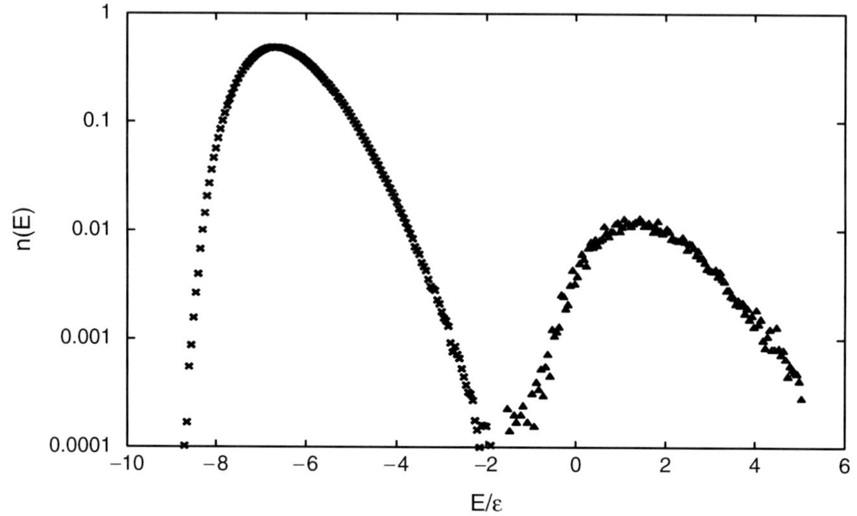

Figure 3 Energy distributions of pentamers. The crosses show the distribution for pentamers in equilibrium at temperature $T = 0.5$, whereas the triangles show the distribution of pentamers that have grown from dimers

discussed above. The capillary approximation results are much smaller than our simulation results. Values are approximately linear on the log–log plot shown, indicating a power law dependence on collision frequency (or equivalently, saturation). However, this dependence is stronger for the capillary approximation results (close to S^{M-2} as expected) than for the simulation ones (which are closer to S^{M-3}). Finally Figure 3 shows the energy distribution of the pentamers that grow from dimers, compared to that for an equilibrium distribution of pentamers at temperature $T = 0.5$. The energies of the grown pentamers are much higher than the average equilibrium energy, indicating that these clusters are highly unstable.

Conclusions

We have presented some preliminary results of simulations of cluster growth. We find that far more clusters grow than predicted by the capillary approximation, but that the energy of these grown clusters is much higher than their mean energy in equilibrium. While these two effects act in opposite directions, they indicate that (not surprisingly) the assumptions of the capillary approximations are questionable for very small clusters. We also find that the saturation dependence of the growth fraction is weaker than expected based on equilibrium considerations. Further work is required to determine the significance of these results for nucleation theory. In particular, we want to incorporate the effects of collisions with non-condensing gas, which should act to stabilize the growing clusters.

An Insight into the Failure of Classical Nucleation Theory

Joonas Merikanto, Antti Lauri, Evgeni Zapadinsky, and Hanna Vehkamäki

Abstract We have calculated the work of formation of Lennard–Jones (LJ) and water clusters with Monte Carlo simulations. A comparison to classical nucleation theory shows that down to very small cluster sizes, the liquid drop model accurately describes the work required to add a monomer to the cluster. The liquid drop model only fails in its description of the smallest of clusters. This results in an erroneous absolute value for the cluster work of formation throughout the whole size range.

Keywords Nucleation theory, cluster

Introduction

Besides its shortcomings, the classical nucleation theory (CNT) is the main tool for calculations of nucleation rates in practically relevant systems. However, its predictions are known to be unreliable; experiments on homogeneous nucleation of water[1] show that CNT succeeds reasonably well in predicting the magnitude of nucleation rate at various supersaturations and temperatures, although the temperature dependency of the measured rates is not correctly predicted. On the other hand, measurements on homogeneous nucleation of argon[2] indicate that CNT completely mispredicts the onset pressure of nucleation. If CNT fails in providing the homogeneous nucleation rates of simple substances, it cannot be viewed as a trustworthy tool in predicting binary or ternary nucleation rates for the compounds found in the atmosphere.

The discussion about the failure of CNT has primarily focused on two possible explanations: CNT intrinsically miscalculates the degrees of freedom of nucleating clusters (the Lothe–Pound paradox), and the use of liquid drop (LD) model assigning macroscopic thermodynamic properties to molecular-size systems is inaccurate.

Department of Atmospheric Sciences, P.O. Box 64, 00014 University of Helsinki, Finland

C.D. O'Dowd and P.E. Wagner (eds.), *Nucleation and Atmospheric Aerosols*, 121–125.

In approaches based on statistical mechanics, such as Monte Carlo simulations, the molecular clusters are properly treated as microscopic objects and the degrees of freedom can be counted correctly. Hence, the Monte Carlo calculations are free from the Lothe–Pound paradox. Therefore, such simulations can be used in solving the origin of the failure of CNT. Our results here are based on this type of study. A full description of our results will be published shortly.[3]

Results

The Work of Cluster Formation

The theoretical basis of our simulation method is presented elsewhere.[3,4] Here, we note that starting from the treatment of the cluster-partition function, one can gain a self-consistent cluster size distribution in a saturated or supersaturated vapor. The work of formation of clusters of each size governs this distribution. This work can be written in a form that is comparable to the expression used in the CNT.

We calculate the change in cluster work of formation with respect to cluster size, $\delta\Delta W_n$, both in LJ and water vapors with Monte Carlo simulations. In our simulations the clusters are defined according to the Stillinger cluster definition. The water clusters are simulated with three different potential models, from which one is polarizable. These results are then compared to the predictions of the LD model applied in the CNT. In the LD model, the change in the formation work of the cluster in equilibrium vapor is given solely by the change in cluster surface area, since the clusters are considered to have a bulk surface tension $\gamma\infty$;

$$\delta\Delta W^e_{n,LD} = \Delta W^e_{n,LD} - \Delta W^e_{n-1,LD} = (A_n - A_{n-1})\gamma_\infty. \tag{1}$$

In the above equation, A_n is the equimolar surface of a cluster containing n molecules.

Figure 1 shows the simulated values of $\delta\Delta W_n$ compared to those obtain with Eq. (1). At small cluster sizes the simulation results deviate from those given by the LD model. However, there is a threshold cluster size after which the simulation results overlap with the LD model results for all larger cluster sizes. For these clusters, the slope of the simulated values of $\delta\Delta W_n$ against the change in surface area is given exactly by the model specific value of $A_1\ \gamma\infty$.

The threshold cluster size after which the LD model correctly predicts the change in cluster work contains between 7 and 9 water molecules, or between 20 and 50 LJ particles, depending on temperature. Simulations with all three applied water models give the same threshold cluster sizes, and the curve below the threshold size has the same form in all cases. For a spherically symmetric LJ potential $\delta\Delta W_n$ is a smooth function of the cluster size, whereas the preferred arrangements of molecules govern $\delta\Delta W_n$ in case of water clusters.

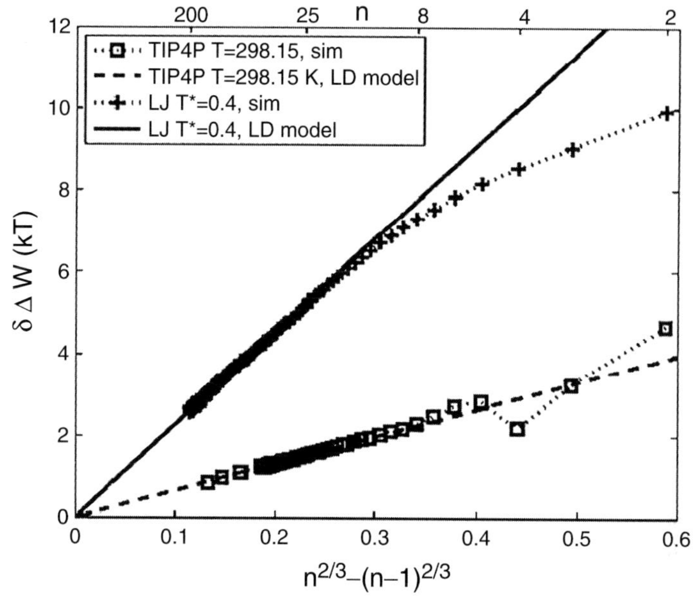

Figure 1 The change in cluster work of formation (y-axis) in a saturated vapor with respect to the change in cluster surface area (lower x-axis). The results from simulations with Lennard–Jones (LJ) clusters and TIP4P water clusters are compared to the predictions of the liquid drop (LD) model. The number of molecules in the cluster is shown in the upper x-axis

The total work of cluster formation of a n-cluster is obtained by summing $\delta\Delta W_n$ over n,

$$ {}_n\Delta W^e_{n,LD} = \sum_{n'=1}^{n} \delta\Delta W_{n'}. \tag{2} $$

In Monte Carlo approaches the work of formation of a monomer cluster is taken as zero, whereas the LD model assigns an unphysical value of $A_1\gamma_\infty$ to a monomer formation. Therefore, the work of formation of a cluster containing less than the threshold number of molecules can be expressed as:

$$ {}_n\Delta W^e_{n,sim} = \Delta W^e_{n,LD} - A_1\gamma_\infty - \sum_{n=2}^{n} b_n, \tag{3} $$

where b_n describes the deviation between the simulated and LD values of $\delta\Delta W_n$. However, above the threshold cluster size b_n terms are zero, and thus

$$ \Delta W^e_{n>n^{trh},sim} = \Delta W^e_{n,LD} - A_1\gamma_\infty - B, \tag{4} $$

where B is given by the summation over b_n and is a constant for all large cluster sizes. Therefore, for these cluster the scaling law proposed by McGraw and Laaksonen[5,6] applies, stating that the correct work of formation differs from the LD model work of formation only by a constant.

The Surface Tension Size Dependence

It is important to realize that while the change of work of formation for clusters larger than the threshold size is governed by the bulk surface tension, these clusters do not actually exhibit a bulk surface tension. The size-dependent surface tension acting on the equimolar surface, $\gamma_n(n)$, can be calculated from

$$\gamma_e(n) = \frac{\Delta W^e_{n,sim}}{A_n} = \gamma_\infty - \frac{A_1\gamma_\infty + \sum_{n'=2}^{n} b_n}{A_n} \tag{5}$$

nn which is in close connection to that proposed by McGraw and Laaksonen,[6] where last term in the equation is given by a constant divided by the surface area of the cluster. Figure 2 shows our calculated surface tension size dependence of LJ clusters at the reduced temperature $T = 0.7$ compared to those proposed by McGraw and Laaksonen,[6] and Tolman.[7] Our calculated surface tension agrees to McGraw and Laaksonen expression above the threshold size, but deviates from it at smaller sizes, tending to zero for a monomer cluster.

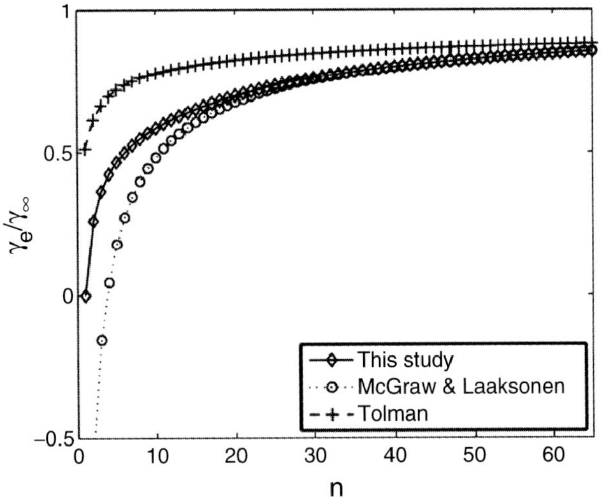

Figure 2 The size dependence of the surface tension on the equimolar surface (*y*-axis) with respect to the cluster size (*x*-axis) for Lennard–Jones clusters at reduced temperature $T^* = 0.7$. The Tolman expression shown for comparison is gained by using a recent estimate 0.16 σ for the Tolman length

Conclusion

Our analysis reveals how the self-consistent CNT, valid for a monomer cluster, evolves to the McGraw and Laaksonen scaling law, valid for a cluster larger than a threshold size. This threshold size appears to be rather small in general, consisting only a few or few tens of molecules depending on the interaction potential and temperature. Our results for water clusters explain why CNT succeeds fairly well in producing the experimental nucleation rates of water, and fails drastically in case of argon. The calculated values of water nucleation rates and argon nucleation onset pressures from our simulation results are in a good comparison to experiments; a more thorough comparison can be found from Ref. 3.

References

1. Wölk, J. and Strey, R., *J. Chem. Phys.*, **105**, 11683 (2001).
2. Fladerer, A. and Strey, R., *J. Chem. Phys.*, **124**, 164710 (2006).
3. Merikanto, J., Zapadinsky, E., Lauri, A., and Vehkamäki, H., *Phys. Rev. Lett.* (in press).
4. Merikanto, J., Vehkamäki, H., and Zapadinsky, E., *J. Chem. Phys.*, **121**, 914 (2004).
5. McGraw, R. and Laaksonen, A., *Phys. Rev. Lett.*, **76**, 2754 (1996).
6. McGraw, R. and Laaksonen, A., *J. Chem. Phys.*, **106**, 5284 (1996).
7. Tolman, R.C., *J. Phys. Chem.*, **17**, 333 (1949).

Nucleation Chemical Physics: From Vapor-phase Clusters to Crystals in Solution

Shawn M. Kathmann

Abstract We discuss some of the relevant issues concerning the chemical physics of vapor-phase and condensed-phase nucleation processes. The presentation will highlight issues involved in going from interaction potentials to thermodynamics and kinetics.

Keywords Nucleation, interaction potentials, statistical mechanics, charge transfer, clusters, crystals, luminescence

Introduction and Background

Both vapor-phase clusters and condensed-phase crystals are important in a wide variety of fundamental and applied problems in chemical physics. Favorable fluctuations in a supersaturated phase generate clusters of the new phase – exactly how one defines these new clusters as distinct from the mother phase represents a continuing challenge in molecular theories of nucleation. These incipient clusters can form homogeneously within the mother phase or heterogeneously on seeds, dust, impurities, ions, or others stability-inducing atomic/molecular structures (e.g., steps, edges, vacancies, etc.). Upon reaching a critical size the clusters may grow to macroscopic dimensions if enough nucleating material is present in surrounding environment or until relaxation processes dominate bringing the phase transformation to completion.

The central quantities of interest in either vapor-phase or condensed-phase nucleation are the free energies and rate constants for the individual reaction steps underlying the phase transformation. Modern computational chemical physics has advanced to such a degree that its accuracy can, at times, match experimental observation. There are many challenges in the accurate treatment of both vapor and crystal nucleation. Calculations of accurate reaction rate constants require reliable

Molecular Interactions and Transformations Group, Fundamental Sciences Directorate, Pacific Northwest National Laboratory, Richland, WA 99352, USA

C.D. O'Dowd and P.E. Wagner (eds.), *Nucleation and Atmospheric Aerosols*, 126–130.

representations of molecular interaction potentials typically beyond what empirical interaction potentials can provide. However, the direct application of high-level *ab initio* methods with appropriate unbiased statistical mechanical sampling of nucleation thermodynamics and dynamics remains computationally intractable due to the large number of relevant degrees of freedom and the sheer volume of phase space to be explored. Fluctuations leading to the formation of clusters of the new phase are often so rare (under common ambient conditions) that one may never "see" a clustering event occur on the time scales of typical molecular dynamics simulations – the majority of time being spent searching irrelevant regions of the configuration space. Methods to bias the ensemble sampling can be employed to bring the clustering event into focus during reasonable simulation times. Understanding clustering thermodynamics requires accurate free energies in the vapor phase and/or potentials of mean force along the relevant reaction coordinates in the condensed phase. There are also quantum effects that influence clustering like zero-point-energy and tunneling, which can be especially important for processes involving light atoms such as hydrogen.

Compared to gas-phase cluster formation, which in many cases can be viewed as isolated encounters of two reactant species, the constant proximity of solvent molecules to the crystallizing solute can profoundly alter reaction kinetics and thermodynamics. Solvents can influence the course of crystallization in many ways – providing an efficient route for energy transfer, initiating reactions that would not occur in the gas phase, stabilizing reaction intermediates, donating or accepting electrons, limiting the mass transfer of reacting species, stabilizing transitions states and thus modifying reaction barriers, and moderating collision dynamics. Moreover, solvents affect solubility of reactants and intermediate species – in order for a reaction in condensed phase to occur the reactants must first dissolve in solution before they can encounter one another. Solvation ultimately influences the electronic structure of the solute species and thus alters their reactivity. Computing reaction barriers and understanding mechanisms in condensed phase is much more complicated than those computed in the gas phase because the reactants are surrounded by solvent molecules and the configurations, energy flow, and electronic structure of the entire statistical assembly must be considered.

Controlled crystallization is currently of great interest to the broader scientific community because crystal properties, such as electronic and optical properties, depend strongly on size, shape, and composition that differ significantly from either bulk or isolated molecules. The desire is to control the crystallization process to "tune" desired characteristics. Nucleation is an extremely sensitive process, and it is exactly this sensitivity that has impeded exploiting it to exercise this control. The physical reason why nucleation is so sensitive can be understood intuitively. Since the nucleation process involves many (typically 10s–100s) mechanistic steps, small changes in the thermodynamics and kinetics of each step can be amplified over the total number of steps required to reach the critical cluster. Another way of looking at nucleation sensitivity is as a measure of the degree to which a system can extend into a region of thermodynamic metastability (e.g., scratching the side of a beaker to induce crystallization). The sensitivity of nucleation [1] has also been exploited partially as a means to detect mercury vapor contamination down to levels of one part in 10^{14}. Subtle variations in interaction potentials, charge transfer, quantum treatment of nuclear degrees of freedom, etc. can alter the nucleation rate by many orders of magnitude and thus motivates the search for relative effects on nucleation in solution, including the consequence of various seeds.

Finally, crystallization of several materials (including NaCl and AgCl) in solution is accompanied by the emission of light, appropriately called *crystalloluminescence*, the origin of which [2] is not currently understood. We are investigating the nature of crystalloluminescence systems with high-level electronic structure calculations. The results of these calculations will be reported for isolated clusters of ions in the gas phase, microsolvated ions in the gas phase, and in the bulk solution phase using time-dependent density functional theory.

Results and Discussion

We have shown [3] how various seed ions can have a profound influence on the ion-induced nucleation thermodynamics of water from the vapor and how a molecular-level description radically departs from a continuum treatment such as that given by Classical Nucleation Theory. This work also addressed the century-old controversy concerning water's sign preference showing that the ion's chemical identity, not just the sign of the charge, plays an essential role in determining cluster thermodynamics. We have also studied [4] the influence of anharmonicity on the chemical potentials of $Na^+(H_2O)_i$ and $Cl^-(H_2O)_i$ clusters. We compare in Figure 1 the chemical potentials using the full anharmonic results using 10s–100s of millions of configurations to the Rigid-Rotor Harmonic Oscillator Approximation (RRHOA) at 298 K. The agreement between experiment and the full anharmonic results is better than 1 kcal/mol. This comparison shows that using the RRHOA without quantifying its effect is unjustified. It is important to point out that we are referring to two different types of anharmonicity: (1) local anharmonicity of the vibrations for a given cluster configuration and 2) global anharmonicity resulting from sampling between the large number of configurations available within the relevant volume of configuration space. Additionally, using multiple minima in the RRHOA partition function (the "superposition" approximation) still underestimates the full anharmonic free energy.

We are currently simulating aqueous salt clusters of various sizes to understand the free energy landscape as ions are added to aqueous clusters (Top: Figure 2). We will study the size- and concentration-dependent thermodynamics for these finite aqueous salt clusters using empirical interaction potentials and test their convergence to the bulk limit. Since the bulk limit of the size-dependent chemical potentials can be used to predict the concentration-dependent surface tension and the partial vapor pressure of water above the solution, these calculations will provide essential benchmarks of the validity of empirical potentials in describing the properties of concentrated salt solutions. Figure 2 (bottom) also shows the difference in charge transfer between NaCl and AgCl.

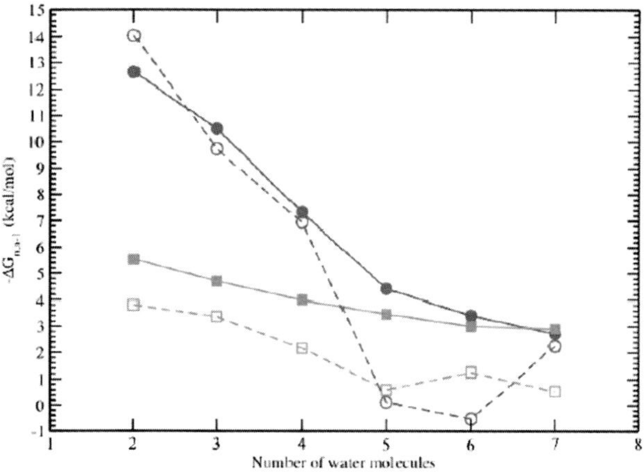

Figure 1 Anharmonic Gibbs free energy differences (**solid lines**) for $Na^{+}(H_2O)_i$ (**circles**) and $Cl^{-}(H_2O)_i$ (**squares**) compared to the RRHOA results (**dashed lines**) at 298 K

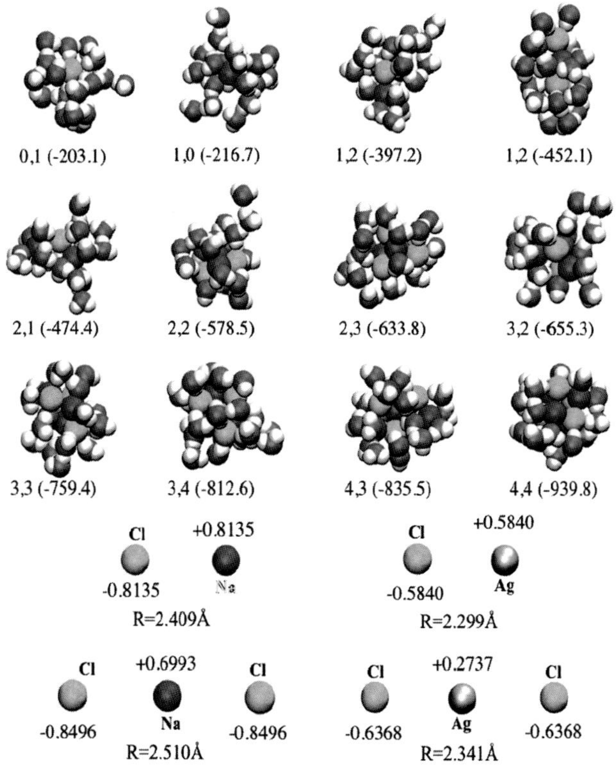

Figure 2 (**Top**) Anharmonic Helmholtz free energies for $Na_iCl_j(H_2O)_{20}$ clusters at 298 K. The clusters are denoted with as: *i*, *j* (A[$Na_iCl_j(H_2O)_{20}$]) where the cluster Helmholtz free energies are in kcal/mol. (**Bottom**) Comparison of *ab initio* charges (Gaussian98/QCISD/aug-cc-pvtz-pp) between NaCl, $NaCl_2$, AgCl, and $AgCl_2$

Acknowledgments This work was supported by the Chemical and Material Sciences Division, Office of Basic Energy Sciences, Department of Energy. The Pacific Northwest National Laboratory is operated by Battelle for the US Department of Energy.

References

1. Ereifej, H., Doster, G., Schmitt, J., and Story, J., Extreme sensitivity in trace element detection, *Appl. Phys. B*, **68**, 141–144 (1999).
2. Gibbon, M.A., Sopp, H., Swanson J., and Walton A.J., Image intensifier studies of crystalloluminescence of NaCl, *J. Phys. C*, **21**, 1921–1934 (1988).
3. Kathmann, S.M., Schenter, G.K., and Garrett, B.C., Ion-induced nucleation: the importance of chemistry, *Phys. Rev. Lett.*, **94**, 116104 (2005).
4. Kathmann, S.M., Schenter, G.K., and Garrett, B.C., The critical role of anharmonicity in the aqueous ionic clusters relevant to nucleation, *J. Phys. Chem. B*, in press (2007).

Homogeneous Nucleation: What I Understand, What I Somewhat Understand, What I do not Understand

Professor Joseph L. Katz

Keywords Homogeneous nucleation

More than one century ago Gibbs laid some of the foundation of our present understanding of nucleation. Since then there has been some progress. We now have a kinetic scheme which I think is valid for all fluid systems which are sufficiently dilute, *when* we know the kinetics of the individual steps in the process. For such systems the nucleation process can be described as the series of kinetic steps which cause the formation of clusters which are sufficiently large that their probability to grow is larger than their probability to become smaller. For low-density gasses, Maxwell and others have provided us with equations for the collision rates of molecules. These we can use to tell us the accretion rate of molecules by the clusters. For somewhat higher-density gasses, Boltzmann's equation, combined with Enskog's approximate solution of his equation, tells us how to calculate the needed collision rates.[1]

I think that the above-mentioned scheme is quite general, i.e., we can use it for all dilute fluid to fluid nucleation processes for which one is able to create a reasonably accurately model for the arrival rates. However, I have no idea how to describe homo-geneous nucleation when the density is so large that the concept of discrete clusters of molecules no longer has meaning.

To be able to make a predictive theory one also needs values for the rates at which molecules leave the various-sized clusters. If we had a kinetic model for the evapor-ation rate, one then would have a quantitative predictive theory which would need only laboratory measurable quantities: the vapor pressure, the surface

Department of Chemical and Biomolecular Engineering, Johns Hopkins University, Baltimore, MD 21218, USA

[1] Enskog theory has not been used at for nucleation higher gas densities, probably because one needs to know the intermolecular potential and then to use it to evaluate complicated integrals which probably are known only for spherical molecules. Note that corrections in arrival rate are exponential, i.e., they change the effective supersaturation and thus can have an enormous effect on the rate of nucleation.

C.D. O'Dowd and P.E. Wagner (eds.), *Nucleation and Atmospheric Aerosols*, 131–133.

tension, etc. and the operating condition variables: temperature, concentration, substance, solvent, etc. to be used.

For gas to liquid nucleation one can use the Kelvin equation as our evaporation rate model. However, the Kelvin equation is derived using equilibrium thermodynamics. In a supersaturated gas there is equilibrium *only* for the clusters of critical size. In principle, using it for all other sizes is incorrect. Evidence that doing so indeed is incorrect is provided by the comparison of its predictions with collection of experimentally measured nucleation rates. The Kelvin equation version of theory results in an equation for the rate of nucleation which is almost identical to the equation for the self-consistent version of the Classical Theory of Nucleation. The predicted rates of nucleation rate differ negligibly. We all know that for all, carefully measured fluids the self-consistent version of classical theory fairly accurately predicts the measured *supersaturation* dependence of the nucleation rate. (The Kelvin equation makes predictions which are identical to well within the accuracy of this statement.) Unfortunately, we all also know that the predictions of this equation (and also of the Kelvin equation) do *not* agree with the measured *temperature* dependence of the nucleation rate.

Farkas (who credits Szilard) created a scheme for obtaining these evaporation rates by equating them to evaporation rates of identically sized cluster in a supersaturated vapor constrained to be in "equilibrium". I have created one which equates the evaporation rates to the evaporation rates from identically sized clusters, but in a truly equilibrium vapor, the saturated vapor at the same temperature. To obtain these evaporation rates one then assumes that the clusters are identical to *small droplets* of liquid density with liquid surface tension, etc. Improvements have consisted of using a variety of thermodynamic or statistical mechanics models, models which try to correct for perceived differences such as entropy effects, a better estimate of the relationship of droplets to clusters, attributing a size dependence to the surface tension, etc. I know these well; my name is on several such papers. Nonetheless, I remain uncertain about the relevance of any of these models to a quantitative description the rate at which molecules leave clusters.

When thinking about the homogeneous nucleation of superheated liquids I am quite puzzled. Homogeneous Nucleation theory very accurately predicts the temperatures at which explosive vaporization occurs. This is true for every substance that has been measured (except for water). But this the same theory of nucleation we discussed above, a theory based on formation of clusters by the accretion and loss of individual molecules (and for some substances of dimers, trimers, etc.). Superheated liquids nucleate by forming a critical-sized void, not a large-enough cluster of "gas" molecules. What is being accreted and lost is free volume. Free volume is continuous, not discrete. If the kinetics is based on the wrong variable why does it work?

I know that I do not understand the nucleation of crystals, neither from a melt (ice from water) nor from solution. I somewhat understand the nucleation of a glass from a supersaturated vapor. But a glass is not a crystal, it is a frozen fluid. Its nucleation is similar to that of a liquid from a vapor.

I think that I do understand the first half of the nucleation process that leads to the nucleation of a crystal from a gas. What first occurs is the nucleation of a supercooled liquid. This then is followed by the nucleation of the crystal from the supercooled liquid. I assert this sequence is necessary because it is impossible to create an ordered cluster of molecules if they arrive one at a time without having first created a stable disordered cluster. For stable or growing cluster, if the mobility of the molecules is large, then the molecules can continuously sample different configurations until they create an ordered one. When the mobility is not large enough one gets a glass.

In subcritical "crystals" there are two variables that relate to the rate of nucleation. One is the number of molecules in the cluster; the other is related to the degree of geometric ordering of molecules. The general interrelationship of order with size is a mystery to me. But in two of the limiting cases, I think I do have some understanding.

In dilute solutions of those substances for which *ordering is easy*, it is the net rate of aggregation of the molecules which determines the rate of nucleation. In other words, because ordering is easy, this process is similar to that of a gas nucleating to a liquid. The only key difference is that arrival rate is controlled by diffusivity through the solvent instead of Maxwell collision rate theory.

In a melt of those substances for which ordering is difficult (i.e., critical sizes are large), it is the net rate of ordering which determines the rate of nucleation. In most melts the molecules which will end up in a crystallite already are almost in their desired locations (but not rotational orientation) since (for most substances) crystal densities are almost the same as liquid densities. For these, the nucleation barrier is getting the molecules to line up correctly. This "barrier" is determined by the rate at which one obtains *useful* fluctuations. The useful fluctuations usually are not ones which at each step create a cluster of lowest energy. Most of these lead to structures for which one would need a step of size many kT to proceed. The low-energy fluctuations of interest are those which need only a series kT-sized energy movements to place the molecules into useful locations. Many of the movements will produce kT, thus furthering the process. The end of the "critical" part of the growth paths is reached when a crystallite has been created which much more than kT lower in energy than the liquid. Note that this path usually will not lead to the lowest energy crystal structure. As further growth occurs, fluctuations and the increasing difference in energy, at high temperatures will cause the crystal structure to jump into a lower-energy structure.

Homogeneous Nucleation Rate in Supersaturated Water Vapor

V. Ždímal and D. Brus

Abstract The rate of homogeneous nucleation in supersaturated vapors of water was studied experimentally using a static diffusion chamber. Helium was used as a carrier gas. Droplets grown by condensation were observed by illuminating the chamber across its whole height with a flattened laser beam and recording digital images of droplets trajectories. Image analysis was used to determine local values of nucleation rate from droplets "starting positions". Rates were studied in the range $3 \times 10^{-1} - 3 \times 10^{2}$ cm^{-3} s^{-1} at four isotherms: 290, 300, 310, and 320 K. Measured isothermal dependencies of nucleation rate of water on supersaturation were compared with prediction of classical theory of homogeneous nucleation.

Keywords Homogeneous nucleation, static diffusion chamber, supersaturation, condensation, water

Introduction

Nucleation is a critical step in vapor to liquid phase transition. In most situations the heterogeneous nucleation is important, that occurs under presence of foreign nuclei, aerosol particles, ions, or surfaces. If they are absent, the process takes place by vapor condensation on its own embryos. This process is termed homogeneous nucleation and in unary vapor it represents the simplest system for both experimental and theoretical investigations of nucleation. The experimental techniques used for this purpose are reviewed by Heist and He (1994). Among them static diffusion chambers are used both for determination of critical supersaturation and for nucleation rate measurements. In these studies, the nucleation rate is usually derived from the integral flux of droplets recorded by an optical counter.

Laboratory of Aerosol Chemistry and Physics, Institute of Chemical Process Fundamentals, AS CR, v.v.i., Rozvojová 135, 16502 Prague, Czech Republic

C.D. O'Dowd and P.E. Wagner (eds.), *Nucleation and Atmospheric Aerosols*, 134–138.

The inherent presence of nonuniformities in temperature and supersaturation, and the way of detection cause difficulties in comparison of results with the results obtained by other techniques. A couple of years ago, the way how nucleation is detected in the static diffusion chamber was modified (Smolík and Ždímal 1994). This approach allowed us to determine the rate of nucleation in dependence on temperature and supersaturation, independently on any nucleation theory. Here this technique is used to measure the homogeneous nucleation rate in supersaturated vapors of water in helium at four temperatures: 290, 300, 310, and 320 K. It enables us to compare our data both with theoretical predictions and with other data obtained using various expansion devices (Allen and Kassner 1969; Miller et al. 1983; Viisanen et al. 1993; Viisanen et al. 2000; Wölk et al. 2000; Schmit et al. 2000; Wölk and Strey 2001) and diffusion chambers (Heist and Reiss 1973; Mikheev et al. 2002).

Experimental

The static diffusion chamber used in this research consists of two duralumin plates separated by a 162 mm i.d. optical glass ring, 25 mm in height, sealed in plates by EPDM sealing. The bottom plate with approximately 1 mm film of liquid investigated is heated, vapor diffuses through stagnant carrier gas and condenses on the cooler top plate. The top plate is slightly conical, so that the condensate flows to its edge and along the glass wall back to the pool. If the amount of carrier gas is properly selected, so that convection is avoided and the chamber operates at the steady state, temperature T and partial vapor pressure P decrease almost linearly from the bottom to the top. Since, the equilibrium vapor pressure Peq is approximately an exponential function of temperature, it decreases with the height of the chamber more rapidly than the actual vapor pressure. The vapor in the chamber therefore becomes supersaturated with supersaturation S (defined as the ratio of the partial vapor pressure to the equilibrium one S = P/Peq) having its maximum close to the top plate of the chamber. By increasing the temperature difference between the plates one can increase supersaturation gradually and arrange the state in which supersaturation is sufficient for homogeneous nucleation to begin. The droplets, once formed by nucleation, grow rapidly to visible size and fall down to the liquid pool.

Calibrated thermocouples are used to measure temperature of both plates; pressure transducer allows to measure total pressure inside the chamber. By solving appropriate transport equations (e.g., Katz 1970), profiles of temperature and partial vapor pressure (and hence supersaturation) are found.

To observe nucleation in the chamber, the interior of the chamber is illuminated from side by a flat laser beam passing through the center of the chamber. Trajectories of the droplets, formed inside this beam, are visible and could be photographed using a camera in which axis is positioned exactly perpendicular to the beam. Each droplet photographed might be characterized by its starting point (point of its origin) and its trajectory ending in the liquid pool at the bottom plate. After evaluating enough particles (starting points) in one experiment, one gets the distribution of homogeneous nucleation rates as a function of height in the chamber. The measured distribution is then related to the corresponding values of temperature and supersaturation. This procedure can be seen in Figure 1. On the left side of this figure the trajectories of droplets as they appear on photographs are shown schematically. In the three windows on the right there are local values of nucleation rate J as determined from photographs, and profiles of temperature T and supersaturation S calculated from the model of transport.

This approach has a major advantage: it allows us to determine local nucleation rate as a function of local temperature and local supersaturation, independently on any nucleation theory (Brus et al. 2005). Major drawback of this approach is the necessity to manually process and evaluate large numbers of photographs.

In order to keep the advantages of the photographic approach and at the same time minimize its drawbacks, the photographic camera was replaced by a digital CCD camera. A 16 bit camera ST-7 (SBIG, USA) widely used in astronomy for imaging of faint objects was chosen for its sensitivity and acceptable resolution (512 × 756 pixels). The camera is provided with a Nikkor lens. Recorded images are downloaded to a PC through its USB port. The camera is software controlled which makes possible to take long series of pictures with predetermined parameters. The electronic noise of the camera is negligible in comparison with the optical noise from the experiment. Quality of pictures taken by this camera is comparable with quality of photographs, but the digital pictures can be evaluated automatically using image analysis.

Starting points of droplet trajectories are detected as local maxima of normalized correlation with a template (Šonka et al. 1998). On a Pentium PC under Linux the

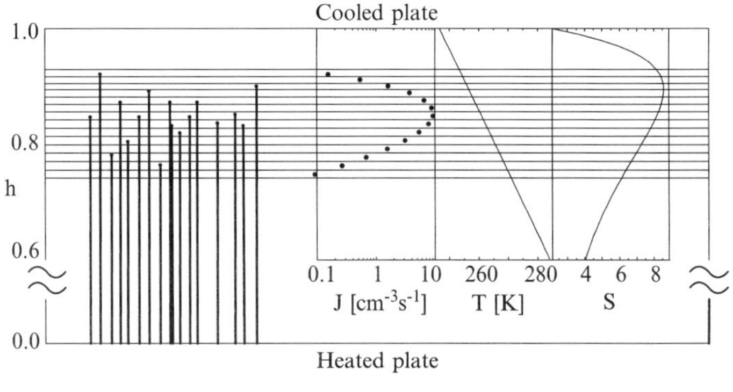

Figure 1 Nucleation rate as function of temperature and supersaturation – way of evaluation

processing is completed within seconds. Position of liquid films, defining the coordinate system, is found by a randomized optimization approach. The criterion function – the sum of intensities along a line – is approximately evaluated from a set of samples (typically 10% of pixels along a line). Exhaustive search is performed in the space of nearly horizontal lines. The estimated error in determining the starting vertical position of a droplet is less than 1 pixel. This approach slightly surpasses in precision former visual evaluation of photographs. The ability to process a series of pictures automatically makes it much more effective (Ždímal et al. 2000).

Results

The technique described above was used to determine the dependence of homogeneous nucleation rate of water on supersaturation at four isotherms. It means, that in all experiments on one isotherm the temperatures of the plates were chosen so, that temperature of the isotherm laid very near to the experimentally found centre of the nucleation zone. Results of these experiments are given in Figure 2. It can be seen, that the experimental data is in reasonable qualitative agreement with the prediction of classical theory of homogeneous nucleation. However, the experimental points are about two orders in magnitude lower than the corresponding theoretical curves.

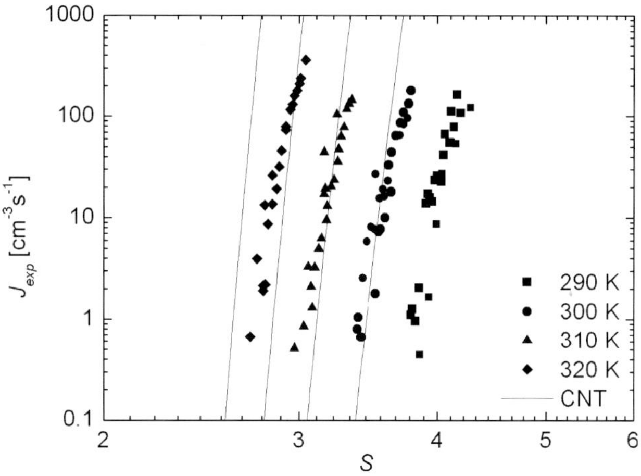

Figure 2 Comparison of the measured nucleation rate with prediction by classical nucleation theory (CNT)

Acknowledgments The authors of this paper gratefully acknowledge the support by Czech Science Foundation, grant No. 101/05/2524. The authors would also like to thank the head of the laboratory, Dr. Jiří Smolík, for his long-term support of nucleation research. Dr. Hermann Uchtmann from Philipps-University Marburg deserves thanks for his advice concerning the surface coating of the chamber plates.

References

Allen, E. and Kassner, J., *J. Colloid Interface Sci.*, **30**, 81 (1969).

Brus, D., Hyvärinen, A.-P., Ždímal, V., Lihavainen, H., *J. Chem. Phys.*, **122**, 214506 (2005).

Heist, R.H. and He, H., *J. Phys. Chem. Ref. Data*, **23**, 781 (1994).

Heist, R.H. and Reiss, H., *J. Chem. Phys.*, **59**, 665 (1973).

Katz, J.L., *J. Chem. Phys.*, **52**, 4733 (1970).

Mikheev, V.B., Irving, P.M., Laulainen, N.S., Barlow, S.E., Pervukhin, V.V., *J. Chem. Phys.*, **113**, 10772 (2002).

Miller, R., Anderson, R.J., Kassner, J.K., and Hagen, D.E., *J. Chem. Phys.*, **78**, 3204 (1983).

Schmitt, J.L., Brunt, K.V., and Doster G.J., in Wölk et al., p. 51 (2000).

Smolík, J. and Ždímal, V., *Aerosol Sci. Technol.*, **20**, 127 (1994).

Šonka, M., Hlaváč, V., and Boyle, R.D., *Image Processing, Analysis and Machine Vision*, Boston, MA: PWS (1998).

Viisanen, Y., Strey, R., and Reiss, H., *J. Chem. Phys.*, **112**, 8205 (2000).

Viisanen, Y., Strey, R., and Reiss, H., *J. Chem. Phys.*, **99**, 4680 (1993).

Wölk, J. and Strey, R., *J. Phys. Chem. B*, **105**, 11683 (2001).

Wölk, J., Viisanen, Y., Strey, R., *Proceedings of the 15th International Conference on Nucleation and Atmospheric Aerosols*, edited by B. Hale and M. Kulmala (American Institute of Physics, Melville, NY, 2000), Vol. 7, p. 7.

Ždímal, V., Smolík, J., Hopke, P.K., Matas, J., in Wölk et al., p. 311 (2000).

Displacement Barrier Heights from Experimental Nucleation Rate Data: Scaling and Universality

Jussi Malila[1], Antti-Pekka Hyvärinen[2], Yrjö Viisanen[2], and Ari Laaksonen[1]

Abstract Experimental nucleation data was analysed in the framework of scaling relations stemming from the nucleation theorem. Differences between the classical nucleation theory and experimental data were given as displacement barrier heights, $D(T)$. For molecular liquids, most of the analysed data gave $D(T)$ with similar magnitude and trend. Also other studied substances follow similar temperature trend, though it is not always clear if deviations from this trend are due to uncertainty in nucleation rate data or assumptions behind the scaling relation.

Keywords Homogeneous nucleation, nucleation experiments, classical nucleation theory, nucleation theorem, displacement barrier height, scaling.

Motivation

Nucleation phenomena are ubiquitous in the nature, and they influence, e.g., many processes in the atmosphere.[1] The basic tool to analyse these phenomena is the classical nucleation theory (CNT),[2] which has been applied to describe the nucleation behaviour in different fields of science more or less successfully.[1,3] However, from vapour to liquid nucleation experiments it has been learned that CNT predicts the supersaturation dependence of nucleation rate correctly, while failing to predict the temperature dependence. To overcome this problem, several phenomenological theories have been presented, which usually can successfully improve CNT predictions for some substances, but fail to describe nucleation behaviour of others. Here we present results of a nucleation theorem-based analysis of experimental nucleation rate data of a variety of substances, and discuss the general nature of corrections needed to bring CNT unison with experiments.

---

[1] *University of Kuopio, Department of Physics, P.O. Box 1627, FI-70211 Kuopio, Finland*

[2] *Finnish Meteorological Institute, P.O. Box 503, FI-00101 Helsinki, Finland*

C.D. O'Dowd and P.E. Wagner (eds.), *Nucleation and Atmospheric Aerosols*, 139–143.

Theory

The observable quantity in nucleation experiments is the nucleation rate, J, giving the number of new stable clusters appearing per time and volume as a function of partial vapour pressure and temperature. This experimental nucleation rate is compared to the classical Becker-Döring nucleation rate expression,[2]

$$J_{\text{CNT}} = \sqrt{\frac{2\sigma}{m\pi}}\, vS \left(\frac{p}{kT}\right)^2 e^{-W_{\text{CNT}}/kT}. \tag{1}$$

Here S is the saturation ratio, v the molecular volume in the (bulk) liquid, m mass of the molecule, p the equilibrium vapour pressure, and $W_{\text{CNT}} = 4\sigma R^2/3$ the work of formation, with σ the (planar) surface tension and R the radius of the critical cluster (assumed spherical). In Eq. (1), the $1/S$ correction[4] that makes the cluster distribution to satisfy the law of mass action is used, and it is assumed that the mass accommodation coefficient is unity[5].

Based on the model independent nucleation theorem (for a review, see Ref. 6) and the Kelvin equation, McGraw and Laaksonen[7] proposed the following scaling for W:

$$-\mathrm{D}(T) = W - W_{\text{CNT}}. \tag{2}$$

Here $D(T)$, which is a function of temperature only, is the correction between the "true" free energy of formation and the CNT prediction and is known as the displacement barrier height. As the kinetic prefactor in Eq. (1) is a slowly varying function of S and T when compared to the exponential Boltzmann factor, $D(T)$ is very nearly given as $\ln (J/J_{\text{CNT}})$. Furthermore, it is known that when measured and calculated nucleation rates are presented in ($\ln J$, $\ln J_{\text{CNT}}$) -coordinates, they appear as straight lines. This implies that $D(T)$ can be presented in functional form $A + BT$ (we note that Strey and co-workers[8] ended up with form $A + B/T$ based on similar arguments and experimental results).

Data Analysis

Experimental Data

A review of experimental methods used to measure nucleation rate is given by Heist and He,[9] including also review of nucleation data prior to 1992. Most of the earlier data concentrates on determination of the critical saturation ratio for given on-set nucleation rate and temperature. Recently, more quantitative nucleation rate measurements have been done on various substances, suiting better to the critical evaluation of various nucleation theories. Because it is impossible to make reference here to all the data-sets used to extract $D(T)$ from experimental data, as well as to parameterisations of thermophysical properties needed to evaluate J_{CNT}, these are given in the supplementary web site at http://physics.uku.fi/~jmalila/nucleation_data.html. Only nucleation experiments, where the total pressure (vapour and carrier gas) was below 1.5 atm were used for data extraction, in order to avoid effects stemming from non-ideal behaviour of carrier gas.[10]

Results and Discussion

Fits for parameters A and B in reduced form for different substances are given in Table 1. In Figure 1, results from nucleation rate measurements of n-pentanol and four n-alkanols are depicted. The data of each compound falls approximately into a single line as predicted. There seems to be general trend with heavier n-alkanes having systematically larger a and smaller b. For n-alkanols, results from some experimental set-ups (laminar flow diffusion chambers, and to lesser extent, from supersonic nozzles and some expansion cloud chambers) tend to have different temperature dependence than data from other experimental set-ups, as already known,[11] part of which may be due to pressure effects at very low total pressures,[10] though the exact nature of the inert carrier gas does not affect the results. Results are very sensitive to the measured T-dependence, as can be seen from the data of n-butanol, where exclusion of inconsistent data-sets leads to better agreement with other n-alkanols (results in parenthesis). In Table 1, also examples of other substances

Table 1 Examples of linear least squares fits of displacement barrier height for different substances: $D(T)/kT = a + bt$, where $t = T/T_c$. N is the number of data points used for each fit

Substance	*a*	*b*	*N*
A. $C_nH_{2n+1}OH$, n = 0–5 (water and n-alcohols)			
–water	59.054	−140.474	353
–methanol	49.267	−133.634	25
–ethanol	72.228	−149.641	78
–n-propanol	32.843	−67.853	53
–n-butanol	14.637 (58.679)	−20.212 (−117.052)	577 (38)
–n-pentanol	82.651	−164.968	1,327
B. C_nH_{2n+2}, n = 7–10 (n-alkanes)			
–n-heptane	56.023	−122.385	16
–n-octane	66.778	−141.610	38
–n-nonane	68.969	−146.194	134
–n-decane	81.553	−164.660	37
C. Metals			
–caesium	−6.289	−166.849	65
–mercury	−1,621.290 (188.162)	11,265.682 (−100.166)	8 (6)
D. Argon	391.620	−893.989	182

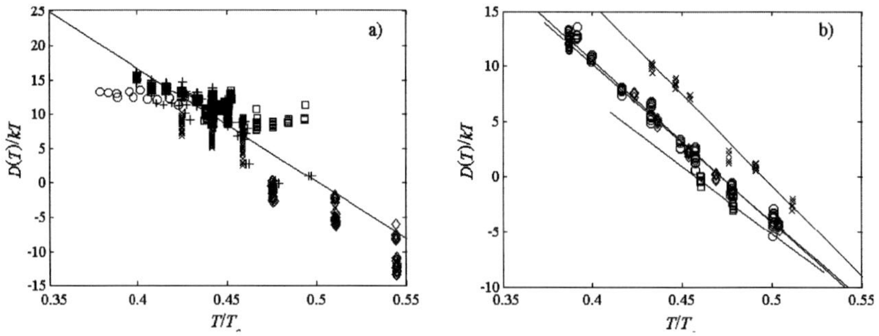

Figure 1 Displacement barrier heights from experimental data for (a) *n*-pentanol and (b) *n*-alkanes as a function of reduced temperature (T_c is the critical temperature). The equations for linear fits are given in Table 1. In (a) different symbols present different experimental techniques: × stands for piston-expansion tube, ϒ for laminar flow diffusion cloud chamber, ◊ for upward static diffusion cloud chamber, o for supersonic nozzle, and + for different expansion cloud chambers, respectively. In (b), different symbols present different alkanes: *n*-heptane (□), *n*-octane (◊), *n*-nonane (o), and *n*-decane (×)

studied are given: for caesium, results are consistent with molecular liquids, but for last two elements in Table 1, Hg and Ar, the method seems to give distinctly different results. For mercury, results agree better with other substances, if measurements from two highest temperatures – which were treated already differently when the raw data was analysed[12] – are excluded (results in parenthesis). For argon, a possible, though not only, reason for dissimilar behaviour can be inaccuracy of equilibrium vapour pressure data at low temperatures. Overall, the suggested fit does provide a suitable scaling and giving coefficients *a* and *b* with the same order of magnitude for all substances studied with some exceptions. It is still unclear if these exceptions are due to inaccuracies in experimental nucleation rate or physicochemical data or assumptions behind the scaling relation (2).

Acknowledgments This work was supported by the Academy of Finland. All the researchers who provided us with their data in electronic form are warmly acknowledged.

References

1. Kulmala, M., *Science*, **302**, 1000–1001 (2003); Cantrell, W. and Heymsfield, A., *Bull. Am. Meteorol. Soc.*, **86**, 797–807 (2005).
2. Becker, R. and Döring, W., *Ann. Phys. (Leipzig)*, **24**, 719–752 (1935).
3. Donn, B. and Nuth, J.A., *Astrophys. J.*, **288**, 187–190 (1985); Kämpfer, B., *Ann. Phys. (Leipzig)*, **9**, 605–635 (2000); Zandi, R., van der Schoot, P., Reguera, D., Kegel, W., and Reiss, H., *Biophys. J.*, **90**, 1939–1948 (2006).
4. Courtney, W.G., *J. Chem. Phys.*, **35**, 2249–2250 (1961).
5. Davis, E.J., *Atmos. Res.*, **82**, 561–578 (2006).
6. Kashchiev, D., *J. Chem. Phys.*, **125**, 014502 (2006).

7. McGraw, R. and Laaksonen, A., *Phys. Rev. Lett.*, **76**, 2754–2757 (1996).
8. Strey, R., Wagner, P.E., and Schmeling, T., *J. Chem. Phys.*, **84**, 2325–2335 (1986); Wölk, J., Strey, R., Heath, C.H., and Wyslouzil, B.E., *J. Chem. Phys.*, **117**, 4954–4960 (2002).
9. Heist, R.H. and He, H., *J. Phys. Chem. Ref. Data*, **23**, 781–804 (1994).
10. Heist, R.H., Ahmed, J., and Janjua, M., *J. Phys. Chem.*, **99**, 375–383 (1995); Brus, D., Ždímal, V., and Stratmann, F., *J. Chem. Phys.*, **124**, 164306 (2006).
11. Hyvärinen, A.-P., Lihavainen, H., Viisanen, Y., and Kulmala, M., *J. Chem. Phys.*, **120**, 11621–11633 (2004).
12. Martens, J., Uchtmann, H., and Hensel, F., *J. Phys. Chem.*, **91**, 2489–2492 (1987).

Arrhenius Temperature Dependence of Homogeneous Nucleation Rates

Robert McGraw

Abstract A simple yet physically based and highly accurate parameterization of the nucleation rate is obtained. In essence, the log nucleation rate is expanded in special coordinates suggested by the first and second nucleation theorems and only linear terms are required. The results support an Arrhenius model of the temperature dependence and are highly accurate over the range of a typical set of measurements – about 3–5 orders of magnitude in nucleation rate. This range is likely sufficient for most applications to atmospheric particle formation, which now seem feasible without the need for bulk properties estimation, as would be required using classical nucleation theory, and without significant higher-order corrections to the linear result.

Keywords Heterogeneous nucleation, nucleation theorem, Arrhenius temperature dependence

The present study is an outgrowth of two new papers that combine nucleation theorems and multivariate analysis to obtain a simple parameterization of nucleation rate – one both physically based and highly accurate [1,2]. In essence, the log nucleation rate is expanded in special coordinates suggested by the first and second nucleation theorems with only linear terms retained. For the most general result, homogeneous nucleation in a multicomponent vapor system, the result is:

$$\ln J \approx \ln J_0 + \sum_i (g^*_i + \delta_i)(\ln n_i - \ln n_i^0) - \frac{\Delta E(g^*_1, g^*_2, \cdots)}{k}\left(\frac{1}{T} - \frac{1}{T_0}\right). \quad (1)$$

$J(cm^{-3}s^{-1})$ is nucleation rate, k is Boltzmann's constant, T is temperature (°K), and $n_i(cm^{-3})$ is the vapor concentration of species i. The label "0" refers to the center of expansion – a reference condition conveniently set by the centroid of the experimental data expressed in nucleation-theorem motivated coordinates, $\{\ln J, \{\ln n_i\}, 1/T\}$, with $n+1$

Environmental Sciences Department, Atmospheric Sciences Division

Brookhaven National Laboratory

C.D. O'Dowd and P.E. Wagner (eds.), *Nucleation and Atmospheric Aerosols*, 144–148.

coordinates for an n-component mixture and temperature dependence. The coefficients in Eq. 1 are readily obtained by applying standard multilinear regression to a set of rate measurements. Especially noteworthy is that these coefficients have direct physical significance: g^*_i is the number of molecules of species i present in the critical nucleus, δ_i is a small kinetic term (its value is between 0 and 1) related to the direction of nucleation flux over the free energy surface. $\Delta E(g^*_1, g^*_2, \ldots)$ is the energy of critical cluster formation from its *vapor-phase* components. Equivalent to Eq. 1:

$$J = J_0 \prod_i \left(\frac{n_i}{n_i^0} \right)^{g^*_i + \delta_i} Exp\left[-\frac{\Delta E(g_1^*, g_2^*, \cdots)}{k} \left(\frac{1}{T} - \frac{1}{T_0} \right) \right] \tag{2}$$

showing the Arrhenius temperature dependence. The latter follows in principle from the second nucleation theorem expressed here in the less conventional form:

$$\left(\frac{\partial \ln J}{\partial T} \right)_{n_1, n_2 \cdots} = \frac{\Delta E(g^*_1, g^*_2, \cdots)}{kT^2} \tag{3}$$

whereby the concentrations of vapor species are held constant in taking the partial derivative [2]. In the more conventional form of the second nucleation theorem the saturation ratio is held constant – and the energy difference appearing on the right side is replaced by the difference between the critical cluster energy and that of the same number of molecules of bulk liquid (see Eq. 5 below).

In the absence of comprehensive multicomponent nucleation rate measurements spanning a significant range of both temperature and vapor composition coordinates, Eqs. 1 and 2 were tested using reported measurements from several different sources. These included aerosol chamber measurements of nucleation rate in ternary organic acid/sulfuric acid/water systems at constant temperature [3], and nucleation pulse chamber measurements varying both temperature and vapor concentration for several single component vapor systems including water [4], methanol and n-hexanol [5], and nonane [6]. Eqs. 1 and 2 were found to be highly accurate over the reported range of measurements, even for this diversity of cases that typically span about 5 orders-of magnitude in nucleation rate. Indeed a variety of techniques are available for homogeneous nucleation measurement, but each of these is typically limited in coverage to a range of 3–5 orders-of-magnitude in nucleation rate (see Figure 1 of Ref. 7 for a comparison of various techniques and their range of rate coverage). This is very encouraging: A good fit to multivariate nucleation data over the range of the aerosol chamber measurements (about 10^{-1} to 10^4 $cm^{-3}s^{-1}$) is probably sufficient for most applications to atmospheric particle formation, and this now seems feasible using Eqs. 1 and 2 without significant higher order curvature corrections required.

As noted above, a local Arrhenius temperature dependence for the nucleation rate is expected from the second nucleation theorem. More surprising is the persistence of Arrhenius behavior over the full temperature range of the studied measurements. This is suggestive of a small difference in heat capacity between the critical cluster and its dissociated vapor form – the larger the difference in heat capacity, the greater

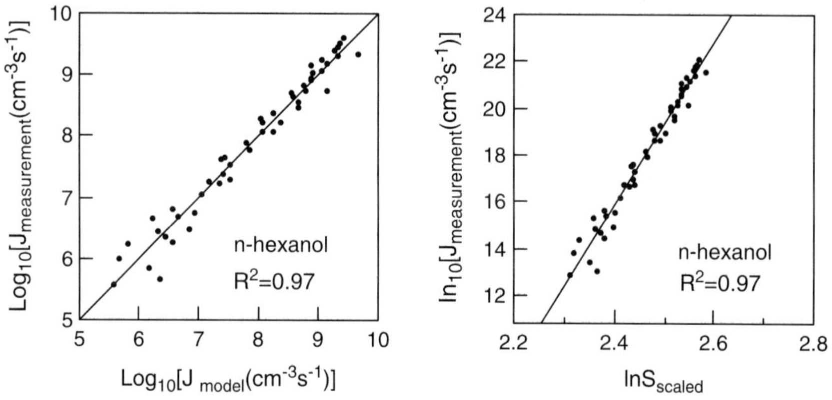

Figure 1 Comparison of measured *n*-hexanol nucleation rate with model predictions using the Arrhenius temperature dependence

the curvature expected in an Arrhenius plot. Related studies note a similar (linear in $1/T$) dependence, but for $Log(J/J_{CNT})$, where J_{CNT} is the classical nucleation rate, rather than for $LogJ$ itself [5]. In a study of nucleation in condensed phase systems, the saturation pressure of water vapor over supercooled liquid at the homogeneous ice nucleation threshold, and at the efflorescence point for heterogeneous nucleation of ammonium sulfate onto small particles of calcium carbonate dispersed in a supersaturated aqueous solution of ammonium sulfate, were each found to follow the Arrhenius temperature dependence. In these last two cases the measurements were of nucleation threshold conditions and not of nucleation rate [8].

Eqs. 1–3 are expressed in terms of vapor concentration, which is most useful for analysis of complex systems where even crude estimates of saturation vapor pressure are unavailable. Nevertheless, because saturation vapor pressures also tend to follow Arrhenius behavior, at least over short temperature spans, one can expect similar quality fits to experimental data with Eqs. 1–2 rewritten in terms of $\ln S$ instead of $\ln n$. Thus comparable quality fits were obtained using the bilinear form:

$$\ln J = a + b/T + c\ln S \qquad (4)$$

to parameterize nucleation rates for the four unary vapor systems mentioned above. Eq. 4 describes a plane in the 3D space of coordinates $\{\ln J, \ln S, 1/T\}$ and to a remarkable extent the experimental points were found to lie in this plane. Values for the quality of fit were obtained as: $R^2 = 0.94$ for water, 0.90 for nonane, 0.89 for methanol, and 0.97 for hexanol.

This paper examines this simple Arrhenius model, its connections to Hale's temperature dependent scaling of the nucleation rate [9], and comparisons between the two temperature functions. There are advantages and disadvantages to each approach: the Arrhenius model does an excellent job over temperature spans of at least 40 K (the maximum range of the reported measurements). And because the

Arrhenius model derives from the nucleation theorems, which are in turn based on the even more fundamental principles of mass action and detailed balance, the methods are not beholden to the usual drop-model assumptions of classical nucleation theory. The Hale scaling function, like the corresponding states temperature correlations that preceded it [10], find application over a much broader temperature range – but carry the disadvantage of being closely linked to the spherical drop model of classical nucleation theory.

The Arrhenius scaling approach is illustrated here for *n*-hexanol vapor, the nucleation properties of which are known to depart strongly from the predictions of classical nucleation theory [5]. Figure 1 (left) illustrates the fit of the Arrhenius model (Eq. 4) to the 50 hexanol measurements reported by Strey et al. (Fit parameters are: $a = 103.425$, $b = -47035.2$, $c = 34.6337$). The right panel of Figure 1 shows the logarithm of the measured nucleation rate plotted versus the new Arrhenius-scaled supersaturation defined as:

$$\ln S_{scaled} = \ln S - \frac{E(g^*) - g^* E_1^b}{kg^*}\left(\frac{1}{T} - \frac{1}{T_0}\right) \qquad (5)$$

where $E_1{}^b$ is the energy per molecule in the *bulk liquid phase*. The slope of the regression line provides an estimate of the number of molecules of hexanol present in the critical nucleus, g*. The numerator of the energy coefficient is the difference between the critical cluster energy and g* molecules of the bulk phase. Its numerical value and g* itself are each available from the fit parameters listed above in connection with Eq. 4. Thus with correction for energy of monomer vaporization [2], $(E(g^*) - g^* E_1{}^b)/k = 39804. = -b -(E_1 - E_1{}^b)/k$, a value grossly overestimated in predictions based on the liquid drop model of classical nucleation theory. On the other hand the critical cluster size, $g^* = 34$ (to nearest integer), is in much closer agreement with $g_{CNT}{}^* = 36$ predicted by the Kelvin relation.

Acknowledgment This research was supported by the DOE Atmospheric Sciences Program.

References

1. McGraw, R. and Zhang, R., *Multivariate Analysis of Homogeneous Nucleation Rate Measurements: I. Nucleation in the p-toluic Acid/Sulfuric Acid/Water System*, submitted for publication (2007).
2. McGraw, R., *Multivariate Analysis of Homogeneous Nucleation Rate Measurements: II. Temperature and Vapor Concentration Dependence*, submitted for publication (2007).
3. Zhang, R., Suh, I., Zhao, J., Zhang, D., Fortner, E.C., Tie, X., Molina, L.T., and Molina, M.J., Atmospheric new particle formation enhanced by organic acids, *Science*, **304**, 1487–1490 (2004).
4. Wölk, J. and Strey, R., Homogeneous nucleation of H_2O and D_2O in comparison: the isotope effect, *J. Phys. Chem. B*, **105**, 11683–11701 (2001).
5. Strey, R., Wagner, P.E., and Schmeling, T., Homogeneous nucleation rates for *n*-alcohol vapors measured in a two-piston expansion chamber, *J. Chem. Phys.*, **84**, 2325–2335 (1986).

6. Wagner, P.E. and Strey, R., Measurements of homogeneous nucleation rate for *n*-nonane vapor using a two-piston expansion chamber, *J. Chem. Phys.*, **80**, 5266–5275 (1984).
7. Iland, K., Wedekind, J., Wölk, J., Wagner, P.E. and Strey, R., Homogeneous nucleation rates of pentanol, *J. Chem. Phys.*, **121**, 12259–12264 (2004).
8. Onasch, T.B., McGraw, R. and Imre, D., Temperature-dependent heterogeneous efflorescence of mixed ammonium sulfate/calcium carbonate particles, *J. Phys. Chem. A*, **104**, 10797–10806 (2000).
9. Hale, B.N., Temperature dependence of homogeneous nucleation rates for water: near equivalence of the empirical fit of Wölk and Strey, and the scaled nucleation model, *J. Chem. Phys.*, **122**, 204509 (1–3) (2005).
10. McGraw, R., A corresponding states correlation of the homogeneous nucleation thresholds of supercooled vapors, *J. Chem. Phys.*, **75**, 5514–5521 (1981).

A Study of Scaled Nucleation in a Model Lennard–Jones System

Barbara N. Hale[1] **and Jerry Kiefer**[2]

Abstract Scaling of the vapor-to-liquid nucleation rate is examined in a model Lennard-Jones system using Monte Carlo derived rate constant ratios for growth and decay of small clusters. The model assumes a dilute vapor system of noninteracting clusters and the steady-state nucleation rate formalism expressed as a summation over products of rate constant ratios. The nucleation rates so obtained are examined in a scaling plot,[1,2] $-\log[J/10^{26}]$ vs. $C_o\,[T_c/T-1]^3/(\ln S)^2$ the form of which has been recently used to test the consistency of nucleation rate data.[3,4]

Keywords Homogeneous nucleation, Lennard-Jones, Monte Carlo, cluster

The Model and Simulations

The model system is a Lennard-Jones (full potential) dilute vapor with volume, V, composed of a noninteracting mixture of ideal gases with each *n*-cluster size constituting an ideal gas of N_n clusters. This permits (with the separable classical Hamiltonian) a law of mass action form for ratios of cluster concentrations in terms of classical configurational canonical partition functions, $Q(n)$:

$$\frac{N_n}{N_{n-1}N_1} = \frac{Q(n)}{Q(n-1)Q(1)n} = \frac{\beta(n-1)}{\mu(n)}. \tag{1}$$

At equilibrium in volume, V, $\beta(n-1)N_{n-1}N_1 = \mu(n)N_n$. The steady-state nucleation rate formalism (which uses steady-state cluster size distributions, N^{ss}_n) assumes

[1] *Physics Department, University of Missouri-Rolla Rolla, MO 65409, USA*

[2] *Physics Department, St. Bonaventure University St. Bonaventure, NY 14778, USA*

C.D. O'Dowd and P.E. Wagner (eds.), *Nucleation and Atmospheric Aerosols*, 149–152.

$\beta(n-1)N^{ss}_{n-1}N^{ss}_{1} = \mu(n)N^{ss}_{n}$ with cluster growth and decay taking place via monomers. In terms of the equilibrium rate constants, β and μ, the steady state nucleation rate, J, is given by:

$$\frac{1}{J} = \frac{1}{\beta(1)N_1N_1} + \sum_{n=2}^{M}\left[\prod_{j=2}^{n}\frac{\beta(j-1)}{\mu(j)}\right]^{-1}\frac{1}{\beta(n)}\left[\frac{1}{N^{ss}_{1}}\right]^{n+1}. \tag{2}$$

With Eq. (1) the above equation becomes,

$$\frac{1}{J} = \frac{1}{\beta(1)N_1N_1} + \sum_{n=2}^{M}\prod_{j=2}^{n}\left[\frac{Q(j)N_1}{Q(j-1)Q(1)j}\right]^{-1}\frac{1}{\beta(n)N_1N_1}\left[\frac{N_1}{N^{ss}_{1}}\right]^{n+1}, \tag{3}$$

where N^{ss}_{1}/N_1 is the monomer supersaturation ratio, S, and M is sufficiently large to ensure convergence of $1/J$. Monte Carlo simulations are used to determine the Helmholtz free energy differences which appear in the first term on the right of the following expression,

$$\ln\left[\frac{Q(n)N_1}{Q(n-1)Q(1)n}\right] = \ln\left[\frac{Q(n)}{Q(n-1)Q(1)\frac{v_n}{V}}\right] - \ln\frac{n/v_n}{N_1/V}. \tag{4}$$

The $Q(n)$ have been normalized with V^n so that $Q(1) = 1$. In this model, the assumed cluster definition is n atoms constrained within a spherical volume, $v_n = \alpha n/\rho_{liq}$. Further details are in Refs. (5)–(8). Formally, the expression in Eq. (4) is independent of α. However, volumes too large or too small place physically unrealistic constraints on the cluster definition. A working range is $5 \leq \alpha \leq 8$. To determine properties of the Lennard–Jones system, the corresponding states approach of Dunikov[9] is used. In particular, the reduced critical temperature, $T_c^* = 1.31$, and the reduced density, $\rho_c^* = 0.310$, are assumed. The simulations are carried out at $T^* = 0.335$, 0.419, and 0.503 corresponding roughly to experimental argon temperatures of T = 40, 50, and 60 K. The Monte Carlo Bennett simulations utilize two independent ensembles: one in which all n atoms interact normally and a second in which the interaction of one of the atoms is turned off. Both ensembles have volume, v_n. Since the simulations produce a sequence of rate constant ratios (and not an energy of formation for the critical sized cluster) it is not necessary to correct for differing $Q(n-1)$ volumes in sequential values of free energy differences, as was done by Lauri et al.[10]

Simulation Data and LJ Model Scaling Properties

In terms of the above formalism, the Bennett Monte Carlo simulation free energy differences, $-\delta_n f(n)$, are

$$-\delta_n f(n) = \ln\left[Q(n)/\left[Q(n-1)Q(1)\frac{v_n}{V}\right]\right] + \ln\alpha. \quad (5)$$

Figure 1 shows a plot of $-\delta_n f(n))/[T^*_c/T^*-1]$ vs. $n^{-1/3}$ for some preliminary data. The predicted intercept is $\ln[\rho_{liqLJ}/\rho_{vapLJ}]/[T^*_c/T^*-1]$ and depends on the LJ full potential system properties. An advantage of this analysis is that one can use the scaled $-\delta_n f(n)$ to predict nucleation rates at temperatures other than those of the simulation.

The $-\delta_n f(n)$ are used in Eq. (3) to determine nucleation rates for the three values of T^* and a range of S. In the final form for J, $\beta(n) = [v/4]\,(36\pi)^{1/3}[\rho_{liqLJ}(T^*)]^{-2/3}n^{2/3}$, $N_1/V = \rho_{vapLJ}(T^*)$, and $N^{ss}_1/N_1 = S$ are used. No calculation is made of n-cluster free energies of formation, or of the critical cluster size. However, the critical cluster size, n^*, can be estimated for a given S and T^* from $-\delta_n f(n^*) = (\ln[\rho_{liqLJ}/\rho_{vapLJ}]) - \ln S$ in Figure 1.

Figure 2 shows thc plot of $-\log[J/10^{26}]$ vs. $C_o\,[T_c^*/T^* -1]^3/(\ln S)^2$. To demonstrate scaling one could use any constant for C_o, as well as any factor $J/10^p$. Here the nearly universal $[J/(\lambda_{thermal})^{-3}]$ » $[J/10^{26}]$ sec^{-1} is used, with J converted to cgs units via the LJ parameters, σ and ε. $C_o = (16\pi/3)\Omega^3/\ln(10)$ and $\Omega = 2.1$. The $[T_c/T -1]^3/(\ln S)^2 = [\ln S_{scaled}]^{-2}$, where $\ln S_{scaled} = \ln S/[T_c/T -1]^{3/2}$ is the scaled supersaturation. This latter form was proposed some time ago as the scaling function for the nucleation rate far below the critical temperature.[11,12] In this simple LJ model study, the scaling appears reasonably well satisfied except at high supersaturation ratios, where the critical cluster size is small. Future work will examine the $-\delta_n f(n)$ over a wider range of n.

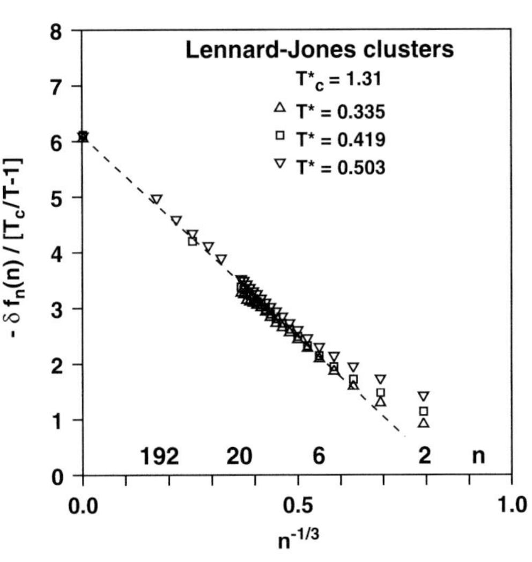

Figure 1 The free energy differences, $-\delta_n f(n)$, vs. $n^{-1/3}$, for three reduced temperatures in the model LJ system. The $-\delta_n f(n)$ have been scaled with $[T^*_c/T^* -1]$, where $T^*_c = 1.31$

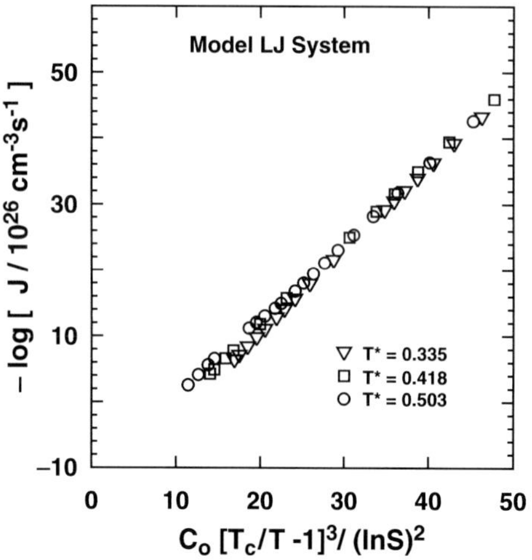

Figure 2 $-\log[J/10^{26}]$ from the LJ Monte Carlo rate constant ratio simulations vs. $C_o\,[T_c/T-1]^3/(\ln S)^2$. The $Co = (16\pi/3)\Omega^3/\ln(10)$, $\Omega = 2.1$, $T_c = 1.31\varepsilon/k$ and $T = T^*\varepsilon/k$. J is converted to cgs units using the LJ parameters, σ and ε, and the atomic mass for argon

References

1. Hale, B.N., *J. Chem. Phys.*, **122**, 204509 (2005).
2. Hale, B.N., Computer simulations, nucleation rate predictions and scaling, in: *Nucleation and Atmospheric Aerosols 2004*, edited by M. Kasahara and M. Kulmala, Kyoto University Press, Kyoto, Japan, pp. 3–14 (2004).
3. Gharibeh, M., Kim, Y., Dieregsweiler, U., Wyslouzil, B., Ghosh, D., and Strey, R., *J. Chem. Phys.*, **122**, 094523 (2005).
4. Brus, D., Zdimal, V., and Stratmann, F., *J. Chem. Phys.*, **124**, 164306 (2006).
5. Hale, B.N. and Ward, R.C., *J. Stat. Phys.*, **28**, 487 (1982).
6. Hale, B.N., *Aust. J. Phys.* **49**, 425 (1996).
7. Hale, B.N. and Kathmann, S.M., *J. Phys. Chem.*, **B105**, 11719 (2001).
8. Hale, B.N. and DiMattio, D.J., *J. Phys. Chem.* B, **108**, 19780 (2004).
9. Dunikov, D.O., Mallyshenko, S.P., and Zhakhovskii, V.V., *J. Chem. Phys.*, **115**, 6623 (2001).
10. Lauri, A., Merikanto, J., Zapadinsky, E., and Vehkamaki, H., *Atm. Res.*, **82**, 489 (2006).
11. Hale, B.N., *Phys. Rev A*, **33**, 4156 (1986).
12. Hale, B.N., *Metallurgical Transactions*, **23A**, 1863 (1992).

Using Small Angle X-ray Scattering to Measure Homogeneous Nucleation Rates of *n*-propanol in a Supersonic Nozzle

David Ghosh[1], Reinhard Strey[1], and Barbara E. Wyslouzil[2]

Abstract In an earlier publication[1] we presented the results of axial pressure measurements characterizing condensation of a series of *n*-alcohols ($C_iH_{2i+1}OH$, $i = 3–5$) in a supersonic nozzle. Although we were able to determine the temperature T_{Jmax}, condensible pressure p_{Jmax} and characteristic time Δt_{Jmax}, associated with the peak nucleation rate, we could only estimate the nucleation rate J_{max}, because we were unable to directly measure the number density N of the aerosol. Here we present the results of our small angle x-ray scattering (SAXS) experiments that characterize *n*-propanol droplets formed in a comparable supersonic nozzle. By fitting the radially averaged scattering spectrum, we obtain the mean radius $\langle r \rangle$, the width of the size distribution, and the particle number density N of the aerosol at a fixed position in the nozzle. Finally, by combining the data from both sets of experiments, we find that our measured nucleation rates are about half an order of magnitude higher than the original estimates.

Keywords Homogeneous nucleation, nozzle, *n*-propanol

Introduction

In an earlier publication,[1] we presented the results of pressure trace measurements in our standard supersonic nozzle (nozzle A) for *n*-propanol, *n*-butanol, and *n*-pentanol. By analyzing the pressure data we determined the temperature T_{Jmax}, partial pressure of condensible p_{Jmax}, and characteristic time Δt_{Jmax}, corresponding to the maximum nucleation rate J_{max}, where J_{max} is defined by

[1] *Institut für Physikalische Chemie, Universität zu Köln, D-50939 Köln, Germany*

[2] *Department of Chemical Engineering, Ohio State University, Columbus, OH 43210, USA*

C.D. O'Dowd and P.E. Wagner (eds.), *Nucleation and Atmospheric Aerosols*, 153–157.

$$J_{max} = \frac{N}{\Delta t_{J\,max}} f. \tag{1}$$

In Eq. (1) N is the number density of the aerosol and f is a factor that corrects for any difference in the gas density between the nucleation zone and the point in the nozzle where N is measured.

In the earlier work, we estimated J_{max} by assuming that N for the alcohol aerosols was in the same range as the values of N that we measured in nozzle A for H_2O, D_2O, and H_2O–D_2O aerosols using small angle neutron scattering (SANS).[2] In this publication we extend our investigation and report the results of our pioneering small angle x-ray scattering (SAXS) experiments from *n*-propanol droplets formed in a similar supersonic nozzle. By fitting the SAXS spectra to scattering from a polydisperse distribution of spheres, we can determine the mean radius $\langle r \rangle$, the width of the distribution function σ, and the particle number density N as a function of the condensible flow rate. By combining data from the pressure trace and SAXS experiments we can now quantify the nucleation rate as a function of T_{Jmax} and p_{Jmax} in our supersonic nozzles.

Experimental Setup

The setup used to maintain supersonic flow in the nozzle during the SAXS measurements is described in detail elsewhere.[3] All of the experiments used N_2 as the carrier gas. The stagnation pressure p_0 was maintained at $p_0 = 30$ kPa by adjusting the flow rate of N_2 as the flow rate of alcohol was varied. Three stagnation temperatures, $T_0 = 35°C$, 45°C, and 50°C, were investigated to increase the range of condensible flow rates. Although there are some minor differences between the flow systems used for SAXS and the pressure measurements,[1] we leave detailed discussions to a later publication.

Our nozzles consist of two contoured nozzle blocks sandwiched between flat sidewalls. For the pressure trace measurements[1] we used our standard nozzle, nozzle A, characterized by a linear expansion rate of $d(A/A^*)/dx = 0.0486\,cm^{-1}$. For the SAXS measurements we used a nozzle (H2) with a slightly faster expansion rate, $d(A/A^*)/dx = 0.054\,cm^{-1}$. The windows in the sidewalls of nozzle H2 consist of 10 mm high × 100mm long sheets of 0.025 mm thick ruby mica glued across 1 mm high × 90 mm long slots machined parallel to the flow. The x-ray beam passes through the mica windows at a right angle to the flow.

All SAXS measurements were conducted using the BESSERC 12-ID beam line at the Advance Photon Source, Argonne National Laboratory, Illinois. We used a 0.2 mm × 0.2 mm beam of 12 keV x-rays and a fixed sample to detector distance of 2 m. The SAXS spectra were all measured 6.5 cm downstream of the nozzle throat. The total integration time for each spectrum was 5 sec for the sample and 5 sec for the empty cell.

Results and Discussion

We analyzed each radially averaged SAXS spectrum by assuming the aerosol is described by a Gaussian distribution of spheres. Figure 1 summarizes the variation in mean radius <*r*>, distribution width σ, and particle number density *N* with the alcohol mass flow rate. As the flow rate increases from 2 to 15 g min^{-1}, <*r*> increases by a factor of ~3, σ increases by a factor of ~2.2, and *N* decreases by a factor of ~5. The rapid increase in <*r*> with condensible flow rate $\dot{m}$ is expected because the concentration of condensible vapor is directly proportional to $\dot{m}$. Indeed, if *N* and the fraction of incoming vapor that condense are constant, a simple volumetric balance shows that <*r*> should be proportional to $\dot{m}^{1/3}$. The solid line in Figure 1 corresponds to this scaling law with $N = 10^{12}\ \text{cm}^{-3}$. Our measured data follow the simple scaling reasonably well, except at the lowest flow rates where the assumption of constant *N* breaks down. Since *N* increases but the condensible concentration is fixed, <*r*> must decrease as $(10^{12}/N_{obs})^{1/3}$. Finally, the droplet size observed at a fixed position in the nozzle also depends on the length of time available for droplet growth. Droplets formed further upstream have a longer time of flight before they reach the observation window, than those formed further downstream. If we increase T_0 at a fixed $\dot{m}$, particle formation occurs further downstream and the particles have less time to grow. This effect can be seen in Figure 1, where the radii for plenum temperatures of T_0= 35°C are slightly higher than the radii for experiments at similar flow rates but higher plenum

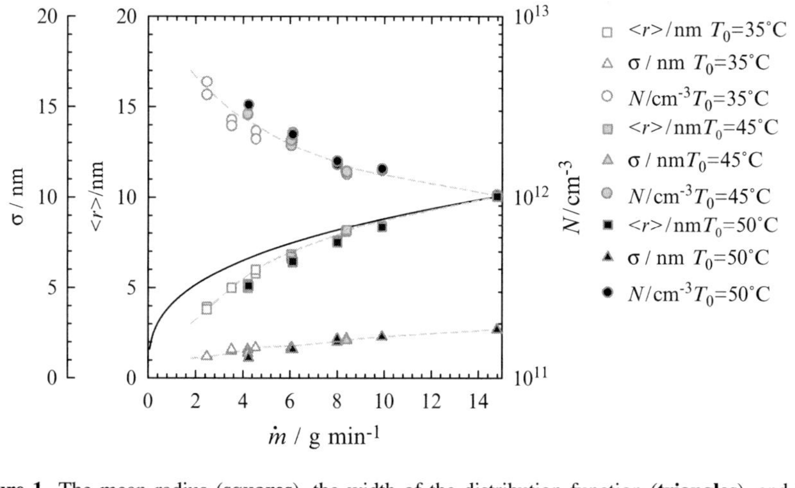

Figure 1 The mean radius (**squares**), the width of the distribution function (**triangles**), and the particle number density (**circles**) as a function of the mass flow rate for *n*-propanol at T_0 = 35°C (white), T_0 = 45°C (gray), and T_0 = 50°C (black). The dashed lines are intended to guide the eye. The solid line indicates the proportionality of the mean radius with the cube root of the flow rate

temperatures. Since the changes in <r> for a given flow rate at changing plenum temperature are minor, we conclude that the time of flight effect is secondary.

A further observation from Figure 1 is that as the alcohol flow rate decreases, the particle number density increases. Decreasing the alcohol flow rate shifts nucleation toward lower temperatures and the characteristic time $\Delta t_{J\max}$ increases. The latter suggests that the self-quenching process is not as effective at low temperatures as at higher nucleation temperatures.

In contrast to the rapid changes in <r> and N with $\dot{m}$, the width of the size distribution σ increases from 1.2 to 2.7 nm over the investigated flow rate range, while the polydispersity index p = <r>/σ decreases from 0.31 to 0.27. Thus, the most "monodisperse" droplets are formed at the highest flow rates because <r> increases much more rapidly with $\dot{m}$ than σ does.

The main purpose of our research is to measure the maximum nucleation rate during condensation in our supersonic nozzles. To combine the SAXS and pressure trace measurements, we first corrected N for the difference in expansion rates between nozzles A and H2 using the relative changes in N that Kim et al.[2] observed for D_2O condensation in nozzles with three different expansion rates. We then correlated the corrected values of N with $\dot{m}$ to account for slight differences in the mass flow rates used in the SAXS and pressure measurements. In our previous work[1] we estimated the particle number density to be $10^{11} < N/\text{cm}^{-3} < 10^{12}$. From the SAXS measurements we find that the actual particle number densities are less than half an order of magnitude higher than our previous estimates. Thus, our earlier estimates for the nucleation rate $J = 5 \times 10^{16}$ cm^{-3} s^{-1} are half an order of magnitude lower than the nucleation rates determined here. Figure 2 illustrates our measured nucleation rates as a function of the supersaturation and temperature.

Our results encourage us to plan future position-resolved SAXS experiments to understand growth behavior during supersonic nozzle expansion better.

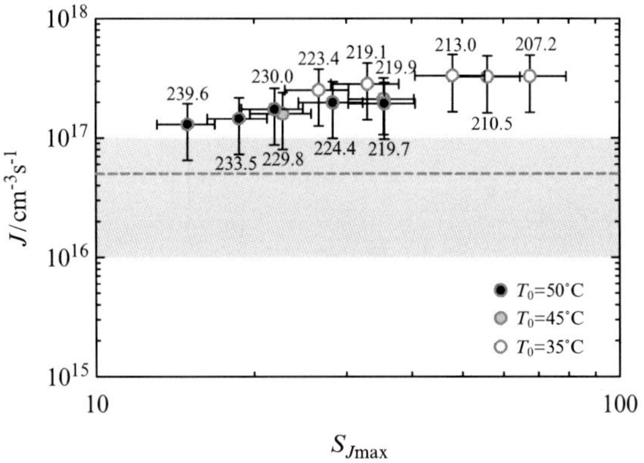

Figure 2 The actual nucleation rate J vs. the supersaturation S for n-propanol. The gray dashed line (**shaded area**) indicates our earlier estimate for the nucleation rate (uncertainty of the estimate)

Acknowledgments This work was supported by the National Science Foundation under Grant numbers CHE-0410045 and CHE-0518042 and by the Donors of the Petroleum Research Fund administered by the American Chemical Society. The use of the Advanced Photon Source was supported by the US Department of Energy, Office of Science, Office of Basic Energy Sciences, under Contract No. DE-AC02-06CH11357. We thank R. Winans and S. Seifert for their help conducting the SAXS experiments.

References

1. Gharibeh, M., Kim, Y., Dieregsweiler, U., Wyslouzil, B.E., Ghosh, D., Strey, R., *J. Chem. Phys.*, **122**, 94512–94521 (2005).
2. Kim, Y.J., Wyslouzil, B.E., Wilemski, G., Wölk, J., Strey, R., *J. Phys. Chem. A*, **108**, 4365–4377 (2004).
3. Tanimura, S., Zvinevich, Y., Wyslouzil, B.E., Zahniser, M., Shorter, J., Nelson, D., McManus, B., *J. Chem. Phys.*, **122**, 194304–194315 (2005).

Condensation of a Supersaturated Lattice Gas

Vitaly A. Shneidman

Abstract A cold supersaturated lattice gas on a square lattice with Glauber/ Metropolis dynamics provides a unique possibility of first-principle evaluation of the steady-state and transient rates for nucleation of two-dimensional crystals. Alternatively, nucleation can be studied using Monte Carlo methods. Due to current advances in computational powers a temperature domain can be identified where both approaches overlap, allowing for an instructive analysis. The Becker-Döring model also can be assessed.

Keywords Homogeneous transient nucleation, Monte Carlo simulations, clusters

The Model

In lattice gas interaction potential between two particles has a value $-\phi$ if particles are nearest neighbors, a value of $+\mu$ if particles attempt to occupy the same lattice site, and is zero otherwise. The gas is stable for any chemical potential $\mu < \mu_0 \equiv -2\phi$. For larger μ the gas is supersaturated and will tend to nucleate and condense. Dimensionless supersaturation is defined as

$$S = (\mu - \mu_0)/|\mu_0| \tag{1}$$

Gas becomes fully unstable (the barrier to nucleation vanishes) for $S \geq 1$, while the interface between the gas and the "liquid" becomes unstable for $S \geq 1/2$. Without restrictions, in the following the value $\phi = 4$ will be used. In that case $T_c = 2.269\ldots$ corresponds to the critical temperature of the gas (this follows from the Onsager equilibrium solution of the Ising model, to which the lattice gas picture is equivalent). Of interest for the present study will be temperatures which are noticeably smaller than T_c and moderate supersaturations $S < 1/2$.

New Jersey Institute of Technology, USA

C.D. O'Dowd and P.E. Wagner (eds.), *Nucleation and Atmospheric Aerosols*, 158–161.

The "work" required to form a given cluster with n spins and perimeter P is given by

$$W = P/2 - 2Sn \tag{2}$$

Dynamics is realized through the Metropolis rule – if an attempt to add a particle at a randomly selected site decreases the value of W, such an attempt is accepted, otherwise it is accepted with probability exp $(-\Delta W/T)$(Boltzmann constant is taken as 1). Glauber dynamics is similar – see Ref. [3].

Becker-Döring Expectations

In this work no attempt will be made to use the interfacial tension (also available from the Onsager solution) in order to calculate the work required to form a critical cluster. This is possible, but can lead to exponentially large errors in the nucleation rate – see discussion in Ref. [5]. However, the classical *kinetic* picture based on the Becker-Döring equation (BDE) can be extremely useful.

The following expressions for the time-dependent flux $j(n,t)$ at a given cluster size $n > n_*$ and the number of nuclei $N(n,t)$ which exceed the size n were obtained in Refs. [1,2] from the matched asymptotic (singular perturbation) solution of the BDE:

$$j(n,t) = j_{st} e^{-e^{-x}}, \quad N(n,t) = \tau j_{st} E_1\left(e^{-x}\right), \quad x \equiv \frac{t - t_i(n)}{\tau} \tag{3}$$

Here j_{st} is the steady-state flux, τ the "relaxation time" (related to properties of near-critical clusters), $t_i(n)$ is the incubation time, which also depends on off-critical decay and growth of clusters. E_1 is the standard first exponential integral. At large times one expects $N(n,t) \sim j_{st}\,[t-t_0(n)]$ with $t_0(n)$ being the "time-lag". Connection to the incubation time is given by [2]

$$t_0(n) = t_i(n) + \gamma\tau \tag{4}$$

Specific form of $t_0(n)$ depends on the particular selection of the Becker-Döring coefficients (see Ref. [2] for explicit relations); otherwise, the above dependences are universal.

Nucleation Rates in Lattice Gas at Small T

The problem can be described from a large system of kinetic equations describing transitions between clusters of various shapes and sizes. Such system of equations represents a generalized version of the BDE in the multidimensional space of cluster configurations. In steady-state the equations can be solved analytically [3,4] or with the help of symbolic computations [7,8], allowing to obtain an accurate expression for the nucleation rate, including and exact (low-temperature) pre-exponential. In the time-dependent case exact matrix expressions for the flux, time-lag, etc. can be obtained [6] but those expressions are formal and have to be evaluated numerically.

Transient flux accurately follows the double-exponential dependence of Eq. (3). However, unlike the case of BDE where all parameters of that dependence can be calculated analytically, the case of lattice gas requires their numerical evaluation. Results relevant to the following discussion ($T=0.8$, $S=0.22$, a 1000×1000 lattice and $n=20$) are given by [6]

$$j_{st} = 1.75,\, \tau = 6.134,\, t_0 = 19.964 \tag{5}$$

(note that the flux reported in Ref.[6], J_{st} is *per site*, i.e., should be multiplied by 10^{-6} compared to the above). Despite the qualitative correctness of the kinetic part of the Becker-Döring picture, there are several features, such as the collapse of transient fluxes [6] which require an alternative description.

Monte Carlo Simulations

There are two possibilities to study the nucleation behavior. If the nucleation rate is not too small, one can consider a large system with many nuclei formed – see Figure 1. In that case the number of nuclei can be directly counted and compared with the theoretical predictions. The other possibility is to consider a "small" system, with typically not more than one nucleus formed. In that case the nucleation process is random with distribution of waiting times given by

$$w_n(t) = j(n,t)\exp\left[-N(n,t)\right] \tag{6}$$

(and the steady-state flux with linear $N(t)$ would correspond to an exponential distribution, typical for Poissonian processes). Note that while formally the theoretical values of $N(n,t)$ are given by the same Eq. (3), one requires $N(n,t) \gg 1$ or $N(n,t) \ll 1$ to justify the former or the latter approach, respectively. For the present parameters both situations could be achieved by using different sizes of the lattice – see Figure 2.

Good correspondence between Monte Carlo simulations and rigorous low-temperature analytical and computational data is observed, and implications of this comparison will be discussed.

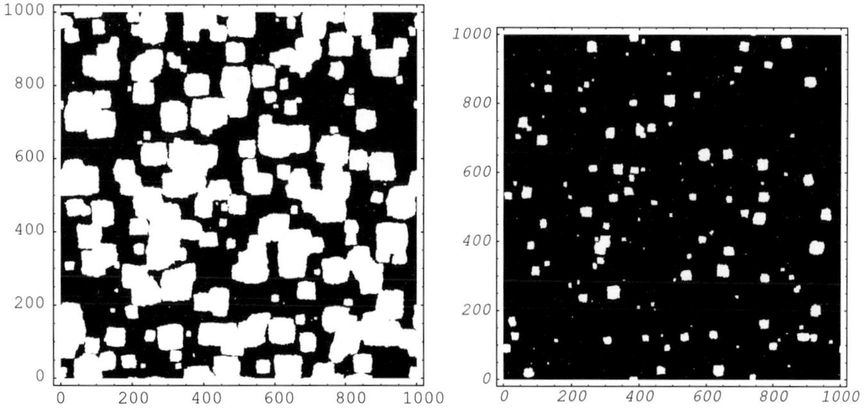

Figure 1 Low-temperature condensation of a supersaturated lattice gas for $T = 0.8$ and $S = 0.22$ at two different times $t = 91$ (**left**) and $t = 191$ (**right**). Note that nuclei ("crystals") are noncircular and are close to square shapes due to low-temperature anisotropy

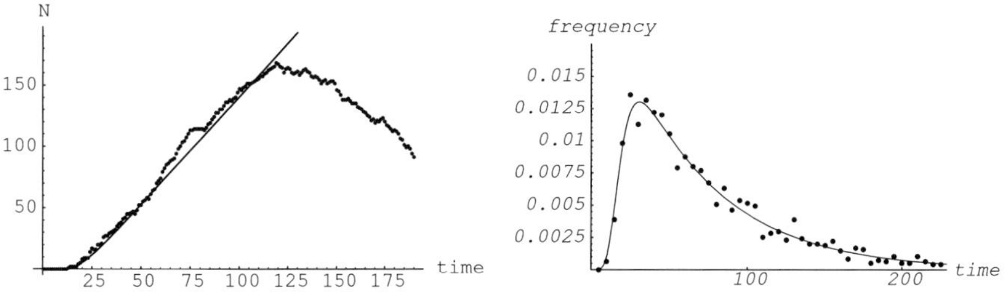

Figure 2 Results of Monte Carlo simulations for a large (**left**) and a small (**right**) systems, respectively. **Left**: Number of nuclei with size exceeding $n = 20$ in a 1000×1000 lattice, as in Figure 1. Points – simulation data, line – Eq. (3) with parameters evaluated from "first principles", Eq. (5). Lowering of the simulation curve at larger times is due to cluster coagulation (see Figure 1, **right**), which is absent in the Becker-Döring picture and in Eq. (3). **Right**: distribution of waiting times to observe the first nucleus with the same parameters, but a smaller, 100×100 lattice. Solid line – Eq. (6) with $j_{st} = 1.75 \cdot 10^{-2}$ (no matching parameters were used)

Acknowledgment The author benefited from discussions and collaboration with G. Nita and E. Goldstein

Reference

1. V.A. Shneidman. *Sov. Phys. Tech. Phys.,* 32:76, 1987.
2. V.A. Shneidman. *Sov. Phys. Tech. Phys.,* 33:1338, 1988.
3. V.A. Shneidman. *J. Stat. Phys.,* 112:293, 2003.
4. V.A. Shneidman. *New Journal of Physics,* 7:12, 2005. doi:10. 1088/1367-2630/7/1/012.
5. V.A. Shneidman, K.A. Jackson, and K.M. Beatty. *J. Chem. Phys,* 111:6932, 1999.
6. V.A. Shneidman and G.M Nita. *Phys Rev. Lett.,* 97:065703, 2006.
7. V.A. Shneidman and G.M Nita. *Phys Rev. Lett.,* 89:25701, 2002.
8. V.A. Shneidman and G.M Nita. *Phys Rev.,* E 68:021605, 2003.

Argon Nucleation in a Cryogenic Nucleation Pulse Chamber

Judith Wölk[1], **Kristina Iland**[1], **Reinhard Strey**[1]

Abstract Homogeneous nucleation of argon has been measured with a cryogenic nucleation pulse chamber. Due to the very fast growth of the argon droplets nucleation and growth could not be decoupled. However, we can present a systematic set of onset data for argon measured in a temperature range from 42 to 58K and for vapor pressures from 0.3 to 10kPa. Using the most recent expressions for temperature-dependent vapor pressures, surface tensions and densities predictions by the classical *Becker-Döring* nucleation theory are calculated. A rather high deviation of up to 26 orders of magnitude between theory and experiment is found. As in the case of other systems (e.g., water, alcohols, and alkanes) classical theory shows a stronger temperature dependence than experimentally observed.

Keywords Homogeneous nucleation, onset measurements

Introduction

Nucleation as the first step of first-order phase transitions is an omnipresent process. It can be found in nature as well as in many scientific fields. Simple examples are condensation, melting, and evaporation. The understanding of homogeneous nucleation kinetics is of high interest for a wide range of applications like meteorological modeling or industrial design. Since the fundamental experiments of Wilson (1897),[1] homogeneous nucleation in supersaturated vapors has been studied with various devices and substances.[2] Parallel to the experimental work there have been many efforts to describe this nucleation process theoretically.[3–6] Unfortunately no theory is able to describe the nucleation process quantitatively. For instance, a comparison with classical nucleation theory (CNT) developed by Becker and Döring[3] and experiments show a weaker temperature dependence what means that

[1] *Institut für Physikalische Chemie, Universität zu Köln, Luxemburger Str. 116, 50939 Köln, Germany*

C.D. O'Dowd and P.E. Wagner (eds.), *Nucleation and Atmospheric Aerosols*, 162–166.

at higher temperatures the predictions of the theory are too high and at lower once they are too low. For the newer theories, like molecular theories,[7–9] the density functional theory[10–12] or for computer simulations[13,14] of the nucleation process the exact intermolecular potentials has to be known in order to perform the calculations. Very often a *Lennard-Jones* potential is chosen, which only accounts for rare gases like argon. In order to test the validity of the theories it is of great interest to study nucleation of such simple atoms experimentally. Up to now experimental work on the homogeneous nucleation of argon does not give a consistent pattern.[15–22] In order to generate a consistent data set on argon nucleation, Fladerer and Strey[23] built a nucleation pulse chamber that is able to work on temperatures below 83K. This chamber is able to expand an argon–helium gas mixture adiabatically and thus induce argon nucleation. Besides the growth measurements on argon droplets they detected the first onset of argon nucleation data with this new device. Unfortunately the growth process of argon was found to be so fast that they were not able to decouple nucleation and growth. Since then we have replaced their expansion volume by a larger one so that now deeper expansions are possible.

Results

Here we report a consistent set of homogeneous nucleation experiments for argon over a wide range of temperatures with helium as carrier gas.[24] Description of the details of the experimental setup and procedure can be found elsewhere[23,24].

Onset Experiments

In Figure 1 our onset data of argon (filled circles)[24] are compared to older data from literature in a *Wilson* plot.

It can be seen that using a bigger expansion volume than Fladerer and Strey[23] results in an extension of the experimental window. The deviations of the *Fladerer* and *Strey* data (circles) at lower temperatures can be explained by the fact that toward the lower end of their measuring window the expansion rate decreased which results in smaller nucleation rates. In contrast the data from Iland et al.[24] (filled circles) were made under comparable expansion rates and thus for the same nucleation rates ($J = 10^7$ cm^{-3} s^{-1}) for the whole experimental window.

The data from Wu et al.[18] (gray hexagons) and data from Stein[17] (gray triangles) were measured with supersonic nozzles. Because of the higher expansion rates and correspondingly higher nucleation rates of supersonic nozzles ($10^{15} < J$/cm^{-3} s$^{-1} < 10^{17}$) these data should lie on the left side of the data measured with the nucleation pulse chamber. This is found for the data of Wu et al. All the other data by Matthew and Steinwandel[19,20] (turned triangles) and by Zahoransky et al.[21,22] (data from 1995: squares, data from 1999: diamonds) were measured with shock tubes $10^4 < J$/cm^{-3} s$^{-1} < 10^{10}$. The data from Iland et al. agree well with the data of

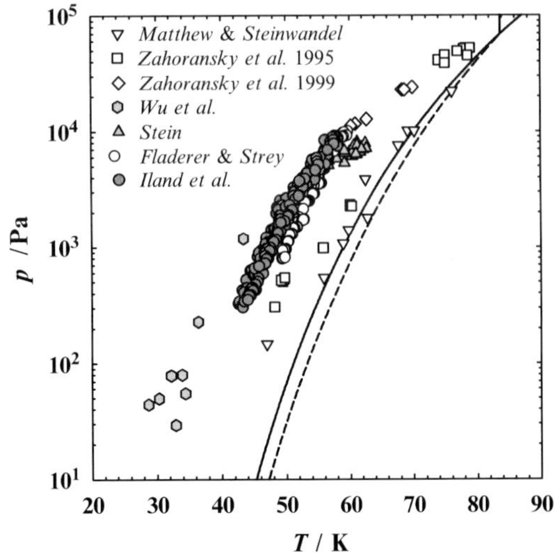

Figure 1 Onset conditions of argon-nucleation determined by Iland et al.[24] (**filled circles**) in comparison to the data of Fladerer and Strey[23] (**empty circles**), Matthew and Steinwandel[19] (**downward triangles**), Zahoransky et al.[21,22]. (Data from 1995: squares, data from 1999: diamonds.) Wu et al.[18] (**hexagons**) and Stein[17] (**upright triangles**). The gas–solid binodal is drawn as dashed line and the extrapolated gas–liquid binodal as solid line

Zahoransky et al. from 1999 and some of the data from 1995. The data of Matthew and Steinwandel[19] are lying right on the binodal of the system, which might be explained by the detection of heterogeneous nucleation.

Comparison with Classical Nucleation Theory

The classical nucleation theory of Becker and Döring[3] is the most commonly used model for quantitative prediction of nucleation phenomena. Due to space limitations we compare here our measured onsets for argon only to the predictions of the classical theory. The theoretical predictions have been calculated using the most recent expressions for vapor pressure, surface tension, and density as function of temperature.[23,24]

The onset data (filled circles) as well as the gas–liquid binodal (bottom solid line) are plotted in Figure 2a as a function of inverse temperature. In this *Volmer* plot[25] they form nearly straight lines so that the comparison with theory is simplified. The theoretical prediction for constant nucleation rates J_{CNT} of 10^{-20} cm^{-3} s^{-1}, 10^{-10} cm^{-3} s^{-1}, 1 cm^{-3} s^{-1}, and 10^{10} cm^{-3} s^{-1} are shown also as solid lines. The region where the theoretical prediction matches the experimental window for nucleation rates from $J_{CNT} = 10^5$ cm^{-3} s^{-1} to $J_{CNT} = 10^9$ cm^{-3} s^{-1}, is marked as hatched region in the figure. If CNT is giving the right prediction this region would be lying on top of the experimental data. As can be seen this is not the case. The hatched region is lying at pressures twice to six times as high as the onset pressures of argon, with growing discrepancies with decreasing temperature. From estimating an nucleation time and the knowledge of the size of the scattering volume it is possible to estimate

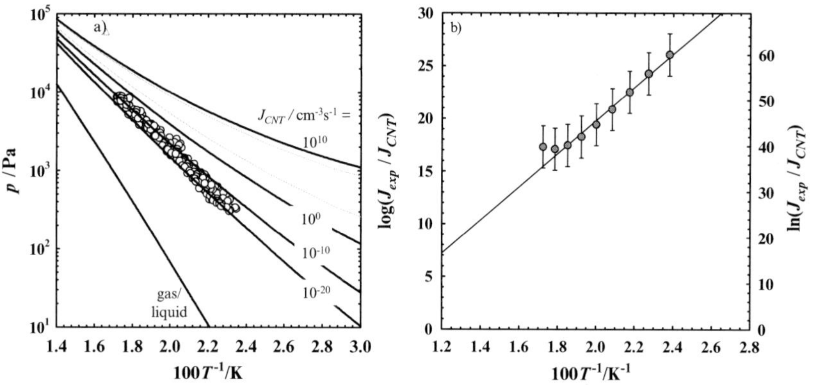

Figure 2 (a) Onset of argon nucleation (**circles**) and the gas–liquid binodal (**bottom solid line**) as a function of inverse temperature T. Also shown is the prediction of classical nucleation theory for constant nucleation rates J_{CNT} of $10^{\ 20}$ cm^{-3} s^{-1}, 10^{-10} cm^{-3} s^{-1}, 1 cm^{-3} s^{-1}, and 10^{10} cm^{-3} s^{-1}. The hatched region represents $J_{CNT} = 10^5$ cm^{-3} s^{-1} to $J_{CNT} = 10^9$ cm^{-3} s^{-1}, which corresponds to our measuring window (b) Ratio of experimental nucleation rates $J_{exp} = 10^7$ cm^{-3} s^{-1} and nucleation rates J_{CNT} calculated with help of classical nucleation theory CNT (**circles**) as a function of inverse temperature T

the nucleation rates corresponding to the onset data as log (J_{exp}/cm^{-3} s^{-1}) = 7 ± 2. The ratio of this average nucleation rate and the rate calculated with help of CNT for the experimental conditions is plotted in Figure 2b as function of inverse temperature (circles). A discrepancy between experiment and theory of about 16 orders of magnitude at high temperatures (around 58K) and of 26 orders of magnitude at low temperatures (42K) is found. Both plots show a weaker temperature dependence of the experiments compared to theory.

Conclusion

New experimental onset data for argon are presented and compared with data from literature as well as to the predictions of classical nucleation theory. The latter predicts rates 16 to 26 orders of magnitude lower than the experimental rates and has a stronger temperature dependence than the experiment.

References

1. Wilson, C.R.T., *Philos. Trans. R. Soc. London Ser. A*, **189**, 871 (1897).
2. Heist, R.H. and He, H. J., *J. Phys. Ref. Data*, **23**, 781 (1994).
3. Becker, R. and Döring, W., *Ann. Phys.*, **24**, 719 (1935).
4. Girshick, S.L. and Chiu, C.-P., *J. Chem. Phys.*, **93**, 1273 (1990).

5. Reiss, H., Kegel, W.K., and Katz, J.L., *Phys. Rev. Lett.*, **78**(23), 4506 (1997).
6. Reiss, H., Kegel, W.K., and Katz, J.L., *J. Phys. Chem. A*, **102**(44), 8548 (1998).
7. Schenter, G.K., Kathmann, S.M., and Garret, B.C., *Phys. Rev. Lett.*, **82**(17), 3484 (1999).
8. Senger, B., Schaaf, P., Corti, D.S., Bowles, R.K., Voegel, J.-C., and Reiss, H., *J. Chem. Phys.*, **110**(13), 6421 (1999).
9. Senger, B., Schaaf, P., Corti, D.S., Bowles, R.K., Pointu, D., Voegel, J.-C., and Reiss, H., *J. Chem. Phys.*, **110**(13), 6438 (1999).
10. Zeng, X. C. and Oxtoby, D.W., *J. Chem. Phys.*, **94**(6), 4472 (1991).
11. Oxtoby, D.W., *J. Phys.: Condensed Matter*, **4**, 7627 (1992).
12. Talanquer, V. and Oxtoby, D.W., *J. Chem. Phys.*, **100**(7), 5190 (1994).
13. Laasonen, K., Wonczak, S., Strey, R., and Laaksonen, A., *J. Chem. Phys.*, **113**(21), 9741–9747 (2000).
14. Toxvaerd, S., *J. Chem. Phys.*, **115**(19), 8913 (2001).
15. Pierce, T., Sherman, P.M., and McBride, D.D., *Astronautica Acta*, **16**, 1 (1971).
16. Lewis, J.W.L., and Williams, W.D., *NTIS-Report No. AD782445* (1974).
17. Stein, G.D., *NTIS-Report No. ADA007357* (1974).
18. Wu, B.J.C., Wegener, P.P., and Stein, G.D., *J. Chem. Phys.*, **69**(4), 1776 (1978).
19. Matthew, M.W. and Steinwandel, J., *J. Aerosol Sci.*, **14**(6), 755 (1983).
20. Steinwandel, J. and Buchholz, T., *Aerosol Science and Technology*, **3**, 71 (1984).
21. Zahoransky, R.A., Höschele, J., and Steinwandel, J., *J. Chem. Phys.*, **103**(20), 9038 (1995).
22. Zahoransky, R.A., Höschele, J., and Steinwandel, J., *J. Chem. Phys.*, **110**(17), 8842 (1999).
23. Fladerer, A. and Strey, R., *J. Chem Phys.*, **124**(16) (2006).
24. Iland, K., Wölk, J., Kashchiev, D., and Strey, R., *J. Chem Phys.*, to be submitted (2007).
25. Volmer, M., *Kinetik der Phasenbildung*, , Dresden, Germany: Steinkopff (1939).

Water Homogeneous Nucleation: Importance of Clustering Thermodynamics

Hua Du, Fangqun Yu, and Alexey B. Nadykto

Abstract Quantum methods, which are developed to accurately predict the physical properties of small clusters, are applied to study the water nucleation phenomenon. The self-consistency corrected (SCC) theory of water homogeneous nucleation, which satisfies the law of mass action and avoids the mismatch in the cluster distribution for monomers, overpredicts the nucleation by several orders of magnitude. In this study, we show that the overprediction is likely due to the inadequate description of thermodynamic properties of small water cluster using capillarity approximation. The Density Functional Theory (DFT) methods – PBEPBE and PW91PW91, were used to calculate the thermodynamic properties of $(H_2O)_n$ (n = 1–10). These data were then used in the kinetic water nucleation model to study the effect of thermochemistries of the first ten water clusters on the prediction of nucleation rate by SCC. A considerable difference between Gibbs free energy changes of the $(H_2O)_n$ formation used in classical nucleation thoery (CNT) and results obtained using quantum methods has been illustrated. We also found that predicted nucleation rates by the SCC model can be improved when the model is constrained by the stepwise Gibbs free energy changes obtained by DFT methods. This study highlights the importance of properly treating the thermochemistry of small cluster formation and demonstrates the feasibility of employing thermodynamic properties of small clusters obtained using quantum methods instead of those derived based on traditional capillarity approximation.

Keywords Water homogeneous nucleation, density functional theory, thermochemistry

---

Atmospheric Sciences Research Center, State University of NewYork at Albany, Albany, NY 12203, USA

C.D. O'Dowd and P.E. Wagner (eds.), *Nucleation and Atmospheric Aerosols*, 167–171.

Comparisons of Stepwise Gibbs Free Energy Change of Formation of Small Water Clusters

A comparison of stepwise dG^0 in Figure 1 shows that dG^0 of formation of water dimer given by PBEPBE and PW91PW91 with aug-cc-pvtz are in good agreement with those predicted by high-accuracy model chemistry methods and MP2/6-311 + + G(2d,2p) with the anharmonic correction. They also fall within the experimental uncertainty range. The stepwise dG^0 for $(H_2O)_n$ (n = 1–10) given by the PBEPBE and PW91PW91 is in agreement within ~0.5 kcal/mol per step. The difference between the DFT-calculated dG^0 and those given by the above three model chemistry methods ranges from 0.06 to 0.17 kcal/mol per step and 0.42–0.65 kcal/mol per step for PBEPBE/aug-cc-pvtz and PW91PW91/aug-cc-pvtz, respectively. Compared to PW91PW91/aug-cc-pvtz, PBEPBE/aug-cc-pvtz yields closer results to the model chemistry methods. In addition, the dG^0 curve given by CNT[1] is much lower than those given by the quantum methods. The Gibbs free energy of 2.77 kcal/mol is assigned by CNT to n = 0®1 transition, which is obviously unphysical. Since the nucleation rate is very sensitive to the dG^0, the significant difference in dG^0 of formation of small water clusters between CNT and quantum methods is expected to lead to a large difference in the predicted nucleation rates.

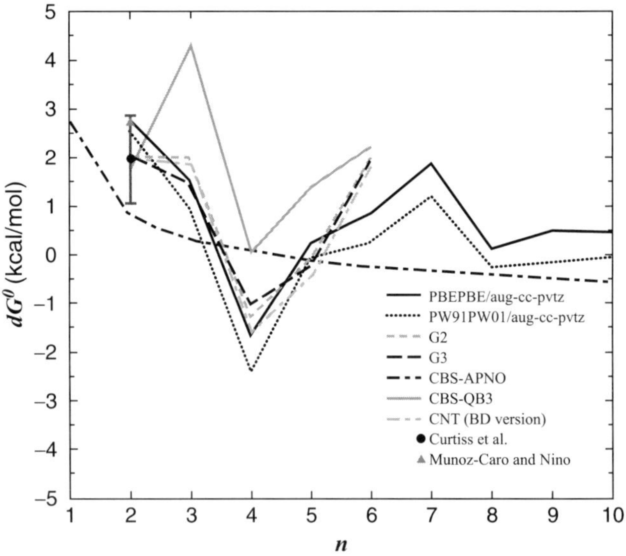

Figure 1 dG^0 for $(H_2O)_{n-1} + H_2O \rightarrow (H_2O)_n$ reaction as a function of the number of molecules in the cluster at standard condition (T = 298.15K and P = 1 atm). The square with the error bar and triangle refer to Curtiss et al.,[2] and Munoz-Caro and Nino,[3] respectively

Comparisons of Nucleation Rates from Various Studies

Figure 2 shows the nucleation rates based on QM-based nucleation model, SCC theory,[4] the Monte Carlo study,[5] and experimental results as a function of supersaturation ratio at four different temperatures. For the QM-based model, the stepwise dG^0 for $(H_2O)_n$ (n = 1–10) are derived from two DFT methods while those for $(H_2O)_n$ (n > 10) are calculated in the same way as in SCC theory.[4] Compared to PW91PW91/aug-cc-pvtz (the long-dashed curves), PBEPBE/aug-cc-pvtz (the dot-dahsed curves) gives much better agreement with the experimental data due to the different prediction of stepwise dG^0 (Figure 1). Under all four temperatures studied here, the nucleation rates predicted by PBEPBE/aug-cc-pvtz are consistently lower than those predicted by both SCC theory and the Monte Carlo study. The only difference between PBEPBE/aug-cc-pvtz and SCC is the different dG^0 for the first ten water clusters, however, the deviations in the predicted nucleation rate range from

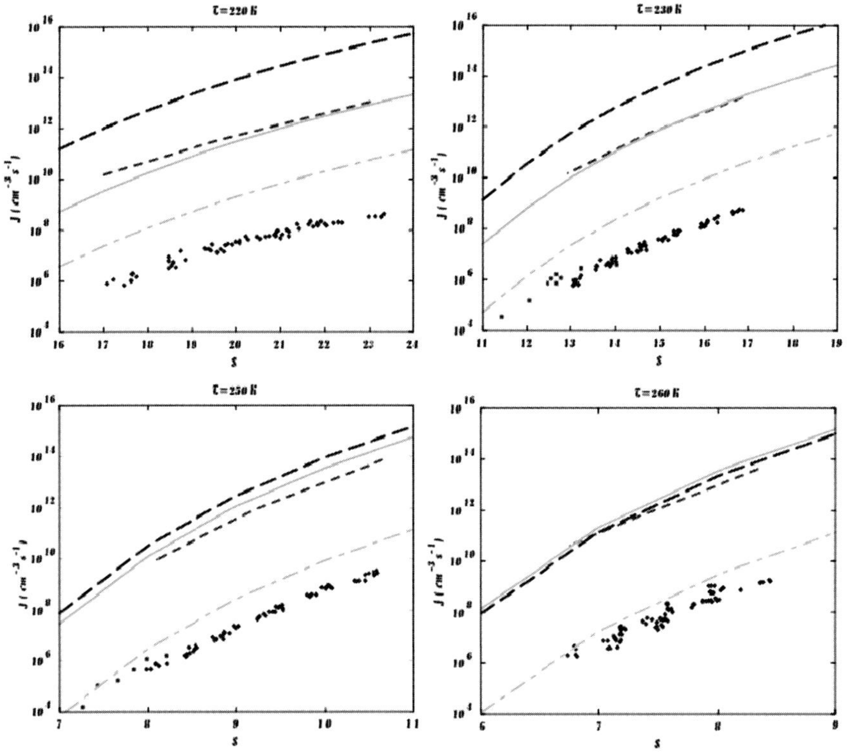

Figure 2 Supersaturation ratio dependence of the water nucleation rates at T = 220, 230, 250, and 260K (as indicated): **Circles** – experimental data of Wolk and Strey[6]; **squares** – experimental data of Mikheev et al.[7]; **solid curves** – SCC prediction; **dashed curves** – Monte Carlo simulations[5]; **long-dashed curves** – PW91PW91/aug-cc-pvtz; **dot-dashed curves** – PBEPBE/aug-cc-pvtz

10^2 at T = 220K to 10^4 at T = 260K. It clearly indicates that the more accurate description of dG^0 for the first ten clusters significantly improves the nucleation rate prediction of SCC nucleation theory. The overprediction of nucleation rate by SCC is likely due to the underestimation of dG^0 for the small clusters which suggests that the disagreement of SCC theory with experimental results may be associated with the error in the capillarity approximation for the small clusters.

Effects of Thermochemistry of $(H_2O)_n$ (n > 10) on Predicted Nucleation Rates

The quantum-derived dG^0 significantly improve the prediction of nucleation rate by SCC theory in terms of magnitude, however, the temperature dependence of its prediction is not satisfying. We speculated this may be due to the use of dG^0 predicted using capillarity approximation for $(H_2O)_n$ (n > 10). Due to the unavailability of quantum-calculated thermodynamic properties of water clusters larger than $(H_2O)_{10}$, we extrapolated the dG^0 from $(H_2O)_{11}$ to the $(H_2O)_{20}$, as shown in Figure 3(a). It is done in such a way that the dG^0 rapidly decreases, and then gradually approaches the classical value of $(H_2O)_{20}$. The choose of $(H_2O)_{20}$ is arbitary as the main purpose is to qualitatively show the importance of thermodynamic properties of larger water clusters on the temperature dependence of the predicted nucleation rate. Figure 3(b) shows the comparison of nucleation rates calculated based on the updated dG^0 as shown in Figure 3(a) with the experimental study. Clearly, the temperature dependence of nucleation rates has been significantly improved. The

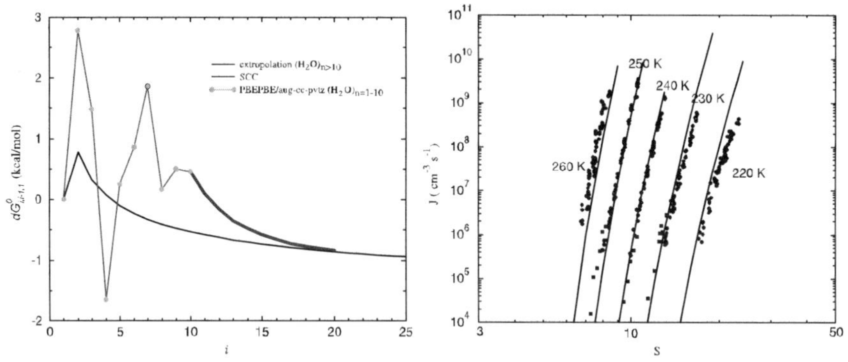

Figure 3 The stepwise Gibbs free energy changes from DFT using PBEPBE/aug-cc-pvtz method for water clusters from monomer to decamer (**circles**) at the standard condition. The dot-dashed curve is the extrapolation, so that the Gibbs free energy changes approach to that predicted by the SCC model. (b) Predicted nucleation rates at five different temperatures (as indicated), calculated with the DFT-based stepwise Gibbs free energy changes and extrapolations as shown in Figure 3(a)

excellent agreement with the experimental values may suggest that the additional dG^0 for clusters bigger than the decamer may be the reason for the insufficiently accurate temperature dependence of predicted nucleation rate in Figure 2. Further study is needed to obtain more accurate values of dG^0 for $i > 10$.

Acknowledgment Support of this work by the US National Science Foundation under grant ATM0618124 is gratefully acknowledged.

References

1. Becker, R. and Doring, W., *Ann. Phys. (Paris)*, **24,** 719 (1935).
2. Curtiss, L.A., Frurip, D.J., and Blander, M., *J. Chem. Phys.*, **71**(6), 2703–2711 (1979).
3. Munoz-Caro, C. and Nino, A., *J. Phys. Chem. A*, **101**(22), 4128–4135 (1997).
4. Girshick, S.L. and Chiu, C.-P., *J. Chem. Phys.*, **93**(2), 1273–1277 (1990).
5. Merikanto, J., Vehkamaki, H., and Zapadinsky, E., *J. Chem. Phys.*, **121**(2), 914–924 (2004).
6. Wolk, J. and Strey, R., *J. Phys. Chem. B*, **105**(47), 11683–11701 (2001).
7. Mikheev, V.B., et al., *J. Chem. Phys.*, **116**(24), 10772–10786 (2002).

Validation and Development of a New Hailstone Formation Theory–Numerical Simulations of a Strong Hailstorm Occurring Over the Qinghai-Tibetan Plateau

Fengqin Kang, Qiang Zhang, and Shihua Lu

Abstract Hailstorms occur frequently over the northeastern border of the Qinghai-Tibetan Plateau and its surroundings because of the combined geographical and meteorological features of this region. Formation and growth of the hailstones in a typical hailstorm are simulated using a three-dimensional (3D) cloud model with hail-bin microphysics developed by the Institute of Atmospheric Physics of the Chinese Academy of Sciences (IAP/CAS). The information of the large-scale circulations for the cloud model was provided by the MM5V3 model. The results show that (1) the water content of each hailstone bin is significantly large in the "cave channels"; (2) at the initial stage of hail formation, there is another high water content region consisting of small ice particles ($D < 1$ mm), graupel and hail embryos ($1\,mm < D < 5\,mm$), as well as small hailstones ($5\,mm < D < 10\,mm$), around the altitude of −30~−50°C above the high water content center associated with the "cave channels"; between them there is a gap of lower water content, which means that the main mechanisms of hail formation are different in those two regions; (3) as the hail and rain fall, the maximum center at higher-level drops until it merges with a lower equivalent; the larger the hail particles are, the earlier the maximum centers merge with each other; (4) during the hailstorm dissipation period the downdraft occurs in the region of "cave channels" and the "cave channels" fade; however, it is still the center of high hail water content, even though all updraft airflow turns to downdraft airflow; (5) "cave channels" are not the only regions of hailstones formation, but are nonetheless effective in the growth of hailstones, so which should be the main region of suppressing hail growth from small to large.

Keywords The Qinghai-Tibetan Plateau, hailstorms, hailstone formation

Key Laboratory of Arid Climatic Change and Reducing Disaster of Gansu Province, Lanzhou Institute of Arid Meteorology, Lanzhou, Gansu,China730020

C.D. O'Dowd and P.E. Wagner (eds.), *Nucleation and Atmospheric Aerosols*, 172–176.

Introduction

Over the past few decades, much effort has been devoted to understanding the physical and dynamic processes taking place in hailstorms, and a number of theoretical models of hailstone formation have been proposed (Barge and Bergwall 1976; Browning 1964; Browning and Foote 1976; Chisholm and Renick 1972; English 1986; Krauss and Marwitz 1984; Lemon and Doswell 1979; Miller et al. 1988) (see also Cotton and Anthes 1989 for an excellent review). Despite the great progress, our understanding and modeling of relevant physical mechanisms are still far from complete, which seriously limits our ability to mitigate the destruction caused by such extreme weather phenomena. For example, existing models of hailstone formation and growth cannot explain many observational phenomena (Smith 2003; Dostalek et al. 2004). Different models often give inconsistent results (Gilmore et al. 2004; van den Heever and Cotton 2004).

Over thc last decade or so, Xu and his colleagues in China have developed a "cave channels" theory (CC theory) on the formation and growth of hailstones (Xu et al. 2004). The CC theory has been applied to explain observations successfully (Tian et al. 2005). However, previous studies regarding the CC theory have been performed by use of a 3D Lagrangian model that is unable to simulate the growth of all the particles, and as a result, it is unclear if the CC theory can be applied to all the hydrometeors in a hailstorm.

CC Theory

We recapitulate the CC theory here because it has not been well known outside China, and it is used to compare and contrast with simulation results. As illustrated by Figure 1 (Xu et al. 2004), this theory has the following main points. First, there is a core of main updrafts (MUD) and an area by this core where the horizontal wind speed relative to the hailstorm equals zero because of the strong convective airflow of the hailstorm. Below this zero line, winds blow toward the core, whereas winds blow away from the core above this zero line. The growth travel trajectories of cloud particles rotate around the zero line. Particles enter the core of MUD circle by circle and grow into hailstones gradually. Second, there is a "cave channel" (CC) close to the core of MUD and below the zero line. Although the cave channel occupies only about 6% or smaller of the total volume of the hailstorm, it acts as a trap to attract particles. Once a particle enters the CC, it cannot escape the attraction of the CC until it becomes a large hailstone and falls from the CC exit, for it has been growing from small to large in CC until the updrafts cannot bear its weight. Therefore hail embryos form in the entrance end of the CC and grow into large hailstones in the exit end of CC. Finally, the existence of CC and its location depend on the airflow. The rate of hailstone growth and the lengths of the growth trajectories also depend on the field of supercooled water. A 3D Euler model is used for macrocloud and microcloud fields, and a 3D Lagrangian growth travel model is used for the behavior of hydrometeor particles (see also Xu et al. 2004). Until now we cannot find an analytic expression to show these characteristics.

Simulations Results and Concluding Remarks

Hailstorms that occurred from 1961 to 2001 over northwestern China are statistically analyzed. It is shown that hailstorms occur frequently over the northeastern border of the Qinghai-Tibetan Plateau and its surroundings because of the combined geographical and meteorological features of this region. A typical hailstorm is simulated by coupling a 3D cloud model with hail-bin microphysics developed by the Institute of Atmospheric Physics of the Chinese Academy of Sciences (IAP/CAS). Simulation results are further analyzed in light of the CC theory that has been recently proposed by Chinese scientists to understand the physical mechanisms for formation and growth of hailstones in hailstorms. The following points can be drawn from this study.

First, the simulation results are largely consistent with the so-called CC theory on hailstone formation. Throughout the life cycle of a hailstorm, the CC region assumes high water contents of hydrometeors of all sizes, including large hailstones. Furthermore, in the early stage of the hailstorm, there is an additional center of high water contents at high altitude (–30~–50°C) for smaller particles (first three classes, i.e., small ice particles (D < 1 mm), graupel and hail embryos (1 mm < D < 5 mm), and small hailstones (5 mm < D < 10 mm). There is a gap of lower water contents between the two centers, suggesting different mechanisms for the formation and growth of the particles in these two zones. Although both zones are all important for the formation and growth of particles of the first three classes, the CC region is necessary for further growth of smaller hail/graupel particles into large hailstones.

Second, as the hailstorm evolves, the high-level water content center gradually decreases in altitude to merge with the lower one, and an earlier merger leads to the

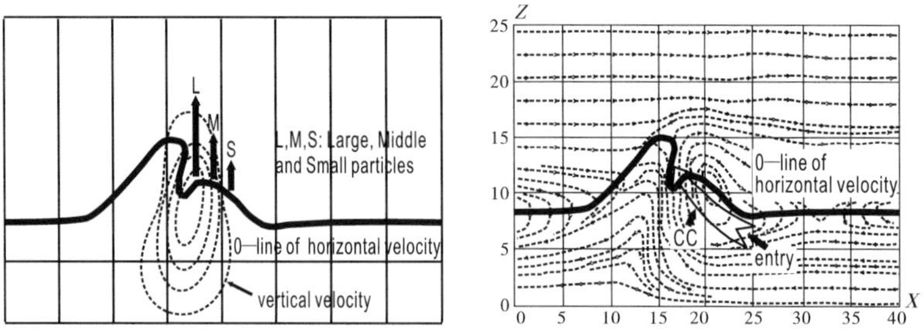

Figure 1 A schematic illustration of the "cave channels" (CC) theory (**right**) Location of CC in a relative updraft velocities. (From Xu et al. 2004.)

formation of larger hailstones. During the dissipation period, downdraft occurs in the CC region. Nevertheless, the CC region still exhibits high hail water contents.

Third, the results show that the "accumulation zone of cloud water" appears only at the initial stage of the hailstorm, and its maximal cloud water content is below the level of 0°C. Therefore it seems that the theory of "accumulation zone of supercooled water" as proposed by Sulakvelidze (Sulakvelidze et al. 1967) does not work in this hailstorm.

Finally, for the purpose of hail suppression this study suggests that artificial seeding of particles into the CC area may inhibit the formation of hailstones, especially large ones, because of the competition between the natural and artificial embryos.

Acknowledgments The authors would like to thank the reviewers for their critical and constructive comments. The authors thank the Institute of Atmospheric Physics of the Chinese Academy of Sciences for their hail-bin cloud model, especially Xueliang Guo. Thanks are also due to Huiming Ji, Wei Zhou, and Yanzhong Liu for their contributions to this paper. This work is supported under the Ministry of Science and Technology of the People's Republic of China grant 2002dib10046.

References

Barge, B.L. and Bergwall, F., *Fine Scale Structure of Convective Storms Associated with Hail Production*, Edmonton Rep. 76-2, Alberta, Canada: Atmospheric Science Division, Alberta Research Council (1976)

Browning, K.A., Airflow and precipitation trajectories within severe local storms which travel to the right of the winds, *J. Atmos. Sci.*, **21**, 634–639 (1964).

Browning, K.A. and Foote, G.B., Airflow and hail growth in supercell storms and some implications for hail suppression, *Q. J. R. Meteorol. Soc.*, **102**, 499– 533 (1976).

Chisholm, A.J., and Renick, J.H., *The Kinematics of Multicell and Supercell Alberta Hailstorms*, Rep. 72-2, Alberta, Canada: Alberta Hail Studies, Research Council of Alberta (1972).

Cotton, W.C. and Anthes, R.A., *Storm and Cloud Dynamics*, New York: Elsevier, 880 pp. (1989).

Dostalek, J.F., Weaver, J.F., and Loren Phillips, G., Aspects of a tornadic left-moving thunderstorm of 25 May 1999, *Weather Forecasting*, **19**, 614–626 (2004).

English, M., The testing of hail suppression hypotheses by the Alberta Hail Project, *Paper Presented at the 10th Conference on Weather Modification*, American Meteorological Society, Arlington, VA (1986).

Fletcher, N.H., *The Physics of Rain Clouds*, New York: Cambridge University Press, 390 pp. (1969).

Gilmore, M.S., Straka, J. M., and Rasmussen, E.N., Precipitation and evolution sensitivity in simulated deep convective storms: Comparisons between liquid-only and simple ice and liquid phase microphysics, *Mon. Weather Rev.*, **8**, 1897–1916 (2004).

Krauss, T.W. and Marwitz, J.D., Precipitation processes within an Alberta supercell hailstorm, *J. Atmos. Sci.*, **41**, 1025–1034 (1984).

Lemon, L.R. and Doswell III, C.A., Severe thunderstorm evolution and mesocyclone structure as related to tornadogenesis, *Mon. Weather Rev.*, **107**, 1184–1197 (1979).

Miller, L.J., Tuttle, J.D., and Knight, C.A., Airflow and hail growth in a severe northern High Plains supercell, *J. Atmos. Sci.*, **45**, 736–762 (1988).

Smith, P.L., "Raindrop size distributions: exponential or gamma – does the difference matter?", *J. Appl. Meteorol.*, **42**, 1031–1034 (2003).

Sulakvelidze, G.K., Bibilashvili, N.S., and Lapcheva, V.F., *Formation of Precipitation and Modification of Hail Processes*, Washington, DC: National Science Foundation (1967).

Tian, L.Q., Xu, H.B., and Wang, A.S., The observation validation and repeat of a new view of hail cloud mechanism, *Plateau Meteorol.*, **24**, 77–83 (2005).

van den Heever, S.C. and Cotton, W.R., The impact of hail size on simulated supercell storms, *J. Atmos. Sci.*, **61**, 1596–1609 (2004).

Xu, H.B., Duan, Y., and Liu, H.Y., *The Physics of Hailstorm and the Principle and Design of Hail Suppression*, Beijing: Meteorology Press (2004).

Scaling of Critical Nuclei Composed of Diatomic and Triatomic Molecules

I. Napari[1] and A. Laaksonen[2,3]

Abstract Density functional theory has been used to investigate scaling properties of critical nuclei composed of two or three Lennard-Jones sites. Varying amount of asymmetry is introduced in the attractive site–site interactions to induce orientational order at liquid–vapour interface. Systems showing little orientation obey the scaling laws of McGraw and Laaksonen (*Phys. Rev. Lett.*, **76**, 2754–2757 (1996)). However, fluids with a layered structure at the interface show considerable deviations from the scaling laws.

Keywords Homogeneous nucleation, scaling, density functional theory

Introduction

Nucleation in atmospheric sciences and technological applications is usually described using classical nucleation theory (CNT), with often ill effects. CNT is based on bulk thermodynamics and it is thus not well suited to small molecular clusters. Various amendments and alternatives have been sought to enhance the performance of CNT or replace it, but the results from years of study to develop a practical yet reliable approach to nucleation has been meager and the original CNT is still commonly used.

An intriguing viewpoint to nucleation is offered by scaling theories. Based on the nucleation theorem, McGraw and Laaksonen[1] developed scaling laws for the number of particles in the critical nucleus and height of the nucleation barrier. Although subject to certain assumptions, the scaling laws have been shown to be valid for simple Lennard-Jones (LJ) fluids[1,2] as well as, for example, *n*-alkanes.[3]

[1] *Department of Physical Sciences, University of Helsinki, P.O. Box 64, FI-00014 Helsinki, Finland*

[2] *Department of Applied Physics, University of Kuopio, P.O. Box 1627, FI-70211 Kuopio, Finland*

[3] *Finnish Meteorological Institute, P.O. Box 503, FI-00101 Helsinki, Finland*

C.D. O'Dowd and P.E. Wagner (eds.), *Nucleation and Atmospheric Aerosols*, 177–180.

In this work we use density functional theory (DFT) to study the validity of the scaling laws for model fluids consisting of diatomic or triatomic molecules. The molecules consist of LJ interaction sites and we vary the strength of the attractive interaction between the sites to induce orientational order at liquid–vapour interface. We show that interfacial structure is indicative of how closely the scaling laws are obeyed: serious departure from the scaling relations is observed if a layered zone exists between the homogeneous liquid and vapour phases.

Scaling

McGraw and Laaksonen[1] developed the following scaling relations for the number of particles in the critical nucleus N^* and the work of formation W^*:

$$N^* = C(T)(\Delta\mu)^{-3} \tag{1}$$

$$\frac{W^*}{N^*\Delta\mu} = \frac{1}{2} - \frac{D(T)}{C(T)}(\Delta\mu)^2 \tag{2}$$

where $\Delta\mu$ is the chemical potential difference between the supersaturated and saturated vapour. Here $C(T)$ and $D(T)$ are functions only of temperature. The scaling laws represent a special solution of a differential equation, but otherwise they do not depend on the microscopic details of the system.

Model and Methods

Our model fluids consist of diatomic ("dimer") and triatomic ("trimer") species with the atomic site–site interaction described by hard sphere repulsion and attractive Lennard-Jones (LJ) tail. The dimer models are numbered according to the asymmetry and behaviour at the liquid–vapour interface: DI (identical sites, no molecular orientation at the interface), DII (modest orientation with the weakly interacting sites pointing toward vapour), DIII (strongly oriented surface layer), DIV (multiple oriented layers). The special feature of these dimer fluids is that they have the same equation of state. We have also studied a trimer model (T) where the molecule consists of a stiff chain of identical LJ sites. The results are compared to a simple monoatomic ("monomer", M) fluid.

The nucleation behaviour is studied using density functional theory (DFT). Density functional method is based on reduction of the microscopic details of the system into spatially varying density distributions and on the division of the Helmholtz free energy into a reference and perturbation part (hard spheres and attractive LJ tail in our case). DFT calculation gives the equation of state, the liquid–vapour surface tension, and the properties of critical nuclei in a consistent manner (e.g., see Ref. 4 for more details).

Results

Figure 1 illustrates the validity of the particle number scaling law for the model fluids. On the horizontal axis a is the relative vapour phase activity defined by $a = exp(\Delta\mu/kT)$. Fluids showing little or no surface structure at liquid–vapour interface (DI, DII, M, T) obey the scaling, attested by the linear dependence between $(\Delta\mu)^{-3}$ and N^*. In contrast, system with oscillating density profiles at interface (DIII, DIV) deviate from the scaling law somewhat when the cluster is small.

The second scaling law is demonstrated in Figure 2. Again, systems DI, DII, T, and M conform to the scaling relation while DIII and DIV fail to do so. Now, however, the failure in the case of models DIII and DIV is of more serious nature, because there is not even approximate correspondence between the scaling theory and DFT calculation (although the scaling seems to hold for large clusters in system DIII).

The sudden jumps in the curves of Figure 2 reflect structural changes in the cluster: The model DIV, for example, shows multiple-oriented layers at the planar liquid–vapour interface. A small cluster may have just one layer and but, as the cluster grows, it can incorporate more layers. The failure of the scaling laws can be attributed to the varying thickness of the layered "phase" in clusters, which should be included in the thermodynamic description of the cluster, although this may be impossible in practice.

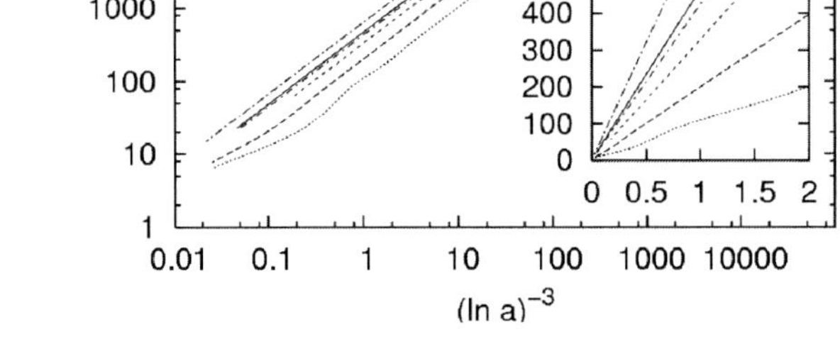

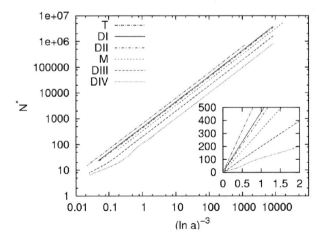

Figure 1 Scaling of the particle number for the studied models. The inset shows the scaling for small clusters in more detail

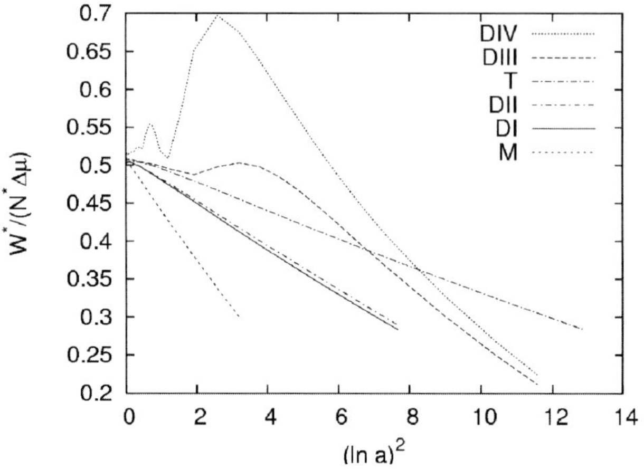

Figure 2 Scaling for $W^*/(N^*\Delta\mu)$

Conclusions

We have shown that the scaling relations of McGraw and Laaksonen[1] are obeyed if the liquid–vapour interface is (almost) structureless, that is the molecules at the interface are randomly oriented. The scaling laws fail for fluids with a layered interface. Such fluids do exist, for example, some metals[5] and diblock copolymers.[6] Fortunately, according to our study, the scaling laws should be valid for most fluid systems studied in nucleation experiments and in atmospheric sciences.

Acknowledgment This work was supported by the Academy of Finland.

References

1. McGraw, R. and Laaksonen, A., *Phys. Rev. Lett.*, **76**, 2754–2757 (1996).
2. Chen, B. and Siepmann, J.I., Oh, K.J. and Klein, M.L., *J. Phys. Chem.*, **115**, 10903–10913 (2001).
3. Chen, B., and Siepmann, J. I., Oh, K. J., and Klein, M. L., *J. Phys. Chem.*, **116**, 4317–4329 (2002).
4. Napari, I. and Laaksonen, A., *J. Phys. Chem.*, **113**, 4480–4487 (2001).
5. DiMasi, E., Tostmann, H., Ocko, B.M., Pershan, P.S., and Deutsch, M., *Phys. Rev. B* **58**, R13419–R13422 (1998).
6. Hariharan, A. and Harris, J.G., *J. Chem. Phys.*, **101**, 4156–4165 (1994).

Heterogeneous Condensation on Nanoparticle

Sergey P. Fisenko[1], **Manabu Shimada**[2], **and Kikuo Okuyama**[2]

Abstract Vapor condensation on nanoparticle with radius smaller than the Kelvin radius is considered as fluctuation or as the heterogeneous nucleation. The expression for steady-state heterogeneous nucleation rate is obtained.

Keywords The Kelvin radius, core-shell structure, heterogeneous nucleation rate

Introduction

From thermodynamic point of view a vapor condensation starts on the spherical nanoparticle if the radius R of the nanoparticle is greater then the Kelvin radius, R_c. The latter radius is determined by means of the expression [1]:

$$R_c = 2\sigma v_a / kTln(S) \tag{1}$$

where σ is the surface tension of the condensed liquid, v_a is the volume per molecule in this liquid, S is the vapor supersaturation, k is the Boltzmann's constant, T is the temperature of the system. If nanoparticle radius is smaller the Kelvin radius the condensation is prohibited from thermodynamic point of view. Nevertheless due to fluctuations the vapor condensation is possible, we call this process as heterogeneous nucleation. Basic parameters of heterogeneous nucleation are the main subject of our report. It is worthy to emphasize that insight into heterogeneous nucleation is practically important for evaluation of the performance of particle size magnifiers (PSM) [2,3].

---

[1] *A.V. Luikov Heat & Mass Transfer Institute, National Academy of Sciences, 15, P. Brovka St., Minsk 220072, Belarus*

[2] *Department of Chemical Engineering, Graduate School of Engineering, Hiroshima University, Higashi Hiroshima, 739–8527 Japan*

C.D. O'Dowd and P.E. Wagner (eds.), *Nucleation and Atmospheric Aerosols*, 181–184.

Free Energy of Wetting Film Formation

The free energy of formation of wetting film $\Delta\Phi(R,g)$, which has g molecules, on the surface of the spherical nanoparticle can be written as:

$$\Delta\Phi(R, g) = -gkTln(S) + 4\pi [R^2 (\sigma_{sl} - \sigma_{sv}) + R_1^2 \sigma] \quad (2)$$

where σ_{sl} and σ_{sv} are correspondingly, the surface tension between the solid nanoparticle and the liquid, and the surface tension between the solid nanoparticle and vapor, R_1 is the outer radius of wetting film, we will call it the shell radius. In particular for system of gold–water, the contact angle is equal to zero, and for calculation σ_{sv} we have the condition:

$$\sigma_{sv} = \sigma + \sigma_{sl}.$$

For core gold nanoparticle with radius $R = 2.5$ nm and water vapor supersaturation $S = 1.4$, the free energy of wetting film formation versus the radius of the shell is displayed in Figure 1. The existence of thermodynamic barrier of wetting film formation is obvious. The height of this barrier (~50 kT) is typical for homogeneous vapor nucleation. The position of the maximum of the thermodynamic barrier exactly corresponds to the Kelvin radius R_c. Qualitatively it is clear from data in Figure 1 that overcoming of thermodynamic barrier is important if the nanoparticle radius is slightly smaller than R_c. Analytic results are presented below.

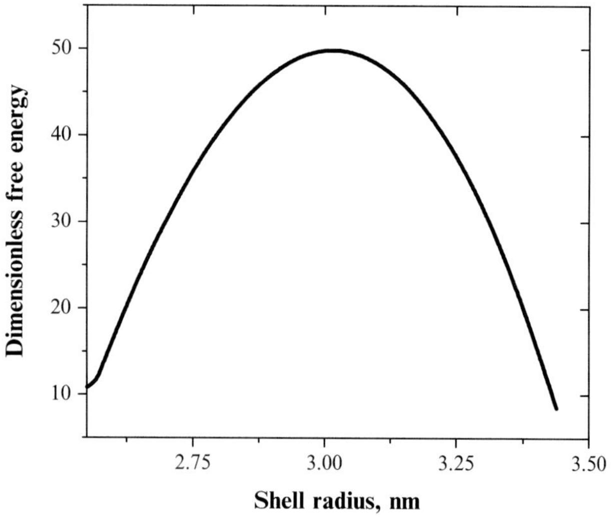

Figure 1 Dimensionless free energy of the liquid shell formation $\Delta\Phi$(2.5 nm,g)/kT versus shell radius R1

Kinetics of Heterogeneous Nucleation

Let us introduce the ensemble of heterogeneous droplet with core nanoparticles of the same size and corresponding to the distribution function *f*(*g*) of the number molecules *g* in the liquid shell over the nanoparticle. We assume below that *g* is the continuous variable.

The kinetic equation for distribution function is the continuity equation, which includes the Brownian diffusion in the space of the number of molecules is the liquid shell. It can be shown that the kinetic equation for heterogeneous nucleation is formally similar to the kinetic equation for homogeneous nucleation [4]:

$$\partial_g\left[f(g)L_{11}\partial_g\left(\ln f(g)+\beta\Delta\Phi(R,g)\right)\right]=-\partial_g\left[J_h\right] \tag{3}$$

where L_{11} is the free molecular flux of vapor molecules on the surface of the critical heterogeneous droplet, which radius coincide with Kelvin's radius; J_h is the rate of heterogeneous nucleation. The physical meaning of J_h is the flux of heterogeneous droplets overcoming thermodynamic barrier of the heterogeneous condensation.

Boundary conditions to the kinetic equation (3) are:

For small *g* we have equilibrium with adsorbed molecules

$$f(g)=C\,exp\left[-\beta\Delta\Phi(R,g)\right]; \tag{4}$$

For large *g* we have the standard condition

$$f(g)=0,$$

where *C* is the normalization constant. The normalization constant *C* is related with the number density of nanoparticles $N_p(g)$ with *g* molecules in liquid core by the relationship:

$$N_p(g)=C\,exp\left(-\beta\Delta\Phi(R,g)\right)dg \tag{5}$$

After approximate integration of the both sides of (5) over *g* we have the expression:

$$C\cong N_t\,exp\left(\beta\Delta\Phi(R,0)\right),$$

where N_t is the total number of nanoparticles in the ensemble.

The heterogeneous nucleation rate J_h is the first integral of the kinetic equation (3), and after calculations we have the formula:

$$J_h=N_tL_{11}\,exp\left[-\beta\left(\Delta\Phi(R,g^*)-\Delta\Phi(R,0)\right)\right]\sqrt{\alpha/\pi} \tag{6}$$

where g^* is the number of molecule at the critical shell, and

$$\alpha=0.5\left|\partial^2_{gg}\Delta\Phi(R,g^*)\right|.$$

To remind that g^* depends on radius of the nanoparticle, temperature, supersaturation, and thermophysical properties of the liquid. In particular for conditions shown in Figure1, heterogeneous nucleation gives substantial effect if the nanoparticle radius is larger $0.95R_c$.

Conclusions

It was shown that vapor condensation can start on the surface of the nanoparticle even if its radius is smaller than the Kelvin radius. This process is stochastic process, and the probability to overcome thermodynamic barrier is smaller than one. The free energy of wetting film formation is given by the expression (2). The flux of heterogeneous droplets, which overcome the thermodynamic barrier, is the rate of the heterogeneous nucleation. For calculation of heterogeneous nucleation rate the formula (6) was obtained. If the radius of core increases the heterogeneous nucleation rate drastically increases because free energy of wetting film formation drops.

The heterogeneous nucleation on charged nanoparticles is also discussed in the frame-developed kinetic description.

References

1. Kubo, R., *Thermodynamics*, Amsterdam: North-Holland (1968).
2. Kim, C.S., Okuyama, K., and de la Mora, J.F., *Aerosol Sci. and Tech.*, **37**, 791–803 (2003).
3. Fisenko, S.P., Wang, W.N., Shimada, M., and Okuyama, K., *Intern. J. Heat & Mass Transfer*, **50** (2007).
4. Frenkel, Y.I., *Kinetic Theory of Liquids,* New York: Dover (1955).

Kinetics of Multicomponent Nucleation in Gas Phase

Sergey P. Fisenko

Abstract Kinetics of n-component nucleation is considered as the Brownian diffusion in the size and $(n-1)$ composition spaces. The formula for nucleation rate is obtained, which has correct transitions to all possible unary cases.

Keywords Nucleation rate, the Brownian diffusion coefficients

Introduction

For the first time the binary nucleation kinetics as the Brownian diffusion in the cluster size and in the composition space was considered in [1]. I present here the generalization of this approach to the nucleation kinetics of n vapors, which has growing scientific and industrial importance. For this case we have to consider one cluster size space and $(n-1)$ composition spaces.

Nucleation Kinetics in Cluster Size and Composition Spaces

We consider below n-component supersaturated mixture of vapors. For description of multicomponent clusters we use the distribution function of clusters, to denote it as $f(g_1, \ldots g_n)$, where g_1 is the number of molecules of the first kind. The steady-state kinetic equation for $f(g_1, \ldots g_n)$ can be written, using the formal analogy with the binary nucleation in gases. This kinetic equation is the n-dimensional Fokker-Planck equation of special kind [2]:

$$\partial_{g_1}\left[fL_{11}\partial_{g_1}\left(ln\, f + \beta\Delta\Phi(g_1 \ldots g_n)\right)\right] + \ldots + \partial_{g_n}\left[fL_{nn}\partial_{g_n}\left(ln\, f + \beta\Delta\Phi(g_1 \ldots g_n)\right)\right] = 0, \qquad (1)$$

A.V. Luikov Heat & Mass Transfer Institute, National Academy of Sciences, 15, P. Brovka St., Minsk 220072, Belarus

C.D. O'Dowd and P.E. Wagner (eds.), *Nucleation and Atmospheric Aerosols*, 185–189.

where $\beta = 1/kT$, k is the Boltzmann's constant, T is the temperature of vapors mixture, $\Delta\Phi$ is the free energy formation of multicomponent cluster; for every $i = 1..n$ parameter L_{ii} is the flux of molecules of i-kind on the critical cluster at the free molecular regime. The critical cluster and its properties are determined from the system of equations

$$\partial_{g_1} \Delta\Phi = ... = \partial_{g_n} \Delta\Phi = 0. \tag{2}$$

Usually the radius of critical cluster is about 1 nm; the total number of molecules in it is about 100. The dimensionless free energy of critical cluster formation $\beta\Delta\Phi$ is about several tens.

The equilibrium solution f_e of the equation (1) has the form:

$$f_e = C\, exp\left[-\beta\Delta\Phi(g_1,...g_n)\right], \tag{3}$$

where C is the normalization "constant" [3], which depends on all the number densities of vapor molecules.

Let us start the transition to the natural variables: the total number of molecules in a cluster, g:

$$g = \sum_{i=1}^{n} g_i;$$

and $(n-1)$ molar compositions of component, x_i: $x_i = g_i / g$. In order to introduce new distribution function of clusters $\varphi(g, x_1, \ldots x_{n-1})$, I apply the condition [2]:

$$\varphi(g, x_1,...x_{n-1})\, dg dx_1 ... dx_{n-1} = f(g_1,...g_n) dg_1 ... dg_n \tag{4}$$

After the calculation of the Jacobean it is easy to show that the equilibrium distribution function $\varphi_e(g, x_1, \ldots x_{n-1})$, based on new variables, is:

$$\varphi_e = C_1 g^{n-1}\, exp\left[-\beta\Delta\Phi\right], \tag{5}$$

where C_1 is new normalization constant. It is worth to note that equilibrium function (5) has wonderful property, $\varphi_e(0, x_1, \ldots x_{n-1}) = 0$. Therefore the problem of calculation of C_1 is not so ambiguous even if we use the capillary approximation for description of thermodynamic properties of small clusters. Even more, this fact gives directions for future development of the nucleation kinetics [4]. To note that at new variables the critical cluster is determined by the system of equations:

$$\partial_g \Delta\Phi = \partial_{x_1} \Delta\Phi = ... = \partial_{x_{n-1}} \Delta\Phi = 0$$

It is useful to introduce a new function $y(g, x_1, \ldots x_{n-1})$ in order to the nonequilibrium distribution function can be written as $\varphi(g, x_1, \ldots x_{n-1}) = \varphi_e(g, x_1, \ldots x_{n-1}) y(g, x_1, \ldots x_{n-1})$. Kinetic equation for new distribution function y is obtained from equation (1), following the transformation rules from [2]:

$$\partial_g \left\{ \varphi_e \left[L\partial_g y + L(n-1)y/g + \sum_{i=1}^{n-1} \partial_{x_i} \left(a_i(g, x_{1,}...x_{n-1})y \right) \right] + \right.$$

$$\sum_{i=1}^{n-1} \partial_{x_i} \left\{ \left\{ \varphi_e \left[\partial_g \left(a_i(g, x_{1,}...x_{n-1})y \right) + a_i y(n-1)/g \right\} + \right.\right.$$

$$\left. \partial_{x_i} \left(b_i(g, x_{1,}...x_{n-1})y \right) + \sum_{s \neq i} \partial_{x_s} \left(c_{is}(g, x_{1,}...x_{n-1})y \right) \right\} \qquad (6)$$

Following [1] the Brownian diffusion coefficient L in the cluster size space is defined:

$$L = \sum_{i=1}^{n} L_{ii}.$$

Correspondingly, $b_i(g, x_1, \ldots x_{n-1})$ is the Brownian diffusion coefficient at i-composition space. In equation (6) terms with a_1 describe cross-flows between the cluster size space and i-composition space; terms with c_{is} describe cross-flow between i- and s-composition spaces. For binary nucleation we have identities $c_{is} \equiv 0$. Explicit expressions for coefficients are:

$$a_i(g, x_1, ...x_{n-1}) = \frac{L_{ii} - x_i L}{g} \, b_i(g, x_1, ...x_{n-1}) = \frac{(L - L_{ii}) x_i^2 + (x_i - 1)^2 L_{ii}}{g^2}, \qquad (7)$$

$$c_{is}(g, x_1, ...x_{n-1}) = \frac{L x_i x_s - x_s L_{ii} - x_i L_{ss}}{g^2}. \qquad (8)$$

The Brownian diffusion coefficients in cluster size and compositions space are always positive. As it follows from expressions above that if a cluster size increases the contribution of cross-flows to the evolution of the distribution function decreases. Diffusion spreading of the distribution function in the compositions spaces is smaller for larger clusters. There are the set of characteristic compositions, x_i^k, when cross-flow between cluster size and i-composition space is equal to zero: $x_i^k = L_{ii} / L$. For x_i^k the value of Brownian diffusion coefficient in i-composition space is minimal one. Additionally, for x_i^k and x_s^k, the value of c_{is} has an extreme value. Physical meaning of characteristic composition is clarified in [1] for binary nucleation; for multicomponent nucleation probably situation is more diverse. After overcoming thermodynamic barrier the molar composition of i component of growing cluster has tendency to reach characteristic composition x_i^k. Thus the simplest case of multicomponent nucleation kinetics is the case when molar compositions in the critical cluster coincide with x_i^k.

The most important flux for steady-state nucleation kinetics is the flux in the cluster size space J_g, which is:

$$J_g = -\varphi_e \left[L\partial_g y + L(n-1)y/g + \sum_{i=1}^{n-1} \partial_{x_i} \left(a_i(g, x_{1,}...x_{n-1})y \right) \right]$$

For multicomponent the boundary conditions are obvious generalization of the boundary conditions for binary nucleation [1]: On every plane in n-dimensional space, where $x_i = 1$ or $x_i = 0$, the corresponding the scalar product $J_{xi}\, n = 0$; where n is the normal to the plane. For small g we put the condition of equilibrium cluster distribution function:

$$\varphi(g, x_1 \cdots, x_{n-1}) = \varphi_e(g, x_1 \cdots, x_{n-1}); \text{ therefore } y = 1$$

for relatively large clusters, which are large than critical one, we have standard condition:

$$\varphi(g, x_1, \ldots x_{n-1}) = y = 0$$

Thus the domain for the distribution function is a parallelepiped in n-dimensional space.

Let us denote the solution of (2), which correspond the lowest value of the free energy of cluster formation as g^* and x_1^*, $\ldots x_{n-1}^*$. Usually this solution corresponds to the saddle point of the surface of free energy of cluster formation.

The total nucleation rate I is the $(n-1)$ dimensional integral of the flux in the cluster size space over all possible compositions; the expression for I at the point g is written below:

$$I(g) = \int_0^1 dx_1 \ldots \int_0^1 dx_{n-1} J_g(g, x_1, \ldots x_{n-1}). \tag{9}$$

The special solution of the kinetic equation (6) in the form

$$\varphi(g, x_1, \ldots x_{n-1}) = y = 0 \tag{10}$$

where δ is the Dirac's δ function, gives the expression for nucleation rate I:

$$I = C_1 \left(g^*\right)^{n-1} L exp\left[-\beta \Delta\Phi(g^*, x_1^*, \ldots x_{n-1}^*\right] \sqrt{\alpha / \pi}, \tag{11}$$

where α is semiwidth of the surface of the free energy cluster formation in cluster size space at the saddle point: $\alpha = 0.5 \partial_{gg} \Delta\Phi(g^*, x_1^* \ldots x_{n-1}^*)$. The expression (11) is the generalization of formulas obtained in [1] for binary nucleation. The deviation of the expression (11) from previous ones for nucleation rate, obtained in [3,4] is obvious. It is worth to emphasize that the expression (11) has the correct transition to the unary nucleation rate for any component if concentrations of all other components go to zero [5].

Conclusions

The expression for multicomponent nucleation rate, which has $n-1$ correct transitions to unary nucleation, is derived. The feature of multicomponent nucleation kinetics is flows between different composition spaces. Explicit expressions are obtained for the Brownian diffusion coefficient in cluster size space and for the Brownian diffusion coefficients in n–1 composition spaces. Brownian diffusion coefficient in compositions space drastically decreases if the cluster size grows.

References

1. Fisenko, S.P. and Wilemski, G., *Phys. Rev. E*, **70** (056119) (2004).
2. Risken, H., *The Fokker-Planck Equations*, New York: Springer (1989).
3. Kurasov, V.B., *Physica A*, ***353***, 159–216 (2005).
4. Kalikmanov, V.I. and van Dongen, M., *Phys. Rev. E*, **55**, 1607 (1997).
5. Lin Zhuo and Wu, D.T., *J. Chem. Phys.*, **125** (104506) (2006).

Simulation of Nucleation Experiments in Laminar Flow Diffusion Chamber

Sergey P. Fisenko and Anton A. Brin

Abstract The positive feedback between growing droplets and thermodynamic and spatial parameters of nucleation zone in laminar flow diffusion chamber is considered. For calculation we use our advanced mathematical model. It is shown that for given temperature parameters the volume of the nucleation zone depends on pressure and flow rate of gaseous mixture.

Keywords Volume of the nucleation zone, droplet growth

Introduction

Recent experiments, devoted to homogeneous nucleation of vapors in diffusion cloud chambers and laminar flow diffusion chambers show that the number of optically detectable droplets, depends on the total pressure at the system [1]. Such experimental results are at the deep contradiction with the classical theory of nucleation kinetics [2], which does not show any such dependence. What are the physical fundamentals of such deviation between theoretical predictions and experimental results? We consider that for laminar flow diffusion chambers main reasons are: the finite size of nucleation zone and simultaneously, the positive feedback between droplets growth and motion and the state of the nucleation zone.

We developed the new mathematical model of the performance of laminar flow diffusion chamber, which take into account a nonlinear positive feedback between the number of droplets and the conditions at nucleation zone. We illustrate our approach by the simulation of the performance of the setup, used for low-pressure experiments at laminar flow diffusion chamber (LFDC) [1]; working vapor is butanol, helium is the carrier gas.

A.V. Luikov Heat & Mass Transfer Institute, National Academy of Sciences,15, P. Brovka St., Minsk 220072, Belarus

C.D. O'Dowd and P.E. Wagner (eds.), *Nucleation and Atmospheric Aerosols*, 190–194.

Simulation Results

Our mathematical mode is described in details in [3]. It has two partial differential equations for the temperature field and the field of the number density of vapor with nonlinear terms for interaction with growing droplets. Additionally it has ordinary differential equations for droplets growth and motion. Growing droplets heat the gaseous mixture around them and simultaneously, decrease the number density of a vapor. One of simulation results, dimensionless free energy of the critical cluster formation, $\Delta\Phi/kT_w$, is shown in Figure 1. The points *A* and *B* mark the boundary of nucleation zone. The length of nucleation zone $L = z(B)-z(A)$; L is about 4R in Figure1, where R is the radius of the condenser. We determine that the volume of nucleation zone Ω is the volume where the dimensionless free energy of the critical cluster formation is not larger than the dimensionless global minimum on two units. In other words, inside this zone at different places nucleation rates have the same order of magnitude.

For steady-state condition in LFDC the number density of droplets $N(r,z)$ and the local nucleation rate $J(r,z)$ are connected by the continuity equation:

$$\partial_z\left[N(r,z)u(r)\right] = J(r,z), \tag{1}$$

where $u(r)$ is the axial velocity of flow. After integration of the equation (1), we have approximate expression for total nucleation rate I:

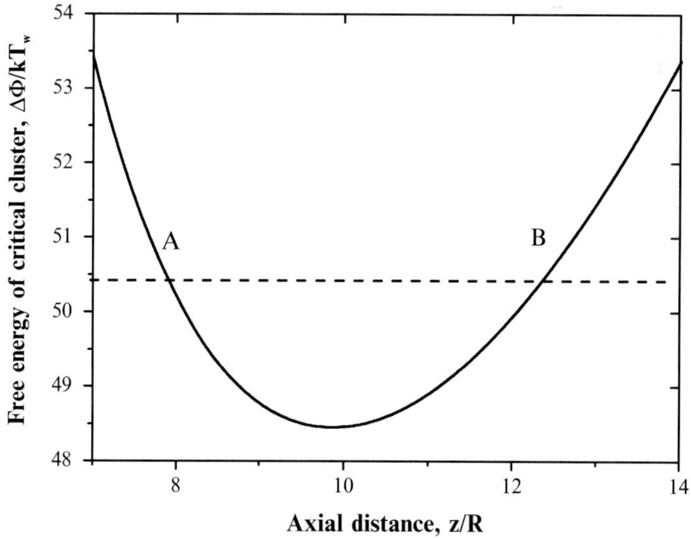

Figure 1 Free energy of the critical cluster formation versus axial position at the centerline of condenser

$$I \approx N_t u(0)\Omega / L, \tag{2}$$

where N_t is already the total number of droplet. Thus we have to know the volume of nucleation zone and its axial size in order to connect nucleation rate and the number of formed droplets.

The analysis of equations of our mathematical model leads to the conclusion that we can neglect effects of the positive feedback if the following inequality is valid:

$$N(r,z) << R^{-2}\lambda^{-1},$$

where λ is the mean free pass of vapor molecules. Simple numerical estimation gives that for atmospheric conditions, $R = 2$ mm and $\lambda = 0.1$ µm, $N(r,z) << 10^{11}$ droplet/m^3. Well known that $\lambda \sim P^{-1}$, where P is the total pressure at the flow chamber. Thus the positive feedback is stronger for relatively small pressure. It is worth to mention that the Lewis number Le is also pressure dependent for small pressures of carrier gas,

$$Le = \lambda / (\rho c D),$$

where λ, ρ, c are correspondingly, the heat conductivity of mixture in LFDC, density and heat capacity; D is the diffusion coefficient of vapor in carrier gas. It was shown in [3] that for different Le numbers the profiles of temperature and supersaturation are different. The dependence of dimensionless volume of the nucleation zone versus the total pressure is shown in Figure 2 for the same flow rate (conditions of experiments in [1]).

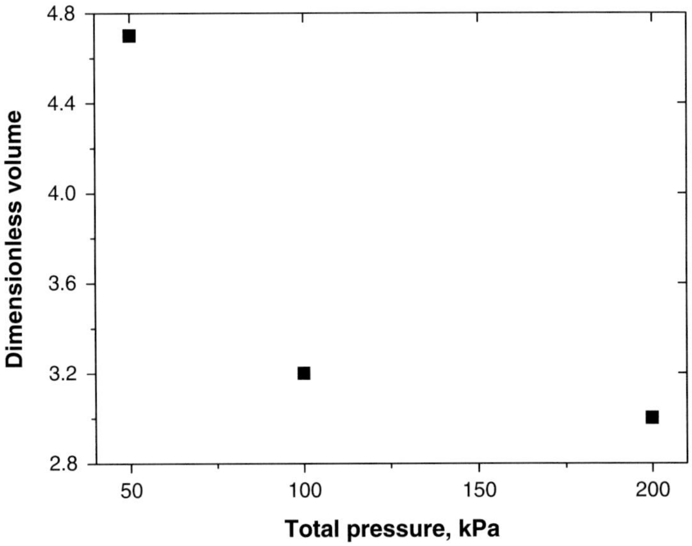

Figure 2 Nucleation zone volume Ω versus total pressure; for conditions of experiments N1 from tables at [1]

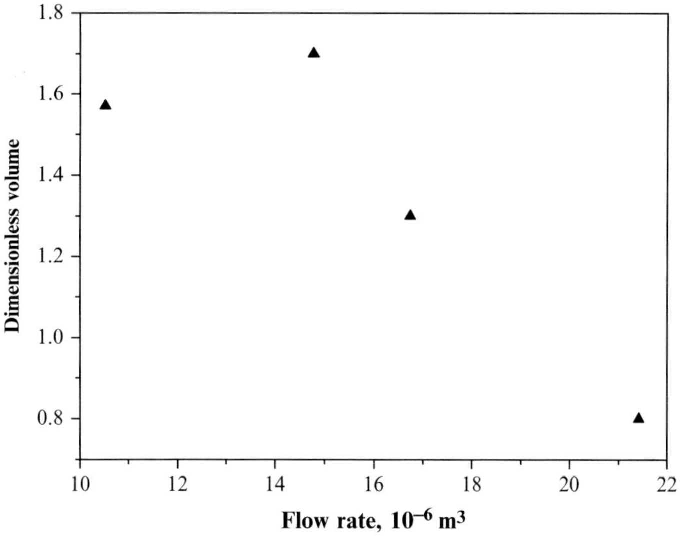

Figure 3 Nucleation zone volume Ω versus flow rate; for conditions of experiments from [1] (constant temperature at nucleation zone, pressure adjusted)

The scale of volume of nucleation zone is πR^3. For relatively high flow rate of gaseous mixture if we increase pressure, as our simulation shows, the volume of the nucleation zone decreases. The obtained dependence of dimensionless volume of nucleation zone versus flow rate (for the same total pressure) is shown in Figure3.

It is not trivial dependence because the value of flow rate affects significantly on temperature and vapor density profiles [3]. Near the global minimum of the free energy of the critical cluster formation nucleation rate is practically constant. Thus the data in Figure 3 presents, in fact, the number of formed droplet in LFDC. The similar results have been reported in [1, Figure 5].

Conclusions

For high nucleation rate, higher than 10^{11} particles/(m^3s) for the setup [1], nonlinear interaction between growing droplets and vapor in nucleation zone of the laminar flow diffusion chamber significantly disturbs the temperature and supersaturation profiles. For such conditions the interpretation of the results of nucleation experiments should be based on more sophisticated mathematical models than standard ones. In particular, for low total pressure LFDC these effects are very profound. We found that there is the significant reducing of axial size of the nucleation zone in compare with the results of standard mathematical models. Another results of our simulations and the comparison with experimental ones are discussed.

For comparison of predictions of the homogeneous nucleation theory and experimental results these nonlinear effects of interaction between droplets and vapor put the limits on the use of data obtained be means of laminar flow diffusion chamber.

References

1. Hyvarinen, A.P., Brus, D., Zdimal, V., et al., *J. Chemical Physics*, **124**, 224304 (2006).
2. Frenkel, Y.I., *Kinetic Theory of Liquids*, New York: Dover (1955).
3. Fisenko, S.P. and Brin, A.A., *Intern. J. Heat and Mass Transfer*, **49**, 1004–1014 (2006).

A Comparative Study in Molecular Dynamics Simulation of Nucleation

Jan Julin, Ismo Napari, and Hanna Vehkamäki

Abstract Gas-liquid nucleation of 1,000 Lennard–Jones atoms is simulated to evaluate temperature regulation methods and methods to obtain nucleation rate. The thermostats compared are the Berendsen and Andersen thermostats, of which the Andersen thermostat yields nucleation rates about two times higher than the Berendsen thermostat. The nucleation rates are obtained both through direct observation of times of nucleation onset and by the method of Yasuoka and Matsumoto. The nucleation rates obtained with the latter method are about three times higher than those obtained with direct observation.

Keywords Nucleation, molecular dynamics, thermostat

Introduction

Molecular dynamics (MD) is the most straightforward molecular level method to investigate gas–liquid nucleation. In the direct MD method a large number of molecules are placed in a simulation box, the system is quenched to a supersaturated state, and the particle trajectories are followed by integrating the equations of motion. The nucleation event and the cluster growth that follows can then readily be observed.

If an inert carrier gas is added to the simulation, the nucleating substance is thermalized by the interaction with the carrier gas and the nucleation process is isothermal. In atmospheric nucleation a carrier gas is always present in the form of "air molecules" (mainly O_2 and N_2). A carrier gas is often missing from MD simulations since the number of carrier gas particles required for effective thermalization multiplies the simulation time and may place an unacceptably low limit to the number of nucleating particles. Without the carrier gas the nucleation occurs non-isothermally and the observed nucleation rates are lower[1]; however, the presence of a carrier gas

Department of Physical Sciences, University of Helsinki P.O. Box 64, FI-00014 Helsinki, Finland

C.D. O'Dowd and P.E. Wagner (eds.), *Nucleation and Atmospheric Aerosols*, 195–199.

can be imitated by coupling the system to an artificial thermostat. While using a thermostat avoids the downsides of a carrier gas, the thermostat may disturb the dynamics of the system in a nonrealistic manner.

In this work we perform gas–liquid nucleation simulations of simple Lennard–Jones particles. We compare two existing methods to obtain the nucleation rate and assess two popular thermostats, the Berendsen[2] and Andersen[3] thermostats.

Results

We performed direct MD simulations with a system consisting of 1,000 Lennard–Jones atoms in a 120×120×120 Å box with periodic boundary conditions. The parameter values for argon, $\sigma = 3.40$ Å and $\varepsilon/k_b = 120$ K, were used. The potential cutoff was at 5σ. Initially the system was at a temperature of 130 K, and was quenched to a target temperature of 85 K after 1 ns of simulation time. The time step used was 6 fs. Depending on which happened first, the simulation was terminated after 10 ns of simulation time or after the largest cluster exceeded the size of 200 atoms.

Two well-known thermostats, the Berendsen and the Andersen thermostats were used. When thermostatting with the Berendsen thermostat the velocities of every particle is scaled by a factor

$$\lambda = \sqrt{1 + \frac{\Delta t}{\tau}\left(\frac{T_0}{T} - 1\right)}, \quad (1)$$

where T_0 is the target temperature, T is the current kinetic temperature of the system, Δt is the time step, and τ is a preset time parameter, for which the value $\tau = 400$ fs was used.

When using the Andersen thermostat a number of particles are randomly selected to be given new velocities drawn from a Maxwell–Boltzmann distribution corresponding to the target temperature. The probability that a particle is given a new velocity is $\nu\Delta t$, where Δt is again the time step and ν is a parameter describing the collision frequency with an imaginary heat bath. In these simulations the value of the parameter was set to $\nu = 5 \times 10^{-4}$ 1/fs.

There is a clear unnaturalness in both these thermostats as they remove heat from particles inside large clusters as well. In order to test the effect this has on the nucleation process we also performed simulations where the thermostats are only applied to free particles, where the free particles are those that do not belong to any clusters when a cluster is defined by the Stillinger[4] criterion with a cutoff at 1.5σ.

To obtain the nucleation rate from the simulations two methods were used. In the method of Yasuoka and Matsumoto[5] the number of clusters that exceed a certain threshold size, is plotted as a function of simulation time. The slope of the linear dependence found is then divided by the volume of the simulation box to obtain the nucleation rate.

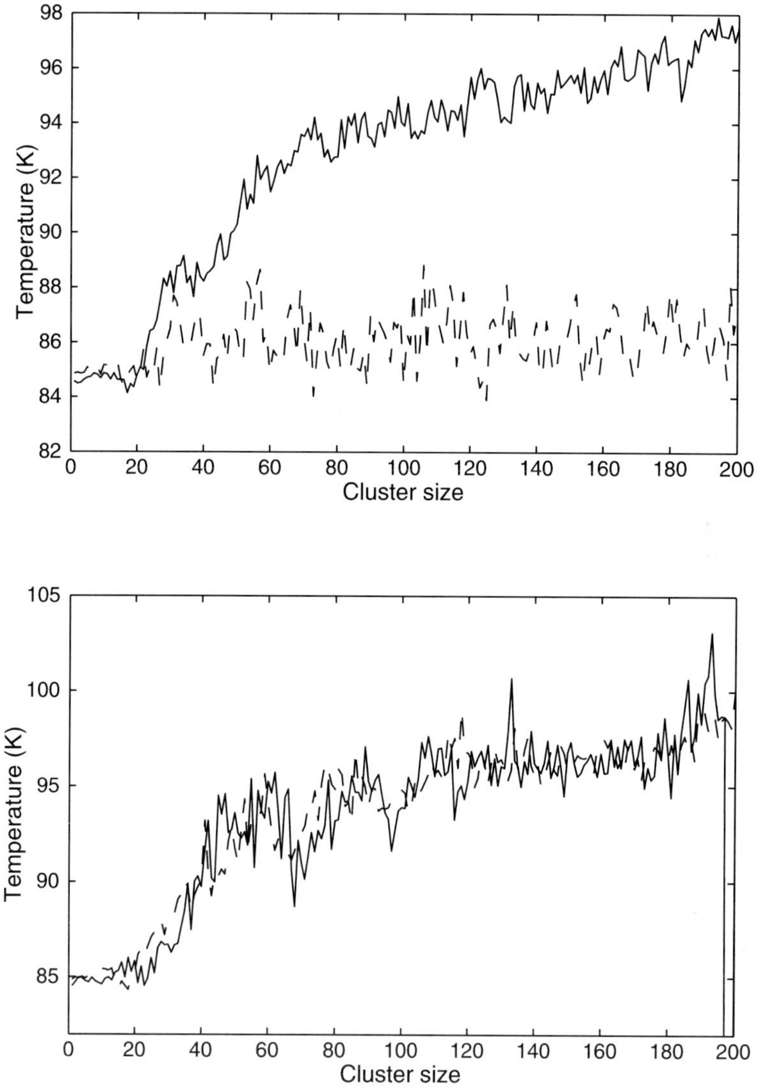

Figure 1 Cluster temperatures averaged over several runs. Top: the Berendsen (solid line) and Andersen (dashed line) thermostatted runs. Bottom: thermostatting only applied to the monomers, solid line represents the Berendsen and dashed line the Andersen thermostat

The nucleation rate was also obtained with a direct observation method, where one observes the time of nucleation onset, the time when the first cluster able to grow appears. The nucleation rate is then estimated as (one cluster)/(time of onset X the volume of the simulation box). The direct observation method has the downside that a relatively high number of simulation runs is required to obtain sufficient statistics, unlike the method of Yasuoka and Matsumoto where even a single run might suffice.

Table 1 Nucleation rates obtained by different methods and thermostats

	Yasuoka–Matsumoto	Direct observation	Number of simulations
Andersen	2.3×10^{-7}	1.0×10^{-7}	99
Berendsen	1.3×10^{-7}	0.5×10^{-7}	97
Andersen to free	1.5×10^{-7}	0.4×10^{-7}	87
Berendsen to free	1.3×10^{-7}	0.4×10^{-7}	84

The thermostats manage to keep the temperature at the desired level with varying success. As can be seen in Figure 1, where the temperatures are collected starting from the quench until the end of the simulation, the deviation from the desired temperature is at most around 3K even for larger clusters when using the Andersen thermostat. However, the Berendsen thermostat manages only to keep the temperature of clusters up to size 25 in check, which is close to the critical cluster size of our simulations. For larger clusters than this the temperature rises rapidly, and the largest clusters are over 10 K hotter than the target temperature.

For the runs where the thermostats are only applied to the free particles the temperature of clusters behaves similarly to the regular Berendsen thermostatted runs. When the Berendsen thermostat is only applied to monomers the temperatures of larger clusters are slightly higher than those in regular Berendsen thermostatted runs, as one should expect. And since both thermostats managed to keep the monomer temperature where desired, it comes as no surprise that when the Andersen thermostat is used only on monomers, the cluster temperatures mimic the behavior seen in corresponding Berendsen thermostatted runs.

Table 1 collects the nucleation rates obtained by the different methods and thermostats along with the number of simulations where a nucleation event was observed. The nucleation rates are given in reduced units where we have taken the Lennard–Jones parameters σ and ε as units of length and energy respectively, and the single particle mass m, 6.63×10^{-26} kg for argon, as the unit of mass. The unit of nucleation rate is then $[J] = \sigma^{-3}\tau^{-1}$, where $\tau = 2.15$ ps is the unit time.

The nucleation rates for the different thermostats are fairly close to each other, the rates from the regular Andersen thermostatted runs being a bit higher than the others. There is little difference in nucleation rates between the runs where only the monomers are thermostatted and the regular thermostatted runs. For the Berendsen thermostat the difference is practically nonexistent. The method of Yasuoka and Matsumoto yields —two to three times higher nucleation rates than direct observation. This is still a fairly good agreement.

Conclusions

We have performed MD simulations of 1,000 Lennard–Jones atoms to assess the effects of simulation methods on nucleation. The Andersen thermostat regulates the temperature of the clusters better than the Berendsen thermostat; however the nucleation rates are still fairly similar for the two thermostats. Applying the thermostats only on monomers instead of all atoms in the system does not appear to have any significant effect on the nucleation rates. It seems that thermostatting method is not a critical issue in obtaining the nucleation rate. However, this conclusion may not apply if the critical cluster is larger than in our simulations.

References

1. Barrett, J., Clement, C., and Ford, I., *J. Phys. A: Math. Gen.*, **26**, 529–548 (1993).
2. Berendsen, H.J.C., Postma, J.P.M., van Gunsteren, W.F., DiNola, A., and Haak, J.R., *J. Chem. Phys.*, **81**, 3684–3690 (1984).
3. Andersen, H.C., *J. Chem. Phys.*, **72**, 2384–2393 (1980).
4. Stillinger, F.H., *J. Chem. Phys.*, **38**, 1486–1494 (1963).
5. Yasuoka, K. and Matsumoto, M., *J. Chem. Phys.*, **109**, 8451–8462 (1998).

The Reactions of Sulfuric Acid with Biogenic Criegee Intermediates or Secondary Ozonides as a Possible Source of Nucleation Precursors

Theo Kurtén, Boris Bonn, Hanna Vehkamäki, and Markku Kulmala

Abstract We have studied the reactions of sulfuric acid with stabilized Criegee Intermediates (sCIs) and secondary ozonides formed in the ozonolysis of biogenic mono- and sesquiterpenes. Activation energy calculations on model species show that the reaction between sulfuric acid and sCIs is almost barrierless. If the lifetime of biogenic sCIs in the atmosphere is sufficiently long for them to undergo bimolecular reactions, then the sCI plus sulfuric acid reaction could account some part of the organically assisted new-particle formation events observed in the atmosphere. If, on the other hand, unimolecular formation of secondary ozonides is the main sink reaction of sCIs, then the reactions between sulfuric acid and the secondary ozonides could be of atmospheric importance. For the secondary ozonide formed from α-pinene, we have found a strongly exothermic reaction which results in the formation of a sulfuric acid monoester plus a water molecule. However, this reaction may have high activation energy.

Keywords Nucleation, sulfuric acid, Criegee intermediate, secondary ozonide, quantum chemistry

The Formation and Reactions of Criegee Intermediates

Recent experimental evidence[1] indicates that new-particle formation from organic vapors in atmospheric conditions may proceed via reactions of stabilized Criegee intermediates (sCIs). Criegee intermediates are biradicals formed (via several steps) in the ozonolysis of alkenes (e.g., terpenes):

$$R_1R_2C = CR_3R_4 + O_3 => R_1R_2CO + R_3R_4COO^{\ddagger} \qquad (1)$$

Where $R_1 \ldots R_4$ are functional groups and ‡ denotes vibrational excitation. Small Criegee biradicals rapidly undergo unimolecular decomposition, but larger compounds are expected to be collisionally stabilized, and can react with, e.g., water, carbonyl compounds, or organic acids. For endocyclic terpenes (i.e., terpenes in

Department of Physical Sciences, University of Helsinki, PL 64, 00014 Helsingin yliopisto, Finland

C.D. O'Dowd and P.E. Wagner (eds.), *Nucleation and Atmospheric Aerosols*, 200–203.

which the double bond is located inside a ring structure), unimolecular formation of secondary ozonides (SOZ) is also possible:

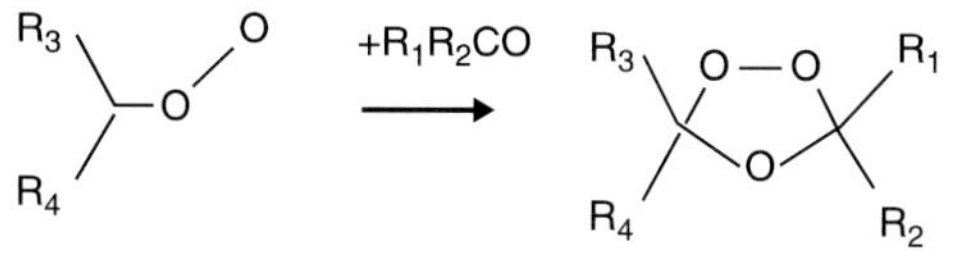

where the two reacting moieties are located on the same molecule. It is unclear whether bimolecular reactions (mainly, that with water vapor) or unimolecular SOZ formation is the main sink reaction for biogenic sCIs in the atmosphere. Chuong et al. postulated,[2] based on quantum chemical and kinetic calculations, that unimolecular SOZ formation is the main sink reaction for sCIs formed from endocyclic sesquiterpenes. However, it is possible[3] that their computed activation barrier for SOZ formation is somewhat too low due to the small basis set used in the calculations.

Reactions of Criegee Intermediates with H_2SO_4

Since sulfuric acid is known to participate in tropospheric nucleation in a wide variety of conditions, a plausible first step for organically assisted nucleation processes is the reaction of some reactive organic species with sulfuric acid.

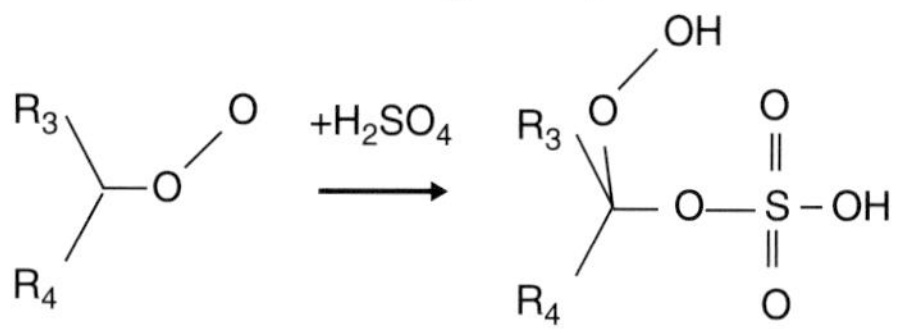

We have recently proposed[3] a mechanism for the reaction of sCIs with sulfuric acid: where the product is a peroxide complex stabilized by OOH•••O = S hydrogen bond.

We have calculated[3] thermochemical parameters for the above reaction for a variety of sCIs. The key results of this study are presented in Table 1. The structures of the reaction products of sulfuric acid and two biogenic sCIs are shown in Figure 1.

It can be seen from Table 1 that the formation of the peroxide complex is strongly exothermal. However, this is unsurprising given the high reactivity of the sCI. For the reaction to be of atmospheric importance, it must have significantly lower activation energy than the competing pathways, such as the bimolecular reaction with water vapor. We have studied the reactions of the model species $(CH_3)_2COO$ with sulfuric

Table 1 Reaction energies (ΔE_0) and standard Gibbs free energies ($\Delta G°$) for the reactions of three sCIs with H_2SO_4, computed at the B3LYP/6–311 + G(2d,p) level, with an extra set of diffuse and polarization on the sulfuric acid hydrogens. All values in kcal mol^{-1}

Reaction	ΔE_0	$\Delta G°$ (298 K, 1 atm)
$(CH_3)_2COO + H_2SO_4 => (CH_3)_2C(OOH)\text{-}O\text{-}SO_3H$	−30.0	−14.0
$C_{10}H_{16}O_3$[a] + $H_2SO_4 => C_{10}H_{18}O_7S$[c]	−23.4	−7.9
$C_{15}H_{24}O_3$[b] + $H_2SO_4 => C_{15}H_{26}O_7S$[c]	−30.0	−13.7

[a]The more stable of the two possible sCIs formed in the ozonolysis of α-pinene.
[b]The more stable of the two possible sCIs formed in the ozonolysis of β-caryophyllene.
[c]See Figure 1 for the structures of the reaction products.

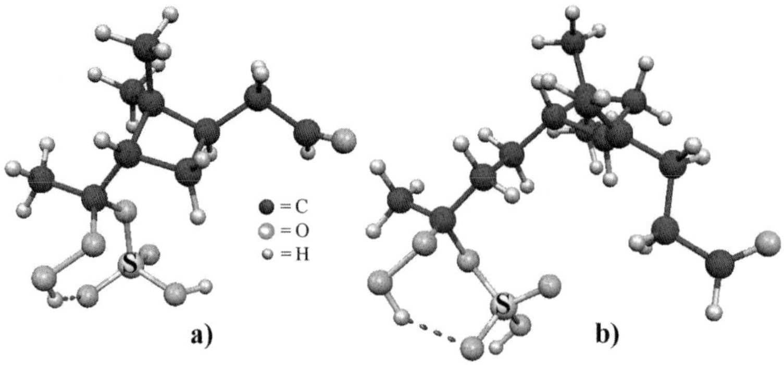

Figure 1 The products of the reaction of sulfuric acid with the sCIs formed in the ozonolysis of **(a)** α-pinene **(b)** β-caryophyllene, optimized at the B3LYP/6–311 + G(2d, p) level with extra diffuse and polarization functions on the sulfuric acid hydrogen atoms

Table 2 Reaction energies (ΔE_0) for the various steps of the reaction of $(CH_3)_2COO$ with sulfuric acid and water, computed using two different quantum chemical methods. ••• denotes the hydrogen-bonded complex and ‡ denotes the transition state structure. The basis set for the B3LYP computations was the same as in Table 1. The basis set for the RI-CC2 calculations was def2-QZVPP. The RI-CC2 transition state geometry corresponds to the maximum bracketed along the B3LYP IRC path with a step size of 0.05 a.u

Reaction	ΔE_0, B3LYP, kcal mol^{-1}	ΔE_0, RI-CC2, kcal mol^{-1}
$(CH_3)_2COO + H_2SO_4$ = > $(CH_3)_2COO{\bullet\bullet\bullet}H_2SO_4$	−20.5	−22.4
$(CH_3)_2COO{\bullet\bullet\bullet}H_2SO_4$ = > $[(CH_3)_2COO{\bullet}H_2SO_4]^‡$	0.1	−0.2
$[(CH_3)_2COO{\bullet}H_2SO_4]^‡$ = > $(CH_3)_2C(OOH)\text{-}O\text{-}SO_3H$	−9.6	−16.5
$(CH_3)_2COO + H_2O$ = > $(CH_3)_2COO{\bullet\bullet\bullet}H_2O$	−9.2	−9.3
$(CH_3)_2COO{\bullet\bullet\bullet}H_2O$ = > $[(CH_3)_2COO{\bullet}H_2O]^‡$	+15.2	+12.9
$[(CH_3)_2COO{\bullet}H_2O]^‡$ = > $(CH_3)_2C(OOH)\text{-}OH$	−36.8	−37.9

acid and water in greater detail using Intrinsic Reaction Co-ordinate (IRC) scans at the B3LYP level[4,5] together with high-level single-point energy evaluations at the RI-CC2 level.[6,7] Both reactions proceed with the same general mechanism. First, a hydrogen-bonded complex is formed without a barrier. Next, a proton is transferred to the COO group, followed by the formation of a new C–O bond. This reaction step has a small or nonexistent barrier in the case of sulfuric acid, and a high barrier in the case of water. The energetics of the individual reaction steps are shown in Table 2.

It can be seen from Table 2 that the difference between activation energies of the reactions of $(CH_3)_2COO$ with water and sulfuric acid is around 12–15 kcal/mol. The result is probably roughly applicable also to the larger biogenic sCIs, since the reacting functional groups are the same. A naive application of conventional transition state theory would imply that the sCI + H_2SO_4 reactions should proceed about 10^{10}–10^{11} times faster than the sCI + H_2O reactions in atmospheric conditions. However, the quantum chemical parameters presented here are not accurate enough for actual quantitative rate constant predictions, and tunneling processes or reactions with water clusters[8] may increase the effective rate constants for the sCI + H_2O reactions significantly. We can nevertheless conclude that the rate constants for the sCI + H_2SO_4 reactions are likely to be at least as large as those determined for the reactions of various sCIs with organic acids,[9] (which are around five orders of magnitude faster than the reactions with water) and might be significantly larger. This would imply that if the biogenic sCIs livelong enough to undergo bimolecular reactions, a significant fraction of the sCIs formed in atmospheric conditions may react with sulfuric acid.

Reactions of Secondary Ozonides (SOZ) with H2SO4

We have also investigated the reaction of sulfuric acid with secondary ozonides. Based on the experimental observations of Surratt et al.,[10] we have assumed that the end product is a sulfuric acid ester. Stoichiometrically, the most probable reaction mechanism would then involve elimination of water, e.g., as follows:

We have investigated the thermodynamics of the above reaction for the secondary ozonide formed in the ozonolysis of α-pinene at the B3LYP/6–311 + + G(2d,2p) level. The standard Gibbs free energy of the reaction was $-67.8\,\text{kcal mol}^{-1}$, indicating that it is very favorable thermodynamically. However, the reaction mechanism is likely to be quite complex, and may involve high-energy barriers due to the breaking of C–O and O–O bonds. Further investigations of the kinetics of the reaction are necessary before any conclusions about its atmospheric relevance can be drawn.

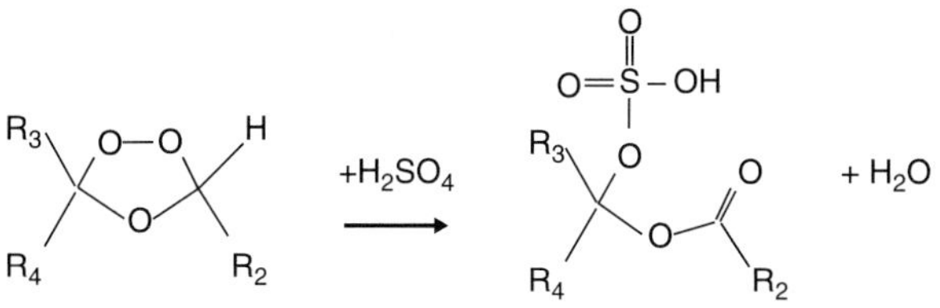

All our computations have been carried out using the Gaussian 03[11] and Turbomole[12] v. 5.8. program suites. The thermal contributions to the free energies have been computed using the standard rigid rotor and harmonic oscillator approximations.

Acknowledgments The authors thank the Scientific Computing Center (CSC) in Espoo, Finland for computing time and the Academy of Finland for funding.

References

1. Bonn, B., Schuster, G., and Moortgat, G.K., *J. Phys. Chem. A*, **106**, 2869–2881 (2002).
2. Chuong, B., Zhang, J.Y., and Donahue, N.M., *J. Am. Chem. Soc.*, **126**, 12363–12373 (2004).
3. Kurtén, T., Bonn, B., Vehkamäki, H., and Kulmala, M., *J. Phys. Chem. A*, in press (2007).
4. Becke, A.D., *J. Chem. Phys.*, **98**, 5648–5652 (1993).
5. Lee, C., Yang, W., and Parr, R.G., *Phys. Rev. B*, **37**, 785–789 (1988).
6. Christiansen, O., Koch, H., and Jørgensen, P., *Chem. Phys. Lett.*, **243**, 409–418 (1995).
7. Hättig, C. and Weigend, F., *J. Chem. Phys.*, **113**, 5154–5161 (2000).
8. Ryzhkov, A.B. and Ariya, P.A., *Chem. Phys. Lett.*, **419**, 479–485 (2006).
9. Tobias, H.J. and Ziemann, P.J., *Phys. Chem. A*, **105**, 6129–6135 (2001).
10. Surratt, J.D., Kroll, J.H., Kleindienst, T.E., Edney, E.O., Claeys, M., Sorooshian, A., Ng, N.L., Offenberg, J.H., Lewandowski, M., Jaoui, M., Flagan, R.C., and Seinfeld, J.H., *Environ. Sci. Technol.*, **41**, 517–527 (2007).
11. Frisch, M.J. et al., Gaussian 03, Revision C.02, Gaussian, Wallingford, CT (2004).
12. Ahlrichs, R., Bär, M., Häser, M., Horn, H., and Kölmel, C., *Chem. Phys. Lett.*, **162**, 165–169 (1989).

Condensation-Relaxation of Supersaturated Vapor and the Possibility of the Experimental Determination of the Nucleation Rate

Naum Kortsensteyn and Eugene Samujlov

Abstract In this paper, we propose a method for determining the nucleation rate immediately at the drop nucleation step. The method was developed by analyzing the results of numerical modeling by moment's method of the bulk condensation of vapor from a vapor–gas mixture after instantaneous creation of a supersaturated state with regard to heat release. The analysis of the numerical modeling results enabled us to make simplifying assumptions and derive scaling relations, which relate the condensation relaxation time and the number density of droplets to the nucleation rate under the initial conditions. It is these relations, who provide a fundamental possibility of experimentally determining the nucleation rate, form the basis for the method proposed.

Keywords Supersaturated vapor, condensation-relaxation, scaling relations, nucleation rate

Introduction

The key role in a first-order phase transition is played by nucleation. Along with classical nucleation theory (CNT), a lot of studies were performed, where either various corrections were made to the classical theory for the calculated results to agree better with experimental data or attempts were made to construct a nucleation theory on a new basis. With regard to this, it is desirable to perform a systematic experimental investigation of nucleation rate in vapors of various substances, similar to previous studies of nucleation in superheated liquids [1]. The nucleation rate in a vapor is difficult to experimentally determine, in particular, because of the fact that droplets at the moment of nucleation have size ~1nm and cannot be detected by optical diagnostic methods. In this context, along with the existing methods [2], it seems important to experimentally determine the nucleation rate by methods

Krzhizhanovsky Power Engineering Institute, Leninskii pr. 19, Moscow, 119991 Russia

C.D. O'Dowd and P.E. Wagner (eds.), *Nucleation and Atmospheric Aerosols*, 204–208.

involving no optical detection of droplets. In this paper, we propose a method for determining the nucleation rate immediately at the droplet nucleation step. This method was developed by analyzing the results of numerical modeling of the bulk condensation in a spatially homogeneous medium with regard to heat release. The analysis of the numerical modeling results enabled us to make simplifying assumptions and derive (with a certain accuracy) scaling relations, which relate the condensation relaxation time and the number density of drops to the nucleation rate under the initial conditions. It is these relations, who provide a fundamental possibility of experimentally determining the nucleation rate, form the basis for the method proposed.

Formulation of the Relaxation Problem

It was assumed that, at the initial moment of time, the immovable vapor–gas mixture contains no droplets and, at $t = 0$, the supersaturation ratio instantaneously takes a value $s = s_0 > 1$. In a supersaturated vapor, as in any metastable system, the relaxation proceeds to the equilibrium state. In the case considered, this is a condensation relaxation in a spatially homogeneous medium including the processes of nucleation and the growth of forming droplets. The time of condensation relaxation τ_c [3] is understood as such a time interval over which the initial supersaturation ratio S_0 due to vapor expansion decreases by e times ($e = 2.718$). Changing of the supersaturation ratio is described by the following equation:

$$\frac{d\ln s}{dt} = -\alpha\pi r_d^2 n_d v_T \left(1 + x_v \frac{L}{C_p T}\left(\frac{L\overline{M}_v}{\overline{R}T} - 1\right)\right), \tag{1}$$

were α is condensation coefficient, υ_T is the thermal velocity of vapor molecules, L is the heat of vaporization; $\overline{M}_v$ is the molar mass of vapor; N_A is Avogadro's number; x_v is the mass concentration of the vapor, number droplet density $n_d = \rho\Omega_0$, $r_d = \Omega_1 / \Omega_0$, C_p and ρ is the specific heat capacity and density of vapor-gas-droplet mixture. Here Ω_i are the distribution function moments:

$$\Omega_i(t) = \int_{r_{cr}}^{\infty} r^i f(r,t)dr. \tag{2}$$

In the kinetic regime of droplet growth from kinetic equation for the distribution function of droplet size follows the set of coupled moment equations [4,5]

$$\frac{d\Omega_i}{dt} = i\dot{r}\Omega_{i-1} + \frac{I}{\rho} r_{cr}^{\ i}. \tag{3}$$

The set of equations (1), (3) presents macro kinetic model of the process of condensation relaxation. When using this model one should control the fulfillment of conditions

$$\tau_c >> \tau_g >> \tau_{lag}, \tag{4}$$

where τ_g is the time of vapor transition to the metastable state and τ_{lag} is the time lag in the theory of nucleation. The first of these conditions makes it possible to simulate the condensation relaxation in a quasi-steady-state approximation, while the second condition allows us to consider the creation of metastable state as a rather fast process and disregard the processes accompanying transition of vapor to the metastable state.

Results and Discussion

Set of equations (1), (3) with the use of CNT for calculating the nucleation rate I and the Hertz–Knudsen formula for determining the rate of droplet growth $\dot{r}$ was employed in this work for the numerical simulation of condensation relaxation in a mixture of cesium vapors and argon in the formulation described above. For integrating of set of Eqs. (1), (3) was used the Kutta–Merson method [6]. When calculating, we varied the initial values of the supersaturation ratio (from 3 to 6), temperature (from 580 to 610 K), and volume vapor concentration in a mixture (from 0.001 to 0.85). Some results of numerical simulation of condensation relaxation are shown in Figure 1.

As is seen from Figure 1, in the course of condensation relaxation we can specify the "induction period" τ_i, during which the formation of droplets proceeds at a constant

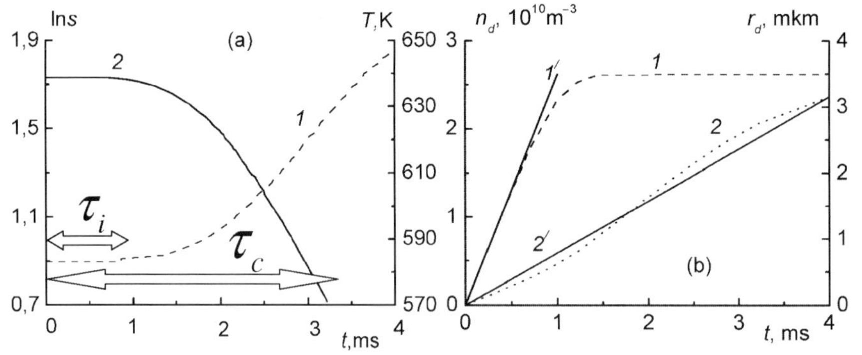

Figure 1 (a) Dependences of temperature (1) and supersaturation ratio (2) on time; **(b)** Dependences of number droplet density (1, 1′) and average droplet radius (2, 2′) on time; (1′, 2′) denote calculation by Eq. (5)

nucleation rate I_0 and almost invariable thermodynamic parameters of a system (temperature, supersaturation ratio). In this case, the forming droplets grow over the entire period of condensation relaxation at a slow varying rate. Hence, the number density of nucleated droplets and their average radii can be presented in the form

$$n_d = I_0 t, 0 < t < \tau_i, n_d = I_0 \tau_i, t \geq \tau_i, r_d = b\dot{r}_0 t, 0 \leq t \leq \tau_c, \tag{5}$$

where I_0 is the nucleation rate determined by the initial values of temperature T_0 and the supersaturation ratio S_0; $\dot{r}_0$ is the rate of droplet growth corresponding to T_0 and S_0; $\tau_i = a\tau_c$; a is the fraction of induction period of the total time of condensation relaxation; and b is constant equals~1.

One can integrate Eq. (1) with allowance for relations (5) within the interval from τ_i to τ_c those results in

$$1 = \frac{\pi a b^2}{3}(1 - a^3) V_T \dot{r}_0^2 A_0 I_0 \tau_c^4. \tag{6}$$

The derived expression (6) turned out to be the basis for the analytical consideration of the process of condensation relaxation. First, relations connecting the time of condensation relaxation and number droplet density (i.e., resultant process characteristics) with nucleation rate at the initial moment of time

$$\tau_c = P_\tau I_0^{-1/4}, n_d = P_n I_0^{3/4}, \tag{7}$$

$$P_\tau = \left(\pi a b^2 \dot{r}_0^2 V_T A_0 (1 - a^3)/3\right)^{-1/4}, A_0 = x_v^0 \frac{L}{C_p T_0}\left(\frac{L\overline{M}_v}{\overline{R}T_0} - 1\right), P_n = aP_\tau. \tag{8}$$

Relation (7) can be presented in the form

$$I_0 = B\tau_i^{-4}, B = 3a^3/[\pi b^2 \dot{r}_0^2 V_T A_0 (1 - a^3)], \tag{9}$$

that allows us to use it for measuring nucleation rate. Indeed, the duration of "induction period" τ_i can be determined by the time dependence of temperature in the volume of condensing vapor, the a value is known, and the remaining parameters in Eq. (10) are the characteristics of the studied substance or they are set by the initial experimental conditions. There is the possibility to estimate the accuracy of the experimental data received by the suggested method as follows. From the general reasons we can write down (see (10)):

$$\text{Accuracy } (I_0) = \text{Accuracy } (B) + 4 \times \text{Accuracy } (\tau_i). \tag{10}$$

Factor B brings an error not less than 10%. The error in determination of duration of the"induction period" τ_i is connected to accuracy of experimental determination of temperature depending on time in concrete experimental installation. As a result we have

$$\text{Accuracy } (I_0) > 4 \times \text{Accuracy } (T(t)) + 10\%. \tag{11}$$

Acknowledgments This work was supported by the Russian Foundation for Basic Research, projects no. 05-08-01512 and 07-08-00082.

References

1. Baidakov, V.G., in Peregrev kriogennykh zhidkostei (Overheating of Cryogenic Liquids), Ural Otd. Ross. Akad. Nauk, Yekaterinburg (1995).
2. Wölk, J., Strey, R., and Wyslouzil, B.E., Homogeneous Nucleation Rates of Water: State of the Art. *Nucleation and Atmospheric Aerosols 2004: 16th Int'l Conf.*, edited by M. Kasahara and M. Kulmala, pp. 101–114.
3. Kortsenshtein, N.M. and Samuilov, E.V., *Dokl. Phys. Chem.*, **397**, part 2, 169–173 (2004).
4. Sternin, L.E., *Osnovy gazodinamiki dvukhfaznykh techenii v soplakh* (Fundamentals of Gas Dynamics of Two-Phase Flows in Nozzles), Moscow: Mashinostroenie (1974).
5. Frenclach, M. and Harris, S.J., *J. Coll. Interface Sci.*, **118**, 252–261 (1987).
6. Ford, B. and Pool, J.C.T. The evolving NAG library service, in: Sources and Development of Mathematical Software, edited by W. Cowell Englewood Cliffs, NJ: Prentice-Hall, pp. 375–397 (1984).

Nanoparticle Detection Using Nucleation Regime of the CPC

Genrik Mordas, Mikko Sipilä, and Markku Kulmala

Abstract In this study we present a method how to improve the particle detection efficiency of the condensation particle counter. The limit of the CPC UF-02proto was explored. Our investigations show that the instrument background is 0.002 cm^{-3} at default regime (temperature difference 32.5°C), and the instrument cut-size (D50%) is 4.4 nm. Increasing the temperature difference, the homogeneous nucleation starts around the temperature difference of 39.0°C. Subsequent temperature difference increasing enhances the number concentration of the nucleated particles. However, the experimental results show that the nucleation process can be prevented by aerosol particles existing in the supersaturated region. When the number concentration of 15 nm silver particles exceeds 4,000 cm^{-3} then the homogeneous nucleation inside CPC disappears. The detection efficiency as a function of the temperature difference is also investigated. The lowest reliable instrument cut-size D50% was 1.8 ± 0.2 nm when the temperature difference was 44.0°C.

Keywords Condensation particle counter, detection efficiency, ultrafine particles

Introduction

Atmospheric aerosol impacts on the natural ecosystems[1] and human health[2] depend strongly on the particle size distribution, which is determined in many respects by new particle formation phenomenon.[3] New particle formation was studied during couple of decades in various locations.[4,5,6] However, it still remains unclear, because newly formed particles have very small size and cannot be measured by the commercial instruments. Thus, the instrument modification and development can lead to better understanding of new particle formation process.

Division of Atmospheric Sciences, Department of Physical Sciences, P. O. Box 64, FI-00014 University of Helsinki, Finland

C.D. O'Dowd and P.E. Wagner (eds.), *Nucleation and Atmospheric Aerosols*, 209–213.

The most available ultrafine particle detectors are condensation particle counters (CPCs).[7] The CPCs are classified by the supersaturation creating methods: an adiabatic expansion CPC,[8] a conductive cooling CPC,[9,10] a mixing-type CPC[11,12] and water diffusivity CPC.[13] The recently developed CPC UF-02proto is a conductive cooling CPC with the swirling flow inside saturator-condenser.[14]

The CPC is characterized by detection efficiency parameter, or more practical parameter – a cut-size D50%. The cut size presents the particles size, which drops the particle detection efficiency at half. It is determined by the particle transport efficiency, optical registration and it is strongly depended on the temperature difference between saturator and condenser. Increasing the temperature difference, the cut-size is improved. However, the main problem in attempts to improve the cut-size is that homogeneous nucleation finally always buries the signal from droplets nucleated heterogeneously on the surfaces of existing particles or clusters. The critical particle size at which the homogeneous nucleation overcomes the particle activation depends on several factors including the detailed design of the CPC, physicochemical properties and supersaturation of condensing vapor, supersaturation profile, and initial particle concentration in the supersaturated regime.

In this paper, we present the new point of view on the operation regimes of the CPC and the method to get lower cut-size using recently developed CPC.[14]

Calibration Methods

The particle detection efficiency as a function of particle size can be determined experimentally.[15] The principles of this method were used in our calibration setup.[14]

Polydisperse silver aerosol was produced by a tube furnace (Carbolite Furnaces MTF 12/388). The generated silver particles were charged in an alpha-active^{241} Am bipolar source after which they were classified with a differential mobility analyzer (DMA VIE-08, Hauke, length 0.109 m). After DMA, the selected monodisperse aerosol was further diluted. Then, the prepared calibration aerosol was divided to a CPC UF-02proto and to a reference electrometer (TSI3068). Since the generated particles in this study were less than 30 nm in diameter, they were most probably not more than singly charged.[16] This justified the use of an electrometer as a reference instrument.

In the experiments where the effect of the background aerosol population on the detection efficiency of the CPC was investigated, a second DMA (DMA VIE-08, Hauke, length 0.109 m) was applied. In the second DMA the monodisperse fraction of larger particles (background aerosol) was extracted from the same generated polydisperse silver aerosol and mixed with the calibration aerosol. Both DMAs were operated with 2.0–3.5 l min^{-1} aerosol and 15–21 l min^{-1} sheath and excess air flows in open-loop arrangements. The flows were measured with a Gillian bubble flow meter and controlled with needle valves.

Experimental Results

The basic idea to use CPCs in the default regime is that the instrument is operated in safe conditions, i.e., that the instrument background is close zero. Therefore the default regime is chosen so that there is no homogeneous nucleation inside the CPC and the displayed value is the number concentration of existing particles larger than determined instrument cut-size. The cut-size of the CPC UF-02proto is 4.4 nm at default regime, when the temperature difference is 32.5°C.[14]

Stolzenburg and McMurry[17] showed that the instrument cut-size depends strongly on the temperature difference. Increasing temperature difference improves the cut-size. However, it also enlarges a probability to start homogeneous nucleation in the supersaturated region. Thereby, the instrument background will be changed too.[3] Thus, at default regime, the CPC UF-02proto background is 0.002 cm^{-3}.[14] Fixing the condenser temperature at 10.0°C and increasing the saturator temperature, the background, caused by homogeneous nucleation, increased slowly. However, when the temperature difference exceeded 39.0°C, the number concentration started to increase significantly. The same process was investigated using the ultrafine CPC TSI 3025.[18]

However, the nucleation process inside the CPC can be affected by the aerosol particles existing in the supersaturated volume. So, operating the CPC in nucleation regime (the temperature difference was 52.0°C), it measured the number concentration of the nucleated particles inside the condenser. Adding the background aerosol of 15 nm silver particles, it showed the number concentration of silver particles plus particles nucleated in the CPC. However, increasing concentration of background aerosol, the number concentration of the nucleated particles decreased. Finally, matching the background aerosol concentration at 4,000 cm^{-3}, and the nucleation of the new particles was completely prevented. Thereby, the large background particles formed a condensation sink for the butanol vapor and thus decreased the supersaturation inside the condenser. So, the condensation sink for the butanol vapor affected the nucleation rate significantly.

Using this homogeneous nucleation preventing method, we were able performed investigations of the CPC detection efficiency, operating it in the nucleation regime. Three examples of calibration curves are presented in Figure 1. They show that increasing the temperature difference, the detection efficiency of instrument amends. However, the shape of the detection efficiency curve is changed. At default regime, the curve decreases rapidly, but it is less steep for larger temperature differences. Investigation results showed that the lowest determined cut-size of silver particles was 1.8 nm with temperature difference of 44.0°C. Further increase of the temperature difference did not anymore improve the cut-size. This can be explained by limiting of the instrument detection efficiency not by particle activation inside the condenser but by the particle transport to the supersaturated region in the condenser. This would also partly explain why the homogeneous nucleation rate increased with increasing temperature difference while the cut-size was not anymore decreasing.

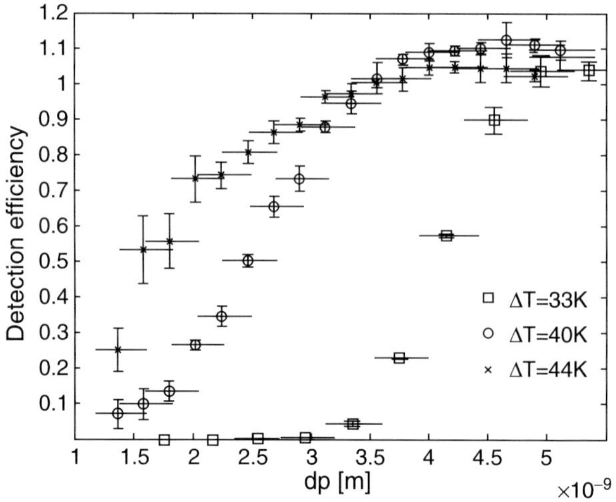

Figure 1 The detection efficiency of the CPC UF-02proto at different instrument regimes

Conclusions

The homogeneous nucleation determines the background of the CPC. In the default regime (the temperature difference is 32.5°C), the background of CPC UF-02proto is 0.002 cm^{-3}. The instrument background increased significantly, when the temperature difference exceeded 39.0 K and the homogeneous nucleation started. However, the nucleation process can be prevented using the background aerosol. Using monodisperse aerosol of 15 nm silver particles, the nucleation inside CPC stops when number concentration of background aerosol concentration is 4,000 cm^{-3}.

The cut-size of the UF-02proto was determined as a function of temperature difference between the saturator and the condenser. At the default regime (temperature difference 32.5 °C), the cut-size was 4.4 nm. Using 44.0 K temperature difference, the cut-size was 1.8 ± 0.2 nm. Further increase of temperature difference did not improve the cut-size but increased the concentration of homogeneously nucleated particles.

Acknowledgments The financial support of the investigations from the Academy of Finland is gratefully acknowledged.

References

Likens, G.E., Driscoll, C.T., and Buso, D.C., *Science*, **272**(5259), 244–246 (1996).
Davidson, C.I., Phalen, R.F., and Solomon, P.A., *Aerosol Sci.Technol.*, **39**, 737–749. (2005).

Kulmala, M., Lehtinen, K.E.J., Laakso, L., Mordas, G., and Hämeri, K., *Boreal Env. Res.*, **10**, 79–87 (2005).

O'Dowd, C.D., McFiggans, G., Creasey, D.J., Pirjola, L., Hoell, C., Smith, M.H., Allan, B.J., Plane, J.M.C., Heard, D.E., Lee, J.D., Pilling, M.J., and Kulmala, M., *Geophys. Res. Lett.*, **26,** 1707–1710 (1999).

Birmili, W., Berresheim, H., Plass-Dülmer, C., Elste, T., Gilge, S., Wiedensohler, A., and Uhrner, U., *Atmos. Chem. Phys.*, **3**, 361–376 (2003).

Kulmala, M., Vehkamäki, H., Petäjä, T., Dal Maso, M., Lauri, A., Kerminen, V.-M., Birmili, W., and McMurry, P.H., *J. Aerosol Sci.*, **35**, 143–176 (2004).

McMurry, P.H., *Aerosol Sci. Technol.*, **33**, 297–322 (2000).

Metnieks, L. and Pollak, L.W., *Introduction for Use of Photo-electric*, Dublin: Institute of Advanced Studies (1959).

Bricard, J.A., Delattre, P. Madeleine, G., and Pourprix, M., Detection of ultra-fine particles by means of a continuous flux condensation nuclei counter, in: *Fine Particles*, edited by B.Y.H. Liu, New York: Academic Press, pp. 565—580 (1976).

Sem, G.J., *Atmos. Res.*, **62**, 267–294 (2002).

Wang, J., McNeill, V.F., Collins, D.R., and Flagan, R.C., *Aerosol Sci. Technol.*, **36**, 678–689 (2002).

Mavliev, R., *Atmos. Res.*, **62**, 302–314 (2002).

Hering, S.V. and Stoltzenburg, M.R., *Aerosol Sci. Technol.*, **39**, 428–436 (2005).

Mordas, G., Kulmala, M., Petäjä, T., Aalto, P.P., Matulevicius, V., Grigoraitis, V., Ulevicius, V., Grauslys, V., Ukkonen, A., and Hämeri, K., *Boreal Env. Res.*, **10**, 543–552 (2005).

Scheibel H.G. and Porstendörfer J., *J. Aerosol Sci.*, **14**(2), 113–126 (1983).

Wiedensohler, A., *J. Aerosol Sci.*, **19**, 387—389 (1988).

Stoltzenburg, M.R. and McMurry, P.H., *Aerosol Sci. Technol.*, **14**, 48–65 (1991).

Hämeri, K., Augustin, J., Kulmala, M., Vesala, T., Mäkelä, J., Aalto P.P., and Krissinel' E.J., *Aerosol Sci.*, **26**, 1003–1008 (1995).

Structure of Sulfuric Acid–Water Clusters

Martta Salonen, Ismo Napari, and Hanna Vehkamäki

Abstract We have studied sulfuric acid–water clusters using molecular dynamics method. We have first simulated clusters containing water and undissociated sulfuric acid molecules. To explore the influence of molecular dissociation on cluster structure, we have also studied clusters containing water and bisulfate-hydronium ion pairs. Simulations show that sulfuric acid tends to lie on the cluster surface whereas bisulfate is inside.

Keywords Homogeneous nucleation, sulfuric acid, cluster structure

Introduction

Sulfuric acid is a key component in atmospheric aerosol formation.[1] Due to its high acidity, sulfuric acid is also a good catalyst for chemical reactions.[2] Atmospheric aerosol particles provide both surface and liquid phase reaction sites for heterogenous chemical reactions. Heterogenous reactions are significant part of the atmospheric chemistry.

In aqueous solution sulfuric acid tends to deprotonate. Dissociation of H_2SO_4 in small water clusters have been studied by several groups for example Ding et al.[3] The quantum chemical calculations show that the sulfuric acid molecules deprotonate easily in a small water cluster. However, experimental studies have shown that sulfuric acid on the liquid surface tends to appear also as a neutral molecule.[4]

In order to study small water–sulfuric acid clusters, we have used molecular dynamics simulation method. Molecular dynamics simulations provide significant molecular level insight that is otherwise unattainable. In this study we concentrate on structure of the critical clusters. Our system consists of one-to-four sulfuric acid and 100 water molecules. We studied two different systems: first we observed

Department of Physical Sciences, P.O. Box 64,
00014 University of Helsinki, Finland

C.D. O'Dowd and P.E. Wagner (eds.), *Nucleation and Atmospheric Aerosols*, 214–217.

clusters where the sulfuric acid molecules are not deprotonated and second a system consisting of bisulfate and hydronium ions, and water.

Computational Details

Our simulations have been done using GROMACS[5] (Groningen Machine for Chemical Simulations) molecular dynamic program. We used a reliable potential model for clusters of sulfuric acid and water constructed by Ding et al.[6] In this model two atoms of different molecules interact via the pair potential

$$u(r_{ij}) = A\frac{q_j q_i}{r_{ij}} + \varepsilon_{ij}\left[\left(\frac{\sigma_{ij}}{r_{ij}}\right)^{12} - \left(\frac{\sigma_{ij}}{r_{ij}}\right)^{6}\right], \tag{1}$$

where A is a constant and r_{ij} is the distance between i and j atoms. Charges q_i and q_j, Lennard–Jones energy parameter ε_{ij}, and distance parameter σ_{ij} are the interaction parameters obtained from the parametrization of quantum chemistry calculations. The intramolecular potentials are simple harmonic potentials.

The simulations have been carried out with periodic boundaries. To speed up the calculations we used a cutoff of 5.5 nm for intermolecular potentials. The temperature of system is fixed at 260 K throughout the present study. The system was coupled to a Nosé–Hoover thermostat. We used Stillinger definition to distinguish between vapor and cluster atoms.

At the beginning of simulations all molecules are in the vapor phase. During the equilibration period, small clusters are formed via nucleation. Coagulation of clusters and condensation of free molecules leads the system to a situation, where we have only one cluster surrounded by vapor phase. Following an equilibration period, compact cluster-vapor systems are simulated. Structure analysis has been done only for the equilibrium cluster.

Results

In the Figures 1A and 1B we illustrate the radial dependences of cluster densities. Figure 1A shows the density profiles for a system containing one sulfuric acid and 100 water molecules. Figure 1B shows the corresponding density profile for ionized system containing one bisulfate, one hydronium ion, and 99 water molecules. To simplify the figures we only consider sulfur atom in the sulfuric acid and bisulfate molecules and oxygen in the water and hydronium molecules.

As Ding et al.[6] show, the bisulfate ion is located inside the cluster and hydronium on the surface. We found that neutral sulfuric acid molecule in the water cluster was found to be on the surface.

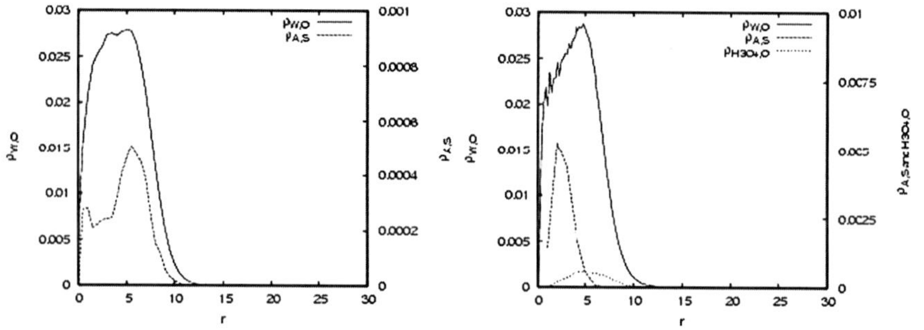

Figure 1 Radial dependence of density. In Figure 1A the simulated system includes only one sulfuric acid and the total number of water molecules is 100. Density profiles in Figure 1B are for the system including one bisulfate hydronium ion pair and 99 water molecules

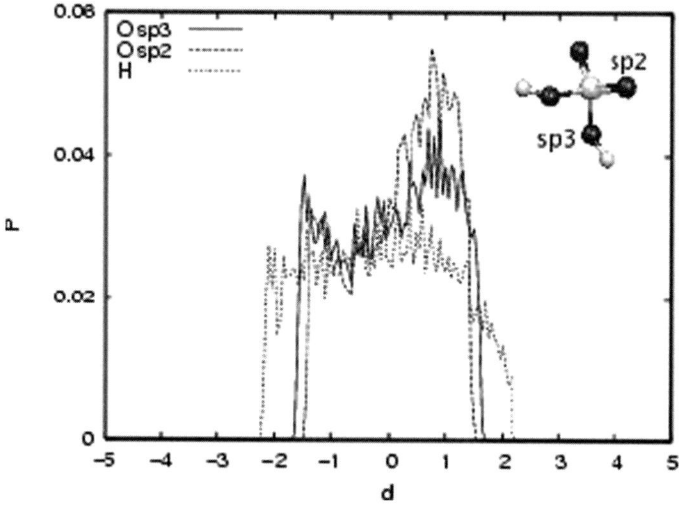

Figure 2 Orientation of the sulfuric acid molecule in the cluster

Figure 2 shows the orientation of the sulfuric acid molecule along the cluster radius with the sulfur atom at the origin. The lines show the position probability for other atoms. As we can see from the figure, sulfuric acid has clear orientation on surface. Position probability of double bounded oxygen (ps2) has a peak away from the center of cluster. Correspondingly, hydrogen probability decreases with the increasing distance.

We have also simulated systems containing two-to-four sulfuric acid molecules or bisulfate hydronium ions pairs. Sulfuric acid molecules nucleate easily and only one big cluster will be formed in the systems. Preliminary observations of clusters indicate that sulfuric acid molecules tend to distribute evenly on the cluster surface. In contrast to results of Ding et al.,[6] in our simulation conditions repulsive interaction between bisulfate ions is too strong to allow coagulation of clusters with bisulfate ions.

Conclusions

In this study we simulated water–sulfuric acid clusters and focused our research to cluster structure. Due to sulfuric acid molecules willingness to donate one proton to a water molecule in the liquid phase, we simulated also systems containing bisulfate and hydronium ions. Our calculations indicate that sulfuric acid molecules (1–4) in the water (100 molecules) cluster appear on the surface. At least in the cluster containing one sulfuric acid molecule we can see orientation of the acid molecule. The orientation reflects the tendency of acid molecule to maximize number of hydrogen bonds. As Ding et al.[6] reported earlier a bisulfate ion is found to be inside the cluster and hydronium ion lies on the surface. In our simulation conditions small clusters containing one ion pair could not coagulate forming a stable cluster with two or more ion pairs. Repulsive interaction between the ions seems to be too strong to allow formation of clusters with several bisulfate ions.

Acknowledgments This work was supported by the Academy of Finland.

References

1. Kulmala, M., Laaksonen, A., and Pirjola L., *J. Geophys. Res.*, **103**, 8301–8307 (1998).
2. Wayne, R.P. *Chemistry of Atmospheres: an Introduction to the Chemistry of the Atmospheres of Earth, the Planets and their Satellites*, Oxford: Oxford University Press (2000).
3. Ding, C.-G. and Laasonen, K., *Chem. Phys. Lett.*, **390**, 307–313 (2004).
4. Bianco, R., Wang, S., and Hynes, T. H., *J. Phys Chem. B*, **109**, 21313–21321 (2005).
5. Berendsen H.J.C., van der Spoel, D., and van Drunen, R., *Comp. Phys. Comm.*, **91**, 43–56 (1995).
6. Ding, C.-G., Taskila, T., Laasonen, K., and Laaksonen, A., *Chem. Phys.*, **287**, 7–19 (2002).

Quantum Chemical Calculations of the Binding Energies of $(H_2SO_4)_2$, $HOSO_2{\bullet}H_2SO_4$, and $HOSO_4{\bullet}H_2SO_4$

Martta Salonen[1], Théo Kurten[1], Markku Sundberg[2], and Hanna Vehkamäki[1]

Abstract We use quantum chemical calculations to attempt to explain why sulfuric acid produced in situ via the reaction of OH radicals with SO_2 in the presence of water vapor nucleates so much better than sulfuric acid taking from liquid phase. We test whether or not selected formation reaction intermediates or alternative reaction products could take part in nucleation. We have calculated binding energies for the $(H_2SO_4)_2$, $HOSO_2{\bullet}H_2SO_4$ and $HOSO_4{\bullet}H_2SO_4$ dimers at the RI-MP2/QZVPP level. Our calculations indicate that the binding between H_2SO_4 and either of these radicals is weaker than the binding between two H_2SO_4 molecules. The binding of these intermediates to sulfuric acid is therefore not likely to explain the observations.

Keywords Homogeneous nucleation, sulfuric acid, quantum chemistry

Introduction

Several experimental and theoretical studies have shown that sulfuric acid is an important species for nucleation in the atmosphere (see e.g., Kusaka et al.[1] and references therein). However, the first steps of cluster formation in the sulfuric acid–water system are still unclear.

A major part of atmospheric sulfur originates from anthropogenic sources.[2] In the atmosphere, emissions of SO_2 are converted to H_2SO_4. The hydroxyl radical seems to initiate the most probable route by addition to SO_2 in the presence of a catalyst (here denoted M). The initially formed $HOSO_2$ radical is further oxidized to H_2SO_4.[3]

$$SO_2 + OH + M \rightarrow HOSO_2 + M \tag{1}$$

[1]*Department of Physical Sciences, P.O. Box 64, 00014 University of Helsinki, Finland*

[2]*Department of Chemistry, P.O. Box 55, 00014 University of Helsinki, Finland*

C.D. O'Dowd and P.E. Wagner (eds.), *Nucleation and Atmospheric Aerosols*, 218–221.

$$HOSO_2 + O_2 \rightarrow HO_2 + SO_3 \tag{2}$$

$$SO_3 + H_2O + M \rightarrow H_2SO_4 + M \tag{3}$$

Berndt et al.[4,5] have experimentally observed the formation of new particles in the H_2SO_4/H_2O system. They found that the threshold H_2SO_4 concentration was 10^7 molecule cm^{-3} if they produced H_2SO_4 in situ via the reaction OH radicals with SO_2 in the presence of water vapor, and 10^{10} molecule cm^{-3} if the H_2SO_4 was taken from a liquid sample. The experimental results give reason to expect that other species than H_2SO_4 molecules alone are responsible for particle formation.

In this study we have looked at how strongly sulfuric acid is bound to the reaction intermediates $HOSO_2$ and $HOSO_4$. The $HOSO_4$ is formed via an alternative path of oxidation of $HOSO_2$.[6]

$$HOSO_2 + O_2 + M \rightarrow HOSO_4 + M \tag{4}$$

If these radicals could bind sulfuric acid strongly, then the collision between two such radical – sulfuric acid clusters could be an alternative method to form a post-critical sulfuric acid cluster without having to go via the critical cluster itself. We calculated the binding energies for $HOSO_2{\bullet}H_2SO_4$ and $HOSO_4{\bullet}H_2SO_4$ using a high-level *ab initio* molecular orbital method, and compared the obtained values with the binding energy of $(H_2SO_4)_2$.

Methods

Calculations have been performed using the TURBOMOL[7] program suite and the computational level RI-MP2/QZVPP[8] employing the default convergence criterion. The guess geometries for the $HOSO_2$ and $HOSO_4$ radicals were taken from calculations of Majundar D. et al.[6] We used three different initial configurations for H_2SO_4 – dimers obtained from Ding et al.[9]

Results

We calculated the binding energies for $HOSO_2{\bullet}H_2SO_4$ and $HOSO_4{\bullet}H_2SO_4$ using an *ab initio* molecular orbital method and compared those with the binding energies of $(H_2SO_4)_2$. The structures of calculated sulfuric acid–radical dimers are shown in Figure 1. These geometries correspond the lowest energy configurations found in our study.

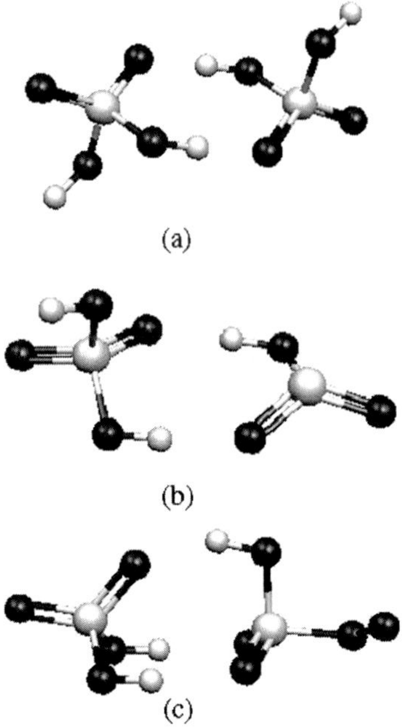

Figure 1 Molecular geometries of **(a)** $(H_2SO_4)_2$ **(b)** $HOSO_2 \cdot H_2SO_4$ **(c)** $HOSO_4 \cdot H_2SO_4$

Table 1 Binding energies ΔE_0

Reaction	ΔE_0 kcal/mol
$H_2SO_4 + H_2SO_4 \Rightarrow (H_2SO_4)_2$	−18.2
$H_2SO_4 + HOSO_2 \Rightarrow HOSO_2 \cdot H_2SO_4$	−16.5
$H_2SO_4 + HOSO_4 \Rightarrow HOSO_4 \cdot H_2SO_4$	−13.9

The binding energies are shown in Table 1. According to our calculations, the binding between two H_2SO_4 molecules is stronger the binding between H_2SO_4 and either of these radicals.

Conclusions

In this work, we have tried to explain the experimental results concerning the differences in nucleation rate when H_2SO_4 is produced in situ instead of being taken from the liquid phase. We calculated the binding energies between H_2SO_4 and two radical intermediates $HOSO_2$, and $HOSO_4$.

The binding between H_2SO_4 and either of the studied radicals is weaker than the binding between two H_2SO_4 molecules. However, we have not yet looked at how the presence of water molecules, which are always present at least in some concentration in the experiments, changes the situation. Also, radical–radical addition products (e.g., $H_2S_2O_8$) may also play a role.[9] More work needed to clarify the mechanism leading to the enhanced particle formation.

Acknowledgments This work was supported by the Academy of Finland. We thank the CSC center for scientific computing for computer time.

References

1. Kusaka, Z.-G.W. and Seinfeld, J.H., *J. Chem. Phys.*, **108**, 6829–6848 (1998).
2. Seinfeld, J.H. and Pandis, S.N., *Athmospheric Chemistry and Physics: From Air Pollution to Climate Cange*, New York: Wiley (1998).
3. Wayne, R.P., *Chemistry of Atmospheres: An Introduction to the Chemistry of the Atmospheres of Earth, the Planets and their Satellites*, Oxford: Oxford University Press (2000).
4. Berndt, T., Böge, O., Stratmann, F., Heintzenberg, J., and Kulmala, M., *Science*, **307(5710)**, 698–700 (2005).
5. Berndt, T., Böge, O., and Stratmann, F., *Geophys. Res. Lett.*, **33**, L15817 (2006).
6. Majumdar, D., Kim, G.-S., Kim, J., Oh, J.Y., Lee, J.Y., Kim, K.S., Choi, W.Y., Lee, S.-H., Kang M.-H., and Mhin, B.J., *J. Phys. Chem.*, **122**, 723–730 (1999).
7. Ahlrichs, R., Bär, M., Häser, M., Horn, H., and Kölmel, C., *Chem. Phys. Lett.*, **162**, 165–169.(1989).
8. Weigend, F. and Häser, M, *Theor. Chem. Acc.*, **97**, 331–340 (1997).
9. Ding, C.-G., Taskila, T., Laasonen, K., and Laaksonen, A., *Chem. Phys.*, **287**, 7–19 (2002).
10. Friend, J.P., Burnes, R.A., and Vasta, R.M., *J. Phys. Chem.*, **84**, 2423–2436 (1980).

Calculation of Cluster-free Energies Using the Jarzynski Relation

Hoi Yu Tang and Ian J Ford

Abstract Knowledge of the free energy of formation of molecular clusters, based on a microscopic model of the intermolecular interactions, forms the basis of the theoretical understanding of aerosol nucleation. A number of methods have been developed to calculate this quantity, but most are rather time consuming. As an alternative, we have sought to exploit the Jarzynski relation, which is a powerful result in nonequilibrium statistical physics, connecting a free energy change to the mechanical work associated with the change. Thus we plan to pull a cluster apart with external forces over the course of a molecular dynamics simulation and use the Jarzynski relation to evaluate the associated free energy of formation, or rather disassembly. As a first step we have tested the procedure on the separation of a dimer.

Keywords Homogeneous nucleation, free energy

Free Energies in Nucleation Theory

Free energy calculations are central to statistical physics, and not least to studies of nucleation processes. In equilibrium, the population of clusters of size *i*, defined in whatever fashion thought appropriate, is proportional to the exponential of a free energy. This is normal canonical statistical mechanics. Nucleation is usually regarded as a perturbation to the equilibrium situation, whereby a drift, or current in size space is brought about by the removal, or simply the absence from the outset, of the very largest clusters [1]. Knowledge of the equilibrium populations therefore underlies the usual description of the nonequilibrium nucleation process, and free energies are central quantities of interest for computation.

Department of Physics and Astronomy and London Centre for Nanotechnology
University College London, Gower St., London WC1E 6BT, UK

C.D. O'Dowd and P.E. Wagner (eds.), *Nucleation and Atmospheric Aerosols*, 222–225.

Even having reduced the complicated nonequilibrium problem to one involving equilibrium quantities, the task before us is still very difficult. A free energy is related to a partition function, which is an integral over (usually) an enormous number of molecular positions and momenta, subject (often) to a complicated and sometimes arbitrary definition of what constitutes a cluster and what does not. The definition issue is not one we shall address here, but instead we shall imagine that the criteria are settled and the remaining job is to evaluate the integral:

$$Z = \int dx^i dp^i \exp(-H/kT) \tag{1}$$

where H is the Hamiltonian, k is Boltzmann's constant, T is the temperature, and the integration over the positions and momenta of the i molecules is represented by $dx^i dp^i$. The prime on the integration indicates that a cluster definition has defined a specific region of phase space for integration.

Jarzynski Relation

Phase space integrals are hard to calculate. However, an exciting recent development in statistical physics suggest a different approach, rather close in spirit to the fundamental character of free energy in thermodynamics. It is the Jarzynski relation [2], which is a connection between the external work done on a system and a free energy change:

$$\langle \exp(-W/kT) \rangle = \exp(-\Delta F/kT) \tag{2}$$

This pleasingly simple result is to be interpreted as follows. External fields are introduced and vary in a specified manner for a specified period of time. The system receives energy from those fields: this is called external work. Since the system may start the process in a variety of microstates, the amount of energy received will vary from trial to trial. The left hand side in the above equation is the average of the exponentiated work done over a large number of trials, each with an initial state chosen from a canonical distribution. Remarkably, this average is exactly related to the free energy change associated with going from the initial state, without fields, to the final state, with fields. It does not depend on the rate at which the fields vary. In thermodynamics we expect $\Delta F = W$ for quasistatic, infinitely slow operations. The Jarzynski relation is an extension to this result.

Strategies for Cluster-Free Energies

Our strategy is clear. We begin with a cluster sampled from a canonical distribution. We then introduce a force to pull one or more molecules out of the cluster over the course of a molecular dynamics simulation. At the end of this pulling period, an amount of external work has been done. Repeating the process with a different initial state allows us to evaluate the average exponentiated work required for use in Eq. (2). If one molecule is extracted, the free energy difference between clusters differing in size by one molecule is calculable. However, this quantity is the focus of attention of other methods.

A rather more ambitious strategy would be to pull all molecules simultaneously, with forces designed to separate the cluster into its individual constituents. The average exponentiated external work done is the free energy of disassembly of the cluster, directly related to equilibrium cluster populations. It is our aim to prove the feasibility of such an approach.

Dimer Disassembly

Cluster disassembly is a future task, but we have attempted to prove the principle by pulling apart a dimer of argon atoms with external forces created by two moving tether points interacting harmonically with each atom. The process is sketched in Figure 1.

In initial calculations, the speed of separation appears to affect properties of the probability distribution for external work. This is expected, but what is unexpected is that the mean exponentiated work is also rate dependent. We hope to clarify this matter with further work

Outlook

The neatness of Jarzynski's relation is very appealing. Free energy ought to be calculable from mechanical work. However, some practical issues need to be resolved before we can conclude whether our ambitious strategy to calculate cluster-free energies is achievable. Time will tell.

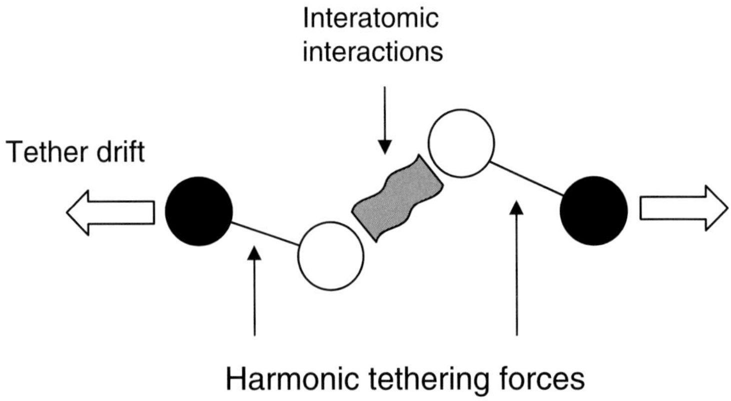

Figure 1 Sketch of the pulling of the atoms (empty circles) of a dimer apart through the gradual separation of harmonic tether points (shown as filled circles)

Acknowledgments This work was made possible by the award of a studentship from the UK Engineering and Physical Science Research Council.

References

1. Ford, I.J., Statistical mechanics of nucleation: a review, *Proc. Instn. Mech. Engrs 218 Part C: J. Mech. Eng. Sci.*, 883–899 (2004).
2. Jarzynski, C., Nonequilibrium equality for free energy differences, *Phys. Rev. Lett.*, **78**, 2690 (1997).

New Particle Formation and Sulphuric Acid Concentrations During 26 Months in Hyytiälä

T. Nieminen, M. Kulmala, I. Riipinen, A. Hirsikko, M. Boy, and L. Laakso

Abstract Over two year time series of calculated sulphuric acid concentrations and air ion size distribution measurements are compared. New particle formation is seen when sulphuric acid concentration is high enough ($>10^6$ cm^{-3}). Negative ions activate typically in lower sulphuric acid concentrations than positive ions.

Keywords Particle formation and growth, sulphuric acid, nucleation, ions

Introduction

Formation of new aerosol particles in the atmosphere has been observed in many locations.[1] This process is currently not completely understood. One of the difficulties is in detecting the new 1–2 nm particles right after they are formed. This is not possible at the moment with commonly used instruments (such as the Differential Mobility Particle Sizer, DMPS) which can measure particles down to 3 nm in diameter at best. Another open question is identifying the nucleating compounds. Studying the very first stages of new particle formation and the initial growth is important in order to get information about the nucleation mechanisms and the vapours participating in them. Using ion spectrometers that measure naturally charged particles as small as 0.5 nm it has become possible to get this information.

Methods

Air ion size distributions have been measured continuously in Hyytiälä SMEAR II station in southern Finland since March 2003 using the Balanced Scanning Mobility Analyzer (BSMA).[2] The BSMA detects ions in the size range 0.4–7.5 nm

Department of Physical Sciences, University of Helsinki
P.O. Box 64, FI-00014 University of Helsinki, Finland

C.D. O'Dowd and P.E. Wagner (eds.), *Nucleation and Atmospheric Aerosols*, 226–229.

in the mass-based Tammet diameter.[3] Also neutral particle size distributions between 3–500 nm are measured in Hyytiälä with a DMPS.[4] To get information about the role of sulphuric acid in new particle formation we have calculated atmospheric sulphuric acid concentrations for 26 months from April 2003 to May 2005 using a pseudo steady-state model.[5] The calculations use measurement data of CO, NO_x, O_3, SO_2, methane, non-methane hydrocarbon concentrations, and solar radiation. Since sulphuric acid is not measured regularly this is the only way to get a long time series of the behaviour of its ambient concentrations. When compared to the available measured sulphuric acid concentrations during the QUEST 2 and 4 campaigns in the spring 2003 and 2005 the calculated and measured data agree quite well.

Results

Negative and positive cluster ions smaller than about 1.5 nm are always present. During new particle formation particles appear first in sizes 1.5–2 nm. Between April 2003 and March 2005 we have seen such formation events with the BSMA on 134 days in negative ions and on 107 days in positive ions.[6] In Figure 1 the mean sulphuric acid concentration during particle formation events and at times when there is no new particle formation are represented as a function of the day of year. The event points are mean values between the start and end times of the new particle appearance in sizes 1.3–2.4 nm. Non-event points are mean values from non-event days during the average time of the particle formation events, which is 0930–1430 for negative ions and 0950–1400 for positive ions. There is a clear difference in the sulphuric acid concentration between the event and non-event days. During most of the events the mean sulphuric acid concentration is higher than about 10^6 cm^{-3} and on the other hand nearly all non-event days are below this limit. Only during summer (June–August) there are non-event days when the sulphuric acid concentrations are high. This could be related to the observed summer minimum in growth rates of 1–3 nm ions.[7] It can also be seen from Figure 1 that new particle formation occurs most often in spring from March to May. This is the time when sulphuric acid concentrations are highest during the studied period.

An example of the typical behaviour of ion concentrations in the size range 1.3–2.4 nm during new particle formation is shown in Figure 2. The negative ions start to appear almost at the same time when the sulphuric acid concentration increases, whereas the positive ion concentration rises somewhat later. There are also notably more negative than positive ions formed. On some days there is new particle formation seen only in negative ions. This indicates that in ion-induced nucleation negative ions are particularly important. By measuring air ion size distributions we can observe the first stages of new particle formation, and see how negative ions are activated before positive ions and neutral particles.

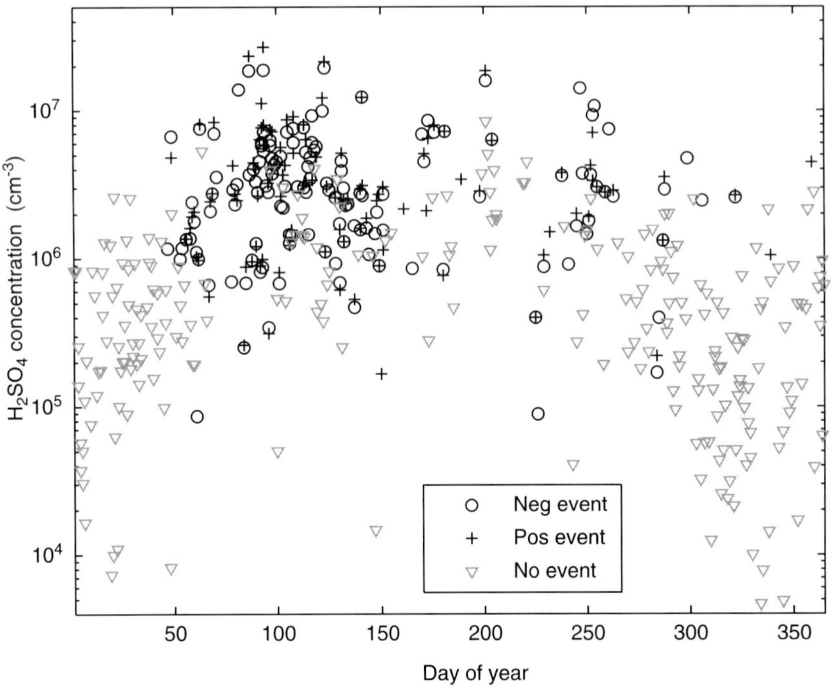

Figure 1 Mean concentration of sulphuric acid during new particle formation event and non-event days as a function of day of year

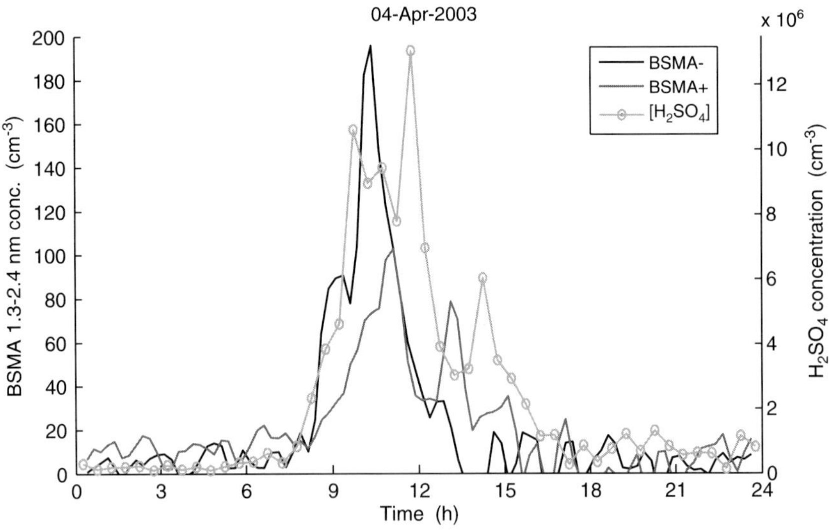

Figure 2 Concentration of ions in size range 1.3–2.4 nm and sulphuric acid concentration on new particle formation day 4 April 2003

References

1. Kulmala M., Vehkamäki H., Petäjä T., Dal Maso M., Lauri A., Kerminen V.-M., Birmili W., and McMurry P.H., Formation and growth rates of ultrafine atmospheric particles: a review of observations, *J. Aerosol Sci.*, **35**, 143–176 (2004).
2. Tammet H., Continuous scanning of the mobility and size distribution of charged clusters and nanometer particles in atmospheric air and the balanced scanning mobility analyzer BSMA, *Atmos. Res.*, **82**, 523–535 (2006).
3. Tammet H., Size and mobility of nanometer particles, clusters and ions, *J. Aerosol Sci.*, **26**, 459–475 (1995).
4. Aalto P., Hämeri K., Becker E., Weber R., Salm J., Mäkelä J.M., Hoell C., O'Dowd C.D., Karlsson H., Hansson H.-C., Väkevä M., Koponen I.K., Buzorius G., and Kulmala M, Physical characterization of aerosol particles during nucleation events, *Tellus*, **53B**, 344–358 (2001).
5. Boy M., Kulmala M., Ruuskanen T.M., Pihlatie M., Reissel A., Aalto P.P., Keronen P., Dal Maso M., Hellen H., Hakola H., Jansson R., Hanke M., and Arnold F, Sulphuric acid closure and contribution to nucleation mode particle growth, *Atmos. Chem. Phys.*, **5**, 863–878 (2005).
6. Hirsikko A., Bergman T., Laakso L., Dal Maso M., Riipinen I., Hõrrak U., and Kulmala M, Identification and classification of the formation of intermediate ions measured in boreal forest, *Atmos. Chem. Phys.*, **7**, 201–210 (2007).
7. Hirsikko A., Laakso L., Hõrrak U., Aalto P.P., Kerminen V.-M., and Kulmala M., Annual and size dependent variation of growth rates and ion concentrations in boreal forest, *Boreal Env. Res.*, **10**, 357–369 (2005).

The First Heterogeneous Nucleation Theorem Including Line Tension: Analysis of Experimental Data

Anca I. Hienola,[1] Hanna Vehkamäki,[1] Antti Lauri,[1] Paul E. Wagner,[2] Paul M. Winkler,[2] and Markku Kulmala[1]

Abstract We have incorporated the concept of line tension in the derivation of the first nucleation theorem in the case of a one-component system. The derivative of the critical cluster formation energy still gives the total number of the molecules in the cluster: the sum of the number of molecules in the bulk liquid and the number of molecules along the three-phase contact line.

Keywords Heterogeneous nucleation, critical cluster, line tension

Introduction

The nucleation theorems[1] give a theory-independent insight to nucleation phenomena beyond experimental limits. Earlier, we have introduced a way to interpret an experimentally achievable quantity in heterogeneous nucleation: the nucleation probability, in terms of the nucleation theorems.[2] In this study we include the concept of line tension in the analysis, and derive the first nucleation theorem for one-component systems. The results are applied for one-component nucleation of water and *n*-propanol.

Theory

Starting from heterogeneous nucleation geometry, the inclusion of line tension produces an extra term to Young's equation:

$$\cos\theta = \frac{\sigma_{sg} - \sigma_{sl}}{\sigma_{lg}} - \frac{\sigma_t}{\sigma_{lg} R \tan\phi}. \quad (1)$$

[1]*Department of Physical Sciences, University of Helsinki, P.O. Box 64, FI-00014 University of Helsinki*

[2]*Institut für Experimentalphysik, Universität Wien, Boltzmanngasse 5, AT-1090 Vienna, Austria*

C.D. O'Dowd and P.E. Wagner (eds.), *Nucleation and Atmospheric Aerosols*, 230–234.

Figure 1 The geometry of heterogeneous nucleation on a spherical aerosol particle

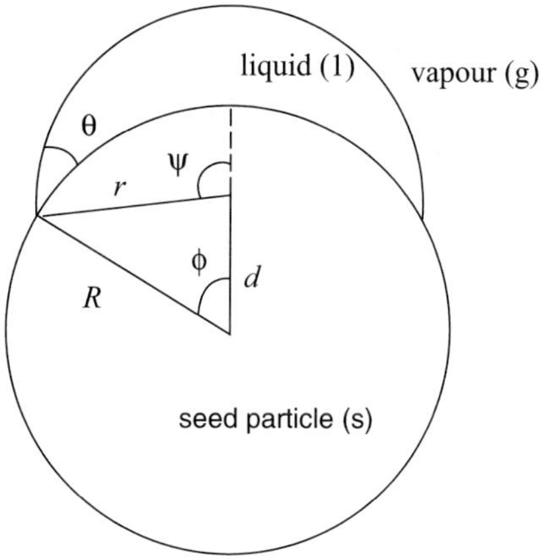

For angles ϕ and θ, see *Figure 1*. σ_{ij} represents the interfacial tension between phases i an j (subindexes g, l and s correspond to vapour, liquid, and substrate phases, respectively); σ_t is line tension, the reversible work needed to expand a unit length of the three-phase contact line isothermally, and R is the radius of the seed particle.

The critical formation free energy in heterogeneous nucleation, including line tension, is given by[3]

$$\Delta G^*_{\mathrm{het}} = f_{\mathrm{g}} \Delta G^*_{\mathrm{hom}} - \frac{\sigma_t A_{sg}}{R\tan\phi} + 2\pi R\sigma_t \sin\phi, \tag{2}$$

where f_g is a geometric factor,[4] and A is surface area. Heterogeneous nucleation rate, i.e., the rate of formation of embyos on the substrate, is given by

$$J = K\exp\left(-\frac{\Delta G^*_{\mathrm{het}}}{k_B T}\right), \tag{3}$$

where K is a kinetic pre-factor, k_B is the Boltzmann constant, and T is temperature.

Assuming that the equimolar surface coincides with the surface of tension, and applying the Gibbs adsorption isotherm in one dimension, the nucleation theorem in terms of critical free energy of formation reads as

$$\left[\frac{\partial\left(\frac{\partial \Delta G^*_{\mathrm{het}}}{k_B T}\right)}{\partial \ln S}\right]_T = \Delta N^*_{\mathrm{het}}, \tag{4}$$

where S is saturation ratio. Similarly as in the homogeneous case, the derivative gives the number of molecules in the critical cluster ΔN^*_{het}. In terms of nucleation probability P (the fraction of pre-existing particles activated for growth through heterogeneous nucleation) the theorem reads as[5]

$$\left[\frac{\partial \ln\left(\ln\frac{1}{1-P}\right)}{\partial \ln S}\right]_T = \Delta N^*_{het} + k, \tag{5}$$

where k corresponds to the effect of the kinetic pre-factor K.

Results

We applied the heterogeneous nucleation theorem Eq. (5) to estimate the number of molecules in the critical cluster in unary heterogeneous nucleation of water and n-propanol on 8 nm monodispersed Ag particles. Using the theorem, we analysed recently obtained experimental data.[6]

The Effect of the Kinetic Pre-factor

Using the kinetic approach of the classical nucleation theory (CNT) the derivative of the kinetic pre-factor $\ln K$ with respect to $\ln S$ can be estimated numerically. For the range of experimental conditions it yields 3–8 for pure water, and 8–16 for pure n-propanol. These estimates were obtained using the macroscopic contact angle 19.1°.[6] We also generated nucleation probability data with the classical theory of heterogeneous nucleation, and applied the heterogeneous nucleation theorem to this data. The aim of this procedure was to check that the resulting numbers of molecules in the critical cluster are equal to the numbers of molecules given by the classical theory.

We found that the kinetic contribution to the first nucleation theorem is not in the range of 1–2 as in the homogeneous case.[7–9] Our calculations showed that the contribution of the pre-factor is sensitive to the critical cluster size, as well as to the contact angle: the values obtained using classical theory for very large clusters are not applicable to the experimental results, which indicate the clusters to be much smaller. Subtraction of the classical values for the kinetic contribution from the results given by the data analysis would lead to negative numbers when the experiments indicate the clusters to be small.

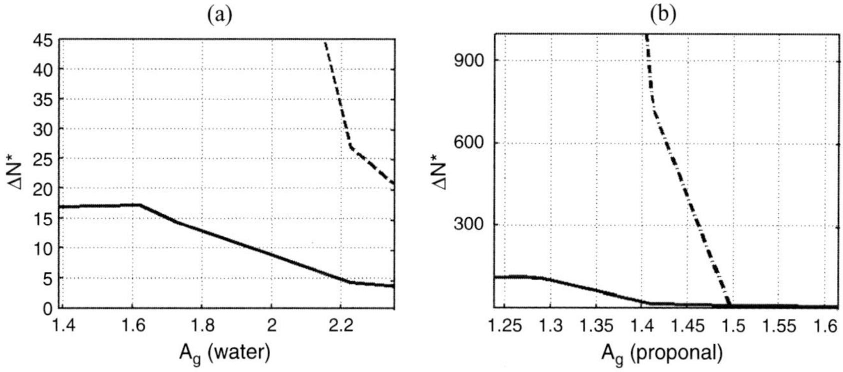

Figure 2 The critical cluster sizes in heterogeneous nucleation of (a) water and (b) *n*-propanol as a function of saturation ratio. The solid lines represent the nucleation theorem analysis of experimental nucleation probability results,[6] and the dashed line (a) and dash-dotted line (b) the classical nucleation theory (CNT) results.

Analysis by the Heterogeneous Nucleation Theorem

Figures 2a (water) and 2b (*n*-propanol) show the results of our analysis. It should be noted that the effect of the kinetic pre-factor was not taken into account in these calculations. A more sophisticated model for the kinetics is needed to estimate the role of the kinetic pre-factor in the heterogeneous nucleation theorem.

According to the theorem, critical water clusters contain less than 20 molecules, and for *n*-propanol the largest clusters have around 110 molecules. These values are considerably smaller than the ones predicted by the classical theory, as shown in Figures 2a and 2b.

Acknowledgment Mr. Kai Ruusuvuori is gratefully acknowledged for his assistance.

References

1. Kashchiev, D., *Nucleation: Basic Theory with Applications*, Oxford: Butterworth-Heinemann (2000).
2. Hienola, A.I., Winkler, P.M., Wagner, P.E., Vehkamäki, H., Lauri, A., Napari, I., and Kulmala, M., *J. Chem. Phys.*, **126**, 094705 (2007).
3. Lazaridis, M., *J. Colloid Interf. Sci.*, **155**, 386–391 (1993).
4. Määttänen, A., Vehkamäki, H., Lauri, A., Merikallio, S., Kauhanen, J., Savijärvi, H., and Kulmala, M., *J. Geophys. Res.*, **110**, E02002 (2005).

5. Vehkamäki, H., Määttänen, A., Lauri, A., Kulmala, M., Winkler, P., Vrtala, A., and Wagner, P.E., *J. Chem. Phys.*, accepted for publication (2007).
6. Wagner, P.E., Kaller, D., Vrtala, A., Lauri, A., Kulmala, M., and Laaksonen, A., *Phys. Rev. E*, **67**, 021605 (2003).
7. Ford, I.J., *J. Chem. Phys.*, **105**, 8324–8332 (1996).
8. Ford, I.J., *Phys. Rev. E*, **56**, 5615–5629 (1997).
9. Vehkamäki, H. and Ford, I.J., *J. Chem. Phys.*, **113**, 3261–3269 (2000).

Heterogeneous Nucleation Theorems for Multicomponent Systems

Hanna Vehkamäki[1], Antti Lauri[1], Anni Määttänen[1], Paul E. Wagner[2], Paul M. Winkler[2], and Markku Kulmala[1]

Abstract We present a new form of the nucleation theorems applicable to heterogeneous nucleation. The heterogeneous nucleation theorems allow direct determination of properties of nanoclusters formed on pre-existing particles from measured heterogeneous nucleation probabilities. The theorems can be used to analyse the size (first theorem) and the energetics (second theorem) of heterogeneous clusters independent of any specific nucleation model.

Keywords Heterogeneous nucleation, critical cluster, nucleation theorems

Introduction

Nanoparticles have received intensive attention in many branches of technology, and heterogeneous nucleation is controlling their formation processes. Atmospheric nanoparticles can affect human health, and when they grow to larger sizes they also reduce visibility and play a role in determining the Earth's radiation budget. We have investigated the size of small clusters formed at the surface of a pre-existing aerosol particle. The number of molecules in a critical cluster – acting as a starting point of phase transition – can be obtained using the first nucleation theorem, which we have derived for heterogeneous nucleation. We have also derived the second heterogeneous nucleation theorem, giving the binding energy of the critical cluster. General forms for the nucleation theorems have been presented earlier,[1] but the specific form applicable to analysis of heterogeneous nucleation probability data has made the use of these powerful analytical tools difficult until now.

[1] *Department of Physical Sciences, University of Helsinki, P.O. Box 64, FI-00014 University of Helsinki*

[2] *Institut für Experimentalphysik, Universität Wien, Boltzmanngasse 5, AT-1090 Vienna, Austria*

C.D. O'Dowd and P.E. Wagner (eds.), *Nucleation and Atmospheric Aerosols*, 235–239.

Derivation of the Nucleation Theorems

The First Theorem

As shown by Kashchiev,[1] in the isothermal case the nucleation theorem for heterogeneous multicomponent nucleation is

$$\left(\frac{\partial \Delta G^{*}_{\mathrm{het}}}{\partial \mu^{g,i}}\right)_{T,\mu^{g,j\neq i}} = -\Delta N^{*}_{\mathrm{het},i}, \quad (1)$$

where $\Delta G^{*}_{\mathrm{het}}$ is the formation free energy of a critical cluster, $\mu^{g,i}$ is the gas-phase chemical potential of component i, T is temperature, and $\Delta \mathrm{N}^{*}_{\mathrm{het},i}$ is the excess number of molecules of component i in the critical cluster. In terms of an experimentally more favourable quantity, gas-phase activity $A_{g,i}$, the theorem is given by

$$\left[\frac{\partial\left(\frac{\partial \Delta G^{*}_{\mathrm{het}}}{kT}\right)}{\partial \ln A_{g,i}}\right]_{T,A_{g,j\neq i}} = \Delta N^{*}_{\mathrm{het},i}, \quad (2)$$

where k is the Boltzmann constant.

The Second Theorem

The formation free energy of a heterogeneous cluster is given by

$$\Delta G^{*}_{\mathrm{het}} = \left(P^{g} - P^{l}\right)V^{l}_{\mathrm{het}} + \Phi, \quad (3)$$

where P^{g} and P^{l} are the pressures in the gas-phase and in the liquid cluster, respectively; Φ is the effective surface energy. Using the generalized Laplace equation, and the Gibbs adsorption equation for each surface phase, assuming a metastable equilibrium throughout the system, and keeping the gas-phase chemical potentials constant the second theorem can be written by

$$\left(\frac{\partial \Delta G^{*}_{\mathrm{het}}}{\partial T}\right)_{\mu^{g,i}} = -\Delta S^{*}, \quad (4)$$

where S stands for entropy. In practical applications it is usually more convenient to keep the gas-phase activities rather than the chemical potentials constant. Applying the Gibbs–Duhem equation and Clausius–Clapeyron equation, assuming that the partial molecular volume in the liquid is negligible compared to the one in the vapour, a simple form of the second theorem reads

$$\left[\frac{\partial\left(-\frac{\Delta G^*_{\text{het}}}{kT}\right)}{\partial T}\right]_{A_{g,i}} = \frac{\Delta\left(\Delta_{\text{pure},l}U^*\right)}{kT^2}, \tag{5}$$

Where U is the total energy, $\Delta_{\text{pure},l}$ refers to the difference compared to pure liquids, and the first Δ to difference between the cluster and the same space occupied by gas phase.

The Classical Formalism: Theorems in Terms of the Nucleation Probability

The classical theory of heterogeneous nucleation treats the critical cluster as a cap-shaped embryo with radius r^* attached on a spherical seed particle of radius R with contact angle θ. The density of the cluster is assumed to be the same as bulk liquid, and the energy bound in the surface of the embryo is the surface area times the liquid surface tension. The measurable quantity in heterogeneous nucleation is usually the nucleation probability P – the fraction of aerosol particles activated for condensational growth through heterogeneous nucleation, given as a function of the heterogeneous nucleation rate $J_{\text{het}} = K\exp[-\Delta G^*_{\text{het}}/(kT)]$, where K is a kinetic pre-factor, by $P = 1 - \exp(-J_{\text{het}}t)$, where t is the time period under consideration.

With the above listed presumptions the first and second nucleation theorems can be written as[2]

$$\left[\frac{\partial\ln\left(\ln\frac{1}{1-P}\right)}{\partial\ln A_{g,i}}\right]_{T,A_{g,j\neq i}} = \Delta N^*_{\text{het}} + \left(\frac{\partial\ln K}{\partial\ln A_{g,i}}\right)_{T,A_{g,j\neq i}}, \tag{6}$$

and

$$\left[\frac{\partial\ln\left(\ln\frac{1}{1-P}\right)}{\partial T}\right]_{A_{g,i}} = \Delta\left(\Delta_{\text{pure},l}U^*\right) + \left(\frac{\partial\ln K}{\partial T}\right)A_{g,i}, \tag{7}$$

respectively.

Results

We applied the first nucleation theorem for the analysis of heterogeneous nucleation experiments of Wagner et al.[3] They have obtained nucleation probabilities for binary heterogeneous nucleation of water and *n*-propanol vapours on nearly monodispersed silver nanoparticles with a geometric mean diameter of 8 nm and geometric standard deviation 1.035. The experiments were carried out with several gas-phase activity fractions of *n*-propanol at $T = 285\,K$. Experiment with each activity fraction involved several single measurements on different saturation ratios. Nucleation probability curves were plotted for each activity fraction. The critical cluster size depends strongly on the steepness of the nucleation probability curve near the heterogeneous nucleation onset ($P = 0.5$).

Figures 1 and 2 show the result of the analysis of experimentally obtained nucleation probabilities for two gas-phase activity fractions of *n*-propanol: $X_g = 0.225$ (Figure 1) and $X_g = 0.541$ (Figure 2). According to the heterogeneous nucleation theorem the critical cluster sizes in the present conditions are typically between 10 and 100, usually smaller than predicted by the classical theory (also shown in the figures). Unfortunately there do not yet exist experimental data to be analysed by the second nucleation theorem.

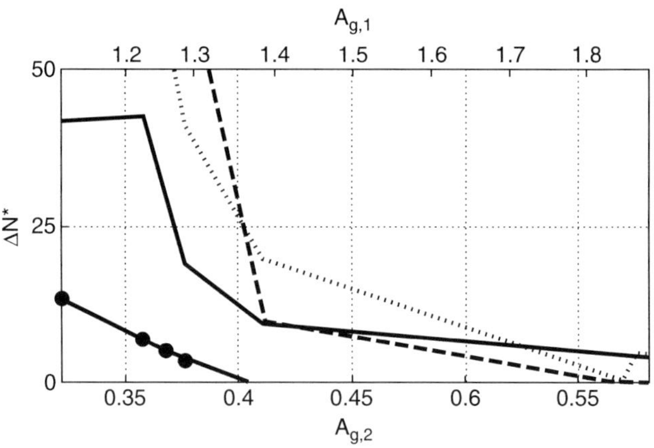

Figure 1 The number of water (solid line) and *n*-propanol (solid line with dots) molecules in the critical cluster according to the heterogeneous nucleation theorem. The classical theory predictions are also shown (dashed line: water, dotted line: *n*-propanol). $A_{g,1}$ refers to water, $A_{g,2}$ to *n*-propanol. $X_g = 0.225$

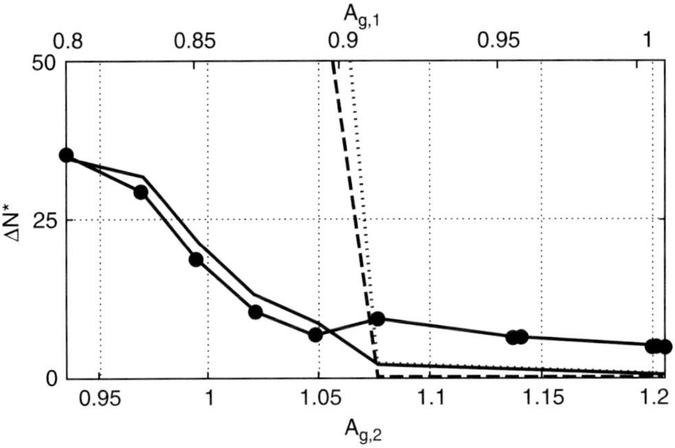

Figure 2 Same as Figure 1, but for $X_g = 0.541$

Acknowledgments Mr. Kai Ruusuvuori is gratefully acknowledged for his assistance.

References

1. Kashchiev, D., *Nucleation: Basic Theory with Applications*, Oxford: Butterworth-Heinemann (2000).
2. Vehkamäki, H., Määttänen, A., Lauri, A., Kulmala, M., Winkler, P., Vrtala, A., and Wagner, P.E., *J. Chem. Phys.*, accepted for publication (2007).
3. Wagner, P.E., Kaller, D., Vrtala, A., Lauri, A., Kulmala, M., and Laaksonen, A., *Phys. Rev. E*, **67**, 021605 (2003).

Nucleation Events in Melpitz, Germany, and Po Valley, Italy: Similarities and Differences

A. Hamed[1], Kari E.J. Lehtinen[1,2], J. Joutsensaari[3], B. Wehner[4], W. Birmili[4], A. Wiedensohler[4], F. Cavalli[5], S. Decesari[5], M. Mircea[5], S. Fuzzi[5], M.C. Facchini[5], and A. Laaksonen[1,6]

Abstract This study has been designed to investigate the factors that influence the occurrence of new particle formation events (particles >3 nm) in two anthropogenically influenced boundary layer regimes in Central Europe (Melpitz, Eastern Germany) and Southern Europe (San Pietro Capofiume "SPC" in the Po Valley, Italy). In particular, we study the similarities and differences of factors driving nucleation events at these two locations. The analysis encompasses three years of data at each observation site, between July 2003 and June 2006. Particle size distribution measurements were carried out using twin DMPS systems (Differential Mobility Particle Sizers) at each site, with particle size ranges of 3–800 nm, and 3–600 nm at Melpitz and SPC, respectively. In addition to particle size measurements, several gas and meteorological parameters are being measured for stations, including SO_2, NO, NO_2, NO_x, O_3, temperature, relative humidity, wind direction, wind speed, global radiation, precipitation, and atmospheric pressure. We utilized these parameters in our analysis of particle formation and growth processes. Our preliminarily results show that for Melpitz station, it is possible to separate between nucleation days and nonnucleation days by using just two variables (a product of radiation and sulphuric dioxide – its increase indicates increase of sulphuric acid concentration – and condensation sink"CS"). This result for Melpitz indicates that low enough CS value is needed together with high enough sulphuric acid production to drive nucleation and vice verse for nonnucleation days. For SPC station the same criterion could not separate nucleation days and nonnucleation days as efficiently as for Melpitz, and an overlap region between nucleation and nonnucleation days remained. Interestingly, however, the nonnucleation days in Po Valley are bound by the same criterion as in Melpitz, and it is the nucleation days that cause the Po Valley overlap.

[1]*Department of Physics, University of Kuopio, P.O. Box 70211 Kuopio, Finland*

[2]*Finnish Meteorological Institute, Kuopio Unit, P.O. Box 1627, FIN-70210 Kuopio, Finland*

[3]*Department of Environmental Sciences, University of Kuopio, FIN-70211 Kuopio, Finland*

[4]*Institute for Tropospheric Research, Leipzig, Germany*

[5]*Instituto di Scienze dell'Atmosfera e del Clima Consiglio Nazionale delle Richerche, Bologna, Italy*

[6]*Finnish Meteorological Institute, P.O. Box 503, 00101 Helsinki, Finland*

C.D. O'Dowd and P.E. Wagner (eds.), *Nucleation and Atmospheric Aerosols*, 240–244.

Keywords Nucleation days, nonnucleation days, driving nucleation factors

Introduction

New particle formation from supersaturated vapors (nucleation) can occur almost everywhere in the atmosphere, in clean areas, rural, coastal, and polluted areas (Kulmala et al., 2004). A direct chemical analysis of particles <5 nm is technically not feasible at the moment; therefore, knowledge on the atmospheric nucleation and particle growth mechanism and its governing circumstances has only been gathered indirectly, for instance by time series analysis of the number concentration of freshly formed particles, gas-phase precursor, and meteorological variables. A comparison of sufficiently long atmospheric data sets in different environments is expected to reveal useful information on the atmospheric conditions governing new particle formation events.

In this presentation we show the results from the long-term analysis made for Melpitz and for SPC stations for the period of three years (July 2003–June2006) in order to reveal the driving nucleation factors for both stations and to highlight the similarities and differences of the criterion for the favorable conditions for new particle formation.

Methods

In a first step, the particle formation days were classified into two categories, i.e., nucleation days and nonnucleation days. The nucleation days were identified by increase of the particle number concentrations in the nucleation mode where for those days the newly formed particles show a clear growth for several hours. In contrast, the days when no particle formation was observed, as seen from the absence of the nucleation mode, were named nonnucleation days. As a second step, we determined the time when nucleation starts on nucleation days. Note that only particles larger than about 3 nm in diameter (that is the minimum detectable size for current aerosol instruments) can be observed (Kulmala et al., 2004). The smallest nucleated clusters, on the other hand, are about 1 nm in diameter, and need time to grow to 3 nm size, this time varying under different atmospheric situations. However, because the exact growth time is not known, the observed start of the particle formation will be used as nucleation start. In a third step we calculated the condensation sink for nucleation days and for nonnucleation days by using the method described by Kulmala et al. (2001). In practice, the vapor was assumed to have very low vapor pressure at the surface of the particle, and molecular properties were assumed similar to those of sulfuric acid.

Particularly, we are interested in finding out what are the factors that could drive nucleation. We therefore studied systematically the values of different parameters

(CS, meteorological parameters, and gas-phase concentrations) at nucleation event start, and during daytime hours of nonnucleation days.

Results and Discussion

Figure (1) illustrates the condensational sink plotted against the product of SO_2 concentration and global radiation, a proxy for sulfuric acid production, for both stations (Melpitz and SPC) respectively. The needed parameters are calculated at the nucleation start for nucleation days. For nonnucleation days, we considered only time range from 6 am to 7 pm as this is the time of day when nucleation is expected to occur.

For Melpitz station, the figure can be divided into two regions. The upper portion of the figure is dominated by nucleation days where CS values were relatively low and sulfuric acid production was high enough. This indicates that low enough CS value is needed together with high enough sulfuric acid production to drive nucleation in Melpitz. In contrast, the lower portion of the figure was dominated by nonnucleation days probably due to high values for CS that in turn inhibited nucleation, in addition to the production of sulfuric acid being too low. Stanier et al. (2004) suggest that favorable conditions for nucleation, in Pittsburgh, Pennsylvania, can be described using a product of UV radiation and sulfur dioxide and the condensation sink. Their results agree with our finding for Melpitz (note, however, that we use global radiation rather than UV).

For SPC station, the same two predictive factors (CS and product of SO_2 concentration and global radiation) were not sufficient to make a clear division between nucleation days and nonnucleation days. It is probably due to an additional parameters should be included, that seem to play an important role for occurrence nucleation in SPC.

However, we transferred exactly the same line for Melpitz to the SPC plot to be able to compare between the stations and to discuss their similarities and differences by using those two predictive parameters. It can be seen that the separation line is a border for the nonnucleation days for Melpitz and for SPC as well, while it is a border just for nucleation days in Melpitz but not in SPC where there is an overlap region, below the separating line, between nucleation days and nonnucleation days. Therefore the problem is actually concerning the nucleation days in the overlap region. At the moment, we can speculate that other factors, such as production of condensable organics capable of speeding up particle growth, are needed to produce a successful parameterization of the occurrence probability of SPC nucleation days. However, that is something we need to learn more about and will discuss in more details in our future work.

Acknowledgments This work was supported by the Academy of Finland (107826), BACCI (Nordic Centre of Excellence on Biosphere-Aerosol-Cloud-Climate Interactions), and Emil Aaltonen foundation. Also by ACCENT (Atmospheric Composition Change: a European Network). Data in Melpitz were collected under the grant Feinstaubmessung by Umweltbundesamt, Dessau.

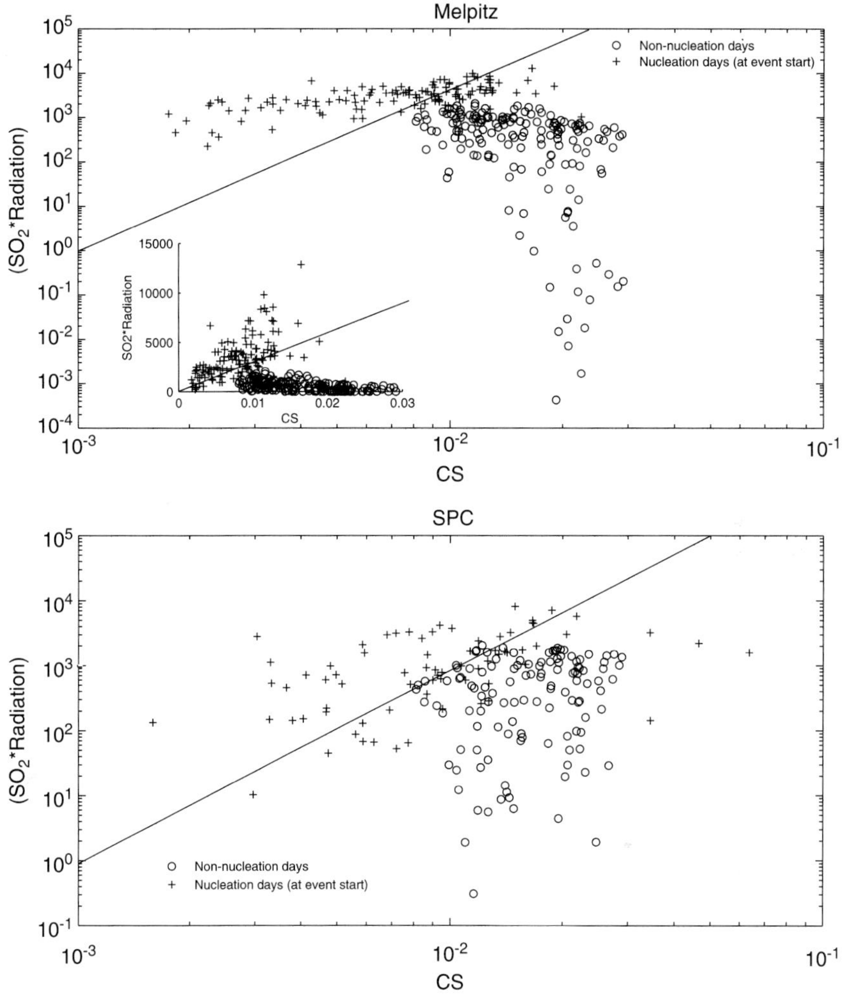

Figure 1 Condensation Sink (CS) versus product of SO_2 and radiation for nucleation days (+ symbols) where the parameters were calculated at nucleation start time and for nonnucleation days (o symbols) where the parameters were calculated from 6 am to 7 pm, to cover the expected time for nucleation to occur. The correlating applied for both stations (Melpitz and SPC) respectively. All times are local winter times (UTC + 1). The inner plot in Melpitz shows the same correlation but for linear scale

References

1. Kulmala, M., Vehkamäki, H., Petäjä, T., Dal Maso, M. Lauri, A., Kerminen, V.-M. W. Birmili, W., and McMurry. P.H., Formation and growth rates of ultrafine atmospheric particles: a review of observations, *J. Aerosol Sci.*, **35**, 143–176 (2004).

2. Kulmala, M., Hämeri, K., Aalto, P.P., Mäkelä, J.M., Pirjola, L., Nilsson, E.D., Buzorius, G., Rannik, U., Dal Maso, M., Seidl, W., Hoffmann, T., Jansson, R., Hansson, H.-C., Viisanen, Y., Laaksonen, A., and O'Dowd C.D., Overview of the international project on biogenic aerosol formation in the boreal forest (BIOFOR), *Tellus*, **53B**, 324–343 (2001).
3. Stanier, C.O., Khlystov, A.Y., and Pandis, S.N., Nucleation events during the pittsburgh air quality study: description and relation to key meteorological, gas phase, and aerosol parameters, *Aerosol Sci. Tech.*, **38(S1)**, 253–264 (2004).

Simulation of an Aerosol Nucleation Burst

J. Salm

Abstract A preliminary attempt has been made to simulate an aerosol nucleation burst recorded at SMEAR II Station, Hyytiälä, Finland, on April 8, 2000. The simulation tool by Tammet and Kulmala has been applied. The time series of the concentration of particles in the diameter interval of 2.8–8.6 nm served as the main characteristic of the burst.

Keywords Tropospheric aerosols, nucleation bursts, simulation

Introduction

The bursts of charged atmospheric nanoparticles were discovered by air ion measurements at Tahkuse Observatory in 1985 [1]. In the 1990s and later, the study of the nucleation bursts of atmospheric nanoparticles has spread over the world because of the potential relationships with the Earth climate [2]. Several different hypotheses and models have been elaborated for the explanation of this phenomenon. In spite of a large number of observed nucleation bursts, and numerous hypotheses or theoretical models, the microphysical mechanism of the phenomenon has remained unclear up till now.

Due to the complexity of the phenomenon, Tammet and Kulmala have developed a numerical tool for the simulation of atmospheric aerosol nucleation bursts [3]. The tool takes into account a number of various processes and relationships, which probably affect the generation and growth of aerosol particles in the atmosphere. This paper presents a preliminary attempt to simulate a measured nucleation burst.

Institute of Environmental Physics, University of Tartu, Tartu, Estonia

C.D. O'Dowd and P.E. Wagner (eds.), *Nucleation and Atmospheric Aerosols*, 245–249.

Measurements

Complex measurements were performed at SMEAR II Station, Hyytiälä, Southern Finland (61 51'N, 24 17'E, 170 m above sea level), from March 20 to April 20, 2000, at the annual maximum of the new particle burst occurrence. The instrumentation and certain results have been published earlier [4,5,6]. Of stationary facilities, mainly atmospheric and aerosol instrumentation was used for this research. The instrumentation was essentially completed with special apparatus for aerosol and air ion measurements. The main novel apparatus was a set of two identical differential mobility particle sizers (DMPS) with different pre-charging conditions. One of these DMPS operated at the very customary mode: the aerosol particles were charged in a bipolar charger. Another DMPS was identical, but without the charger; thus it measured only naturally charged aerosol particles. In both DMPS, negatively charged particles were analyzed by means of differential mobility analyzers (DMA) according to their mobility.

The main measurement campaign lasted from March 29 to April 14, 2000, when the maximum number of instruments was operating. The database of the measurements consists of 91 measured parameters with 10 min or 30 min resolution. Intense or medium nucleation bursts of aerosol particles occurred on 7 days. The main earlier results of that campaign can be summarized as follows [4,5,6]:

(1) The mobility distributions measured by two mentioned spectrometers with bipolar pre-charging and without charging coincide well. Significant deviations from steady state charge distribution towards the excess or deficit occurred only in the fractions of smallest particles during the nucleation events.
(2) The nucleation bursts were intensive at low concentration of pre-existing background aerosols.
(3) The correlation of the concentration of nanometer size aerosols with the concentrations of several measured polluting trace gases was insignificant.

A remarkable nucleation burst occurred on April 8, 2000. This was a medium nucleation burst around noon. The graph of the concentration of nanoparticles is presented in Figure 1. A significant excess charge of naturally charged particles is clearly presented, despite the fact that the neutralization of charged particles is much faster process than the charging of neutral particles. Obviously this excess charge indicates ion-induced nucleation. That burst was chosen as an object of simulation.

Simulation

The simulation tool has 95 control lines; among them are 48 physical input parameters [3]. The values of the physical parameters are varied, in order to achieve conformity between the output diagrams and the measured graphs for April 8, 2000.

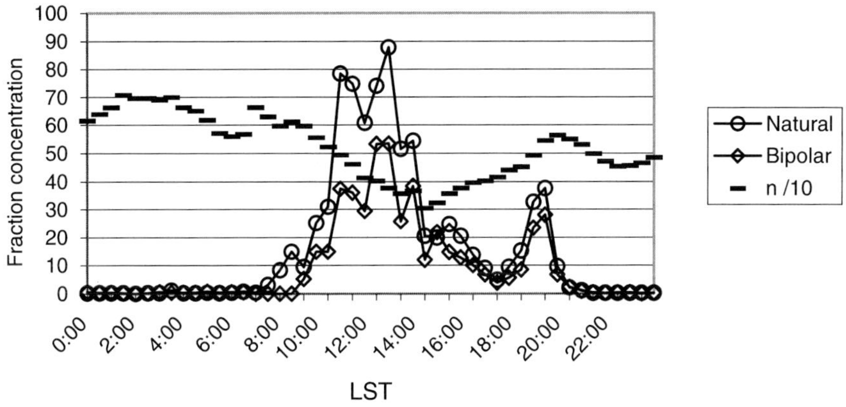

Figure 1 Time series of the fraction concentrations (cm^{-3}) of negatively charged aerosol particles and air ions at Hyytiälä, Finland, on April 8, 2000. "Natural" denotes naturally charged particles in a diameter range of 2.8–8.6 nm, "Bipolar" particles in the same size range charged in the bipolar charger, and "n" cluster air ions measured by means of air ion counters

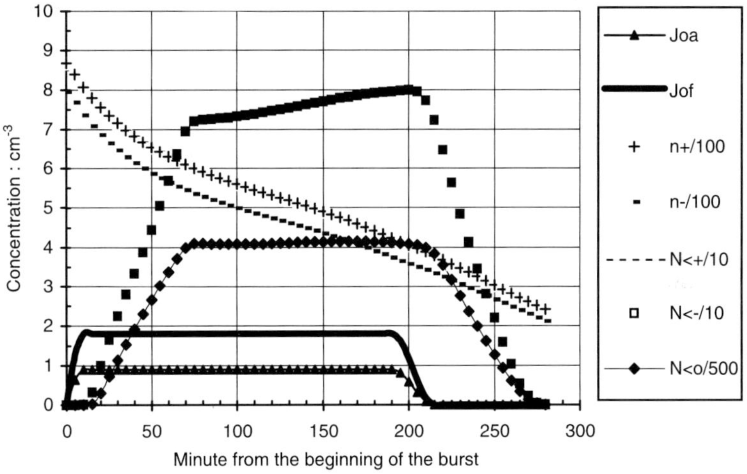

Figure 2 Simulation results for the fraction concentrations of charged and neutral aerosol particles and of cluster ions, on the basis of measurements at Hyytiälä, Finland, on April 8, 2000. "Joa" and "Jof" denote the maximum nucleation rate for neutral nucleation in free air and in forest, respectively, "n +" and "n-" are the concentrations of cluster ions, "N < +", "N < -", and "N < o" are the concentrations of positive, negative, and neutral particles in the diameter interval of 2.8–8.6 nm. The nucleation rate for neutral particles in open air "Joa" and in forest "Jof" in units cm^{-3} s^{-1} are also presented

The meteorological and the background aerosol parameters are taken from the measurements. Many parameters are kept as in the user manual version 20060912, e.g., the electric mobility of cluster ions, density, and diameter of growth units, polarizability etc. Slightly altered parameters: ionization rate 4 cm^{-3} s^{-1}; birth size of particles 1.6 nm; conifer needle length 50 m^{-2}. The low end and the high end of the particle presentation range are taken equal to those of the first fraction of the DMPS, 2.8 nm, and 8.6 nm.

Many parameters are varied in a wider range: maximum nucleation rates for neutral and charged nucleation; time of steady nucleation activity; growth rates of neutral particles. As the main output parameters, the concentrations of charged and neutral particles in the size interval of 2.8–8.6 nm are kept into consideration. This interval corresponds to the first size fraction of DMPS. More than 60 output diagrams with different sets of input parameters are analyzed. An output diagram of the simulation is presented in Figure 2.

The set of varied input parameters corresponding to Figure 2: initial, half-time, and final ionization rate is 4 cm^{-3} s^{-1} both in open air and in forest; maximum nucleation rate for positive and negative ion nucleation is 0.06 cm^{-3} s^{-1} both in open air and in forest; maximum nucleation rate for neutral nucleation is 0.9 cm^{-3} s^{-1} in open air and 1.8 cm^{-3} s^{-1} in forest; time of steady nucleation activity is 180 min; initial, half-time, and final plain Knudsen growth rate of neutral particles is 3 nm/h both for the first and for the second condensing substance.

Discussion

As a result of the simulation, the diagram in Figure 2 has been obtained. The width and the height of the curve for charged particles in the diameter interval of 2.8–8.6 nm are comparable with those of the measured curves in Figure 1. The average percentage of charged particles in the above diameter interval would be about 2% in steady state. However, according to Figure 1, the measured concentration "Natural" was two times higher than the "Bipolar" or steady-state concentration. Thus the percentage of charged particles, 4%, in Figure 2, is also in accordance with measured values.

The measurement range of DMPS begins at a diameter of 2.8 nm, but the growth of aerosol particles begins at 1.6 nm. Thus the growth rate of particles is an especially significant parameter, when we try to measure the electrical state of nanoparticles. The charging steady state for nanoparticles is attained in about 1 h. If the growth rate is too low, about 1 nm/h or lower, the particles can attain charging steady state before they will be measured by DMPS.

Acknowledgments This research has in part been supported by the Estonian Science Foundation through grants 6223 and 6988. The author thanks Dr. J. M. Mäkelä for the organization of measurements in 2000.

References

1. Tammet, H., Salm, J., and Iher, H., in: *Atmospheric Aerosols and Nucleation. Lecture Notes in Physics*, Vol. 309, Vienna: Springer, pp. 239–240 (1988).
2. Kulmala, M., Vehkamäki, H., Petäjä, T., Dal Maso, M., Lauri, A., Kerminen, V.-M., Birmili, W., and McMurry, P.H., *J. Aerosol Sci.*, **35**, 143–176 (2004).
3. Tammet, H. and Kulmala, M., *J. Aerosol Sci.*, **36**, 173–196 (2005).
4. Mäkelä, J.M., Salm, J., Smirnov, V.V. et al., in *Proceedings of 12th International Conference on Atmospheric Electricity*, Versailles, France, 2003, pp. 793–796.
5. Smirnov, V.V., Salm, J., Mäkelä, J.M., Paatero, J., Dynamics of atmospheric aerosol, ions and trace gases at invasion of the arctic air masses (translated by AGU), *Atmos. Oceanic Opt.*, **17**, 61–69 (2004).
6. Smirnov, V.V., Salm, J., and Mäkelä, J.M., in: *Nucleation and Atmospheric Aerosols 2004, 16th International Conference*, edited by M. Kasahara and M. Kulmala, Kyoto, Japan, pp. 316–319 (2004).

Heterogeneous Nucleation of Nitrogen

Guram Chkonia[1], Jan Wedekind[2], Judith Wölk[1], and Reinhard Strey[2]

Abstract Homogeneous and heterogeneous vapor–liquid nucleation experiments of nitrogen are carried out using a cryogenic nucleation pulse chamber. For the latter water ice crystals are used as heterogeneous seeds. The onset of nucleation is detected by constant angle *Mie* scattering at pressures from 2 to 25 kPa and temperatures from 50 to 70 K. The measured onset data are compared to experimental results from literature.

Keywords Homogeneous and heterogeneous nucleation, nitrogen, triple point

Introduction

Since nitrogen is the main component of natural air, its condensation behavior is of fundamental interest for all techniques where air is cooled down rapidly. During such a cooling process the system gets supersaturated and a first-order phase transition is initialized by nucleation, followed by growth and aging. Of all these processes, nucleation is the least understood. Therefore many efforts are undertaken to clarify the nucleation process experimentally[1] as well as theoretically[2–14].

The first nucleation experiments of nitrogen from the gas-phase date back to 1952, when Faro et al.[15] made the first experiments with a wind tunnel. In the same year Willmarth and Nagamatsu[16] used a supersonic nozzle to observe the onset of nitrogen nucleation. Since then several other groups used supersonic nozzles and shock tubes to investigate the nucleation kinetics of nitrogen. Very recently Iland[17] measured the onset conditions for nitrogen in a cryogenic nucleation pulse chamber for the first time. The techniques mentioned above differ in their respective measuring window by reaching different supersaturations of the gas phase. Therefore the typical nucleation rates of nucleation pulse chambers range in $10^5 < J/\mathrm{cm}^{-3}\ \mathrm{s}^{-1} < 10^9$,

[1]*Institut für Physikalische Chemie, Universität zu Köln, Luxemburger Str. 116, 50939 Köln, Germany*

[2]*Departament de Física Fonamental, Universitat de Barcelona, Martí i Franquès, 1, 08028 Barcelona, Spain*

C.D. O'Dowd and P.E. Wagner (eds.), *Nucleation and Atmospheric Aerosols*, 250–254.

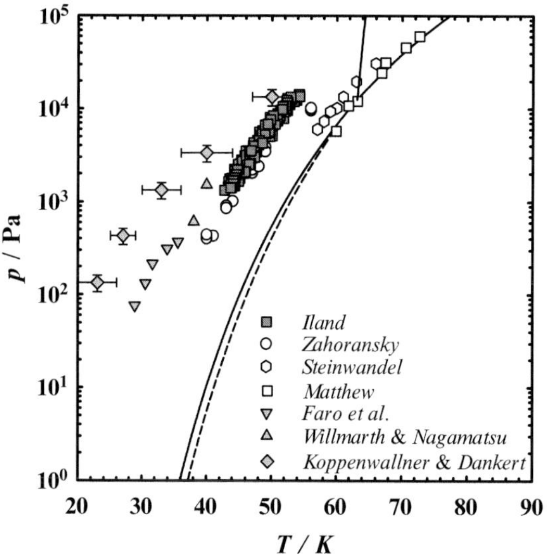

Figure 1 Wilson-plot containing the nitrogen nucleation onset data from Iland determined with the cryogenic chamber (dark filled squares)[15,16,18] in comparison with supersonic nozzle data (grey filled data points)[19–21] and data measured with shock tubes (empty data points). The solid lines are the binodals for liquid–solid and gas–liquid phase transition, the dashed line is the binodal for gas–solid phase transition

supersonic nozzles yield rates of the order of $J = 10^{17}$ cm^{-3} s^{-1} and shock tubes measure about $10^7 < J$/cm^{-3} s^{-1} $< 10^9$. In Figure 1, we present the onset data found in literature. The nozzle data should lie on the left side of the data measured with the nucleation pulse chamber because of the higher expansion rates and correspondingly higher nucleation rates of supersonic nozzles. This is found for the measurements from Koppenwallner and Dankert[18]. The data by Faro et al.[15] and Willmarth and Nagamatsu[16] have been measured at lower temperatures than the data from Iland[17], but they still show a consistent pattern.

The empty symbols in Figure 1 are measured with shock tubes. Since in shock tubes the onset of nucleation is usually detected at similar supersaturations as in expansion chambers, the data should lie close to each other. This is the case for the data measured by Zahoransky[19] in 1986 (empty circles), while the data measured by Steinwandel[20] (empty hexagons) are lying at much higher temperatures for the same onset-pressure. The data of Matthew[21] (empty squares) are shifted even to higher temperatures and are lying on top of the gas–liquid binodal. The latter might be explained by the detection of heterogeneous nucleation.

In general, two sorts of nucleation have to be distinguished: homogeneous and heterogeneous nucleation. While homogeneous nucleation occurs solely due to density fluctuations in the system, heterogeneous nucleation is taking place on preexisting surfaces. By adding water vapor into the cryogenic nucleation pulse chamber, the vapor resublimes and the formed crystals act as heterogeneous nuclei.

Heterogeneous nuclei lower the nucleation barrier substantially. Therefore, compared to homogeneous nucleation, much smaller values of supersaturation are sufficient to start heterogeneous nucleation.

Experiment

Here we report a consistent set of homogeneous as well as heterogeneous nucleation experiments of nitrogen measured over a wide range of temperatures with helium as carrier gas. The experiments are carried out in a cryogenic nucleation pulse chamber developed by Fladerer and Strey[22] and modified by Iland et al.[17] For the heterogeneous nucleation experiments the chamber has been provided with an extra mixing volume by Mensah[23]. This modification enables the injection of a water–helium mixture directly into the pulse chamber prior to the nucleation pulse experiment. A detailed description of the experimental setup and procedure can be found in the above mentioned papers.

Results and Discussion

In Figure 2 our homogeneous (crosses) and heterogeneous (grey diamonds) onset data of nitrogen are shown in a Wilson-plot (left diagram) as well as in a Volmer-plot. We were able to reproduce the homogeneous data of Iland[17] (squares) very well. As expected the heterogeneous onsets occur at lower supersaturations and

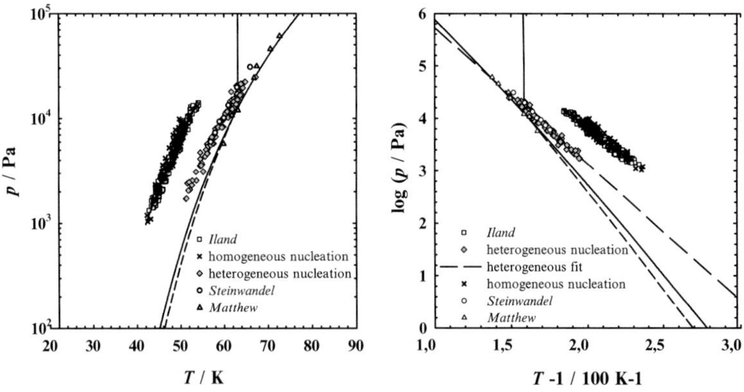

Figure 2 Onset data shown in a Wilson- and a Volmer-plot. The crosses are homogeneous measurements, the diamonds are heterogeneous measurements. Our results are compared to data obtained by Iland[17], Matthew[21], and Steinwandel[20]. The solid lines are the gas–liquid and the solid–liquid binodals, the short dashed lines are gas–solid binodals. The long dashed line in the Volmer-plot is fitted to heterogeneous measurements

therefore the data lie closer to the binodal than the homogeneous ones. It is noteworthy that the shock tube measurements from Matthew[21] and from Steinwandel[20] are in agreement with the heterogeneous results. Therefore the assumption that their experiments may be contaminated and that they detected also heterogeneous nucleation seems reasonable.

The measured heterogeneous data do not converge to the triple point as expected but are shifted to slightly lower temperatures. This temperature shift arises for experiments with high nitrogen partial pressures. For high condensable fractions latent heat-effects might no longer be negligible. Since the onset-temperature is computed from the measured pressure, latent heat-effects cause an underestimation which could possibly lead to the observed shift.

The effect might also be caused by gas–solid nucleation. Volmer[24] described Ostwald's step rule as a consequence of the change in the phase behavior for small systems. According to Volmer, the phase that forms first is determined by the position of the "metastability boundary curves", which correspond to the formation rate of the new phase. Consequently, the energy barrier for gas–solid nucleation is usually higher than the barrier for gas–liquid nucleation and the liquid phase builds up faster than the solid one. Since his comments are based rather on theoretical concepts than on experimental results, the applicability of this model to every single system needs to be revised possibly. In this case gas–solid nucleation could become energetically preferred.

The gas–liquid binodal has to be extrapolated below the triple point, where no experimental results are available. This information, however, is crucial for the calculation of the supersaturation. Thus, an experimental vapor pressure curve would greatly improve the accuracy of the onset evaluation. Assuming that our heterogeneous particle radius is larger than 20 nm[25] the critical supersaturation S_{crit} needed to induce heterogeneous nucleation should be smaller than 1.1, which is very close to the actual equilibrium value at the binodal. Therefore the results of this work could be used to derive an experimental vapor–pressure curve.

Acknowledgments We would like to thank Amewu Mensah for many fruitful discussions and support with the experiments.

References

1. Heist, R.H. and He, H.H., *J. Phys. Chem. Ref. Data*, **23**(5), 781–805 (1994).
2. Becker, R. and Döring, W., *Ann. Phys.*, **24**, 719 (1935).
3. Girshick, S.L. and Chiu, C.P., *J. Chem. Phys.*, **93**(2), 1273–1277 (1990).
4. Zeng, X.C. and Oxtoby, D.W., *J. Chem. Phys.*, **94**(6), 4472–4478 (1991).
5. Oxtoby, D.W., *J. Phys.: Condens. Matter*, **4**, 7627–7650 (1992).
6. Talanquer, V. and Oxtoby, D.W., *J. Chem. Phys.*, **100**(7), 5190–5200 (1994).
7. Reiss, H., Kegel, W.K., and Katz, J.L., *Phys. Rev. Lett.*, **78**(23), 4506–4509 (1997).
8. Reiss, H., Kegel, W.K., and Katz, J.L., *J. Phys. Chem. A*, **102**(44), 8548–8555 (1998).
9. Senger, B., Schaaf, P., Corti, D.S., Bowles, R., Voegel, J.C., and Reiss, H., *J. Chem. Phys.*, **110**(13), 6421–6437 (1999).

10. Senger, B., Schaaf, P., Corti, D.S., Bowles, R., Pointu, D., Voegel, J.C., and Reiss, H., *J. Chem. Phys.*, **110**(13), 6438–6450 (1999).
11. Schenter, G.K., Kathmann, S.M., and Garret, B.C., *Phys. Rev. Lett.*, **82**(17), 3484–3487 (1999).
12. Reguera, D. and Reiss, H., *Phys. Rev. Lett.*, **93**(16) (2004).
13. Reguera, D. and Reiss, H., *J. Phys. Chem. B*, **108**(51), 19831–19842 (2004).
14. Zandi, R., Reguera, D., and Reiss, H., *J. Phys. Chem. B*, **110**(44), 22251–22260 (2006).
15. Faro, I., Small, T.R., and Hill, F.K., *J. Appl. Phys.*, **23**(1), 40–43 (1952).
16. Willmarth, W.W., and Nagamatsu, H.T., *J. Appl. Phys.*, **23**(10), 1089–1095 (1952).
17. Iland, K., Experimente zur homogenen Keimbildung von Argon und Stickstoff, *Dissertation*, Universität zu Köln (2004).
18. Koppenwallner, G. and Dankert, C., *J. Phys. Chem.*, **91**, 2482–2486 (1987).
19. Zahoransky, R.A., *Zeitschrift für Flugwissenschaften und Weltraumforschung*, **10**(1), 34 (1986).
20. Steinwandel, J., *Ber. Bunsenges. Phys. Chem.*, **89**, 481 (1985).
21. Matthew, M.W., Condensation of argon and nitrogen in cryogenic shock tubes, *Dissertation*, Yale University (1982).
22. Fladerer, A. and Strey, R., *J. Chem. Phys.* **124**(16) (2006).
23. Mensah, A., First experimental data on the gas-liquid vapor pressure curve of liquid argon between 52 K and 72 K, Universität zu Köln (2005).
24. Volmer, M., *Kinetik der Phasenbildung*, Dresden: Steinkopff (1939).
25. Chen, C.C., Guo, M.S., Tsai, Y.J., and Huang, C.C., *J. Colloid. Interf. Sci.*, **198**(2), 354–367 (1998).

Homogeneous Nucleation of a Homologous Series of *n*-Alkanes in a Supersonic Nozzle

Dirk Bergmann[1], David Ghosh[1], Judith Wölk[1], Reinhard Strey[1], Shinobu Tanimura[2], and Barbara E. Wyslouzil[2]

Abstract We conducted static pressure measurements along the axis of a supersonic nozzle to follow condensation of a homologous series of *n*-alkanes (C_iH_{2i+2}; $i = 7$–10) from dilute mixtures of vapor in nitrogen or argon. We determined the pressure $p_{J\max}$, temperature $T_{J\max}$, and supersaturation $S_{J\max}$ corresponding to the maximum nucleation rate $J_{\max}$ as a function of the condensable flow rate. As the condensable flow rate increases, the onset of condensation moves further upstream in the nozzle and the latent heat released to the flow due to condensation increases. The release of latent heat is detected as a deviation between the pressure measured for the condensing flow and the pressure expected for an isentropic expansion of the same gas mixture. In a Volmer-plot of the maximum nucleation pressure $p_{J\max}$ versus the inverse temperature $1/T_{J\max}$, the data points for each *n*-alkane lie on straight lines parallel to each other. At a fixed temperature $T_{J\max}$, the pressure $p_{J\max}$ decreases as the carbon chain length increases while the supersaturation $S_{J\max}$ increases as the carbon chain length increases.

Keywords Homogeneous nucleation, *n*-alkanes, supersonic nozzle

Introduction

In earlier studies we investigated the nucleation behavior of H_2O,[1] D_2O,[1], and a homologous series of *n*-alcohols[2] in supersonic nozzles. Here, we extend our investigations to a homologous series of *n*-alkanes (C_iH_{2i+2}; $i = 7$–10). Most of these compounds are constituents of raw natural gas and must be separated from CH_4 prior to transport. Thus, understanding their condensation behavior is important in order to develop environmentally friendly separation and purification processes for natural gas. In this work we perform static pressure trace measurements along the center

[1] *Institut für Physikalische Chemie, Universität zu Köln, D-50939 Köln, Germany*

[2] *Department of Chemical Engineering, Ohio State University, Columbus, OH 43210, USA*

C.D. O'Dowd and P.E. Wagner (eds.), *Nucleation and Atmospheric Aerosols*, 255–259.

line of a supersonic nozzle and determine the temperature $T_{J\max}$, pressure $p_{J\max}$, and supersaturation $S_{J\max}$ corresponding to the maximum nucleation rate $J_{\max}$.

Experimental Setup

The principle of the experimental setup has been described in detail elsewhere,[1] and only a basic description and a simplified schematic, Figure 1, are given here.

Depending on the n-alkane the carrier gas, nitrogen or argon, is drawn from two liquid cylinders. The gas leaving the cylinders is warmed by electrical heaters and the flow rate is controlled and monitored by two MKS mass flow controllers. One of the carrier gas streams is heated further, and enters the vapor generator where part of the flow disperses the condensable liquid into a fine spray while the remaining flow provides the energy to evaporate the droplets and dilute the flow. The n-alkanes are pumped from a flask into the vapor generator using a peristaltic pump. The vapor-rich gas stream is combined with the main carrier gas stream and the gas temperature is adjusted in a heat exchanger placed in a temperature controlled circulating water bath. This water bath also controls the temperature in the plenum while the plenum pressure is controlled by the MKS flow controllers. Both the plenum and nozzle are located in a large evacuated metal box with a lid to reduce the heat transfer to the surrounding. The gas mixture passes through the plenum and continues to flow through the supersonic nozzle. Here, a movable pressure probe measures the static pressure as a function of axial position along the center line of the nozzle. Finally, the flow is exhausted to atmospheric pressure by two vacuum pumps.

Results and Discussion

Nucleation and growth is accompanied by the release of latent heat. Static pressure trace measurements can detect this latent heat as a deviation between the condensing pressure trace and the theoretical isentropic expansion of a gas with identical

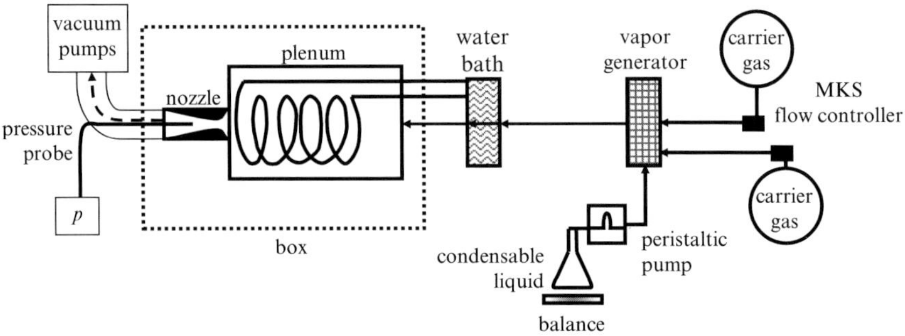

Figure 1 A schematic diagram of the experimental setup.

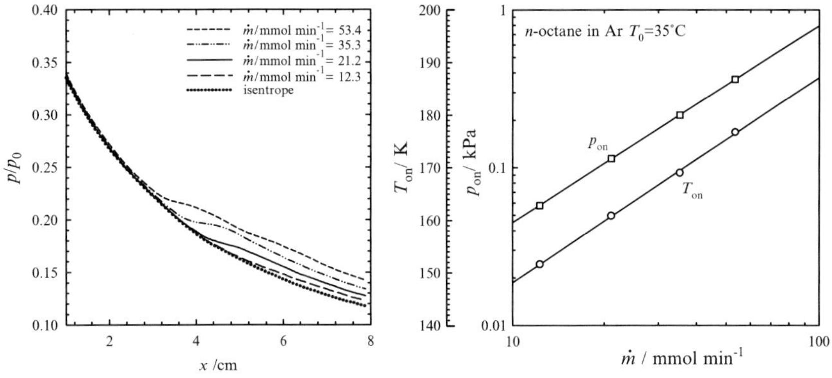

Figure 2 (a) The pressure ratio p/p0 is plotted as a function of the distance x from the throat of the nozzle for n-octane in Ar at T0 = 35°C. The wet isentrope corresponding to an n-octane flow rate of 12.3 mmol min^{-1} is also shown. (b) The onset pressure of the condensable pon and the onset temperature Ton are shown as a function of the condensable flow rate for the n-octane experiments

thermophysical properties as the condensing flow that does not condense, the theoretical wet isentrope. As in our previous work[3] we define the onset of condensation in terms of a temperature difference between the condensing flow curve and the wet isentrope. Our onset criterion is a temperature difference of 0.5 K. Ideally, the condensing flow and the wet isentrope agree up to the onset of condensation. For some n-alkanes, however, we observed deviations between the condensing flow and the wet isentrope prior to the onset that may arise from uncertainties in the temperature dependent heat capacities $c_p(T)$ in the temperature range of 140–250 K. This deviation is minimized if we introduce correction factors, on the order of 10%, for the heat capacities $c_p(T)$. Figure 2 (a) illustrates the deviation of the condensing pressure traces from the wet isentrope corresponding to the lowest flow rate. In these experiments the stagnation temperature T_0 was 35°C and the stagnation pressure p_0 was 30 kPa. As the amount of condensable increases, the onset of condensation moves further upstream to higher pressures and temperatures, Figure 2 (b), and the deviation from the wet isentrope increases.

Even though the onset of condensation is a convenient parameter to characterize the limit of stability in an expanding flow, the definition of this term strongly depends on the experimental device used. Furthermore, in supersonic nozzle experiments, where nucleation and growth are not decoupled, this parameter depends upon both nucleation and droplet growth. The more physically meaningful parameters are the temperature $T_{J\max}$ and pressure $p_{J\max}$ corresponding to the maximum nucleation rate $J_{\max}$. To determine these parameters, we calculate the nucleation rate as a function of position using the experimentally derived pressure $p(x)$ and temperature $T(x)$ profiles and the classical nucleation rate[4] expression

$$J_{CNT} = \sqrt{\frac{2\sigma}{\pi\mu_v}} v_m \left(\frac{p_v}{kT}\right)^2 \exp\left(\frac{-16\pi v_m^2 \sigma^3}{3(kT)^3 (\ln S)^2}\right), \tag{1}$$

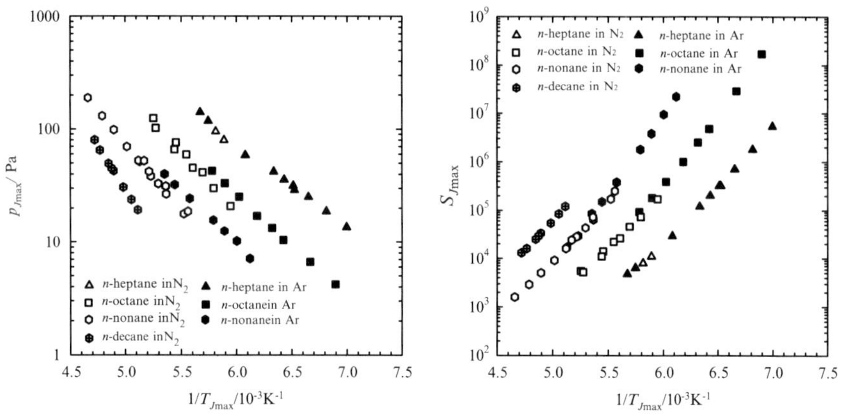

Figure 3 (a) A Volmer[6]-plot of the pressure p_{Jmax} versus the inverse temperature $1/T_{Jmax}$ corresponding to the maximum nucleation rate for the *n*-alkanes. (b) The critical supersaturation S_{Jmax} versus the inverse temperature $1/T_{Jmax}$

where σ is the surface tension of the macroscopic fluid–vapor interface, v_m the molecular volume, μ_v is the molecular mass of the condensable vapor, k the Boltzmann constant, and the supersaturation S is given by

$$S = \frac{p_v(x)}{p_{eq}\left(T(x)\right)}. \tag{2}$$

Here, $p_{eq}(T(x))$ is the equilibrium vapor pressure[5] of the condensable species and $p_v(x)$ the partial pressure of the *n*-alkane at the position x. From the nucleation rate profile it is easy to determine the values of S_{Jmax} and T_{Jmax} that correspond to the maximum nucleation rate, and hence determine the corresponding partial pressure p_{Jmax} of the *n*-alkane. Plotting the pressure p_{Jmax} as a function of the inverse temperature $1/T_{Jmax}$, Figure 3 (a),[6] the data for each *n*-alkane lie on straight lines that are quite parallel to each other, and, at a fixed temperature T_{Jmax}, the pressure p_{Jmax} increases as the carbon chain length of the *n*-alkane decreases. The data from experiments using Ar as the carrier gas lie above those where N_2 is the carrier gas, because the cooling rates are higher in the former case and, thus, we can probe the metastable region more deeply. When plotting the supersaturation S_{Jmax} as a function of the inverse temperature $1/T_{Jmax}$, Figure 3 (b), the data points for each *n*-alkane also lie on straight lines reasonably parallel to each other. Here, however, at a fixed temperature T_{Jmax} the supersaturation S_{Jmax} increases with increasing *n*-alkane chain length.

Acknowledgment This work was supported by the National Science Foundation under Grant numbers CHE-0410045 and CHE-0518042 and by the Donors of the Petroleum Research Fund administered by the American Chemical Society.

References

1. Kim, Y.J., Wyslouzil, B.E., Wilemski, G., Wölk, J., and Strey, R., *J. Phys. Chem. A*, **108,** 4365–4377 (2004).
2. Gharibeh, M., Kim, Y., Dieregsweiler, U., Wyslouzil, B.E., Ghosh, D., and Strey, R., *J. Chem. Phys.*, **122**, 094512–094518 (2005).
3. Streletzky, K.A., Zvinevich, Y., Wyslouzil, B.E., and Strey, R., *J. Chem. Phys.*, **116**, 4058–4070 (2002).
4. Becker, R. and Döring, W., *Ann. Phys.*, **24**, 719–752 (1935).
5. Katz, J.L., *J. Chem. Phys.*, **52**, 4733–4748 (1970).
6. Volmer, M., *Kinetik der Phasenbildung*, Dresden: Steinkopff (1939).

Preliminary Results on Homogeneous Nucleation of Water: A Novel Measurement Technique using the Two-valve Expansion Chamber

Alexandra Manka, Dirk Bergmann, David Ghosh, Judith Wölk, and Reinhard Strey

Abstract Using the expansion chamber by performing simply an expansion the onset conditions for homogeneous water nucleation were measured. Thereby the latent heat released into the system during nucleation is detected as a deviation in pressure between the expansion of the water–argon mixture and the expansion of pure argon under identical conditions. The onset of nucleation is defined by the detection of a light scattering signal. Comparing the onset conditions for temperature and pressure with data from Wölk and Strey (Wölk, J. and Strey, R., *J. Phys. Chem. B*, **105**, 11683–11701 (2001)) and Heath et al. (Heath, C.H., Streletzky, K., Wyslouzil, B.E., Wölk, J., and Strey, R., *J. Chem. Phys.*, **117**, 6176–6185 (2002)) enables us to estimate the nucleation rate during our experiment to $J = (10^{13} \pm 10^{2})\ \mathrm{cm^{-3}s^{-1}}$. Further, comparing our estimated nucleation rates with the predictions of classical nucleation theory reveals that for $T = 240\,\mathrm{K}$ a rather good agreement is found.

Keywords Homogeneous nucleation, water, expansion chamber, onset

Introduction

Water which covers our planet with roughly 71% can be found in nature in three phases, e.g., gaseous steam, liquid water, and solid ice. Despite its common appearance in all three phases it is up to date not fully understood how phase transitions between these phases take place. To resolve the uncertainties during these phase transitions considerable research efforts have been undertaken. First experimental studies of homogeneous water nucleation date back to 1897 when Wilson[3] conducted the first quantitative study of water nucleation. Since then, numerous experimental devices[4] for the investigation of nucleation have been designed.

Institut für Physikalische Chemie, Universität zu Köln, Luxemburger St. *116, 50939 Köln, Germany*

C.D. O'Dowd and P.E. Wagner (eds.), *Nucleation and Atmospheric Aerosols*, 260–264.

In this work we use the two-valve expansion chamber to investigate the nucleation behavior of water by simply carrying out an expansion. We compare our measured data with those from Wölk and Strey[1] also measured in the two-valve expansion chamber ($10^5 < J/\text{cm}^{-3}\,\text{s}^{-1} < 10^9$) and Heath et al.[2] measured in a supersonic nozzle ($J \sim 5.10^{16}\text{cm}^{-3}\,\text{s}^{-1}$).

Experimental Setup

A well-defined water–argon gas mixture with a partial pressure $p_{0\,H2O}$ of water as condensable and a partial pressure p_{Ar} of argon as carrier gas is prepared in a mixing receptacle. The prepared mixture is drawn into the measurement volume of the two-valve expansion chamber so that a given total pressure p_0 from 50–100 kPa is reached. The chamber temperature is set at $T_0 = 25°\text{C}$. In the expansion volume we set a pressure p_{set} of 14–25 kPa, respectively. The water–argon gas mixture is adiabatically expanded to a pressure p_{exp} by opening the valve connecting the measurement and the expansion volume. During the expansion the temperature of the system falls rapidly to lower values and the former thermodynamically stable condensable vapor changes its state to a metastable vapor. At a given pressure and temperature the metastable vapor relaxes into its thermodynamically stable state, the liquid state. Nuclei are formed from the supersaturated vapor phase and grow to a certain size. These droplets scatter the transmitted laser light. In this work we define the point at which a light scattering signal is observed as the onset of condensation.

The general experimental setup used here has been described in detail elsewhere.[5] However, compared to the given literature[5] reference, we do not carry out a recompression after the expansion; thus particle formation and growth are not decoupled.

Results and Discussion

We have carried out expansions of water–argon mixtures with varying initial water pressure $p_{0\,H2O}$ as well as expansions of pure argon. Figure 1 shows the results for two different expansions with $p_{0\,H2O} = 1.027$ kPa (left) and $p_{0\,H2O} = 2.263$ kPa (right).

Nucleation is accompanied by the release of latent heat. This release in latent heat can be observed in the deviation of pressure between the water–argon expansion (solid line) and the pure argon expansion (dashed line) under identical conditions (Figure 1). We defined the onset in nucleation as the point where a light scattering signal (dotted line) is observed. As the initial partial pressure $p_{0\,H2O}$ of water in the vapor phase increases, the temperature corresponding to the activating supersaturation S^* shifts toward higher values. Consequently, the light scattering signal is detected earlier (Figure 1). Likewise, the deviation in pressure between the

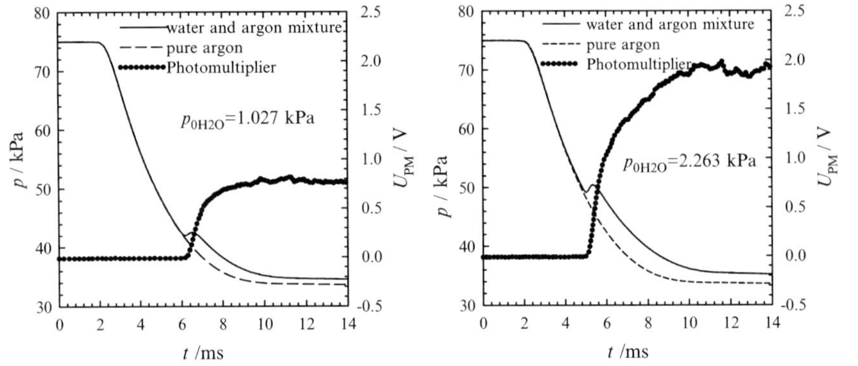

Figure 1 The pressure during the expansion for $p_{0\,H2O} = 1.027\,kPa$ (**left diagram solid line**) and $p_{0\,H2O} = 2.263\,kPa$ (**right diagram solid line**), the pressure for the pure argon expansion (dashed line) and the light scattering signal of the photomultiplier (dotted line) versus the time

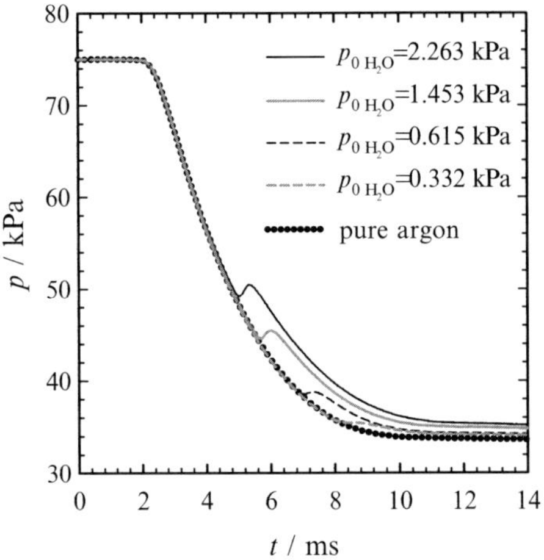

Figure 2 The pressure during expansion for various initial partial pressures of water $p_{0\,H2O}$ versus the time. All experiments are carried out at $T_0 = 25°C$ and $p_0 = 75\,kPa$. The dotted line indicates the pure argon expansion under identical conditions

water–argon mixture and the pure argon expansion shifts toward higher pressures. Further, the intensity in scattered light is amplified as the initial partial pressure $p_{0\,H2O}$ of water is increased (Figure 1). This is expected because as the amount of condensable in the vapor phase increases, the potential amount of material for condensation is amplified. Thus, more nuclei are enabled to be formed. Additionally, the existing nuclei have the ability to grow larger. Likewise, the deviation in pressure between the water–argon expansion and the pure argon expansion is expected to be propagated as the initial partial pressure $p_{0\,H2O}$ of water in the vapor phase is increased (Figure 2), because the amount of latent heat released during nucleation is directly proportional to the amount of matter condensed.

The isentropic relations hold up to the first point of nucleation (onset). Thus, we determine the onset temperature T_{on} with the Poisson equation

$$T_{on} = T_0 \left(\frac{p_{on}}{p_0} \right)^{\frac{\gamma_m - 1}{\gamma_m}} . \tag{1}$$

Here T_0 is the chamber temperature, p_0 is the initial chamber pressure, p_{on} is the onset pressure, and γ_m is the ratio of the heat capacities at constant pressure and temperature of the mixture. In Figure 3(a) we compare our measured onset pressures and temperatures with existing literature data.[1–2] The data measured by Wölk and Strey[1] lie in the nucleation rate regime of $10^5 < J/\text{cm}^{-3}\ \text{s}^{-1} < 10^9$ (grey and black solid line, respectively). The Heath et al.[2] data, measured with a supersonic nozzle, correspond to a nucleation rate $J \sim 5.10^{16}\text{cm}^{-3}\ \text{s}^{-1}$ and lie at a given temperature at higher pressure compared to the Wölk and Strey[1] data points. The onset data points from this work lie between the two sets of literature data. Thus, we conclude that the nucleation rates during this work lie between $10^9 < J/\text{cm}^{-3}\ \text{s}^{-1} <\sim 5.10^{16}$. Estimating a nucleation rate of $J = (10^{13} \pm 10^2)\ \text{cm}^{-3}\ \text{s}^{-1}$ for our onset data points (Figure 3(b)) and comparing these to the predictions of classical nucleation theory (CNT)[7] reveals that the predictions of CNT and experiment coincide at $T = 240\,\text{K}$. A good agreement for water between CNT[7] and experiment at $T = 240\,\text{K}$ was already shown by Wölk and Strey[1] in literature. Thus we conclude that our estimated nucleation rates exhibit a satisfying accuracy. Furthermore, we learn that if no recompression is carried out it is possible to reach higher nucleation rates in our chamber than with our typical measurement technique[1,5] ($10^5 < J/\text{cm}^{-3}\ \text{s}^{-1} < 10^9$).

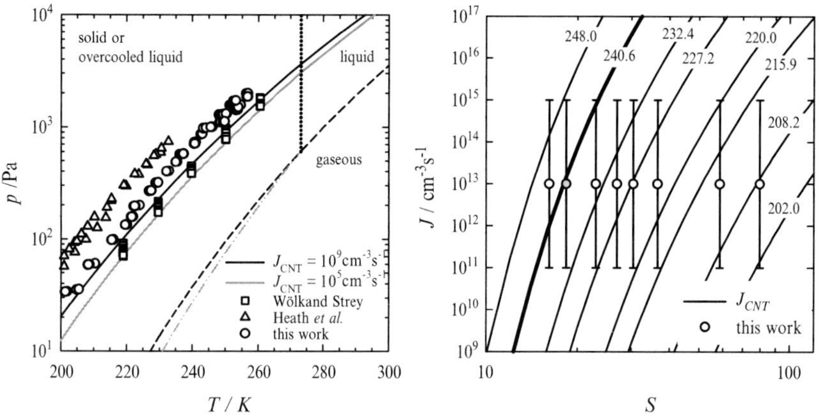

Figure 3 **(a)** A phase diagram for water containing the onset conditions determined in this work in comparison with Wölk and Strey[1], and Heath et al.[2] The gas–solid binodal[6] is drawn as a grey dashed dotted line, the extrapolated gas–liquid binodal[1] as a black dashed line, and the solid–liquid phase boundary as a dotted line.[6] The black and grey solid lines indicates the nucleation rate JCNT = 105 cm^{-3} s^{-1} and JCNT = 109 cm^{-3} s^{-1} according to classic nucleation theory (CNT)[7]; respectively. **(b)** The estimated nucleation rates *J* (circles) versus supersaturation *S*. The solid lines indicate isothermal nucleation rates JCNT according to classic nucleation theory. The T = 240 K measurement point and the corresponding CNT isotherm have been highlighted as a grey filled circle and a bolt line, respectively

References

1. Wölk, J. and Strey, R., *J. Phys. Chem. B*, **105**, 11683–11701 (2001).
2. Heath, C.H., Streletzky, K., Wyslouzil, B.E., Wölk, J., and Strey, R., *J. Chem. Phys.*, **117**, 6176–6185 (2002).
3. Wilson, C.R.T., *Philos. Trans. R. Soc. London, Ser. A*, **189**, 265 (1897).
4. Heist, R.H. and He, H., *J. Phys. Chem. Ref. Data*, **23**, 781–805 (1994).
5. Strey, R., Wagner, P.E., and Viisanen, Y., *J. Chem. Phys.*, **98**, 7748–7758 (1994).
6. Landold, H. and Börnstein, R., *Zahlenwerte und Funktionen aus Physik, Chemie, Astronomie, Geophysik und Technik, 6. Auflage, Band II/2a*, Berlin: Springer, pp. 32 and 224 (1960).
7. Becker, R. and Döring, W., *Ann. Phys.*, **24**, 719–752 (1935).

Part II
Nucleation: Binary, Homogeneous, and Heterogeneous

Homogeneous Binary Nucleation Theory and the Structure of Binary Nanodroplets

Gerald Wilemski

Abstract The structure of a critical nucleus or a larger post-critical droplet refers to the spatial distributions of the different chemical species within the droplet. Structure directly affects the free energy of formation of droplets. This has a strong effect on the nucleation rate as well as on the rates of droplet growth and evaporation. Structure also affects how aerosol droplets interact with their environment, influencing phenomena such as heterogeneous chemistry, trace gas adsorbtion and uptake, and the radiative properties of larger droplets. This paper briefly reviews recent advances in understanding the structure of critical nuclei in highly nonideal binary systems. It then focuses on recent theoretical work to predict the structure of larger binary nanodroplets using density functional theory (DFT) and lattice Monte Carlo (MC) simulations of model aqueous-organic mixtures.

Keywords Homogeneous nucleation, binary nucleation, nanodroplets, clusters

Introduction

Evidence for multicomponent homogeneous nucleation of particles in the atmosphere is considerable and increasing.[1–3] Under the right conditions, these particles will continue to grow, passing through the several nanometer-size range until they become Aitken nuclei. With further growth, these particles may become cloud condensation nuclei. While these processes are broadly understood, many of the details are still hazy, and will depend on the local thermodynamic conditions and chemical composition of the atmosphere, among other factors. It is well-known that temperature and the partial pressures and chemical identities of the participating molecular species have a strong influence on particle nucleation and growth rates. These factors also strongly influence the structure of critical nuclei and of the larger post-critical droplets, and this paper will summarize some recent developments in this

Department of Physics and Cloud and Aerosol Sciences Laboratory, University of Missouri-Rolla, Rolla Missouri USA 65409-0640

C.D. O'Dowd and P.E. Wagner (eds.), *Nucleation and Atmospheric Aerosols*, 267–277.

area. By structure, I mean the average spatial distributions of the various chemical species within the droplet. This notion is most useful for droplets containing many thousands down to several tens of molecules, and its utility rapidly diminishes with further decreases in cluster size. Hence, my considerations will not apply to the structure of small molecular clusters with fewer than, say, ~30 molecules. This review is not intended to be comprehensive, and it will concentrate mainly on my own studies on binary nanodroplets of 1–30 nm radius.

Small Critical Nuclei

For typical experimental conditions, binary critical nuclei generally contain 30–100 molecules. The role of structure in these cases is easiest to appreciate by considering the well-known effect of surface enrichment in alcohol–water systems. Alcohol molecules preferentially concentrate at the vapor–liquid interfaces of small droplets and bulk systems, lowering their interfacial tensions. In binary nucleation, this greatly reduces the work of forming the critical nucleus, thereby leading to the well-known mutual enhancement of nucleation.[4–7] Bulk thermodynamic models[8–10] of this phenomenon lead to unphysical results,[10] but progress has been made using phenomenological models for the surface composition.[11–13] In particular, the Laaksonen and Kulmala model[12] has been quite successful at describing structure in terms of compositional variables.[14,15] Models of this type provide only a simple characterization of the cluster structure. To obtain detailed spatial resolution, microscopic treatments based on statistical mechanics or molecular simulations are required.

Early microscopic treatments of surface enrichment in binary nucleation were made with density functional theory (DFT) for simple binary Lennard–Jones (LJ) clusters[16,17] and more complex models of amphiphiles.[18] More sophisticated DFT models of amphiphilic systems with LJ dimers and trimers have been used recently to demonstrate interesting lamellar structures in the critical nucleus, as well as the mutual enhancement of nucleation.[19,20]

The clearest demonstrations of surface enrichment are from computer simulations using molecular dynamics[21,22] (MD) or Monte Carlo[21,23] (MC) techniques, which produce computer-generated images of the clusters. Beyond this visualization capability, these techniques allow the full effect of cluster structure on the nucleation process to be studied quantitatively for quite realistic model systems. Recent innovations in MC sampling techniques by Chen et al.[23] and by McKenzie and Chen[24] now allow the construction of the free energy surface (FES) on which nucleation takes place for systems governed by complex and realistic intermolecular potentials. This capability permits detailed studies of nucleation pathways and mechanisms under conditions comparable to those studied experimentally. Recent studies[23, 24] of water/ethanol nucleation reproduced the mutual enhancement effect with results that were in excellent qualitative and reasonable quantitative accord with the experimental critical activities.[14]

Surface enrichment in the miscible water–alcohol systems is a clear manifestation of nonideal thermodynamic behavior.[25] Even more interesting structural effects arise when nonideality is extreme, i.e., in partially miscible and nearly immiscible systems. Using classical nucleation theory (CNT), Ray et al.[26] found that the FES of a partially miscible system could have double saddle points (DSP), i.e., two types of mixed critical nuclei may occur for the same vapor state. Although CNT is useful in studying how the DSP affects the kinetics of nucleation,[26,27] it does not reveal the structures of the critical nuclei. In a notable MC/MD study on small binary LJ clusters at low temperatures, Clarke et al.[21] identified numerous different cluster structures that could arise depending on the relative strengths of the different LJ energy parameters. Among the types found were well-mixed (WM) droplets, surface-enriched droplets, and nonspherical, phase-separated droplets. Although these insights into the energetics of cluster structure are valuable, they are less useful from a nucleation perspective since it is not clear when or if these structures may occur as critical nuclei. A step in this direction was provided by Talanquer and Oxtoby,[28] who used DFT in the square-gradient approximation to study binary nucleation of partially miscible LJ systems. In addition to finding DSP behavior, they also found nonspherical, phase-separated droplets, similar to those of Clarke et al.,[21] that appeared to afford an alternative nucleation pathway. Subsequent MC simulations by ten Wolde and Frenkel[29] and more sophisticated DFT calculations by Napari and Laaksonen[30] for binary LJ clusters showed that the phase-separated clusters were not critical nuclei but corresponded to local maxima on the FES, and hence were not important for nucleation. Note that the work of Talanquer and Oxtoby,[28] ten Wolde and Frenkel,[29] and Napari and Laaksonen[30] was conducted at a single, relatively high temperature, roughly twice as high as that used in the earlier study of Clarke et al.[21] Thus, while nonspherical, phase-separated LJ droplets may not serve as critical nuclei at temperatures near the triple point, this possibility remains an open question at lower temperatures.

Recent work by McKenzie and Chen[24] bears on this issue. One of the systems they studied, nonane–ethanol, exhibits quite unusual behavior at $T = 230$K, the low temperature used in the experiments of Viisanen et al.[31] For a small range of vapor conditions, the saddle region on the FES was broad and flat, rendering ambiguous the notion of a single critical nucleus with a well-defined composition. This feature persisted at 300K and, to a lesser extent, at 360K as well. They analyzed a subset of clusters at 230K with similar numbers (20 ± 2) of n-nonane and ethanol molecules. These compositions lie on the flat part of the saddle. They found that these clusters had a structure resembling that of the nonspherical, phase-separated LJ droplets studied by Clarke et al.[21] and Talanquer and Oxtoby.[28] They did not discuss the structure of the high temperature critical nuclei. Thus, in a system decidedly different from a binary LJ mixture, there seems to be strong evidence for the role of highly asymmetric critical nuclei at low temperatures. As I will discuss in the next section, the key to the stability of much larger phase-separated droplets also appears to be low temperature.

Large Nanodroplets

Recent experimental and theoretical efforts to understand the structure of aqueous-organic nanodroplets are largely motivated by the atmospheric importance of these systems.[32,33] Typically, the organic species are only sparingly soluble in water, and the bulk mixtures have miscibility gaps between coexisting water-rich and organic-rich phases whose compositions depend on the nature of the organic species. This phase behavior leads one to anticipate a nonuniform distribution of organic matter in water-rich aqueous-organic aerosol droplets, e.g., in the form of an aqueous core surrounded by a shell of organic material.[34] Indeed, there is experimental support for such core-shell (CS) structures from surface analyses of atmospheric aerosol particles[35] and even more evidence from small-angle neutron scattering measurements on nanodroplets containing the partially miscible species water and *n*-butanol.[36] Two recent theoretical studies explore the structures of aqueous-organic nanodroplets that can occur under various conditions. Both DFT[37] and lattice MC[38] (LMC) techniques were used to treat binary fluid models that emulated the characteristics of partially miscible aqueous-organic systems.

The DFT model of Li and Wilemski[37] is a binary mixture of hard spheres with attractive Yukawa forces. This approach is an extension to spherical droplets of Sullivan's theory[39] of interfaces in the binary van der Waals fluid. It generalizes the work of Oxtoby and Evans[40] for unary systems and is closely related to the DFT of Zeng and Oxtoby[16] for nucleation in binary fluids. In these theories, the grand potential is a functional of the species densities. The nature of the functional determines the complexity of the resulting Euler–Lagrange equations that must be solved to find the equilibrium droplet-density profiles. The earliest and simplest functionals[16,39] employ the random phase and local density approximations and produce integral Euler–Lagrange equations of the form

$$\mu_i = \mu_{ih}[\rho_1(\mathbf{r}),\rho_2(\mathbf{r})] + \sum_{j=1}^{2}\int d\mathbf{r}'\phi_{ij}\left(|\mathbf{r}-\mathbf{r}'|\right)\rho_j(\mathbf{r}'), \tag{1}$$

where μ_i is the chemical potential of the *i*th component in the system, μ_{ih} is the local chemical potential of species *i* in a hard sphere fluid mixture,[41] $\rho_i(\mathbf{r})$ is the average number density of species *i* at point **r**, and ϕ_{ij} is the attractive part of the pair potential between molecules of species *i* and species *j*. Although more sophisticated models of amphiphilic systems have been developed,[18–20] we adopted the simpler, pseudopotential approach of Sullivan[39] by retaining Eq. (1) and choosing ϕ_{ij} as the Yukawa potential

$$\phi_{ij}(r) = -\alpha_{ij}\lambda^3\exp(-\lambda r)/(4\pi\lambda r). \tag{2}$$

Here, α_{ij} controls the strength of the attractive intermolecular potential, and λ is an inverse range parameter that is assumed to be the same for all pair interactions. For this choice of ϕ_{ij}, acting with ∇^2 on the coupled integral Euler–Lagrange Eq. (1), produces two coupled differential equations that resemble nonlinear diffusion equations,

$$^2\mu_{ih} = \lambda^2(\mu_{ih}(\rho_1,\rho_2) - \mu_i - \sum_{j=1}^{2}\alpha_{ij}\rho_j). \tag{3}$$

The cross interaction term is assumed to obey the so-called Bertholet (geometric mean) mixing rule, $\alpha_{12} = \sqrt{\alpha_{11} + \alpha_{22}}$. This leaves five independent parameters (λ, α_{11}, α_{22}, and two hard sphere diameters) whose values are chosen to fit the density of pure water and the surface tensions and vapor pressures of both pure fluids at 250K.

The mean-field van der Waals equation of state for this pseudo water–pentanol mixture captures most of the features of real water–pentanol mixtures. As illustrated in the published phase diagram,[37] the model correctly predicts bulk liquid–liquid phase separation at small values of $x_{p\text{-Pentanol}}$, the *p*-pentanol mole fraction.[42] The model bulk surface tension is also realistic: there is a steep drop at low values of $x_{p\text{-Pentanol}}$ in the water-rich (L1) phase and a slow variation in the alcohol-rich (L2) phase.[37]

Two principal types of droplet structures were evident in the extensive DFT calculations. These are illustrated in Figure 1 for two moderately large nanodroplets (radius ~ 4–5 nm). In the WM droplet structure, Figure 1(a), the density profiles are fairly flat throughout most of the droplet. As the vapor–liquid interface is approached, the *p*-water density tends to decay more quickly leaving a *p*-pentanol-rich coating on the droplet surface. These structures resemble the bulk L2-liquid–vapor interface. Similarly the CS structure in Figure 1(b) resembles the bulk L1-vapor interface. Furthermore, the CS structure is consistent with that inferred from the recent SANS measurements of Wyslouzil et al.[36] the core is very rich in *p*-water, and the *p*-pentanol concentration remains small until the vapor–liquid interface is reached. In the interfacial region, the *p*-water density falls rapidly, while the *p*-pentanol density profile is a roughly Gaussian shaped adsorbed layer. As shown elsewhere,[37] the thickness and density of this adsorbed layer varies with the *p*-pentanol vapor concentration.

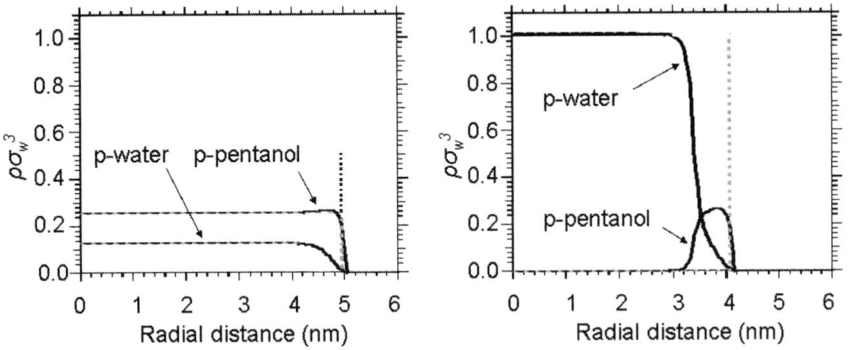

Figure 1 Density profiles of critical droplets at vapor state activities, $a_P = 1.002$, $a_W = 1.178$ ($a_i = p_i/p_i^0$, where p_i and p_i^0 are the actual and equilibrium vapor pressures of species *i*, respectively); the mol percentage of *p*-pentanol is 2.64% in the bulk gas phase. The droplet size is well described by the vertical dashed line, which is the Gibbs dividing surface of Buff.[43] (a) Left: a well-mixed structure. (b) Right: a core-shell structure. Note that both structures occur at the same vapor state

Despite the relative simplicity of our model, the interfacial density profiles are quite similar to those found in MC simulations of bulk butanol–water interfaces with realistic intermolecular potentials.[44] The similarity of the two sets of results supports the utility and efficacy of this semiempirical model in these types of investigations.

Figure 2 is a phase diagram for the different droplet structures illustrated in Figure 1 calculated using DFT.[37] The coordinates of the diagram are the vapor phase activities of the *p*-water and *p*-pentanol species, as defined in the Figure 1 caption. The theory predicts that there is a bistructural region where both types of droplet structures may occur simultaneously at the same vapor state. Binary CNT[45] also predicts a bi-structural region (shown in Ref. 37) whose lower boundary agrees fairly well with the DFT results, although there is a growing discrepancy as the water activity increases. In contrast there is a major discrepancy in the upper boundary: the coexistence region predicted by CNT is much broader than that predicted by DFT.[37] These discrepancies with CNT highlight the need for a non-classical theory, such as DFT. Note that the droplet structures shown in Figure 1 are calculated for the same vapor phase state in the bistructural region of Figure 2. The bistructural region of the phase diagram is consistent with the DSP behavior found in earlier studies of vapor phase nucleation using CNT[26,27] and DFT,[28] as discussed earlier.

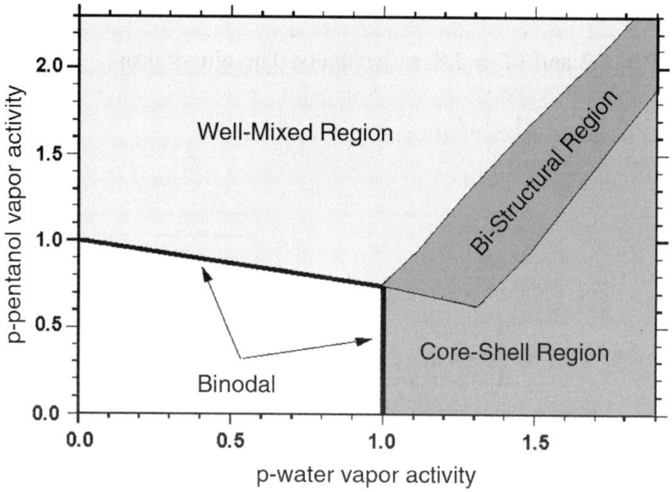

Figure 2 A structural phase diagram for aerosol nanodroplets. The heavy solid lines are the vapor binodal compositions. The shaded regions are supersaturated vapor states. In the bistructural region, both types of droplet structures can occur at the same vapor state, although with different free energies. In the upper region, only well-mixed droplets, as shown in Figure 1a, are found. In the lower region, only core-shell droplet structures, as shown in Figure 1b, are found. Droplet radii along the upper and right boundaries of this diagram are roughly 1–2 nm; the largest droplets studied, with 30 nm radii, are found close to the binodal lines

Phase diagrams of the sort shown in Figure 2 may play an important role in future models of the radiative effects of atmospheric aerosols, because the internal structure of a particle directly affects its radiative properties and strongly influences the size and number density of droplets that may form in clouds. The latter, so-called indirect effect, has a major influence on the earth's albedo and hydrological cycles, but is still poorly quantified.[33] Progress on this problem will depend in part on a better understanding of aerosol-particle structure. For example, the bistructural region in Figure 2 implies that aerosol structural transitions will exhibit hysteresis in the same way that salt-containing aerosol particles do when undergoing deliquescence and efflorescence.[46,47] Finally, it should be noted that the phase diagram of Figure 2 may be incomplete since the theory underlying it considered only spherically symmetric droplet structures.

Recently, Ning and Wilemski have begun to address the effect of temperature on nanodroplet structure with MC simulations of model aqueous-organic nanodroplets containing roughly 4,000 particles.[37] They used a simple fcc lattice model with nearest neighbor interactions. Their approach generalizes the earlier LMC work of Cordeiro and Pakula[48] who simulated unary droplets. For the results shown here, a 32,000 site fcc lattice was partially filled by 1,400 type 1 and 3,400 type 2 beads. The beads interact repulsively with vacant sites, i.e., E_1 and $E_2 > 0$, and the interaction between different bead types E_3 may be repulsive (>0) or attractive (<0), depending on the desired type of mixture behavior. The energy scale is set by taking the type 1 bead-vacancy interaction to be unity, $E_1 = 1$. Simulation temperatures are given in terms of kT scaled by E_1. For a system of pure type 1 beads, the estimates of Cordeiro and Pakula[48] for the critical temperature and the "triple point" temperature are $kT_c = 5.3$ and $kT_t = 2.8$, respectively. The binary droplet simulations were made for $kT < 2.75$, and are, thus, at temperatures subcooled with respect to the pure type 1 "liquid." The basic MC move is the interchange of a randomly chosen bead with a randomly selected nearest neighbor. The usual Metropolis acceptance rule is then applied. One MC step equals N exchange attempts, where N is the total number of beads on the lattice.

By varying E_2 and E_3, various types of droplet structures can be realized. These include WM, surface-enriched, CS, and a nonspherical, phase-separated structure, termed a Russian doll (RD). Snapshots of the latter two structures are shown in Figure 3. The RD structure is similar to that found by Clarke et al.[21] and Talanquer and Oxtoby[28] for much smaller droplets. It occurs at low temperatures for a highly unfavorable cross interaction, $E_3 = 0.8$. A remarkable feature of this structure is that it gradually converts into a CS structure as the temperature increases. This can be understood as a wetting transition that occurs at a relatively well-defined temperature (for the droplet size studied). Young's equation[49] provides a simple macroscopic interpretation of the transition in the idealized case of a nondeformable type 1 substrate coated by type 2 fluid. The RD structure corresponds to the imperfect wetting by type 2 fluid of the type 1 substrate with a nonzero contact angle θ. Young's equation for the balance of interfacial tensions then reads, $\gamma_{bv} = \gamma_{rb} + \gamma_{rv} \cos \theta$, from which it follows that there is a surface free energy penalty associated with spreading fluid 2 over substrate 1 ($\gamma_{bv} < \gamma_{rb} + \gamma_{rv}$). As T increases, the interfacial

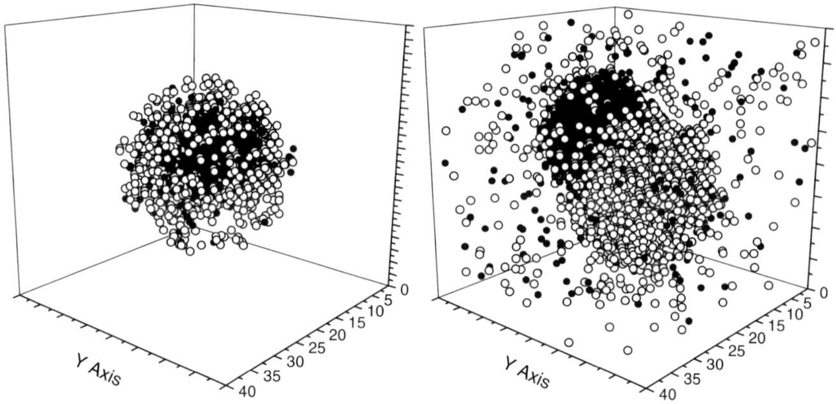

Figure 3 Snapshots of binary lattice droplet structures formed for $E_2 = 2/3$ and $E_3 = 0.8$: (**Left**) Core-shell structure at $kT = 2.6$ with vapor removed. (**Right**) Russian doll structure at $kT = 2.0$ showing vapor. Type 1 beads (water-like) are shown in black, while type 2 beads (organic-like) are open circles

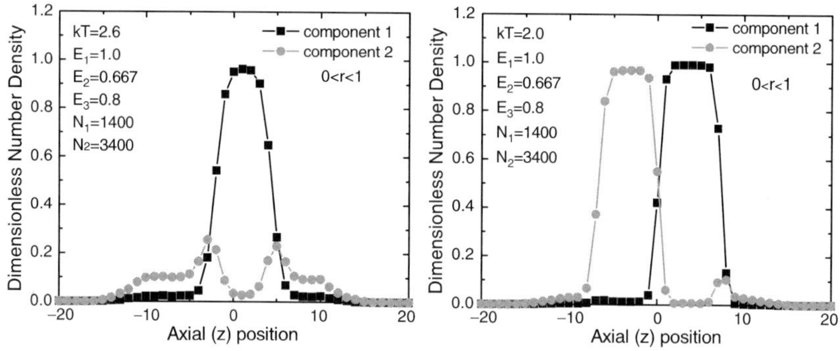

Figure 4 Axial density profiles for core-shell and Russian doll structures averaged over 10^7 Monte Carlo steps. The density is the fraction of sites within a small region occupied by beads of type 1 or type 2. Note that the CS droplet profile (**left**) at $kT = 2.6$ is slightly asymmetric because of dewetting. The RD droplet profile (**right**) at $kT = 2.0$ is highly asymmetric due to dewetting and phase separation

tensions decrease in such a way as to reach the condition $\theta = 0$. Antonow's rule for perfect wetting is then satisfied, $\gamma_{bv} = \gamma_{rb} + \gamma_{rv}$, the free energy penalty is absent, and fluid 2 spreads spontaneously over the substrate 1 resulting in the CS droplet structure.

The transition is reversible in the sense that changing the temperature in either direction causes a change in structure to occur without apparent hysteresis. The RD structures are stable over runs as long as 10^7 MC time steps, and statistically similar results were found from several different initial conditions. The radial density profiles for the CS droplet are similar to those of Figure 1b.

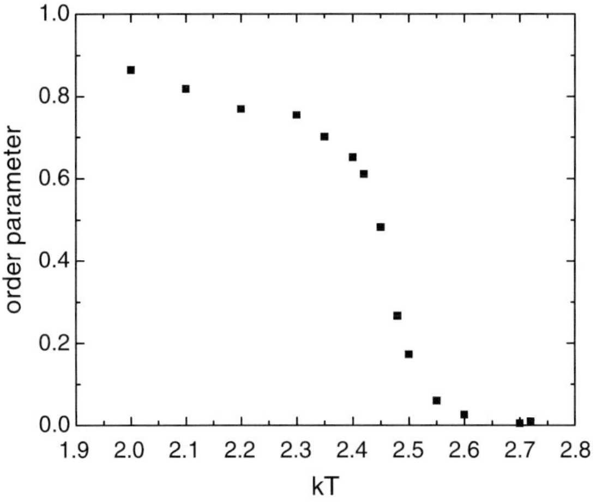

Figure 5 The microscopic order parameter η for the wetting transition is plotted versus T; η is a simple measure of the asymmetry in the type 2 bead number density profile with respect to $z = 0$

Density profiles along an internal z-axis for the CS and RD structures are illustrated in Figure 4. This z-axis is a line passing through the centers of "mass" of the type 1 and type 2 beads with $z = 0$ at the overall center of mass. The wetting/dewetting transition may be characterized by a microscopic order parameter η, defined here as the difference in the peak heights of the two maxima in the type 2 bead axial density profile. A plot of this order parameter versus T is shown in Figure 5. For CS structures η is quite small, but not zero due to fluctuations. Below the dewetting transition temperature, η increases monotonically as the RD structure grows increasingly anisotropic.

Indirect support for imperfect wetting of organics on water droplets comes from recent droplet growth measurements by Peeters et al.[50] in ternary water–nonane–methane mixtures at high pressure and low temperature (242 and 247K). In summary, the key to the stability of the RD droplets appears to be low temperature. Given the range of temperatures in the atmosphere and the ubiquity of aqueous-organic droplets, structural transitions of this sort could be a common occurrence there.

References

1. Spracklen, D.V., Carslaw, K.S., Kulmala, M., Kerminen, V.-M., Mann, G.W., and Sihto, S.-L., The contribution of boundary layer nucleation events to total particle concentrations on regional and global scales, *Atmos. Chem. Phys.*, **6**, 5631–5648 (2006).
2. O'Dowd, C.D., and Hoffman, T., Coastal new particle formation: a review of the current state-of-the-art, *Environ. Chem.*, **2**, 245–255 (2005).

3. Dunn, M.J., Jimenez, J.L., Baumgardner, D., Castro, T., McMurry, P.H., and Smith, J.N., Measurements of Mexico City nanoparticle size distributions: observations of new particle formation and growth, *Geophys. Res. Lett.*, **33**, L15802/1–4, (2006).
4. Schmitt, J.L., Whitten, J., Adams, G.W., and Zalabsky, R.A., Binary nucleation of ethanol and water, *J. Chem. Phys.*, **92**, 3693–3699 (1990).
5. Strey, R., Viisanen, Y., and Wagner, P.E., Measurement of the molecular content of binary nuclei. III. Use of the nucleation rate surfaces for the water-*n*-alcohol series, *J.Chem.Phys.*, **103**, 4333–4345 (1995).
6. Rodemann, T. and Peters, F., Experimental investigation of binary nucleation rates of water-*n*-propanol and water-*n*-butanol vapors by means of a pex-tube, *J.Chem.Phys.*, **105**, 5168–5176 (1996).
7. Wyslouzil, B.E., Heath, C.H., Cheung, J.L., and Wilemski, G., Binary condensation in a supersonic nozzle, *J.Chem.Phys.*, **113**, 7317–7329 (2000).
8. Flood, H., Formation of droplets in mixtures of the vapors of ethyl alcohol and water, *Z. Physik. Chem. A*, **170**, 286–294 (1934).
9. Neumann, K. and Döring, W., Droplet formation in supersaturated mixtures of the vapors of two completely miscible liquids, *Z. Physik. Chem. A*, **186**, 203–226 (1940).
10. Wilemski, G., Revised classical binary nucleation theory for aqueous alcohol and acetone vapors, *J. Phys. Chem.*, **91**, 2492–2498 (1987).
11. Flageollet-Daniel, C., Garnier, J.P., and Mirabel, P., Microscopic surface tension and binary nucleation, *J. Chem. Phys.*, **78**, 2600–2606 (1983).
12. Laaksonen, A. and Kulmala, M., An explicit cluster model for binary nuclei in water-alcohol systems, *J. Chem. Phys.*, **95**, 6745–6748 (1991).
13. Laaksonen, A., Nucleation of binary water-*n*-alcohol vapors, *J. Chem. Phys.*, **97**, 1983–1989 (1992).
14. Viisanen, Y., Strey, R., Laaksonen, A., and Kulmala, M., Measurement of the molecular content of binary nuclei. II. Use of the nucleation rate surface for water-ethanol, *J. Chem. Phys.*, **100**, 6062–6072, (1994).
15. Salonen, M., Napari, I., and Laaksonen, A., Modeling of critical cluster structure in surface active systems, in: *Nucleation and Atmospheric Aerosols, 2004*, edited by M. Kasahara and M. Kulmala, Kyoto: Kyoto University Press (2004), pp. 277–280.
16. Zeng, X.C. and Oxtoby, D.W., Binary homogeneous nucleation theory for the gas-liquid transition: A nonclassical approach, *J. Chem. Phys.*, **95**, 5940–5947 (1991).
17. Laaksonen, A. and Oxtoby, D.W., Gas-liquid nucleation of nonideal binary mixtures. I. A density functional study, *J. Chem. Phys.*, **102**, 5803–5810 (1995).
18. Talanquer, V. and Oxtoby, D.W., Nucleation in the presence of an amphiphile: a density functional approach, *J. Chem. Phys.*, **106**, 3673–3680 (1997).
19. Napari, I. and Laaksonen, A., Surfactant effects and an order-disorder transition in binary gas-liquid nucleation, *Phys. Rev. Lett.*, **84**, 2184–2187 (2000).
20. Napari, I., Laaksonen, A., and Strey, R., Density-functional studies of amphiphilic binary mixtures. II. Gas-liquid nucleation, *J. Chem. Phys.*, **113**, 4480–4487 (2000).
21. Clarke, A.S., Kapral, R., and Patey, G.N., Structure of two-component clusters, *J. Chem. Phys.*, **101**, 2432–2445 (1994).
22. Tarek, M. and Klein, M.L., Molecular dynamics study of two-component systems: The shape and surface structure of water/ethanol droplets, *J. Phys. Chem. A*, **101**, 8639–8642 (1997).
23. Chen, B., Siepmann, J.I., and Klein, M.L., Simulating the nucleation of water/ethanol and water/*n*-nonane mixtures: mutual enhancement and two-pathway mechanism, *J. Am. Chem. Soc.*, **125**, 3113–3118 (2003).
24. McKenzie, M.E. and Chen, B., Unravelling the peculiar nucleation mechanisms for non-ideal binary mixtures with atomistic simulations, *J. Phys. Chem. B*, **110**, 3511–3516 (2006).
25. Osborne, M.J., and Lacks, D.J., Surface segregation in liquid mixtures with strong interspecies attraction, *Phys. Rev. E*, **70**, 010501(R)/1–4, (2004).
26. Ray, A.K., Chalam, M., and Peters, L.K., Homogeneous nucleation of binary vapors partially miscible in liquid state, *J. Chem. Phys.*, **85**, 2161–2168 (1986).

27. Wyslouzil, B.E. and Chen, S., Binary nucleation kinetics. 6. Partially miscible systems, *J. Phys. Chem. B*, **105**, 11566–11573 (2001).
28. Talanquer, V. and Oxtoby, D.W., Critical clusters in binary mixtures: a density functional approach, *J. Chem. Phys.*, **104**, 1993–1999 (1996).
29. ten Wolde, P.R. and Frenkel, D., Numerical study of gas-liquid nucleation in partially miscible binary mixtures, *J. Chem. Phys.*, **109**, 9919–9927 (1998).
30. Napari, I. and Laaksonen, A., Gas-liquid nucleation in partially miscible systems: Free-energy surfaces and structures of nuclei from density functional calculations, *J. Chem. Phys.*, **111**, 5485–5490 (1999).
31. Viisanen, Y., Wagner, P.E., and Strey, R., Measurement of the molecular content of binary nuclei. IV. Use of the nucleation rate surfaces for the *n*-nonane-*n*-alcohol series, *J.Chem. Phys.*, **108**, 4257–4266 (1998).
32. Charlson, R.J., Seinfeld, J.H, Nenes, A., Kulmala M., Laaksonen, A., and Facchini, M.C., Atmospheric science. Reshaping the theory of cloud formation, *Science*, **292**, 2025–2026 (2001).
33. Schwartz, S.E., Aerosols, clouds, and climate change, in: *Nucleation and Atmospheric Aerosols, 2004*, edited by M. Kasahara and M. Kulmala, Kyoto: Kyoto University Press , pp. 323–338 (2004).
34. Ellison, G.B., Tuck, A.F., and Vaida, V., Atmospheric processing of organic aerosols, *J. Geophys. Res. Atmos.*, **104**, 11633–11641 (1999).
35. Tervahattu, H., Juhanoja, J., Vaida, V., Tuck, A.F., Niemi, J.V., Kupiainen, K., Kulmala, M., and Vehkamaki, H, Fatty acids on continental sulfate aerosol particles, *J. Geophys. Res., Atmos.*, **110**, D06207/1–9 (2005).
36. Wyslouzil, B.E., Wilemski, G., Strey, R., Heath, C.H., and Dieregsweiler, U., Experimental evidence for internal structure in aqueous-organic nanodroplets, *Phys. Chem. Chem. Phys.*, **8**, 54–57 (2006).
37. Li, J.S. and Wilemski, G., A structural phase diagram for model aqueous organic nanodroplets, *Phys. Chem. Chem. Phys.*, **8**, 1266–1270 (2006).
38. Ning, H.X. and Wilemski, G., Structure in binary nanodroplets, presented at the 2006 APS March Meeting, Baltimore, MD, 16 March 2006; *Bull. Am. Phys. Soc.*, **51**, 1227 (2006).
39. Sullivan, D.E., Interfacial density profiles of a binary van der Waals fluid, *J. Chem. Phys.*, **77**, 2632–2638 (1982).
40. Oxtoby, D.W. and Evans, R. Nonclassical nucleation theory for the gas-liquid transition, *J. Chem. Phys.*, **89**, 7521–7530 (1988).
41. Mansoori, G.A., Carnahan, N.F., Starling, K.E., and Leland, T.W., Equilibrium thermodynamic properties of mixture of hard spheres, *J. Chem. Phys.*, **54**, 1523–1525 (1971).
42. Gross, J. and Sadowski, G., Application of the perturbed-chain SAFT equation of state to associating systems, *Ind. Eng. Chem. Res.*, **41**, 5510–5515 (2002).
43. Buff, F.P., Some considerations of surface tension, *Z. Elektrochem.*, **56**, 311–313 (1952).
44. Chen, B., Siepmann, J.I., and Klein, M.L., Vapor-liquid interfacial properties of mutually saturated water/1-butanol solutions, *J. Am. Chem. Soc.*, **124**, 12232–12237 (2002).
45. Nishioka, K. and Kusaka, I., Thermodynamic formulas of liquid-phase nucleation from vapor in multi-component systems, *J. Chem. Phys.*, **96**, 5370–5376 (1992).
46. Randles, C.A., Russell, L.M., and Ramaswamy, V., Hygroscopic and optical properties of organic sea salt aerosol and consequences for climate forcing, *Geophys. Res. Lett.*, **31**, L16108/1–4, (2004).
47. Tang, I., Thermodynamic and optical properties of mixed-salt aerosols of atmospheric importance, *J. Geophys. Res.*, **102**, 1883–1893 (1997).
48. Cordeiro, R. M., and Pakula, T., Behavior of evaporating droplets at nonsoluble and soluble surfaces: Modeling with molecular resolution, *J. Phys. Chem. B*, **109**, 4152–4161 (2005).
49. Rowlinson, J.S. and Widom, B., *Molecular Theory of Capillarity*, Oxford: Oxford University Press , p. 9 (1982).
50. Peeters, P., Pieterse, G., Hrubý, J., and van Dongen, M.E.H., Multi-component droplet growth. I. Experiments with supersaturated *n*-nonane vapor and water vapor in methane, *Phys. Fluids*, **16**, 2567–2574 (2004).

Nucleation Versus Spinodal Decomposition in Confined Binary Solutions

Alexander S. Abyzov[1] and Jürn W.P. Schmelzer[2]

Abstract Basic features of spinodal decomposition, on one side, and nucleation, on the other side, and the transition between both mechanisms are analyzed within the framework of the thermodynamic cluster model based on the generalized Gibbs approach. Hereby the clusters, representing the density or composition variations in the system, may change with time both in size and intensive state parameters. The effect of changes of the state parameters in confined systems (depletion effect) on cluster evolution is analyzed.

Keywords Nucleation, spinodal decomposition, crystallization, multicomponent systems, scenario of first-order phase transitions

Introduction

Nucleation and spinodal decomposition are two major mechanisms how first-order phase transitions may proceed in a variety of systems. Which one of the mentioned mechanisms dominates in the decomposition process is commonly assumed to depend on the degree of instability of the initial state of a phase-separating system: nucleation and growth for metastable systems, and spinodal decomposition for thermodynamically unstable systems. In nucleation a nucleus of initially small size is supposed to be formed stochastically with state parameters widely similar to the properties of the newly evolving macroscopic phases. In contrast, spinodal decomposition is characterized by initially smooth changes of the state parameters of the system (composition, density, etc.) extended, in general, over large regions in space [1]. In the simplest formulation, in the initial stages of spinodal decomposition the change of density and composition is determined for a more or less fixed size of the new

[1]*National Science Center, Kharkov Institute of Physics and Technology, Academician st. 1, 61108 Kharkov, Ukraine*

[2]*Institut für Physik der Universität Rostock, Universitätsplatz, 18051 Rostock, Germany*

C.D. O'Dowd and P.E. Wagner (eds.), *Nucleation and Atmospheric Aerosols*, 278–281.

phase regions, while nucleation-growth models draw the attention to a change of the size of the clusters at given values of their intensive state parameters.

In the present contribution, basic features of spinodal decomposition, on one side, and nucleation, on the other side, and the transition between both mechanisms are analyzed within the framework of a thermodynamic cluster model based on the generalized Gibbs approach [2]. Hereby the clusters, representing the density or composition variations in the system, may change with time both in size and intensive state parameters. As a model system for the analysis, we consider phase separation in a binary regular solution.

Phase Separation in Finite Domains

We consider new phase formation in a binary solid or liquid solution in a finite size domain. The solution is considered as a regular one representing one of the simplest models of a system consisting of two kinds of interacting molecules. The domain, where the processes of nucleation and/or spinodal decomposition are assumed to proceed, is considered as a sphere of radius R_0. The limiting situation of an infinite system is thus reached for $R_0 \to \infty$, while finite-size effects take place for finite values of R_0.

Cluster formation in a binary solution results from a redistribution of molecules. Following Gibbs' model approach, we consider a cluster as a spatially homogeneous part of the domain volume with a composition different from the ambient phase. Both size and composition of the cluster may vary in a wide range. As the dividing surface, separating the cluster from the ambient phase in the thermodynamic description underlying the method of analysis, we always employ here the surface of tension. The effect of the finite size is taken into account only by the conservation laws for the numbers of particles of the different components in the cluster ($n_{1\alpha}$, $n_{2\alpha}$) and in the ambient phase ($n_{1\beta}$, $n_{2\beta}$): $n_{1\alpha} + n_{2\alpha} = \text{const}$, $n_{1\beta} + n_{2\beta} = \text{const}$. The molar fractions of the second component in the ambient phase (x_β) and the cluster (x_α) are defined as $x_\beta = n_{2\beta}/n_\beta$ and $x_\alpha = n_{2\alpha}/n_\alpha$, where $n_\beta = n_{1\beta} + n_{2\beta}$, $n_\alpha = n_{1\alpha} + n_{2\alpha}$. The initial state is either a metastable or unstable homogeneous state, characterized by $x_\alpha(0) = x_\beta(0) \equiv x$. The cluster is considered as a sphere of radius $R = (3n_\alpha\omega_\alpha/4\pi)^{1/3}$, where n_β is the volume per particle. The change of the Gibbs free energy, ΔG, connected with the formation of one cluster in the initially homogeneous ambient phase can be written in a commonly good approximation as

$$\Delta G = \sigma A + \sum_j n_{j\alpha}\left(\mu_{j\alpha} - \mu_{j\beta}\right) + \sum_j n_j\left(\mu_{j\beta} - \mu_{j0}\right), \qquad (1)$$

where σ is the interfacial tension, A is the surface area of the cluster, the chemical potential per particle in the cluster is denoted by $\mu_{j\alpha}$, in the ambient phase in the initial state by μ_{j0} and in the state once a cluster has been formed by $\mu_{j\beta}$, correspondingly. The surface tension is given, according to Becker ([3], see also [2]) by $\sigma = \tilde{\sigma}\,(x_\alpha - x_\beta)^2$.

The most probable trajectory of evolution is determined by the macroscopic growth equations, which can be written in the form

$$\frac{dn_{1\alpha}}{dt} = -D_1(1-x_\beta)\Theta(n_{1\alpha},n_{2\alpha})\frac{d\Delta G}{dn_{1\alpha}},$$
$$\frac{dn_{2\alpha}}{dt} = -D_2 x_\beta\Theta(n_{1\alpha},n_{2\alpha})\frac{d\Delta G}{dn_{2\alpha}}, \qquad (2)$$

where $\Theta(n_{1\alpha}, n_{2\alpha}) = \Theta_0 n_\alpha^{\kappa}$, D_1 and D_2 are the partial diffusion coefficients of the different components in the ambient phase, the parameter κ has the value $\kappa = 2/3$ for kinetic-limited growth, and $\kappa = 1/3$ for diffusion-limited growth, and Θ_0 is a parameter depending only on temperature (we suppose $T = 0.7T_c$, T_c is the critical temperature).

Dependence of ΔG along the evolution path is shown on Figure 1a for $x = 0.3$. For large domains, $R>6.83R_\sigma$, spinodal decomposition is possible, while for $R < 6.83R_\sigma$ nucleation barrier arises and the system transforms to a metastable one. For $R = 4.87R_\sigma$, system becomes bistable (homogeneous and heterogeneous states have identical energies), and in the interval $4.7R_\sigma < R < 4.87R_\sigma$ system is metastable but the final state has a higher Gibbs free energy than the initial one (here, $R_\sigma = (3n_\sigma\omega_\alpha/4\pi)^{1/3}$, $n_\sigma = (2\tilde{\sigma}/k_BT)(4\pi/3)^{1/3}\omega_\alpha^{2/3}$, k_B is the Boltzmann constant). And only for $R<4.7\ R_\sigma$ the system is stable. The metastable state in a wide sense exists for $4.7\ R_\sigma < R < 6.83\ R_\sigma$, however the possibility for phase transition of an initial state exists only within the interval $6.83\ R_\sigma > R > 4.87\ R_\sigma$. Dependence of the reduced critical radius R_c/R_σ on the initial solute concentration, x, for different values of the domain size R_0 is illustrated on Figure 1b. In the unstable region, two critical radii exist: the smaller one is determined by the balance between volume energy gain and the loss during surface formation (as for infinite domain), and larger one is determined by the effect of changes of the state parameters (depletion effect). Dependence of the minimum value of the work of critical cluster formation $\Delta G_c/n_\sigma k_BT$ on the initial solute concentration, x, for different values of the domain size, R_0, is shown on Figure 1c. In the region $x<x_{sp}$ with domain size reduction ΔG_c increases insignificantly, while for $x>x_{sp}$ $\Delta G_c = 0$ for $R_0 \rightarrow \infty$ and nonzero values of ΔG_c arise only for finite values of R_0 (here $x_b \approx 0.086$ is binodal and $x_{sp} \approx 0.226$ is spinodal). On Figure 1d the Cahn plot, i.e., growth increment $\gamma(k)/k^2$ versus k^2, for $x = 0.45$ and various domain sizes R_0 is shown (full curves, $k \sim 1/R$). Note, that this plot is different from the linear classic one [1], and is very similar to the experimental data for spinodal decomposition in the glass SiO_2-12.5 Na_2O [4] (circles in Figure 1d).

Discussion

Basic features of spinodal decomposition, on one side, and nucleation, on the other side, and the transition between both mechanisms are analyzed within the framework of the uniform model based on the generalized Gibbs approach [2]. As shown,

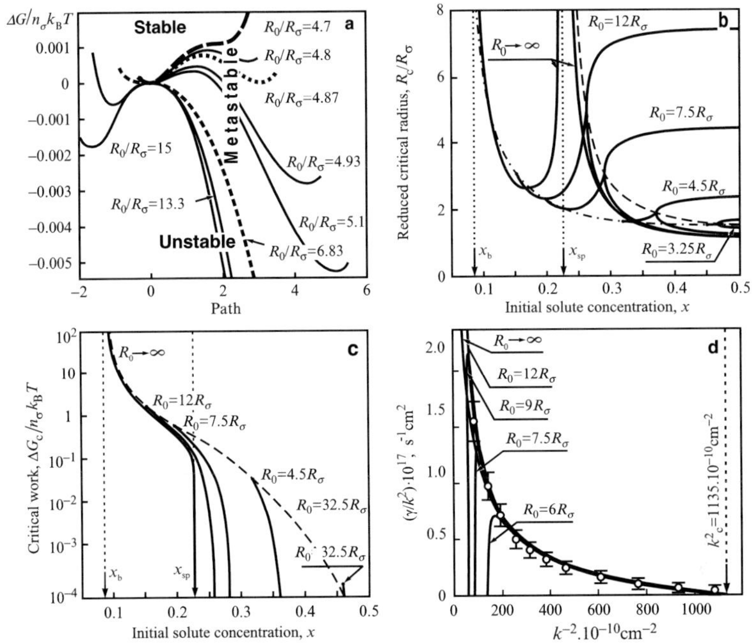

Figure 1 (a) Gibbs free energy along the path for $x = 0.3$, and for different values of the domain size, R_0; (b) dependence of the reduced critical radius, R_c / R_σ, on the initial solute concentration, x, for different values of the domain size; (c) dependence of the minimum value of the work of critical cluster formation $\Delta G_c/n_\sigma k_B T$ on the initial solute concentration, x, for different values of the domain size; (d) γ/k^2 via k^2: solid curves – calculation for $x = 0.45$, circles – experimental data [4]

there is no qualitative difference between nucleation and spinodal decomposition with respect to the basic mechanism of system evolution. Nucleation processes, starting from thermodynamically metastable initial states, proceed qualitatively widely similar as compared with processes of phase formation governed by spinodal decomposition. This similarity is especially noticeable, if we consider unstable system in the finite domain: for large domains spinodal decomposition is possible, while with the domain size diminishing the nucleation barrier arises and the system transforms to a metastable one.

Acknowledgments The authors would like to express their gratitude to the Deutsche Forschungs gemeinschaft (DFG) for financial support.

References

1. Cahn, J.W. and Hilliard, J.E., *J. Chem. Phys.*, **28**, 258 (1958); **31**, 688 (1959).
2. Schmelzer, J.W.P., Abyzov, A.S., and Möller, J., *J. Chem. Phys.*, **121**, 6900 (2004).
3. Becker, R., *Ann. Phys.*, **32**, 128 (1938).
4. Andreev, N.S., Boiko, G.G., and Bokov, N.A., *J. Non-Cryst. Solids*, **5**, 41 (1970).

Stabilization of H_2SO_4–H_2O Clusters by Organic Acids

A.B. Nadykto[1] and F. Yu[1]

Abstract While atmospheric aerosols are known to contain a significant fraction of organic substances, the influence of organics on the formation of H_2SO_4–H_2O clusters and subsequent nucleation in the atmosphere is poorly understood. In the present study, bonding among formic, acetic and benzoic acids, sulfuric acid, ammonia and water is studied using Density Functional Theory (DFT) at PW91PW91/6-311 + + G(3df,3pd) level. The stabilizing effect of formic, acetic, and benzoic acids is found to be close that of ammonia that indicates that the common organic substances may efficiently stabilize small H_2SO_4–H_2O clusters and their involvement, alongside with or without ammonia, in the atmospheric nucleation should be studied further

Keywords Organics enhanced nucleation, binary clusters, ternary clusters, nucleation precursors

Introduction

New particle formation frequently observed in various locations in the atmosphere is an important source of atmospheric aerosols that are responsible for climate change and human health impacts. Although nucleation phenomena have been intensively studied in the past, there are still major uncertainties concerning nucleation mechanisms and species involved in the atmospheric nucleation. Nucleation in the Earth's atmosphere is essentially multicomponent. Nucleation rates are very sensitive to the concentration of H_2SO_4 vapor, which is likely a principal nucleation agent. However; binary homogeneous nucleation of H_2SO_4 and H_2O cannot explain some of the nucleation events observed in the lower atmosphere. Ternary homogeneous nucleation (THN) involving H_2SO_4 and H_2O as the key nucleation agents and NH_3 as a principal stabilizer of H_2SO_4–H_2O clusters has been considered as an

[1]*Atmospheric Sciences Research Center, State University of New York at Albany, 251 Fuller Rd., Albany 12203, NY, USA*

C.D. O'Dowd and P.E. Wagner (eds.), *Nucleation and Atmospheric Aerosols*, 282–286.

alternative nucleation mechanism in the lower atmosphere since 1990s. While the original THN [1] predicts that NH_3 at ppt level can increase the H_2SO_4–H_2O nucleation rate by up to ~30 orders of magnitude, the laboratory experiments [2–4] indicate that the presence of NH_3 at ppb—ppm levels enhances the H_2SO_4–H_2O nucleation by up to ~2 orders of magnitude only [5]. Other species that may be involved in atmospheric nucleation are ions, iodine-containing vapors, and organics.

In this paper, the thermodynamic stability of hydrogen-bonded complexes of formic, acetic, and benzoic acids with free and hydrated sulfuric acid has been studied. This study has been carried out using the quantum theory at DFT PW91PW91/6-311 + + G(3df,3pd) level.

Results and Discussion

Tables 1 and 2 present changes in the Gibbs free energy associated with the cluster formation. Calculations have been performed at temperature of 298.15K and pressure of 101.3 kPa.

As seen from Table 1, hydrates of the sulfuric acid are stable thermodynamically and hydration free energies obtained in the present study are consistent with the experiments [6]. Another important detail is that the presence of ammonia is unlikely to enhance the affinity of water to H_2SO_4–H_2O clusters. As seen from Table 2, this quantity depends on the ammonia content only weakly. The affinity of ammonia to the monomer of the sulfuric acid obtained in the present study (−7.77 kcal mole^{-1}) is in agreement with experimental data [7] (~ −8.5 kcal mole^{-1}) and its value is ~3.2 kcal mole^{-1} more negative than that given in the B3LYP study [8]. The affinity of ammonia to binary sulfuric acid–water clusters gradually decreases as the water content grows. This finding is consistent with observation showing that the effect of ammonia decreases as the saturation ratio for water and, consequently, water content in the cluster, increases [5]. Gibbs free energy associated with addition of sulfuric acid to $(H_2SO_4)_2(H_2O)_nNH_3$ clusters (−11.5:−13.8 kcal mole^{-1}) is much larger than sulfuric acid dimerization (−5.58 kcal mole^{-1}) and $(H_2SO_4) + NH_3 = (H_2SO_4)(NH_3)$ (−7.77 kcal mole^{-1}). The stabilizing effect of ammonia on the formation of small H_2SO_4–H_2O clusters is likely to increase with the number of sulfuric acid molecules in the cluster. Thermochemistry of H_2SO_4 and NH_3 in small H_2SO_4-H_2O-NH_3 clusters is either virtually independent of or depends weakly on the water content. This suggests that the stabilizing effect of ammonia at initial steps of the cluster growth is associated mainly with the sulfuric acid.

As seen from Table 2, the interaction of formic acid, acetic and benzoic acids with sulfuric acid and water leads to the formation of stable hydrogen-bonded complexes. The stability of such complexes is close to the stability of complexes of sulfuric acid with ammonia. In addition to the formation of strong hydrogen-bonded complexes with sulfuric acid and water, formic, acetic, and benzoic acids form thermodynamically stable complexes with ammonia.

Table 1 Enthalpies, entropies, and Gibbs free energy changes associated with reactions among clusters composed of atmospheric precursors (sulfuric acid, water, and ammonia) calculated at $T = 298.15$K and $P = 101.3$kPa. Superscripts (a), (b), (c), (d), (e), and (f) refer to studies [9], [10], [6], [8], [7], and [11], respectively. Superscript* refers to experimental data

Reaction	ΔH	ΔS	ΔG
$H_2SO_4 + H_2O \Leftrightarrow (H_2SO_4)(H_2O)$	−11.76	−31.80	−2.28 (−2.5)[a] (−0.6)[b] (−3.6 ± 1.0)[c*]
$(H_2SO_4)(H_2O) + H_2O \Leftrightarrow (H_2SO_4)(H_2O)_2$	−12.57	−32.08	−3.00 (−1.8)[a] (−0.1)[b] (−2.3 ± 0.3)[c*]
$(H_2SO_4)(H_2O)_2 + H_2O \Leftrightarrow (H_2SO_4)(H_2O)_3$	−11.34	−31.71	−1.89 (−0.5)[b]
$(H_2SO_4)(NH_3) + H_2O \Leftrightarrow (H_2SO_4)(NH_3)(H_2O)$	−10.96 (−8.87)[d]	−32.03 (−30.70)[d]	−1.41 (0.62)[d]
$(H_2SO_4)(NH_3)(H_2O) + H_2O \Leftrightarrow (H_2SO_4)(NH_3)(H_2O)_2$	−11.92 (−9.0)[d]	−32.34(−30.12)[d]	−2.28 (−0.02)[d]
$(H_2SO_4)(NH_3)(H_2O)_2 + H_2O \Leftrightarrow (H_2SO_4)(NH_3)(H_2O)_3$	−11.51(−9.45)[d]	−33.51(−32.47)[d]	−1.52(0.23)[d]
$(H_2SO_4)_2(NH_3) + H_2O \Leftrightarrow (H_2SO_4)_2(NH_3)(H_2O)$	−11.68	−31.32	−2.31
$(H_2SO_4)_2(NH_3)(H_2O) + H_2O \Leftrightarrow (H_2SO_4)_2(NH_3)(H_2O)_2$	−11.17	−33.40	−1.21
$(H_2SO_4)_2(NH_3)(H_2O)_2 + H_2O \Leftrightarrow (H_2SO_4)_2(NH_3)(H_2O)_3$	−12.18	−34.00	−2.04
$H_2SO_4 + NH_3 \Leftrightarrow (H_2SO_4)(NH_3)$	−16.72 (−13.76)[d]	−30.01 (−30.91)[d]	−7.77 (−4.54)[d](−8.5)[e*]
$(H_2SO_4)(H_2O) + NH_3 \Leftrightarrow (H_2SO_4)(H_2O)(NH_3)$	−15.91 (−12.91)[d]	−30.23 (−31.09)[d]	−6.90 (−3.64)[d]
$(H_2SO_4)(H_2O)_2 + NH_3 \Leftrightarrow (H_2SO_4)(H_2O)_2(NH_3)$	−15.27	−30.49	−6.18
$(H_2SO_4)(H_2O)_3 + NH_3 \Leftrightarrow (H_2SO_4)(H_2O)_3(NH_3)$	−15.44	−32.30	−5.81
$H_2SO_4 + H_2SO_4 \Leftrightarrow (H_2SO_4)_2$	−16.16 (−13.2)[f]	−35.46(−35.6)[f]	−5.59 (−3.1)[a](−2.5)[f]
$(H_2SO_4)(NH_3) + (H_2SO_4) \Leftrightarrow (H_2SO_4)_2(NH_3)$	−25.11	−45.14	−11.65
$(H_2SO_4)(NH_3)(H_2O) + (H_2SO_4) \Leftrightarrow (H_2SO_4)_2(NH_3)(H_2O)$	−25.83	−44.42	−12.59
$(H_2SO_4)(NH_3)(H_2O)_2 + (H_2SO_4) \Leftrightarrow (H_2SO_4)_2(NH_3)(H_2O)_2$	−25.08	−45.49	−11.52
$(H_2SO_4)(NH_3)(H_2O)_3 + (H_2SO_4) \Leftrightarrow (H_2SO_4)_2(NH_3)(H_2O)_3$	−25.75	−45.98	−12.04
$(H_2SO_4)_2 + NH_3 \Leftrightarrow (H_2SO_4)_2(NH_3)$	−25.67	−39.68	−13.83
$NH_3 + H_2O \Leftrightarrow (NH_3)(H_2O)$	−5.81	−19.87	0.11

This indicates that formic, acetic, and benzoic acids may efficiently stabilize small sulfuric acid–water complexes and they can interact actively with ammonia. This suggests that the involvement of these acids, with or without ammonia, in

Table 2 Enthalpy, entropy, and Gibbs free energy changes associated with reactions among clusters composed of atmospheric precursors (sulfuric acid, water, and ammonia) and formic, acetic, and benzoic acids calculated at temperature of 298.15K and pressure of 101.3 kPa

Reaction	ΔH (kcal mole^{-1})	ΔS (cal mole^{-1} K^{-1})	ΔG (kcal mole^{-1})
$CH_2O_2 + CH_2O_2 \Leftrightarrow (CH_2O_2)_2$	−17.02	−37.99	−5.69
$C_2H_4O_2 + C_2H_4O_2 \Leftrightarrow (C_2H_4O_2)_2$	−17.15	−36.54	−6.26
$C_2H_4O_2 + CH_2O_2 \Leftrightarrow (C_2H_4O_2)(CH_2O_2)$	−17.33	−37.10	−6.27
$H_2SO_4 + CH_2O_2 \Leftrightarrow (H_2SO_4)(CH_2O_2)$	−17.67	−37.67	−6.44
$H_2SO_4 + C_2H_4O_2 \Leftrightarrow (H_2SO_4)(C_2H_4O_2)$	−18.71	−37.71	−7.46
$CH_2O_2 + H_2O \Leftrightarrow (CH_2O_2)(H_2O)$	−10.03	−32.12	−0.45
$C_2H_4O_2 + H_2O \Leftrightarrow (C_2H_4O_2)(H_2O)$	−10.08	−32.06	−0.52
$(CH_2O_2)(H_2O) + H_2O \Leftrightarrow (CH_2O_2)(H_2O)_2$	−11.32	−31.96	−1.79
$(C_2H_4O_2)(H_2O) + H_2O \Leftrightarrow (C_2H_4O_2)(H_2O)_2$	−10.74	−31.65	−1.31
$(H_2SO_4)(C_2H_4O_2) + H_2O \Leftrightarrow (H_2SO_4)(C_2H_4O_2)(H_2O)$	−11.43	−30.10	−2.45
$(H_2SO_4)(CH_2O_2) + H_2O \Leftrightarrow (H_2SO_4)(CH_2O_2)(H_2O)$	−11.68	−31.21	−2.37
$(H_2SO_4)(H_2O) + CH_2O_2 \Leftrightarrow (H_2SO_4)(H_2O)(CH_2O_2)$	−17.58	−11.06	−6.53
$(H_2SO_4)(H_2O) + C_2H_4O_2 \Leftrightarrow (H_2SO_4)(H_2O)(C_2H_4O_2)$	−18.37	−36.01	−7.64
$H_2SO_4 + (CH_2O_2)(H_2O) \Leftrightarrow (CH_2O_2)(H_2O)(H_2SO_4)$	−19.32	−36.76	−8.36
$H_2SO_4 + (C_2H_4O_2)(H_2O) \Leftrightarrow (C_2H_4O_2)(H_2O)(H_2SO_4)$	−20.06	−35.75	−9.40
$NH_3 + CH_2O_2 \Leftrightarrow (NH_3)(CH_2O_2)$	−11.63	−29.53	−2.82
$NH_3 + C_2H_4O_2 \Leftrightarrow (NH_3)(C_2H_4O_2)$	−10.78	−28.29	−2.35

clustering and subsequent nucleation of sulfuric acid and water should be studied in further detail.

Acknowledgments NSF funded this work under grant ATM0618124 and NOAA under grant NA05OAR4310103.

References

1. Napari, I., Noppel, M., Vehkamäki, H., and Kulmala, M., *J. Geophys. Res.*, **107**(D19), 4381–4387, doi:10.1029/2002JD002132 (2002).
2. Ball, S.M., Hanson, D.R., Eisele, F.L., and McMurry, P.H., *J. Geophys. Res.*, **104**(D19), 23709–23721, 10.1029/1999JD900411 (1999).
3. Kim, T.O., Ishida, T., Adachi, M., Okuyama, K., and Seinfeld, J.H., *Aerosol Sci. Techn.*, **29**, 112–119 (1998).
4. Christensen, P.S., Wedel, S., and Livbjerg, H., *Chem. Eng. Sci.*, **49**, 4605–4617 (1999).

5. Yu, F., *J. Geophys. Res.*, 111, D01204, doi:10.1029/2005JD005968 (2006).
6. Hanson, D.R. and Eisele, F.L., *J. Phys. Chem. A*, **104**, 1715–1723 (2000).
7. Hanson, D.R. and Eisele, F.L., *J.Geophys.Res.*, **107**(12), 4158 10.1029/2001JD001100 (2002).
8. Ianni, J.C. and Bandy, A.R., *J. Phys. Chem. A*, **103**, 2801–2815 (1999).
9. Ding, C.-G., Laasonen, K., and Laaksonen, A., *J. Phys. Chem. A*, **107**(41) 8648–8658 (2003).
10. Bandy, A.R. and Ianni, J.C., *J. Phys. Chem. A*, **102**(32) , 6533–6546 (1998).
11. Ianni, J.C. and Bandy, A.R., *J. Mol. Structure (Theochem)*, **497**, 19–30 (2000).

Critical Cluster Content in High-pressure Binary Nucleation: Compensation Pressure Effect

Vitaly I. Kalikmanov[1] and Dzmitry G. Labetski[2]

Abstract Nucleation experiments in binary (*a-b*) mixtures, when component *a* is supersaturated and *b* (carrier gas) is undersaturated, reveal that for some mixtures at high pressures the *a*-content of the critical cluster dramatically decreases with the pressure contrary to the expectations based on the Classical Nucleation Theory. We show that this phenomenon is a manifestation of the dominant role of the unlike interactions at high pressures resulting in the negative partial molar volume of component *a* in the vapor phase beyond the compensation pressure. The analysis is based on the Pressure Nucleation Theorem (PNT) for multicomponent systems which is invariant to a nucleation model.

Keywords Binary nucleation, Nucleation theorem, cluster composition, compensation pressure effect

Introduction

Vapor–liquid nucleation in binary (*a-b*) mixtures at high pressures attracts considerable experimental attention during recent years.[1–2] An important fundamental issue here is the role of real gas effects, and in particular, the interactions between molecules of different species: for some mixtures these effects become very pronounced at high pressures thereby substantially influencing nucleation behavior. From the point of view of applications understanding high pressure binary (or, more generally, multicomponent) nucleation is important for chemical plants and natural gas processing systems where nucleation takes place in the presence of high-pressure carrier gas.

[1]*Twister Supersonic Gas Solutions, Einsteinlaan 10, 2289 CC, Rijswijk, The Netherlands, e-mail: Vitaly.Kalikmanov@twisterbv.com*

[2]*Department of Applied Physics, Eindhoven University of Technology, P.O.Box 513, 5600 MB, Eindhoven, The Netherlands, e-mail: D.G.Labetski@tue.nl*

C.D. O'Dowd and P.E. Wagner (eds.), *Nucleation and Atmospheric Aerosols*, 287–292.

The aim of the present study is to analyze the effect of total pressure on the critical cluster content using the available experimental data on nucleation rate without making a priori assumptions about the excess numbers Δn_a^*, Δn_b^* of molecules of both components in the critical cluster. We show that the anomalous behavior of a critical cluster discovered in experiments[2] can be attributed to a phenomenon, termed the *compensation pressure effect*, which is a manifestation of the dominant role played in some mixtures by unlike interactions at high pressures.

Pressure Nucleation Theorem for Multicomponent Mixtures

The Nucleation Theorem for a multicomponent mixture[3]

$$\left.\frac{\partial W^*}{\partial \mu_i^v}\right|_{T,\{\mu_j^v\},j\neq i} = -\Delta n_i^* \tag{1}$$

provides a general, model-independent tool for analysis of nucleation phenomena. Here W^* is the work of the critical cluster formation, μ_j^v is the chemical potential of component j in the vapor, T is the temperature. However, in experiments the directly measurable quantity is the nucleation rate $J = J_0 \exp(-\beta W^*)$ and not W^* (here $\beta = 1/k_B T$, k_B is the Boltzmann constant). Using (1) and the Maxwell relations, we find:

$$\left.\frac{\partial \ln(J/J_0)}{\partial \ln p^v}\right|_{\{y_k\},T} = \frac{p^v}{k_B T}\sum_i \upsilon_i^v \Delta n_i^*, \tag{2}$$

where υ_i^v is the partial molar volume of component i in the vapor phase

$$\upsilon_i^v = \frac{1}{\rho^v}\left[1+\sum_{j\neq i} y_j \frac{\partial \ln \rho^v}{\partial y_j}\right], \tag{3}$$

ρ^v is the bulk vapor density. Eqs. (2) and (3) represent the general form of the *Pressure Nucleation Theorem* (PNT) for a multicomponent mixture. Note, that both NT and PNT result from Hill's treatment of thermodynamics of small systems.[4]

Consider now a *binary a-b* mixture in which nucleation occurs inside the coexistence region. We characterize the nonequilibrium state of the mixture by the *metastability parameters* $S_i = y_i\, p^v/y_{i,eq}(p^v,T)\, p^v = y_i\,/y_{i,eq}$, $i = a,b$, where y_i and $y_{i,eq}(p^v,T)$ are, respectively, the actual and the equilibrium vapor molar fraction of component i at the same pressure p^v and temperature T. If $S_a > 1$, then the normalization condition $\Sigma_i y_i = \Sigma_i y_{i,eq} = 1$ implies $S_b > 1$. We thus call a the *supersaturated component*, while b is the *undersaturated* one. Using the single-component isomorphism for binary nucleation[5] we find from (1)

$$\Delta n_a^* = \left(\frac{\partial \ln J}{\partial \ln S_a}\right)_{p^v,T} - 1 \tag{4}$$

while the prefactor[6] $J_0 \sim S_a y_{a,eq}^2 (p^v)^2$. Then PNT takes the form

$$\frac{\partial \ln J}{\partial \ln p^v} = \frac{p^v}{k_B T}\left(\upsilon_a^v \Delta n_a^* + \upsilon_b^v \Delta n_b^*\right) + h + 2, h\left(p^v, T\right) \equiv \frac{\partial \ln y_{a,eq}}{\partial \ln p^v}\Big|_T, \tag{5}$$

Equations (4) and (5) do not refer to a particular nucleation model, they fully determine the critical cluster given the measured $J(y_a, p^v, T)$ and an equation of state (EoS) for a mixture.

Compensation Pressure Effect

Consider the partial molar volume of component a of a binary mixture. From Eq. (3):

$$\upsilon_a^v = \frac{1}{\rho^v}\left(1 - y_b \frac{\partial \ln \rho^v}{\partial y_a}\right). \tag{6}$$

One can identify two competing factors when an a-molecule is introduced into the binary vapor at a fixed p^v. The first tendency is the increase of V^v to preserve p^v. The opposite tendency, manifested by the second term in Eq. (6), is the reduction of V^v: at a sufficiently high p^v, the separation between b molecules can become of the order of the range of unlike (a-b) interactions, so that a certain number of b molecules move in the direction of the a molecule thereby decreasing the volume occupied by them. At a certain pressure p_{comp}, satisfying

$$\left(\frac{\partial \ln \rho^v}{\partial y_a}\right)\Big|_{p_{comp},T} = \frac{1}{y_b}, \tag{7}$$

the two trends compensate each other resulting in $v_a^v = 0$. At $p^v > p_{comp}$: v_a^v becomes negative implying that the "squeezing tendency" prevails and a number of b-molecules find themselves attached to an a-molecule. Since component a is supersaturated, the a-molecules tend to form a liquid-like cluster, "entraining" the attached b-molecules into it. The presence of the more volatile component decreases the specific surface free energy of a cluster. These features imply that the a-content of the cluster at sufficiently high p^v becomes small – in accordance with the experimental observations. We stress that this *compensation pressure effect* is mainly due to unlike interactions which manifest themselves strongly at $p^v > p_{comp}$. The simplest way to estimate p_{comp} (T) is the virial expansion[7]

$$\rho^v = \frac{p^v}{k_B T}\left(1 - b_2\right), b_2 \equiv \frac{B_2 p^v}{k_B T}, \tag{8}$$

which is valid when $|b_2| << 1$. Here $B_2 = \Sigma_i \Sigma_j y_i y_j B_{2,ij}$ is the second virial coefficient of the mixture; $B_{2,aa}(T)$, $B_{2,bb}(T)$ are the second virial coefficients of the pure substances a and b respectively, and the cross term $B_{2,ab}$ is constructed according to the combination rules.[7] We obtain from Eqs. (7) and (8), linearizing in b_2:

$$p_{comp}(T) = \frac{1}{2}\frac{k_B T}{\left[B_{2,bb}(T) - B_{2,ab}(T)\right]} \tag{9}$$

(here we set $y_a \approx 0$, $y_a \approx 1$). This result demonstrates the leading role of the a-b interactions, giving rise to $B_{2,ab}$, which usually satisfies $|B_{2,ab}| > |B_{2,bb}|$.

n-Nonane/Methane Nucleation

Figure 1 shows the application of the proposed model to n-nonane/methane nucleation. The symbols correspond to the nucleation rate measurements at $p^v = 10$, 25, 33, and 40 bar[2] and the connecting lines serve the purpose of guiding the eye. The analysis is based on Eqs. (4) and (5) and the Redlich-Kwong-Soave (RKS) EoS which is most suitable for alkanes.[7] At the lowest pressure, $p^v = 1$ bar, the effect of carrier gas on J is negligible – we deal here with pure n-nonane nucleation. In the absence of 1 bar measurements we estimate this point using the Mean-field Kinetic Nucleation Theory (MKNT)[6] which is in good agreement with low pressure n-nonane nucleation experiments.[8] From the RKS-EoS we find $p_{comp} \approx 17.8$ bar. At $p^v < 10$ bar Δn_a increases with p^v. From the presented considerations one can expect the growth of Δn_a up to p_{comp} and its decrease beyond p_{comp}. However, since p_{comp} is located in between the experimental points, the decrease of Δn_a on Figure 1 starts at 10 bar. The b-content of the cluster also grows at sufficiently low p^v; due to the same reasons the point of maximum growth is expected to be at p_{comp}. Further experiments are desirable to verify this conjecture. The dramatic decrease of Δn_a at higher pressures (down to $\Delta n_a \approx 5$ at 40 bar) limits the growth of Δn_b since the

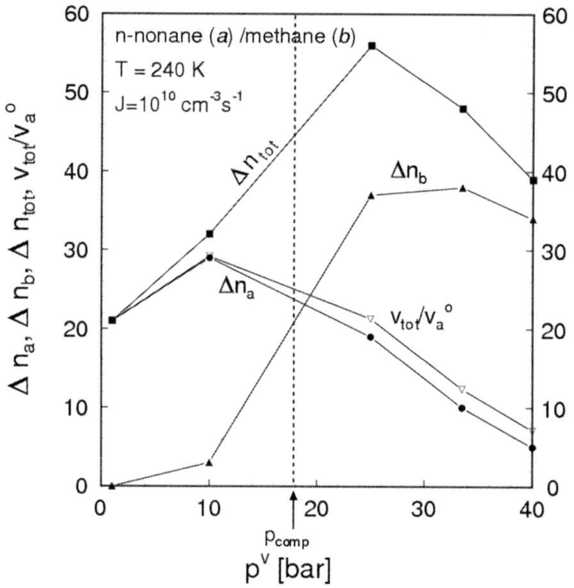

Figure 1 Content of the critical cluster in the binary *n*-nonane/methane nucleation at $T = 240\text{K}$. The symbols correspond to experiments[2], the 1 bar point is the theoretical MKNT prediction.[6] Solid lines are to guide the eye. Maximum of Δn_a and the inflection point of Δn_b are predicted to be located at the vertical dashed line $p_{comp}(T = 240\text{K}) \approx 17.8\,\text{bar}$. Also shown are the total number of molecules $\Delta n_{tot} = \Delta n_a + \Delta n_b$ and the total volume v_{tot} of the critical cluster in the units of the volume $v_a^{\,0}$ of the *n*-nonane molecule

b-molecules can only enter the cluster being entrained by the molecules of the supersaturated component *a*.

Studying $v_a^{\,v}(p^v)$ at different T we find scaling: $p_{comp}(T) \sim T^{3.1}$ showing that p_{comp} is a strongly increasing function of the temperature.

References

1. Heist, R.H. and He, H., *J. Phys. Chem. Ref. Data*, **23**, 781 (1994); Katz, J.L., Fisk, J.L., and Chakarov, V., in: *Nucleation and Atmospheric Aerosols*, edited by N. Fukuta and P.E. Wagner, Hampton: Deepak, p.11 (1992); Looijmans, K.N.H., Kriesels, P.C., and van Dongen, M.E.H., *Exp. Fluids*, **15**, 61 (1993); Looijmans, K.N.H., Ph.D. thesis, Eindhoven University (1995); Looijmans, K.N.H. and van Dongen, M.E.H., *Exp. Fluids*, **23**, 54 (1997).
2. Looijmans, K.N.H., Luijten, C.C.M., and van Dongen, M.E.H., *J. Chem. Phys.*, **103**, 1714 (1995); Luijten, C.C.M., Ph.D. thesis, Eindhoven University (1999); Peeters, P., Ph.D. thesis, Eindhoven University (2002); Luijten, C.C.M., Peeters, P., and van Dongen, M.E.H., *J. Chem. Phys.*, **111**, 8535 (1999); Labetski, D., Ph.D. thesis Eindhoven University (2007).
3. Oxtoby, D.W. and Kashchiev, D., *J. Chem. Phys.*, **100**, 7665 (1994).

4. Hill, T.L., *Thermodynamics of Small Systems*, Dover, NY.(1994); Kashchiev, D., *J. Chem. Phys.*, **125**, 014502 (2006).
5. Kalikmanov, V.I. and Labetski, D.G., *Phys. Rev. Lett.*, **98**, 085701 (2007).
6. Kalikmanov, V.I., *J. Chem. Phys.*, **124**, 124505 (2006).
7. Poling, B.E., Prausnitz, J.M., and O'Connell, J.P., *The Properties of Gases and Liquids*, 5th edn. NY: McGraw-Hill (2001).
8. Viisanen, Y., Wagner, P.E., and Strey, R., *J. Chem. Phys.*, **108**, 4257 (1998).

The Effect of Total Pressure on Nucleation in a Laminar Flow Diffusion Chamber: *n*-Pentanol + Helium

A.-P. Hyvärinen[1], D. Brus[1,2], V. Ždímal[2], J. Smolík[2], M. Kulmala[3], Y. Viisanen[1], and H. Lihavainen[1]

Abstract Homogeneous nucleation rates of *n*-pentanol were measured in a laminar flow diffusion chamber (LFDC) using helium as carrier gas at total pressure ranging from 50 to 400 kPa. Total pressure was observed to decrease the nucleation rate at temperatures above 280K, but increase it at temperatures below 280K

Keywords Homogeneous nucleation, pressure effect, laminar flow diffusion chamber, *n*-pentanol

Introduction

Homogenous vapor-to-liquid nucleation refers to the formation of liquid particles from condensable vapors without preexisting surfaces. As nucleation has many applications in science and technology, it has received a considerable amount of interest both from theoretical and experimental points of view.

In vapor–liquid nucleation experiments, the condensable vapor is typically dispersed in a large background of inert carrier gas. Its main function is to serve as a heat-bath for the latent heat release during condensation. It was generally assumed that the presence of a noncondensable carrier gas does not influence the clustering process, but several experimental and theoretical studies implicate that both the type and pressure of the carrier gas can affect the measured nucleation kinetics. The pressure effect has been mostly observed in diffusion-based devices such as the thermal diffusion cloud chamber (TDCC) and the laminar flow diffusion chamber

[1] *Finnish Meteorological Institute, Erik Palménin aukio, P.O. Box 503, FI-00101 Helsinki, Finland*

[2] *Laboratory Aerosol Chemistry and Physics, Institute of Chemical Process Fundamentals, Academy of Sciences of the Czech Republic, Rozvojová 135, 165 02 Prague 6, Czech Republic*

[3] *University of Helsinki, Dept. Physical Sciences, P.O. Box 64, 00014 Univ. of Helsinki, Finland*

C.D. O'Dowd and P.E. Wagner (eds.), *Nucleation and Atmospheric Aerosols*, 293–296.

(LFDC). In these measurements, nucleation rate has been repeatedly observed to decrease as a function of increasing pressure.[1–3] In pressure-dependent nucleation devices, such as expansion cloud chamber, only negligible carrier gas effects have been observed,e.g.[4]

Comparison between measurements is still difficult due to different compounds and thermodynamic conditions applied in the experiments. In this work, we have measured the homogeneous nucleation rates of *n*-pentanol in a LFDC using helium as a carrier gas at total pressures ranging from 50 to 400 kPa.

Experimental Methods

The LFDC used in this work is based on design of Lihavainen and Viisanen[5] and was modified to sustain both under- and overpressure in the range of 50–400 kPa.[3]

The LFDC consists of three main parts, each separately temperature-controlled. A flow of a carrier gas is first brought into a saturator, a horizontal tube half filled with the nucleating substance. The carrier gas gets fully saturated with the vapor of the substance in the saturator. The vapor–gas mixture then continues to flow through the preheater and the condenser, two coaxial vertical tubes with the same inner diameter. The preheater is kept at a higher temperature than the saturator It assures that the flow becomes laminar with a known velocity profile. It also defines the boundary and initial temperature for the mixture before it enters the condenser. The condenser is at a much lower temperature than the saturator. The nearly stepwise temperature drop leads to an increase in the saturation ratio of the vapor, and nucleation is observed if critical saturation ratio is exceeded. The mixture remains supersaturated long enough to allow nucleated particles to grow to an optically detectable size. A mathematical model is used to determine the theoretical temperature saturation ratio and nucleation rate profiles in the tube. The correct and stable operation of the device at different pressures was rigorously tested.[3]

Results

Figure 1 presents nucleation rates measured as a function of pressure at constant saturation ratio. For the first time, the pressure effect was observed to be both negative and positive with the same substance, depending on temperature. A transition concerning the pressure dependence occurs at 280K. A positive pressure effect is observed at temperatures below 280K and a negative effect above it. At 280K the effect is ambiguous. The positive pressure effect below 280K is in accordance with results presented by Anisimov et al.[6] The pressure effect is always more notable at pressures under 200 kPa.

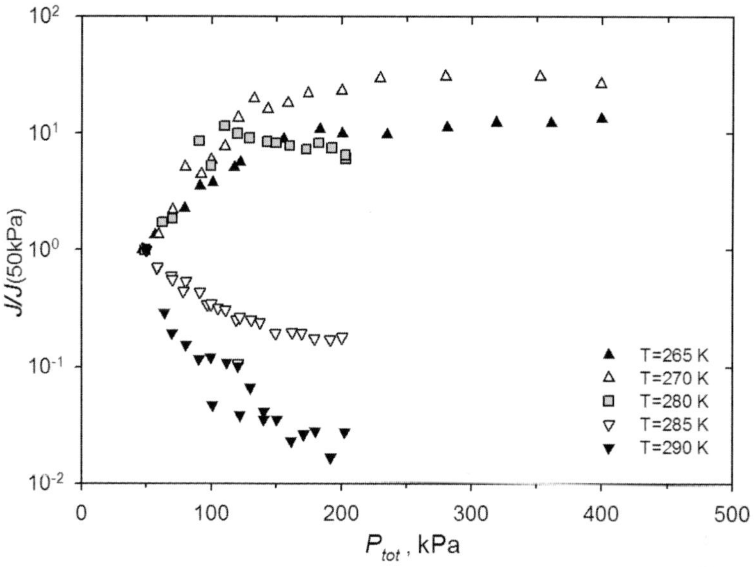

Figure 1 Nucleation rates as a function of pressure at constant saturation ratio. Results at each temperature level are normalized by setting the nucleation rate at 50 kPa to unity

Nucleation rates were also measured as a function of saturation ratio at constant temperature and pressure. The critical cluster sizes were determined from these measurements with the nucleation theorem[7]:

$$\left(\frac{\partial \ln J}{\partial \ln S}\right)_T \approx n^* \qquad (1)$$

In Eq. (1) J is nucleation rate, S is saturation ratio and n^* is the molecular content of critical cluster. Increasing pressure in the measurements was observed to decrease the number of molecules in a critical cluster. Similar pressure dependence has also been observed with *n*-butanol.[3]

Conclusions

Increasing the total pressure can either increase or decrease the nucleation rate of *n*-pentanol in a LFDC. This observation is surprising. However, the temperature dependence of the pressure effect is systematic when compared to earlier measurements with *n*-butanol.

Acknowledgments This work was supported by Maj and Tor Nessling foundation and the foundation of Vilho, Yrjö and Kalle Väisälä.

References

1. Katz, J.L., Hung, C.-H., and Krasnopoler, M., in: *Proceedings of the 12th ICNAA*, Vienna; Berlin: Springer , p. 356 (1988).
2. Brus, D., Ždímal, V., and Stratmann, F., *J. Chem. Phys.*, **124**, 164306 (2006).
3. Hyvärinen, A.-P., Brus, D., Ždímal, V., Smolík, J., Kulmala, M., Viisanen, Y., and Lihavainen, H. *J. Chem. Phys.*, **124**, 224304 (2006).
4. Van Remoortere, P., Heath, C., Wagner, P.E., and Strey, R., in: *Proceedings of the 14th ICNAA,* Helsinki, Oxford: Pergamon, edited by M. Kulmala and P.E. Wagner, p. 256 (1996).
5. Lihavainen, H. and Viisanen, Y. *J. Phys. Chem. B*, **105**, 11619 (2001).
6. Anisimov, M.P., Hopke, P.K., Shandakov, S.D. and Shvets, I.I., *J. Chem. Phys.*, **113**, 1971 (2000).
7. Kashchiev, D., *J. Chem. Phys.*, **104**, 8671 (1996).

Thermochemistry of $(H_2SO_4)_m$ $(H_2O)_n$ $(NH_3)_k$: A DFT Study

A.B. Nadykto[1] and F. Yu[1]

Abstract H_2SO_4–H_2O and H_2SO_4-H_2O-NH_3 nucleation rates in earth's atmosphere are very sensitive to thermodynamics of initial growth steps. However, reliable data on thermochemistry of small H_2SO_4–H_2O and H_2SO_4-H_2O-NH_3 clusters are lacking. This theoretical study is dedicated to the quantum modeling of H_2SO_4–H_2O and H_2SO_4-H_2O-NH_3 clusters using Density Functional Theory (DFT) at PW91PW91 level with the largest Pople 6–311 + + G(3df,3pd) basis set. We have obtained equilibrium geometries, and computed enthalpies, entropies, and Gibbs free energies of $(H_2SO_4)_m$ $(H_2O)_n$ $(NH_3)_k$ (n = 1,2; m = 1–5; k = 0,1) formation. The obtained results have been compared with earlier theoretical studies. It has been pointed out that the application of PW91PW91 coupled with large 6–311 + + G(3df,3pd) basis set allows to improve the quality of the model predictions, and helps to avoid serious problems with H_2SO_4 thermochemistry found in earlier studies.

Keywords Binary clusters, ternary clusters, Density Functional Theory, atmospheric nucleation

Introduction

Quantum methods have been used to study H_2SO_4–H_2O and H_2SO_4-H_2O-NH_3 clusters in the past. However, the quality of quantum predictions was not always satisfactory. The pioneering studies of Ianni and Bandy (1998), (1999) [1,2], in which H_2SO_4–H_2O and H_2SO_4-H_2O-NH_3 structures were identified for the first time, were unable to predict the stability of gas-phase hydrates of sulfuric acid and ammonia bisulfate complex in agreement with experiments. More recent studies of Al Natsheh et al. (2004) [3], Kurten et al. (2006) [4], Nadykto and Yu (2007) [5] showed that insufficient accuracy of water and ammonia affinities to the sulfuric

[1] *Atmospheric Sciences Research Center, State University of New York at Albany, 251 Fuller Rd., Albany 12203, NY, USA*

C.D. O'Dowd and P.E. Wagner (eds.), *Nucleation and Atmospheric Aerosols*, 297–301.

acid is caused by deficiencies of the B3LYP in describing intermolecular interactions in hydrogen-bonded complexes. According to Tsuzuki and Luthi (2001) [6], Natsheh et al. (2004, 2006), Kurten et al. (2006), Nadykto et al. (2006) [7], Nadykto and Yu (2007) [5], Nadykto et al. (2007) [8], PW91PW91 functional is a better choice. A comprehensive structural analysis of H_2SO_4–H_2O clusters has been carried out in Ding et al. (2003) [9] using PW91PW91 method with the DNP basis set. The same combination was used in a recent study by Kurten et al. (2007), where a limited number of H_2SO_4-H_2O-NH_3 structures were identified. The DNP is similar to 6–31G** basis in Gaussian, but the basis functions are numerical atomic orbitals augmented by polarization functions. This basis does not include diffuse functions. While PW91PW91/DNP adequately describes first two hydration steps, the PW91PW91/DNP dimerization free energy for the sulfuric acid deviates from PW91PW91/6–311 + + G(3df,3pd) results by > 2.5 kcal mol^{-1} [5]. Such a big deviation, which is probably caused by the basis set quality, may lead to erroneous conclusions about the thermochemistry of the sulfuric acid, which is a key substance controlling the atmospheric nucleation. Thus, the application of a more appropriate, presumably larger, basis set is necessary.

In this paper, the structure, thermochemical properties, and dipole moments of $(H_2SO_4)_m$ $(H_2O)_n$ $(NH_3)_k$ (n = 1, 2; m = 1–5; k = 0,1) clusters have been studied using DFT at PW91PW91 level with large 6–311 + + G(3df,3pd) basis set. The hydration of H_2SO_4 monomer and dimer has been investigated so as the effect of ammonia on stability of $(H_2SO_4)_m$ $(H_2O)_n$ clusters.

Results and Discussion

At present time, the experimental data are available for (H_2SO_4) (H_2O) only. Table 1 presents a comparison of optimized (H_2SO_4) (H_2O) geometries at different levels of theory with experimental data. As seen from Table 1, PW91PW91/6–311

Table 1 Comparison of (H_2SO_4) (H_2O) geometry obtained using ab initio and DFT methods with experiments

	MP2*	Re	Ianni Bandy	Ding	Natsheh	Nadykto Yu	exp.**
R(H1-O2)	0.967	0.975		0.978	0.98	0.976	0.95
R(O2-S)	1.579	1.636	1.611	1.634	1.63	1.614	1.578
R(O3-S)	1.557	1.603		1.604	1.59	1.576	1.567
R(O4-S)	1.43	1.466	1.439	1.463	1.46	1.446	1.464
R(O5-S)	1.42	1.458	1.43	1.453	1.44	1.435	1.41
R(H6-O3)	0.997	1.009	0.999	1.023	1.03	1.022	1.04
R(H6-O8)	1.654	1.651	1.682	1.621	1.61	1.61	1.645
R(H9-O8)	0.976					0.98	0.98
R(H10-O8)	0.96					0.97	0.98
R(H9-O4)	2.15	2.23	2.207	2.141		2.05	2.04

* MP2(full)/6–311 + + G(3df,3pd), present study; ** Fiacco et al. (2003) [10].

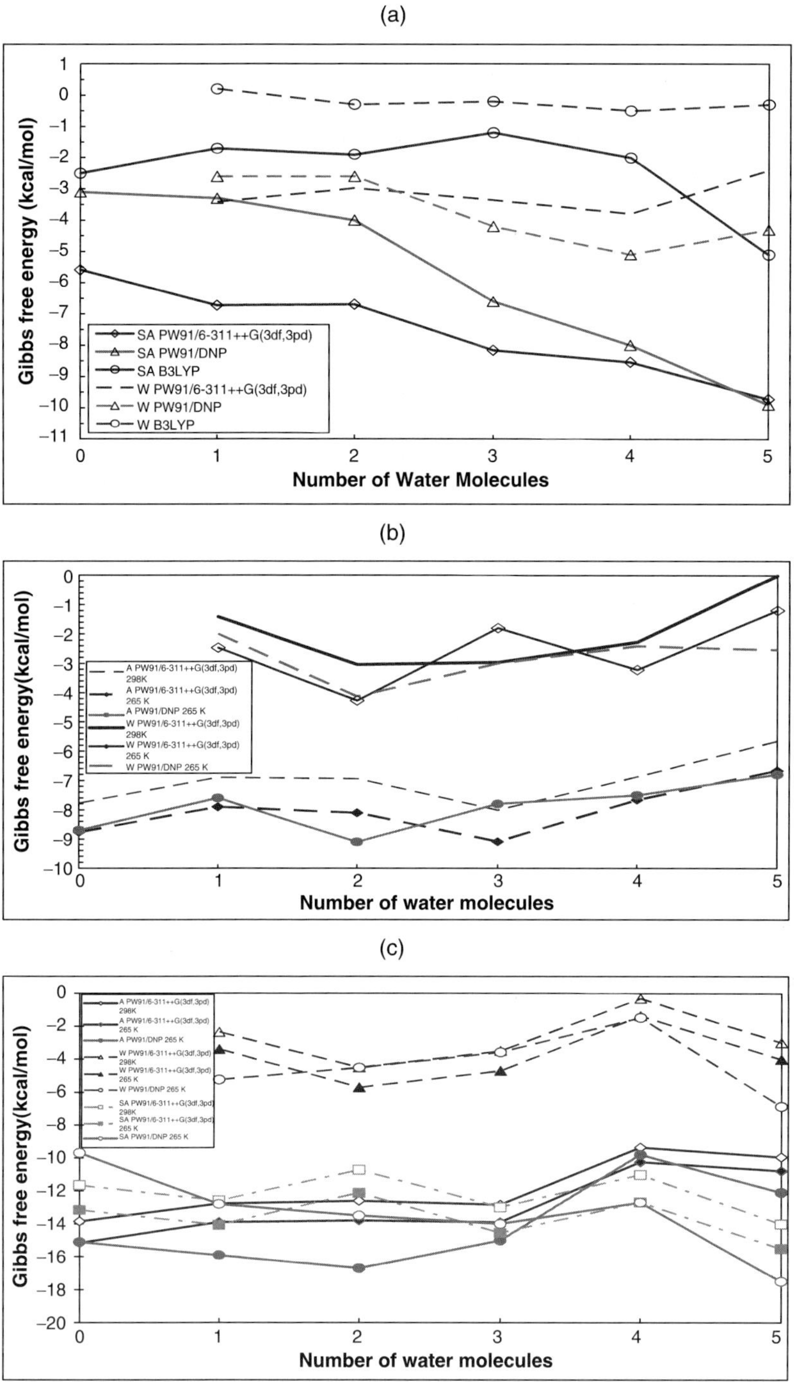

Figure 1 Changes of Gibbs free energies associated with the formation of (a) $(H_2SO_4)_2\ (H_2O)_n$; (b) $(H_2SO_4)\ (NH_3)\ (H_2O)_n$; (c) $(H_2SO_4)_2\ (NH_3)\ (H_2O)$. Abbreviations B3LYP, PW91/DNP, and PW91/6–311 + + G(3df,3pd) refers to studies Ianni and Bandy(2000), Ding et al.(2003) (a) and Kurten et al. (2007)(b,c), and the present study, respectively. Gibbs free energy changes were calculated at $P = 101.3$ kPa. $T = 298.15$K unless specified

+ + G(3df,3pd) predictions are in best overall agreement with experiments. Another important detail is that PW91PW91/6–311 + + G(3df,3pd) is the only method that predicts longest R(H9-O4) bond length controlling the (H_2SO_4) (H_2O) stability in agreement with experiment data. The difference in the (H9-O4) bond length between PW91PW91/6–311 + + G(3df,3pd) and PW91PW91/DNP is considerable, and may be attributed to the difference in the basis set quality.

Another important detail is that a vast majority of clusters studied here is highly polar. Due to high polarity, they may be involved in the dipole-charge interaction with airborne ions enhancing the ion-mediated nucleation rates.

Figure1 presents the comparison of Gibbs free energy changes associated with the formation of $(H_2SO_4)_m$ $(H_2O)_n$ $(NH_3)_k$ complexes by addition of water (W), sulfuric acid (S), and ammonia (A). As seen from Figure1(a), both Ianni and Bandy (2000) [11], and Ding et al. (2003) underestimate the affinity of sulfuric acid to (H_2SO_4) $(H_2O)_n$ complexes by 1–5 kcal mol^{-1} per step.

While B3LYP/6–311 + + G(2d,2p) predictions of Ianni and Bandy (2000) [11] consistently deviate from our results, the PW91/DNP energies at n = 4,5 are quite close to ours. This suggests that PW91/DNP results could probably used to approximate the thermochemistry of larger (H_2SO_4) $(H_2O)_n$ complexes, for which a more rigorous treatment would be prohibitively expensive.

The comparison of theoretical affinities with average experimental value of –8.9 kcal mol^{-1} for $(H_2SO_4)_2$ $(H_2O)_n$ (n = 1–10; T = 242K) leads us conclude that the present results are in better agreement with experiments than the previous studies. The difference between hydration free energies obtained using B3LYP/6–311 + + G(2d,2p) and those obtained using PW91 with DNP and 6–311 + + G(3df,3pd) basis sets is very large. The difference between PW91/DNP and PW91/6–311 + + G(3df,3pd) is a lot smaller, yet it exceeds ~ 1 kcal mol^{-1} per step on average. As seen from Figure 1(b), affinities of ammonia to hydrated sulfuric acid and H_2O affinities to (H_2SO_4) $(H_2O)_n$ (NH_3) predicted by two PW91 studies are quite close. However, thermochemical properties of $(H_2SO_4)_2$ $(H_2O)_n$ (NH_3) (Figure1(c)), especially NH_3 and H_2SO_4 affinities, are predicted by PW91/DNP with much less success. The performance of DNP basis set is quite different from that of 6–311 + + G(3df,3pd), and it is clear that the application of more accurate 6–311 + + G(3df,3pd) basis set leads to a significant improvement in the quality of the model predictions.

Acknowledgment This work was funded by NSF under grant ATM0618124, and NOAA under grant NA05OAR4310103.

References

1. Bandy, A.R. and Ianni, J.C., *J. Phys. Chem. A,* **102**(32), 6533 (1998).
2. Ianni, J.C. and Bandy, A.R. *J. Phys. Chem. A,* **103**, 2801 (1999).
3. Al Natsheh, A. et al., *J. Phys. Chem. A,* **108**(41), 8914–8928 (2004).
4. Kurten, T. et al., *J. Phys. Chem. A,* **110**, 7178.
5. Nadykto, A.B. and Yu, F., *Chem.Phys.Lett.*, in press (2007).

6. Tsuzuki, S. and Lüthi, H.P., *J. Chem. Phys.*, **114**, 3949 (2001).
7. Nadykto, A.B. et al., *Phys. Rev. Lett.* **96**, 125701, 1–4 (2006).
8. Nadykto, A.B., Du, H., and Yu, F., *Vibr.Spectroscopy*, in press (2007).
9. Ding, C.-G., Laasonen, K., and Laaksonen, A., *J. Phys. Chem. A.*, **107**(41), 8648–8658 (2003).
10. Fiacco, D.L., Hunt, S.W., and Leopold, K.R., *J. Am. Chem. Soc.* **124**(16), 4504–4513 (2002).
11. Ianni, J.C. and Bandy, A.R., *J. Mol. Structure (Theochem)*, **497**,19–30 (2000).

Monte Carlo Simulations on Heterogeneous Nucleation II: Line Tension

Antti Lauri, Evgeni Zapadinsky, Anca I. Hienola, Hanna Vehkamäki, and Markku Kulmala

Abstract We have interpreted the results of molecular Monte Carlo simulations from the perspective of line tension in heterogeneous nucleation. Our method enables a numerical estimate of the line tension from the Monte Carlo data. For our test case, Lennard–Jones argon nucleating on platinum substrate the obtained line tension value is $\sigma_t = 1.2 \times 10^{-12}$ N, which is in the range of earlier reported line tension values for other substances.

Keywords Heterogeneous nucleation, Monte Carlo simulations, line tension

Introduction

Line tension is a quantity describing the effect of the energy bound in the contact line between the three phases usually present in the heterogeneous nucleation phenomenon: vapour, nucleating embryo and insoluble substrate surface. Line tension as a concept in heterogeneous nucleation has been a matter of discussion during the past decades. Gretz[1] was the first to notice the significance of the three-phase interface in theoretical considerations of heterogeneous nucleation. Since then, it has been commonly agreed that line tension should be taken into account in order to have a thermodynamically correct description of heterogeneous nucleation, and the line tension concept has been successfully used to explain experimental results.[2]

It has, however, proved to be very difficult to make quantitative estimates of the magnitude, or even the sign of the line tension to be utilized in computational model calculations. In this study we present a new approach for the calculation of an estimate of line tension from molecular simulations.

Department of Physical Sciences, University of Helsinki, P.O.Box 64, FI-00014 University of Helsinki, Finland

C.D. O'Dowd and P.E. Wagner (eds.), *Nucleation and Atmospheric Aerosols*, 302–305.

Theory

If line tension is not considered, the formation free energy in heterogeneous nucleation is given by

$$\Delta G_{\text{het}} = \rho_l(\mu_l - \mu_v)V_{\text{het}} + \sigma_{lg}A_{lg} + (\sigma_{sl} - \sigma_{sg})A_{sl}, \tag{1}$$

where ρ is density, μ is chemical potential, σ is surface tension, V is volume, and A is surface area. Subindexes g, l, and s correspond to vapour, liquid, and substrate phases, respectively. Including the effect of line tension, an extra term is added to the free energy formula[3]:

$$\Delta G^t_{\text{het}} = \Delta G_{\text{het}} + 2\pi\sigma_t \sin\theta, \tag{2}$$

where r is the radius of the cluster, and θ is the contact angle between the surfaces of the nucleating embryo and a flat substrate.

Earlier we have derived the difference between the formation free energy of an n-cluster in heterogeneous and homogeneous nucleation.[4] It is straightforward to include the extra term including the line tension in the derivation:

$$\Delta G^t_{\text{het}}(n) - \Delta G_{\text{hom}}(n) = \alpha\left(f^{1/3} - 1\right)n^{2/3} + \left[\frac{24\pi^2}{\rho_l}\frac{(1+m)^2}{(2+m)\left(1-m^2\right)^{1/2}}\right]^{1/3}\sigma_t n^{1/3}, \tag{3}$$

where $\alpha = (36\,\pi)^{1/3}\,\rho_l^{-2/3}\,\sigma_{lg}$, f is a geometrical factor, and $m = \cos\,\theta$.

Simulations

Earlier, starting from Boltzmann equilibrium cluster distribution, we have reported the heterogeneous formation free energy of an n-cluster.[5] This can be easily converted to the form where the line tension is included:

$$\Delta G^t_{\text{het}}(n) - \Delta G_{hom}(n) = [F_{\text{het}}(n) - F_{\text{hom}}(n)] - [F_{\text{het}}(1) - F_{\text{hom}}(1)], \tag{4}$$

where $F(n)$ is the Helmholtz free energy of an n-cluster.

The overlapping distribution Monte Carlo method[6] is a powerful tool in the evaluation of the difference in formation free energy of a heterogeneous and homogeneous n-cluster. We applied the method for calculating the formation free energy differences from molecular Monte Carlo simulations.

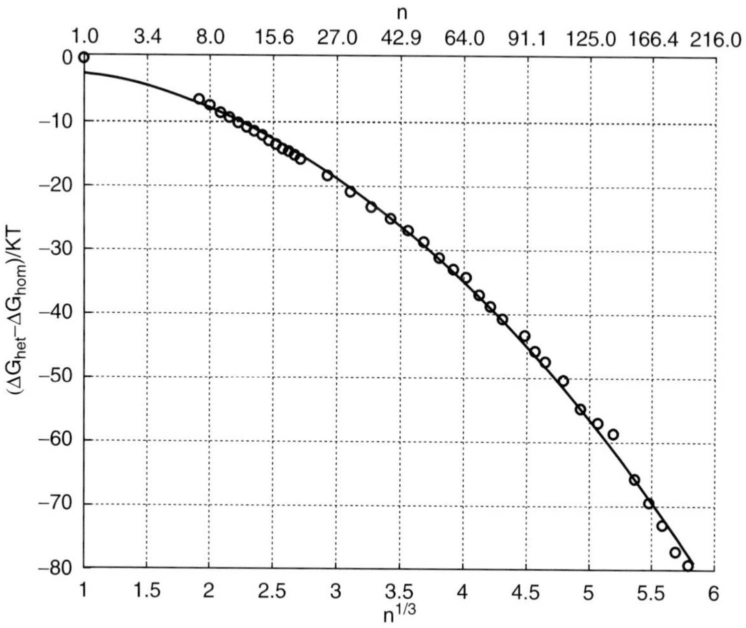

Figure 1 The difference between the heterogeneous and homogeneous formation free energies at $T = 60$K. Circles represent results from single simulation runs, and the solid curve is a second-order polynomial fit to the data

Results

We simulated molecular clusters consisting of 1–200 argon molecules in the presence of an fcc(111) monolayer of platinum atoms at $T = 60$K. Both argon–argon and argon–platinum interactions were represented by truncated and shifted Lennard–Jones potentials with cutoff distance 2.5 σ, where σ is the Lennard–Jones distance parameter.

The formation free energy differences obtained from the simulations are plotted as a function of $n^{1/3}$ in Figure 1. A second-order polynomial function fit is also shown. The sign and magnitude of line tension can be calculated from the first-order term in the fit [see Eq. (3)], resulting in a positive value of the line tension, $\sigma_t = 1.2 \times 10^{-12}$ N.

Conclusion

We have interpreted our molecular Monte Carlo simulation results in the viewpoint of the line tension in heterogeneous nucleation. Our first analysis results with Lennard–Jones argon show that it is possible to get a numerical value for the line tension from molecular Monte Carlo simulations. The obtained value for 2.5 σ cutoff Lennard–Jones argon nucleating on platinum substrate, $\sigma_t = 1.2 \times 10^{-12}$ N, is in the range of earlier estimations for the line tension value for other substances. The presented method is applicable for other substances and cluster–substrate interaction models as well.

References

1. Gretz, R.D., *Surf. Sci.*, **5**, 239–251 (1966).
2. Hienola, A.I., Winkler, P.M., Wagner, P.E., Vehkamäki, H., Lauri, A., Napari, I., and Kulmala, M., *J. Chem. Phys.*, **126**, 094705 (2007).
3. Lazaridis, M., *J. Colloid Interf. Sci.*, **155**, 386–391 (1993).
4. Lauri, A., Zapadinsky, E., Vehkamäki, H., and Kulmala, M., *J. Chem. Phys.*, **125**, 164712.
5. Zapadinsky, E., Lauri, A., and Kulmala, M., *J. Chem. Phys.*, **122**, 114709 (2005).
6. Bennett, C.H., *J. Comput. Phys.*, **22**, 245–268 (1976).

Insights from the Atomistic Simulations of a Ternary Nucleating System

Ricky B. Nellas[1], Bin Chen[2], and J. Ilja Siepmann[3]

Abstract The development of a few advanced simulation techniques in our group has enabled us to venture on the very first atomistic simulations of a three-component nucleating system, water/*n*-nonane/1-butanol, that provided results which are in good agreement with experimental data. This study has practical implications in aerosol formation, separation process, nano-materials engineering, and micro-emulsion research. In this investigation, we found that the three nonideal nucleation behaviors we observed in the three binary mixtures (mutual nucleation enhancement for water/1-butanol, reluctant co-nucleation for *n*-nonane/1-butanol, and independent nucleation for water/*n*-nonane) were carried forward and combined in this ternary mixture at all compositions. Such diverse spectrum of nonideal behavior implies that changing the gas-phase compositions could lead to various nucleation pathways. In addition, the "micro-phase diagrams" (or nucleation free energy landscape) constructed at intermediate vapor-phase compositions show coexistence of multiple "micro-phases" (clusters with distinct compositions). Visual inspection of the clusters that contain equal molar amounts of each component further reveals the existence of an internal phase separation into a multilayered structure. In particular, water and 1-butanol adopt a core/shell pattern with the nonpolar butyl chains pointing outwards that favors the deposition of *n*-nonane compared to the bare water surface. This explains the enhanced miscibility between water and *n*-nonane with the presence of 1-butanol. However, the tendency for *n*-nonane to deposit on one end of (rather than wrapping around) the core/shell structure suggests that 1-butanol is just too short to act effectively as a surfactant. Simulations for ternary mixtures involving either shorter or longer alcohols are currently under investigation to examine this speculation and to search for possible ways of improving the amphiphile. We will also present results for those mixtures at the meeting.

[1] *Department of Chemistry, Louisiana State University, Baton Rouge, LA 70803–1804, USA*

[2] *Department of Chemistry, Louisiana State University, Baton Rouge, LA 70803–1804, USA*

[3] *Department of Chemistry and of Chemical Engineering and Materials Science, University of Minnesota, 207 Pleasant Street S.E., Minneapolis, Minnesota 55455–0431, USA*

C.D. O'Dowd and P.E. Wagner (eds.), *Nucleation and Atmospheric Aerosols*, 306–309.

Keywords Ternary nucleation, nonideal nucleation behaviors, simulation, phase separation, micro-phase diagram, critical cluster, surfactant

Onset Activities

The onset activity plot in Figure 1 exhibits one of the richest nucleation behaviors for a given system obtained by both simulations and experiments.[1] We found that the three nonideal nucleation behaviors observed in the three binary mixtures (mutual nucleation enhancement for water/1-butanol, reluctant co-nucleation for *n*-nonane/1-butanol, and independent nucleation for water/*n*-nonane)[2–4] were carried forward and combined in this ternary mixture at all compositions. At higher *n*-nonane activity fraction, the plot resembles to that observed for the binary *n*-nonane/1-butanol mixture. In contrast, at low *n*-nonane activity, the onset activities curved to the opposite side, and looks like the binary water/1-butanol mixture onset activity plot. At intermediate *n*-nonane activity, both features are present. Our simulation results are in good agreement with the experimental data.[1]

"Micro-Phase Diagram"

The "micro-phase diagram" elaborates the microscopic miscibility details of the ternary system. As depicted in Figure 2, the most probable distribution is widely spread out, extending from pure *n*-nonane end to the edge that connects the water

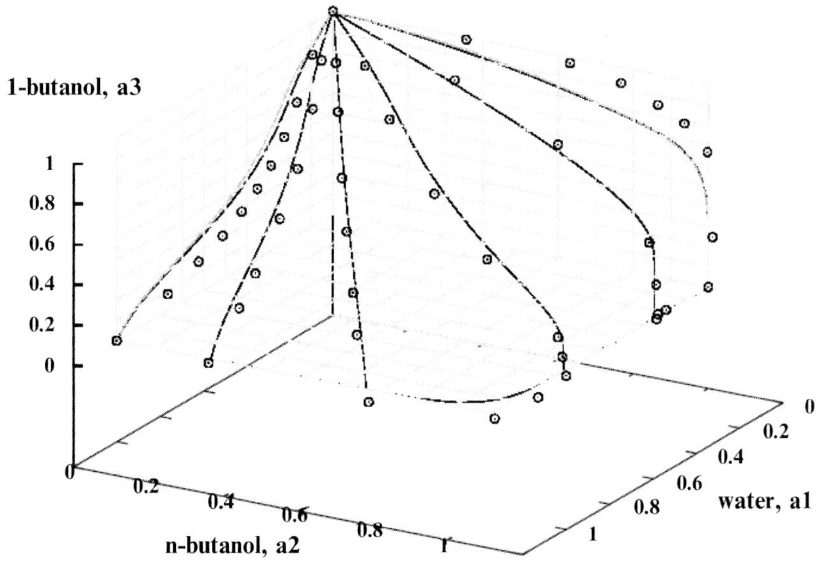

Figure 1 Onset activity plot for simulation at 230K (**dark lines**), the experiment at 240K (**circles**), and simulations results interpreted from histogram-reweighting technique at 240K (**faded lines**)

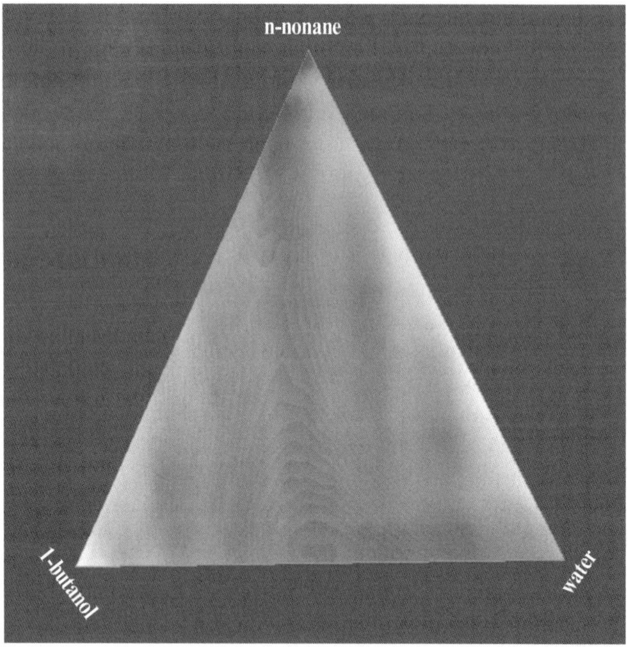

Figure 2 Thermodynamic "micro-phase diagram" for cluster consisting of 45 molecules

and 1-butanol end. Although, improved miscibility between water and *n*-nonane with addition of 1-butanol is evident, a small dip centered in the middle of this triangular diagram implies that the system still tends to demix toward either *n*-nonane-enriched or water/1-butanol-enriched "micro-phases."

Critical Cluster Configuration

The aforementioned dip in Figure 2 implies that a full miscibility is still unfavorable. Visual inspection of clusters containing equal molar amounts of the three components indicates an internal phase separation into a multilayered structure. Specifically, as shown in Figure 3, the water core is wrapped by a layer of 1-butanol molecules. This core/shell structure favors the deposition of *n*-nonane compared to the bare water surface.[5] This leads to the enhanced miscibility of water with *n*-nonane in the presence of 1-butanol. However, the tendency for *n*-nonane to stay on one side (rather than wrapping around) may suggest that 1-butanol is just too short to effectively act as an amphiphile to form a more symmetrical spherical structure.

Acknowledgments We thank Professor Les Butler and Heath Barnett for the assistance provided in the use of the Amira software. Part of the computer resources were provided by CCT and OCS at LSU (SuperMike and Pelican). Financial support from LSU start-up fund, the NSF

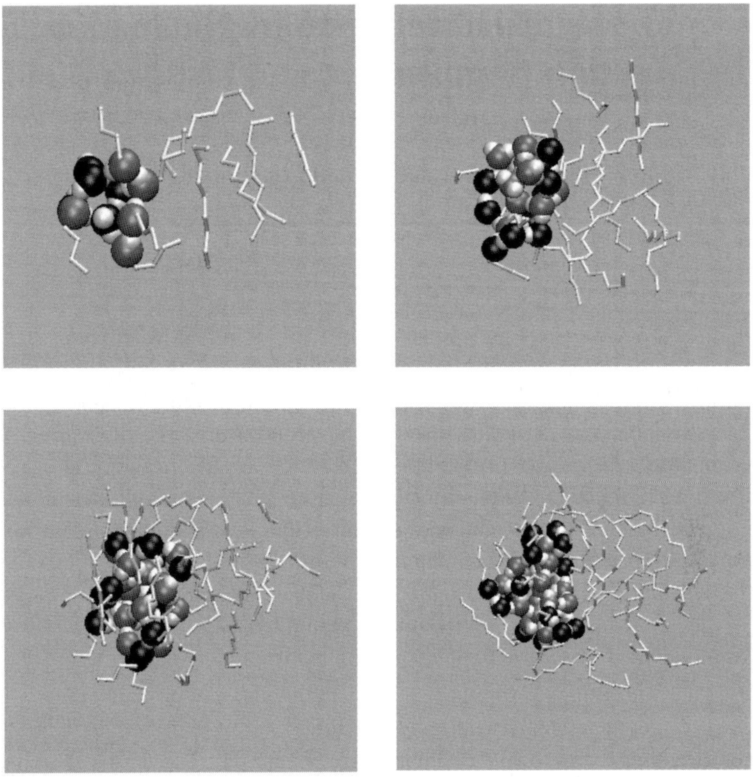

Figure 3 Representative cluster configurations consisting of 5, 10, 15, and 20 molecules of each component. Color code: water oxygen (**gray**); 1-butanol oxygen (**black**); hydrogen (**white**); and alkyl tails (**white stick**)

(CHE/MCB-0448918), the ACS-PRF (Grant No. 41933-G9), and Louisiana Board of Regents Support Fund (LEQSF(2005–08)-RD-A-02) is gratefully acknowledged.

References

1. Viisanen, Y. and Strey, R. *J. Chem. Phys.*, 105, 8293 (1996).
2. Chen, B., Siepmann, J.I., and Klein, M.L., *J. Am. Chem. Soc.*, 125, 3113 (2003).
3. McKenzie, M.E. and Chen, B., *J. Phys. Chem. B.*, 110, 3511 (2006).
4. Nellas, R.B., McKenzie, M.E., and Chen, B., *J. Phys. Chem. B*, 110, 18619 (2006).
5. Chen, B., Siepmann, I.J., and Klein, M.L., *J. Am. Chem. Soc.*, 124, 12232 (2002).

Two-component Heterogeneous Nucleation in the Martian Atmosphere

Anni Määttänen[1], Hanna Vehkamäki[2], Antti Lauri[2], Ismo Napari[2], and Markku Kulmala[2]

Abstract We have been modeling binary nucleation in the Martian atmosphere, which has not been investigated before. The properties of the nucleating water–carbon dioxide system are unknown, which brings up the need for assumptions. The binary nucleation model shows interesting results for binary nucleation at the initial stages of unary water nucleation, but only when assuming ideal mixture. The nonideal mixture is less favorable for the process.

Keywords Heterogeneous nucleation

Introduction

In the Martian atmosphere dust and other types of aerosol particles (most importantly carbon dioxide and water ices) scatter and absorb both short- and long-wave radiation. This affects the thermal structure of the atmosphere and flow fields therein. Flow patterns affect the aerosol particle distribution by saltation and advection, which again modifies its thermal and flow structure. This feedback between aerosols and flow patterns may, have a great influence in giving rise to Martian dust storms. For understanding this feedback, both the boundary layer processes[1,2] being involved in lifting and advecting dust particles and creating atmospheric circumstances favorable to the onset of nucleation, and the aerosol processes themselves[3,4,5] should be studied in detail.

On one hand, Martian aerosol processes are simpler than the Terrestrial ones, since there are practically no organic vapors or anthropogenic emissions influencing the processes. On the other hand, the Martian atmosphere is nearly pure CO_2 vapor, which takes part in the processes of nucleation and condensation. In a near-pure

[1] *Division of Atmospheric Sciences, University of Helsinki/Space Research, Finnish Meteorological Institute, Helsinki, Finland*

[2] *Division of Atmospheric Sciences, University of Helsinki, Helsinki, Finland*

C.D. O'Dowd and P.E. Wagner (eds.), *Nucleation and Atmospheric Aerosols*, 310–313.

situation, nucleation becomes nonisothermal and strong coupling of fluxes arises in multicomponent condensation. These effects need to be taken into account in the theories used.

So far studies on Martian clouds have been focusing on clouds formed of only one component, either water or carbon dioxide. Our work has been focusing on testing the possibilities for binary nucleation.

Two-Component Nucleation

We have used a two-component heterogeneous nucleation model[6] to model binary nucleation in the Martian atmosphere.[7]

The water–carbon dioxide system has not been modelled before. The Martian atmosphere is 95.3% carbon dioxide, and there is no liquid phase for the nucleating substances. Clouds are presumed to form from either water or CO_2 ice crystals, and this is the first study of binary nucleation on Mars. There is no thermodynamic data available for the ice mixture properties. The non-isothermal coefficient (important in nucleation occurring in a near-pure substance) is calculated using the formulation[8] as was also done in.[5]

We calculated, as a first approximation, the thermodynamic data (surface energy, ice density) for the mixture assuming ideal mixing. For ideal mixture, the solid-phase activities equal the mole fractions of the respective species. We tested the behavior of the system by using the activity coefficients of water–*n*-propanol mixture for the water–carbon dioxide mixture. The activities of the water–*n*-propanol system are thought to realistically mimic the H_2O–CO_2 system since both CO_2 and *n*-propanol are nonpolar molecules. The results show (Figure 1) that the ideal mixture is more favorable for nucleation compared to the nonideal nonpolar system of water and *n*-propanol.

We tested the possibility of nucleation in the Martian near-surface conditions (temperature range and pressure, condensation nuclei, CN, size) taking realistic temperature and vapor amount values, and our results show (Figure 2) that the onset of binary nucleation happens at slightly lower activities than the onset of unary nucleation of water.

We looked at the numbers of molecules in the critical cluster in these cases, and noticed that the cluster is mainly composed of water molecules. In some cases with low amount of water vapor (1 ppm), the number of CO_2 molecules stays fairly large (tens of molecules) in a larger range of temperatures (about one degree), but for 300 ppm of water even in the first critical (binary) cluster the number of CO_2 molecules is only 2–3 and after that drops to less than one (which implies unary water nucleation). So theoretically at the onset of water nucleation on Mars it seems that binary nucleation might have a role in facilitating the process, but the number of CO_2 molecules in the critical clusters is so small, that particle formation process is nearly pure water nucleation.

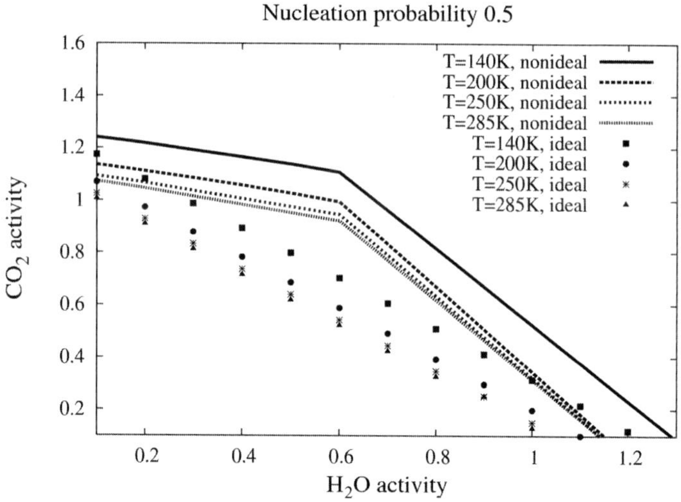

Figure 1 Activity plot for the Martian binary system: the critical activities required for nucleation probability to reach or exceed 0.5. CO_2 activity on the *y*-axis and H_2O activity on the *x*-axis. The dots describe the behavior for the ideal mixture and the lines the nonideal mixture

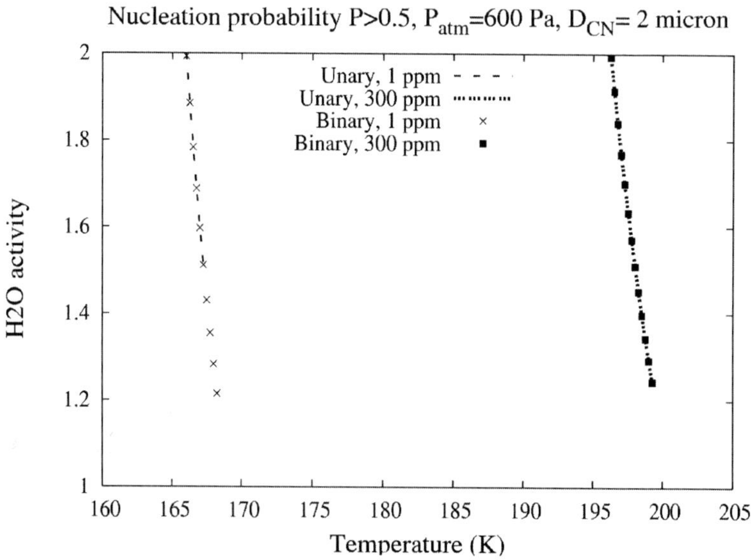

Figure 2 Critical water activity for nucleation probability of P = 0.5 in unary and binary nucleation. The two cases represented are for 1 and 300 ppm of water. The dots are for binary nucleation and the lines unary nucleation. Especially in the leftmost case of 1 ppm of water it can be clearly seen that binary nucleation requires lower activities than unary nucleation to start

Summary

We have modeled binary nucleation in the Martian water–CO_2 system. According to our results we can presume that in the present Martian atmosphere ice clouds do form one component at a time, as described in our earlier paper[5] and several other authors.[9–12] However, we cannot completely rule out the possibility of binary nucleation of the CO_2–H_2O mixture, possibly at the very first stages of unary water nucleation. This result is naturally dependent on the assumptions made for the system (for example, the ideal mixture assumption) and validating the results will have to wait until there are data or experiments for the real thermodynamic properties of the system.

Acknowledgments AM would like to thank financial support from the Kordelin foundation and the Graduate school of astronomy and space physics. We also like to thank the Academy of Finland for funding.

References

1. Savijärvi, H., *Q. J. Roy. Meteorol. Soc.*, **125**, 483–493 (1999).
2. Savijärvi, H., Määttänen, A., Kauhanen, J., and Harri, A.-M., *Q. J. Roy. Meteorol. Soc.* (2004).
3. Noppel, M., Vehkamäki, H., and Kulmala, M., *J. Chem. Phys.*, **116**, 218–228, (2002).
4. Korhonen, H., Lehtinen, K.E.J., and Kulmala, M., *Atmos. Chem. Phys.*, **4**, 757–771 (2004).
5. Määttänen, A., Vehkamäki, H., Lauri, A., Merikallio, S., Kauhanen, J., Savijärvi, H. and Kulmala, M., *J. Geophys. Res.*, **110**, E02002 (2005).
6. Määttänen, A., Vehkamäki, H., Lauri, A., Napari, I., and Kulmala, M., A kinetically correct and an approximative model of heterogeneous nucleation. 17th ICNAA conference abstract (2007).
7. Määttänen, A., Vehkamäki, H., Lauri, A., Napari, I., and Kulmala, M., *J. Chem. Phys.*, Submitted (2007).
8. Feder, J., Russell, K.C., Lothe, J., and Pound, G.M., *Adv. Phys.*, **15**, 111–178 (1966).
9. Michelangeli, D.V., Toon, O.B., Haberle, R.M., and Pollack, J.B., *Icarus*, **100**, 261–285 (1993).
10. Colaprete, A., Toon, O.B., and Magalhaes, J.A., *J. Geophys. Res.*, **104**(E4), 9043–9054 (1999).
11. Colaprete, A. and Toon, O.B., *J. Geophys. Res.*, **107**(E7), 5051, doi:10.1029/ 2001JE001758 (2002).
12. Montmessin, F., Rannou, P., and Cabane, M., *J. Geophys. Res.*, **107**(E6), 5037, doi:10.1029/ 2001JE001520 (2002).

Binary Homogenous Nucleation of Sulfuric Acid and Water Mixture: Experimental Device and Setup

David Brus[1], Antti-Pekka Hyvärinen[1], Heikki Lihavainen[1], Yrjö Viisanen[1], and Markku Kulmala[2]

Abstract New particle formation in the atmosphere still attracts both atmospheric scientists and aerosol researchers. New laminar flow chamber was recently built in Finnish Meteorological Institute to study particle formation of binary and ternary compounds such as sulfuric acid–water and sulfuric acid–ammonia–water, respectively. The laminar flow chamber and its experimental setup are presented here.

Keywords Homogeneous nucleation, particle formation, sulfuric acid–water

Introduction

New particle formation in the atmosphere has received considerable attention lately both from atmospheric scientists and aerosol researchers. Atmospheric new particles have been observed to form by self-condensing or nucleating homogeneously in events lasting a couple of hours nearly all around the world.[1]

The first step of new particle formation or any first-order phase transition is nucleation. It has been calculated that in the atmosphere the equilibrium vapor pressure of sulfuric acid is low enough for it to be a likely candidate to nucleate homogeneously.[2]

All recent experiments on homogeneous sulfuric acid and water nucleation have relied on a flow-based measurement technique.[3–7] In general the results are in fair agreement with each other, although somehow dependent on the method of generating the sulfuric acid vapor.

A new laminar flow chamber built recently in Finnish Meteorological Institute is presented here. It is designed for homogeneous nucleation experiments of binary and ternary compounds such as sulfuric acid–water and sulfuric acid–ammonia–water, respectively.

[1] *Finnish Meteorological Institute, Erik Palménin aukio 1, P.O. Box 503, FIN-00100 Helsinki, Finland*

[2] *Department of Physical Sciences, University of Helsinki, P.O. Box 64, FIN-00014 Helsinki, Finland*

C.D. O'Dowd and P.E. Wagner (eds.), *Nucleation and Atmospheric Aerosols*, 314–316.

Experimental Device and Setup

The laminar flow chamber is positioned vertically and experimental setup consists of an atomizer, a furnace, a mixing unit, a nucleation chamber, and a measurement unit.

A known amount of studied solution is introduced to furnace with HPLC Pump through a ruby micro-orifice (20 µm). The dispersion is then vaporized in a Pyrex glass tube wrapped with resistant heating wires. The furnace temperature is kept at 330°C, and the temperature of vapor inside furnace is about 130°C. After furnace, the vapor is filtered with Teflon filter, introduced to Teflon mixing unit and cooled by turbulent mixing with particle free air to 60°C. The vapor gas mixture is then cooled to wanted nucleation temperature in nucleating chamber which is kept at constant temperature with two liquid circulating baths. The nucleation chamber is made of stainless steel and its whole length is 200 cm. Temperature of the stream is registered along the nucleation chamber using six PT100 probes.

For nucleation at room temperature, 25°C is achieved at a distance approximately 115 cm from the mixing unit. The measured temperature profile is shown in Figure 1.

Concentration of water vapor is measured at the end of nucleation chamber with a dew point monitor. The relative humidity might be changed in two ways. First by changing the flow rate of solution from HPLC Pump, second by sucking humid air just after the furnace while keeping overall flow rate in the chamber unchanged.

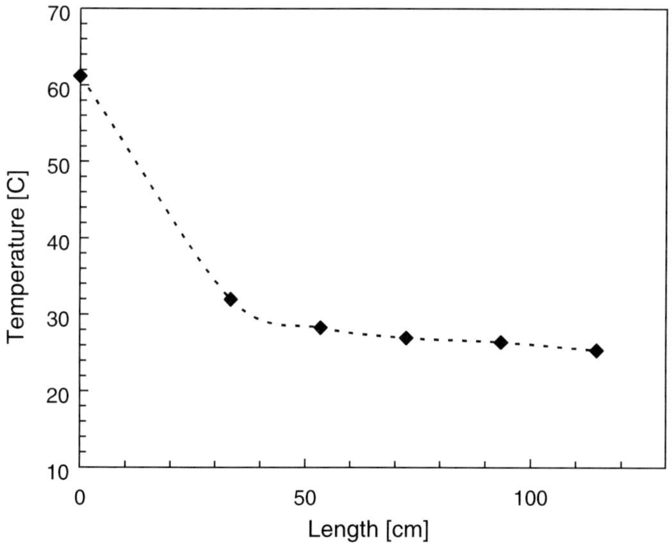

Figure 1 Temperature profile in the axis of laminar flow tube as a function of distance from mixing unit

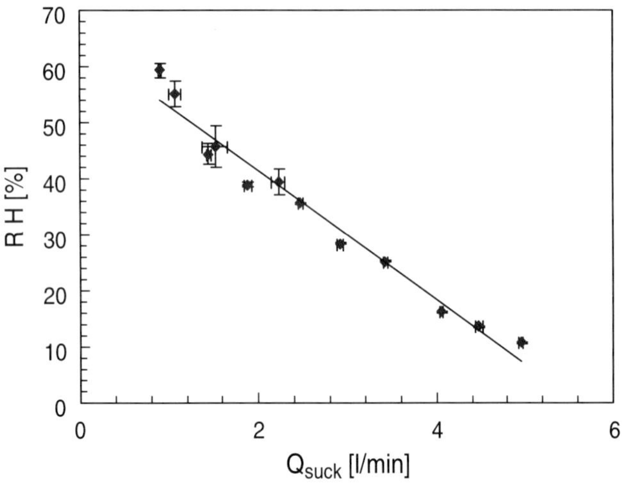

Figure 2 Relative humidity as a function of humid air flow sucked after furnace, error bars are standard deviations from mean values, and solid line is linear fit to data

Second case is displayed in Figure 2. Our applicable range of relative humidity is from 10% to 65%.

Sulfuric acid concentration is determined from samples taken inside the tube. The aerosol number concentration is measured with an ultrafine condensation nucleus counter (TSI 3025).

References

1. Kulmala, M., Vehkamäki, H., Petäjä, T., Dal Maso, M., Lauri, A., Kerminen, V.-M., Birmili, W., and McMurry, P.H., *J. Aerosol Sci.*, **35**, 143 (2004).
2. Seinfeld, J.H. and Pandis, S.N., *Atmospheric Chemistry and Physics: From Air Pollution to Climate Change*, New York: Wiley (1998).
3. Wyslouzil, B.E., Sienfeld, J.H., Flagan, R.C., and Okuyama, K., *J. Chem. Phys.*, **94**, 6842–6868 (1991).
4. Viisanen, Y., Kulmala, M., and Laaksonen, A., *J. Chem. Phys.*, **107**, 920 (1997).
5. Ball, S.M., Hanson, D.R., and Eisele, F.L., and McMurry, P.H., *J. Geophys. Res.*, **104**, D19, 23.709–23.718 (1999).
6. Zhang, R., Suh, I., Zhao, J., Zhang, D., Fortner, E.C., Tie, X., Molina, L.T., and Molina, M.J., *Science*, **304**, 1487 (2004).
7. Berndt, T., Böge, O., and Stratmann, F., *Geophys. Res. Lett.*, **33**, L15817 (2006).

Monte Carlo Simulations on Heterogeneous Nucleation I: The Point Where the Classical Theory Fails

Antti Lauri, Evgeni Zapadinsky, Joonas Merikanto, Hanna Vehkamäki, and Markku Kulmala

Abstract We have performed Monte Carlo simulations of homogeneous and heterogeneous nucleations of Lennard–Jones argon clusters. We interpreted the simulation results using the major concept posing a difference between the homogeneous and heterogeneous classical nucleation theories, the contact parameter. Our results show that the concept describes the cluster–substrate interaction surprisingly well, even for very small molecular clusters. The use of the classical theory concept results in an underestimation of the heterogeneous nucleation rate by 2–3 orders of magnitude. The main contribution is induced by the failure of the classical theory of homogeneous nucleation to predict the energy involved in bringing one molecule from the vapor to the cluster for clusters containing less than approximately 15 molecules.

Keywords Heterogeneous nucleation, Monte Carlo simulations, argon

Introduction

Heterogeneous nucleation is a candidate for triggering atmospheric aerosol activation processes.[1] For practical atmospheric applications it is important to predict the heterogeneous nucleation rate theoretically. Usually the predictions are made using the classical heterogeneous nucleation theory (CHNT).[2] It is based on the same principles as the classical nucleation theory (CNT)[3]: molecular clusters are treated as liquid droplets with bulk liquid values for density and surface tension. When the interaction of a droplet with a substrate surface is taken into account, only one macroscopic parameter, namely the contact angle between the nucleating droplet and the surface is needed. There has been an attempt to substitute the macroscopic contact angle with a microscopic fit.[4]

Department of Physical Sciences, University of Helsinki, P.O. Box 64, FI-00014 University of Helsinki, Finland

C.D. O'Dowd and P.E. Wagner (eds.), *Nucleation and Atmospheric Aerosols*, 317–321.

Besides the classical theory, there exist a number of other methods available for the calculation of the nucleation rate and free energy of formation. With the development of computers, the statistical mechanics (molecular) approach[5] to nucleation has become possible. It allows one to calculate the nucleation rate for homogeneous nucleation with Monte Carlo (MC) simulations. However, the practical application of the molecular approach is limited because of the complex nature of the intermolecular interaction. Most of MC simulations related to nucleation are performed for argon because of the simplicity of the intermolecular potential. On the other hand, experiment on the nucleation of argon is rather difficult, and comparison of the theory and experiment does not allow conclusions on the validity of various nucleation theories. A recent experimental study by Fladerer and Strey,[6] suggesting a vast 30 orders of magnitude difference between experimental and theoretical nucleation rates, gives an extensive data set for comparison.

We have developed a molecular approach to heterogeneous nucleation,[7] and utilized our simulation approach to compare the results to the CHNT concept of the multiplication of the homogeneous nucleation barrier height by the contact parameter. The simulated model system is an argon cluster on platinum substrate. The main question is whether it is enough to know just the contact angle between the cluster and the substrate and the homogeneous nucleation rate in order to calculate the heterogeneous nucleation rate.

Theory

The Classical Approach

The MC simulation method, which we are using, produces the derivative of the free energy of formation with respect to cluster size. To be able to compare the results of the classical approach with the statistical mechanical results, the differential form of the free energy of formation in the CNT is presented here.

To find the relation between the difference in the work of formation between an n- and an $(n-1)$-cluster we will need to express the work of formation in terms of the number of molecules n. The free energy of formation in homogeneous nucleation can be written as

$$\Delta G_{\mathrm{hom}} = -\, nkT \ln S + \alpha n^{2/3}, \tag{1}$$

where $\alpha = (36\pi)^{1/3}\rho_l^{-2/3}\sigma_{lg}$. Differentiation of the work of formation over the number of molecules leads to

$$\frac{\partial \Delta G_{\mathrm{hom}}}{\partial n} = -kT \ln S + \frac{2}{3}\alpha n^{-1/3}. \tag{2}$$

The approximated difference between the work of formation of an n- and an $(n-1)$ -cluster is now given by

$$\Delta G_{\mathrm{hom}}(n)-\Delta G_{\mathrm{hom}}(n-1)\approx\frac{\partial\Delta G_{\mathrm{hom}}}{\partial n}\Delta n, \tag{3}$$

where $\Delta n = 1$. The discrete approach requires rather the work of bringing one molecule from the vapor to the cluster than the derivative, utilizing the concept of the increment of the work of formation:

$$\Delta G_{\mathrm{hom}}(n)-\Delta G_{\mathrm{hom}}(n-1)=\frac{3}{2}s_{\mathrm{hom}}\left[n^{2/3}-(n-1)^{2/3}\right]-kT\ln S,\ (n\geq 2) \tag{4}$$

where s_{hom} is the slope of the linear derivative function.

The Statistical Mechanical Approach

Starting from the law of mass action and Boltzmann-type distribution of clusters of different sizes we have derived the expression for the work of formation of an n-cluster (a cluster containing n molecules) in homogeneous nucleation[8]:

$$\Delta G_{\mathrm{hom}}(n)=\sum_{i=2}^{n}\left(F^{A}_{\mathrm{hom}}(i)-F^{B}_{\mathrm{hom}}(i)-kT\ln\left[i+\delta(i)\right]+kT\ln\frac{i}{\rho_{\infty}\langle v_{\mathrm{free}}\rangle}-kT\ln S\right), \tag{5}$$

where F is the Helmholtz free energy, δ is a term arising from the boundary condition of the cluster definition, and ρ_{∞} is the bulk density of the cluster. Superscripts A and B correspond to two closely related systems, namely an n-cluster and an $(n-1)$-cluster with one noninteracting (free) molecule, and $\langle v_{\mathrm{free}}\rangle$ is the canonically averaged volume available for the free molecule in system B.

Starting from similar starting points as in the homogeneous case, the work of n-cluster formation in heterogeneous nucleation is given by[7]

$$\Delta G_{\mathrm{het}}(n)=\left[F_{\mathrm{het}}(n)-F_{\mathrm{hom}}(n)\right]-\left[F_{\mathrm{het}}(1)-F_{\mathrm{hom}}(1)\right]+\Delta G_{\mathrm{hom}}(n). \tag{6}$$

In this formulation the equilibrium cluster distribution is normalized with respect to the distribution near the substrate surface.

The Simulations

We performed molecular simulations using the MC discrete summation method[9] together with the overlapping distribution method.[10,11] The modeled substance was Lennard–Jones argon (cutoff distance 2.5σ), in heterogeneous nucleation on platinum surface.

We applied the cluster definition of Stillinger[12] with 1.5σ connectivity distance. The simulations were carried out at T = 60K.

Results

One set of simulations was run to obtain the work of formation in homogeneous nucleation for clusters of size of 2–900, and another one for calculating the difference between the work of formation of the clusters in homogeneous nucleation and that in heterogeneous nucleation for clusters of size of 1–195.

Figure 1(a) presents the surface energy difference over a large range of cluster sizes. The results of the simulation runs can be interpreted using the differential form of the classical free energy of formation (Eq. (2)). The slope of the linear fit equals 2/3α, which can be further utilized to obtain a numerical value for the surface tension. A single linear fit is not the proper choice, but use of two separate fits produces a fair agreement with the simulation results. For both cases, the linear dependence changes at n = 16 (n–1/3 ≈ 0.4), producing two different slopes for the linear fit to the data. As expected, in both cases the fits approach zero for infinitely large cluster sizes. The calculated value for the surface tension of Lennard–Jones argon of 2.5σ cutoff is σ_{lg} = 0.0168 N/m.[13]

Figure 1(b) shows that for the free energy difference between heterogeneous and homogeneous nucleation a single fit is a feasible solution – the slopes of the lines are almost identical with and without the 15 smallest clusters. These results suggest that at least in this case the problem of the CNT to describe the properties of the smallest clusters correctly is related to the homogeneous nucleation theory, whereas the CHNT concept of the contact parameter works well. However, the use of the classical homogeneous nucleation theory to obtain the free energy of formation leads to 2–3 orders of magnitude underestimation of the heterogeneous nucleation rate.[13]

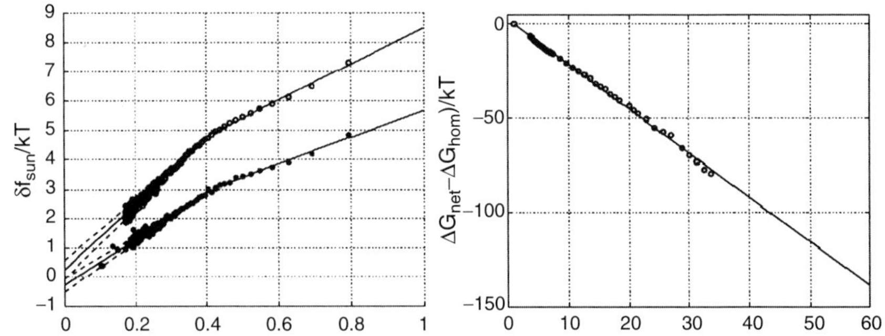

Figure 1 (a) The homogeneous formation free energy difference between an n- and an (n–1)-cluster at T = 60K, S = 1. Dots (potential cutoff 2.5σ) and circles (full potential; shown for reference) correspond to single simulation results; lines represent linear fits to the data. (b) The formation free energy between heterogeneous and homogeneous nucleation at T = 60K

References

1. Kulmala, M., Lehtinen, K.E.J., and Laaksonen, A., *Atmos. Chem. Phys.*, **6**, 787–793 (2006).
2. Volmer, M., *Kinetik der Phasenbildung*, Dresden: Steinkopff (1939).
3. Volmer, M. and Weber, A., *Z. Phys. Chem.*, **119**, 277–301 (1925).
4. Wagner, P.E., Kaller, D., Vrtala, A., Lauri, A., Kulmala, M., and Laaksonen, A., *Phys. Rev. E*, **67**, 021605 (2003).
5. Lee, J.K., Barker, J.A., and Abraham, F.F., *J. Chem. Phys.*, **58**, 3166–3180 (1973).
6. Fladerer, A. and Strey, R., *J. Chem. Phys.*, **124**, 164710 (2006).
7. Zapadinsky, E., Lauri, A., and Kulmala, M., *J. Chem. Phys.*, **122**, 114709 (2005).
8. Lauri, A., Merikanto, J., Zapadinsky, E., and Vehkamäki, H., *Atmos. Res.*, **82**, 489–502 (2006).
9. Bennett, C.H., *J. Comput. Phys.*, **22**, 245–268 (1976).
10. Hale, B.N., *Aust. J. Phys.*, **49**, 425–434 (1996).
11. Hale, B.N. and Ward, R., *J. Stat. Phys.*, **28**, 487–495 (1982).
12. Stillinger, F.H., *J. Chem. Phys.*, **38**, 1486–1494 (1963).
13. Lauri, A., Zapadinsky, E., Vehkamäki, H., and Kulmala, M., *J. Chem. Phys.*, **125**, 164712.

A Kinetically Correct and an Approximate Model of Heterogeneous Nucleation

Anni Määttänen[1,2], Antti Lauri[2], Hanna Vehkamäki[2], Ismo Napari[2], and Markku Kulmala[2]

Abstract We have developed a two-component heterogeneous nucleation model which includes the exact calculation of the kinetic pre-factor using the correct heterogeneous Zeldovich factor for a two-component system. The model is tested by comparing its predictions with a simplified model and with experimental data for water–*n*-propanol nucleating on silver particles.

Keywords Heterogeneous nucleation, cluster

Introduction

Formation of new aerosol particles via gas-to-particle conversion plays an important role in the atmosphere.[1] For example, binary nucleation of water and sulphuric acid,[2] and ternary nucleation of water, sulphuric acid, and ammonia[3] occur more easily than one-component nucleation of water. It has been speculated that in the atmosphere also the nucleation of organics provides new particle formation routes.[4]

In heterogeneous nucleation clusters form on the surfaces of pre-existing particles, facilitating the start of condensational growth. Nucleation rate, or in heterogeneous nucleation probability, is the quantity accessible to laboratory and field measurements. The formation free energy is the key quantity in predicting particle formation rate, which is proportional to the exponential of the formation free energy. The kinetics of the cluster growth is described by two pre-exponential factors: the average growth rate of a critical cluster and the Zeldovich factor, correcting, e.g., for the fact that some clusters that have reached the critical size decay.

[1] *Space Research, Finnish Meteorological Institute, Helsinki, Finland*

[2] *Department of Physical Sciences, University of Helsinki, P.O.Box 64, FI-00014 University of Helsinki, Finland*

C.D. O'Dowd and P.E. Wagner (eds.), *Nucleation and Atmospheric Aerosols*, 322–326.

Theory

The nucleation rate is given by[5]

$$J = \frac{|\lambda_1|}{2\pi kT} F^e \exp\left(-\frac{\Delta G^*_{\text{het}}}{kT}\right)\frac{1}{\sqrt{\left|\det \frac{\mathbf{W}^*}{2\pi kT}\right|}} = R_{\text{av}} F^e \exp\left(-\frac{\Delta G^*_{\text{het}}}{kT}\right) Z, \tag{1}$$

where λ_1 is the negative eigenvalue of the matrix product $\mathbf{R}^*.\mathbf{W}^*$ ($\mathbf{W}^*$ is formed from the second derivatives of formation free energy with respect to cluster sizes, and $\mathbf{R}^*$ is the heterogeneous growth matrix); k is the Boltzmann constant, T is temperature, F^e is the total number of molecules adsorbed per surface area, ΔG^*_{het} is the formation free energy, R_{av} is the average growth rate, and Z is the Zeldovich factor, by definition given by

$$Z_{\text{het}} = \left[\frac{-1}{2\pi kT}\left(\frac{\partial^2 \Delta G_{\text{het}}}{\partial N^2_{\text{het}}}\right)^*\right]^{1/2}. \tag{2}$$

It should be noted that in binary systems N_{het} corresponds to the total number of molecules in the cluster. To obtain an explicit form for the Zeldovich factor we have to take the second derivative of the formation free energy

$$\begin{aligned}\Delta G_{\text{het}} &= \Delta G^{\text{s}}_{\text{het}} + \Delta G^{\text{V}}_{\text{het}} \\ &= \sigma_{gl} 2\pi r^2 (1-\cos\Psi) + (\sigma_{sl} - \sigma_{sg}) 2\pi R^2 (1-\cos\phi) + \Delta\mu N_{\text{het}},\end{aligned} \tag{3}$$

where σ_{gl} is the gas–liquid surface tension, σ_{sl} and σ_{sg} are the liquid–solid and gas–solid surface energies, respectively; r is the radius of the cluster and R the radius of the pre-existing particle. $\Delta\mu$ is the chemical potential difference between gas and liquid, both at the gas pressure. Consult Figure 1 for angles ϕ and Ψ.

The Exact Model

We have used the equimolar surface condition for all the phase interfaces, which means that the total number of molecules in the cluster equals the number of molecules in the hypothetical bulk liquid, since surface excess contributions are set to zero. We have derived the exact Zeldovich factor for one-component heterogeneous nucleation in the case of curved pre-existing surface. In this case the cluster surface area is not proportional to $N^{2/3}$. Our earlier formulation[6] is well applicable for large pre-existing particles acting as CN, but for small CN the differences may be large. For one-component systems the exact formula for the Zeldovich factor in heterogeneous nucleation reads as[7]

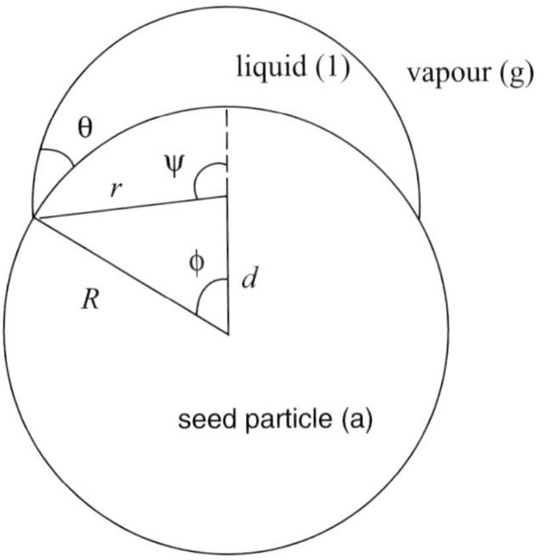

Figure 1 A critical cluster of radius r* on the surface of a pre-existing particle of radius R

$$Z_{\text{het}} = \frac{v_l}{\pi r^{*2}}\left(\frac{\sigma_{lg}}{kT}\right)^{1/2}\left\{2+\frac{(1-mX)\left[2-4mX-\left(m^2-3\right)X^2\right]}{\left(1-2mX+X^2\right)^{3/2}}\right\}^{-1/2}$$
$$= 2Z_{\text{hom}}\left\{2+\frac{(1-mX)\left[2-4mX-\left(m^2-3\right)X^2\right]}{\left(1-2mX+X^2\right)^{3/2}}\right\}^{-1/2}. \quad (4)$$

For binary systems the Zeldovich factor is given by[8]

$$Z_{\text{het}} = \frac{-\left(\mathbf{W}_{11}^{*}+2\mathbf{W}_{12}^{*}\tan\Psi+\mathbf{W}_{22}^{*}\tan\Psi\right)}{1+\tan\Psi}\frac{1}{\sqrt{\left|\det\mathbf{W}^{*}\right|}}, \quad (5)$$

where Ψ is the direction angle of the growth vector in the (N_1, N_2) coordinate system, given by the eigenvector related to the eigenvalue λ_1.

The Simplified Model

In the simplified model,[9] the nucleation rate is calculated by Eq. (1), but with the following two approximations. First, the average growth rate R_{av} is calculated by approximating the direction angle of the growth vector as $\Psi=\arctan[x/(1 - x)]$ where x is the mole fraction of the solvent in the solution. Second, for the Zeldovich factor Z the one-component homogeneous expression

$$Z = \sqrt{\frac{\sigma_{lg}}{kT}\frac{v_m}{2\pi r^{*2}}} \tag{6}$$

is extended to the binary case using the virtual monomer volume $v_m = xv_1 + (1 - x)v_2$, where v_i are the bulk liquid partial molecular volumes of species i. These approximations make it possible to avoid the calculation of the second derivatives of the formation energy, which is numerically the most demanding task in the application of heterogeneous classical nucleation theory.

Results

We tested the models by a comparison to three experiments conducted for water–*n*-propanol mixture nucleating on oxidized Ag particles with an average diameter of 8 nm at T = 285K.[10] The experiments were done for several propanol liquid mass fractions X_l. We chose three cases for the comparison with the models (X_l = 0.653, 0.763, and 0.926).

Figures 2(a) and 2(b) show the kinetic pre-factors and the Zeldovich factors, respectively, of both models for conditions at which nucleation probability P = 0.5 as functions of water activity. The difference in the kinetic pre-factor between the models originates mainly from the difference in the Zeldovich factors. The approximate growth angle Ψ plays a minor role. The Zeldovich factors, and thus the kinetic pre-factors differ 1–5 orders of magnitude, but this does not affect the prediction of the onset conditions.

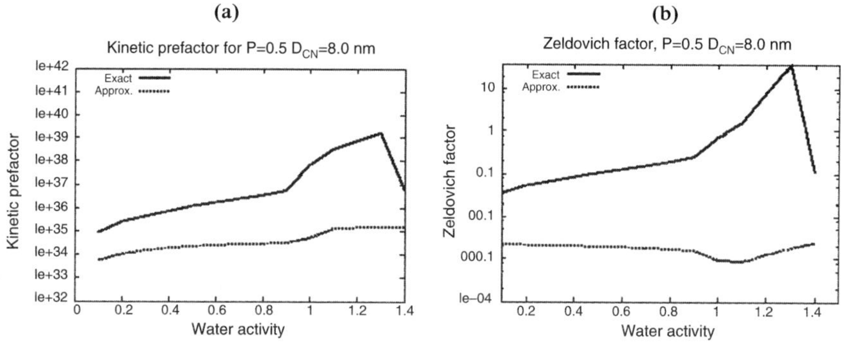

Figure 2 The kinetic pre-factor (a) and Zeldovich factor (b) for conditions at which nucleation probability P=0.5 for binary nucleation with both models

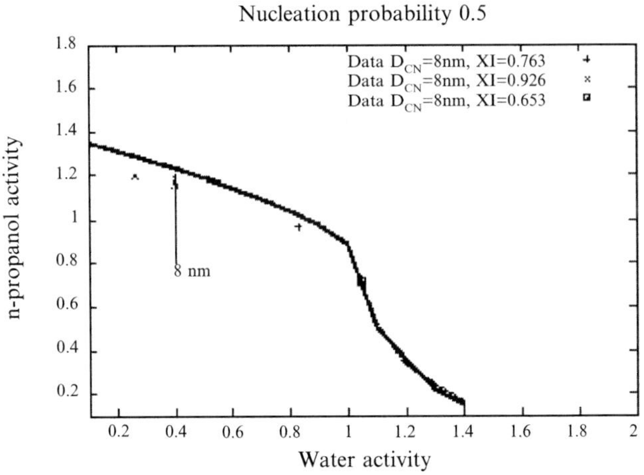

Figure 3 Activity plot for conditions at which nucleation probability P=0.5 for binary nucleation

Figure 3 shows an activity plot for binary nucleation of water and *n*-propanol. Both models are able to reproduce the experimental results fairly well.

References

1. Kulmala, M., Lehtinen, K.E.J., and Laaksonen, A., *Atmos. Chem. Phys.*, **6**, 787–793 (2006).
2. Noppel, M., Vehkamäki, H., and Kulmala, M., *J. Chem. Phys.*, **116**, 218–228 (2002).
3. Napari, I., Noppel, M., Vehkamäki, H., and Kulmala, M., *J. Geophys. Res.*, **107**, 10.1029/2002JD002132 (2002).
4. Gaman, A.I., Kulmala, M., Vehkamäki, H., Napari, I., Mircea, M., and Facchini, M.C., *J. Chem. Phys.*, **120**, 282–291 (2004).
5. Trinkaus, H., *Phys. Rev. B*, **27**, 7372–7378 (1983).
6. Määttänen, A., Vehkamäki, H., Lauri, A., Merikallio, S., Kauhanen, J., Savijärvi, H., and Kulmala, M., *J. Geophys. Res.*, **110**, E02002 (2005).
7. Vehkamäki, H., Määttänen, A., Lauri, A., Napari, I., and Kulmala, M., *Atmos. Chem. Phys.*, **7**, 309–313 (2006).
8. Määttänen, A., Vehkamäki, H., Lauri, A., Napari, I., and Kulmala, M., *J. Phys. Chem.*, submitted (2007).
9. Kulmala, M., Lauri, A., Vehkamäki, H., Laaksonen, A., Petersen, D., and Wagner, P.E., *J. Phys. Chem. B*, **105**, 11800–11808 (2001).
10. Wagner, P.E., Kaller, D., Vrtala, A., Lauri, A., Kulmala, M., and Laaksonen, A., *Phys. Rev. E*, **67**, 021605 (2003).

Reversible Work of the Heterogeneous Formation of an Embryo of a New Phase on a Spherical Charged Conductor Within a Uniform Multicomponent Macroscopic Mother Phase

M. Noppel[1], S. Mirme[1], A. Hienola[2], H. Vehkamäki[2], M. Kulmala[2], and P.E. Wagner[3]

Abstract A thermodynamically consistent formalism is applied to calculate the reversible work needed to form an embryo of a new phase on a charged insoluble conducting sphere within a uniform macroscopic mother phase. An approximate procedure for the calculation of formation free energy of a spherical cap like embryo is described.

Keywords Heterogeneous nucleation, ion-induced nucleation

Introduction

Condensational growth of insoluble aerosol particles is often initiated by heterogeneous nucleation on the surface of these particles. The classical theory of heterogeneous nucleation was developed by Fletcher (1958). The theory was extended to binary systems using the capillarity approximation by Lazaridis et al. (1991). The effect of electrical charges on nucleation is basically considered for the case of ion-induced nucleation where a layer of new phase forms on a perfectly wetted spherical ion or, in case of nonspherical symmetry, the effect is still treated in the frame of this theory and not always thermodynamic consistency is followed.

In this paper we present an expression for the work of embryo formation obtained in a thermodynamically consistent way (Debenedetti and Reiss, 1998). The embryo is formed on a charged seed particle modeled as a spherical conductor. The conditions of equilibrium between the critical nucleus and the mother phase are considered.

[1] *Institute of Environmental Physics, University of Tartu, 18 Ülikooli Str. 50090 Tartu, Estonia*

[2] *Department of Physical Sciences, P.O. Box 64, FIN-00014 University of Helsinki, Finland*

[3] *Fakultaet fuer Physik, Universitaet Wien, Boltzmanngasse 5, A-1090 Wien, Austria*

C.D. O'Dowd and P.E. Wagner (eds.), *Nucleation and Atmospheric Aerosols*, 327–331.

Reversible Work of Formation

The formation of an embryo on a seed particle is considered to take place isothermally in two stages. First an embryo is formed on neutral seed particle and then the seed particle with an embryo on it is charged keeping thermodynamic parameters that determine the state of neutral embryo unchanged. Critical embryos are fragments that are in unstable equilibrium with the mother phase whereas arbitrary size embryos are not. It is assumed implicitly that there exist constraints, which keep also the embryos of arbitrary noncritical size in equilibrium. The reversible work to form a general-size cluster is given by

$$W^{rev} = (p_g^0 - p_l^0)V_l + \sigma_{lg}A_{lg} + (\sigma_{lg} - \sigma_{cg})A_{cl} + \sum_i(\mu_{i,l} - \mu_{i,g})N_{i,l}$$
$$+\sum_s\sum_i(\mu_{i,s} - \mu_{i,g})N_{i,s} + \tfrac{1}{2}\left(\int \mathbf{ED}\big|_{clg}dV - \int \mathbf{ED}\big|_{cg}dV\right), \tag{1}$$

where p_l^0 is the pressure in the electric field-free reference liquid which is in chemical equilibrium with an embryo and p_g^0 is the pressure of vapor phase far away from an embryo, where electric field is absent; V_l is the volume of an embryo, A_{lg} and σ_{lg} are the area and surface tension of the surface between the embryo and vapor phase, A_{cl} and σ_{cl} are the area and surface tension of the surface between the embryo and the seed particle, $N_{l,i}$ and $N_{s,i}$, are the number of molecules of species i in the bulk liquid phase of an embryo and the number of molecules on the sth surface ($s = cl$ or $s = lg$), respectively; $\mu_{i,l}$, $\mu_{i,s}$, and $\mu_{i,g}$ are corresponding chemical potentials of this species in liquid, surface, and vapor phases, respectively; **D** and **E** are the dielectric displacement and the electric field vectors, respectively; subindexes *clg* and *cg* refer to the systems of seed particle–embryo–vapor and seed particle–vapor, respectively. Integration goes over all the volume where electric field is nonzero.

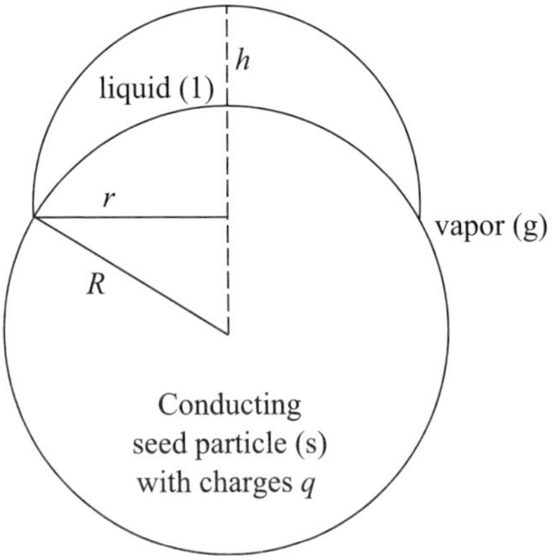

Figure 1 The geometry of an embryo on a spherical seed particle. Radius of the seed particle is R, the height of the top of the liquid embryo from the seed particle surface is h, and the radius of the embryo is r

The last term of the Eq. (1) can be represented also as

$$\tfrac{1}{2}\left(\int \mathbf{ED}\Big|_{clg} dV - \int \mathbf{ED}\Big|_{cg} dV\right) = \frac{q^2}{2}\left(\frac{1}{C_{cl}} - \frac{1}{C_{cg}}\right) = -\tfrac{1}{2}\left(\int_{V_l} (\mathbf{D} - \varepsilon_g \varepsilon_0\, \mathbf{E})\mathbf{E}_0\, dV\right) \quad (2)$$

where q is the electrical charge of a seed particle; C_{cl} and $C_{cg} = 4\pi\varepsilon_0\varepsilon_g R$ are the electrical capacitances of a seed particle with and without an embryo on it, respectively; ε_0 is the permittivity of vacuum; ε_g is the dielectric constant of vapor; R is the seed particle radius, $\mathbf{E}_0$ is the electric field vector around seed particle in a vacuum. The last equality holds for the linear relationship $\mathbf{D} = \varepsilon\varepsilon_0\mathbf{E}$.

In case of nucleation on neutral seed particle, an embryo has a shape of spherical cap. The electric field causes stresses inside an embryo that depend on field strength and direction and, therefore, the shape of an embryo is not any more strictly spherical cap. The search of minimum of Eq. (1) involves solving of a variation problem where the shape of an embryo, and strength of electric field as functions of space coordinates should be varied to find the minimum. This makes the problem complicated, and thus as a first approximation, we assume that the embryo preserves approximately the shape of spherical cap also in electric field and look for the minimum among these type embryos.

The Condition of a Critical Embryo

The conditions of extrema of Eq. (1) yield the conditions of equilibrium between the critical embryo and the mother phase:

$$\mu_{i,l} = \mu_{i,s} = \mu_{i,g} \quad (3)$$

$$(p_l^0 - p_g^0) = \sigma_{lg}\frac{\partial A_{lg}}{\partial V_l} + (\sigma_{cl} - \sigma_{cg})\frac{\partial A_{cl}}{\partial V_l} + \frac{q_2}{2}\frac{\partial}{\partial V_l}\left(\frac{1}{C_{cl}}\right)$$

$$= \frac{2\sigma_{lg}}{r} - \frac{1}{2}\frac{\partial}{\partial V_l}\left(\int_{V_l} \mathbf{PE}_0 dV\right), \quad (4)$$

where r is the radius of spherical surface of liquid–vapor interface of an embryo, Fletcher (1958) has given the relations that determine the relationship of surface areas A_{lg} and A_{cl} versus volume of an embryo V_l. $\mathbf{P}$ is the polarization vector. In deriving the last equality of Eq. (4), the polarization of vapor phase is ignored ($\varepsilon_g = 1$) and the relation $\mathbf{D} = \varepsilon_0\mathbf{E} + \mathbf{P}$ is used in Eq. (2). The integral in the Eq. (4) can be approximately calculated as explained in the following.

For a conducting embryo an external field inside embryo is screened by induced surface charges. These charges create an electric field, which is equal to external field by strength but opposite by direction. Similar behavior is expected for a dielectric embryo. The electric field $\mathbf{E}$ in a dielectric sphere placed into homogeneous external electric field $\mathbf{E}_0$ is given by $\mathbf{E} = 3\mathbf{E}_0/(\varepsilon + 2)$. Supposing that this relationship holds also for an embryo inside the nonhomogeneous field of the seed particle, the polarization can be expressed as $P = E_0 3\varepsilon_0(\varepsilon_l - 1)/(\varepsilon_l + 2)$. The electric field of a seed particle in a vacuum is given by $E_0 = q/(4\pi\varepsilon_0 R^2)$. The volume of an embryo as spherical cap is divided into spherical layers, each layer parallel to the surface of a seed particle. The volume of each layer is multiplied with characteristic value of E_0^2 of that layer and, the integrand is summed up over layers. For the integral of the last term in Eq. (4) an upper and lower bound can be found by assigning E_0^2 its maximum value (at the surface of the seed particle) and minimum value (at the top of the cap shaped embryo)

$$\frac{3(\varepsilon_l - 1)q^2}{16\pi^2\varepsilon_0(\varepsilon_l + 2)} \frac{V_l}{(R+h)^4} < \int_{V_l} \mathbf{PE}_0 dV < \frac{3(\varepsilon_l - 1)q^2}{16\pi^2\varepsilon_0(\varepsilon_l + 2)} \frac{V_l}{R^4}, \tag{5}$$

where h is the distance from the top of an embryo to seed particle surface (the maximal distance of separation).

Assuming incompressibility of an embryo the Eq. (3) can be expressed as

$$k_b T \ln\left(\frac{A_{i,g}}{A_{i,l}(x_{i,l})}\right) = v_i \left(\frac{2\sigma_{\mathbf{lg}}}{r} - \frac{1}{2}\frac{\partial}{\partial V_l}\left(\int_{V_l} \mathbf{PE}_0 \, dV\right)\right), \tag{6}$$

where $A_{i,g} = p^0_{i,g} / p^{pure}_{i,sat}$ is the gas-phase activity of species i, $p^{pure}_{i,sat}$ is the saturated vapor of species i over flat surface of pure liquid, $p^0_{i,g}$ is the partial pressure of species i in the nucleation vapor, $A_i(T,x_{i,l}) = p^0_{i,sat} / p^{pure}_{i,sat}$ is the liquid-phase activity, $p^0_{i,sat}$ is the partial pressure of free molecules of component i in the equilibrium vapor above a flat surface of a solution which composition is given by mole fractions $x_{i,l}$, and v_i is the partial molecular volume of species i.

Set of Eqs. (6), one for each component i, determine the radius r and the composition $x_{i,l}$ of a critical embryo. The first term in Eq. (1) can be expressed as

$$(p_g^0 - p_l^0)V_l = -\frac{V_l}{v} k_b T \ln(S), \tag{7}$$

where

$$\ln(S) = \sum_i x_{i,l} \ln\left(A_{i,g} / A_{i,l}(x_{i,l})\right), \tag{8}$$

$$v = \sum_i x_{i,l} v_i. \tag{9}$$

Eq. (1) with the first term expressed by Eq. (7) gives the formation free energy of a critical embryo.

$$\Delta G = -\frac{V_l}{v} k_b T \ln(S) + \sigma_{cl} A_{lg} + (\sigma_{cl} - \sigma_{cg}) A_{cl} + \tfrac{1}{2}\left(\int \mathbf{ED}\big|_{clg} dV - \int \mathbf{ED}\big|_{clg} dV \right)$$

$$= -\frac{V_l}{v} k_b T \ln(S) + \sigma_{lg} A_{lg} + (\sigma_{cl} - \sigma_{cg}) A_{cl} - \tfrac{1}{2} \frac{3\varepsilon_0(\varepsilon_0 - 1)}{(\varepsilon_l + 2)} \int_{V_l} E_0^2 dV \qquad (10)$$

Acknowledgment This work was supported by the Estonian Science Foundation Grants 5387 and 6223 and the Academy of Finland.

References

Debenedetti, P.G., and Reiss, H., *J. Chem. Phys.*, **108**, 5498–5505 (1998).
Fletcher, N.H., *J Chem. Phys.*, **29**, 572–578 (1958).
Lazaridis, M., Kulmala, M., and Laaksonen, A.B., *J. Aerosol Sci.*, **22**, 823–830, (1991).

Stochastic Birth and Death Equations to Treat Chemistry and Nucleation in Small Systems

Christiane Maria Losert-Valiente Kroon and Ian J. Ford

Abstract We explore a new framework for describing the kinetics of simple chemical reactions which can also be applied to describe the kinetics of heterogeneous nucleation. Traditional treatments neglect the effect of statistical fluctuations in populations. We employ techniques in a manner analogous to the treatment of quantum systems to develop a stochastic description of processes beyond the mean field approximation.

Keywords Nucleation, fundamental aerosol physics, aerosol chemistry, Monte Carlo simulations

Introduction

Chemical and nucleation processes taking place in very small open systems do not necessarily proceed at the scaled down rates of similar processes taking place in large systems. For example, reactions taking place on the surface of a particle of area 1 nm^2 will not proceed at one millionth of the rate of the same process on a particle of area 1 μm^2. In making this claim, we are not appealing to differences in surface properties, such as curvature. The crucial difference is that the populations of reactants in the smaller system are lower, and therefore more susceptible to statistical fluctuations which can alter the reaction rate in an important way. Normal classical chemical and activation properties of ultrafine aerosols can therefore differ from expectations. The effect of this small system stochasticity was explored by Lushnikov, Bhatt, and Ford (2003) and Bhatt and Ford (2003).

The reaction $A + A \rightarrow C$ can provide a simple illustration of the reason for the failure of traditional kinetics. The rate for such a two-body process is proportional to the square of the population reactant A. The mean reaction rate is therefore proportional to the time-average of the squared reactant population. However,

Department of Physics and Astron omy, University College London, Gower Street, London, WC1E 6BT, UK

C.D. O'Dowd and P.E. Wagner (eds.), *Nucleation and Atmospheric Aerosols*, 332–336.

traditional kinetic equations employ the square of the mean reactant population, rather than the mean of the square, and these quantities differ for small systems. For nucleation, the growth rate of a molecular cluster is normally taken to be proportional to the product of mean populations of monomer and cluster, and this assumption is similarly flawed.

In order to treat population fluctuations correctly, we can replace mean population dynamics with a description using master equations. These describe the evolution of the full probability distribution of the reactant population. The methods used are similar to treatments of quantum mechanical systems, where the fluctuations are due to quantum uncertainty, rather than the population uncertainty associated with a small, open, classical system. However, except in some special cases, the equations are too complicated to be solved exactly, and numerical methods are very cumbersome.

Mathematical Methods

A general master equation can be written in the following form

$$\frac{d}{dt}P(m) = \sum_n T(n\,|\,m)P(n) - \sum_n T(m\,|\,n)P(m), \tag{1}$$

where $T(n|m)$ represents the transition amplitude or propagator from a microstate n to a microstate m and $P(m)$ is the probability to find the system in state m. Considering a d-dimensional lattice L, the microstates correspond to the occupation numbers $\{N_i\}$ $= \{N_1,N_2,\ldots,N_f\}$ at each lattice site $i \in L$. We allow multiple occupancy of the lattice sites (bosonic representation). The master equation can be mapped to a second quantized operator description (Doi 1976). We construct a Fock space by introducing creation and annihilation operators at each lattice site which satisfy certain commutation relations. In these terms it can be shown that the master equation is equivalent to a Schroedinger equation with imaginary time, namely

$$\frac{d}{dt}|\Psi(t)\rangle = -H\,|\Psi(t)\rangle, \tag{2}$$

where H is the time-evolution operator and $|\Psi(t)\rangle$ is the many-body wave function. The initial state is chosen corresponding to a Poisson distribution

$$P(\{N_i\};t=0) = e^{n_0}\prod_i \frac{n_0^{N_i}}{N_i!}, \tag{3}$$

at each lattice site with n_0 being the total initial particle density. The second quantized operator description can in turn be recast into a path integral formalism via the coherent state representation (Peliti 1985).

We want to derive the functional integral representation for an average particle distribution. The expectation value of an observable O is given by

$$\langle O\rangle := \sum_{\{N_i\}} O(\{N_i\})P(\{N_i\};t). \tag{4}$$

We define projection states which are left eigenstates of all creation operators with unit eigenvalue. By means of techniques originally used to treat quantum systems one is now able to recast the above average (Täuber, Howard, and Vollmayr-Lee 2005) in the following way

$$\langle O \rangle = C \int \prod_i D\Phi_i D\Phi_i^* O e^{-S[\{\Phi\},\{\Phi^*\}]}, \tag{5}$$

where C is the normalization constant, S denotes the action, Φ is a complex, fluctuating field and Φ^* its shifted complex conjugate.

Chemical Reaction $A + A \rightarrow C$

The chemical reaction between two reaction partners of species A producing a chemical compound C is to take place on the surface of a third particle, for example, two hydrogen atoms H reacting on a space dust particle forming H_2. The evolution of the probability distribution P for N_A, the number of reaction partners A, and N_C, the number of reaction products C, in time t is given by a master equation of the form

$$\begin{aligned}\frac{d}{dt} P(\{N_A\}, \{N_C\}; t) &= \text{absorption of a molecule of species A or C} \\ &+ \text{binary reaction} \\ &+ \text{evaporation of a molecule of species A or C.}\end{aligned} \tag{6}$$

The above master equation is equivalent to a Schroedinger-like wave equation (2) with the Hamiltonian $H = H[\Phi_A(t),\Phi_C(t),\Phi_A^*(t),\Phi_C^*(t)]$. The functions $\Phi(t)$ are fluctuating fields associated with the various particle species. Performing a Gaussian transformation brings in a white Gaussian noise $\eta(t)$, that is the stochastic noise has zero mean value and a correlation given by

$$\langle \eta(t)\eta(t') \rangle = \delta(t - t'). \tag{7}$$

This leads to the Hamiltonian $\bar{H} = \bar{H}[\Phi_A(t),\Phi_C(t),\Phi_A^*(t),\Phi_C^*(t),\eta(t)]$. Taking into account the definition of a functional Dirac δ-distribution and integrating out over the fluctuating fields one obtains the following form for the average particle density of the molecules of species A

$$\langle \Phi_A(t) \rangle = C \int D\eta \Phi_A(t) e^{-\frac{1}{2}\int_0^t \eta^2(s)ds}, \tag{8}$$

with C a normalization constant. The above expression for the averaged particle number is valid only if two constraint equations are satisfied by the fluctuation fields, namely

$$\frac{d}{dt}\Phi_A(t) + 2\kappa\Phi_A^2(t) + \lambda_A\Phi_A(t) - j_A - i\sqrt{2\kappa}\Phi_A(t)\eta(t) = 0, \tag{9}$$

$$\frac{d}{dt}\Phi_C(t) - \kappa\Phi_A^2(t) + \lambda_C\Phi_C(t) - j_C = 0, \tag{10}$$

where κ is the reaction rate to form a reaction product C, λ is the evaporation rate, and j the source rate. The stochastic differential equation (9) for the field $\Phi_A(t)$ resembles the evolution equation for the mean population of A molecules $n_A(t)$ in an astonishing way:

$$\frac{d}{dt}n_A(t) + 2\kappa n_A^2(t) + \lambda_A n_A(t) - j_A = 0. \tag{11}$$

This correspondence suggests that the solution of Eq. (9), $\Phi_A(t)$, averaged over all noise histories corresponds to the mean population $n_A(t)$.

Monte Carlo Simulation

By means of the Milstein scheme we are able to generate a solution to the stochastic equation (9) for the field $\Phi_A(t)$. In Figure 1, the real part of a random walk according to Eq. (9) is plotted for the reaction $H + H \rightarrow H_2$. The values of the rate coefficients at a temperature $T = 10K$, with $\kappa = 9.28 \times 10^4$ $1/s$ and $\lambda_H = 1.88 \times 10^{-3}$ $1/s$, were taken from a paper by Stantcheva, Caselli, and Herbst (2001)

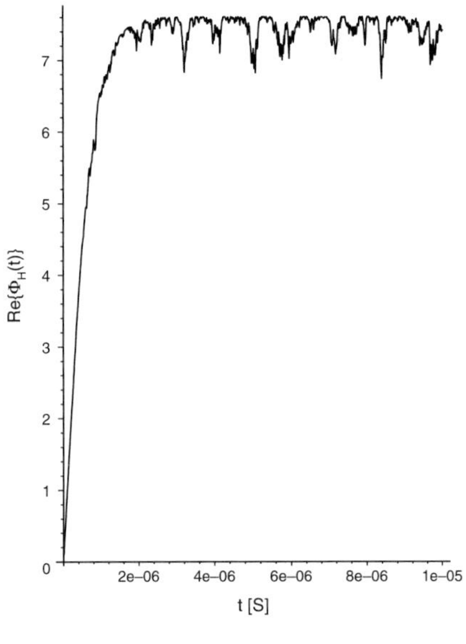

Figure 1 Real part of the solution to the stochastic constraint equation

who investigated the use of rate equations to describe diffusive chemistry occurring on the surfaces of interstellar grains. The source rate was taken to be $j_H = 10^7\ 1/s$, which corresponds to a situation where the classical solution is still accurate. For the above values of the rate coefficients the stochastic approach undertaken by Lushnikov, Bhatt, and Ford (2003) gives an average particle density of $\langle \Phi_H \rangle = 7.47$ for the steady state, whereas the average classical particle number for the steady state is $n_A = 7.34$. One observes from Figure 1 that the real part of the random walk solution to Eq. (9) fluctuates, after a relaxation time, around the steady state value.

This model will be extended to deal with the Becker–Döring birth and death equations central to nucleation theory.

Acknowledgment This work was supported by the Leverhulme Trust under grant F/07134/BV.

References

1. Bhatt, J.S. and Ford, I.J., *J.Chem.Phys.*, **118**, 3166–3176 (2003).
2. Doi, M., *J. Phys. A: Math. Gen.*, **9**, 1465–1477 (1976).
3. Lushnikov, A.A., Bhatt, J.S., and Ford, I.J., *J. Aerosol Sci.*, **34**, 1117 (2003).
4. Peliti, L., *J. Phys.*, **46**, 1469 (1985).
5. Stantcheva, T. Caselli, P., and Herbst, E., *A&A*, **375**, 673–679 (2001).
6. Täuber, U.C., Howard, M., and Vollmayr-Lee, B.P., *J. Phys. A: Math. Gen.*, **38**, R79-R131 (2005).

Part III
Nucleation: Ion and Cluster Properties

Surface Nucleation in Freezing Nanoparticles

Richard K. Bowles and Eduardo Mendez-Villuendas

Abstract Monte Carlo simulation techniques are used to calculate the nucleation free energy barrier for freezing in a 456 atom gold nanoparticle. Our results show that the solid embryo forms at the surface of the particle and that the usual classical nucleation model, with the embryo growing in the core of the cluster, is unable to predict the shape of the free energy barrier. We use a simple partial wetting model that treats the crystal as a lens-shaped nucleus at the liquid–vapour interface and find that it is necessary to introduce a negative line tension associated with the three phase contact line to obtain a good fit to the calculated nucleation barrier.

Keywords Nanoparticle, freezing, pseudo-heterogeneous nucleation, line tension

Introduction

It is well known that, as particles become smaller, surface effects become more important so it should not be surprising that surface phenomena are important in the nucleation behaviour of nanoscale systems. For example, molecular dynamics simulations used to study the freezing of a 561 atom gold cluster, cooled at a constant rate from above the melting temperature, show the formation of the icosahedral structure is initiated by ordering at the surface of the particle rather than in the core.[1] On the other hand, Lennard–Jones clusters initially freeze to a core-ordered icosahedron with a disordered surface,[2] which points to a core nucleation mechanism. This suggests that the preferred location for the formation of the growing embryo will be determined by the wetting behaviour of the liquid–solid interface. In particular, Djikaev et al.[3] used a simple phenomenological model to show that the pseudo-heterogeneous nucleation process, with the solid embryo growing at the liquid–vapour interface, is thermodynamically favoured compared to nucleation in the bulk, if the

Department of Chemistry, University of Saskatchewan, Saskatoon, Saskatchewan, Canada S7N 5C9

C.D. O'Dowd and P.E. Wagner (eds.), *Nucleation and Atmospheric Aerosols*, 339–343.

liquid partially wets the crystal. Understanding where nucleation takes place within the nanoparticle is important because core and surface nucleation rates can differ by orders of magnitude. Furthermore, surface tensions or surface free energy densities in the case of crystals are often obtained by fitting phenomenological models to nucleation data and we need to know the details of the proposed model.

Simulation Methods

The use of umbrella Monte Carlo (MC) simulation techniques combined with parallel tempering to calculate the free energy barrier associated with forming an n-sized embryo is now well established.[4,5] The key step is to develop an embryo criteria that is able to identify solid-like particles within the liquid and determine which of these atoms are locally connected such that they belong to the same embryo. We follow the work of Frenkel[5] and use the correlation of the Steinhardt q_6 order parameter between two neighbouring atoms to designate that two particles are "bonded", using a threshold value for $q_6 \cdot q_6 > 0.65$ which was obtained by studying distribution of the liquid and various solid structures of the cluster. We then label a particle as being solid when it is bonded to at least half its neighbours. Finally, two solid atoms are considered to be in the same embryo if they are bonded.

We use the semi-empirical embedded-atom method (EAM) potential for gold[6] and our cluster consists of N = 456 atoms. The simulations are carried out using periodic boundary conditions with a cell volume V = 1,500 A^0.[3] At each temperature, we divide our embryo size coordinate, n, into eight windows each with a parabolic biasing potential, $w(n_{max}) = 0.0005\,(n_{max} - n_0)^2$, which forces the system sample states where the largest embryo in the system, n_{max}, is near n_0. We choose $n_0 = 0,10,20,30,\ldots,70$ and study temperatures $T = 750,730,710,690,680,670,660,650$. Our tempering scheme allows exchanges between neighbouring windows in both n_0 and T and these exchanges have acceptance ratios of about 0.4 and 0.6, respectively. The embryo criterion is computationally expensive to apply so we use trajectories that consist of 10 normal MC moves for every particle in the cluster sampling the potential, followed by a test against $w(n_{max})$. If the final move is rejected, the system is returned to state at the beginning of the trajectory. Our results were obtained as an average of four independent simulations, with each sampling 436,000 trajectories in each window at every temperature, after the system reached equilibrium.

Results and Discussion

To determine if the solid embryo is forming in the core of the nanoparticle or at the surface, we simply examine a number of configurations and count the number of atoms that are both in the embryo and on the surface of the particle. To identify surface atoms, we use a "cone" algorithm with an apex angle of

120°. Figure 1 shows that approximately 46% of the atoms in embryos larger than 20 atoms are on the surface of the cluster, clearly suggesting the solid is only partially wet by the liquid. For smaller embryos, we see that this percentage increases to 63%. The degree of partial wetting also appears to be independent of temperature.

The work of forming an n-sized embryo is obtained from the simulation by calculating the equilibrium number of n-sized embryos, and using the relation[4,5]

$$N_n / N \approx \exp[-\Delta W(n)/kT], \tag{1}$$

where the approximation holds for rare embryos, i.e., cases of low to moderate undercooling. Figure 2 shows the resulting free energy barrier at $T = 710\,K$, along with a fit for the data using a simple capillarity-type model where we have assumed complete wetting (core nucleation) and a spherical embryo (dotted line). While the model gives a fair estimate of the height of the barrier, it clearly fails to predict the shape and is particularly poor in the region of small embryos.

To account for the partial wetting of the embryo we assume the solid embryo grows at a planar liquid–vapour interface in the shape of a lens[7] (see Figure 3). The work of forming such an n-sized embryo is given by

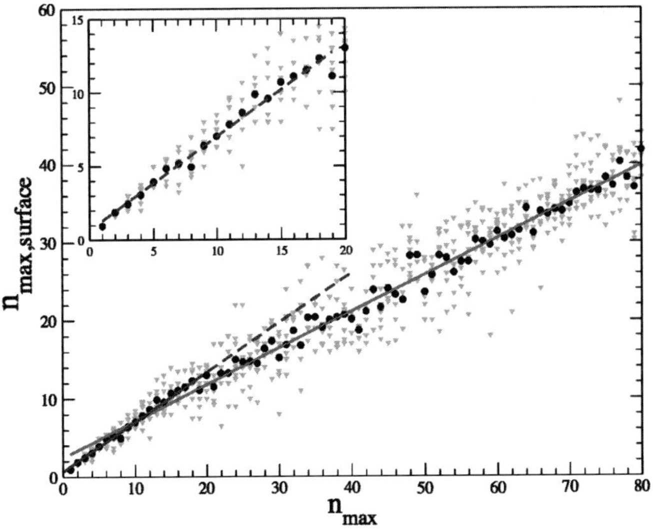

Figure 1 The average number of atoms in an n_{max}-sized embryo that appear on the surface of the nanoparticle as a function of the size of the embryo. Dark circles represent the average over all temperatures studied. The light triangles are the averages obtained for each independent temperature. The dotted and solid lines represent linear best fits to the data for small ($n_{max} < 20$) and large ($n_{max} > 20$) embryos. **Inset**: Expansion of small embryo region of plot. (Reprinted with permission from Ref. 7.)

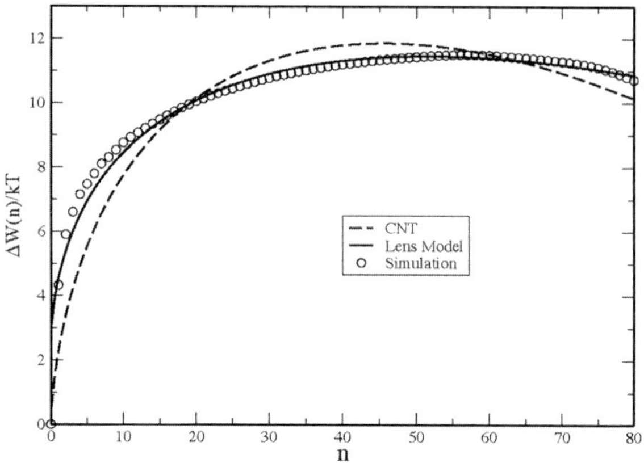

Figure 2 Phenomenological model data fits to the calculated free energy barrier at T = 710 K. The lens model with line tension included (**solid line**) and CNT, assuming core nucleation (**dotted line**). (Reprinted with permission from Ref. 7.)

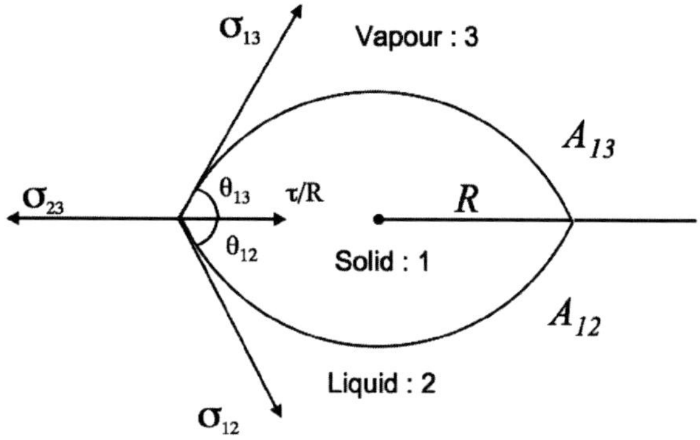

Figure 3 Schematic cross section of a solid (**phase 1**) lens nucleus forming at the liquid (**phase 2**)–vapour (**phase 3**) interface. A_{12} and A_{13} are the solid–liquid and solid–vapour interfacial surface areas respectively and R is the radius of the lens. The four arrows originating from the 3-phase contact are the force vectors of the surface tensions σ_{ij} and the line tension τ/R. (Reprinted with permission from Ref. 7.)

$$\frac{\Delta \mathrm{W(n)}}{\mathrm{kT}} = n\Delta\mu + R^2\left[\sum_{i=2,3}\frac{2\pi\sigma_{1i}}{1+\cos\theta_{1i}} - \pi\sigma_{23}\right] + 2\pi R\left[\tau + \frac{\tau_0}{R}\right], \tag{2}$$

where $\Delta\mu$ is the difference in chemical potential between the liquid and solid phase, the terms in R^2 arise from the surfaces A_{12} and A_{13}, while the final term is due to the line tension τ, and its curvature correction τ_0, associated with the three phase contact line. It should be noted that, by including the line tension in the force balance at the contact line, θ_{12} and θ_{13} become size dependent.

Figure 2 shows a fit of our model to our calculated data where we have used $\Delta\mu$, σ_{12}, τ and τ_0 as fit parameters and the known values of σ_{13} and σ_{23}. This is a clear improvement over the core nucleation model. Data fits that do not include both τ and τ_0 only do as well as the core model, suggesting the line tension is a key element in the work of forming solid-like embryos at t he liquid–vapour interface. Finally, we note that τ was negative and this may help explain the size dependency of the number of embryo atoms in the surface. A negative line tension acts to stretch the lens of a given size, pulling more atoms into the surface. However, τ is linear in R, and so is only important at small sizes. As the embryo increases in size, the surface terms dominate, causing the lens to become more spherical.

Acknowledgments We would like to thank Ivan Saika-Voivod and Peter Poole for helpful discussions. We acknowledge NSERC for funding support and WestGrid for computing resources.

References

1. Nam H.-S., Hwang, N.M., Yu, B.D. and Yoon, J.-K., *Phys. Rev. Lett.*, **89**, 275502 (2002).
2. Noya, G. and Doye, J.P.K., *J. Chem. Phys.*, **124**, 104503 (2006).
3. Djikaev, Y.S., Tabazadeh, A., Hamill, P., and Reiss, H., *J. Phys. Chem.*, **106**, 10247 (2002).
4. ten Wolde, P.R. and Frenkel, D., *J. Chem. Phys.*, **109**, 9901 (1998).
5. Auer, A. and Frenkel, D., *Phys. Rev. Lett.*, **91**, 015703 (2003).
6. Foiles, S.M., Baskes M.I., and Daw, M.S., *Phys. Rev. B*, **33**, 7983 (1986).
7. Mendez-Villuendas, E. and Bowles, R.K., *Phys. Rev. Lett.*, **98**, 185503 (2007).

Biogenic Sesquiterpenes and Atmospheric New Particle Formation: A Boreal Forest Site Investigation

Boris Bonn[1,2], Michael Boy[1], Miikka Dal Maso, Hannele Hakola[3], Anne Hirsikko[1], Markku Kulmala[1], Theo Kurtén[1], Lauri Laakso[1], Jyrki Mäkelä[4], Ilona Riipinen[1], Üllar Rannik[1], Sanna-Liisa Sihto[1], and Taina M. Ruuskanen[1]

Abstract The link between ambient biogenic volatile organic compounds (VOCs) and secondary new particle formation has been hypothesized for a long time. However no clear indication for the participation in the very first steps has been found yet. In this study we link atmospheric air ion concentrations between 0.56 and 0.75 nm in mobility-derived diameter to a first stage oxidation product of sesquiterpenes, the C_{15}-stabilized Criegee intermediates. These intermediates, which are obtained from air ion measurements at Hyytiälä (Finland), react subsequently with either sulphuric acid to form organo sulfates or with pinonaldehyde producing secondary ozonides to initiate new particle formation. Assuming activation to occur by reactive sulphuric acid uptake, we calculated new particle formation rates observed during the BACCI/QUEST IV campaign and found a good match between the calculated and the measured ones. However, the height of new particle formation is dependent on the stability of the boundary layer and the resulting vertical profile of sesquiterpenes. The nice agreement between measurements and calculations provide a first indication about the importance of the biosphere with respect to nucleation and a further feedback within the climate system.

Keywords Heterogeneous nucleation, sesquiterpenes, sulphuric acid, boreal forest

[1]*Department of Physical Sciences, Helsinki University, P.O. Box 64, FI-00014 Helsinki, Finland*

[2]*Now at: Department of Plant Physiology, Estonian University of Life Sciences, Tartu, Estonia*

[3]*Finish Meteorological Institute, Helsinki, Finland*

[4]*Institute of Physics, Tampere University of Technology, Tampere, Finland*

C.D. O'Dowd and P.E. Wagner (eds.), *Nucleation and Atmospheric Aerosols*, 344–349.

Ambient Sesquiterpene Concentrations at a Boreal Forest Site

About 50 years ago Went (1960) postulated the importance of biogenic VOCs emitted by forests for atmospheric new particle formation in his "blue haze" article. However, neither a proof nor the importance of different VOCs with respect to new particle formation has been gained so far. This is mainly caused by the lack of understanding of the very first steps of organic nucleation. Bonn and Moortgat (2003) hypothesized that sesquiterpenes ($C_{15}H_{24}$), which are predominantly reacting with ozone, might be capable in doing so. It has been shown in laboratory studies (Bonn et al. 2002; Bonn and Moortgat 2003) that this organic nucleation is anticorrelated with water vapour, which reacts with the stabilized Criegee Intermediates (sCIs), preventing the formation of nucleation initiating species.

This anticorrelation of nucleation events and water vapour has been found at ambient conditions too (Hyvönen et al. 2005). By contrast, Laakso et al. (2004) and Kulmala et al. (2006) found strong correlations of nucleation rates with ambient sulphuric acid concentrations at the boreal forest site in Hyytiälä (Finland, 61°51′ N, 24°17′ E, 180 m a.s.l.). So which one is right or are all observations explainable by a formation mechanism including both, the biogenic VOCs and sulphuric acid? The major problem to measure ambient sesquiterpene concentrations and to test hypotheses is the high reactivity with ozone, leading to an atmospheric lifetime of less than 2 min, which is too short to gain sufficient material for chemical analysis.

Here we use an indirect approach (Bonn et al. 2006) to gain ambient sesquiterpene concentrations from air ion measurements performed continuously with the Balanced Scanning Mobility Analyzer (BSMA (Tammet 2004)) and the Air Ion Spectrometer (AIS, Airel Ltd.) at the boreal forest station in Hyytiälä. For this purpose we assume the measured positive and negative air ions between 0.56 and 0.75 nm in diameter to be charged sCIs formed by the reaction of β-caryophyllene (sesquiterpene) with ozone. Calculations of electron and proton affinities have shown that the sCIs are highly favourable in receiving ions of both polarities and that the amount of ions is remarkably anticorrelated with ambient water vapour. This is expected because sesquiterpene derived sCIs originate from the reaction of sesquiterpenes with ozone and are destroyed mainly by reaction with water vapour due its high abundance. For more details please refer to Bonn et al. (2006).

This resulted in a seasonal behaviour of sesquiterpene volume mixing ratios as shown in Figure 1. As expected sesquiterpene mixing ratio and thus concentration is highest during summer, but it was found to be significant even in early spring, e.g., sometimes exceeding 10 pptv.

Vertical Distribution of Sesquiterpenes

When investigating new particle formation from sesquiterpene reactions a further aspect needs to be taken into account, i.e., the remarkable variation of sesquiterpene concentrations with altitude, caused by the emission in the canopy and the high

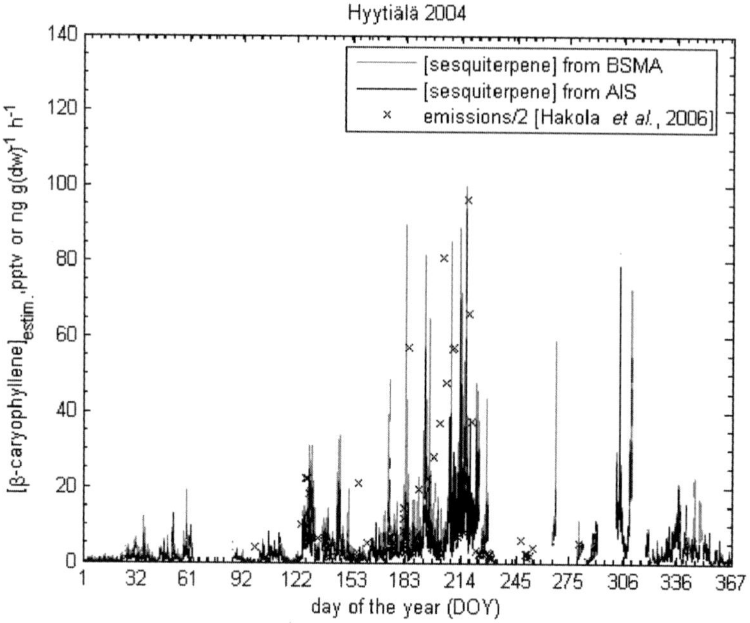

Figure 1 Shown are derived sesquiterpene volume mixing ratios in ppt at cottage level from BSMA and AIS measurements. Data gaps in August and September are caused by a gap in water vapour or ozone measurements. Spotwise branch emission measurements (x) by Hakola et al. (2006) are shown for intercomparison

reactivity of the VOC with ozone. To investigate this, we performed Lagrangian back trajectories for the air sampled at 2 m in height (cottage) until reaching the canopy. Several thousands of back trajectories have been calculated for different stability conditions of the meteorological boundary layer and have been averaged for 13 different stability cases. With this the destruction by ozone was taken into account. According to the actually measured stability the different cases were applied to obtain sesquiterpene concentrations at the top of the canopy.

This resulted in a severe gradient during the nocturnal stable boundary layer and a weaker profile during noon. This can be seen in Figure 2 for an exemplary time period in 2004. During daytime it can occur that surface layer sesquiterpene concentration exceed the canopy ones because of the fast transport in the canopy compared to the surface layer.

The Link of Biogenic Terpenes to New Particle Formation

It has been found by Laakso et al. (2004) and Kulmala et al. (2006) that new particle formation at the boreal forest site of Hyytiälä is linked to the concentration of sulphuric acid either linearly or squared. Such kind of behaviour can be explained

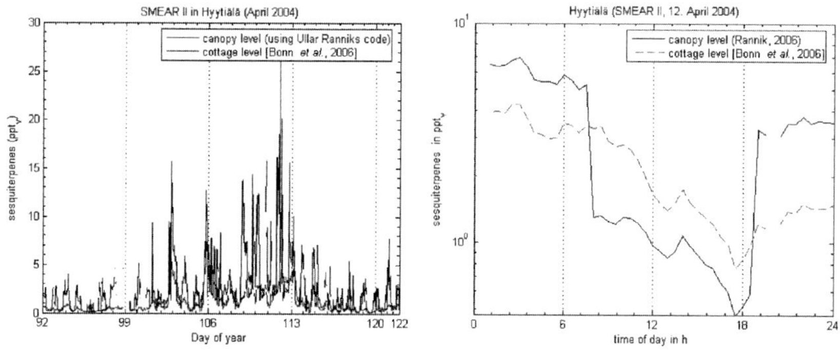

Figure 2 Displayed are calculated sesquiterpene volume mixing ratios at cottage level and at the top of the canopy. Shown is a typical daily pattern in April 2004. Especially during night-time the canopy concentration is much higher than the surface layer one at 2 m in height

by the participation of sulphuric acid in the formation process of the initiating molecule (sCI + H_2SO_4 → organo sulfates (Kurtén et al. 2007)) as well as in the activation.

Participation of sulphuric acid in both leads to a squared relationship, while participation in the activation process only results in a linear relationship. The linear relationship assumes secondary ozonides, such as formed by sesquiterpene derived sCIs and pinonaldehyde (oxidation product of the monoterpene α-pinene), to be the initial molecule to be activated. Both initiating molecules (i) organo sulfates and (ii) secondary ozonides have the advantage to be reactive and will bind further sulphuric acid molecules by reactions rather than by condensation (see contribution of Kurtén et al. 2007b).

Therefore, we do not assume a homogeneous nucleation of a single compound, but an activation type process similar to the cloud condensation nuclei replacing the condensation process by reactions in the very first stage of activation. The formation rate at about 1.2 nm in diameter is formulated like this (Bonn et al. 2007):

$$J_{1.2,linear} = k_{sCI+aldehyde} \cdot [sCI] \cdot [pinonaldehyde] \cdot \tau_{SOZ} \cdot [H_2SO_4] \cdot collision \quad (1a)$$

$$= A \cdot [H_2SO_4] \text{ and} \quad (1b)$$

$$J_{1.2,squared} = k_{sCI+H_2SO_4} \cdot [sCI] \cdot [H_2SO_4] \cdot \tau_{OS} \cdot [H_2SO_4] \cdot collision \quad (2a)$$

$$= K \cdot [H_2SO_4]^2. \quad (2b)$$

$k_{sCI+aldehyde}$ is the reaction rate constant between sCIs and pinonaldehyde (here: 10^{-14} cm^3 molecule^{-1} s^{-1}) and $k_{sCI+H2SO4}$ the one of sCIs and sulphuric acid (here: 5×10^{-13} cm^3 molecules^{-1} s^{-1}). τ abbreviates the lifetime of secondary ozonides (SOZ) or organo sulfates (OS). *Collision* is the collision rate of sulphuric acid (H_2SO_4) molecules

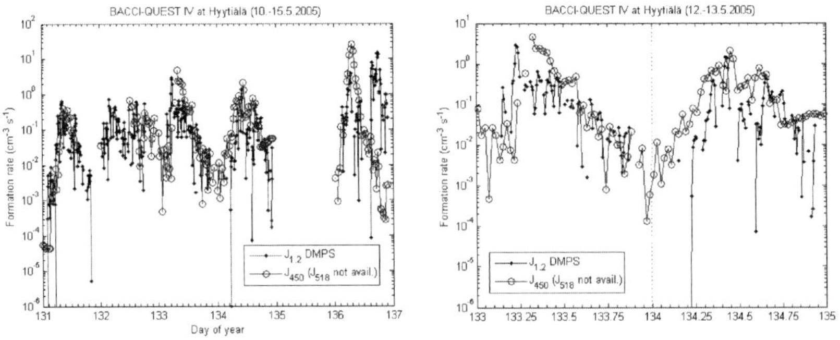

Figure 3 Displayed are nucleation rates obtained from DMPS measurements, condensation sink, and growth rate and the formation rate of organo sulfates activated by a single sulphuric acid molecule (squared relationship of sulphuric acid concentration and formation rate) for an exemplary week (**left picture**) and for two selected days (**right plot**)

and SOZ or OS. And finally $[i]$ is the concentration of compound i. The lifetime of the nucleation initiating molecules (*ca.* 1.2 nm) is been given by the inverse condensation sink (CS) in s^{-1} for sulphuric acid molecules.

To check the formulation given above by real atmospheric conditions, we applied the formulas to the conditions found during the BACCI/QUEST IV campaign at Hyytiälä in spring 2006 and compared the results with air ion and DMPS measurements. Therefore we used ozone, water vapour, sulphuric acid, temperature, and pinonaldehyde measurements as well as condensation sink values calculated from measured particle size distributions. Sesquiterpene concentrations at cottage level (2 m in height) have been obtained by the method of Bonn et al. (2006). An exemplary intercomparison is shown in Figure 3.

Taking into account the uncertainty in sesquiterpene concentrations because of the height of initial cluster formation the calculated values agree fairly well with the formation rates obtained from measurements. Thus, this approach can be considered as a first indication for the link of sesquiterpenes and sulphuric acid to new particle formation events observed at Hyytiälä. However, much more information about the emission characteristics of sesquiterpenes and about the individual steps of new particle formation needs to be known to prove this relation and to extend the present results to other conditions and different sites on the Earth's surface.

Acknowledgments We acknowledge the support from the staff of the Hyytiälä field forest station as well as by all the collaborating partners within the BACCI and QUEST projects. The BACCI project is kindly acknowledged for financial support.

References

1. Bonn, B., Schuster, G., and Moortgat, G.K., *J. Phys. Chem. A*, **106**, 2869–2881 (2002).
2. Bonn, B. and Moortgat, G., *Geophys. Res. Lett.*, **30**, 1585, doi: 10.1029/2003GL017000 (2003).

3. Bonn, B., Hirsikko, A., Hakola, H., Kurtén, T., Laakso, L., Boy, M., Dal Maso, M., Mäkelä, M., and Kulmala, M., *Atmos. Chem. Phys. Diss.*, **6**, 13165–13224 (2006).
4. Bonn, B., Kulmala, M., Riipinen, I., Sihto, S.-L., and Ruuskanen, T.M., *J. Aerosol Sci.*, submitted (2007).
5. Hyvönen, S., Junninen, H., Laakso, L., Dal Maso, M., Grönholm, T., Bonn, B., Keronen, P., Aalto, P., Hiltunen, V., Pohja, T., Launiainen, S., Hari, P., Mannila, H., and Kulmala, M., *Atmos. Chem. Phys.*, **5**, 3345–3356 (2005).
6. Laakso, L., Anttila, T., Lehtinen, K.E.J., Aalto, P.P., Kulmala, M., Hõrrak, U., Paatero, J., Hanke, M., and Arnold, F., *Atmos. Chem. Phys.*, **4**, 2353–2366 (2004).
7. Kulmala, M., Lehtinen, K.E.J., and Laaksonen, A., *Atmos. Chem. Phys.*, **6**, 787–793 (2006).
8. Kurtén, T., Bonn, B., Vehkamäki, H., and Kulmala, M., *J. Phys. Chem. A*, in press (2007).
9. Kurtén, T., Bonn, B., Vehkamäki, H., and Kulmala, M., submitted to *Proceedings of 17th International Conference on Nucleation and Atmospheric Aerosols*, Galway (2007b).
10. Tammet, H., in *Proceedings of 16th International Conference on Nucleation and Atmospheric Aerosols*, Kyoto (2004).
11. Went, F.W., *Nature*, **187**, 641–643 (1960).

Using Charging State in Estimating the Relative Importance of Neutral and Ion-induced Nucleation

Veli-Matti Kerminen[1], Tatu Anttila[1], Tuukka Petäjä[2], Lauri Laakso[2], Stephanie Gagne[2], Kari E. J. Lehtinen[3], and Markku Kulmala[2]

Abstract In this work we investigate which quantities or parameters determine the charging state of a growing nucleation mode under different atmospheric conditions. We further investigate whether and when measured charging states can be used to get information about the relative importance of neutral and ion-induced nucleation.

Keywords Atmospheric nucleation, air ion measurements, analytical expression, fitting

Introduction

The role of ions in atmospheric aerosol formation is a controversial topic. The main reason for this is that ion-induced nucleation rates have traditionally been estimated using models only (e.g., Yu et al. 2006). An alternative way to approach the problem is to use available measurements from the 3–20 nm size range and, by using the data from these measurements, to try to interpret what has been taking place at cluster diameters below 3 nm.

One promising way of estimating the role of ion-induced nucleation is to use the concept of a charging state (Iida et al. 2006; Vana et al. 2006; Laakso et al. 2007). The charging state is defined as the ratio between two quantities: the charged fraction of particles at a certain diameter and the corresponding equilibrium charged fraction.

[1] *Finnish Meteorological Institute, Research and Development, P.O. Box 503, FI-00101 Helsinki, Finland*

[2] *University of Helsinki, Department of Physics, P.O. Box 64, FI-00014 Helsinki, Finland*

[3] *Finnish Meteorological Institute, Kuopio Unit and University of Kuopio, Department of Physics, P.O. Box 1624 FI-70811 Kuopio, Finland*

C.D. O'Dowd and P.E. Wagner (eds.), *Nucleation and Atmospheric Aerosols*, 350–353.

The aim of this work is to find out which quantities, or parameters, determine the charging state of a growing nucleation mode under different atmospheric conditions, and to illustrate under which conditions the measured charging state could tell us something about the original nucleation mechanism. We approach the problem using a simple theoretical framework.

Theoretical Framework

The charging state of a growing nucleation mode is a dynamical variable that depends in a complicated way on its initial value determined by the nucleation mechanism (neutral versus ion-induced) and the rates at which air ions collide with each other and with neutral aerosol particles (see, e.g., Hoppel and Frick 1986).

By making a number of simplifying assumptions, it can be shown that the size dependency of the charging state (S) of a growing nucleation mode can be described with the following differential equation (manuscript in preparation):

$$\frac{dS(d_p)}{dd_p} = K - \left(K + \lambda/d_p\right)S. \tag{1}$$

Here d_p is the mean diameter of the nucleation mode, λ is a parameter related to the aerosol-ion attachment rate (typically in the range 1–1.5), and K is a parameter given by

$$K = \frac{\alpha N_{\pm}^{C}}{\mathrm{GR}(d_p)}, \tag{2}$$

where α is the ion–ion recombination coefficient (about 1.6×10^{-6} cm^3 s^{-1}), $N_{\pm}^{C}$ is the concentration of negative or position cluster ions, and GR is the nuclei growth rate.

Equation (1) can be solved analytically only if the particle growth rate is independent of particles size, i.e., $\mathrm{GR}(d_p) = \mathrm{GR} =$ constant. Assuming further that $\lambda = 1$ we obtain:

$$S(d_p) = 1 - \frac{1}{Kd_p} + \frac{1+(S_0-1)Kd_0}{Kd_p}\exp\left[-K(d_p-d_0)\right], \tag{3}$$

where S_0 is the nuclei initial charging state at the diameter of d_0. Comparisons to detailed simulations with full ion balance equations showed that the error S predicted by equation (3) is usually in the range 10–20% in moderately polluted environments. Larger errors may occur if the total number concentration of nucleation mode particles is much larger than the cluster ion concentration, if the value of K varies with time or size, or if the initial charging state of the nuclei population is larger than about 50–100, i.e., the nuclei have been borne dominantly by ion-induced nucleation.

Results and Discussion

The analysis made in the previous section suggests that the charging state of a growing nucleation mode depends essentially on two parameters: the initial charging state of the mode (S_0) and the parameter K given by equation (2). The value of S_0 is expected to depend on the nucleation mechanisms, being larger when the contribution of ion-induced nucleation to the overall nucleation rate is higher. The parameter K tells essentially how rapidly the growing nucleation mode approaches charge equilibrium. Since K is directly proportional to the number concentration of ion clusters and inversely proportional to the nuclei growth rate, charge equilibrium should be achieved faster for slowly growing nuclei populations and when the number concentration of cluster ions is high.

Typical values of K in the lower troposphere are in the range 0.1–10 nm^{-1}. By relying on our theoretical framework, it can be shown that when the value of K is larger than a certain threshold (2–4 nm^{-1}), practically all information about the initial charging state of the growing nuclei population will be lost by the time these nuclei reach measurable sizes (>3 nm diameter). In such cases there is no way of making interpretation about the potential importance of ion-induced nucleation using current instrumental techniques.

When the value of K is smaller than about 2–4 nm^{-1}, measured charging states can in principle be extrapolated down to the sizes at which nuclei were initially born in the atmosphere. We tested this in practice by fitting the function given by equation (3) to real measurement data obtained from Ion-DMPS measurements. The Ion-DMPS is a new instrument which measures the nuclei charging state directly at a few selected sizes in the range 3–15 nm (Laakso et al. 2007). Our tests demonstrated that equation (3) can be fitted to such data to obtain the values of K and $S_0(d_0)$. This means when combined with suitable measurements, our simple theoretical framework can be used to get information about the relative importance of neutral and ion-induced nucleation in the atmosphere.

Conclusions

By using a theoretical approach, we found that the time evolution of the nuclei charging state is governed by two parameters: the initial nuclei charging state determined by the nucleation mechanism, and the parameter K that is directly proportional to the cluster ion concentration and inversely proportional to the nuclei growth rate. We demonstrated that if the value K is larger than a certain threshold value, any information about the initial charging state of the nuclei, and thereby about the nucleation mechanism, will be lost by the time the nuclei grow into the measurement size range (>3 nm). By making a few simplifying assumptions, we derived an analytical expression for the functional dependence of the nuclei charging state on the nuclei size. We demonstrated that the derived expression can usually be fitted into experimental data on nuclei charging states. When the value of K is small enough, the obtained fitting can be extrapolated successfully down to sizes where the nucleation has taken place to obtain information about the relative importance of neutral and ion-induced nucleation.

Acknowledgments This work was funded by the Academy of Finland and by European Union.

References

Hoppel, W.A. and Frick, G.M., *Aerosol Sci. Technol.*, **5**, 1–21 (1986).Yu, F., *Atmos. Chem. Phys.*, **6**, 5193–5211 (2006).Iida, K., Stolzenburg, M., McMurry, P., Dunn, M., Smith, J., Eisele, F., and Keady, P., *J. Geophys. Res.*, **111**, D23201, doi:10.1029/2006JD007167 (2006).

Laakso, L., Gagne, S., Petäjä, T., Hirsikko, A., Aalto, P.P., Kulmala, M., and Kerminen, V.-M., *Atmos. Chem. Phys.*, **7**, 1333–1345 (2007).

Vana, M., Tamm, E., Hõrrak, U., Mirme, E., Tammet, H., Laakso, L., Aalto, P.P., and Kulmala, M., *Atmos. Res.*, **82**, 536–546 (2006).

The Effect of Clouds on Ion Clusters and Aerosol Nucleation at 1,465 m a.s.l.

Hervé Venzac, Karine Sellegri, and Paolo Laj

Abstract Aerosol and ion size distributions were measured at the top of the puy de Dôme (1,465 m a.s.l) for a three-month period. The goals were to investigate (1) the vertical extent of nucleation in the atmosphere and (2) the effect of clouds on nucleation. Nucleation and new particle formation events were classified in four classes: burst of cluster ions (class 1), large ions formation starting from 10 nm (class 2), burst of cluster ions followed by large ions formation with a gap of intermediate ions (class 3), and burst of ions with continuous growth to the >10 nm size (class 4). All together these events were observed during nearly half of the analyzed days. Large concentrations of intermediate ions (1.4–6 nm) seem to be appropriate to detect the occurrence of most of nucleation events. In cloud, cluster ions (<1.4 nm) concentrations are lower than during clear sky conditions, presumably being scavenged by cloud droplets, but the intermediate ions concentrations remain unchanged. We observed that large ions formation starting at 10 nm (class 2) and continuous growth of ions (class 4) occur preferably under clear sky conditions, but that all event classes except class 2 can occur in cloudy conditions.

Keywords Ion clusters, cloud droplets, scavenging

Introduction

Our ability to predict the aerosol size and concentration in time and space is dependent on our knowledge of the aerosol sources. The secondary aerosol formation processes are complex and extremely variable according to the environment. Ultrafine aerosols formations, i.e., nucleation events were observed in many environments (marine, boreal forest, Antarctica) (Kulmala et al. 2004) but very few data have been reported at high-elevation sites.

Laboratoire de Météorologie Physique, Observatoire de Physique du Globe de Clermont-Ferrand, Université Blaise Pascal, France

C.D. O'Dowd and P.E. Wagner (eds.), *Nucleation and Atmospheric Aerosols*, 354–357.

We propose to present here a unique set of measurements of ions and charged aerosol particles performed by an Air Ion Spectrometer (AIS) (0.4–44 nm ion size distribution) at the summit of the puy de Dôme Research Station during spring 2006. This device provides us with the measurements of air ions at sizes which witness the very initial steps of aerosol formation. The goals were to investigate (1) the vertical extent of nucleation in the atmosphere and (2) the effect of clouds on nucleation.

Site and Methods

The puy de Dôme station is located at 1,465 m a.s.l. and lies in a region where both the upper part of the boundary layer (BL) and the free troposphere (FT) can be sampled. Because the puy de Dôme station is in-cloud for more than 50% of the time, it offers a valuable opportunity to study the interaction between new particle formation processes and clouds. The mobility distributions of atmospheric positive and negative ions are measured with the AIS (Air Ion Spectrometer, AIREL Ltd., Estonia), providing ion size distribution, 0.4–44 nm (mobility ranges: 3.162–0.0013 $cm^2\ V^{-1}\ s^{-1}$). The AIS sampling principle is based on the simultaneous selection of 21 different sizes of atmospheric ions of each polarity (negative and positive) along two differential mobility analysers and their subsequent simultaneous detection using electrometers in parallel.

Results

All together, nucleation events were observed at the site more than one third of the days during the spring period. They were classified into four different classes of events (Figure 1): (1) burst of cluster ions, (2) large ions formation starting from 10 nm, (3) burst of cluster ions followed by large ions formation with a gap of intermediate ions, and (4) burst of ions with continuous growth to the >10 nm size.

Then, in order to relate ion size distributions and other meteorological parameters such as the RH, we reduced the AIS data set in three size ranges: ions clusters (0.4–1.4 nm), intermediate ions (1.4–6 nm), and large ions (6–44 nm). Cluster ions concentrations vary typically between 100 and 1,000 ions cm^{-3}, intermediate ions concentrations are usually lower than 500 ions cm^{-3} but can exceed 3,000 ions cm^{-3} during nucleation events. Large concentrations of intermediate ions seem to be appropriate to detect the occurrence of most of nucleation events. A statistical analysis of the concurrent presence of ion clusters and cloud droplets on the one hand, and intermediate ions and cloud droplets on the other hand, brings new information regarding our current understanding of atmospheric aerosol dynamics. In cloud, the cluster ions concentrations are significantly lower than during clear sky periods, presumably because they are efficiently scavenged by cloud droplets, but the intermediate

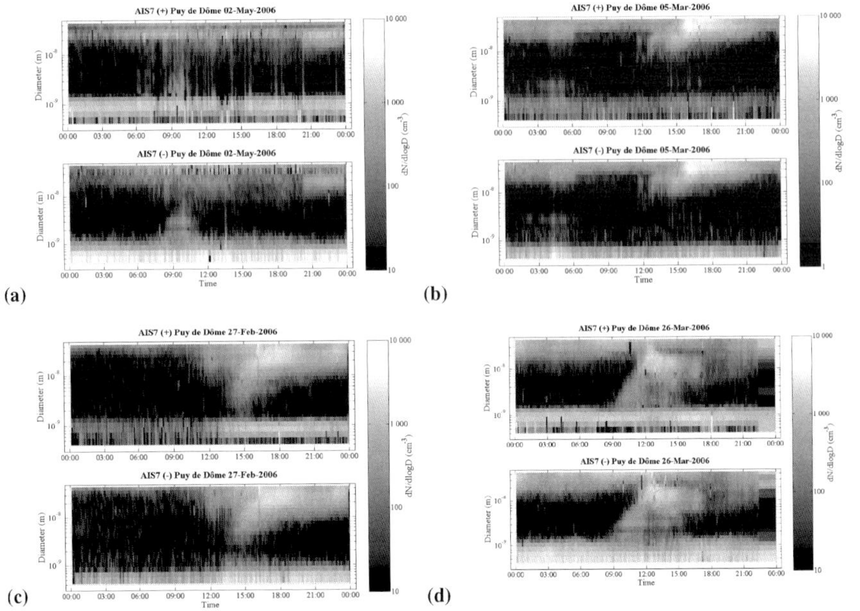

Figure 1 Different nucleation classes were observed at the puy de Dôme station with the Air Ion Spectrometer. Examples are shown for (a) class 1, (b) class 2, (c) class 3, and (d) class 4

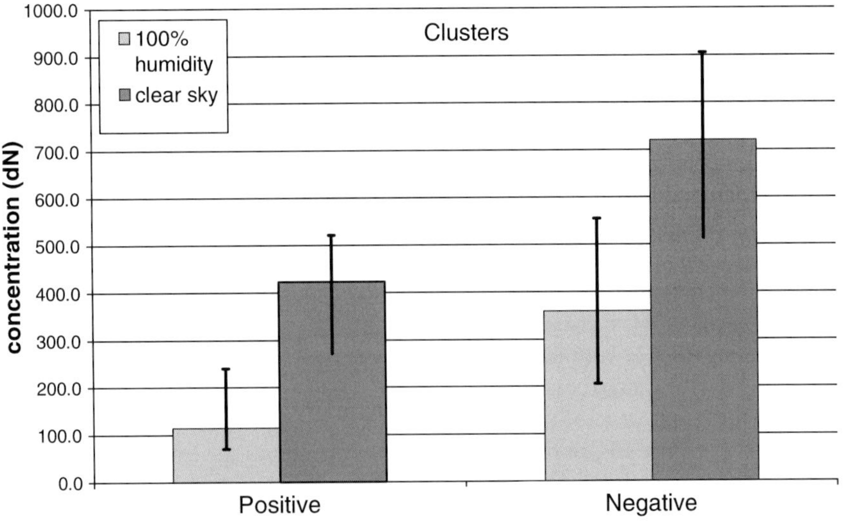

Figure 2 Median cluster ion concentration for in-cloud and clear sky conditions, both for positive and negative ions

ions concentrations remain unchanged (Figure 2). This would indicate that cluster ions can still grow into intermediate ions within clouds and that the condensing gas responsible for the growth has a limited solubility and is not scavenged by cloud droplets. This result would lead to the conclusion that the limiting factor for new particle formation is not the amount of precursors (clusters) but other parameters such as nonsoluble condensing gases concentration.

A statistical analysis of the RH percentiles measured during each different event class and nonevent class show significant differences. We observed that large ions formation starting at 10 nm (class 2) and continuous growth of ions (class 4) occur preferably under clear sky conditions, but that all event classes except class 2 can occur in cloudy conditions.

Acknowledgments This work was supported by INSU, within the frame of the French National Program on Atmospheric Chemistry (PNCA).

References

1. Kulmala, M., Vehkamäki, H., Petäjä, T., Dal Maso, M., Lauri, A., Kerminen, V.-M., Birmili, W., and McMurry, P.H., *J. Aerosol Sci.*, **35**, 143–176 (2004).

The Effect of Seed Particle Charge on Heterogeneous Nucleation

P.M. Winkler[1], G.W. Steiner[1], G.P. Reischl[1], A. Vrtala[1], P.E. Wagner[1], and M. Kulmala[2]

Abstract Heterogeneous nucleation of n-propanol vapor was measured using charged and uncharged particles in the diameter range from 1.4 to 4 nm as condensation nuclei. Neutral particles were obtained by neutralization of electrostatically classified particles by means of a bipolar diffusion charger whereas for charged particles the charger was replaced by a corresponding dummy. Depending on the sign of the voltage supplied to the classifier electrode, positively or negatively charged particles were obtained, respectively. Heterogeneous nucleation measurements were performed using a fast expansion chamber. In this paper we illustrate how the charging state of equally sized particles influences the nucleation behavior.

Keywords Heterogeneous nucleation, charging state, bipolar diffusion charger

Introduction

Heterogeneous nucleation is not necessarily restricted to neutral seed particles but may also take place on particles carrying an electrical charge. In such a case the electrostatic forces due to the particle charge can be expected to influence the nucleation process significantly. In addition, not only the charge itself, but also the sign may have an impact on the nucleation process. Quite recently, Nadykto et al.[1] have shown that such a sign preference is essentially of quantum nature depending on the electronic structure of individual ions. This finding is especially important for correctly modeling the effect of seed particle charge and heterogeneous nucleation experiments under well-defined conditions are crucial for testing theoretical predictions. In this paper we present new experimental findings showing that charge effects become increasingly important below a certain seed particle size and that sign effects indeed have a significant impact on the nucleation process.

[1] *Fakultät für Physik, Universität Wien, Boltzmanngasse 5, A-1090 Wien, Austria*

[2] *Department of Physical Sciences, University of Helsinki, P.O. Box 64, FIN-00014 Helsinki, Finland*

C.D. O'Dowd and P.E. Wagner (eds.), *Nucleation and Atmospheric Aerosols*, 358–362.

Experimental Method

The experimental system – shown schematically in Figure 1 – includes a source of monodisperse particles and a vapor generation unit. Vapor supersaturation is achieved by adiabatic expansion in a computer-controlled thermostated expansion chamber, the size analyzing nuclei counter (SANC). Droplet growth is observed using the constant-angle Mie scattering (CAMS) detection method.[2] Details of the experimental system are presented elsewhere.[3]

Generation of seed particles was performed with a newly developed WO_x-Generator (Grimm Aerosoltechnik, Modell 7860). To this end, tungsten oxide sublimates into a purified carrier gas passed through the generator. Primary produced molecules subsequently grow to larger clusters by coagulation. Using a newly developed differential mobility analyzer (Nano-DMA) well-defined unipolarly charged monodisperse particle fractions were extracted. Typically, after passing the Nano-DMA, particles are brought into charge equilibrium by means of a radioactive source in a bipolar diffusion charger. However, by replacing the charger with a corresponding dummy, particles undergo practically the same diffusion losses but

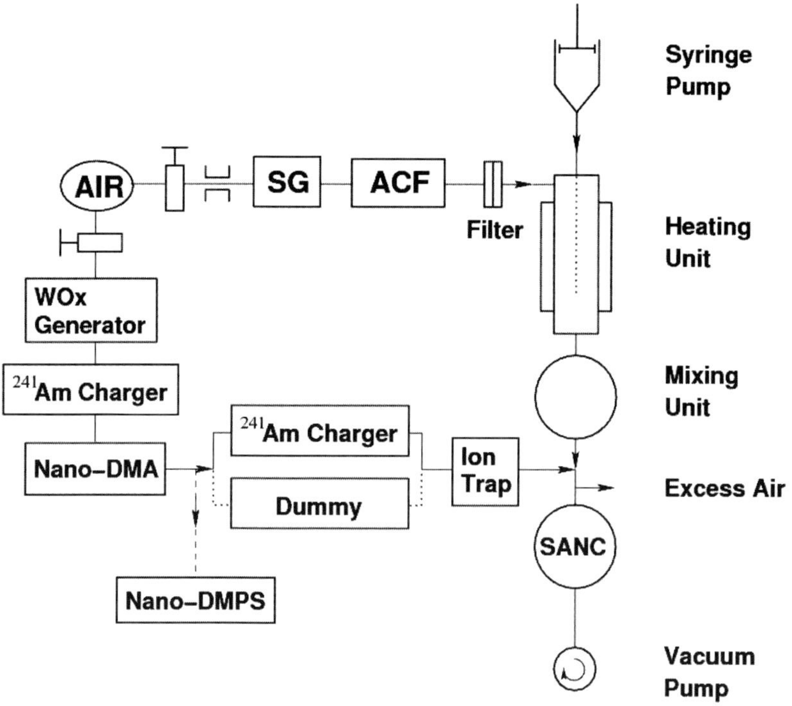

Figure 1 Experimental arrangement for measuring heterogeneous nucleation on charged and uncharged particles. SG: silicagel gas dryer, ACF: active carbon filter.

stay charged. By switching the polarity of the classifier electrode we obtained negatively and positively charged particles, respectively. The size distribution of the particle fractions obtained was monitored by a newly developed differential mobility particle spectrometer (Nano-DMPS, Tapcon & Analysesysteme GmbH, EMS VIE-10) allowing to analyze in the diameter range from 0.5 to 40 nm. Corresponding size distributions for positively and negatively charged particles are shown in Figure 2. As can be seen, excellent agreement between the sizes of positive and negative particles is observed. For positive particles a slightly higher geometric standard deviation is found resulting in a higher number concentration. It should be noted that neutral particles are obtained by classification of negatively charged particles and subsequent neutralization. Accordingly, the size distribution showing negative particles is also representative for neutral particles.

n-Propanol vapor was added to the system by controlled injection of a liquid beam and subsequent quantitative evaporation in a heating unit (see Figure 1). In order to achieve the desired vapor supersaturations, a well-defined binary vapor–air mixture together with the seed particles were passed into a thermostated expansion chamber. During a computer-controlled measurement cycle, the chamber was subsequently connected to a low-pressure buffer tank resulting in adiabatic expansion with expansion times around 8 msec and well-defined vapor supersaturations were obtained.

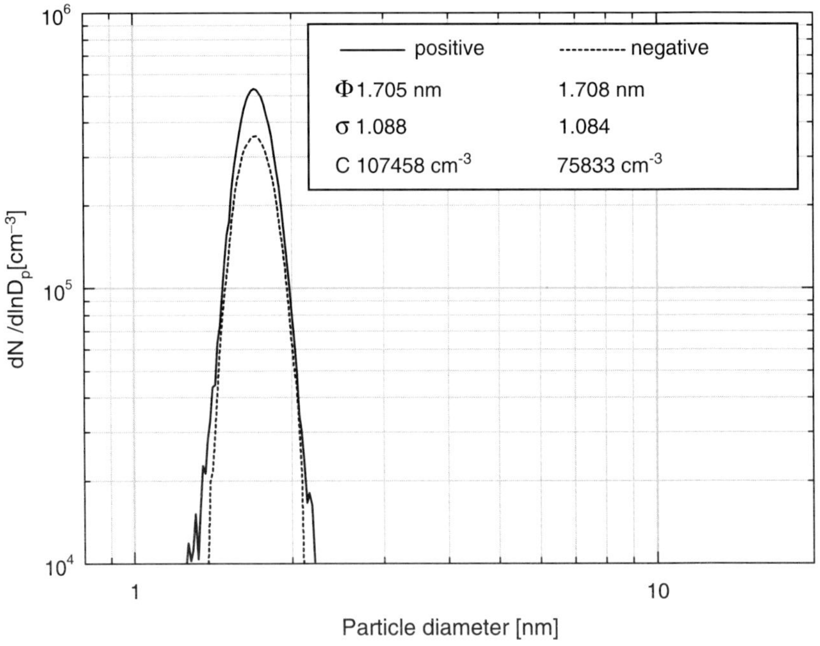

Figure 2 Size distribution of negatively and positively charged WO_x particles with 1.7 nm geometric mean particle diameter

Those particles inducing heterogeneous nucleation at the vapor saturation ratios considered lead to the formation and growth of liquid droplets. Using the CAMS detection method,[2] the radius and number concentration of the growing droplets can be determined simultaneously at various times during the growth process without any external calibration. In the present study the droplet concentrations were determined immediately after expansion from the height of the first Mie maximum.

Results and Conclusions

Using the SANC we have measured the number concentration of droplets heterogeneously nucleated by charged and uncharged WO_x particles at various saturation ratios and at constant nucleation temperature. At low supersaturation ratios no particles are activated, whereas with increasing supersaturation more and more particles will be activated and grow to detectable sizes. Beyond a certain saturation ratio all particles are activated and no further increase of the particle number concentration is observed. Thereby we obtained the heterogeneous nucleation probability, i.e., the number concentration of activated droplets normalized relative to the total concentration of seed particles. In Figure 3, corresponding heterogeneous nucleation

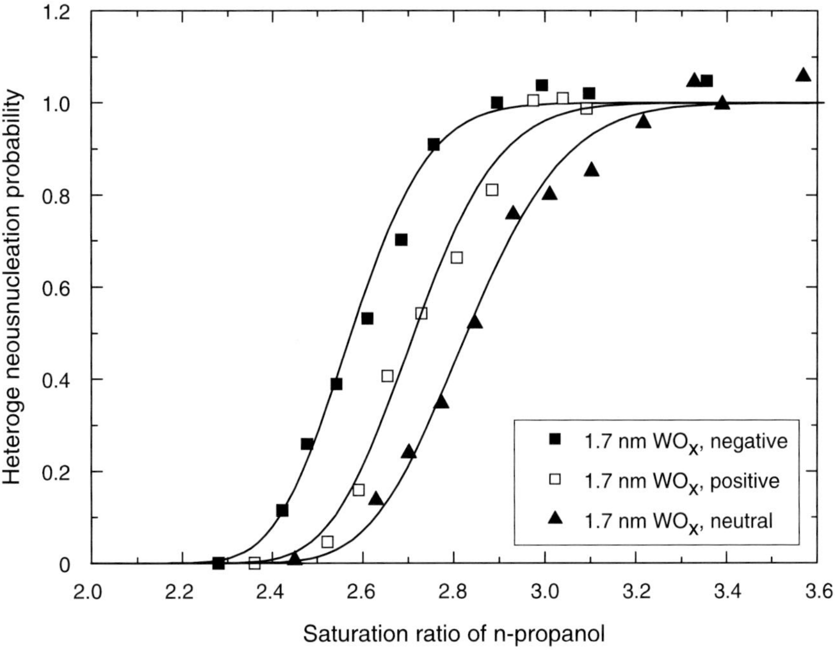

Figure 3 Heterogeneous nucleation probabilities for WO_x seed particles with 1.7 nm geometric mean particle diameter and different charging state

probabilities are shown as functions of the vapor saturation ratio for negatively and positively charged particles as well as for uncharged particles, respectively. Solid lines are fitted to the data to guide the eye. As can be seen from this figure, particles of same size but different charging state exhibit different nucleation behavior. Negatively charged WO_x particles appear to be more easily nucleated to growth than positively charged particles and neutral ones. Smaller saturation ratios are generally required to see charged particles being nucleated. This behavior was observed for a set of measurements covering the size range from 1.4 to 4 nm particle diameter. In conclusion we can state that for *n*-propanol vapor the charge effect on nucleation becomes evident considering seed particle diameters below 4 nm. According to our results, a negative sign preference is clearly observed.

Acknowledgment This work was supported by the Austrian Science Foundation (FWF, Project No. P16958-N02), and the Academy of Finland.

References

1. Nadykto A.B., Natsheh, A.A., Yu, F., Mikkelsen, K.V., and Ruuskanen, J., *Phys. Rev. Lett.*, **96**, 125701 (2006).
2. Wagner, P.E., *J. Colloid Interface Sci.*, **105**, 456–467 (1985).
3. Wagner, P.E., Kaller, D., Vrtala, A., Lauri, A., and Kulmala, M., *Phys. Rev.*, **E67**, 021605 (2003).

Experimental Observation of Heterogeneous Nucleation Probabilities for Ion-induced Nucleation

P.M. Winkler[1], G.W. Steiner[1], G.P. Reischl[1], A. Vrtala[1], P.E. Wagner[1], and M. Kulmala

Abstract Heterogeneous nucleation probabilities for nucleation of *n*-propanol vapor on positive and negative ions around 1 nm electrical mobility diameter were experimentally determined using a fast expansion chamber. Ions were obtained from a ^{241}Am bipolar diffusion charger and subsequently classified by means of a newly developed Differential Mobility Analyzer (DMA) especially designed for operation in the few nanometer and even subnanometer size range. Although positive and negative ions exhibit clearly different mobility diameters, the corresponding onset saturation ratios were found to be surprisingly similar.

Keywords Ion-induced nucleation, heterogeneous nucleation probability, DMA

Introduction

In the past two decades atmospheric new particle formation due to vapor nucleation has been observed in a variety of locations all over the world[1] indicating the importance of this mechanism for the global aerosol budget and consequently global climate. Detailed understanding of the relevant nucleation processes describing atmospheric new particle formation properly is thus essential for the further development of climate models. Besides field studies modeling and laboratory experiments under well-defined conditions are of crucial importance. Presently, the influence of charged molecular clusters and ionizing radiation on the formation of atmospheric nanoparticles and subsequently on cloud formation and climate change is intensively discussed.[2,3] Additionally, condensable organic vapors probably responsible for the condensational growth of nanoparticles to cloud condensation nuclei have become subject of intensive studies.[4,5,6]

[1] *Fakultät für Physik, Universität Wien, Boltzmanngasse 5, A-1090 Wien, Austria*

[2] *Department of Physical Sciences, University of Helsinki, P.O. Box 64, FIN-00014 Helsinki, Finland*

C.D. O'Dowd and P.E. Wagner (eds.), *Nucleation and Atmospheric Aerosols*, 363–367.

In this paper we present first quantitative nucleation probabilities obtained from laboratory experiments on heterogeneous nucleation using positive and negative ions as seed particles. Both ion species are obtained from a radioactive source in a bipolar diffusion charger and subsequent electrostatic classification in a Differential Mobility Analyzer (DMA). *n*-Propanol being a model substance for organic vapors was used as the working fluid.

Experimental Method

A schematic diagram of the experimental arrangement can be seen in Figure 1. The experimental system mainly consists of an ion generation unit, a vapor generation unit, and an expansion chamber for measuring droplet number concentration. A main feature of this arrangement is the generation of unipolar ions. To this end, a highly dried and filtered well-defined air flow passes a ^{241}Am bipolar diffusion charger where the radioactive material produces high number concentrations of positive and negative ions. Subsequently, the bipolar ions are passed through a newly developed Differential Mobility Analyzer (UDMA) where ions are separated depending on the polarity supplied to the electrode of the UDMA. Positive voltage yields negative ions and vice versa. The UDMA was especially designed for operation in the few nanometer down to even subnanometer size range. The channel

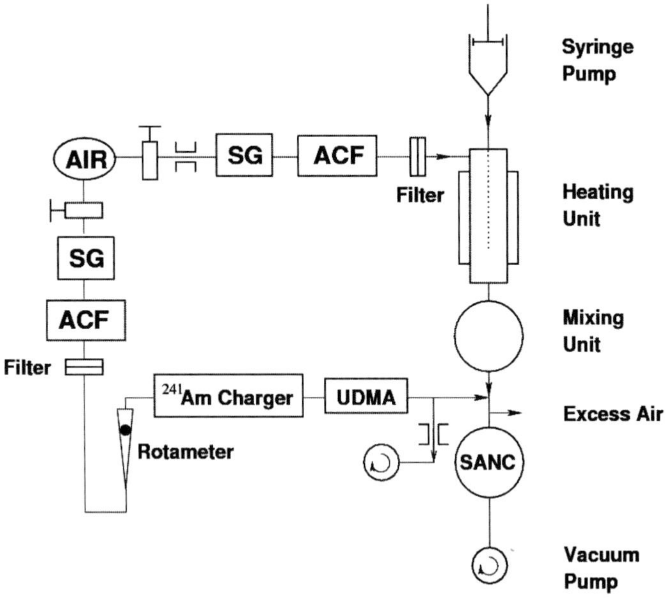

Figure 1 Experimental arrangement for heterogeneous nucleation measurements on unipolar ions. SG: silicagel gas dryer, ACF: active carbon filter

length of only 6.5 mm as well as the sheath air flow rate of 288 L/min allows to minimize diffusion losses and provide excellent conditions for classification of ion sizes even below 1 nm mobility diameter. Vapor generation is performed by means of a high-precision syringe pump injecting a well-defined liquid beam into a heating unit where the liquid is quantitatively vaporized and mixed with a highly dried and purified well-defined air flow acting as carrier gas flow. In the present study *n*-propanol was used as the working fluid. Finally, the ions are mixed with the vapor and passed into the expansion chamber (SANC). Growth rates and number densities of the condensing droplets were measured using the CAMS detection method.[7] In order to reduce diffusion losses on the way to the expansion chamber,combined flow rates of ion and vapor flow up to nearly 10 L/min were chosen from which 7 L/min were passed through the SANC. Sampling lines had to be kept as short as possible.

Results and Discussion

As heterogeneous nucleation is strongly size dependent, we have first of all determined the size distributions of positive and negative ions. Therefore the voltage in the UDMA was stepwise varied and the corresponding number concentrations were measured with the SANC. As can be seen from Figure 2 positive and negative ions exhibit clearly different mean mobility diameters. For negative ions the peak is found around 0.93 nm whereas for positive ions the peak is found at 1.1 nm mobility diameter. It should be noted that the size resolution of the UDMA is of the order of 0.02 nm. In order to extract monodisperse ion fractions with as high-number concentrations as possible, voltages were supplied to the UDMA corresponding to the peaks of the respective primary ion size distributions.

Using the SANC we have measured the number concentration of droplets heterogeneously nucleated by negative and positive ions at various vapor saturation ratios. With increasing supersaturation more and more ions will be activated and grow to detectable sizes. Beyond a certain saturation ratio all ions are activated and no further increase of the ion number concentration is observed. The heterogeneous nucleation probability was obtained by normalizing the number concentration of activated droplets relative to the total concentration of seed particles (ions). In Figure 3, corresponding heterogeneous nucleation probabilities are plotted for negative and positive ions. As can be seen from this figure, positive ions are activated only at slightly smaller saturation ratios than negative ions, despite the considerable difference in ion mobility diameter. This indicates a possible sign effect for the nucleation process. It is important to note that nucleation of such small particles still occurs well below the onset of homogeneous nucleation, which is indicated by the shaded area in Figure 3. Clearly a plateau is reached, where all ions are activated allowing determination of total ion number concentration and thus heterogeneous nucleation probability. Future work will concentrate on water as condensable vapor. Besides, determination of the chemical composition of the ions considered will allow a more detailed interpretation of ion behavior.

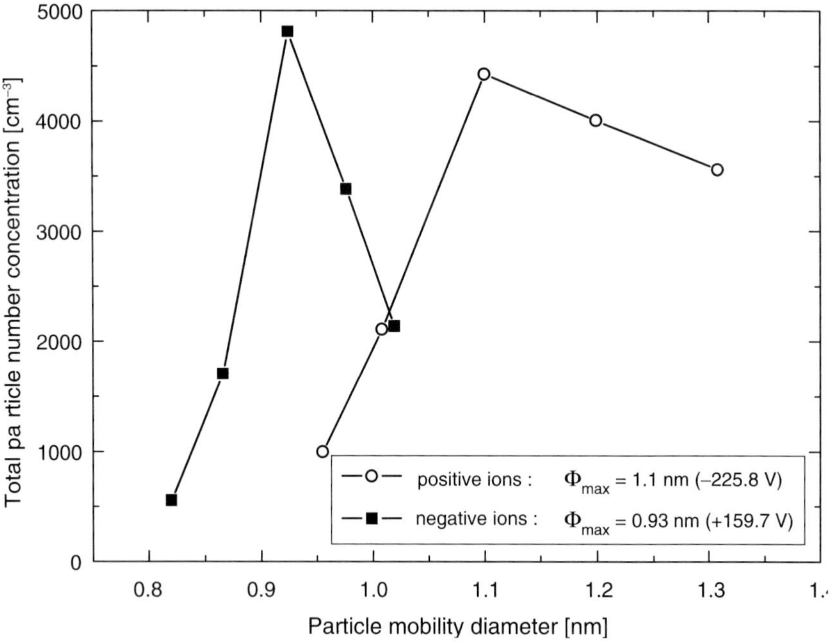

Figure 2 Size distribution of negative and positive ions. Geometric mean diameters and corresponding classifier voltages are shown for comparison

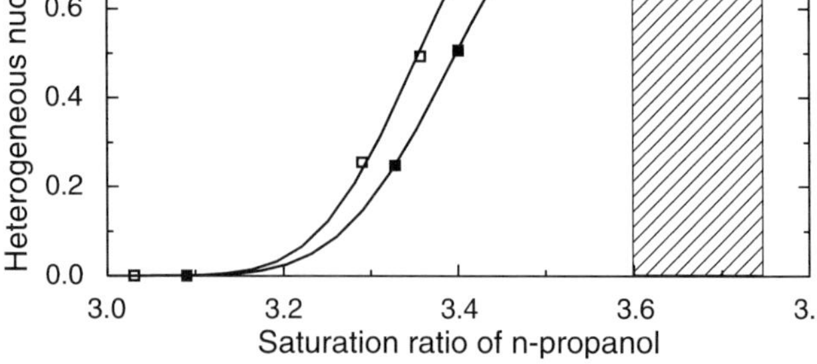

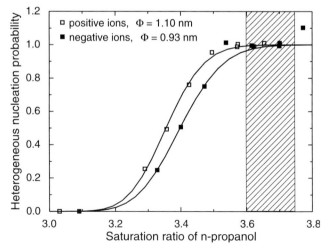

Figure 3 Heterogeneous nucleation probabilities for positive and negative ions. The shaded area indicates the onset of homogeneous nucleation of *n*-propanol according to Wedekind et al.[8]

Acknowledgment This work was supported by the Austrian Science Foundation (FWF, Project No. P16958-N02), and by the Academy of Finland.

References

1. Kulmala, M., Vehkamäki, H., Petäjä, T., Dal Maso, M., Lauri, A., Kerminen, V.-M., Birmili, W., McMurry, P.H., *J. Aerosol Sci.*, **35**, 143 (2004).
2. Kanipe, J., *Nature*, **443**, No. 7108, 141–143 (2006).
3. Laakso, L., Anttila, T., Lehtinen, K.E.J., Aalto, P.P., Kulmala, M., Horrak, U., Paatero, J., Hanke, M., Arnold, F., *Atmos. Chem. Phys.*, **4**, 2353, (2004).
4. Anttila, T., Kerminen, V.-M., Kulmala, M., Laaksonen, A., O'Dowd, C., *Atmos. Chem. Phys.*, **4**, 1071 (2004).
5. Kulmala, M., Kerminen, V.-M., Anttila, T., Laaksonen, A., O'Dowd, C., *J. Geophys. Res.*, **109**, D04205 (2004).
6. Kulmala, M., Laakso, L., Lehtinen, K.E.J., Riipinen, I., Dal Maso, M., Anttila, T., Kerminen, V.-M., Horrak, U., Vana, M., Tammet, H., *Atmos. Chem. Phys.*, **4**, 2553 (2004).
7. Wagner, P. E., *J. Colloid Interface Sci.*, **105**, 456–467 (1985).
8. Wedekind, J., Iland, K., Wagner, P.E., and Strey, R., "Homogeneous nucleation of 1-Alcohol vapors", in *Nucleation and Atmospheric Aerosols 2004,* edited by M. Kasahara and M. Kulmala, *16th ICNAA Conference Proceedings*, Kyoto University Press, pp. 49–52, (2004).

Air Ion Measurements During a Cruise from Europe to Antarctica

Marko Vana[1,3], Aki Virkkula[2], Anne Hirsikko[1], Pasi Aalto[1], Markku Kulmala[1], and Risto Hillamo[2]

Abstract Latitudinal air ion mobility distribution variations were characterized in the marine boundary layer during a cruise over the Eastern Atlantic and Southern Ocean. We measured concentrations of cluster ions and charged aerosol particles in the size range of 0.4–40 nm. The measured ion mobility distributions were characterized and classified according to meteorological conditions and air mass back-trajectories. New particle formation episodes (bursts of intermediate air ions) were observed above the pack-ice region of the Southern Ocean and several times during showers in the tropical marine boundary layer.

Keywords Atmospheric ions, marine aerosols, particle formation, field measurements

Introduction

Formation of secondary aerosol particles and their further growth have been observed in numerous locations over the world.[1] As the oceans cover about 70% of the Earth's surface, they are significant source of aerosol particles and play an important role in climate processes. Importance of air ions (naturally charged clusters and aerosol particles) in new particle formation and the role of ion-induced nucleation in the marine boundary layer are poorly known yet. Cosmic rays are the main source of ionization above the oceans. Numerous aerosol size distribution measurement campaigns have been conducted in the marine boundary layer. However, air ion measurements over the oceans are rare so far.[2]

Currently the lower limit of the diameter measurement range of the particle size spectrometers is approximately 3 nm. To understand initial steps of particle nuclea-

[1] *Department of Physical Sciences, P.O. Box 64, FI-00014 University of Helsinki, Finland*

[2] *Finnish Meteorological Institute, Research and Development, P.O. Box 503, FI-00101 Helsinki, Finland*

[3] *Institute of Environmental Physics, University of Tartu, Ülikooli 18, 50090 Tartu, Estonia*

C.D. O'Dowd and P.E. Wagner (eds.), *Nucleation and Atmospheric Aerosols*, 368–372.

tion, measurements of smaller particles are important. Air ion mobility spectrometers are able to measure mobility (size) distribution of charged particles even for the smallest thermodynamically stable particles with diameter about 1–2 nm.[3,4] Simultaneous measurements of air ion mobility and aerosol particle size distributions can be used for assessing the importance of air ions in the new particle formation. The aim of the present study is to characterize air ions measured over the Eastern Atlantic and Southern Ocean.

Data Acquisition

An Air Ion Spectrometer (AIS), designed in the University of Tartu and manufactured by AIREL Ltd., Estonia, was used for measuring the mobility distribution of air ions.[5] The AIS is a multichannel, parallel-principle device, measuring simultaneously ion concentrations in 27 mobility fractions of both positive and negative ions. The spectrometer consists of two identical differential mobility analyzers, one for measuring positive ions and the second for negative ions. The mobility range of single-charged particles is 0.0013–3.2 $cm^2\ V^{-1}\ s^{-1}$, which corresponds to the particle mass diameter range from 0.4 to 40 nm. Particle diameter and mobility are uniquely related through the modified Millikan formula.[5]

The air ion data were partly accompanied by the measurements of aerosol size distribution and by the measurements of black carbon. Aerosol particles larger than about 10 nm were measured using a differential mobility particle sizer (DMPS) that consisted of a medium-size Hauke-type differential mobility analyzer (DMA) and a TSI 3010 CPC. The closed-loop sheath-flow arrangement was used. A two-wavelength Magee Scientific Aethalometer (AE-20) was used for monitoring black carbon and thus discarding data that were collected during smoke from the ship stack. All the three instruments were installed above the bridge, about 30 m above sea level.

Figure 1 illustrates the route of the expedition. Air ion mobility and aerosol particle size distributions were measured onboard the Russian research vessel Akademik Fedorov between Europe and the coast of Queen Maud Land, Antarctica. The measurements were part of the FINNARP-2004 expedition. The measurements started on November 3, 2004, at the harbor of Bremerhaven, Germany, continued during the cruise to Cape Town, South Africa, where the ship arrived on November 22, restarted on November 27 at Cape Town and continued until the vessel reached the Antarctic shelf ice on December 9, 2004.

Results

Several interesting results have been obtained from these measurements. We classified the measured air ion mobility distributions according to air mass back-trajectories and meteorological conditions. Latitudinal variations of different

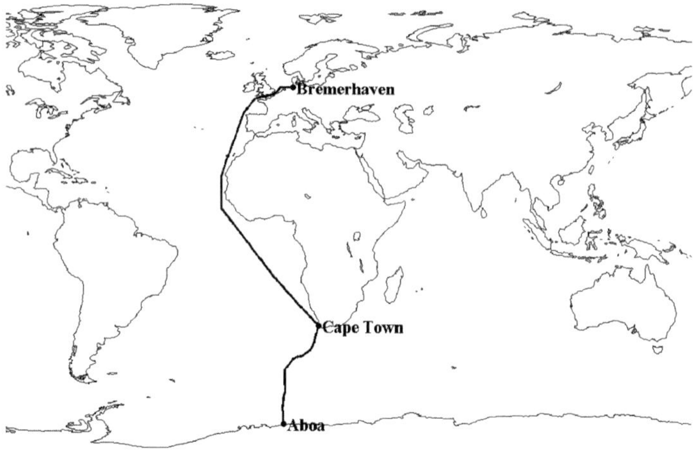

Figure 1 The route of the research vessel Akademik Fedorov during November 3–December 9, 2004

classes of air ions were characterized. We also identified and separated the data, which were contaminated, either by direct emissions from the chimney of ship or by other activities on board the ship.

High-mobility small ions are mainly responsible for atmospheric electric conductivity over the oceans. Therefore, it is important to know their concentrations. Small ions (0.4–1.6 nm) were found to present all the time and our measurements show that their concentrations in clean marine conditions were typically between 100–600 cm^{-3} (see Figure 2). The concentrations of intermediate ions (1.6–7 nm) with one polarity were most of the time below 20 cm^{-3} in clean air. The concentrations of large ions (7–22 nm) with one polarity were mostly below 50 cm^{-3} in clean air. Figure 2 shows the latitudinal variation of small ion concentrations and air temperature during the cruise from Europe to Antarctica. We can see similarity in variations of these parameters, especially in lower latitudes.

In the air ion size distributions a peak at sizes smaller than about 1 nm was present during all time of the cruise. On the Atlantic Ocean between Europe and South Africa no obvious nucleation events were observed in the clean air, only some weak bursts during the presence of polluted air masses. During an earlier cruise onboard the same research vessel Koponen et al.[7] observed weak particle formation episodes above the pack-ice region of the Southern Ocean. During the present cruise similar observation was made on December 3, 2004. Figure 3 shows the temporal variation of the air ion size distribution and intermediate air ions during the nucleation event day. The nucleation episode lasted approximately 8 hours and accompanied by intensive solar radiation. The concentrations of negatively charged nanometer particles were twice higher than the concentrations of positively charged particles during first hours of the nucleation event.

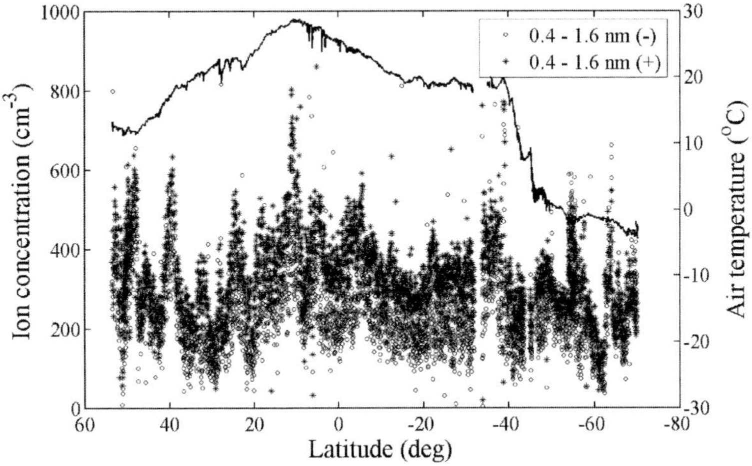

Figure 2 Latitudinal variation of small ion concentrations and air temperature during the expedition from Bremerhaven to Aboa

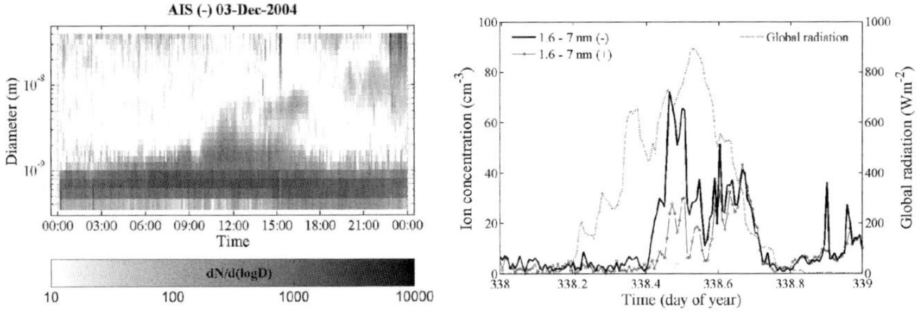

Figure 3 A particle formation episode above the pack-ice region of the Southern Ocean (about 55°S; 5.5°E): time variation of the size distribution of negative air ions (**left plot**), time variation of intermediate air ions and global radiation (**right plot**) on December 3, 2004

One special effect, which we observed, is the formation of intermediate ions during rain. The formation of intermediate air ions is accompanied by sudden decrease in temperature during short-time showers in the tropical marine boundary layer. In general, the concentrations of positively charged nanometer particles were many times smaller than the concentrations of negatively charged ones. The lifetime of particles is relatively small, their concentration decreasing quickly after the rain due to the coagulation with larger particles. A potential way to obtain charged particles from a waterfall is described by Laakso et al.[8] The formation of charged nanometer particles during rainfall may possibly be caused by the ion evaporation from charged water droplets. The effect of rain has also been observed in a boreal forest and in urban environments.

Acknowledgment This work was supported by the Academy of Finland (Finnish Antarctic Research Program, contract no. 53669).

References

1. Kulmala, M. et al., *J. Aerosol Sci.*, **35**, 143–176 (2004).
2. Siingh, D. et al.,. *J. Geophys. Res.*, **110**, doi:10.1029/2005JD005765 (2005).
3. Kulmala, M. et al., *Nature* **404**, 66–69 (2000).
4. Hõrrak, U. et al., *J. Geophys. Res.*, **105**, 9291–9302 (2000).
5. Mirme, A. et al., *Boreal Env. Res.*, **12** (2007).
6. Tammet, H., *J. Aerosol Sci.*, **26**, 459–475 (1995).
7. Koponen, I.K. et al., *J. Geophys. Res.*, **107**, doi:10.1029/2002JD002533 (2002).
8. Laakso, L. et al., *Atm. Chem. Phys. Discuss.*, **6**, 9297–9314 (2006).

Air Ion Measurements at Mace Head on the West Coast of Ireland

Marko Vana[1,4], Mikael Ehn[1], Tuukka Petäjä[1], Henri Vuollekoski[1], Pasi Aalto[1], Gerrit de Leeuw[1,2], Darius Ceburnis[3], Colin D. O'Dowd[3], and Markku Kulmala[1]

Abstract Coastal nucleation events and behavior of cluster ions were characterized through the measurements of air ion mobility distributions at the Mace Head research station on the west coast of Ireland in 2006. We measured concentrations of cluster ions and charged aerosol particles in the size range of 0.4–40 nm. These measurements allow us to characterize freshly nucleated charged particles with diameter below 3 nm. The analysis shows that bursts of intermediate ions (1.6–7 nm) are a frequent phenomenon in marine coastal environment. Nucleation evens occurred during most of the measurement days. We classified the nucleation burst events. Particle formation and growth events mostly coincided with the presence of low tide. Small ions concentrations appear to be strongly dependent on the variations of meteorological parameters including wind speed and direction.

Keywords Nucleation, atmospheric ions, marine aerosols, particle formation and growth

Introduction

Coastal regions are places where new particle formation occurs frequently.[1,2] Therefore, coastal aerosols can significantly contribute to the natural background aerosol population. Observations have shown that these nucleation events usually coincide with the occurrence of low tide and solar irradiation.[3] Though the concentrations of sulfuric acid and ammonia are found to be sufficient for nucleation of

[1] *Department of Physical Sciences, P.O. Box 64, FI-00014 University of Helsinki, Finland*

[2] *Finnish Meteorological Institute, Research and Development, P. O. Box 503, FI-00101 Helsinki, Finland*

[3] *Department of Experimental Physics and Environmental Change Institute, National University of Ireland, Galway, Ireland*

[4] *Institute of Environmental Physics, University of Tartu, Ülikooli 18, 50090 Tartu, Estonia*

C.D. O'Dowd and P.E. Wagner (eds.), *Nucleation and Atmospheric Aerosols*, 373–377.

thermodynamically stable clusters,[4] this does not explain the rapid growth of the freshly formed particles. O'Dowd et al.[1] showed that biogenic iodine oxides can participate in the coastal new particle production and growth. However, the detailed nucleation mechanisms and the role of ion-induced nucleation in the coastal aerosol formation are still unknown.

Currently the lower limit of the diameter measurement range of the particle size spectrometers is approximately 3 nm. To understand particle nucleation mechanisms, measurements of smaller particles are important. In this work we applied an air ion mobility spectrometer capable of measuring the mobility (size) distribution of charged particles even for the smallest thermodynamically stable particles. One of our aims was to get more information on the behavior of particles with diameter below 3 nm and also to detect possible seasonal variation in the particle formation.

Data Collection

We measured air ion mobility distributions with an Air Ion Spectrometer (AIS), designed in the University of Tartu, and built by Airel Ltd., Estonia.[5] The AIS is a multichannel, parallel-principle device, measuring simultaneously ion concentrations in 27 mobility fractions of both positive and negative ions. The measurement range of ion mobility was from 0.0013 to 3.2 $cm^2\ V^{-1}\ s^{-1}$, which corresponds to the particle mass diameter range from 0.4 to 40 nm. Particle diameter and mobility were uniquely related through the modified Millikan formula.[6]

Data were collected as a part of the EU project MAP (Marine Aerosol Production) at the Mace Head Atmospheric Research Station (53°19′ N, 9°54′′ W) on the west coast of Ireland in 2006. The location of the monitoring station provides a good opportunity to study particle formation events at different distances from the tidal source regions. Currently the database includes continuous measurements for more than 1 year.

Results and Discussion

Classification of Nucleation Events

Nucleation burst events usually coincide with the occurrence of low tide and have been observed to occur during different air mass trajectory regimes which correspond to different distances from the tidal source regions. Two main types of events can occur in the coastal environment. One is similar to the banana-type events, which have been observed in many places around the world (Figure 1, November 2, 2006). In this case polluted air advects over the tidal regions far from measurement station or corresponds to the polluted air with no advection over tidal regions. The second

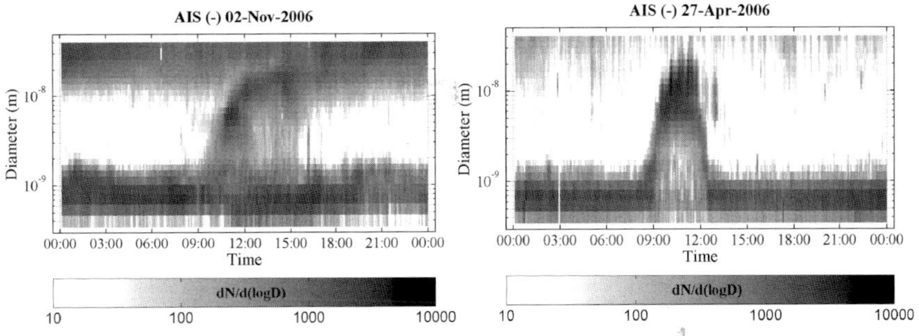

Figure 1 Surface plots of the size distributions of two clear nucleation days at the Mace Head research station

type corresponds to clean marine air, which encountered the tidal regions closer to measurement station (Figure 1, April 27, 2006). The surface plots in Figure 1 show the size distributions of negative air ions.

We analyzed the 1-year data set of air ion size distributions. We divided all measurement days into six groups depending on whether new particle formation occurred or not. In this classification we have one nonevent category, three event categories, and two "undefined" categories. The group "undefined" includes the days when the data contained high noise levels, high background concentrations, or other uncertainties were detected.

We compared days with and without events separately for each month. All months, except December, showed particle formation during more than half of the days. During the period from June to September it was around 75% and during April–May it was around 85%. The "undefined" and missing measurement days add uncertainty to this analysis. Due to this reason, we had to discard up to half of the days for some months. According to our classification, 83 nonevent days, 193 event days, 67 "undefined" days, and 22 missing measurement days occurred at the Mace Head research station in the year 2006.

Characteristics of Air Ions in the Marine Coastal Environment

Our measurements show that the concentrations of small ions (0.4–1.6 nm) with one polarity were typically between 200 and 800 cm^{-3}. The concentrations of intermediate ions (1.6–7 nm) with one polarity were on average about 40 cm^{-3} reaching 500–1,500 cm^{-3} during nucleation events. The concentrations of light large ions (7–22 nm) with one polarity were on average approximately 70 cm^{-3}.

Figure 2 illustrates the temporal variation of the concentration of particles with diameter below 7 nm for one of the coastal nucleation episodes. During the nucleation event, the concentration of small ions decreased considerably. The concentration of intermediate ions with diameter of 1.6–3 nm increased in the first stage of

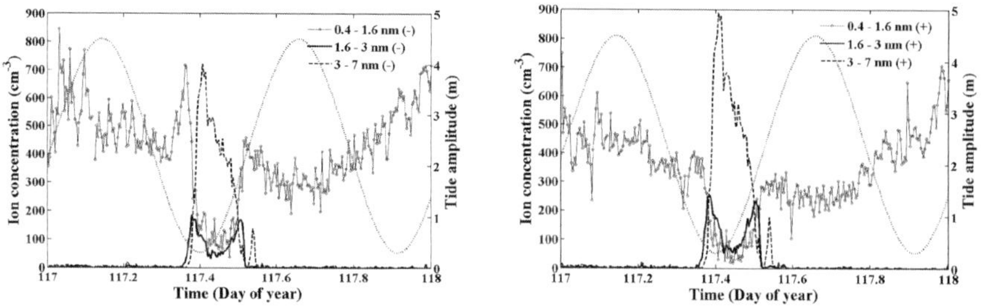

Figure 2 Small ion and intermediate ion concentrations and tide amplitude (**gray dashed line**) during a coastal particle production event, April 2006

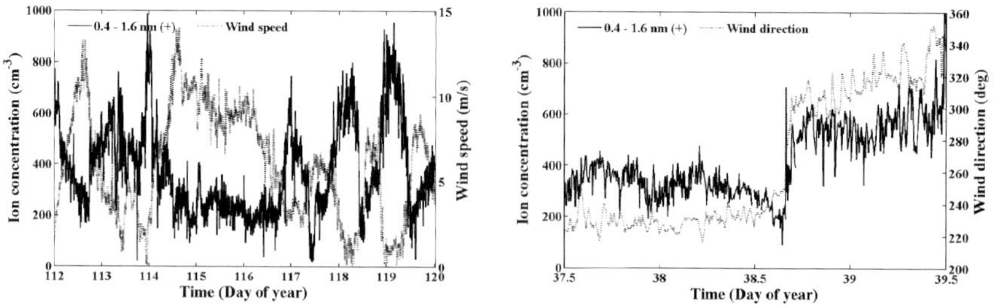

Figure 3 Variation of the concentration of small ions with wind speed (**left**), and with wind direction (**right**)

the nucleation event and started to follow the pattern of the small ion concentrations at some point where the concentration of intermediate ions with diameter of 3–7 nm was approximately 200 cm^{-3}. If we have deep depletion in the concentration of small ions during a nucleation event, then two possible mechanisms for nucleation can occur. First, homogeneous nucleation occurred and cluster ions were attached to the nucleated particles; second, ion-induced nucleation took place and cluster ions actively participated in the new particle formation.

We also studied the dependence of the concentration of small ions on meteorological parameters. Figure 3 illustrates the time variation of the concentration of small ions in relation to the wind speed (left plot). These two parameters are clearly anticorrelated, the concentration of small ions had highest values during periods with low wind speeds. The cause of high concentrations can be the accumulation of radon near the ground during low wind speeds. It can also be that the wind speed relationship has probably nothing to do with wind speed but more direction as when the air is from polluted sectors, typically we have a high pressure and lower wind speeds at Mace Head. Figure 3 also shows that in some cases the concentration of small ions is dependent on wind direction (right plot). The concentration of small ions increased considerably when the wind direction turned from the clean ocean sector (180–300°) to the sector where air was advected over sparsely populated land in the northwest-to-north direction.

Acknowledgment This work was supported by EU (FP6, project number 018332).

References

1. O'Dowd, C.D. et al., *Nature*, **417**, 632–636 (2002).
2. Wen, J., Zhao, Y., and Wexler, A.J., *Geophys.Res.*, **111**, doi:10129/2005JD006210 (2006).
3. O'Dowd, C.D. et al., *J. Geophys. Res.*, **107**, 10.1029/2001JD000206 (2002).
4. Kulmala, M., Pirjola, L., and Mäkelä, J.M., *Nature*, **404**, 66–69 (2000).
5. Mirme, A. et al., *Boreal Env. Res.*, **12** (2007).
6. Tammet, H., *J. Aerosol Sci.*, **26**, 459–475 (1995).

Improving Particle Detection Efficiency of a Condensation Particle Counter by Means of Pulse Height Analysis

Mikko Sipilä[1], Pasi Aalto[1], Heikki Junninen[1], Hanna Manninen[1], Tuukka Petäjä[1], Mikael Ehn[1], Ilona Riipinen[1], Colin O'Dowd[2], and Markku Kulmala[1]

Abstract Pulse height analysis of white light scattered from the droplets formed in the CPC condenser was applied to distinguish between homogeneous and heterogeneous nucleation with high butanol supersaturation. With this method the activation of calibration aerosol particles down to ca. 1.5 nm was clearly observable. Experimental results show that the instrument described here is a powerful tool in atmospheric sub-3 nm aerosol particle measurements.

Keywords Condensation particle counter, pulse height analysis, detection efficiency, clusters

Introduction

Atmospheric new particle formation from gaseous precursors by nucleation has been observed in a numerous locations all over the world. Even though several nucleation mechanisms have been suggested[1] none has been experimentally verified and the first steps of nucleation remain completely unclear. Main reason for that is that atmospheric nucleation takes place in the particle sizes below 3 nm which is the lowest detection limit of any commercial instrument. Therefore the measurements in the sub-3 nm range are necessary in order to solve the detailed pathways of the nucleation process.

Condensation particle counters (CPC) are most commonly used instruments in aerosol nanoparticle detection. Operation of a CPC is based on particle activation and growth in supersaturated (typically alcohol) atmosphere and subsequent optical detection of formed droplet. By increasing the supersaturation smaller particles are

[1]*Department of Physical Sciences, P.O. Box 64, 00014 University of Helsinki, Finland*

[2]*Department of Experimental Physics, National University of Ireland, University Road, Galway, Ireland*

C.D. O'Dowd and P.E. Wagner (eds.), *Nucleation and Atmospheric Aerosols*, 378–381.

activated. However, with increasing supersaturation homogeneous nucleation of supersaturated vapour starts leading to false counts in the optics. In conductive cooling type CPCs the supersaturation field has radial and axial gradient. Therefore the different sizes of particles activate at different axial positions inside the CPC condenser. Bigger particles activate earlier and they have more time for growth thus yielding to larger final droplets. By measuring the amount of white light scattered by the droplet the initial particle size can be concluded.

This pulse height analysis method has been used in size distribution measurements between 3 and 10 nm[2] as well as to determine the composition of freshly nucleated nanoparticles.[3] Here we report on the application of PHA-technique together with a modified commercial CPC (model TSI-3025A) to extend the measurable particle size range below 2 nm.

Experiment and Results

Pulse height CPC (PH-CPC) used in this study is TSI-3025A conductive cooling type CPC with modified optics. Amount of light scattered by the droplet and thus the pulse height in the photodetector depends on the droplet size. However, the relation is not monotonic for monochromatic light and thus the optical detection system of the CPC was modified[4]. Laser, for example, was replaced by a white light source.

In normal operation conditions the pulse height is size dependent approximately from 3 nm (instruments lower detection limit) to 10 nm.[2] We, however, increased the supersaturation inside the condenser by increasing the saturator temperature from 37°C up to 44°C. As expected, homogeneous nucleation of butanol vapour started inside the condenser. However, homogeneously nucleated droplets were clearly distinguishable from heterogeneously nucleated droplets in PH-spectra (Figure 1) even at sizes approaching 1 nm (negatively charged WOx-particles). In ordinary CPC, in which only the total number of pulses is recorded, homogeneous nucleation would cover the signal from heterogeneously nucleated droplets.

By this method the detection efficiency of PH-CPC was investigated as a function of particle size. In case of negatively charged WOx-particles the detection efficiency is mainly determined by particle transport losses in the CPC capillary and practically all particles are activated down to ca. 1.5 nm (Figure 2).

Besides the physical size of entering aerosol particle the supersaturation required to activate the particle for growth depends also on particle charge and composition. Bigger particles are activated at lower supersaturations than smaller ones. Also charged particles and particles soluble in condensing vapour yield larger final droplets than neutral particles or insoluble particles initially of the same size. Moreover, the total condensation sink formed by the aerosol entering the condenser affects the pulse height spectrum (Figure 3). All this makes the interpretation of measurement data very complicated without information on physicochemical properties of sampled particles.

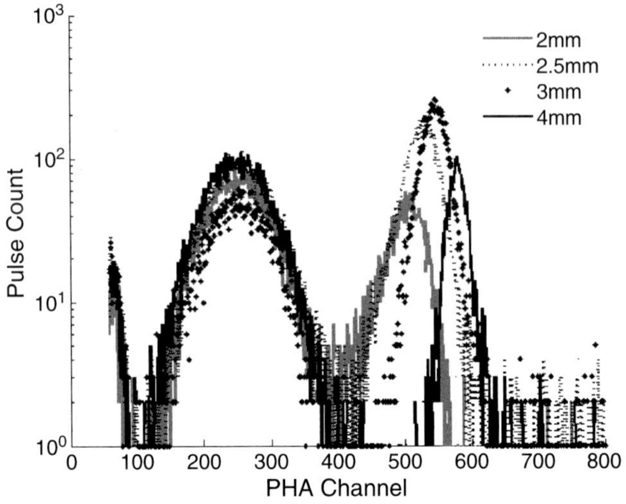

Figure 1 PHA spectra for DMA-classified WOx aerosol. Homogeneously nucleated droplets (**mode on the left hand side**) are clearly distinguishable from heterogeneously nucleated droplets

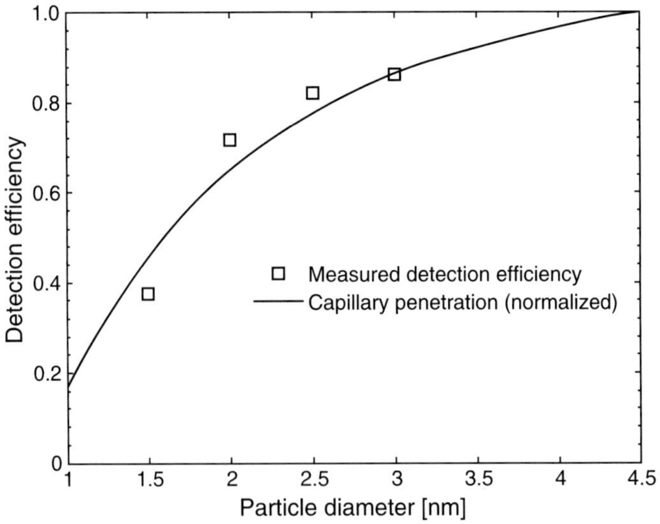

Figure 2 Detection efficiency of PH-CPC is mainly determined by particle penetration to the condenser. Capillary penetration is calculated[5] and because of inaccuracy of the aerosol flow normalized to match the data.

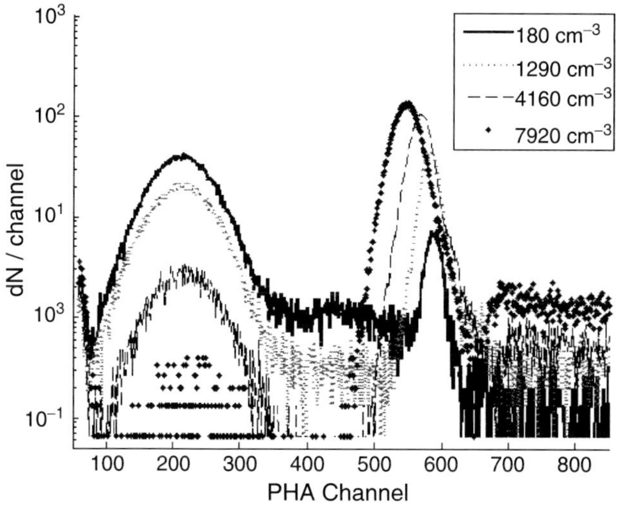

Figure 3 The dependence of pulse height on initial particle concentration (13 nm WO_x particles). Size of heterogeneously nucleated droplets decreases with increasing particle concentration whereas the size of the homogeneously nucleated droplets is not affected. However, the concentration of homogeneously nucleated droplets decreases with increasing condensation sink.

Conclusions

Experimental results show the huge potential of the PH-CPC in nanocluster detection. However, there are still several problems associated with this method, including the affect of particle chemical composition and charge, which need to be solved. Also the transport efficiency of nanometer-sized clusters should be improved in order to increase the sensitivity for low particle concentrations.

References

1. Kulmala, M., *Science*, **302**, 1000–1001 (2003).
2. Weber, R.J., Stolzenburg, M.R., Pandis, S.N., and McMurry, P.H., *J. Aerosol Sci.*, **29**, 601–615 (1998).
3. O'Dowd, C.D., Aalto, P., Hämeri, K., Kulmala, M., and Hoffmann, T., *Nature*, **416**, 497–498 (2002).
4. Dick, W.D., McMurry, P.H., Weber, R.J., and Quant, F.R., *J. Nanoparticle Res.*, **2**, 85–90 (2000).
5. Stolzenburg, M.R. and McMurry, P.H., *Aerosol Sci. Tech.*, **14**, 48–65 (1991).

The Effect of Temperature on Ion Cluster Formation of 2-propanol

A.-K. Viitanen[1], T. Mattila[2], J.M. Mäkelä[1], M. Marjamäki[1], and J. Keskinen[1]

Abstract Ion-mobility spectrometry offers a robust and effective technique to detect ion clusters. Here, influence of temperature on the ion cluster formation has been measured. Especially temperatures below 0°C favor formation of dimers and trimers for the 2-propanol used as a trace compound.

Keywords Ion clusters, ion-mobility spectrometry, time-of-flight

Introduction

Ions can be measured directly from atmosphere using sophisticated sampling, sequential pressure drop units, and mass spectrometer [1]. Also, long-time monitoring of the ambient ion-mobility distribution has been performed using aspiration type mobility analyzers [2]. Recently, ions have been observed to be linked with new particle formation process in the atmosphere [3].

Ion-mobility spectrometry (IMS) is a conventional in situ technique for detecting and identifying low-concentration gas phase compounds [4]. Most common commercial IMS-applications are utilized to detect warfare agents, drugs, and explosives [4,5]. IMS has been implemented also for the environmental applications such as monitoring ammonia in the atmosphere [6]. Ewing et al. [7] have measured 2-propanol in cold conditions using membrane. In this study, IMS is used to examine cluster ion formation without membrane. Purified air with trace concentrations of 2-propanol is ionized, and the ion mobility distribution is measured in different temperatures. We hope to gain information on ion mobility distributions in cold ambient conditions.

[1] *Aerosol physics laboratory, Tampere University of Technology, P.O. Box 692, 33101 Tampere, Finland*

[2] *Environics Oy, P.O. Box 1199, FI-70211 Kuopio, Finland*

C.D. O'Dowd and P.E. Wagner (eds.), *Nucleation and Atmospheric Aerosols*, 382–386.

Experimental

The commercial ion mobility spectrometer RAID I (Bruker Daltonics GmbH) was used for the measurements. In commercial IMS-devices the membrane is usually used before the drift-tube to attain better sensitivity and to remove moisture and impurities from the sample [4]. However, the device used here has been tuned for direct use by removing the membrane, enabling exact studies of the actual ion phenomena in environmental applications.

The RAID I is based on the time-of-light method (Figure 1). The sample gas flow of 0.42 L/min is being aspirated automatically and ionized in atmospheric pressure by β-source (^{63}Ni). Pulse of ions, with length of 0.3 msec, is introduced into the drift region where different ions are separated by electric field. At the end of the drift chamber the collector plate detects the current of arriving ions, forming a time-of-flight distribution of the sample.

The device was placed into a thermally controlled chamber. Temperature of the sample gas was measured from the outside wall of the drift tube with an accuracy of about 1 °C. It is assumed that the gas temperature is the same as the wall temperature. Due to internal structures and materials of the drift tube, the drift gas temperature may differ from the wall temperature and is also likely to vary throughout the drift tube [8]. Thus by implementing the temperature control in this way a certain inaccuracy is to be accepted.

In the measurement setup (Figure 1), PTFE (Teflon) tubing was used for connecting the parts throughout the system. Pressurized air, used as a carrier gas, was lead through silica gel and active charcoal to remove moisture and hydrocarbons. The primary flow was distributed into sample and sheath flow using rotameters. Sample flow was routed to the device or into exhaust through a magnetic valve. The liquid 2-propanol sample was injected using a 100 μl glass syringe (Hamilton) with the assistance of syringe pump (Cole-Parmer). Different injection rates were

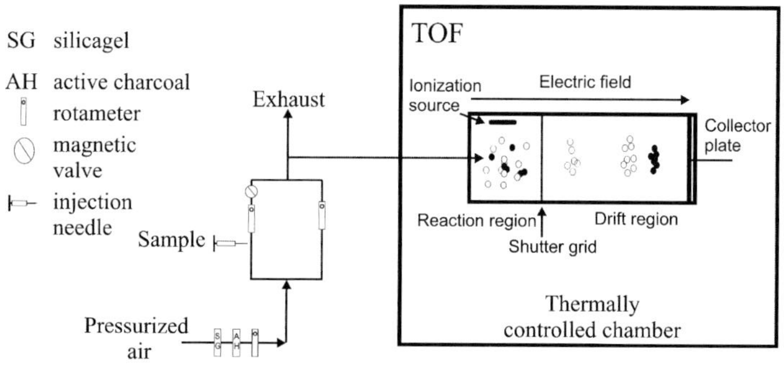

Figure 1 The measurement setup and the operation principle of the IMS based on time-of-flight method

used to vary the concentration. On the grounds of the injection rate, the sample properties and the dilution flow the theoretical concentration was determined. The actual concentration was not recorded but it might be smaller than the theoretical concentration [9].

The time-of-flight of the ions is measured, then the mobility values (Z) are first derived in the gas temperature prevailing in the drift tube. According to a procedure common within the IMS community, the mobility values Z are modified to standard conditions to obtain a reduced mobility Z_0, using:

$$Z_0 = Z\frac{273.15K}{T}\frac{P}{101.325kPa} \tag{1}$$

This mobility reduction to 273.15 K is performed due to consistency and comparability of different measurements. Note that our true temperature scale being −10°C to +80°C can generate 30% differences in the actual measured mobility values Z.

Results and Discussion

Mobility spectra of 2-propanol, a common solvent, was measured on different concentrations in the temperature range of −10°C to +80°C. Reduced mobility distributions are presented in Figure 2. The peaks are identified according to their reduced mobility as a monomer (1.8 cm^2/Vs), dimer (1.6 cm^2/Vs) and trimer (1.4 cm^2/Vs) the corresponding particle sizes according to [10] being in the range of 0.6 nm. Despite the reducing the reduced mobility values have some variation depending on humidity and temperature of the measurement which occurs from the varying amount of water molecules attached to a multimer. This phenomenon has the strongest effect on monomers. Thus the reduced mobility of the monomer has more temperature dependency than the dimer or the trimer (this can be seen from Figure 2).

The results show a strong temperature dependence for 2-propanol. At highest temperature there are no trimers at all. Near the room temperature some trimers are detected, but they are unstable and some of them break down during the time-of-flight measurement. At the room temperature more trimers are detected when the concentration is high. At the lowest measured temperature (−10°C) trimer-ions are stable being the most intensive peak at concentrations 10 ppm and 30 ppm. Even at the lowest measured concentration (3 ppm) there are trimers more than monomers.

In this study, we have presented the thermal dependence of 2-propanol and its highly increased tendency to cluster while the temperature decreases. As a conclusion: Effect of the temperature is more important on the appearance of 2-propanol trimers than the concentration in the ppm range.

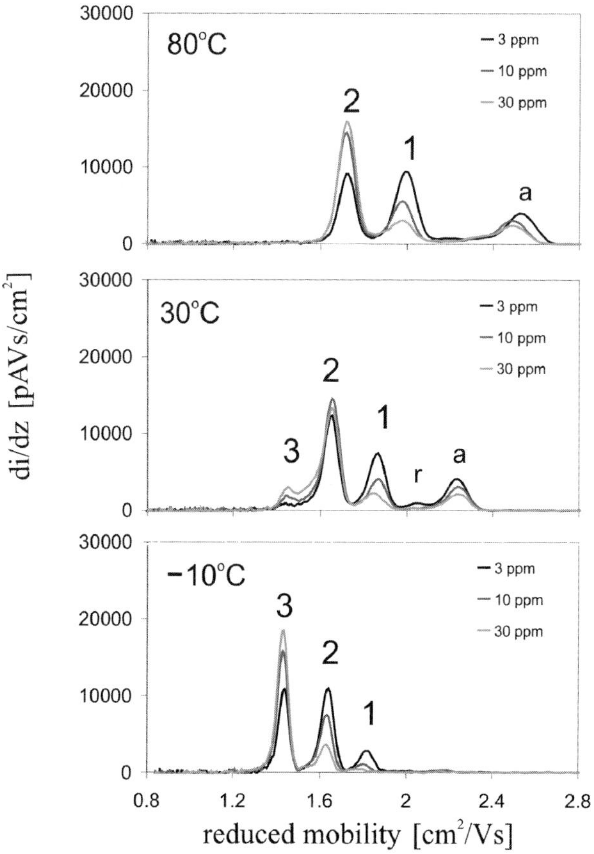

Figure 2 Mobility distribution for 2-propanol in different temperatures and with different concentrations. The numbers 1–3 refers to the number of multimer, the letter *r* to reaction ion peak and the letter *a* to ammonia

Acknowledgments The authors would like to acknowledge Finnish Defense Forces for funding the project.

References

1. Eisele F.L., First tandem mass spectrometric measurement of tropospheric ions, in *J. Geophys. Res.*, **93**, 716–724 (1988).
2. Horrak, U., Salm, J., Tammet, H., Bursts of intermediate ions in atmosphere, in *J. Geophys. Res.*, **103**, 13909–13915 (1998).
3. Kulmala, M., Vehkamäki, H., Petäjä, T., Dal Maso, M., Lauri, A., Kerminen, V.-M., Birmili, W., McMurry, P.H., Formation and growth rates of ultrafine atmospheric particles: a review of observations, in *Aerosol Sci.*, **35**, 143–176 (2004).

4. Eiceman, G.A. and Karpas, Z., *Ion Mobility Spectrometry*, 2nd edn. Boca Raton, FL: CRC Press, Taylor & Francis (2005).
5. Eiceman, G.A. Ion-mobility spectrometry as a fast monitor of chemical composition, in *Trends in Analytical Chemistry*, **21**(4) 259–275 (2002).
6. Myles, L., Meyers, P., Robinson, L., Atmospheric ammonia measurement with an ion mobility spectrometer, in *Atmos. Environ.*, **40**, 5745–5752 (2006).
7. Ewing, E.G., Eiceman, G.A., Stone, J.A., Proton-bound cluster ions in ion mobility spectrometry, in *Int. J. Mass Spectrom.*, **193**, 57–68.
8. Thomas, C.L., Rezgui, N.D., Kanu, A.B., Munro, W.A., Measuring the temperature of the drift gas in an ion mobility spectrometer: a technical note, in *IJIMS 2002*, **5**, 31–36.
9. Nousiainen, M., Peräkorpi, K., Sillanpää, M., Determination of gas-phase produced ethyl parathion and toluene 2,4-diisocyanate by ion mobility spectrometry, gas chromatography and liquid chromatography, in *Talanta, 2007*, article in press.
10. Tammet, H., "Size and mobility of nanometer particles, clusters and ions", in *J. Aerosol Sci.*, **26**, 459–475 (1995).

Heterogeneous Ion-induced Nucleation: Condensation of Supersaturated Vapors on Neutral and Charged Nanoparticles

Victor Abdelsayed and M. Samy El-Shall

Abstract A new method is described to study the condensation of supersaturated vapors on nanoparticles under well-defined conditions of vapor supersaturation, temperature, and carrier gas pressure. The nanoparticles can be activated to act as condensation nuclei at supersaturations significantly lower than those required for homogeneous nucleation. A small difference is observed in the number of droplets formed on positively and negatively charged nanoparticles. The charge preference effect depends on the nature of the nucleating vapor molecules and the degree of vapor supersaturation.

Keywords Heterogeneous nucleation, ion nucleation, nanoparticles, sign effect

Introduction

In many natural processes the condensation of supersaturated vapors occurs on charged nanoparticles, where both ion-induced and heterogeneous nucleation mechanisms are simultaneously involved.[1–4] Because of the strong ion-dipole and ion-induced dipole forces relative to van der Waals dispersion forces, ion nucleation occurs preferentially with respect to homogeneous nucleation.[5] In addition, the presence of nanoparticles may serve to lower the thermodynamic barrier to nucleation by providing the interface required for nucleation since a surface already exists.[5] For charged nanoparticles, the size of the particles, the surface properties, the contact angle between the condensing molecules and the particle, as well as the sign and the magnitude of the charge play the most important roles in determining the thermodynamics and kinetics of the process. The molecular properties of the condensing vapor molecules such as dipole moment and polarizability can also affect the nucleation process.

Department of Chemistry Virginia Commonwealth University Richmond, VA 23284–2006, USA

C.D. O'Dowd and P.E. Wagner (eds.), *Nucleation and Atmospheric Aerosols*, 387–391.

In this work, we demonstrate the application of a new technique to study the condensation of supersaturated vapors on metal nanoparticles generated by pulsed laser vaporization inside a diffusion cloud chamber (DCC). The systems investigated include the condensation of supersaturated vapors of polar (acetonitrile and trifluoroethanol, TFE) and nonpolar (*n*-hexane) molecules on Mg and Al nanoparticles. The results provide new insights on the condensation of supersaturated vapors on nanoparticles including the sign effect of the charged nuclei. The results presented here are focused on the condensation of supersaturated TFE vapor on Mg nanoparticles.

Experimental

Detailed descriptions of the DCC and the principles of its operation are available in the literature.[6,7] In the current experiments; the vapor supersaturation of TFE is adjusted below the value required for homogeneous nucleation. A single laser pulse (1.6×10^8 W/cm^2) is fired from a Quanta ray Nd:YAG laser (532 nm, 2 ns pulse width) focused on the Mg target placed on the bottom plate of the DCC. The nucleating droplets are counted by observing the forward scattering of light from the droplets falling through a horizontal He-Ne laser beam traversing the chamber using a photomultiplier connected to counting electronics. The He-Ne laser beam is set above that of the YAG laser to detect droplets of the TFE-coated Mg nanoparticles. The number of droplets is counted within a well-defined volume.

Results and Discussion

Under the laser power used in the current experiments a single laser pulse generates about 10^{13}–10^{14} metal atoms including about 10^5–10^6 atomic ions.[8] These atoms immediately condense to form neutral nanoparticles with a small fraction of charged ones. The nanoparticles diffuse through the DCC where they may become activated in the nucleation zone (~0.7 reduced height of the DCC) thus resulting in the condensation of the supersaturated TFE vapor. The activation process depends on the size and charge of the Mg nanoparticles generated in the DCC as well as the degree of the supersaturation of the TFE vapor as discussed in the following sections.

Effect of Vapor Supersaturation

Figure 1(a) displays the nucleation time profiles measured at various TFE supersaturations (1.387–1.829, at a total pressure of 675 Torr) following the laser vaporization of a Mg target using a single laser pulse. It is clear that the total droplets' count (TDC), Figure 1(b), decreases by decreasing the supersaturation of the TFE vapor.

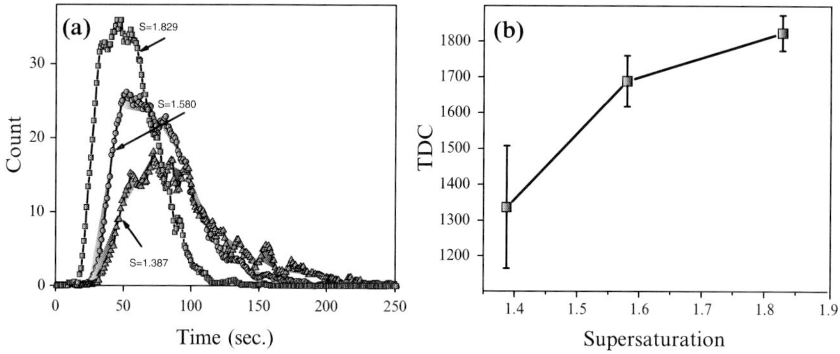

Figure 1 (a) Time profiles showing the number of TFE droplets condensed on Mg nanoparticles after a single laser pulse (1.6×10^8 W/cm^2) at different supersaturations. (b) Total number of droplets (integrated time profile) as a function of supersaturation

The nucleation peak is also shifted to longer times as the supersaturation ratio of the TFE vapor is decreased. At higher supersaturations most of the Mg nanoparticles are activated and the nucleation rate reaches its maximum at early times. On the other hand, at lower supersaturations only large Mg nanoparticles can be activated, and since these particles diffuse slowly the nucleation rate reaches its maximum at longer times.

Effect of Applied Electric Field

Since laser vaporization of metal targets generates a fraction of atomic ions,[8] it is expected to see a significant effect on the nucleation profiles by applying an electric field between the chamber plates. Figure 2(a) displays the nucleation time profiles in the presence of different electric fields (0.15–15.38 kV/m) at a constant TFE supersaturation ratio of 1.819 and a total pressure of 675 Torr.

Figure 2(a) shows that the time width of the nucleation cloud as well as the total intensity decreases as the applied electric field is increased. The inset in Figure 2(a) clearly shows that the total number of droplets decreases as the applied field increases. This behavior can be attributed to the short residence time that the charged nanoparticles spend in the nucleation zone in the presence of an electric field. In the absence of the applied field, the motion of the nanoparticles is described in terms of diffusion, and the measured nucleation profile reflects the longer time of diffusion as compared to the drift velocity of the charged particles.

The number of droplets forming on the positively charged (applying the negative field on the top plate) and negatively charged (applying the positive field on the top plate) nanoparticles appears to be similar at lower applied fields. However, at higher fields the number of droplets forming on the negatively charged particles is more than the number formed on the positively charged particles. This effect is also observed in the total number of droplets integrated over the nucleation time profile as shown in

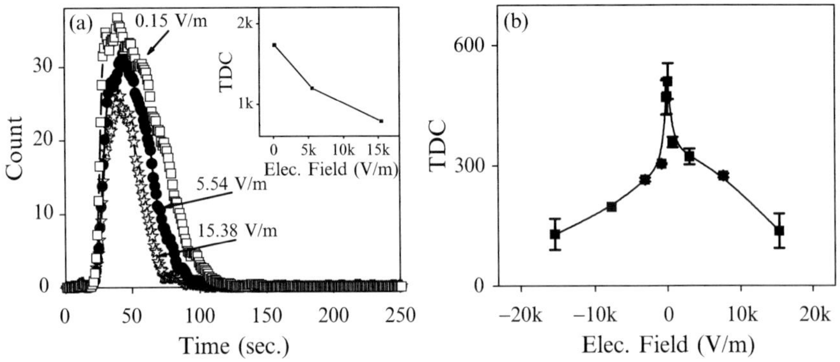

Figure 2 (a) Time profiles showing the number of TFE droplets at (S_{max} = 1.819, T_{max} = 278.6 K, P = 675.0 Torr) after a single laser pulse (1.6×10^8 W/cm^2) under different fields applied between the top and bottom plates of the DCC (5 cm). Inset shows the total number of droplets as a function of applied electric field. (b) Total number of TFE droplets as a function of the magnitude and sign (on the top plate) of the applied fields (S_{max} = 2.39, T_{max} = 269.0 K, P = 644.2 Torr)

Figure 2(b). The number of droplets decreases by similar amounts for the positively and negatively charged particles in the presence of small applied fields. This is due to the removal of ions from the nucleation zone before condensing the supersaturated vapor. This result suggests that both the positive and negative particles have almost the same ability to induce condensation of the TFE vapor. However, Figure 2(b) shows that at higher applied fields the number of droplets formed on the negatively charged particles appears to be higher than that formed on the positively charged particles (the decrease in the number of droplets by applying positive and negative fields is not symmetric). This could be explained by the low mobility of the negatively charged particles. In this case, the increase in the number of droplets would be due to the longer residence time of the negatively charged particles in the nucleation zone as compared to the more mobile positively charged particles. It is clear that definitive answers regarding the sign preference in ion nucleation is still lacking.

Conclusions

In conclusion, condensation of supersaturated TFE vapor on Mg nanoparticles generated by laser vaporization has been studied at supersaturations significantly lower than those required for homogeneous nucleation. The number of nanoparticles that can be activated increases with increasing the vapor supersaturation. The small difference observed in the number of droplets formed on positively and negatively charged nanoparticles is attributed to the difference in the mobilities of these nanoparticles. Therefore, no significant charge preference is observed for the condensation of TFE vapor on the Mg nanoparticles.

Acknowledgment Acknowledgement is made to the National Science Foundation (CHE-0414613) and to NASA (NNG04GH45G) for the support of this research.

References

1. Kulmala, M., Vehkamaki, H., Petaja, T., Dal Maso, M., Lauri, A., Kerminen, V. M., W. Birmili, W., and McMurry, P.J., *Aerosol Sci.*, **35**, 143 (2004).
2. Laakso, L., Anttila, T., Lehtinen, K.E.J., Aalto, P.P., Kulmala, M., Horrak, U., Paatero, J., Hanke, M., and Arnold, F., *Atmos. Chem. Phys.*, **4**, 2353 (2004).
3. Chen, C.C., Tao, C.J., and Cheng, H.C., *J. Colloid Interface Sci.*, **255**, 158 (2002).
4. Mavliev, R., Hopke, P.K., Wang, H.C., and Lee, D.W., *J. Phys. Chem. B*, **108**, 4558 (2004).
5. Kashchiev, D., *Nucleation: Basic Theory with Applications*, Oxford: Butterworth-Heinemann (2000).
6. Rusyniak, M., Abdelsayed, V., Campbell, J., and El-Shall, M.S., *J. Phys. Chem. B*, **105**, 11866 (2001).
7. Abdelsayed, V. and El-Shall, M.S., *J. Chem. Phys.*, **126**, 024706 (2007).
8. El-Shall, M.S. and Li, S., Synthesis and characterization of metal and semiconductor nanoparticles, in: *Advances in Metal and Semiconductor Clusters*, Vol. 4, edited by M.A. Duncan, London: JAI Press, pp.115–177 (1998).

Charged State of Freshly Nucleated Particles: Implications for Nucleation Mechanisms

Fangqun Yu[1] **and Richard P. Turco**[2]

Abstract Aerosol nucleation events have been observed at a variety of locations worldwide, and may have significant climatic and health implications. Despite extensive studies in the past, the mechanisms of particle formation remain illusive. While ions have long been suggested as favorable nucleation embryos, their significance as a source of atmospheric particles has remained uncertain. An analysis of recent measurements of the electrical charge on freshly nucleated particles can shed new light on the underlying nucleation mechanisms. Here, we demonstrate, based on a conservative analytical analysis and detailed kinetic simulations, that ion-mediated nucleation (IMN) can consistently explain the charged states of new particles, and particularly the excess charge (or overcharging) most frequently detected. We contrast our results to other analyses of these measurements, which conclude that the ion contribution to particle formation is relatively small, and demonstrate that the ion contribution may in fact be dominant.

Keywords Nucleation theory, ion-mediated nucleation, electrical charge state, charged fraction

Mechanisms of Atmospheric Particle Formation

Measurements indicate that H_2SO_4 and H_2O are clearly involved in many, if not most, nucleation events observed in the atmosphere. Binary homogeneous nucleation of sulfuric acid and water is now generally taken to be negligible in the lower troposphere.[1] Recent analyses of laboratory measurements also show that the contribution of ternary homogeneous nucleation, which involves ammonia, to the formation of new particles is likely to be very limited.[2] While certain organics have been shown to enhance H_2SO_4–H_2O nucleation rates in the laboratory,[3] the level of

[1]*Atmospheric Sciences Research Center, State University of New York, Albany, New York, USA*

[2]*Department of Atmospheric and Oceanic Sciences, University of California, Los Angeles, California, USA*

C.D. O'Dowd and P.E. Wagner (eds.), *Nucleation and Atmospheric Aerosols*, 392–396.

enhancement is similar to that of ammonia, and the significance of organics in initiating nucleation is probably small. On the other hand, organics clearly contribute to the growth of freshly nucleated particles.

Ions, which are generated continuously and ubiquitously throughout the atmosphere by cosmic radiation and radioactive decay, have long been known to promote nucleation. To study nucleation processes involving ion clusters, Yu and Turco[4, 5] developed a kinetic model that explicitly treats the complex interactions among small air ions, neutral and charged clusters of various sizes, precursor vapor molecules, and the preexisting aerosol. Yu and Turco[5] refer to the coupled formation and evolution of aerosol size distributions, including both charged and neutral clusters – under the influence of ionization, recombination, neutralization, condensation, evaporation, coagulation, and scavenging – as ion-mediated nucleation (IMN). This IMN model has been updated by incorporating recently available thermodynamic data and algorithms.[6]

The multiple-year ion charge and mobility measurements taken in Hyytiälä, Finland indicate that ions are involved in more than 90% of the particle formation events that can be clearly identified.[7, 8] Nevertheless, the relative importance of IMN versus neutral nucleation in the atmosphere remains unclear. Laakso et al.[7] conclude that the average contribution of ion nucleation to the overall nucleation rate is relatively small. Yu[9] points out, however, that the same measurements presented by Laakso et al.[7] may actually indicate the dominance of IMN in the observed particle formation. Accordingly, a careful interpretation of the measured charged fractions of freshly nucleated particles is needed to identify the underlying nucleation processes.

Overcharging of Freshly Nucleated Particles

We assume that IMN, and classical homogeneous (neutral) nucleation, create thermodynamically stable particles at rates, J^{IMN} and J^{HOM}, respectively, with an initial diameter, $d = d_0$. It should be noted that the formation rates of neutral particles will generally be enhanced by IMN through ion–ion neutralization processes involving sub-critical clusters, which is ignored here. Thus, the analysis below is conservative with regard to the contribution of IMN to the overall nucleation rate. Under these assumptions, the fraction of new particles that are initially charged (i.e., the charge fraction, CF) would be approximated as,

$$CF(d_0) = J^{IMN}/(J^{IMN}+J^{HOM}) \quad (1)$$

This fraction applies only to particles nucleated during a specific time interval that is short relative to variations in the nucleation rates. Generally, these particles will grow into larger stable aerosols within a definite size range.

The particles nucleated on ions are quickly neutralized due to recombination during their initial growth. Thus, as the particles increase in size from d_0 to d_1, the fraction originally nucleated on ions (i.e., at a rate, J^{IMN}) that remain charged at $d = d_1$ is roughly,

$$X^C = e^{-\alpha C \Delta t} \quad (2)$$

where α is the ion–ion recombination coefficient for a small ion with a charged nanoparticle of opposite sign, and C is total concentration of small (negative or positive) ions. Here, $\Delta t = (d_1 - d_0)/GR$ is the time needed to grow particles from d_0 to d_1 at a fixed growth rate, GR.

An aerosol immersed in a steady-state ion-plasma achieves an *equilibrium* (or steady-state) charge distribution. Hence, the attachment of small ions to neutral particles also contributes to the charged fraction, and we identify CF'(d_1) as the charged fraction at $d=d_1$ due to this contribution. CF'(d_1) is always smaller than the corresponding equilibrium charged fraction $CF^0(d_1)$, because charging of a neutral aerosol approaches equilibrium from lower charge values. Under transient conditions, the ratio of an instantaneous CF to the equilibrium CF^0 at the same size is defined as the overcharge ratio, OR. The values of OR for particles at $d = d_1$ can be estimated as follows,

$$OR \leq [CF' \times (J^{IMN} + J^{HOM}) + X^C \; J^{IMN}]/[CF^0 \; (J^{IMN} + J^{HOM})] < 1 + e^{-\alpha C \Delta t} \times J^{IMN}/[CF^0 \times (J^{IMN} + J^{HOM})] \quad (3)$$

The particles are overcharged if OR>1, and undercharged if OR<1. Under typical conditions, $\alpha = \sim 1.5 \times 10^{-6}$ cm^3/s, $C = \sim 750$/cm^3. The growth rates of sub-3 nm intermediate ions at Hyytiälä have been estimated from ion mobility spectra to be in the range of ~0–4 nm/hour with a mean value of 1 nm/hr.[10] If we assume d_1=3 nm, d_0=1.5 nm, and GR = 1.5 nm/hr (for sub-3 nm particles), we determine,

$$OR(3\text{ nm}) < 1 + 0.017 \, J^{IMN} / [CF^0 \; (J^{IMN} + J^{HOM})] \quad (4)$$

The equilibrium charged fraction $CF^0(d)$ at d = 3 nm is around 1%. From equation (4) we can then estimate OR for particles of 3 nm for several scenarios:

Case 1: 100% IMN nucleation, $J^{IMN}/(J^{IMN}+J^{HOM}) = 1.0$: OR < 2.7
Case 2: 50% IMN nucleation, 50% homogeneous nucleation: OR < 1.85
Case 3: 10% IMN nucleation, 90% homogeneous nucleation: OR < 1.17
Case 4: 100% homogeneous nucleation: OR < 1

For the 27 nucleation event-days described by Laakso et al. (2006), in which OR(3 nm) values are given, the following data apply: OR(3 nm) > 2 for 20 days (74%); 1 < OR(3 nm) < 2 for 5 days (18.5%); and OR(3 nm) < 1 for 2 days (7.5%). Thus, based on a simple, and conservative, interpretation, these measurements[7] indicate that IMN dominates particle formation on most of the days sampled. Of course, the analysis is subject to uncertainties associated with the values of α, C and GR. In real situations, α and GR depend on the size and type of particles, and α, C and GR also vary with time during nucleation events. Accordingly, full kinetic modeling with resolved microphysics is required to resolve the effects of such variations.

We simulated this competitive, highly nonlinear homogeneous/ion nucleation regime using a model that explicitly resolves positive, negative, and neutral particles ranging in size from molecular scales to several micrometers, while also treating the co-condensation of sulfuric acid and organic compounds.[6] Figure 1 provides a comparison of simulated and observed size-dependent OR values. The predicted rapid decrease in OR as particle sizes increase from ~2–3 nm to 7 nm is consistent with measurements. It appears that most of the observed behavior in OR can be explained in a straightforward way by variations in the concentrations of the key precursor gases (sulfuric acid and low volatility organics), and by sensitivity to the particle size at which organic vapors begin to condense (the activation size).

An overcharge on freshly nucleated particles is an unambiguous signature of an IMN mechanism. Based on our simulations, competing neutral H_2SO_4–H_2O and ternary nucleation are negligible under the conditions encountered. Thus, practically all of the neutral particles between 1–3 nm are formed via the neutralization of charged particles, while activation and growth of these 1–3 nm particles provides the major source (>90%) of the aerosol detected at sizes ≥3 nm. Laakso et al.[7] interpreted the

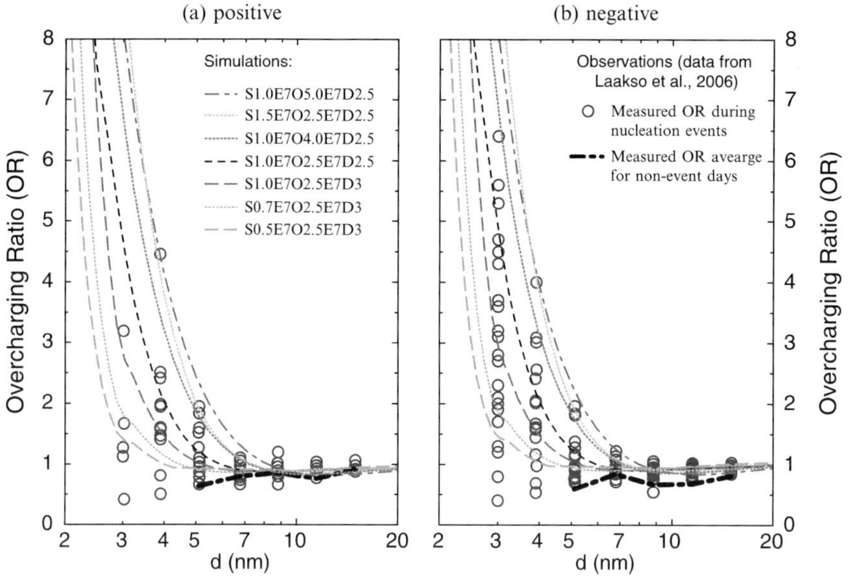

Figure 1 Simulated and observed size-dependent overcharging ratios (ORs) are compared. Each curve represents calculated OR values averaged over nucleation events corresponding to specific atmospheric conditions (**below**). The open circles are observed OR values for 34 nucleation events sampled with an ion-DMPS during an intensive field campaign reported by Laakso et al.[7]; the thick dashed lines are mean ORs seen on 5 nonevent days. In the simulations, temperature (T), relative humidity (RH), and precursor gas concentrations (sulfuric acid [S], condensable organics [O]) were parameterized as sine functions of local time to mimic diurnal variations. In the legend, read S1.0E7O5.0E7D2.5 as [S]max =1.0 ×107 cm^{-3}, [O]max = 5.0 × 107 cm^{-3}, and Dact = 2.5 nm (the particle activation diameter for organic condensation). The daily mean T and RH (i.e., T0 and RH0) are 270 K and 60%, respectively

same overcharge measurements and concluded that the average contribution of IMN to the total nucleation rate is relatively small. Here, both a conservative analytical analysis and a detailed kinetic modeling assessment, confirm that the contribution of ion nucleation is dominant. We suggest that the analysis of particle electrical overcharging data requires a comprehensive analytical interpretation accounting for the range of realistic conditions encountered in the field.

Acknowledgment This study is supported by NSF under grant 0618124 and the NOAA/DOC under grant NA05OAR4310103.

References

1. Yu, F., *J. Geophys. Res.*, **111**, D04201, doi:10.1029/2005JD006358 (2006).
2. Yu, F., *J. Geophys. Res.*, **111**, D01204, doi:10.1029/2005JD005968 (2006).
3. Zhang, R. et al., *Science*, **304**, 1487–1490 (2004).
4. Yu, F., Turco, R.P., *Geophys. Res. Lett.*, **24**, 1927–1930 (1997).
5. Yu, F. and Turco, R.P., *Geophys. Res. Lett.*, **27**, 883–886 (2000).
6. Yu, F., *Atmos. Chem. Phys.*, **6**, 5193–5211 (2006).
7. Laakso, L. et al., *Atmos. Chem. Phys. Discuss.*, **7**, 1333–1345 (2007).
8. Hirsikko, A. et al., *Atmos. Chem. Phys. Discuss.*, **7**, 201–210 (2007).
9. Yu, F., *Atmos. Chem. Phys. Discuss.*, **6**, S4727–S4734 (2006).
10. Kulmala, M. et al., *Atmos. Chem. Phys.*, **4**, 2553–2560 (2004).

Measurement of the Charging State with an Ion-DMPS to Estimate the Contribution of Ion-induced Nucleation

Stéphanie Gagné[1], Lauri Laakso[1], Tuukka Petäjä[1], Veli-Matti Kerminen[2], Pasi P. Aalto[1], and Markku Kulmala[1]

Abstract In this work, the Ion-DMPS is presented and the analyzed data from over one year of measurements in Hyytiälä, Finland (April 13, 2005–September 21, 2006), is discussed. The Ion-DMPS allows calculation of the charging state of particles taking part in a nucleation event. A physically sound equation that describes that charging state of particles during a nucleation event is presented and fitted to the Ion-DMPS data. The behavior predicted by the equation of the charging state fitted very well to the experimental data obtained from the Ion-DMPS. According to this analysis, an overwhelming majority of the new particle formation events appeared to include at least some participation of ion-induced nucleation. The analysis of a year of data showed that the contribution of ion-induced nucleation varied between 0% and slightly less than 40%. The median ion-induced participation for the new particle formation events analyzed was between 5% and 10%.

Keywords Ion-induced nucleation, atmospheric aerosols, field measurements, Ion-DMPS

Introduction

Aerosol particle formation takes place frequently and in many different types of environments (Kulmala et al. 2004). Many different nucleation mechanisms have been proposed to explain particle formation. However, despite many observations, the exact mechanisms are not well known. Amongst the mechanisms ion-induced, ternary, binary, and kinetic nucleation, have been proposed, but the relative importance of these processes in different environments is still unknown. Ternary and binary nucleation refers to the number of vapors taking part in the nucleation

[1]*University of Helsinki, Department of Physical Sciences, P.O. Box 64, FI-00014 Helsinki, Finland*

[2]*Finnish Meteorological Institute, Climate and Global Change, Erik Palmenin Aukio 1, P.O. Box 503, FI-00101 Helsinki, Finland*

C.D. O'Dowd and P.E. Wagner (eds.), *Nucleation and Atmospheric Aerosols*, 397–401.

process, whereas ion-induced nucleation refers to the presence of charged particles in the nucleation event (Weber at al. 1996; Laakso et al. 2002). Thus, both binary and ternary nucleation can be either ion-induced or neutral. Some proposed mechanisms also involve both ion-induced and neutral nucleation together (e.g. Kulmala et al. 2006).

In a nucleation event, the distribution of electrical charges onto the particles carries information about the nucleation mechanisms. Indeed, it is possible to detect if ion-induced nucleation was involved or not in the formation of new aerosol particles from the evolution charged fraction of particles during nucleation. The Ion-DMPS, a relatively new instrument (Laakso et al. 2007) allowing retrieving the charging state of particles during a nucleation event.

Methods

The Ion-DMPS is based on the well-known DMPS system except for a modification in the bipolar charger which can be switched on or off. The DMA can also be switched polarity in such a way that four modes of measurement are possible: (1) ambient sample, negative ions; (2) ambient sample, positive ions; (3) neutralized sample, negative particles; (4) neutralized sample, positive particles. The ratio of ambient mode over the neutralized mode for the same charge sign for each diameter and time gives the charging state of the particles at this particular diameter and time. The Ion-DMPS measures diameters from 3 nm to 13 nm with 7 size bins, it measures in the four modes within around 13 min.

A physically sound equation that describes the behavior of the charging state as a function of the diameter has been developed (Kerminen et al. 2007). This equation corresponds very well to the behavior as observed with the Ion-DMPS and it is shown here:

$$S(d_p) = 1 - \frac{1}{kd_p} + \frac{(S_0 - 1)kd_0 + 1}{kd_p}\exp(-k(d_p - d_0)) \tag{1}$$

where $S(d_p)$ and S_0 are the charging states at sizes d_p and d_0, where d_0 is the diameter at which the new particle form. The parameter k is defined as:

$$k = \frac{\alpha N_{\pm}^{C}}{GR} \tag{2}$$

where α is the recombination coefficient (here $\alpha = 1.6 \cdot 10^{-12}$ m^3/s), $N_{\pm}^{c}$ is the number of cluster ions and GR is the growth rate of particles. The parameter k as fitted to the Ion-DMPS data can be verified against values from other measurement devices. Moreover, the parameter k defines the degree to which a result can be trusted. This is the "memory effect". When the value of k is too high, even if the charging state of particles was very high at smaller particle sizes, the charging state would decrease so rapidly with respect to size that it would have reached equilibrium (charging state = 1) already at 3 nm, size at which the Ion-DMPS starts to detect particles (see Figure 1).

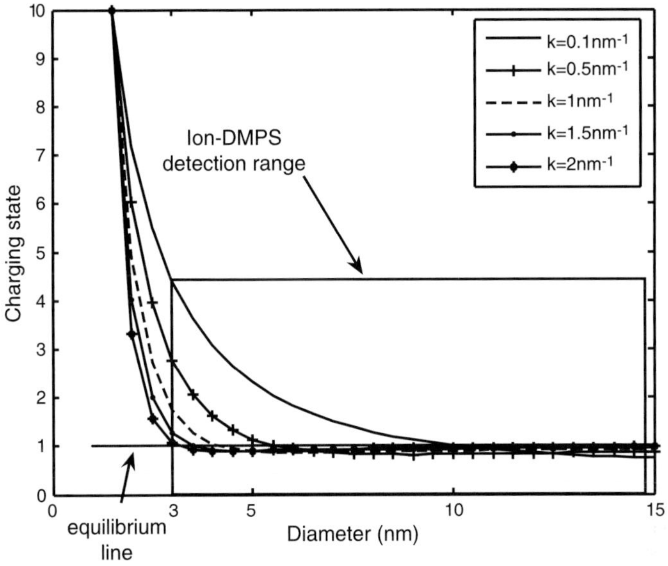

Figure 1 In this figure Eq. 1 is ploted with $S_0 = 10$ for a varying value of *k*. As the value of *k* is growing, it becomes more difficult tell precisely at what charging state the particles were for a smaller size.

Results

The median charging state for each size over the period for which the nucleation mode was in the said size bin is calculated and placed onto a plot of the median charging state as a function of the diameter (see example Figure 2). A fit of Eq. 1 to the experimental data was performed for all days with a nucleation event. The equation agreed very well with the measured data and the fit values of *k* that were verified with data available through other measurement devices (BSMA). Within lower values of the fitted and the calculated *k* and by eliminating value which differed by more than the maximum uncertainty for the calculated *k*, the values agreed rather well.

According to Eq. 1, the charging state of particles at 1.5 nm, S0 has been extrapolated. The charging states observed at 1.5 nm varied between –1.0 and 42.6 for negatively charged particles while it varied between –0.5 and 51.8 for positively charged particles. The negative values of S0 are physically impossible but it can be explained by the sensitivity of the fit when the charging state is below equilibrium or because the diameter d0 at which nucleation starts is too small. One can however be sure in these cases that the nucleation had no ion-induced nucleation involved. A clear majority of the nucleation events involved at least some ion-induced nucleation (around 3/4 of the events). The median of the particle flux resulting from ion-induced nucleation was slightly less than 10% while the maximum was slightly less than 40%.

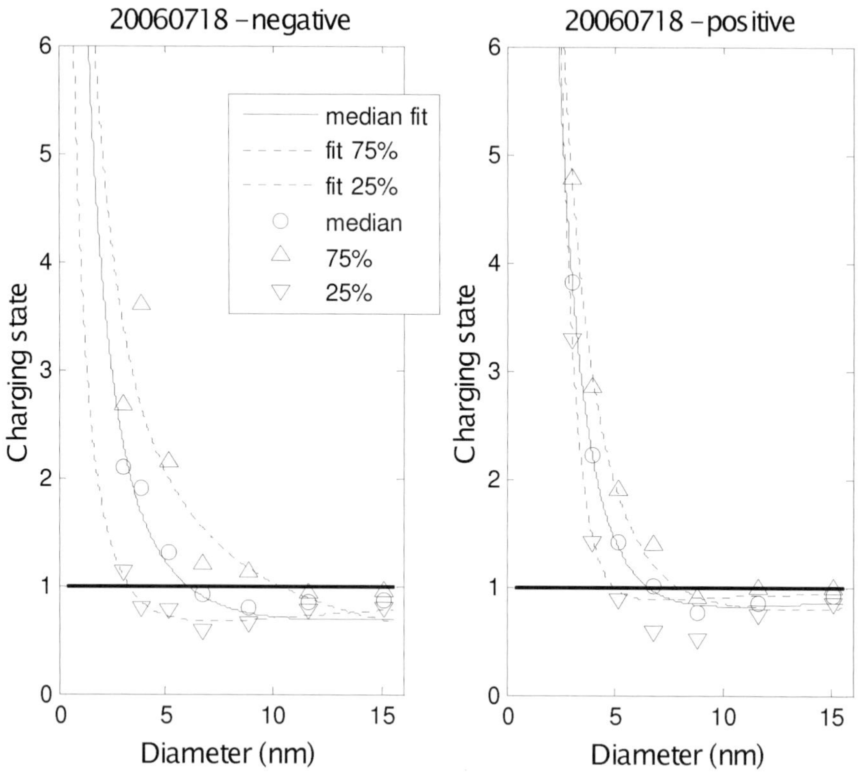

Figure 2 The median charging state of the particles as a function of the diameter is presented for the example day (July 18, 2006). The median charging state is represented by circles while the 25th and 75th percentiles are represented by triangles. The fit to Eq. 1 of the median charging state is represented by a full line while the dotted lines show the fits for the 25th and 75th percentiles. The negatively and positively charged particles are shown on the left and right subplot respectively. The parameters are $k = 0.2\,nm^{-1}$ and $S_0 = 5.6$ on the negative side and $k = 0.4\,nm^{-1}$ and $S_0 = 13.5$ on the positive side

Acknowledgments Erkki Siivola and Pekka Pihkala are acknowledged for their work on the Ion-DMPS. Veijo Hiltunen and Heikki Laakso are acknowledged for their efforts maintaining the instrument during the field measurements. Anne Hirsikko is acknowledged for providing the BSMA data. This work was supported by Biosphere-Aerosol-Cloud-Climate Interactions (BACCI).

References

1. Kulmala, M., Vehkamäki, H., Petäjä, T., Dal Maso, M., Lauri, A., Kerminen, V.-M., Birmili, W., and McMurry, P.H., *J. Aerosol Sci.*, **35**, 143–176 (2004).
2. Weber, R.J., Marti J., McMurry, P.H., Eisele, F.L., Tanner, D.J., and Jefferson, A., *Chem. Eng. Comm.*, **151**, 53–64 (1996).

3. Laakso, L., Mäkelä, J.M., Pirjola, L., and Kulmala, M., *J. Geophys. Res. D*, **107**(D20), 4427, doi:10.1029/2002JD002140 (2002).
4. Kulmala, M., Lehtinen, K.E.J., and Laaksonen, A., *Atmos. Chem. Phys.*, **6**, 787–793 (2006).
5. Laakso L.,Gagné, S., Petäjä, T., Hirsikko, A., Aalto, P.P., Kulmala, M., and Kerminen, V.-M., *Atmos. Chem. Phys.*, **7**, 1333–1345 (2007).
6. Kerminen et al., manuscript in preparation (2007).

Part IV
Ice Nucleation

Progress and Issues in Quantifying Ice Nucleation Involving Atmospheric Aerosols

Paul J. DeMott

Abstract A brief review/update is provided on recent studies of ice nucleation occurring within or as a consequence of the properties of atmospheric aerosol particles. Laboratory and atmospheric studies of homogeneous freezing and heterogeneous ice nucleation are considered. Significant progress is noted in some areas and key issues for future research are highlighted.

Keywords Heterogeneous ice nucleation, homogeneous ice nucleation, aerosol indirect effects

Introduction

It is important to identify the ice formation processes involving different types of atmospheric aerosol particles to understand and constrain predictions of their role in affecting cloud microphysics, cloud chemistry, precipitation, and climate. This paper b riefly reviews some recent progress and identifies some key and evolving issues in ice formation involving atmospheric aerosol particles. Significant theoretical and experimental progress in understanding homogeneous freezing of pure water droplets and the analogous process for solute aerosols at temperatures below the pure water freezing limit is noted, with the roles of surface versus volume freezing and the effect of organic aerosols remaining to be resolved. Increased numbers of laboratory studies are providing inferences regarding the roles of common atmospheric particle types as ice nuclei (IN) and the role of secondary phase transitions within aerosols as enablers of heterogeneous ice nucleation. Field measurements of IN made in the last decade have provided insight into the concentrations and variability of natural IN populations, and their relation to aerosol properties. While natural IN concentrations sometimes correlate with the concentrations of ice

Department of Atmospheric Science, Colorado State University, Fort Collins, CO, USA 80523–1371

C.D. O'Dowd and P.E. Wagner (eds.), *Nucleation and Atmospheric Aerosols*, 405–417.

particles exceeding 100 μm, discrepancy remains compared to the high concentrations of smaller ice crystals measured in clouds that in some cases reflects unresolved ice formation mechanisms. New measurement methods are needed and many are in progress.

Homogeneous Freezing Nucleation

Homogeneous ice formation by freezing is a vigorous and important process for ice crystal formation relevant to upper tropospheric cloud conditions. It may occur in dilute droplets lofted below about −35°C in cumulus clouds and it may occur for solution (haze) droplets or any unfrozen particles containing solutions at lower temperatures. Some of the recent progress in quantifying homogeneous freezing has been described in other review articles devoted to this topic specifically[1] or as part of a review of atmospheric ice formation processes.[2] Benz et al.[3] confirm the consistency of recent data sets concerning the homogeneous freezing rates of pure water drops and suggest only modest adjustments to any parametric descriptions of these data based on expansion cloud chamber studies. Stöckel et al.[4] used levitation methods to confirm the warmer homogeneous freezing temperature of D_2O versus H_2O observed previously in emulsion studies and attribute this to the greater degree of intermolecular association in D_2O.

Knowledge concerning the homogeneous freezing of solution drops, as occurs below about −38°C, has also evolved to an apparent point of high understanding. The relation between freezing temperatures and water activity suggested by Koop et al.[5] has undergone scrutiny in a number of regards. Baker and Baker[6] used a thermodynamic model for water to demonstrate that compressibility is maximized (as would be fluctuations) at the homogeneous freezing temperature over a wide range of pressures and then compared compressibility and freezing data for NaCl solutions to show that solutions having the same compressibility as pure water as a function of pressure also freeze homogeneously at the same temperature. This result appears to confirm the suggestion of Koop et al.[5] that the underpinning of the general relation between freezing and water activity is due to the similar impacts of pressure and solute on the hydrogen-bonding network of water. Khvorostyanov and Curry[7] demonstrated that a classical theoretical approach when constrained with proper physical data can and does predict these same solute and pressure dependencies. Abbatt et al.[8] note that most measurements of the homogeneous freezing conditions of sulfuric acid aerosol particles are consistent at low temperatures, but suggest that discrepancies for ammonium sulfate relate to realization in some experiments of the apparent heterogeneous ice nucleating behavior of crystalline ammonium sulfate (next section). Larson and Swanson[9] found more ideal freezing behavior for freely falling ammonium sulfate drops and added nucleation rate data for these. Wise et al.[10] found that mixtures of ammonium sulfate and dicarboxylic acids froze for the same water activity condition at low temperatures although their data did not quantitatively agree with the parameterization of Koop et al.[5]

Koop[1] demonstrated that some substances such as PEG that supercool a great deal more than ideal solutes such as sulfuric acid do so because of the strong temperature dependence of their water activity.

The first observations of homogeneous freezing in natural particles generally confirmed the relative humidity conditions for the onset of homogeneous freezing found in laboratory studies,[11] but also suggested organic aerosol influences on freezing onset. Cziczo et al.[12] also showed that aerosols containing organic compounds remain preferentially unfrozen in cirrus until cloud formation is strongly forced, supporting the inferences from nucleation studies. Kärcher and Koop[13] framed the possible mechanisms at play in these observations: the lowered rate of water uptake due to lowered hygroscopicity and lowering of the mass accommodation coefficient of organics. Beaver et al.[14] show that the carbon chain length of aldehydes and ketones added to sulfuric acid aerosols can impact their freezing temperatures in either direction from the homogeneous freezing conditions. Very short chains (C_3) inhibited freezing, intermediate chain length organics had no impact, and longer chain species ($>C_8$) with lower solubilities led to freezing at warmer temperatures that suggests heterogeneous nucleation. More and higher quality field measurements on homogeneous freezing in concert with particle composition data are needed to resolve organic aerosol impacts.

An issue that remains for quantitative understanding of homogeneous freezing regards whether freezing ensues in the volume or preferentially close to the surface–air interface of drops as proposed by Tabazadeh et al.[15] Koop[5] points out that existing data of sufficient quantitative accuracy to confirm whether surface or volumetric nucleation rates at different drop sizes. Duft and Leisner[16] and Stöckel et al.[4] show data supporting that volumetric freezing dominates for larger droplets. New methods are needed to explore a broad droplet size regime. New molecular dynamics simulations[17] suggest that homogeneous freezing starts in the droplet subsurface rather than the bulk. An ameliorating factor in deciding if this issue is critically relevant for quantifying real atmospheric ice formation is the fact that competition for water vapor between condensation and consumption by nucleated crystals is an important feature of freezing at cirrus altitudes and temperatures.[18] Thus, nucleation is usually rapidly quenched.

Heterogeneous Ice Nucleation

Theoretical Developments

Khvorostyanov and Curry[19] consider the properties of the critical radius, activation energy and nucleation rate in an evolved classical theoretical model for heterogeneous nucleation by freezing of deliquescent CCN. Khvorostyanov and Curry[20] implement this theory within a cloud parcel model framework to demonstrate that a Bergeron-like process limits the fraction of CCN that can freeze heterogeneously.

Their apparent assumption that this same behavior is reflected in observations of the concentration of such particles active as IN in the atmosphere is not supported by measurements described in the next section. Thus, suggestions by these authors[21] that primary nucleation provides a potential explanation of past observations of ice crystal concentrations far exceeding IN concentrations are unconfirmed. Although a classical theoretical model may yet describe heterogeneous ice nucleation by realistic mixed atmospheric aerosols with the same success as apparent for homogeneous freezing, some work remains to properly constrain aerosol-related nucleation parameters.

Important Ice Nucleus Aerosol Types

Special studies have provided a paradigm for gathering new information on the composition of atmospheric IN in recent years and this has guided recent laboratory studies of heterogeneous ice nucleation. In one method, glaciated cloud is sampled with a counterflow virtual impactor and the compositions are assessed on a single particle basis using electron microscopy[22] or mass spectrometry.[23] Alternately, collection of freshly activated IN[24,25] offers direct insights into the composition of these particles. A primary method for processing the concentrations of atmospheric IN as a function of temperature and relative humidity in recent years is the Continuous Flow Diffusion Chamber (CFDC).[26] Figure 1 summarizes the compositions of IN measured for particles activated by CFDC instruments in a number of studies while processing atmospheric particles in the heterogeneous ice nucleation environmental regime (i.e., warmer than about −36°C or at lower relative humidity than require for activation of homogeneous freezing at lower temperatures), and without regard to the actual processing conditions. The Particle Ablation by Laser Mass Spectrometry (PALMS) instrument was linked to a CFDC in two studies[11,27] (referred to by the

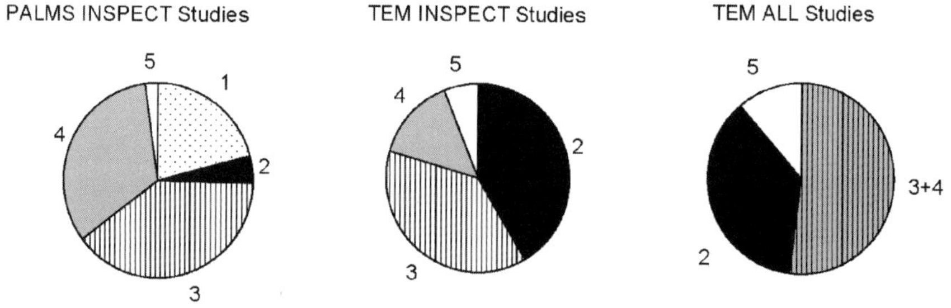

Figure 1 Pie charts of the predominant compositions of individual apparent ice nuclei (IN) processed by the CFDC technique, as measured using the Particle Ablation by Laser Mass Spectrometry (PALMS) instrument and Transmission Electron Microscopy (TEM). Temperature and supersaturation dependencies are not considered. Categories of particle types are: 1 (sulfate/organic), 2 (C for TEM; K, C, organic for PALMS), 3 (mineral dust), 4 (metallic), and 5 (Other – includes some sulfates for TEM data)

INSPECT acronym). Transmission Electron Microscopy (TEM) analyses of IN collected as described by Chen et al.[24] have been done in several more studies. Nevertheless, the TEM data from the INSPECT studies are quite similar to the collective project results in indicating the importance of mineral dust and carbonaceous aerosols as major contributors to IN populations for particles in the 0.05 to 1 μm size range. The C particles cannot be easily distinguished with regard to their sources or nature (e.g., organic or inorganic). Only modest amounts of sulfates are ever indicated in the TEM studies. In slight contrast, PALMS, measuring IN compositions larger than about 0.2 μm, finds more apparent sulfate/organic particles as IN and much smaller contributions from primarily C particles.

Mineral Dusts

The role of mineral dusts as IN has been known for over 50 years. DeMott et al.[28] measured extremely strong response of Saharan dusts as IN at great distances from their source and even studies in remote locations.[11,27] A large number of laboratory studies have occurred in the last few years to investigate the ice nucleating activity of surrogate and natural mineral dusts representative of those in the atmosphere in order to improve fundamental understanding and support parameterization development for cloud and climate modeling. Techniques employed have included chilled microscope stages,[29–31] CFDC,[32–34] expansion cloud chamber,[35,36] aerosol flow tube,[37] and acoustic levitation.[38] Typical activation behaviors of mineral dusts in their pure state are as displayed in Figure 2. Excellent qualitative agreement has been obtained on the tendency for relatively constant ice formation onset (and presumably nucleation rate) at constant ice supersaturation at low temperatures, but water saturation is often found to be required for ice formation at warmer supercooled temperatures. Quantitative differences between studies are likely related to the role particle size effects, chemical compositional differences, and sample preparation and generation. Few studies have yet validated ice formation by natural dusts as warm as suggested by cloud lidar studies,[39] but studies using supermicron dusts[34,36] clearly show warmer temperature onsets than those using submicron particles.[32,33] Coating of mineral dusts with solutes leads to modest[32] or no change[30] in ice nucleation activation conditions, although the freezing process does show the same apparent relation to water activity that exists for homogeneous freezing of solutions.[31,32] Studies are underway to examine the role of dust coating by other pollutants. Recent studies[30,36] differ on confirming previously noted[40] effects of preactivation on enhancing the ice forming ability of dusts.

Carbonaceous Particles from Combustion

The potential ice nucleation behavior of primary carbonaceous particles from biomass and fossil fuel combustion processes is poorly understood despite a great deal of recent research. This situation may reflect that surface and water interaction characteristics

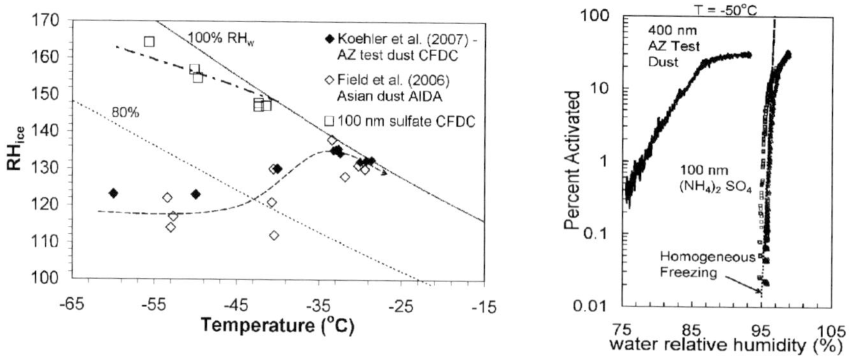

Figure 2 **Left panel**: Similarity of conditions where approximately 1% by number of noted dust particles nucleate ice heterogeneously in studies using a CFDC versus the AIDA chamber. The dashed arrow is to draw the eye toward the characteristic ice formation behaviors. Conditions for homogeneously freezing 1% of ammonium sulfate in the CFDC are also shown in comparison to the parameterization of Koop et al.[5] implemented in a Köhler theory calculation of haze growth for the CFDC processing conditions. **Right panel**: the fractional ice activation measured during slow relative humidity changes in the CFDC studies are shown (replicated sulfate experiments shown). The expected homogeneous freezing fraction for the sulfate drops is shown as the diagonal line

that influence the ice nucleation behavior of these particles are also strongly influenced by physical and chemical processing, factors that are difficult to simulate or control. Consequently, investigators have conducted measurements using both surrogates for carbonaceous aerosols (e.g., certain manufactured black carbons or flame-generated particles) that can be processed in different ways and combustion particles from real sources. Much laboratory study focus has been on upper tropospheric conditions where black carbon acting as an ice nucleus could perturb cirrus cloud processes[41] and where a direct source is present in the form of aircraft fuel combustion.[42] Results from recent CFDC studies for both model and real soot particles at –55°C are shown in Figure 3. Many biomass combustion particles contain varying amounts of soluble matter and are found to freeze nearly indistinguishably from pure haze particles at low temperatures, suggesting no apparent heterogeneous IN ability. More hygroscopic ones containing more inorganic matter appear to have a slight advantage in the timescale of the CFDC measurement. Real aircraft combustor soot[43] containing modest overall soluble content either act similarly to stimulate homogeneous freezing or act as heterogeneous IN for conditions exceeding the onset relative humidity for homogeneous freezing of soluble aerosols. The behavior of hydrophobic diesel exhaust particles as apparent centers for a homogeneous-like freezing is unexpected, but may reflect the conversion of semi-volatile matter in the gas phase to hydrophilic condensed matter during cooling. Interestingly, only very hydrophobic graphitized carbon particles demonstrated clear heterogeneous IN ability, albeit less efficiently than for mineral dust. We have not found this to be a behavior unique to all hydrophobic particles, and in any case, a processed particle seems the more likely model for atmospheric soot.

Other studies have found lower relative humidity thresholds for ice formation by pure and coated soot,[44,45] but there appears a tendency for coating to push the freezing

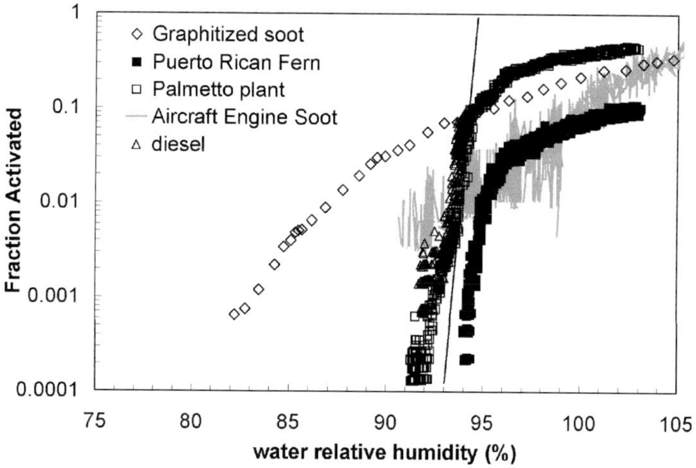

Figure 3 Ice formation by various combustion produced particles measured in the Colorado State University "laboratory" CFDC operating at −55 ± 2°C. The graphitized soot sample and the aircraft engine soot sample[43] were generated dry by the fluidized bed technique and were size selected at 200–250 nm, respectively. The diesel soot particles were collected from burning a low sulfur (~15 ppm) fuel in a tractor engine and then size selecting the exhaust at 200 nm. Of these types, only the aircraft soot had critical diameters for CCN activation that were less than their Kelvin (wettable) values. The plant samples were burned in a large chamber at the US Fire Sciences Laboratory (Missoula, MT), then size selected at 100 nm. The palmetto burned particles contained lower organic content and were much more hygroscopic (based on CCN and water uptake measurements) than the fern-produced particles. The solid diagonal line denotes the predicted homogeneous freezing fraction for pure 100 nm solutes as described previously

behavior of particles toward the homogeneous freezing condition of the solute. Ice formation may be delayed even to water saturation for high organic carbon fractions.[46]

There is a great need for additional studies of ice nucleation associated with carbonaceous particles in the regime of mixed-phase stratiform and other precipitating clouds. Past cloud chamber and supercooled cloud tunnel studies suggest modest[45] to strong ice activation[47,48] by certain black carbons in supercooled cloud conditions at temperatures above −30°C. Yet these results have not been reproduced in recent expansion chamber studies[44] or in diffusion chamber studies.[49] This may relate to the nature of the experiments and/or the role of different surface functional groups.[2]

Biological Particles and Other Organic IN

The capability of biological particles to act as IN at modest supercooling suggests their important role in cold cloud processes.[50] These particles may have sources as biogenic particles produced from the gas phase or from bacteria, pollens and decaying plant matter lofted by surface winds. Unlike the high selectivity of IN activity expression for bacteria, ice nucleating activity appears to be a general characteristic of pollen populations.[51] The subject area of biological ice nucleation is receiving revived interest as reflected by recent workshops. No comprehensive review is provided here.

The capability of certain insoluble organics to serve as IN has been known for some time, but few atmospherically relevant species are known. Zobrist et al.[55] demonstrate how the homogeneous freezing of haze drops containing oxalic acid, a common atmospheric organic, can lead to the formation of a hydrate that catalyzes ice formation during subsequent aerosol cooling and humidification cycles. Potentially relevant amphiphilic molecules have also been studied recently to explore fundamental aspects of heterogeneous ice nucleation[59] and efficacy as IN for solution droplets.[60]

Apparent IN Activity of Crystalline Sulfates and Salts

Recent optical microscopy observations of particles collected on surfaces and processed in controlled environments have demonstrated apparent ice nucleation behavior for crystalline ammonium sulfate at low temperatures.[29,56] This behavior is fundamentally unexplained to this point, but makes sulfate particles suspect as heterogeneous IN for cirrus formation.[8] Such heterogeneous IN activity of crystalline ammonium sulfate was not noted for aerosols in earlier CFDC studies[57] and is not apparent in the refined experiments shown in Figure 2. However, only small fractions of all sulfate particles (1 in 10^4–10^5) appear to express this behavior following efflorescence and there may be unresolved size effects that have prevented the ready observation of this process in CFDC instruments to this point.

Ice Nucleation Mechanisms

Ice nucleation mechanisms[52] have not been directly discussed here. Many of the above-mentioned studies have examined ice nucleation in the deposition, condensation-freezing and immersion-freezing regimes and may be so interpreted for developing quantitative understanding. Some studies of freezing of solute-coated dust particles have presented freezing nucleation rate data.[29,32,37] The validity of applying these nucleation rates in stochastic nucleation models to simulate cloud parcels has not been proven. Caution is suggested based on the cloud chamber deposition nucleation experiments of Möhler et al.[35] that demonstrate dependence on ice saturation ratio and not cooling rate. More data is needed on the time dependence of nucleation at constant conditions.

The discovery that contact freezing nucleation may occur just as readily due to particle collisions with drop surfaces from the inside as from the outside supports the importance of surface nucleation phenomena in homogeneous and heterogeneous freezing.[53,54] Confirmation of the suggestion[54] that this process offers and explanation for strong ice enhancement during cloud evaporation is dependent on demonstrating the existence of a population of contact freezing nuclei that are not represented in freezing nuclei populations already measured by other means.

New studies noted in the previous section concerning the role of crystallization or chemical-phase transitions within aerosols in ice formation are complex mechanisms which follow pathways that are not directly measurable except in a laboratory setting.

Atmospheric Measurements and Relation to Ice Formation

Data on free-tropospheric concentrations of IN have been collected with the CFDC technique in a large number of studies over the last decade. These data give inferences to the dependencies of atmospheric IN concentrations active by deposition, condensation and (possibly) immersion freezing processes on temperature, supersaturation, and characteristics of the sampled air masses. A selection of these results is shown here (Figure 4). More extensive analyses will be presented at a later date. The smaller increases in IN at lower temperatures compared to the 0 to –20°C regime in Figure 4 is a consistent feature of measurements. Data reflect seasonal influences such as mineral dust transport and the relation between IN concentrations and concentrations of larger aerosol particles (not shown). The Arctic region sees strong seasonal changes in IN and these are hypothesized to regulate clouds and climate there.[60]

In examining the role of IN in cloud ice formation, direct comparison to ice crystal concentrations is sometimes insufficient since ice concentrations very often

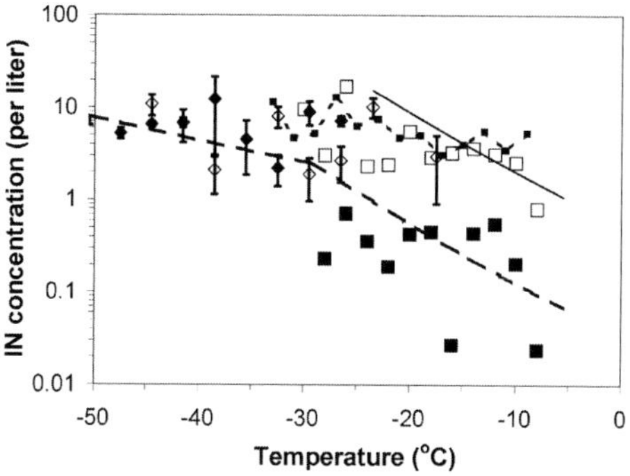

Figure 4 Ice nuclei (IN) concentration data binned versus temperature from mid-latitude Spring[27] (⊇), mid-latitude Fall[11] (filled diamonds), Arctic Spring[64] (γ), and Arctic Fall[60] (□) experiments. Variable ice supersaturation conditions are present in the data. Error bars show 95% confidence limits about mean values. The solid line is the parameterization suggested by Meyers et al.[65] The dashed line is a parameterization suggested by Phillips et al.[66] τηατ αδφυστσ ωιτη αεροσολ τψπε ανδ μασσ λοαδινγ ανδ ισ ρεπρεσεντατιϖε ηερε οφ τηειρ δεφινιτιον οφ □λοω δυστ□ χονδιτιονσ. Βιννεδ δατα φορ χονχεντ ρατιονσ οφ ιχε χρψσταλσ λαργερ τηαν 100μιχρονσ ιν τηε Αρχτιχ Σπρινγ προφεχτ χλουδσ61 ισ αλσο σηοων (♠)

exceed the concentrations attributable to primary IN by a great deal. While in some cases this is clearly because of known secondary ice formation mechanisms, even climatological assessments of ice concentrations in clouds find widespread high concentrations of ice crystals in developed stratiform clouds.[61,62] Nevertheless, Figure 4 shows that temperature-binned ice concentrations measured in clouds at sizes above about 100 μm agree remarkably well with the IN concentrations during a common study period. Total ice concentrations, including small crystals measured by an FSSP instrument in glaciated clouds, exceed IN concentrations by a factor of 100 in this case. Similar findings from other locales provide a conundrum for explaining small ice crystal concentrations that suggest the existence of cloud particle measurement issues and/or unidentified ice formation mechanisms such as noted in some cases.[63] More focused data on IN versus ice concentrations in clouds for circumstances where only primary ice nucleation is expected is needed.

Modeling Aerosol Impacts on Atmospheric Ice Formation

Much impetus exists for including ice nucleation processes in numerical models at a range of scales from process to global scale due to the importance of ice formation processes in determining cloud properties and precipitation, and thereby potential impacts of present and changing aerosol populations on climate. The transfer of new information on ice nucleation into cloud and climate models is usually achieved via parameterization. This may involve direct empirical parameterizations or constrained theory applied within process to cloud-scale models, followed by development of parameterizations suitable for use to the global scale. Examples of such developments already exist in the literature, but are not reviewed here due to space limitations.

Future Outlook

It is tempting to suggest that we are entering a new area of rapid progress in understanding ice formation by atmospheric aerosols. Whether or not this is borne out in fact depends on continued progress on all fronts from theoretical, to experimental (laboratory and field), to modeling, and to the utility of global aerosol and cloud-phase data sets (e.g., from satellite studies) which continue to motivate new interest and measurement needs. Significant issues remain, especially for resolving the role of carbonaceous particles as IN and collecting atmospheric data on IN concentrations representative of all mechanisms, including pre-activation. The need for new instruments is paramount. Further, a program of reference atmospheric IN sampling seems recommended in order to isolate instrumental influences from real aerosol influences on ice nucleation as measured by a growing community of researchers. A first step in this direction will occur during 2007 in the form of the first ice nucleation workshop in over 30 years, to occur at the AIDA facility in Karlsruhe, Germany.

Acknowledgments Portions of the material presented is based upon work supported by NASA under grant numbers NNG06GF00G and NNG06GB60G, the US-NSF under grant numbers ATM-0334228, ATM-0436196 and ATM-0611936, the US DOE under grant numbers DE-FG02-06ER64176, and National Institute for Climatic Change Research grant MPC35TA-A. I am indebted to Ottmar Möhler, Vitaly Khvorostyanov, and Thomas Koop for their discussions. I gratefully acknowledge Sonia Kreidenweis, Daniel Cziczo, Markus Petters, Kirsten Koehler, Anthony Prenni, Matt Richardson, David Rogers, and Olga Popivicheva for providing data for analyses and presentation.

References

1. Koop, T., *Z. Phys. Chem.*, **218**, 1231–1258 (2004).
2. Cantrell, W. and Heymsfield, A.J., *Bull. Amer. Meteorol. Soc.*, **86**, 795–807 (2005).
3. Benz, S., Megahed, K., Möhler, O., Saathoff, H., Wagner, R., and Schurath, U., *J.Photochem. Photobio. A: Chemistry*, **176,** 208–217 (2005).
4. Stöckel, P., Weidinger, I.W., Baumgärtel, H., and Leisner, T., *J. Phys. Chem. A*, **109**, 2540–2546 (2005).
5. Koop, T., Luo, B.P., Tsias, A., and Peter, T., *Nature*, **406**(6796), 611–614 (2000).
6. Baker, M.B. and Baker, M., *Geophys. Res. Lett.*, **31**, L19102, doi:10.1029/2004GL020483 (2004).
7. Khvorostyanov, V.I. and Curry, J.A., *J. Phys. Chem. A*, **108**, 11073–11085 (2004).
8. Abbatt, J.P.D., Benz, S., Cziczo, D.J., Kanji, Z., Lohmann, U., and Möhler, O., *Science*, **313**, 1770–1773 (2006).
9. Larson, B.H. and Swanson, B.D., *J. Phys. Chem. A*, **110**, 1907–1916 (2006).
10. Wise, M.E., Garland, R.M., and Tolbert, M.A., *J. Geophys. Res.*, **109**, D19203, doi:10.1029/2003JD004313 (2004).
11. DeMott, P.J., Cziczo, D.J., Prenni, A.J., Murphy, D.M., Kreidenweis, S.M., Thomson, D.S., Borys, R., and Rogers, D.C., *Proc. Natnl. Acad. Sci.*, **100**, 14655–14660 (2003).
12. Cziczo, D.J., DeMott, P.J., Brooks, S.D., Prenni, A.J., Thomson, D.S., Baumgardner, D., Wilson, J.C., Kreidenweis, S.M., and Murphy, D.M., *Geophys. Res. Lett.*, **31**, 12116, doi:10.1029/2004GL019822.
13. Kärcher, B. and Koop, T., *Atmos. Chem. Phys.*, **5**, 703–714 (2005).
14. Beaver, M.R., Elrod, M.J., Garland, R.M., and Tolbert, M.A., *Atmos. Chem. Phys.*, **6**, 3231–3242 (2006).
15. Tabazadeh, A., Djikaev, Y.S., and Reiss H., *Proc. Natl. Acad. Sci.*, **99**, 15873–15878 (2002).
16. Duft, D. and Leisner, T., *Atmos. Chem. Phys.*, **4**, 1997–2000 (2004).
17. Vrbka, L. and Jungwirth, P., *J. Phys. Chem. B*, **110**, 18126–18129 (2006).
18. Lin, R-F., Starr, D.O.C, DeMott, P.J., Cotton, R., Sassen, K., Jensen, E., Karcher, B., and Liu, X., *J. Atmos. Sci.*, **59**, 2305–2329 (2002).
19. Khvorostyanov, V.I. and Curry, J.A., *J. Atmos. Sci.*, **61**, 2676–2691 (2004).
20. Khvorostyanov, V.I. and Curry, J.A., *J. Atmos. Sci.*, **62**, 261–285 (2005).
21. Hobbs, P.V. and Rangno, A.L., *J. Atmos. Sci.*, **47**, 2710–2722 (1990).
22. Twohy, C.H. and Poellot, M.R., *Atmos. Chem. Phys.*, **5**, 2289–2297 (2005).
23. Cziczo, D.J., Murphy, D.M., Hudson, P.K., and Thomson, D.S., *J. Geophys. Res.*, **109**, D04201, doi:10.1029/2003JD004032 (2004).
24. Chen, Y., Kreidenweis, S.M., McInnes, L.M., Rogers, D.C., and DeMott, P.J., *Geophys. Res. Lett.*, **25**, 1391–1394 (1998).

25. Cziczo, D.J., DeMott, P.J., Brock, C., Hudson, P.K., Jesse, B., Kreidenweis, S.M., Prenni, A.J., Schreiner, J., Thomson, D.S., and Murphy, D.M., *Aerosol Sci. and Technol.*, **37**, 460–470 (2003).
26. Rogers, D.C., DeMott, P.J., Kreidenweis, S.M., and Chen, Y., *J. Atmos. Oceanic Technol.*, **18**, 725–741 (2001).
27. Richardson, M.S., DeMott, P.J., Kreidenweis, S.M., Cziczo, D.J., Dunlea, E., Jimenez, J.L., Thompson, D.S., Ashbaugh, L.L., Borys, R.D., Westphal, D.S., Cassucio, G.S., and Lersch, T.L., *J. Geophys. Res.*, **112**, D02209, doi:10.1029/2006JD007500 (2007).
28. DeMott, P.J., Sassen, K., Poellot, M.R., Baumgardner, D., Rogers, D.C., Brooks, S.D., Prenni, A.J., and Kreidenweis, S.M., *Geophys. Res. Lett.*, **30**, 1732, doi:10.1029/2003GL017410 (2003).
29. Kanji, Z.A. and Abbatt, J.P.D., *J. Geophys. Res.*, **111**, D16204, doi:10.1029/2005JD006766 (2006).
30. Knopf, D.A. and Koop,T., *J. Geophys. Res.*, **111**, 111, D12201, doi:10.1029/2005JD006894 (2006).
31. Zuberi, B., Bertram, A.K., Cassa, C.A., Molina, L.T., and Molina, M.J., *Geophys. Res. Lett.*, **29**, 1504, doi: 10.1029/2001GL014289 (2002).
32. Archuleta, C.A., DeMott, P.J., and Kreidenweis, S.M., *Atmos. Chem. Phys.*, **5**, 2617–2634 (2005).
33. Koehler, K.A., Kreidenweis, S.M., DeMott, P.J., Prenni, A.J., and Petters, M.D., *J. Geophys. Res.*, **112**, accepted (2006).
34. Salam, A., Lohmann, U., Crenna, B., Lesins, G., Klages, P., Rogers, D., Irani, R., MacGillivray, A., and Coffin, M., *Aerosol Sci. and Technol.*, **40**, 134–143 (2006).
35. Möhler, O., Field, P.R., Connolly, P., Benz, S., Saathoff, H., Schnaiter, M., Wagner, R., Cotton, R., Krämer, M., Mangold, A., and Heymfield, A.J., *Atmos. Chem. Phys.*, **6**, 3007–3021 (2006).
36. Field, P.R., Möhler, O., Connolly, P., Krämer, M., Cotton, R., Heymsfield, A.J., Saathoff, H., and Schnaiter, M., *Atmos. Chem. Phys.*, **6**, 2991–3006 (2006).
37. Hung, H.M., Malinowski, A., and Martin, S.T., *J. Phys. Chem. A*, **107**, 1296–1306 (2003).
38. Ettner, M., Mitra, S.K., and Borrmann, S., *Atmos. Chem. Phys.*, **4**, 1925–1932 (2004).
39. Sassen, K., DeMott, P.J., Prospero, J., and Poellot, M.R., *Geophys. Res. Lett.*, **30**, 1633, doi:10.1029/2003GL017371 (2003).
40. Roberts, P. and Hallett, J., *Quart. J. Royal Meteorol. Soc.*, **94**, 25–34 (1968).
41. Kärcher, B. and Lohmann, U., *J. Geophys. Res.*, **108**, 4402, doi:10.1029/2002JD003220 (2003).
42. Popovicheva, O.B., Persiantseva, N.M., Lukhovitskaya, E.E, Shonija, N.K., Zubareva, N.A., Demirdjian, B., Ferry, D., and Suzanne, J., *Geophys. Res. Lett.*, **31**, L11104, doi:10.1029/2003GL018888 (2004).
43. Popovicheva O.B., Persiantseva N.M., Shonija N.K., DeMott, P.J., Koehler, K., Petters, M.D., Kreidenweis, S.M., Tishkova, V., Demirdjian, B., and Suzanne, J., Submitted to *Phys. Chem. Chem. Phys.* (2007).
44. Möhler, O., Büttner, S., Linke, C., Schnaiter, M., Saathoff, H., Stetzer, O., Wagner, R. Krämer, M., Mangold, A., Ebert, V., and Schurath, U., *J. Geophys. Res.*, **110**, D11210, doi:10.1029/2004JD005169 (2005).
45. Möhler, O., Linke, C., Saathoff, H., Schnaiter, M., Wagner, R., Mangold, A., Krämer, M., and Schurath, U., *Meteorol. Z.*, **14**, 477–484 (2005).
46. DeMott, P.J., *J. Appl. Meteor.*, **29,** 1072–1079 (1990).
47. Gobunov, B., Baklanov, A., Kakutkina, N., Windsor, H.L., and Toumi, R., *Aerosol Sci.*, **32** 199–215 (2001).
48. Diehl, K. and Mitra, S.K., *Atmos. Environ.*, **32**, 3145–3151 (1998).
49. Dymarska, M., Murray, B.J., Sun, L., and Eastwood, M.L., *J. Geophys. Res.*, **111**, D04204 doi:10.1029/2005JD006627 (2006).
50. Morris, C.E., Georgakopoulos, D.G., and Sands, D.C., *Journal De Physique IV*, **121**, 87–103 (2004).

51. Von Blohn, N., Mitra, S.K., Diehl, K., and Borrmann, S., *Atmos. Res.*, **78,** 182– 189 (2005).
52. Vali, G. and Vali, G., *J. Rech. Atmos.*, **19**, 105–115 (1985).
53. Shaw, R.A., Durant, A.J., and Mi, Y., *J. Phys. Chem. B.*, **109**, 9865–9868 (2005).
54. Durant, A.J. and Shaw, R.A., *Geophys. Res. Lett.*, **32**, L20814, doi:10.1029/2005GL024175 (2005).
55. Zobrist, B., Marcolli, C., Koop, T., Luo B.P., Murphy D.M., Lohmann, U., Zardini, A.A., Krieger, U.K., Corti T., Cziczo, D.J., Fueglistaler, S., Hudson, P.K., Thomson, D.S., and Peter, T., *Atmos. Chem. Phys.*, **6**, 3115–3129 (2006).
56. Shilling, J.E., Fortin, T.J., and Tolbert, M.A., *J. Geophys. Res.*, **111**, D12204, doi:10.1029/2005JD006664 (2006).
57. Chen, Y, DeMott, P.J., Kreidenweis, S.M., Rogers, D.C., and Sherman, D.E., *J. Atmos. Sci.*, **57**, 3752–3766 (2000).
58. Zobrist, B., Koop, T., Luo, B.P., Marcolli, C., and Peter, T., *J. Phys. Chem. C*, **111**, 2149–2155 (2007).
59. Cantrell, W.A. and Robinson, C., *Geophys. Res. Lett.*, **33**, L07802, doi:10.1029/2005GL024945 (2006).
60. Prenni, A.J., Harrington, J.Y., Tjernström, M., DeMott, P.J., Avramov, A., Long, C.N., Kreidenweis, S.M., Olsson, P.Q., and Verlinde, J., *Bull. Amer. Meteor. Soc.*, **88**, in press (2007).
61. Gultepe, I., Isaac, G.A., and Cober, S.G., *Int. J. Climatol.*, **21**, 1281–1302 (2001).
62. Korolev, A., Isaac, G.A., Cober, S.G., Strapp, J.W., and Hallett, J., *Q. J. R. Meteorol. Soc.*, **129**, 39–65 (2003).
63. Cotton, R.J. and Field, P.R., *Q. J. R. Meteorol. Soc.*, **128**, 2417–2437 (2002).
64. Rogers D.C., DeMott, P.J., and Kreidenweis, S.M., *J. Geophys. Res.*, **106**, 15053–15063 (2001).
65. Meyers M.P., DeMott, P.J., and Cotton, W.R., *J. Appl. Meteor.*, **31**, 708–721 (1992).
66. Phillips V.T.J., DeMott, P.J., and Andronache, C., *J. Atmos. Sci.*, submitted (2007).

Immersion Freezing Efficiency of Soot Particles with Various Physico-Chemical Properties

O.B. Popovicheva[1], N.M. Persiantseva[1], E.D. Kireeva[1], T.D. Khokhlova[2], N.K. Shonija[2], and E.V. Vlasenko[2]

Abstract 2,000 water droplets with soot particles immersed are monitored individually to establish the link between physico-chemical properties and freezing efficiency. Soot behaviour in water (relating with number density and agglomerate size), wetting, and water soluble fraction (WSF) are found as key parameters defining the freezing temperature. The maximum freezing efficiency, 6°C, is reported for homogeneously distributed low size soot agglomerates of low number density and with hydrophilic coverage.

Keywords Heterogeneous freezing, soot aerosols, wetting, hydration

Introduction

The major source of uncertainties in assessing aerosol indirect effects on climate is a strong impact on ice cloud formation. One of the reasons is that soot particles emitted into atmosphere from a great variety of combustion sources have a wide range of a natural variability. Field measurements provide evidence for the deduced freezing threshold of cirrus cloud formation at the northern hemisphere, suggesting that soot aerosol exhibits signatures of pollution. The high concentration of non-volatile particles points to black carbon (soot) as a source of ice nuclei (IN) in clouds, especially in cirrus caused by aircraft exhaust (Petzold et al., 1998). Lack of characterization data for each kind of soot, from one side, and absence of knowledge about their freezing efficiency, from the other side, limit our ability to establish the link between physico-chemical properties of soot particles and their ice nucleation behaviour in clouds.

The situation is complicated by a limit of possibilities for in situ exhaust measurement of main soot characteristics such as microstructure, surface chemistry,

[1] *Institute of Nuclear Physics, Moscow State University, Moscow, Russia*

[2] *Chemical Department, Moscow State University, Moscow, Russia*

C.D. O'Dowd and P.E. Wagner (eds.), *Nucleation and Atmospheric Aerosols*, 418–422.

wetting and hygroscopicity: the sparse documentation exists on the properties of aircraft soot particles measured at contrail and cirrus levels. Laboratory studies can provide relevant information, but the current laboratory data on surrogate soot from different sources demonstrate the great variety of nucleation properties, frequently without established connection with soot physico-chemical characteristics.

This report presents the elaboration of an advanced approach to address the immersion freezing efficiency of soot particles in connection with its behaviour in water, wetting, and water soluble fraction. The unique set of soot samples from many sources, commercial, and original ones, is proposed for examination. The *list* of main physico-chemical characteristics required for obliging analysis has elaborated. Measurements of median freezing temperature for the population of soot seeding water droplets demonstrates a clear impact of different physical (density, particle size) and chemical (wetting, hygroscopicity) soot properties. Results for original aircraft engine combustor soot allow some atmospheric application.

Set of Soots

Four commercial available soots and original aircraft engine combustor (AEC) soot chosen for analysis are presented in Table 1. AEC soot was collected behind the combustor of D30-KU engine operating at cruise conditions (Popovicheva et al. 2004).

Freezing Measurements

We chose the approach focusing on the freezing of the droplet population supported on a cold stage (Zuberi et al. 2002); it allows monitoring the droplets individually and verifying the properties of soot in a wide range. The apparatus consists from a cell for freezing droplets and the controlled atmospheric chamber providing a stable cooling rate. Droplets of a controllable size (of 0.13 cm radius) are rested on the perfectly polished hydrophobic surface to minimize the effects of contact and achieve the spherical shape of droplets. The droplet temperature is continuously

Table 1 Set of soots for immersion freezing experiments and specific properties

Soots, name	ρ, g/cm^3	θ, degree	Distribution in droplets	S, m^2/g	WSF, %	ΔT_f,C
acetylene (AS)	0.09	47	●	86	0	4.4
FW2	0.3	40	◒	420	1.1	1.9
Furnace (FS)	0.28	28	◒	100		3.7
Lamp	0,3	28	◒	22	0.43	0.8
aircraft engine combustor (AEC)	0,35		◒	12.6	13.5	0.9

measured by the thermometer having the resistance temperature sensor that touches a droplet 10 mkm under its support. Fully automated temperature control is accomplished starting from a preparation temperature of 2°C with a cooling rate standardized to 1.5 ± 0.2°C min^{-1}. The time moment of the sharp change of the temperature profile (latent heat pulse) indicates the freezing onset (nucleation) and the freezing temperature, T_f.

The suspensions with the soot mass ratio ~2.5 wt% in water are prepared by gentle stirring. To address the kinetics of wetting the soot suspensions freshly prepared and settling for 2 days are analysed. The measure of the freezing efficiency is a difference between the median freezing temperatures, ΔT_f, for the ensembles of 100 soot suspensions and of 100 pure water droplets. The median freezing temperature for pure water droplets was obtained 11.5 ± 2.1°C in the given experiments.

The most important physical characteristic impacting the freezing efficiency is the soot particle behaviour in water. In dependence on the number density of the soot bed, and the wetting characteristics the homogeneous distribution of particles over droplet volume (sedimentation stability) or heterogeneity leading to sedimentation of particles on the bottom (sedimentation instability) are observed. Figure 1 shows the typical images of water droplets with immersed soot of the different spatial distribution. AES particles are separated on two fractions; more light particles are accumulated on the top.

The number density of the soot bed, ρ, was measured by a routine method. Table 1 presents the configuration of the soot spatial distribution in the droplets as well as the values of ρ and ΔT_f for settled soot suspensions. The best freezing efficiency is found for AS soot forming the homogeneously distributed suspension, ΔT_f = 4.4°C. This configuration is perfect for treating the heterogeneously freezing systems in the classical theory of heterogeneous nucleation. The particles of other soots sediment on the droplet bottom and demonstrate less freezing efficiency. Freshly prepared suspensions give the lower freezing temperatures than settled ones: for AS soot ΔT_f = 1.8°C. This result may be explained by slow kinetics of wetting of large soot agglomerates which may approach as much as 1 mm of diameter in original soot powder. Experiments with AS soot agglomerates less then 80 μkm show the absence of the dependence on the settling time; ΔT_f = 4.7°C is obtained. Figure 2 plots histograms of the freezing droplet observation for ensemble of pure water droplets and of droplets with immersed AS soot agglomerate.

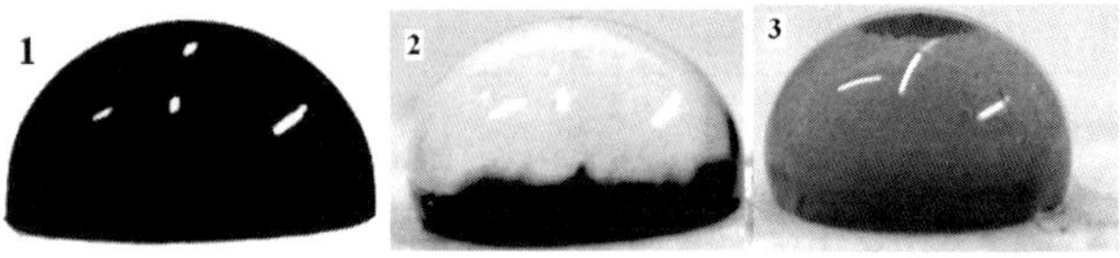

Figure 1 Images of macroscopic water droplets with immersed (1) AS, (2) GTS, and (3) AEC soots

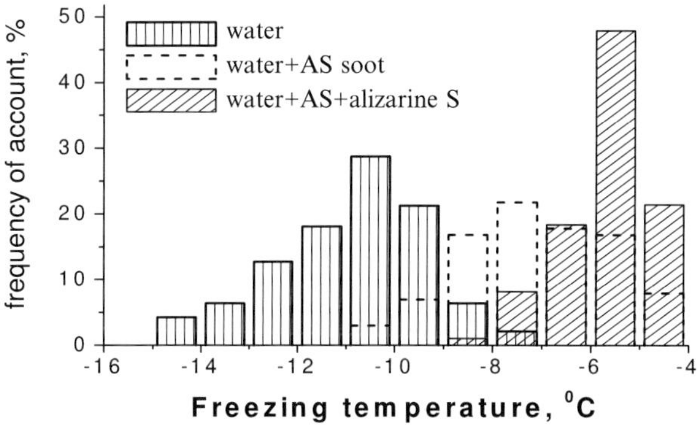

Figure 2 Freezing histogrames less then 80 μkm

Impact of Wetting, Solubles, and Hygroscopicity

The classical theory of heterogeneous nucleation operates by a wetting parameter. The contact angle of water droplet on the soot surface, θ, determines the rate of heterogeneous freezing of ice germ originating on the substrate active sites. In this work θ was obtained by "sessile droplet" method (Popovicheva et al. 2003) and is presented in Table1. However, a water contact angle is the macroscopic measure, which can not reveal the amount of active sites. But it correlates with the level of soot hygroscopicity, which may be obtained by measurements of water uptake from vapour phase (Persiantseva et al. 2004). Figure 3 plots the amount of water molecules per nanometre square of the surface of AS, LS, FW2, and FS soots, measured as a function of the relative humidity (RH) by a gravimetrical method. The specific soot surface areas, S, obtained by BET single point method was taken into account (see Table1). AS soot is revealed to be really hydrophobic substances because negligible water uptake in comparison with one statistic water monolayer (1 ML), while AEC soot demonstrates the features of highly hydrophilic soot with water uptake >5 ML at low RH. The level of hygroscopicity, defined at low RH, is increased from LS, FW2 to FS soot. According to fundamentals of water adsorption on carbonaceous adsorbents it correlates with the increasing amount of active sites. And we found the highest $\Delta T_f = 3.7$ for FS soot (see Table 1). But hydrophobic AS soot demonstrates the displacement even 1° more, in contrast to other soots. This effect takes place because the homogeneous distributed AS soot particles over the droplet volumes, probably the distribution is a dominated factor in heterogeneous water freezing. However, the temperature of freezing may be additionally increased if AS soot surface may become more hydrophilic. To prove the hypothesis that the hydrophilic surface with high amount of active sites impacts the freezing efficiency laboratory-made soot samples were produced by adsorption modification of hydrophilic organic compounds. The impact of the 0.2 ML ionic modifier (sodium salt of sulfoasid – alizarin S) deposited on AS soot is demonstrated by higher uptake than on original AS soot, see Figure 3, and ΔT_f for particles less then 80 μkm is found 5.9°C. Additionally, θ for AS soot

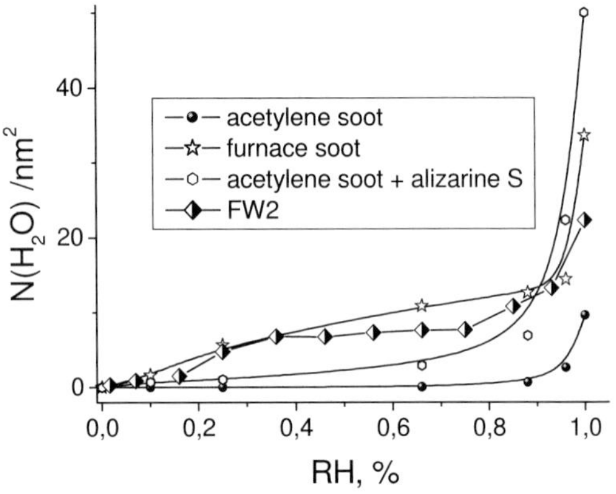

Figure 3 Water uptake by set of soots

with alizarin S is found 36°, less then for original AS soot, 47°, that stresses the connection between wetting and hygroscopicity.

Finally, we note the lowest ΔT_f, only 0.9°C, is found for AEC soot. Its high hygroscopicity is determined by extremely large WSF, accumulated at high pressure and temperature combustion in AEC (Popovicheva et al. 2004). Comparison of WSFs and ΔT_f, reported in Table 1, allows the conclusion about the dissolution of WSF from the AEC soot surface in water, in confirmation with insolubility requirement that, in general, IN are highly water insoluble (Pruppacher and Klett 1978).

Conclusions

The highest impact of soot immersion on heterogeneous freezing is observed for the droplets with homogeneous distributed soot agglomerates of low size. Sedimentation instability limits the freezing efficiency. More wettable and hydrophilic soots demonstrate higher freezing temperatures but soluble compounds may be easy dissolved in water leading to loosing the potential freezing efficiency.

Acknowledgments This work is supported by EC QUANTIFY-TTC Project, contract no., 3893 and by the grant of President of Russian Federation SS-7101.2006.02.

References

Persiantseva, N.M., Popovicheva, O.B., and Shonija, N.K., *J. Environ. Monit.*, **6**, 939–945 (2004).

Petzold, A., Strom, J., Ohlsson, S., and Schroder, F.P., *Atmos. Res.*, **49**, 21–34 (1998).

Popovicheva, O.B., Persiantseva et al., *J. Geophys. Res. Lett.*, 107 (D23): Art. no. 4734 (2004).

Pruppacher, H.R. and Klett, J.D., *Microphysics and Cloud Precipitation*, Hingham MA: D. Reidel (1978).

Zuberi, B., Bertram, A.K., Cassa, C.A., et al., *Geophys. Res. Lett.*, **29**(10), 1504 (2002).

Ice Nucleation Characteristics of Atmospheric Trace Gas Aged Mineral Dust Aerosols with a Continuous Flow Diffusion Chamber

Abdus Salam[1,2,*], Glen Lesins[2], and Ulrike Lohmann[3]

Abstract Ice nucleation characteristics of montmorillonite mineral dust aerosol particless with and without surface modification by atmospheric trace gas (ammonia and sulfur dioxide) were studied at different temperatures and saturation conditions with a continuous flow diffusion chamber (CFDC) at Dalhousie University, Canada. There was no significant change in the size distributions due to ammonia and sulfur dioxide exposure to the mineral dust particles. The percentage of activation of ammonia exposed montmorillonite attains a saturation level when the exposure times reach 120 min. Ammonia aged montmorillonite aerosols are more efficient as ice nuclei (IN) with increasing relative humidities and decreasing temperatures. Ammonia gas enhanced the ice nucleation efficiency of montmorillonite mineral dust particles from 4 to 11 times at our experimental conditions. The activation temperature of ammonia exposed montmorillonite was higher than for pure montmorillonite particles. Sulfur dioxide gas aged montmorillonite particles were also efficient in ice nucleation. But there was no significant difference in the ice nucleation efficiency of pure montmorillonite and sulfur dioxide exposed montmorillonite.

Keywords Ice nucleation, montmorillonite, ammonia, sulfur dioxide

Introduction

Understanding ice nucleation and cloud properties is important because it affects both climate and air quality. Aerosol radiative forcing cools the climate and even small changes in the aerosol induced cloud amount or properties can

[1]*Department of Chemistry, University of Dhaka, Faculty of Science, Dhaka – 1000, Bangladesh*

[2]*Department of Physics and Atmospheric Science, Dalhousie University, Halifax, Nova Scotia, B3H 3J5, Canada*

[3]*Institute of Atmospheric and Climate Science, ETH Zurich, Switzerland*

* *Corresponding author: E-mail: asalam@gmail.com; Tel.: + 88-01817061160; Fax: + 88-02-8615583*

C.D. O'Dowd and P.E. Wagner (eds.), *Nucleation and Atmospheric Aerosols*, 423–426.

partly offset the anthropogenic greenhouse gas radiative forcing. Cloud lifetime and radiative transfer are the key elements contributing to the indirect aerosol radiative effect.[1] Atmospheric dust aerosol particles are also important for the direct climate forcing due to their effect on scattering and absorption of solar radiation. However, it is important to understand the chemical interactions between atmospheric trace gases and aerosol particles serving as ice nuclei (IN). Laboratory studies have also indicated a strong ice nucleating behavior associated with dust in the atmosphere,[2] especially of clay minerals[3,4] and many of the metal oxide components of dust.[5] In this study we use a continuous flow diffusion chamber (CFDC) to demonstrate the effect of atmospheric trace gases (ammonia and sulfur dioxide) on the ice nucleation characteristics of montmorillonite mineral dust aerosols at different temperatures and saturation conditions.

Methods and Experiments

Mineral dust aerosols were exposed to trace gases (ammonia and sulfur dioxide) in a stainless steel cylindrical chamber. About 5 g of mineral dust particles were placed into the chamber, and trace gases was allowed to pass into the chamber for different exposure times with occasional stirring. A transmission electron microscope (TEM), Model Philips EM 210 was used to characterize the surface of trace gas aged montmorillonite dust particles before and after exposure. Aged mineral dust particles were placed into the aerosol generator of the CFDC. The aerosol generator is an airtight reservoir with a vibrating membrane at its base. The aerosol particles produced from the generator were characterized for the size distributions with an Aerodynamic Particle Sizer (APS), Model TSI 3321 and introduced in the flow of the CFDC. Before entering the chamber the aerosol particles passed through impactors to cut off particles larger than 5 μm. The CFDC is a vertically oriented flow chamber consisting of two concentric circular copper cylinders. The inside cylinder walls of the chamber are ice covered and are held at two different temperatures to expose the aerosol particles to supersaturation with respect to ice within the annular region. Typical operating conditions of the chamber are temperatures between −5°C and −45°C; temperature differences between the two walls of 0–20°C; supersaturation with respect to water of −30–+10%; supersaturation with respect to ice of 0–+49%; total air flow 2.83 L min^{-1}, and a residence time of the aerosol particles into the chamber is 20 sec. Aerosol particles smaller than 5 μm in diameter produced from the generator were injected into the center of the gap near the location of the maximum supersaturation. Ice crystals were activated and grew in the chamber and were identified with a commercially available MetOne optical particle counter, Model 278B at the outlet of the chamber.

Result and Discussions

Ice nucleation characteristics of montmorillonite mineral dust aerosols with and without surface modification by ammonia (25 ppm), sulfur dioxide (100% pure and 44.5 ppm) gases at 1 atm pressure are studied at different temperatures and saturation conditions with a CFDC at Dalhousie University, Canada. The size distributions between 0.5 and 20 µm of the mineral dust particles with and without trace gas exposed montmorillonite are measured. The total number concentration of aerosol particles smaller than 5 µm in diameter that are input into the CFDC is almost the same for both exposed and nonexposed montmorillonite.

Ice nucleation experiments of aged montmorillonite dust particles are studied by varying ammonia exposure times from 0 to 150 min before injecting them into the CFDC at different relative humidity conditions at a temperature of −20°C. The fraction of activated IN (IC > 5/APS2-5) was calculated from the ratio of the total number of ice crystals larger than 5 µm measured by the MetOne (IC > 5) to the average of the total number of aerosol particles between 2 and 5 µm measured by the APS (APS2-5) before and after of each ice nucleation experiment. About 2.5% of the aged montmorillonite particles were activated as IN at 90% relative humidity with respect to water (RHw), and 7% at 100% RHw for one minute of ammonia gas exposure. The higher values in the case of aged montmorillonite at 100% RHw are caused by either condensation freezing or enhanced deposition nucleation.

Ice nucleation experiments using montmorillonite mineral dust particles with and without ammonia aging were carried out at temperatures between −5°C and −35°C at different relative humidity conditions in 5°C temperature intervals. The efficiency of montmorillonite dust particles to act as IN increased 4 to 11 times due to the modification of the dust surface by ammonia gas at 90% RHw, and 6 to 11 times at 100% RHw within the temperatures between −150°C and −350°C. Ammonia aged montmorillonite aerosols are more efficient as IN with increasing relative humidities and decreasing temperatures. Salam et al.[4] also reported similar results for nonaged kaolinite and montmorillonite mineral dust aerosols as IN. Ice nucleation activity was observed at a temperature between −5°C and −150°C for the ammonia-aged montmorillonite mineral dust aerosols at 100% and 90% RHw, respectively; whereas no ice nucleation activity was observed at temperatures above −15°C at 100% RHw and above −27°C at 90% RHw for pure montmorillonite dust particles. It is also possible that the addition of ammonia enables liquid water uptake so that the ammonia-aged dust particles are activated in condensation freezing mode instead of the deposition mode, especially at 100% RHw.

The influence of sulfur dioxide on the ice nucleation of montmorillonite mineral dust aerosols was investigated. Two concentrations of sulfur dioxide were chosen – pure SO_2 (100%), and a trace concentration of SO_2 (44.5 ppm in N_2 gas medium). The active IN (%) versus temperature of montmorillonite after exposure to sulfur dioxide at 100% RHw were studied. The exposure times for these experiments are 2.5 h for pure SO_2 and 18 h for 44.5 ppm SO_2. The sulfur dioxide exposed montmorillonite acts as efficient IN within current experimental conditions at both concentration levels. There was no significant difference in ice nucleation efficiency between unexposed montmorillonite and 100% pure SO_2-aged montmorillonite, and 44.5 ppm SO_2-aged montmorillonite. Knopf and Koop[6] also studied the ice nucleation efficiency of pure Arizona test dust (ATD) and sulfuric acid coated ATD. They found that pure and sulfuric acid coated ATD particles nucleate ice at considerably lower relative humidities than required for homogeneous ice nucleation in liquid aerosols. No significant difference in the ice nucleation ability of pure and sulfuric acid coated ATD particles was observed.

Conclusions

Ice nucleation characteristics of montmorillonite mineral dust aerosols with and without surface modification by ammonia and sulfur dioxide gas were studied at different temperatures and saturation conditions with a CFDC at Dalhousie University, Canada. There was no significant change in the unactivated size distribution due to the ammonia and sulfur dioxide gas exposure to the mineral dust aerosols. Ammonia-aged montmorillonite aerosols are more efficient as IN with increasing relative humidities and decreasing temperatures. Ammonia gas enhanced the ice nucleation efficiency of montmorillonite mineral dust particles from 4 to 11 times at our experimental conditions. The activation temperature of ammonia exposed montmorillonite was higher than for pure montmorillonite particles. The sulfur dioxide exposed montmorillonite acts as efficient IN within current experimental conditions at both concentration levels. However, there was no significant difference in the ice nucleation efficiency of pure montmorillonite and sulfur dioxide exposed montmorillonite.

Acknowledgments We are grateful for support from the Canadian Foundation for Climate and Atmospheric Sciences (CFCAS) and the National Science and Engineering Research Council (NSERC) of Canada. We thank Brian Crenna for his excellent work in constructing and testing the CFDC. We also acknowledge Kevin Borgel and Andy George for their technical help.

References

1. Lohmann, U., Feichter, J., Penner, J.E., and Leaitch, R., *J. Geophy. Res.*, **105**, 12193–12206 (2000).
2. Isono, K., Komabayasi, M., and Ono, A., *J. Meteoro. Soc. Japan*, **37**, 211–233 (1959).
3. Zuberi, B., Bertram, A.K., Cassa, C.A., Molina, L.T., and Molina, M.J., *Geophy. Res. Letter*, **29**(10), 1504, doi: 10.1029/2001GL014289 (2002).
4. Salam, A., Lohmann, U., Crenna, B., Lesins, G., Klages, P., Rogers, D., Irani, R., MacGillivray, A., and Coffin, M., *Aero. Sci. Techn.*, 40, 134–143 (2006).
5. Hung, H.-M., Malinowski, A., and Martin, S.T., *J. Phy. Chem. A*, **107**, 1296–1306 (2003).
6. Knopf, D.A. and Koop, T., *J. Geophys. Res.*, **111**, D12201, doi: 10.1029/2005JD006894 (2006).

Ice Initiation for Various Ice Nuclei Types and its Influence on Precipitation Formation in Convective Clouds

Martin Simmel[1], **Harald Heinrich**[1], **and Karoline Diehl**[2]

Abstract A spectral bin microphysics model is applied to a convective cloud to investigate the influence of varying ice nuclei (IN) type on ice initiation, evolution, and, finally, precipitation formation as well as cloud dynamics. Sensitivity studies were made using an air parcel model and a cylinder-symmetric model of the Asai–Kasahara type. In the air parcel model the vertical cloud evolution highly depends on IN type resulting in a bimodal behavior: either a sufficient amount of ice forms, then the cloud reaches about 3 km higher than the warm cloud, or ice formation is not effective enough which restricts the mixed cloud dynamics to the warm cloud one. The Asai–Kasahara model does not show this strong difference in cloud dynamics. However, cloud lifetime as well as precipitation is enhanced for effective ice formation depending on IN type.

Keywords Heterogeneous ice nucleation, primary ice nucleation, contact freezing, immersion freezing, precipitation formation

Introduction

In the atmosphere homogeneous ice nucleation is restricted to temperatures of about minus 35°C and lower. Since ice is observed in clouds at much higher temperatures it must have been formed by another process. Primary atmospheric ice formation at warmer temperatures typically occurs via heterogeneous nucleation, i.e., with participation of an ice nucleus (IN). Possible IN are insoluble aerosol particles such as biological particles, minerals or soot. Four heterogeneous freezing processes are known to be relevant (e.g., Pruppacher and Klett 1997): deposition freezing, condensation freezing, immersion freezing, and contact freezing. Deposition and condensation freezing are closely connected to an interaction with the vapor phase, whereas for immersion

[1]*Leibniz Institute for Tropospheric Research, Leipzig, Germany*

[2]*Institute of Atmospheric Physics, University of Mainz, Mainz, Germany*

C.D. O'Dowd and P.E. Wagner (eds.), *Nucleation and Atmospheric Aerosols*, 427–431.

and contact freezing supercooled droplets are involved. These two processes are focused on during the present study. Several laboratory experiments (summarized in Diehl et al. 2006) showed that the freezing temperatures may largely differ for various IN types and for different freezing processes.

Model Description

For the present study the spectral bin microphysics model of Simmel and Wurzler (2006) with the ice microphysics described in Diehl et al. (2006) was used. The model includes warm microphysics (drop activation, condensation/evaporation, collision/coalescence, break-up) as well as mixed-phase microphysics (heterogeneous ice nucleation via immersion and contact freezing, water vapor deposition/sublimation on ice particles, interaction of drops and ice particles via collision). In contrast to Diehl et al. (2006) in the present study, the microphysics is used in a one-dimensional discretization, i.e., the hydrometeors are resolved spectrally only according to their (liquid and/or frozen) water mass (e.g., Simmel and Wurzler 2006). Compared to Diehl et al. (2006) some extensions have been applied to the ice-phase microphysics, namely (a) the introduction of a spectrally resolved insoluble particle mode serving as contact IN, (b) the introduction of a liquid water shell for the ice particles as well as the corresponding processes of freezing (of the liquid shell for T < 0°C) and melting (of the frozen core for T > 0°C) and, (c) the possibility to describe immersion freezing for a population consisting of different externally mixed IN types. The change of the number of frozen drops N_f is proportional to the number of unfrozen drops N_u, the drop volume V_d and the temperature change dT:

$$\frac{dN_f}{dt} = -N_u V_d \sum_i f_i a_i B_i \exp(-a_i T) \frac{dT}{dt} \qquad (1)$$

i describes the immersion IN type (e.g., bacteria, montmorillonite, soot), a_i and B_i are IN type specific constants (Diehl and Wurzler 2004), and f_i is the respective fraction of drops with the specific IN type included. Homogeneous freezing is treated formally equivalent as immersion freezing using alternative values for a_i and B_i. Typically, most of the drops do not include an immersion IN resulting in, e.g., $i = 1$ and a fraction f_1 close to 1 (homogeneous freezing). Only small fractions $f_i << 1$ of the drops include immersion IN ($i > 1$).

Diehl et al. (2006, 2007) discussed model sensitivity studies in the frame of an air parcel model with entrainment which, however, could not give any information about precipitation. Therefore, in the present investigations the spectral bin microphysics was additionally implemented into a vertically resolved (200 m) cylinder-symmetric model of Asai–Kasahara type (Asai and Kasahara 1967) consisting of two concentric cylinders. The cloud evolves in the updraft of the inner cylinder (with 2 km radius) whereas the outer cylinder (with 20 km radius) is responsible for the compensating downward motion. This allows for a more realistic description of the convective dynamics, cloud lifetime, and, especially, for a quantitative representation of precipitation.

Sensitivity Studies: Results and Conclusions

As thermodynamic background a strongly convective profile obtained by the regional model REMO is used (Langmann, personal communication). For the sensitivity studies a rural background aerosol distribution (Jaenicke 1988) is investigated. Different mixed-phase runs with varying IN type for combined immersion and contact freezing is compared to the respective warm cases.

Air Parcel Model

Figure 1 shows the vertical velocity of the air parcel (left) and the ice particle number (right) for different IN types (bacteria, montmorillonite, soot). Soot is the least effective IN, forming only a negligible number of ice particles. The minor amount of ice formed and latent heat released are not able to enhance cloud dynamics substantially compared to the warm case. In contrast to soot, montmorillonite and bacteria are more efficient IN. Considerable amounts of ice particles are formed and sufficient latent heat is released intensifying vertical dynamics. Ice formation at about 6–7 km height is due to contact freezing for all 3 IN types. For the bacteria run, immersion freezing is responsible for the unrealistically high ice particle numbers at about 9 km and, finally, for the higher vertical cloud extension caused by more vapor consumption. A bimodal behavior can be observed: Either sufficient ice is formed to enhance cloud dynamics then cloud reaches significantly

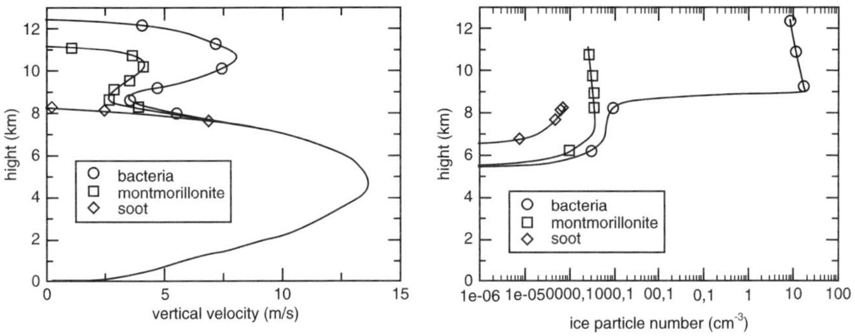

Figure 1 Vertical velocity (**left**) and ice particle number (**right**) for different IN types and ratios

higher than in the warm case or too little ice forms then cloud height is restricted to the same value as for the warm case. In the present case the model switches from one mode to the other when montmorillonite is used as IN instead of soot. Generally, the crucial point for switching from one mode to the other depends on IN type and number, warm microphysics (drop number and size) as well as on the thermodynamics.

Asai–Kasahara Model

Figure 2 compares the liquid water mixing ratio of the warm case (upper left) with the liquid water mixing ratio (lower left) and the ice-phase mixing ratio (lower right) of the mixed-phase run, using montmorillonite as IN. One can observe that both runs are almost identical up to about 20 min simulation time. Then, ice forms above 4,000 m which leads to an earlier onset of precipitation and to a longer cloud life time, especially in the upper part. Precipitation (upper right) is enhanced by about 10% for bacteria, 18% for montmorillonite, and 27% for soot. This means that even though less ice is formed for soot (not shown) than for montmorillonite

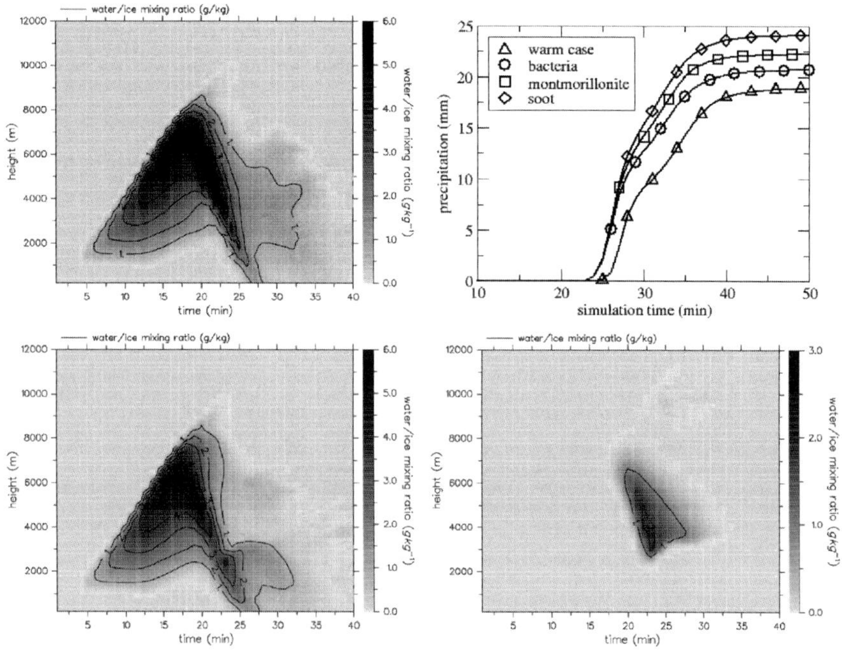

Figure 2 Time evolution of vertical profiles of hydrometer mixing ratios. Liquid phase for the warm cloud case (**upper left**), liquid (**lower left**), and ice phase (**lower right**) for the mixed-phase cloud using montmorillonite as IN. Precipitation for several IN types (upper right)

or bacteria precipitation is intensified even more. This could be due to the smaller ice particle number in case of soot which can grow faster by the Bergeron–Findeisen process and by riming resulting in higher vertical velocities. The strong enhancement of vertical dynamics as observed in the parcel model could not be confirmed. This indicates an overestimation by the air parcel model possibly due to the disregard of precipitation.

Acknowledgments The study was funded by the DFG under contract RE 1333/3-1. For the use of the thermodynamic profile Dr. B. Langmann (MPI Hamburg, Germany) is acknowledged.

References

1. Asai, T. and Kasahara, A., *J. Atmos. Sci.*, **24**, 487–496 (1967).
2. Diehl, K. and Wurzler, S., *J. Atmos. Sci.*, **61**, 2063–2072 (2004).
3. Diehl, K., Simmel, M., and Wurzler, S., *J. Geophys. Res.*, **111**, D07202 (2006).
4. Diehl, K., Simmel, M., and Wurzler, S., *Atmos. Env.*, **41**, 303–314 (2007).
5. Jaenicke, R., Aerosol physics and chemistry, in: Landolt-Bernstein, *Zahlenwerte und Funktionen aus Naturwissenschaften und Technik* V4B, New York: Springer, pp. 391–457 (1988).
6. Pruppacher, H.R. and Klett, J.D., *Microphysics of Clouds and Precipitation*, Dordrecht: Kluwer (1997).
7. Simmel, M. and Wurzler, S., *Atmos. Res.*, **80**, 218–236 (2006).

Strong Dependence of Cubic Ice Formation on Aqueous Droplet Ammonium to Sulphate Ratio

Benjamin J. Murray[1,2] **and Allan K. Bertram**[1]

Abstract The homogeneous freezing of aqueous solution droplets in the Earth's upper troposphere (UT) and tropopause region is thought to be an important process in the formation of cold ice clouds. We have shown that the phase of ice that crystallizes in solution droplets, under conditions relevant for the UT and tropopause region (typically >188 K), is strongly dependent on the ammonium to sulphate ratio (ASR) of the solute. Droplets of aqueous $(NH_4)_3H(SO_4)_2$ (ASR = 1.5) freeze dominantly to cubic ice below 200 K; whereas aqueous NH_4HSO_4 (ASR = 1.0) droplets only freeze dominantly to cubic ice below 180 K; a temperature not typical of the UT or tropopause region.

Keywords Homogeneous nucleation, ice clouds, cubic ice, hexagonal ice, ice, tropospheric aerosols

Introduction

Ice clouds that occur in the Earth's upper troposphere (UT) and tropopause region play an important role in the Earth's climate by scattering and absorbing radiation. However, the sign and magnitude of their net climatic impact remains uncertain.[1] In addition, they influence the distribution of water vapour in the troposphere as well as the amount that enters the stratosphere, where changes in humidity may have consequences for polar stratospheric ozone.[2]

An important mechanism of ice cloud formation in the UT and tropopause region is thought to be homogeneous freezing of submicron aqueous droplets.[3] Recently, it was shown that the metastable cubic crystalline phase of ice (ice I_c) was produced when a range of solution droplets froze homogeneously under conditions that are relevant for the UT and tropopause.[4] Previous to this work, it was often

[1]*Department of Chemistry, University of British Columbia, 2036 Main Mall, Vancouver, British Columbia V6T 1Z1, Canada*

[2]*School of Chemistry, University of Leeds, Woodhouse Lane, Leeds LS2 9JT, UK*

C.D. O'Dowd and P.E. Wagner (eds.), *Nucleation and Atmospheric Aerosols*, 432–435.

assumed that only the stable hexagonal phase (ice I_h) would form under conditions relevant for the UT and tropopause.[5] Under the extreme conditions of the polar high-latitude summer mesosphere, where temperatures regularly fall below 150 K, cubic ice is also thought to exist.[6]

Recent laboratory measurements have shown clouds composed of ice I_c will have a saturation vapour pressure 11 ± 3% larger than that of a cloud composed of ice I_h.[7] In fact, the atmosphere within cold ice clouds is often found to be more humid than expected if the ice clouds were composed of hexagonal ice particles in equilibrium with their environment.[7,8] Hence, the formation of cubic ice in the UT and tropopause region may have some important implications: Murphy[9] has shown that the difference in vapour pressure between ice I_c and ice I_h may strongly influence cloud properties and lifetimes through a process analogous to the Bergeron–Findeisen process in which hexagonal ice crystals grow at the expense of metastable cubic ice crystals. Jensen and Pfister[2] examined the impact of persistent high ice supersaturations in cold cirrus clouds and found that transport of water into the stratosphere is enhanced and cloud properties were altered. In addition, particles composed of cubic ice may have different shapes to hexagonal ice particles; forming crystals with cubic, octahedron or cubo-octahedron habits, and possibly also crystals of threefold symmetry.[10] In addition, a recent laboratory investigation revealed that cubic and hexagonal ice have different heterogeneous chemistries,[11] which may have important implications for the processing of gas-phase species by ice clouds.

As mentioned, recently it was shown that ice I_c was produced when a range of solution droplets froze homogeneously under conditions that are relevant for the Earth's atmosphere.[4] However, the effect of the solute type on the amount of I_c produced was not studied in detail. We have focused this study on solutions of $(NH_4)_3H(SO_4)_2/H_2O$ (ammonium to sulphate ratios, ASR, = 1.5) and NH_4HSO_4/H_2O (ASR = 1.0) because initial tests showed that they exhibit strongly contrasting behaviour and also because these compositions are relevant for the UT and tropopause region. These solutions are referred to as LET and AHS for the remainder of this abstract; indicating that the overall stoichiometry of the solute in the droplets corresponds to letovicite and ammonium bisulphate, respectively.

Experimental

The experimental technique has been described previously.[4,12] In these experiments aqueous solution droplets (volume median diameters of between 2 and 20 μm with geometric standard deviations between 1.3 and 1.8) were suspended in an oil matrix, by emulsification, and the crystalline phases that formed when the droplets froze were monitored using x-ray diffraction.

In a typical experiment, emulsified droplets were cooled at a rate of 10 K min^{-1} from room temperature to 173 K, while recording the diffraction pattern to determine the temperature range over which the droplets froze, these temperature were in good agreement with a parameterization relating water activity to freezing temperature.[13] When at 173 K (or 163 K in a few cases where the freezing temperature was particularly low) a diffraction pattern was recorded by signal averaging for 30–40 min. These low-temperature diffraction patterns were then used to determine the phase of ice that precipitated in the solution droplets.

Results

Examples of the diffraction patterns of frozen LET, and AHS solution droplets, which froze at similar temperatures (Mean freezing temperature = 194 K), are illustrated in Figure 1. It is clear from this figure that the diffraction patterns of the resulting frozen droplets are very different, even though the mean freezing temperatures of the droplets were almost identical (and their water activates at freezing were therefore almost identical[13]). The patterns in Figure 1 are complicated by crystalline solute-phase peaks, which have been identified by comparing the patterns in Figure 1 to those of pure frozen water droplets.[12] The solute-phase peaks are labelled “S”. Fortunately, the peaks which indicate bulk hexagonal ice at 33.5–44.5°C (labelled “h”), as well as those common to cubic and hexagonal ice at 4047°C (labelled “h + c”), do not overlap with the major solute-phase peaks. Inspection of these peaks reveals that AHS droplets freeze to a significant amount of ice I_h, since the hexagonal peaks at 33.543.5°C are prominent. In fact, this diffraction pattern indicates ice I_h is the dominant product in these AHS droplets. Whereas, the hexagonal peaks at 33.543.5°C are absent in the frozen LET droplets, indicating that cubic ice was the dominant product in these droplets. However, this pattern does have a strong peak at the position of the ice I_h (100) reflection at 23°. In addition, the region between $2\theta = 22.526.5$°C is raised above the background. These features indicate that the cubic ice contains “hexagonal like” stacking faults. In fact, stacking faults are though to be inherent in ice I_c.[4]

Diffraction patterns similar to those illustrated in Figure 1, but for varying concentrations of LET and AHS (and therefore varying freezing temperatures), reveal that the freezing temperature below which cubic ice is the dominant product is strongly dependent on the chemical make-up of the solute. For LET droplets, cubic ice is the dominant product below 200 K, whereas for AHS droplets cubic ice is not the dominant product until below a freezing temperature of about 180 K.

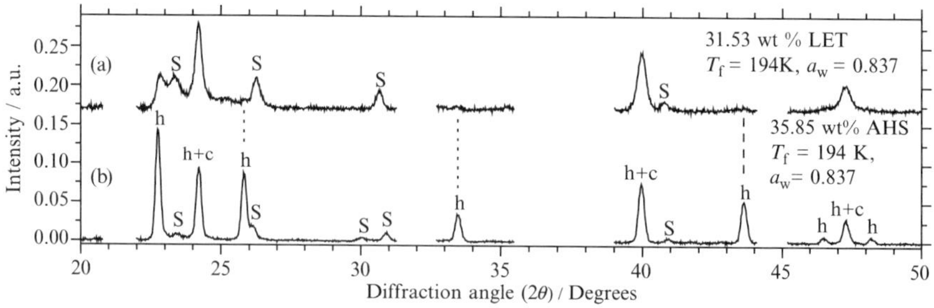

Figure 1 Diffractions patterns of frozen 31.53 wt% $(NH_4)_3H(SO_4)_2$ (a) and 35.85 wt% NH_4HSO_4 (b) droplets, which both froze at around 194 K, while being cooled at a rate of 10 K min^{-1}. These patterns were collected after the droplets had been cooled to 173 K

Conclusions

In the UT and tropopause region the lower temperature limit is approximately 188 K. Furthermore, the temperature in the topical tropopause region is typically between 188 and 200 K.[14] Measurements of the ice phase that forms in LET and AHS solution droplets of varying concentration indicate that the phase of ice in the UT and tropopause region will strongly depend on the ASR of the sulphate aerosols from which ice clouds form. Our results show that droplets with ASR = 1.5 (LET) will freeze to ice I_c below ~200 K, which is in the temperature range relevant for the UT and tropopause. However, droplets with ASR = 1.0 (AHS) will not freeze to ice I_c under temperatures relevant for the UT and tropopause region. Unfortunately our knowledge of upper tropospheric aerosol composition is very poor and we therefore can not predict where and when cubic ice will be the main product in the atmosphere. However, it is intriguing to note that the emissions of ammonia, largely from livestock, have risen faster than emissions of sulphate, which may suggest tropospheric aerosols now have a larger ASR than they did in pre-industrial times. We speculate that this may have led to more ice I_c formation in the UT, which may have altered cloud properties and may therefore have influenced climate.

Acknowledgments The authors thank A. Lam and B. Patrick for assistance with x-ray diffraction measurements and interpretation of the diffraction patterns. This work was funded by the Natural Science and Engineering Research Council of Canada, NSERC, the Canadian Foundation for Climate and Atmospheric Sciences, CFCAS, and the Canada Foundation for Innovation, CFI. BJM acknowledges the Natural Environmental Research Council, NERC, for a fellowship held at the School of Chemistry (University of Leeds) which commenced September 2006.

References

1. Lohmann, U. and J. Feichter, *Atmos. Chem. Phys.*, **5**, 715–737 (2005).
2. Jensen, E. and L. Pfister, *Geophys. Res. Lett.*, **32**, D03208, doi:10.1029/2004GL021125 (2005).
3. DeMott, P.J., Laboratory studies of cirrus cloud processes, in: *Cirrus*, edited by D.K. Lynch, et al., Oxford: Oxford University Press, pp. 102–135, (2002).
4. Murray, B.J., Knopf, D.A., and Bertram, A.K., *Nature*, **434**, 202–205 (2005).
5. Pruppacher, H.R. and Klett, J.D., *Microphysics of Clouds and Precipitation*, Dordrecht: Kluwer (1997).
6. Murray, B.J. and Plane, J.M.C., Phys. Chem. *Chem. Phys.*, **7**, 3970–3979 (2005).
7. Shilling, J.E., et al., *Geophys. Res. Lett.*, **33**, L17801, doi:10.1029/2006gl026671 (2006).
8. Peter, T., et al., *Science*, **314**, 1399–1402 (2006).
9. Murphy, D.M., *Geophys. Res. Lett.*, **30**, 2230, doi:10.1029/2003GL018566 (2003).
10. Hallett, J., et al., Ice crystals in cirrus, in: *Cirrus*, edited by D.K. Lynch, et al., Oxfrd: Oxford University Press, pp. 102–135 (2002).
11. Behr, P., Terziyski, A., and Zellner, R., *J. Phys. Chem. A*, **110**, 8098–8107 (2006).
12. Murray, B.J. and Bertram, A.K., *Phys. Chem. Chem. Phys.*, **8**, 186–192 (2006).
13. Koop, T., et al., *Nature*, **406**, 611–614 (2000).
14. Zhou, X.L., Geller, M.A., and Zhang, M.H., *J. Climate*, **17**, 2901–2908 (2004).

Ice Cloud Formation Mechanisms Inferred from in situ Measurements of Particle Number-size Distribution

Y. Tobo[1], **D. Zhang**[2], **Y. Iwasaka**[3], **and G.-Y. Shi**[4]

Abstract Ice crystals in cirrus clouds are formed via homogeneous or heterogeneous pathway. This work addresses in situ measurements of ice crystal concentrations in the region that is thought to be in a favorable condition for ice nucleation. Based on a series of measurements by a balloon-borne optical particle counter, possible scenarios of ice cloud formations will be discussed.

Keywords Balloon-borne measurements, optical particle counter, ice crystals, homogeneous and heterogeneous ice nucleation

Introduction

The indirect effects of aerosols on global climate are a subject of current scientific debate due in part to an incomplete understanding of ice cloud formation mechanisms.[1] Freezing of ice in aerosols can take place via homogeneous or heterogeneous nucleation. If relative humidity with respect to ice (RH_{ice}) in the upper troposphere (UT) or tropopause region exceeds the threshold of ~140–160%, homogeneous nucleation occurs and no liquid aerosols can exist.[2] On the other hand, heterogeneous nucleation occurs on a fraction of the background aerosols, so that this pathway is expected to lead to low number density of ice crystals in cirrus clouds (e.g., subvisible or thin cirrus clouds) as compared to the pathway via homogeneous freezing.[3] Therefore, understanding the distribution and variability of ice crystal concentrations is important for the validation of cirrus cloud formation mechanisms and the climate prediction.

[1] *Graduate School of Natural Science and Technology, Kanazawa University, Kanazawa 920–1192, Japan*

[2] *Faculty of Environmental and Symbiotic Sciences, Prefectural University of Kumamoto, Kumamoto 862–8502, Japan*

[3] *Institute of Natural and Environmental Technology, Kanazawa University, Kanazawa 920–1192, Japan*

[4] *Institute of Atmospheric Physics, Chinese Academy of Science, Beijing 100029, China*

C.D. O'Dowd and P.E. Wagner (eds.), *Nucleation and Atmospheric Aerosols*, 436–439.

In this study, we present measurements of particle number-size distributions in the UT and tropopause region. A series of measurements in the tropics and subtropics provide evidence that there are relatively large particles that are most likely ice crystals. It is noteworthy that there are occasionally measurable differences between the concentrations of ice crystals in different observational cases, and these differences may partly depend upon differences in their formation mechanisms. On the basis of these measurements, possible scenarios of ice formations in a cirrus cloud formation region will be discussed.

Instrumentation

Vertical profiles of size-resolved particle concentrations are obtained using a balloon-borne optical particle counter (OPC). The OPC instrument measures particle number concentrations by detecting forward scatter of laser beams at 780 or 810 nm at a sample flow rate of 3 L min^{-1}. It counts particles by equivalent optical diameters (D_p) and the available size ranges are $D_p > 0.3$, 0.5, 0.8, 1.2, and 3.6 μm. The counting uncertainties about the concentrations of 10^{-1}, 10^{-2}, and 10^{-3} cm^{-3} are about ±10%, 32%, and 100%, respectively. The detailed characteristics of the OPC instrument used here had been described previously.[4] In addition to the particle number concentrations, the vertical structures of air temperature and pressure are measured by an onboard Vaisala radiosonde. The vertical structures of them are usually measured during the balloon ascent with a vertical resolution of about 100–110 m (every 20 sec) from ground to an altitude of ~10 hPa.

The background aerosols in the upper troposphere and lower stratosphere are mainly composed of submicron-sized particles, such as solutions (H_2SO_4, HNO_3, NH_3, H_2O, or organics), followed by soot, mineral dust, fly ash, and so on.[5] On the other hand, there are typically few particles larger than 3.6 μm above the UT (e.g., see Figure 1). As for Figure 1, two peaks are evident: (1) peaks extending from the ground surface to 400 hPa, and (2) peaks at ~150 hPa measured only on August 17, 1999. The former probably attribute to convective transport of continental large particles (e.g., mineral dust). However, this possibility should be ruled out at higher altitudes. Considering that ice crystals are usually much larger than submicron-sized aerosols, the presence of these large particles in the UT is presumably regarded as ice crystals that are formed by freezing of ice in aerosols.

Results and Discussion

Once homogeneous ice nucleation occurs, freezing in almost all particles can occur very rapidly (within a few seconds or minutes) and no liquid aerosols can exist. Vertical profiles of aerosols obtained from balloon-borne particle measurements in the tropics frequently show that there are significant number concentrations of

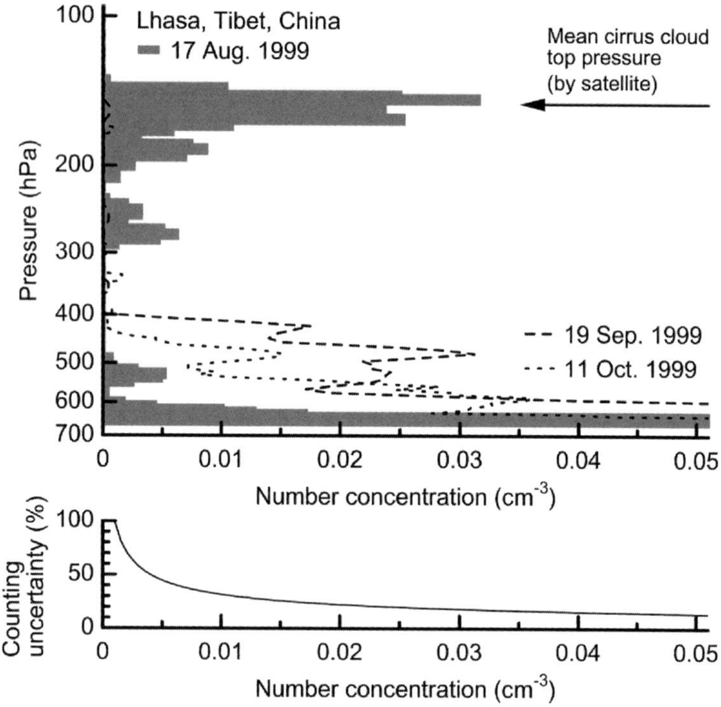

Figure 1 (**Top**) Vertical profiles of number concentrations of particles with $D_p > 3.6\,\mu m$ over Lhasa (29.4°N, 91.1°E, 3,650 m a.s.l., or ~650 hPa), China, on August 17, 1999, September 19, 1999, and October 11, 1999. The positions of the tropopause are ~100 hPa. Note that data smoothing is made by averaging per 1 min during the balloon ascent. (**Bottom**) Changes in counting uncertainties of the OPC measurements depending on particle number concentrations

particles larger than 3.6 μm (>1 cm^{-3}) in the cold-point tropopause region.[6] The observed tropopause temperatures in the tropics are very cold (<~195 K) and therefore probably well below freezing. Thus, the observed large particles are expected to be ice crystals that are formed by homogeneous freezing.

In contrast, a balloon-borne measurement at a specific location on the Tibetan Plateau shows that there are low number concentrations of large particles (<0.04 cm^{-3}) in the upper troposphere (Figure 1). This measurement also indicate that submicron-sized particles ($D_p = 0.3$–$1.2\,\mu m$) with the concentration of >0.2 cm^{-3} coexist in the same vertical layer.[7] One possible interpretation is that freezing of aerosols occur on a fraction of the background aerosols rather than the total background aerosols. This also raises the question of whether other processes than homogeneous freezing may contribute to the observed ice clouds, such as heterogeneous ice nucleation on a small fraction of the background submicron-sized aerosols.

Satellite observations have shown the frequent occurrence of ice clouds in the upper troposphere (mean cloud top pressure of ~150 hPa) over the Tibetan Plateau,[8,9] so that this altitude is likely in a relatively favorable condition for ice nucleation.

A chemical-transport model simulated that convectively lifted continental aerosols from South Asia frequently extend to ~150 hPa and then are entrained by the upper tropospheric anticyclone over the Tibetan Plateau for several days.[8]

Recent laboratory studies with a cloud chamber showed that ice nucleation occurs on continental aerosols including solid-phased sulfate (e.g., ammonium sulfate), soot, mineral dust, fly ash, and so on, via deposition or immersion nucleation at low ice supersaturation (~100–140% RH_{ice}) at the upper tropospheric temperatures (~210–220 K).[10,11,12] The heterogeneous pathway is expected to become more important under the influence of deep convection.[3] Considering that continental aerosols reaching the upper troposphere could act as heterogeneous ice nuclei under the influence of larger-scale upwelling or low cooling rate, the observed ice crystals at ~150 hPa might be formed by heterogeneous nucleation rather than homogeneous nucleation.

Thus, in-situ measurements of particle number–size distributions above the upper troposphere are useful to provide some insight into ice cloud formation mechanisms. In this presentation, we will report and discuss some examples of ice clouds, which are likely formed via homogeneous or heterogeneous pathway based on balloon-borne measurements and other approaches.

Acknowledgments This research was funded by the Japan Ministry of Education, Culture, Sports, Science and Technology (PI: Y. Iwasaka, No.10144104 & No.10041115), National Natural Science Foundation of China (PI: G.-Y. Shi, No.49775275), and 21st-Century COE programs of Kanazawa University (PI: K. Hayakawa).

References

1. Lohmann, U. and Feichter, J., *Atmos. Chem. Phys.*, **5**, 715–737 (2005).
2. Koop, T., Luo, B., Tsias, A., and Peter, T., *Nature*, **406**, 611–614 (2000).
3. DeMott, P.J., Cziczo, D.J., Prenni, A.J., Murphy, D.M., Kreidenweis, S.M., Thomson, D.S., Borys, R., and Rogers, D.C., *Natl. Acad. Sci. U.S.A.*, **100**, 14655–14660 (2003).
4. Hayashi, M., Iwasaka, Y., Watanabe, M., Shibata, T., Fujiwara, M., Adachi, H., Sakai, T., Nagatani, M., Gernandt, H., Neuber, R., and Tsuchiya, M., J. Meteorol. *Soc. Jpn.*, **76**, 549–560 (1998).
5. Murphy, D.M., Thomson, D.S., and Mahoney, M.J., *Science*, **282**, 1664–1669 (1998).
6. Matsumura, T., Hayashi, M., Fujiwara, M., Matsunaga, K., Yanai, M., Saraspriya, S., Manik, T., and Suripto, A., *J. Meteorol. Soc. Jpn.*, **79**, 709–718 (2001).
7. Tobo, Y., Iwasaka, Y., Shi, G.-Y., Kim, Y.-S., Ohashi, T., Tamura, K., and Zhang, D., *Atmos. Res.*, doi:10.1016/j.atmosres.2006.08.003 (2006).
8. Li, Q., Jiang, J.H., Wu, D.L., Read, W.G., Livesey, N.J., Waters, J.W., Zhang, Y., Wang, B., Filipiak, M.J., Davis, C.P., Turquety, S., Wu, S., Park, R.J., Yantosca, R. M., and Jacob, D.J., *Geophys. Res. Lett.*, **32**, L14826, doi:10.1029/2005GL022762 (2005).
9. Fu, R., Hu, Y., Wright, J.S., Jiang, J.H., Dickinson, R.E., Chen, M., Filipiak, M., Read, W.G., Waters, J.W., and Wu, D.L., *Proc. Natl. Acad. Sci. U. S. A.*, **103**, 5664–5669 (2006).
10. Abbatt, J.P.D., Benz, S., Cziczo, D.J., Kanji, Z., Lohmann, U., and Möhler, O., *Science*, **313**, 1770–1773 (2006).
11. Möhler, O., Büttner, S., Linke, C., Schnaiter, M., Saathoff, H., Stetzer, O., Wagner, R., Krämer, M., Mangold, A., Ebert, V., and Schurath, U., *J. Geophys. Res.*, **110**, D11210, doi:10.1029/2004JD005169 (2005).
12. Möhler, O., Field, P.R., Connolly, P., Benz, S., Saathoff, H., Schnaiter, M., Wagner, R., Cotton, R., Krämer, M., Mangold, A., and Heymsfield, A.J., *Atmos. Chem. Phys.*, **6**, 3007–3021 (2006).

The FINCH (Fast Ice Nucleus Chamber counter)

U. Bundke[1], H. Bingemer[1], B. Nillius[2], R. Jaenicke[2], and T. Wetter[1]

Abstract We present first results of our new developed ice nucleus (IN) counter FINCH at the sixth Cloud and Aerosol Characterisation Experiment (CLACE 6) campaign at Jungfraujoch station 3,571 m a.s.l. Measurements were made at the total and the ICE CVI inlet.

Keywords Heterogeneous nucleation, ice nucleus counter, cluster (times 10 pt)

Introduction

Among the cloud condensation nuclei (CCN) in the troposphere ice nuclei (IN) play a special role for the formation of precipitation by the Bergeron–Findeisen process (Findeisen 1938). The number concentration of the IN, the temperature and the ice supersaturation to activate the IN are considered as key information for the understanding of ice formation in clouds. In the past, comparative measurements with different instruments showed large differences of deposition mode freezing IN concentrations (Vali 1975) due to different measuring methods. Furthermore long sampling times are needed to obtain a sufficient statistic. During long sampling periods considerable concentration fluctuations may occur. These points complicate the interpretation of measured data.

Method

Therefore a Fast Ice Nuclei Counter FINCH (Fast Ice Nucleus Chamber) (Bundke 2006a) was developed at the Institute for Atmosphere and Environment Frankfurt together with the Institute for Atmospheric Physics in Mainz (Figure 1). In particles

[1]*Institute for Atmosphere and Environment, J.W. Goethe Universit, Frankfurt, Germany*

[2]*Institute for Atmospheric Physics, J. Gutenberg University, Mainz Germany*
Contact: bundke@meteor.uni-frankfurt.de

C.D. O'Dowd and P.E. Wagner (eds.), *Nucleation and Atmospheric Aerosols*, 440–444.

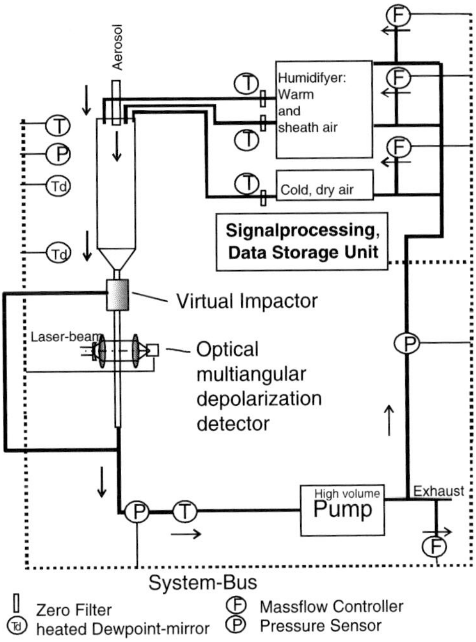

Figure 1 Schematic flow diagram of the FINCH counter

are activated within the chamber at certain ice supersaturation and temperature by mixing three gas flows, a particle free warm moist, a particle free cold dry, and an aerosol flow. Temperature and supersaturation are actively controlled using PID (Pro-portional Integral Differential) algorithms with measurements of temperature and frost point as process variables. After activation the particles will grow in a processing chamber of about 1 m length to macroscopic ice crystals or supercooled water droplets. With a new developed optical detector it is possible to distinguish between supercooled droplets and ice particles to obtain the number concentration of CCN and IN. FINCH is called a fast IN-Counter because of the high flow rate of the aerosol probe of 5–10 L min^{-1}. Considering a number concentrations of 10 IN L^{-1}, a measuring period of 12 minutes is sufficient to obtain good statistics.

Model Calculations and Characterization

FLUENT three-dimensional (3D) calculations were performed for the actual chamber geometry with an aerosol flow of about 10 L min^{-1} to determine the conditions within the chamber corresponding to different flow rates. The flow rates where adjusted for optimum residence time for particle growth and minimization of aerosol losses.

The calculations show aerosol residence times of about 4.5 sec in the development section of the chamber at an aerosol flow rate of 10 L min^{-1} combined with a sheath air flow and a cold flow rate of 20 L min^{-1} and a warm flow rate of 10 L min^{-1}.

Higher flow rates of cold and sheath flow up to 50 L min^{-1} are possible without changing the principle flow characteristics, but with considerable shorter residence times. Considering this result we are able to achieve a wide range of possible supersaturations and temperatures for IN and CCN activation.

These calculations were supported by measurements made with a newly developed small heated dew-point-mirror (Bundke et al. 2006b) at different temperatures and saturations in the chamber which is also used to control supersaturation within the mixing process. The result shows a very good consistency of measurement and calculation. Furthermore at 0.4 m distance nearly constant minimum temperature and maximum saturation conditions were reached. Thus it is possible to deduce at which temperature and saturation the IN were activated with measurement of temperature and frost point at one location.

Detection of IN

For the determination of supercooled droplets and ice crystals, as a result of activation of existent IN and CCN an own new optical sensor was developed, shown in Figure 2.

A Laser beam passes first a circular polarizer to before entering the detection volume. Droplets and ice crystals scatter this incident beam of radiation. Backscattered light is detected and analyzed for the circular polarization under an angle of 115°. Therefore the scattered light passes a λ/4 retarder (quarter wave plate) and a polarizing beam splitting cube. The quarter wave plate converts the circularly polarized light into linear polarized light that is polarized at either 45° or

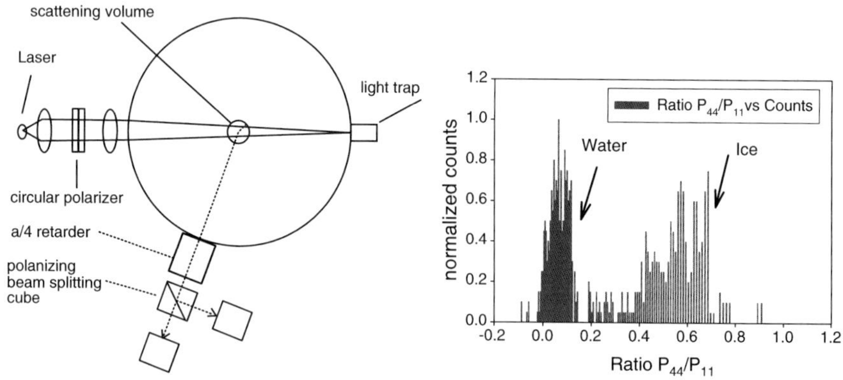

Figure 2 Optical sensor for detection and discrimination of supercooled droplets and ice crystals (**left**) and Histogram (normalized frequency of occurrence) of FINCH measurements of the P44/P11 ratio for detected single particles (**right**)

135° from the principal axis, depending on the direction of rotation of the incident circular polarization. The quarter wave plate is followed by a polarizing beam splitter oriented to separate the components of linear polarized light. Separate detectors measure the two perpendicular linear polarized beams emanating from the polarizing beam splitter. Thus it is possible to determine the ratio of P44/P11 of the scattering matrix. Following Hu et al. (2003) this ratio shows a maximal difference for water droplets and different types of ice crystals as function of scattering angle at a ± 15° rage centered at 115 degree scattering angle.

Thus it is possible to discriminate between droplets and crystals with high accuracy (see Figure 2 (right))

First Measurements during CLACE 6

FINCH first shows his capabilities during the CLACE 6 experiment as a quick look. Figure 3 shows IN concentration just before a cloud reaches the station.

Conclusion and Outlook

FLUENT calculation and the measurement of the temperature and saturation ratio in the FINCH IN counter were made. The results show a very good consistency of measurement and calculation. The temperature and saturation is controlled

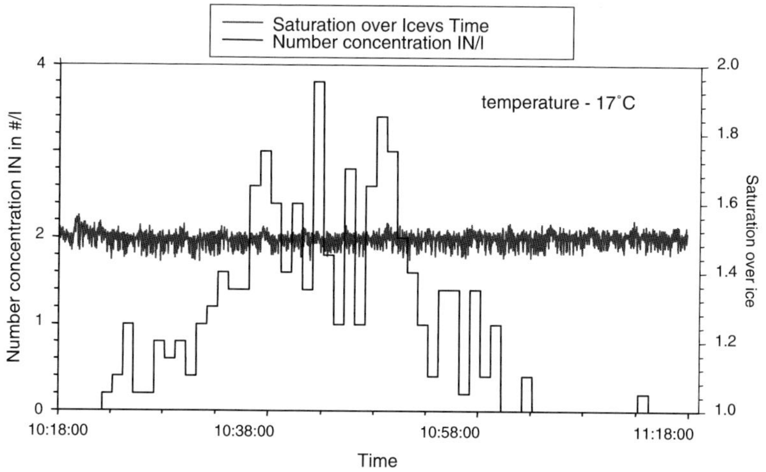

Figure 3 As an example this figure shows the number concentration of IN as minute mean values during a cloud event. Boundary layer air reaches Jungfraujoch station at 1030, where IN concentration rises until at 1050 a cloud reaches the station. IN concentration falls rapidly, because IN were incorporated into hydrometeors, which cannot enter the total inlet

automatically. So it is possible to activate IN at well-known conditions and with a high-time resolution. It is also possible to determine 3D IN spectra with temperature and supersaturation as independent variables. The presentation will show the first results of the CLACE 6 field measurement campaign at Jungraujoch in February/March 2007 and furthermore the delisquescense curves of caolinite, silveriodid, and different dust samples to characterize the instrument.

Acknowledgments This work has been performed within Project A1 of the Collaborative Research Centre (SFB) 641 "The Tropospheric Ice Phase", funded by the funded by the German Science Foundation

References

Vali, G., Ice nucleation workshop, *Bull. Amer.Meteorol. Soc.*, **56**, 1180–1184 (1975).

Bundke, U., Bingemer, H., Wetter, T., Nillius, B., Jaenicke, R., The FINCH (Frankfurt Ice Nuclei Chamber) counter – New Developments and First Measurements. *7th International Aerosol Conference September 10–15*, MN, USA: St. Paul, p. 1350 (2006a).

Bundke et al., A new, fast Frost-/ Due point hygrometer with an active Sensor area of 4 mm^2, *Proceedings of the EGU* (2006b).

Hu et al., Discriminating between spherical and non-spherical scatterers with lidar using circular polarization: a theoretical study, *J. Quant. Spectrosc. Radiat. Transf.*, **79–80**, 757–764 (2003).

Heterogeneous Ice Nucleation on Soot Particles

Ottmar Möhler

Abstract The AIDA (Aerosol Interaction and Dynamics in the Atmosphere) facility of Forschungszentrum Karlsruhe was used to investigate the heterogeneous ice nucleation efficiency of various soot samples in the temperature range 240–195 K. Flame soot (FS) with little amount of organic carbon (OC) showed significant heterogeneous ice nucleation below the threshold supersatuaration for homogeneous freezing of solution droplets, similar to untreated and sulphuric acid coated graphite spark generator soot (GS) investigated in earlier studies. An increasing amount of OC almost completely suppressed the ice nucleation activity of FS below water saturation. The results are presented and discussed together with a review of literature data on soot ice nucleation from other laboratory studies.

Keywords Heterogeneous ice nucleation, soot particles, organic coating

Introduction

The consumption of fossil fuel in vehicle motors or aircraft engines induces the emission of large numbers of soot particles internally mixed with variable amounts of organic compounds and sulphuric acid. If these particles are involved in cloud formation processes, the insoluble soot core may induce heterogeneous ice nucleation and therefore affect the formation, life cycle, and radiative properties of mixed-phase and cirrus clouds. The heterogeneous ice nucleation of soot particles was investigated in different laboratory studies at temperatures between 268 and 188 K, but the topic is far from a good understanding as will be discussed in the following paragraphs. New results from recent experiments in the AIDA (Aerosol Interaction and Dymamics in the Atmosphere) chamber will be discussed together with previous AIDA and literature results.

Institute for Meteorology and Climate Research, Atmospheric Aerosol Research (IMK-AAF) Forschungszentrum Karlsruhe, Germany

C.D. O'Dowd and P.E. Wagner (eds.), *Nucleation and Atmospheric Aerosols*, 445–449.

At warmer temperatures, above about 235 K, recent laboratory studies report little, if any, ice activity of different soot samples.[1–3] Only a minor fraction of soot particles acted as immersion freezing nuclei after water condensed to the same particles. These results contradict earlier studies[4,5] which reported relatively large number fractions of soot containing droplets to freeze at temperatures as high as 253 and 268 K, respectively. The work by Gorbunov et al.[5] suffers from a lack of aerosol characterization as well as ice particle measurements during the mixing-type cloud chamber experiments. The results from this study at relatively warm temperatures should not be extrapolated to cold cirrus temperatures. Diehl and Mitra[4] investigated the freezing of large water droplets with diameters between 0.3 and 0.8 mm suspended in a vertical wind tunnel. Each droplet probably contained a large number of soot particles which was not specified. Therefore it is not possible to relate the fraction of frozen droplets to the soot particle number or mass in the droplets. Freezing could be induced by a minor fraction of the soot particles immersed in the droplet. Again the results should not be extrapolated to cirrus conditions. Laboratory experiments at cirrus cloud temperatures will be discussed in the results and discussion section.

Experimental Methods

The experimental methods of AIDA cloud expansion experiments and techniques of soot generation and characterization are carefully described by Möhler et al.[6,7] Figure 1 shows data time series of a typical experiment starting at about 970 hPa and a uniform temperature of 229 K (first panel). The large aerosol vessel (volume 84 m^3) is filled with almost ice saturated synthetic air (solid line in panel 2) and soot aerosol from a propane flame burner with a number concentration of about 200 cm^{-3} (dashed line in panel 3). The organic carbon (OC) content of the burner soot can be varied by the fuel-to-air ratio in the diffusion flame in about the range of 15–40% by weight.[7]

Pumping expansion starts at t = 0 (vertical line in Figure 1), lowers the gas temperature (dashed line in panel 1), and thereby rises the relative humidity with respect to ice (RHi, solid line in panel 3) and accordingly with respect to water (dashed line in panel 3). Ice particles measured with an optical particle counter (solid line in panel 4) start to nucleate during this experiment after 60 s of pumping at a relative ice humidity of about 120%. The ice active number fraction of the soot aerosol (bottom panel) steadily increases to a maximum value of 0.2 until RHi reaches its peak value of 127%.

Results and Discussion

The relation between the ice-active aerosol number fraction f_{ice} and the ice saturation ratio S_{ice} was determined at different temperatures and for FS of different OC content. Figure 2 shows examples for two experiments started at the same temperature

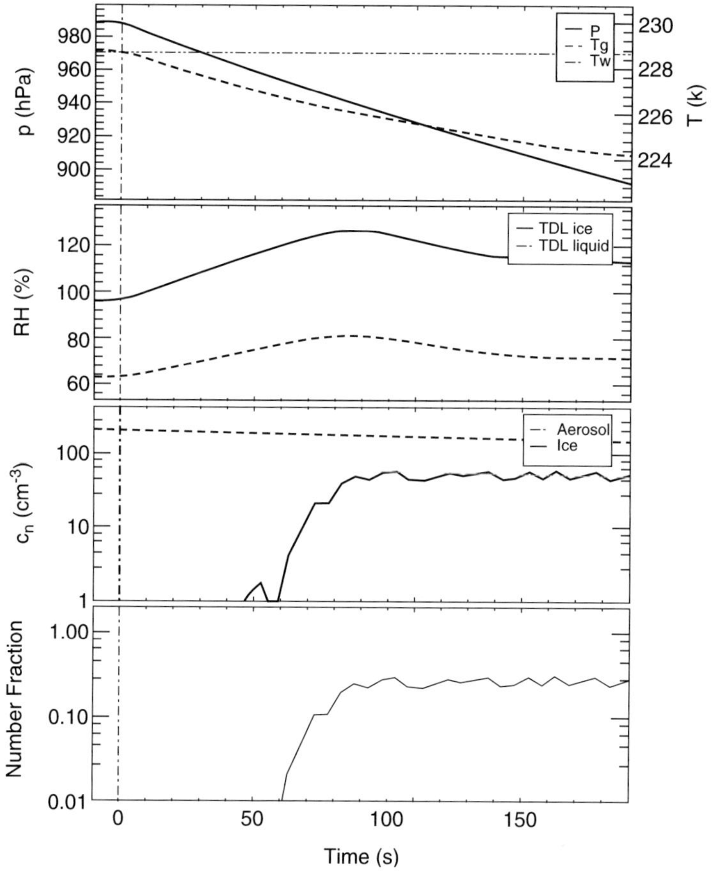

Figure 1 Data time series of an Aerosol Interaction and Dynamics in the Atmosphere (AIDA) cloud expansion experiment starting at a temperature of 229 K. Flame soot (FS) aerosol with minimum OC content was added at a number concentration of about 200 cm^{-3}

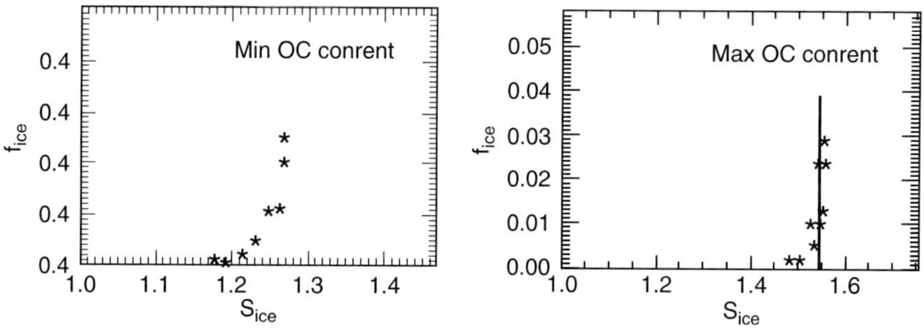

Figure 2 Relation between ice-activated particle fraction and the ice saturation ratio S_{ice} for two experiments with minimum OC soot (left) and maximum OC soot (right), both starting at the same temperature of 229 K and using the same expansion and cooling rate. The different scales of the x- and y-axis should be noted. The vertical line in the right figure indicates water saturated conditions at 226 K

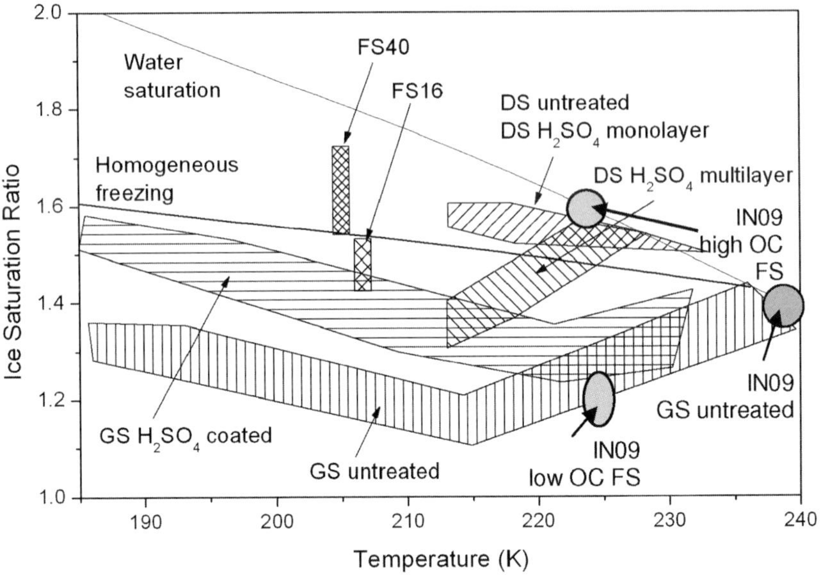

Figure 3 Summary of threshold conditions for ice nucleation by a particle fraction of about 1% in different soot aerosols measured in the temperature range 240–185 K

of 229 K. Soot aerosol with minimum (left panel) and maximum (right panel) OC content was exposed to the same rates of cooling and increase of S_{ice}. The minimum OC soot showed significant ice nucleation at S_{ice} between 1.20 and 1.27 (see also Figure 1), whereas the ice nucleation is markedly suppressed in the soot aerosol with maximum OC content. Water saturation (vertical line in the right panel) needs to be reached to activate a few percent of the particles.

Figure 3 depicts a summary of ice nucleation onsets measured for various soot aerosols in the temperature range 195–240 K. The areas show the range of S_{ice} where about 1% of the respective soot particles nucleated ice at the given temperature. More particles may be ice-active at higher S_{ice} as shown in Figure 2 above. Here we compare only the threshold values for the activation of 1% of all particles. Results from the present experiments (AIDA campaign IN09) with FS are indicated with light shaded ovals. Earlier studies with FS at lower temperatures[7] showed a similar trend of strong ice nucleation suppression with increasing OC content. The ice nucleation of untreated graphite spark generator soot (GS) was also investigated (darker grey-shaded oval) and compared well with earlier AIDA studies using the same soot type[6]. The present results are further compared in Figure 3 with literature data from laboratory studies using sulphuric acid coated GS[6] as well as monolayer and multilayer coated Degussa soot (DS).[8] It is obvious from Figure 3 that there is a huge variability of ice nucleation efficiency of different soot samples. The reason for that is unclear and more experiments are needed to better understand the relation between specific soot particle properties and their ice nucleation efficiency.

Acknowledgments Contributions by the AIDA team are gratefully acknowledged. This work is part of the HGF (Helmholtz-Gemeinschaft Deutscher Forschungszentren) project "Particles and Cirrus Clouds (PAZI-II)". Part of this work was funded within the European Union Integrated Project SCOUT-O3.

References

1. DeMott, P.J., *J. Appl. Meteorol.*, **29**, 1072 (1990).
2. Dymarska, M., Murray, B.J., Sun, L.M., Eastwood, M.L., Knopf, D.A., and Bertram, A.K., *J. Geophys. Res.*, **111** (2006).
3. Kanji, Z.A. and Abbatt, J.P.D., *J. Geophys. Res.*, **111**, D16204, doi:10.1029/2005JD006766 (2006).
4. Diehl, K. and Mitra, S.K., *Atmos. Environ.*, **32**, 3145 (1998).
5. Gorbunov, B., Baklanov, A., Kakutkina, N., Windsor, H.L., and Toumi, R., *J. Aerosol Sci.*, **32**, 199 (2001).
6. Möhler, O., Büttner, S., Linke, C., Schnaiter, M., Saathoff, S., Stetzer, O., Wagner, R., Krämer, M., Mangold, A., Ebert, V., and Schurath, U., *J. Geophys. Res.*, **110**, D11210, doi:10.1029/2004JD005169 (2005).
7. Möhler, O., Linke, C., Saathoff, H., Schnaiter, M., Wagner, R., Mangold, A., Krämer, M., and Schurath, U., *Meteorol. Z.*, **14**, 477 (2005).
8. DeMott, P.J., Chen, Y., Kreidenweis, S.M., Rogers, D.C., and Sherman, D.E., *Geophys. Res. Lett.*, **26**, 2429 (1999).

An Examination of a Continuous Flow Diffusion Chamber's Performance: Implications for Field Measurements of Ice Nuclei

Mathews S. Richardson, Anthony J. Prenni, Paul J. DeMott, and Sonia M. Kreidenweis

Abstract Recent field studies using the aircraft version of the Colorado State University Continuous Flow Diffusion Chamber (CFDC) indicate that the kinetic aspects of diffusional growth (both of water and ice) and ice nucleation, coupled with the limited aerosol residence time in the chamber, may result in a delay in the detection of the onset of homogeneous freezing. Through a series of controlled laboratory studies, we confirmed that the onset of homogeneous freezing of ammonium sulfate particles larger than ~100 nm was not detected at conditions consistent with expectations from theory. Current work involves modeling of the fluid dynamical and thermodynamical fields through the chamber, isolating particle trajectories from these fields, and running a microphysical model along these trajectories. The microphysical model is initialized with a distribution of dry particles of specified composition and the processes simulated include deliquescence, diffusional growth of droplets and crystals, and homogeneous freezing.

Keywords: Ice nucleation, homogeneous freezing, tropospheric aerosols, continuous flow diffusion chamber

Introduction

Ice plays an important role in the atmosphere. Ice crystals affect the radiative properties of clouds, play a role in convective cloud dynamics, and impact precipitation. Despite its importance, relatively little is known about the initiation of ice formation in the atmosphere. One instrument used extensively to study atmospheric ice nuclei (IN) concentrations in situ is the aircraft version of the Colorado State University Continuous Flow Diffusion Chamber (CFDC)[1]; the aircraft version has a shorter growth column than the laboratory unit owing to limitations on allowable instrument height. Depending on the configuration of refrigeration system, the aircraft CFDC can measure both heterogeneous IN and homogeneously freezing particle

Department of Atmospheric Science, Colorado State University, Fort Collins, CO, USA

C.D. O'Dowd and P.E. Wagner (eds.), *Nucleation and Atmospheric Aerosols*, 450–454.

concentrations. Data from two recent field campaigns (INSPECT-I and -II)[2, 3] employing this CFDC indicate that the conditions for the onset of homogeneous freezing were closer to water saturation than expected based on the parameterization proposed by Koop et al.[4] To understand these results, we undertook laboratory and computational studies to ascertain whether and how ice formation onset conditions measured by the field instrument might be impacted by the kinetics of ice nucleation, diffusional growth of droplets, and diffusional growth of crystals.

Laboratory Studies

In the laboratory experiments, an atomizer (TSI 3076; Minneapolis, MN) and differential mobility analyzer (DMA, TSI 3080) generated a near-monodisperse distribution of ammonium sulfate aerosol. A condensation particle counter (CPC, TSI 3010) measured concentrations of particles at the outlet of the DMA. The resulting aerosol stream was diluted with dry air and further dried with two diffusion driers prior to entering the chamber. Rogers[5] describes the theory of CFDC operation and the calculation of the aerosol lamina conditions. In previous versions of the aircraft CFDC, the outer wall of the evaporation region consisted of PVC, a material considered a good thermal insulator and barrier to water vapor diffusion. However, laboratory and computational studies suggested that the PVC was not an ideal insulator, resulting in partial evaporation of the ice. This PVC section has since been replaced with copper which is actively cooled to cold wall temperatures and ice-coated thereby leading to greater confidence in the predicted evaporation in this region. Laboratory results presented here are for measurements with the actively cooled evaporation section.

Figure 1 summarizes results from the laboratory tests. Shown in the figure are measurements of freezing onset conditions (0.1–1% activated) for ammonium sulfate at –42 °C. These results suggest that higher supersaturations are required for freezing with increasing particle size. This is in contrast to the homogeneous freezing parameterization of Koop et al., shown for ammonium sulfate particles for 0.1–1.0% freezing (dashed lines). These measurements are in keeping with the data from the INSPECT campaigns. Additional laboratory studies (not shown here) suggest that this deviation may result due to insufficient processing time in the CFDC for homogeneous freezing temperatures. To examine the role of time dependent nucleation and growth processes, we modeled conditions in the CFDC and the particle response to its evolving environment. These simulations are described below.

Computational Studies

The computational model consists of two parts run individually. Fluent (ANSYS, Inc., Lebanon, NH) simulated the thermodynamic and fluid dynamic properties of the chamber using a mesh generated in Gambit (ANSYS, Inc.). The model domain

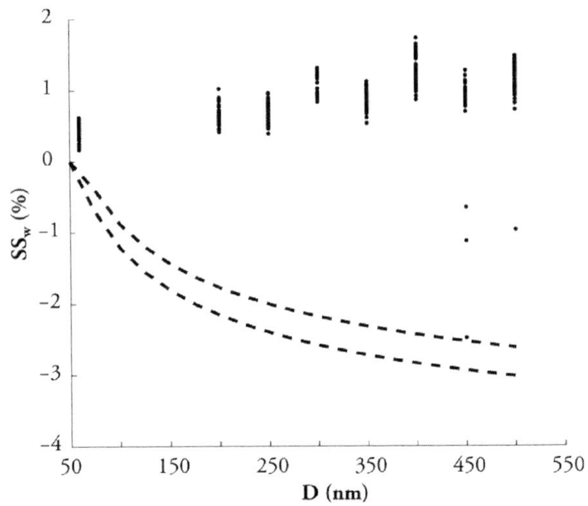

Figure 1 Supersaturation with respect to water (SS_w) required for the onset of homogeneous freezing of ammonium sulfate particles at −42 °C, as measured by the CFDC (filled circles). The range of measurements cover 0.1–1% of the particles activated. For comparison, dashed lines show the predicted onset conditions for 0.1% (**lower**) and 1% (**upper**) of the particles activated based on Koop et al.

consisted of a three dimensional, 5° wedge capturing flow from one of the 72 pairs of sheath holes and a portion of the sample inlet. The fluid domain is subdivided into three distinct regions: the inlet, the annular gap and the outlet (Figure 2a). In addition, solid regions such as the outlet cone and the inlet blades are modeled so as to capture heat transfer through these elements.

In all simulations the flow is laminar. Wall temperatures in the inlet region, at the collar and in the evaporation regions are not actively controlled and are governed by solutions to the energy equation. In the growth region, the chamber is operated such that the outer wall is warmer than the inner wall whereas in the evaporation region the outer wall temperature is set to that of the inner wall temperature. At the boundaries in these regions, the water vapor mixing ratio is set to that of ice saturation at the wall temperature. The water vapor mixing ratios in the domain are solved through the solution of the convective-diffusion equations. Particles are modeled using the Fluent Lagrangian particle model. In these simulations, particles are released at various points of the sample inlet. A thermodynamic trajectory for a particle is shown in Figure 2b.

In the microphysical model which uses the trajectories produced by Fluent, ammonium sulfate particles are initially dry but deliquesce to their equilibrium size at their deliquescence relative humidity[7] using the water activity parameterization of Clegg et al.[8] Water droplets then grow diffusionally[9] and ice nucleates from the aqueous phase via Koop et al.[4] and proceed to grow or evaporate diffusionally. The

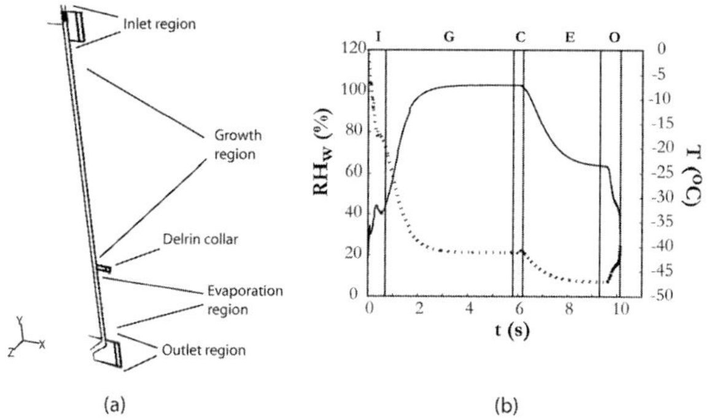

Figure 2 (a) The computational domain for CFDC fluid dynamic simulations. (b) A sample output trajectory from the Fluent Lagrangian trajectory particle model of relative humidity (solid line) and lamina temperature (dashed line). The relative humidity is calculated using Buck.[6] The letters indicate the domain region for which that particular portion of the trajectory is located (I = inlet, G = growth, etc.). The total residence time in the growth region at the steady state conditions is approximately 3–4 sec

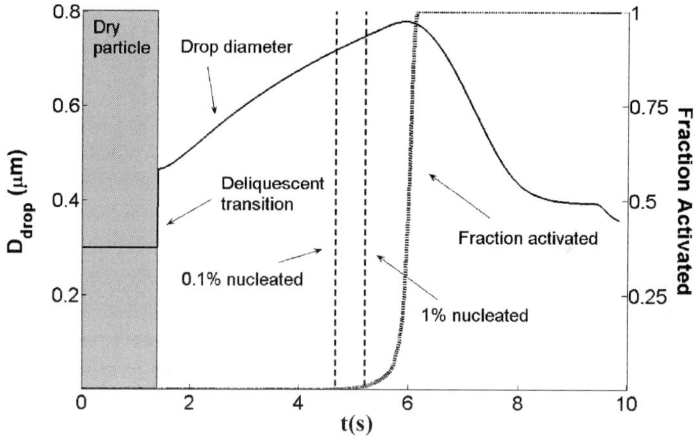

Figure 3 Simulated droplet growth and the calculated cumulative activated fraction at each time step for 300 nm ammonium sulfate particles. These conditions are for a trajectory with a steady state lamina RH_w of ~101% and temperature of −42.2°C (**Figure 2**b). The condensation coefficient was 0.1 (similar to that recommended by Shaw and Lamb[10]) and the initial particle population was 1,000 cm^{-3}. All particles nucleated as predicted by Koop et al. and the maximum crystal size was 2.02 µm; however, due to the delayed onset of nucleation, the mean crystal diameter was ~1 µm

impact of diffusional growth kinetics is illustrated in the figure below. From this figure, it is evident that the kinetics play an important role in CFDC measurements. Although all particles may nucleate ice at the given conditions, most do not form ice until reaching the end of the growth region, limiting their growth and detection in the instrument. We are undertaking further laboratory and computational studies to examine these effects more closely.

Acknowledgments We thank D. Straub, D. Rogers, D. Dandy, and M. Petters for helpful discussions. This research was supported through NSF grants ATM-0334228 and ATM-0436196.

References

1. Rogers, D.C., DeMott, P.J., Kreidenweis, S.M., and Chen, Y., *J. Atmos. Oceanic Technol.*, **18**, 725–741 (2001).
2. DeMott, P.J., Cziczo, D.J., Prenni, A.J., Murphy, D.M., Kreidenweis, S.M., Thomson, D.S., Borys, R., and Rogers, D.C., *Proc. Natl. Acad. Sci.*, **100**, 14655–14660 (2003).
3. Richardson, M.S., DeMott, P.J., Kreidenweis, S.M., Cziczo, D.J., Dunlea, E.J., Jimenez, J.L., Thomson, D.S., Ashbaugh, L.L., Borys, R.D., Westphal, D.L., Casuccio, G.S., and Lersch, T.L., *J. Geophys. Res.*, **112**, D02209 (2007).
4. Koop, T., Luo, B., Tsias, A., and Peter, T., *Nature*, **406**, 611–614 (2000).
5. Rogers, D.C., *Atmos. Res.*, **22**, 149–181 (1988).
6. Buck, A.L., *J. Appl. Meteor.*, **20**, 1527–1532 (1981).
7. Onasch, T.B., Siefert, R.L., Brooks, S.D., Prenni, A.J., Murray, B., Wilson, M.A., and Tolbert, M.A., *J. Geophys. Res.*, **104**, 21317 (1999).
8. Clegg, S.L., Brimblecombe, P., and Wexler, A.S., *J. Chem. Phys. A*, **102**, 2137–2154 (1998).
9. Pruppacher, H.R., and Klett, J.D., *Microphysics of Clouds and Precipitation*, Boston: Kluwer (1997).
10. Shaw, R.A. and D. Lamb, *J. Chem. Phys.*, **111**, 10659–10663 (1999).

Activation of Artificial Ice Nuclei

N.S. Kim[1], A.G. Shilin[1], J. Backmann[2], and I.A. Garaba[3]

Abstract This paper presents new methods to modify artificial ice nuclei (IN) in order to increase the activity of the latter. Based on theoretical and experimental data we demonstrated that the activity of particles can be essentially changed if hygroscopic compounds are introduced or if the particles are treated with gaseous haloids. Application of complex compounds such as xAgJ–yMeJ increases the efficiency of pyrotechnic mixtures by several orders of magnitude. At the same time, the temperature threshold of aerosol activity is about −5°C.

Keywords Heterogeneous nucleation, ice nuclei, pyrotechnic mixtures

Introduction

Usually, active influence on supercooled clouds is carried out by means of aerosols of ice forming compounds. To this kind of compounds belongs also AgJ. This reagent is the most effective one. Its advantages consist in the relative ecological safety and in the easiness by which it converts into aerosol state. To generate the aerosols of AgJ, one uses thermo-condensation during the burning of pyrotechnic mixtures and combustible solvents of this compound.

Recent studies on optimizing physicochemical conditions for converting of AgJ into a highly dispersed state allow creating very effective pyrotechnic mixtures. These compositions are used in anti-hail rockets and in pyro-cartridges of different calibers.

Pyrotechnic mixtures like "AU" which contain 10.5% of AgJ and its modifications, developed by S.A.M.-Vienna, Austria, also the Russian composition AD-1 which contains 8% of AgJ (Chernikov et al. 2003). At the temperature of a supercooled cloud which is about −6°C these compositions give about 5×10^{12} ice

[1] *Russia, adk@obninsk.ru*

[2] *Republic of Bulgaria, "Grandotech"*

[3] *Republic of Moldova*

C.D. O'Dowd and P.E. Wagner (eds.), *Nucleation and Atmospheric Aerosols*, 455–460.

nuclei per gram, and at the temperature of $-10°C$ – even 1.5×10^{13} to 2×10^{13} nuclei per gram. At present, the anti-hail rockets and pyro-cartridges, loaded with these mixtures are applied worldwide in different programs including the active influence of hail-forming processes and the increasing of precipitations on specific territories.

However, the results of recent theoretical and experimental studies show, that the potential of AgJ as a source of artificial ice nuclei (IN) is not exhausted yet.

Activation of Ice Nuclei

The burning of pyrotechnical mixtures and solutions containing an ice forming compound, results in highly dispersed aerosol. The size distribution of particles in such an aerosol can be described by a function f (r). The particle-forming process takes place at relatively high temperature (about 1,000°C) and is the result of complex physical and chemical interactions between combustion products. Therefore, artificial ice-forming particles have a complex spatial structure and well developed surface containing active centers of ice nucleation (Kim and Shkodkin 1988). The number of active ice particles in this aerosol can be calculated using the following general formula:

$$N = N_1 \int_0^{\infty} \theta(r-r^*)\left\{1-\exp\left[-4\pi r^2 e^{k_0(\Delta T-\Delta T_{50})}\cdot\frac{(1-e^{\beta\cdot\tau})ln2}{1-e^{-\beta\tau_{\Pi}}\cdot S_{\Pi}}\right]\right\} f(r)dr.$$

where $S_э$ – is the total surface of particles, $\boldsymbol{\tau}_э$ – fixed time, ΔT_{50} – threshold temperature and β – parameter characterizing the solubility of active centers.

According to this equation, the activity of ice-forming aerosols depends on several factors:

1. Humidity of air. The humidity effect is described by the function $\theta(r - r^*)$
2. Adsorbing and crystallizing characteristics of the particle surface
3. The temperature of supercooling, which defines the probability of ice nucleation
4. Dispersion characteristics of the formed aerosol

Corresponding to this equation, we can activate the aerosols by modifying the surface of ice-forming particles. For example, the presence of hygroscopic substances on the surface of the particles can essentially increase their nucleation activity.

The surface of ice-forming particles can be modified by chemical adsorption of gases from the atmosphere or of combustion products. Thus, the commonly used methods to generate aerosols by burning of pyrotechnical mixtures are based on the release of a large amount of chlorine and iodine into the gas phase.

Our studies with different substances demonstrated that haloids can strongly influence the activity of ice-forming particles. In this case the concentration of haloids was close to the mass concentration of ice-forming aerosols, this is about 1–10 mol per liter.

Figure 1 shows the efficiency of IN in the presence of chlorine and gaseous iodine.

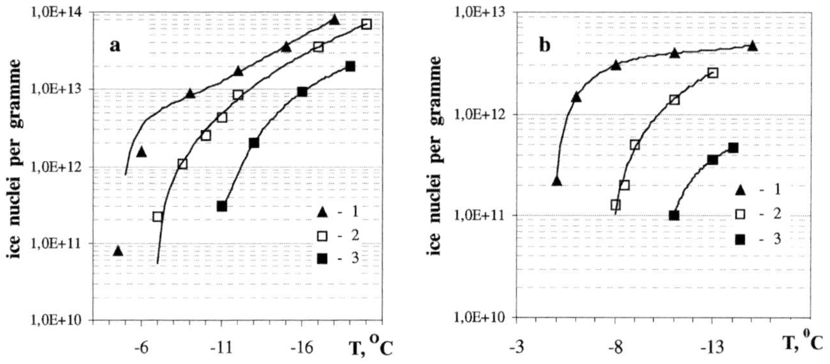

Figure 1 Efficiency of ice nuclei (IN) in the presence of chlorine and gaseous iodine. (a) Specific yield of IN for copper acetyl acetonate after addition of a halogen: 1 – iodine, 2 – without additives, 3 – chlorine. (b) Specific yield of IN for pyro-composition containing 0.4% of AgJ: 1 – iodine, 2 – without influencing combustion products, 3 – binding of iodine with KOH granules

Corresponding to these data, the ice-forming activity of aerosols in the presence of a halogen depends on the temperature. Figure 1a shows data obtained for aerosols of copper acetyl acetonate, Figure 1b shows data obtained for pyrotechnic mixture containing 0.4% AgJ but no metal oxides. It is clear, that the introduction of iodine increases the yield significantly, especially for the region of slight supercooling. The lower curve on the Figure 1b was obtained under condition when the iodine released upon the burning of pyrotechnic mixture was bound by means of potassium hydroxide (KOH) granules. The granules were placed into the aerosol chamber where the burning took place.

Complex Compounds

Another possibility to increase the ice-forming activity of artificial aerosols consists in the application of complex compounds of AgJ as a reagent. Passarely et al. (1974) found out that drops of colloid solution of the complex reagent 3AgJ–CuI freeze in already at the temperature of −1.5°C. Likewise, analogous compounds on the particle surface can increase its nuclear activity too. However, this kind of complex compounds has not found application for active influences on supercooled clouds yet. The reasons for this are firstly, insufficiently explored physicochemical properties of these complexes and secondly, technological obstacles to obtain aerosols of these reagents.

For common studies on ice-forming properties, aerosol particles of complexes like xAgJ–y$\mathbf{Me_1J}$ seem to be the most interesting. Below, the results of our studies on complexes are presented.

In order to prove the efficiency of complexes, we used a special aerodynamic stand (Kim N. 2003) and the standard method to determine the efficiency of small (up to 0.1 g) pyrotechnic species. Sublimation of regents was carried out during the burning process. At the same time, the recipe of the mixture was close to the composition of regular mixtures used in anti-hail rocket and pyro-cartridges.

Figure 2 shows the efficiency of ice formation for two different mixtures containing complex compounds with iodide of the metals Me_1 and Me_2. The content of the complex in the pyrotechnic mixture was about 15%. The ordinate axis indicates the yield of ice-forming particles, the abscissa axis shows the ratio of x to y (x:y)

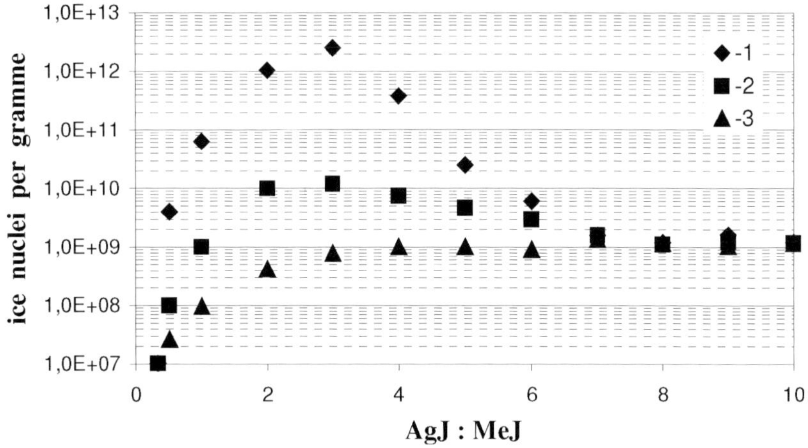

Figure 2 Dependence of the yield of active particles on the ratio of AgJ and MeJ in the complex. 1–complex xAgJ:yMe_1J, 2–complex xAgJ:yMe_2J, 3–mechanical mixture of AgJ and Me_1J

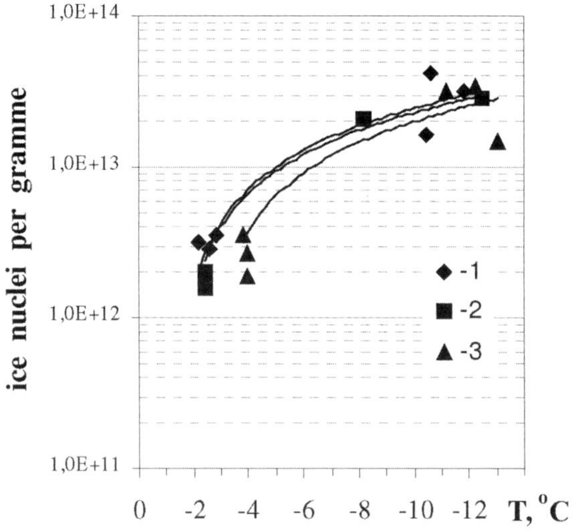

Figure 3 Temperature dependency of the yield for mixtures with different content of the complex xAgJ:y**Me_1J** (total mass 0.1 g): 1–5%, 2–10%, 3–15%

in the complex. The temperature of the supercooled fog in cloud chambers was about – 5°C.

Corresponding to these results, the activation at the temperature of –5°C can be increased by several orders of magnitude.

The next figure (Figure3) shows the dependence of the yield of active particles on the temperature. The yield is calculated for 1 g of pyrotechnic mixture. In this experiments we used 0.1 g of pyrotechnic mixtures containing different amount of the complex xAgJ:y$\mathbf{Me_1J}$. According to these data, increasing the content of the complex compound xAgJ:y$\mathbf{Me_1J}$ in the mixture gives a considerably higher yield of active particles, especially at high temperature. Here, we would like to stress, that application of this complex in pyrotechnic mixtures increases the threshold temperature essentially. For the composition containing 15% of xAgJ:y$\mathbf{Me_1J}$, the threshold temperature is just –2°C. In the temperature range from –2°C to –3°C, the yield of the active particles is about 10^{12} per gram. These values are typical only for the cooling agents like solid carbon dioxide and liquid nitrogen.

Figure 4 shows the efficiency of the ice-forming mixture measured in the wind tunnel according to the standard method. For this experiment, we used pyrotechnic devices like the pyro-cartridge PV-26. The weight of the applied mixture was 30 g and the air flow rate in the wind tunnel was about 30 ms^{-1}. If we compare the results shown on Figures 5 and 6, it is obvious that the efficiency of these mixtures produced by the industrial process corresponds to the results of lab experiment obtained for small species. We would like to stress that the investigated mixtures excel the conventional mixtures AU and AD-1 (Ад-1) in efficiency, and this applies for the whole temperature range tested in our experiments.

Based on these results we hope to develop more efficient pyrotechnic mixtures for active influences. Application of these mixtures can broaden the supercooled

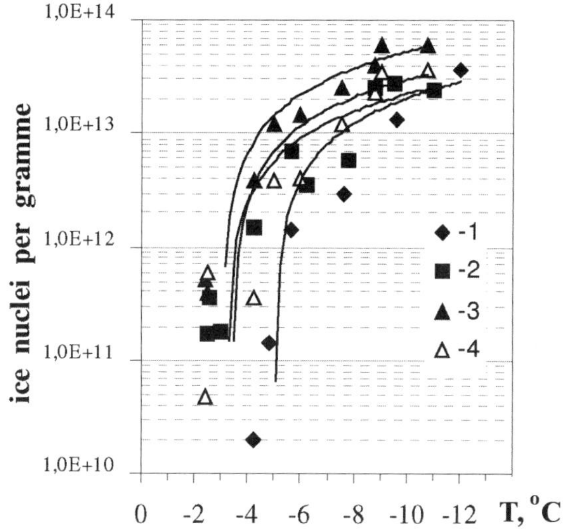

Figure 4 Temperature dependency of the yield for mixtures containing the complex xAgJ:y$\mathbf{Me_1J}$ (total mass 30 g): 1 – AD-1, 2 – 10%, 3 – 15%, 4 – 10%

areas in a cloud. Moreover, these mixtures can replace inconvenient cooling agents such as solid carbon dioxide, which are still used to increase precipitations and dispersion of clouds.

References

1. Chernikov A.A., Plaude N.O., Kim N.S., Korneev V.P., Nesmeyanov P.A., Dubinin B.N., and Sidorov A.I., New Russian pyrotechnic aids for seeding supercooled clouds. Proc. VIII WMO *Scintific Conf.on Weather Modification* (cosponsored by IAMAS), Casablanca, Morocco, 479–480 (2003).
2. Kim N.S. and Shkodkin A.V., Stochastic ice nucleation on nonhomogeneous surface, in: *Atmospheric Aerosols and Nucleation*, edited by P.E.Wagner and G.Vali, Berlin: Springer, pp.705–708 (1988).
3. Passarely R.E., Chessin H., and Vonnegut B., Ice nucleation in supercooled cloud by CuJ-3AgJ and AgJ aerosols, *J. Appl. Meteorol.*, **13**(8), 946–948 (1974).
4. Kim N.S., Equipment and testing results of crystallizing efficiency of anti-hail rockets. *Abstract of Presentations on Meeting of experts of WMO on Hail Suppression*. Nalchik, Russia (2003).
5. Passarely R.E., Chessin H., and Vonnegut B. Ice nucleation in supercooled cloud by CuJ-3AgJ and AgJ aerosols, *J. Appl. Meteorol.*, **13**(8), 946–948 (1974).

Heterogeneous Ice Nucleation of Aqueous Solutions with Immersed Mineral Dust Particles

B. Zobrist[1,2], C. Marcolli[1], T. Koop[2], and T. Peter[1]

Abstract Heterogeneous ice freezing experiments with emulsified samples made of suspensions containing Arizona Test Dust (ATD) together with different solutes have been investigated with a differential scanning calorimeter (DSC). The measurements indicate that ATD particles act as highly potent ice nuclei (IN) in the immersion mode, whereas the heterogeneous freezing temperature depends on the water activity of the solution. A reduced water activity of the sample leads to a lowered freezing temperature. Water-activity-based nucleation theory was applied to the data, indicating that a constant offset ($\Delta a_{w,het} = 0.203$) reproduces the measurements well. Such a constant offset for a specific IN was found in previous studies,[1] indicating the water-activity-based nucleation theory can describe heterogeneous ice nucleation in the immersion mode.

Keywords Heterogeneous ice nucleation, mineral dust, water activity, differential scanning calorimeter

Introduction

Cirrus clouds cover about 30% of the Earth's surface, affecting chemical and physical processes of the atmosphere. Their presence increases the scattering and absorption of solar radiation as well as the absorption of long wave terrestrial radiation. A change in cirrus cloud coverage may significantly alter the global radiation balance and hence the Earth's climate. So far, the exact mechanisms of cirrus cloud formation are largely unknown. Ice particles in cirrus clouds can form via homogeneous ice nucle ation from liquid aerosols or by heterogeneous ice nucleation on solid ice nuclei (IN). It seems likely that solid IN may appear immersed in liquid aerosols in the upper troposphere. It was

[1]*Institute for Atmospheric and Climate Science, ETH Zurich, Switzerland*

[2]*Department of Chemistry, Bielefeld University, Bielefeld, Germany*

C.D. O'Dowd and P.E. Wagner (eds.), *Nucleation and Atmospheric Aerosols*, 461–465.

found, that those liquid aerosols usually consist of inorganic/organic mixtures, i.e., 30% to over 80% of the aerosol mass in the free troposphere is carbonaceous material.[2]

Field measurements have found in ice residuals a notable fraction of mineral dust particles,[3] which can act as IN far from the source region.[4] Mineral dust is considered to be one of the most important IN. Arizona Test Dust (ATD, Powder Technology Inc.) is a commercially available mineral dust composed of various mineral species, but its composition is similar to dusts originating from deserts.[5] So far, it has been shown that ATD is a potent IN in the deposition mode.[6,7] Here, the heterogeneous ice freezing ability of ATD immersed in various aqueous solutions (organic, inorganic, and organic/inorganic mixtures) is investigated. The obtained results were analysed with water-activity-based nucleation theory.

Experimental

A commercial differential scanning calorimeter (DSC, TA Q10) with an external cooling device (LNCS) is used in this study. The DSC temperature calibration was performed with the melting point of ice and the ferroelectric phase transition of $(NH_4)_2SO_4$ at 223.1 K. The accuracy of the reported ice freezing and ice melting points is ±0.5 K and ±0.3 K, respectively. Fine Arizona Test Dust (ATD 0-3 μm particles diameter, Powder Technology Inc.) is used to prepare the suspensions without any further modification. Cooling and heating rates of 10 $Kmin^{-1}$ and 1 $Kmin^{-1}$ were used in all experiments, respectively.

Typical DSC Experiment

Figure 1 shows a typical freezing and melting thermogram of an emulsion containing ATD particles. Two exothermal peaks at two different temperatures are observed in the cooling cycle, whereas the first one exhibits a distinct larger size than the second one. The peak at higher temperature is attributed to heterogeneous ice freezing on the ATD particles. The other peak is due to homogeneous ice nucleation, indicating that not all emulsion droplets contain ATD particles. During the warming cycle two endothermic peaks appear. The first one corresponds to the eutectic melting of solid NaCl and ice, whereas the second one arises from the ice melting point of the solution. Similar experiments have been performed with ATD suspensions containing either 4.83, 10.18, 19.60 or 30.00 wt% $(NH_4)_2SO_4$, 5.05 wt% H_2SO_4, 5.94 wt% NaCl, 10.52 or 25.05 wt% malonic acid, 11 wt% poly(ethylene glycol) with molar mass of ~300 $gmol^{-1}$, pure water or a 4.1/8.3 wt% NaCl/Malonic acid solution. ATD always represents ~5% of the total mass.

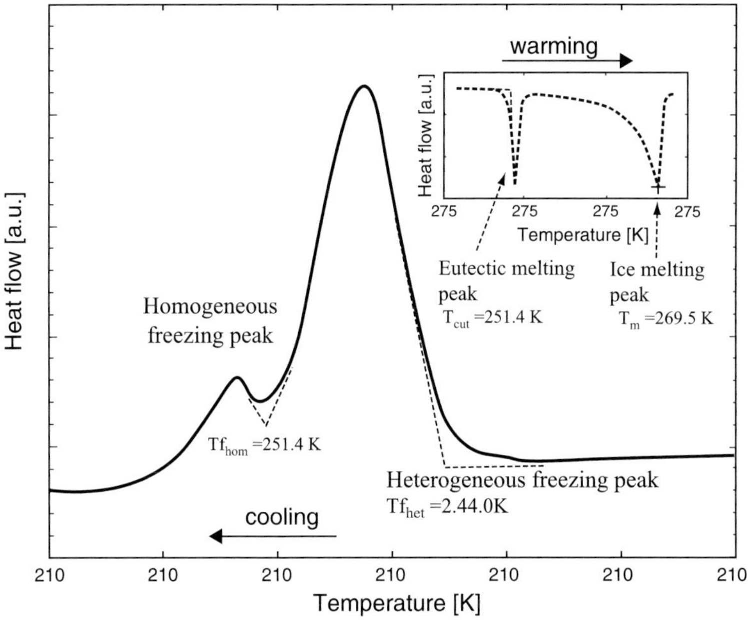

Figure 1 Typical DSC thermograms of an emulsion made with a 5.94 wt% NaCl/5 wt% ATD aqueous suspension as a function of temperature. Solid line: freezing thermogram recorded with 10 Kmin^{-1}. Dashed line in the insert: a part of the melting thermogram recorded with 1 Kmin^{-1}. The homogeneous (Tf_{hom}) and heterogeneous (Tf_{het}) ice freezing as well as the eutectic melting temperature (T_{eut}) are determined at the onset of the peak. The ice melting point (T_m) is derived form the minima of the heating curve (see the large cross in the insert)

Results and Discussion

Since it was found that homogeneous ice nucleation depends only on the water activity of the sample, i.e., independently of the nature of the solute,[8] the data are evaluated in the same manner for heterogeneous nucleation. Figure 2 depicts the measured homogeneous and heterogeneous ice freezing points as a function of the water activity of the solution, which can be derived from the ice melting point. The homogeneous as well as the heterogeneous freezing points decrease with decreasing water activity of the sample. The homogeneous freezing points are typically somewhat lower than predicted by Koop et al.,[8] which may have two reasons. Firstly, the used water activity estimation may not hold for all solutes. Secondly, homogeneous ice nucleation will mainly occur in the smallest droplets of the emulsion, since those are more likely not to contain an ATD particle (Marcolli et al., manuscript in preparation and Marcolli et al. and abstract submitted to the ICNAA 2007). However, the difference between the prediction and the measurements is at most 4 K. The heterogeneous freezing points at similar water activity exhibit similar freezing temperatures. Figure 2 includes also a horizontally shifted curve form

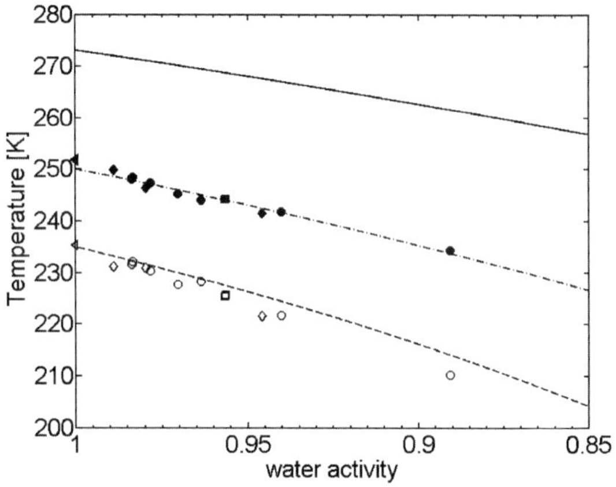

Figure 2 Heterogeneous freezing points of emulsified samples containing all 5 wt% ATD particles as a function of water activity. Open and filled symbols: Heterogeneous and homogeneous freezing points, respectively. Solutes: Circles: inorganic; diamonds: Organics; triangle: water; squares: inorganic/organic mixtures. Solid and dashed line: Ice melting and homogeneous freezing point line.[8] Dashed dotted line: Horizontally shifted curve form the solid line with a constant offset of $\Delta a_{w,het} = 0.203$

the ice melting point line with a constant offset of $\Delta a_{w,het} = 0.203$. There is maximum deviation of the curve and the measurements of ~1 K. The good agreement between the measurements and the prediction indicates that the water-activity-based nucleation can describe heterogeneous ice nucleation in the immersion mode. Such simple parameterization can be included in either microphysical or even global models.[1]

Conclusion and Outlook

DSC freezing experiments of emulsified samples containing an aqueous suspension of ATD particles and additional solute(s) have been investigated. Inorganic, organic as well as a mixture of organic/inorganic aqueous ATD suspension were examined. A lowered heterogeneous freezing temperature is observed for reduced water activity of the sample. Water-activity-based nucleation theory was used to evaluate the data, indicating that a constant offset ($\Delta a_{w,het}$) of 0.203 describes the measurements extremely accurate. Since for all experiments the same ATD concentration was used, it remains unclear whether a distinct ATD concentration change may also alter the heterogeneous freezing temperature. Such a behavior was at least observed in pure water ATD suspensions (Marcolli et al., manuscript in preparation). Another interesting part would be the ice nucleation influence of ATD under strong acidic or alkaline conditions. It seems likely that anthropogenic emissions may lead to surface changes on natural mineral dust particles, which then may change the ice nucleation efficiency of such particles.

Acknowledgments We are grateful for support by the Swiss National Fund in various projects and by the European Commission through the integrated project SCOUT-O3.

References

1. Zobrist, B., Marcolli, C., Koop, T., Luo, B.P., Murphy, D.M., Lohmann, U., Zardini, A.A., Krieger, U.K., Corti, T., Cziczo, D.J., Fueglistaler, S., Hudson, P.K., Thomson, D.S., and Peter T., *Atmos. Chem. Phys.*, **6**, 3115–3129 (2006).
2. Murphy, D.M., Cziczo, D.J., Froyd, K.D., Hudson, P.K., Matthew, B.M., Middlebrook, A.M., Peltier, R.E., Sullivan, A., Thomson, D.S., and Weber, R.J., *J. Geophys. Res.*, **111**, D23S32, doi:10.1029/2006JD007340 (2006).
3. DeMott, P.J., Cziczo, D.J., Prenni, A.J., Murphy, D.M., Kreidenweis, S.M., Thomson, D.S., Borys, R., and Rogers, D.C., *Proc. Natl. Acad. Sci. U. S. A.*, **100**, 14655–14660 (2003).
4. Sassen, K., DeMott, P.J., Prospero, J.M., and Poellot, M.R, *Geophys. Res. Lett.*, **30**, 1633, doi:10.1029/2003GL017371 (2003).
5. Krueger, B. J., Grassian, V.H., Cowin, J.P., and Laskin A., *Atmos. Environ.*, **39**(2), 395 (2005).
6. Knopf, D.A. and Koop T., *J. Geophys. Res.*, **111**, D12201, doi:10.1029/2005JD006894 (2006).
7. Möhler, O., Field, P.R., Connolly, P., Benz, S., Saathoff, H., Schnaiter, M., Wagner, R., Cotton, R., Krämer, M., Mangold, A., and Heymsfield, A.J., *Atmos. Chem. Phys.*, **6**, 3007–3021, (2006).
8. Koop, T., Luo, B.P. Tsias, A. and Peter, T., *Nature*, **406**, 611–614 (2000).

Immersion Freezing in Emulsified Aqueous Sulfuric Acid Solutions Containing AgI Particles

Markus Böttcher and Thomas Koop

Abstract We present a method to investigate heterogeneous immersion freezing in aqueous sulfuric acid solutions. This was achieved using water-in-oil emulsions of aqueous solution droplets in an inert oil matrix. Silver iodide (AgI) particles have been employed as ice nuclei (IN). Homogeneous and heterogeneous freezing is detected by means of differential scanning calorimetry. We observed that homogeneous ice nucleation as well as heterogeneous immersion freezing induced by AgI particles both occurs at lower temperatures with increasing sulfuric acid concentrations.

Keywords Heterogeneous ice nucleation, sulfuric acid, silver iodide

Introduction

Cirrus clouds influence atmospheric chemistry by catalyzing heterogeneous surface reactions. They also affect the Earth's radiation balance by reflecting incoming solar radiation and absorbing terrestrial infrared radiation. Furthermore, formation of cirrus clouds leads to dehydration of upper tropospheric air through sedimentation of ice particles. All these processes depend on the number and size of ice crystals, which in turn depend on the formation mechanism of the particular cirrus cloud [1]. Ice particles in cirrus clouds can form via homogeneous ice nucleation from preexisting liquid aerosol droplets, or by various heterogeneous ice nucleation processes induced by ice nuclei (IN). Heterogeneous nucleation normally leads to fewer but larger ice crystals at a lower ice saturation ratio when compared to homogeneous nucleation. However, heterogeneous ice nucleation pathways for cirrus clouds are only poorly understood. Laboratory studies of heterogeneous ice nucleation are, thus, a useful tool to reduce the associated uncertainties. Here we present laboratory experiments of heterogeneous ice nucleation in aqueous sulfuric acid

Department of Chemistry, Bielefeld University, Germany

C.D. O'Dowd and P.E. Wagner (eds.), *Nucleation and Atmospheric Aerosols*, 466–470.

solutions. Silver iodide (AgI) was employed for inducing heterogeneous ice nucleation, because it is known to be a very potent IN [2]. We use differential scanning calorimetry to determine the freezing temperatures of emulsified aqueous solution droplets.

Methods and Results

Sample Preparation

Homogeneous ice nucleation in micrometer-sized aqueous droplets can be studied using water-in-oil emulsion [3]. In this method the aqueous phase is immersed in an inert oil phase by homogenizing the phases with high-speed stirrer. The resulting emulsion is stabilized by a surfactant and consists of ~10^5 aqueous droplets which can be studied in a single measurement, thereby allowing statistically significant results. Furthermore, the large majority of the droplets are free of dust particles originating from laboratory air, thus allowing homogeneous ice nucleation to be investigated. We use a mixture of Methylcylopentane and Methylcyclohexane (1:1 by volume) as the organic phase, in which the surfactant Span 65 is dissolved (7 wt%). All three components are insoluble in water and are stable with respect to sulfuric acid.

Here, we want to study heterogeneous ice nucleation in the immersion mode. Therefore, a method to introduce IN deliberately into a large fraction of the droplets is required. We have done this by in-situ precipitation of insoluble AgI within individual droplets. The procedure is shown schematically in Figure 1.

0.5 ml of a dilute aqueous silver nitrate, $AgNO_3$, solution (0.16 wt%), and 1 ml of an aqueous sulfuric acid solution of twice the desired final concentration were placed into a test tube (black in Figure 1a). Next, 2 ml of the organic phase was added (white). Then 0.5 ml of a dilute aqueous potassium iodide, KI, solution was carefully placed on top of the organic phase (0.16 wt%; grey). Immediately thereafter, all components are mixed with a high-speed homogenizer for 10 minutes. During the initial phase of the stirring process, emulsified droplets form, that contains either of the two aqueous phases (light grey and black circles in Figure 1b). Later, droplets collide and coalesce during stirring. When two droplets of different type coalesce, Ag^+ ions and I^- ions react to form solid AgI, which immediately precipitates because of its extremely low solubility in aqueous solutions (see dark grey circles with white squares in Figure 1c). The potassium and nitrate ions remain in solution. However, because of their very low concentration (~0.08 wt%) they do not affect the ice nucleation measurements in any significant way, see below. Because the collision between different droplets is a stochastic process, some pure potassium iodide and silver nitrate droplets without any precipitated AgI remain at the end of the stirring (light grey and black circles in Figure 1c). The typical diameter of the aqueous droplets in the emulsion is about 3–9 μm, as confirmed by optical microscopy.

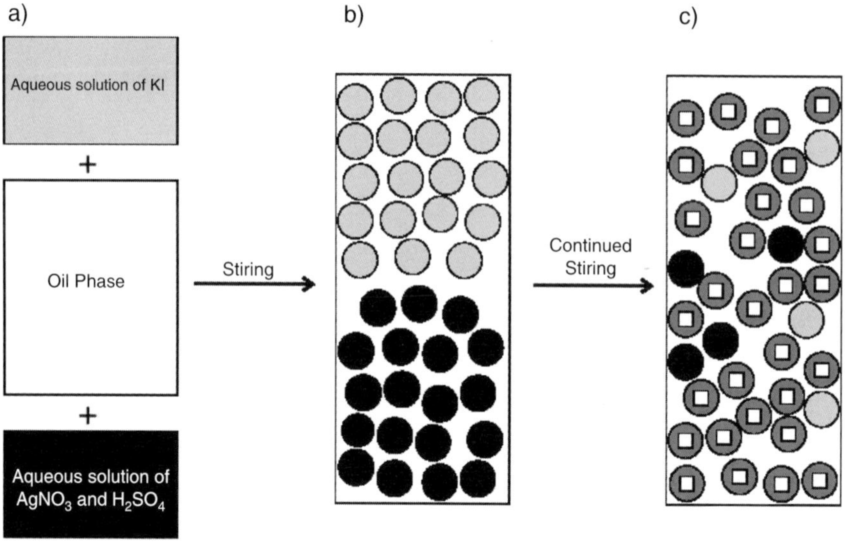

Figure 1 Preparation of water-in-oil emulsions containing silver iodide, AgI. Two aqueous phases are mixed with an oil phase in order to produce an emulsion consisting of droplets that contain either a potassium iodide, KI, solution (light grey circles) or droplets containing silver nitrate, $AgNO_3$ and sulfuric acid, H_2SO_4 (black circles). When these droplets collied and coalesce, silver iodide (AgI, white squares) precipitates because of its extremely low solubility

Ice Nucleation Experiments

Freezing of the emulsified samples was measured with a commercial differential scanning calorimeter (DSC, TA-Instruments Q 100). About 10 µl of an emulsion sample was cooled at a rate of $10\,K\,min^{-1}$ from room temperature to 170 K. The resulting DSC thermograms for various samples are shown in Figure 2. In these thermograms the heat flow measured by the DSC is shown as a function of sample temperature. A positive signal indicates an exothermic process such as ice crystallization from a supercooled liquid.

The bottom thermogram shows the freezing of a pure water emulsion that does not contain any AgI particles. Only one peak is observable which can be attributed to homogeneous ice nucleation. The homogeneous freezing temperature was determined as the onset of the signals and was found to be 236 K, in good agreement with literature values for homogeneous ice nucleation in micrometer sized droplets [3]. In contrast, the second thermogram shows results of an emulsion, in which most water droplets do contain AgI particles. Such samples do reveal two freezing signals, one corresponding to homogeneous ice nucleation (lower T signal) and one corresponding to heterogeneous ice nucleation induced by AgI (higher T signal). We note that the homogeneous ice nucleation temperature of

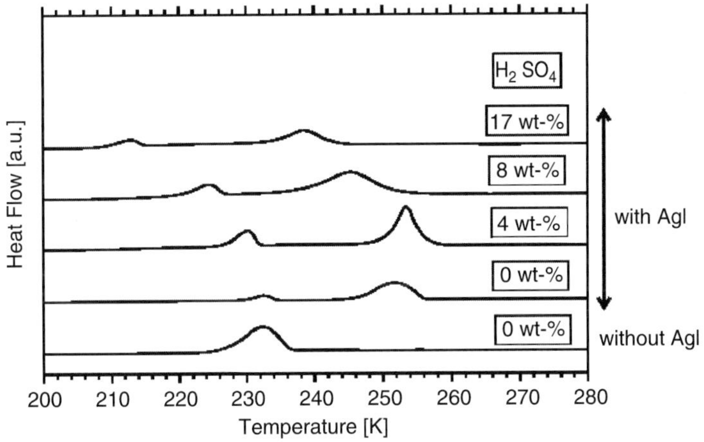

Figure 2 DSC thermograms of aqueous sulfuric acid emulsion samples. Only the cooling experiments are shown. The sulfuric acid concentration is indicated at each thermogram. The different thermograms have been shifted vertically for clarity. Note, that in case of the 4 wt% solution a cooling rate of 1 K min^{-1} is used

both samples is almost identical, showing that the traces of dissolved $AgNO_3$ or KI do not affect the homogeneous ice nucleation temperature in any significant way. The larger signal at about 257 K indicates that a large fraction of the droplets contain AgI particle and, thus, freeze heterogeneously at a much higher temperature. The other thermograms are measurements of droplets containing AgI particles and a varying concentration of sulfuric acid. Both the homogeneous ice nucleation temperature as well as the heterogeneous freezing temperature of the droplets decreases with increasing sulfuric acid concentration. We have also observed the formation of several sulfuric acid hydrates in the temperature range between 181and 193 K (not shown here).

Conclusion

A laboratory method to investigate heterogeneous freezing of aqueous sulfuric acid solutions is presented. It is shown that solid AgI IN can be produced by in situ precipitation in individual micrometer-sized aqueous droplets. These are stabilized in water-in-oil emulsions. Using differential scanning calorimetry, ice nucleation was studied as a function of sulfuric acid concentration. The homogeneous ice nucleation temperature decreased with increasing sulfuric acid concentration. In addition, heterogeneous ice nucleation on AgI particles occurred in the immersion mode. The observed immersion freezing temperatures also decreased with increasing sulfuric acid concentration.

References

1. Baker, M.B., *Science*, **276**, 1072–1078 (1997).
2. Vonnegut, B., *J Appl. Phys.*, **18**, 593–595 (1947).
3. Koop, T., *Z. Phys. Chem.*, **218**, 1231–1258 (2004).

Part V
Air Quality – Climate

Global Dimming by Air Pollution and Global Warming by Greenhouse Gases: Global and Regional Perspectives

V. Ramanathan

Abstract The global build up of greenhouse gases (GHGs), is the most vexing global environmental issue facing the planet. GHGs warm the surface and the atmosphere with significant implications for, rainfall, retreat of glaciers and sea ice, sea level, among other factors. What is less recognized, however, is a comparably major global problem dealing with air pollution. Until about 10 years ago, air pollution was thought to be just an urban or a local problem. But new data have revealed that, due to fast long-range transport, air pollution is transported across continents and ocean basins, resulting in transoceanic and transcontinental plumes of atmospheric brown clouds (ABCs) containing submicron-size particles, i.e., aerosols. ABCs intercept sunlight by absorbing as well as reflecting it, both of which lead to a large surface dimming. The dimming effect is enhanced further because aerosols nucleate more cloud drops which makes the clouds reflect more solar radiation. The surface cooling from this dimming effect has masked the warming due to GHGs. ABCs are concentrated in regional and megacity hot spots. Long-range transport from these hot spots gives rise to widespread plumes over the adjacent oceans. Such a pattern of regionally concentrated surface dimming and atmospheric solar heating, accompanied by widespread dimming over the oceans, gives rise to large regional effects. Only during the last decade, we have begun to comprehend the surprisingly large regional impacts. The large north–south gradient in the ABC dimming has altered the north–south gradients in sea surface temperatures, which in turn has been shown by models to decrease rainfall over the continents. In addition to their climate effects, ABCs lead to acidification of rain and also result in over few million fatalities worldwide. The surface cooling effect of ABCs may have masked as much 50% of the global warming. This presents a dilemma since efforts to curb air pollution may unmask the ABC cooling effect and enhance the surface warming. Thus efforts to curb GHGs and air pollution should be done under one framework. The uncertainties in our understanding of the ABC effects are large, but we are discovering new ways in which human activities are changing the climate and the environment.

Keywords Global dimming, global warming

Scripps Institution of Oceanography, University of California at San Diego, USA

C.D. O'Dowd and P.E. Wagner (eds.), *Nucleation and Atmospheric Aerosols*, 473–483.

Introduction

Human activities are releasing tiny particles, aerosols, into the atmosphere in large quantities. Our understanding of the impact of these aerosols has undergone a major revision, due to new experimental findings from field observations such as the Indian Ocean Experiment (Ramanathan et al. 2001) and ACE-Asia (Huebert et al. 2003), among others and global modeling studies. Aerosols enhance scattering and absorption of solar radiation and also produce brighter clouds that are less efficient at releasing precipitation. These in turn lead to large reductions in the amount of solar radiation reaching Earth's surface, a corresponding increase in atmospheric solar heating, changes in atmospheric thermal structure, surface cooling, disruption of regional circulation systems such as the monsoons, suppression of rainfall, and less efficient removal of pollutants. Black carbon plays a major role in the dimming of the surface. Man-made aerosols have dimmed the surface of the planet, while making it brighter at the top of the atmosphere.

Together the aerosol radiation and microphysical effects can lead to a weaker hydrological cycle and drying of the planet which connects aerosols directly to availability of fresh water, a major environmental issue of the twenty-first century. For example, the Sahelian drought during the last century is attributed by models to aerosols. In addition, new coupled-ocean atmosphere model studies suggest that aerosols may be the major source for some of the observed drying of the land regions of the planet during the last 50 years. Regionally aerosol-induced radiative changes (forcing) are an order of magnitude larger than that of the greenhouse gases (GHGs), but because of the global nature of the greenhouse forcing, its global climate effects are still more important. However, there is one important distinction to be made. While the warming due to the GHGs will make the planet wetter, i.e., more rainfall, the large reduction in surface solar radiation due to absorbing aerosols will make the planet drier.

Without a proper treatment of the regional aerosol effects in climate models, it is nearly impossible to reliably interpret the causal factors for observed regional as well as global climate changes during the last century. Until the 1950s, the extratropical regions played a dominant role in emissions of aerosols, but since the 1970s the tropical regions have become major contributors to aerosol emissions, particularly black carbon. The chemistry and hence the radiative effects of aerosols emitted in the extratropics are very different (even possibly in the sign) from that of the aerosols emitted in the tropics. The paper presents the following recent results and findings:

1. Identification of regional and megacity aerosol hotspots
2. Global distribution of the surface dimming and atmospheric heating by ABCs
3. The magnitude of the masking effect of global warming by ABCs
4. Reconciliation of the large surface dimming with the observed global warming
5. Effects of ABCs on the Asian monsoon
6. The combined effects of ABCs and GHGs on regional climate

We conclude with the following outstanding issues that can influence the response of our climate to GHGs increases during the next several decades:

1. Why is the planetary albedo 29% and how is it regulated?
2. Will a warmer planet also become more cloudy?
3. What is the role of the biosphere in determining the aerosol chemistry and nuclei for cloud formation?
4. What is the role of ABCs in nucleating mixed-phase and ice-phase clouds in the mid and upper troposphere?

ABCs: Regional and Megacity Hotspots

It is important to first recognize that ABCs are a worldwide problem. Monthly mean aerosol optical depth (AOD) analyzed from thc MODIS instrument (Kaufman et al. 2002) on NASA's TERRA satellite that monthly mean AODS exceeding 0.2(indicative of strong pollution) occur even over large areas of industrialized

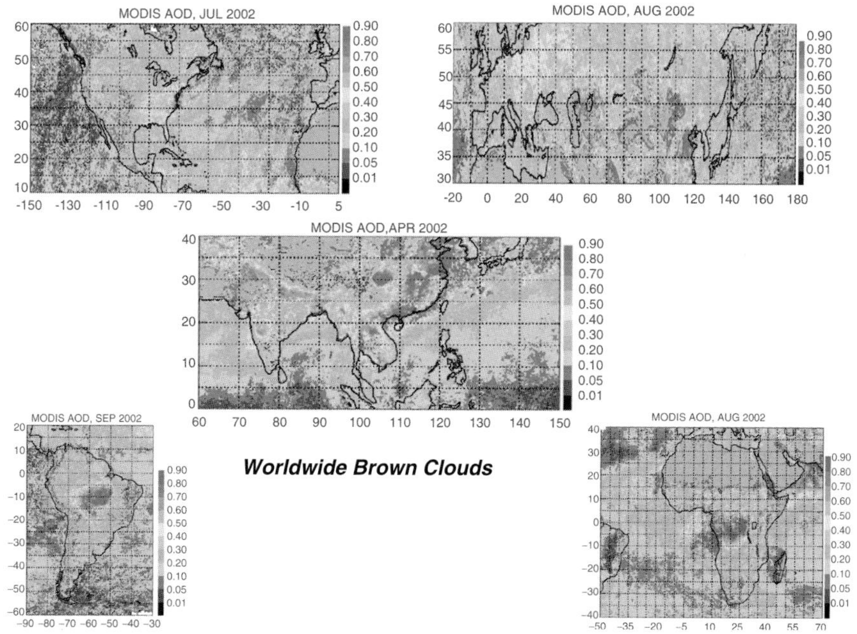

Figure 1 Monthly mean aerosol optical depths derived from MODIS aerosol instrument on NASA's TERRA satellite. The color shading is dark blue for AODs smaller than 0.05 (clean marine background); green for 0.2 (visible brown clouds), yellow for 0.4 to 0.5 (very hazy) and red for AODs > 0.6 (heavily polluted)

countries. The figure (Figure 1) also indicates how, due to long-range transport, ABCs from east Asia spreads across the Pacific and like wise, pollution from North America spreads across the Atlantic.

Using a combination of satellite observations, aerosol transport models and field campaign data, Ramanathan et al. (2007) recently identified five *regional hot spots* around the world: (1) East Asia (eastern China, Thailand, Vietnam, and Cambodia); (2) Indo-Gangetic Plains in South Asia (the northwest to northeast region extending from eastern Pakistan, across India to Bangladesh and Myanmar); (3) Indonesia; (4) Southern Africa extending southwards from sub-Saharan Africa into Angola and Zambia and Zimbabwe; (5) The Amazon basin in South America. In addition, the following 13 *megacity hot spots* have been identified: Bangkok, Beijing, Cairo, Dhaka, Karachi, Kolkata, Lagos, Mumbai (Bombay), New Delhi, Seoul, Shanghai, Shenzen, and Tehran. Over these hotspots, the annual mean AODs exceed 0.3 and the absorption optical depth is about 10% of the AOD, indicative of the presence of strongly absorbing soot accounting for about 10% of the anthropogenic aerosol amount.

Global Dimming by ABCs

Is the Planet dimmer now than it was during the early twentieth century? Solar radiometers around the world are indicating that surface solar radiation in the extra tropics was less by as much as 510% during the mid-twentieth century (e.g., see Stanhill and Cohen and Wild et al.), while in the tropics, such dimming trends have been reported to extend into the twenty-first century. But many of these radiometers are close to urban areas and it is unclear if the published trends are representative of true regional averages. The Indian Ocean Experiment (Ramanathan et al. 2001) used a variety of chemical, physical and optical measurements to convincingly demonstrate (Satheesh and Ramanathan 2000) that ABCs with AODs of about 0.2 and absorption of AODs of about 0.02 can lead to dimming as large as 510% (i.e., decrease in annual mean absorbed solar radiation of about 15 Wm^{-2}). In order to get a handle on the global average dimming, recently we integrated such field observations with satellite data and aerosol transport models to retrieve an observationally constrained estimate.

As seen from Figure 2, over large regions the reduction of solar absorption at the surface exceeds 12 Wm^{-2} (>5%), which is consistent with the dimming reported from surface observations. The global annual average dimming (for 2002), however, is −3.5 Wm^{-2}. *Thus great care should be exercised to extrapolate surface measurements over land areas to global averages*. The global dimming of −3.5 Wm^{-2} has been compared with the GHGs forcing of 3 Wm^{-2} from 1850 to present, i.e., 2005, (IPCC 2007). Such comparisons, without a proper context could be misleading, since the dimming at the surface is not the complete forcing. It does not account for the atmospheric solar heating by ABCs, discussed next.

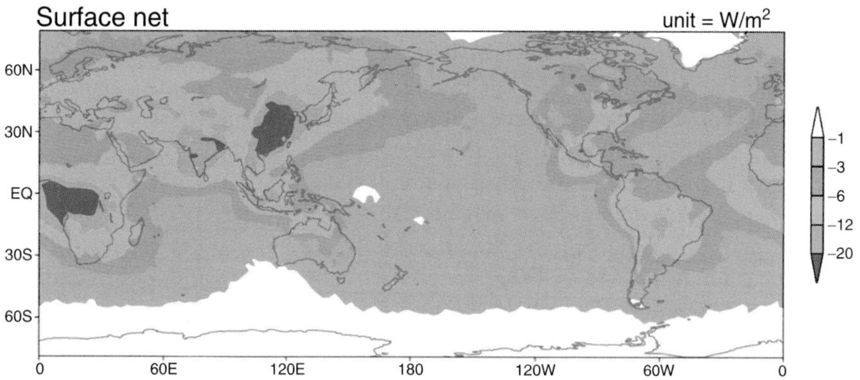

Figure 2 Integrated and observationally constrained estimate of annual mean global dimming by ABCs around the world for 2001–2003. (Ref. Chunget al. 2005.)

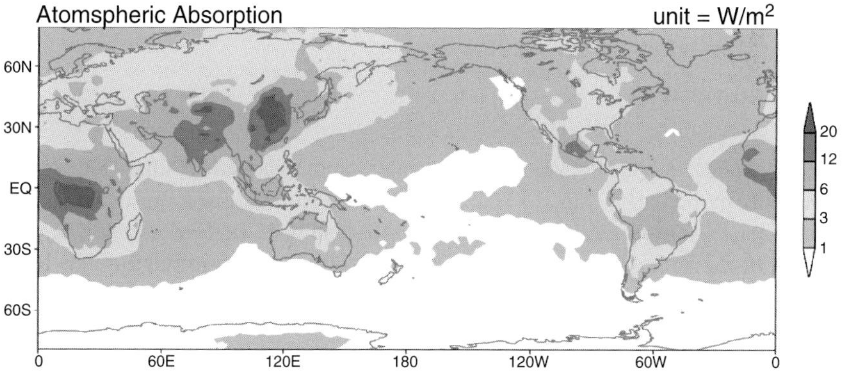

Figure 3 Integrated and observationally constrained estimate of annual mean atmospheric solar heating by ABCs for 2001–2003. (Ref. Chung, Ramanathan, Podgorny and Kim 2005.)

Global Solar Heating of Atmosphere by ABCs

There is an important distinction in the forcing by scattering aerosols, like sulfates, and that due to absorbing aerosols like soot (see Ramanathan et al. 2001 for a detailed elaboration of the points noted later). For sulfates, the dimming at the surface is nearly the same as the net radiative forcing due to aerosol since there is no compensatory heating of atmosphere, and hence a direct comparison of the surface dimming with GHGs forcing is appropriate. For soot, however, the dimming at the surface is mostly by the increase in atmospheric solar absorption, and hence the dimming does not necessarily reflect a cooling effect. It should also be noted that the dimming at the surface due to soot solar absorption can be factor of 3 larger than the dimming due to reflection of solar (a cooling effect). Figure 3 shows our recent estimates of the global distribution of the atmospheric solar heating by man-made aerosols for the period 2001–2003.

Masking of Global Warming

The global mean estimates shown in Figure 4 underscores the relative contributions of aerosols and GHGs at the surface, the atmosphere and the surface. While at the surface, the aerosol dimming (negative forcing of $-4.4\,Wm^{-2}$) is much larger than the GHGs forcing of 1.6, the positive atmospheric forcing of $3\,Wm^{-2}$ within the atmosphere by aerosols (ABCs) enhances the GHGs forcing of $+1.4\,Wm^{-2}$, such that the sum of the surface and the atmospheric forcing, i.e., forcing at TOA, is $-1.4\,Wm^{-2}$ for ABCs and $+3\,Wm^{-2}$ for GHGs. Thus the net anthropogenic forcing by anthropogenic modification of the radiative forcing is positive. We also note that, globally, the ABC forcing ($-1.4\,Wm^{-2}$) has masked about 50% of the GHGs forcing ($+3\,Wm^{-2}$). The implications of this masking to global warming and climate sensitivity is discussed in Andreae et al. (2005).

Interactions Between GHGs and ABCs on Regional Scales: The Asian Monsoon as an Example

The fundamental driver of evaporation of water vapor is absorbed solar radiation at the surface, particularly, over the sea surface. The precipitation over land is driven by two major source terms: evaporation from the land surface and long-range transport of moisture from the oceans and its subsequent convergence over the land regions. It is then logical to posit that the large reduction of absorbed solar radiation

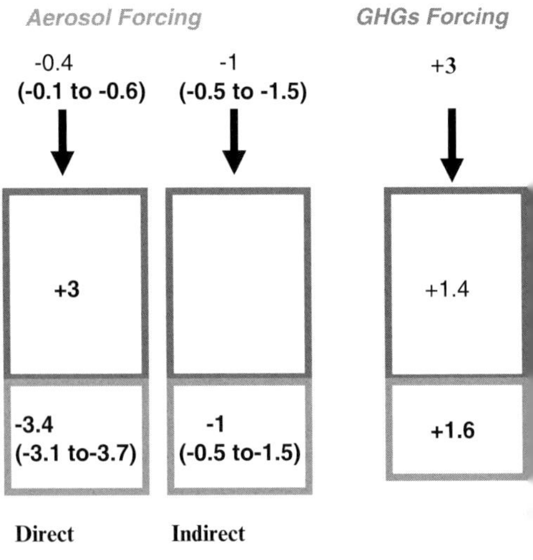

Figure 4 Global mean anthropogenic forcing due to aerosols (the two panels on the right) and due to greenhouse gases. The blue boxes show the atmospheric forcing; the brown box at the bottom shows the surface forcing; the sum of the two is the forcing at top-of-atmosphere (TOA). The aerosol forcing is for 2001–2003 and is from the present study; while the GHGs at TOA is from IPCC (2007). Also see Crutzen and Ramanathan (2003) on the *parasol* effect of aerosols

by the land and sea surface due to interception of sunlight by ABCs (Figure 2) should lead to an overall reduction of rainfall. The observed precipitation trends over the last 50 years reveal major regions which experienced an overall reduction of rainfall (Sahel and the Indian monsoon) as well as a shift in the rainfall patterns (Figure 5). Numerous climate model studies have been published, which suggest that inclusion of the aerosol dimming can help explain the Sahelian drought (Rotstayn and Lohmann 2002); the decrease in Indian monsoon rainfall (Chung et al. 2002; Ramanathan et al. 2005; Meehl et al. 2007; Lau et al. 2007); and the north–south shift in east Asian rainfall (Menon et al. 2002). Ramanathan et al. (2005) conducted a coupled ocean atmosphere model study with prescribed greenhouse forcing and ABC forcing (Figures 2 and 3) over South Asia from 1930 to 2002. For the time-dependent ABC forcing, they scaled the observationally constrained forcing (for 2001–2003) with history of SO_2 and soot emission for South Asia from 1930 to 2002. Their model simulations led to the following findings

1. *Dimming Trends*: The simulated trend in dimming of about 7% over India was consistent with the observed trends obtained from radiometer stations (12 stations) in India, thus providing evidence for large dimming due to ABCs
2. *Atmosphere*: heated by absorption and scattering of solar radiation
   – Warmer atmosphere is more stable: ***less precipitation***
3. *Surface*: less solar radiation ("dimming"), thus more cooling (offset GHG warming)
   – Reduced solar radiation over Northern Indian Ocean (NIO): less evaporation, ***less precipitation***

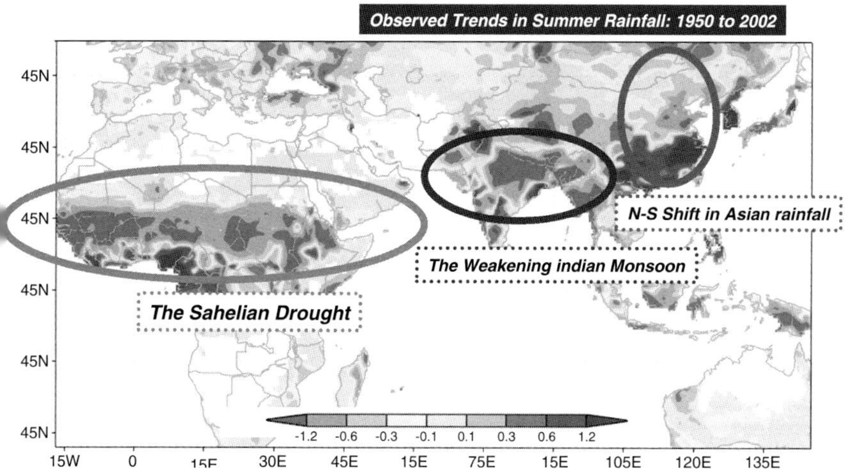

Figure 5 Trend in observed rainfall from 1950 to 2002. The figure shows the change in rainfall between 2002 and 1950; It was obtained by multiplying the year linear trend in mm/day/year by 52 years. The precipitation data is the Hadley center CRU data. (Ref. Mitchell and Jones 2005.)

– Pollution is greater over NIO than SIO, which weakens the summertime sea surface temperature gradient: less circulation, weaker monsoon, ***less precipitation***

4. *Monsoon Impact*: The resulting deceleration of the summer monsoonal circulation, the decrease in evaporation, and the increase in stability are the primary mechanisms for the reduction in the summer monsoon rainfall

These recent findings have catalyzed the creation of an international program for a better understanding of aerosol effects on the Asian monsoon (Lau et al. 2007).

Outstanding Issues

How Will Clouds and Their Radiative Forcing Respond to and Feedback on Global Warming?

Figure 6 shows the net radiative forcing (in Wm^{-2}) due to clouds obtained from cloud forcing observations (Ramanathan et al. 1989) from the Earth Radiation Budget Experiment (Harrison et al. 1990). Globally, clouds enhance the albedo (percent reflection to space of incident solar radiation) of the planet from about 14% in clear skies to the observed albedo of about 29%. The resulting shortwave radiative forcing is about -48 (uncertainty is about ± 5) Wm^{-2}. Clouds also enhance the greenhouse effect and this positive forcing is about $+30\,Wm^{-2}$, such that the net effect of clouds is to exert a net negative forcing of about $-18\,Wm^{-2}$. The global distribution of the net cloud forcing reveals the strong regional variations.

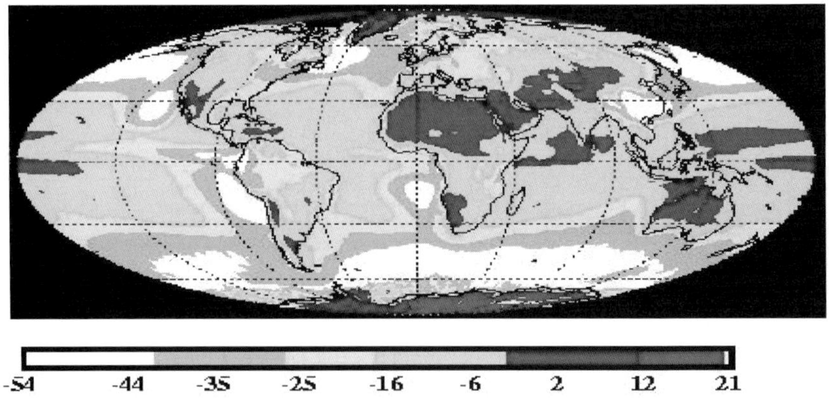

Figure 6 Net (sum of the short wave forcing, i.e., reflection of solar, and greenhouse effect of clouds) radiative forcing due to clouds for the 1985 to1989 period. (See Ramanathan et al. 1989 for how cloud forcing data were obtained.)

Essentially the extratropical storm track cloud systems (ET_STCS), are the dominant cooling agents of the whole planet (Weaver and Ramanathan 1999), without which the entire planet would be warmer by more than 10°C. Thus a small change in the cloud forcing, say a 2% decrease or increase in cloud shortwave forcing, is enough to amplify or ameliorate the 0.8 K global warming of the last 100 years by about 70%. Currently we neither have any theoretical constraint nor observational constraint to rule out such a possibility. The other major source of concern is that the ET_STCS cloud systems are susceptible to anthropogenic effects because: (1) the fast retreat of sea ice is reducing the equator to pole temperature gradient which is one of the major factors that regulate the latitudinal extent of the extra tropical cloud systems; and (2) advection of ABCs from the surrounding continents (east Asia for the Pacific; and North America for the Atlantic) as shown in Figures 2 and 3, can significantly perturb the nucleation of water drops and ice crystals in these systems.

Biospheric Influences on Aerosols and Clouds and its Feedback Effects

Biogenic emissions of aerosol precursor gases by terrestrial and marine micro-organisms, DMS (Charlson et al. 1989), Isoprenes (Meskidze and Nenes 2006), and other organic vapors (O'Dowd et al. 2002) are now well known to be important biogenic sources of aerosols and cloud condensation nuclei. There have been several suggestions that the biogenic emission sources respond to temperature changes and thus possibly feedback on climate changes (e.g., Charlson et al. 1989; O'Dowd et al. 2002). The climate models have not been able to incorporate these biogenic effects largely because the details of the aerosol formation nor the emission strengths are largely unknown. Yet the potential of these biogenic aerosols to influence climate can be appreciable and we need observational constraints for their effects.

Emission Sources for Elemental and Organic and Carbon

These carbonaceous aerosols are the major source of dimming by ABCs (Ramanathan et al. 2001, 2007) but our ability to model their effects in climate models are severely limited. One of the main reasons is the factor of 2 or more uncertainty in the current estimates of the emission of the organic (OC) and elemental carbon (EC). Further more, that biomass and biofuel burning contribute over 50% of the OC and EC and the historical trends (during the last 100 years) in these emissions are unknown and models currently resort to scaling the current day emissions with past trends in population.

Why is the Planetary Albedo 29%?

One of the fundamental unanswered (and perhaps unasked) questions in climate is: Why is the planetary albedo 29% (± 1%)? Let us consider first why this question is important. A global albedo of 32% would plunge the Earth into a climate similar to that of the last ice age; while an albedo of 26% would be comparable to a sixfold increase in the CO_2 concentration from about 300–1800 ppm. Is it sheer chance that the planet just settled into the 29% albedo? I am not aware of any theory that has contemplated this issue seriously, except for the untestable Gaia theory (Lovelock and Margulis 1974). It is remarkable that our general circulation climate models are able to explain the observed temperature trends during the last century solely through variations in GHGs, aerosols, and solar irradiance. This implies that the role of clouds in planetary albedo has not changed during the last 100 years by more than ±0.3% (out of 29%). Clearly we need theoretical and observational verification for this remarkable model result. Most discussions of pale climate changes also implicitly assume clouds have not changed much over time scales ranging from 10^3 to 10^8 years (see http://www-c4.ucsd.edu/gap/ for additional discussions of this topic).

Acknowledgments The research reported here was funded and supported by NSF (J Fein); NOAA (Koblinsky) and NASA (Langley Res Center). Some of the material in this paper is taken from a lecture given at the Pontifical Academy of Sciences (2006) and the Bjerkenes lecture given at the AGU (2006).

References

1. Andreae, M.O., Jones, C.D., and Cox, P.M., Strong present-day aerosol cooling implies a hot future, *Nature*, **435**, 1187–1190 (2005).
2. Charlson, R.J., Lovelock, J.E., Andreae, M.O., and Warren, S.G.,Correspondence re. sulphate aerosols & climate, *Nature*, **340**, 437–438 (1989)
3. Chung, C.E., Ramanathan, V., and Jeffrey T. Kiehl, Effects of the South Asian absorbing haze on the northeast monsoon and surface-air heat exchange, *J. Climate*, 2462–2476 (2002).
4. Chung, C.E., Ramanathan, V., Kim, D., and Podgorny, I.A., Global anthropogenic aerosol direct forcing derived from satellite and ground-based observations, *J. Geophys. Res.*, **110** D24207, doi:10.1029/2005JD00635 (2005).
5. Crutzen, P.J. and Ramanathan, V., The parasol effect in climate, *Science*, **302**, 1679–168 (2003).
6. Harrison, E.F., Minnis, P., Barkstrom, B.R., Ramanathan, V., Cess, R.D., and Gibson, G.G. Seasonal variation of cloud radiative forcing derived from the earth radiation budget experiment, *J. Geophys. Res. Atmospheres*, **95**(18), 687–18,703 (1990).
7. IPCC Climate Change 2007: The Physical Science Basis. Summary for Policy Makers. Contribution of WG1 to the fourth assessment report. IPCC Secretariat, Geneva.
8. Huebert, B.J., Bates, T., Russell, P.B., Shi, G., Kim, Y.J., Kawamura, K., Carmichael, G., and T (2007).
9. Nakajima, An overview of ACE-Asia: strategies for quantifying the relationships between Asian aerosols and their climatic impacts, *J. Geophys. Res.*, **108**(D23), 8633, doi:10.1029 2003JD003550 (2003).

10. Kaufman, Y.J., Tanré, D., and Boucher, O., A satellite view of aerosols in the climate system, *Nature*, **419**, 215–223 (2002).
11. Lau, W.M et al, Aerosol-hydrological cycle research: a new challenge for monsoon climate research. Submitted to *Bulletin of the American Meteorological Society* (2007).
12. Lovelock, J.E. and Margulis, L., Atmospheric homeostasis by and for the biosphere: the Gaia hypothesis, *Tellus*, **26**, 2–10 (1974).
13. Meehl, G.A., Arblaster, J.M., and Collins, W.D., Effects of black carbon aerosols on the Indian monsoon, *J. Climate*, accepted (2007).
14. Menon, S., Hansen, J.E., Nazarenko, L., and Luo, Y., Climate effects of black carbon aerosols in China and India, *Science*, **297**, 2250–2253 (2002).
15. Mitchell, T.D. and Jones, P.D., An improved method of constructing a database of monthly climate observations and associated high-resolution grids, *Int. J. Climatology*, **25**, 693–712 (2005).
16. O'Dowd, C.D., Aalto, P., Hameri, K., Kulmala, M., and Hoffman, T., Atmospheric particles from organic vapors, *Nature*, **406**, 49–498 (2002).
17. Ramanathan, V., Cess, R.D., Harrison, E.F., Minnis, P., Barkstrom, B.R., Ahmad, E., and Hartmann, D., Cloud-radiative forcing and climate: results from the earth radiation budget experiment, *Science*, **243**: 57–63 (1989).
18. Ramanathan, V. et al., The Indian Ocean experiment: an integrated assessment of the climate forcing and effects of the Great Indo-Asian Haze, *J. Geophys. Res. Atmospheres*, **106**, (D 22), 28,371–28,399 (2001).
19. Ramanathan, V., Chung, C., Kim, D., Bettge, T., Buja, L., Kiehl, J.T., Washington, W.M., Fu, Q., Sikka, D.R., and Wild, M., Atmospheric brown clouds: impacts on South Asian climate and hydrological cycle. *PNAS*, Vol. 102, No. 15, 5326–5333 (2005).
20. Ramanathan et al, Atmospheric brown clouds: hemispherical and regional variations in long range transport, absorption and radiative forcing, Accepted in *J Geophys Res.* (2007).
21. Rotstayn, L.D. and Lohmann, U., Tropical rainfall trends and the indirect aerosol effect, *J. Climate*, **15**, 2103–2116 (2002).
22. Satheesh, S.K. and Ramanathan, V., Large differences in tropical aerosol forcing at the top of the atmosphere and earth's surface, *Nature*, **405**, 60–63 (2000).
23. Stanhill, G. and Cohen, S., Global dimming: a review of the evidence for a widespread and significant reduction in global radiation with discussion of its probable causes and possible agricultural consequences, *Agric. Forest Meteorol.*, **107**, 255–278 (2001).
24. Wild, M., Gilgen, H., Roesch, A., Ohmura, A., Long, C., Dutton, E., Forgan, B., Kallis, A., Russak, V., and Tsvetkov A., From dimming to brightening: decadal changes in solar radiation at the earth's surface, *Science*, **308**, 847–850 (2005).

Freshly Formed Aerosol Particles: Connections to Precipitation

Binod Pokharel, Jefferson R. Snider, and David Leon

Abstract Aerosol particle measurements made onboard the Wyoming King Air research aircraft during the summer and winter of 2006 are examined in the context of aerosol particle source and sink processes. These measurements were obtained over the North-Eastern Pacific (off the coast of northern California) and over a high-elevation valley in southeastern Wyoming. Our focus is on measurements of the concentration of recently formed particles with diameters between 0.003 and 0.015 μm; these are supplemented with airborne radar observations of precipitation. The summer/marine measurements suggest that recently formed particles appear in response to removal of larger particles by precipitation. The winter/continental measurements demonstrate that the recently formed particles are scavenged via their interaction with precipitation. It appears that attachment to ice crystals is the dominating sink process in this context. Emphasis on these recently formed particles, and their interaction with precipitation, is motivated by recognition that both their genesis and removal are linked to the hydrologic cycle.

Introduction

Particle production in the remote atmosphere occurs sporadically but is thought to be a dominating source which balances, on average, aerosol removal processes. Some studies have demonstrated that the genesis of new particles (hereafter ultrafine nuclei or UFN) is indirectly coupled to precipitation via the reduction in aerosol surface area (SA) by precipitation (Covert et al. 1992). The direct impact of precipitation (and cloud) on the UFN via their irreversible attachment to hydrometeors (Martin et al. 1980) has been less frequently studied. This study reports on airborne measurements which shed new light on both processes.

University of Wyoming Department of Atmospheric Science

C.D. O'Dowd and P.E. Wagner (eds.), *Nucleation and Atmospheric Aerosols*, 484–488.

Measurements

Aerosol were measured on the Wyoming King Air with an ultrafine condensation particle counter (UCPC), a condensation particle counter (CPC), and a passive cavity aerosol spectrometer probe (PCASP). Laboratory tests show that these instruments measure particles of diameter larger than 0.003, 0.015, and 0.12 µm, respectively. The UFN are defined operationally as the difference between concentrations measurements made by the UCPC and CPC. These measurements were complimented by observations from the Wyoming Cloud Radar, which used an upward- and a downward-looking antenna to obtain vertical cross sections of radar reflectivity and Doppler velocity through the depth of the cloud layer with the exception of a 200 m "dead-zone" centered on the aircraft. The minimum detectible signal for the WCR, roughly –25 dBZ at 1 km, is sufficient to detect cloud droplets at liquid water contents down to ~ 0.1 g m^{-3} at typical cloud-droplet number concentrations, or to detect drizzle size (~100 µm radius) droplets at concentrations of 0.1 L^{-1}.

Aerosol Scavenging by Precipitation

Vertical cross sections of radar reflectivity are shown in Figures 1 and 2. The summer/marine case (Figure 1) was observed 200 km west of the northern California coast; heavy drizzle is evident in the drizzle core observed at ~19:47 UTC. The Wyoming King Air made four passes through the drizzle cell at a range of altitudes (all within the marine boundary layer). Aerosol measurements were obtained both at 100 m altitude, when using the radar to acquire the data presented in Figure 1, and in the same air mass at 400 m about 10 min later. The aerosol measurements made at 400 m were taken to be representative of air that had been processed by the drizzle core. To avoid artifacts due to drizzle splashing we selected the aerosol data where the cloud droplet concentration was less than 1 cm^{-3} and where larger hydrometeor concentrations were less than 1 L^{-1}. We refer to the region at 400 m altitude as "outflow" and contrast PCASP measurements made there to measurements made at 100 m altitude (the "inflow" region). The degree of scavenging between the inflow and outflow regions is presented in the lower-right panel of Figure 1. The lower-right panel of Figure 2 also portrays precipitation affecting the aerosol, but in this case it is UFN which are depleted between the upwind and downwind sides of a winter orographic cloud.

Vertical Profiles

Representative vertical profiles of conserved temperatures and aerosol (expressed as a count mixing ratio) are presented in Figures 3 and 4 for the summer/marine and winter/continental campaigns, respectively. The marine profile starts at 30 m above

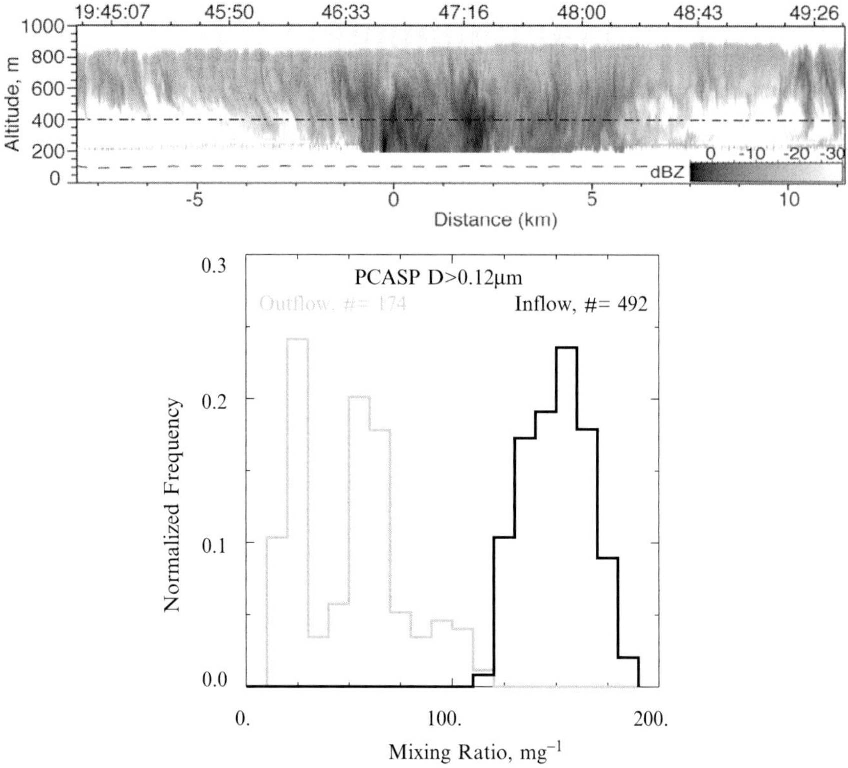

Figure 1 **(Upper)**–Radar reflectivity cross section from the marine/summer campaign (20060629). A drizzle cell is centered at ~19:47 UTC. Dashed-dot line indicates the aircraft track at 400 m (outflow level). The flight leg at the "inflow" level is indicated by the dashed line. **(Lower)**–Frequency distributions of cumulative PCASP count mixing ratio measured in the inflow air entering the drizzle core at 100 m altitude and in outflow from the drizzle core measured at 400 m altitude

the eastern Pacific Ocean and the continental profile starts at 2700 m; approximately 600 m above the Laramie River Valley. Evident in both the marine and continental profiles is a well-mixed boundary layer and an overlying free tropospheric air mass. Both soundings were conducted during daytime while avoiding clouds. The marine profile (Figure 3) shows abundant UFN at ~550 m, within the boundary layer, and values of SA < 5 μm^2 cm^{-3} throughout the boundary layer. In the free tropospheric portion of the continental profile (Figure 4), the UFN coexisted with values of SA that ranged between 10 and 80 μm^2 cm^{-3}. Within the continental boundary layer there were intermediate values of SA (~25 μm^2 cm^{-3}) and no UFN. Of the 150 profiles available from the two campaigns, 18 exhibited UFN in the boundary layer and 15 exhibited UFN in the free troposphere.

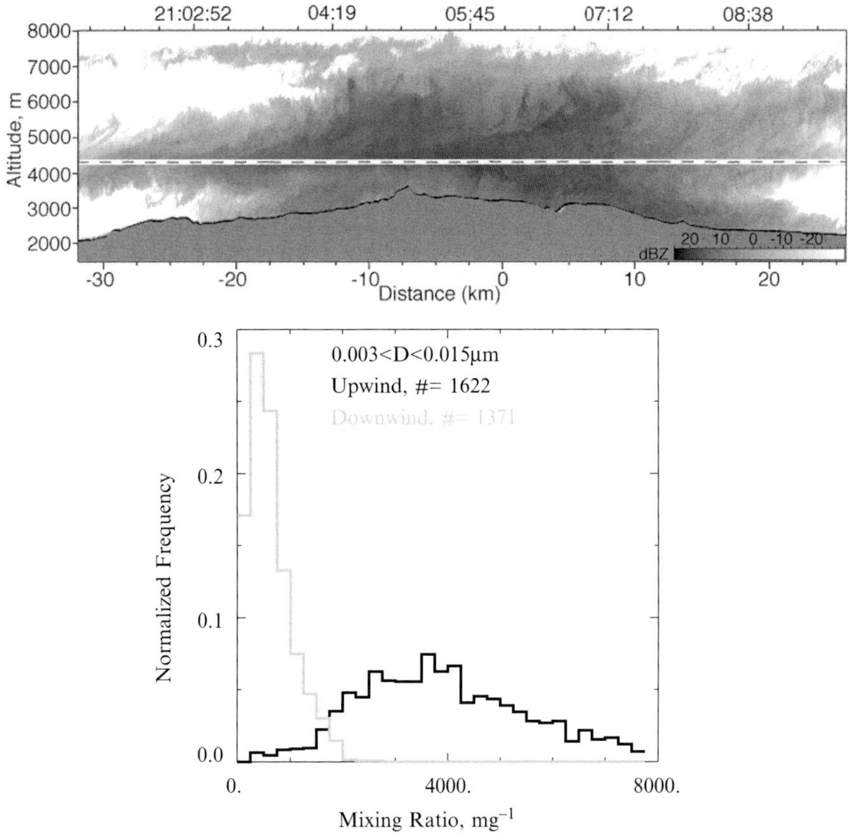

Figure 2 (Upper)–Radar reflect-ivity cross section from the continental/winter campaign (20060118). Dashed line indicates the flight level. Gap in data just above/below the flight level is the result of the ~200 m dead zone of the WCR. Temperature along the flight track is −14°C. The wind is from left to right at 22 m/s; **(Lower)**–Frequency distributions of UFN count mixing ratio measured upwind and downwind of the orographic cloud during flight legs conducted parallel to the mountain barrier

Summary

We present observations of UFN within both marine and continental boundary layers. Preliminary analyses suggest that precipitation-scavenging of boundary-layer accumulation mode and coarse mode aerosol is a necessary condition for UFN formation. Using an airborne radar we also present views of precipitation structures, and their boundaries. These data are enabling refined analyses of aerosol scavenging as a function of aerosol size, precipitation depth, and precipitation rate.

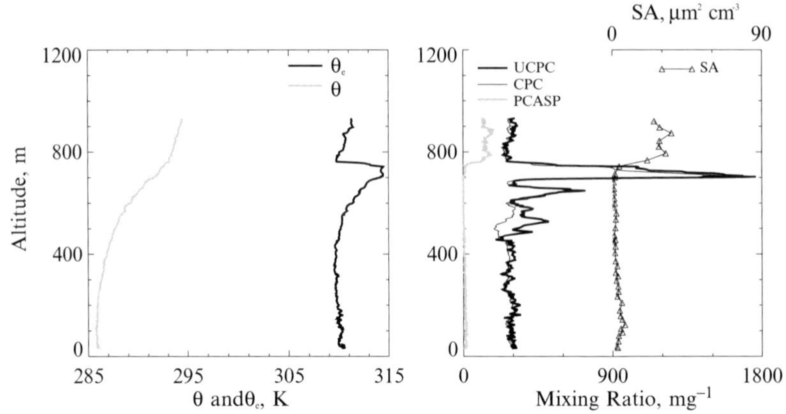

Figure 3 Summer/marine profile (20060604), **(left)** conserved temperatures indicating marine boundary layer top at ~700 m, **(right)** aerosol showing UFN at ~550 m and depletion of aerosol surface area throughout the boundary layer

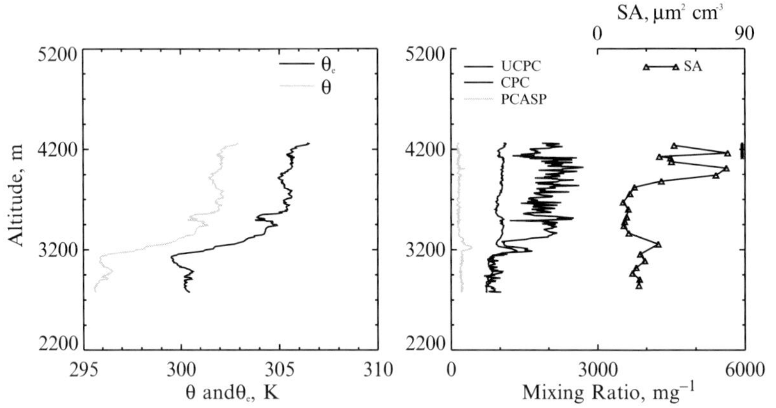

Figure 4 Winter/continental profile (20060126), (**left**) conserved temperatures indicating boundary layer top at ~3100 m, (**right**) aerosol showing UFN in the free troposphere

References

Covert, D.S., Kapustin, V.N., Quinn, P.K., and Bates, T.S., New particle formation in the marine boundary layer, *J. Geophys. Res.*, **97**, 20,581–20,589 (1992).

Martin, J.J., Wang, P.K., and Pruppacher, H.R., A theoretical determination of the efficiency with which aerosol particles are collected by simple ice crystal plates, *J. Atmos. Sci.*, **37**, 1628–1638 (1980).

Relative Humidity Dependence of Aerosol Optical Properties and Direct Radiative Forcing in the Surface Boundary Layer of Southeastern China

Y.F. Cheng[1,2], A. Wiedensohler[1], H. Eichler[1,3], Y.H. Zhang[2], W. Birmili[1], E. Brüggemann[1], T. Gnauk[1], M. Tesche[1], H. Herrmann[1], A. Ansmann[1], D. Althausen[1], R. Englemann[1], M. Wendisch[1,3], H. Su[2], and M. Hu[2]

Abstract In a numerical study, influences of relative humidity (RH) on particle optical properties and direct radiative forcing of the surface aerosol at Xinken (XK, in Pearl River Delta of China, 2004) are investigated based on the observed particle hygroscopic growth ($f_g(D_p,RH)$) and relevant physical and chemical properties. The model is validated by comparing the simulated extinction coefficient with the one measured by a Raman LIDAR. At a wavelength (λ) of 550 nm, the scattering coefficient increases with a factor of 1.54 ± 0.06 and 2.31 ± 0.20 from RH 30% to 80% and 90%, respectively. This ratio is mainly controlled by $f_g(D_p,RH)$, and generally increases with increasing λ and decreases with increasing particle effective diameter. From RH 30% to 90%, the average single scattering albedo changes from 0.77 to 0.88, 0.91 to 0.97, and 0.86 to 0.94 for internal, respectively external mixtures, and the retrieved mixing state of elemental carbon (EC). The assumption that the absorption coefficient does not change upon humidification alone can introduce an overestimation of ω_0 of 0.02 (~ 5%) at RH 90% for an internal mixture, however, accounting for the EC mixing state at XK, this error can only be about 0.3%. The upscatter fraction is estimated as 0.233 at RH 30% with a 30% decrease to 0.191 at RH 90%. The estimation of aerosol direct radiative forcing (ΔF_R) in the surface boundary layer at XK is strongly dependent on RH as well as the EC mixing state. The critical single scattering albedo at XK is estimated to be about 0.77–0.80. Assuming an internal or coated mixing state of EC, no pronounced ΔF_R is observed for RH < 60%, whereas a cooling effect is observed for RH > 60%. Accounting for the EC mixing state at XK, the effect of ΔF_R is cooling. Over 40% of this cooling effect is contributed by water at RH 80%, and ΔF_R at RH 90% exceeds that at RH 30% by about a factor of 2.7. This RH dependence is increased with increasing λ.

Keywords Aerosol optical property, direct radiative forcing, relative humidity, PRD of China

[1] *Leibniz Institute for Tropospheric Research, Leipzig, Germany*

[2] *College of Environmental Sciences, Peking University, Beijing, China*

[3] *Now at Institute for Atmospheric Physics, Johannes Gutenberg-University, Mainz, Germany*

C.D. O'Dowd and P.E. Wagner (eds.), *Nucleation and Atmospheric Aerosols*, 489–493.

Introduction

One key parameter governing the aerosol optical properties (AOPs) is the relative humidity (RH), because most of the parameters, which control the AOPs, such as particle size distribution and relative refractive index, are functions of RH. The RH dependence of AOPs needs to be addressed to better understand the aerosol climate forcing at ambient condition.

Methodology

As part of the Observation Experiment for Regional Air Quality in Pearl River Delta (PRD) of China, ground-based measurements of aerosol optical, chemical, and physical properties were conducted from 4 Oct to 5 Nov of 2004 (278–310 DOY (Day of Year)) at Xinken (XK, 22.6° N, 113.6° E), which is located downstream (near the mouth) of the Pearl River on the northwest–southeast axis between Canton and Hong Kong. The measured aerosol properties include number size distributions and scattering and absorption coefficients of the dry aerosol, particle number size distributions at controlled RHs, as well as the size-segregated aerosol mass and black carbon concentrations from the chemical analyzed impactor samples.

In the numerical study, a three-component aerosol optical model was assumed for the simulation of the particle optical properties. Except water, the dry particle material was classified into light absorbing (elemental carbon (EC)) and non-light-absorbing components (such as sulfate, ammonium, nitrate, organic carbon (OC), and the chemical undetermined). Four kinds of EC mixing states were used to describe the particles at XK. The first two are the completely external and internal mixtures. The EC mixing states retrieved from a dry optical closure study was also applied, as well as the core-shell mixture. A Mie code was used to calculate the AOPs at a series of RHs. The uncertainties of all the modeling results have been evaluated with Monte Carlo simulations.

Size-segregated aerosol hygroscopic growth factors have been determined with the measured particle number size distributions at dry conditions and controlled RHs. These hygroscopic data can be used to calculate the water uptake and the particle number size distributions at any defined RH. The input parameters for the Mie code were derived according to the volume fractions according to the size-segregated aerosol mass and black carbon concentrations from the chemical analyzed impactor samples and the hygroscopic growth factors.

In our case, the humidity dependence of AOPs can be characterized by defining an humidification factor $\xi_{AOP}(RH)$ for the same aerosol mixing state:

$$\xi_{AOP}(RH) = AOP^{wet}(RH) / AOP^{dry}(RH_0). \quad (1)$$

Here, $RH_0 = 30\%$ is set as dry condition.

The aerosol direct radiative forcing at the surface boundary layer has been evaluated with one simple thin layer aerosol radiative model.

Results and Discussion

The model is validated by comparing the simulated extinction coefficient (σ_{ep}) with the measured one by a Raman LIDAR. As shown in Figure 1, they generally agree within 10%, except for an outlier on 298.96 DOY, when a high concentration of pollutants accumulated at the ground due to calm wind during the whole night.

Figure 2 represents the humidification factors of extinction and scattering coefficients ($\xi_{ep}(RH)$ and $\xi_{sp}(RH)$). At a wavelength λ of 550 nm, the scattering coefficient increases with a factor of 1.54 ± 0.06 and 2.31 ± 0.20 from RH 30% to 80% and 90%, respectively. This ratio is mainly controlled by $f_g(D_p, RH)$, and generally increases with increasing λ and decreases with increasing particle effective diameter.

From RH 30% to 90%, the average single scattering albedo (ω_0) changes from 0.77 to 0.88, 0.91 to 0.97, and 0.86 to 0.94 for internal and external mixture, and the retrieved mixing state of the EC at XK, respectively (see Figure 3). The assumption that the absorption coefficient does not change upon humidification alone can introduce an overestimation of ω_0 of 0.02 (~5%) at RH 90% for an internal mixture, however, accounting for the EC mixing state at XK, this error can only be about 0.3%.

The upscatter fraction is estimated as 0.233 at RH 30% with a 30% decrease to 0.191 at RH 90%. The estimation of aerosol direct radiative forcing (ΔF_R) in the surface boundary layer at XK is strongly dependent on RH as well as the EC mixing

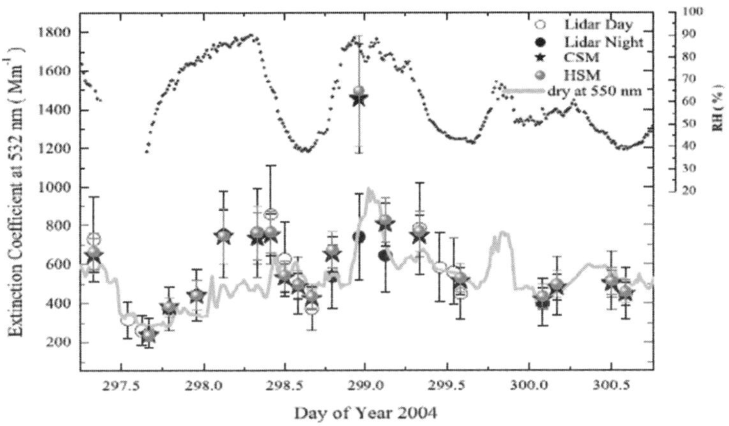

Figure 1 Time series of modeled and LIDAR-derived particle extinction coefficients (σ_{ep}) at 532 nm. CSM and HSM refer to σ_{ep} calculated with the coated spherical model and the homogeneous spherical model, respectively. Error bars were applied according to the model and measurement uncertainties (99% confidence level). The gray line represents σ_{ep} derived from measurements at dry conditions. In addition, the relative humidity is shown

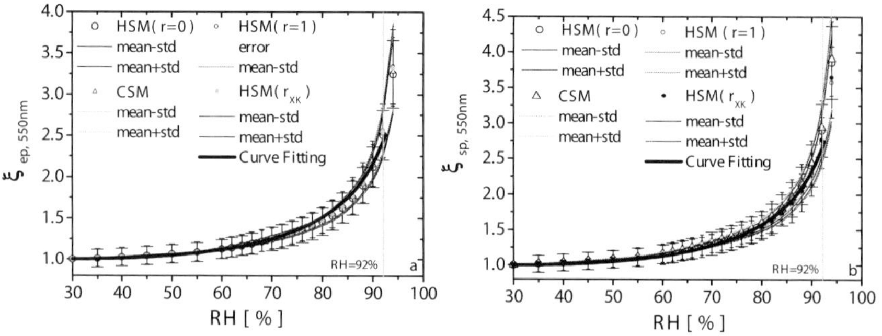

Figure 2 The humidification factors of σ_{ep} and scattering coefficient σ_{sp} at wavelength of 550 nm with four aerosol mixing states. *r* is defined as the external mixing ratio of elemental carbon (EC) in the total mass of EC. *r* = *0* refers to the completely internal mixture whereas *r* = *1* is the external mixture. r_{XK} refers to the retrieved mixing ratio of EC at Xinken with the dry optical closure study

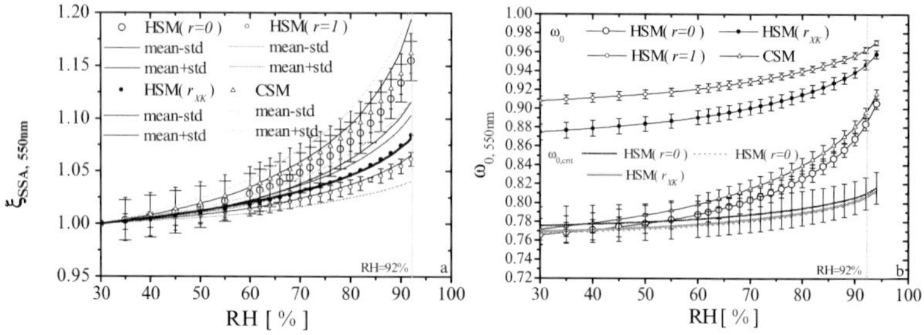

Figure 3 Humidification factor of the single scattering albedo (*SSA*, ω_0) at wavelength of 550 nm with four aerosol mixing states. The point is the mean value, and the standard deviations are presented by the shot dot or solid line. The bar indicates the uncertainties with one relative standard deviation (s) at 68% confidence level derived from the Monte Carlo simulation. In (b) the average critical single scattering albedo ($\omega_{0,crit,550nm}$) are also presented

state. The critical single scattering albedo ($\omega_{0,crit,550nm}$) at XK is estimated to be about 0.77–0.80 with an assumed surface albedo of 0.23. Assuming an internal or coated mixing state of EC, no pronounced ΔF_R is observed for RH < 60%, whereas a cooling effect is observed for RH > 60%. Accounting for the EC mixing state at XK, the effect of ΔF_R in the surface boundary layer is cooling. Over 40% of this cooling effect is contributed by water at RH 80%, and ΔF_R at RH 90% exceeds that at RH 30% by about a factor of 2.7. This RH dependence increases with increasing λ.

Acknowledgments This study is mainly supported by China National Basic Research and Development Programs 2002CB410801 and 2002CB211605. We also acknowledge Gottlieb Daimler and Karl Benz Foundation for the scholarship supporting the study of Y.F. Cheng in Leibniz Institute for Tropospheric Research (*IfT*, Leipzig, Germany).

References

1. Charlson, R.J., Schwartz, S.E., Hales, J.M. Cess, R.D., Jr., J.A.C., Hansen, J.E., Hofmann, D.J., *Science*, **255**, 423–430 (1992).
2. Cheng, Y.F., Eichler, H., Wiedensohler, A., Heintzenberg, J., Zhang, Y.H., Hu, M., Herrmann, H., Zeng, L.M., Liu, S., Gnauk, T., Brüggemann, E., and He, L.Y., *J. Geophys., Res.*, **111**, D20204 (2006).

The Effect of Including Aerosol Nucleation and Coagulation in a Global Model

Minghuai Wang and Joyce E. Penner

Abstract In this study, we evaluate how the predicted aerosol size information and the mixing states between sulfate aerosol and non-sulfate aerosols from our aerosol microphysics model affect cloud properties and aerosol indirect effects. Our results show that the use of the predicted sulfate aerosol number decreases cloud droplet number concentrations over regions with high DMS emission. The treatment of mixing between sulfate aerosol and non-sulfate aerosol causes a decrease in the cloud droplet number concentration over remote marine regions but an increase in the cloud droplet number concentrations over continental regions. Overall, the effect of anthropogenic aerosols is increased when the mixing between sulfate and non-sulfate aerosols is included

Keywords Aerosol microphysics, aerosol indirect effect, cloud droplet nucleation

Introduction

Atmospheric aerosol is an important component of the global climate system. One of the important ways for aerosol to affect climate is to modify cloud properties by acting as cloud condensation nuclei (CCN). This "aerosol indirect effect," is one of the largest uncertainties in our understanding of climate. One major challenge is to determine number concentration and ability of aerosol particles to act as CCN. Size information and the mixing state of aerosol components are needed to decide whether a given aerosol particle will act as a CCN.[1,2] The assumed size distribution used in models of the aerosol indirect effect contributes a large fraction of the total uncertainty associated with estimates of the first indirect forcing or "albedo effect".[3] In the past, global aerosol models did not predict the size-resolved aerosol composition and assumed size distributions were used to calculate the nucleated

Department of Atmospheric, Oceanic, and Space Science, University of Michigan, Ann Arbor, MI, 48109-2143, USA

C.D. O'Dowd and P.E. Wagner (eds.), *Nucleation and Atmospheric Aerosols*, 494–498.

cloud droplet number. In addition, most global aerosol models either assume that different aerosol components are externally mixed or are completely internally mixed. These assumptions cannot capture the complexity of the aerosol system. In recent years, however, detailed aerosol microphysics models have been added to global aerosol models,[4,5] which enable the prediction of more realistic size distributions and mixing states for cloud droplet nucleation parameterizations. As of yet, however, there are no studies that evaluate how these predicted aerosol properties affect cloud properties and aerosol forcing.

We have recently developed a global aerosol model that includes the prediction of the size distribution for pure sulfate aerosol as well as the interaction of sulfate and the non-sulfate aerosol components.[5] In this study, we investigate how the size and mixing state of the aerosol affect the cloud droplet number, cloud optical depth, and anthropogenic aerosol indirect effect.

Models and Methods

Aerosol fields and cloud fields from a coupled climate and aerosol model are used here (Wang et al., in preparation). The coupled model has two components: the NCAR CAM3 general circulation model and the IMPACT aerosol model. The latter includes an aerosol microphysics module to simulate sulfate aerosol dynamics (nucleation, condensation, coagulation) and its interactions with primary emitted non-sulfate aerosols. Two modes (nucleation and accumulation) and two moments (mass and number) are used to describe the pure sulfate aerosol. Non-sulfate aerosols are assumed to follow predefined background size distributions (Table 1 in Liu et al.[5]). Sulfuric acid gas is produced from the gas-phase oxidation of sulfur dioxide. It can nucleate to form new sulfate particles in the nucleation mode or condense on preexisting sulfate or non-sulfate particles. The hydrophilic and hydrophobic properties of non-sulfate aerosols are determined by the amount of sulfate coating. The aqueous production of sulfate is equally distributed among hygroscopic aerosol particles that are larger than 0.05 μm in radius. Aerosol fields from the coupled model reproduce well the results from the IMPACT standalone version driven by reanalysis data.[5]

We use an offline calculation similar to that used by Chen and Penner[3] to investigate how the size and mixing state of the aerosol affects the cloud properties at the top of the cloud. A cloud droplet nucleation parameterization based on Abdul-Razzak and coauthor's work[6] is used in this study. It combines the treatment of multiple aerosol types and a sectional representation to deal with arbitrary external mixing and arbitrary size distributions. Five categories of aerosols are externally mixed in our model: pure sulfate, biomass burning OM/BC with coated sulfate, fossil fuel OM/BC with coated sulfate, sea salt with coated sulfate and dust with coated sulfate. Within each category of aerosols, the coated sulfate is treated as an internal mixture. The size spectrum for each aerosol category is divided into 30 bins. The size distribution for the non-sulfate aerosol is that assumed in our previous study.[5] We limit the effects of aerosols on clouds to low and warm clouds only (T > 273 K, and P > 640 mb).

In our reference case ("REF"), the size distribution of pure sulfate aerosol is that prescribed in Chuang et al.[7] and the number of sulfate aerosol particles is calculated based on the predicted mass concentration. In our second case ("SULN"), the number of sulfate particles is that predicted from the model. The difference between "REF" and "SULN" indicates the effects of predicting the sulfate aerosol number on cloud properties. In order to study the effects of the mixing state between sulfate and non-sulfate aerosols, we included a third case ("NOMIX"), in which we remove all sulfate aerosol coated on non-sulfate aerosol and assumed this sulfate mass is part of the pure sulfate aerosol. Each case included two calculations: one for aerosol from present day emissions, and one for aerosol from preindustrial emissions. Below we examine the effect in January.

Results and Discussion

First we examine how the predicted sulfate aerosol number affects the cloud droplet number and cloud optical properties. For aerosols from present day emissions, in the "REF" case, large cloud droplet number concentrations correspond to regions with high sulfate production, such as regions with high pollution and high DMS (Dimethyl Sulfide) emission. But in the "SULN" case, we do not always get large droplet number concentrations in regions with high sulfate production. In high DMS emission regions, nucleation is not a favored process[5] and the sulfate gas condenses onto preexisting aerosols. So in those regions most of the sulfate is condensed on sea salt particles. Our model predicts less-pure sulfate aerosol number in the accumulation mode in "SULN" case than the one in "REF" case, which leads to smaller cloud droplet number concentrations over high latitude regions of the southern hemisphere.

For aerosols from the preindustrial emissions, the difference between the SULN and REF cases is similar to that for the present day calculation. The difference between the present day calculation and preindustrial calculation determines the effects on cloud properties from anthropogenic aerosol. The "SULN" case produces a larger anthropogenic effect than does the "REF" case. The global average change in the effective radius at cloud top is −1.17 μm in the "SULN" case, but it is only −.96 μm in the "REF" case. Figure 1a) shows the spatial pattern of the difference between the change in effective radius caused by anthropogenic aerosols from the "SULN" and "REF" cases (SULN–REF). As shown there, over most continental regions, anthropogenic aerosol produces a larger decrease in droplet effective radius in "SULN" than in REF. But over the east coast of East Asia and over the east coast of North America, the opposite is true. For low and warm clouds, anthropogenic aerosols cause a global average increase in the cloud optical depth by 3.39 in the "SULN" case and by 1.86 in the "REF" case. The larger increase in cloud optical depth in the "SULN" case implies a larger first indirect forcing.

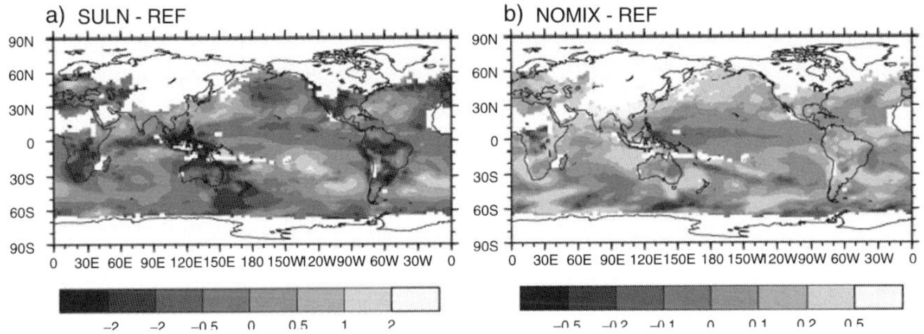

Figure 1 Difference in the change in cloud droplet effective radius caused by anthropogenic aerosol, a) SULN–REF, and b) NOMIX–REF (unit: μm)

Next we examined the effects of the predicted mixing state between sulfate and non-sulfate aerosols. The mixing between sulfate and non-sulfate aerosol has two effects. One is to increase the hygroscopic properties of non-sulfate aerosol, which causes an increase in the cloud droplet number. The other is to decrease the amount of pure sulfate aerosols, which causes a decrease in the cloud droplet number. Our calculations show that, over most regions with high pollution, the first effect dominates, and the "NOMIX" case produces a smaller cloud droplet number concentration than the "REF" case. But over remote marine regions, the second effect dominates and the NOMIX case produces a larger cloud droplet number concentration than the "REF" case. This is because sea salt aerosol in these regions is active as CCN so that the removal of the mixed sulfate on these particles causes only a small effect. If we examine the difference between the present day and preindustrial droplet concentrations, the effect of anthropogenic aerosols is smaller in the NOMIX case than that in the REF case. The change in the global average droplet effective radius is −0.80 μm in the NOMIX case compared to −0.96 μm in the REF case. Figure 1b) shows the spatial pattern of the difference in effective radius caused by anthropogenic aerosol between the NOMIX and REF cases (NOMIX–REF). The effect of anthropogenic aerosol is to increase the low and warm cloud optical depth by 1.64 in the "NOMIX" case and by 1.86 in the "REF" case.

Our results show that the effect of including the predicted sulfate aerosol number is to decrease the cloud droplet number concentration over regions with high DMS emissions and to increase the effect of anthropogenic aerosol on cloud optical depth. The treatment of mixing between sulfate aerosol and non-sulfate aerosol causes a decrease in the cloud droplet number concentration over remote marine regions but an increase in the cloud droplet number concentrations over continental regions. Overall, the effect of anthropogenic aerosols is increased when the mixing between sulfate and non-sulfate aerosols is included. The results reported here will be affected by the choice of assumed size distributions for the aerosols in our model.

Acknowledgments We thank Dr. Xiaohong Liu at PNNL for helpful discussions regarding the IMPACT aerosol model, and Dr. Yang Chen at JPL for help in implementing the cloud nucleation parameterization. NASA grant NNG04GC01G provided funding for this project.

References

1. Dusek, U., Frank, G.P., Hildebrandt, L., Curtius, J., Schneider, J., Walter, S., Chand, D., Drewnick, F., Hings, S., Jung, D., Borrmann, S., and Andreae, M.O., *Science*, **312**, 1375–1378 (2006).
2. Nenes, A., Charlson., R.J., Facchini, M.C., Kulmala, M., Laaksonen, A., and Seinfeld, J.H., *Geophys. Rev. Lett.*, **29**, doi:10.1029/2002GL015295 (2002).
3. Chen, Y., and Penner, J.E., *Atmos. Chem. Phys.*, **5**, 2935–2948 (2005).
4. Adams, P.J. and Seinfeld, J.H., J. *Geophys. Rev.*, **107**(D19), 4370–4392(2002).
5. Liu, X., Penner, J.E., and Herzog, M., *J. Geophys. Rev.*, **110**, D18206, doi:10.1029/2004JD005674 (2005).
6. Abdul-Razzak, H. and Ghan, S.J., *J. Geophys. Rev.*, **107**, doi:10.1029/2001JD000483 (2002).
7. Chuang, C.C., J.E. Penner, K.E. Taylor, A.S. Grossman, and J.J. Walton, An assessment of the radiative effects of anthropogenic sulfate, *J. Geophys. Res.*, **102**, 3761–3778 (1997).

On the Estimated Impact of Anthropogenic Cloud Condensation Nuclei (CCN) in the Region of Mexico City on the Development of Precipitation

M.L. Frias-Cisneros and D.G. Baumgardner

During an intensive measurement campaign in March 2006 at the Cortez Pass (4,000 m, a.s.l.), 60 km SE of Mexico City, cloud condensation nuclei (CCN) were measured with a thermal gradient CCN counter at operated at supersaturations that ranged from 0.2% to 0.7%. Aerosol size distributions were measured with a scanning mobility particle sizer (SMPS). The source of the air masses at the research site was estimated from wind trajectories computed from NCEP analysis. The CCN concentrations depend on the direction of the air masses with the highest concentration related to Mexico City when winds were from the NW and lowest when coming from Cuernavaca in the SW. The average maximum CCN concentration when air is coming from the direction of Mexico City is 1,500 cm^{-3} at a supersaturation of 0.5%.

In this work, in order to understand the influence of anthropogenic emissions from Mexico City, we are using a cloud model with explicit microphysics to look at the sensitivity of droplet formation and the development of precipitation to the properties of the CCN that were measured at our research site. The presentation will discuss the results of these simulations and the general characteristics of the CCN in the Mexico City region.

Centro de Ciencias de la Atmósfera, UNAM, Ciudad Universitaria, C.P. 04510, México D.F.

C.D. O'Dowd and P.E. Wagner (eds.), *Nucleation and Atmospheric Aerosols*, 499.

Emission of Aerosol Black Carbon in the Atmosphere of Chhattisgarh

N.K. Jaiswal[1], K.S. Patel[1], Saathoff[2], and U. Schurath[2]

Aerosol Black Carbon (ABC), is the major anthropogenic component of atmospheric aerosols. Being produced by all combustion processes, it is chemically inert and hence has a significantly long lifetime. Real time measurements of the mass concentration of ambient ABC have been made of exists over Chhattisgarh. Annual variations show highest concentration during dry winter months of December–February and lowest values during the monsoon months of June–September. Apportionment to total aerosol mass concentration shows that about 8–10% share for ABC. The fine particulate ($PM_{2.5}$) and coarse particulate (PM_{10}) samples were collected on preheated quartz fibre filters with Sequential Speciation Sampler (Model 2300 USA).ABC concentration is found to be in the range 5–81 $\mu g cm^{-3}$.The details and implications will be presented.

[1] *School of Studies in Chemistry, Ravishankar Shukla University Raipur, Chhattisgarh*

[2] *Institute of Meteorology and Climate Research (IMK), Atmoshpheric Aerosol Research Division(IMK-AAF), Forschungszentrum Krlshruhe, Germany*

C.D. O'Dowd and P.E. Wagner (eds.), *Nucleation and Atmospheric Aerosols*, 500.

Impact of Saharan Dust on Tropical Cyclogenesis

N. Christina Hsu, Si-Chee Tsay, and K.M. Lau

Abstract Monsoon rainfalls sustain the livelihood of more than half of the world's population. Understanding the mechanism that drives the water cycle and fresh water distribution is highlighted as one of the major near-term goals in NASA's Earth Science Strategy, and the interaction between natural/anthropogenic aerosols, clouds, and precipitation is a critical component of that mechanism. Analyses of the long-term trend of July–August precipitation anomaly for the last 50 years in the twentieth century depict that the largest regional precipitation deficit occurs over the Sahel, where the monsoon water cycle plays an important role. Thus, it is of paramount importance to study how dust aerosols, as well as air pollution and smoke, influence monsoon variability.

The Saharan Air Layer (SAL) occurs during the late spring through early fall and originates as a result of low-level convergence induced by heat lows over the Sahara that lifts hot, dry, dust-laden air aloft into a well-mixed layer that extends up to 500 mb. During years of normal rainfall in Sahel, the midlevel easterly jet (MLEJ) increases positive vorticity, cyclonic shear, and baroclinic instability south of this jet, which favors the development of tropical cyclone (TC) in the eastern Atlantic. This response of the MLEJ is partly due to interaction with SAL. During dry years the SAL is more pronounced and through inhibition of convection and widening the MLEJ decreases shear, and baroclinic and barotropic instability, which are negative influences on TC development. A number of studies have examined the SAL but its impact on cyclogenesis remains inconclusive!

The NASA African Monsoon Multidisciplinary Activities (NAMMA) was conducted during the international AMMA Special Observation Period (SOP-3) of September 2006 to better comprehend the key attributes of the SAL and how they evolve from the source regions to the Atlantic Ocean. This is crucial for understanding its impact on the key processes such as the MLEJ, AEWs, and mesoscale convective systems that determine precipitation over west Africa and TC genesis. Preliminary results, obtained from the synergy of satellite (Deep-Blue) and surface

NASA Goddard Space Flight Center, Greenbelt, MD 20771 USA

C.D. O'Dowd and P.E. Wagner (eds.), *Nucleation and Atmospheric Aerosols*, 501–502.

(SMART-COMMIT) observations, will be presented. The following questions will be discussed: What is the radiative forcing associated with the SAL and possible thermodynamic and dynamic consequences such as strengthening of the marine boundary layer or the impact on AEWs? How do the optical and physical characteristics of the dust in the SAL evolve from the continent to the marine environment? What is the possible impact of dust on microphysical processes (e.g., nucleation)?

Aerosol Distributions over Europe: A Regional Model Evaluation

Bärbel Langmann[1], Saji Varghese[1], Colin O'Dowd[1], and Elina Marmer[2]

Abstract This paper summarizes an evaluation of the regional scale atmospheric climate-chemistry/aerosol model REMOTE which has been extended recently by a microphysical aerosol module. Applications over Europe will be presented in comparison with available measurements (e.g., from the EMEP network) focusing on the European distribution and variability of primary and secondary aerosols.

Keywords Tropospheric aerosols, regional modeling

Introduction

Tropospheric aerosols have significant effects on human health, environment, and climate. An improved understanding of anthropogenic and natural emission sources, secondary aerosol formation, modification of the aerosol chemical composition, and size distribution is essential for efficient emission reduction policies to improve air quality. Moreover, aerosol effects on climate due to direct and indirect aerosol radiative forcing are mainly determined by the atmospheric aerosol chemical composition and size distribution. Due to the relative short residence times, troposheric aerosol distribution shows considerable spatial and temporal variability with inter-annual variability strongly dependent on the prevailing meteorological conditions.

Major efforts have been made in recent years to improve atmospheric aerosol modeling on the global and regional scale. The early three-dimensional models determined only bulk aerosol distributions whereas several new approaches are available today to simulate aerosol chemical composition and size distribution. We have extended a regional online atmosphere-photochemistry model by an aerosol

[1] *Department of Experimental Physics, National University of Ireland, Galway, Ireland*

[2] *JRC Ispra, Italy*

C.D. O'Dowd and P.E. Wagner (eds.), *Nucleation and Atmospheric Aerosols*, 503–506.

microphysical module and utilized it to simulate European wide distributions of sulfate, sea salt (SS), black carbon (BC), primary organic carbon (POC) as well as secondary organic carbon (SOC) over Europe.

Model Setup

We use the regional three-dimensional online climate-chemistry/aerosol model REMOTE (Regional Model with Tracer Extension, http://www.mpimet.mpg.de/en/wissenschaft/modelle/remote.html) [1] which is one of the few regional climate models that determines the physical, photochemical, and aerosol microphysical state of the model atmosphere at every model time step. The physical parameterizations of the global ECHAM-4 model [2] are used for the current study. After emission, photochemical and aerosol trace species undergo transport processes (horizontal and vertical advection, transport in convective clouds, vertical turbulent diffusion) and are removed from the atmosphere by sedimentation, dry and wet deposition. For the determination of aerosol dynamics and thermodynamics we use the M7 module [3,4]. The five aerosol components considered in M7 are sulphate (SO4), BC, organic carbon (OC), SS, and mineral dust (DU). These components have either negligible or low solubility or are treated as an internal mixture of insoluble and soluble compounds. The aerosol size spectrum is subdivided into nucleation, aitken, accumulation, and coarse modes. Each mode can be described by three moments: aerosol number N, number median radius r, and standard deviation σ. Standard deviations are prescribed in M7, so that the median radius of each mode can be calculated from the corresponding aerosol number and aerosol mass, which are transported as 25 tracers. Thus, the total number of prognostic trace species in REMOTE is 63, where 38 of these are included to describe photochemical transformations [1].

Temporally variable anthropogenic emissions of SOx, NOx, NH_3, CO, VOCs, BC, and POC are obtained from the EMEP emission inventory. For primary anthropogenic aerosol emissions, number mean radius and number concentration of the respective size mode is related to the mass concentration based on [4]. In addition to terrestrial biogenic terpene and isoprene emissions from forests [1] we consider isoprene emissions from the ocean based on [5]. For coarse mode SS, we use the same lookup table as described in [4]. The net accumulation sea-spray flux is based on [6] and is used as an organic–inorganic source function for the mixture of POC and SS aerosols. Recent measurements at the Mace Head station [7] have shown that OC contributes a considerable fraction to sea-spray during periods of increased biological activity of the ocean. Further details are given in [8].

REMOTE is applied with 20 vertical layers of increasing thickness between the earth's surface and the 10 hPa pressure level using terrain following hybrid pressure-sigma coordinates. The model domain covers Europe and the North-east Atlantic Ocean. REMOTE is initialized at the first time step using meteorological analysis data of the European Centre for Medium Range Weather Forecast (ECMWF), which are updated at the lateral boundaries every 6 h. A previous version of REMOTE with a bulk aerosol approach has been evaluated extensively [9]. Here we analyze results from spring 2002 to summer 2003 focusing on aerosol chemical composition and size distribution.

Results

Preliminary model results of near surface accumulation mode POC distributions over Europe during winter and summer show significant differences over both the continent and the ocean (Figure 1). During winter the highest concentrations are determined over eastern Europe due to domestic heating when accumulation under high-pressure conditions in the planetary boundary layer takes place. Even higher near surface POC concentrations are found during summer in the marine boundary layer where POC is released together with SS dependent on wind speed [8]. Transport of SS and marine POC from the ocean to continental areas is small. However coastal areas are affected.

Comparisons of measured concentration at the Mace Head station show reasonable agreement with simulated accumulation mode SS and POC concentration. This also holds for the comparison of modeled precipitation with monthly GPCP data and EMEP station data (Figure 2), and measurement data of wind speed and direction.

Outlook

A careful evaluation of the quality of the meteorological and aerosol simulation results is necessary for future application of REMOTE, e.g., SOC formation from terrestrial and marine sources, aerosol–cloud interaction studies, coupled ocean-atmosphere

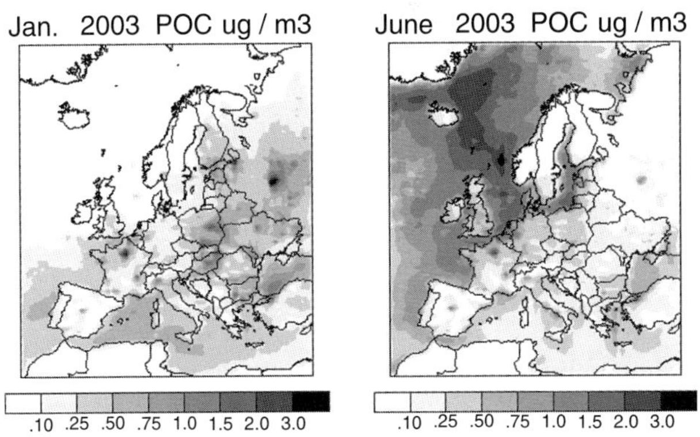

Figure 1 Near surface POC distribution in the accumulation mode as simulated by REMOTE for January and June 2003

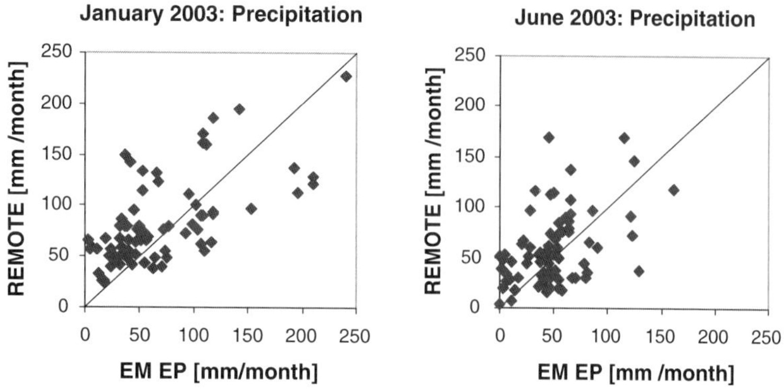

Figure 2 Measured and simulated precipitation during January and June 2003

biogeochemical modeling, or applications over other regions of the Earth. Further evaluation of the REMOTE/M7 simulations over Europe is going on and will be presented on the conference.

References

1. Langmann, B., *Atmos. Environ.*, **34**, 3585–3598 (2000).
2. Roeckner, E., Arpe, K., Bengtsson, L., Christoph, M., Claussen, M., Dümenil, L., Esch, M., Giorgetta, M., Schlese, U., and Schulzweida, U., Report No. 218, Hamburg, Germany: Max Planck Institute for Meteorology (1996).
3. Vignati, E., Wilson, J., and Stier, P., *J. Geophys. Res.*, **109**, doi:10.1029/2003JD004485 (2004).
4. Stier, P., Feichter, J., Kinne, S., Kloster, S., Vignati, E., Wilson, J., Ganzeveld, L., Tegen, I., Werner, M., Balkanski, Y., Schulz, M., Boucher, O., Minikin, A., and Petzold, A., *Atmos. Chem. Phys.*, **5**, 1125–1156 (2005).
5. Meskhidze and Nennes (2006).
6. Geever, M., O'Dowd, C., van Ekeren, S., Flanagan, R., Nilsson, E.D., de Leeuw, G., and Rannik, Ü., *Geophys. Res. Lett.*, **32**, doi: 10.1029/2005GL023081 (2005).
7. Yoon, Y.J., Ceburnis, D., Cavalli, F., Jourdan, O., Putaid, J.P., Facchini, M.C., Decesari, S., Fuzzi, S., Jennings, S.G., and O'Dowd, C., *J. Geophys. Res.*, **112**, doi:10.1029/2005JD007044 (2007).
8. O'Dowd, C., Langmann, B., Varghese, S., Scannell, C., Ceburnis, D., and Facchini, M.C., to be submitted to *Geophys. Res. Lett.* (2007).
9. Marmer, E. and Langmann, B., accepted, *J. Geophys. Res.* (2007).

An Analysis on Aerosol Variable Tendency and Clouds Physical Effect for Decrease of Natural Precipitation over Hebei Area

Duan Ying, Wu Zhihui, Shi Lixin, and Yang Baodong

Abstract In this paper the variable rule of mean annual air temperature were analyzed in Hebei province area. Secondary, the interaction between the variation tendency of mean annual precipitation and the variation of annual mean air temperature was studied. ON the basis of above study, using the observational data of the low clouds frequency, the decrease of drizzle days number and analyzed results of aerosol concentration with height in different decade, preliminary researched results was obtained.The results show that there is good correlation between air temperature rising and the decrease of natural precipitation and above mentioned elements varieties. Furthermore the possible physical effect for aerosol varieties was analyzed.

Keywords Mean annual air temperature, mean annual precipitation, aerosols, variable tendency, cloud physical effect

Introduction

Hebei Province locates in the north of China, with continental monsoon climate. In China, since the 1990s decade of the twentieth century, the study on the variable tendency of air temperature and natural precipitation were more researched, and in this field also have obtained some important study results. Chen Longxun (1998), Huang Ronghui et al. (1999) utilized many kinds of meteorological essential factors and synthetically analyzed the climate variation near 50 years in China in order to do some research about the variable tendency of air temperature and precipitation.[1–2] The results show that the annual mean air temperature is raising and the climatic tend to warmer in North China and Northeast China. At the same time the mean annual precipitation reduces clearly. Recently, the research about clouds physical effect of the climate variation is increasing. Some investigators also start to attach

Weather Modification Office of Hebei Province, Shijiazhuang, 050021, China

C.D. O'Dowd and P.E. Wagner (eds.), *Nucleation and Atmospheric Aerosols*, 507–511.

to the research about aerosol environment effect. Shi Guangyu, Xu Li et al. (1993) using observed data by upper air balloon over the Xianghe and Yutian area in Hebei province, the analyzed observation from 1984 to 1993 in this area, the results show that the real number density of aerosol in 1993 is larger than that in 1984, especially the value near surface layer increased more.[3] The same observational result was obtained by the aircraft PMS system over Hebei province area in summer of 2005.

Analysis of the Air Temperature Variation Tendency

Based on the statistic analysis of air temperature data from 1950 to 1999 in Hebei province 40 weather stations, the analysis results of annual mean air temperature and mean air temperature in winter and summer given in Table 1.

From Table 1, in the last 50 years, the results show that the tendency of annual mean air temperature in Hebei province increased as time goes by. It increased gradually beginning with the 1960s to the maximum value in the 1990s; the increased value was 0.7°C and increases 0.23°C every 10 years. The mean air temperature was the minimum in winter in 1960s. Then it moved up from the late 1970s and got to the maximum in 1990s. In the 1990s it was 1.9°C higher than that in the 1960s. Every 10 years it increases 0.63°C. Namely, the range of mean air temperature tending to warmer is larger in winter. But in summer it changes a little with the minimum value in the 1970s and lesser rising up in the 1980s. Namely the variation of mean air temperature with decade is not obviously in summer. Thus it can be seen that in the last 50 years, annual mean air temperature stepping up and climate warming are mainly in winter.

Correlation Analysis Precipitation and Air Temperature Varition

Based on the statistic analysis of 50 years precipitation data in Hebei province, the main tendency of annual mean precipitation was decreased gradually from 1951 to 2000. The mean annual precipitation was 585.1 mm in the 1950s and it was 512.3 mm in the 1990s. It lessened 72.8 mm from the 1950s to the 1990s. But with time passing by, corresponding annual mean air temperature variation climbed up. This result shows that natural precipitation is tending to decrease along with the

Table 1 The variable tendency of the mean air temperature (°C)

Content	1950s	1960s	1970s	1980s	1990s
Annual	9.9	9.6	9.8	10.0	10.3
Winter	−5.2	−5.8	−5.1	−5.1	−3.9
Summer	23.7	23.5	23.1	23.3	23.6

increasing annual mean air temperature from the 1950s to the 1990s in Hebei province. So the decreasing of annual natural precipitation and the rising of mean annual air temperature are anti-correlation.

Variation of Drizzle Days and Low Cloud Days

Using historical observation data of three routine stations near Bohai gulf (Leting, Huanghua, and Tanghai) in Hebei area, the variation of mean annual air temperature, annual number of drizzle days and low clouds days were analyzed synthetically in this area. The result shows that the annual mean air temperature increases 0.85°C from the 1960s to the 1990s.

At the same time the tendency of precipitation took on obvious descent and precipitation decreased 133 mm from the 1960s to the 1990s. The decreased rate gets to 19%. The range of air temperature increase and rainfall decrease are larger in this area than that in the whole province. Figure 2 shows that the variation of low clouds days and drizzle days number (other figures are omitted). The results show that the number of drizzle days of each station reduced obviously in the area, the number of low clouds days increased clearly, and corresponding annual mean precipitation trends to reduce.

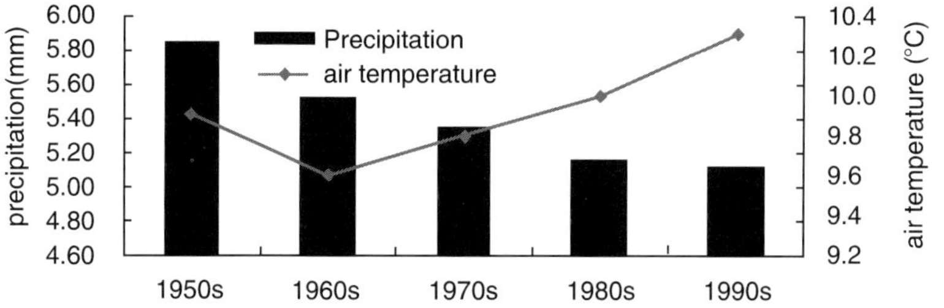

Figure 1 The variation tendency of the precipitation and the air temperature with decade in Hebei Province

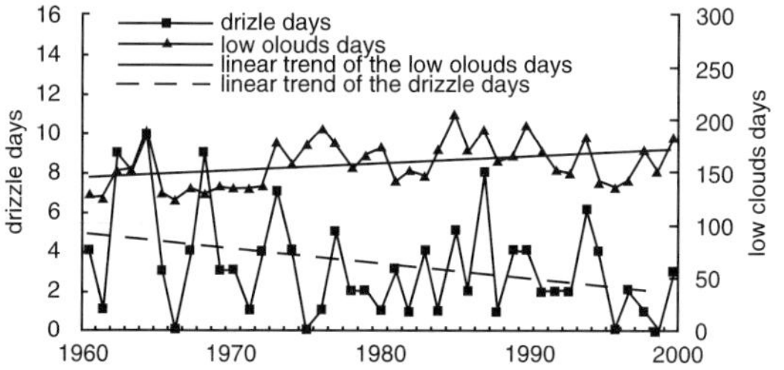

Figure 2 Variation tendency low clouds days and number of drizzle days in the Huanghua weather station

Aerosol Observation Results over the Region

The vertical distribution of aerosol density is show in Figure 3. It is find that the aerosol concentration near surface layer increases with year to year quickly from 1984 to 1993 over this area (Shi Guangyu et al. 1993). This is possibly because of the development of the industry and the society.

It is a new detecting result in the same area over Bohai Sea Gulf of China on 21 June in 2005. It is accomplished by this paper's authors using the aircraft with aerosol probe (PCASP-100X) of PMS (Figure 4). It indicates that the aerosol concentration obviously augments more than previous value. It shows that the aerosol over this area is persistently increasing.

Discussion on Aerosol Cloud Physical Effect

Based on above results, it is primarily guessed that the effect causing by the increase of atmospheric aerosol particle directly affects the balance of atmospheric radiation in this area. Its indirect effect is that the increase of aerosol can obviously increase the number of cloud condensation nuclei (CCN). In clouds, it will results

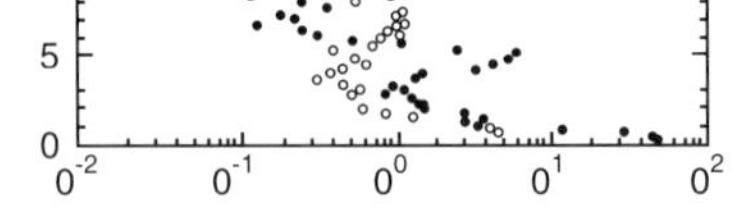

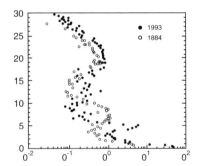

Figure 3 The comparison of observation results of aerosol density with height (1980s to 1990s)

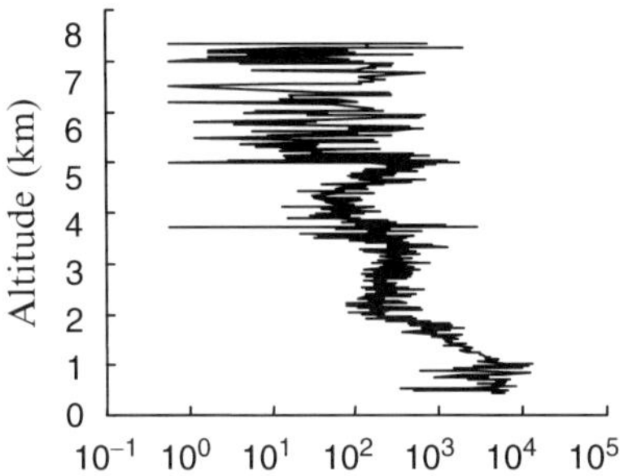

Figure 4 The observational results of aerosol concentration with height in 2005

in the decrease of mean radius of cloud drop when the water content is constant. Therefore, it will affect the microstructure characteristics of cloud. According to the observations of Huanghua and Laoting weather stations, it seems that this mechanism exists in this area.[4] The analysis indicates that there is some correlation between the increase of aerosol concentration and the changing rule of the low cloud and the drizzle. The rising of aerosol concentration resulted directly in the increase of CCN, furthermore, it possibly affected the the occurrence of the low cloud days and the drizzle days.

The low clouds precipitation belongs to warm cloud precipitation process. Cloud droplet grows up gradually through condensation and until certain scale condensation will be very slow. After this period, it is difficult that cloud droplet grows up to raindrop through condensation process. So precipitation is mainly because of collision and coalescence growth process. The gravity collision and coalescence is the most important form in all kinds of collision and coalescence form. The condition lies in some cloud drops can become to bigger drops. If there are enough CCN in the atmosphere, they would form many cloud droplets. These adequate cloud droplets compete the limit in cloud each other. As a result, the big drop forms difficultly and precipitation appears not easily.

Summarily, the research results show that there are good correlativity between the rising of annual mean air temperature and decreasing of mean annual precipitation in certain region. There is good correlativity between low clouds days increasing and number of drizzle days decreasing. Increase of aerosol in the atmosphere in which reflects on adding to CCN and ice nucleus. Therefore increasing of aerosol in the atmosphere is one of possible physical effect which influence the variation of air temperature rising and precipitation decreasing.

Acknowledgment This research was supported by National Natural Science Foundation of China (No.40475003).

References

1. Chen Longxun et al., The research of climatic change over China in the past 45 years. *ACTA Meteorologica Sinica*, **56**(3), 257–270 (In Chinese) (1998).
2. Huang Ronghui et al., The rainfall change with year-to-year in summer and the droughty tendency in North China, *Plateau Meteoro.*, **18**(4), 465–476 (In Chinese) (1999).
3. Shi Guangyu et al., The vertical distribution sounding of atmospheric ozone and aerosol by air-balloon, *Atmospheric Science*, **20**(4), 401–407 (In Chinese) (1996).
4. An Yuegai, Guo Jinping, Duan Ying et al., An analysis of the evolution characteristics and physical cause of formation about forty years natural rainfall in Hebei Province. Proceeding of 13th International Conference on Clouds and Precipitation, Reno, Nevada, 395–397 (2000).

An Assessment of the Direct Radiative Forcing of the PM_{10}-Study Case

Sabina Stefan[1], Anca Nemuc[2], C. Talianu[2], and C. Necula[1]

Abstract Today the interest in aerosols is high mainly because of their effect on human health and their role in climate change. They have also a determining effect on visibility and contribute to the soiling of monuments. Observations and model calculations show that the increase in the atmospheric aerosol burden is delaying the global warming expected from the increase in greenhouse gasses [1]. The aim of present paper is to analyze the direct effect of the aerosol and to compute the radiative forcing of the aerosol in different sites and meteorological conditions. The calculation of PM_{10} levels and radiative forcing must necessarily be based on a description of the emissions of the individual chemical species and how they transform and mix in the atmosphere. We performed a statistical analysis of the emission of the aerosol for one year, amplitude of the concentrations and the temporal variability as is described in Section 1. The optical parameters and the equations used to compute the radiative forcing are presented in Section 2. Discussion of the results and a few concluding remarks are made in Section 3

Keywords Radiative forcing, tropospheric aerosols, PM_{10}

Temporal Variability of the PM_{10} Concentrations

Daily PM_{10} concentrations of samples collected at two sites, urban and regional background, were used to estimate aerosol direct radiative forcing. The measurements were performed at 2 m above the ground level. Statistical analysis of the concentration values was made with the HSpec software which plot Hilbert energy spectrum using Maximal Overlap Discrete Wavelet Transform method [2,3]. This method was developed in order to analyze nonstationary and nonlinear multicomponent signals and it

[1] *University of Bucharest, Faculty of Physics, Atmospheric Physics Dept, P.O.BOX, MG-11, 077125 Magurele, Romania*

[2] *National Institute for R&D of Optoelectronics INOE, Magurele, Bucharest, Romania*

C.D. O'Dowd and P.E. Wagner (eds.), *Nucleation and Atmospheric Aerosols*, 512–516 is based both on filtration of signals with wavelet filters and on partition the frequency axis into 2^j equal width frequency bands. We used a Fejer–Korovkin filter in 22 points in this study, and we splinted the frequency axis into 2^7 bands, giving thus a good

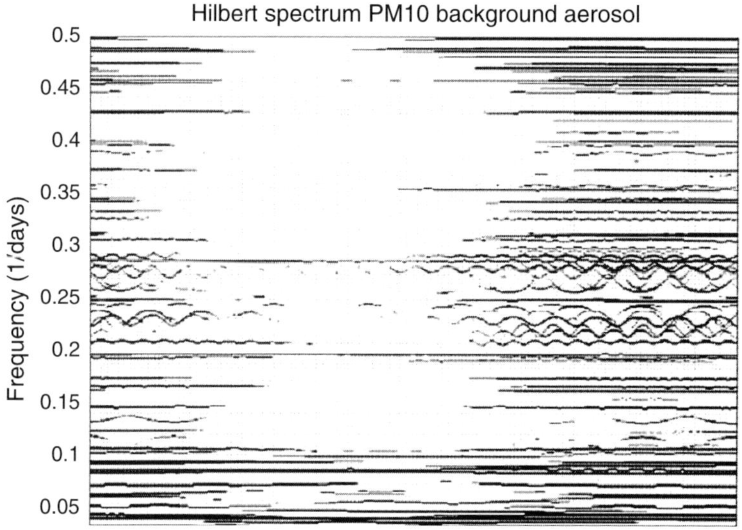

Figure 1 Hilbert spectrum of the concentration values of PM_{10} for regional background location

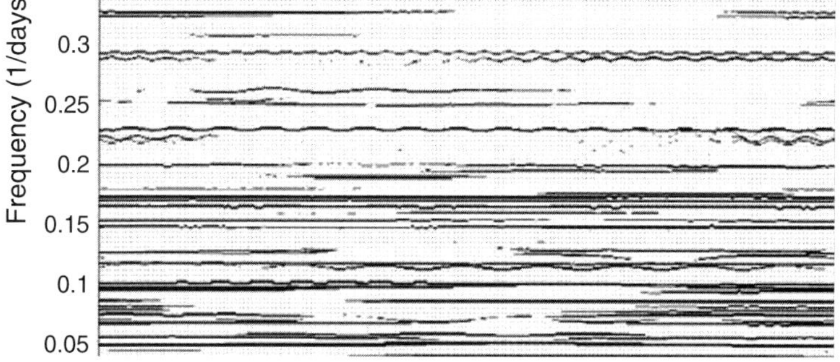

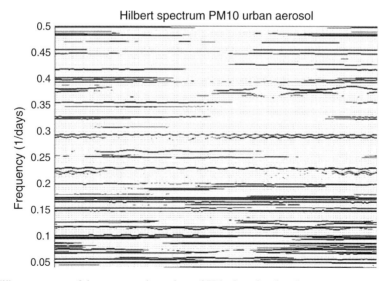

Figure 2 Hilbert spectrum of the concentration values of PM_{10} for urban location

frequency resolution, enough to separate the main frequencies of our time series. Spectral analysis for the both sites shows clear seasonal cycle (Figure 1 and 2) with maxima in winter.

Optical Properties of the Aerosol

The aerosol properties needed to estimate the magnitude and sign of direct aerosol radiative forcing: the single scattering albedo, the aerosol optical depth and the aerosol up-scatter fraction, were computed using OPAC [4]. The aerosol type chosen is valid for the mixing layer (the first atmospheric layer above the ground). The optical depth of the boundary layer is calculated with the height profiles by using for aerosol scale height 1 km in winter and 2 km in summer. Aerosol direct radiative forcing was computed for the wavelengths in the range 0.3–10.0 µm and cloud free conditions [5]:

$$\Delta F(\lambda) = F_0(1 - A_c)T_a^2\left[\left(r + \frac{t^2 R_s}{1 - R_s r}\right) - R_s\right] \quad (1)$$

The quantity $\Delta F > 0$ if ΔR_p, the quantity in square brackets, is grater than zero When interpreted in terms of climate, the net change in forcing is $-\Delta F$. ΔF depends on the following parameters:

$F_0 = \frac{S_0}{4}$ incident solar flux (Wm^{-2}); Solar constant is $S_0 = 1368.3\ Wm^{-2}$
A_c fraction of the surface covered by clouds
T_a fractional transmittance of the atmosphere
R_s albedo of underlying Earth's surface
ϖ single-scattering albedo of the aerosol
β up-scatter fraction of the aerosol (0.21 for regional background aerosol and 0.22 for urban aerosol [1].
τ aerosol optical depth

The single-scattering albedo, ϖ depends on the aerosol size distribution and chemical composition and it is wavelength dependent. The up-scatter fraction of the aerosol depends on aerosol size and composition, as well as on the solar zenith angle. Aerosol optical depth depends largely on the mass concentration of aerosol The key parameter governing the amount of cooling versus heating is the single-scattering albedo. The boundary between cooling and heating, that is, at $\Delta R_p = 0$ occurs for values of ϖ_{crit}:

$$\varpi_{crit} = \frac{2R_s}{2R_s + \beta(1 - R_s)^2}$$ The values of $\varpi > \varpi_{crit}$ lead to cooling.

The single most important parameter in determining direct aerosol forcing is relative humidity, and the most important process is the increase of aerosol mass as a result of water uptake. The coarse mode is estimated to contribute less than 10% of the total radiative forcing for all RHs of interest [6]. The surface albedo is assumed equal to 0.2 for urban site and 0.06 for regional site for all wavelengths.

Results and Discussions

The radiative budget in an urban atmosphere appears strongly modified in comparison with background sites due to atmospheric pollution and the changes to the surface of the land caused by urban development, The radiative forcing for the

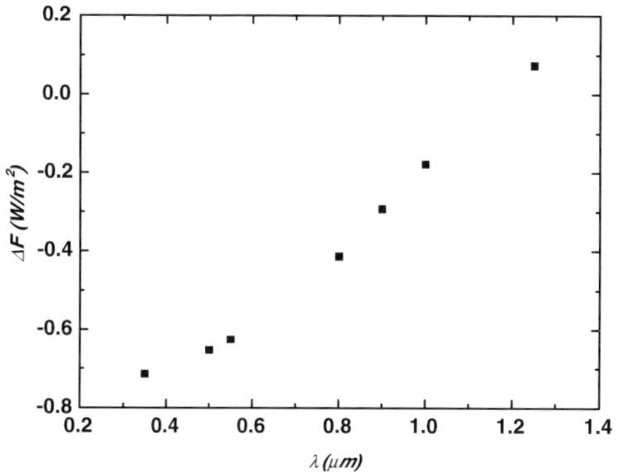

Figure 3 The forcing for the two sites during winter. Radiative total direct forcing values are for weekly concentration values of PM_{10} and relative humidity 50%

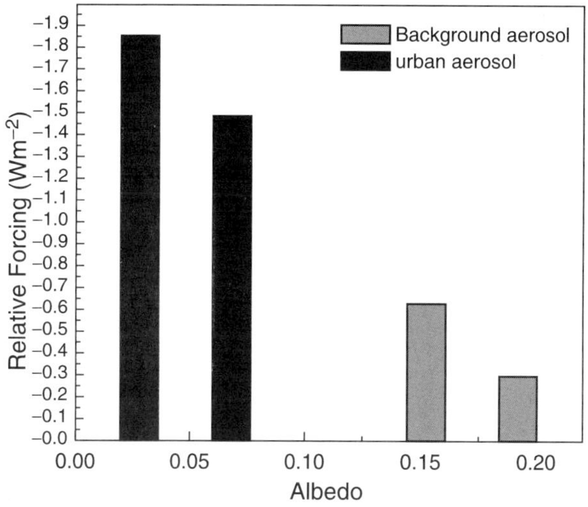

Figure 4 Radiative forcing for urban and background sites versus albedo. Radiative total direct forcing values are for weekly concentration values of PM_{10} and relative humidity of 50%

winter season for weekly concentrations of PM_{10} values, relative humidity 50% can be observed in Figure 3. The dependence of radiative forcing on wavelengths shows the influence of the aerosol on single scattering albedo.

The differences observed in the values for the two sites emphasize the importance of knowledge of number concentration and composition of aerosol (Figure 4). Our results are in agreement with those of Pilinis [6] confirming that coarse aerosol particles induce the small radiative forcing values.

Acknowledgments This study was supported by AMCSIT Program, contract CEEX 112/2005. The authors thank to National Institute for Environment for providing PM_{10} data.

References

1. Intergovernmental Panel on Climate Change, IPCC2001, New York.
2. Olhede, S. and Walden, A.T., The Hilbert spectrum via wavelet projections, *Proc.R. Soc. Lond A*, **460**, 955–975 (2004).
3. Olhede, S. and Walden, A., A generalized demodulation approach to time-frequency projections for multicomponent signals, *Proc. R. Soc. A*, **461**, 2159–2179 (2005).
4. Hess, M., Koepke, P., and Schult I., Optical properties of aerosol and clouds – the software package OPAC, *Bull. Amer. Meteorol. Soc.*, **79** (5), 831–844 (1998).
5. Charlson, R.J., Schwartz, S.E., hales, J.M., Cess, R.D., Jr. Coakley, J.A., Hansen, J.E., and Hofmann, D.J., Climate forcing by antropogenic aerosols, *Science*, **255**, 423–430 (1992).
6. Pilinis, C., Pandis, S.N., and Seinfeld, J.H., Sensitivity of direct climate forcing by atmospheric aerosol to aerosol size composition, *J. Geophys. Res.*, **100** (D9), 18739–18754 (1995).

Measurements of Hydroxylated Polycyclic Aromatic Hydrocarbons in Atmospheric Aerosols from an Urban Site of Madrid (Spain)

Ana I. Barrado, Susana García, Rosa M. Pérez, and Oscar Pindado

Abstract This paper describes an analytical methodology for the separation and quantitative determination by HPLC with fluorescence detection of hydroxylated polycyclic aromatic hydrocarbons (OH-PAHs) and their parents PAHs in atmospheric urban aerosol. The sampling site is located in the outskirts of Madrid and can be considered an open urban area. The aim of this work was the optimization of analytical methods, applied in several samples and characterize the concentration levels of selected PAHs-OH and PAHs. The preliminary results show concentrations between 2.5–53.2 pg/m^3 for 2-OHPhenathrene (2-OHPH) and 27.8–34.5 pg/m^3 for 1-OHPyrene (1-OHPYR).

Keywords Atmospheric urban aerosol, PAHs, OH-PAHs, HPLC/Fluorescence

Introduction

Hydroxylated polycyclic aromatic hydrocarbons (OH-PAHs) are known as the metabolites of the PAHs and commonly used to assess human exposure to PAHs from atmospheric air. These compounds are more toxic than the corresponding PAHs due to their direct carcinogenicity, thus OH-PAHs have showed a very big cito-toxicity because can form adducts with the DNA. For this reason, there are many paper related with the determination of the OH-PAHs in the urine. Regarding OH-PAHs, few literatures exist to measure these compounds in air samples, and in particular, HPLC measurements with fluorescence detection are rarely applied. Therefore, a sensitive and reliable method is necessary for monitoring of their concentrations in atmospheric as well as in biological fluid. OH-PAHs in atmosphere can be the result of direct emission from combustion sources, such as diesel engines, and/or formation in the atmosphere by the hydroxyl radical oxidation of PAHs.

Chemistry Division, Department of Technology, CIEMAT Avd. Complutense 22, 28040 Madrid (Spain)

C.D. O'Dowd and P.E. Wagner (eds.), *Nucleation and Atmospheric Aerosols*, 517–521.

The present work shows the preliminary results of measurements of select OH-PAHs and their parent PAHs in order to characterize their concentrations levels and study their potential relationship.

Objective

The aim of this work was the optimization of analytical methods to determine the trace levels of 1-Hydroxypyrene (1-OHPYR) and 2-Hydroxyphenanthrene (2-OHPH) in airborne particulates. The analytical methods were applied to several ambient air samples.

Experimental Part

Air Sampling

Sampling site was located at CIEMAT, situated 2 km from the urban center of Madrid in the northwest of the city. The location is characterized by open areas, and can be considerate representative of urban background.

High volume MCV sampler equipped with a Whatman glass fibre filter previously heated, was used to collect 16 ambient air samples from 15 March 2006 to 02 August 2006. The volume of ambient air draw through filter during 24 h sampling was about 720 m^3.

Chromatographic Analysis

After collection, filters were stored in a refrigerator until analysis. Samples were cut into a four pieces; two pieces extracted ultrasonically during 30 min with methanol to determine 1-OHPYR and 2-OHPH, and the others were ultrasonic extracted with dichloromethane, to determine their corresponding PAHs. Extracts were filtrated and evaporated under nitrogen flow, finally were refilled to 1.0 ml of methanol and dichloromethane respectively. Due to the good resolution of chromatographic peaks of selected OH-PAHs obtained a clean-up step was not necessary.

An Aliquot of 25 µl of these solutions were analyzed using a Hewlett-Packard series 1050 liquid chromatograph with a C18 Supelcosil thermostated column and an Agilent 1100 series fluorescence detector.

The mobile phase was a mixture of HPLC grade of acetonitrile/water (45:55) programed up to 100% of acetonitrile in 23 min and keeping it there for 10 min; the mobile phase flow rate was 1.5 mL min^{-1}. An equilibration delay of 6 min was applied to the next injection and the experimental conditions of fluorescence detector were obtained from literature.

Result and Discussion

Optimization of Extraction Conditions

In a preliminary study, extraction by ultrasonication was examined with several solvents such as, methanol, dichloromethane, hexane, acetone, and different mixtures of them. The best results were obtained with methanol. Figure 1 shows the recovery percentage of several PAHs and OH-PAHs with methanol (MEOH) and dichloromethane (DCM), being the maximum extractability obtained for 1-OHPYR with MEOH. For this reason 1-OHPYR and 2-OHPH will be extracted with MEOH.

It has been proved that an additional extraction with MEOH and DCM did not improve the extractability of these compounds, so there were not necessary re-extraction step.

Determination of OHPAHs and PAHs in Airborne Particulates

The development method was applied to determination of PAHs and OH-PAHs in 16 samples colleted from 15 March 2006 to 02 August 2006.

Figures 2a and 2b illustrate concentrations levels of ambient air samples, measured for OHPAHs and their parent PAHs, respectively.

As it can be seen, there is no significant linear correlation between the concentrations of the OH-PAHs and their parent PAHs. A possible explanation would be that their abundances in the air are determinated by many factors, such as temperature, oxidant concentrations, meteorological conditions, and their ambient lifetimes.

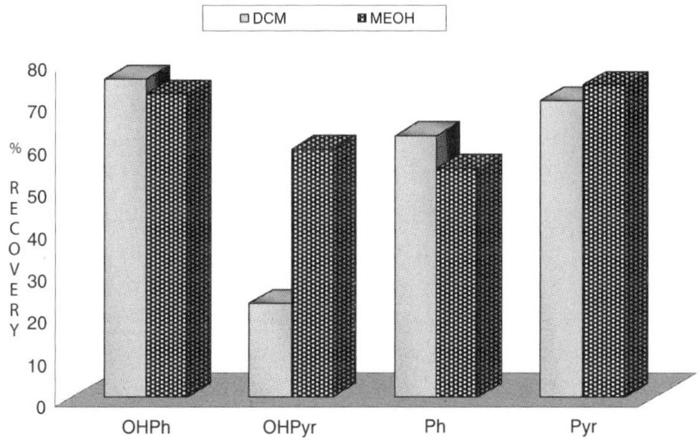

Figure 1 % Recovery of OHPAHs and their parents PAHs

Table 1 Shows minimum, maximum, and mean values of concentrations of OH-PAHs and their parents PAHs

	2-OHPH	1-OHPYR	PHEN	PYR
Mean values pg/m^3	15.5	30.4	41.7	84.1
Min–max pg/m^3	2.5–53.1	27.8–34.5	6.0–109.0	24.7–458.5

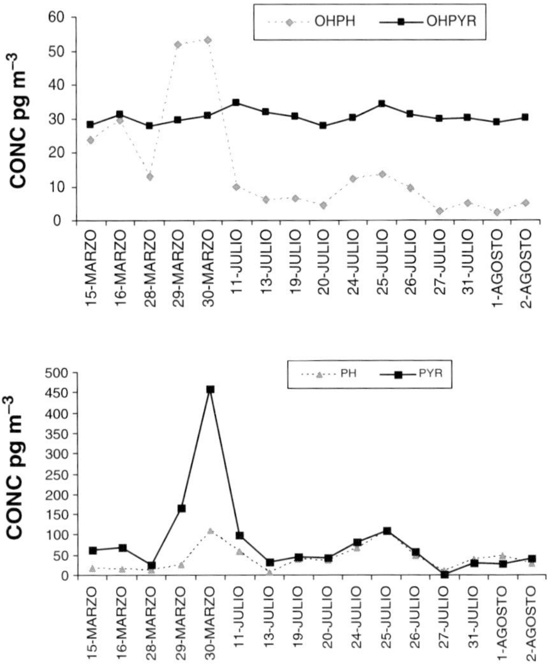

Figure 2 (a) Concentration of OHP-AHs (pg/m^3) (b) Concentration of their due PAHs (pg/m^3)

The preliminary results show that the concentration of 1-OHPYR has been kept constant in the sampling period, meanwhile the concentration of 2-OHPH shows an increment in a few days of March. This similar variation was obtained for particles concentration (Figure 3), so this would be consistent with the particle dependence of 1-OHPYR, but not 2-OHPH. In this sense, more concentration data is required to complete these observations and make a correlation study between selected OH-PAHs, and particle concentration.

Although a very few studies on OH-PAHs in the ambient air have been documented, our preliminary results agree with the previously values.

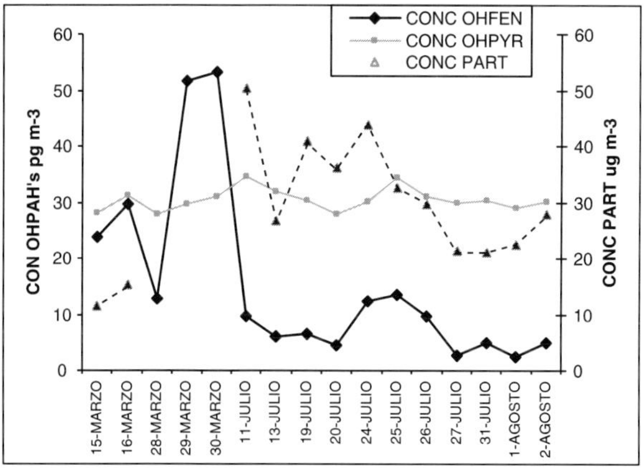

Figure 3 Concentration of OHPAHs and particles

References

1. Kishikawa, N., Morita, S., Wada, M., Ohba, Y., Nakashima, K., and Kuroda, N., *Analytical-Sciences*, **20**, 129–132 (2004).
2. Wang, G., Kawamura, K., Zhao, X., Li, Q., Dai, z., and Niu, H., *Atmospheric Environment*, **41**, 407–416 (2007).
3. García-Alonso, S., Pérez-Pastor, R., Sevillano-Castañon, M.L., *Environmental Science and Pollution Research*, **6**, N° 3 (1999).

The Simulation of Aerosol Transport over East Asia Region Using CMAQ

Youn-Seo Koo and Sung-Tae Kim

Abstract The modeling system using MM5, SMOKE, and CMAQ was developed to investigate the aerosol transport from China to Korea. The model prediction was validated with measurements in Korea during the high PM_{10} episode days. The simulation showed that the significant amount of PM_{10} in Korea was transported from China, especially from Beijing Area

Keywords Aerosol transport, PM_{10}, simulation, CMAQ, East Asia

Introduction

The East Asian Region is world's most populous area with a rapid growing economy resulting in the large air pollutant emissions. China in this region is especially major source to determine the background level of air quality in Korea and Japan.

The model system using US EPA's Models-3/CMAQ was developed to assess the aerosol transport from China to Korea. The meteorological model was MM5 and emission data were processed using SMOKE. The emission data from TRACE-P and ACE-Asia experiments were used in this study. The model simulation was compared with measurements from ambient air quality stations in Seoul Metropolitan Area and the impact of China sources to Korea in high PM_{10} period was evaluated.

Model Description

The modeling system using SMOKE, MM5, and CMAQ was developed with a configuration covering East Asia region in Figure 1. $27 \times 27\,km^2$ (D1) grid resolution for Asia region as well as $9 \times 9\,km^2$ (D2) and $3 \times 3\,km^2$ for Korea peninsula set up as the modeling domains. The number of vertical layers of MM5 and CMAQ were 22.

Department of Environmental Engineering, Anyang University, Korea

C.D. O'Dowd and P.E. Wagner (eds.), *Nucleation and Atmospheric Aerosols*, 522–525.

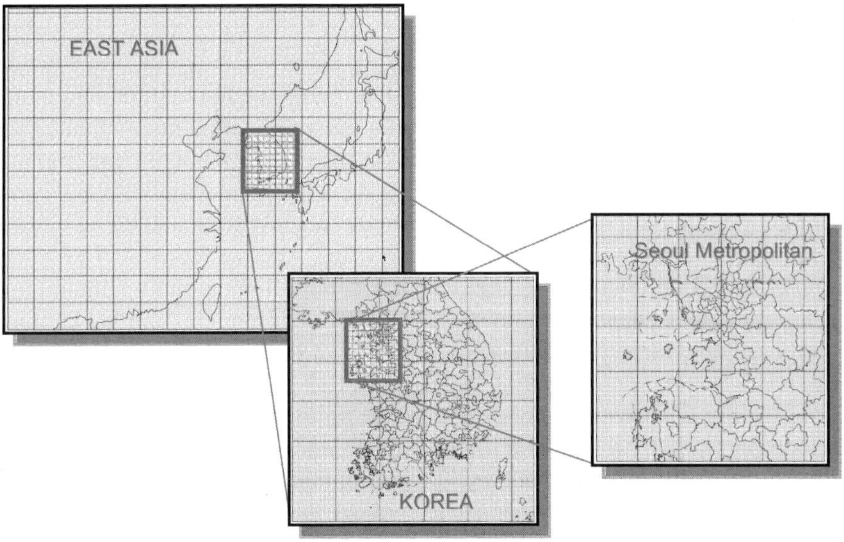

Figure 1 The modeling domains to simulate the air quality in East Asia region

Table 1 Emission inventory of ACE-Asia and CAPSS

Country	SO2	NOx	CO	CH4	VOC	NH3	BC	OC	$PM_{2.5}$	PM_{10}
China	20,385	11,347	115,749	38,356	17,432	13,570	1,049	3,385	12,243	14,345
Japan	801	2,198	6,806	1,143	1,920	352	53	74	324	332
N. Korea	227	273	3,556	1,345	234	98	22	106	326	366
S. Korea	829	1,322	2,824	1,433	1,161	172	22	28	136	171
CAPSS (2003)	499	1,186	920		824	236				67.4

The emission rate for 27 km domain was estimated based on the ACE-Asia emission data of year 2000 (Streets et al. 2003) and detailed emission rates are in Table 1. The emission data of 9 and 3 km domains were from the Korea emission data of CAPSS (Clean Air Policy Support System) of year 2003. Those included point, mobile and area emission sources.

The VOC and PM-10 emission data were speciated to 11 CB-IV chemical species and eight particulate species by using SMOKE system. The hourly biogenic emission data were prepared using GloBeis 2 and MM5 output data.

The simulation was run over 20 days of January 10–30, 2007 and was compared with measurements to test its performance. The hourly PM10 concentration in Seoul Metropolitan Area had reached up to over 150 ug/m^3 on 17th and 23rd of January in 2007.

Results and Discussions

The predicted distributions of wind filed and PM_{10} over East Asia region were shown in Figure 2. In order to identify the influence from China, the model simulations were run with Korean emission and without Korean emission. The Case 1 represented the model run with Korean emission while Case 2 was the simulation without Korean emission data. Model predictions were compared with ambient air quality monitoring stations located in Seoul Metropolitan Area in Figure 3. The model simulation of Case 1 was in good agreement with measurements which implied that the model could depict PM_{10} transport in East Asia region. The background concentration which was represented by Case 2 in Figure 4 showed that the PM_{10} transport from China was significant. The contribution from China to Seoul Metropolitan Area in Figure 4 could be up to about 80% in this particular episode days.

In order to assess more extensively the aerosol transport in East Asia region, the study to run the model in full year and to compare the results with speciated measurements of aerosols at super sites such as $PM_{2.5}$, sulfate, nitrate, organic carbon, element carbon, and particle size distributions is underway.

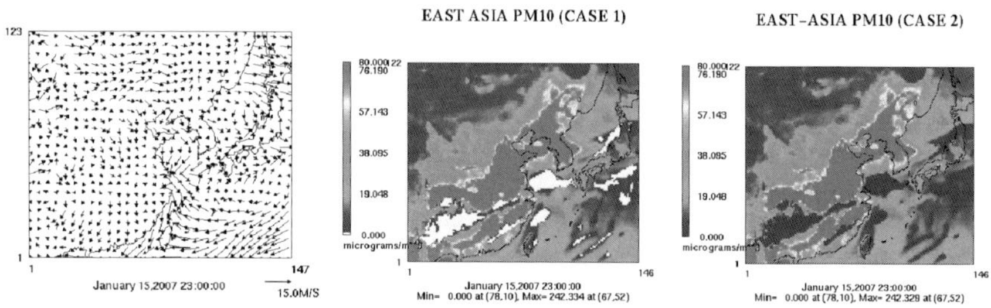

Figure 2 The predicted wind field and PM_{10} distributions for Case 1 and 2

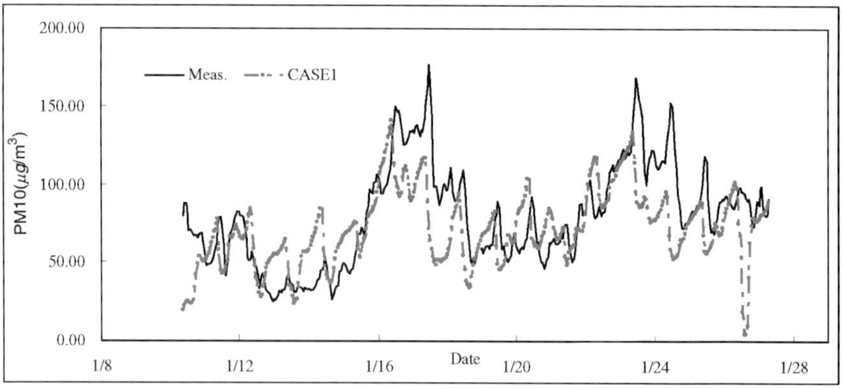

Figure 3 The comparison of measured PM_{10} with predictions for Case 1

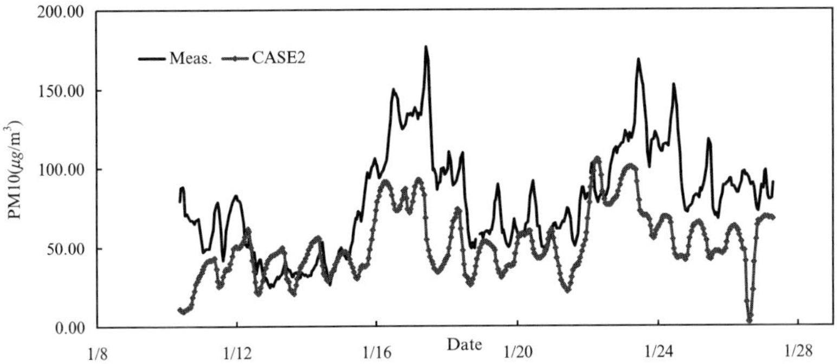

Figure 4 The comparison of measured PM_{10} and predicted PM_{10} in Seoul Metropolitan Area without Korea emission (CASE 2)

Conclusions

The modeling system was developed to simulate aerosol transport in East Asia region focusing on Korean peninsula. The system was tested for PM_{10} episode day in January, 2007. The predictions by the model were compared with measurements in Korea. They were in good agreement with measurements. The results showed that the PM_{10} transport from China was significant.

In order to assess more extensively the aerosol transport in East Asia region, the study to run the model in full year with speciated measurements of aerosols at super sites is underway.

Acknowledgements We, the authors, would like to thank the financial support of Korean Ministry of Environment and KIEST.

References

1. Streets, D.G., Bond, T.C., Carmichael, G.R., Fernandes, S.D., Fu, Q., He, D., Klimont, Z., Nelson, S.M., Tsai, N.Y., Wang, M.Q., Woo, J.-H., and Yarber, K.F., *J. Geophys. Res.*, **108**(D21), 8809, doi:10.1029/2002JD003093 (2003).
2. CMAS, *Community Modeling and Analysis System*, http://www.cmascenter.org (2006).
3. MM5, *MM5 Community Model*, http://www.mmm.ucar.edu/mm5 (2005).

Part VI
Aerosol–Cloud Interactions

Current Understanding of Aerosol–Climate Effects

Surabi Menon

Abstract The climate effects of aerosols are based on aerosol impacts on radiation budgets via the direct scattering and absorption of solar radiation and the modification to cloud radiative properties via the interactions between aerosols and cloud properties that affect the cloud droplet/ice crystal number, cloud/ice particle size, and condensate distribution. These radiative effects include changes both to the top of the atmosphere and surface fluxes. Other important aerosol climate effects include the semidirect effect that affects cloud distributions, the thermodynamical aerosol effects that affect atmospheric stability and circulation and changes to the hydrological cycle. These aerosol climate effects can have profound implications on future temperature changes, water supplies and air quality. However, the climate effects of aerosols remain subject to a large uncertainty and realistic future predictions of climate change requires a better understanding of aerosol effects on climate through integrated modeling and observational studies.

Keywords Tropospheric aerosols, cloud microphysics, radiative forcing

Aerosol–Climate Interactions

The largest uncertainty in IPCC projections of climate change is the forcing associated with aerosols and aerosol–cloud interactions. Besides the radiative forcings associated with aerosols, changes to the thermodynamical fluxes and to the hydrological cycle due to aerosol impacts (e.g., aerosol effects on liquid-phase clouds can suppress precipitation formation and aerosol effects on ice-phase clouds can modify snowfall rates via an increase in precipitation efficiency) are of critical importance. While the direct scattering and absorption of radiation by aerosols (sulfates, organic carbon (OC), black carbon (BC), nitrates, sea salt and dust) are considered to be within +0.8 to −1.0 Wm^{-2}

Lawrence Berkeley National Laboratory, 1 Cyclotron Road, Berkeley, CA, USA

C.D. O'Dowd and P.E. Wagner (eds.), *Nucleation and Atmospheric Aerosols*, 529–539.

(Hansen et al. 2005), the forcing associated with aerosol–cloud interactions encompass a broader range −1.0 to −4.0 Wm^{-2} (Menon 2004; Lohmann and Feichter 2005; Hansen et al. 2005). Thus, the negative forcings associated with aerosol–cloud–radiation interactions are relatively large compared to GHG forcings, that range from 3–4 Wm^{-2} (Hansen et al. 2005), but given the large uncertainty associated with aerosol forcing (due to uncertainties in aerosol emissions, aerosol chemical and physical transformation in the atmosphere, deposition on snow surfaces, optical properties of internally mixed aerosol particles, ability to participate in cloud processes especially for cold clouds, treatment of aerosol–cloud–precipitation processes, boundary layer representation and low cloud formation in climate models) considerable progress in climate model parameterizations is needed to constrain these anthropogenic forcings within observed temperature changes as well as in predicting their climate impacts. Based on surface temperature changes observed over the last century or so (1880–2006) and more recently over the last 25 years (1980–2006), as shown in Figure 1, the rapid rate of warming over the recent years may be related to increased GHG emissions (including effects from accumulated GHG emissions) or/and declining aerosol emissions. To project future temperature changes, accurate quantification of aerosol climate effects are clearly needed. Observational constraints on aerosol properties and associated changes in clouds remain uncertain with the available suite of satellite measurements and thus the overall radiative forcing effects of aerosols and their climate impacts remain subject to a large uncertainty.

Differences between observationally based and model-derived assessments of aerosol forcings remain large. The clear-sky direct present-day global radiative forcing effects of aerosols have been estimated via satellite-based aerosol measurements and surface wind speeds to be $-1.9 \pm 0.3\,Wm^{-2}$, which is much larger than model estimates that range from −0.5 to −0.9 Wm^{-2} (Bellouin et al. 2005). However, a global assessment of differences in the indirect effect estimate is more challenging since dynamical effects and cloud feedbacks have to be isolated from the aerosol indirect effects to get a true sense of the changes to cloud properties from aerosols. Aerosol effects on low-level stratus clouds have been the focus of most prior aerosol–cloud interaction studies. The role of cumulus clouds have become a more recent focus given the large amount of latent heat released from convective towers, corresponding changes in precipitation, especially in biomass regions due to convective heating effects and associated circulation changes (Graf et al. 2004; Nober et al. 2003; Menon and Rotstayn 2006). Relevant questions regarding the impact of biomass aerosols on convective cloud properties include the effects of vertical transport of aerosols, spatial and temporal distribution of rainfall, vertical shift in latent heat release, phase shift of precipitation, circulation and their impacts on radiation (Menon and Del Genio 2006). Besides changes to top of the atmosphere (TOA) radiative budgets, surface heating or cooling effects of aerosols can lead to changes in atmospheric stability and thermodynamics that in turn modify cloud macro/micro properties (Ramanathan et al. 2001; Menon et al. 2002a; Chung et al. 2002).

These surface effects include recent studies on "global dimming trends" over some land surfaces based on observations, which suggest that changing aerosol emissions over different regions may be responsible for the decline and the increase in

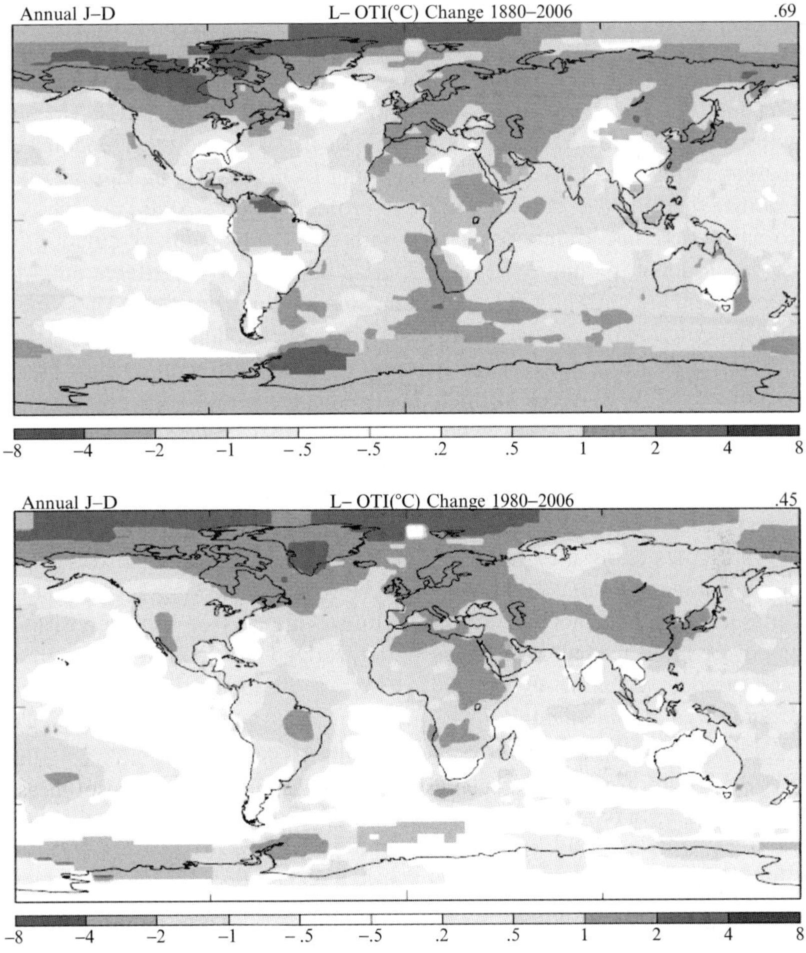

Figure 1 Observed global annual temperature (Ts) changes for 1880–2006 and 1980–2006 based on local linear trends from the GISTEMP analysis by Hansen et al. (http://data.giss.nasa.gov/gistemp). Global mean values are listed on r.h.s of the graph

surface solar radiation at several locations (Wild et al. 2005; Ramanthan et al. 2005; Nazarenko and Menon 2005). Over land surfaces, this decrease in surface shortwave radiation (~3–6 Wm^{-2} per decade) has been observed between 1960 and 1990, whereas, over the same period surface temperatures actually increased by 0.4 K. These in turn are thought to affect evaporation and precipitation (Wild et al. 2005).

Besides these direct, semidirect, cloud microphysical and thermodynamical impacts from aerosols, alteration of surface albedos, especially snow and ice-covered surfaces, due to absorbing aerosols could contribute to melting of ice and permafrost and the early advent of Spring (Hansen and Nazarenko 2004). These aerosol effects on the Arctic snow surfaces and recent melting and retreat of glaciers worldwide have resulted in an increased focus on aerosol and other climate impacts on ice-covered surfaces since much is still unknown regarding the sign of the total surface forcing that may be associated either with aerosols or other GHGs. Upcoming activities related to the International Polar Year in 2008, and other ongoing field campaigns in polar regions (e.g., the upcoming DOE ARM campaign in Alaska) will further shed light on aerosol and cloud properties in the Arctic region.

To narrow the uncertainty associated with aerosol and aerosol–cloud interactions, development of improved parameterizations of aerosol effects on climate are needed and are currently being implemented in climate models. These include the implementation of improved aerosol microphysics schemes (Stier et al. 2005; Koch et al. 2007) that account for number, mass and different species of internally mixed aerosols; cloud droplet nucleation schemes (such as those from Abdul-Razak and Ghan 2002; Nenes et al. 2003) that replace older empirically based representation of cloud droplet number (Boucher and Lohmann 1995; Menon et al. 2002b); improved representation of ice nucleation processes (Lohmann et al. 2007) and the reformulation of the standard bulk cloud microphysics schemes using a two-moment cloud microphysics scheme (Morrison et al. 2005; Lohmann et al. 2007) in GCMs that are able to account for both number and mass of the hydrometeor spectra. Climate impacts of soot deposition on snow-covered surfaces are still being parameterized via snow and ice radiative transfer models (Flanner et al. 2007) and are in the early stages of being incorporated in climate models.

Figure 2 presents a simple schematic of the steps used to represent aerosol effects on climate. The advent of elaborate treatments of aerosol physical processes and their interactions with clouds in climate models require observations of aerosol properties, cloud microphysical and radiative properties from both satellites and field measurements that can help constrain the physical processes associated with aerosols and their impacts on cloud properties and radiation than were available previously. Availability of newer satellite measurements (e.g., NASA "A" train based measurements) on aerosol and cloud vertical distributions, aerosol chemical composition, cloud phase in remote polar regions, etc., would further be valuable in evaluating model aerosol and cloud representation. Thus, progress in these areas via integrated modeling and observational analyses (satellite, field and laboratory-based) will clearly improve current predictions of aerosol climate effects.

Here, we outline aerosol effects on climate via their direct and indirect radiative effects, and changes to the hydrological cycle from aerosol surface forcings. While some of our assessments are based on modeling studies with the NASA Goddard Institute for Space Studies (GISS) climate model, we also compare our results with those from other climate models to provide an overview on the forcing ranges obtained from different models. Finally, we present an assessment of observational requriements that may help constrain future predictions of aerosol climate effects.

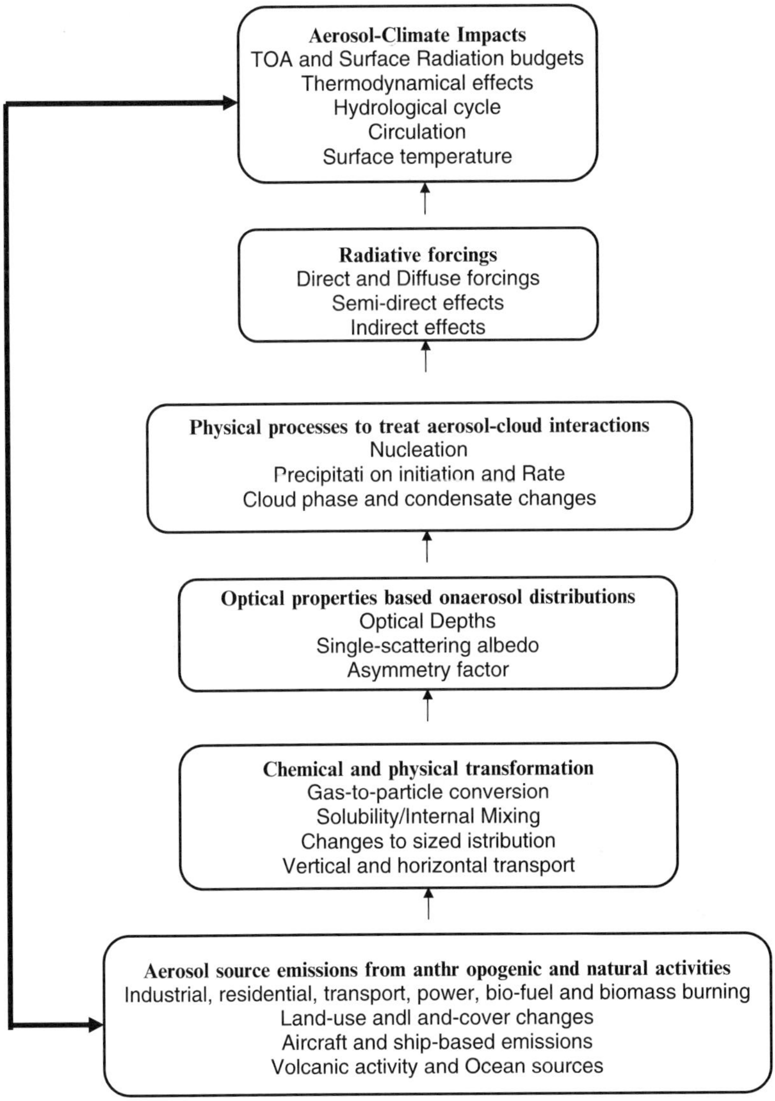

Figure 2 Simple schematic of the representation of aerosols in a climate model

Model Simulations of Aerosol–Climate Effects

Since forward-based model predictions of aerosol–climate effects from pre-industrial (PI) to present-day (PD) tend to overestimate the negative forcing associated with aerosols (some of which may be related to the uncertainty in PI aerosol emissions), examining aerosol effects on climate over the time period of global observations may be useful, especially when attempting to predict future aerosol effects on climate.

We present climate effects of aerosols based on changes between PD and PI aerosols emisisons, as well as changes over the last 40 years (1960–2004). Future projections of aerosol effects on climate are also examined to understand regions that may be subject to adverse aerosol effects and will be presented during the lecture. Our simulations are based on the GISS climate model (ModelE) (Schmidt et al. 2005; Hansen et al. 2005) that includes a microphysics based cumulus scheme (Del Genio et al. 2005) coupled to an online aerosol chemistry/transport model (Koch et al. 2006). Aerosols simulated include sulfates, organic matter (OM = 1.3*OC to account for other organic species), BC and sea salt, with prescribed dust (Hansen et al. 2005). A description of the aerosol emissions, processes treated and schemes used to couple the aerosols with the clouds is in Koch et al. (2006) and Menon and Del Genio (2006). The aerosol–cloud interactions we simulate for liquid-phase stratus and cumulus clouds, include (a) the first aerosol indirect effect: increased (decreased) cloud reflectivity due to an increase (decrease) in aerosols and cloud droplet number concentrations (CDNC) and reduced (increased) droplet sizes (Twomey 1991) and (b) the second aerosol indirect effect: change in cloud cover, cloud liquid water path (LWP) and precipitation due to smaller droplet sizes that inhibit precipitation processes, thereby increasing cloud cover and LWP (and thus cloud optical depths) (Albrecht 1989).

Aerosol emissions that we use come from a variety of sources (Dentener et al. 2006) and we present results based on the emissions specified by AEROCOM (An aerosol model intercomparison project, http://nansen.ipsl.jussieu.fr/AEROCOM). Figure 3 shows the global annual column burden change (for PD versus PI aerosol

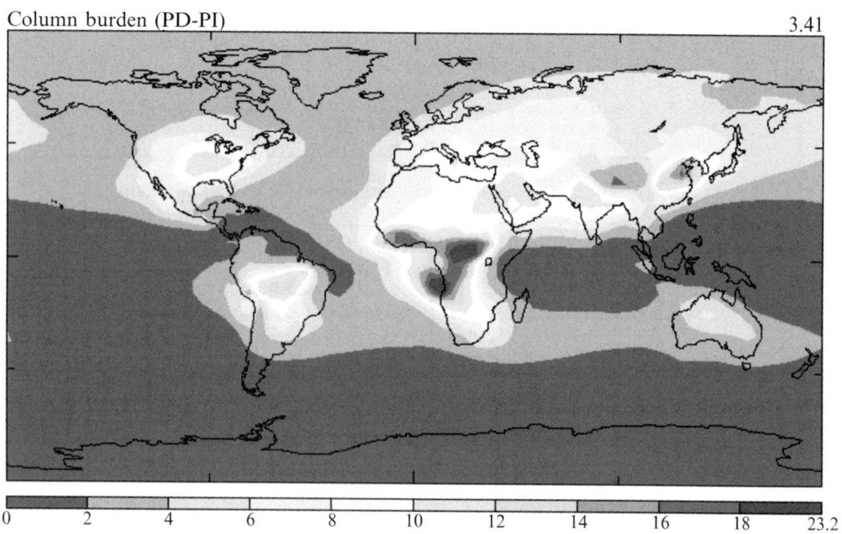

Figure 3 Change in global annual total aerosol column burden (mgm^{-2}) for present-day (PD) versus pre-industrial (PI) aerosol emissions simulated by the GISS GCM. Aerosols simulated include sulfates, organic matter and black carbon. Global average value is shown on the r.h.s. of the figure

emisisons) for aerosols (sulfates, OM and BC) as simulated by the GISS GCM. As can be seen larger changes are obtained for land locations where anthropogenic emisisons from fossil-fuel use or from biomass burning are large.

Our aerosol indirect effect paramterization is based on a semiprognostic treatment of CDNC, based on the empirical relationships between aerosol number and CDNC from field studies with separate equations being used to treat aerosol effects on liquid-phase stratus and cumulus clouds (Menon and Rotstayn 2006). At present we do not treat aerosol effects on ice clouds, but ongoing work includes a fully prognostic treatment of nucleation schemes for both liquid and ice-phase clouds and their climate impacts and will be discussed in the future (during the plenary lecture).

Aerosol–Direct and Indirect Effects

Here, we present an overview of the range of direct forcings obtained by nine different models that participated in the AEROCOM study. Since all models use similar emissions as specified by AEROCOM, the range provides an assessment on the types of differences that may be expected from differences in models and model treatment of aerosol processes. Results are summarized in detail in Schulz et al. (2006) and we mainly report on the general range for the different species. For sulfate aerosols, the direct forcing ranges from –0.16 to –0.58 Wm^{-2}, with an average value of –0.35 ± 0.15 Wm^{-2}. For carbonaceous aerosols, the direct forcing for OM ranges from –0.06 to –0.23 Wm^{-2}, with an average value of –0.14 ± 0.05 Wm^{-2}. For BC, the direct forcing ranges from 0.08 to 0.36 Wm^{-2} with an average value of 0.10 ± 0.09 Wm^{-2}. The average values for TOA direct radiative forcing for all-skies is –0.22 ± 0.16 Wm^{-2} [0.04 to –0.41 Wm^{-2}], for surface forcing is –1.02 ± 0.23 Wm^{-2} [–0.81 to –1.49 Wm^{-2}], and for the atmospheric forcing values obtained were 0.82 ± 0.17 Wm^{-2} [0.61 to 1.14 Wm^{-2}].

For similar emissions, values we simulate with the GISS GCM for the aerosol direct, semidirect and indirect effects are –0.18, –0.08, and –0.65 Wm^{-2}, respectively (Menon and Del Genio 2006). The regional climate response to such forcings are described in more detail in Menon and Del Genio (2006) who found an increase in precipitation in regions with positive atmospheric forcings for the India/Indian Ocean region and for China. Similar results have also been reported by Ramanathan et al. (2001). For example, over the northern (34–42° N, 90–120° E) and southern (18–30° N, 90–120° E) parts of China for the summer season (June to August) differences in atmospheric forcing (for PD to PI aerosol emissions) were found to be –1.08 and 2.08 Wm^{-2}, that resulted in precipitation changes of –0.39 and 0.05 mm/d, respectively. For the India/Indian Ocean (0–20° N, 40–100° E) region an increase in atmospheric forcing of 4.36 Wm^{-2} resulted in a 0.35 mm/day increase in precipitation. Thus, besides the cooling effects of aerosols on surface forcing, aerosol impacts on atmospheric forcing can impact surface energy fluxes and precipitation through a feedback loop. Further examples of these regional climate impacts from aerosols will be presented during the talk.

In terms of more recent changes, Nazarenko and Menon (2005) in their analysis of the influence of aerosols on surface radiation trends (the "dimming" between 1960–1990 and reversal of "dimming" after 1990 phenomena) find that transient GISS GCM simulations (for the 1960–2004 time period) with anthropogenic aerosols match observed trends in surface radiation at the relevant time intervals as well as observed surface temperature changes and simulations without anthropogenic aerosols tend to overpredict surface warming. Thus, without anthropogenic aerosols, warming trends would be far larger than were observed over the years when aerosol emissions were large.

With regard to the indirect effect, Penner et al. (2006) present a range of values obtained from three models that participated in the AEROCOM exercise. Given the even larger uncertainty in indirect effect estimates, sensitivity tests were conducted to understand the range of values that may result from assumptions used to treat different indirect effect processes. Differences in cloud schemes and condensate treatment in the model were found to produce large differences in forcing estimates and are therefore more difficult to evaluate. Penner et al. (2006) provide a range of values from 0.06 to $-1.4\,Wm^{-2}$ for the three models for six different sensitivity tests. The first indirect effect alone was found to be -0.21 to $-0.66\,Wm^{-2}$ and with the second indirect effect, values for a change in cloud forcing for PD versus PI aerosol emissions were about -0.34 to $-1.13\,Wm^{-2}$, more on the lower range of previously reported values.

Here, we have mainly detailed results from simulations that treat liquid-phase clouds. Global model results on the effects of aerosols on all cloud phases and types are mainly available from Lohmann et al. (2007) that indicate a net indirect effect of $-1.8\,Wm^{-2}$ with changes of 6.5 and $0.2\,gm^{-2}$ in the liquid and ice water paths respectively. The longwave component is about $0.2\,Wm^{-2}$. Larger values of the indirect effect were obtained for differences in cloud schemes ($-2.9\,Wm^{-2}$) and aerosol emissions ($-2.8\,Wm^{-2}$). These results were obtained with the ECHAM model that had an improved aerosol microphysics and a two-moment cloud microphysics scheme. Use of improved ice nucleation schemes and climate effects of aerosol on cold clouds from different models are not easily available at present but should be available in a few months. An important focus then would be the change to the longwave component and the surface forcing in polar regions that can have a profound climate impact.

Menon (2004) and Lohmann and Feichter (2005) provide an overview of uncertainties pertaining to the aerosol–climate effect and the aerosol indirect effect, respectively. Menon (2004) finds a total net aerosol forcing of -1.6 to $-2.7\,Wm^{-2}$ (for an indirect effect ranging from -1.0 to $-2.1\,Wm^{-2}$). Lohmann and Feichter (2005) find that for the large range in the indirect effects, the climate sensitivity parameter (ratio of surface temperature change to the radiative forcing) λ, varies from 0.45 to $0.87\,Km^2\,W^{-1}$, based on results from different models and for different aerosols (either for sulfates only or for all aerosols). The climate sensitivity for all aerosol effects was found to be between 0.49 and $0.62\,Km^2W^{-1}$, compared to that for GHGs that are about 0.48 to $1.01\,Km^2\,W^{-1}$ for $2 \times CO_2$.

Thus, the relatively large values of forcing and the climate sensitivity parameter obtained for aerosols compared to that for GHGs indicate the relatively large contribution of aerosol effects to climate and that realistic future predictions can only be obtained if aerosol forcings are constrained with observations to a narrower range.

Constraining Aerosol–Climate Effects

Global or regional observational constrains on the effects of aerosols on climate may be obtained by comparing specific products with available satellite retrievals. A general schematic of various satellite sensors in orbit or being launched are shown in Figure 4 (from http://www-calipso.larc.nasa.gov/about/atrain.php). For the direct effect, common parameters evaluated include the aerosol optical depth, fine fraction mode, Angstrom exponent, clear-sky radiative forcings, absorption efficiencies, etc. These have been summarized in various studies and a review of the range across models (from AEROCOM) with several satellite retrievals are in Kinne et al. (2006). They find that while models have somewhat similar values for aerosol optical depths, the species-specific aerosol optical depths vary considerably across models. Similar comparisons of simulations with the indirect effects to satellite-based observations are being used to evaluate and constrain aerosol indirect effects. More details on such comparisons will be presented at the meeting. One of the major barriers towards a conclusive constraint on the aerosol direct or indirect effect is the vertical distribution of aerosols and clouds. Model fields of e.g., liquid water path are notoriously difficult to evaluate with satellite products of liquid water path that may obscure cloud thickness issues that tend to be overpredicted by climate models. Also species-specific aerosol optical depths

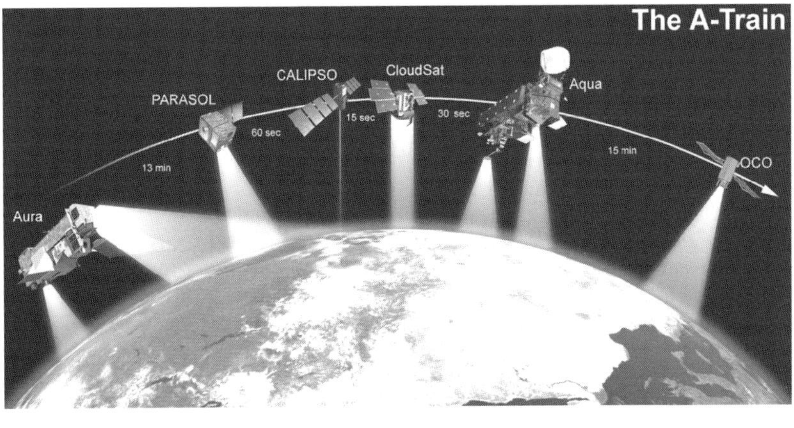

Figure 4 Series of satellites in or to be in orbit that will provide a more quantitative understanding of the Earth's climate system.

are currently not available via satellite measurements, but future missions (e.g., the GLORY mission with the Aerosol Polarimetry Sensor and the POLDER instrument on PARASOL) that can differentiate between aerosol species can provide a better evaluation on aerosol optical depths for the different aerosol types. Further information on aerosol and cloud vertical distribution through CloudSat and CALIPSO will be helpful in evaluating model aerosol and cloud distributions that affect transport mechanisms, atmospheric stability features, indirect effects and thus net radiative forcings.

Improved model representation of aerosol climate effects in turn dictate better observational capabilities that are not always available via satellite retrievals. Clearly, field campaigns are necessary since they provide a better evaluation at the process level. Laboratory measurements that can guide parameterization development are also essential. An integrated approach that combines as many observational products and improved physical process representation in climate models can ultimately pave the way for us to obtain a meaningful evaluation of aerosol climate effects.

Acknowledgment S. Menon graciously acknowledges support from the NASA Modeling, Analysis and Prediction Program for climate simulations used in this study, the DOE Integrated Climate Assessment Program and data resources from the AEROCOM project. Contributions from D. Koch, A.D. Del Genio, L. Nazarenko, and J. Hansen for the climate modeling work reported here are greatly appreciated.

References

Abdul-Razak, H. and Ghan, S.J., *J. Geophys. Res.*, **107**, doi:10.1029/2001JD000483 (2002)
Albrecht, A., *Science*, **245**, 1227–1230 (1989).
Bellouin, N., Boucher, O., Haywood, J., and Reddy M.S., *Nature*, **438,** 1138–1141 (2005).
Boucher, O. and Lohmann, U., *Tellus*, **47**, 281–300 (1995).
Chung, C.E., Ramanathan, V., and Kiehl, J.T., *J. Clim.*, **15**, 2462–2476 (2002).
Del Genio, A.D., Kovari, W., Yao, M.S., and Jonas, J., *J. Clim.*, **18**, 2376–2387 (2005).
Dentenr, F. et al. *Atmos. Chem.Phys.*, **6**, 4321–4344 (2006).
Flanner, M.G., Zender, C.S., Randerson, J.T., and Rasch, P.J., *J. Geophys. Res.*, in review (2007).
Graf, H.F., *Science*, **303**, 1309–1311 (2004).
Hansen, J. et al., *J. Geophys. Res.*, **108**, doi:10.1029/2002JD002911 (2005).
Hansen, J. and Nazarenko, L., *Proc. Nat. Acad. Sci.*, **101**, 423–428 (2004).
Kinne, S. et al., *Atmos. Chem.Phys.*, **6**, 1815–183 (2006).
Koch, D., Menon, S., McGraw, R., Bauer, S., Wright, D., *NASA Modeling, Analysis and Prediction Program's Science team Meeting*, March 7–9, Baltimore, MD (2007).
Koch, D., Schmidt, G.A., and Field, C.V., *J. Geophys. Res.*, **111**, doi:10.1029/2004JD005550 (2006).
Lohamnn, U. and Feichter, *J. Atmos. Chem. Phys.*, **5**, 715–737 (2005).
Lohmann, U., Stier, P., Hoose, C., Ferrachat, S., Roeckner, E., and Xhang, J., *Atmos. Chem. Phys.*, **7**, 73719–73761 (2007).
Menon, S. and Del Genio, A.D., Evaluating the impacts of carbonaceous aerosols on clouds and climate, In: *Human-induced Climate Change: An Interdisciplinary Assessment*, edited by Schlesinger et al., Cambridge University Press (2006).

Menon, S. and Rotstayn, L., *Climate Dynamics*, **27**, 345–356 (2006).
Menon, S., *Ann. Rev.*, **29**, 1–30 (2004).
Menon, S., Hansen, J., Nazarenko, L., and Luo, Y., *Science*, **297**, 2250–2253 (2002a).
Menon, S., Del Genio, A.D., Koch, D., Tselioudis, G., *J. Atmos. Sci.*, **59**, 692–713 (2002b).
Morrison, H., Curry, J.A., and Khvorostyanov, V.I., *J. Atmos. Sci.*, **62**, 1665–1677 (2005).
Nazarenko, L. and Menon, S., *Geophys. Res. Lett.*, **32**, doi:10.1029/2005GL024089 (2005).
Nenes, A. and Seinfeld, J.H., *J. Geophys. Res.*, **108**, doi:10.1029/2002JD002911 (2003).
Nober, F.J., Graf, H.F., and Rosenfeld, D., *Glob. Planetary Change*, **37**, 57–80 (2003).
Penner, J.E., Quaas, J., Storelvmo, T., Takemura, T., Boucher, O., Guo, H., Kirkevag, A., Kristjansson, J.E., and Seland, O., *Atmos. Chem.Phys.*, **6**, 3391–3405 (2006).
Ramanathan, V. et al., *Proc. Nat. Acad. Sci.*, **102**, 5326–5333 (2005).
Ramanathan, V. et al., *J. Geophys. Res.*, **106**, 28371–28398 (2001).
Schmidt, G. et al., *J. Climate*, **19**, 153–192 (2006).
Schulz, M. et al., *Atmos. Chem.Phys.*, **6**, 5225–5246 (2006).
Stier, P. et al., *Atmos. Chem. Phys.*, **5**(4), 1125–1156 (2005).
Takemura, T., Nozawa, T., Emori, S., Nakajima, T.Y., Kawamoto, K., and Nakajima, T., *J. Geophys. Res.*, **110**, doi10.1029/2004JD00502 (2005).
Twomey,S., *J. Atmos. Sci.*, **34**, 1149–1152 (1977).
Wild, M. et al., *Science*, **308**, 847–850 (2005).

Mass Spectral Evidence that Small Changes in Composition Caused by Oxidative Aging Processes Alter Aerosol CCN

J.E. Shilling[1], S.M. King[1], M. Mochida[1,2], D.R. Worsnop[3], and S.T. Martin[1,*]

Abstract Oxidative processing (i.e., "aging") of organic aerosol particles in the troposphere affects their cloud condensation nuclei (CCN) activity, yet the chemical mechanisms remain poorly understood. In this study, oleic acid aerosol particles were reacted with ozone while particle chemical composition and CCN activity were simultaneously monitored. The CCN activated fraction at 0.66 ± 0.06% supersaturation was zero for 200 nm mobility diameter particles exposed to 565 to 8,320 ppmv O3 for less than 30 s. For greater exposure times, however, the particles became CCN active. The corresponding chemical change shown in the particle mass spectra was the oxidation of aldehyde groups to form carboxylic acid groups. Specifically, 9-oxononanoic acid was oxidized to azelaic acid, although the azelaic acid remained a minor component, comprising 3–5% of the mass in the CCN-inactive particles compared to 4–6% in the CCN-active particles. Similarly, the aldehyde groups of α-acyloxyalkyl-hydroperoxide (AAHP) products were also oxidized to carboxylic acid groups. On a mass basis, this conversion was at least as important as the increased azelaic acid yield. Analysis of our results with Köhler theory suggests that an increase in the water-soluble material brought about by the aldehyde-to-carboxylic acid conversion is an insufficient explanation for the increased CCN activity. An increased concentration of surface-active species, which decreases the surface tension of the aqueous droplet during activation, is an interpretation consistent with the chemical composition observations and Köhler theory. These results suggest that small changes in particle chemical composition caused by oxidation could increase the CCN activity of tropospheric aerosol particles during their atmospheric residence time.

Keywords Aging, CCN activity, oleic acid, AMS, Köhler theory

[1]*School of Engineering and Applied Sciences, Harvard University, Cambridge, MA 02138, USA*

[2]*Institute for Advanced Research, Nagoya University, Nagoya 464–8601, Japan*

[3]*Aerodyne Research, Inc., Billerica, MA 08121, USA*

C.D. O'Dowd and P.E. Wagner (eds.), *Nucleation and Atmospheric Aerosols*, 540–544.

Introduction

Organic aerosol particles in the atmosphere can be transformed from a hydrophobic to a hydrophilic state by several processes, which are collectively referred to as aging.[1,2] At present, many climate and atmospheric chemistry models[3–5] include particle aging processes as a simple conversion from a hydrophobic to a hydrophilic state after 1–2 days in the atmosphere.[2] Before a more accurate representation can be included in models, however, a more detailed understanding of the hydrophobic-to-hydrophilic aging mechanisms of organic aerosol particles, as well as consequent effects on CCN activity, is necessary.

The ozonolysis of oleic acid has served as a model system to investigate aerosol aging.[6,7] Broekhuizen et al. showed that exposure to high ozone levels transforms oleic acid aerosol particles from CCN-inactive to CCN-active; however, the chemical changes and mechanisms responsible for CCN-activation remain unclear.[8] In this study, oleic acid particles are subjected to increasing ozone exposure while CCN activity and chemical composition are simultaneously monitored. With this information, a detailed chemical mechanism is developed and discussed within the framework of Köhler theory to rationalize the increase in CCN activity.

Experimental

Oleic acid aerosol was generated by homogeneous nucleation and by atomizing solutions of oleic acid in ethyl acetate. The oleic acid aerosol particles were introduced through a movable injector into the center of a laminar flow tube at 298 K, atmospheric pressure, and 0% relative humidity. During experiments, the particles were exposed to variable mixing ratios of ozone (565–8,320 ppmv) for adjustable lengths of time (5–50 sec) while simultaneously monitoring the aerosol chemical composition with an Aerodyne AMS and CCN activity with a Droplet Measurement Technologies CCN counter.

Results and Discussion

The activated fractions (F_a) of reacted oleic acid aerosol particles are shown in Figure 1 for increasing ozone exposure. For a particular ozone mixing ratio, the exposure was varied by increasing the ozonolysis time. As seen in Figure 1, the oleic acid particles were transformed from CCN-inactive to CCN-active by increasing exposure to ozone. The exposures required to activate 181 nm particles atomized from ethyl acetate (Figure 1A) were slightly lower than those required to activate 200 nm particles generated by homogeneous nucleation (Figure 1B).

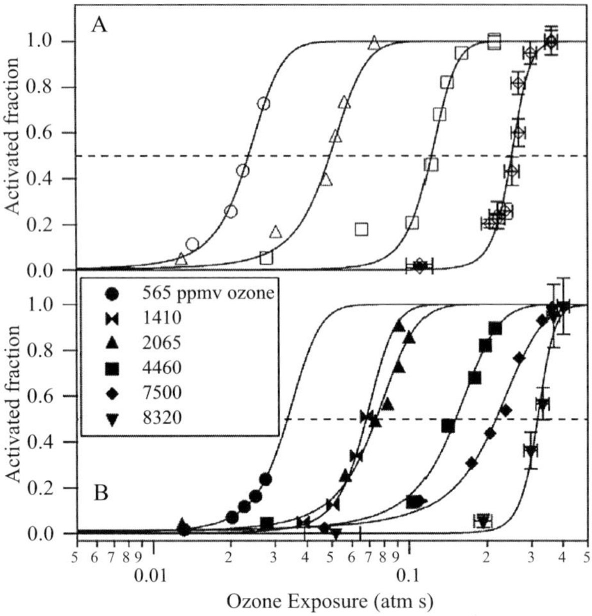

Figure 1 CCN-activated fraction at 0.66 ± 0.06% supersaturation for 181 nm particles generated by atomization (panel A) and for 200 nm particles generated by homogeneous nucleation (panel B) as a function of ozone exposure

The mass spectra of reacted but CCN-inactive particles and reacted, CCN-active particles are similar, indicating that the differences in the chemical compositions of inactive and active particles are small. Small changes, however, can be highlighted by a fractional difference plot between groups of averaged spectra (Figure 2). Figure 2 shows that, when the particles became CCN-active, the signal intensities having contributions from azelaic acid and from AAHP-1* and AAHP-2* increased while those having contributions from 9-oxononanoic acid and from AAHP-1 and AAHP-2 decreased. Therefore, an increase in carboxylic acid functionalities (i.e., gain of AA, AAHP-1*, and AAHP-2*) and a decrease in aldehyde functionalities (i.e., loss of OA, AAHP-1, and AAHP-2) were associated with the increase in CCN activity.

The conversion of aldehydes to carboxylic acids is a slow but known reaction under high ozone exposure.[9,10] The proposed mechanism is shown in Figure 3. A two-step evolution therefore leads to CCN activation. The products reported in the literature form in step 1 with their reported yields,[11,12] but, upon further exposure, some of the aldehydes begin to react with ozone.

By calibrating the AMS with the pure compounds, we are able to determine that azelaic acid increased from 3–5% to 4–6% by mass upon activation. The corresponding 6% loss of 9-oxononanoic acid relative to a base yield of 30% by mass[11,12] was consistent with the increase in AA, suggesting stoichiometric conversion of OA to AA. Because AAHP-1 and AAHP-2 both contain 9-oxononanoic acid as a monomer unit (Figure 3), we believe that the conversions of AAHP-1 to

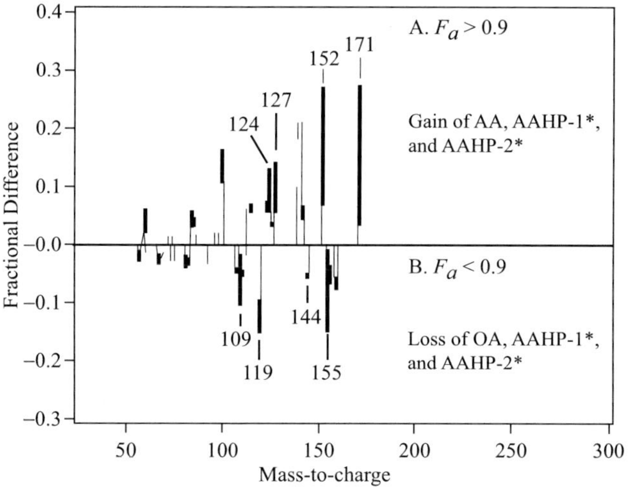

Figure 2 Fractional difference ($\Delta_{m/z}$) calculated as: $\Delta_{m/z} = (I^A_{m/z} - I^B_{m/z})/ I^B_{m/z}$, where $I^A_{m/z}$ and $I^B_{m/z}$ are the signal intensities of the averaged mass spectra for $F_a > 0.9$ and $F_a < 0.1$, respectively, for the data of Figure 3B. Signals having signal-to-noise ratios less than 40 are excluded from the analysis. Error bars (3σ) are shown as white bar overlays so that only statistically significant $\Delta_{m/z}$ values can be seen.

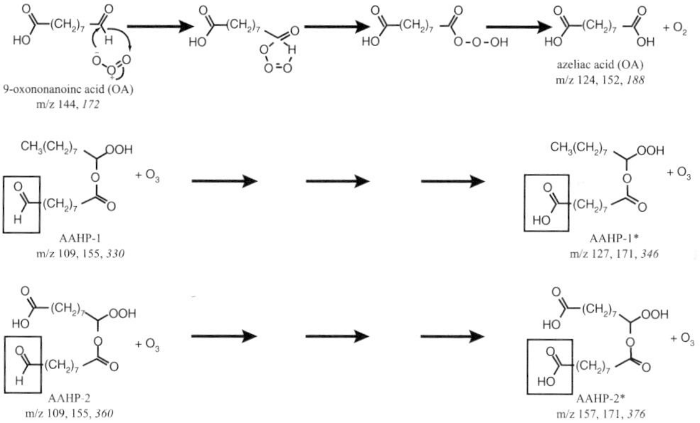

Figure 3 Chemical mechanism showing the oxidation of aldehydes to carboxylic acids by ozone. The conversion of aldehyde to acid functionalities is associated with the increase in CCN activity observed for oleic acid aerosol particles at high ozone exposures

AAHP-1* and AAHP-2 to AAHP-2* were at least as important on a mass basis as the conversion of OA to AA.

We have used Köhler theory to determine whether the increased CCN activity is driven by production of water-soluble material or by production of surface-active species in the particles. These modeling results indicate that an increased content of more-soluble material cannot alone explain the experimentally observed activation at 0.66 ± 0.06% supersaturation. Instead, decreased surface tension accompanying the conversion of aldehydes to carboxylic acids is a chemically reasonable explanation of the observations. Specifically, we hypothesize that molecules such as AAHP-1*, which is present in large quantities only in the CCN-active particles, lower the surface tension of the activating droplets.

Conclusions and Atmospheric Implications

These laboratory findings begin to provide a mechanistic and quantitative basis for further thinking on the oxidative chemical aging reactions active in the troposphere, and two main implications are apparent. (1) Organic compounds bearing aldehyde functionalities occur in diverse populations of atmospheric aerosol particles.[13,14] Aldehyde-to-carboxylic acid conversion by ozonolysis may therefore be a chemical aging mechanism having widespread applicability. (2) Only small changes in chemical composition are necessary to alter the aerosol CCN properties. Thus, the slow conversion of aldehydes to carboxylic acids over the lifetime of atmospheric particles may plausibly transform them from CCN-inactive to CCN-active, especially in the case that surface-active products form.

Acknowledgment This material is based upon work supported by the National Science Foundation under Grant No. ATM-0513463. SMK acknowledges support from the EPA STAR fellowship program. The authors thank P. Ziemann for helpful discussions.

References

1. Rudich, Y., *Chem. Rev.*, **103**, 5097 (2003).
2. Kanakidou, M. et al., *Atmos. Chem. Phys.*, **5**, 1053 (2005).
3. Koch, D., *J. Geophys. Res.*, **106**, 20311 (2001).
4. Cooke, W.F., Wilson, J.J.N., *J. Geophys. Res.*, **101**, 19395 (1996).
5. Chung, S.H., Seinfeld, J.H., *J. Geophys. Res.*, **107**, 4407 (2002).
6. Rudich, Y., Donahue, N.M., Mentel, T.F., *Annu. Rev. Phys. Chem.*, **58**, 321 (2007).
7. Zahardis, J., Petrucci, G.A., *Atmos. Chem. Phys.*, in press (2006).
8. Broekhuizen, K.E., Thornberry, T., Kumar, P.P., Abbatt, J.P.D., *J. Geophys. Res.*, **109**, D24206 (2004).
9. Bailey, P.S., Ozonation in organic chemistry, in: *Nonolefinic Compounds*, Vol. II, edited by H.H. Wasserman, New York: Academic Press (1982).
10. Story, P.R., Alford, J.A., Burgess, J.R., Ray, W.C., *J. Am. Chem. Soc.*, **93**, 3042 (1971).
11. Katrib, Y. et al., *J. Phys. Chem. A*, **108**, 6686 (2004).
12. Ziemann, P.J., *Faraday Discuss*, **130**, 469 (2005).
13. Saxena, P., Hildemann, L.M., *J. Atmos. Chem.*, **24**, 57 (1996).
14. Rogge, W.F., Mazurek, M.A., Hildemann, L.M., Cass, G.R., Simoneit, B.R.T., *Atmos. Environ.*, **27A**, 1309 (1993).

Iodine Speciation in Rain and Snow

B.S. Gilfedder[1], M. Petri[2], and H. Beister[1]

Abstract Iodine has been of interest recently due its role in new aerosol formation. However, it is also a vital nutrient for all mammals, and iodine deficiency disorders are though to affect 30% of the global population. Here we show that, in contrast to current models, organo-I is the most abundant species in rain and snow from Australia, New Zealand, Ireland, Germany and Switzerland. Moreover, at least one unknown iodine species was consistently observed in IC-ICP-MS chromatograms. It is proposed that these species are formed during breakdown of larger organo-I molecules derived either from the ocean during bubble bursting or reactions in the aerosol during reduction of iodate and/or oxidation of iodide.

Keywords Iodine speciation, organic iodine, iodide, iodate, IC-ICP-MS

Introduction

The speciation of iodine in aerosols and precipitation has been of interest recently in both atmospheric research and more applied environmental geochemistry and health. First, gas to particle conversion has been suggests as a novel processes leading to new particle formation in the troposphere,[1] which may, in turn, influence climate through scattering of incoming solar radiation. So far, the new particles are thought to form through polymerization of IO_2 dimmers and to initially consist of purely oxidized forms of iodine such as I_2O_4 or I_2O_5.[2] These species should decay to IO_3^- shortly after formation or may even be taken up as HIO_3 from the gas phase.[3] However, field aerosol and precipitation measurements fail to find significant quantities of iodate, with iodide and organo-I generally being the dominant species, particularly in submicron particles.[4] On a more applied level, 30% of the world's

[1] Institute for Environmental Geochemistry, Heidelberg University Germany

[2] Bodensee-Wasserversorgung, Sipplingen Laboratory Germany

C.D. O'Dowd and P.E. Wagner (eds.), *Nucleation and Atmospheric Aerosols*, 545–550.

population suffers from insufficient iodine intake. This is not unilaterally confined to less developed countries either; many of the more industrialized countries also have insufficient iodine intake despite iodine supplements added to salt. Iodine deficiency disorders are particularly prevalent in mountainous terrains such as the Himalayas,[5] which suggests some relationship between altitude and iodine concentrations in precipitation. As such, the sources and species of natural iodine in precipitation (which is the dominant iodine contributor to soils[6]) are of concern to a wider population. The aim of this paper is to show that organically bound iodine is the dominant iodine species in precipitation from Germany, Ireland, Australia, New Zealand, and Switzerland.

Methods

Rain samples were obtained from various locations around the world: two samples from a rural location in Australia (Barkers Vale, N.S.W.); one sample from a costal rural location in New Zealand (Oakura); four samples from Patagonia Chile (from Biester et al.[7] as given in Gilfedder et al.[8]); eight samples from Mace Head (four from the station and four from the residence cottage), Ireland; one sample from rural East Germany (Lauchhammer); one sample from the Black Forest Germany; 26 samples from Lake Constance, Germany from[8]; and two samples from different parts of the Alps (Sedrun and Interlarken), Switzerland. Hail was also collected during a summer storm in the Alps near Interlaken. Snow samples were obtained from Lake Constance, and the Black Forest (from Ref. 3), both Germany, and Fiescherhorn Glacier ice, the Alps (Interlarken), Switzerland. All rain and snow samples were filtered through 0.45 µm filters and analysed for iodine by ion chromatography-inductively coupled plasma mass spectrometry (IC-ICP-MS) by the methods outlined in Gilfedder et al.[8] Total iodine was measured by normal mode ICP-MS. Organically bound iodine was calculated as total I - Σinorganic species. Unknown species identified in IC-ICP-MS chromatograms were quantified using the iodide calibration curve. Total iodine and iodide calibrations with the IC-ICP-MS were checked periodically with standard reference material BCR611, which has a recommended total iodine concentration of 9.1 µg/L. Concentrations were always within the standard deviation given in the certificate (Figure 1).

Results

Total iodine concentrations in rain were surprisingly uniform given the various sampling locations, which ranged from continental to costal to mountainous. Total iodine ranged from an average of 5.2 nmol in Australia to an average of 32 nmol at

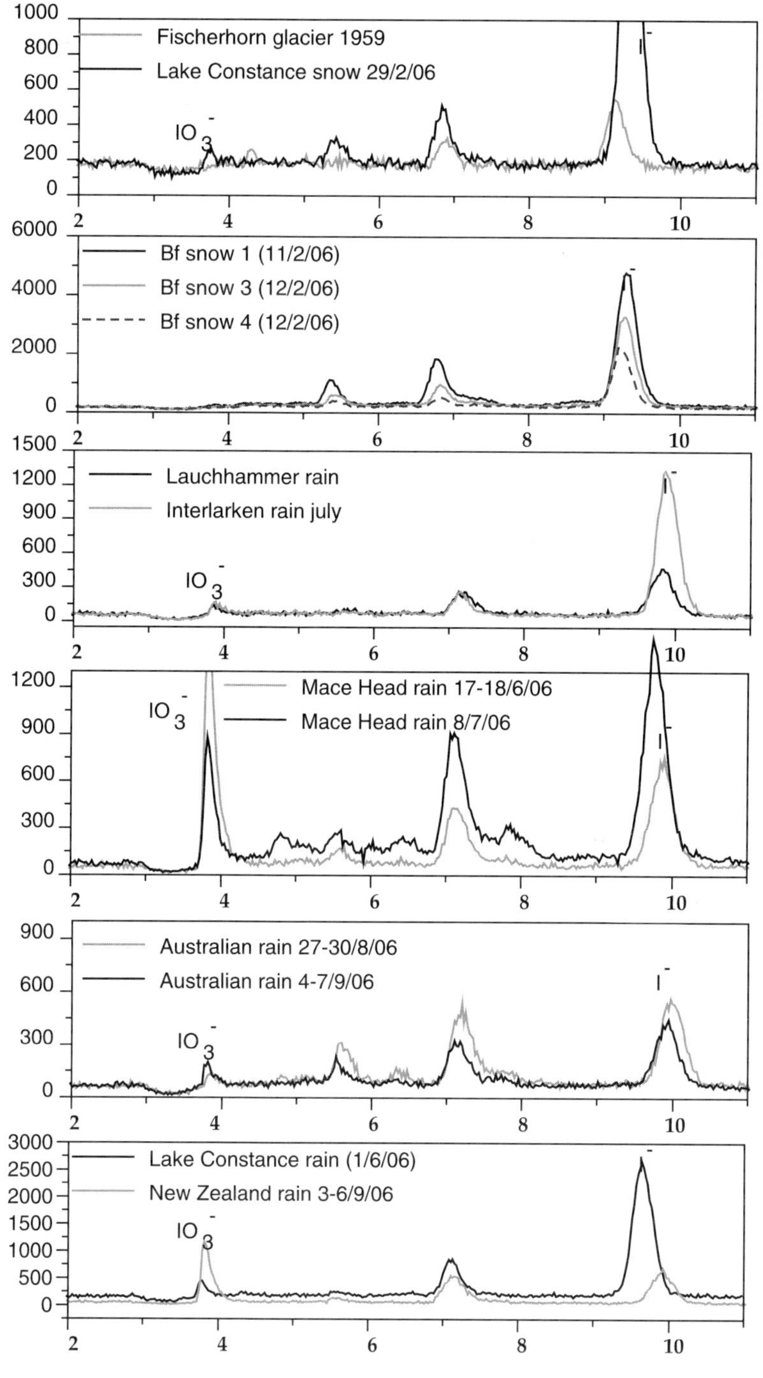

igure 1 Selected IC-ICP-MS chromatograms of rain samples from various locations

Mace Head, Ireland (Table 1). Iodine concentrations in snow where generally lower than in rain, except for samples taken during February in the Black Forest, where total iodine levels where dependent on altitude. In contrast to current models for aqueous phase iodine chemistry,[9–11] organically bound iodine was the most abundant species in all rain and snow samples, followed by iodide and then iodate (when it was above the detection limit). Iodate to iodide ratios were quite variable, ranging from 0 when iodate was below the detection limit to 1.64 in rain from the Mace Head Cottage. In the majority of cases the IO_3^-/I^- ratio was less than 1. However this ratio relatively insignificant given that >50% of the iodine was associated with organic material.

A number of unidentified iodine compounds were also observed in the majority of rain and snow samples. The largest of these eluted at ca.7 min depending on the age of the ion exchange column. A smaller peak was often, although less consistently than the large unknown peak, found in rain and snow samples. This smaller peak eluted at ca.5.5 min.

Table 1 Iodine speciation in rain and snow

Sample location (n)	Total iodine (nmol)	Iodide (nmol)	Iodate nmol	Org-I (nmol
Rain				
Lake Constance	11.9 ± 6.7	4.1 ± 2.6	1.3 ± 0.86	7.2 ± 4.6
Germany (26)	4.7–23	0.9–9.2	<0.2–3.1	0.6–17
Australia (2)	5.2	1.8	0.47	3
	2–3.2	1.3–2.4	0.4–0.55	2.3–3.7
New Zealand (1)	13.7	2.6	1.8	8.7
East Germany (1)	–	2	0.3	–
Mace head station (4)	24.8 ± 8.7	4.8 ± 1.6	3.3 ± 1.1	16.7 ± 6.3
	18–37	3–6.8	2.4–5.8	
Mace head cottage (4)	32 ± 11	5.8 ± 2.7	3.5 ± 1.2	23 ± 10.7
Switzerland (Sedrun) (1)	10.1	4.1	0.17	5.7
The Black Forest (1)	14	3.31	1.9	8.8
Patagonia (3)	31.6	2.7	<0.8	1.34
Snow				
Lake Constance (16)	5.3 ± 2	2.7 ± 1	0.4 ± 0.2 (only 4 samples above detection)	2.6 ± 1.25
	0.2 – 9.8			0.7–5.3
Black Forest (19)*	18.7	5.7	0.4 (only 2 samples above detection)	12.6
Fiescherhorn glacier, Switzerland (20)	3.2 ± 1.6	0.9 ± 0.94 +	<0.2	–
	1.6–5.8			
Interlaken hail (1)	1.0	0.55	Nd	0.45

Concentrations in these samples are highly dependent on elevation and so no ± is given. Se Gilfedder et al.[12] for further details. +Only four samples analysed.

Discussion

Field observations of iodine speciation in aerosols and rain currently differ significantly from laboratory results and current models of atmospheric iodine chemistry. In particular, laboratory studies on the nucleation of iodine suggest that iodine oxide is the dominant product of the gas to particle conversion process. Moreover, models suggest that due to the abundance of oxidizing species in the atmosphere iodide should be rapidly (at a diffusion-controlled rate) oxidized to I_2 or other volatile species and escape into the gas phase.[3,10] In these reaction sequences iodate is the only sink species and should accumulate in the particles. However, Baker[4] has found during extended ship cruises that organo-I is the dominant species in submicron aerosols, whereas iodate is more prevalent in larger particles.

It is evident from our results that iodate is the least abundant species in rain and snow, and that organically bound iodine is the most important sink for iodine. Moreover, it is also obvious that there are a number of species making up this pool of "organo-I" as seen by the unidentified peaks in the chromatograms. It is thought that these unidentified species, which makes up 5–20% of the total iodine, are a relatively small ionic organic molecules. However, to date no available standards have matched this peak. We have tested iodoacetic acid, diiodoacetic acid, I_3^-, and a range of aromatic iodine species. It is possible that this species is formed during the breakdown of large organo-I compounds that are ejected into the atmosphere from the ocean microlayer during bubble bursting.[13] This may also be a source of the excess iodide observed in our rain samples and the aerosol measurements of Baker,[4] as Wong and Cheng[14] have shown that organo-I decomposes to iodide in near shore sea water on exposure to light. Alternatively, iodine may bind to organics during reduction of iodate as HOI, a predicted iodine intermediate species, is highly nucleophilic.

Acknowledgment This project was funded by the Deutsche Forschungsgemeinschaft BI 734/5-1/2 to Harald Biester. We would like to thank Joelle Buxmann for taking the rain samples from Mace Head.

References

1. O'Dowd, C.D., Jimenez, J.L., Bahreini, R. et al., *Nature*, **417** (6 June), 632 (2002).
2. Saunders R.W. and Plane, J.M.C., *Environ. Chem.*, **2**, 299 (2005).
3. Pechtl, S., Schmitz, G., and von Glasow, R., *Atmos. Chem. Phys. Discuss.*, **6**, 10959 (2006).
4. Baker, A.R., *Environ. Chem.*, **2**, 295 (2005).
5. Stewart, A., *British Medical Journal*, **300** (9 June), 1507 (1990).
6. Fuge, R. and Johnson, C.C., *Environ. Geochem. Health*, **8** (2), 31 (1986).
7. Biester, H., Keppler, F., Putschew, A. et al., *Environ. Sci. Technol.ogy* **38**, 1984 (2003).
8. Gilfedder, B.S., Petri, M., and Biester, H., *J. Geophys. Res.*, accepted (2006).
9. Chatfield, R.B. and Crutzen, P.J., *J. Geophys. Res.*, **95** (D13), 22319 (1990).

10. Vogt, R., Sander, R., and von Glasow, R. et al., *J. Atmos. Chem.*, **32**, 375 (1999).
11. McFiggans, G., Plane, J.M.C., and Allan, B.J. et al., *J. Geophys. Res.*, **105** (D11), 14371 (2000).
12. Gilfedder, B.S., Petri, M., and Biester, H., *Atmos. Chem. Phys. Discuss.* (2006).
13. O'Dowd, C.D., Facchini, M.C., Cavalli, F. et al., *Nature*, **431** (October), 676 (2004).
14. Wong, G.T.F. and Cheng, X.H., *Marine Chem.*, **74**, 53 (2001).

Using Aerosol Number to Volume Ratio in Predicting Cloud Droplet Number Concentration

Niku Kivekäs, Veli-Matti Kerminen, Tatu Anttila, Hannele Korhonen, Mika Komppula, and Heikki Lihavainen

Abstract In this work we present a parameterization for estimating cloud droplet number concentration from the submicron particle volume concentration, soluble fraction, updraft velocity, and a size distribution shape parameter $R(0.1\,\mu m)$. The parameterization will also be tested against measured values.

Keywords Cloud droplet activation, parameterization

Introduction

In atmospheric aerosol systems, the number of particles >70–120 nm in diameter is related closely to the number of particles able to act as cloud condensation nuclei (CCN). Many large-scale atmospheric models, however, have the particle mass or volume concentration as the only prognostic variable.

The aim of this study is to investigate how accurate the estimation of cloud droplet number concentration (CDNC) would be if an empirical relation between the number concentration of CCN-size particles and particle volume (or mass) concentration was available. The relation used here is

$$R(d_c) = \frac{N(d > d_c)}{V_{tot}}, \tag{1}$$

where $N(d > d_c)$ is the number concentration of submicron particles with diameter larger than a cutoff diameter d_c and V_{tot} is the total volume concentration of all submicron particles.

The ratio R has been found to vary little in marine air (Hegg and Russell 2000; Van Dingenen et al. 2000), but it seems to depend on the air mass type and ambient

Finnish Meteorological Institute, Research and Development, P.O. Box 503, FI00101 Helsinki, Finland

C.D. O'Dowd and P.E. Wagner (eds.), *Nucleation and Atmospheric Aerosols*, 551–555.

temperature (Dusek et al. 2004; Kivekäs et al.; 2007). If the value of R was known or could be parameterized, it would significantly simplify the estimation of CDNC in models where atmospheric particles are represented as volume or mass concentration.

In this study we have parameterized the number concentration of cloud droplets as a function of submicron volume concentration (V_{tot}), $R(0.1\,\mu m)$, updraft velocity (υ_{up}) and the soluble fraction of particle mass (ε). The parameterization has been made by fitting the parameters to a set of simulations made with an adiabatic air parcel model (Korhonen et al. 2005; Anttila et al. 2002). The parameterization has been tested by comparing the results to real cloud values measured at Pallas in northern Finland.

Methods

Cloud droplet activation was modeled using an adiabatic cloud model. The model used here calculates the number concentration of cloud droplets in an adiabatically rising air parcel having a fixed updraft velocity.

The number size distribution parameters are given to the model as lognormal modes, the parameters being the number concentration of a mode (N_p in cm^{-3}), the mean diameter of the mode (d_p in µm) and the geometric standard deviation of the mode (σ). Also the soluble (mass) fraction (ε) of the particles and the updraft velocity (υ_{up} in m/s) are given. The model is sectional, converting a given modal size distribution into 500 size bins over the diameter range 0.001–3.0 µm.

For the particle size distribution variables, particle soluble fraction of the particle mass and updraft velocity, we chose a set of values that are typical for background air in northern Europe. All combinations of values of different variables were simulated. The values had also a weight factor attached to each value, being 1 in the more extreme cases and 2 or 3 for the more typical conditions. The weight factor of a simulation was the product of weight factors of each variable. By this way the weight factors of a simulation varied from 1–12.

V_{tot} and $R(0.1\,\mu m)$ were calculated for each dry particle number size distribution. In the simulations the activation diameter D50 was defined as the dry diameter of the smallest size bin that activated into cloud droplets.

To calculate the number concentration of cloud droplets (CDNC), we need $R(d_c)$ where d_c is equal to the actual D50 diameter. From Kivekäs et al. (2007) we see that $R(d_c)$ can be expressed in the form $a \times d_c^{-1} + b$ in the size range $50\,nm < d_c < 200\,nm$ at different locations. Therefore we decided to link $R(d_c)$ and $R(0.1\,\mu m)$ in the form

$$R(d_c) = \alpha \times \left(\frac{R(0.1\mu m)}{d_c} \right) + \beta \qquad (2)$$

Parameterization

The parameterization linking V_{tot} (µm³/cm³) $R(0.1\,\mu m)$ (µm⁻³), ε and υ_{up} (m/s) to CDNC (cm⁻³) consists of two parts. The first part (Eq. 3) estimates the D50-diameter (µm) and the second part (eq 4) calculates the CDNC based on Eq. 2 using the result of the first part as d_c:

$$D50 = \left(0.014 \times \ln V_{tot} + 0.008 \times \ln R(0.1\mu m) + 0.016\right) \times \varepsilon^{-0.19} \times v_{up}^{-0.32} \quad (3)$$

$$CDNC = \left(0.10 \times \frac{R(0.1\mu m)}{D50}\right) \times V_{tot} - 6cm^{-3}. \quad (4)$$

The correlation coefficient between the simulated and parameterized values of D50 was 0.96, as was also the correlation coefficient between the simulated and parameterized values of CDNC. The relation between simulated and parameterized values of D50 and CDNC for a set of 1,158 individual cases is illustrated in Figure 1.

Testing the Parameterization

The parameterization was tested against measured cloud data from the Pallas GAW station. The station has several measurement sites, of which those considered here are the Sammaltunturi (67°58′N, 24°07′E, 560 m above sea level) and Matorova sites (68°00′N, 24°14′E, 340 m above sea level) (Hatakka et al. 2003). There are periods of time when the higher altitude site (Sammaltunturi) is inside cloud whereas the lower altitude site (Matorova) is below the cloud. During these cloud

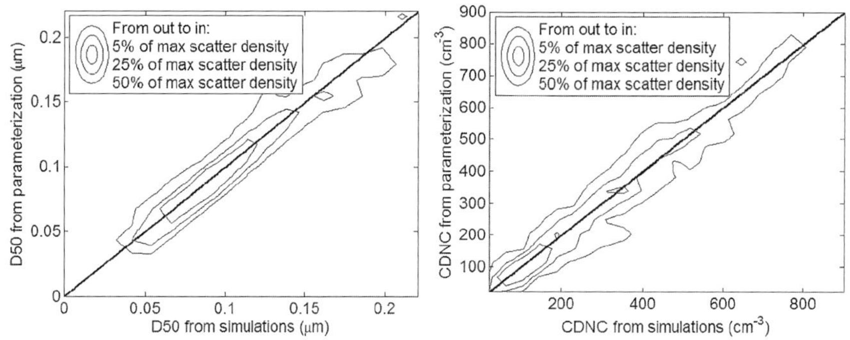

Figure 1 Parameterized versus simulated values of D50 (**left**) and CDNC (**right**). The contours represent the scatter point density, being (from out to in) 5%, 25%, and 50% of the maximum point density in the picture. The thick straight line is 1:1

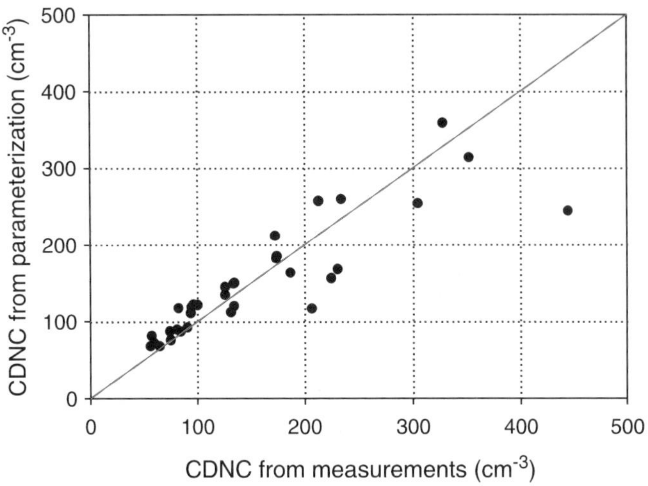

Figure 2 The parameterized versus measured cloud droplet number concentrations at Pallas during 33 cloud events. The straight line is 1:1

events the activated fraction of each size class can be calculated by subtracting the in-cloud number size distribution from the ones measured below the cloud. (Komppula et al. 2005)

For the 33 cloud events analyzed here, V_{tot} and $R(0.1\,\mu m)$ were calculated from the particle size distributions measured below the cloud. The soluble fraction was estimated to be 50% of the particle mass and updraft velocity was set equal to 0.43 m/s for all the events.

The parameterized versus measured values of cloud droplet number concentration are shown in Figure 2. The correlation coefficient between those two data sets was 0.76 and the largest offsets produced by the parameterization were within 50% of the measured value.

It should be noted that two parameters, ε and υ_{up}, were fixed in this analysis. If additional information on their values in individual cloud events were available, the correlation between the measured and parameterized cloud droplet number concentrations would probably have been higher than that shown by Figure 2.

Conclusions

The concept of parameterizing cloud droplet number concentrations (CDNC) with number-to-volume concentration ratios ($R(d_c)$) works well in Pallas, which is a station representing continental background air. This kind of a parameterization would be a computationally efficient way to estimate the aerosol impact on cloud formation in large scale models. This demonstrates the benefit achievable from investigating parameterizations for $R(0.1\,\mu m)$ or for some other $R(d_c)$.

References

1. Anttila, T. and Kerminen, V.-M-, Influence of organic compounds on the cloud droplet activation: a model investigation concidering the volatility, water solubility, and surface activity of organic matter, *J. Geophys. Res.*, **107**(D22), 4662, doi:10.1029/2001JD001482 (2002).
2. Dusek, U., Covert, D.S., Wiedensohler, A., Neususs, C., and Weise, D., Aerosol number to volume ratios in Southwest Portugal during ACE-2, *Tellus*, **56B**, 477–491
3. Hatakka, J., Aalto, T., Aaltonen, V., Aurela, M., Hakola, H., Komppula, M., Laurila, T., Lihavainen, H., Paatero, J., Salminen, K., and Viisanen, Y., Overview of the atmospheric research activities and results at Pallas GAW station, *Boreal Environ. Res.*, **8**, 365–384.
4. Hegg, D.A. and Russell, L.M. Analysis of processes determining the number-to-volume relationship for submicron aerosol in the eastern Atlantic, *J. Geophys. Res.*, **105**, 15321–15328.
5. Kivekäs, N., Kerminen, V.-M., Engler, C., Lihavainen, H., Komppula, M., Viisanen, Y., and Kulmala, M., Particle number to volume concentration ratios at two measurement sites in Finland, *J. Geophys. Res.*, accepted for publication (2007).
6. Komppula, M., Lihavainen, H., Hatakka, J., Aalto, P.P., Kulmala, M., and Viisanen, Y., Observations of new particle formation and size distribution at two different heights and surroundings in subarctic area in northern Finland, *J. Geophys. Res.*, **108**(D9), 4295, doi:10.1029/2002JD002939 (2003).
7. Komppula, M., Lihavainen, H., and Kerminen, V.-M., Measurements of cloud droplet activation of aerosol particles at a clean subarctic background, *J. Geophys. Res.*, **110**(D0), 6204, doi:10.1029/2004JD005200 (2005).
8. Korhonen, H., Kerminen, V. -M., Lehtinen, K.E.J., and Kulmala, M., CCN activation and cloud processing in sectional aerosol models with low size resolution, *Atmos. Chem. Phys.*, **5**, 2561–2570.
9. Van Dingenen, R., Virkkula, A.O., Raes, F., Bates, T.S., and Wiedensohler, A., A simple nonlinear analytical relationship between aerosol accumulation number and sub-micron volume, explaining their observed ratio in the clean and polluted marine boundary layer, *Tellus*, **52B**, 439–451.

Influence of Surface Tension on the Connection Between Hygroscopic Growth and Activation

H. Wex[1], T. Hennig[1], D. Niedermeier[1], E. Nilsson[1,2], R. Ocskay[3], D. Rose[4], I. Salma[3], M. Ziese[1], and F. Stratmann[1]

Abstract The hygroscopic growth and the critical supersaturation needed for the activation of different kinds of particles were measured with LACIS (Leipzig Aerosol Cloud Interaction Simulator), with a HH-TDMA (high humidity tandem differential analyzer) and with the CCNc (Cloud Condensation Nuclei counter) from DMT (Droplet Measurement Techniques) during different measurement campaigns. Among the examined substances were particles generated from different seawater samples, soot particles coated with ammonium sulfate or levoglucosan, and particles from two different HULIS (HUmic LIke Substances) samples. The attempt to connect the hygroscopic growth of these particles with their activation by using a simple form of the Köhler equation and by assuming a surface tension of water was successful for the seawater samples and for the coated soot particles. However, for the HULIS particles, the value used for the surface tension had to be lower than that of water to achieve closure between hygroscopic growth and activation. This is in accordance with the fact that HULIS is known to reduce the surface tension. The surface tension lowering that had to be taken into account increased towards more concentrated solutions. For dry sizes above about 90 nm, the surface tension of water could be used even in the case of the HULIS particles. For smaller dry sizes, the activation could be modeled correctly by using surface tensions between 60 and 65 mN/m. The surface tension decreased further for the modeled hygroscopic growth. This tendency in the surface tension is consistent with the HULIS concentrations in the particles/droplets. Particles grown to their equilibrium diameters at subsaturation conditions are more concentrated than particles at activation. Of the activated particles, the ones with the larger dry diameters have larger growth factors at the point of activation and thus build on more dilute solutions, making the influence of the surface tension less pronounced with increasing dry particle size.

[1] *Leibniz-Institute for Tropospheric Research, Permoser Str. 15, 04318 Leipzig, Germany*

[2] *Lund University, P.O. Box 118, S-221 00 Lund, Sweden*

[3] *Eötvös University, Institute of Chemistry, Budapest, Hungary*

[4] *Max Planck Institute for Chemistry, Biogeochemistry Department, P.O. Box 3060, 55020, Mainz, Germany*

C.D. O'Dowd and P.E. Wagner (eds.), *Nucleation and Atmospheric Aerosols*, 556–560.

Keywords Hygroscopic growth, activation, surface tension

Introduction

In the past, it often had been tried to connect the hygroscopic growth of aerosol particles with the activation of these particles to cloud droplets (e.g., [1]–[3]). The connecting equation is the Köhler equation, describing the water vapor saturation over the surface of a solution droplet. In general, connecting the hygroscopic growth of particles to their activation by using the Köhler equation was successful more often when relatively simple substances (e.g., ammonium sulfates) were examined, while it more often did not work out for atmospheric aerosol particles or for mixtures of substances including organic compounds.

During several measurement campaigns at the ACCENT infrastructure site LACIS (Leipzig Aerosol Cloud Interaction Simulator) [4], hygroscopic growth and activation were measured for different types of aerosol particles. This abstract comprises an overview of the results obtained for the data collected during these campaigns with respect to the connection between hygroscopic growth and activation for the different particle types.

Measurements

The following substances were used to generate particles: (a) NaCl and three different seawater samples; (b) soot particles from a spark-generator coated with either ammonium sulfate or levoglucosan (during the ACCENT campaign LExNo (LACIS Experiment in November)); (c) two different HULIS (HUmic Like Substances) samples, collected and prepared in Budapest.

Besides LACIS, which measured both, hygroscopic growth and activation, for the samples given in (a) and (c), an HH-TDMA (High Humidity Tandem Differential Mobility Analyzer) [5] and a continuous-flow streamwise thermal-gradient CCNc (Cloud Condensation Nucleus counter) [6] measured hygroscopic growth and activation, respectively, during LExNo. Both, LACIS and the HH-TDMA, measured the hygroscopic growth up to high relative humidities (RHs) above 95%, with the HH-TDMA measuring up to 98% RH and LACIS measuring up to 99.5% RH.

Modeling

The modeling followed the approach described in [7], with a parameter ρ_{ion} defined as: $\rho_{ion} = (\Phi\nu\, \rho_{sol})/M_{sol}$ (with the osmotic coefficient Φ, ν being the number of ions the substance dissociates to in solution, and with the density ρ_{sol} and the molecular weight M_{sol} of the solute). This parameter is internally included in the water activity term in one of the possible formulations of the Köhler theory, which will not be depicted here due to the limited length of this abstract. The description is given in detail in [7].

ρ_{ion} includes all the parameters of the solute which occur in the water activity term and for which values possibly are unknown. With this, there are two unknowns in the Köhler equation, ρ_{ion} in the water activity term and the surface tension σ in the Kelvin term.

Considering the data, there are two different sets of measurements, those of the hygroscopic growth and those of the activation. In the following, for each of the measured substances the two free parameters ρ_{ion} and σ were adjusted such, that both, the measured hygroscopic growth and the measured activation could be modeled with the Köhler equation. It is shown, that this adjustment leads to reasonable values for ρ_{ion} and σ for all substances investigated and reveals different behavior of the sea-water and the coated soot particles compared to HULIS.

Particles Generated from NaCl and Seawater and Coated Soot Particles

For these substances it was found, that the surface tension of water (σ = 72.8 mN/m) could be used to describe both, hygroscopic growth and activation. Using σ of water in the Köhler equation, ρ_{ion} was adjusted such, that the equation reproduced the measured particle sizes at the respective RHs. This was done separately for the different substances at all the different RHs.

For the seawater samples, the values for ρ_{ion} were about constant at RHs above 95%. This suggests an ideal behavior of the solutions in this concentration range, which, at these RHs, is at a molality below 1.5 approximately. The average values of ρ_{ion} derived at RHs between 96% and 99% are given in Figure 1 (left panel). In case of NaCl, the ρ_{ion} derived from the measurements is close to the one that is obtained with the values of Φ, ν, ρ_{sol}, and M_{sol} for NaCl (using Φ = 1, ν = 2). Using these values of ρ_{ion} together with σ of water, the Köhler equation was used to obtain critical diameters for the activation at supersaturations of 0.1% and 0.3%. These results are compared with the measurements in Figure 1 (right panel). Except for two data points, agreement between measurements and calculations within uncertainty was found.

For the coated soot particles that were examined during LExNo, the hygroscopic growth at 98% RH was taken to adjust ρ_{ion}. Then, again, the obtained values of ρ_{ion}, were used to model the critical super-saturation for the particles. A comparison of critical super-saturations measured with the DMT-CCNc and modeled values is given in Figure 2, showing an excellent agreement (R^2 = 0.98).

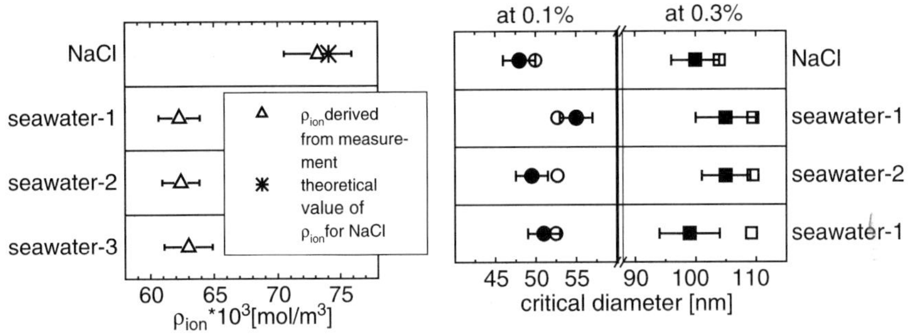

Figure 1 The left panel shows ρ_{ion} derived from the measurements. The right panel shows the critical diameters at super-saturations of 0.1% and 0.3% which were calculated based on ρ_{ion} (open symbols) and the ones which were measured with LACIS (filled symbols)

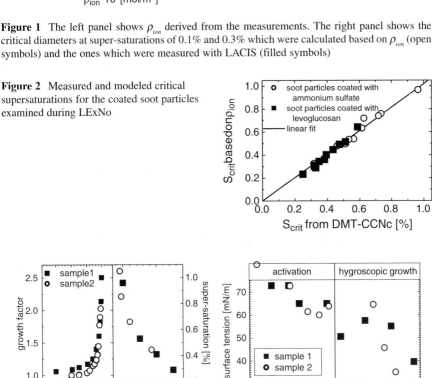

Figure 2 Measured and modeled critical supersaturations for the coated soot particles examined during LExNo

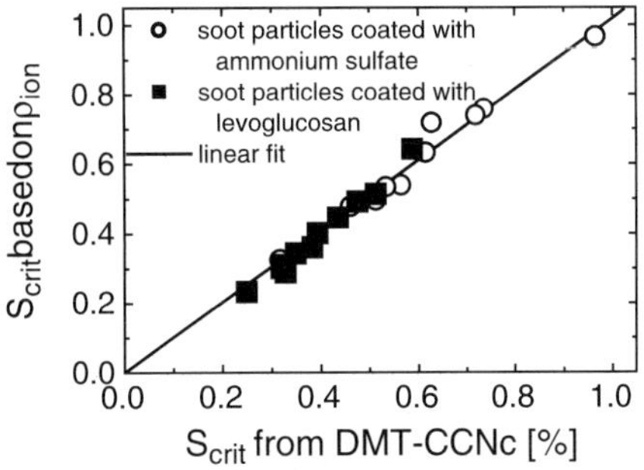

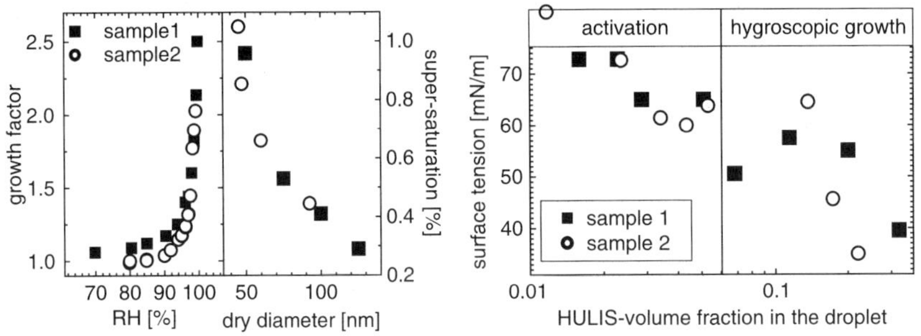

Figure 3 **Left panel**: measured hygroscopic growth and activation of two different HULIS samples. **Right panel**: surface tensions needed to reproduce these measured data

Hulis Particles

Two different HULIS samples were used, both extracted from urban aerosol samples from Budapest. The hygroscopic growth of the two samples differed at RHs below about 95%, while hygroscopic growth at larger RHs and the activation behavior was similar (see Figure 3, left panel). To reproduce the measured activation data for these samples, it was necessary to vary σ, (assuming a constant ρ_{ion}, as before) as depicted in Figure 3 (right panel). The resulting tendency in σ is an increase toward σ of water for more dilute droplet solutions. For activation on particles with diameters above about 90 nm, a lowering of the surface tension due to HULIS was not observed, being consistent with the fact that these particles consist of the most diluted solutions that were examined during these studies.

Acknowledgments We thank ACCENT for support to the guest scientists during LExNo. Sampling and preparation of the HULIS samples was supported by the Hungarian Scientific Research Fund under grant K061193.

References

1. Covert, D.S., Gras, J.L., Wiedensohler, A., and Stratmann, F., *J. Geophys. Res.*, **103**(D13), 16597–16608 (1998).
2. Brechtel, F.J. and Kreidenweis, S.M., *J. Atmos. Sci.*, **57**, 1854–1871 (2000).
3. Svenningsson, B. et al., *Atmos. Chem. Phys.*, **6**, 1937–1952 (2006).
4. Stratmann, F. et al., *J. Atmos. Oceanic Technol.*, 2**1**, 876–887 (2004).
5. Hennig, T., Massling, A., Brechtel, F., and Wiedensohler, A., *J. Aerosol Sci.*, **36**, 1210–1223 (2005).
6. Roberts, G. and Nenes, A., *Aerosol Sci. Tech.*, **39**, 206–221 (2005).
7. Wex, H. et al., *Geophys. Res. Lett.*, **34**, L02818, doi:10.1029/2006GL028260 (2007).

On Aerosol-cloud Studies at the Puijo Semiurban Measurement Station

A. Leskinen[1], T. Raatikainen[3], J. Hirvonen[1], A.-P. Hyvärinen[3], A. Kortelainen[2], P. Miettinen[2], N. Pietikäinen[1], H. Portin[1], J. Rautiainen[2], R. Sorjamaa[2], P. Tiitta[2], P. Vaattovaara[2], A. Laaksonen[2,3], K.E.J. Lehtinen[1,2], H. Lihavainen[3], and Y. Viisanen[3]

Abstract A measurement site for aerosol-cloud interaction studies is described. The newly established measurement site at Puijo observation tower produces data from semiurban environment for climatic models and particle formation studies. In addition to the basic measurements, intensive yearly measurement campaigns are organized at the station. The first aerosol-cloud interaction experiment (PUCE1) took place in October–November 2006. Some results of the measurement campaing are presented and discussed.

Keywords Aerosol-cloud interaction, aerosol mass spectrometry, urban aerosols

Introduction

Atmospheric fine particles affect the climate directly by scattering and absorbing energy, and indirectly through cloud formation and cloud optical properties, which are reported to have a cooling effect on the climate, with great uncertainty.[1] In order to make climatic model estimations more accurate, more experimental data are needed, e.g., from measurement stations, which are at times in cloud. These kinds of stations are, e.g., the GAW (Global Atmospheric Watch) stations at Jungfraujoch in Switzerland and at Pallas in Finland.

Similar particle and cloud research was recently started in Kuopio, Finland at the Puijo measurement station, which is located on the top of an observation tower. Measurements at Puijo produce data from a semiurban environment for climatic models and particle formation studies. The data from Puijo are compared with those measured at the Pallas background station.[2]

[1] *Finnish Meteorological Institute, Kuopio Unit, P.O. Box 1627, FI-70211 Kuopio, Finland*

[2] *Department of Physics, University of Kuopio, P.O. Box 1627, FI-70211 Kuopio, Finland*

[3] *Finnish Meteorological Institute, Research and Development, P.O. Box 503, FI-00101 Helsinki, Finland*

C.D. O'Dowd and P.E. Wagner (eds.), *Nucleation and Atmospheric Aerosols*, 561–564.

Experimental

Site Description

The Puijo measurement station is located on the top of an observation tower (62≡54϶32” N, 27≡39϶31” E). The 75 m high tower is located on the Puijo hill (232 m a.s.l.), 2 km from the centre of Kuopio (population ~90,000). The station has been planned and instrumented by the Finnish Meteorological Institute and the University of Kuopio for aerosol, cloud droplet, weather, and pollutant gas measurements (Table 1).

The weather parameters have been recorded continuously since October 2005. The aerosol inlets and size distribution and total concentration instruments were installed in June 2006. The cloud droplet size and aerosol optical properties measurements started in August 2006 and the pollutant gas measurements in October 2006.

The First Aerosol-Cloud Experiment

In addition to the permanent measurements at Puijo, intensive measurement campaigns with a larger set of instruments are carried out yearly. The first aerosol-cloud experiment (PUCE1) was arranged from October 16, 2006 to November 17, 2006.

Table 1 Summary of permanent measurements at the Puijo measurement site

Component	Measurement method/instrument
Aerosol light absorbing coefficient	Multiangle absorption photometer (Thermo MAAP 5012)
Aerosol number concentration	Condensation particle counter (TSI 3010 CPC, 3785 WCPC)
Aerosol light scattering coefficient	Three wavelength integrating nephelometer (TSI 3563)
Aerosol size distribution (cloud interstitial)	Differential mobility particle sizer (DMPS, 7–900 nm)
Aerosol size distribution (total)	DMPS (7–900 nm), dust monitor (Grimm #190, 0.25–32 μm)
Atmospheric pressure	Capasitive absolute pressure sensor (Vaisala BAROCAP)
Cloud droplet size distribution	Optical cloud droplet spectrometer (DMT, 2–50 μm)
Icing conditions	Ice detector
Nitrogen oxide concentration	Chemiluminescent NO–NO_x analyzer (Thermo 42i)
Ozone concentration	UV photometric O_3 analyzer (Thermo 49i)
Sulfur dioxide concentration	UV fluorescence SO_2 analyzer (Thermo 43i)
Temperature and relative humidity	Pt100 and Vaisala HUMICAP
Visibility, present weather, and precipitation	Present weather sensor (Vaisala FD12P)
Wind speed and direction	Ultrasonic two-dimensional anemometer (Thies Ultrasonic)

Cloud events with no rain were characterized by a sudden drop in visibility below 200 m and changes in aerosol scattering coefficient and cloud interstitial particle size distribution, onset of cloud droplets, and precipitation intensity less than 0.2 mm/h. The activated fraction was also measured with a cloud condensation nuclei (CCN) counter in 0.1–2% supersaturations. Aerosol mass spectrometer was used to get the aerosol chemical composition. The cloud interstitial and total particle size distributions were compared with each other and with the cloud droplet data to get the activated fraction of the particles.[3] The activated fraction was compared to the CCN-counter data. The aerosol and gas compound measurement data collected during cloudless days were utilized in particle formation studies.

Results and Discussion

The particle concentrations measured at Puijo are typical for an urban environment. The average concentration of 10–500 nm particles during June–August 2006 was 2,150 cm^{-3} (190–13,200 cm^{-3}). For comparison, at the Pallas station the long-time hourly average concentration of 7–500 nm particles has been 700 cm^{-3} (10–9,300 cm^{-3}).[4] The fraction of the particle concentration at Puijo emerging from local traffic and energy production is under further studies.

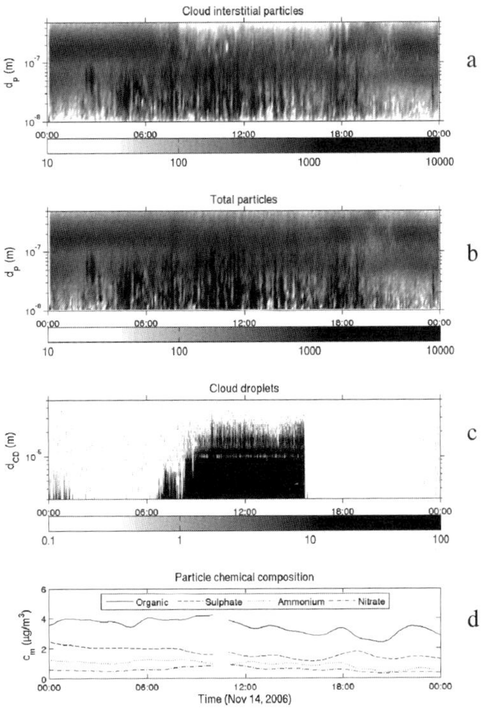

Figure 1 Cloud interstitial particle size distribution (a), total particle size distribution (b), cloud droplet size distribution (c), and particle chemical composition (d) at the Puijo measurement station on November 14, 2006. The horizontal bars denote particle and cloud droplet concentration (L/cm^3)

The weather data were analyzed for possible cloud events. The percentage when the top of the tower was in cloud (visibility below 200 m) was 15%, of which 3/5 were categorized as icing clouds. Condidering this, the suitability of the Puijo measurement site for icing studies is under discussion.

During PUCE1, a dozen of cloud events, lasting 1–8 h, were observed. As an example, cloud interstitial and total particle size distributions, cloud droplet size distribution and concentrations of organics, sulfate, ammonium, and nitrate on one day with a cloud event are shown in Figure 1. During the cloud event the wind was blowing from southeast/south, where the center of Kuopio town and major roads are located. As the cloud event started at around 8 am, a part of the accumulation mode particles were seen to disappear as they were activated into cloud droplets. The onset of the cloud event can also be seen as an increase in cloud droplet concentration. In this event the cloud droplet size ranged up to 30 µm. During the cloud event the mass concentrations of organics and sulphate in the submicron range particles were seen to decrease.

Acknowledgment The instrumentation was supported financially by the European Regional Development Fund (ERDF).

References

1. IPCC, climate change 2007: the physical science basis. Summary for policymakers. *A Contribution of Working Group I to the Fourth Assessment Report of the Intergovernmental Panel on Climate Change*. Available at http://www.ipcc.ch/SPM2feb07.pdf (20.3.2007).
2. Hatakka, J., Aalto, T., Aaltonen, V., Aurela, M., Hakola, H., Komppula, M., Laurila, T., Lihavainen, H., Paatero, J., Salminen, K., and Viisanen, Y., *Boreal Env. Res.*, **8**, 365–383 (2003).
3. Komppula, M., Lihavainen, H., Hatakka, J., Paatero, J., Aalto, P., Kulmala, M., and Viisanen, Y., *J. Geophys. Res.*, **108**(D9), 4295, doi:10.1029/2002JD002939 (2003).
4. Komppula, M., Lihavainen, H., Kerminen, V.-M., Kulmala, M., and Viisanen, Y. (2005), *J. Geophys. Res.*, **110**(D6), 6204, doi:10.1029/2004JD005200 (2005).

Partitioning of Aerosol Particles in Mixed-phase Clouds at a High Alpine Site

J. Cozic[1], B. Verheggen[1,*], E. Weingartner[1], U. Baltensperger[1], S. Mertes[2], K.N. Bower[3], I. Crawford[3], M. Flynn[3], P. Connolly[3], M. Gallagher[3], S. Walter[4], J. Schneider[4], J. Curtius[5], and A. Petzold[6]

Abstract The partitioning of aerosol particles between the cloud and the interstitial phase (i.e., unactivated aerosol) has been investigated during several Cloud and Aerosol Characterization Experiments (CLACE) conducted in winter and summer 2004 and winter 2005 at the high alpine research station Jungfraujoch (3,580 m altitude, Switzerland). Ambient air was sampled using different inlets in order to determine the scavenged fraction of aerosol particles, F_{Scav}, and of black carbon, $F_{Scav,BC}$. They denote the fraction of the aerosol volume concentration and of the black carbon (BC) mass, respectively, that has been incorporated into cloud droplets and ice crystals. They are both found to increase with increasing liquid water content, and to decrease with increasing particle number concentration. The scavenged fraction also decreases with increasing cloud ice mass fraction and with decreasing temperature from 0 to −25°C. This can be explained by the Wegener-Bergeron-Findeisen process, which describes the effect of a water vapour flux from liquid droplets to ice crystals, thus releasing the formerly activated particles back into the interstitial phase. The presence of ice could also have prevented additional particles from activating. BC was found to be scavenged into the cloud phase to the same extent as the bulk aerosol, which suggests that BC was covered with soluble material through aging processes, rendering it more hygroscopic. However, BC was found to be enriched in small ice crystals compared to the bulk aerosol, indicating that BC containing particles preferentially act as ice nuclei. If this finding is representative, it would mean that in addition to an indirect effect on liquid cloud

[1] *Laboratory of Atmospheric Chemistry, Paul Scherrer Institut, CH-5232, Villigen PSI, Switzerland*

[2] *Leibniz-Institute for Tropospheric Research, D-04318, Leipzig, Germany*

[3] *University of Manchester, M60 1QD, Manchester, UK*

[4] *Max Planck Institute for Chemistry, D-55128, Mainz, Germany*

[5] *Johannes Gutenberg University, D-55099, Mainz, Germany*

[6] *German Aerospace Centre, D-82234, Wessling, Germany*

* *Now at Institute for Atmospheric and Climate Science, ETH Zürich, 8092 Zürich, Switzerland*

C.D. O'Dowd and P.E. Wagner (eds.), *Nucleation and Atmospheric Aerosols*, 565–569.

formation, there is an indirect aerosol effect via glaciation of clouds. The observed partitioning behaviour has substantial implications for our understanding of the indirect effect of aerosols on climate.

Keywords CLACE, scavenging, CCN, ice nuclei, black carbon

Introduction

A series of international field campaigns were carried out at the Jungfraujoch station (3,580m a.s.l.) in Switzerland under the name CLACE (Cloud and Aerosol Characterization Experiment). A main focus of CLACE is the investigation of aerosol-cloud interactions in mixed-phase clouds. The Jungfraujoch station is well suited to study these processes since it is situated in the free troposphere with only minor boundary layer influence and is within clouds about 40% of the year. Studying the partitioning of particles into cloud hydrometeors (i.e., droplets and ice crystals) is important because the microphysical and optical properties of the cloud can be altered (indirect aerosol effect). This aerosol indirect effect has been recognized as the greatest source of uncertainty in assessing human impact on climate.[1]

Material and Methods

The aerosol was sampled by three well-characterized inlets: A total inlet (tot) heated to 25°C designed to evaporate cloud constituents at an early stage of sampling (i.e., sampling both cloud residuals and interstitial particles), an interstitial inlet (int) using a PM2 cyclone and collecting only unactivated aerosol particles ($d_p > 2\,\mu m$) and an Ice-Counterflow Virtual Impactor (Ice-CVI)[2] designed to sample residual particles of small ice crystals (i.e., particles that served as ice nuclei). Differencing the response downstream of the different inlets provides insight into the partitioning of the aerosol particles into cloud droplets and ice crystals.

A wide variety of physical and chemical parameters was determined downstream of these inlets and were complemented by in-situ measurements of cloud microphysical parameters. Two Scanning Mobility Particle Sizers (SMPS, TSI 3934) were used to measure the particle size distribution between 17 and 900nm (dry) diameter (one switching between the total and interstitial inlet and another one behind the Ice-CVI inlet). Two Aerosol Mass Spectrometers (AMS, Aerodyne) were operated in parallel to the two SMPS and enabled the determination of the size segregated mass loading of non-refractory chemical components (e.g., sulfate, nitrate, ammonium and organic components) in the size range of 50–1,000nm. Two optical particle counters (OPC, Grimm Dustmonitor 1.108) measured the size distribution in the diameter range $d_p = 0.3$–$20\,\mu m$. Three different instrument types measured the aerosol light absorption from which the black carbon (BC) mass concentration was deduced: a multiwavelength Aethalometer (AE31, Magee Scientific), two Multi-Angle Absorption Photometers (MAAP 5012, Thermo Electron Cooperation) and two Particle Soot Absorption Photometers (PSAP, Radiance Research, USA). Cloud droplet size distributions were measured in situ by means of a Forward Scattering Spectrometer Probe (FSSP; modified Model SPP100). A Cloud Particle Imager (CPI; SPEC Inc. Model 230X) was deployed to observe and record real-time CCD images of the ice particles and supercooled droplets with $d_p = 10\text{–}2{,}300\,\mu m$ present in the clouds. From these images the ice crystal number and mass concentration was determined. The cloud liquid water content (LWC) was continuously measured with two particulate volume monitors (PVM-100, Gerber Scientific).

Results

Partitioning of Aerosol Particles and BC

A result from the latest CLACE campaigns is that the partitioning of aerosol particles to the cloud phase is strongly dependent on the relative fraction of ice in the cloud. Figure 1a shows that the scavenged volume fraction (derived from the size distribution measurements downstream of the total and interstitial inlets and defined as $(V_{tot} - V_{int})/V_{tot}$) is about 60% in liquid clouds. The fraction of scavenged particles decreases with increasing cloud ice mass fraction (IMF)[3] to reach $F_{Scav} < 10\%$ in mixed-phase clouds with IMF>0.2. This can be explained by the Wegener–Bergeron–Findeisen process, which describes the effect of a water vapour flux from liquid droplets to ice crystals. The formation of ice during the early stages of cloud development could have prevented additional particles from activating by quickly lowering the supersaturation. This is also due to the difference in vapour pressure over ice and liquid. Figure 1b shows that black carbon (BC) mass is scavenged into the cloud phase to the same extent as the bulk aerosol. Such

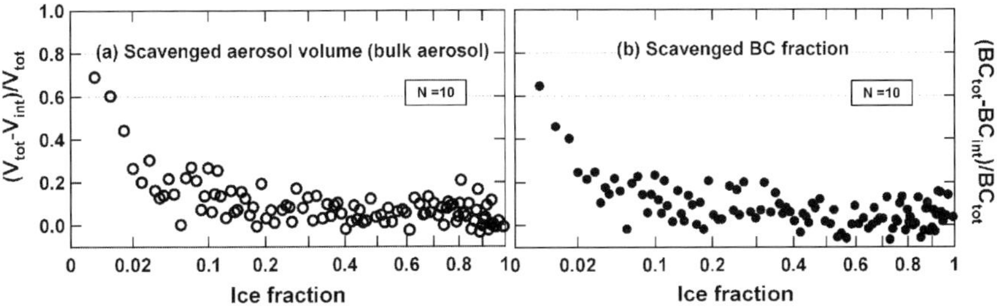

Figure 1 Scavenged fraction of aerosol volume (a) and black carbon mass (b) versus the ice mass fraction of mixed-phase clouds. Each point represents an average of 100 min of measurement

behaviour is not expected for freshly emitted soot particles because they are hydrophobic.[4] Most soot particles on the Jungfraujoch experienced aging processes which transformed them into an internally mixed hygroscopic aerosol.[5] The scavenged fraction was increased in liquid cloud with increasing liquid water content (LWC) up to a plateau of 60% and decreased with increasing particle (or BC) concentration since there is an increased competition for the available water vapour.

Enrichment of Black Carbon in Ice Residuals

The Ice-CVI allowed for the sampling and subsequent analysis of residual particles in small ice crystals (ice residuals). The chemical composition of ice residuals was remarkably different from the total aerosol (Figure 2). Comparison of SMPS and AMS data confirms the findings of Krivacsy et al.[6] that this aerosol is composed to about 95% of non-refractory material (vapourized at 600°C). Ice residuals show a significantly different signature: Ice crystal residuals sampled by the Ice-CVI show a negligible mass concentration of non-refractory material as measured by the aerosol mass spectrometer compared to the SMPS derived mass, indicating that preferably refractory (i.e., non-volatile, such as BC or mineral dust) particles act as ice nuclei. An analysis of the size resolved mass size distributions shows that the ice residuals experience a relatively larger mass contribution from particles larger than 300 nm, suggesting that larger particles (e.g., mineral dust) preferentially act as ice nuclei.

The BC mass fraction behind the total inlet ($BC_{tot}/(V_{tot}\cdot\rho)$, assuming an aerosol density of $\rho = 1.5\,g/cm^3$) was compared to the BC mass fraction behind the Ice-CVI ($BC_{cvi}/(V_{cvi}\cdot\rho)$, assuming an aerosol density of $\rho = 2\,g/cm^3$). It can be observed that

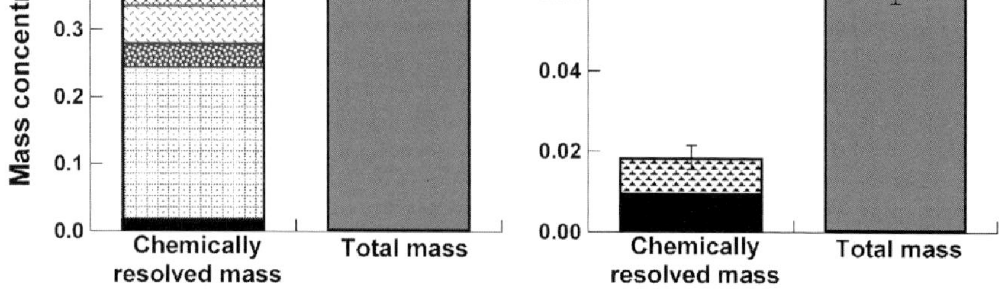

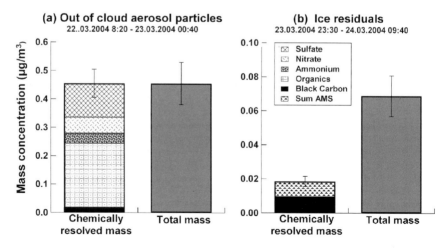

Figure 2 Example of submicrometer chemical composition of (a) the out-of-cloud aerosol and (b) ice residuals

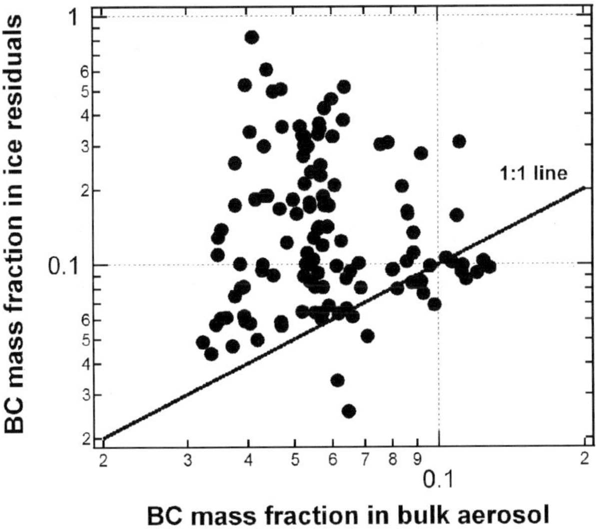

Figure 3 Comparison of the BC mass fraction in the ice residual phase with the corresponding fraction in the bulk aerosol phase in cloud. Each point represents an average of 30 min of measurement. Note that the scales are logarithmic

BC behind the total inlet represents between 3 to 10% of the total aerosol mass whereas in ice residuals it represents from 4% up to 60% of the aerosol mass. Points above the 1:1 line have a larger BC fraction in the ice residuals than in the total aerosol, while below it is the reverse is true. Figure 3 thus shows that most of the time, BC is enriched in the ice residuals compared to the bulk aerosol. On average 20% of the ice residuals is composed of BC.

Besides dust, BC also acts as potential ice nuclei. If generally true, this means that in addition to an indirect effect on liquid cloud formation, there is an indirect aerosol effect via glaciation of clouds. This result is highly important for climate since BC has a predominately anthropogenic origin.

Acknowledgment This work was supported by MeteoSwiss in the framework of the Global Atmosphere Watch program, the German Research Foundation DFG, and the U.K. Natural Environment Research Council NERC.

References

1. Intergovernmental Panel on Climate Change (IPCC) (2001).
2. Mertes, S. et al., *Environ. Sci. Technol.*, submitted (2007).
3. Verheggen, B. et al., *J. Geoph. Res.*, submitted (2007).
4. Weingartner, E. et al., *Environ. Sci. Technol.*, **36**, 55–62 (2002).
5. Cozic, J. et al., *Atmos. Chem. Phys. Discuss.*, **6**, 11877–11912 (2006).
6. Krivacsy, Z. et al., *J. Atmos. Chem.*, **39**, 235–259 (2001).

The Role of Aerosol Characteristics in the Evolution of Clouds and Precipitation

Verena Grützun, Oswald Knoth, Martin Simmel, and Ralf Wolke

Abstract A spectral microphysics scheme is implemented into the three-dimensional Lokalmodell of the German Weather Service in order to investigate the role of aerosol characteristics and cloud droplet nucleation for the evolution of clouds and the formation of precipitation. The scheme includes a combined spectral description of wetted aerosols including cloud condensation nuclei as well as cloud droplets and rain drops. An initial spectrum is prescribed which can freely evolve afterwards. Sensitivity studies on an artificial test case are undertaken showing the impact of various aerosol characteristics, such as particle number distribution, on the cloud properties. The studies show, e.g., that an increased initial particle number with sufficiently large particles leads to diminished cloud droplet sizes and a relative humidity near 100%, while fewer particles lead to supersaturations up to 15% and larger droplets. This results in a larger amount of precipitation for fewer but not too small particles.

Keywords Cloud condensation nuclei, cloud droplet nucleation, cloud aerosol interaction, precipitation formation, mesoscale modeling, precipitation forecast

Introduction

In weather forecasting, the correct quantitative prediction of precipitation is one of the main issues. Unfortunately it is only poorly done (Hense 2003). Especially extreme events are seldom predicted correctly. Most models relay on simple descriptions such as one or two moment bulk schemes parameterizing conversion rates between water vapor, cloud water, and rain water without looking at specific aerosol parameters. Yet it is clear and known for years that certain cloud properties and thus the evolution of clouds are strongly dependent on the available aerosol particles in the atmosphere. This is illustrated, e.g., by the differences between

Leibniz Institute for Tropospheric Research, Leipzig, Germany

C.D. O'Dowd and P.E. Wagner (eds.), *Nucleation and Atmospheric Aerosols*, 570–575.

clouds evolving in a maritime regime with only few aerosol particles present and continental ones with much more available particles. Another example was given by Rosenfeld (2000), who showed the influence of polluted plumes (i.e., plumes with an increased particle number) on the size of cloud droplets (CDs) as well as on the amount of the resulting precipitation. The nucleation of droplets hereby serves as the direct link between the cloud condensation nuclei (CCN) and the CDs. Various theoretical and experimental studies (e.g., Kogan 1991; Khain et al. 1999; Martinsson 1999) show that, depending on the size, composition and number of available aerosol particles, CDs of differing size distribution will evolve in the cloud. Those droplets grow by condensation and collision and eventually form precipitation.

To investigate the link between aerosol characteristics, cloud evolution, and precipitation, we combine the mesoscale Lokalmodell (Doms and Schättler 1999) with a spectral microphysics model (Simmel et al. 2002; Simmel and Wurzler 2006). Currently, we concentrate on warm microphysics, although the scheme is capable of describing mixed phase microphysics as well (Diehl et al. 2006).

Model Description

The Lokalmodell (LM) (Doms and Schättler 1999) is the regional part of the operational model system of the German Weather Service (DWD). It is nonhydrostatic and fully elastic and operates on scales in the order of 1 km. The description of grid-scale precipitation is based on a Kessler-type formulation. In our approach, the original bulk microphysics of the LM is replaced by the spectral bin microphysics creating the model system LM-SPECS. The microphysics combines wetted aerosol particles, CDs and rain drops in one joint spectrum using 66 size bins, whereby the particle/droplet water mass doubles from bin to bin. The spectrum starts at a radius of 1 nm and reaches up to several mm. In each size bin, particle number mixing ratio and water mass mixing ratio as well as soluble and insoluble aerosol mass mixing ratio are considered. A dry aerosol spectrum is prescribed and brought into equilibrium with the initial atmosphere thus being transformed into a wetted aerosol spectrum. It can evolve freely afterwards. Condensation/evaporation, collision/coalescence, and spontaneous drop breakup are the underlying microphysical processes for the changes in the hydrometeor spectrum. Since this study concentrates on warm microphysics only, the mixed phase will not be considered further.

The coupled model system works with two time steps. In a large time step of 10–100 sec, the dynamics of the winds, pressure, and temperature is computed. The dynamical tendencies of these quantities are used to drive the microphysics in the small time step of 1 sec or smaller, whereby they are linearly approximated in time. The microphysical tendencies as well as the transport of moisture-related quantities are calculated in the small time step. Since the microphysics is very sensitive to small changes in the masses, it is crucial to use a mass conserving advection scheme (Knoth and Wolke 1998).

Results

The model system LM-SPECS is applied to an artificial heat bubble. The model domain is $80 \times 80\,km^2$ with a horizontal resolution of 1 km. The vertical resolution is about 90 m near ground up to 600 m near top, whereby 48 height levels are used. A heat bubble with a maximum temperature disturbance of 2 K in the center, a horizontal radius of 10 km, and a vertical extension of 2.8 km was placed in the middle of the domain at a height of 1.4 km. The initial wind was set to zero. This bubble induces the development of a deep convective cloud. As initial particle spectrum, a unimodal lognormal distribution was used for most model runs. Presented is a sensitivity study with respect to the initial number concentration of the aerosol particles. Four cases are shown, one clean case with an initial number concentration of $100\,cm^{-3}$ (N0100), the Kreidenweis case ($566\,cm^{-3}$, N0566, Kreidenweis 2003), and a polluted one ($1{,}000\,cm^{-3}$, N1000). An additional run with a trimodal distribution (trimodal) was performed in order to investigate the influence of the shape of the distribution. The initial dry aerosol particle distributions are shown in Figure 1 (left).

The middle picture shows the particle number mixing ratio after 12 min at a height of 1.2 km. In all cases, cloud droplets are already activated, but no rain drops are present. More particles lead to an increased number of CDs which are smaller in size, though. This causes strong differences in the subsequent cloud evolution. For the small droplets, the onset of collision is delayed, so that rain drops form at

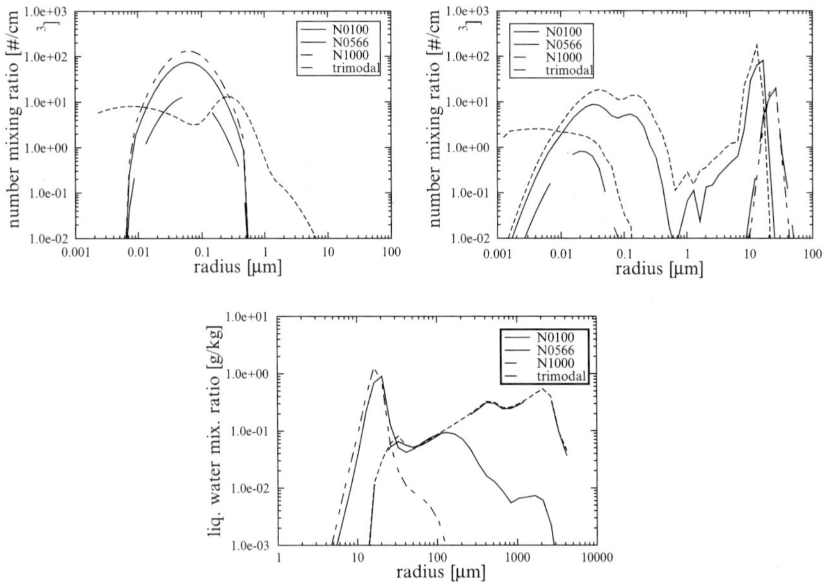

Figure 1 **Left**: Initial dry particle number mixing ratio per bin, middle: number mixing ratio per bin after 12 min, **Right**: liquid water mixing ratio per bin after 24 min and a height of 1.2 km

later times. This can be seen in the right picture showing the liquid water mixing ratio after 24 min. While cases N0100 and N0566 already have quite a few drops with radii larger than 100 μm and thus have a large amount of liquid water in the corresponding part of the spectrum, N1000 still has only small drops. In the activated part of the spectrum, the trimodal case stays near N0100 because a similar number of CDs has been activated at the beginning.

Vertical profiles of total liquid water mixing ratio, i.e., the sum over all spectral bins, (left), relative humidity (middle) and total number mixing ratio (right) after 24 min in the cloud center are shown in Figure 2. The cloud base for all cases is at almost the same height at 1 km which can be seen from the drop number mixing ratio and the relative humidity. Cloud top is at about 8 km. The maximum liquid water contents are located at different heights, though, whereby a smaller droplet number leads to a greater height. N0100 and the trimodal case already precipitate at 24 min resulting in a reduction of liquid water mixing ratio inside the cloud and an increase of liquid water mixing ratio below the cloud down to the ground due to falling rain drops.

The relative humidity shows a strong sensitivity on the initial particle number density. Whereas case N1000 stays close to saturation, the supersaturation reaches up to about 15% for N0100. This can be explained with a depletion of particles

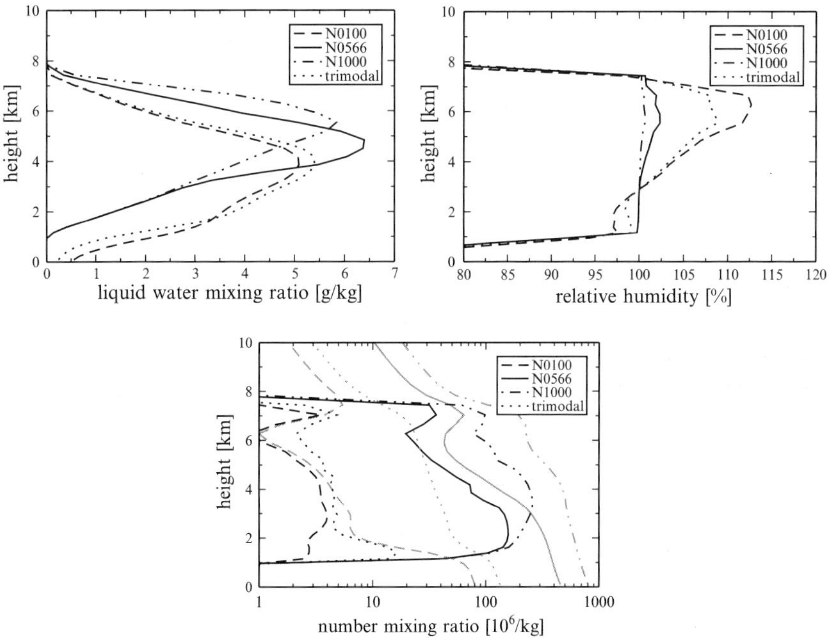

Figure 2 Vertical profiles of liquid water mixing ratio (**left**), relative humidity (**middle**) and total number mixing ratio (**right**) at center point after 24 min. Gray lines in the number concentration denote total number concentration of particles and drops, black ones only drops (radius > 1 μm)

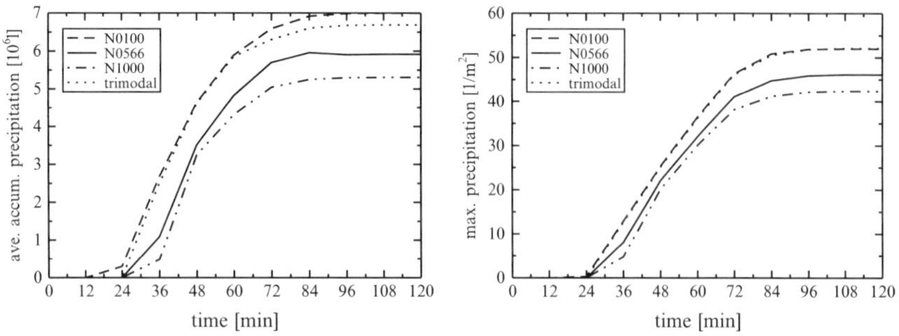

Figure 3 Average rain in precipitation area (**left**) and maximum rain at single grid point (**right**)

illustrated in the right graph. While the trimodal case shows great similarities with N0100, the depletion cannot be seen here, although the supersaturation reaches up to 10%. This follows from the large amount of particles which are too small to be activated, even at such high supersaturations, as the trimodal spectrum starts at much smaller sizes.

Figure 3 shows the resulting average accumulated precipitation of the various cases (left). It can be clearly seen that a smaller initial number mixing ratio of aerosol particles results in a larger amount of precipitation. Here again, the trimodal case stays close to the N0100. This becomes even more evident in the maximum precipitation at single grid points (right). Here, the difference between the two extremes N0100 and N1000 reaches almost 10 L/m^2, which might hint to a potential reason for the poor prediction of extreme precipitation events.

Conclusions and Outlook

A spectral microphysics scheme was coupled to the mesoscale weather forecast model LM of the DWD. The focus was on the link of aerosol particles and cloud characteristics and the subsequent development of precipitation. In a sensitivity study, it was shown that aerosol characteristics such as the initial particle size distribution play an important role in the cloud evolution and for the resulting accumulated precipitation. A smaller initial particle number concentration as in case N0100 leads to larger CDs, to an increase of supersaturation in the cloud, and to an enhanced amount of rain. Further studies will be done on various parameters, e.g., the particle chemical composition. Also, mixed phase microphysics will be investigated. Finally, we will focus on offering new approaches of considering aerosol characteristics in mesoscale modeling in a more explicit way in order to improve the precipitation forecast.

Acknowledgment This study is part of the Priority Program 1167 "Quantitative Precipitation Forecast" funded by the German Research Foundation (DFG).

References

Diehl, K., Simmel, M., and Wurzler, S., *J. Geophys. Res.-Atmos.*, **111**(D7) (2006).
Doms, G. and Schättler, U., *Techn. Rep. Part I: Scientific Documentation*, DWD (1999).
Hense, A., Adrian, G., Kottmeier, C., Simmer, C., and Wulfmeyer, V., *Initial proposal of PP 1167* (2003).
Khain, A.P., Pokrovsky, A., and Sednev, I.L., *Atmos. Res.*, **52**, 195–220 (1999).
Knoth, O. and Wolke, R., *Atmos. Environ.*, **32**, 1785–1797 (1998).
Kogan, Y.L., *J. Atmos. Sci.*, **48**, 1160–1189 (1991).
Kreidenweis, S.M. and Coauthors, *J. Geophys. Res.-Atmos.*, **108**(D7) (2003).
Martinsson, B.C. and Coauthors, *Atmos. Res.*, **50**, 289–315 (1999).
Rosenfeld, D., *Science*, **287**, 1793–1796 (2000).
Simmel, M. and Wurzler, S., *Atmos. Res.*, **80**, 218–236 (2006).
Simmel, M., Trautmann, T., and Tetzlaff, G., *Atmos. Res.*, **61**, 137–150 (2002).

Maritime CCN Drizzle Relationships

Subhashree Mishra, James G. Hudson, and C.F. Rogers

Abstract Variations in CCN concentrations in clean maritime air modulated not only cloud droplet concentrations (N_c) but also large cloud droplet concentrations (N_{LD}).

Keywords Drizzle, CCN

Introduction

During the Rain in Cumulus over the Ocean (RICO) field project in December–January (2004–2005) in the eastern Caribbean differences in horizontal wind velocity (v) produced significant differences in giant nucleus (GN) concentrations.[1] But N_{LD} concentrations were negatively associated with GN.[1] Since a good correlation between v and vertical wind (w) and between w and N_c, and higher N_{LD} when w and N_c were low, it appeared that lower w, which should produce lower N_c, enhanced precipitation.[1]

Measurements

Flight-averaged CCN measurements at 100 m altitude by the Desert Research Institute (DRI) CCN spectrometers[2] are presented in Table 1 for 12 RICO flights that were previously analyzed[1] and two additional flights (RF1 and 2; D 7 and 8). Criteria for flight-averaged N_c from the NCAR PMS Forward Scattering Spectrometer Probe (FSSP) are 600–900 m altitude, LWC > 0.25 g m^{-3}, and w > 0.5 m s^{-1}.[1]

Desert Research Institute Nevada System of Higher Education Reno, NV 89512–1095, USA

C.D. O'Dowd and P.E. Wagner (eds.), *Nucleation and Atmospheric Aerosols*, 576–579.

Table 1 Flight number, date, flight-averaged N_c (cm^{-3}), CCN, and w ($m\ s^{-1}$). Adjacent columns show flight rankings of these quantities for the latter 12 flights

Flight	Date	N_c	N_c rank	w	w rank	CCN	CCN rank
RF01	D07	201		1.77		200	
RF02	D08	123		1.64		133	
RF03	D09	74	10	2.20	3	98	7
RF05	D13	71	11	1.89	9	91	9
RF06	D16	111	5	2.17	6	131	2
RF07	D17	46	12	1.85	10	55	12
RF08	D19	98	7	2.18	5	79	11
RF09	D20	74	9	1.64	12	94	8
RF10	J05	80	8	2.00	8	108	6
RF12	J11	135	2	2.19	4	90	10
RF13	J12	114	3	2.50	1	119	4
RF14	J14	207	1	2.48	2	142	1
RF15	J16	112	4	2.15	7	108	5
RF18	J23	104	6	1.73	11	126	3
ave	111			2.05		112	
sd	46			0.29		35	

Results

Figure 1 indicates that CCN concentration variations had as much influence on N_c as w – correlation coefficient (R) 0.80. For only the 12 flights previously considered[1] R=0.70 versus 0.66 for w-N_c.[1] A crude demonstration of the combined effects of N_c and w is a plot of the product of CCN and w against N_c. This brings R for the 12 flights up to 0.82 and for the 14 flights to 0.86 (Figure 2). Since this exercise is only meant to estimate the relative influences, the products were normalized by dividing by the RICO average w (2.03 m/s).

There was a trend of increasing N_{LD} with decreasing v and w.[1] In Figure 3A, where the flight numbers are used as data points, CCN are somewhat correlated with v, R = 0.37. Two exceptions to the N_{LD}-v and w inverse relationship – RF6 and 18 (D16 and J23) were reported.[1] These did not conform because they had the 2nd and 3rd lowest v, 6th and 11th ranked w and 5th and 6th ranked N_c, but low N_{LD}. The explanation for this may be 2nd and 3rd highest CCN concentrations (Table 1 and Figure 3A where they lie out to the upper left)! When these two flights are excluded R is 0.82 (Figure 3B)! Therefore, since CCN are so well correlated with v for the ten flights for which N_{LD} is inversely related to v and w, it would seem that N_{LD} is inversely related to CCN concentrations. Those two exceptions to the N_{LD}-v and N_{LD}-w inverse relationship with 2nd and 3rd ranked CCN concentrations provide further testimony of the positive effect of CCN on N_c and the negative effect on N_{LD}. For these flights N_c was 5th and 6th ranked and N_{LD} was low. The average of the CCN and w ranks for these flights is 4 and 7, which closely matches the N_c ranks of these flights.

Five of the flights that followed the N_{LD}-v-w inverse trend are displayed in Figure 3 of Colon-Robles et al.[1] and Table 2 here in N_{LD} order – highest to lowest – rows 1 to 5.

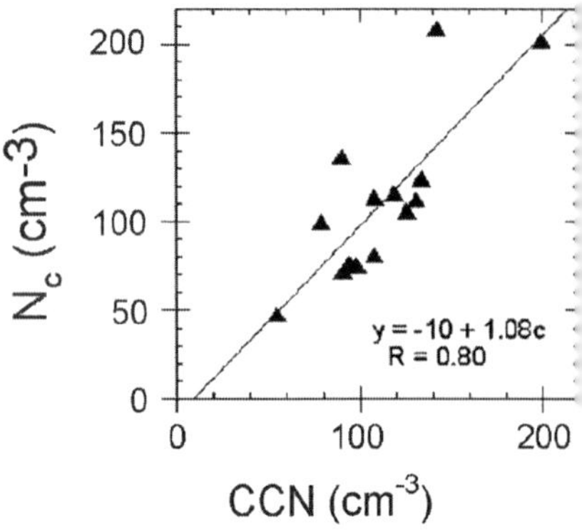

Figure 1 Flight-averaged cloud droplet concentrations (Nc) against flight-averaged CCN

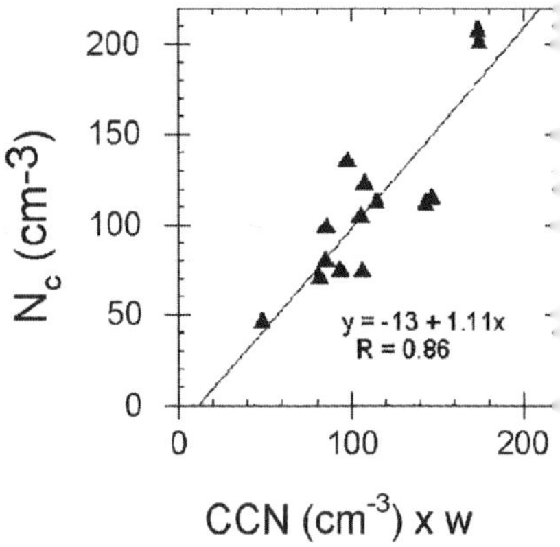

Figure 2 As Figure 1 but Nc versus the product of CCN concentrations and normalized updraft velocity (w)

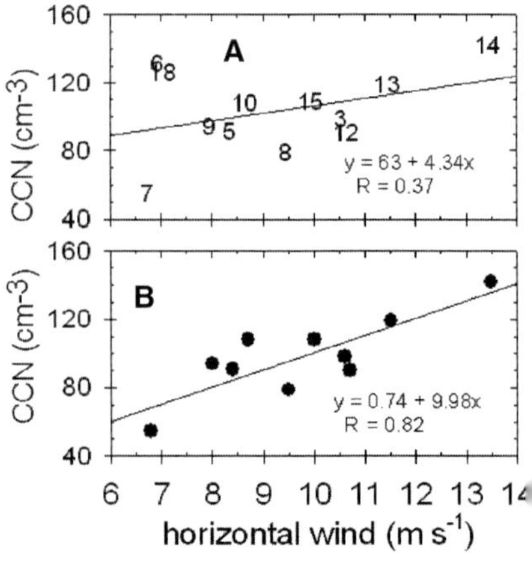

Figure 3 (a) CCN at 1% S versus horizontal wind speed, data points are flight numbers. (b) As A but excluding RF6 and RF18 (D16 and J23)

Table 2 Flight ranking of large cloud droplet concentrations (NLD) for five of the 10 flights that ollowed the NLD-v inverse trend1, flight number, date, horizontal wind (v) (m s[-1]), average Nc cm[−3]) and overall Nc flight ranks, average CCN concentrations (cm[-3]), and overall CCN flight anks, overall flight ranks of updraft velocity (w), and average of the CCN and w flight ranks

N_{LD} ank	Flight	Date	v	N_c	N_c rank	CCN	CCN rank	w rank	Ave CCN and w rank
[illegible]	7	D17	6.8	46	12	55	12	10	11
[illegible]	9	D20	8.0	74	9	94	8	12	10
[illegible]	8	D19	9.5	98	7	79	11	5	8
[illegible]	12	J11	10.7	135	2	90	10	4	7
[illegible]	14	J14	13.5	207	1	142	1	2	1.5

The order of v and N_c is of course completely reversed, while CCN and w are nostly reversed. The flight with the highest N_{LD} (D17) had the lowest N_c and CCN out only the 3rd lowest w. The flight with the lowest N_{LD} (J14) had the highest N_c and CCN but only the 2nd highest w. The last column, which combines CCN and w, is in complete reverse order.

Conclusions

We have shown that the factor of four variability in flight-averaged CCN concentrations observed during RICO were related to both the total cloud droplet concentrations (N_c) and the concentrations of large cloud droplets (N_{LD}). In combination with previous results[1] this analyses shows that the aerosol with the most nfluence on N_c and N_{LD} (precipitation) is CCN and not GN. The present results lso demonstrate that CCN variability may modulate precipitation in clean maritime air masses.

Acknowledgment Support was from the US National Science Foundation Grant ATM-0342618. Measurements other than CCN were provided by the Research Aviation Facility of NCAR, which rovided the measurement platform, the C-130 airplane.

References

. Colon-Robles, M., Rauber, R.M., and Jensen, J.B., *Geophys. Res. Lett.*, **33**, L20814, (2006).
. Hudson, J.G., *J. Atmos. & Ocean. Technol.* **6**, 1055–1065 (1989).

Effect of Nucleation and Secondary Organic Aerosol Formation on Cloud Droplet Number Concentrations

Risto Makkonen[1], Ari Asmi[1], Hannele Korhonen[2], Harri Kokkola[3], Simo Järvenoja[4], Petri Räisänen[4], Kari Lehtinen[5], Ari Laaksonen[5], Veli-Matti Kerminen[4], Heikki Järvinen[4], Ulrike Lohmann[6], Johann Feichter[7], and Markku Kulmala[1]

Abstract The global general circulation model ECHAM5 is used together wit HAM aerosol module to investigate the effect of the nucleation scheme on clou droplet number concentrations. It is shown that nucleation can have a significan role on indirect aerosol effect. Also an efficient SOA formation scheme is intro duced, and results are compared with original ECHAM5-HAM.

Keywords Nucleation, SOA, atmospheric aerosols, organic aerosols

Introduction

The continental biosphere emits large quantities of volatile organic compound (VOCs) capable of forming secondary organic aerosols (SOA). By acting effec tively as cloud condensation nuclei [1], these aerosols have the potential to influ ence climate. The climatic effects of biogenic SOA are, however, extremely poorl known [2]. Biogenic VOC emissions may increase substantially in the future as result of global warming, in addition to which there are potentially complex feed back mechanisms between biogenic VOC emission, aerosols, clouds and climat [3]. As a result, the climatic effects of biogenic VOC emissions cannot be quanti fied without an Earth System Model that couples the biosphere and atmosphere an explicitly treats the microphysical processes associated biogenic SOA formatior

[1] *Department of Physical Sciences, University of Helsinki, FI-00014, Helsinki, Finland*

[2] *School of Earth and Environment, University of Leeds, LS2 9JT, Leeds, UK*

[3] *Finnish Meteorological Institute, FI-70211, Kuopio, Finland*

[4] *Finnish Meteorological Institute, FI-00101, Helsinki, Finland*

[5] *Department of Physics, University of Kuopio, FI-70211, Kuopio, Finland*

[6] *Institute of Atmospheric and Climate Science, ETH Zurich, CH-8092, Zurich, Switzerland*

[7] *Max Planck Institute for Meteorology, D-20146, Hamburg, Germany*

C.D. O'Dowd and P.E. Wagner (eds.), *Nucleation and Atmospheric Aerosols*, 580–584.

Nucleation is often speculated as the main source of atmospheric aerosol particles in terms of their number concentration, especially in remote areas. The current models are based on binary nucleation schemes, which have been found to be insufficient for explaining observed particle formation in the lower troposphere. New formulations for atmospheric nucleation exist, but their strength and effect on climatically important aerosol properties have not been quantified.

Here we introduce the implementation of a new, computationally effective secondary organic aerosol (SOA) formation scheme into a global climate model, together with new nucleation mechanisms.

Methods

Modelling Framework

Aerosol modelling is done with HAM-module within the ECHAM5 atmospheric general circulation model (GCM) [4]. HAM uses M7 microphysics [5] to describe the aerosol populations which are defined by seven modes: a soluble nucleation mode, and both soluble and insoluble Aitken, accumulation, and coarse modes. The chemical compounds in aerosol particles include sulfate, primary organic carbon, black carbon, mineral dust, and sea salt. Aerosol dynamics with log-normal distributions in M7 have proven to be relatively fast and accurate, hence applicable for a GCM. We use a cloud droplet nucleation model [6] to investigate aerosol-cloud interaction.

Nucleation

The standard setup of the ECHAM5-HAM uses a binary nucleation parameterization [7] to account for the production of new aerosol particles. This scheme produces results that are comparable with measurements made in the upper troposphere and stratosphere but, in contrast to observations, there is a lack of nucleation mode particles in the lower troposphere. The ECHAM5-HAM also includes an option for another nucleation parameterization [8], but this parameterization results in even lower particle number concentrations.

We have included two new nucleation mechanisms in ECHAM5-HAM: kinetic and activation-type nucleation, both of which have been used for modelling particle formation events in boreal forests [9]. Here we present results for the activation-type nucleation only.

Nucleated particles were placed in the soluble nucleation mode by calculating the formation rate of 3 nm particles using the following parameterization [10]

$$J_{3nm} = A \cdot [H_2SO_4] \cdot \exp\left\{\gamma \cdot \left(\frac{1}{3} - 1\right) \cdot \frac{CS'}{GR}\right\} \qquad (1)$$

Here $A = 2 \cdot 10^{-6}$ (s^{-1}) is the so-called activation coefficient for nucleation, [H_2SO_4] is the concentration of gaseous sulphuric acid (cm^{-3}), *CS* is proportional to the condensation sink, *GR* is particle growth rate and γ is a coefficient that accounts for the ambient temperature and pre-existing aerosol particle population. The growth rate was estimated for sulphuric acid condensation only.

SOA Formation from Biogenic Sources

Originally the HAM-module handles the biogenic SOA formation by distributing the soluble part (assumed to be 65%) of biogenic organic matter evenly between the soluble Aitken and accumulation mode, and the insoluble part into insoluble Aitken mode.

Our approach distributes the biogenic organic matter by calculating condensation on existing aerosol particle population. The amount of condensing organic matter is estimated by assuming an aerosol yield of 5–20% from biogenic VOC emissions. Due to the rapid photochemistry of biogenic VOCs, this fraction is condensed into aerosol particles within the same grid box as in which VOCs are emitted. Organic matter is allowed to condense on all seven modes, and its mass is processed in coagulation and condensation calculations. The mass from insoluble modes is transferred to soluble modes during particle aging.

Results

The vertical profiles of cloud droplet number concentrations (CDNC) obtained in the two simulations (activation-type and binary nucleation) are shown in Figure 1. The enhanced particle production in the lower troposphere, typical for the activation type nucleation, seems to have a significant effect on CDNC.

The results for a one-year simulation with the new SOA scheme are presented in Table 1, including a comparison with the original setup. It can be noted that most of biogenic organic carbon is distributed to the largest modes with the new scheme. The total amount of organic carbon is not exactly the same in the two simulations, since the production of condensing organic matter is dependent on the yield, and the strength of sink processes vary between the different aerosol modes.

Discussion

It is shown that nucleation can have a significant effect on cloud properties by modifying cloud droplet number concentrations. The choice of a nucleation scheme is therefore important in climate simulations.

Mechanistic treatment of biogenic SOA formation changes global averages of organic carbon size distributions. For example, the significance of organic mass in the coarse mode increases. We are currently performing simulations to find out the

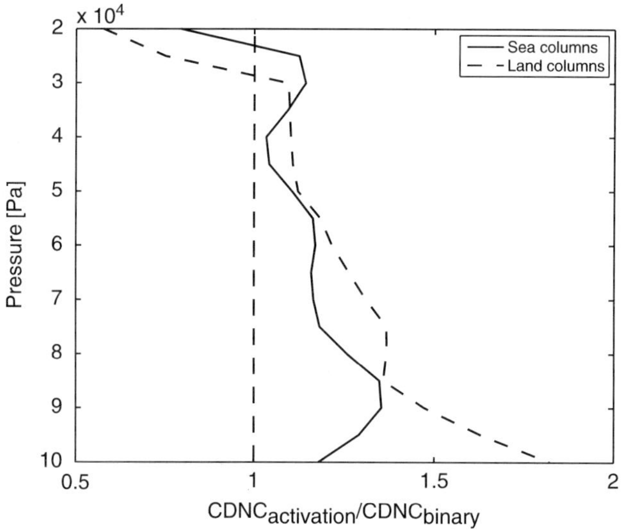

Figure 1 Ratio of CDNC between simulations with activation-type and binary nucleation. Values are yearly averages from northern hemisphere, separately for land and sea areas

Table 1 Distribution of biogenic organic carbon into different aerosol modes

Mode	Mechanistic treatment (%)	Implicit treatment (%)
Nucleation – soluble	0.02	NA
Aitken – soluble	0.72	32.50
Accumulation – soluble	7.35	32.50
Coarse – soluble	51.14	0.00
Aitken – insoluble	1.81	35.00
Accumulation – insoluble	11.69	NA
Coarse – insoluble	27.27	NA

simultaneous effect of the newly implemented nucleation schemes and explicit SOA formation. It is expected that CDNC increases even more, since there are more particles available for organic matter to grow.

References

1. Hartz, K.E.H., Rosenorn, T., Ferchak, S.R., Raymond, T.M., Bilde, M., Donahue, N.M., and Pandis, S.N., Cloud condensation nuclei activation of monoterpene and sesquiterpene secondary organic aerosol, *J. Geophys. Res.*, **110**, D14208, doi:10.1029/2004JD005754 (2005).
2. Tsigaridis, K., Lathiere, J., Kanakidou, M., and Hauglustaine, D.A., Naturally driven variability in the global secondary organic aerosol over a decade, *Atmos. Chem. Phys.*, **5**, 1891–1904 (2005).

3. Kulmala, M., Suni, T., Lehtinen, K.E.J., Dal Maso, M., Boy, M., Reissell, A., Rannik, U., Aalto, P., Keronen, P., Hakola, H., Bäck, J., Hoffmann, T., Vesala, T., and Hari, P., N new feedback mechanism linking forests, aerosols, and climate, *Atmos. Chem. Phys.*, **4**, 557–562 (2004).
4. Stier, P., Feichter, J., Kinne, S., Kloster, S., Vignati, E., Wilson, J., Genzeverld, L., Tegen, I., Werner, M., Balkanski, Y., Schultz, M., Boucher, O., Minikin, A., and Petzold, A., The aerosol-climate model ECHAM-HAM, *Atmos. Chem. Phys.*, **5**, 1125–1156 (2005).
5. Vignati, E., Wilson, J., and Stier, P., M7: an efficient size-resolved aerosol microphysics module for large-scale aerosol transport models, *J. Geophys. Res.*, **109**, D22202, doi:10.1029/2003JD004485 (2004).
6. Lohmann, U., Stier, P., Hoose, C., Ferrachat, S., Roeckner, E., Zhang, J., Cloud microphysics and aerosol indirect effects in the global climate model ECHAM5-HAM, *Atmos. Chem. Phys. Discuss.*, **7**, 3719–3761 (2007).
7. Vehkamäki, H., Kulmala, M., Napari, I., Lehtinen, K.E.J., Timmreck, C., Noppel, M., and Laaksonen, A., An improved parameterization for sulfuric acid/water nucleation rates for tropospheric and stratospheric conditions, *J. Geophysical Res.*, **107**(D22), 4622, doi:10.1029/2002JD002184 (2002).
8. Kulmala, M., Laaksonen, A., Pirjola, L., Parameterization for sulphuric acid/water nucleation rates, *J. Geophys. Res.*, **103**, 8301–8307 (1998).
9. Kulmala, M., Lehtinen, K.E.J., and Laaksonen, A., Why formation rate of 3 nm particles depends linearly on sulphuric acid concentration?, *Atmos. Chem. Phys. Discuss.*, **5**, 11277–11293 (2005).
10. Kerminen, V.-M., Anttila, T., Lehtinen, K.E.J., and Kulmala, M., Parametrization for atmospheric new-particle formation: application to a system involving sulfuric acid and condensable water-soluble organic vapors, *Aerosol Sci. Technol.*, **38**, 1001–1008 (2004).

Measurements of the Rate of Cloud Droplet Formation on Atmospheric Particles

Chris Ruehl[1], Patrick Chuang[1], and Athanasios Nenes[2]

Abstract The influence of aerosols on cloud properties is an important modulator of the climate system, however, it is not known to what extent particles exist in the atmosphere that may be prevented from acting as CCN by kinetic limitations. We measured the rate of cloud droplet formation on atmospheric particles sampled at four sites across the United States during the summer of 2006, and found that on 7 of 16 days at 3 of 4 sites, ambient aerosols that grew significantly more slowly than ammonium sulfate aerosol were present. These results suggest that for some air masses, accurate quantification of CCN concentrations may need to account for kinetic limitations.

Keywords Cloud condensation nuclei, tropospheric aerosols, indirect effects

Introduction

After several decades of research on the influence of human activities on the Earth's climate, the largest source of uncertainty in anthropogenic radiative forcing of the atmosphere remains the effect of atmospheric particles on cloud properties. The ability to predict the size distribution of cloud droplets given a size distribution of suspended particles is essential if the magnitudes of aerosol indirect effects are to be quantified. In the atmosphere, cloud droplets form on preexisting aerosol particles when the water vapor supersaturation (S) exceeds some threshold. For almost a century, Köhler theory[1] has been used to determine the minimum (critical) supersaturation (S_C) required to activate a particle of known size and soluble fraction, allowing the particle to grow into a cloud droplet via condensation of water vapor. Classical Köhler theory, however, predicts only the equilibrium S_C of a particle, and thus does not incorporate any potential kinetic limitations to cloud droplet formation. Failure

[1] Earth & Planetary Sciences Department, University of California, Santa Cruz, CA

[2] Chemical and Biomolecular Engineering, Georgia Institute of Technology

C.D. O'Dowd and P.E. Wagner (eds.), *Nucleation and Atmospheric Aerosols*, 585–590.

to take into account kinetic limitations could result in errors in calculated radiative forcing similar to that of anthropogenic greenhouse gases (~2 W/m^2).[2] It is therefore of interest if particles exist in the atmosphere that, under typical atmospheric supersaturations (~0.1–1%), would form cloud droplets at equilibrium, but not within atmospherically relevant time scales. The kinetics of cloud droplet growth are often parameterized with the mass accommodation coefficient (α), which conceptually is the probability that a water vapor molecule colliding with a growing droplet will be incorporated into the liquid phase. In this study, we refer to the "apparent" mass accommodation coefficient (α_{app}) because we cannot distinguish between different mechanisms of kinetic limitation to droplet growth. The purpose of this study is to measure α_{app} for various ambient aerosols (urban, regional polluted, and background), to see if aerosols with potential kinetic limitations to droplet formation exist in the atmosphere.

Experimental

We measured the condensational growth rate of particles that had passed through a conditioner with RH ~80% just upstream of a differential mobility analyzer (DMA) (Figure 1). The DMA selected a quasi-monodisperse particle population with a mean diameter in the range of 100–250 nm. This flow was split between a condensation

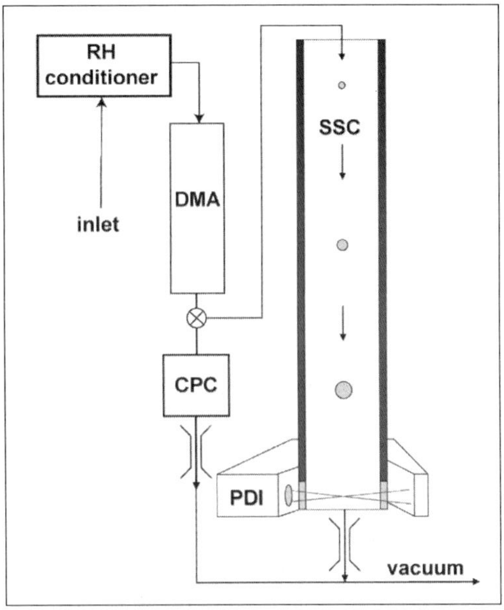

Figure 1 Experimental setup

particle counter (CPC) and a supersaturating column-phase Doppler interferometer (SSC-PDI). The SSC produced a water vapor supersaturation (S) along its centerline when a temperature gradient was applied to the wetted walls of the column. S was calibrated with laboratory-generated ammonium sulfate particles, which have well-known droplet activation properties. After flowing for 10 s through an isothermal entrance length with RH~100%, particles were exposed to a known S in the SSC for 30 sec. The velocity and diameter of the activated droplets was then measured with the PDI while still subject to S (i.e., before the particles exited the SSC).

We parameterized droplet growth rates by transforming observed droplet diameter (D_d) distributions into α_{app} distributions with a fully coupled numerical flow model that simulated conditions in the SSC.[3] This model solves for the time-dependent droplet condensational growth, correcting for non-continuum effects.[4] Figure 2 shows how modeled α varies with S and D_d when a dry ammonium sulfate particle with D = 100 nm is exposed to S for 30 sec. We group PDI measurements taken at a range of S and determined the percentage of droplets with α_{app} below two arbitrary thresholds (10^{-3} and $10^{-2.5}$). Using this approach, laboratory-generated ammonium sulfate particles with D = 100 nm had $\alpha_{app} \sim 10^{-2}$.

All equipment was housed in a trailer, and was deployed at four sites across the United States (Figure 3) during the summer of 2006. The sites were selected to sample a variety of general air mass types: urban (HOU – Houston, Texas), polluted regional (GSM – Great Smoky Mountain National Park, Tennessee), and background continental (BON – Bondville, Illinois, and SGP – Lamont, Oklahoma).

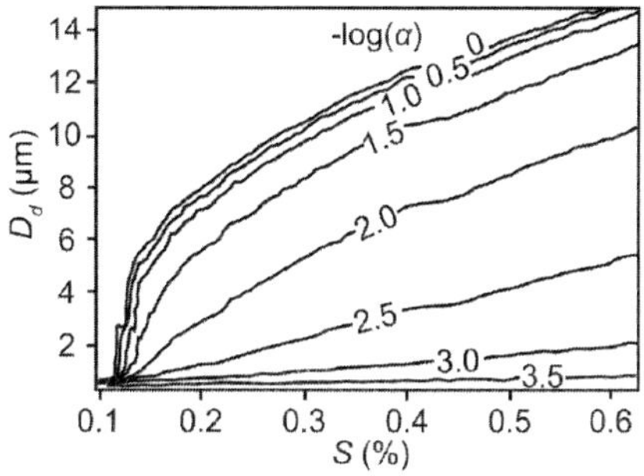

Figure 2 $\alpha(D_d, S)$ for $NH_4(SO_4)_2$ particles (Ddry = 100 nm)

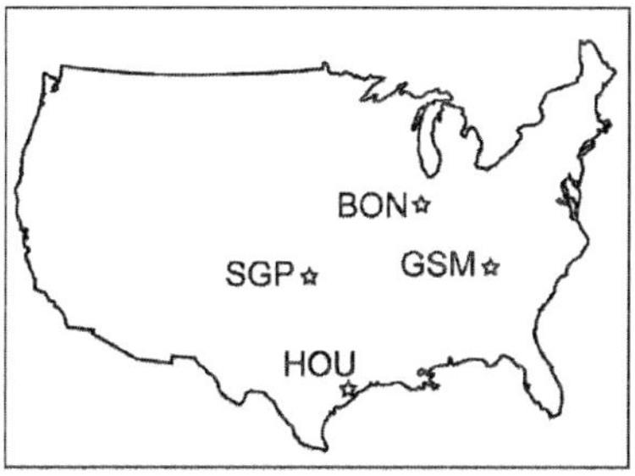

Figure 3 Map of field sites

Ambient aerosol was analyzed in the same way as laboratory-generated ammonium sulfate particles.

Results/Conclusions

The growth rate of droplets was measured as described above on a total of 16 days between August 10 and September 22, 2006 (Table 1). On days when multiple scans yielded distinct results, the scans are listed separately. We broadly classify these 16 days into three groups: (1) little/no, (2) moderate, or (3) highest-observed slowly growing droplets. Nine of the 16 days had little/no low-α_{app} droplets (<0.9% and <21.4% with $\alpha_{app} < 10^{-3}$ and $10^{-2.5}$, respectively), 5 days had moderate low-α_{app} droplets (2.7 – 4.9% and 19.6 – 48.0% with $\alpha_{app} < 10^{-3}$ and $10^{-2.5}$, respectively), and 2 days had one scan with the highest observed low-α_{app} droplets (>10.8% and >48.5% with $\alpha_{app} < 10^{-3}$ and $10^{-2.5}$, respectively). Of the 7 days in which moderate or highest-observed low-α_{app} droplets were observed, 4 were at the urban site (HOU), 1 was at the polluted regional site (GSM), and 2 were at one of the continental background sites (BON).

Table 1 Results of all droplet growth rate measurements

Date	Time	Site	n	$\alpha_{app}<10^{-3}$ (%)	$\alpha_{app}<10^{-2.5}$ (%)
August 10, 2006[1]	10:35–15:32	GSM	1867	0.7	8.8
August 12, 2006[1]	12:12–19:02	GSM	481	0.1	5.0
August 13, 2006[2]	12:58–15:22	GSM	101	2.7	21.9
August 15, 2006[1]	21:27–21:32	GSM	19	0	0
August 19, 2006[1]	13:58–14:10	BON	58	0	4.7
August 23, 2006[1]	13:17–15:26	BON	33	0	5.2
August 24, 2006[3]	12:24–13:02	BON	420	1.2	26.2
	13:06–13:47	BON	202	10.9	48.5
	17:02–17:29	BON	95	0	16.8
August 25, 2006[3]	11:49–12:27	BON	219	0.9	8.2
	13:06–14:22	BON	166	10.8	53.0
	14:38–17:18	BON	485	0.4	9.7
August 28, 2006[1]	13:40–14:33	BON	85	0	0
September 4, 2006[2]	16:27–16:50	HOU	43	3.8	19.6
September 6, 2006[2]	17:29–19:39	HOU	123	4.9	28.5
September 7, 2006[2]	11:47–12:16	HOU	683	3.3	48.0
	16:58–17:40	HOU	274	2.2	18.4
September 8, 2006[1]	11:24–11:32	HOU	164	0	1.8
September 11, 2006[2]	12:23–21:17	HOU	466	2.8	26.8
September 16, 2006[1]	16:01–16:36	SGP	97	0.9	21.4
September 22, 2006[1]	14:39–14:57	SGP	43	0	13.5

We compared NOAA HYSPLIT back-trajectories for BON, and according to these analyses, air from aloft (>1,000 m elevation) arrived at BON throughout both the days in which low-α_{app} droplets were highest-observed (24 and 25 August), but was absent on other days (Figure 4). A similar pattern was seen at GSM: Air from aloft arrived on 13 August, when low-α_{app} droplets were moderate, but was absent on other days.

These results suggest that aerosols containing particles with α_{app} lower than that observed for laboratory-generated ammonium sulfate by at least a factor of ten are common in the atmosphere, as they were observed on seven out of sixteen days at three of four field sites. Kinetic limitations of this magnitude could keep these particles from forming cloud droplets. These particles might also have a longer atmospheric lifetime due to less efficient removal by wet deposition. This could explain why slowly growing particles seem to be more prevalent in air masses arriving from aloft, if these particles have already been subject to one or more cycles of cloud-forming conditions.

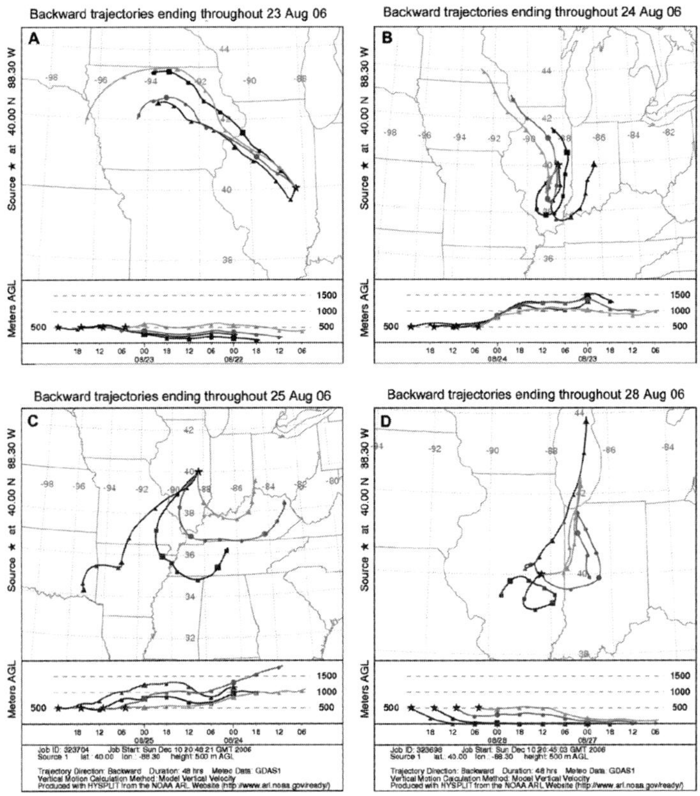

Figure 4 Back-trajectories from BON

Acknowledgments The authors would like to acknowledge funding from the Nation Science Foundation Graduate Research Fellowship Program, as well as the National Aeronautics and Space Administration.

References

1. Köhler, H., *Trans. Far. Soc.*, **32**, 1152–1161 (1936).
2. Chuang, P.Y., Charlson, R.J., and Seinfeld, J.H., *Nature*, **390**, 594–596 (1997).
3. Lance, S., Medina, J., Smith, J.N., and Nenes, A., *Aer. Sci. Tech.*, **40**, 242–254 (2006).
4. Fukuta, N. and Walter, N.A., *J. Atmos. Sci.*, **27**, 1160–1172 (1970).

Aerosol Particle Formation in Different Types of Air Masses in Hyytiälä, Southern Finland

L. Sogacheva[1], L. Saukkonen[2], M. Dal Maso[1], and M. Kulmala[1]

Abstract We analyzed synoptic weather maps and satellite images over Scandinavia for the period of 2003–2005 in order to reveal in which types of air masses the new particles formation occur more frequently in Hyytiälä, southern Finland. We calculated monthly number of different types of air masses (Arctic, Polar, and Subtropical) and the frequency of different types of air masses for the days classified according the intensity of new particle formation. The results show that for all seasons Arctic air masses bring most favorable condition for aerosol particle formation in boreal forest. In about 40% of arctic air surges new particles are registered at Hyytiälä.

We also calculated the median particle concentration for three different modes (nucleation, <25 nm, Aitken, 25–90 nm, and accumulation, >90 nm) in Arctic, Polar, and Subtropical air masses during the days when no frontal passages have been observed over Hyytiälä.

Keywords Particle formation, air mass, particle concentration

Introduction

Climatological aspects of new particle formation are of great interest last decades because of incontestable effect of aerosol particles on the Earth heat balance vie direct and indirect effects. The role of the solar radiation has been emphasized.[1,2] Hellmuth[3] investigated nucleation burst evolution in convective boundary layer and using model calculations confirmed the conclusion made by Hyvönen et al.[4] that the probability of having event day is anticorrelated with condensation sink and relative humidity. The effects of air masses and synoptic weather on aerosol

[1]*Department of Physical Sciences, University of Helsinki, P.O. Box 64 FIN-00014, Helsinki, Finland*

[2]*Finnish Meteorological Institute, Erik Palménin aukio, FI-00560 Helsinki, Finland*

C.D. O'Dowd and P.E. Wagner (eds.), *Nucleation and Atmospheric Aerosols*, 591–595.

particle formation, the influence of physical and meteorological parameters on nucleation events have been investigated in Hyytiälä, southern Finland. Nilsson et al.[5] reported that during the BIOFOR 3 experiment (March–April 1999) nucleation have occurred in Arctic and to some extent in Polar air masses. Sogacheva et al.[6] by means of trajectories analysis confirmed that new particle formation occur more frequently in the air masses originated and transported to Hyytiälä from over the Arctic and North Atlantic.

In this presentation we show the results from the long-term air masses analysis made for Hyytiälä for the period of 2003–2005 in order to reveal how the monthly distribution of air masses types coincide with nucleation event monthly frequency, which types of air masses are most typical for event and nonevent days, and what is the mean monthly concentration of particles of different sizes in different types of air masses.

Methods

To investigate type of the air masses arrived at Hyytiälä and the passages of the atmospheric fronts we analyzed the synoptic maps of the surface and 850 hPa layers ("Berliner Wetterkarte") together with the satellite images form the AVHRR five channel scanning radiometer. We have used the air mass classification made daily at Berliner Wetterkarte for 00 UTC (Finnish wintertime). The air mass classification recognize Arctic, Polar, and Subtropical air masses, each divided into marine and continental and transition case in between.

The particle number concentration and size distribution has been measured at SMEAR II station in Hyytiälä, southern Finland. Days were classified either as event days, nonevent days or undefined days. Event days were further classified into separate classes (class Ia, class Ib, and class II) regarding the possibility to drive the characteristics from size distribution.[7] For class I days the growth rate could be determined with good confidence level; while for class II the derivation of these parameters was not possible or the accuracy of the results was questionable.

Results and Discussion

It is logical that Arctic air masses, which are considered to be the most favorable for new particle formation, are observed more frequently during winter period (November–March), in Hyytiälä (Figure1a). However, November, January and February are the months with the lowest frequency of new particle formation (5, 4, and 8 events for three years of measurements, respectively). Considering the solar radiation as one of the key parameters in particle formation, this can be explained be the fact that about 50% of the days during that period are characterized by opaque clouds (>4 octas), Figure 1b.

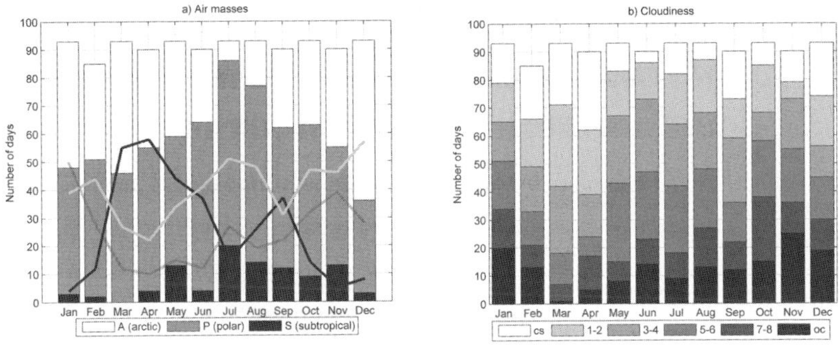

Figure 1 Monthly distributions of different types of air masses (a) for Hyytiälä, years 2003–2005 (**bar**), event (**black line**), nonevent (**gray line**) and undefined days (**light gray line**) and cloud fraction (b, in octas, cs = clear sky conditions, oc = overcast)

In March and April, when the frequency of Arctic air is still high and cloudiness is the lowest (for more than 80% of the days the cloudiness was less than 5 octas), the maximum frequency of aerosol particle formation is observed. Second maximum in particle formation in September can be explained (in the view of air mass type and cloudiness) by the increasing, comparing to summer period, frequency of Arctic air intrusion and lower, comparing to next months, cloudiness. Convection, which gives rise to Cumulus clouds often in summer, becomes weaker, and frontogenesis on the line between Arctic and Polar air masses is still weak because of the insignificant difference in temperature between the air masses mentioned above.

More than 70% of new particle formation classified as class Ia, occurred in Arctic air masses (Figure 2). For the rest days from class Ia the Polar air masses have been observed at the station. For event classes Ia and II the fraction of Arctic and Polar air was about 47% and 50% respectively. About 3% of events classified as class Ia and II have occurred in Subtropical air masses. However, our results slightly contradict to Nilsson et al. (2001), who conclude that nucleation never occurred in Subtropical air. We can explain this in two different ways. First, out studies cover the period of tree years, wherever Nilsson et al.[5] analyzed the synoptic situation for the campaign measurements. Second, in the present analysis midlatitude Subtropical air masses (Sp in air mass classification), have been included to Subtropical type, even if their properties (temperature and relative humidity) are more typical for Polar, than to Subtropical air. We followed the statement that the air mass origin influences the new particle formation as well.[6]
Number concentration of particles of nucleation (<25 nm) and accumulation (>90 nm) modes have strongly pronounced difference with respect to air mass type. The concentration of nucleation mode particles (Figure 3a) is several times higher in March and September in Arctic air masses, whereas the difference in concentration in Polar and Subtropical air is not much significant for the whole year. Concentration of the accumulation mode particles is higher in Subtropical air mass for the whole year (Figure 3c). The highest concentration of accumulation

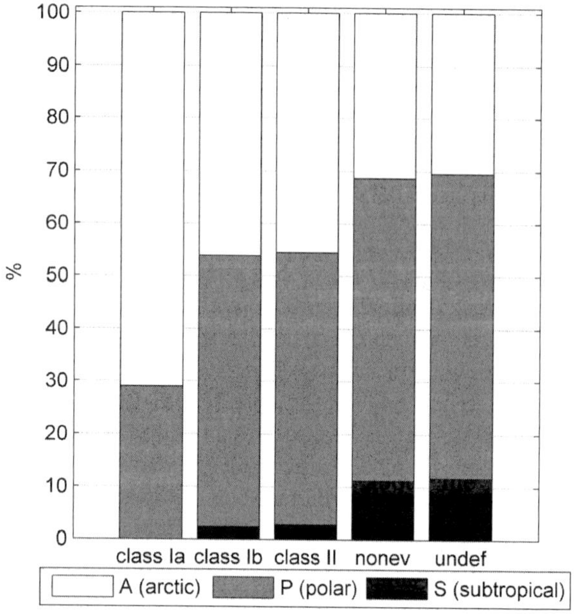

Figure 2 Fraction of different types of air masses for the days classified according to the intensity of new particle formation in Hyytiälä (2003–2005)

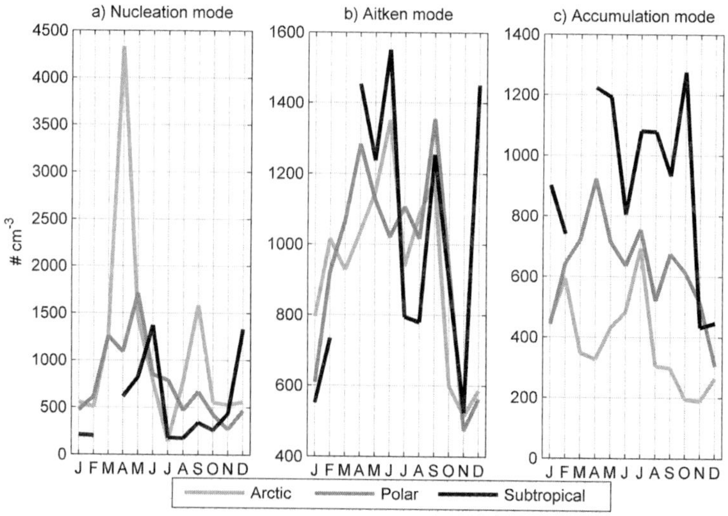

Figure 3 Mean monthly concentration of particles in Nucleation (a), Aitken (b), and Accumulation (c) modes in different types of air masses in Hyytiälä (2003–2005)

particles in Arctic air mass is observed in February and in July. Number concentration of Aitken mode particles (25–90 nm) is in the same order for most of the months in all the air mass types (Figure 3b).

References

1. Clement, C.F., Pirjola, L., Dal Maso, M., Mäkelä, J.M., and Kulmala, M., Analysis of particle formation bursts observed in Finland, *J. Aeros. Sci.*, **32**, 217–236 (2001).
2. Boy, M. and Kulmala, M., Nucleation events in the continental boundary layer: Influence of physical and meteorological parameters, *Atmos. Chem. Phys.*, **2**, 1–16 (2002).
3. Hellmuth, O., Columnar modeling of nucleation burst evolution in convective boundary layer – first results from feasibility study, Parts I–IV, *Atmos. Chem. Phys.*, **6**, 4175–4274 (2006).
4. Hyvönen, S., Junninen., H., Laakso, L., Dal Maso, M., Grönholm, T., Bonn, B., Keronen, P., Aalto, P., Hiltunen, V., Pohja, T., Launainen, S., Hari, P., Mannila, H., and Kulmala, M., A look at aerosol formation using data mining technique, *Atmos. Chem. Phys.*, **5**, 3345–3356 (2005).
5. Nilsson, E.D., Paatero, J., and Boy, M., Effect of air masses and synoptic weather on aerosol formation in the continental boundary layer, *Tellus*, **53B**, 462–478 (2001).
6. Sogacheva, L., Dal Maso, M., Kerminen, V.-M., and Kulmala, M., Probability of nucleation events and aerosol particle concentration in different air mass types arriving at Hyytiälä, southern Finland, based on back trajectory analysis, *Boreal Env. Res.*, **10**, 479–491 (2005).
7. Dal Maso, M., Kulmala, M., Riipinen, I., Wagner, R., Hussein, T., Aalto, P.P., and Lehtinen, K.E.J., *Boreal Env. Res.*, **10**, 323–336 (2005).

Model Studies of Nitric Acid Condensation in Mixed-phase Clouds

Joni-Pekka Pietikäinen[1], J. Hienola[2], H. Kokkola[3], S. Romakkaniemi[4], K. Lehtinen[3], M. Kulmala[2], and A. Laaksonen[1]

Abstract In the upper troposphere, clouds are mainly composed of ice particles which work as a platform for heterogeneous chemistry. They also have a crucial role in distributing compounds, like nitric acid, between different altitudes in troposphere. In this study we use air parcel model to investigate the adsorption of HNO_3 and the possibility that some of HNO_3 remains in the ice particles during the freezing of supercooled liquid particles. We compare the results from our model simulations to airborne measurements of nitric acid partitioning between ice particles and gas phase. The results shows that it is likely that atleast some of HNO_3 is captured to particles during freezing, but the results are highly dependent on the form of Langmuir's isotherm used.

Keywords Adsorption, ice clouds, ice nuclei, modelling, nitric acid

Introduction

Cirrus clouds play an important role in the Earth's energy balance. The ability to scatter and absorb both solar (shortwave) and Earth (longwave) radiation makes cirrus clouds interesting part of climate. Moreover, cirrus clouds play an important role in redistributing HNO_3 via uptake and sedimentation in the upper troposphere. Also, the ice particles provide platform for heterogeneous processes which can influence the chemical composition of mid and upper troposphere compounds (Ullerstam et al. 2005). This is why it is important to understand the physical

[1] *Department of Physics, University of Kuopio, P.O. Box 1627, FI-70211, Kuopio, Finland*

[2] *Department of Physical Sciences, University of Helsinki, P.O. Box 64, FI-00014, Helsinki, Finland*

[3] *Finnish Meteorological Institute, University of Kuopio, P.O. Box 1627, FI-70211, Kuopio, Finland*

[4] *Centre for Atmospheric Science, School of Earth, Atmospheric and Environmental Science, University of Manchester, P.O. Box 88 M60 1QD, Manchester, UK*

C.D. O'Dowd and P.E. Wagner (eds.), *Nucleation and Atmospheric Aerosols*, 596–600.

processes governing the uptake of trace gases on the ice crystals and snowflakes (Kärcher and Basko 2004).

The uptake of HNO_3 has been studied widely during recent years. It is unclear whether nitric acid actually freezes or stays in the particles due to their complicated structure. Some studies propose that part of nitric acid stays in the ice particles during the freezing process (Krämer et al. 2006), some that HNO_3 is adsorbed only on the surface of ice particles (Gao et al. 2004) and it has been also proposed that adsorbed HNO_3 could trap in to the ice particles during the growth phase (Kärcher and Voigt 2006).

The adsorption of nitric acid is one of the most studied mechanisms for HNO_3 uptake (Popp et al. 2005; Ullerstam et al. 2005; Gao et al. 2004). Adsorption models are equilibrium (steady state) models, and different kind of isotherms for the adsorption is available. Adsorbed nitric acid can change the properties of ice particles, for example, it can affect the condensation of water, thus preventing the ice/vapour system to gain equilibrium (Gao et al. 2004).

It is still unclear which mechanism for HNO_3 uptake is the appropriate approach. Freezing of nitric acid is difficult to measure because the measurement methods used only give the total amount of nitric acid and do not precisely tell how nitric acid is distributed in the particles. This is why it is important to develop and test different mechanisms. Adsorption has been used widely but its weakness is the assumption of equilibrium state. Due to the fluctuations in temperature, the ice particles undergo a series of growing and evaporating cycles and this means that the particles are not always in the steady state conditions (Kärcher and Basko 2004). Trapping models seem to have solved the equilibrium problem but they are not included in our model yet (Kärcher and Basko 2004).

Using a microphysical cloud model we have compared different mechanisms for HNO_3 uptake to ice particles. Furthermore, different forms of Langmuir's isotherm for adsorption have been tested. We compare results from Gao et al. (2004) to the results from our model. Two mechanisms for HNO_3 uptake had been used. Nitric acid either stays in the particles during freezing or adsorbs to the surface of the particles. We have found that, depending on the form of the isotherm, some nitric acid can freeze or for example, trap inside the ice particles.

The Cloud Model

The model used in this work is a cloud microphysics model, which is based on a condensation model by Kokkola (2003). The model is a moving sectional model which describes non-equilibrium growth of an aerosol population. The model solves time-varying differential equations describing condensation, evaporation, chemical reactions in the liquid and gas phase, adiabatic variables and the emissions of gas-phase species. The differential equations are solved using the VODE ode-solver (Brown et al. 1989). Thermodynamical equilibrium model used in this work is the aerosol inorganic model (AIM) (Clegg et al. 1998).

A new module, describing the freezing process, was made and attached to existing cloud model. The ice module calculates the probability of freezing P_i for every liquid-phase size class i on every time step $^{\Delta t}$:

$$P_i = 1 - e^{-I_{fi} V_i \Delta t} \tag{1}$$

where I_{fi} is freezing rate and V_i is the volume of liquid-phase size class i. If freezing (homogeneous) occurs, the frozen liquid-phase particles are moved to the ice particle distribution. Equally to liquid phase, ice particles are divided to separate size classes which uses moving sectional structure.

Homogeneous freezing is done in the model by Koop's parameterization (Koop et al. 2000). In the Koop''s parameterization, ice nucleation rate is a function of water activity and pressure. After the ice particles have formed, the adsorption of nitric acid is calculated based on the Langmuir's isotherm:

$$\Theta = \frac{(K_{eq}P)^{v}}{1+(K_{eq}P)^{v}}, \tag{2}$$

where Θ is the surface coverage, K_{eq} is the equilibrium adsorption constant and P is the partial pressure on HNO_3 and v is the heterogeneity parameter (Ullerstam et al. 2005; Gao et al. 2004). If the occupation of neighbouring molecyles in the surface does not affect to the ability of a particle to bind to the surface this condition is typically to be a non-dissociative Langmuir isotherm. The dissociative form assumes that adsorbed molecyle consumes two sites on the surface. In Eq. (2) the heterogeneity parameter has the value of $v = 1$ if non-dissociative form is used and $v = 1/2$ if dissociative form is used (Ullerstam et al. 2005).

Results and Discussion

Figures 1 and 2 show preliminary results of the difference between dissociative and non-dissociative form of Langmuir isotherm for cases with and without nitric acid trapping during freezing. Also, Figures 1 and 2 represent the measured values for the amount of nitric as a function of surface area of ice particles from Gao et al. (2004). The results show that if the dissociative form of Langmuir's isotherm is used, the adsorption of nitric acid can not explain the measured values. However, if the non-dissociative form is used, adsorption by itself can explain, and actually overestimates, the measured values.

There are some uncertainties which involve in these calculations. The number of HNO_3 molecules taken by a monolayer is somewhat uncertain. Also, the equilibrium adsorption constant has at least two different temperature varying parameterizations. The effects of these uncertainties are not represented in this study.

In the future, the difference between these two forms of Langmuir's isotherm will be studied further. Also, simulations will be made for different kind of

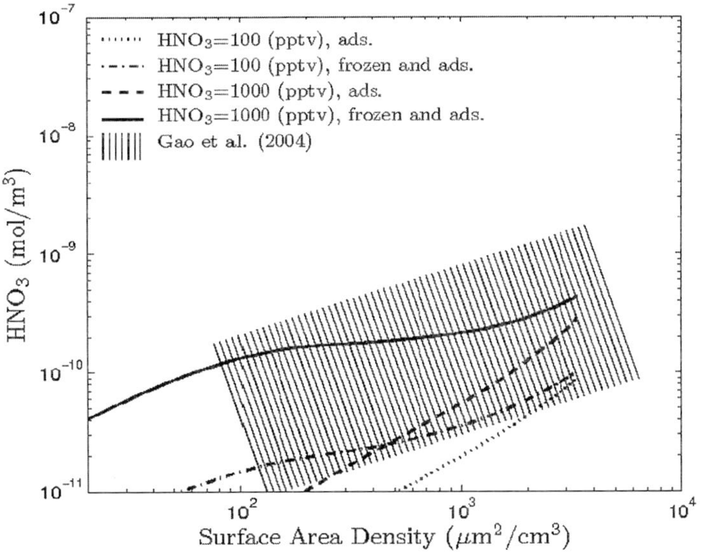

Figure 1 Concentration of nitric acid HNO_3 as a function of surface area density. Different lines represent the lower (100 ppt) and upper (1,000 ppt) limits for initial gas-phase nitric acid volume mixing ratio for cases where nitric acid is frozen or evaporated. The dissociative form for nitric acid adsorption has been used

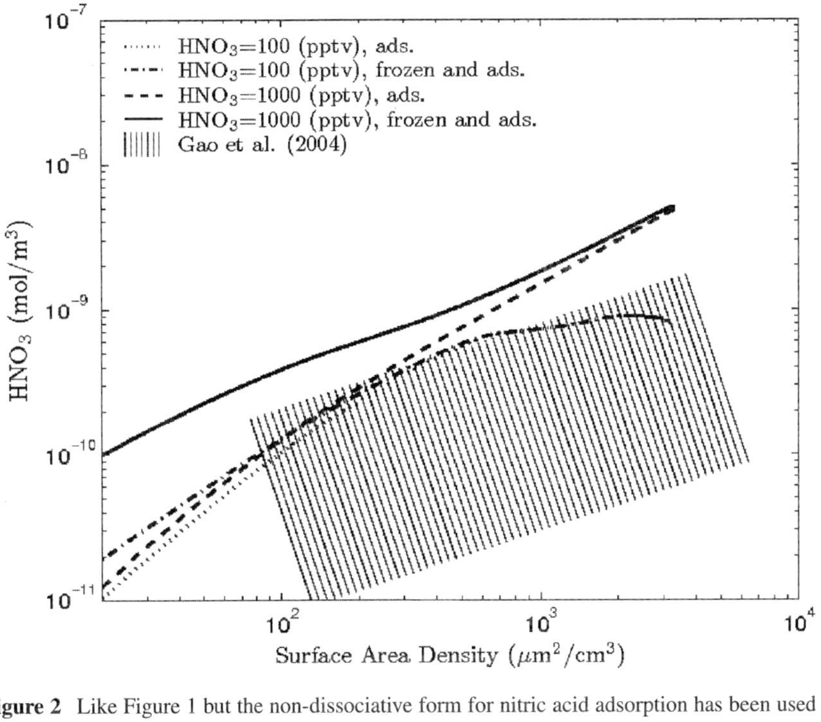

Figure 2 Like Figure 1 but the non-dissociative form for nitric acid adsorption has been used

parameterizations for adsorption equilibrium constants and the effect of monolayer approach will be studied.

Acknowledgment This works was supported by Academy of Finland (Center of Excellence program).

References

1. Brown, P.N., Byrne, G.D., and Hindmarsh, A.C, *J. Sci. Stat. Comput.*, **10**:1038–1051 (1989).
2. Clegg, S.L., Brimblecombe, P., and Wexler, A.S., *J. Phys. Chem. A*, **102**:2137–2154 (1998).
3. Gao et al. *Science*, **303**, 516–520 (2004).
4. Kokkola, H., Ph.D. thesis, University of Kuopio (2003).
5. Koop, T., Luo, B., Tsias, A., and Peter, T., *Nature*, **406**, 611–614 (2000).
6. Krämer, M., Sciller, C., Ziereis, H., Ovarlez, J., and Bunz, H., *Tellus*, **58B**, 141–147 (2006).
7. Kärcher, B. and Basko, M. M., *J. Geophys. Res.*, **109**, D22204 (2004).
8. Kärcher, B. and Voigt, C., *Geophys. Res. Lett.*, **33**, L08806 (2006).
9. Popp et al. *J. Geophys. Res.*, **10**, D06302 (2004).
10. Ullerstam, M., Thornberry, T., and Abbatt, J.P.D., *Faraday Discuss.*,**130**, 211–226 (2005).

Modelling Cirrus Cloud Fields for Climate and Atmospheric Chemistry Studies

Andrew Horseman, A. Rob MacKenzie, and Chuansen Ren

Abstract The occurrence and properties of cirrus are not well captured by current global climate and chemistry models. At Lancaster, we have developed a strategy for modelling cirrus that incorporates recent advances in the theory of ice particle formation but is computationally efficient. We have successfully used this model (LACM) in recent tropical field campaigns (TroCCiNOx) and are now developing the model further and expanding its application to other parts of the globe, with a view to incorporating the model into a climate-chemistry model and so assessing the global radiative and chemical impact of cirrus. Assessing model performance in other regions will require detailed comparison of model results with satellite and airborne observations of cirrus, and interpretation of observations and model results using analysis and forecast fields supplied by Meteorological Agencies.

Keywords Cirrus, tropopause transition layer, heterogeneous nucleation, tropospheric aerosols

Introduction

Cirrus clouds (including cirrostratus, cirrocumulus, etc.) are ice clouds that form near the tropopause, covering about 30% of the Earth's surface on average.[5] These clouds are important in climate since they contribute to the reflection of sunlight, and to the absorption/re-emission of infrared terrestrial radiation. Cirrus clouds also have indirect climate and atmospheric chemistry effects by redistributing water and other compounds in the atmosphere. There may also be direct chemical effects of cirrus clouds due to reactions on the surface of the cloud particles – leading to destruction of ozone for instance.

Environmental Science Deptartment, Lancaster University, UK

C.D. O'Dowd and P.E. Wagner (eds.), *Nucleation and Atmospheric Aerosols*, 601–604.

Cirrus Modelling

The ice particles that make up cirrus clouds form from nucleation of existing aerosols either homogeneously or heterogeneously. A new parameterisation of the homogeneous mechanism by Kärcher and Lohmann[2,3] and subsequently improved by Ren and MacKenzie[1] have made detailed modelling of the homogeneous nucleation practical. Figure 1, produced using the parameterisation of Kärcher and Lohmann, illustrates the temperature dependence of those cirrus ice crystal properties important for the modelling of factors such as their radiative potential and sedimentation rate. The enhanced parameterisation[1] has been incorporated into a Lagrangian air-parcel cirrus model (LACM). LACM has been used to model the conservation of total water in the tropical tropopause layer (TTL). Ren et al.[4] discusses the successful use of LACM to model the total amount of water in the TTL measured along the flight tracks of the research aircraft over Brazil during the TroCCiNOx (Tropical Convection, Cirrus and Nitrogen Oxides Experiment) campaign. This work reconstructed the total water mixing ratios measured along the tracks using ECMWF analyses and matched them to observations (Figure 2), i.e., successfully predicting the dehydration effects of cirrus formation.

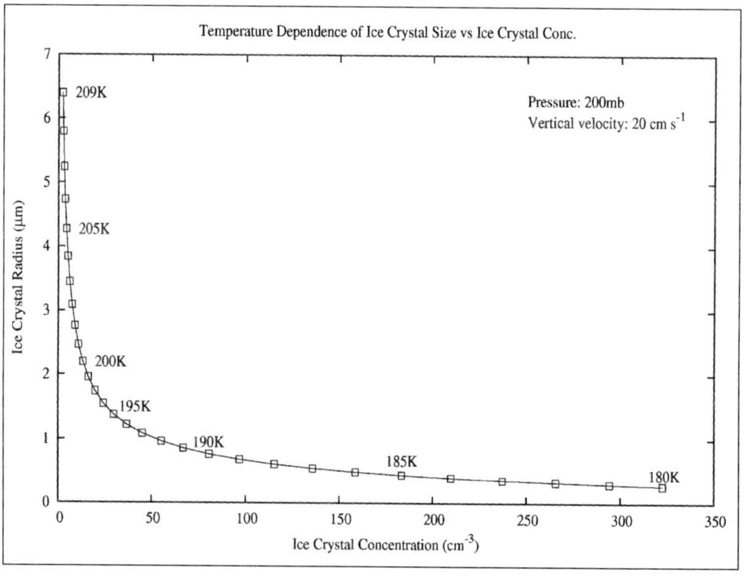

Figure 1 The effects of temperature on the size and number of ice crystals formed by homogeneous freezing generated from the parameterisation. The pressure and vertical velocity were held constant, only the temperature was varied from 180–240 K in steps of 1 K. Note these values are for 1 min of growth

The results from LACM give a good approximation to observations and also indicate the need for a process that re-hydrates the air-parcel. The physical mechanism is most likely responsible for this re-hydration is the injection of moist air by small-scale deep convection. LACM incorporates the main processes contributing to the dehydration and re-hydration of the TTL, i.e., (i) transport, (ii) freezing and drying, and (iii) moistening by injection. To date work with LACM has concentrated on the tropical troposphere.

The Next Step

Figure 2 demonstrates the improvement over current analyses, and so a second stage of work to extend the use of this parameterisation to cirrus prediction on a global scale is now underway. The strategy for this stage is to incorporate the homogeneous nucleation parameterisation into the SLIMCAT chemical transport model (CTM). This will provide us with a 3D Eulerian model with global scope and extend the vertical coverage up into the stratosphere enabling us to test the parameterisation in regions of the Brewer-Dobson circulation. The road map for the work includes initial validation of the enhanced CTM against the already proven LACM model and a second stage of validation against satellite data. The successful

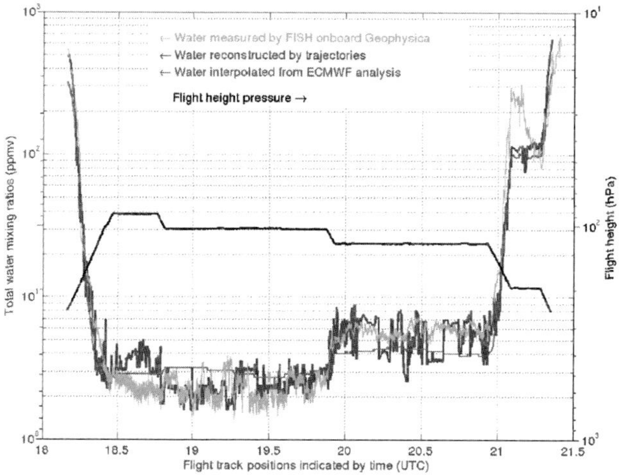

Figure 2 Comparison of measured water vapour mixing ratios (μmol/mol) from the FISH (Fast In situ Stratospheric Hygrometer) instrument with that produced by a global model (ECMWF) and by using the Lancaster parameterisation calculated along air trajectories. With two notable exceptions, at 18.7 UTC and 21.2 UTC, the Lancaster model is in better agreement with the observations. This is generally the case in regions away from deep convection (near convection neither model does well)

completion of this will provide a tool for the further investigation of processes such as the dehydration of the stratosphere where recent work suggests an increase in the amount of water present[7] or possible problems like radiative measurements that are currently incorrectly attributed to CO_2 as the reflection/radiation may be from sub-visual cirrus.

Acknowledgments This work was carried out with the support of the Natural and Environment Research Council (NER/S/A/2006/14022 and NER/T/S/2000/00977). Thanks are also due to Martyn Chipperfield for providing the SLIMCAT software and Bernd Kärcher for an implementation of the nucleation parameterisation.

References

1. Ren, C. and MacKenzie, A.R., Cirrus parameterization and the role of ice nuclei, *Quart. J. Roy. Meteorol. Soc.*, **131**, 1585–1605 (2005).
2. Kärcher, B. and Lohmann, U., A Parameterization of cirrus cloud formation: homogeneous freezing including effects of aerosol size, *J. Geophys. Res.*, **107** (D23), 9–11 (2002a).
3. Kärcher, B. and Lohmann, U., A Parameterization of cirrus cloud formation: Homogeneous freezing of supercooled aerosols, *J. Geophys. Res.*, **107**(D2), AAC 4-2 (2002b).
4. Ren, C., MacKenzie, A.R., Schiller, C., Shur, G., and Yushkov, V., Diagnosis of processes controlling water vapour in the tropical tropopause layer by a Lagrangian cirrus model, *Atmos. Chem. Phys.*, in press (2007).
5. Wylie D.P., Menzel W.P., Woolf, H.M., Strabala, K.L., Four years of global cirrus cloud statistics using HIRS, *J. Climate*, **7**, 315–335 (1994).
6. Chipperfield, M.P., New version of the TOMCAT/SLIMCAT off-line chemical transport model: intercomparison of stratospheric tracer experiments, *Quart. J. Roy. Meteorol. Soc.*, **132**, 1179 (2006).
7. Kley, D., Russell III, J. M., and Phillips, C., *SPARC Assessment of Upper Tropospheric and Stratospheric Water vapour*. SPARC Report No. 2, **312** (2000).

A Modelling Study of the Effect of Increasing Background Ozone on PM Production in Clouds

Robert J. Flanagan and Colin D. O'Dowd

Abstract The Answers to Urbino Questions (ACCENT policy-driven synthesis report, 2006) highlighted the fact that background ozone concentrations are increasing over Europe. Measurements have shown that levels have increased by up to 5 ppbv/decade over the last 20–30 years and a similar trend is envisioned to be likely in the coming decades. Much of the focus of this increase is on the possible direct health effects of higher levels of ozone. Here, we have investigated the effects that this increase will have on in-cloud PM production. Aerosol-cloud-chemistry process model simulations show that this increase in ozone will also lead to a significant increase of in-cloud sulphate production. Indeed, given such a trend by 2,100 almost 20% more sulphate mass will be produced in the simulated cloud cycle than present day.

Keywords PM, cloud processing, ozone

Introduction

As highlighted in the Answers to Urbino Questions (ACCENT policy-driven synthesis report, 2006),[1] background levels of ozone in Europe have been increasing over the last century. The main cause of this increase is the growth in ozone precursor emissions from industry, road, air, and ship transport, households, and agriculture. Increased levels of background ozone may pose a serious problem as ozone is a toxic gas and can have adverse effects on plant, animal, and human life.[2]

Simmonds et al.[3] showed that over a period of 16 years from 1987 to 2003 background ozone measured at the Mace Head Atmospheric Research Station on the west coast of Ireland increased by about 8 ppb. This is an average increase of 0.5 ppb per year which is at the higher end of the average increase over Europe and

Department of Experimental Physics and Environmental Change Institute, National University of Ireland, Galway, University Road, Galway, Ireland

C.D. O'Dowd and P.E. Wagner (eds.), *Nucleation and Atmospheric Aerosols*, 605–609.

could lead to background ozone levels of in excess of 80 ppb by 2,100. The possible health effects of such increases have already been mentioned, but here we investigate the effect they will have on some in-cloud processes.

Given that most of the production of aerosol sulphate relates to in-cloud aqueous phase processes and that the ozone-driven oxidation pathway is a key process for sulphate production in the marine environment model simulations have been performed to study the impact of fluctuating ozone concentrations. We use an aerosol-cloud-chemistry process model[4] to simulate sulphate production in a rising air parcel in conditions representative of the North Atlantic clean marine environment.

Model Details

The model used for the simulations is a size resolved aerosol-cloud-chemistry process model. The model consists of two basic units; a physical cloud module and a chemistry module. Atmospheric aerosols form solutions which are too concentrated to be considered ideal and hence the activity coefficients need to be calculated. The activity coefficients are calculated using a Pitzer model method.[5] The Pitzer model method involves generating a virial expansion of the Gibbs free energy of the solutions. The minimum in this relationship is then related to the activity coefficients. The major aqueous phase oxidants of sulphur dioxide are ozone, hydrogen peroxide and also oxygen when in the presence of iron and manganese. However in the maritime atmosphere concentrations of iron and manganese are very low and thus only the ozone and hydrogen peroxide driven pathways are included in the model. Elementary thermodynamic equations were used to describe the dynamic development of the air parcel.[6]

In order to accurately describe the growth of real aerosols in the atmosphere, one must account for non-equilibrium growth effects. The magnitude of these effects will vary for different vapour species[7] and different meteorological conditions. A sophisticated multi-component mass and heat flux module[8,9] developed at the University of Helsinki has been incorporated into the model. The mass and heat flux module requires the calculation of latent heats, thermal conductivities, saturation vapour pressures and many other parameters which can depend on the composition, temperature, and pressure of the aqueous aerosol mixture.

Simulation Scenarios

Simulations were chosen to be representative of the clean marine environment. A full range of simulations were performed with ozone levels from 15 to 80 ppb (pre-industrial – ~2,100) Heterogeneous sulphate production on sulphate and sea-salt aerosol populations and an external mixture of both was simulated. Both the sea-salt and sulphate distributions were represented in the model by 30 size bins. Some of the initial environmental conditions and trace gas concentrations for the model runs are given in Table 1.

Table 1 Simulations initial conditions

Variable	Value
Relative humidity	97.5%
Temperature	5°C
Pressure	920 mb
Updraft velocity	0.4 ms^{-1}
CO_2	350 ppmv
NH_3	1.0 ppbv
SO_2	1.5 ppbv
H_2O_2	1.5 ppbv

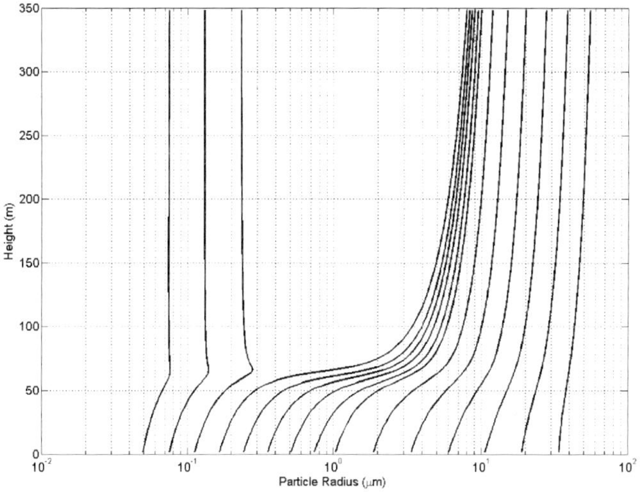

Figure 1 A typical growth pattern for a selection of the aerosol size classes

Results

Figure 1 shows a typical growth pattern of a selection of the aerosol size classes. For this simulation cloud base was 56 m above the initialization level. Only those aerosols over a certain critical size will be activated and become cloud droplets.

Figure 2 shows the results of simulations performed to study the effect of increased background on heterogeneous sulphate production on a sulphate aerosol population during a cloud cycle.

Over 17% more sulphate is produced at projected 2,100 ozone levels compared to present day. Simulations were also carried out on sea-salt aerosol and sea salt, sulphate external mixtures and results are currently being finalized.

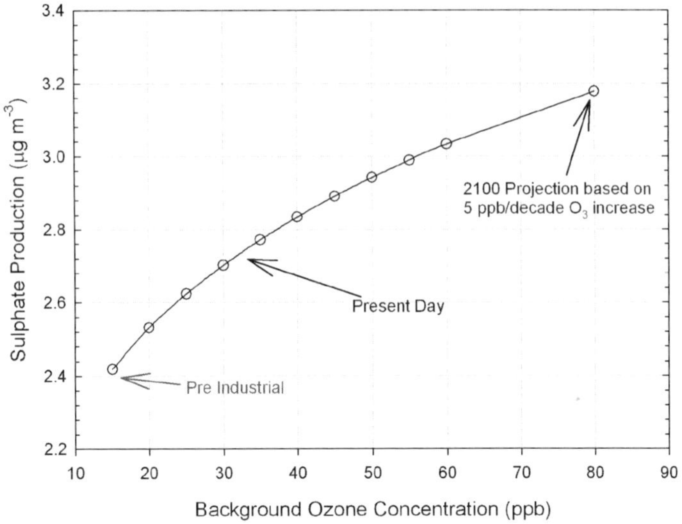

Figure 2 Sulphate production versus background ozone concentration. A range of ozone values was chosen to represent pre-industrial to 2,100 levels based on a 5 ppb/decade increase

Conclusions

Model simulations have shown that increasing levels of background ozone will have a significant impact on in-cloud sulphate production. Further simulations are being performed to study scenarios other than the clean marine environment. Results of these simulations will be incorporated into a regional climate model to better understand and quantify PM levels across Europe and assess its effect on climate.

Acknowledgment This work was supported by the Irish Environmental Protection Agency.

References

1. Answers to Urbino Questions (ACCENT policy-driven synthesis report, 2006).
2. WHO, *Air Quality Guidelines for Europe*, European Series No. 23, World Health Organization Copenhagen, Denmark: WHO.
3. Simmonds, P.G., Derwent, R.G., Manning, A.L., and Spain, G., *Atmos. Environ.* **38**(28) 4769–4778 (2004).
4. O'Dowd, C.D, Lowe, J.A., Clegg, N., Clegg, S.L. and Smith, M.H., *J. Geophys. Res.*, **105** 7143–7160 (2000).
5. Pitzer, K.S., Ion interaction approach: theory and data correlation, in *Activity Coefficients in Electrolyte Solutions*, Boca Raton, FL: CRC Press, pp. 75–154 (1991).

6. Pruppacher, H.R. and Klett, J.D., *Microphysics of Clouds and Precipitation*, Dordrecht, The Netherlands: Kluwer (1997).
7. Meng, Z. and Seinfeld. J., *Atmos. Environ.*, **30**(16), 2889–2900, (1996).
8. Vesala, T., Kulmala, M., Rudolf, R., Vrtala, A., and Wagner, P.E., *J. Aerosol. Sci.*, **28**(4), 565–598 (1997).
9. Hienola, J., Kulmala, M., and Laaksonen, A., *J. Aerosol Sci.*, **32**, 351–374 (2001).

Part VII
Aerosol Characterisation and Properties

Stratospheric Aerosol: Measurements, Importance, Life Cycle, Anomalous Aerosol

Terry Deshler

Abstract Stratospheric aerosol, noted after large volcanic eruptions since at least the late 1800s, were first measured in the late 1950s, with the modern continuous record beginning in the 1970s. Stratospheric aerosol, both volcanic and nonvolcanic are sulfuric acid droplets with radii (concentrations) on the order of 0.02–0.5 μm (0.5–0.005 cm^{-3}), increasing by factors of 2–4 (10–10^3) after large volcanic eruptions. The source of the sulfur for the aerosol is either through direct injection from sulfur-rich volcanic eruptions, or from tropical injection of tropospheric air containing OCS, SO_2, and sulfate particles. The life cycle of nonvolcanic stratospheric aerosol (SA) (photodissociation and oxidation of sulfur source gases, nucleation/condensation in the tropics, transport poleward and downward in the global planetary wave driven tropical pump) leads to a quasi steady-state relative maximum in particle number concentration at around 20 km in the mid latitudes. SA have significant impacts on the Earth's radiation balance for several years following volcanic eruptions. Away from large eruptions, the direct radiation impact is small and well characterized; however, these particles also may play a role in the nucleation of near tropopause cirrus, and thus indirectly affect radiation. SA play a larger role in the chemical, particularly ozone, balance of the stratosphere. In the midlatitudes they interact with both nitrous oxides and chlorine reservoirs, thus indirectly ozone. In the polar regions they provide condensation sites for polar stratospheric clouds (PSCs) which then provide the surfaces necessary to convert inactive to active chlorine leading to polar ozone loss. Until the mid 1990s the modern record has been dominated by three large sulfur-rich eruptions: Fuego (1974), El Chichon (1982) and Pinatubo (1991), thus definitive conclusions concerning the trend of nonvolcanic SA could only recently be made. Although anthropogenic emissions of SO_2 have changed somewhat over the last 30 years, the measurements during volcanically quiescent periods indicate no long term trend in nonvolcanic stratospheric aerosol.

Keywords Stratospheric aerosol, volcanic aerosol, stratospheric aerosol – measurements, trends, polar stratospheric clouds, stratospheric chemistry

---

Department of Atmospheric Science, University of Wyoming, Laramie, Wyoming, USA

C.D. O'Dowd and P.E. Wagner (eds.), *Nucleation and Atmospheric Aerosols*, 613–624.

Stratospheric Aerosol – Measurement Record

The existence of persistent atmospheric aerosol after volcanic eruptions must have been apparent to careful observers from early times. The reddish diffraction ring around the sun following large volcanic eruptions was first described from Hawaii by Sereno Bishop nine days after the eruption of Krakatoa, in late August 1883, and led to the name Bishop's ring. Less certain is that early observers suspected that aerosol persisted in the stratosphere during volcanically quiescent periods. The first published report suggesting a persistent aerosol layer in the stratosphere used purple twilight observations.[1]

Quantitative observations of stratospheric aerosol (SA) characterizing the altitudes, sizes, masses, and compositions were first provided by balloon-borne impactor measurements in the late 1950s.[2] Subsequent measurements by both aircraft and balloon-borne impactors suggested the global distribution of SA. While Aitken nuclei concentrations decreased through the stratosphere, particles > 0.1 μm had a maximum in concentration near 20 km, suggesting an aerosol source in this region. These initial stratospheric measurements were followed by a suite of instruments to measure SA including passive extinction measurements,[3] active scattering measurements,[4] and counting and sizing single particles.[5]

Surface based measurements, balloon-borne, and lidar, include several fairly continuous records of measurements of SA which began in the early 1970s,[6] about a decade after Junge's initial measurements. The balloon-borne measurements have been limited primarily to Laramie, Wy, with sporadic measurements from Lauder, NZ, and both polar regions. These measurements provide profiles of size resolved aerosol number concentration. The first lidar (light detection and ranging) measurements of SA were completed shortly after Junge et al.'s initial measurements.[4] Lidar sites investigating SA now range in latitude from 90° S to 80° N, with a number of sites in northern midlatitudes, and a few stations in the subtropics and southern midlatitudes. Lidars provide remote, vertically resolved measurements of atmospheric backscatter from both molecules and aerosols at one or more wavelengths. Both lidar and in situ measurements provide aerosol profiles (Figure 1). The peak in SA >0.1 μm near 20 km is clear. The concentration of smaller particles, >0.01 μm, is flat or decreasing.

The balloon-borne measurements from Laramie[6–8] along with lidar measurements from two midlatitude sites[9–10] and two tropical sites[11–12] provide the longest SA record available (Figure 2). Three eruptions have dominated the modern SA record; however this high level of volcanic activity is a bit anomalous. Solar and stellar extinction data show that since 1880 the stratosphere has been dominated by eight major eruptions.[13–1] Four of these occurred between 1880 and 1910 and four since 1960. The long-term SA measurements, which began in the 1970s, have captured the complete cycle for three major eruptions with a global stratospheric impact: Fuego (14° N, October 1974, 3–6 Tg of aerosol), El Chichón (17° N, April 1982, 12 Tg), and Pinatubo (15° N, June 1991, 30 Tg). Within this record there have been four periods when volcanic influences were at a minimum, 1974, 1978–1980, 1988–1991, and 1997–present.

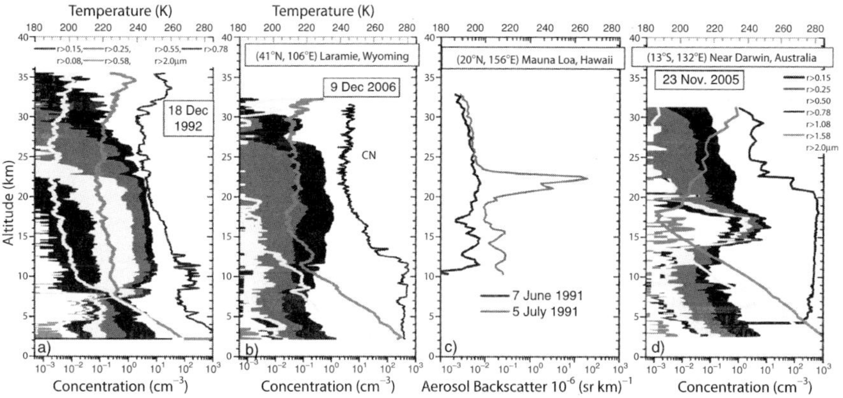

Figure 1 Vertical profiles of aerosol concentration at seven sizes, 0.15, 0.25, 0.5, 0.78, 1.08, 1.58, 2.0 µm, condensation nuclei (CN) and temperature and frost point above Laramie, Wy for measurements (a) 1.5 and (b) 15 years after Pinatubo. (c) Lidar measurements just before and after the Pinatubo eruption from Mauna Loa, Hawaii. (d) Aerosol concentration profiles above Darwin, Australia in November 2005

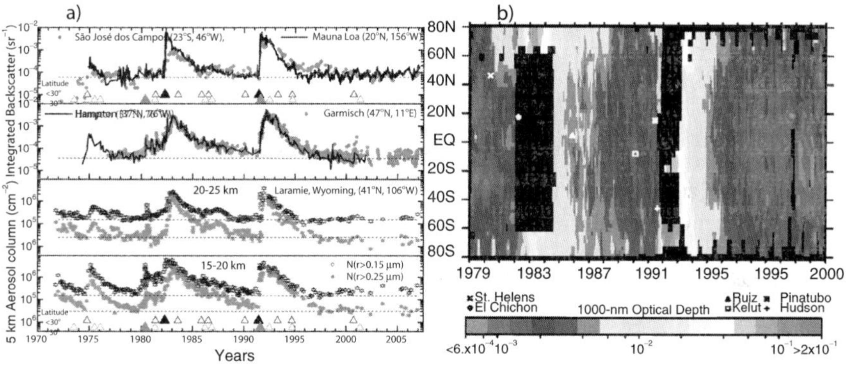

Figure 2 (a) History of SA from two tropical lidar sites, two midlatitude lidar sites, and one set of midlatitude in situ measurements. The lidar measurements show integrated SA backscatter and represents the measurements above the tropopause at each site. The in situ data are presented as integral columns of number between 15–20 and 20–25 km at two sizes. The time of volcanic eruptions for VEI = 4 (**open triangle**) and VEI = 5 (**closed triangle**) and for latitudes <30° and >30° are shown in the bottom of the top and bottom panels. (b)[27]: SAM II, SAGE, and SAGE II SA optical depth at 1.02 µm, 1979–2003. Profiles which do not extend to the tropopause are excluded, leading to a significant amount of missing data after Pinatubo. The gap between 1982 and 1984 was between SAGE and SAGE II. The loss of the high-latitude measurements after 1991 is due to the loss of SAM II

Beginning in the late 1970s satellite instruments: SAM II[15] (1979–1994), SAGE (1979–1981) and SAGE II[16] (1984–2005), added considerably to the SA record. These initial instruments which extended the local measurements to a global picture (see Figure 2),[31] have been supplemented with a number of additional instruments, e.g., HALOE (80°S–80° N, 1991–2005), POAM II, III (63–88° S, 54–71° N, 1993–1996, 1998–2005), SAGE III (2003–2006). The current satellite instruments, ACE-FTS, MAESTRO, CALYPSO are all challenged by the low level of SA during background conditions.

Importance of Stratospheric Aerosol

SA play a role in the global radiation and chemical balance of the atmosphere.SA scatter solar radiation and must be considered in global climate models. Their impact is small during nonvolcanic, or background, periods, optical depths of ~0.01, but is significant following large volcanic eruptions. Optical depths reached 0.1 following El Chichon and Pinatubo. This is high enough to cause measurable impacts on global climate. Both stratospheric warming,[18] and tropospheric cooling[19] have been documented. These effects are observable for 2–3 years following the eruption, after which the optical depths have fallen by at least a factor of two.

SA play a relatively larger role in the chemical balance of the atmosphere. During volcanically quiescent periods SA affect the budget of several trace gases, in particular NO_x, while after volcanic activity SA can directly increase the abundance of activated chlorine.[20] Decreases in midlatitude ozone have been observed following both El Chichon[21] and Pinatubo.[22] NO_2 columns were reduced after both El Chichón and Pinatubo[23] from the hydrolysis of N_2O_5 on water bearing volcanic SA.[24] At low aerosol loading NO_x increases and induces ozone loss from the nitrogen catalytic cycle.[25] There are direct measurements of the anticorrelation of aerosol surface area and the NO_x/NO_y ratio.[26] The hydrolysis of N_2O_5 saturates as aerosol surface area increases above ~10 μm^2 cm^{-3}, thus SA control NO_x only during periods of low aerosol loading.

In the polar regions SA provide the nucleation/condensation sites for the growth of polar stratospheric clouds (PSCs), providing the surfaces necessary to convert reservoir chlorine molecules to a photo-labile species easily photolyzed with the return of the sun in spring. This leads to the annual ozone hole above Antarctica and to occasional Arctic ozone loss. The stability of the Antarctic polar vortex leads regularly to temperatures conducive for PSCs, and thus ozone loss. In contrast the northern hemisphere topography forces large amplitude planetary waves leading to a warmer and less stable polar vortex. Thus PSCs, and polar ozone loss, are much less frequent.

Life Cycle of Stratospheric Aerosol

The initial impactor measurements by Junge allowed for estimates of the chemical composition of the particles using an electron microprobe analyzer.[2] The only element to appear in appreciable quantities was sulfur, leading to a suggestion that the aerosol were primarily composed of sulfate. Further balloon-borne measurements established the boiling point of SA,[5] which was consistent with a composition of 75% sulfuric acid and 25% water. Additional filter measurements also confirmed the dominance of sulfur, although other species such as silicates are present particularly shortly after large volcanic eruptions. SA are always under saturated with respect to sulfuric acid, because of the very low vapor pressure of sulfuric acid. Thus any free sulfuric acid molecules are quickly captured by a particle which then condenses water, with an abundance about 10^7 times that of sulfuric acid, to maintain equilibrium with the environment. Slight decreases (increases) in temperature lead to uptake (release) of water. Thus the size and composition of SA is controlled by temperature and the abundance of gas-phase sulfuric acid. The initial heated inlet in situ measurements,[8] repeated following El Chichon and Pinatubo,[27] and impactor measurements following Pinatubo, were all consistent. Lingering questions concerning the possibilities of contamination from such impactor measurements were laid to rest by direct aerosol composition measurements in 1995, while the Pinatubo aerosol was still decaying[28] and in 1998 when the Pinatubo influence was minimal.[29] Based on these and other measurements it is clear that SA is composed primarily of sulfuric acid and water during both volcanically active and quiescent periods.

Origin of Stratospheric Aerosol

For a volcanic eruption to significantly perturb the stratosphere it must have enough energy to penetrate the tropopause and it must be rich in sulfur. Generally eruptions with a volcanic explosivity index (VEI) ≥ 4, will penetrate into the stratosphere.[30] In the period of modern measurements (1960–2006) there have been about 33 eruptions with such a stratospheric impact.[31] Of these five, three tropical and two high latitude, had a VEI ≥ 5. One of these, Mt. Saint Helens, released less than 1 Tg of SO_2 and the silicates and other large particles quickly fell out. The impact of Mt. St. Helens was gone in about a year (see Figure 2). Cerro Hudson, the other high latitude eruption, released ~3 Tg of SO_2, but occurred in 1991 just after Pinatubo (~30 Tg of SO_2), so the signal from Cerro Hudson was quickly lost in the Pinatubo aerosol. The other three eruptions, Agung (1963), El Chichon (1982), and Pinatubo (1991) were all at low latitudes and the latter two clearly produced a stratospheric impact lasting five or more years (see Figure 2). The impact of Agung is unknown, due to a lack of measurements; however, this eruption, along with a number of eruptions in the late 1960s with a VEI = 4 were responsible for the rather large amount of SA observed in the early 1970s when measurements began. The one eruption with a significant stratospheric impact but a VEI < 5 is Fuego (1974) at 14°N. The lower VEI produces an impact primarily below 20 km, as indicated in Figure 2.

Tropical eruptions rich in sulfur directly inject particles and gases to altitudes of >20 km (VEI = 4) and to >25 km (VEI = 5) in the tropical stratospheric reservoir.[32] The sulfur is oxidized and transformed into sulfuric acid within about a month. The high concentrations of gas-phase sulfuric acid lead to nucleation of new binary particles. The nucleation continues until the sulfuric acid is condensed into either new particles or onto preexisting particles. The aerosol population then further evolves through coagulation and meridional mixing. The meridional transport of material out of the tropical reservoir depends on season and the phase of the quasi biennial oscillation (QBO).[32] After Pinatubo, filaments showing both condensation on preexisting particles and new particles were observed about 1 month after the eruption.[27] The injection of the precursor gases for SA directly into the tropical reservoir, and the confinement of this reservoir to the tropics prolong the stratospheric impact of tropical eruptions. Material from the tropical reservoir is generally transported out of the tropical reservoir primarily during winter for a westerly shear QBO, while easterly phase QBOs are more confining.[32] Thus after Pinatubo aerosol was primarily transported into the northern hemisphere. Although it is assumed that the impact of a high-latitude eruption would be much shorter lived, since the aerosol would be fairly quickly dispersed poleward and downward, there has been no opportunity to test this suggestion.

The fact that nonvolcanic SA is also primarily sulfuric acid and water[5,27,29] implies that sulfur source gases must be transported across the tropopause, and transformed into sulfuric acid which can then either nucleate new particles or condense on preexisting particles. In addition sulfate aerosol can also be deposited in the stratosphere as remnants of overshooting convection. Scavenging of particles larger than 0.1 μm leads to a particle deficit near the tropopause. Carbonyl sulfide (OCS), an inert sulfur bearing molecule was first suggested as the source of stratospheric SO_2[33]; however, it has been shown to be insufficient. The most recent comparison of SA models with observations of background aerosol indicates that sulfur source gases for SA must be about equally divided between OCS and SO_2, with direct injection of sulfate particles providing a source equivalent to the sum of the gas-phase precursors.[34] This conclusion is arrived at through sensitivity analysis for the SO_2 and sulfate inputs, while using available measurements of free tropospheric OCS.[35] The input of sulfate aerosol is based on tropical tropopause aerosol measurements[34]; however SO_2 measurements at tropopause level are not available. SO_2 has a short, highly variable lifetime (10–20 days) with sources that are intense yet regionally limited, SO_2 mixing ratios in the free troposphere vary by several orders of magnitude from peaks of 1,000 pptv in air from urban centers and in continental outflow to less than 10 pptv in the upper troposphere above remote regions.[35] The only stratospheric measurements of SO_2 were completed in 1985 above 30 km, capturing the SO_2 released from the evaporation of SA, >10 pptv. The models require inputs of 80 pptv SO_2 to achieve agreement with satellite extinction measurements below 25 km. Above 25 km OCS is the dominant source of sulfur.

Loss of Stratospheric Aerosol

The removal of aerosol from the stratosphere follows similar paths for either volcanic or background aerosol. Both types of particles originate from the condensation of sulfuric acid and water onto new or preexisting aerosol, coagulation, and growth in the presence of additional sulfuric acid vapor. The nucleation of new particles and coagulation occur over short timescales. Even in the presence of high concentrations of particles and sulfuric acid vapor, following large volcanic eruptions, the evolution of the aerosol size distribution through nucleation and coagulation will continue for less than a few months.[36] After this time, diffusional growth, mixing, and gravitational settling lead to further changes in size distribution. Gravitational settling is an important factor in controlling the rather constant nature of the background aerosol population, but is a small contributor to the removal of SA. Gravity has the effect of moving the larger particles closer to the tropopause where they are removed through stratosphere troposphere exchange processes, which are rapid compared with gravitational settling.

The exchange of air between the troposphere and stratosphere occurs, in order of importance, through the extratropical wave driven equator to pole circulation, through tropopause folds, and through subsidence in the polar vortex.[37] The extratropical tropospheric planetary wave driven circulation creates equatorial upwelling and extratropical downwelling. The peak activity thus occurs in the winter hemisphere, which depending on the phase of the QBO,[32] has the effect of moving air meridionally and downward from the tropical reservoir. Estimates of the flux of air created by this circulation, which is in agreement with tracer measurements,[37] suggests stratospheric residence times of approximately 2 years.[36] This is significantly faster than either tropopause folds or polar vortex subsidence. The latter is estimated at about 8 years.[36]

The initial SA observations[2] concluded that there was a source of SA near 20 km, now known as the Junge layer, because of the local maximum in aerosol concentration that did, and does, appear there in mid latitude measurements (see Figure 1). We now know that the source of these aerosol are a complex interaction between equatorial upwelling, transporting sulfur source gases and nascent particles into the stratosphere, nucleation, condensation, and coagulation to develop a relatively static size distribution, and then gravitational settling, and extratropical wave driven downwelling causing removal. The interplay of these various sources, sinks, and evoluting forces create a relatively static aerosol profile during volcanically quiescent conditions, thus the only real opportunity to observe stratospheric decay processes is after large volcanic eruptions (see Figure 2).

For each of the stratospheric events observed in Figure 2, e-folding times can be calculated. Generally after eruptions e-folding times vary between 0.8 and 1.5 years depending on the observation/geophysical parameter considered. Aerosol volume/extinction will have somewhat faster e-folding times than aerosol number; however, these differences are not large since gravity plays a relatively small roll in SA removal. For comparison of decay of the major eruptions in Figure 2, the exponential growth/decay coefficient calculated at two year intervals following the three major eruptions is shown in Figure 3. After El Chichon there was still significant decay observed even after 5–7 years, whereas the Pinatubo decay rate was nearly zero at 5–7 years.

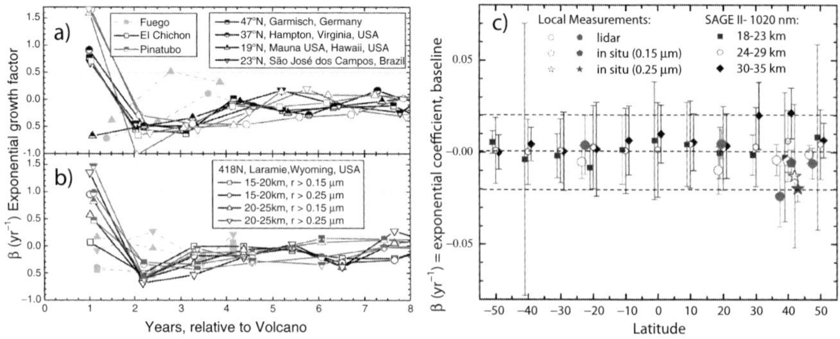

Figure 3 Exponential growth rate, assuming $N(t) = N_0 \exp(\beta t)$, calculated in 2-year intervals for the time after Fuego (1974), El Chichon (1982), and Pinatubo (1991) for (a) the four long-term lidar sites and (b) in situ measurements at two sizes and two altitude intervals (see Figure 2 a). (c)[31]: β (yr^{-1}) ± 95% confidence interval, as a function of latitude for ±5° zonal averages of 5 km optical depths at 1.02 µm from SAGE II for three altitude intervals, for integrated backscatter from São José dos Campos, Mauna Loa, Hampton, and Garmisch, and for 15–30 km column integrals of aerosol concentration for particles with radius ≥0.15 and 0.25 µm. For the lidar and in situ data the open symbol represents β based on the analysis of quiescent periods, while the closed symbol represents β obtained from the parametric model. In all cases the values of β are plotted around the central latitude for clarity

For the reasons stated above it is difficult to assign a lifetime to the nonvolcanic component of SA. This problem was made doubly difficult by the fact that from the time that the stratospheric measurement record began until 1995 the stratosphere was primarily dominated by volcanic eruptions, and volcanic interludes were brief (see Figure 2). Early work concluding that nonvolcanic SA was increasing may have been misled by uncertainty as to whether the volcanic influence was really minimal. Only the long post-Pinatubo quiescent period has allowed us to confidently address the question of changes in background SA[31] utilizing four lidar and one balloon-borne measurement records, spanning 30 years, and one satellite (SAGE II) record spanning 20 years. The records were investigated by comparing the volcanically quiescent periods and by fitting a parametric model to each record to account for, and remove, volcanic aerosol. Figure 3 summarizes the estimates of the exponential growth/decay coefficient for each record and each method as a function of latitude. The conclusion is that background SA has not displayed a long-term trend over the period 1970–2005.[31]

Anomalous Stratospheric Aerosol

Stratospheric clouds have been documented in the polar regions since the late 1800s. These clouds became the subject of intense interest after the discovery of their role in converting inactive to active chlorine and in denitrifying the stratosphere. Observational, laboratory, and theoretical work have since shown that PSC particles above the ice point are either: nitric acid trihydrate (NAT); or liquid ternary aerosol (LTA) droplets. Composition and phase of the particles determines threshold temperatures at which they occur and persist. NAT will exist at temperatures 3–5 K above LTA temperatures, and LTA begin appearing about 3 K above the ice point for polar stratospheric levels of gaseous nitric acid and water. Particle phase also controls development of particle size which plays a large role in the extent and rate of dehydration/denitrification, thus determining the time frame during which chlorine will remain active.

The major remaining question concerning the formation of PSCs in the atmosphere is the process by which NAT forms. LTA growth is strictly a function of gas-phase mixing ratios and temperature, as the vapor pressures of sulfuric acid, nitric acid, and water adjust to environmental temperature and molecular collisions. LTA begins to grow on the SA as soon as temperatures reach about 3 K above the ice point. At these temperatures the concentration of NAT particles is often at best only 0.001 of the aerosol population[38–39] indicating clearly a nucleation barrier which is not well understood. Laboratory investigations suggest that LTA must be significantly supercooled for NAT nucleation. Arctic PSC observations in the early 2000s suggest the importance of a small population of NAT particles[38] and that such a population forms at temperatures just a few degrees below the NAT point in both lee waves and synoptic situations.[39] These latter observations are the clearest evidence yet indicating that NAT forms readily on a small fraction (10^{-4}–10^{-5}) of SA through heterogeneous nucleation.

The midlatitude stratosphere has been the focus of many long-term measurement series, and aside from an occasional PSC straying over Europe, there have been no reported observations of anomalous aerosol, until recent work on wildfire generated convection, PyroCBs,[40] indicating the injection of smoke from massive forest fires to altitudes of up to 7 km above the tropopause.

In contrast observations in the tropics have been limited to two lidar sites, and occasional aircraft and balloon measurements. After the polar stratosphere the tropical tropopause is the next coldest region of the lower atmosphere. Suspicions of the existence of tropical nitric acid and water clouds have been documented by satellite[41] and aircraft measurements.[42] Evidence from the satellite (HALOE) measurements[41] indicate observations of NAT, LTA, and ice, all very near the equator and between 3 km below to 1 km above the tropopause. The aircraft observations,[42] also near the equator, are of very low concentration ($<10^{-4}$ cm^{-3}) particles ($>0.8\,\mu m$ radius) which contain nitric acid at 1 km above the tropopause. There is evidence for both sets of observation to suggest that the air at and just above the tropopause could be supersaturated with respect to NAT. The two observations; however, are not consistent with each other. The estimates of extinction from the aircraft measurements are a factor of 1,000 below the satellite observations. The aircraft observations are consistent with Arctic PSC observations indicating heterogeneous nucleation of NAT on a small fraction of SA. In addition to these observations there are balloon-borne measurements of particle layers at higher concentrations and warmer temperatures both spanning the tropopause and 4–5 km into the stratosphere.[43–44] An example of such a profile is shown in Figure 1d.

Considering the size of the particles (>1.0 μm), the high number concentrations ($>10^{-2}$ cm^{-3}), and the thickness of these layers (6 km), these particles are perhaps the nonvolatile remnants of overshooting thunderstorms; however, a definitive explanation of these observations has yet to be offered.

Acknowledgments The measurement records presented here were supported by a number of funding agencies in the USA, Germany, and Brazil. These include the US National Science Foundation and National Aeronautics and Space Administration, the Deutsche Forschungsgemeinschaft and Bundesministerium für Bildung und Forschung, and the European Union. John Barnes, Dale Simonich, and Thomas Trickl are acknowledged for supplying recent lidar measurements.

References

1. Gruner, P. and Kleinert, H., Die Dämmerungserscheinungen, *Probl. Kosm. Phys.*, **10**, 1–113 (1927).
2. Junge, C.E., Changnon, C.W., and Manson, J.E., Stratospheric aerosols, *J. Meteor.*, **18**, 81–108 (1961).
3. Volz, F.E. and Goody, R.M., The intensity of the twilight and upper atmospheric dust, *J. Atmos. Sci.*, **19**, 385–406 (1962).
4. Fiocco G. and Grams, G., Observation of aerosol layer of 20 km by optical radar, *J. Atmos. Sci.*, **21**, 323–324 (1964).
5. Rosen, J.M., The vertical distribution of dust to 30 km, *J. Geophys. Res.*, **69**, 4673–4676 (1964).
6. Hofmann, D.J., Rosen, J.M., Pepin, T.J., and Pinnick, R.G., Stratospheric aerosol measurements, I, time variations at northern midlatitudes, *J. Atmos. Sci.*, **32**, 1446–1456 (1975).
7. Hofmann, D.J., Increase in the stratospheric background sulfuric acid aerosol mass in the past 10 Years, *Science*, **248**, 996–1000 (1990).
8. Deshler, T., Hervig, M.E., Hofmann, D.J., Rosen, J.M., and Liley, J.B., Thirty years of *in situ* stratospheric aerosol size distribution measurements from Laramie, Wyoming (41°N), using balloon-borne instruments, *J. Geophys. Res.*, **108**(D5.), 4167, doi:10.1029/2002JD002514 (2003).
9. Jäger, H., Long-term record of lidar observations of the stratospheric aerosol layer at Garmisch-Partenkirchen, *J. Geophys. Res.*, **110**, D08106, doi:10.1029/2004JD005506 (2005).
10. Osborn, M.T., DeCoursey, R.J., Trepte, C.R., Winder, D.M., and woods, D.C., Evolution of the Pinatubo volcanic cloud over Hampton, Virginia, *Geophys. Res. Lett.*, **22**, 1101–1104 (1995).
11. Barnes, J.E. and Hofmann, D.J., Lidar measurements of stratospheric aerosol over Mauna Loa, *Geophys. Res. Lett.*, **24**, 1923–1926 (1997).
12. Simonich, D.M. and Clemesha, B.R., A history of aerosol measurements at São José dos Campos, Brazil (23 S, 46 W) from 1972 to 1995, *Advances in Atmospheric Remote Sensing with Lidar – Selected Papers of the 18th International Laser Radar Conference*, Springer, Berlin, Germany, 481–484 (1997).
13. Sato, M., Hansen, J.E., McCormick, M.P., and Pollack, J.B., Stratospheric aerosol optical depths, 1850–1990 *J. Geophys. Res.*, **98**, 22987–22994 (1993).
14. Stothers, R.B., Major optical depth perturbations to the stratosphere from volcanic eruptions: Pyrheliometric period, 1881–1960, *J. Geophys. Res.*, **101**, 3901–3920 (1996).
15. McCormick, M.P., et al., High-latitude stratospheric aerosols measured by the SAM II satellite system in 1978 and 1979, *Science*, **214**, 328–331 (1981).

16. Thomason, L.W., Poole, L.R., and Deshler, T., A global climatology of stratospheric aerosol surface area density deduced from stratospheric aerosol and gas experiment II measurements: 1984–1994, *J. Geophys. Res.*, **102**, 8967–8976 (1997).
17. Hamill, P. and Brogniez, Stratospheric record and climatology, Chapter 4 of the *SPARC Assessment of Stratospheric Aerosol Properties* (L.W. Thomason and Th. Peter, editors), WCRP-124, WMO/TD-No. 1295, SPARC Report No. 4.
18. Labitzke, K. and McCormick, M.P., Stratospheric temperature increases due to Pinatubo aerosols, *Geophys. Res. Lett.*, **19**, 207–210 (1992).
19. Dutton, E.G. and Christy, J.R., Solar radiative forcing at selected locations and evidence for global lower tropospheric cooling following the eruptions of El Chichón and Pinatubo, *Geophys. Res. Lett.*, **19**, 2313–2316 (1992).
20. Hanson, D.R. and Lovejoy, E.R., The reaction of ClONO2 with submicrometer sulfuric acid aerosol, *Science*, **267**, 1326–1328 (1995).
21. Hofmann, D.J. and Solomon, S., Ozone destruction through heterogeneous chemistry following the eruption of El Chichón, *J. Geophys. Res.*, **94**, 5029–5041 (1989).
22. Gleason, J.F., et al., Record low global ozone in 1992, *Science*, **260**, 523–526 (1993).
23. Johnston, P.V., McKenzie, R.L., Keys, J.G., and Matthews, W.A., Observations of depleted stratospheric NO_2 following the Pinatubo volcanic eruption, *Geophys. Res. Lett.*, **19**, 211–213 (1992).
24. Rowland, F.S., Sato, H., Khwaja, H., and Elliott, S.M., The hydrolysis of chlorine nitrate and its possible atmospheric significance, *J. Phys. Chem.*, **90**, 1985–1988 (1986).
25. Crutzen, P.J., The possible importance of CSO for the sulfate layer of the stratosphere, *Geophys. Res. Lett.*, **3**, 73–76 (1976).
26. Fahey, D.W., et al. *In situ* measurements constraining the role of sulphate aerosols in mid-latitude ozone depletion, *Nature*, **363**, 509–514 (1993).
27. Deshler, T., Johnson, B.J., and Rozier, W.R., Balloonborne measurements of Pinatubo aerosol during 1991 and 1992 at 41°N, vertical profiles, size distribution, and volatility, *Geophys. Res. Lett.*, **20**, 1435–1438 (1993).
28. Arnold, F., Curtius, J., Spreng, S., and Deshler, T., Stratospheric aerosol sulfuric acid: first direct in situ measurements using a novel balloon-based mass spectrometer apparatus, *J. Atmos. Chem.*, **30**, 3–10 (1998).
29. Murphy, D.M., Thomson, D.S., and Mahoney, M.J., In situ measurements of organics, meteoritic material, mercury, and other elements in aerosols at 5 to 19 kilometers, *Science*, **282**, 1664–1669 (1998).
30. Newhall, C.G. and Self, S., The volcanic explosivity index (VEI): an estimate of explosive magnitude for historical volcanism, *J. Geophys. Res.*, **87**, 1231–1238 (1982).
31. Deshler, T. and Anderson-Sprecher, R., Non-volcanic stratospheric aerosol trends: 1971–2004, Chapter 5 of the *SPARC Assessment of Stratospheric Aerosol Properties* (L.W. Thomason and Th. Peter editors), WCRP-124, WMO/TD-No. 1295, SPARC Report No. 4.
32. Trepte, C.R. and Hitchman, M.H., Tropical stratospheric circulation deduced from satellite aerosol data, *Nature*, **355**, 626–628 (1992).
33. Crutzen, P.J., The possible importance of CSO for the sulfate layer of the stratosphere, *Geophys. Res. Lett.*, **3**, 73–76 (1976).
34. Weisenstein, D. and Bekki, S., Modeling of stratospheric aerosols, Chapter 6 of the *SPARC Assessment of Stratospheric Aerosol Properties* (L.W. Thomason and Th. Peter editors), WCRP-124, WMO/TD-No. 1295, SPARC Report No. 4.
35. Notholt, J. and Bingemer, H., Precurso Gas Measurements, Chapter 2 of the *SPARC Assessment of Stratospheric Aerosol Properties* (L.W. Thomason and Th. Peter editors), WCRP-124, WMO/TD-No. 1295, SPARC Report No. 4.
36. Hamill, P., Jensen, E.J., Russell, P.B., and Bauman, J.J., The life cycle of stratospheric aerosol particles, *Bull. Amer. Met. Soc.*, **7**, 1395–1410 (1997).
37. Holton, J.R., Haynes, P.H., McIntyre, M.E., Douglass, A.R., and Rood, R.B., Stratosphere-Troposphere exchange, *Rev. Geophys.*, **33**, 403–439 (1995).

38. Fahey, D.W., et al., The detection of large HNO_3-containing particles in the winter arctic stratosphere, *Science*, **291**, 1026–1031 (2001).
39. Larsen, N., Knudsen, B.M., Svendsen, S.H., Deshler, T., Rosen, J.M., Kivi, R., Weisser, C., Schreiner, J., Mauersberger, K., Cairo, F., Ovarlez, J., Oelhaf, H., and Schmidt, A., Formation of solid particles in synoptic-scale arctic PSCs in early winter 2002/2003, *Atmos. Chem. Physics.*, **4**, 2001–2013 (2004).
40. Fromm, M., Bevilacqua, R., Servranckx, R., Rosen, J., Thayer, J.P., Herman, J., and Larko, D., Pyro-cumulonimbus injection of smoke to the stratosphere: observations and impact of a super blowup in northwestern Canada on 3–4 August 1998, *J. Geophys. Res.*, **110**, D08205, doi:10.1029/2004JD005350 (2005).
41. Hervig, M. and McHugh, M., Tropical nitric acid clouds, *Geophys. Res. Lett.*, **29**(7), 1125, doi:10.1029/2001GL014271 (2002).
42. Popp, P.J., et al., The observation of nitric acid-containing particles in the tropical lower stratosphere, *Atmos. Chem. Phys.*, **6**, 601–611 (2006).
43. Rosen, J.M., Morales, R.M., Kjome, N.T., Kirchhoff, V.W.J.H., and Silva, F.R.da, Equatorial aerosol-ozone structure and variations as observed by balloon-borne backscattersondes since 1995 at Natal, Brazil (6°S), *J. Geophys. Res.*, **109**, D03201, doi:10.1029/2003JD003715 (2004).
44. T. Deshler and Mercer, J.L., Large Particles in the Tropical Lower Stratosphere, In Situ Size Distributions, Oral presentation, *AGU fall meeting*, San Francisco, *Amer. Geophys. Union* (December 2006).

Highlights of Fifty Years of Atmospheric Aerosol Research at Mace Head

T.C. O'Connor, S.G. Jennings, and C.D. O'Dowd

Abstract This paper summarises the results of 50 years of research on aerosols in the marine atmosphere at the Mace Head Atmospheric Research Station on the west coast of Ireland. It concentrates on the key sources, physico-chemical properties, number and mass concentrations, size range, volatility and chemical composition of aerosols in different air masses. It also examines key optical properties of the aerosols and their long-range transport.

Keywords Mace head, marine aerosols, sources, volatility, optical properties

The Development of the Mace Head Station

The development of the Atmospheric Research Station on Mace Head began in 1957 with the search for a suitable site to undertake research on air from the Atlantic Ocean that was not complicated by nearby sources of man made pollutants. The site finally chosen was close to the centre of the Atlantic coast of Ireland and of Europe. Its coordinates are 53° 19.5′ N, 9° 54′ W. Winds from a direction between 180° and 300° from North reach it without passing over any inhabited area. In the last 20 years the facilities at the station have been considerably expanded and it has hosted several important international research programmes, field campaigns and workshops. Since 1994 it has been a baseline station for the Global Atmospheric Watch (GAW) programme of the World Meteorological Organisation. The aim of this paper is to summarise the development of the station and to provide a brief account of the milestones of the atmospheric research programme carried out at Mace Head over the years, with the emphasis on aerosol particulate matter.

Department of Experimental Physics & Environmental Change Institute, National University of Ireland, Galway, University Road, Galway, Ireland

C.D. O'Dowd and P.E. Wagner (eds.), *Nucleation and Atmospheric Aerosols*, 625–629.

The initial research was undertaken with two manual Nolan-Pollak (NP) condensation nucleus counters (CNC) which were used to measure the concentration and mean size of nuclei in various air masses and the fraction of them that are charged. This was to test if Boltzmann ionisation equilibrium exists in the air over the ocean.[1] The measurements were occasionally interrupted by the production of large concentrations of very small nuclei in the littoral zone. Investigations of the natural sources of these nuclei at that time were inconclusive but recent work by O'Dowd et al.[2] has identified iodine oxides as being the primary player in the process. The highlights of other research undertaken by personnel of the Atmospheric Research Group in NUI, Galway are outlined below.

Aerosol Physical and Chemical Characteristics

Aerosol number concentrations at Mace Head were first measured[3] by using manual photoelectric nucleus counters and showed baseline levels of around 700 cm^{-3} for marine air conditions. Further measurements[4,5] since 1989 with automatic NP and TSI CNCs have given a range from 50 to 600 per cm^3 for clean marine air conditions. The particle concentrations in bursts of the ultrafine particles from the littoral zone frequently exceeded 10^5 cm^{-3}

Aerosol particle size and mass distribution (diameter range from 0.09 to 3.0 μm) measurements using an optical particle counter have been carried out at Mace Head since 1987.[6] More recently, use is being made of differential mobility analyser/ scanning mobility particle sizers over a particle diameter range from 10 nm to about 0.4 μm. An increase in accumulation mode diameter from 0.1 μm in winter to 0.177 μm in summer has been shown.[7] This is attributed to increased biological activity associated with North Atlantic phytoplankton blooms during the spring-summer seasons. Direct aerosol mass distributions measurements[4] using multistage Berner impactors show an accumulation mode with a geometric mean diameter between 0.4–0.5 μm and a coarse mode centred between 4–5 μm.

Aerosol volatile properties of marine air aerosols have been measured,[6] with the operating scanning temperature range extended from 40°C to 850°C for the first time. The work indicated the presence of sodium chloride particles in the submicron mode – which previously had been ascribed only to the coarse fraction. The volatility technique was extended later[8] to infer for the first time elemental carbon mass.

Aerosol chemical composition of aerosol particles was made to help to identify their origin. Total suspended particulate measurements of the mass concentration of the inorganic fraction at Mace Head were carried out through the Atmosphere/ Ocean Chemistry Experiment (AEROCE) programme [1988–1994]. Some of the results of the work[9] showed that the concentration of sea salt has a seasonal pattern, with a minimum in summer and a maximum in winter because sea salt content in the marine aerosol is dependent on wind speed over the North Atlantic. By contrast, the nss-sulphate mass concentration showed an opposite seasonal pattern with lower values during winter and higher values during mid summer. Seasonal variation of nss-sulphate concentration and of the submicron organic fraction of the aerosol at Mace Head is attributed[25,7] to the cycle of marine biota which is active during the North Atlantic phytoplankton blooming seasons – from spring to autumn. Size segregated chemical composition of clean marine air at Mace Head has been quantified[10,11] and it was concluded that within the majority of the clean marine aerosol masses, inorganic sea salt and organic matter, linked to air bubble mediated aerosol production, significantly dominates the submicron aerosol concentration and mass.

Primary aerosol flux measurements confirmed a submicron source of sea-spray. In terms of primary marine aerosol production, the 1st oceanic eddy-covarience measurements over major oceans[12] confirmed the ocean as a significant source of submicron sea-spray aerosol and found that approximately 50% of the flux (2×10^6 m^{-2} s^{-1} at 10 m s^{-1}) resided in sizes from 10–100 nm and 50% from 100–500 nm. Simultaneously, a dominant organic fraction of submicron aerosol mass was quantified and linked to biological activity in surface waters.[11,13] Very recent chemical-flux measurements confirmed a primary aerosol source of water insoluble organic carbon while the water soluble organic carbon appeared to result from secondary aerosol formation processes.[14]

Aerosol Radiative Properties

Aerosol absorption coefficient measurements have been made since February 1989 via light attenuation through aerosol laden quartz fibre filters using the Magee Scientific Aethalometer Models AE-8 and AE-9. These measurements have been made on a continuous basis with occasional unavoidable breaks. A rigorous time series analysis [17] of monthly averaged black carbon aerosol levels of both marine and continental sector aerosol based on continuous measurements from 1989 to 2003 at Mace Head, shows a significant increase (7.7% $annum^{-1}$ & 13.3%/annum for marine and continental sector aerosol) from 1989 to 1997, but no significant trend thereafter.

Aerosol scattering coefficient measurements by means of nephelometry started at Mace Head in January 1997 using a three-wavelength integrating nephelometer (TSI Model 3563) and a single-wavelength integrating nephelometer (TSI Model 3551). Overall annual mean scattering coefficient amounts to about 20.0 Mm^{-1}. This scattering level can be regarded as representative for the marine boundary layer.[18] Time series analysis[19] indicates a seasonal cycle with up to a factor of 3 higher values in winter compared to summer. The local sea salt production contributes about 20% to the variability of the scattering coefficient, as derived from its correlation with wind speed.[20]

Aerosol Optical Depth (AOD), since March 2000, has been measured using column-integrated light-extinction at Mace Head in order to derive AOD data. The measurements are performed with a Precision Filter Radiometer (PFR), developed at the Physikalisch-Meteorologisches Observatorium Davos/World Radiation

Centre, (PMOD/WRC), Switzerland. Overall mean AOD (500 nm) are of order 0.1 and is in good agreement with baseline AOD values of 0.07 over the Atlantic Ocean[21] and of 0.11 over the North Atlantic Ocean[22]; however, the most recent results[15] illustrate AOD values of 0.4 in wind speeds of 18 m s^{-1} and suggest that the natural primary marine AOD may rival AOD associated with polluted air advecting out over the ocean. Similarly, cloud condensation nuclei[16] are observed to increase with increasing wind speed, pointing to a significant influence of primary marine aerosol on aerosol radiative properties.

Sources of New Particles

The elucidation of new particle formation processes is incredibly difficult due the small sizes of particles involved and the minute mass associated with these particles. Nevertheless, coastal production events at Mace Head were so strong that it enabled sufficient particles of 6–8 nm in size to be electrically extracted from air samples in order to perform high resolution STEM EDX analysis.[23] From this analysis, iodine was found in all analysed particles and suggested an iodine source driving the nucleation and formation mechanism. This was confirmed through a series of chamber experiments on algae emissions and CH_2I_2 and I_2 photo-oxidation experiments.[24,25] The current consensus is that I_2 emissions drive the events and oxidation in the presence of ozone leads to the formation of iodine oxide clusters and aerosol particles. Further growth of particles is driven by sulphuric acid and organic vapour condensation.[26] Sulphuric acid is not present in high enough concentrations to account for growth of clusters into aerosol particles.

Long-range Transport

Hemispheric transport of air-borne pollutants can also be detected at Mace Head and this can contribute to international efforts to quantify both the import and export of pollution to and from Europe. Chemical analysis of aerosol samples show that man's activities and transport of sulphate aerosol from anthropogenic sources have had a major impact on the chemistry of the atmosphere over a large area of the North Atlantic.[9] In addition there is evidence[27] of transport of boreal forest fire emissions from Canada to Mace Head.

Conclusions to Date

Mace Head has proved to be an excellent marine aerosol research station over the years. Significant new insights into aerosol formation and transformation mechanisms have been gained, along with detailed quantification of aerosol physical, chemical and optical properties. Mace Head continues to play an important national and international role in basic aerosol research and research into atmospheric composition and pollution transport. More detail can be found on http://macehead.nuigalway.ie.

References

1. O'Connor, T.C. and Sharkey, W.P., *Proc. Roy. Irish Acad.*, **61 A**, 15–27 (1960).
2. O'Dowd, C.D., et al., *Nature*, **417**, 632–636 (2002).
3. O'Connor, T.C., et al., *Q. Jour. Roy. Met. Soc*, **87**, 105–108 (1961).
4. Jennings, S.G., et al., *Atmos. Environ.* 31, 2795–2808 (1997).
5. McGovern, F.M., *Atmos. Environ.*, **33**, 1711–1722 (1999).
6. Jennings, S.G. and O'Dowd, C.D., *J. Geophys. Res.*, **95**, 13937–13948 (1990).
7. Yoon, Y.J., et al., *J. Geophys. Res.*, **112**, doi:10.1029/2005JD007044 (2007).
8. Jennings, S.G. et al., *Geophys. Res. Letts.*, **21**, 1719–1722 (1994).
9. Savoie, D.L., et al., *J. Geophys. Res.*, **107**, 4356. doi:1029/2001JD000970 (2002).
10. Cavalli, F., et al., *J. Geophys. Res.*, **109**, doi:10.1029/2004JD005137 (2004).
11. O'Dowd, C.D., et al., *Nature*, doi:10 1038/nature02959 (2004).
12. Geever, M., et al., *Geophys. Res. Letts*, doi:10.1029/2001GL02081 (2005).
13. O'Dowd, C.D. and de Leeuw, G., *Phil. Trans. R. Soc. A*, doi:10.1098/rsta.2007.2043 (2007).
14. Ceburnis, D., et al., *Proceedings of the 17th International Conference on Nucleation and Atmospheric Aerosols*, eds. C.D. O'Dowd and P. Wagner, Springer, this issue (2007).
15. Mulcahy, J.P., et al., *Proceedings of the 17th International Conference on Nucleation and Atmospheric Aerosols*, eds. C.D. O'Dowd and P. Wagner, Springer, this issue (2007).
16. Jennings, S.G., et al., *Atmos. Res.*, **46**, 1661–1664 (1998).
17. Junker, C., et al., *Atmos. Chem. Phys.*, **6**, 1913–1925 (2006).
18. Quinn, P.K., et al., *J. Geophys. Res*, **106**, 20783–20810 (2001).
19. Jennings, S.G., et al, *Boreal Environ. Res.*, **8**, 303–314 (2003).
20. Kleefeld, C., et al., *J. Geophys. Res.*, **107**, doi:10 1029/2000JD000262 (2002).
21. Kaufman, Y.J., et al., *Geophys. Res. Letts.*, **28**, 3251–3254 (2001).
22. Villevalde, Y.V. et al., *J. Geophys. Res.*, **99**, 20983–20988 (1994).
23. Mäkelä, J.M., et al., *J. Geophys. Res.*, **107**, doi:10.1029/2001JD000580 (2002).
24. Hoffmann, T., et al., *Geophys. Res. Letts.*, **28**, 1949–1952 (2001).
25. O'Dowd, C.D., et al., *J. Geophys. Res.*, **107**, doi:10.1029/2000JD000206 (2002).
26. Vaattovaara, P., et al., *Atmos. Chem. Phys.*, **6**, 4601–4616 (2006).
27. Forster, C., et al., *J. Geophys. Res.*, **106**, 22887–22906 (2001).

Transformation of Nitrates during the Transport of Air from China to Okinawa, Japan

Shiro Hatakeyama[1,2], Akinori Takami[1], and Yoshihiro Takiguchi[2]

Abstract Comprehensive measurements of chemical, physical, and radiative properties of aerosols have been carried out at a new Japanese super site for aerosol measurements (Cape Hedo Atmosphere and Aerosol Monitoring Station: CHAAMS). Particulate nitrate concentrations were measured both with an aerosol mass spectrometer (PM_1) and a nitrate monitor (PM_{10}) simultaneously. NO_3^- in Okinawa exists mostly in coarse mode. This is a big difference from that measured in China, where NO_3^- exists mainly in fine mode. Chemical transformation takes place during the long range transport.

Keywords Aerosol chemical composition, nitrate, Okinawa, CHAAMS

Introduction

Anthropogenic emission in East Asia has been increasing due to the rapid economic growth.[1] Particularly, the emission of NOx in China is increasing drastically. Satellite monitoring showed a large increase of column abundance of NO_2 over East Asia in these 6 years.[2] After aerosol is emitted and/or is formed from the gaseous species in the source region, then the aerosol is transported. During the transport, the chemical composition of the aerosol is changed due to the gas–aerosol interaction.[3] The change of the chemical composition affects the regional climate change.

A new station (Cape Hedo Atmosphere and Aerosol Monitoring Station: CHAAMS) has been established in Okinawa at the northern tip of the main island. We have carried out the aerosol measurement in order to understand the chemical composition and the chemical transformation of aerosol in East Asia. In this presentation the results of NO_3^- measurements made at Cape Hedo in the year of 2006 are shown.

---

[1]*National Institute for Environmental Studies, Tsukuba, Ibaraki, Japan*

[2]*Graduate School of Emvironmental Sciences, Tsukuba University, Tsukuba, Ibaraki, Japan*

C.D. O'Dowd and P.E. Wagner (eds.), *Nucleation and Atmospheric Aerosols*, 630–634.

Experiments

Observations were all carried out at CHAAMS which is situated at the north end of Okinawa Island as shown in Figure 1 (128.25E, 26.87N, 60 m a.s.l.).[4,5] There is no large industrial area near the observation site. The average temperature is 21°C and the relative humidity is about 75%. The station was renewed and expanded in June, 2005 and in January, 2007.

Measurements of aerosol chemical composition (sulfate, nitrate, ammonium, and organics are main targets) were made by use of an aerosol mass spectrometer (AMS, Aerodyne Research Inc.). AMS can measure both mass spectra and size distribution of ambient aerosol. Aerosols are separated from gaseous species by an aerodynamic lens and vaporized at 600°C on a vaporizer. The vaporized molecules are ionized by the standard 70 eV electron impact ionization and ions are analyzed by a quadrupole mass spectrometer, which gives mass spectra of aerosol components. The size distribution, expressed by the vacuum aerodynamic diameter measured in the free molecule regime, is calculated from the flight time of aerosol. By use of AMS aerosols of PM_1 size range can be measured.

PM_{10} NO_3^- was measured with a nitrate monitor (R & P 8400) equipped with a PM_{10} impactor for removal of particles larger than 10 µm. Here, nitrate measured with AMS is dealt with as nitrate in fine mode, and the difference between PM_{10} nitrate and AMS nitrate is dealt with as nitrate in coarse mode. The data taken simultaneously at Cape Hedo such as total NOy and gaseous HNO_3 concentrations (data provided by Professor Bandow of Osaka Prefecture University) and NOx concentrations (data provided by Ministry of Environment, Japan) were also used for analyses.

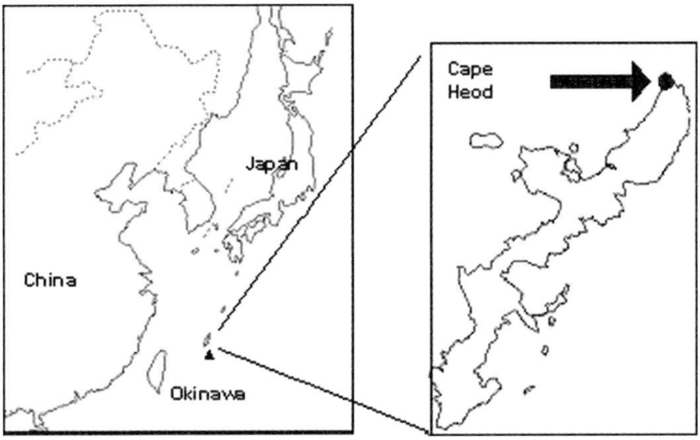

Figure 1 Location of Cape Hedo Atmosphere and Aerosol Monitoring Station (CHAAMS)

Back trajectory analyses were made by use of NOAA HYSPLIT back trajectory analysis system on the website (http://www.arl.noaa.gov/ready/hysplit4.html). The trajectory of air mass going back up to 96h before was analyzed.

Results and Discussion

Concentration of Gaseous and Particulate Nitrogen Oxides

Figure 2 shows the variation of total NOy, gaseous nitrogen (NOx + HNO_3g) and particulate nitrate (PM_{10} NO_3^-) measured in March–November, 2006 at CHAAMS. High-concentration peaks of NOy were seen in spring. Concentration of NO_3^- also showed peaks simultaneously. Back trajectories clearly suggested that those high concentrations of pollutants were transported mainly from China.

As shown in Figure 2 NOy and NO_3 are relatively high in March and April, whereas they are low in summer. This is also due to the difference of air mass origin. Back trajectory analyses showed that air masses come from the Pacific Ocean in summer, while they come from the continent in spring. Origins of pollutants are clearly different in two seasons.

Distribution of Oxygenated Nitrogen between Gas and Particle and Fraction of Coarse Particles and Fine Particles in Particulate NO_3

Oxygenated nitrogen is distributed between gases and particles. In this work gases are defined as NOx (= NO + NO_2) and gaseous HNO_3. Changes of the fraction of gases and particles depend on the temperature following the thermal equilibrium between

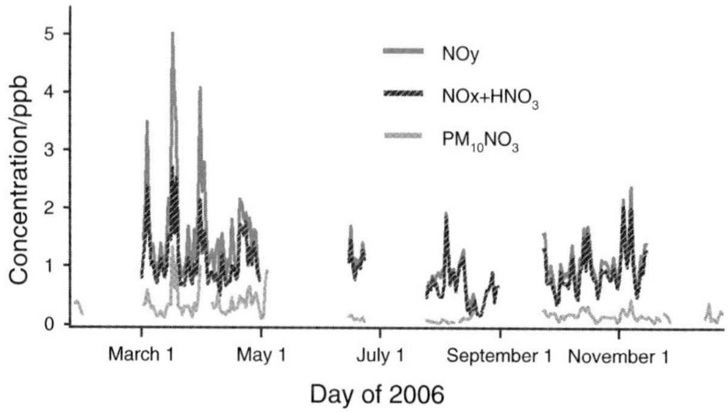

Figure 2 Variation of NOy, NOx + HNO_3g, and NO_3 (PM_{10})

ammonium nitrate and gaseous (NH_3 + HNO_3), i.e., $NH_4NO_3 \rightleftarrows NH_3 + HNO_3$. As shown in Figure 3 the fraction of particles is relatively high in spring. In summer the fraction of particles is low possibly due to the decomposition of NH_4NO_3 to NH_3 and HNO_3 during the long-range transport.

When decomposition of ammonium nitrate takes place, resulting HNO_3 gas can be adsorbed on coarse particles such as sea salts and dust particles. Therefore, the longer the transport at relatively high temperature the larger the fraction of particulate NO_3^- becomes and the larger the fraction of NO_3^- contained in coarse particles becomes. Accordingly, the smaller the fraction of gaseous nitrogen oxides and the smaller the fraction of NO_3^- contained in fine particles becomes. Such situation can be clearly seen in Figure 4. In the figure, the data of time = 0 are those obtained in China in winter, 2002 and in spring, 2003.

Nitrate is reported to exist in a coarse mode in aerosols in relatively clean air,[3,6,7] because gaseous nitric acid quickly deposits on preexisting large particles. The same situation was observed at Cape Hedo, Okinawa.

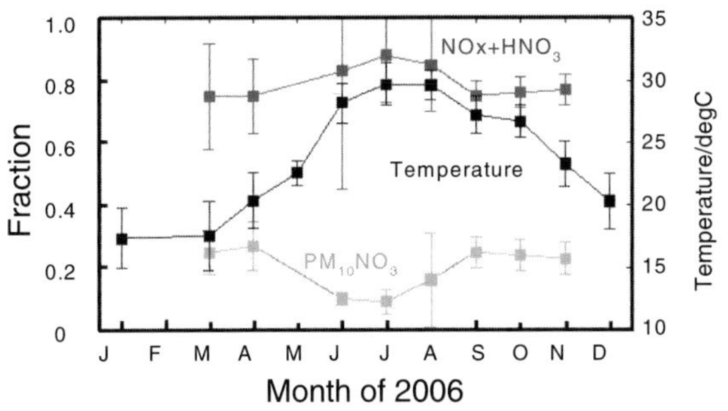

Figure 3 Average fraction of gases (**filled circles**) and particles (**filled triangles**) in oxygenated nitrogen species as well as average temperature (filled squares)

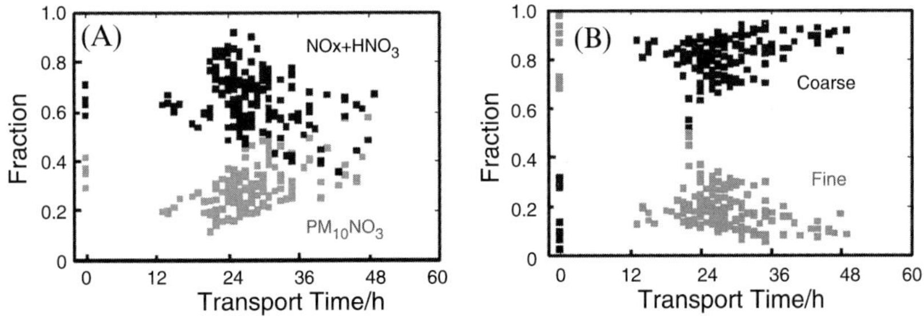

Figure 4 (a) Fraction of gaseous and particulate nitrogenous species. (b) Fraction of NO_3^- contained in coarse and fine particles

Acknowledgment This research has been supported by Global Environment Research Fund (C-051) from Ministry of Environment, Japan.

References

1. Streets, D.G., et al., *J. Geophys. Res.*, **108**, D21 8809, doi:10.1029/2002JD003093 (2003).
2. Richter, A., Burrows, J.P., Nuss, H., Grainer, C., and Niemeier, U., *Nature (London)*, **437**, 129–132, doi:10.1038/nature04092 (2005).
3. Song, C.-H. and Carmichael, G.R., *Atmos. Environ.*, **33**, 2203–2218 (1999).
4. Kato, S., Kajii, Y., Itokazu, R., Hirokawa, J., Koda, S., and Kinjo, Y., *Atmos. Environ.*, **38**, 2975–2981 (2004).
5. Jaffe, D.J., Prestbo, E.M., Swartzendruber, P., Weiss-Penzias, P., Kato, S., Takami, A., Hatakeyama, S., and Kajii, Y., *Atmos. Environ.*, **39**, 3029–3038 (2005).
6. Shimohara, T., Oishi, O., Utsunomiya, A., Mukai, H., Hatakeyama, S., Jang, E.-S., Uno, I., and Murano, K., *Atmos. Environ.*, **35**, 667–681 (2001).
7. Jordan, C.E., Dibb, J.E., Anderson, B.E., and Fuelberg, H.E., *J. Geophys. Res.*, **108**(D20), 8817, doi:10.1029/2002JD003101 (2003).

Major Factors Affecting the Ambient Particulate Nitrate Level at Gosan, Korea

N.K. Kim[1], Y.P. Kim[1]*, and C.H. Kang[2]

Abstract Major factors that are affecting the ambient particulate nitrate concentration levels at Gosan, Korea, were identified based on the inorganic ions data in TSP and PM2.5 measured at Gosan, Jeju Island, Korea between March 1998 and February 2003. According to the correlation coefficient analysis and factor analysis to the TSP and PM2.5 data, it was found that nitrate in TSP is highly correlated with crustal species such as nss-Ca^{2+} or nss-Mg^{2+}, while nitrate in PM2.5 is highly correlated with anthropogenic species such as nss-SO_4^{2-} or NH_4^+. Also, backward trajectory analysis is carried out to find out the difference in trajectories of air parcel movement between two cases. For the cases of good correlation between nitrate and crustal species, air parcels mainly moved from Mongolia, and for the cases of good correlation between nitrate and anthropogenic species, air parcels mainly moved from China. High-nitrate concentration cases occurred most often in spring (59.8%) and when the air parcels moved from Mongolia or China (80.0%).

Keywords Nitrate concentration, correlation, factor analysis, backward trajectory analysis

Introduction

Northeast Asia including Korea, China, and Japan can be characterized by high emissions of anthropogenic species. Among them, China emits an overwhelming fraction (Akimoto and Narita 1994; van Aardenne et al. 1999; Streets et al. 2000).

Especially, the emissions of NO_x from China are increasing continuously. Streets and Waldhoff (2000) suggested the NO_x emission from China increase 122% from 1995 to 2020. Thus, it is important to understand the effects of the

[1] *Department of Environmental Science and Engineering, Ewha Womans University, Korea*

[2] *Department of Chemistry, Cheju National University, Korea*

C.D. O'Dowd and P.E. Wagner (eds.), *Nucleation and Atmospheric Aerosols*, 635–638.

emissions and transport of ambient trace species from northeast Asia, especially China, on the regional air quality and global environment.

For example, Park and Lee (2003) estimated that about half of nitrogen deposition between 1994 and 1998 in Korea was originated from outside. Among them, nitrate aerosols accounted for 80% of them. Therefore, it is important to identify the factors affecting ambient particulate nitrate level in the region.

According to the correlation coefficient analysis and factor analysis to the TSP and PM2.5 data at Gosan between 1998 and 2002, it was found that nitrate in TSP was highly correlated with crystal species such as nss-Ca^{2+} or nss-Mg^{2+}, while nitrate in PM2.5 was highly correlated with anthropogenic species such as nss-SO_4^{2-} or NH_4^+ (Kim et al. 2003, 2004).

In this work, based on the inorganic ions data in TSP and PM2.5 measured at Gosan, Jeju Island, Korea between March 1998 and February 2003, major factors that are related with the nitrate concentration are identified. Backward trajectory analysis results and various statistical tools were used.

Data

The measurement site, Gosan is located on the western tip of Jeju Island (126° 10′E, 33°17′N). Jeju Island is located at about 100 km south of Korean Peninsula, about 500 km west of China (Jiangsu province), and about 200 km east of Japanese Islands (Kyushu). Gosan is located at a 78 m high hill, the far western edge of the island, the least developed area in the island on the grounds of a meteorological station. Several intensive and routine measurements studies including PEM-West A and B, ACE-Asia, and ABC have been carried out at Gosan.

TSP particles were collected by a high volume tape sampler (Kimoto Electric Co., Model 195A). Particles were collected for either 6 or 24 h, and 24 h averaged data are used. The flow rate was about 170 LPM. PM2.5 particles were collected by a low-volume sampler. The sampler consists of a Teflon-coated aluminum cyclone with a cut size 2.5 m at a flow rate of 16.71 LPM (URG, USA), a Teflon filter holder for 47 mm filters (Sarvillex, USA), a critical orifice (BGI, USA), and a pump (Dayton, USA). Daily sampling started at 0900 LST and lasted for 24 h.

Eight ions were analyzed. NH_4^+ was analyzed by the indophenol method with a UV-Visible spectrophotometer and Na^+, K^+, Ca^{2+}, and Mg^{2+} by an atomic absorption spectroscopy. Anions (SO_4^{2-}, NO_3^-, and Cl^-) were analyzed by an ion chromatography. Non-sea-salt (nss)-K^+, nss-Ca^{2+}, nss-Mg^{2+}, and nss-SO_4^{2-} concentrations are estimated by assuming all Na^+ were from sea salt and subtracting the sea water composition. Details on the sampling and analysis were given in Park et al. (2004).

To ensure the quality of the data, three steps of quality control procedures were taken. First, instrument quality check had been carried out. Second, the sampling and analysis QA/QC procedures were checked. Third, ion balance was used to check the validity of the data. The data with the ratio of the sum of the equivalent cation concentrations to the equivalent anion concentrations being within 30% are used for further data analysis. In case of TSP, among the total data sets of 636, 26 data (about 4.1%) are discarded. In case of PM2.5, among the total data sets of 375, 68 data (about 18.1%) are discarded based on these three quality checks. In this work, remaining 610 data sets of TSP and 307 data sets of PM2.5 are used for further analysis.

Results

The data with the highest 10% and the lowest 10% of the nitrate concentrations were selected and analyzed. The highest 10% cases have occurred mostly in spring, and the lowest cases have occurred mostly in summer. Also, the highest 10% cases have occurred when the air parcels moved from Mongolia or China, and the lowest cases have occurred when the air parcels moved through the Pacific Ocean. However, in the samples with the similar trajectories, crustal species was the major factor that determined the nitrate concentration in TSP, while NH_4^+ was the factor that determined the nitrate concentration in PM2.5.

To further identify the characteristics of the high-nitrate concentration cases, the data with the nitrate concentration higher than the sum of the mean and one standard deviation were selected. These cases in TSP were correlated well with crustal species such as nss-Ca^{2+} or nss-Mg^{2+}, while these cases in PM2.5 were correlated well with anthropogenic species such as NH_4^+ (Table 1).

For the cases of good correlation between nitrate and anthropogenic species, backward trajectory analysis result showed that air parcels mainly moved from China. For the cases of good correlation between nitrate and crustal species, air

Table 1 Factor analysis result for ions in TSP and PM2.5 at Gosan

	Component				Component		
TSP	1	2	3	PM2.5	1	2	3
Na^+	0.990	0.008	0.100	ss-SO_4^{2-}	0.986	−0.011	0.133
ss-SO_4^{2-}	0.989	0.008	0.097	Na^+	0.986	−0.011	0.132
ss-Ca^{2+}	0.984	0.015	0.101	ss-K^+	0.984	−0.009	0.137
ss-Mg^{2+}	0.969	0.038	0.169	ss-Mg^{2+}	0.974	0.007	0.167
ss-K^+	0.966	0.046	0.124	ss-Ca^{2+}	0.968	0.015	0.149
Cl^-	0.881	−0.187	0.229	Cl^-	0.903	−0.106	−0.021
NH_4^+	−0.097	0.947	0.008	NH_4^+	−0.093	0.965	0.031
nss-SO_4^{2-}	0.010	0.915	0.282	nss-SO_4^{2-}	−0.081	0.852	0.264
nss-K^+	0.053	0.814	0.395	NO_3^-	0.042	0.754	−0.036
nss-Ca^{2+}	0.170	0.120	0.917	nss-K^+	0.027	0.722	0.368
nss-Mg^{2+}	0.078	0.183	0.874	nss-Ca^{2+}	0.106	0.148	0.891
NO_3^-	0.279	0.324	0.705	nss-Mg^{2+}	0.268	0.197	0.814
Eigen Value	6.143	3.330	1.280	Eigen Value	5.944	3.202	1.154
% of Variance	51.194	27.747	10.664	% of Variance	49.534	26.684	9.620
Cumulative %	51.194	78.941	89.605	Cumulative %	49.534	76.217	85.837

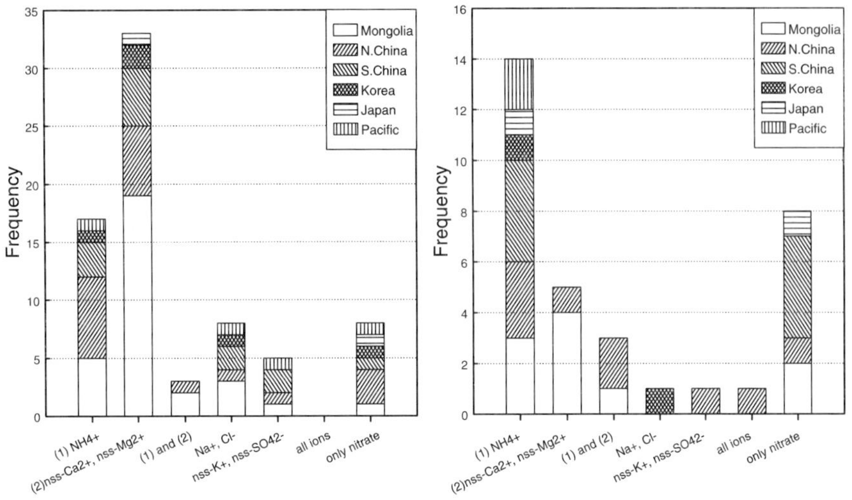

(a) TSP (b) PM2.5

Figure 1 Number of days of high nitrate concentration cases with simultaneous increase of other major ions and the sectors of air parcel trajectories

parcels mainly moved from Mongolia (Figure 1). High nitrate concentration cases occurred most often in spring (59.8%) and when the air parcels moved from Mongolia or China (80.0%).

Acknowledgments The authors gratefully acknowledge the NOAA Air Resources Laboratory (ARL) for the provision of the HYSPLIT transport and dispersion model and/or READY website (http://www.arl.noaa.gov/ready.html) used in this work. This work was supported by the Korea Science and Engineering Foundation (KOSEF) through the National Research Lab. Program funded by the Ministry of Science and Technology (No. M10600000221-06J0000-22110).

References

1. Akimoto, H. and Narita, H., *Atmos. Environ.*, **28**, 213–225 (1994).
2. Kim, N.K., Kim, Y.P., Kang, C.H., and Moon, K.C., *J. Kor. Soc. for Atmos. Environ.* (in Korean), **19**, 333–343 (2003).
3. Kim, N.K., Kim, Y.P., Kang, C.H. and Moon, K.C., *J. Kor. Soc. for Atmos. Environ.* (in Korean), **20**, 119–128 (2004).
4. Kim, Y.P., Moon, K.-C., Shim, S.-G., Lee, J.H., Kim, J.Y., Fung, K., Carmichael, G.R., Song, C.H., Kang, C.H., Kim, H.-K., and Lee, C.B., *Atmos. Environ.*, **34**, 5053–5060 (2000).
5. Park M.H., Kim, Y.P., Kang, C.H., and Shim, S.G., *J. Geophys. Res.*, **109**, D19S13, doi:10,1029/2003JD0014110 (2004).
6. Park, S.U. and Lee, E.H., *Atmos. Environ.*, **37**, 3967–3980 (2003).
7. Streets, D.G. and Waldhoff, S.T., *Atmos. Environ.*, **34**, 363–374 (2000).
8. Streets, D.G., Tsai, N.Y., Akimoto, H., and Oka, K., *Atmos. Environ.*, **34**, 4413–4424 (2000).
9. van Aardenne, J.A., Carmichael, G.R., Levy II, H., Streets, D., and Hordijk, L., *Atmos. Environ.*, **33**, 633–646 (1999).

Functional Group Analysis Using Tandem Mass Spectrometry, Method Development, and Application to Particulate Organic Matter

J. Dron, N. Marchand, and H. Wortham

Abstract The present study describes the development of new analytical techniques for the determination of oxygenated functional groups using atmospheric pressure chemical ionization – tandem mass spectrometry (APCI-MS/MS) operated in the constant neutral loss scanning (CNLS) mode. The carboxylic acid and carbonyl moieties were investigated. Applying this method together with an innovative statistical approach demonstrated its suitability for quantitative measurements. Measurements of carboxylic concentrations inside atmospheric aerosol samples also proved its applicability to particulate organic matter (POM).

Keywords Functional group analysis, particulate organic matter, chemical composition, mass spectrometry

Introduction

The particulate organic matter (POM) in atmospheric aerosols contributes to approximately 20–50% of the total fine aerosol mass at continental midlatitudes and is known to have a high potential impact on both human health and climate change.[1] Considering the degree of complexity of POM, conventional analytical methods lead to the characterization of at best 20% of the organic carbon total mass.[2] Functional group analysis is a complementary approach enabling the characterization of a larger fraction of the aerosol mass and providing valuable chemical composition data. This technique being besides the best suited for modelling purposes, it appears imperative to improve its performances.[3,4] The most popular techniques for functional group analysis are Fourier-Transform Infra-Red (FT-IR) and proton Nuclear Magnetic Resonance (H-NMR) spectroscopy. However, both techniques have shown limitations due to poor robustness in quantitative determination,[5] heavy instrumentation and relatively low sensitivity.[6,7]

LCE, Université de Provence (case 29), 3 place Victor Hugo, 13331 Marseille Cédex 3, France

C.D. O'Dowd and P.E. Wagner (eds.), *Nucleation and Atmospheric Aerosols*, 639–643.

The aim of this work is to propose a new method for the analysis of functional groups and to highlight its relevance for atmospheric chemistry purposes. The analytical method is based on tandem MS operating in the constant neutral loss scanning (CNLS) mode. The quantification of the functional groups relies on the ability of the compounds bearing the same functional group to loose an identical and characteristic neutral fragment into the MS collision cell. Compared to FT-IR and H-NMR spectroscopy, this analytical strategy offers major benefits: (i) molecular weight distribution of the detected compounds (ii) high accuracy quantification of the functional groups (iii) detection limits allowing its application to environmental measurements.

The targeted functional groups were carboxylic acids and carbonyl compounds, which originate from primary emissions as well as in situ formation through oxidative processes.[8–11] Real sample analysis proved that the presented procedure enables the determination of functional groups in atmospheric POM with high sensitivity and robustness.

Experimental Section

Twenty-nine carboxylic acids and 23 carbonyls were selected in order to study a wide panel of molecules having different molecular environment around the targeted functional group and because of their similitude with other compounds detected[11,12] in atmospheric POM. Twenty-five reference mixtures containing all the 29 or 23 compounds at different and randomly determined concentrations were prepared in methanol. Before analysis, the carboxylic acids were methylated into their corresponding methyl esters with boron trifluoride (BF_3/Methanol, 14%, v/v) and the carbonyl compounds were derivatized with pentafluorophenylhydrazine (PFPH).

The analyses were performed on a triple quadrupole mass spectrometer equipped with an atmospheric pressure chemical ionization (APCI) chamber. The standard solutions and the samples were injected directly into the ionization chamber, no HPLC column was used in this work.

Atmospheric particles were collected downstream a charcoal denuder on teflon-coated glass fiber filters in July 2006 in the center of the city of Marseille, France. The filters were extracted by means of an Accelerated Solvent Extraction (ASE) device with methanol/hexane (1 + 1, v/v) at 100°C and 100 bars during 5 min.

Results and Discussion

Analysis of Standard Material

The carboxylic acids were first individually analyzed after derivatization and by neutral loss of 32 arbitrary mass units (amu) corresponding to the loss of a methanol neutral fragment. The responses in terms of signal to noise (S/N) atios could be divided into four categories. The diacids and oxoacids presented he highest S/N ratios (from 30% to 180% taking hexanedioic acid as reference), ıitro and aminoacids had S/N ratios of 15% and phenyl, hydroxy groups or ılkanoic acids ranged from 1% to 5%. Finally, branched and linear alkanoic ıcids were not detected.

Carbonyl compounds were detected by neutral loss 183 amu corresponding to he loss of a pentafluoroaniline fragment. As for carboxylic acids, a significant ıeterogeneity could be observed. Ketones presented S/N ratios approximately 5 to .0 times higher than aldehydes. Also, the general tendency was the opposite as 'or carboxylic acids, linear ketones having the highest S/N ratios while diones or lialdehydes having the lowest ones.

The results above show that great disparities in terms of signal intensities are ›bserved due the nature of the molecular environment around the targeted 'unctional groups. For that reason and for a better consideration of the realistic :onditions in the POM, the study of reference mixtures of standard compounds vas a prerequisite.

Figure 1 shows the S/N ratio distribution obtained for study of the carboxylic ıcid reference mixtures, each of them containing each carboxylic acid at lifferent randomly determined concentrations. It has a Gaussian shape and :orresponds to a calculated relative standard deviation (RSD) of 12% over the 25 reference mixtures. Also, the RSD was 5% for seven injections of the same 'eference mixture. Similar results were obtained in the case of the carbonyl 'unctionality with respectively, 14% and 5%. Considering reference mixture naterial analysis, the detection and quantification limits were respectively).005 mM and 0.02 mM.

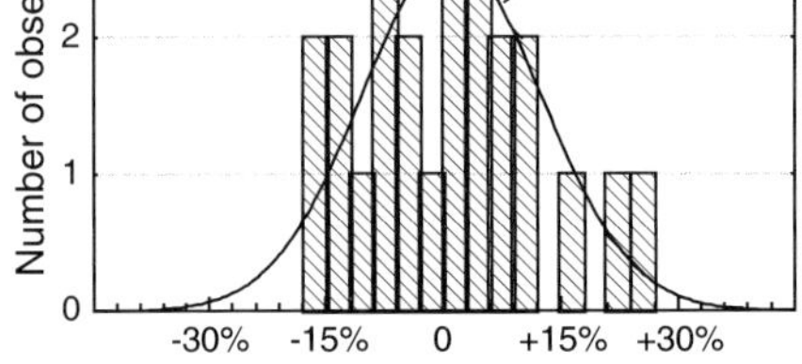

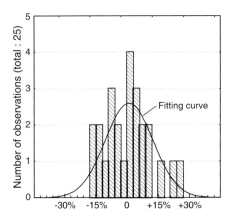

'igure 1 Distribution of the S/N ratios obtained for the 25 reference mixtures. X-axis, divided n 3% intervals, represents the variation in the S/N ratios obtained for all 25 different composition 'eference mixtures compared to the average calculated value for which X = 0. The number of ›bservations corresponds to the number of mixtures for which the S/N ratio was included inside ›ne interval

Application to Atmospheric Particulate Matter

Only carboxylic functional groups were to date determined inside an atmospheric sample following the analytical method described earlier. The results of a daytime filter analysis before and after derivatization are presented in Figure 2. The measured intensity is clearly higher when analyzing the sample after derivatization and represents the addition of initially present methyl ester and methoxy functional groups together with derivatized carboxylic moieties. As expected, we can observe a molecular weight distribution of the carboxylic acids inside the sample, with a carboxylic functional group background ranging from approximately m/z 150 to 380 with a maximum located around m/z 220.

The carboxylic acid molar concentrations were 12.00 ± 2.48 nmol/m^3 and 7.94 ± 2.39 nmol/m^3 for respectively day and nighttime samplings. They correspond to around 10% of the PM10 total mass according to an average molecular weight of 250 g/mol. These results are consistent with what found elsewhere[7] in terms of

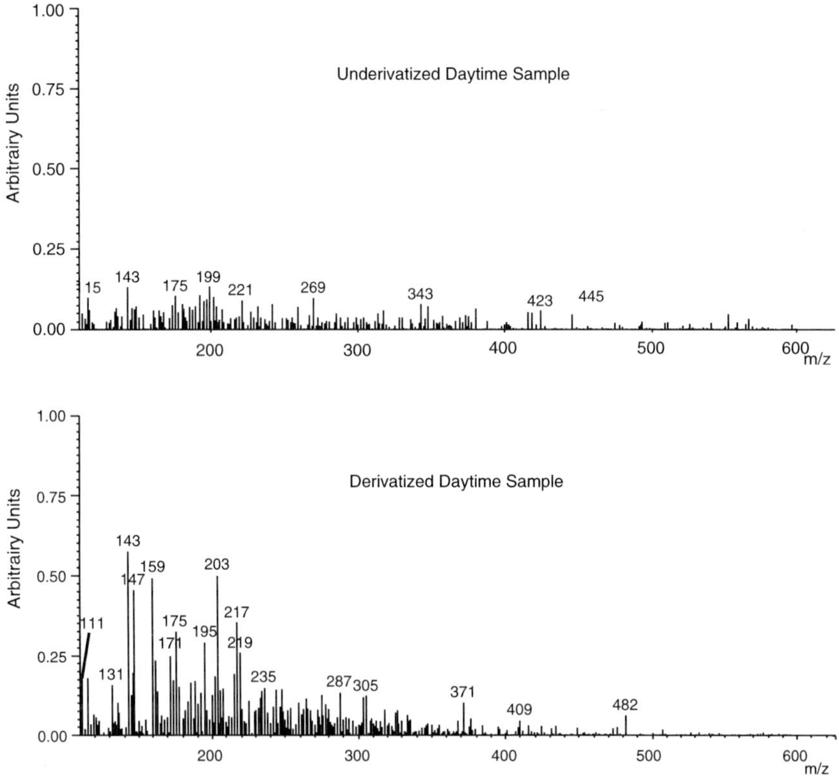

Figure 2 Average of ten mass spectra obtained for the neutral loss analysis of methanol on the same sample (urban atmospheric aerosol collected during daytime in July 2006) before and after derivatization

concentrations and prove the applicability of the neutral loss tandem mass spectrometry method to the analysis of POM samples.

Conclusions

An innovative method using MS/MS for the quantitative functional group analysis of POM was successfully developed and is presented here. The results obtained on standard material illustrate the higher robustness and the lower detection limits achieved by this method compared to the existing ones, down to 0.8 nmol/m^3 for the carboxylic acid complete procedure including sampling and extraction. The instrumentation employed here is also lighter and more universal as it is also used for identification an online measurement purposes. Moreover, it provides an approximate molecular weight distribution of the molecules contained in the sample and bearing the targeted functional group. The comparison of such distributions could be of great interest for source apportioning and reaction modeling in the particulate organic matter issued from atmospheric aerosols.

Comparative experiments between measurements carried out using the presented procedure and modeling calculation results for *n*-decane photooxidation will also be presented.

References

1. Kanakidou, M. et al., *Atmos. Chem. Phys.*, **5**, 1053–1123 (2005).
2. Jacobson, C.M. et al., *Rev. Geophys.*, **38**, 267–294 (2000).
3. Fuzzi, S. et al., *Atmos. Chem. Phys.*, **6**, 2017–2038 (2006).
4. Decesari, S. et al., *J. Geophys. Res.*, **105**, 1481–1489 (2000).
5. Blando, J.D. et al., *Environ. Sci. Technol.*, **32**, 604–613 (1998).
6. Suzuki, Y. et al., *Environ. Sci. Technol.*, **35**, 2656–2664 (2001).
7. Tagliavini, E. et al., *Atmos. Chem. Phys.*, **6**, 1003–1019 (2006).
8. Edney, E.O. et al., *Atmos. Environ.*, **37**, 3947–3965 (2003).
9. Kawamura, K. et al., *Atmos. Environ.*, **30**, 1709–1722 (1996).
10. Jaoui, M. et al., *Anal. Chem.*, **39**, 5661–5673 (2004).
11. Li, Y.-C. et al., *Environ. Sci. Technol.*, **39**, 7616–7624 (2005).
12. Schauer, J.J. et al., *Environ. Sci. Technol.*, **36**, 1169–1180 (2002).

Study on the Distribution and Variation Trends of Atmospheric Aerosol Optical Depth over the Yangtze River Delta in China

Duan Jing and Mao Jietai

Abstract Characteristics of distribution and seasonal variation trends o aerosol optical depth (AOD, on 550 nm) over the Yangtze River Delta (YRD were analyzed in this paper with the NASA MODIS Level-2 products fror 2000 to 2005. The results indicate that the areas of regions with larger than 1. increased most quickly. The AOD over main big cities in the YRD increase gradually year by year according to the variation characteristics of AOD pea values in the whole year over the cities in plain regions appear in summer bu those over the cities in mountainous areas appear in spring. The yearly increasin trend of AOD over the plain cities is larger than those over the mountainou area cities.

Keywords MODIS, aerosol optical depth, Yangtze River Delta in China

Introduction

Recently the research and the observation show that atmospheric aerosol play a important part on the global average radiation forcing. The radiation forcing produce by atmospheric aerosol is comparative with the radiation forcing produced b greenhouse gases.[1] AOD is a primary parameter which can reflect the aerosol impa to the radiation. Also the reliability of research about atmospheric aerosol is ver low now.[2] The main reason is we do not know aerosol accurate characters we (including physical and chemical characters of aerosol; distribution and optic characters of aerosol; particle size distribution).

The YRD is the important economic central region in coastland of easter China. It is also one of the areas in which aerosol is also increased the mo quickly. To study the climate and environment effects of aerosol will have ver

Department of Atmospheric Sciences, School of Physics, Peking University, Beijing, China, 100871

C.D. O'Dowd and P.E. Wagner (eds.), *Nucleation and Atmospheric Aerosols*, 644–648.

mportant meaning. In this paper, the AOD retrieved from MODIS data was analyzed in order to get the region and seasonal distribution and the variation endency of the YRD in recent year.

Data and Methods

MODIS (Moderate Resolution Imaging Spectroradiometer) is a sensor aboard the TERRA and AQUA satellites of NASA with 36 channels from visible to infrared ight, a highest visible light resolution of 250 m and a scanning width of 2,330 km. The spatial resolution of MODIS aerosol products is 10 km. This paper uses the AOD land data of February 2000 to July 2005 to summarize the AOD seasonal listribution and variation tendency over the YRD in China.

The AOD from the MODIS products is not exactly accurate. It is because of uncertainty of albedo of underlying surface. Wherefore ground-channel sun-photometer remote sensing is generally used to verify satellite remote sensing esults. Because of the previous analysis results between ground observations and satellite remote sensing,[3,4] the AOD products from MODIS data had have the need quality. It can be used in this study.

Results and Discussion

The research region is 112°~122° E, 26°~34° N. The season divisiory standard is rom December to February as winter; from March to May as spring; from June to August as summer; from September to November as autumn. Thirty-one cities are hosen in the YRD in order to analyses the seasonal variation of these cities.

Seasonal Distribution Character of AOD in YRD

Figure 1 shows that seasonal average spatial distribution of MODIS Level-2 AOD product from 2000 to 2005. The result is terrain and density of population impact the istribution of the AOD. There are three central areas in this region in which the AOD s higher than other areas. They are Lianghu plain, Boyanghu plain, and YRD plain. The ig value region in YRD plain is mostly in the south of the Yangtze River. The AOD of he plain in the south of Shanghai is bigger than 0.8; the biggest value appear in summer. The big humidity is one of the reasons. Because of hygroscopic aerosol the AOD will e increased by sorbing the water. The actinism in summer also works on it.

In order to explain the AOD variation year by year, Figure 2 shows the classified tatistic of the given areas from by value of the AOD from 2000 to 2005 (the AOD s disparted into four species). The area with less than 0.6 of AOD decreases year

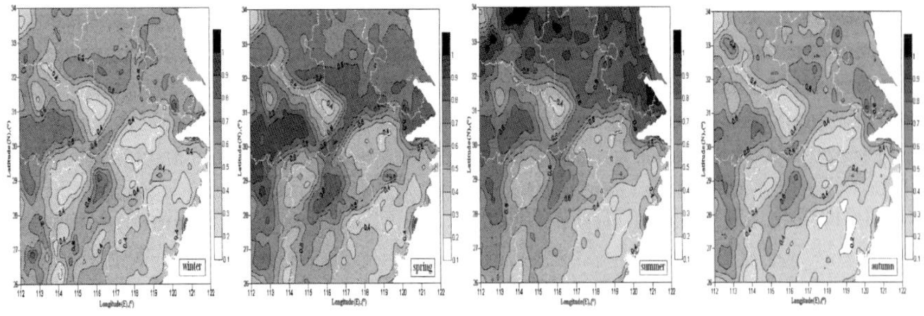

Figure 1 Seasonal average spatial distribution of MODIS Level-2 AOD product (2000~2005)

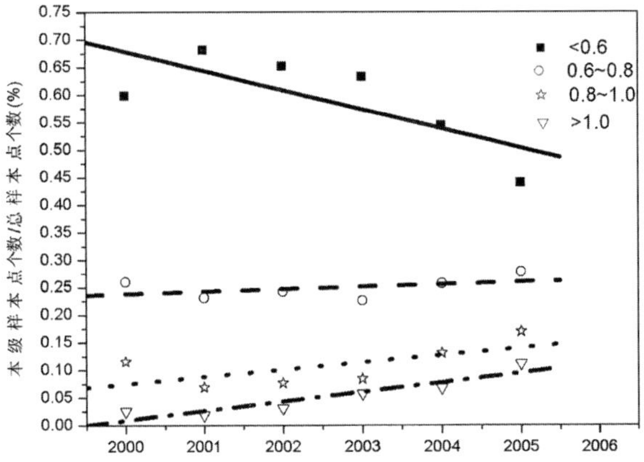

Figure 2 The classified statistic of the given areas from by value of the AOD (2000~2005)

by year. The area with larger than 1.0 of AOD increase most quickly. It shows tha the pollution gradually expands in 2000 ~ 2005.

Variation Characters of AOD in Cities of the YRD

Seasonal Variation of the Cities

Thirty-one cities chosen in the YRD are analyzed. According to the differen season which is the big value of the AOD, two kinds of cities are separated in a cities. They are cities in plain (Figure 3 (a); Figures 4 and 5 sample 1) and cities i mountain (Figure 3 (b); Figures 4 and 5 sample 2). Sample 1 is the big value of th AOD in summer. Sample 2 is not observes the big value of the AOD in summe The AOD in plain is less than in mountain. The AOD increases in summer in bot of them. But the AOD of plain cities increases in summer observably.

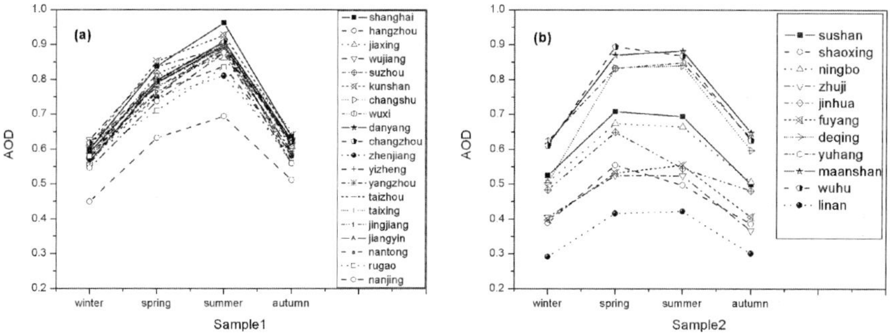

Figure 3 The sorts and characters of seasonal variation of 31 stations

Figure 4 AOD seasonal variation of two samples (2000~2005)

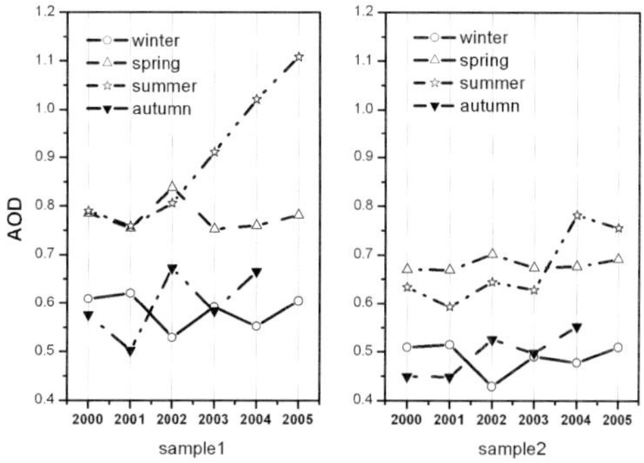

Figure 5 AOD time series of two samples (2000~2005)

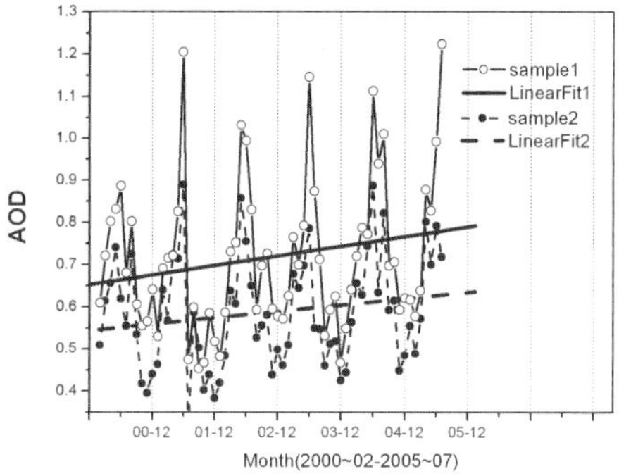

Monthly Average Variation Characters of the Cities in All 6 Years

According to the sorts of cities, monthly variation of two kinds of cities is shown in Figure 5. The AOD in both of the two kinds of cities increased year by year. The increasing tendency of sample 1 is faster than sample 2. This kind of tendency closely contacts with the development of the industry.

Summary

The results shows that bigger value area of AOD over the YRD is increasing year by year, especially the area of the AOD bigger than one increases the most quickly. The polluted range expanded continuously in 2000 ~ 2005. The big value season of the AOD change from spring to summer. The decrease of dust weather in spring maybe is the reason. On average the atmospheric AOD of each main city gradually increased in 2000 ~ 2005. The yearly increasing trend of AOD over the plain cities is larger than those over the mountainous area cities.

Acknowledgment This research was supported by National Natural Science Foundation of China (No. 40475003).

References

1. Mao J.T., Zhang J.H., and Wang M.H., Summary comment on research of atmospheric aerosol in China[J], *ACTA Meteorologica Sinica*, **60**(5), 625–634 (2002).
2. IPCC, Climate Change 2001: Synthesis Report[R]. IPCC.156–157.
3. Mao J.T., Li C.C., and Zhang J.H., et al., The comparison of remote sensing aerosol optical depth from MODIS Data and ground sun-photometer observations [J], *J. Appl. Meteorol. Sci.*, **13**, 127–135 (2002).
4. Li C.C, Mao J.T, and Lau K.H, et al., Research on the air pollution in beijing and its surroundings with MODIS AOD products [J], *Chinese Atmos. Sci.* (in Chinese), **27**(5), 869–878 (2003).

Detecting Below 3 nm Particles Using Ethylene Glycol-based Ultrafine Condensation Particle Counter

Kenjiro Iida, Mark R. Stolzenburg, and Peter H. McMurry

Abstract An ethylene glycol based ultrafine condensation particle counter (EG-UCPC) was developed by modifying the laminar flow ultrafine condensation particle counter which was originally developed by Stolzenburg and McMurry (1991) [1]. Prior to experiments, theoretical analyses were performed using different working fluids to predict activation efficiency and condensational growth after activation. Ethylene glycol was chosen as the working fluid since it is less toxic than other compounds investigated and it has a predicted 50% cutoff diameter is below 2 nm. Experiments were performed using sodium chloride, ammonium sulfate, and silver as test aerosols. To obtain the activation efficiency from the experimental data we have accounted the size dependence of input particle size distribution and particle penetration within the diffusion-broadened mobility window of transfer function of the DMA. The 50% cutoff sizes of negatively charged particles are 1.3, 1.3, and 1.5 nm mobility diameter for sodium chloride, ammonium sulfate, and silver, respectively. The lower cutoff size for the first two compounds suggests that ethylene glycol vapor heterogeneously nucleates more easily on hygroscopic material.

Keywords Condensation particle counter

Theoretical Analysis

Activation efficiencies for the UCPC were simulated for different working fluids using the theoretical approach of Stolzenburg and McMurry (1991) [1]. The condenser temperature was held at 3°C, and the saturator temperature was changed until the maximum classical homogenous nucleation rate along the axis was equal to 1 cm^{-3} s^{-1}. The capillary flow rate, total flow rate, and the geometry of saturator/condenser of the original design [1] were used in the simulation. Figure 1 shows the theoretical activation efficiencies using formamide, ethylene glycol, and *n*-butanol as the

Particle Technology Laboratory, University of Minnesota, Minneapolis, Minnesota, USA.

C.D. O'Dowd and P.E. Wagner (eds.), *Nucleation and Atmospheric Aerosols*, 649–653.

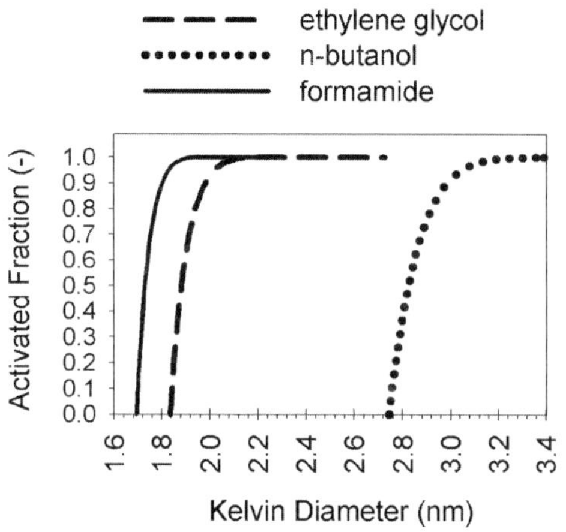

Figure 1 Simulated activation curve of ultrafine CPC using formamide, ethylene glycol and n-butanol as working fluid

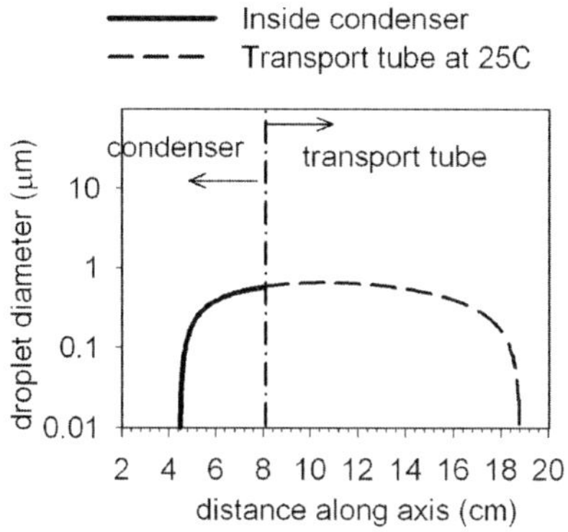

Figure 2 Simulated growth of activated particles along the axis inside the condenser of UG-UCPC, a transport tube after the condenser

working fluid. The calculated saturator temperature for formamide, ethylene glycol, and *n*-butanol are 58.0°C, 54.7°C, and 42.9°C, respectively. We have chosen ethylene glycol (EG) as working fluid since it is less toxic than formamide. Additionally, Kim et al.[2] detected near 100% of 1.6 nm (mobility diameter) NaCl particles using their ethylene glycol based turbulent-mixing type CPC. We have chosen to modify the original design of Stolzenburg and McMurry (1991) since its commercial version has been used for many applications for more than a decade. However, since the vapor pressure of EG is comparatively low a second CPC is needed downstream to grow particles to a size that can be detected optically (>1 μm). Figure 2 shows the calculated diameter of activated particles along the condenser axis and a transport tube downstream. Depletion of vapor inside the transport tube is likely to cause evaporation of the EG droplets if the transport tube is too long.

Experiments

The detection efficiency of the EG-UCPC was evaluated experimentally. In our experiments, the evaporation–condensation method is used to generate nanoparticles consisting of sodium chloride, ammonium sulfate, or silver. A small fraction of nanoparticles are singly and negatively charged as they go through the unipolar charger [3], and these nanoparticles are classified by the nano-DMA [4] operating sheath and aerosol flowrates of 15 LPM and 1.35 LPM, respectively. Figure 3 show the experimental setup downstream of the nano-DMA.

The aerosol flow is symmetrically separated into equal proportions (1.5 LPM each) between the EG-UCPC and Faraday cage electrometer (FCE) at a tee after they pass through a mixing orifice (Ø = 1.5 mm). The flow control scheme of the EG-UCPC is similar to the original design [1]. The core portion of the total flow is extracted at 0.3 LPM, and capillary flow is introduced along the axis of the saturator extension at 1.5 $cm^3 s^{-1}$. The condenser temperature was set at 3°C and the saturator temperature was increased up to the point where false counts remain below 1–2 counts min^{-1}. As theoretical analysis suggested, the final sizes at the condenser exit were too small to be detected by the optics of UCPC; therefore, an additional CPC (TSI 3010) having a 50% cutoff size at 15 nm is used to complete the growth up to optically detectable sizes.

Results

Theoretical inversion of experimental data was performed to obtain the activation efficiency. The equation that establishes the relationship between the activation efficiency, $\eta(D_p)$, and concentration measurements made with the CPC and FCE at a given DMA voltage, V_{DMA}, is:

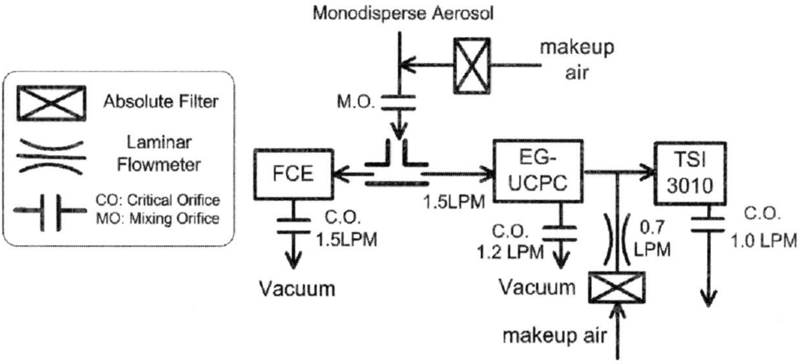

Figure 3 Experimental setup for evaluating the detection efficiency of ethylene glycol based Ultrafine Condensation Particle Counter (EG-UCPC)

$$\left.\frac{N_{CPC}}{N_{FCE}}\right|_{V_{DMA}} = \frac{\int P_{CPC}(D_P)\eta_{CPC}(D_P)\Omega(V_{DMA},D_P)n(D_P)dD_P}{\int P_{FCE}(D_P)\Omega(V_{DMA},D_P)n(D_P)dD_P} \quad (1)$$

where, P_{CPC} and P_{FCE} are theoretically estimated penetration from the nano-DMA exit slit to the end of the UCPC capillary and to Faraday cage, respectively, $\Omega(V_{DMA}, D_p)$ is the transfer function of nano-DMA including the effects of Brownian broadening [5]. The difference between the theoretical and actual performance of the DMA was empirically evaluated. An empirical factor is adjusted until the theoretical diffusion-broadened transfer function fit the mobility spectra of fullerene (C_{60}) which is known to be monodisperse around 1 nm [6]. The particle size distribution, $n(D_p) \equiv dN / dD_p$, of singly charged particles at the DMA inlet slit is obtained by inverting the concentration measured by FCE. Figure 4(a) shows an inverted $n(D_p)$ that had the sharpest decrease toward the lower size end among the datasets and the transfer function evaluated at 4.7 V (1.3 nm). The activation efficiency of EG-UCPC, $\eta(D_p)$, that gave the best fit to data at all values of V_{DMA} are found by solving

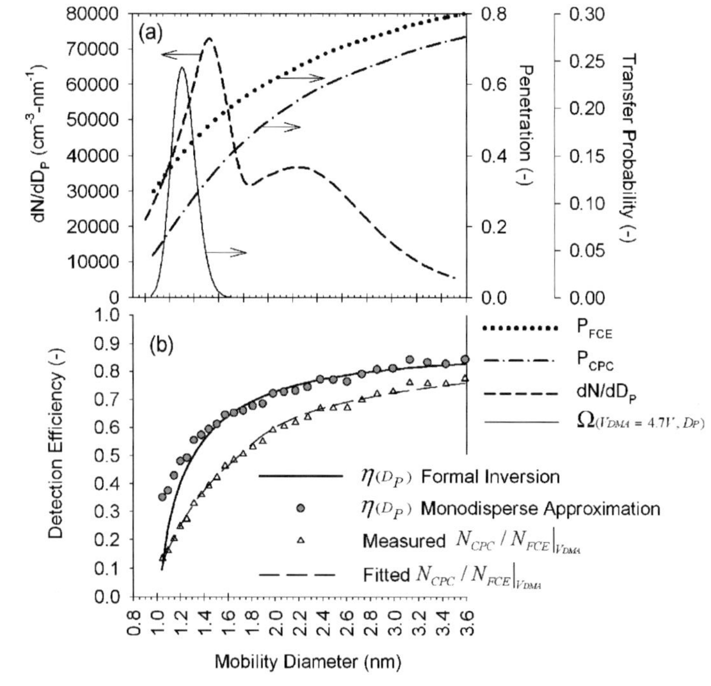

Figure 4 (a) Inverted input distribution, dN/dD_p, of singly charged ammonium sulfate particles at DMA inlet slit, transfer function evaluated at 4.7 V (1.3 nm), and theoretically estimated penetration from DMA exit slit to Faraday Cage. P_{FCE}, and to the end of CPC capillary, P_{CPC}. (b) Measured ratio of N_{CPC}/ N_{FCE}, fitted curve after inversion, and inverted activation efficiencies accounting for size dependence of P_{FCE}, P_{CPC}, $n(D_p)$, and $\eta(D_p)$, shown in (a), within the mobility window of $\Omega(D_p, V_{DMA})$ (**solid line**) and assumes classified particles are monodisperse (dots)

Table 1 50% Cutoff size (nm), D_{P50}, of negatively charged particles

Test Aerosol	Stokes-Einstein	Tammet
Sodium Chloride	1.30	0.91
Ammonium Sulfate	1.31	0.93
Silver	1.54	1.19

Eq. 1. Figure 4(b) shows inverted results $\eta(D_P)$ obtained using P_{CPC}, P_{FCE}, and $n(D_P)$ shown in Figure 4(a). These results obtained by inverting Eq. 1 are compared in Figure 4(b) with values obtained by assuming that particles leaving the DMA were perfectly monodisperse. The difference in the 50% cutoff size between two cases is only about 0.05 nm indicating the transfer function of the nano-DMA is sufficiently narrow to assume that classified particles were monodisperse. Table 1 shows experimentally measured values for D_{P50} for three types of the test aerosol defined using the standard Stokes-Einstein relation and semi-empirical approach of *Tammet* (1995) [7]. The cutoff sizes are lower for sodium chloride and ammonium sulfate than silver suggesting that ethylene glycol vapor heterogeneously nucleates more easily on hygroscopic material.

Acknowledgments This research was supported by NSF Award No. ATM-0506674, "NIRT-Clusters to Nanoparticles:Implications for Atmospheric Nucleation."

References

1. Stolzenburg, M. R. and P. H. McMurry, *Aerosol Sci. Technol.,* **14**, 1, 48–65 (1991).
2. Kim, C. S., K. Okuyama and J. Fernãndez de la Mora, *Aerosol Sci. Technol.,* **37**, 1–13 (2003).
3. Chen, D.-R. and D. Y. H. Pui, *J. Nanoparticle Res.,* **1**, 1, 115–126 (1999).
4. Chen, D.-R., D. Y. H. Pui, D. Hummes, H. Fissan, F. R. Quant and G. J. Sem, *J. Aerosol Sci.,* **29**, 5/6, 497–509 (1998).
5. Stolzenburg, M. R., (1988), Ph.D. Thesis, Dept. of Mech. Eng., Univ. Minnesota, Minneapolis, Minnesota, USA.
6. Seol, K. S., J. Yabumoto and K. Takeuchi, *J. Aerosol Sci.,* **33**, 1481–1492 (2002).
7. Tammet, H., *J. Aerosol Sci.,* **26**, 3, 459–75 (1995).

Chlorine-phase Partitioning at Melpitz near Leipzig

Detlev Möller and Karin Acker

Abstract Hydrochloric acid (HCl) in the gas phase, chloride, and sodium in particle phase were measured first time with high time-resolution and simultaneously with a number of other atmospheric components (in gas and particulate phase) as well meteorological parameters during a campaign at the research station Melpitz (Germany) in summer 2006 to study the Cl partitioning. On most of the 19 measurement days the HCl concentration showed a broad maximum around noon/afternoon (on average 0.1 μg m^{-3}) and much lower concentrations during night (0.01 μg m^{-3}) with high correlation to HNO_3. The data support that (1) HNO_3 is responsible for Cl depletion, (2) there is an increase in the Na/Cl ratio due to faster HCl removal during continental air mass transport, and (3) on average 50% of total Cl is being as gas-phase HCl.

Keywords Atmospheric chemistry, chlorine, degassing, multiphase chemistry, particulate matter, partitioning, sea salt

Introduction

Sea salt aerosol (SSA) has multiple impacts (beside other particulate matter categories) on atmospheric properties: response to climate by optical properties, providing cloud condensation nuclei, being a heterogeneous surface for multiphase chemical reactions, e.g., SO_2 oxidation, and a source for reactive chlorine (Finlayson-Pitts 2003).

Recently it has been found evidence that SSA is also occurring in the very fine fraction below 250 nm (down to 10 nm) depending from sea state (e.g., Clarke et al. 2006). Loss of chlorine (as HCl) from marine particulate matter (PM) have been observed more than 50 years ago and attributed to surface reactions (acidification) by acids produced from gaseous SO_2 and its oxidation onto sea salt particles (e.g.,

Brandenburgische Technische Universität Cottbus, Lehrstuhl für Luftchemie und Luftreinhaltung (Germany)

C.D. O'Dowd and P.E. Wagner (eds.), *Nucleation and Atmospheric Aerosols*, 654–658.

Junge 1954). The degree of Cl loss depends from many parameters and can be up to 100% for sub-μm particles and even 20–80% for the fraction of 7–15 μm (Kerminen et al. 1998).

Only in the last decade Cl degassing have been observed also from continental PM, mainly believed by gaseous HNO_3 sticking onto the particles. As a consequence, the PM is enriched with sodium (Na). Gaseous HCl will be quantitatively scavenged by precipitation and, in turns, "combined" again with Na (at least partly) in rain water as shown by numerous studies. However, the (mass) Na/Cl ratio found in sea water (R_{sea} = 0.56) is not appropriate to adopt for continental PM due to disproportioning between Cl and Na and have been derived between 0.8 and 1.4 from Scandinavian coastal rain water composition (Möller 1990).

This paper presents simultaneous measurements of HCl/Cl^-, HNO_3/NO_3^- and Na^+ (among many other related chemical parameters) in high-time resolution in gas and PM phase, respectively.

Experimental

Ground-based measurements were performed in early summer of 2006 (12–30 June) at the research station Melpitz (51°32N, 12°54 E; 87 m a.s.l.), hence providing a platform for detailed research studies (Spindler et al. 2004) and featuring the monitoring of a range of standard meteorological and atmospheric chemical parameters for example, within the European Monitoring and Evolution Program (EMEP). The place is located about 40 km NE of the city Leipzig (Germany) and surrounded by agricultural land, mainly grassland for hay harvest, and mixed forests.

Reactive trace gases (HCl, HNO_3,) were measured by a coupled wet denuder sampling/ion chromatography analysis technique. The diffusion-based separation of gaseous species from their particulate counterparts system was extended by a steam chamber for simultaneous sampling the soluble fraction of the total suspended matter. The analytical system provided data continuously with 30 min time resolution, concentrations of 0.010 μg m^{-3} could be reliable recorded (limit of quantification; sample airflow 10 L min^{-1}). A detailed description is given in Acker et al. (2005), but in this experiment for the first time also cations were analysed in a sample aliquot by a second ion chromatograph. The whole system (beside a very short Teflon inlet (20 cm), ~3 m above ground) was operated in an air-conditioned container at ~20°C.

Sampling of PM (24 h) was also done using high-volume sampler (Digitel) with different inlets and 5-stage Berner impactor (EMEP study). 96 h backtrajectories (NOAA) are used to determine the air mass origins.

Results

The air masses reaching the site were on average medium polluted, especially when transported mainly over land: NO_2 up to 30 μg m^{-3} (30 min averages, mean 6 μg m^{-3}); SO_2 up to 11 μg m^{-3} (30 min averages, mean 3 μg m^{-3}), the PM_{10} mass

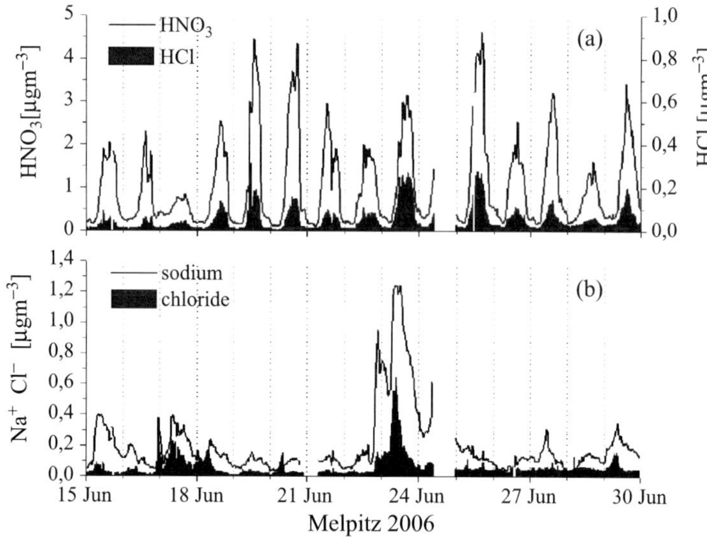

Figure 1 Time series of HNO_3, HCl, particulate sodium and chloride during the Melpitz campaign

concentration ranged from 14.6 to 30 µg m^{-3} (24 h averages).The sampling period generally was very sunny, dry (no rain), and warm (e.g., daytime temperatures 20°C–34°C).

The time series for HCl and HNO_3 are shown in Figure 1a, exhibiting a pronounced diurnal variation with usually low nocturnal values and high values at daytime reaching a maximum around noon. Remarkable, that even the fine structure in temporal variation is identical for HCl and HNO_3, resulting in high correlation:

$$[HCl] = 0.001 + 0.05[HNO_3];\ r^2 = 0.79;\ N = 800.$$

Such cycle is typical for many compounds having direct or indirect photochemical sources and which are mainly produced in the free troposphere, but also processes (e.g., surface related, at interfaces) in the boundary layer (BL) can be a chemical source. During night-time adsorption and deposition are the dominant processes. Diurnal but less pronounced patterns were found also for the corresponding ions in the particle phase, NO_3^- (not shown here) and Cl^-, but also Na^+, see Figure 1b, as a result of vertical and horizontal mixing and deposition. The semi-volatile ammonium nitrate has to be considered as a daytime source of HNO_3 during this experiment due to high air temperatures and low relative humidity (on average 27°C and 48% between 0900 and 1700 CET), despite poor correlation between nitrate and HNO_3 also during that time of the day.

Sea salt contributes to the measured water soluble inorganic PM constituents between 2% and 12% (continental and marine influenced air, respectively). The very good correlation between sodium and magnesium with nearly identical bulk seawater mass Mg/Na ratio (0.12), suggests the main origin of sea salt for Mg:

$$[Mg] = 0.01 + 0.11[Na]; r^2 = 0.86, N = 782.$$

No correlation between sodium and potassium was found when using all data, meaning that soil dust emission was the main source of K in Melpitz during the study time. Only for the period from 22 June (1900 CET) to 24 June (0900 CET), when strong maritime influenced air masses reached the site the high correlation coefficient ($r^2 = 0.89$; $N = 78$) indicates the origin of K from sea salt, although also these samples contain much more K as can be expected from the pure mass K/Na seawater ratio. Similar results were found for the relation between calcium and sodium. It suggests common source process characteristics (despite correlation with wind is poor), here by soil derived K and Ca (latter likely re-suspended).

Highest Na concentrations were found in these maritime air masses from Atlantic (0.5–1.2 µg m^{-3}, 22–24 June), smaller in those from North and Baltic Sea (0.3–0.4 µg m^{-3}, e.g., 15–17 June) and smallest ones in air travelling continental SW Europe (~0.1 µg m^{-3}, e.g., 19–21 June). The overall correlation between Na and Cl is less significant than between Na and Mg due to different distributions. The slope of the function indicates a significant loss of chlorine in the PM reaching the Melpitz site:

$$[Na] = 0.05 + 2.4\,[Cl]; r^2 = 0.65, N = 761.$$

The mass Na/Cl ratio (R_{sample}) varies between 1 and 11 (corresponding to 44–95% Cl loss), not showing a significant dependency from the air mass sector and transport percentage above continent (25–100%). On average, and with only very few significant deviations, mainly given by diurnal cycle, 83% Cl loss have been observed in PM calculated by $Cl_{loss} = 1 - R_{sea}/R_{sample}$.

We assume that gaseous HCl is mostly a result of acidic Cl replacement in SSA by nitrate concerning the reaction

$$NaCl + HNO_3 \rightarrow HCl + NaNO_3.$$

SO_2 seems to play only a minor role in acidic replacement of Cl (poor correlation with HCl) due to drastically reduced emissions in Europe since about 20 years compared to NO_2. If we further assume that there is no other important Cl source beside sea salt, the marine "reference" Na/Cl ratio (in terms of $[Na]/[Cl^- + HCl]$) amounts between 1 and 6 (much higher than expected earlier by Möller 1990), so we can conclude that (a) most of sea salt Cl degassing already did occur above sea and (b) that HCl during continental transport has been removed faster relatively to sodium and chloride in PM, most likely only by dry deposition (however, precipitation would scavenge almost all HCl in contrast to particulate Cl), resulting in an increase of $(Na/Cl)_{total}$. Unfortunately, no rain events did appear during the campaign to support this hypothesis. Due to the distribution of Na (and Mg) over the particle size (derived from impactor data: ~15% in the size range 0.05–1.2 μm; ~55% in the size range 1.2–3.2 μm and ~30% in the size range 3.2–10 μm (aerodynamic diameter), the existence of SSA in the PM fine mode at Melpitz – far from coastal site – is confirmed. Our field data clearly indicate, that HNO_3 is responsible for Cl degassing and that not only the surface of SSA is involved in the HCl acid replacement process, as observed also in several laboratory studies and have been explained by re-crystallisation among nitrate and chloride (Finlayson-Pitts 2003 and references therein).

Acknowledgments This work was supported by Deutsche Forschungsgemeinschaft (DFG). The authors wish to thank the Leibniz-Institut für Troposphärenforschung e.V., Leipzig for valuable assistance.

References

1. Acker, K., Möller, D., Auel, R., Wieprecht, W., and Kalaß, D., *Atmos. Res.*, **74**(1–4), 507–524 (2005).
2. Clarke, A., Owens, S.R., and Zhou, J., *J. Geophys. Res.*, **111**, D06202, doi:10.1029/2005JD006565 (2006).
3. Finlayson-Pitts, B., *J. Chem. Rev.*, **103**, 4801–4822 (2003).
4. Junge, C.E., *J. Meteor.*, **11**, 323–333 (1954).
5. Kerminen, V.-M, Teinilä, K., Hillamo, R., and Pakkanen T., *J. Aerosol Sci.*, **29**, 929–942 (1998).
6. Möller, D., *Tellus*, **42B**, 254–262 (1990).
7. Spindler, G., Müller, K., Brüggemann, E., Gnauk, T., and Herrmann, H., *Atmos. Environ.*, **38**, 5333–5347 (2004).

Measured Neutral and Charged Aerosol Particle Number Size Distributions in Russia

E. Vartiainen[1], M. Kulmala[1], M. Ehn[1], A. Hirsikko[1], H. Junninen[1], T. Petäjä[1], L. Sogacheva[1], S. Kuokka[2], R. Hillamo[2], A. Skorokhod[3], I. Belikov[3], N. Elansky[3], and V.-M. Kerminen[2]

Abstract The properties of aerosol particles in Russia are yet relatively unknown. Thus, we made an intensive 2-week measurement campaign from a train moving along the Trans-Siberian railroad. The measurements included, e.g., aerosol particle and air ion number size distributions, concentrations of gaseous compounds, meteorological variables, and GPS measurements. The results revealed a huge variability of the aerosol concentrations in Russia. The highest concentrations (10,000–40,000 cm^{-3}) were observed near towns or in the vicinity of strong local sources, such as forest fires. The lowest concentrations (500–1,000 cm^{-3}) were detected in remote areas in central and eastern Siberia. Occasionally we observed extremely high cluster ion concentrations. These were mainly due to high radon concentrations. Ion pair production from ^{222}Rn correlated quite well with cluster ion concentrations. In addition, we detected two nucleation events. The aerosol particle growth rates during the events were few nanometers per hour. The moving measurement platform, however, somewhat complicated the analysis of the events.

Keywords Size distributions, air ions, atmospheric Aerosols

Introduction

Atmospheric aerosol particles have direct and indirect effects to the earth's radiation budget and further to the climate change patterns [1]. To model the future climate development, knowledge on aerosol particle concentrations, sources, and atmospheric processes is in major importance.

[1]*Department of Physical Sciences, Univ. of Helsinki, P.O. BOX 64, FIN-00014 Helsinki, Finland*

[2]*Finnish Meteorological Institute, Erik Palmenin aukio, FI-00560, Helsinki, Finland*

[3]*Obukhov Institute of Atmospheric Physics, Pyzhevsky 3, 119017, Moscow, Russia*

C.D. O'Dowd and P.E. Wagner (eds.), *Nucleation and Atmospheric Aerosols*, 659–663.

Aerosol particle number-size distribution measurements have been made in a broad area on the Earth [2]. Measurements have revealed typical features of size distributions in different geographical areas and also shown that particle formation events, which are important sources of particles, exist nearly everywhere measured. Although the level of knowledge has improved significantly with the increasing number of observations and theories the particle formation process is still poorly understood.

Despite the large amount of studies made and the importance of the topic, Russia is still one of the least studied areas on the Earth. To better understand, also in a global scale, the aerosol particle formation and particle sources and transport patterns it would be of great concern to have more knowledge on Russian aerosols.

The foundation of the present work was to improve the knowledge on aerosol particle properties in Russia. The focus was in examining the features of particle number size distributions in a large geographical area and to resolve the sources and transportation pathways of particles in the measured points. New particle formation events, and the properties of air ions that can affect the formation and growth processes, were also of interest.

Methods

Measurements were done during a 2-week campaign between 4 and 18 October 2005. We used a train, travelling along the Trans-Siberian railroad, as a measurement platform and performed the measurements from a special built measurement carriage. The route started from Moscow and went across Ural and southern parts of Siberia trough Siberian boreal forest and steppe zones to Vladivostok on the coast of the Pacific Ocean. Part of the time the train travelled in very remote areas in central and eastern Siberia while it sometimes passed trough big (>1 milj. inhabitants) towns. Track was 9,242 km one-way which took about a week to travel with the train. The stay in Vladivostok was only 11 hours and the travel time back to Moscow again 1 week.

We measured aerosol particle number size distributions along the railroad with a twin-DMPS (Differential Mobility Particle Sizer) instrument. The twin-DMPS consisted of two HAUKE-type differential mobility analysers (DMA) followed by two TSI condensation particle counters (CPC). The first DMPS measured particles from 3 to 50 nm in 20 size sections. Parallel the second DMPS measured particles from 10 to 950 nm in 30 size section. Thus, the instrument gave one size distribution each 10 min.

The number mobility distributions of positively and negatively charged particles (air ions) were measured with an air ion spectrometer (AIS) manufactured by AIREL Ltd., Estonia. The mobility distributions were converted into size distributions [3] and correspondingly the 28 sectioned distributions of the positive and negative ions between diameters 0.4 and 40 nm were achieved every 5 min.

We had also a number of other instruments to study, e.g., the meteorological variables, trace gas concentrations, concentrations of aerosol ionic compounds, and navigational parameters. In addition, we marked extraordinary events, such as forest fires and oncoming trains, observed along the railroad to electronic diaries.

Results and Discussions

Aerosol Particle Concentrations

The measured total aerosol particle number concentration varied from few hundreds to some tens of thousands particles per cubic centimeter. The average particle number concentration during the forward way from Moscow to Vladivostok was 10,016 cm^{-3} (Table 1). During the return from Vladivostok to Moscow the overall average concentration was lower (7,676 cm^{-3}). In both ways, the most of the particles were in the Aitken mode (diameters between 25 and 90 nm).

The highest concentrations were measured when the train passed trough towns and industrial areas in European Russian site, close to Ural Mountains, and as approaching to Vladivostok [4]. On the way from Moscow to Vladivostok, the concentrations in remote areas in central and eastern Siberia were affected by many forest fires which increased the particle concentration especially in the accumulation mode (90–950 nm). The lowest concentrations were detected in these remote areas on the return trip when the existence of forest fires was notably smaller. In addition to the local sources detected, the particle concentrations were affected by the local weather and the origin of the air-masses. Precipitation scavenged particles during some days mainly during the return trip [4]. Close to Vladivostok the particle concentrations increased while air-masses were coming from industrial areas in China.

Ion Concentrations

The air ion number size distribution was divided into representative size ranges. Cluster ions (0.4–1.8 nm) were always present. The cluster ion concentration varied approximately from 100 to 1,500 cm^{-3} (positive) and from 200 to 4,000 cm^{-3} (negative) while average concentrations were around 600 cm^{-3} and 1,300 cm^{-3} respectively. Comparing the average cluster ion concentrations to those observed in Finnish boreal forest on October 2003, which is around 900 cm^{-3} [5], we see they are of the same order. However, in Finnish boreal forest the concentrations of positive and negative cluster ions are closer to each others, detected maximum concentrations are clearly lower, and also the variation of concentrations is smaller.

Table 1 Average aerosol particle concentrations and standard deviations in different size ranges measured between Moscow and Vladivostok during forward and backward ways

Size Range	Average (Std) Moscow–Vladivostok [cm^{-3}]	Average (Std) Moscow–Vladivostok [cm^{-3}]
3–950 nm	10,016 (8,703)	7,676 (10,026)
3–25 nm	2,381 (4,225)	2,130 (4,595)
25–90 nm	4,477 (4,951)	3,315 (5,563)
90–950 nm	3,208 (2,889)	2,267 (2,358)

The concentration of intermediate ions (1.8–7.5 nm) varied the most. Intermediate ions mainly occurred during nucleation or precipitation. Then their concentrations could reach a maximum value of around 200 cm^{-3} (positive) and 250 cm^{-3} (negative).

Cluster ions in the lower troposphere are mainly formed when natural radiation and cosmic rays charge neutral molecules. We calculated the number of ion pairs produced due to ^{222}Rn decay process assuming an average energy needed for formation of an ion pair was 34 eV. Comparing the cluster ion concentrations and ion pair production from ^{222}Rn we noticed quite a good correlation. Sometimes we detected very high ion production rates from ^{222}Rn decay and then also the cluster ion concentrations were extremely high (Figure 1). Comparing the maximum cluster ion concentrations detected in the Finnish boreal forest site [5] we detected approximately twice as high concentrations. Ions are removed due to recombination and coagulation with the preexisting particles. Particles are thus an ion sink. The cluster ion concentrations did not show as clear dependence with the ion sink as with the radon source. Even though, when the highest cluster ion concentrations were detected, also the ion sink was small.

Nucleation

We detected two cases of new particle formation. The growth rates of the nucleated particles, based on the DMPS data, were in both cases in the order of few nanometers per hour. The ion distributions indicated that very small (<3 nm) particles had a lower growth rate. However, the moving measurement platform complicated the calculation of growth rates due to fast changing air masses. Thus, these results contain quite high-error estimates.

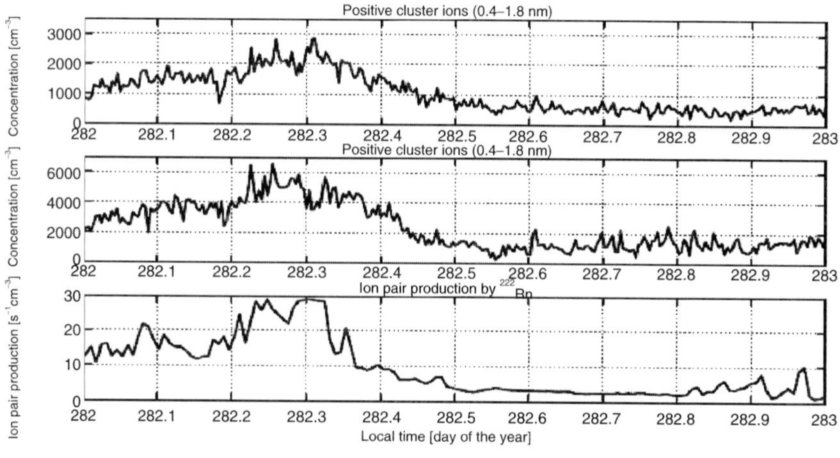

Figure 1 Measured positive (**upper panel**) and negative (middle panel) cluster ion concentration and ion pair production from ^{222}Rn (**lower panel**) on 9 October

In addition, during nucleation the number of negative intermediate ions was higher as compared with the number of positive ions of the same size. The difference could be even in the order of ten.

Acknowledgment This study was funded by the Academy of Finland under grant no. 208208. The TROICA expedition was partly funded by the ISTC Project #2773.

References

1. Karl, T.R. and Trenberth, K.E., *Science*, **302**, 1719–1723 (2003).
2. Kulmala, M., Vehkamäki, H., Petäjä, T., Dal Maso, M., Lauri, A., Kerminen V.-M., Birmili, W., and McMurry, P.H., J. *Aerosol Sci.*, **35**, 143–176 (2004).
3. Tammet, H, J. *Geophys. Res.*, **103**, 13933–13937 (1998).
4. Vartiainen, E., Kulmala, M., Ehn, M., Hirsikko, A., Junninen, H., Petäjä, T., Sogacheva, L., Kuokka, S., Hillamo, R., Skorokhod, A., Belikov, I., Elansky, N., and Kerminen, V.-M., Submitted to *Boreal Environment Research.*
5. Hirsikko, A., Laakso, L., Hõrrak, U., Aalto, P.P., Kerminen, V.-M., and Kulmala M., *Boreal Environ. Res.*, **10**, 357–369 (2005).

Size Distributions, Charging State, and Hygroscopicity of Aerosol Particles in Antarctica

E. Vartiainen[1], M. Ehn[1], P.P. Aalto[1], A. Frey[2], A. Virkkula[2], R. Hillamo[2], A. Arneth[3], and M. Kulmala[1]

Abstract Antarctica is the most isolated continent on Earth. Thus if offers a good background to study aerosol particle properties when only a very limited amount of natural, and nearly no anthropogenic sources affect them. We measured aerosol and ion number size distributions and hygroscopic properties of particles in Antarctica during Antarctic summer 2006/2007. Our preliminary results showed that new particle formation events occurred, during which the particle growth was very slow. The particles typically showed one modal hygroscopic growth (GF ~1.8). Thus, they were internally mixed and also able to act as cloud condensation nuclei (CCN). The cloud droplet formation was one potential reason for the occasionally observed very low-particle concentrations.

Keywords Size distributions, air ions, antarctic aerosols, hygroscopicity

Introduction

The estimates of the effect of aerosol particles to the Earths radiation budget contain many uncertainties [1, 2]. Especially of concern is how the particles influence on the cloud optical properties via acting as cloud condensation nuclei (CCN). In addition, the radiative forcing due to radiation scattering and absorbing properties of particles is of interest.

The total radiative forcing by aerosol particles depends on their physical and chemical properties. Only particles that are large enough in size can act as nuclei for cloud formation. It has been shows that hygroscopic particles are more favorable to act as CCNs [3]. The role of sulfate aerosols is also an important factor.

[1] *Department of Physical Sciences, University of Helsinki, P.O. BOX 64, FIN-00014 Helsinki, Finland*

[2] *Finnish Meteorological Institute, Erik Palmenin aukio, FI-00560, Helsinki, Finland*

[3] *Department of Physical Geography and Ecosystems Analysis, Lund University, Sölvegatan 12, 223 62, Lund, Sweden*

C.D. O'Dowd and P.E. Wagner (eds.), *Nucleation and Atmospheric Aerosols*, 664–668.

Sulfates, that are very abundant in the Earths atmosphere, can cool the climate due to their strong radiation scattering properties.

Aerosol particle chemical and physical properties in different regions of the Earth are characteristic to the environment. Sources of particles are both natural and anthropogenic. However, nearly everywhere, the particle properties are influenced by anthropogenic particle sources, while the air-masses transport particles even to the most remote locations. A major particle source is the secondary particle formation in the atmosphere. The formation of small particles via gas-to-particle conversion is, however, not yet a well understood process and there are several mechanisms presented that may cause the widely detected nucleation events [4].

Measurements all around the Earth are needed to be able to estimate the effects of particles to the Earths climate. An interesting environment, in this respect, is the most remote continent on Earth, Antarctica. In Antarctica, the influences of human activities do not substantially affect the particle properties. In addition, natural sources, such as vegetation or volcanic activities, are small and so the natural particle processes are very distinguishable.

Thus, we wanted to study particle properties, such as their number size distributions together with ion size distributions, particle hygroscopic properties, chemical composition, and optical properties, in Antarctica. Our objectives were to examine particle processes, such as new particle formation, to get some estimates on the chemistry that happens during nucleation and also what is the role of ions into nucleation, study the connections between particle properties and meteorological conditions, and define which are the major sources and transport pathways of particles in Antarctica.

Methods

We measured aerosol particle and air ion number size distributions and hygroscopic properties of the particles in Queen Maud Land, Antarctica. The measurements were conducted at a Finnish research station Aboa (73°**03′**Σ, 13°**25′**Ω) during the Antarctic summer 2006/2007 from 28 December to 29 January. The station was located about 130 km from the coast on a nunatak Basen in the Vestfjella Mountains. We had a separate measurement container, about 250 m from the station, at an altitude of about 500 m above the sea level. In the context of the container an automatic weather station (MILOS) gave data on the alternating meteorological conditions.

The aerosol particle number size distributions were measured with a twin-differential mobility particle sizer (DMPS). The twin-DMPS consisted of two HAUKE-type differential mobility analysers (DMA) followed by two TSI condensation particle counters (CPC). The first DMPS measured particles from 3 to 25 nm in 10 size sections while the second DMPS measured from 10 to 1,000 nm in 30 size sections. The two DMPS's operated parallel with a 10 min time resolution.

The air ion spectrometer (AIS, manufactured by AIREL Ltd., Estonia) measured the number mobility distributions of positive and negative ions with a time resolution of five minutes. The AIS classifies the ions according to their electrical mobility's, which are later converted to corresponding ion diameters. The corresponding size range of the detected ions is from 0.4 to 40 nm, where the ions are divided into 28 size sections.

The hygroscopic growth factors of 10, 25, 50, and 90 nm particles were measured with a hygroscopicity-tandem differential mobility analyser (H-TMDA). The relative humidity in the second DMA, after the moistening, was 90%. Due to very small particle concentrations, the scanning time of the second DMA was long, around 15 min and thus, the growth factors of all scanned particle sizes were achieved in an hour. Calculation of particles was done with a TSI CPC model 3010.

Results and Discussions

Aerosol Particle Number Size Distributions

The total aerosol particle number concentration varied from 50 to around 5,000 particles cm^{-3}. During most of the days, the north-eastern winds blew from the direction of the coast and typical particle concentrations were in the order of around 300 cm^{-3}. Then, the particles were detected in Aitken and accumulation modes.

Particle concentration decreased few times clearly below 100 cm^{-3}. One possible reason for this could be very clean air-masses. Another, more probable, reason was that the particles formed cloud droplets. However, in this very early stage of analysis we can not conclude for certain feather this was the case. This demands the combining of data from different instruments and only then, we may look at the process in more detail.

Particle formation

The particle concentrations occasionally increased due to long-range transport of Aitken and accumulation mode particles. Then the total concentration could increase near 1,000 cm^{-3}. The highest concentration, however, occurred during nucleation events (Figure 1). The small, newly formed, particles then grew to Aitken mode sizes due to condensing vapors. A part of them also coagulated to accumulation mode. The observed growth rates (<1 nm/h) were relatively slow.

Hygroscopicity of particles

Particles showed clear hygroscopic behavior. We observed one growing hygroscopic mode with a growth factor around 1.8, decreasing with decreasing particle size. Thus, it appeared that the particles were internally mixed and their hygroscopic properties favored their acting as CCN.

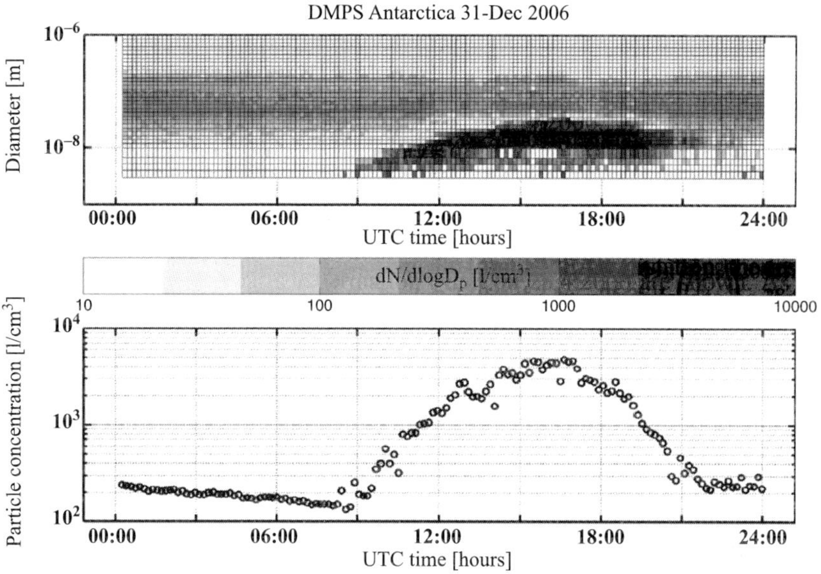

Figure 1 Measured aerosol particle number size distributions on December, 31 2006 (**upper panel**). The logarithmic y-axis shows the particle diameter and the x-axis shows the time. The shade of gray indicates the number of particles. In the **lower panel**, the simultaneous integrated total aerosol particle number concentration is presented

Ion Concentrations

Both very low (close to zero) and very high (thousands of ions cm^{-3}) ion concentrations were detected in all ion size ranges. The concentration of intermediate (~1.8–7.5 nm) ions was mainly close to zero, but during nucleation or some special weather conditions (rain, wind) their concentration increased remarkably. The concentration of cluster (~0.4–1.8 nm) ions was often quite stable. However, in some occasions we detected that cluster ions nearly totally disappeared. What were the reasons for this will be an issue of the detailed analysis.

Conclusions

The measurements were done in the beginning of the year and the analysis is in a very preliminary phase. Thus, the results presented will not yet be able to answer many of the proposed hypotheses. However, the preliminary data shows some very interesting features.

The detected nucleation events will be one of the focuses in the analysis. Especially hygroscopic properties of newly formed particles together with positive and negative ion concentrations, which will provide information of the role of ions in nucleation, are of interest. Hygroscopicity of particles has not, in our knowledge, measured before in Antarctica. Thus, examining the hygroscopic behavior of particles is also one priority. In addition, the occasionally observed very low particle concentrations, which might have been due to cloud droplet formation, will need a more detailed analysis of combined instrument data.

Acknowledgments We acknowledge the logistical support of FINNARP. This study was funded by the Academy of Finland (Finnish Antarctic Research Program, "Dynamics, seasonal variation and chemistry of the Antarctic aerosol").

References

1. IPCC, the fourth assessment report: Climate Change 2007 (2007).
2. Karl, T.R. and Trenberth, K.E., *Science*, **302**, 1719–1723 (2003).
3. Bilde, M. and Svenningsson, B., *Tellus* ***56B***, 128–134 (2004).
4. Kulmala, M., Vehkamäki, H., Petäjä, T., Dal Maso, M., Lauri, A., Kerminen V.-M., Birmili, W., and McMurry, P.H., *J. Aerosol Sci.*, **35**, 143–176 (2004).

Dry Deposition Fluxes and Timescales of Turbulent Deposition over Tall Vegetative Canopies

Fredi Birsan and S.C. Pryor

Abstract Dry deposition of atmospheric aerosols is strongly influenced by turbulence particularly close to the ground where the roughness elements present on the surface considerably amplify the atmosphere surface exchange fluxes. Vegetation elements induce turbulence in the canopy flow which extends vertically above the canopy creating a roughness layer in which turbulent eddies are present. Aerosol particles entering this layer are brought rapidly in contact with vegetation elements increasing deposition. We present a new approach to modelling dry deposition to high-roughness vegetated surfaces. As an initial step, the domain of analysis is one-dimensional (1D), extending in the vertical direction from the ground up to five times the height of the canopy top. In this domain, a time dependent conservation law is solved by constructing several flux components. The turbulent flux generated by vegetation elements is constructed using multiple scales in accordance with the size of vegetation elements. The removal flux is computed as a fraction of the turbulent flux using perfect collection efficiency. The gravitational flux is then calculated based on particle size and turbulence intensity. Once all the fluxes are known the conservation law is advanced in time. Multiple runs are necessary for completing the concentration profile for a given forest canopy structure and wind intensity. Once the concentration profile is determined, deposition velocities are obtained. Extending this method to two space dimensions represents the next step which requires a more complete characterization of the canopy, to include the horizontal vegetation. Several other improvements are under consideration for instance; less than perfect capturing efficiency and particle rebound, inclusion of other fluxes resulting from particle dynamic or chemical processes.

Keywords Turbulent flux, roughness sub-layer, aerosol dry deposition, tall vegetation

---

Atmospheric Science Program, Department of Geography, Indiana University, Bloomington, IN 47405, USA

C.D. O'Dowd and P.E. Wagner (eds.), *Nucleation and Atmospheric Aerosols*, 669–673.

Introduction

Aerosol particles deposition velocities (v_d) exhibit a pronounced dependence on surface type, due in part to the turbulence structure derived from the interaction of the prevailing flow with the surface roughness elements. Modelled deposition velocities exhibit minima of approximately 0.01 and 0.2 cm s^{-1} over grasslands and 0.1 to 1 cm s^{-1} over forests. However, observations over forests generally do not support the pronounced minimum of v_d for particle diameters of 0.1–1 μm manifest in many deposition models [1]. This discrepancy may derive from the fact that all vegetative canopies presents several particular characteristics such as; multiple length scale turbulent flow, three dimensional character, variable domain with complex interaction, which can not be captured by simplified depictions of atmosphere-surface interaction that underpin these models.

Dry Deposition Over Vegetative Canopies

The governing equations for incompressible flow written in non conservative vector form are:

$$\frac{\partial U}{\partial t} + (U \cdot \nabla) \cdot U = -\frac{1}{\rho} \Delta p + \nu \cdot \Delta U + f; \nabla \cdot U = 0; \quad (1)$$

where

U – represents the velocity vector field
p – the scalar pressure field
ρ – fluid density
ν – kinematic viscosity
f – body force (usually gravity but can also Coriolis force or any other force field acting on the fluid)
t – time
∇ – is the divergence operator

A solution of (1) is represented by the velocity vector U, and the scalar field p for a given body force f and for a given fluid with constant density ρ at any point within the specified domain.

Finding a solution to (1) is closely related to the problem of determining the aerosol flux in vegetation which can be formulated as follows: Given a plume of aerosols of known size distribution and number or mass concentration moving with the air over a vegetative canopy, what will be the flux of aerosols impacting vegetation. Our goal is to build a model system such that if certain characteristics of the canopy are specified along with the initial air motion, the aerosol flux can be determined.

Governing Equations for Aerosol Particles

Solid particles much larger in size than the air molecule are suspended due to molecular collisions. The particles with the surrounding air form a continuous fluid for which macromolecular properties such as density or pressure can be defined however; the aerosol particles form a discrete subsystem within the fluid with distinct physical and chemical properties. The motion of the fluid as a whole is described by (1), but the aerosol particles being larger in size and heavier than the air molecules are subjected to larger friction and inertial forces. Moreover, the pressure gradient force which causes the fluid motion is not directly causing aerosol particle motion since they form a discrete system. The consequence of this observation is that the statement of conservation of momentum for aerosol particles is given by (2)

$$\frac{\partial U_p}{\partial t} + \left(U_p \cdot \nabla\right) \cdot U_p = v_p \cdot \Delta U_p; \qquad (2)$$

where

U_p – represents the velocity of aerosol particles
v_p – diffusivity of particles in air

Note: $-U_p$ the solution of (2), is generally different than U the solution of (1), even though both are the result of momentum conservation.

- U_p does not satisfy $\nabla . U_p \neq 0$

The mass conservation for particles in absence of macromolecular motion is given by the diffusion equation. Let c denote the particle density, satisfying:

$$\frac{\partial c}{\partial t} = v_p \cdot \Delta c \qquad (3)$$

If the velocity of particles is given by (2) and their mass density by (3), then:

$$U_p = -2 \cdot v_p \nabla \cdot (\ln c); \qquad (4)$$

The transformation (4), [2] establishes a link between the density of aerosol particles (c) and their velocity (U_p). The mass conservation for aerosol particles then becomes:

$$\frac{\partial c}{\partial t} + \nabla\left(\frac{1}{2} \cdot U_p \cdot c\right) = 0; \qquad (5)$$

Eq. (5) can be generalized to include more particle fluxes present for instance: deposition flux by interception, impaction etc. (F_{di}); gravitational settling flux (F_g)

$$\frac{\partial c}{\partial t} + \nabla \cdot \left(F_t + F_{di} + F_g\right) = 0; \qquad (6)$$

Furthermore, (6) may also include source terms emerging from particle dynamics.

1D Flux Solution

The main difficulty for solving (6) is finding the turbulent flux term F_t which requires finding a solution to (2). In one dimension this process is considerably simplified, the vegetation canopy is parameterized as a vertical distribution of roughness elements (leaves, branches, tree coronas etc.) generating various spatial scales eddies. In this case, the only component of particle velocity (U_p) is w and (2) becomes:

$$\frac{\partial w}{\partial t} + w \cdot \frac{\partial w}{\partial z} = \nu_p \cdot \frac{\partial^2 w}{\partial z^2}; \qquad (7)$$

The time dependent particle velocity w can be constructed by solving (7) and the turbulent flux F_t computed at every time step and introduced in Eq. (6) which becomes:

$$\frac{\partial c}{\partial t} + \frac{\partial}{\partial z}\left(\underbrace{\frac{1}{2} w \cdot c}_{F_t} + F_{di} \right) = 0 \qquad (8)$$

Eq. (8) can be solved iteratively and the aerosol density or concentration profile obtained at every time t starting from a given initial plume distribution for a specific forest canopy given as described above as a distribution of roughness elements. The second factor leading to the concentration profile is wind velocity. The influence of wind intensity U_x is captured by the solution to (7) as having a direct effect on the propagation speed of the solution w.

$$w(z,t) = \sum_{i(z)} \frac{\dfrac{((z - z_0) \pm t \cdot U_x)}{t}}{1 \pm t^{\frac{1}{2}} \cdot \exp\left(\dfrac{((z - z_0) \pm t \cdot U_x)^2}{4 \cdot \nu_p \cdot t} \right)}; \qquad (9)$$

In (9) the sum corresponds to the number of leaves present at the time in the vertical direction throughout canopy height.

The deposition flux F_{di} depends on F_t and the location of receiving surface.

Concluding Remarks

This study proposes a new methodology to quantify dry deposition flux over various vegetative canopies and determine timescales of the dry deposition process. The approach is based on first observing that the momentum conservation equation governing the particle motion are the Burger's equations (2) which balance the inertial and friction forces, and the mass conservation law is given by (6). Next ne concentration profile in 1D was obtained (8) using the turbulent flux (9). The emaining part is to complete the additional flux terms in (6) and improve the 1D olution, then move on to 2D by first formulating the turbulent flux and then ollowing the same methodology.

cknowledgments This work is supported by NSF as part of Project number ATM 0334321.

eferences

. Pryor, S.C. et al., A review of measurement and modelling tools for quantifying particle atmosphere-surface exchange. *Tellus Part B (in review)* (2007).
. Hopf, E., The Partial Differential Equation $u_t + uu_x = u_{xx}$, *Prepared under Navy Contract N6-onr with Indiana University* (1948).

Observations of Particle Nucleation and Growth Events in the Lower Free Troposphere

B. Verheggen[1,2], E. Weingartner[1], J. Cozic[1], M. Vana[3], J. Balzani[2], E. Fries[4], G. Legreid[5], A. Hirsikko[3], M. Kulmala[3], and U. Baltensperger[1]

Abstract Atmospheric nucleation events were regularly observed during the fourth Cloud and Aerosol Characterization Experiment (CLACE 4), conducted at the high alpine research station Jungfraujoch in winter 2005. Nucleation was observed to occur both during the absence and presence of clouds, and was sometimes accompanied by an overcharging of the small particles. We analysed the measured size distributions with the PARGAN procedure to determine empirical particle growth and nucleation rates, taking into account the effect of coagulation. Though nucleation rates were found to be small ($<1\,cm^{-3}\,s^{-1}$), the regular occurrence of nucleation events suggests that they could contribute significantly to the tropospheric aerosol number budget.

Keywords Nucleation rate, growth rate, free tropospheric aerosol, aerosol dynamics

Introduction

Nucleation can occur in almost any environment, subject to a favourable set of conditions. These conditions include a strong source of condensable vapour, high UV radiation intensity, low aerosol surface area, high relative humidity, low temperature, and atmospheric mixing processes. Numerous observations of nucleation events in a variety of locations over the globe have indeed been reported.[1]

[1] *Laboratory for Atmospheric Chemistry, Paul Scherrer Institut, 5232 Villigen PSI, Switzerland*

[2] *Institute for Atmospheric and Climate Science, ETH Zürich, 8092 Zürich, Switzerland*

[3] *Physics Department, University of Helsinki, Helsinki, Finland*

[4] *Institute of Environmental Systems Research, University of Osnabrück, 49076 Osnabrück, Germany*

[5] *Swiss Federal Laboratory for Materials Testing & Research, 8600 Dübendorf, Switzerland*

C.D. O'Dowd and P.E. Wagner (eds.), *Nucleation and Atmospheric Aerosols*, 674–678.

Measurements were made at the Jungfraujoch alpine site in Switzerland during he CLACE 4 campaign (Cloud and Aerosol Characterization Experiment), which ook place from 17 February to 17 March, 2005. Due to its altitude (3,580 m), the ite is far away from significant pollution sources. In winter, the Jungfraujoch can e regarded as representative for the continental lower free troposphere, with only ninor influence from the boundary layer.[2]

Bursts of small particles were regularly observed, but their number concentra-on was lower than is typical for boundary layer nucleation events, where the ource strength of condensable vapour is usually much greater. Selected nucleation vents are analysed to determine empirical nucleation and growth rates from ne measurements of the aerosol size distribution. The relation between nuclea-on events and charged particles, organic compounds, and cloud presence is nvestigated.

Measurements

he aerosol size distribution was measured both indoors under dry conditions (due heating to room temperature) and outdoors under ambient conditions (tempera-ures between −28 and −12°C) for approximately 5 weeks. The outdoor Scanning Mobility Particle Sizer (SMPS) measured the size range from 4 to 100 nm (ambi-nt) diameter, while the indoor spectrum covered the range from 18 to 800 nm (dry) iameters. The indoor SMPS used in this analysis sampled air via a heated inlet 25°C), designed to evaporate cloud hydrometeors at an early stage of sampling to nclude their residual particles in the measurements. A nano-SMPS, on loan from SI, was also operated downstream of the heated inlet for 10 days during the cam-aign. SMPS spectra were also measured downstream of other inlet systems, and sed to quantify the partitioning to the condensed phase during cloud presence.[3] All neasured distributions were corrected for size dependent particle losses. The particle iameters of the spectra measured indoors were corrected towards ambient condi-ons using the empirical relationship $GF = (1-RH)^{\gamma(d)}$, where RH is the measured mbient relative humidity expressed as a fraction, and $\gamma(d)$ is a (weakly size ependent) parameter. Based on the average measured growth factor of 1.4 for an H of 0.85, $\gamma(d) = -0.18$ for a 100 nm diameter particle.

The hygroscopicity correction improved the agreement between the two spectra n terms of sizing, but the ambient SMPS measured a substantially larger number oncentration, independent of particle size. The reason for this discrepancy is not nown, but is likely caused by an imprecisely known aerosol flow rate, due to the arsh conditions of operation. Therefore the number concentrations of the ambient pectra were decreased by a factor 1.3 to improve the agreement with the measured istributions from the other three available SMPS systems. Composite size distri-utions were made from the corrected spectra. The total particle number concentra-on ($D_p > 10$ nm) was measured both in- and outdoors by TSI 3010 condensation article counters (CPC). Correcting the concentrations of the ambient spectra improved the agreement between the total aerosol number concentration determine from the composite distribution ($D_p > 10\,nm$) and the CPC concentration.

An Air Ion Spectrometer (AIS) was operated indoors to measure charged clus ters and particles from 0.9 to 40 nm diameter. It sampled ambient air via a shor inlet (~40 cm) at a high flow rate (~80 lpm) to minimize sampling losses. Clou liquid water content (LWC) was measured with a Particulate Volume Monito (PVM-100, Gerber Scientific). Ice water content (IWC) was derived from image made by a Cloud Particle Imager (CPI) using a novel method described by Connoll et al.[4] The response of the PVM to the presence of ice was quantified and correcte by a combined analysis of the Forward Scattering Spectrometer Probe (FSSP PVM, and CPI data.[3]

Results

Nucleation events usually occurred during relatively clean and clear sky condition often soon after cloud evaporation. On March 6, 2005, we observed the growth c recently nucleated particles during the presence of a mixed-phase cloud. The com bination of a high actinic flux, high relative humidity, low temperature (–22 t –25°C), and evaporating droplets may have created a favourable environment fc nucleation.[5] The presence of ice probably contributed to the evaporation of liqui droplets via the Wegener–Bergeron–Findeisen mechanism. This leads to th release of semi-volatile material into the gas phase, where it could contribute t particle nucleation and growth. Most aerosol particles were present in the interst tial (i.e., unactivated) phase as opposed to the condensed phase, so the cloud drop lets could not have been very numerous so as to present an instantaneous sink fc small aerosol particles. Figure 1 shows a contourplot of composite size distribu tions from the nano and regular SMPS (both operated indoors), as well as LWC an IWC measured during this event on March 6, 2005. During this event, the aeros

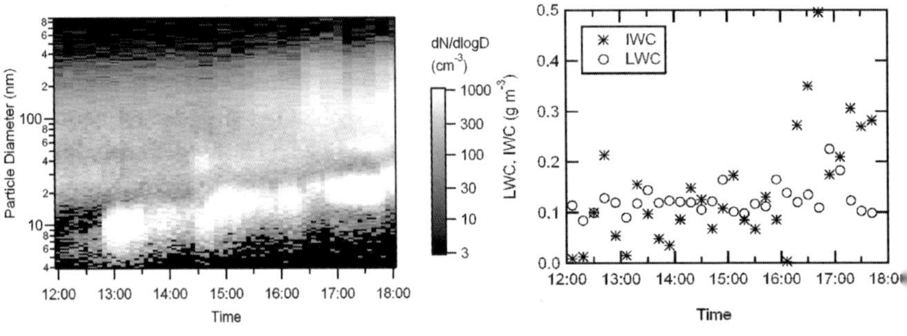

Figure 1 Measured particle size distributions (**left panel**) and liquid and ice water content (LW and IWC, respectively; **right panel**) during nucleation and growth event on March 6, 2005

vas found to be overcharged compared to Boltzmann equilibrium, suggesting that ons may have played a role in the nucleation process.

Five nucleation events were chosen for further analysis (including that of March 6 lescribed above), based on a clear evolution of the spectra over multiple hours, and noderate to low levels of pollutants (CO, NO_x). These were analysed to derive mpirical particle nucleation and growth rates using a recently developed inverse nodelling procedure called PARGAN.[6] The growth rate is determined by non-inear regression analysis of the General Dynamic Equation using only the meas-ıred aerosol size distributions as input. This allows the growth rate to be determined vith a higher time-resolution than can be deduced from inspecting contour plots "banana-plots"). These growth rates are used to estimate the time of formation of ecently nucleated particles (where the critical cluster is assumed to be 1 nm diam-ter). The nucleation rate is determined by correcting for the particle losses that ave occurred between time of nucleation and time of measurement due to coagula-ion. Applying this procedure implicitly assumes that the same air mass is ampled.

The resulting nucleation rates are shown in Figure 2 as a function of the particle rowth rate and the equivalent mixing ratio of condensing vapour. A similar rela-ion between the nucleation rate and the growth rate (and thus vapour concentra-ion) was found for all these events. This relation appears to be close to linear, as een from the linear fit through the average values (black line). If the same (group f) species was responsible for nucleation and growth, this is more consistent with n activation mechanism or kinetic (barrierless) nucleation than with themody-amically limited nucleation. The nucleation rates are relatively low (<1 cm^{-3} s^{-1})

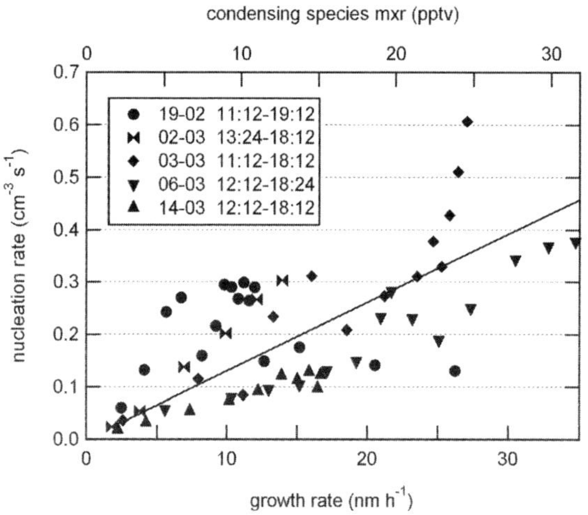

igure 2 Empirical nucleation and growth rates (and equivalent condensing vapour concentra-on) determined for five different events. Both are determined from the measured size distribu-ons using the inverse modelling procedure PARGAN. Each symbol is an average based on pproximately the same amount of data (~90 size bins, each of which is tracked backwards to its me of formation using the growth rate)

for substantial amounts of condensable vapour. This suggests that H_2SO_4 can no be the main contributor to condensation, since it has a very high nucleation poten tial. Some of the heavier aromatic compounds exhibit higher concentrations durin nucleation events than during the absence of such events, suggesting that they ma have contributed to particle growth.

Acknowledgments This work was supported by the Swiss Meteorological Institute (Globa Atmosphere Watch program). We thank the International Foundation High Altitude Researc Stations Jungfraujoch and Gornergrat (HFSJG) for the opportunity to perform experiments at th Jungfraujoch.

References

1. Kulmala, M. et al., *J. Aerosol Sci.*, **35**, 143–176 (2004).
2. Baltensperger, U. et al., *J. Geophys. Res.*, **102**, 19707–19715 (1997).
3. Verheggen, B. et al., submitted (2007).
4. Connolly, P.J. et al., *J. Atmos. Oceanic Technol.*, in press, (2007).
5. Clarke, A.D. et al., *Geophys. Res. Lett.*, **26**, 2425–2428 (1999).
6. Verheggen, B. and Mozurkewich, M., *Atmos. Chem Phys.*, **6**, 2927–2942 (2006).

Size-resolved Aerosol Chemical Concentrations at Rural and Urban Sites in Central California, USA

Judith Chow[1], John Watson[1], Douglas Lowenthal[1], and Karen Magliano[2]

Abstract Aerosol size distributions were measured with MOUDI (Micro-Orifice Uniform Deposit Impactor) cascade impactors at Angiola and Fresno, rural and urban sites in California's San Joaquin Valley (SJV) during the California Regional $PM_{10}/PM_{2.5}$ Air Quality Study (CRPAQS) winter intensive study from December 15, 2000–February 3, 2001. Nitrate size distributions at Angiola were uni-modal and narrow, while nitrate distributions in Fresno appeared multimodal and wider than those at Angiola. The Fresno distributions reflect a mixture of local and distant source contributions. Organic and elemental carbon (OC and EC) appeared on smaller particles than did nitrate at both sites. Relatively more OC was found on the smallest MOUDI stage (0.056 µm) at Angiola than at Fresno. The nitrate geometric mean diameter (GMD) increased from 0.97 to 1.02 µm as the nitrate concentration at Angiola increased from 43 to 66 µg/m^3 during an episode from January 4–7, 2001. There was a direct relationship between nitrate concentration and GMD at Angiola but no such relationship for OC. This demonstrates that secondary aerosol formation increases both concentration and particle size.

Keywords MOUDI, size distribution, ammonium nitrate, gas-to-particle conversion

Introduction

The California Regional $PM_{10}/PM_{2.5}$ Air Quality Study (CRPAQS) was conducted from December 2, 1999 through February 3, 2001 to understand high concentrations of $PM_{2.5}$ and PM_{10} in California's San Joaquin Valley (SJV).[1] Elevated PM concentrations in the SJV are comprised of ammonium nitrate (NH_4NO_3) and

[1] *Desert Research Institute, Division of Atmospheric Sciences, Reno, NV, USA*

[2] *California Air Resources Board, Sacramento, CA, USA*

C.D. O'Dowd and P.E. Wagner (eds.), *Nucleation and Atmospheric Aerosols*, 679–683.

organic carbon (OC) and occur mainly during winter.[2,3] Size-resolved measurements of aerosol chemical concentrations have been used to study aerosol evolution, sources and transport, and light extinction.[4–7] Watson and Chow[8] proposed that wintertime PM episodes in the SJV arise from conversion of urban emissions as they are transported above the boundary layer and subsequently mixed to the surface downwind.[9] Direct relationships between particle size and concentration at remote US National Parks were ascribed to gas-to-particle conversion in-cloud during transport from emissions sources. This paper discusses size-resolved aerosol chemistry at Angiola and Fresno, rural and urban sites, respectively, during CRPAQS. Factors responsible for variations in the observed size distributions are examined.

Methods

MOUDI (Micro-Orifice Uniform Deposit Impactor) samples were collected four times a day at 0000, 0500, 1000, and 1600 PST (GMT-8) at Fresno and Angiola, which is located about 100 km south of Fresno and 90 km north of Bakersfield, CA. Each MOUDI sampler contained eight impaction stages with D_{50}'s (the diameter at which 50% of the particle mass is retained) ranging from 0.056 to 10 µm. Impaction substrates from selected samples were analyzed for chloride (Cl^-), nitrate (NO_3^-), and sulfate ($SO_4^=$) by ion chromatography, ammonium (NH_4^+) by automated colorimetry, water-soluble sodium (Na^+), and potassium (K^+) by atomic absorption, elements by x-ray fluorescence, and organic and elemental carbon (OC and EC) by thermal–optical reflectance.

Results and Conclusions

Figure 1 shows average NO_3^- and OC concentrations (normalized to the sum of the concentrations on the MOUDI stages) at Angiola and Fresno. The smooth curves in the figures were fitted using Twomey's nonlinear least-squares algorithm.[10] The NO_3^- size distributions peaked at around 0.7 µm at both sites but were much narrower at the rural Angiola site. OC mass was concentrated at smaller diameters than NO_3^- at both sites, suggesting that organic material was less hygroscopic than ammonium nitrate (NH_4NO_3). The average relative humidity at Fresno was 72% (60% and 81% during the daytime and nighttime hours, respectively). There was also relatively more OC present in smaller particles at Angiola than in Fresno. This may suggest that organics at Fresno were more hygroscopic than more-aged organic aerosols at Angiola. The presence of a significant amount of OC on the smallest MOUDI stage (0.056 µm) at Angiola could be evidence for organic particle nucleation.

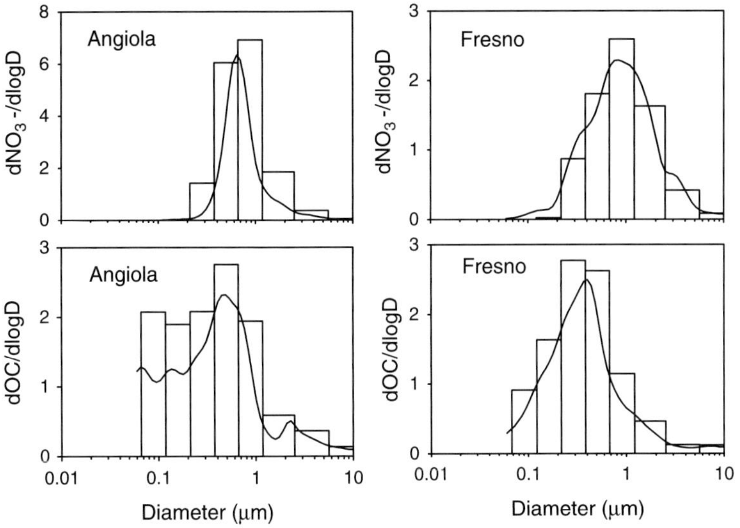

Figure 1 Average nitrate (NO_3^-) and organic carbon (OC) MOUDI size distributions at Angiola and Fresno during CRPAQS

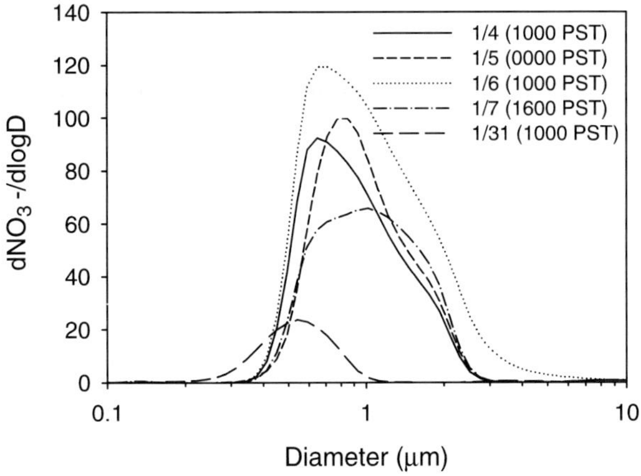

Figure 2 Evolution of the nitrate (NO_3^-) size distribution during the CRPAQS winter intensive study period from January 4–7, 2001. Also shown is a size distribution for January 31, 2001, 1,000 PST, for comparison

Figure 2 represents the evolution of the NO_3^- size distribution at Angiola during an episode January 4–7, 2001, when daily-average NO_3^- concentrations were 43, 43, 56, and 38 μg/m^3, respectively. Over this period, the geometric mean diameter (GMD) increased from 0.90 μm on January 4, 2001 to 1.02 μm at the peak of the episode on January 6, 2001. For comparison, Figure 2 presents a NO_3^- size distribution on

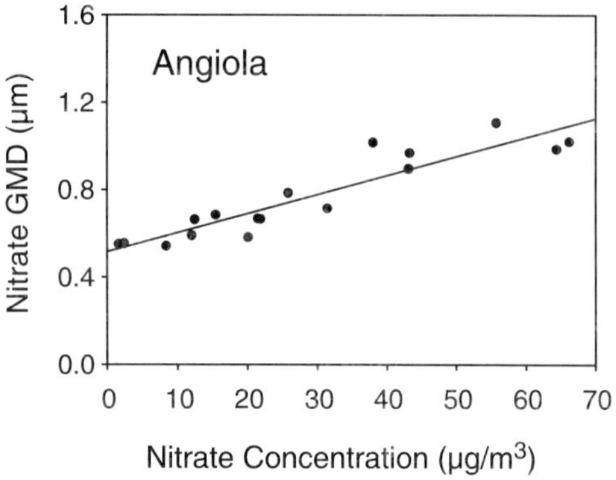

Figure 3 Relationship between NO_3^- and geometric mean diameter (GMD) at Angiola and Fresno

January 31, 2001, when the average NO_3^- concentration had decreased to 8.4 µg/m and the GMD had decreased to 0.54 µm.

The relationship between NO_3^- particle size and concentration at Angiola is apparent as shown in Figure 3. This relationship was not evidence for NO_3^- in Fresno, nor did this relationship exist for OC at either site. This supports the contention that secondary organic aerosol production is limited during winter in the SJV.[7] The results for NO_3^- at Angiola suggest that as air masses are transported from urban source areas, concentrations increase through gas-to-particle conversion, which also increases the particle size. These results are consistent with previous observations at remote US National Parks.[9]

Acknowledgments This work was supported by the California Regional $PM_{10}/PM_{2.5}$ Air Quality Study (CRPAQS) Agency under the management of the California Air Resources Board and by the US Environmental Protection Agency under Contract #R-82805701. Any mention of commercially available products and supplies does not constitute an endorsement of those products and supplies.

References

1. Watson, J.G., DuBois, D.W., DeMandel, R., Kaduwela, A.P., Magliano, K.L., McDade, C. Mueller, P.K., Ranzieri, A.J., Roth, P.M., and Tanrikulu, S. *Field program plan for the California Regional $PM_{2.5}/PM_{10}$ Air Quality Study (CRPAQS)*. Desert Research Institute, Reno, NV (1998).
2. Chow, J.C., Watson, J.G., Lowenthal, D.H., Solomon, P.A., Magliano, K.L., Ziman, S.D., and Richards, L.W., *Aerosol Sci. Technol.*, **18**, 105–128 (1993).
3. Magliano, K.L., Hughes, V.M., Chinkin, L.R., Coe, D.L., Haste, T.L., Kumar, N., and Lurmann, F.W., *Atmos. Environ.*, **33**, 4757–4773 (1999).
4. Li, S.-M. and Winchester, J.W., *J. Geophys. Res.*, **95**(D9), 13897–13908 (1990).
5. Malm, W.C. and Pitchford, M.L., *Atmos. Environ.*, **31**, 1315–1325 (1997).

6. Hand, J.L., Kreidenweis, S.M., Kreiseberg, N., Hering, S., Stolzenburg, Dick, W., and McMurry, P.H., *Atmos. Environ.*, **36**, 1853–1861 (2002).
7. Herner, J.D., Ying, Q., Aw, J., Gao, O., Chang, D.P.Y., and Kleeman, M.J., *Aerosol Sci. Technol.*, **40**, 827–844 (2006).
8. Watson, J.G. and Chow, J.C., A wintertime PM2.5 episode at the Fresno, CA, supersite, *Atmos. Environ.*, 36465–36475 (2002).
9. Lowenthal, D.H. and Kumar, N., *J. Air & Waste Manage. Assoc.*, **54**, 926–934 (2004).
10. Winklmayr W., Wang H.-C., and John W., *Aerosol Sci. Technol.*, **13**, 322–331 (1990).

Measurement of Ultrafine and Fine Particle Black Carbon and its Optical Properties

John Watson[1], Judith Chow[1], Douglas Lowenthal[1], and Nehzat Motallebi[2]

Abstract Carbon aerosols were generated from diesel, acetylene flame, electric arc soot, and wood smoke combustions sources. Elemental carbon (EC; or black carbon [BC]) fractions measured by the IMPROVE_A, STN and the French two-step protocols were within 5% for the sources tested, except for wood smoke (differed by >45%). The French two-step protocol, which is operated in pure oxygen without charring corrections, was more influenced by the presence of sodium chloride in the aerosol mixture than were the IMPROVE_A and STN protocols for source samples on quartz-fiber filters. The EC absorption efficiency (σ_{abs}) at 1,047 nm varied (>50%) between sources, in the range of 2.7–5.3 m^2/g. The Angstrom absorption exponent (α) differed from unity, a value that is typically used in the literature to scale the particle light absorption coefficient (b_{abs}) to different wavelengths. These findings indicate the need for a more complex aerosol optical model beyond a fixed BC/b_{abs} conversion factor.

Keywords Light absorption, black carbon, aethalometer, thermal-optical, photo-acoustic

Introduction

Black carbon (BC) aerosols contribute to warm forcing (0.2–1.0 W/m^2) and enhance evaporation of tropical cumulus.[1] Estimates of these effects require accurate BC emission inventories and mass absorption efficiencies (σ_{abs}, m^2/g) that convert BC concentration into light absorption coefficients (b_{abs}) at different wavelengths (λ). The objective of this study is to better characterize BC, elemental carbon (EC), and b_{abs} measurement methods. There is no single, universally accepted standard for BC

[1] *Desert Research Institute, Division of Atmospheric Sciences, Reno, NV, USA*

[2] *California Air Resources Board, Research Division, Sacramento, CA 95812, USA*

C.D. O'Dowd and P.E. Wagner (eds.), *Nucleation and Atmospheric Aerosols*, 684–688.

or EC measurement, and the available thermal and optical methods can vary by one order of magnitude.[2] Neither are there widely accepted methods to relate BC or EC to b_{abs}, the relevant observable for radiative transfer. Simplified optical theory for calculating σ_{abs} and the single scattering albedo of BC may not be applicable to BC from various sources emissions characterized by different size, morphology, and internal mixing.

Methods

Pure and externally mixed (with NaCl) aerosols from diesel engine, acetylene flame, electric arc, and wood-combustion and carbon black aerosols were generated and sampled in the laboratory under controlled conditions. Continuous b_{abs} and BC measurements were made using the photoacoustic analyzer (PA, 1047 nm) and a seven-color aethalometer (7-AE, 370, 470, 520, 590, 660, 880, 950 nm), along with sample collection on Teflon-membrane and quartz-fiber filters. Organic and elemental carbon (OC and EC) were measured with three thermal-optical methods: (1) IMPROVE_A TOR; (2) STN TOT; and (3) French two-step.[3–5] Corrections for the charring of OC are done by monitoring laser reflectance (TOR) by or transmittance (TOT) through the sample filter during the analysis.

Results and Conclusions

Comparisons of OC and EC concentrations measured with the three thermal methods are shown in Table 1. EC/TC ratios measured by thermal/optical methods were consistent within but variable between source types. The STN and French two-step protocols yielded EC/TC ratios similar to those of the IMPROVE_A protocol for diesel soot (EC/TC ~60%), acetylene flame soot (~96%), and electric arc soot (~50%). The French two-step and STN protocols were lower for EC (86% and 46%, respectively) in wood smoke compared to the IMPROVE_A protocol. Adding NaCl did not affect the OC/EC split in the IMPROVE_A and STN protocols. The French two-step protocol that operates in pure O_2 without a charring correction reported >60–90% lower EC than IMPROVE_A_TOR for all 19 samples.

The mass absorption efficiency (σ_{abs}, m^2/g) at 1,047 nm, calculated for each source from PA b_{abs} (1,047 nm) and EC filter concentration, is presented in Table 2. For all of the experiments using diesel (~5.1 m^2/g), acetylene flame (~3.3 m^2/g), electric arc (~3.3 m^2/g), and carbon black (~5.0 m^2/g), the σ_{abs} were consistent across the different thermal protocols and were similar for each source, with the exception of the source + NaCl samples analyzed by the French two-step protocol. As discussed earlier, the French two-step protocol was the only protocol where the presence of NaCl influenced the amount of EC. The σ_{abs} for the wood-smoke samples were highly variable within and between the thermal protocols. This is probably due

Table 1 OC, EC, and TC measurements, by source and by thermal protocol

Source	Protocol[a]	# of Samples	OC Avg. ($\mu g/m^3$)	EC Avg. ($\mu g/m^3$)	TC Avg. ($\mu g/m^3$)	EC/TC Avg. ± 1σ	
Diesel	IMP	7	125.8	175.2	301	0.590	0.078
	STN	7	119.8	183.1	302.9	0.606	0.044
	FM	7	126.4	174.5	301	0.587	0.058
Acetylene flame	IMP	6	7.6	356.2	358.8	0.982	0.030
	STN	6	18.4	352.3	366.3	0.958	0.074
	FM	6	29.3	329.6	358.8	0.921	0.050
Electric arc	IMP	4	532.4	476	1008.4	0.506	0.097
	STN	4	562.7	422.5	985.2	0.458	0.088
	FM	4	557.5	450.9	1008.4	0.497	0.142
Wood smoke	IMP	6	188.5	52	240.5	0.266	0.118
	STN	6	225.9	27.7	253.6	0.143	0.084
	FM	6	234.2	6.3	240.5	0.036	0.024
Carbon black	IMP	3	1348.5	22415.7	23764.2	0.905	0.072
	STN	3	1004.9	22625.6	23630.4	0.94	0.04
	FM	3	1785.9	21978.3	23764.2	0.855	0.12
Graphite	IMP	3	840.2	14899.7	15739.9	0.929	0.054
	STN	3	711.9	14584.9	15296.8	0.843	0.229
	FM	3	453.7	15286.2	15739.9	0.98	0.023

[a] IMP: IMPROVE_A_TOR protocol; STN: STN_TOT protocol;
FM: French two-step protocol

Table 2 The mass absorption efficiency (σabs) for pure and mixed sources

Source condition	IMPROVE TOR	STN TOT	French two-step
Diesel	5.12 ± 0.41	5.07 ± 0.37	5.28 ± 0.21
Diesel + NaCl	5.26 ± 0.29	4.95 ± 0.14	167.02 ± 110.95
Acetylene flame	3.33 ± 0.16	3.35 ± 0.10	3.58 ± 0.10
Acetylene flame + NaCl	3.28 ±0.37	3.21 ± 0.09	10.44 ± 2.18
Electric arc	3.37 ± 0.35	3.65 ± 0.34	3.31 ± 0.24
Electric arc + NaCl	3.28 ± 0.48	3.53 ± 0.17	106.31 ± 84.37
Wood smoke	2.72 ± 1.25	5.00 ± 0.89	27.04 ± 20.10
Wood smoke + NaCl	4.47 ± 1.44	6.20 ± 1.46	9.07 ± 6.63

to the highly variable EC emissions during the course of the wood combustion, as well as the lower EC fractions compared to the other sources. In addition, the matrix effect due to presence of salts such as K^+ and Cl^- in wood smoke may also contribute to this discrepancy.

Both b_{abs} and σ_{abs} are wavelength-dependent. Absorption by spherical particles is expected to vary inversely with wavelength (λ^{-1}).[6] The presence of organic compounds may enhance absorption at visible and UV wavelengths, thus changing the relationship between b_{abs} and λ.[7] This spectral dependence is described by the

Table 3 Comparison of the Ångstrom exponent (α) for pure and mixed sources

Source condition	A	±	Stdev	R	±	stdev	N
Diesel	0.79	±	0.09	0.983	±	0.029	35
Diesel + NaCl	0.86	±	0.12	0.964	±	0.046	9
Acetylene flame	0.80	±	0.03	0.990	±	0.002	10
Acetylene Flame + NaCl	0.85	±	0.04	0.989	±	0.005	9
Electric arc	1.38	±	0.20	0.945	±	0.031	13
Electric arc + NaCl	1.36	±	0.21	0.888	±	0.065	9
Wood smoke	1.15	±	0.51	0.981	±	0.039	23
Wood smoke + NaCl	1.22	±	0.29	0.995	±	0.004	14
Carbon black	0.53	±	0.01	0.989	±	0.001	3

Angstrom Power Law ($\sigma_{abs} = K\,\lambda^{-\alpha}$), where "α" is the Angstrom absorption exponent and K is a constant. The value of α is derived as follows:

$$\ln (b_{abs}) = -\alpha \ln \lambda + \ln K$$

Multi-wavelength b_{abs} measurements by the 7-AE were used to estimate α for each source. The assumption that these efficiencies vary inversely with wavelength is tested for each source (pure and mixture) and α is presented in Table 3. The correlations are high ($r > 0.96$) for all sources except the electric arc + NaCl samples ($r = 0.89$). The addition of NaCl did not alter many of the α values. Both diesel and diesel + NaCl, and acetylene and acetylene + NaCl samples have α between 0.79 and 0.86. The electric arc and electric arc + NaCl and wood smoke and wood smoke + NaCl samples both have α values greater than 1.15. The lower α values in diesel and acetylene samples also corresponded to higher (>0.5) EC/TC ratios, while the electric arc and wood smoke samples corresponded to lower (≤0.5) EC/TC ratios. Samples with a higher fraction of BC are expected to have α closer to unity, while the presence of organics is expected to increase α greater than unity.[7]

The results indicate that the IMPROVE_A and STN protocols measured similar EC concentrations for different combustion source types except for wood smoke. The presence of an externally mixed scattering component such as NaCl changed the abundances of the EC fractions, but not the OC/EC split in the IMPROVE_A and STN protocols. However, the French two-step protocol was influenced greatly by the aerosol matrix. A single value of σ_{abs} does not exist. Moreover, the Angstrom exponent (α), commonly assumed to be one, was found to vary from 0.5 to 1.4. These observations may be explained by more complex aerosol optical models that consider particle size distributions, morphology, and internal/external mixing characteristics.

Acknowledgments This work was supported by the California Air Resources Board and by the US Environmental Protection Agency under Contract #R-82805701. Any mention of commercially available products and supplies does not constitute an endorsement of those products and supplies.

References

1. Ackerman, A.S., Toon, O.B., Stevens, D.E., Heymsfield, A.J., Ramanathan, V., and Welton, E.J., *Science*, **288**, 1042–1047 (2000).
2. Watson, J.G. and Chow, J.C., A wintertime PM2.5 episode at the Fresno, CA, supersite, *Atmos. Environ.*, **36**,465–475 (2002).
3. Chow, J.C., Watson, J.G., Pritchett, L.C., Pierson, W.R., Frazier, C.A., and Purcell, R.G., *Atmos. Environ.*, **27A**, 1185–1201 (1993).
4. Cachier, H., Bremond, M.P., and Buat-Ménard, P., *Tellus*, **41B**, 379–390 (1989).
5. Peterson, M.R. and Richards, M.H. "Thermal-optical-transmittance analysis for organic, elemental, carbonate, total carbon, and OCX2 in PM2.5 by the EPA/NIOSH method" in Proceedings, Symposium on Air Quality Measurement Methods and Technology-2002, edited by E.D. Winegar and R.J. Tropp, Air & Waste Management Association, pp. 83–1–83–19, Pittsburgh, PA (2002).
6. Bond, T.C. and Bergstrom, R.W., *Aerosol Sci. Technol.*, **40**, 27–67 (2001).
7. Kirchstetter, T.W., Novakov, T., and Hobbs, P.V., *J. Geophys. Res.*, **109**(D21), D21208. ISI: 000225190500010 (2004).

Hygroscopic and Volatile Properties of Ultrafine Particles in the Eucalypt Forests: Comparison with Chamber Experiments and the Role of Sulphates in New Particle Formation

Zoran Ristovski[1], Tanja Suni[2], Nic Meyer[1], Graham Johnson[1], Lidia Morawska[1], Jonathan Duplissy[3], Ernest Weingartner[3], Urs Baltenpserger[3], and Andrew Turnipseed[4]

Abstract Simultaneous measurements of the volatile and hygroscopic properties of ultrafine particles were conducted in a Eucalypt forest in Tumbarumba, South-east Australia, in November 2006. These measurements were part of an intensive field campaign EUCAP 2006 (Eucalypt Forest Aerosols and Precursors). The particles exhibited a 2 step volatilisation with the first component starting to evaporate at temperatures above 50°C. With the onset of evaporation of the first component the hygroscopic growth factor increased. This indicated that the particle was composed of a less volatile, but more hygroscopic core, which was coated with a more volatile, but less hygroscopic, coating. The fraction of the more hygroscopic component was proportional to the measured maximum SO_2 concentration indicating the role of gaseous H_2SO_4 in new particle formation. As the volatilisation temperature of the second more hygroscopic component was above that for H_2SO_4 it is likely that this component is partially or fully neutralised H_2SO_4. Comparison with α-pinene smog chamber experiments shows an excellent agreement with the first step volatilisation indicating its origin in the photooxidation of a monoterpene precursor.

Keywords VH-TDMA, volatilisation, hygroscopic growth, *a*-pinene, eucalypt forest

[1] *ILAQH, Queensland University of Technology, 2 George Street, Brisbane 4000 QLD, Australia*

[2] *CSIRO Marine and Atmospheric Research, GPO Box 1666, Canberra ACT 2601, Australia*

[3] *Laboratory of Atmospheric Chemistry, PSI, CH-5232 Villigen PSI, Switzerland*

[4] *National Center for Atmospheric Research, PO Box 3000, Boulder, Colorado 80305, USA*

C.D. O'Dowd and P.E. Wagner (eds.), *Nucleation and Atmospheric Aerosols*, 689–693.

Introduction

A number of recent studies have shown that formation of new particles is a frequent phenomenon in the atmosphere (see, e.g., a review by Kulmala et al. 2004[1]). In order to understand the mechanisms underlying these new particle formation events it is of utmost importance to determine the chemical composition of the particles. However, to estimate the composition of particles below 20 nm remains a complex task. Very frequently indirect methods such as the measurement of particle hygroscopic properties and/or volatility are used to infer the composition of even the smallest particles below 20 nm.

Recently a new technique, the VH-TDMA has been developed that simultaneously measures the particle volatile and hygroscopic properties.[2] Although the VH-TDMA system measures physical properties such as the volatilisation temperature and hygroscopic growth, chemical properties can be inferred from the measurements, as well as information about the mixing state of the aerosol.

The task of identifying the mechanisms responsible for aerosol formation and quantifying their contribution to aerosol production is accomplished by comparing the physicochemical behaviours observed in field measurements with that of reference aerosols generated in the laboratory.

Experimental Setup

Measurements were conducted at the Tumbarumba flux station that is located in a tall open Eucalypt forest in south-eastern New South Wales in November 2006 (5–12.11.06). The dominant species are *E. delegatensis* (Alpine Ash) and *E. dalrympleana* (Mountain Gum), and the average tree height is 40 m. The instruments were located in a shed on the ground with inlet at 2 m height. The size distribution was measured with an Air Ion Spectrometer (AIS; Airel Ltd, Estonia) (0.34–40 nm) and with an SMPS (TSI 3936) (5–160 nm). During the 1 week period a number of new particle formation events were observed.

The volatile and hygroscopic properties were measured using the VH-TDMA.[2] Once the nucleation events was observed and the newly formed particles reached a relatively stable size of 20–30 nm, this mode was selected and analysed by the VH-TDMA. The thermodenuder temperature was scanned from room temperature until the particles completely volatilised. The maximum thermodenuder temperature necessary for complete volatilisation ranged from 150°C to 180°C. The change in the hygroscopic growth factor was measured simultaneously with the increasing temperature. The relative humidity during temperature scans was kept constant at RH = 90 ± 0.2%.

Experimental Results and Discussion

Figure 1 presents the VH-TDMA spectra of the three nucleation events observed on the 8th, 9th, and 10th of November. The analysed particle size was around 25 nm in all three cases. The particle volatilisation is presented through the volume fraction remaining $V/V_0 = D_v/D_{v0}$, where D_v is the particle diameter at temperature T, and D_{v0} particle diameter at room temperature (left *y*-axis). The change in the hygroscopic properties after volatilisation is presented through the dependence of the growth factor $G_h = D_h/D_v$, on the thermodenuder temperature (T_d), where D_h is the particle diameter after volatilisation and subsequent humidification (right *y*-axis). Open and closed symbols correspond to V/V_0 and G_h respectively.

The first step volatilisation starts at around 50°C and finishes at around 80°C. 30–50% of the volume is lost in this first step volatilisation. During this first step volatilisation the hygroscopic growth factors remains constant. With the removal of the first volatile species, above 80°C, the hygroscopic growth factors start to increase. This indicates that the particles are coated with a more volatile but less hygroscopic compound that suppresses the particle growth. Similar observations were made with the VH-TDMA in the marine environment.[3] The second less volatile but more hygroscopic component volatilises at temperatures above 100°C. For

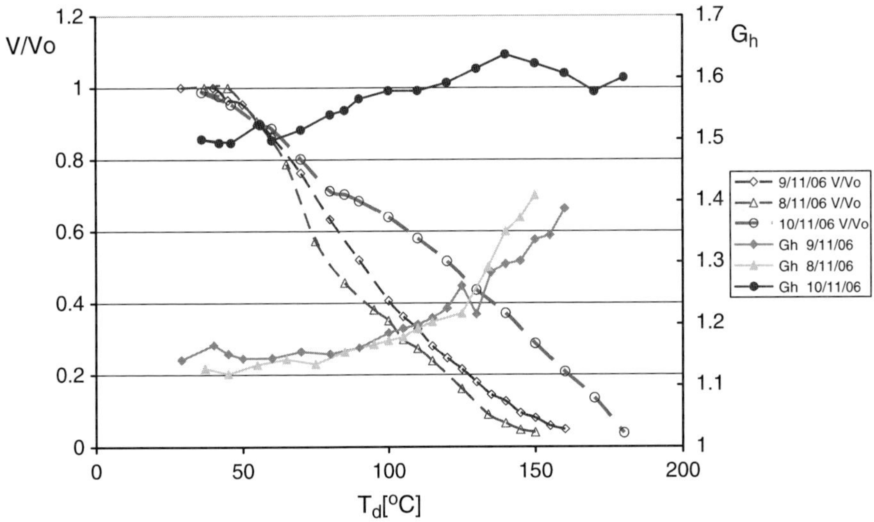

Figure 1 VH-TDMA spectra of the three nucleation events observed on the 8th, 9th, and 10th of November. Initial particle size was 25 nm and the RH was kept constant at 90%. Empty symbols and dashed lines represent volume fraction remaining (V/Vo), and full symbols and full lines hygroscopic growth factors (G_h)

the measurements on the 8th and the 9th, the particle size drops down to below 10 nm at temperatures of around 150°C. This still does not indicate that the second less volatile component had fully evaporated as 10 nm was the lowest particle size detectable by the CPCs used (3010). It is likely that the first more volatile component is of organic origin and a product of photo oxidation of monoterpenes which are relatively abundant in this environment. The second compound is most likely in the form of sulphate.

There is a distinct difference in both volatilisation and hygroscopic growth for the nucleation event observed on the 10th of November. During this event the observed hygroscopic growth factor was significantly higher at room temperature than on the other dates. The presence of the more hygroscopic but less volatile species was much more pronounced on this day.

The reason for this could be in the role that the sulphuric acid plays in the new particle formation events. It has been recently documented[4] that the measured G_h, in boreal forests, correlates positively with the gaseous sulphuric acid concentration. The maximum measured SO_2 concentration on the 8th and 9th of November did not exceed 200 pptv while on the 10th of November the maximum concentration was around 800 pptv. If we take the SO_2 levels as a proxy for the presence of gaseous sulphuric acid than one can assume that the concentration of sulphuric acid

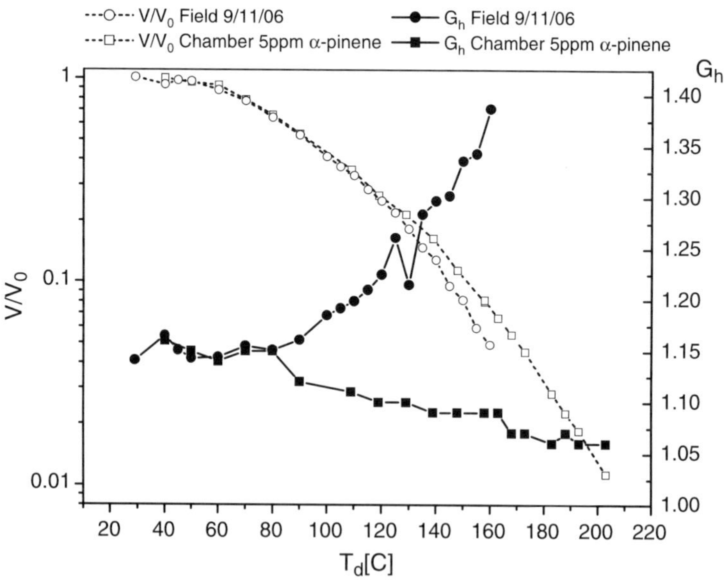

Figure 2 VH-TDMA spectra of the nucleation event observed on the 9th November (**circles**) and smog chamber measurements with 5 ppm α-pinene (**squares**). The volume fraction remaining is presented with open symbols on the left axis, while the hygroscopic growth factors with full symbols on the right axis. In both measurements the relative humidity in the VH-TDMA was kept constant at RH~90%

on the 10th was much higher than on the other two days. This is in clear agreement with the observed increase in the G_h and furthermore supports the idea that the less volatile but more hygroscopic compound observed in the core of the particles is ammonia neutralised sulphuric acid.

Figure 2 presents VH-TDMA spectra of a nucleation event observed on the 9th of November and from chamber measurements conducted at the smog chamber at LAC, PSI in August 2006. In the smog chamber experiments, particles were generated via the photooxidation of *a*-pinene at a concentration of 5 ppm, in a 27-m^3 Teflon chamber at 20°C.[5] The relative humidity for the hygroscopic growth measurements in both field and chamber experiments was kept constant at $90 \pm 0.2\%$. There is an excellent agreement for both the volatility curves and the hygroscopic growth data up to around 80°C which corresponds to the first step volatilisation. At temperatures above 80°C the hygroscopic growth factors start to deviate as the particles measured in the field start to loose the more hydrophilic organic coating. In chamber experiments, care has been taken not to introduce any gaseous sulphuric acid and therefore the particles do not have a hygroscopic sulphate core which would cause an increase in the G_h at higher thermodenuder temperatures. In addition the chamber measurements do not exhibit a two-step volatilisation. This clearly indicates that the more volatile but less hygroscopic particle coating observed in the field measurements is of organic origin and the product of photooxidation of either *a*-pinene or some similar monoterpene.

Acknowledgments The smog chamber measurements were supported by the European Science Foundation (ESF) within the Interdisciplinary Troposphere Research: from the Laboratory to Global Change (INTROP). We also acknowledge funding from the European Network of Excellence ACCENT.

References

1. Kulmala, M. et al., *J. Aerosol. Sci.*, **35**, 143–176 (2004).
2. Johnson, G.R. et al., *J. Aerosol Sci.*, **35**(4), 443–455 (2004).
3. Johnson, G. et al., *J. Geophys. Res.*, **110**, D20203 (2005).
4. Ehn, M. et al., *Atmos. Chem. Phys.*, **7**, 211–222 (2007).
5. Paulsen, D. et al., *Environ. Sci. Technol.*, **39**(8), 2668–2678 (2005).

New Particle Formation in Clean Savannah Environment

L. Laakso[1], T. Petäjä[1], H. Laakso[1], P.P. Aalto[1], T. Pohja[1], E. Siivola[1], P. Keronen[1], S. Haapanala[1], M. Kulmala[1], H. Hakola[2], N. Kgabi[3], M. Molefe[3], D. Mabaso[3], K. Pienaar[3], E. Sjöberg[4], and M. Jokinen[4]

Abstract Southern African savanna background environment lacks previous, continuous long-term combined sub-micrometer aerosol number concentration and gas measurements. We have build a mobile measurement trailer, which contains measurements of aerosol number size distributions 10–840 nm, positive and negative ion size distributions 0.4–40 nm, aerosol mass PM_{10}, $PM_{2.5}$, PM_1, inorganic compound in PM_{10} and $PM_{2.5}$, gases like SO_2, NO, NOx, CO, O_3 and basic meteorology. In addition, we measure periodically VOCs. The trailer is well-protected against thunder storms, electricity breaks and other typical problems. It is connected to internet via GPRS-modem. Preliminary results show that in a clean savanna environment, new particle formation takes place every sunny day, with relatively high nucleation rates and very high growth rates (up to 15 nm/h).

Keywords Particle formation and growth, aerosol measurement, atmospheric aerosol, air pollution

Background

Industrial countries have invested heavily on the air quality monitoring during the last decades. Significant emission and air quality regulations have been set and continuous monitoring in urban and industrial centres has been carried out. In the Third World, these issues are not yet at the same level (Laakso et al. 2006) despite the fact that the poor air quality affects the lives of hundreds of millions of people, in particularly in the developing world.

---

[1]*Department of Physics, University of Helsinki, PB 64, FI-00014, University of Helsinki, Finland*

[2]*Finnish Meteorological Institute, PB 503, FI-00101, Helsinki, Finland*

[3]*The North-West University, Republic of South Africa*

[4]*Department of Agriculture, Conservation and Environment, Mafikeng, Republic of South Africa*

C.D. O'Dowd and P.E. Wagner (eds.), *Nucleation and Atmospheric Aerosols*, 694–697.

In order to assess the role of different particulate sources in the air quality in the developing world, one has to know also the natural background conditions. Thus it is essential to conduct measurements in areas far enough from large industrial sites and cities, which is a challenging task due to very small infrastructure at these locations. In order to reach this goal, a mobile instrument trailer was constructed in Finland. Prior shipping a short comparison campaign was conducted at Hyytiälä, Finland, where the performance of the trailer instruments was evaluated against data obtained from a state-of-the-art atmospheric field station in Hyytiälä, Finland.

Trailer Instrumentation

The trailer was designed and built in Finland. The trailer instrumentation include a Differential Mobility Particle Sizer (DMPS, Aalto et al. 2001), which measures aerosol number size distribution from 10 to 800 nm with 10 min time resolution. Mass concentration of particulate matter is monitored with a TEOM 1400a Ambient Particulate Monitor (Rupprecht and Patashnick). An automatic custom made inlet switch allows measurement of PM_1, $PM_{2.5}$, and PM_{10} with the same instrument in 20 min sampling interval for each mass fraction. Positive and negative air ions between 0.4 and 40 nm are detected with an Air Ion Spectrometer (AIS, Airel Ltd, Tartu, Estonia). Gaseous pollutants (SO_2, O_3, CO, NOx) and local meteorological parameters (temperature, relative humidity, wind speed, wind direction, precipitation rate, and photosynthetically active radiation) are logged continuously. The trailer is

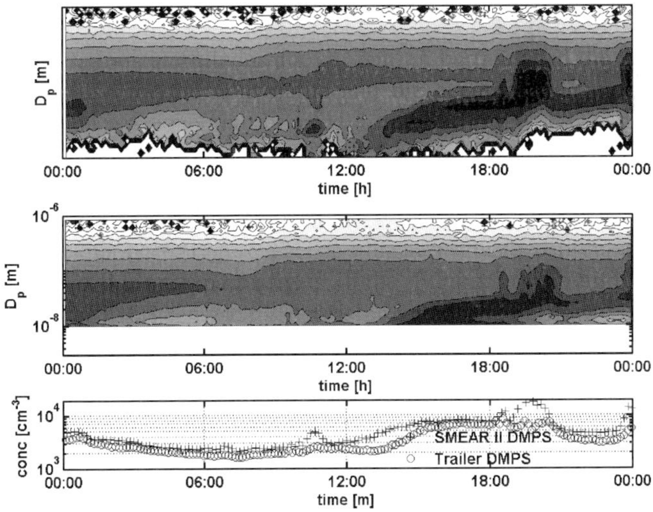

Figure 1 Number size distribution of sub-micron particles measured with SMEAR DMPS (**top-most panel**) and Trailer DMPS (**middle panel**) were in qualitative agreement on 26 April 2006. Due to different lower size detection limits, DMPS in trailer is showing lower values

equipped with a GPS-receiver to pinpoint the measurement location. The measurement data is copied to a server via GPRS-modem wirelessly. This enables remote monitoring of data quality on everyday basis.

During spring 2006, test measurements with the trailer setup were conducted side by side with aerosol and gas-phase equipment of SMEAR II (Vesala et al. 1998) station located at Hyytiälä, Finland. An exemplary data of aerosol number size distribution and total number concentration measurements are presented in Figure 1. New particle formation and subsequent growth was detected by the trailer DMPS and at the station.

Measurements in South Africa

The trailer has been measuring in background savannah since July 2006 and these measurements will continue until end of 2007. During this period, we are measuring VOC-concentrations periodically (~1 week/season) with Tenax-tubes. In addition, inorganics in $PM_{2.5}$ and PM_{10} are monitored weekly. After background savannah measurement, the trailer is located to pollute mining background region for the year 2008. At a later time, measurements will continue at industrial region (2009) and urban background (2010).

Together, these measurements will enable assessment of human activities on the air quality of rapidly developing Republic of South Africa.

Preliminary Results From The Background Savannah

A preliminary data analysis from a period 23 July–15 August 2006 indicated that

- New particle formation took place on every sunny day
- New particle formation produced $1–3 \cdot 10^4$ cm^{-3} new particles in sizes above 10 nm. J_{10} was typically approximately 1 cm^{-3} s^{-1}
- Typical night-time concentration of 10–840 nm particles was approximately 1,000 cm^{-3}
- Particle growth rates were high, up to 15 nm/h
- Average mass concentrations were 12.8, 14.7, and 24.4 µg m^{-3} for PM_1, $PM_{2.5}$, and PM_{10}, respectively
- The VOC concentrations were generally quite low. The sum of monoterpene concentrations varied from 0.3 to 1.6 µg m^{-3}. Typical VOCs observed were α-Pinene, β-pinene, benzene, nonane, isoprene, and camphene

During the period discussed here, there were some plumes from distant field fires. Apart from these periods, air was very clean and there were no significant traces of anthropogenic influence.

Table 1 Median gas concentrations in Botsalano game reserve, 23 July–15 August 2006

Gas	Concentration [ppb]
O_3	34.0
SO_2	1.0
NO	0.1
NOx	0.9
CO	122.4

Figure 2 Our mobile measurement trailer in background savannah in Botsalano game reserve (300 km west of Johannesburg), Republic of South Africa

Acknowledgment The authors want to thank several people who have contributed to this project.

References

Aalto, P. et al., Physical characterization of aerosol particles during nucleation events, *Tellus*, **53B**, 344–358 (2001).

Laakso, L. et al., Aerosol particles in the developing world: a comparison between New Delhi in India and Beijing in China, *Water Air Soil Pollut.*, **173**, 5–20 (2006).

Vesala, T. et al., Long-term field measurements of atmosphere-surface interactions in boreal forest combining forest ecology, micrometeorology, aerosol physics and atmospheric chemistry, *Trends in Heat, Mass & Momentum Transfer*, 4:17–35 (1998).

Hot-air Balloon Measurements of Vertical Variation of Boundary Layer New Particle Formation

L. Laakso[1], T. Grönholm[1], S. Haapanala[1], A. Hirsikko[1], T. Kurtén[1], M. Boy[1], A. Sogachev[1], I. Riipinen[1], M. Kulmala[1], L. Kulmala[2], E.R. Lovejoy[3], J. Kazil[4], D. Nilsson[5], and F. Stratmann[1,6]

Abstract In this study, we used a hot-air balloon as a platform for boundary layer particle and cluster measurements. We did altogether 11 flights during the spring of 2005 and 2006. During the spring of 2006, we observed five new particle formation days. During all days, new particle formation took place in the mixed boundary layer. During one of the days, we observed particle formation in the free troposphere, separate from that of the mixed layer. The observations showed that the concentration of freshly-formed 1.5–2 nm negative ions was several times higher than the concentration of positive ions. We also clearly observed that nucleation during one of the days, 13 March 2006, was a combination of neutral and ion-induced nucleation. During some of the days, particle growth stopped at around 3 nm, probably due to lack of condensable organic vapours. Simulations of boundary layer dynamics showed that particles are formed either throughout the mixed layer or in the lower part of it, not at the top of the layer.

Keywords New particle formation and growth, aerosol measurement, atmospheric aerosol, clusters

Background and Measurements

Boundary layer particle formation has been observed in diverse environments (Kulmala et al. 2004). Since exact location of particle formation is not known, we placed several devices in a hot-air balloon during spring 2006. The instrumentation

[1]*Department of Physical Sciences, University of Helsinki, P.O. Box 64, FI-00014 Finland*

[2]*Department of Forest Ecology, University of Helsinki, P.O. Box 27, FI-00014 University of Helsinki, Finland*

[3]*NOAA Aeronomy Laboratory 325 Broadway, Boulder, CO 80303, USA*

[4]*NOAA Earth System Research Laboratory, 325 Broadway, Boulder, CO 80305, USA*

[5]*Department of Applied Environmental Research, Stockholm University, S-10691 Stockholm, Sweden*

[6]*Institute for Tropospheric Research, Permoserstrasse 15, D-04318 Leipzig, Germany*

C.D. O'Dowd and P.E. Wagner (eds.), *Nucleation and Atmospheric Aerosols*, 698–701.

included: air ion spectrometer (positive and negative ions 0.4–40 nm), CPC (particle number concentration >10 nm), VOC- sampling with bottles, T, RH, p, and CO2-concentration (Laakso et al. 2007).

Results

Together with the ground-based measurements from SMEAR II-station, the results provide new insights in a new particle formation in the boundary layer.

The main characteristics for the spring 2006 are shown in Table 1.

Table 1 Characteristics of spring 2006 flights

	10 March 2006	12 March 2006	13 March 2006	14 March 2006	17 March 2006
1 Temperature at 67 m [°C]	−10.7	−5.4	−0.8	−3.9	4.1
2 Global radiation [Wm^{-2}]	336.6	279.2	346.0	396.6	207.9
3 RH [%]	78.3	65.3	56.4	65.0	54.2
4 Wind speed [ms^{-1}]	3.3	3.8	3.0	5.5	2.9
5 Wind direction [°]	48	51	71	122	271
6 Condensation sink * 1e3 [s^{-1}]	5.27	2.33	2.39	3.14	1.89
7 SO_2 [ppb]	4.78	1.07	0.87	0.78	0.20
8 NO_x [ppb]	4.00	1.82	2.79	1.89	2.95
9 GR (−) 1.3–3 nm [nm/h]	–	1.6	1.8	1	–
10 GR (+) 1.3–3 nm [nm/h]	–	–	–	–	–
11 H_2SO_4 · 1e—6 [cm^{-3}] (Boy et al. 2005)	2.9	2.1	2.6	2.0	0.8
12 BSMA event classification (Hirsikko et al. 2006)	Ib.1	Ib.1	I	Ib.1	II
13 Origin of Air Mass	Baltic countries	Baltic countries	Belarus	S-Russia	Balticsea
14 Production rate of condensable organics in BL [$10^4 \times cm^{-3}s^{-1}$]	0.5	0.4	0.7	1.4	0.7
15 Boundary layer height [m]	200	500	600	500	600
16 Residual layer [m]	200–600	500–1000	600–1200	500–900	600→
17 First decent	13:35–13:52	14:50–15:03	13:56–14:10	11:36–11:52	14:13–14:28
18 Second decent	14:00–14:16	15:10–15:20	14:20–14:30	12:00–12:10	14:44–14:54

Conclusions

We observed boundary layer particle, ion and meteorological profiles during several nucleation event days in a boreal forest. The observations showed that the particle formation observed at the surface was relatively homogeneous throughout the whole boundary layer. However, we also observed separate particle formation inside the residual layer during one day, which is in agreement with earlier observations by Stratmann et al. (2003). Particle concentration profiles from the mixed layer showed that surface measurements usually represent concentrations throughout the mixed layer relatively well.

During our measurements, there was a clear difference between negative and positive ions: the concentration of negative 1.5–2 nm ions was several times larger than that of the positive ions throughout the mixed layer. This is a clear indication of ion-induced nucleation, where negative ions play a significant role. In addition to that, also neutral nucleation/activation was important.

On 13 March 2006, combined data from the balloon and surface measurements showed that at the onset and end of the nucleation event, negative ions activate, and the particle population is negatively overcharged. In the middle of the burst, also positive and neutral clusters start to activate, or neutral homogeneous nucleation starts.

We also noticed that during several days, the activation of small negative ions without growth to sizes observable with DMPS occurred. We suppose that the reason for the lack of growth was low organic vapour concentrations (Haapanala et al. 2006).

If we assume that sulphuric acid plays an important role in new particle formation, the qualitative differences between negative, and positive and neutral nucleation can be related to simple acid-base chemistry. However, they also indicate that ammonia is unlikely to stabilize small negatively charged water–sulphuric acid clusters. When we calculated the negative ion-induced nucleation rates based on the method by Lovejoy et al. (2004), we noticed that pure water– sulphuric acid nucleation is not

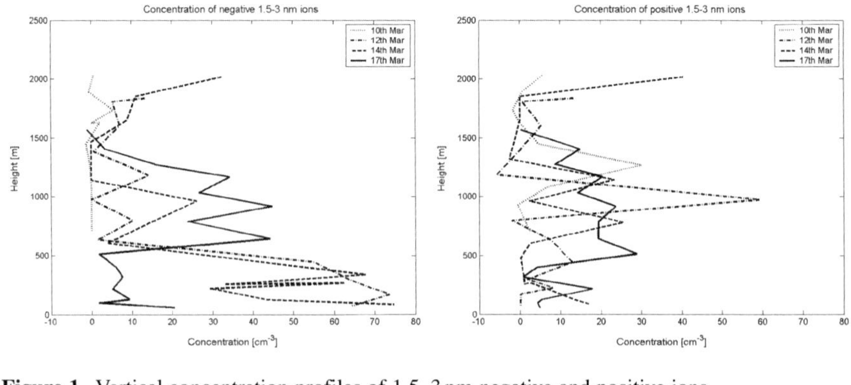

Figure 1 Vertical concentration profiles of 1.5–3 nm negative and positive ions

able to explain the results without assuming some additional stabilizing component as suggested in that article.

To investigate the potential location of nucleation, we carried out boundary layer mixing simulations assuming that nucleation takes place in four different locations. Based on these results, we assume that nucleation did not take place at the top of mixed layer, but rather throughout the layer.

As a final conclusion, our measurements showed that a hot-air balloon is a suitable platform for boundary layer measurements, and it can be used to observe ion dynamics and particle formation.

Acknowledgments The study was in part supported by the Academy of Finland and the Nessling Foundation. Thanks are due to Markku Sipinen and Reijo Lampinen from Aeronaut ballooning company, Tikkakoski, Finland for the balloon flights. We are grateful to Taina Ruuskanen, Erkki Siivola, Heikki Laakso, and Andreas Kürten for their practical help with the measurements.

References

Haapanala, S., Rinne, J., Hakola, H., Hellén, H., Laakso, L., Lihavainen, H., Janson R., Kulmala, M., Boundary layer concentrations and landscape scale emissions of volatile organic compounds in early spring, *Atmos. Chem. Phys. Discuss.*, **6**, 10567–10589 (2006).

Kulmala, M., Vehkamäki, H., Petäjä, T., Dal Maso, M., Lauri, A., Kerminen, V.-M., Birmili, W. and McMurry, P.H., Formation and growth rates of ultrafine atmospheric particles: A review of observations, *J. Aerosol Sci.*, **35**: 143–176 (2004).

Laakso, L., Grönholm, T., Kulmala, L., Haapanala, S., Hirsikko, A., Lovejoy, E.R., Kazil, J., Kurtén, T., Boy, M., Nilsson, E.D., Sogachev, A., Riipinen, I., Stratmann, F., and Kulmala, M. Hot-air balloon measurements of vertical variation of boundary layer new particle formation, Accepted to *Boreal Environment Research* (2007).

Lovejoy, E.R., Curtius, J., and Froyd, K.D., Atmospheric ion-induced nucleation of sulfuric acid and water, *J. Geophys. Res.*, **109**, 10.1029/2003JD004460 (2004).

Stratmann, F., Siebert, H., Spindler, G., Wehner, B., Althausen, D., Heinzenberg, J., Hellmuth, O., Rinke, R., Schmiedler, U., Seidel, C., Tuch, T., Uhrner, U., Wiedensohler, A., Wandinger, U., Wendisch, M., Schell, D., and Stohl, A., New-particle formation events in a continental boundary layer: first reseults from the SATURN experiment, *Atmos. Chem. Phys.*, **3**, 1445–1459 (2003).

Optical Properties and Radiative Effects of Aerosols in a Coastal Zone

Auromeet Saha[1], Marc Mallet[2], Jean Claude Roger[3], Philippe Dubuisson[4], Jacques Piazzola[1], and Serge Despiau[1]

Abstract Physical and optical properties of atmospheric aerosols were measured in the French Mediterranean coastal zone for over one year between October 2005 and 2006. BC mass concentration, absorption and scattering coefficients, number-size distributions of fine and coarse particles were continuously measured at ground level, along with the meteorological parameters. In addition to these surface based in situ measurements, a co-located Cimel Sun photometer continuously measured the columnar aerosol optical depth (AOD), Angstroms coefficient (α) and the other microphysical parameters. Large surface BC concentrations and high values of scattering coefficients occurred during winter, followed by lower values during spring and summer season. However, the columnar aerosols showed different seasonal behaviour, with high AOD values occurring mostly during the summer months, and low to moderate values prevailed during the rest of the period. Monthly mean AOD at mid-visible wavelengths ranged between ~0.1 and 0.34. The Angstroms coefficient (α) remained high (>1.2) during the entire study period, thereby indicating the relative dominance of fine particles. Monthly mean BC concentrations varied between 303 and 1026 ng m^{-3}. The surface single scattering albedo at 525 nm was in the range from 0.7 to 0.8, thereby indicating the dominance of absorbing aerosols over this region. Mean reduction in the diurnally averaged solar flux (as estimated using discrete ordinate radiative transfer model GAME) was found to be 24.9 Wm^{-2} and 61.8 Wm^{-2} for pollution and dust event respectively in June 2006.

Keywords Aerosol optical properties, radiative forcing, coastal zone

[1] *LSEET-LEPI,* UMR 6017, *Université du Sud Toulon-Var, La Garde, France*

[2] *LA, Université Paul Sabatier, Toulouse, France*

[3] *LAMP, OPGC, Clermont Ferrand, France*

[4] *LOCL/MREN, ELICO, Wimereux, France*

C.D. O'Dowd and P.E. Wagner (eds.), *Nucleation and Atmospheric Aerosols*, 702–706.

Site Description

The monitoring station is located in the Toulon University campus in La Valette, southeast of France. It is situated ~8 km towards the east of the urban city of Toulon. The district of Toulon has 22 cantons and 34 communes with a total population of over half a million and population density of >400/km^2. The measurement station is located ~5 km away from the Mediterranean Sea Coast. Other than the traffic, there are no other significant anthropogenic sources of aerosols nearby the measurement site.

Instrumentation and Data

The directly measured aerosol and radiometric quantities included scattering and absorption coefficients, Black Carbon mass concentration, number-size distributions of fine and coarse particles, columnar aerosol optical depth (AOD) and global broadband shortwave fluxes at the surface.

Measurements of Black Carbon (BC) mass concentration and absorption coefficients (β_{ab}) are made using Magee Scientific Aethalometer (model AE-31). The Aethalometer measures the attenuation of light beam at seven wavelengths viz. 370, 470, 520, 590, 660, 880, and 950 nm. The 880 nm is considered as the standard channel for BC measurements, because BC is the principle absorber of light at this wavelength and other aerosol components have negligible absorption at this wavelength. More details on the instrument and the principal of operation are given elsewhere (Hansen et al. 1984, Saha and Despiau 2007). Scattering coefficient of aerosols (β_{sc}) is measured using an Ecotech Nephelometer (Model M9004). Nephelometer draws the ambient air through an inlet tube, which is then illuminated by a flash lamp and the scattered light intensity is measured at 525 nm by a photomultiplier tube. Temperature and pressure measurements made in the scattering chamber are used to calculate the scattering by air molecules, which are then subtracted from the total scattering to get the scattering due to aerosols. The instrument was regularly calibrated in the laboratory, in which the Nephelometer is adjusted to read zero by passing particle free air and the span calibration was done using CO_2 gas. A co-located Cimel sun photometer operated under the PHOTONS/AERONET network has also been used for the present study for the measurements of AOD at 440, 675, 870, and 1020 nm and total columnar water content. The global broadband radiative flux was obtained using CM21 Kipp and Zonen pyranometer. Meteorological parameters such as air temperature, relative humidity, wind speed, wind direction, pressure, and rain intensity were measured using instruments and sensors onboard a tower. All these instruments were part of the monitoring station proposed under the ACCENT program (Despiau et al. 2005).

The measurement campaign was initiated from 17 October 2005, and since then continuous measurements were being made until 31 October 2006. The data collected during the period this period (spanning for 377 days) has been used for the present study.

Results and Discussions

Figure 1 shows the monthly-mean temporal variation of AOD at two representativ wavelengths (440 and 1020 nm), columnar water vapour content and Angstrom coefficient (α). AOD ranges between ~0.1 and 0.34 and has significant monthl variability, with quite high values (>0.2) observed during the summer and autum season, while low values were encountered during the winter and spring months Monthly variations of columnar water vapour showed similar variations as that o AOD, thereby indicating the impact of water vapour on the aerosol optica properties. The Angstroms coefficient (α) was estimated from the AODs at 440 t 870 nm wavelength range. It can be seen from the Figure 1 that, α remained fairl high (>1.2) during the entire study period, thereby indicating the relative dominanc of fine particles.

Figure 2 shows the temporal variation of near surface BC mass concentration and scattering coefficients respectively. Monthly mean BC concentrations varie between 303 and 1026 ng m^{-3}. High values of BC and scattering coefficient wer observed during winter months (November through February) and low value during spring and summer months (March through September). Although th patterns of seasonal variation of BC (and β_{ab}) and scattering coefficient evolv almost identically (as indicated in Figure 2), small differences in the percentag variation of the individual parameters (β_{ab} and β_{sc}) from one month to other woul lead to the observed pattern of single scattering albedo (SSA, the ratio of scatterin

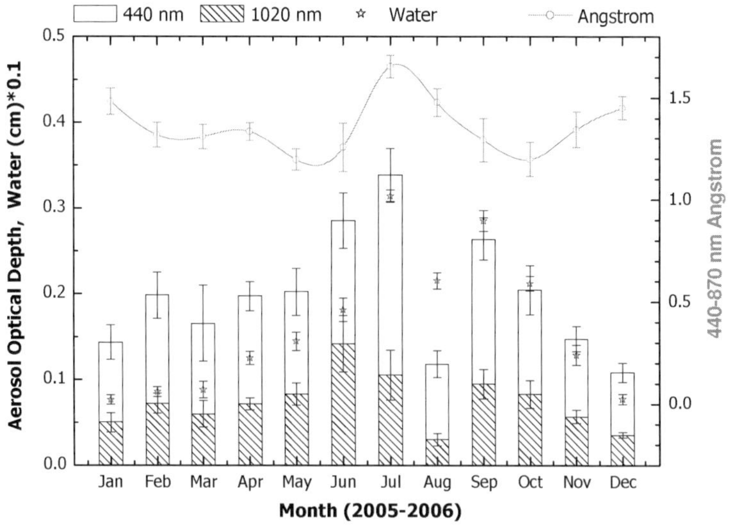

Figure 1 Temporal variation of monthly mean AOD at 440 and 1020 nm, columnar water vapou content (left axis) and Angstrom coefficient (right axis). The error bars represent the standard erro of the monthly ensembles.

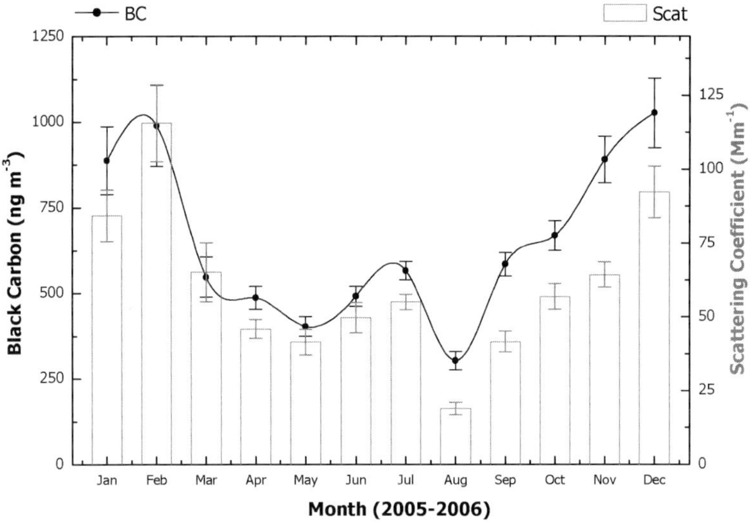

'igure 2 Temporal variation of monthly mean BC mass concentrations (left axis) and scattering oefficient (right axis). The error bars represent the standard error of the monthly ensembles

o extinction coefficient). In the present study, the SSA was computed from β_{sc} at 525 nm measured using Nephelometer and β_{ab} measured using Aethalometer 520 nm channel). The seasonal averaged SSA at 525 nm was found to be 0.76 ± .07, 0.79 ± 0.07, 0.78 ± 0.07, and 0.73 ± 0.08 during winter (DJF), spring (MAM), ummer (JJA), and autumn (SON) seasons, respectively. The significantly low values of SSA indicate the dominance of absorbing aerosols over this region, which nay have implication on the regional radiative forcing.

In order to quantify the implication of aerosols in the regional aerosol radiative orcing, discrete ordinate radiative transfer model GAME (Dubuisson et al. 1996) as been used to perform the radiative transfer calculations. Following the approach roposed in Roger et al. (2006), the daily direct radiative forcing was estimated for hree pollution events (7, 11, and 22 June 2006) and one dust event (19 June 2006). 'or the polluted events (AOD = 0.34 at 440 nm), our simulations indicate a significant reduction of solar radiation reaching the surface (mean $\Delta F_{BOA} = -24.9\,Wm^{-2}$) y reflection to space (mean $\Delta F_{TOA} = -5.1\,Wm^{-2}$), but predominantly by absorption f the solar radiation into the atmosphere (mean $\Delta F_{ATM} = +19.8\,Wm^{-2}$). For the dust vent (AOD = 0.80 at 440 nm), the direct radiative forcing was significantly larger, vith a mean ΔF_{BOA} ($-61.8\,Wm^{-2}$) more than two times larger than for the polluted ases, and the associated ΔF_{TOA} and ΔF_{ATM} were $-7.7\,Wm^{-2}$ and $+54.1\,Wm^{-2}$, espectively.

cknowledgments The present work is supported by French CNRS, INSU, and PNTS Projects.)ne of the authors (AS) would like to acknowledge the French Ministry for Education and Research nd the Université du Sud Toulon-Var for providing the financial support.

References

Despiau, S., Piazzola, J., and Missamou, T., A French Mediterranean site for atmospheric studie and surveillance, First ACCENT Symposium, Urbino (Italy) (2005).

Dubuisson, P., Buriez, J.C., and Fouquart, Y., *J. Quant. Spectrosc. Radiat. Transfer*, **55**, 103–12 (1996).

Hansen, A.D.A., Rosen, H., and Novakov, T., *Science Total Environ.*, **36**, 191–196 (1984).

Roger, J.C., Mallet, M., Dubuisson, P., Cachier, H., Vermote, E., Dubovik, O., and Despiau, S. *J. Geophys. Res.*, **111**, D13208, doi:10.1029/2005JD006361 (2006).

Saha, A. and Despiau, S., *J. Geophys. Res.* (under communication) (2007).

)n Water Condensation Particle ;ounters and their Applicability .o Field Measurements

:. Petäjä[1], H.E. Manninen[1], F. Stratmann[1,2], M. Sipilä[1], G. Mordas[1],
'.P. Aalto[1],H. Vehkamäki[1], W. Birmili[1,2], K. Hämeri[1,3], and M. Kulmala[1]

.bstract The operation of Water Condensation Particle Counters (WCPCs) was erified under laboratory conditions. Variation in the detection efficiency as a ınction of supersaturation is addressed to differences in heterogeneous nucleation ate and critical cluster size. Experimental results were reproduced with a fluid ynamics simulation, when the heterogeneous nucleation process was applied) the activation and growth to detectable sizes. The sensitivity of the WCPC to hemical composition was utilized in a Condensation Particle Counter Battery in eld conditions.

:eywords Condensation Particle Counters, heterogeneous nucleation, atmospheric erosols

ntroduction

ı recent years there has been a growing interest in the health effects of atmospheric ltrafine particles. Increases in aerosol number concentration and in particulate ıatter mass (PM_{10}) had a comparable enhancement factor to hospital readmission f survivors of myocardial infarction.[1] In an urban environment the majority of the article number resides below 100 nm in diameter. The recently developed water ondensation particle counters (WCPC[2,3]) are good candidates for monitoring urposes and they have been shown to perform rather well in comparison with utanol-based counters.[4] A main concern regarding the applicability of a WCPC to otal aerosol particle measurements in particularly in an urban environment arises om their sensitivity to particle composition. This is a major drawback in the case, here the majority of the particles are close to the detection limit of the CPCs. This

)ivision of Atmospheric Sciences, PO BOX 64, FI-00014 University of Helsinki, Finland
nstitute for Tropospheric Research, Permoserstrasse 15, D-04318 Leipzig, Germany
'innish Institute of Occupational Health, FI-00250 Helsinki, Finland

D. O'Dowd and P.E. Wagner (eds.), *Nucleation and Atmospheric Aerosols*, 707–710.
Springer 2007

means that concurrent measurements with a butanol CPC are still needed to secur the results.

The aim of this study is to summarize recent laboratory verifications of water-base CPCs as well as to discuss the theoretical basis of the sensitivity of the WCP detection efficiency to particle composition, i.e., heterogeneous nucleation probabilit This theoretical framework is applied into Computational Fluid Dynamics (CFD Code FLUENT[5] with the Fine Particle Model (FPM[6]) to simulate the heterogeneou nucleation rate inside a TSI 3785 WCPC. Both the laboratory verifications and th computer simulations give rise to a Condensation Particle Counter Battery technique which utilizes the sensitivity of the WCPCs to ultrafine particle composition. This se of instruments was then applied to field measurements to indirectly give an insight int chemical composition of newly formed atmospheric particles.

Methods

The operation of the WCPCs was verified in the laboratory with silver, ammoniu sulfate, and sodium chloride particles. A TSI 3068 electrometer was used as a refe ence instrument.

For modeling fluid flow, heat/mass transfer and particle dynamics inside th WCPC, the FPM,[5] together with the CFD Code FLUENT was used.

A Condensation Particle Counter Battery[7] consists of water and butanol-base CPCs in parallel. The detection efficiencies of butanol and water CPCs were matche in the laboratory for silver particles. As the water CPC is more sensitive to the wat soluble aerosol particles, it either measures higher or lower concentrations than butanol-based particle counter. The concentration difference between the calibrate set of CPCs shows indirectly the solubility of atmospheric nanoparticles to water an butanol. The solubility is, in turn, a function of particle chemical composition.

Results and Discussion

The performance of the TSI 3785 WCPC was verified in the laboratory for silve ammonium sulphate and sodium chloride particles.[8] Similar experiments we performed for TSI 3786 Ultrafine WCPC as well.[9] For both WCPCs, the detectic efficiency is better for water-soluble salts in contrast to inert silver particles. A an example, the TSI 3785 operated at nominal conditions, was able to detect ha of the silver particles (D50) at 5.8 nm. For sodium chloride particles the D50 w 3.6 nm. In comparison, Liu and co-workers[10] obtained D50 values of 5.6 and 3.1 f silver and sodium chloride particles, respectively.

Based on the laboratory experiments,[8,9] the detection efficiency of the WCP can be modified by changing the temperatures of the saturator and the growth tub This is due to the fact that as the supersaturation increases inside the instrumer heterogeneous nucleation rate increases and critical cluster size decreases.[7] Th

:nables tuning of the WCPCs to a desired cutoff size for a given chemical :omposition. The heterogeneous nucleation scheme was implemented into coupled luid and particle dynamics simulations using FLUENT together with the FPM 'or the TSI 3785 WCPC geometry.[11] The measured detection efficiency was eproduced well when a contact angle of 43.5° was assumed (see Figure 1).

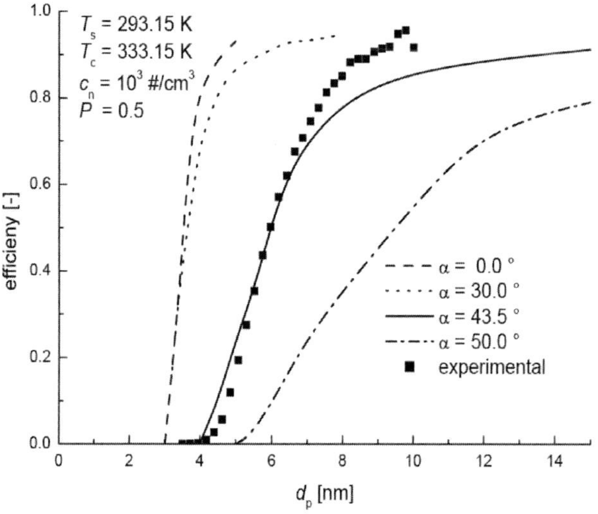

'igure 1 Experimental and modeled counting efficiencies as function of particle size for differ- nt contact angles

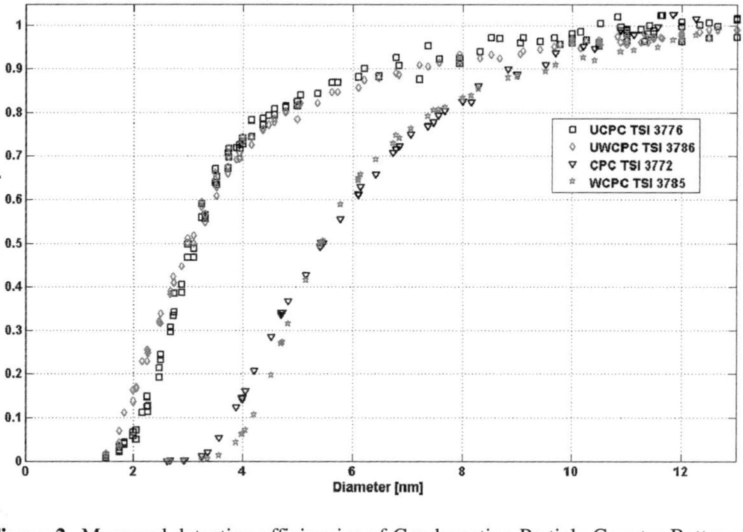

'igure 2 Measured detection efficiencies of Condensation Particle Counter Battery instruments or silver particles

A set of CPCs based on both water (TSI 3785 and TSI 3786) and butanol (TSI 3772 and TSI 3776) were calibrated in the laboratory. The instruments were tuned so that their cutoff sizes matched for silver particles at 3 nm and at 5.5 nm for ultrafine and normal counters, respectively (Figure 2). Previously the CPCB results have shown, that the newly formed particles in the Boreal environment were more soluble to water.[6,12] In contrast, in an urban environment, the CPCB data indicates tentatively that the during a new particle formation the new particles were more soluble to butanol than water.[13]

Conclusions

Detection efficiencies of several water-based CPCs were verified in the laboratory using particles of different chemical composition. Water-soluble components decreased the cutoff size of the WCPCs. This property can be utilized using WCPCs in parallel with butanol based CPCs to study chemical composition of atmospheric aerosol particles.

Furthermore, investigations showed that CPC counting efficiencies are directly related to activation processes. Heterogeneous nucleation is the most important of those processes and can roughly explain experimental observations using a certain contact angle.

Acknowledgments Financial support from the Academy of Finland and 6th framework program project EUCAARI (No 036833–2) are acknowledged.

References

1. Von Klot et al., *Circulation*, **112**, 3073–3079 (2005).
2. Hering, S. and Stolzenburg, M.R., *Aerosol Sci. Technol.*, **39**, 428–436 (2005).
3. Hering et al., *Aerosol Sci. Technol.*, **39**, 659–672 (2005).
4. Biswas et al., *Aerosol Sci. Technol.*, **39**, 419–427 (2005).
5. Fluent, *FLUENT 6.2 User's Guide*, Lebanon, NY: Fluent (2005).
6. Fine Particle Model (FPM) for FLUENT, *Particle Dynamics GmbH*, Leipzig, Germany (www.particle-dynamics.de).
7. Kulmala, M. et al., *J. Aerosol Sci.*, **38**, 289–304 (2007).
8. Petäjä, T. et al., *Aerosol Sci. Technol.*, **40**, 1090–1097 (2006).
9. Mordas, G. et al., *Aerosol Sci. Technol.* (in preparation).
10. Liu, W. et al., *J. Air & Water Manage. Assoc.*, **56**, 444–455 (2006).
11. Stratmann, F. et al., Submitted to EAC 2007 (2007).
12. Manninen, H.E. et al. (this issue).
13. Birmili, W. et al., *7th International Aerosol Conference, Abstract Book, Vol 1*, Minnesota, St. Paul, 1428–1429 (2006).

Relationship of Aerosol Microphysical Properties and Chemical Composition of Aerosol in the Baltic Sea Region

Anna Pugatshova[1], Ülle Kikas[1], Margit Prüssel[1], Aivo Reinart[1], Eduard Tamm[1], and Vidmantas Ulevicius[2]

Abstract The chemical composition and modal structure of aerosol size distribution were studied at the coastal research station Preila, Lithuania (55° 33′ N, 21° 00′ E). The five-modal structure of size distribution was revealed, and the characteristics of the log-normal modes were assessed for different aerosol classes. The highest concentrations of SO_4 and NO_3 (1.49±0.78 μg/m^3 and 0.97± 0.54 μg/m^3, respectively) were found in the polluted continental aerosol carried from Eastern Europe and Russia. The highest concentrations of NH_4 (1.11± 0.75 μg/m^3) were found in the polluted marine aerosol, carried from the Central Atlantics or the North Sea over Western Europe and Southern Scandinavia.

Keywords Tropospheric aerosol, size distribution, chemical composition

Introduction

The microphysical and chemical properties of an aerosol strongly depend on aerosol prehistory, shaped by the geographic regions and emission sources it has passed over.

An important property of the atmospheric aerosol particle size distribution (SD) is its multimodal character. The particle SD can be characterized by means of three modes: nucleation (2–100 nm), accumulation (100–1,000 nm) and coarse (more than 1,000 nm) mode. Further investigations have shown that the nucleation mode, in it's turn, can be split into the pure nucleation (2–20 nm) and Aitken (20–100 nm) modes; the accumulation mode can be split into the condensation (100–600 nm) and droplet (600–1,000 nm) modes (Kulmala et al. 2004). The modes can be associated with different particle sources and stages of aerosol evolution in the

[1] *University of Tartu, Institute of Environmental Physics, Ülikooli 18, 50090 Tartu, Estonia, + 3727375555, + 3727375556*

[2] *Institute of Physics, Environmental Physics and Chemistry Laboratory, Savanoriu 231, LT-02300 Vilnius, Lithuania*

C.D. O'Dowd and P.E. Wagner (eds.), *Nucleation and Atmospheric Aerosols*, 711–715.

atmosphere. Variation of the modal structure of the aerosol by air mass type will give useful information on the prevailing processes in different aerosol types. The particle SD and the chemical composition of the aerosol strongly influence the aerosol-cloud interaction and the optical properties of aerosol. Better knowledge of those characteristics in each geographic region is required for the modeling of aerosol in global climate models.

Measurements and Methods

The research was based on the measurements performed in May to August 2006 at the Air Monitoring Station Preila, locating at the eastern coast of the Baltic Sea, Lithuania (55° 33′ N, 21° 00′ E).

The aerosol SD was measured in the diameter range from 10 to 10,000 nm, using the electric aerosol spectrometer (EAS) (Tammet et al. 2002). The mean SD for each 10 min were obtained as a set of 12 size fraction concentrations, whereas the fraction boundaries were uniformly distributed in a logarithmic scale of the particle diameter.

The modal structure of the particle number size distributions was revealed using an approach introduced by Pugatšova et al. 2007. It bases on the statistical Factor Analysis method. The Factor Analysis enables to group the SD fractions which show strongly correlated time variation. The identified groups were interpreted as the log-normally distributed modes of the particle size distribution. The parameters of the lognormal components were then specified by means of the iterative least squares method. The number concentration N_i, the geometric mean diameter σ_{gi} and the geometric standard deviation D_{gi} were calculated for each (ith) aerosol mode.

The inorganic ions (SO_4, NO_3, Na, Cl, and NH_4) were measured in the 24 h aerosol samples. The sample collection and the chemical analysis were performed according to the EMEP standard procedures.

The surface UV radiation spectra (for wavelength 300–400 nm) were also measured during the campaign in Preila, in order to investigate the impact of aerosol on UV radiative transfer. However, the radiative effect of aerosol is only briefly considered in this work.

Air Masses and Trajectories

Classification of air masses was done by means of the 96-h back-calculated air trajectories for the heights of 250, 750, 1250, 1750, and 2500 m. The trajectories were calculated by means of the HYSPLIT model and the archive of the FNL meteorological database http://www.arl.noaa.gov/ready/hysplit4.html. At first, the air masses were divided into the maritime (M) and continental (C) classes. As a rule, the M air masses had been formed over the Atlantic Ocean, and arrived to Preila over Western Europe, Scandinavia, the Baltic Sea or the Baltic States. The C air masses had been formed over Central or Eastern Europe and the western part of

Russia, and arrived to Preila over the continent. The second classification was made on the basis of the pollution sources the air mass trajectory overpasses. In this manner, the air masses were classified as clean (CL), moderate (MO) or polluted (P). The considered aerosol was divided into the following classes:

M_CL Maritime Clean
M_P Maritime Polluted
C_MO Continental Moderate
C_P Continental Polluted

Results and Discussion

The analysis of SD modes was carried out for 27 cloudless days, when the aerosol–cloud interaction was excluded and the surface UV irradiance was affected mainly by atmospheric aerosol and ozone. The modal structure of aerosol and the characteristics of the lognormal modes for all identified aerosol classes are presented in Table 1. The chemical composition of the respective classes is presented in Table 2.

Figure 1 demonstrates the modal structure for the M_CL aerosol. In this case, the five-modal structure definitely exists. We can clearly see the nucleation mode, split into the pure nucleation and Aitken modes; the large accumulation mode; and the coarse mode, split into two submodes. Compared to other classes, the M_CL aerosol showed the smallest particle number concentration in the accumulation mode and the highest number concentration in the nucleation and coarse modes (Table 1). Also, the highest content of Na and Cl ions was found in the M_CL aerosol. It enables to associate the coarse modes 1 and 2 with sea salt particles. The high share of sea salt particles in clean marine aerosol can be justified with the observed high wind speeds.

The content of SO_4 and NO_3 in aerosol increased with the increasing pollution level in marine and continental air, as was expected. The increase in pollution was accompanied with the diminishing of the daily UV radiation exposures.

The accumulation mode of both continental aerosol classes showed higher concentration and smaller mean diameter of particles, compared to the marine aerosol. Comparison of the ion concentrations and the mode' volume concentrations enabled to assess the share of other substances (organics, water) in different aerosol classes, provided that majority of the aerosol mass presented by SO_4 and NO_3 ions was included in the accumulation mode (Coe et al. 2006). Our investigation showed

Table 1 Inorganic ions in aerosol samples, and daily UV exposures in Preila. Mean concentrations and standard deviations for the air mass classes, and the 12 h UV exposures

	SO_4	NO_3	NH_4	Na	Cl	UV
Aerosol class	µg/m³	µg/m³	µg/m³	µg/m³	µg/m³	J/m²
M_CL	0.36 ± 0.10	0.46 ± 0.29	0.73 ± 0.12	0.60 ± 0.27	1.35 ± 0.66	4100.15
M_P	0.59 ± 0.57	0.61 ± 0.54	1.11 ± 0.75	0.44 ± 0.28	1.07 ± 0.69	3285.39
C_MO	0.45 ± 0.09	0.41 ± 0.09	0.96 ± 0.07	0.31 ± 0.20	0.71 ± 0.27	3726.49
C_P	1.49 ± 0.78	0.97 ± 0.54	0.78 ± 0.34	0.50 ± 0.18	1.02 ± 0.34	3033.59

Table 2 The aerosol modes and modal parameters for aerosol classes

	Marine Clean			Marine Polluted			C_MO			C_P		
Mode	N, cm^{-3}	σ_g	D_g, nm	N, cm^{-3}	σ_g	D_g, nm	N, cm^{-3}	σ_g	D_g, nm	N, cm^{-3}	σ_g	D_g, nm
Pure nucleation	1161	1.48	20	450	1.52	16	500	1.52	16	188	1.20	16
Aitken	3177	1.54	58	2974	1.38	53	1900	1.52	48	1937	1.52	48
Accumulation	60	2.05	250	723	1.71	235	2500	1.55	110	2865	1.55	104
Coarse1	1.00	1.61	1780	0.20	4.00	2400	0.50	1.65	1600	0.20	1.65	2400
Coarse2	0.20	4.00	3200	0.10	5.00	5600	0.20	4.00	2400	0.20	4.00	2400

N – Particle number concentration; σ_g – geometric standard deviation D_g – geometric mean diameter

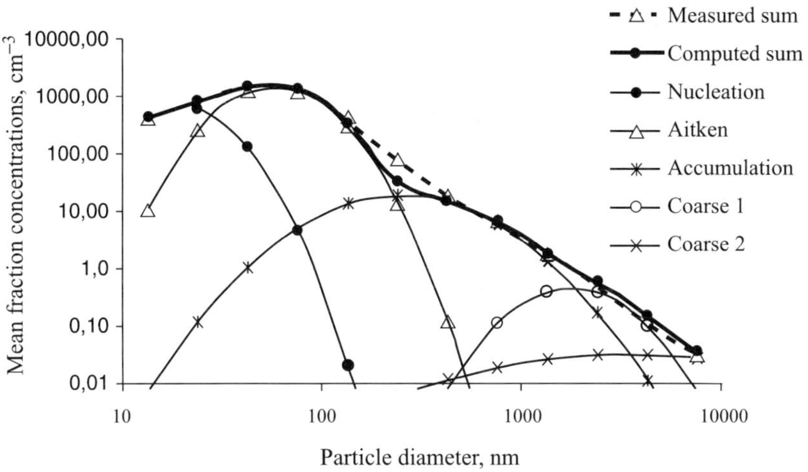

Figure 1 The model structure of the M_CL aerosol type

presence of 16–20% of other substances in the M_P and C_MO aerosol, and almost zero concentration of other substances in M_CL aerosol.

Conclusions

The measurement data were classified into four classes according to the geographic regions and pollution sources the air mass has passed over. The chemical composition, the modal structure of the particle number size distributions and the characteristics of the log-normal aerosol modes were obtained for the clean and polluted marine air, and for the moderately polluted and polluted continental air.

Acknowledgments This work was supported by the Estonian Science Foundation grant 5387 and the ACCENT grant FS8 (3rd call).

References

Coe, H., Allan, J.D., Alfarra, M.R., Bower, K.N. et al., Chemical and physical characteristics of aerosol particles at a remote coastal location, Mace Head, Ireland, during NAMBLEX, *Atmos. Chem. Phys.*, **6**, 3289–3301 (2006).

Kulmala, M., Vehkamäki, H., Petäjä, L., Dal Maso, T., M., Lauri, A., Kerminen, V.-M., Birmili, W., and McMurry, P.H., Formation and growth rates of ultrafine atmospheric particles: a review of observations, *J. Aerosol Sci.*, **35**, 143–176 (2004).

Pugatšova, A., Reinart, A., and Tamm, E., Features of the multimodal aerosol size distribution depending on the air mass origin in the Baltic region, *Atmos. Environ*, in press. doi:10.1016/j.atmosenv.2007.01.044 (2007).

Tammet, H., Mirme, A., and Tamm, E., Electrical aerosol spectrometer of Tartu University, *Atmos. Res.*, **62**, 315–324 (2002).

Hygroscopic Properties of Sub-micrometer Atmospheric Aerosol Particles Measured with H-TDMA Instruments in Various Environments – A Review

K. Hämeri[1], E.Swietlicki[2], H.-C. Hansson[3], A. Massling[4], T. Petäjä[1], P. Tunved[3], E. Weingartner[5], U. Baltensperger[5], P.H. McMurry[6], G. McFiggans[7], B. Svenningsson[8], A. Wiedensohler[4], and M. Kulmala[1]

Abstract A Hygroscopic Tandem Differential Mobility Analyser (H-TDMA) has been used in several field campaigns over the about last 25 years. The investigations were focused on the solubility and state of mixing of atmospheric aerosols. This paper gives a summary of the results of particle hygroscopic properties from different measurement sites. The data can be classified in hygroscopic growth modes indicating the origin and processes of the aerosol particles

Keywords Atmospheric aerosols, hygroscopicity, relative humidity, H-TDMA

Introduction

The hygroscopic properties of atmospheric sub-micrometer aerosol particles are vital for a proper description of how the particles interact with water vapour at sub- and super-saturated conditions, and are thus of major importance in describing the life cycle of the aerosol and the direct and indirect effects of aerosols on climate. The hygroscopic properties can be measured in great detail using Hygroscopic Tandem Differential Mobility Analyzers instruments (H-TDMA).

[1] *Division of Atmospheric Sciences, P.O. Box 64, FI-00014 University of Helsinki, Finland*

[2] *Division of Nuclear Physics, Lund University, P.O. Box 118, S-22100 Lund, Sweden*

[3] *Department of Applied Environmental Science, Stockholm University, S-10691 Stockholm, Sweden*

[4] *Department of Physics, Leibniz-Institute for Tropospheric Research, D-04318 Leipzig, Germany*

[5] *Laboratory of Atmospheric Chemistry, Paul Scherrer Institut, 5232, Villigen PSI, Switzerland*

[6] *University of Minnesota, Department of Mechanical Engineering, 111 Church St SE, Minneapolis, MN 55455, USA*

[7] *Atmospheric Sciences Group, SEAES, University of Manchester, P.O. Box 88, Manchester, M601QD, UK*

[8] *Department of Physical Geography and Ecosystems Analysis, Lund University, P.O. Box 118, S-221 00 Lund, Sweden*

C.D. O'Dowd and P.E. Wagner (eds.), *Nucleation and Atmospheric Aerosols*, 716–720.

The H-TDMA is one of the few instruments capable of providing online and in-situ information regarding the extent of external versus internal mixing of the atmospheric aerosol, since the H-TDMA determines the hygroscopic growth of individual aerosol particles. Although this state of mixing strictly refers only to the hygroscopic properties, it nevertheless implies the extent of chemical mixing of the aerosol.

The primary parameters measured with an H-TDMA as a function of dry particle diameter are: (i) the ratio between humidified and dry particle diameter at a well-defined relative humidity RH – often denoted hygroscopic growth factor; (ii) the number fraction of particles belonging to each of the observed and separable groups of hygroscopic growth; and often also (iii) the spread of diameter growth factors around the arithmetic mean value. Alternatively, the H-TDMA measurements can also be presented as distributions of hygroscopic growth factors for each given dry particle diameter.

While the bulk of H-TDMA data is available for a nominal and high RH (often between 80% and 90%), some field studies have performed scans in RH to explore the aerosol deliquescence and efflorescence behaviour as well as water uptake at low RH.

This work reviews and summarizes the existing H-TDMA data sets, with an emphasis on those published so far in peer-reviewed journals. The aim is to present the data in a way that will make it useful in evaluating models on various spatial and temporal scales incorporating a more detailed aerosol description than simply aerosol mass. To facilitate comparison between sites, growth factors are recalculated to an RH of 90% whenever possible, and classified according to the air mass properties and geographical location.

Experimental Data

Over the last 25 years, H-TDMA instruments have been used in several field measurements in various air masses around the globe. Summary of the measurement sites is presented in a global map (Figure 1).

The measurement sites include remote marine environments (Atlantic Ocean, Pacific Ocean, Indian Ocean, Arctic Ocean), background continental sites (Amazon rain forest, Nordic boreal forest, the Alps, North America), polluted continental sites (Italy, United Kingdom, Germany), and polluted urban sites (Germany, North America, Mexico, Asia). Air masses influenced by fresh as well as aged biomass burning have been studied in North and South America. Emission studies include fresh diesel vehicle and jet engine exhaust and flue gases from biomass combustion. Secondary organic aerosol has been investigated in a variety of smog chamber experiments. Atmospheric new particle formation events have been observed frequently in several locations over the globe. Hygroscopic properties have been measured during intensive campaigns in boreal forest, in coastal site and in an urban environment.

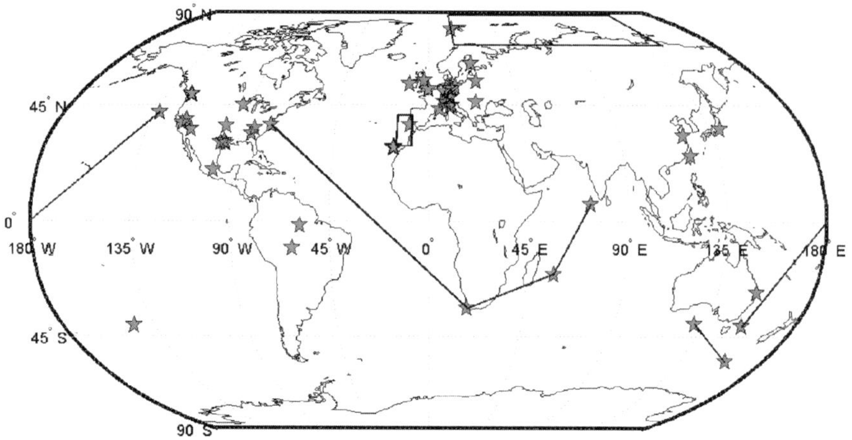

Figure 1 Global map showing locations of H-TDMA experiments

Typical Growth Factors for Different Aerosols

Hygroscopic properties of ambient aerosol vary strongly depending on the origin of the air masses and the location. In continental, polluted air masses the aerosol is often separated to two modes denoted as less and more hygroscopic particles. Clean marine aerosol shows typically hygroscopic growth modes for more hygroscopic particles and sea salt. In urban air, hydrophobic particles originated from combustion are observed mixed with transported background aerosol.

Hygroscopic properties of freshly formed particles depend also on the location and the origin of the air masses. Particles produced during nucleation events in boreal forest revealed diurnal behaviour of hygroscopic growth, with high growth factors at day time and lower during night. The coastal measurements were strongly affected by air mass history. Nucleated particles were hygroscopic in marine background air while particles produced during clean nucleation burst periods were hydrophobic.

Parameterizations of Growth Factor as Function of RH

The various H-TDMA data field data need to be parameterized in order to be incorporated into models describing the life cycle of the atmospheric aerosol. Several alternatives exist. We here propose a straightforward approach that can make full use of existing H-TDMA data sets and that is capable of accounting for hygroscopic growth at subsaturation as well as CCN activation at supersaturation.

For cases when RH scans do not display any obvious step-like deliquescent behaviour, the growth factor as function of RH can be fitted with a continuous one-parameter function (Rissler et al. 2006):

$$Gf = \left[1 + A \cdot \left(\frac{RH/100}{1 - RH/100}\right)\right]^{\frac{1}{3}} \tag{1}$$

Here, A is often the only fitted parameter needed to account for the observed growth as a function of RH (in %). The formula is based on classical Köhler theory and is derived and described in Kreidenweis et al. (2005), in a form originally using three fitted parameters. In Köhler theory, more strictly, water activity (a_w) should replace RH in Eq. 1. To further facilitate the use of H-TDMA data, A can be parameterized as a function of particle dry size, d_p, using relationships such as

$$A(d_p) = \log(d_p) \cdot B + d_p \cdot C + D \tag{2}$$

where B, C, and D are fitted parameters. The parameter A can be written as

$$A = \kappa \cdot M_w / \rho_w \tag{3}$$

where M_w and ρ_w are the molecular weight and density of pure water. The parameter κ represents the number of soluble moles of ions or molecules per unit volume dry particles, and can also be estimated as

$$\kappa = \varepsilon \cdot \upsilon \cdot \rho_s / M_s. \tag{4}$$

Here, ρ_s is the density of the model salt, v the number of ions per molecule of the model salt, and M_s the salt molecular weight. The soluble volume fraction ε – often using ammonium sulfate as reference salt; ε_{AS} – has been widely used to relate H-TDMA growth factor data to the chemical composition of the aerosol particles, and is calculated from the measured growth factor Gf as;

$$\varepsilon_{AS} = \frac{Gf^3 - 1}{Gf_{AS}^{\ 3} - 1}. \tag{5}$$

Here, Gf_{AS} is the growth factor of a fully soluble particle composed entirely of the same solute material (in this case ammonium sulphate) at the same humidified size as the observed particle.

RH-scans

Figure 2 shows an example of this parameterization for an Amazonian biomass burning aerosol (Risser et al 2006).

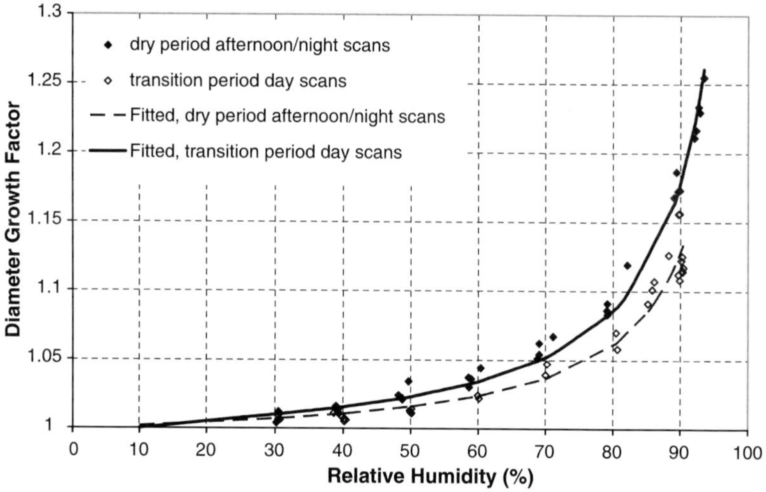

Figure 2 Example of growth factor parameterization *Gf(RH)* according to Equation 1 for a biomass burning aerosol in the Amazon. The dry particle size is 165 nm. Modified from Rissler et al. (2006)

Acknowledgments This work was carried out within the frameworks of the EU FP6 Integrated Project EUCAARI, the EU FP6 Infrastructure Project EUSAAR as well as the NOS-N Nordic Centre of Excellence BACCI.

References

Kreidenweis, S. M., Köhler, K., DeMott, P.J., Prenni, A.J., Carrico, C., and Ervens, B., *Atmospheric Chemistry and Physics Discussion*, **5**, 287–323 (2005).

Rissler, J., Vestin, A., Swietlicki, E., Fisch, G., Zhou, J., Artaxo, P., and Andreae, M.O., *Atmospheric Chemistry and Physics*, **6**, 471–491 (2006).

Contrasting Organic Aerosol Behaviour in Continental Polluted, Biomass Burning and Pristine Tropical Forest Environments

Hugh Coe[1], James Allan[1], Keith Bower[1], Gerard Capes[1], Jonathan Crosier[1], Jim Haywood[2], Simon Osborne[2], Andreas Minnikin[5], Jennifer Murphy[3,4], Andreas Petzold[5], C. Reeves[4], and Paul Williams[1]

Abstract In this paper we will present new measurements of organic aerosol mass taken during the African Monsoon Multivariate Analysis (AMMA) project which took place in West Africa during 2006. The organic mass loading of biomass burning aerosols will be shown to be highly correlated with CO over a wide range of distance scales from the source region to the background. This demonstrates that there is little secondary production of organic aerosol compared to the primary emission over the lifetime of the plume. However, we will show that chemical functionality of the organic mass changes with time, indicating that the aerosol themselves are processed. Aerosol measurements in pristine tropical environments show little evidence for enhanced aerosol loadings despite very large enhancements in possible precursors, placing upper limits on the secondary organic aerosol yield in these environments. This will be compared with global model predictions. This behaviour will be will be compared with previous measurements taken in continental polluted environments.

Keywords Organic aerosols, biomass burning, tropical aerosols, secondary formation

Introduction

Recent measurements have shown that organic aerosol mass loadings in anthropogenically influenced continental regions are significantly enhanced over those predicted by global or regional models (Heald et al. 2005; Johnson et al. 2006). The extent of this enhancement appears to increase with photochemical processing time in the atmosphere and leads to substantial secondary organic aerosol formation

[1] *University of Manchester, SEAES, Manchester, UK*

[2] *The Met Office, Fitzroy Road, Exeter, UK*

[3] *The University of East Anglia, Norwich, UK*

[4] *Now at: University of Toronto, Canada*

[5] *DLR, Oberpfaffenhofen, Germany*

C.D. O'Dowd and P.E. Wagner (eds.), *Nucleation and Atmospheric Aerosols*, 721–725.

in submicron aerosol in these environments (e.g., Volkamer et al. 2006). Global model predictions have extremely poor ability to predict secondary organic aerosol mass, largely because of a lack of a sound physical model of secondary organic aerosol formation as demonstrated by Tsigaridis and Kanakidou (2003). Such models are tuned to yields from measurements of model systems in large chamber studies, which have often been conducted at concentrations and under oxidizing conditions that are significantly different from those observed in the atmosphere.

Unlike continental environments, the formation and transformation of secondary organic aerosol in biomass burning and in pristine tropical forest environments have received rather less attention. Laboratory chamber experiments show that high aerosol yields are observed from monoterpene and isoprene precursor systems. Global models that have been tuned to these studies predict between 1 and 3 $\mu g\ m^{-3}$ of organic aerosol mass produced above tropical forested regions.

That said, studies such as those by Tsigaridis and Kanakidou (2003) have shown that by making different assumptions about the secondary organic aerosol production the model fields can be changed substantially, illustrating again that the physical bases on which the model predictions are made at present are not sound. There are also very few data of organic aerosol mass in pristine tropical regions with which to constrain global models. Whilst organic aerosols clearly dominate during biomass burning, their transformation in the atmosphere and the change in optical properties that ensues from that is a matter of uncertainty at the present time.

There is therefore a real need for in situ measurements of the aerosol organic mass across these environments to establish the magnitude of the observed mass concentration fields and to elucidate if the transformations are similar in nature and magnitude to those observed in mid-latitude polluted continental environments.

Measurement Methodology

Throughout 2006, a large field campaign took place in West Africa, forming part of the international AMMA project. Several ground-based sites and five aircrafts were involved in the project. This paper presents aerosol composition information findings from the aerosol particle measurements made on board the UK Facility for Airborne Atmospheric Measurements (FAAM), a BAe146 research aircraft. The BAe146 was operational at Niamey Airport, Niger during January and February 2006, and again during July and August 2006. Following each of these periods the aircraft was immediately stationed in Dakar, Senegal to investigate the outflow from West Africa. The operating region therefore covers a large area of West Africa, and some of the Atlantic Ocean off the coast of Senegal. Data was collected during both altitude profiles and straight and level runs. The aerosol instrumentation onboard included: a TSI 3025 UCPC measuring number concentrations of particle greater than 3 nm; a Passive Cavity Aerosol Sizing Probe (PCASP) for measurement of the ambient size of particles in the size range 0.3–3 µm); a three wavelength TSI nephelometer to measure the aerosol scattering coefficient; a single wavelength absorption photometer to determine the absorption coefficient; a DMT Cloud Droplet Probe to determine the ambient size of particles in the 3–30 μm diameter range; a GRIMM 1.109 Optical Particle Counter (OPC) to determine the dry sampled size of particles in the 0.3–3 μm size range; and an Aerodyne Aerosol Mass Spectrometer (AMS) to determine the online, size resolved mass loading of submicron, non refractory particulate components at one minute resolution. Limited filter collections were also made. During the January and February period the aircraft was fitted with a suite of large radiometers to investigate the influence of biomass burning on the radiative properties of the column, whilst during the summer period the aircraft was fitted with a wide suite of instruments for studying gas phase photochemistry. Most notably in this paper is the addition of a Proton Transfer Reaction Mass Spectrometer (PTRMS), which can sample isoprene at high time resolution.

Results

Dry Season Biomass Burning Aerosol

During January and February West Africa is in the late part of the dry season. At this time, winds from North East, advect dry air at low level across the Sahara into the Sahel region. This air is often heavily loaded with dust. To the south of Niamey, extensive biomass burning takes place throughout the entire region. This haze layer covers much of the southern Sahel and extends up to an altitude of 2–3 km. As the air moves northwards it lifts the haze above the dry surface layer as it advected westwards towards the Atlantic Ocean.

The BAe-146 sampled biomass burning throughout all these environments: to the south directly in the fires; in regions of burning but away from the main fire sources; to the north in the aged, elevated layers; and in biomass burning haze layers off the west coast of Africa over the Atlantic Ocean. This therefore covers a wide range of chemical ages of the aerosol particles in the atmosphere. Figure 1 shows the relationship between the submicron organic aerosol mass and carbon monoxide, a tracer for the biomass burning in the region for all the flights conducted. There is a clear, linear relationship between the organic mass and CO over all environments, demonstrating a high degree of correlation. This implies that the source ratio is to a first order constant over many days in the atmosphere. The emission ratios we calculate from these data compare well with those of Andreae and Merlet (2001), which were determined close to single fire sources.

This figure also serves to demonstrate a key point: there is no evidence of a substantial contribution from secondary organic aerosol during ageing in the atmosphere. However, what we also observe is that the chemical fingerprint of the organic fraction, as determined from the mass spectra derived from the AMS, changes with time in the atmosphere, from large alkyl fragments and mono-acid moieties close to the source, to more oxygenated multifunctional acids downwind. There appears to be some transformation of the aerosol with time in the atmosphere,

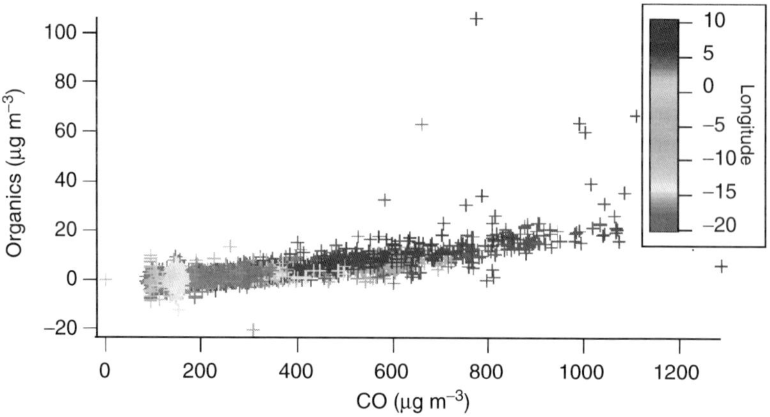

Figure 1 Scatter plot of organic aerosol mass and carbon monoxide, shaded according to longitude, indicating distance from the source region. The close correlation over all regimes demonstrates the lack of secondary organic aerosol production in these air masses

though no significant mass addition occurs. The impact of the aerosol on the optical properties of the layer will also be shown.

Wet Season Pristine Tropical Organic Aerosol

During the summer season, the African monsoon system extends to the north of the region. Large mesoscale convective systems developed on a timescale of 3–4 days, leading to substantial convection and rainfall. Between these periods the vegetation was well watered and rapidly growing. A number of low level flights were conducted to assess vegetation growth, water vapour fluxes and response to rainfall and also to understand the volatile organic carbon flux from the region. This provided an excellent opportunity to examine the organic aerosol formation from natural tropical forest systems. Isoprene concentrations were observed to be often above 1 ppb and peaked between 4 and 5 ppb. Their spatial pattern followed the underlying vegetation closely, showing that the surface emissions were coupled to the sampled air above and substantial volatile organic carbon emissions occurred. Whilst it is debatable whether isoprene is a precursor to secondary organic aerosol formation, it does provide a highly time resolved analogue of other volatile carbon emissions such as terpenes. Bottle samples taken on the ground and in the air show that monoterpenes were also observed at elevated concentrations. Despite these elevated loadings of potential precursors, our measurements show only weak enhancements in organic carbon aerosol mass in these regions, barely above instrumental detection limit. These measurements place limits on the secondary organic aerosol yield in these environments. A contrast with continental polluted environments will be made.

Acknowledgments We acknowledge the staff at FAAM, DirectFlight and Avalon Engineering. GC is in receipt of an NERC studentship. The work was supported by NERC and by the EU.

References

Andreae, M.O. and Merlet, P., *Global Biogeochem. Cycles*, **15**, 955–966 (2001).
Heald, C.L. et al., *Geophys. Res. Letts.*, **32**, L18809 (2005).
Johnson, D. et al., *Atmos. Chem. Phys.*, **6**, 403–418 (2006).
Tsigaridis, K. and Kanakidou, M., *Atmos. Chem. Phys.*, **3**, 1849–1869 (2003).
Volkamer et al., *Geophys. Res. Letts.*, **33**, L17811 (2006).

Airborne Measurements of Tropospheric Aerosol up to 12 km over West Africa during the Monsoon Season in August 2006

Andreas Minikin, Thomas Hamburger, Hans Schlager, Markus Fiebig, and Andreas Petzold

Abstract The German DLR Falcon 20 research aircraft was deployed during the monsoon season from August 1 until August 18, 2006, in Ouagadougou (Burkina Faso) with a scientific payload focusing on in situ measurements of trace gases and aerosol properties. In this contribution observations of aerosol properties made in the close vicinity of several active mesoscale convective systems (MCS) and smaller convective cells as well as in the aged outflow of MCSs are discussed. Particle nucleation is frequently observed in the "fresh" outflow of MCSs (in general occurring only above 8 km altitude) if the air processed by the convective system has aged at least for a few hours.

Keywords Tropospheric aerosols, aerosol formation, monsoon outflow, AMMA

Introduction

One of the many objectives of the international project African Monsoon Multidisciplinary Analyses (AMMA) is to investigate the role of the monsoon related atmospheric circulation, in particular deep convection, in the transport and processing of chemical and particulate atmospheric constituents and to understand, vice versa, the role of chemistry and aerosols in the dynamics of the West African Monsoon (WAM).

The German DLR Falcon 20 (D-F20) was deployed during the monsoon season from August 1 until August 18, 2006, in Ouagadougou (Burkina Faso) with a scientific payload focusing on in situ measurements of trace gases and aerosol properties. Measurements were taken during eight local flights from Ouagadougou as well as during the ferry flights from Germany to Burkina Faso and back. A number of different scientific mission objectives were addressed with these

Deutsches Zentrum für Luft- und Raumfahrt (DLR) Institut für Physik der Atmosphäre Oberpfaffenhofen, 82234 Wessling, Germany

C.D. O'Dowd and P.E. Wagner (eds.), *Nucleation and Atmospheric Aerosols*, 726–730.

flights: observations were made in the close vicinity of several active mesoscale convective systems (MCS) and smaller convective cells as well as in the aged outflow of MCSs, mainly in the Sahel region. Furthermore, measurements were taken during "background" surveys in the middle/upper troposphere (8–12 km altitude, with 12 km being the maximum flight altitude of the D-F20). The aerosol instrumentation of the D-F20 consisted of a combination of a multichannel condensation particle counter system, an optical particle counter, a differential mobility analyzer and a 3-wavelength particle absorption photometer, all connected to the near-isokinetic aerosol inlet of the D-F20 as well as two optical aerosol spectrometer probes mounted under the wings. Some of the cabin instruments were operated behind a thermodenuder heated to 250°C. With this payload the complete size distribution of the aerosol particles, both for the total of all and the nonvolatile particles, can be derived.

Results and Discussion

During ascends and descends of the D-F20 a haze layer in the lower troposphere (typically at 1–5 km altitude) was observed, which proved to be persistent over the measurement period and appears to be associated with the low level African Easterly Jet, probably carrying Sahelian dust. In Figure 1, illustrating the vertical aerosol distribution as found during all the AMMA flights, the haze layer can be clearly identified in the accumulation mode size range.

The analysis of the measurements taken in the vicinity of convective systems indicates that aerosol in all size modes gets efficiently removed in the very fresh outflow of a MCS. This explains the occurrence of some very low particle number concentrations in altitudes above 8 km, in particular in the Aitken mode size range, but also for the nonvolatile and the accumulation mode particles. Particle nucleation is frequently observed in the outflow if the air processed by the convective system has aged at least for a few hours, while the aerosol surface area density is still very low. This can be illustrated for example by the case of August 11, 2006, where the Falcon did fly downstream of a fairly strong MCS system over Burkina Faso (Figure 2). The time series of aerosol number concentrations measured during a part of this flight (Figure 3) leading eastward and downstream of the MCS represents increasing age of the outflow with respect to the processing of air by the convective system (sections 1 to 4 marked in Figure 3). While in the very fresh outflow aerosol concentrations are low in all size ranges, new particle formation sets in after a while, and even later concentration of particles in the Aitken mode range have recovered, indicating further growth of the particles from condensable gases. This occurs on a time scale of hours of outflow age.

In cases of "old" MCS outflow (not shown) and in upper tropospheric "background" the observed number concentrations in particular of nonvolatile particles are higher than in fresh MCS outflow. This indicates that convective processes not

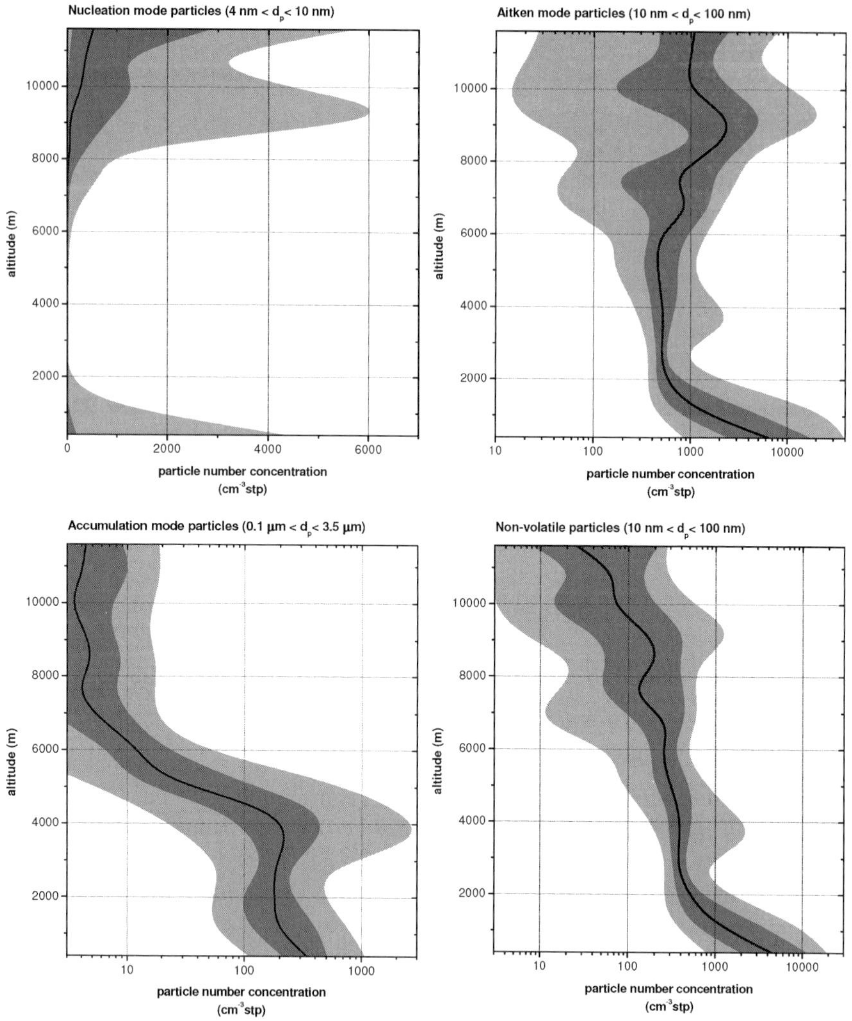

Figure 1 Vertical profiles of nucleation mode (**upper left**), Aitken mode (**upper right**), accumulation mode (**lower left**) and nonvolatile (**lower right**) particle number concentrations and variability during AMMA from all Falcon measurements in August 2006 south of 20° N (cloud-free data only). 5- and 95-percentiles are represented by the shaded area in light gray, the 25- and 75-percentiles in dark gray, the median is the black line. Number concentrations are given for standard conditions (stp, 273.15 K, 1013 hPa)

causing a strong wet removal of aerosol particles must be relevant for the aerosol budget in this altitude range.

A biomass burning plume encountered below 6 km during two flights to the south of Ghana proved to be strongly absorbing. The aerosol size distribution

Figure 2 Meteosat (MSG) image, IR10.8 channel, of 16:30 UT at August 11, 2006 with the DLR Falcon flight track overlayed. Satellite Image provided by the AMMA AOC web page (http://aoc.amma-international.org/). In the first part of the flight, the DLR Falcon moved eastward (indicated by **arrow**) inside the outflow of a few hours old MCS system (the left of the flight track in this image). With increasing distance from the MCS center, the outflow air is of increasing age since being processed by the MCS

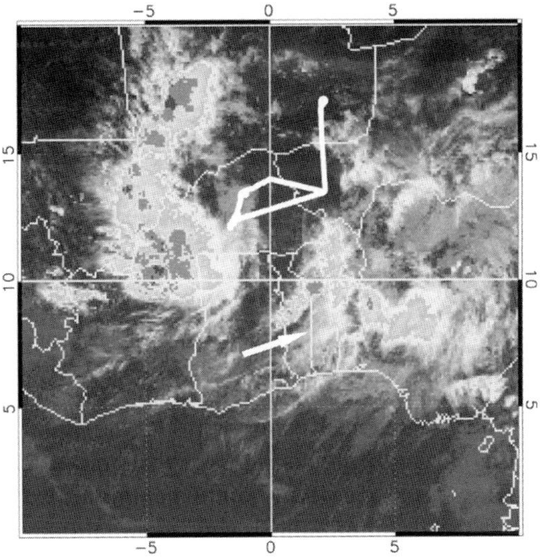

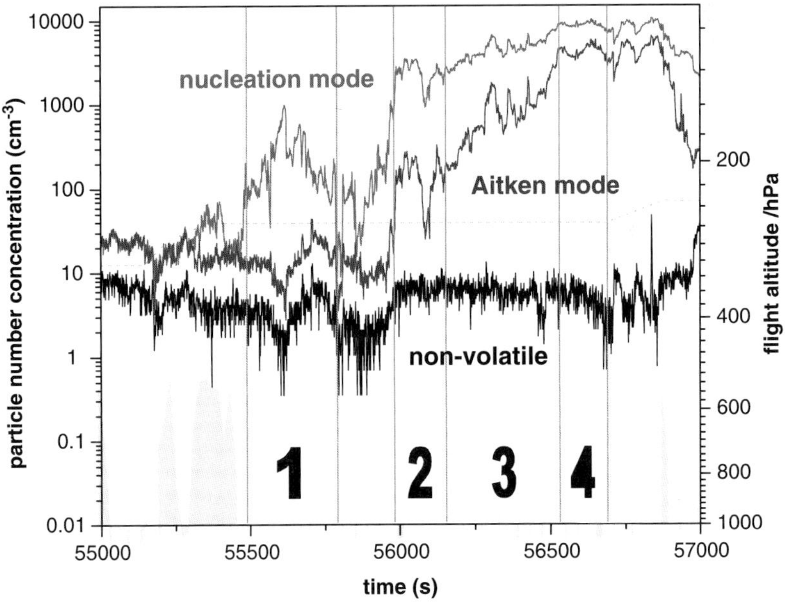

Figure 3 Time series of number concentrations of nucleation mode (**top curve**), Aitken mode (**middle curve**) and nonvolatile (**lower curve**) particle number concentrations measured downstream in the outflow of an MCS during the Falcon flight on 11 August 2006. Four sections of the flight are highlighted corresponding to an increasing age of the measured air mass with respect to being processed in the MCS a couple of hours before

shows a very strong accumulation mode, which gives indication that the plume is considerably aged. Particles in the nucleation mode were not observed in this plume. In general, particle nucleation, if occurring at all, was confined to the upper troposphere and boundary layer (Figure 1).

Acknowledgments Part of this work was supported by the European Union within the AMMA-EU project. We gratefully acknowledge the excellent support during the aircraft campaign by Hans Rüba and the DLR flight department operating the Falcon.

Closure Between Chemical Composition and Hygroscopic Growth of Aerosol Particles During TORCH2

M. Gysel[1,2], J. Crosier[1], D.O. Topping[1], J.D. Whitehead[1], K.N. Bower[1], M.J. Cubison[1], P.I. Williams[1], M.J. Flynn[1], G.B. McFiggans[1], and H. Coe[1]

Abstract Size-resolved chemical composition and hygroscopic growth factors (HGF) of aerosol particles in aged polluted and clean marine air masses were measured at the North Norfolk coastline (UK) using an Aerodyne quadrupole mass spectrometer (Q-AMS) and a hygroscopicity tandem differential mobility analyser (HTDMA), respectively. Particles were found to be sulphate dominated with accordingly high HGFs in the clean marine air masses. Sulphate, nitrate and organics were the dominating compounds in the aged polluted air masses, whereas the respective mass fractions varied considerably with particle size and time. The Zdanovskii-Stokes-Robinson (ZSR) mixing rule was used to predict HGFs using the chemical composition data from the Q-AMS and the HGFs of the components in pure form. The HGFs of the inorganic salts were taken from thermodynamic models, while assuming a bulk HGF of ~1.2 at 90% relative humidity (RH) for the organics resulted in best results for the closure between measured and predicted HGFs. Good closure was achieved in nitrate free periods, whereas systematically smaller measurements were observed when nitrate was a major constituent. An evaporation artefact of ammonium nitrate in the HTDMA instrument, favoured by a long residence time in the instrument, is most likely the reason for this discrepancy. The ZSR mixing rule was found to be an adequate tool for HGF predictions in aged air masses. A critical point for successful closure is matching the variability of composition with particle size and time, and the inorganic speciation was shown to be more important than exact HGF of the organic fraction.

Keywords hygroscopicity, HTDMA, AMS, online measurements

[1] *Atmospheric Sciences Group, SEAES, Univ. of Manchester, P.O. Box 88, Manchester, M60 1QD, UK*

[2] *Now at Laboratory of Atmospheric Chemistry, Paul Scherrer Institut, 5232 Villigen PSI, Switzerland*

C.D. O'Dowd and P.E. Wagner (eds.), *Nucleation and Atmospheric Aerosols*, 731–735.

Introduction

Atmospheric aerosols have many effects on the environment, such as visibility degradation, direct aerosol effect on the earth's climate or aerosol/cloud interactions. The magnitude of these effects is also influenced by the hygroscopic properties of the aerosol particles.

In this study[1] chemical composition and HGFs of atmospheric particles were measured at the North Norfolk coastline during the Tropospheric ORganic CHemistry project (TORCH2). These data are used for a "hygroscopic closure", i.e., prediction of the HGFs of ambient particles based on their chemical composition.

Experimental and Theory

HGFs of particles with different dry diameters (D_0) have been measured using a Hygroscopicity Tandem Differential Mobility Analyser (HTDMA[2]). An Aerodyne aerosol mass spectrometer (Q-AMS[3]) was employed to measure highly time and size-resolved chemical composition of the non-refractory compounds in PM_1 (NR-PM_1), which includes among others the inorganic ions SO_4, NO_3, NH_4 as well as total organics.

A simplified approached was used to predict HGFs of the mixed ambient particles based on their composition as measured by the Q-AMS. The Zdanovskii-Stokes-Robinson mixing rule[4] (ZSR) allows to calculate the HGF of a mixed particle, g_{mixed}, from the volume fractions ε_i of every compound and their respective HGFs g_i in pure form:

$$g_{mixed} \approx \left(\sum_i \varepsilon_i g_i^{\,3} \right)^{1/3} \tag{1}$$

HGFs of pure inorganic salts were taken from a thermodynamic model,[5] while a bulk HGF of 1.2 was assumed for the organic fraction. Volume fractions ε_i of all compounds in the mixed particles were calculated from the mass fractions delivered by the AMS assuming a density of ~1,400 kg/m^3 for the organics.

Results and Discussion

Two distinct air mass types were encountered during the TORCH2 experiment. From 11–13 and 21–23 May clean marine air originating from the Norwegian Sea or the Northeastern Atlantic Ocean without land contact for several days and from 14–20 May aged polluted air originating from the North-eastern Atlantic Ocean and transported across Ireland and North England or the English Midlands was encountered. In the clean marine air the aerosol composition was fairly stable and dominated by ~ 80% sulphate, with only about 20% organics (see Figure 1).

A distinct diurnal pattern with a considerable ammonium nitrate contribution during the night and a relatively higher ammonium sulphate contribution during the day was observed for the aged polluted air mass. This is attributed to the diurnal cycle of back trajectories caused by local wind direction patterns during this period. In the aged polluted air mass organic mass fractions were in the order of 20–40%, 40–60%, and 50–90% for particle diameters of 301, 141, and 66 nm, respectively. Q-AMS data do not allow for a detailed speciation of the organic compounds, but the fragmentation pattern indicated rather oxidised than hydrocarbon like organics as typically found in aged air masses.

A time series of mean HGFs for three different dry diameters is shown in Figure 2 along with two different AMS/ZSR prediction approaches. Observed HGFs were high and fairly constant in the clean marine air mass, as to be expected for a sulphate dominated aerosol. More variability, a clear size dependence and generally smaller

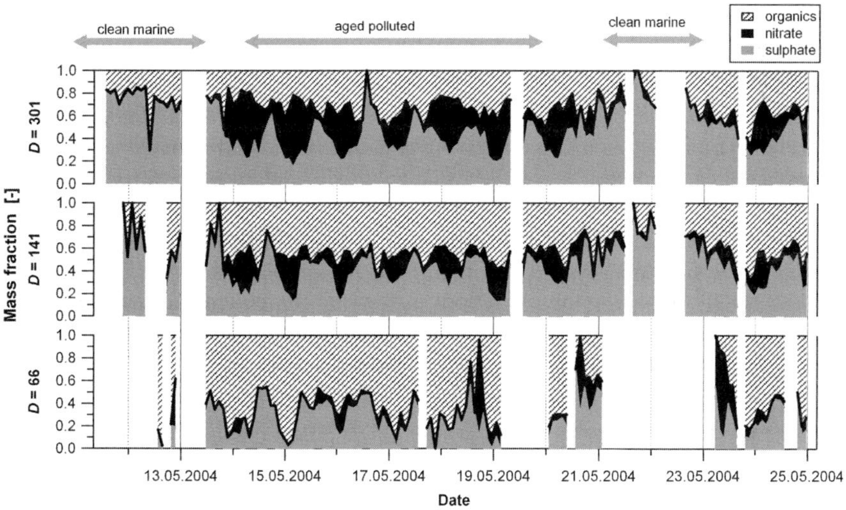

Figure 1 Size-resolved chemical composition as obtained by the Q-AMS

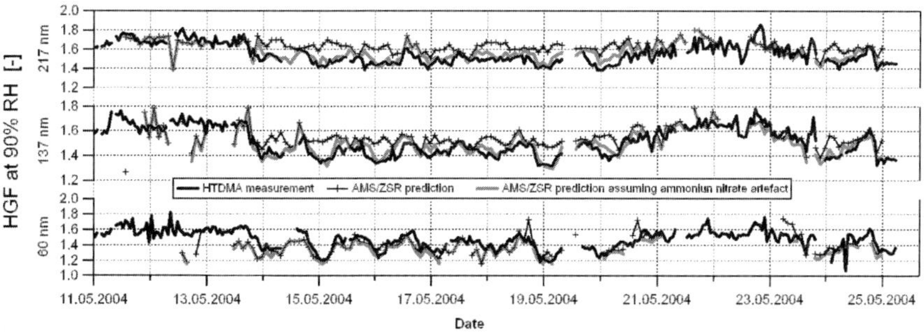

Figure 2 Hygroscopic growth factors from HTDMA measurements and AMS/ZSR predictions

HGFs were seen in the aged polluted air mass, which shows that high resolution in both time and size is often critical for performing a valid hygroscopic closure. In the clean marine events hygroscopic closure is successfully achieved with the basic AMS/ZSR prediction. Successful closure at all times is achieved for the smallest dry size (D = 60 nm), where the organic mass fraction is highest. This shows that the HGF of the organics in aged pollution particles must be of the order of ~1.2 at 90% RH, as assumed for the ZSR predictions. Nevertheless, the inorganic fraction dominates the hygroscopic water uptake at high RH with a fractional contribution of more than 80% at D > 137 nm. Exact inorganic speciation is also more important than the exact organic HGF for precise AMS/ZSR predictions, because of the weighing of the pure compound HGFs by the power of three (Eq. 1).

In the aged polluted event measured HGFs are systematically smaller than predicted, whenever the nitrate mass fraction is high. Similar discrepancies between HTDMA measurements and AMS/ZSR predictions in the presence of nitrate have been reported in a recent study.[6] They have speculated that either the ZSR mixing rule does not hold for mixtures of ammonium nitrate with organics or that some nitrate measured may have been of organic nitrate instead of inorganic nitrate. However, in our study the amount of ammonium measured as well as the ratio of fragments m/z 30 to m/z 46 in the mass spectrum suggest that the nitrate measured was indeed from ammonium nitrate. Such strong deviations of the ZSR mixing rule as required to explain the discrepancies found in our hygroscopic closure are also not to be expected.[7] The Q-AMS does not measure black carbon, but we can indirectly exclude black carbon as the major cause of the discrepancies. After careful consideration of all possibilities we conclude that most likely an ammonium nitrate evaporation artefact occurred in the HTDMA instrument, resulting in measured HGFs being smaller than predicted. The line "AMS/ZSR prediction assuming ammonium nitrate artefact" in Figure 2 shows that good closure at all times an for all sizes is achieved when assuming the ammonium nitrate artefact. This artefact may have been favoured by the relatively long residence time of ~60 s in the HTDMA, which was chosen to assure full equilibration of the HGF at high RH.

Conclusions

The ZSR mixing rule has been shown to be a suitable approach to predict HGFs of aged ambient pollution particles from chemical composition data obtained by a Q-AMS, when there is a mixture of inorganic and organic material. This approach allows for high resolution in both time and size, which is often critical for a valid hygroscopicity closure. The closure results indicate a mean HGF of ~1.2 at 90% RH for the organic compounds. Nevertheless the overall water uptake is clearly dominated by the inorganic salts. Therefore HGF predictions are also more sensitive to the inorganic speciation than to knowing the HGF of the organic compounds precisely. Systematically smaller HGFs than predicted were found in the presence of nitrate, which can most likely be attributed to an evaporation artefact of ammonium nitrate in the HTDMA.

Acknowledgments This work was supported by the British Natural Environment Research Council through grant number NER/T/S/2002/00494 and by the Swiss National Science Foundation.

References

1. Gysel, M., Crosier, J. et al., *Atmos. Chem. Phys. Discuss.*, **6**, 12503–12548 (2006).
2. Cubison, M., Coe, H., and Gysel, M., *J. Aerosol Sci.*, **36**, 846–865 (2005).
3. Jayne, J.T., Leard, D.C. et al., *Aerosol Sci. Technol.*, **33**, 49–70 (2000).
4. Zdanovskii, A.B., *Zhur. Fiz. Khim.*, **22**, 1475–1485 (1948).
5. Topping, D.O., McFiggans, G.B., and Coe, H., *Atmos. Chem. Phys.*, **5**, 1205–1222 (2005).
6. Aklilu, Y., Mozurkewich, M. et al., *Atmos. Env.*, **40**, 2650–2661 (2006).
7. Marcolli, C. and Krieger, U.K., *J. Phys. Chem.*, **A110**, 1881–1893 (2006).

Heterogeneous Oxidation of Saturated Organic Particles by OH

Ingrid George, Alexander Vlasenko, Jay Slowik, and Jonathan Abbatt

Abstract The kinetics and reaction mechanism for the heterogeneous oxidation of saturated organic aerosol by gas-phase OH radicals were investigated. The reaction of dioctyl sebacate (DOS) with OH was studied as a proxy for chemical aging of atmospheric aerosols containing saturated organic matter. We studied the influence of this reaction on the physical and chemical properties of DOS particles using online particle analysis techniques. This reaction system was studied with an aerosol flow tube coupled to an Aerodyne time-of-flight aerosol mass spectrometer (ToF-AMS) and a scanning mobility particle sizer. The heterogeneous kinetics of DOS oxidation by OH were studied by monitoring the loss of DOS with OH exposure with the ToF-AMS. The measured initial reactive uptake coefficient for this reaction is γ_0 = 1.26 (±0.04) indicating that the reaction is very efficient. The density of DOS particles increased with OH exposure by up to 20%, suggesting that the particles are becoming more oxidized. Condensed phase products were identified by electrospray mass spectrometry and consisted of multifunctional carbonyls and alcohols with molecular weights higher than DOS. While we also observed evidence of volatilization of oxidized organics from reacted DOS particles, this appears to not be the major reaction pathway.

Keywords Chemical aging, hydroxyl radicals, organic aerosol

Introduction

Chemical aging of atmospheric organic aerosols by heterogeneous reaction with gas-phase oxidants, i.e., O_3, and OH, can alter the physico-chemical properties of atmospheric aerosols, and thus, influence the role of aerosols on climate. Despite

Department of Chemistry, University of Toronto, 80 St. George Street, Toronto, ON M5S 3H6

C.D. O'Dowd and P.E. Wagner (eds.), *Nucleation and Atmospheric Aerosols*, 736–740.

the potential importance of this process on climate, our current understanding of chemical aging of atmospheric organic aerosols is limited. Most research in this area has focused on the reaction of O_3 with unsaturated organics as a proxy for chemical aging,[1] yet OH may be an important oxidant of saturated atmospheric organic matter. In a recent study, Molina et al.[2] observed complete volatilization of a saturated organic monolayer surface when exposed to OH. These results suggest that heterogeneous oxidation of organic aerosol by OH may lead to the release of oxygenated organics into the troposphere, and that this process may be a significant atmospheric sink for saturated organic aerosol. Thus far, the release of organics has not been yet been observed from OH reaction with saturated liquid surfaces or organic aerosols. Therefore, the reaction mechanism for OH oxidation of saturated organic aerosols is still not clear, especially for liquid particles. Furthermore, the effect of heterogeneous oxidation of saturated organic aerosol by OH on particle properties, such as size, density, and chemical composition, has not been fully explored.

In this study, we sought to address these gaps in knowledge by investigating the heterogeneous reaction of liquid saturated organic aerosol with OH using online particle analysis techniques. Dioctyl sebacate (DOS) particles were used as a proxy for tropospheric aerosol containing saturated organic matter. This research has three major goals: (a) determine the kinetics of the heterogeneous reaction of DOS particles with OH, (b) study the effect of this reaction on the physical properties of DOS particles, and (c) characterize particle-phase products and elucidate the reaction mechanism.

Experimental

The aerosol flow tube setup is shown in Figure 1. DOS particles were produced by homogeneous nucleation. DOS particles with diameters of 150 nm were selected, and the particles were mixed with flows containing ozone and humidified N_2 in the mixing flow tube. Hydroxyl radicals were produced in the reactor flow tube from the photolysis of ozone with 254 nm light from an Hg lamp in the presence of water vapour. OH concentrations were calculated using a photochemical model. The modeled OH concentrations were verified experimentally with a chemical ionization mass spectrometer, which measured changes in the concentrations of SO_2 from its reaction with OH. DOS particles were exposed to OH in the reactor flow tube and were analysed with a scanning mobility particle sizer (SMPS) for size distributions and an Aerodyne time-of-flight aerosol mass spectrometer (ToF-AMS) for particle chemical composition. The ToF-AMS was used to monitor the loss of condensed-phase DOS as a function of OH exposure to calculate the reactive uptake coefficient. Condensed-phase products were identified by analysis of reacted DOS particles with the ToF-AMS and electrospray ionization (ESI) mass spectrometry.

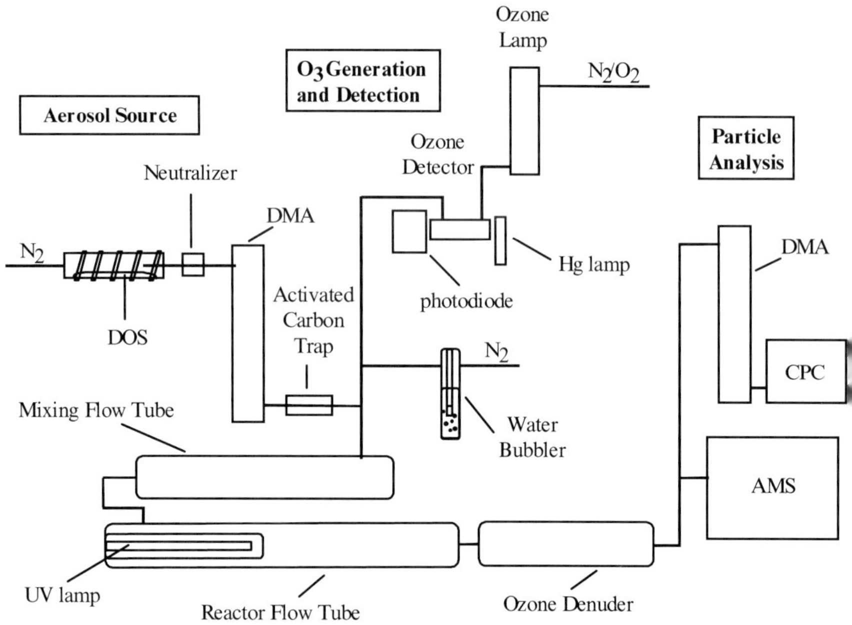

Figure 1 Reaction flow tube setup for the study of heterogeneous oxidation of organic aerosols

Results and Discussion

Kinetics and Particle Properties

The kinetic loss of DOS particles with OH exposure is shown in Figure 2(a). Due to the absence of the DOS molecular peak in the ToF-AMS mass spectrum, the loss of DOS was monitored by a characteristic mass fragment at m/z 297. We calculated the initial reactive uptake coefficient (γ_0) for OH with DOS particles by taking the initial slope as a pseudo first-order rate constant. We calculated the initial reactive uptake coefficient to be $\gamma_0 = 1.26(\pm0.04)$, which is in the range of calculated values in other studies.[2,3,4,5]

The SMPS and ToF-AMS were run simultaneously to gain information on changes in particle volume and density from reaction with OH. Figure 2(b) displays the changes in particle volume and density, where values for reacted particles are compared to those of unreacted particles. Particles were assumed to be spherical after reaction. Particle volume was calculated from the SMPS-measured mobility diameter and particle density (ρ_p) was calculated for spherical particles using Eq. (1)[6]:

$$\rho_p = \frac{D_{va}}{D_m}\rho_0 \tag{1}$$

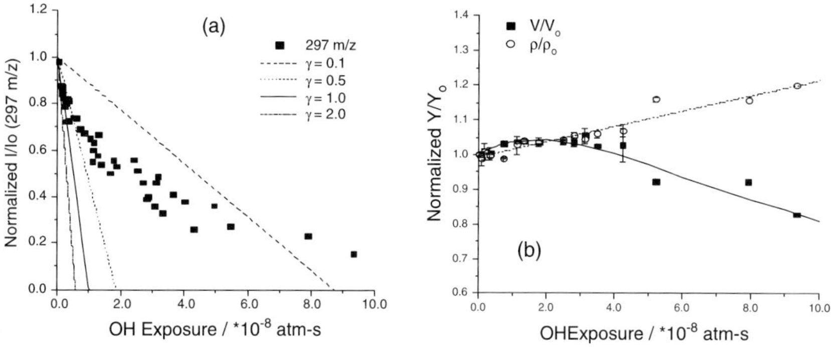

Figure 2 (a) Kinetic loss of m/z 297 as a function of OH exposure. Lines represent calculated slopes for different γ_0 values. (b) Particle density (**open circles**) and volume (**filled squares**) relative to initial values as a function of OH exposure. Lines are fits to guide the eye

Here D_{va} is the vacuum aerodynamic diameter measured by the ToF-AMS, D_m is the mobility diameter, and ρ_0 is a standard density ($\rho_0 = 1.0\,g\ cm^{-3}$). Figure 2(b) shows that the mean particle density for reacted DOS increases linearly with OH exposure by up to 20%, suggesting that the DOS particles are becoming more oxidized. In contrast, the mean particle volume increased slightly at low OH exposure and decreased by up to 17% at high OH exposures. These results suggest that short-chained oxidized organics volatilize from the particles, but that significant evaporation losses occur only at OH exposures above 2.0×10^{-8} atm-s.

Reaction Products

Online and offline mass spectrometry techniques were applied to identify condensed-phase reaction products from OH oxidation of DOS particles. The primary mass fragments of condensed-phase reaction products were observed in the ToF-AMS mass spectra, including m/z 127, 153, and 181. The presence of secondary product fragments such as m/z 44 were also observed suggesting that carboxylic acids are forming as secondary oxidation products.

Figure 3 shows the ESI mass spectra of unreacted DOS particles (Figure 3(a)) and reacted DOS particles (Figure 3(b) and 3(c)). The mass spectrum of reacted DOS clearly shows the presence of masses at higher molecular weights than DOS. These masses are highlighted in Figure 3(c), clearly showing a pattern of product peaks. Closer inspection of the exact masses from the ESI spectrum reveals that these masses correspond to addition of carbonyl and alcohol groups.

Acknowledgments The authors would like to thank Keith Broekhuizen for experimental assistance. This research was funded by NSERC and CFCAS.

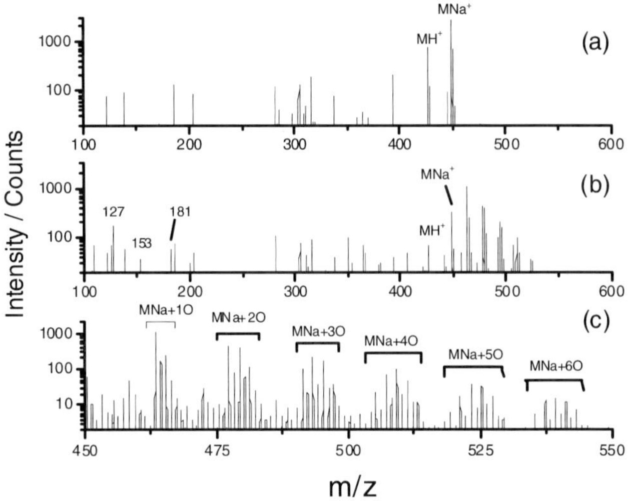

Figure 3 ESI mass spectra: (a) Unreacted DOS particles (b) Reacted DOS particles (c) Same a (b) with mass range 450–550 m/z magnified. MH^+ and MNa^+ denote the protonated and sodiate parent ion, respectively

References

1. Rudich, Y., *Chem. Rev.*, **103**, 5097–5124 (2003).
2. Molina, M.J., Ivanov, A.V., Trakhtenberg, S., and Molina, L.T., *Geophys. Res. Lett.*, **3** L22104 (2004).
3. Cooper, P.L. and Abbatt J.P.D., *J. Phys. Chem.*, **100**, 2249–2254 (1996).
4. Bertram, A.K., Ivanov, A.V., Hunter, M., Molina, L.T., and Molina, M.J., *J. Phys. Chem.* **105**, 9415–9421 (2001).
5. Hearn, J.D. and Smith G.D., *Geophys. Res. Lett.*, **33**, L17805 (2006).
6. DeCarlo, P., Slowik, J.G., Worsnop, D.R., Davidovits, P., and Jimenez, J.L., *Aerosol Sc Technol.*, **38**, 1185–1205 (2004).

Tropospheric Bioaerosols of Southwestern Siberia: Their Concentrations and Variability, Distributions and Long-term Dynamics

A.S. Safatov[1], I.S. Andreeva[1], G.A. Buryak[1], V.V. Marchenko[1], S.E. Ol'kin[1], I.K. Reznikova[1], V.E. Repin[1], A.N. Sergeev[1], B.D. Belan[2], and M.V. Panchenko[2]

Abstract The paper is devoted to the results of the eight-year study of the biogenic components of tropospheric aerosol in Southwestern Siberia at the altitudes up to 7,000 m. The most important among them are: the total protein as an indicator of all substances of biological origin and culturable microorganisms as a component, which is most harmful to man. The eight-year dynamics of mean annual values of the above concentrations, their annual variations and altitude profiles are presented. Possible sources of the observed bioaerosols are discussed.

Keywords Bioaerosols, tropospheric aerosols, long-term trends

Introduction

The knowledge of characteristics of the biogenic components of atmospheric aerosol is required for solving the problems of ecology, meteorology, climatology and applied microbiology. Biogenic components traced by two main components of bioaerosols: the total protein and culturable microorganisms at the attitudes up to 7,000 m [1–3]. The work presents the results of the eight-year study of atmospheric bioaerosols in Southwestern Siberia.

Materials and Methods

Sampling was performed at 8 altitudes from 7,000 to 500 m in the daytime over the forest at the distance of 50 km to the south from Novosibirsk with "Optic-E" laboratory mounted on an Antonov-30 airplane during the last 10 days of each

[1] FSRI SRC VB Vector, 630559, Koltsovo, Novosibirsk region, Russia

[2] IOC SB RAS, Akademichesky Avenue, 1, 634055, Tomsk, Russia

C.D. O'Dowd and P.E. Wagner (eds.), *Nucleation and Atmospheric Aerosols*, 741–745.

month. Air samples were collected on filters of AFA-HA type at a flow rate of approximately 250 L/min and impingers at a flow rate of 50 L/min. On-land samples were collected on a day in the middle of the month to reveal daily variations of the measured values.

The total protein concentration in samples on filters was analyzed in a laboratory using a fluorescent dye described in [4]; the method sensitivity was 0.1 μg/ml of the sample, the determination error did not exceed 20%.

The following nutrient media were used to detect culturable microorganisms on Petri dishes: LB [5] and depleted LB (medium diluted 1:10 to prevent the influence of neighboring colonies on microorganism growth) were used to detect saprophyte bacteria; starch-ammoniac medium [6] was used to detect actinomycetes; soil agar and Sabourau medium [6] were used for low fungi and yeast. The seedings were incubated in thermostats at 30°C for 3–14 days. Morphological characteristics of the grown microorganisms were studied visually and with light microscopy. For this purpose, Gram-stained fixed and live preparations of cell suspensions were prepared and observed with the phase contrast method. Taxonomic groups the microorganisms referred to were determined up to the genus [6–8]. Microorganism concentrations in the sample were calculated with standard methods [9] by averaging 2–3 parallel seedings on 4–5 different media. The number and representation of microorganisms varied in different samples. More detailed analysis of detected bacteria was performed with biochemical methods and genotyping.

Results and Discussion

Altitude profiles and annual dynamics of variations of the total protein and culturable microorganisms concentrations at the altitudes up to 7 km were revealed. Figure presents the data on these concentrations standardized for each flight (to compensate for their time dynamics). The total protein and culturable microorganisms concentrations nearly do not decrease with the altitude, whereas the concentration of aerosol with the diameter of more than 0.4 μm decreases at the altitude of 7 km by more than an order of magnitude as compared with that in the near-ground layer [10]. Such profiles of atmospheric pollutants can be formed by very powerful remote sources such as large woodlands, water surfaces and soil. Aerosols from such sources rising to considerable altitudes are mixed up and transported for long distances, creating the observed profiles as the particles deposit.

Weak dependence of the observed concentrations on the altitude allows us to sum up data for all the altitudes for constructing annual dynamics of the total protein and culturable microorganisms concentrations [11]. An expressed annual variation of the total protein and culturable microorganisms concentrations in the atmosphere of Southwestern Siberia was revealed. Mean annual concentrations of these values in the region at the altitude of 500–7000 m were also determined Figure 2. It presents the observed tendency to a decrease in bioaerosol mass in the atmosphere, but, however, it is not statistically significant.

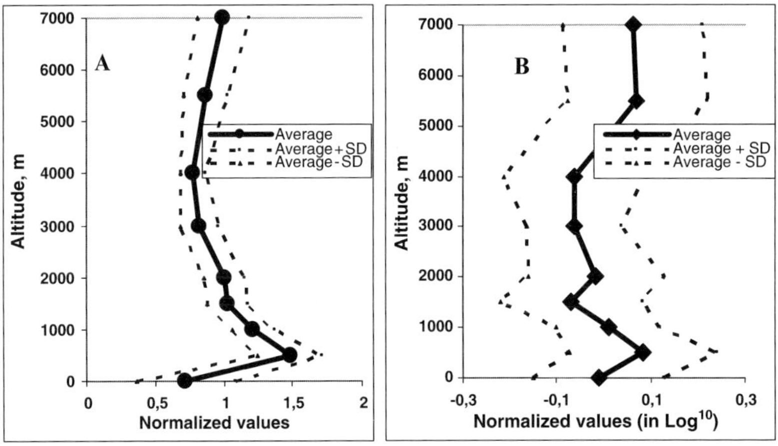

Figure 1 Standardized altitude profiles of the total protein (A) and culturable microorganisms (B) concentrations in Southwestern Siberia

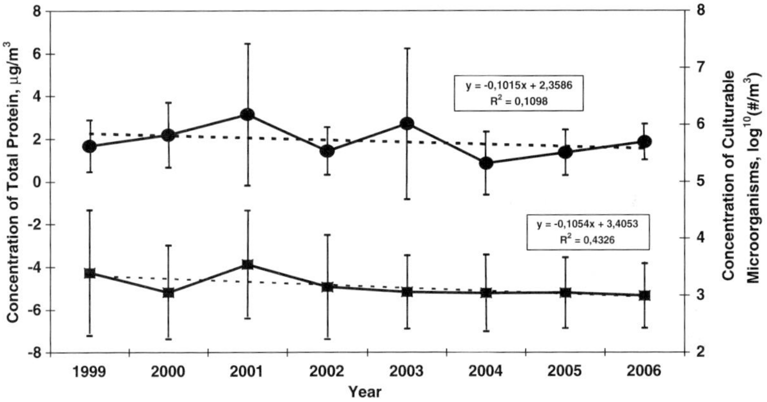

Figure 2 Eight-year variations of mean annual concentrations of the total protein (•) and culturable microorganisms (■) in the atmosphere of Southwestern Siberia at the altitudes of 500–7000 m

No statistically significant differences were revealed in the daily variations of on-land concentrations of the total protein and culturable microorganisms in the atmosphere. No considerable variation in the representation of different culturable microorganisms in samples collected during 24 h was revealed either.

An important characteristic of bioaerosol is its portion in the total mass of atmospheric aerosol. The data of [11] for different seasons revealed the portion of the total protein making up 0.1%–10% of the mass of aerosol particles determined with gravimetrical method though, according to the data of [12], the mass portion of bioaerosol in the vicinity of Lake Baikal makes up from 10%–80% of the total aerosol mass during the year, and according to [13], the portion of the biogenic particles can reach 95%. Estimates performed taking into account that the portion of culturable microorganisms in their total number makes up 0.02–10.6% according to [14] point to the considerable contribution of the detected number of culturable microorganisms to the observed total protein concentrations.

Previously it was shown that the maximal number of protein molecules is in the fraction of particles with aerodynamic diameters 0.16–0.4 μm [11]. The mass portion of the total protein in all particles is maximal in the fraction of 2.1–10 μm. These results agree with the data of [15], which showed that the fraction of atmospheric aerosol with particle diameters exceeding 2 μm is most rich in microorganisms (always containing a considerable amount of protein). Note that mechanical destruction of non-living biological material (remains of plants and animals and their individual cells) to submicron-sized particles requires considerable energy expenses. That's why, naturally, particles with diameters exceeding 0.1 μm are most rich in biogenic components.

Microorganisms referring to a wide range of genera and species were detected in the analyzed samples. Microorganism strains, which cannot be classified under any of the known taxons by their phenotypic characteristics, were also detected. On the whole, the percentage of microorganisms in samples varies with the altitude and between measurements (both at the ground level and at different altitudes). However, at present the obtained data are insufficient for constructing seasonal and altitude dependences for individual microorganisms as was done for their total number.

Conclusion

The performed research work allowed us to determine the dynamics of the mean annual values of the total protein and culturable microorganisms concentrations, their annual variation, altitude profiles and long-term trends. However, the obtained data is insufficient for predicting the variations of these concentrations. Also, they are insufficient for reliable determination of possible sources of the observed atmospheric bioaerosols. That is why observations of atmospheric bioaerosols in the region should be continued.

Acknowledgment The work was partially supported by ISTC, project #3275.

References

1. Andreeva, I.S. et al., *Chem. Sustainable Develop.*, **10**, 547–561 (2002).
2. Borodulin, A.I. et al., *Aerosol Sci.*, **36**, 785–800 (2005).
3. Safatov, A.S. et al., *Chemical Engineering Transactions*, **10**, 401–406 (2006).
4. You, W.W., et al., *Annal. Biochem.*, **244**, 277–282 (1997).
5. Miller, G., *Experiments in molecular genetics*, Moscow: Mir publishers , p. 435 (1976).
6. Saggie, E., *The methods of soil microbiology*, Moscow: Mir publishers, p. 295 (1983).

7. *The Prokaryotes. A Handbook on Habitats, Isolation, and Identification of Bacteria*, edited by M.P. Starr et al., Berlin, Heidelberg, New York, Tokyo: Springer, p. 2596 (1981).
8. *Methods of Experimental Mycology. Reference Book*, Kiev: Naukova dumka, p. 550 (1982).
9. *Statistical Methods in Microbiological Studies*, edited by I.P. Ashmarin and A.A. Vorobyov, Leningrad: Medgiz, p. 180 (1962).
10. Panchenko, M.V. and Pol'kin, V.V., *Atmos. Ocean. Opt.*, **14**, 478–488 (2001).
11. Safatov, A.S. et al., *Atmos. Ocean. Opt.*, **16**, 532–536 (2003).
12. Mattias-Maser, S. et al., *Atmos. Environ.*, **34**, 3805–3811 (2000).
13. Artaxo, P. et al., *J. Geophys. Res.*, **95 D**, 16971–16985 (1990).
14. Lighthart, B., *FEMS Microbiol. Ecology*, **23**, 263–274 (1997).
15. Tong, Y. and Lighthart, B., *Aerosol Sci. Technol.*, **32**, 393–403 (2000).

Assessment of Refractive Index and Microphysical Parameters of Spherical Aerosols from Data of Dual-Polarization Nephelometer

Christophe Verhaege[1], **Pascal Personne**[1], **and Valery Shcherbakov**[1,2]

Abstract It was performed sensitivity tests of dual-polarization polar nephelometer (D2PN) data to optical and microphysical parameters of population of spherical aerosol particles. Measurement errors were modeled as Gaussian random variables with zero mean and a standard deviation of 10%. It is shown that data of the D2PN enable to retrieve microphysical parameters along with the assessment of the refractive index.

Keywords Light scattering, complex refractive index, size distribution, inverse problem, aerosols

Introduction

The knowledge of microphysical and optical characteristics of aerosols is of importance for modeling the radiative balance of Earth's atmosphere, understanding the cloud life cycle, and remote sensing of tropospheric aerosols. Aerosol properties vary widely among various aerosol types (e.g., dust, biomass smoke, urban pollution, sea salt) and geographical regions. A feasible approach to parameterize characteristics of aerosols consists in elaboration of databases. Nowadays a number of light scattering codes are available, often organized in databases. Moreover, a database of experimental results is developed with the light scattering facility in Amsterdam [1].

The long-term objective of the D2PN designed at LaMP is to develop a database of optical and microphysical characteristics of aerosols and to test inverse codes against it. The aim of this work is to present results of sensitivity tests of D2PN data to optical and microphysical parameters of populations of spherical aerosol particles.

[1] *Laboratoire de Météorologie Physique, UMR/CNRS 6016, Université Blaise Pascal, 24 avenue des Landais, 63177 Aubière cedex, France*

[2] *Institute of Physics, Minsk, Belarus*

C.D. O'Dowd and P.E. Wagner (eds.), *Nucleation and Atmospheric Aerosols*, 746–749.

Laboratory Dual-polarization Polar Nephelometer (D2pn)

A diode laser is used as a source of collimated and unpolarized light at the wavelength of 800 nm and with a power of about 1 W. Two detectors allow measurement of polarized scattered intensities perpendicular $I_\perp$ and parallel $I_\parallel$ to the scattering plane with the angular resolution of 1° from 10° through 169°. The measured light power can range from 10 pW up to 3 μW. The measurements are quasi-continuous functions of the scattering angle θ. Thus, the degree of linear polarization and the unnormalized phase function can be obtained as functions of θ within the range between 10° and 169°. The accuracy of the measurements is estimated to be better than 10%.

Sensitivity Tests

In our investigations, we use the flexible inverse code [2] and the software package with kernels [3], which were developed by O. Dubovik and colleagues. The software package enables to calculate phase matrix, extinction, and absorption for polydisperse spheres and/or randomly oriented oblate and prolate spheroids. The inverse code can be easily adapted to particularities of an experimental setup. It is built on the principles of statistical estimation, and provides possibility to retrieve particle size distribution along with the assessment of the refractive index value.

Recall that the information content of data, measured by an instrument, has to be sufficient for accurate retrieval of required parameters. In other words, the measured characteristics have to be sensitive enough to variations of the parameters of interest. Otherwise the retrieved values are only a direct consequence of used a priori constraints. Thus, sensitivity tests are indispensable when the inverse code is adapted to a new experimental setup.

As the first step, we performed the sensitivity tests for the case of populations of spherical aerosol particles. Special attention was paid to the sensitivity of the data of the dual-polarization polar nephelometer to simultaneous variations of the refractive index and microphysical parameters. In the following, the results are analyzed in terms of root mean squared relative errors:

$$RMS = \frac{1}{n}\sqrt{\sum_{i=1}^{n}\frac{\left(\varphi_i(m)-\varphi_{i,meas}\right)^2}{\varphi_{i,meas}^2}} \qquad (1)$$

where n is the number of angles θ, φ is the phase function and m complex refractive index, φ_{meas} corresponds to a synthetic measurement, $\varphi(m)$ is the phase function for the variable refractive index $m = n + \chi i$.

When measurement errors were considered, they were modeled as Gaussian random variables ε with zero mean and a standard deviation of 0.1, i.e., of 10%:

$$\varphi_{meas} = \varphi_{comp}(1+\varepsilon) \tag{2}$$

where φ_{comp} was computed with input model parameters. The *RMS* values were computed as the average over 20 realizations.

Size distributions of particles were modeled by a monomodal lognormal function

$$\frac{dV(r)}{d\ln r} = \frac{C_V}{\sqrt{2\pi}\sigma}\exp\left[-\frac{(\ln r - \ln r_V)^2}{2\sigma^2}\right] \tag{3}$$

with standard deviation σ and median radius r_v. In these simulations, the median radius r_v and the standard deviation σ were varied from 0.25 to 25.0 µm and from 0.3 to 0.7, respectively.

Figure 1 shows the results obtained for the case of nonabsorbing particles. The synthetic measurements were computed for the refractive index value $m_{meas} = 1.3 + 10^{-8}\,i$, $r_v = 5$ µm and $\sigma = 0.5$. The *RMS* values are plotted as 3D surface, which depends on values of the real and imaginary parts of the variable refractive index *m*. The left panel corresponds to $\varepsilon = 0$ and the right panel is for $\varepsilon = 0.1$.

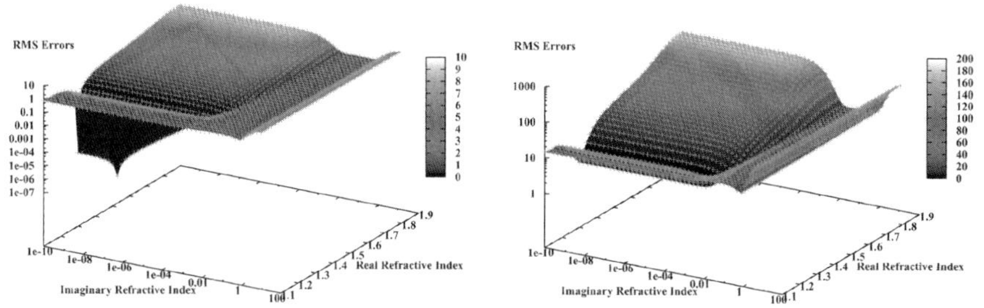

Figure 1 RMS errors functions plotted in the case of nonabsorbing particles $m = 1.3 + 10^{-8}\,i$. Left panel is the test without measurement errors, right panel is the test with measurement errors $\varepsilon = 0.1$

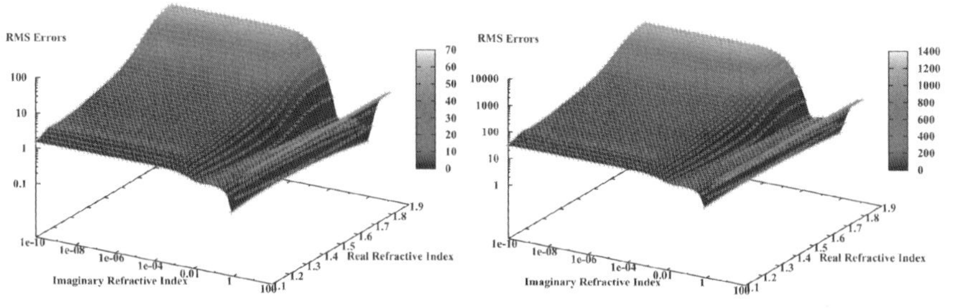

Figure 2 RMS errors functions plotted in the case of absorbing particles $m = 1.3 + 1i$. Left panel is the test without measurement errors, right panel is the test with measurement errors $\varepsilon = 0.1$

In the case without error of measurements, the data of the D2PN are very sensitive to the refractive index values. The sharp minimum is seen exactly at $m = 1.3 + 10^{-8}\, i$. Measurement errors substantially affect the sensitivity. Only the range of variation of the imaginary part can be estimated, i.e., $\chi \leq 10^{-4}$. On the other hand, the data remain sensitive to the real part value.

In the case when the absorption of particles is very high, i.e., $\chi = 1$, the data of the D2PN are not sensitive to the real part of the refractive index (see Figure 2). To the contrary, the imaginary part can be retrieved with good accuracy.

Sensitivity tests of the same kind were performed for the real and the imaginary parts ranged from 1.1 through 1.9 and from 10^{-3} through 10^{-1}, respectively. It follows that in such cases the data of the D2PN provide possibility to estimate the both parts of the refractive index with quite good accuracy (measurement errors $\varepsilon = 0.1$).

The data of the D2PN are sensitive to the microphysical parameters σ and r_v as well. That is, σ and r_v can be retrieved along with the refractive index.

Our calculations, performed for polarized scattered intensities $I_{\perp}$ and $I_{\parallel}$ showed that, compared to the phase function case, the sensitivity increased, but not essentially. At the same time, it is expected that measurements of $I_{\perp}$ and $I_{\parallel}$ will be substantial for characterization of nonspherical particles.

Conclusion

1. The data of the D2PN enable to retrieve microphysical parameters of spherical aerosols along with the assessment of the refractive index.
2. In the case when the absorption of particles is very high, the data of the D2PN are not sensitive to the real part of the refractive index. The imaginary part can be retrieved with good accuracy.
3. Measurement errors substantially reduce the sensitivity to the imaginary part in the case of nonabsorbing particles.

Acknowledgments The authors are grateful to O. Dubovik for providing the flexible inverse code and the software package with kernels.

References

1. Volten, H., Muñoz, O., Hovenier, J.W., de Haan, J.F., Vassen, W., van der Zande, W.J., and Waters, L.B.F.M., WWW scattering matrix database for small mineral particles at 441.6 and 632.8 nm., *JQSRT*, **90**, pp. 191–206 (2005).
2. Dubovik O., Optimization of numerical inversion in photopolarimetric remote sensing, in: *Photopolarimetry in Remote Sensing*, edited by G. Videen, Y. Yatskiv, and M. Mishchenko, Dordrecht, Netherlands: Kluwer Academic Publishers, pp. 65–106 (2004).
3. Dubovik, O., Sinyuk, A., Lapyonok, T., Holben, B.N., Mishchenko, M., Yang, P., Eck, T.F., Volten, H., Muñoz, O., Veihelmann, B., van der Zande, W.J., Leon, J.-F., Sorokin, M., and Slutsker, I. Application of spheroid models to account for aerosol particle nonsphericity in remote sensing of desert dust, *J. Geophys. Res.*, **111**, D11208 (2006).

Number Density and Carbon Concentration of Accumulation Mode Particles over the North Pacific Ocean

Kiyoshi Matsumoto[1] **and Mitsuo Uematsu**[2]

Abstract Measurements of number densities and carbon concentrations of accumulation mode particles were conducted over the widespread areas of the North Pacific Ocean. Higher number densities were found over the coastal seas of the East Asian industrial areas such as the East China Sea, the Japan Sea, and the seas close by Japan due to the continental outflows. The densities decreased to lower values, which depended on the distance from the East Asian industrial area. Similar geographical distribution was found in the concentrations of particulate elemental carbon (EC). The densities of the particles in the summer showed the similar levels with those in the spring over the northwestern North Pacific despite of lower concentrations of particulate EC in the summer. Gas to particle conversion processes via photochemical reactions would have significant impact on the densities over these seas in the summer.

Keywords Particle number density, carbonaceous aerosols, North Pacific Ocean

Introduction

Particle number density is a controlling factor for aerosol optical properties and cloud nucleating processes. Global distribution of their density should be investigated to estimate the incoming solar radiation budget. In particular, accumulation mode particles have a significant impact on radiation field in the atmosphere, since they could scatter the visible rays most efficiently and supplies the largest number of cloud condensation nuclei (CCN). Global distributions of particle number density in the accumulation mode range, however, have been rarely reported in the past researches.

[1] *Faculty of Engineering, Kanagawa University*

[2] *Ocean Research Institute, The University of Tokyo*

C.D. O'Dowd and P.E. Wagner (eds.), *Nucleation and Atmospheric Aerosols*, 750–754.

As well as particle number density, particulate carbonaceous substances including organic carbon (OC) and black carbon (BC) also have large contributions to radiative properties, since OC could effectively scatter solar radiation, contribute to the formation of cloud droplets, and BC is the principal light-absorbing species in the atmosphere.

From 1998 to 2002, seven research cruises have been conducted over the widespread areas of the North Pacific Ocean to investigate physical and chemical properties of marine aerosols. In this paper, we report on number densities and carbon concentrations of accumulation mode particles, and discuss the spatial distributions and seasonal trends of these species over the North Pacific Ocean. Recently, the North Pacific region has been paid to great attention from the viewpoint of atmospheric radiation budget due to outbreaks of anthropogenic and mineral aerosol from Asian continental region.

Observations

The cruises of KH98-3, KH99-3, and MR00-K4 were conducted in the summers over the marginal seas around Japan, northern remote Pacific Ocean, and tropical/subtropical western North Pacific Ocean, respectively. The cruises of MR98-K01, MR99-K02, MR00-K03, and KH02-3 were conducted during the spring or autumn seasons in the continentally influenced air masses over the northwestern North Pacific Ocean, the Japan Sea, and the East China Sea. Tracks of seven research cruises are shown in Figure 1.

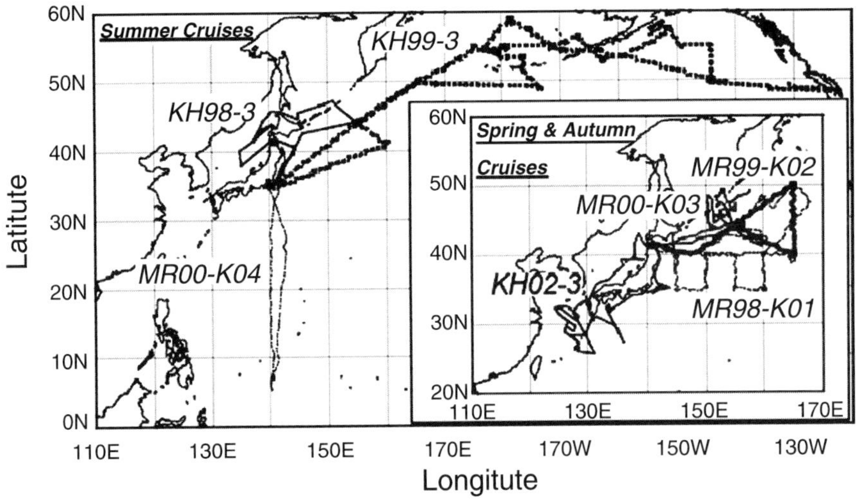

Figure 1 Tracks of seven cruises

Measurements of particle number density and particulate carbon concentration in accumulation mode range (0.10–2.5 μm) were conducted by using an optical particle counter (OPC) (Rion Co. Ltd., KC-18) and an ambient particulate carbon monitor (ACPM) (Rupprecht and Patashnik Co., Inc., Series 5400). Details of the instruments have been reported in our previous papers.[1] The ACPM can automatically measure the concentrations of carbonaceous substances in aerosols by thermal analysis. In this study, OC and total carbon (TC) were defined as the carbonaceous substances evolved below 340°C and below 750°C, respectively. And then, the difference between the amounts of TC and OC was defined as the amount of elemental carbon (EC). Data from the OPC and the ACPM affected by ship's exhaust were discarded in the following discussion.

Results and Discussion

EC Concentrations from the ACPM

Recent studies have pointed out that the EC concentrations from the thermal analysis could be largely affected by pyrolysis OC.[2] Our preliminary experiment conducted at a remote island (Hachijo Island, 300 km south from Tokyo), however, demonstrated that the EC concentrations from the ACPM show a good agreement with the concentrations of BC from the Aethalometer (Magee Scientific Co., AE-16U) that can measure BC in aerosols by an optical absorption, as shown in Figure 2. This means that the EC from the ACPM could be not significantly affected by the pyrolyses OC.

Our preliminary experiment also found that the ACPM with an absolute filter in front of the inlet detected significant concentrations of OC as shown in Table 1, which would be caused by adsorption of organic gases and lead to a significant overestimation of OC. The carbonaceous substances allocated in the EC fraction, however, were not detected from the ACPM with the absolute filter (see Table 1). EC concentrations from the ACPM were not affected by any adsorption of gaseous species.

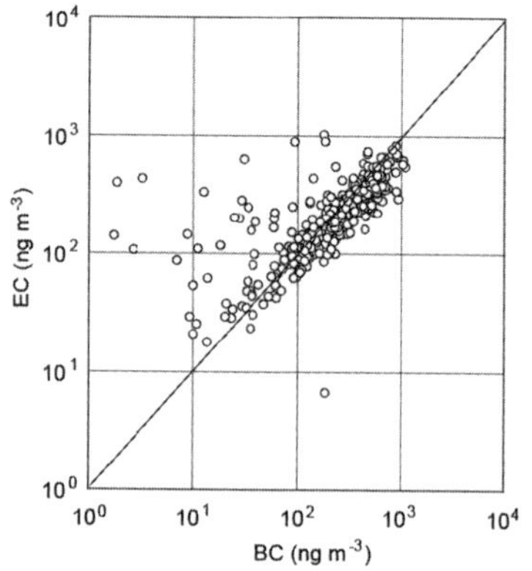

Figure 2 Relationship between the concentrations of EC and BC at Hachijo

Table 1 The ratios of carbon concentrations from the ACPM with an absolute filter to those without the filter

Combustion Temperature	Ratios
Below 200°C	47.7
200–250°C	30.1
250–340°C	14.3
Below 340°C (OC)	36.2
Above 340°C (EC)	4.3

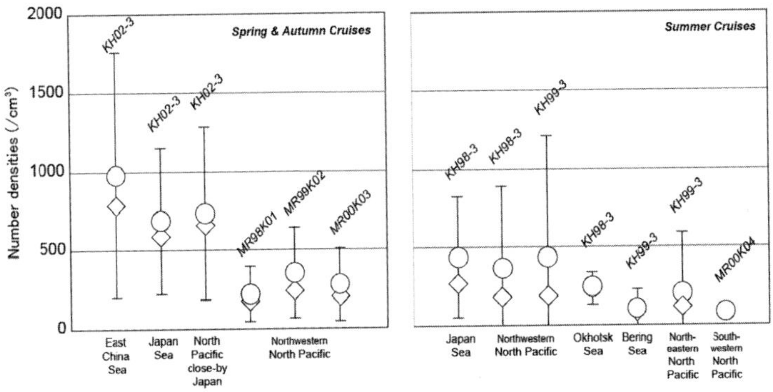

Figure 3 Number densities of the particles with the diameter of 0.1–2.5 μm. Open circles and open diamonds indicate means and medians, respectively, and error bars indicate standard deviations

Measurements of Particle Number Density

Measurements of particle number densities in accumulation mode range (0.10–2.5 μm) are summarized in Figure 3. Higher densities were found over the coastal seas of the East Asian industrial areas such as the East China Sea, the Japan Sea, and the seas close by Japan from the KH02-3 cruise due to the continental outflows. The densities decreased to lower values, which depended on the distance from the East Asian industrial area. Similar geographical distribution was found in the concentrations of particulate EC.

Despite of predominance of the westerlies, however, number densities in the spring and autumn showed the similar levels with those in the summer over the northwestern North Pacific Ocean. In the summer, marine air masses were covered the area and particulate EC showed lower concentrations compared with the case of the spring. Gas to particle conversion processes via photochemical reactions would have significant impact on the densities over these seas in the summer. The densities with the diameters of 0.1–2.5 μm were not affected by wind speed, although wind speed has significant impact on those with the diameter of 0.5–2.5 μm.

Acknowledgments We would like to extend special thanks to Dr. Kaneyasu for the measurement of BC, and the captains and crews of R/V Mirai and R/V Hakuho for their kind supports during our observations. This study was financially supported by Core Research for Evolutional Science and Technology (CREST) of Japan Science and Technology Agency (JST).

References

1. Matsumoto, K., Uyama, Y., Hayano, T., Tanimoto, H., Uno, I., and Uematsu, M., J. *Geophys. Res.*, **108**, doi:10.1029/2003JD003426 (2003).
2. Chow, J.C., Watson, J.G., Crow, D., Lowenthal, D.H., and Merrifield, T., *Aerosol Sci. and Technol.*, **34**, 23–34 (2001).

Comparative Effect of Airborne Pollutants on 3 Ångstrom Turbidity Coefficients

Chih-Chung Wen and Chin-Hsiang Luo

In this paper, the concentration of the atmospheric pollutants (SO_2, NO_2, O_3, and PM_{10}) were used to investigate the Ångstrom turbidity coefficients β (i.e., β_{Lou}, β_{Pin}, and β_{Vis}) at the Taichung Harbor site near Taiwan Strait. The local weather conditions were collected during 2004–2005 at the Wuchi weather station of Taiwan Central Weather Bureau near Taiwan Strait. Furthermore, Pearson correlation analysis is used to investigate the Ångstrom turbidity coefficients (i.e., β_{Lou}, β_{Pin}, and β_{Vis}) correlated with the atmospheric pollutants. The analysis result of the Ångstrom turbidity coefficients and the atmospheric pollutants (SO_2, NO_2, O_3, and PM_{10}) are summarized in Table 1. According to the Pearson correlation, Ångstrom turbidity coefficient, β_{Lou}, is positively correlated with the concentration of SO_2, O_3, and PM_{10} with correlation coefficients r_{sp} = 0.587, 0.624, and 0.625, respectively. Moreover, the results show that β_{Pin} is only correlated with the concentration of SO_2 with a correlation coefficient r_{sp} = 0.597. However, β_{Vis} is higher positively correlated with the concentration of NO_2, O_3, and PM_{10} with correlation coefficients r_{sp} = 0.898, 0.765, and 0.764. In conclusion, the β_{Vis} inherits a stronger environmental impact from characteristics of the atmospheric pollutants by this model. Therefore, visibility should be a direct index to judge the effect of anthropogenic air pollutants.

Table 1 Pearson correlation coefficients of Ångstrom turbidity coefficients and atmospheric pollutants

	β_{Lou}	β_{Pin}	β_{Vis}	SO_2	NO_2	O_3	PM_{10}
β_{Lou}	1.0						
β_{Pin}	0.638*	1.0					
β_{Vis}	0.429	0.144	1.0				
SO_2	0.587	0.597	0.512	1.0			
NO_2	0.407	0.156	0.898**	0.645*	1.0		
O_3	0.624*	0.388	0.765**	0.569	0.645*	1.0	
PM_{10}	0.625*	0.390	0.764**	0.571	0.852**	0.982**	1.0

* the value >0.600

** the value >0.750

Department of Environmental Engineering, Hungkuang University, Sha-Lu, Taichung 43302, Taiwan

C.D. O'Dowd and P.E. Wagner (eds.), *Nucleation and Atmospheric Aerosols*, 755.

Weekly Distribution of the Aerosol Pollution of the Atmosphere in Tbilisi

A.G. Amiranashvili[1], V.A. Amiranashvili[1], D.D. Kirkitadze[1], and K.A. Tavartkiladze[2]

Abstract The influence of the work of transport and industrial factories on the weekly distribution of the aerosol pollution in the atmosphere in Tbilisi is studied. As the parameter of air pollution the aerosol optical depth of the atmosphere (AOD) is used. The effect of an increase of the air pollution during the weekdays in comparison with the weekends is revealed.

Keywords Atmospheric aerosol optical depth

Introduction

The aerosol optical depth of the atmosphere is one of the most important parameters, which characterizes the air pollution. At the same time AOD characterizes the pollution of the entire vertical thickness of the atmosphere and therefore it is the sufficiently inertia parameter. Thus it is important to study reaction AOD to air pollution by transport and industrial factories. For this purpose we carried out the analysis of the data about AOD in Tbilisi during the weekdays and weekends.

Methods

AOD values were calculated by the method [1,2] using to standard actinometric observations of intensity of clear sky integral direct solar radiation. This method for AOD estimation is not so accurate as that founded on sun radiation spectral measurements. And yet it allows to estimate long-term variations of AOD with the help of quite simple and relatively cheap standard actinometric measurements.

[1]*Mikheil Nodia Institute of Geophysics, 1, M. Aleksidze Str., 0193 Tbilisi, Georgia*

[2]*Vakhushti Bagrationi Institute of Geography, 8, M. Aleksidze Str., 0193 Tbilisi, Georgia*

C.D. O'Dowd and P.E. Wagner (eds.), *Nucleation and Atmospheric Aerosols*, 756–760.

Values of the AOD are given for the wavelength $\lambda = 1\,\mu m$ and midday hours. The data of the daily values of AOD in 1980–1990 are used [3].

The following designations will be used below (besides those pointed out above and well known): σ – standard error, Cv – coefficient of variation, A – coefficient of skewness, K – coefficient of kurtosis, α – the level of significance. The estimation of difference between the investigated parameters was evaluated according to Student's criterion with significance level not worse than 0.2.

Results

The results are presented in Table 1 and Figures 1 and 2. Table 1 presents the statistical characteristics of the daily values of atmospheric aerosol optical depth in Tbilisi in 1980–1990 on the days of week.

As follows from Table 1 as a whole the daily values of AOD in weekdays the air pollution is higher than in weekends. Average values of AOD during the weekdays and weekends respectively comprise: year – 0.166 and 0.160 (the difference is significant, $\alpha = 0.25$), warm period – 0.192 and 0.184 (the difference is significant, $\alpha = 0.20$), cold period 0.129 and 0.125 (the difference is low significant, $\alpha = 0.55$). For the separate ranges AOD, and also for series of prolonged everyday measurements AOD this effect appears more clearly.

Figure 1 presents the occurrence of daily values of AOD in weekdays and weekends for year, warm and cold seasons (the width of each range composes 0.05). As follows from Figure 1 the occurrence of low values of AOD (cleaner atmosphere) in weekends is higher than in weekdays, whereas the occurrence of high values of AOD (pollution atmosphere) in weekends is lower than in weekdays. For example, according to the data of the year (curves 1 and 2) the occurrence of AOD in the range from 0.05–0.2 in weekends is 22.5%, and in weekdays – 20.1%–20.9%. At the same time the occurrence of AOD from 0.2 to 0.4 in weekends varies from 0.3%–15.9%, and in weekdays – from 1.4%–17.6%. Values of AOD from 0.4–0.5 in weekends is not observed, whereas in weekdays the occurrence of these values composes 0.2%–0.3%. The similar pattern is observed for the warm and cold seasons.

The several cases of everyday measurements AOD during the week and more presents Figure 2. This figure also clearly demonstrates a considerable increase of air pollution during the weekdays in comparison with the weekend.

Conclusion

Average values of the aerosol optical depth of the atmosphere in the weekdays and weekends in Tbilisi respectively comprise: year – 0.166 and 0.160 (the difference is significant), warm season – 0.192 and 0.184 (the difference is significant), cold season – 0.129 and 0.125 (the difference is low significant). This effect appears

Table 1 Statistical characteristics of the aerosol optical depth of the atmosphere in Tbilisi in 1980–1990 on the days of week

	1	2	3	4	5	6	7
Days of week				Year			
Average	0.166	0.163	0.166	0.171	0.163	0.160	0.159
Min	0.020	0.019	0.034	0.027	0.022	0.032	0.028
Max	0.399	0.476	0.488	0.450	0.408	0.382	0.341
Range	0.379	0.457	0.454	0.423	0.386	0.350	0.313
Median	0.153	0.159	0.162	0.168	0.153	0.157	0.149
Mode	0.202	0.173	0.129	0.168	0.153	0.080	0.181
St Dev	0.082	0.084	0.087	0.080	0.082	0.075	0.070
σ	0.0061	0.0066	0.0067	0.0059	0.0060	0.0059	0.0054
C_v (%)	49.5	51.7	52.6	46.6	50.2	47.2	44.2
A	0.49	0.55	0.69	0.39	0.52	0.36	0.40
K	−0.20	0.20	0.47	−0.22	−0.09	−0.50	−0.64
Count	181	162	169	183	186	165	168
			Warm season				
Average	0.193	0.186	0.191	0.197	0.189	0.184	0.183
Min	0.076	0.033	0.048	0.054	0.024	0.032	0.048
Max	0.399	0.476	0.488	0.374	0.408	0.344	0.341
Range	0.323	0.443	0.440	0.320	0.384	0.312	0.293
Median	0.192	0.195	0.187	0.201	0.176	0.182	0.181
Mode	0.079	0.107	0.080	0.168	0.106	0.104	0.181
Standard Deviation	0.071	0.086	0.083	0.068	0.082	0.071	0.069
σ	0.0069	0.0088	0.0084	0.0066	0.0079	0.0072	0.0068
C_v (%)	36.8	46.1	43.5	34.6	43.1	38.3	37.5
A	0.54	0.41	0.66	−0.11	0.38	0.05	0.17
K	−0.03	0.23	1.26	−0.45	−0.27	−0.68	−0.68
Count	107	96	99	107	107	95	102
			Cold season				
Average	0.127	0.129	0.130	0.133	0.128	0.127	0.122
Min	0.020	0.019	0.034	0.027	0.022	0.033	0.028
Max	0.387	0.313	0.342	0.450	0.351	0.382	0.271
Range	0.367	0.294	0.308	0.423	0.329	0.349	0.243
Median	0.109	0.123	0.110	0.115	0.127	0.119	0.111
Mode	0.037	0.160	0.054	0.096	0.153	0.142	0.084
St Dev	0.082	0.069	0.080	0.080	0.068	0.070	0.056
σ	0.0095	0.0085	0.0096	0.0092	0.0077	0.0083	0.0069
C_v (%)	64.4	53.8	61.8	59.9	53.6	54.8	45.7
A	1.13	0.58	1.07	1.48	0.66	1.06	0.73
K	1.02	−0.17	0.33	2.76	0.46	1.49	−0.22
Count	74	66	70	76	79	70	66

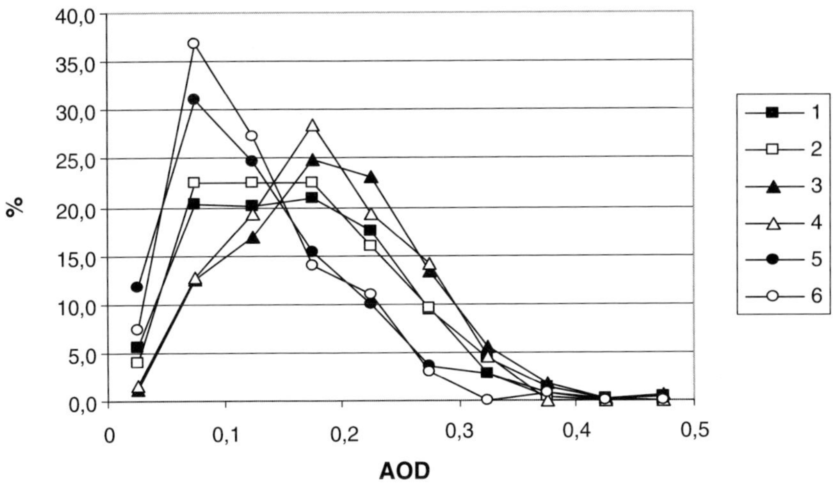

Figure 1 Occurrence of daily AOD values in Tbilisi in 1980–1990. 1 – year, week-days; 2 – year, weekends; 3 – warm season, weekdays; 4 – warm season, weekends; 5 – cold season, weekdays; 6 – cold season, weekends

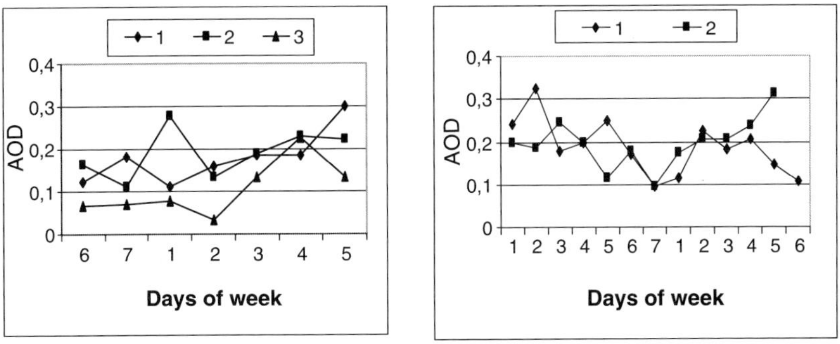

Figure 2 Weekly distribution of AOD in Tbilisi. Left: 1 – 1982 (December 18–24), 2 – 1983 (December 17–23), 3 – 1985 (September 21–27). Right: 1 – 1986 (September 8–20), 2 – 1990 (September 17–28)

more clearly for the separate ranges AOD, and also for series of prolonged every-day measurements AOD. Thus AOD is sensitive to weekly variations of air pollution by transport and industrial factories.

References

1. Tavartkiladze, K.A., *Modelling of the Aerosol Attenuation of Radiation and Atmospheric Pollution Control Methods*, Tbilisi: Metsniereba, pp. 1–204, (in Russian) (1989,).

2. Amiranashvili A.G., Amiranashvili V.A., Gzirishvili T.G., Kharchilava J.F., and Tavartkiladze K.A., Modern climate change in Georgia. radiatively active small atmospheric admixtures, institute of geophysics, Monograph, *Trans. of M.Nodia Institute of Geophysics of Georgian Acad. of Sci.*, ISSN 1512–1135, vol. LIX, 1–128 (2005).
3. Amiranashvili A.G., Amiranashvili V.A., Kirkitadze D.D, and Tavartkiladze K.A., Some Results of Investigation of Variations of the Atmospheric Aerosol Optical Depth in Tbilisi, *Proceedings of 16th Int. Conf. on Nucleation&Atmospheric Aerosols*, Kyoto, Japan, 26–30 July 2004, pp. 416–419.

The Influence of Relative Humidity on the Changeability of the Atmospheric Aerosol Optical Depth

K.A. Tavartkiladze[1] and A.G. Amiranashvili[2]

Abstract Some results of the investigation of relations between atmospheric aerosol optical depth with the assigned level of value and air relative humidity are presented. With an increase of the air relative humidity the probability that the value of the atmospheric aerosol optical depth will not exceed the assigned level at first decreases, reaches the minimum and then increases. The corresponding empirical regression equations between the studied parameters are obtained.

Keywords Atmospheric aerosol optical depth, air relative humidity

Introduction

Variations of the radiatively active small atmospheric admixtures have the determining effect on the formation of the energy balance of system the earth–atmosphere. From these admixtures the influence of the optical properties of atmospheric aerosols on the energy state of the atmosphere is studied thus far insufficiently. The changeability of the content of aerosols in the atmosphere and also their optical properties they can cause both the increase and decrease of the energy level of the earth's surface and atmosphere. On the other hand, the physical state of the atmosphere (especially the water vapor content in the atmosphere) can substantially change both the concentration of aerosols and their different optical properties. The aerosol optical depth of the atmosphere (AOD) is one of the main characteristics of atmospheric aerosols. Earlier we studied the relations between AOD and surface relative humidity for anthropogenic and natural aerosols [1]. This work presents the results of the investigations of relations between AOD with the assigned level of value (P) and the air relative humidity (F).

[1] *Vakhushti Bagrationi Institute of Geography, 8, M. Aleksidze Str., 0193 Tbilisi, Georgia*

[2] *Mikheil Nodia Institute of Geophysics, 1, M. Aleksidze Str., 0193 Tbilisi, Georgia*

C.D. O'Dowd and P.E. Wagner (eds.), *Nucleation and Atmospheric Aerosols*, 761–765.

Methods and Experimental Data

The mentioned relations were established using the data of numerous experiments. The aerosol optical depth is determined in three spectrum range with centers $\lambda_1 = 0.702$, $\lambda_2 = 0.711$ and $\lambda_3 = 0.739\,\mu m$. The AOD was determined by the Sun radiation attenuation and using interferential light filters, whose transmissive ability was within a range of $0.01\,\mu m$ [2]. λ_1 and λ_3 were taken as a range where aerosols only diffuse, while in λ_2 together with diffusion an unintensive absorption of the sun radiation by water vapor took place. The experiments were carried out during four years (the summer of 1979–1982) in the eastern part of the coast of Black sea, which is characterized by the relatively high level of the aerosol pollution of the atmosphere and by significant variations of the air relative humidity. The experiments were carried out, when Sun angle was more than 20°. The total number of measurements was more than 750.

For determining the connection between P and F the following actions were carried out. All data of measurements of AOD were ranked in the correspondence with an increase of the air relative humidity in five ranges. In each range there was an approximately identical quantity of data. The average value of the air relative humidity in these ranges comprised: 0.44, 0.62, 0.69, 0.74, and 0.81. For each range was determined probability that AOD ≤ 0.1, AOD ≤ 0.2, AOD ≤ 0.3, AOD ≤ 0.4, AOD ≤ 0.6 and AOD ≤ 0.8. Then were defined the graphs and the corresponding empirical regression equations of the dependence between P and F. The accuracy of approximation was evaluated by the determination coefficient R^2.

Results

The results of the measurements are presented in Tables 1, 2 and Figure 1. The data about the statistical characteristics of AOD for each year of measurements (average, standard deviation, and coefficient of variation Cv) are represented in Table 1. In particular, as follows from this table the average values of AOD for λ_1 and λ_3 insignificantly differ from each other, whereas values AOD for λ_2 are substantially higher than for λ_1 and λ_3. The greatest changeability AOD is observed for λ_1 (Cv changes from 37.1%–48.7%), smallest – for λ_2 (Cv changes from 25.1%–28.5%).

Figure 1 present the graphs of the relationships between P and F for λ_1 and λ_2. It should be noted that the relationship between P and F for λ_1 practically is the same as for λ_3. The following regularity follows from Figure 1. With an increase of the air relative humidity the probability that the value of the atmospheric aerosol optical depth will not exceed the assigned level at first decreases, reaches the minimum and then increases. With an increase of the atmospheric pollution the dependence between P and F weakens. For example, when $\lambda = 0.702\,\mu m$ the ratio P_{max}/P_{min}

Table 1 The statistical characteristics of the atmospheric aerosol optical depth in different years of measurements

		Years			
Λ, μm	Parameter	1979	1980	1981	1982
0.702	Average	0.318	0.224	0.319	0.251
	Standard Deviation	0.130	0.109	0.130	0.093
	Cv, %	40.9	48.7	40.8	37.1
0.711	Average	0.530	0.452	0.539	0.456
	St Dev	0.133	0.122	0.136	0.130
	Cv, %	25.1	27.0	25.2	28.5
0.739	Average	0.325	0.219	0.327	0.288
	Standard Deviation	0.118	0.098	0.123	0.091
	Cv, %	36.3	44.7	37.6	31.6

Table 2 The empirical regression equations between P and F. Values of regression coefficients

Parameter	Equation	a	B	c	D	R^2
AOD		$\lambda = 0.702\,\mu m$				
≤0.1	$P = a{\cdot}F^3 + b{\cdot}F^2 + c{\cdot}F + d$	25.678	−43.442	22.987	−3.460	0.95
≤0.2	$P = a{\cdot}F^3 + b{\cdot}F^2 + c{\cdot}F + d$	26.076	−44.165	23.375	−3.266	0.99
≤0.3	$P = a + b/\lg F + c/\lg F^2$	1.1107	0.2412	0.0368		0.93
≤0.4	$P = a + b/\lg F + c/\lg F^2 + d/\lg F^3$	1.0518	0.0355	−0.0144	−0.0035	0.75
≤0.6	Dependence is weak					
≤0.8	Dependence is weak					
AOD		$\lambda = 0.711\,\mu m$				
≤0.1	$P = a{\cdot}F^3 + b{\cdot}F^2 + c{\cdot}F + d$	4.344	−4.680	0.336	0.688	0.99
≤0.2	$P = a{\cdot}F^3 + b{\cdot}F^2 + c{\cdot}F + d$	29.644	−52.373	29.148	−4.632	0.99
≤0.3	$P = a + b/\lg F + c/\lg F^2 + d/\lg F^3$	0.5014	−0.4015	−0.2296	−0.0312	0.96
≤0.4	$P = a + b/\lg F + c/\lg F^2$	1.0727	0.2271	0.0275		0.82
≤0.6	$P = a + b/\lg F + c/\lg F^2$	1.0610	0.1455	0.0149		0.98
≤0.8	$P = a + b/\lg F + c/\lg F^2$	1.0521	0.0927	0.0114		0.94

decreases from 374% for AOD ≤0.1 to 111% for AOD ≤0.4. If the pollution of the atmosphere is high the connection between P and F practically is absent. When λ = 0.711 the ratio P_{max}/P_{min} decreases from 259% for AOD ≤0.1 to 111% for AOD ≤0.8. In this case the connection between P and F for all levels of AOD is observed.

The corresponding empirical regression equations between P and F, and also the values of regression coefficients are represented in Table 2. For different levels of the atmospheric pollution and values of wavelength the connection between P and

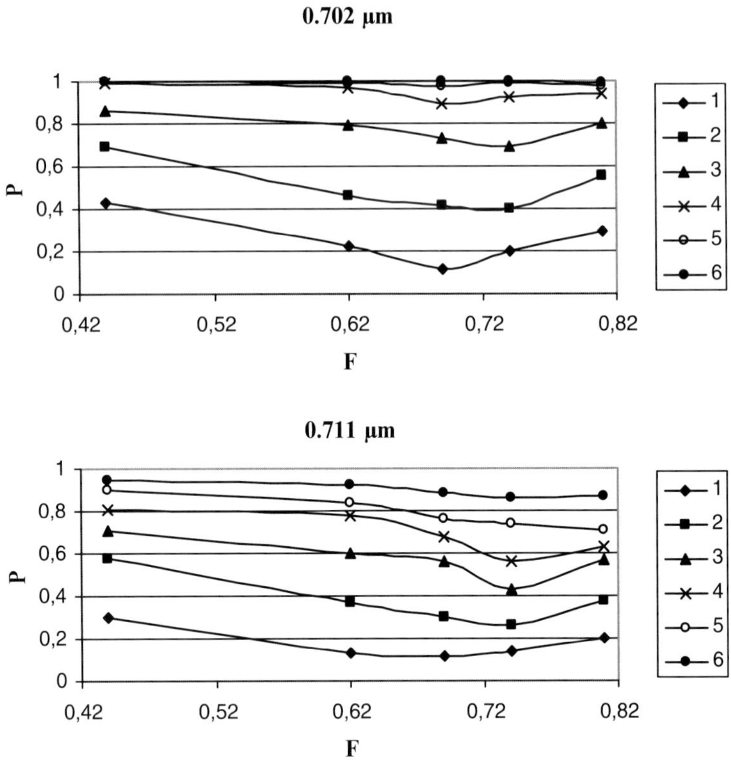

Figure 1 Connection of the probability of the AOD with the assigned level of value with the air relative humidity. 1 – AOD ≤0.1, 2 – AOD ≤0.2, 3 – AOD ≤0.3, 4 – AOD ≤0.4, 5 – AOD ≤0.6, 6 – AOD ≤0.8. Upper – $\lambda = 0.702\,\mu m$, lower – $\lambda = 0.711\,\mu m$

F takes the third-rder polynomial, third-order inverse logarithm and second-order inverse logarithm forms.

Conclusion

The connection of the probability of the atmospheric aerosol optical depth with the assigned level of value with the air relative humidity is studied. The aerosol optical depth is determined in three spectrum range with centers $\lambda_1 = 0.702$, $\lambda_2 = 0.711$, and $\lambda_3 = 0.739\,\mu m$. With an increase of the air relative humidity the probability that the value of the atmospheric aerosol optical depth will not exceed the assigned level at first decreases, reaches the minimum and then increases. For different levels of the atmospheric pollution and values of wavelength the connection between P and F takes the third-order polynomial, third-order inverse logarithm and second-order inverse logarithm forms.

References

1. Tavartkiladze, K., Shengelia, I., Amiranashvili, A., and Amiranashvili, V., The influence of relative humidity on the optical properties of atmospheric aerosols, *J.Aerosol Sci.*, Pergamon, **30**(1), S639–S640 (1999).
2. Tavartkiladze, K.A., *Modelling of the Aerosol Attenuation of Radiation and Atmospheric Pollution Control Methods*, Tbilisi: Metsniereba, pp. 1–204, (in Russian) (1989).

Ground-based Observations of the Chemical Composition of Asian Outflow Aerosols

Charles C.-K. Chou[1], S.J. Chen[2,*], M.T. Cheng[3,*], W.C. Hsu[4,*], C.T. Lee[5,*], Y.L. Wu[6,*], C.S. Yuan[7,*], S.-C. Hsu[1], C.S.-C. Lung[1], and Shaw C. Liu[1]

Abstract To improve our understandings of the characteristics and environmental impacts of the aerosols transported from the Asian Continent to the western North Pacific region, an aerosol observation network has been established on Taiwan and PengHu Islands since the spring of 2002. Aerosol sampling is conducted regularly in the last week of each month. Mass concentration, size distribution, and chemical composition of aerosol samples are measured with a resolution of 12 h. From the measurements of 2003–2005, it was found that the aerosol concentrations and compositions at rural and off-coast stations are rather consistent with each other in the winter monsoon periods. Furthermore they are much higher than those of remote areas, apparently due to long-range transport of air pollutants and dust from the Asian Continent. This is supported by the high percentages of sulfate, ammonium ions and organic carbon which are mostly from anthropogenic sources. It was also found that the Asian outflow aerosols were transformed chemically as they mixing with air pollutants originating from Taiwan. This finding indicates that the Asian outflow aerosols can be important carriers of reactive gases, particularly nitric acid and ammonia, from Taiwan to the downwind areas of the Asian winter monsoons.

Keywords Asian outflow, tropospheric aerosols, aerosol compositions, western North Pacific

---

* authors listed in alphabetical order

[1] *Research Center for Environmental Changes, Academia Sinica, Taipei 115, Taiwan*

[2] *Department of Environmental Science and Engineering, National Pingtung University of Science and Technology, Pingtung 912, Taiwan*

[3] *Department of Environmental Engineering, National Chung Hsing University, Taichung 402, Taiwan*

[4] *Department of Environmental Resources Management, Dahan Institute of Technology, Hualien 971, Taiwan*

[5] *Institute of Environmental Engineering, National Central University, Chungli 320, Taiwan*

[6] *Department of Environmental Engineering, National Cheng Kung University, Tainan 701, Taiwan*

[7] *Institute of Environmental Engineering, National Sun Yat-Sen University, Kaohsiung 804, Taiwan*

C.D. O'Dowd and P.E. Wagner (eds.), *Nucleation and Atmospheric Aerosols*, 766–770.

Introduction

The East China is undergoing the most rapid industrialization in the world. More and more industrial products are made in China in recent years. While those huge international companies moving their manufacturing departments into the mainland China, we can reasonably expect significant increases in the air pollutant emissions over there. Thus, in the meantime of the formation of the Factory of the World, East China is also becoming the most polluted area on the Earth.

There have been a lot of studies focusing at the transport of air pollutants from the Asia Continent to the North Pacific or North America in the last decade: PEM-WEST A/B in 1991–1994, TRACE-P, and ACE-Asia in 2001, for instance.[1–4] Those experiments indeed provided valuable data to improve our understanding of the physical/chemical properties of Asian outflow aerosols. However, because of the short lifetime of aerosols in the atmosphere there are substantial variations in the aerosol properties over different spatial and temporal scales. Hence, the conclusions based on those intensive but fragmented experiments are unavoidably subject to significant uncertainties. To overcome the drawbacks in those intensive campaigns, we hope to establish an accurate, representative, long-term data base for aerosols in Taiwan. With such a database, we can certainly improve our understanding of the properties of aerosols associated with Asian outflows. The data can be useful for characterization of aerosols from local sources. With the accurate data, we can study the transport path of Asian outflows, and the mixing and transformation processes along with the transport path. Furthermore, the data can also be helpful to the investigations on the influences of aerosols on the local air quality, and the probable implications for regional climate.

Experimentals

Taiwan Aerosol Observation Network consists of seven ground-based stations over the Taiwan and PengHu Islands. The geographical locations of the stations are shown in Figure 1. Aerosol samples (PM_{10} and $PM_{2.5}$) are regularly collected in the last week of each month. The sampling period of each sample is 12 h: daytime samples are collected from 00:00 to 12:00 (GMT), whereas nighttime sampling is carried out from 12:00 to 00:00 (GMT). In addition to mass concentration, chemical species (organic carbon, elemental carbon, soluble ions, crustal elements) in the aerosol samples are measured either.

Results and Discussion

Taiwan is under the lee of the Asian winter monsoons which originating in the Asian Continent. In 2003–2005, there were 38 half-day samples successfully collected by the Network during the episodes of prevailing northeasterly monsoons.

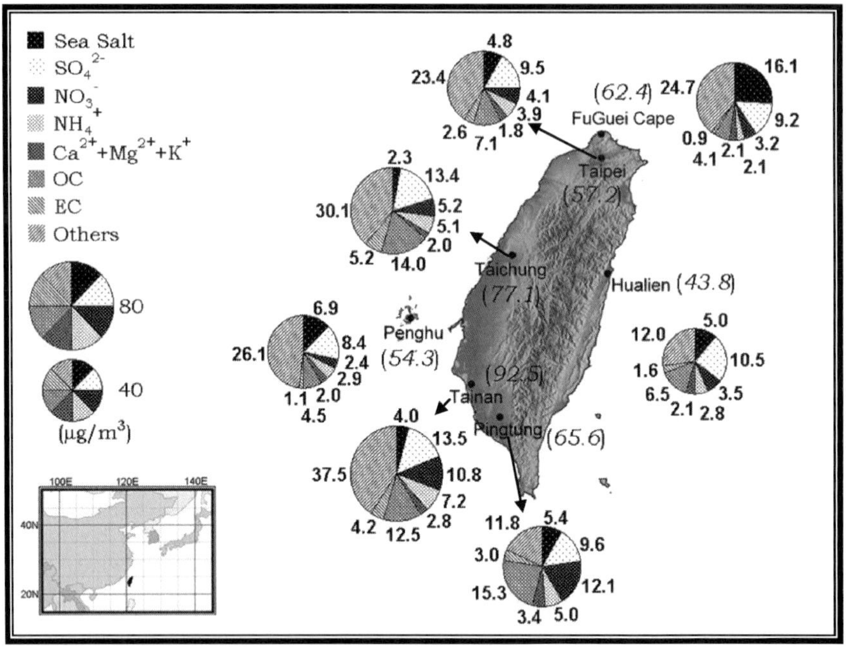

Figure 1 Average mass concentrations and chemical compositions of PM_{10} measured at respective stations of the Taiwan Aerosol Observation Network during the Asian outflow episodes in 2003–2005

Figure 1 illustrates the averages of mass concentrations and chemical compositions of PM_{10} samples at the seven stations, respectively, for the Asian outflow episodes The measurements made at the FuGuei Cape, the northern tip of Taiwan, are ideal to characterize the Asian outflow aerosols. In addition to the sea salt particles generated by the strong northeasters, there are also a substantial amount of anthropogenic species in the aerosols. Furthermore, it was found that the ambient levels of the anthropogenic aerosols (SO_4^{2-}, NO_3^-, OC, EC) are rather consistent among the three rural and off-coast stations: FuGuei Cape, Hualien, and PengHu. In this context, the Asian outflows indeed contained a significant amount of air pollutants and should have important influences upon the air quality as well as atmospheric chemistry in this region.

While the air parcels of Asian outflows passing through Taiwan, the long-range transported aerosols can react with the reactive pollutants emitted locally. It was revealed that the chlorine deficient in aerosols increased from 12% at FuGuei Cape to ~30% at Hualien, and PengHu. Figure 2 shows that the sodium in aerosols correlates linearly with the sum of chlorine and nitrate ions. Furthermore the slope of the line, 1.18, agrees well with the average composition of sea water, implying that the chlorine deficient in aerosols can be almost totally attributed to the reaction between NaCl particles and the nitric acid. Hence, the increases in the chlorine deficient from FuGuei Cape to

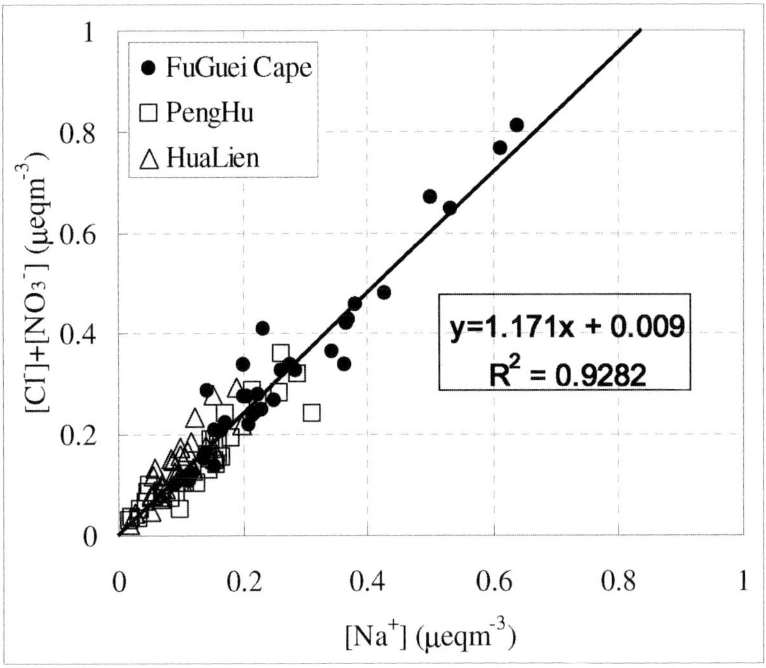

Figure 2 Linear correlation between the sodium and the sum of chlorine and nitrate ions shows that the chlorine deficient in aerosols is a result of the reaction of nitric acid and sea salt particles

the southern stations should be a result of chemical transformation of aerosols by local pollution. Besides, examining the molar ratio of sulfate and ammonium in the aerosols showed that the ammonium bisulfate in Asian outflows could have been neutralized by the locally emitted ammonia as they are passing through Taiwan.

Conclusions

The measurements of aerosol compositions in Taiwan during Asian outflow episodes revealed that the anthropogenic aerosols emitted in the East China have significant influences on the regional atmospheric chemistry and air quality. As the air parcel passing through Taiwan the aerosols can react with reactive species emitted there. Given that the Asian winter monsoons can bring the aerosols further southerly, the aerosols in the Asian outflows indeed play the role of carrier of the reactive gases from Taiwan to the downwind areas, e.g., the South China Sea and Southeastern Asia.

Acknowledgments The authors are grateful to the National Science Council and Environmental Protection Administration of Taiwan for the financial supports.

References

1. Hoell, J.M., Davis, D.D., Liu, S.C., Newell, R., Shipham, M., Akimoto, H., McNeal, R.J. Bendura, R.J., Drewry, J.W., *J. Geophys. Res.*, **101**(D1), 1641–1654 (1996).
2. Hoell, J.M., Davis, D.D., Liu, S.C., Newell, R.E., Akimoto, H., McNeal, R.J., Bendura, R.J., *J. Geophys. Res.*, **102**(D23), 28223–28240 (1997).
3. Huebert, B.J., Bates, T., Russell, P.B., Shi, G., Kim, Y.J., Kawamura, K., Carmichael, G. Nakajima, T., *J. Geophys. Res.*, **108** (D23), 8633 (2003).
4. Jacob, D.J., Crawford, J.H., Kleb, M.M., Connors, V.S., Bendura, R.J., Raper, J.L., Sachse G.W., Gille, J.C., Emmons, L., Heald, C.L., *J. Geophys. Res.*, **108** (D20), 9000 (2003).

The Observations of Cloud Condensation Nuclei over the Bohai Gulf

Lixin Shi, Ying Duan, and Xiaobo Dong

Abstract The aircraft observations of cloud condensation nuclei (CCN) were conducted over the Bohai Gulf during the autumn of 2006. Analyses show that the CCN over nearshore gulf came from two different resources. The low-level nearshore maritime CCN were greatly influenced by anthropogenic emissions.

Keywords Cloud condensation nuclei, anthropogenic emission

ntroduction

ı recent years, the interest of cloud physics has shifted emphasis away from preipitation toward radiative properties as the increasing of global climate studies.[1] 'loud Condensation Nuclei (CCN) are the aerosol particles that can form cloud roplets. There is "indirect effect" arises from the possible influence of anthropoenic CCN. The first is Twomey effect that the increase in cloud droplets number oncentration could increase the multiple scattering within clouds thereby increasıg cloud-top albedo.[2,3] The second is Albrecht effect that the increase in cloud roplet concentration may also inhibit precipitation development, enhancing cloud fetime and resulting in an increase in planetary shortwave albedo[4] and possibly so in the atmospheric absorption of longwave radiation by the resultant increased mospheric loading of liquid water and water vapor.[5]

CCN studies have been made for decades all over the world. In China, observaons on CCN only were performed in 1983–1985 using a static thermal gradient CN counter. The Bohai Gulf locates in the north of China. Around the gulf there e the North China industrial areas. The purpose of this paper is to present the bservations of CCN over Bohai Gulf in the autumn of 2006.

eather Modification Office of Hebei Province, China,050021

D. O'Dowd and P.E. Wagner (eds.), *Nucleation and Atmospheric Aerosols*, 771–776.

Instrumentation

A Piper Cheyenne IIIA twin turbo-prop aircraft was used for the observations. CC were measured by a DMT (Droplet Measurement Technologies, USA) continuou flow streamwise thermal gradient CCN counter.[6] It has a cylindrical continuous flo thermal gradient diffusion chamber employing a novel technique by establishing constant streamwise temperature gradient so that the difference in water vapor an thermal diffusivity yielding a quasi-uniform centerline supersaturation (S). A optical particle counter at the outlet of the chamber counts droplets with diamete larger than 0.75 μm. Those particles larger than 0.75 μm are considered activate CCN and comprise the CCN concentration. The temperature gradient and the flo though the column control the S and may be modified to retrieve CCN spectra. Th instrument can operate between $S = 0.1\%$ and $S = 2\%$ at sampling rata of 1 Hz. I this study, CCN measurements were made at $S = 0.3\%$. CCN spectra were obtaine from 0.1%, 0.2%, 0.3%, and 0.5% saturation cycling measurements.

Case Studies

Two research flights were performed on 27 September and 22 October 2006. Th maritime data were collected about 20 km off coast over the gulf (Figure 1). On th 700 hPa charts for 08:00 BST (omitted), the North China was both dominated b northwesterly winds in the two days. There were little small cumulus clouds and litt stratocumulus clouds over the gulf on 27 September and 22 October, respectively.

27 September 2006

Two CCN layers and an inversion characterized the vertical profile (Figure 2). Th 1,700–2,100 m a.s.l. inversion resulted in the cumulation of high CCN concentr tions below this layer. The CCN concentrations ($S = 0.3\%$) averaged 1,332 cm between 600 and 2,100 m and 59 cm^{-3} between 2,100 and 7,000 m. It seems th CCN over the gulf came from different resources. The back trajectories of the a mass for that day produced by the NOAA HYSPLIT model (http://www.arl.noa gov/ready/hysplit4.html, NOAA Air Resources Laboratory, Silver Spring, MD) a shown in Figure 3. It suggests that the air masses in the boundary layer (top is abo 2,100 m) passed from northern continent of the gulf where that is the heavy indu trial areas (around Tangshan city) of North China and carried regional anthrop genic emissions out over the Bohai Sea. Back trajectories show the air mass abo boundary layer was from Inner Mongolia, and the air mass was not influenced b these cities of Beijing and Tianjin.

Figure 4 gives the horizontal leg measurements at 3,000 m from 13:38 to 13: BST. The averaged concentration was 142 cm^{-3} over sea and 135 cm^{-3} over lan

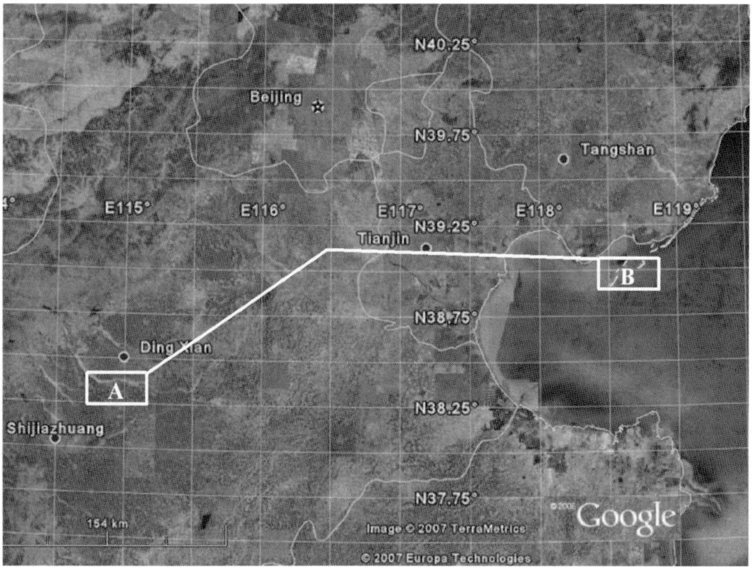

Figure 1 Flights track for CCN observations on 27 September and 22 October 2006. Rectangle A shows the ascent/decent area over the airport. Rectangle B shows the ascent/descent area over Bohai Gulf. Dog-leg path shows the horizontal leg

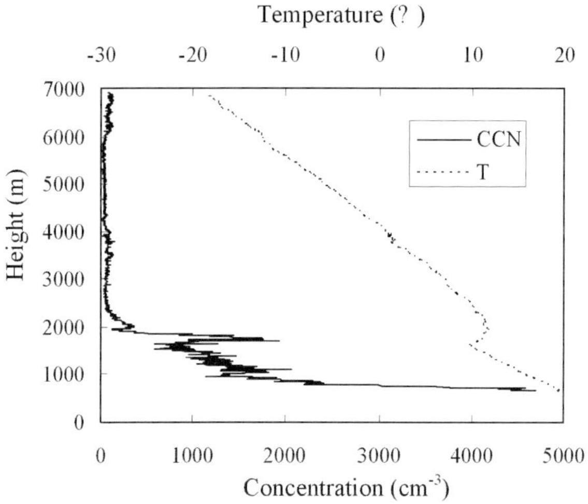

Figure 2 Vertical profile for CCN concentration ($S = 0.3\%$) and temperature over the Bohai Gulf during flight on 27 September 2006

At 13:50 it was over Tianjin City. The inland CCN concentration has the same magnitude even more than 100 km from the coastline (data are not shown here). There was no difference between nearshore and continental conditions above the boundary layer.

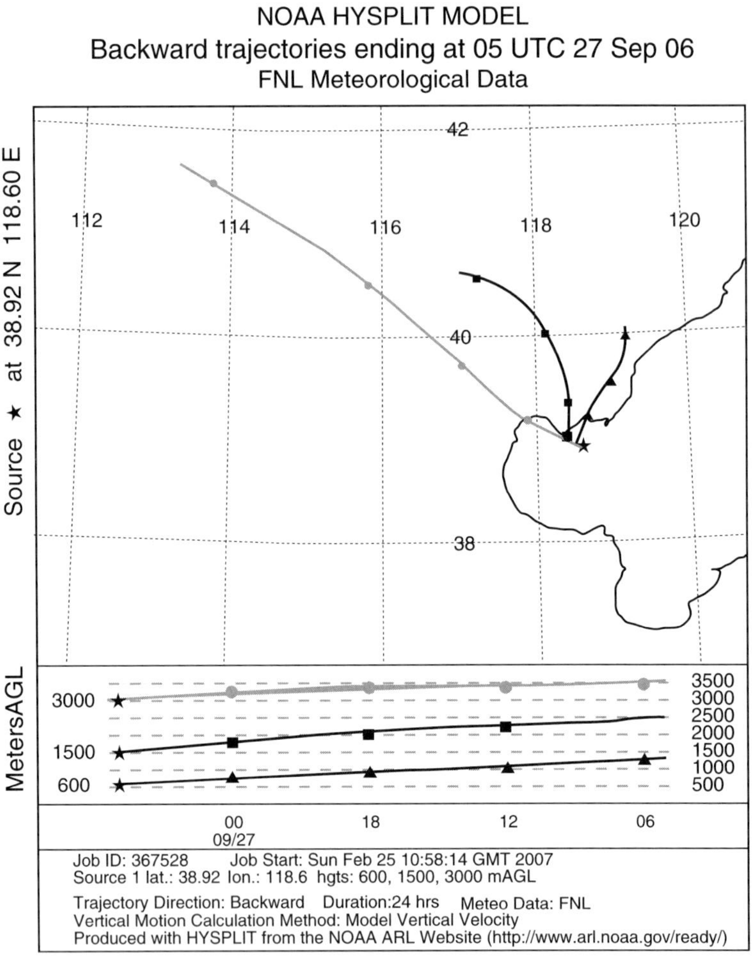

Figure 3 Simulated 24-h back trajectories for the air masses on 27 September 2006

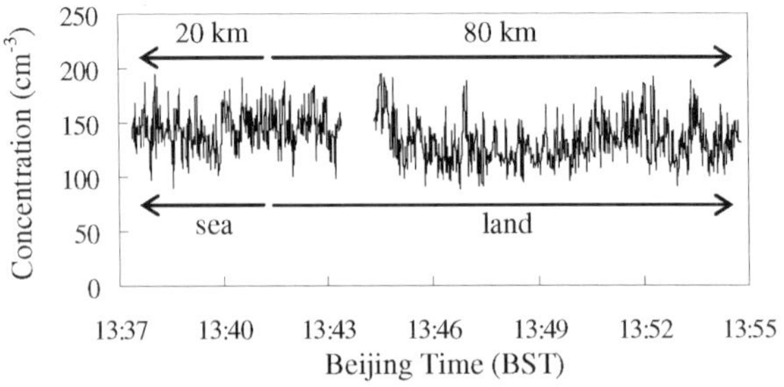

Figure 4 Time series of CCN concentration ($S = 0.3\%$) during horizontal leg at 3,000 m a.s.l. on 27 September 2006. The brief gap is due to occasional instability in the instrument

Figure 5 As Figure 2, but for 22 October 2006

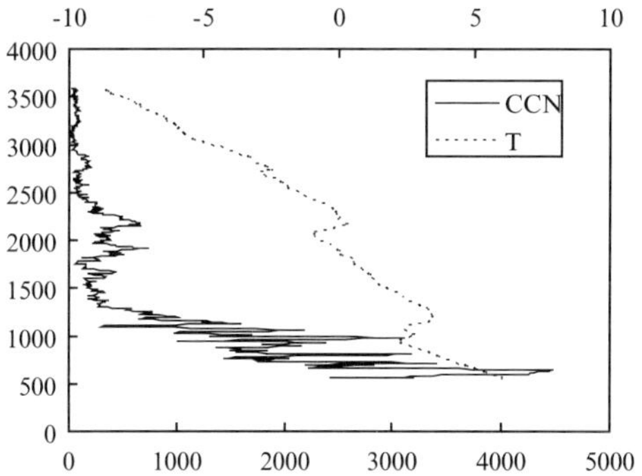

CCN spectra can be described by the expression[7] $N = CS^k$, where N is the number of CCN activated at supersaturation S, C is the number activated at $S = 1\%$, and k is a constant. At 600 m, the calculated CCN spectrum is $N = 6942S^{0.43}$. Twomey and Wojciechowski found values of k ranging from 0.2 to about 1.0, with marine air typically producing values near 0.5 and continental air values near 0.7. Values of C changes greatly, continental air has typically an order of magnitude more CCN than marine air.[8] The high C value in this study shows that the nearshore CCN have been greatly influenced by continental air.

22 October 2006

Three CCN layers and two main inversion layers represented the vertical profile (Figure 5). The 1,100–1,300 m inversion still resulted in the cumulation of high CCN concentration in boundary layer. The averaged CCN concentration ($S = 0.3\%$) was 1,890 cm^{-3} between 600 and 1,300 m, 319 cm^{-3} between 1,300 and 2,200 m, and 115 cm^{-3} between 2,200 and 3,600 m. Back trajectories (omitted) show the similar result as 27 September, the air masses in boundary layer (top about 2,200 m) also passed over Tangshan industrial areas, the air mass above boundary layer was from Mongolia. The CCN spectrum at 600 m is $N = 3841S^{0.65}$. The high C value also shows the low-level air mass was influenced by anthropogenic emissions.

Discussion

Table 1 summarizes the measurements of CCN during horizontal leg of the two flights over the gulf. It can be seen that the CCN concentrations were high in the boundary layer. The vertical profile, back trajectories and CCN spectra analyses show the CCN in boundary layer over nearshore Bohai Gulf belonged to modified maritime type. Additional horizontal measurements at 3,000 m on 27 September

Table 1 CCN data summary for horizontal measurements over the Bohai gulf

			Averaged Concentration (cm^{-3}) at Different *S*			
Date	Time (BST)	Altitude (m)	0.1%	0.2%	0.3%	0.5%
27 September 2006	1245–1257	600	2443	3793	4100	4970
22 October 2006	1306–1317	600	771	1305	1488	2265
22 October 2006	1235–1248	1200	728	1092	1208	1218

suggest the CCN above bounder layer were from free troposphere. However, there also exists the possibility of sea–land-breeze circulation producing mixing with continental polluted air.[9] Because there was no observation at that time, the investigation for this detail mechanism needs to do in the future.

Acknowledgments This research was supported by the National Natural Sciences Foundation Contract 40475003 and the Hebei Province Natural Sciences Foundation Contract D2005000593 We thank the technicians of Yupeng Deng, Yan Jiang, and Zhijun Zhao, also Xinjian Huang and other members of the Cheyenne IIIA crew for help in collecting data, and Professor Yu Qin for insightful advice on this research.

References

1. Hudson, J.G., *J. Appl. Meteor.*, **32**, 596–607 (1993).
2. Twomey, S., Piepgrass M., and Wolfe T.L., *Tellus*, **36B**, 356–366 (1984).
3. Platnick, S.E., and Twomey S., *J. Appl. Meteor.*, **33**, 334–347 (1994).
4. Albrecht, B.A., *Science*, **245**, 1227–1230 (1989).
5. Schwartz, S.E., Cloud droplet nucleation and its connection to aerosol properties, in: *Nucleation and Atmospheric Aerosols 1996, Proceedings of the 14th Int. Conf. Nucleation and Atmospheric Aerosols*, edited by M. Kulmala et al., Oxford: Elsevier Science , pp. 770–779 (1996).
6. Roberts, G.C. and Nenes, *Aeros. Sci. Tech.*, **39**, 206–221 (2005).
7. Twomey, S., *Geofis. Pura Appl.*, **43**, 243–249 (1959).
8. Twomey, S. and Wojciechowski, T.A., *J. Atmos. Sci.*, **26**, 684–688 (1969).
9. Yoshikado, H. and Tsuchida, M.A., *J. Appl. Meteor.*, **35**, 1804–1813 (1996).

Study of Aerosol Emission at Continental Area of Russia

V.V. Smirnov

Abstract During 2005–2006 in the Centre of European territory of Russia the inflow, emission and sink new aerosols (ENA) had regular character in various synoptic situations. The ENA intensity is defined mainly by the mixing conditions in low troposphere, air masses prehistory and rain–snow characteristics. Dependences of ENA from local wind speed and direction as well as the minor gases concentrations, except on ozone are less significant.

Keywords Low troposphere, aerosol transport, nanometer particles

Introduction

In the spring and autumn at few station in Northern Europe [1–3] were observed intensive emissions of aerosols by diameter D > 1 nm (further ENA). During of 5–8 h after the ENA starts can be formed the cloud condensation nuclei by sizes D = 50–150 nm. The meteorological importance of the phenomenon stimulates researches of ENA in various geography regions. In given work the preliminary data of the continuous ENA monitoring is briefly generalized. Measurements were started in February 2005 at forestry site 1 and country site 2. Sites were located in the vicinity of university city Obninsk, placed on 100 km to southwest from Moscow. The primary results allow us to speak about earlier unknown peculiarities of ENA in continental atmosphere.

Long-term aerosol measurement for interval D = 4–1000 nm are provided by high-resolution electric analyzers with a unipolar charger and aerosol electrometer similarly to the model 3030, Thermo-Systems Inc., USA. Input resistance of electrometer is 10^{12} Ohm, noise current is 10^{-16}A, and relaxation time is 60 s. The air samples aspirated from 4 m height with volume velocity of 100 cm^3 s^{-1}. During the daytime ENA were used the small and intermediate air ion mobility spectrometer ALMIS-2 and photoelectric aerosol counter model 218, Royco Inc., USA [2].

Institute of Experimental Meteorology, Obninsk, Russia 249031

C.D. O'Dowd and P.E. Wagner (eds.), *Nucleation and Atmospheric Aerosols*, 777–781.

For studying of the low troposphere structure were used ground-based sonar M300C-IEM (http://sodar.obninsk.org) and sensors of temperature, wind direction and speed at levels of meteorological tower 4, 8, 25, 73, 121, 217, and 301 m (http://typhoon-tower.obninsk.org). Site 1 is located in rarefied coniferous forest at removal ~1 km from the tower and sonar position. Removal of site 2 is 13 km. Local concentrations of ozone and other minor gases were supervised at height of 4 m.

Below we shall generalize the main ENA relations with an atmosphere condition then discuss the ENA features in air masses of high and low atmospheric pressure.

The Basic Parameters of Continental ENA

1. The regular ENA starts in daytime in form of the fast appearance of aerosol particles by sizes D = 4–20 nm and concentration of N = 200–100,000 cm^{-3}. Any appreciable correlations ENA with wind speed and direction on sites 1 and 2 are noted.
2. In 2005–2006 the regular ENA was observed in 70–80% calendar days. The start of the regular ENA coincided with destruction the morning temperature inversion. The ENA reduction coincided with formation of nighttime inversion. The most intensive ENA was associated with winter anticyclones. In sunny days the total number N and surface area S concentrations of new particles reached $N = 3.10^4–3.10^5\ cm^{-3}$, $S = 100–200\ \mu m^2\ cm^{-3}$, that in 10–100 times exceeded nighttime values.
3. Maximal values of N and S did not correlated strong with maximum of sunlight intensity and air temperature. Minimal N and S are observed at nighttime after achievement of the minimal temperature and depth of inversion layer.
4. Character, intensity and duration of ENA were defined mainly by air mass lifetime T_a. So, when air masses were transported from Northern Atlantic (in 2006, T_a = 2–4 days in 60% of episodes,), the ENA appeared already as enriched by particles of D = 10–100 nm. In those conditions the generation of intermediate air ions (effective size of D = 1–2 nm) and nanoparticles (D = 4–10 nm) also took place. When air masses were transported from the Arctic (30% of episodes ENA in 2006, T_a < 15–25 h), the ENA character was approximately as at Finland boreal forest [1–3]. Namely, in the first time there was formed the intermediate air ions and nanoparticles. Then took place the gradual rising for all size fractions and occurrence of cloud condensation nuclei D > 50–100 nm.
5. Unlike observations in Northern Europe [1, 2], the regular ENA at sites 1 and 2 observed not only at the high atmospheric pressure and the cloudless, but also in permanent cloudiness. It is interesting also that in 60% of episodes under developed convection; the ENA intensity was decreased at noon. Concrete illustrations see below.

Regular ENA in Anticyclones

Most long-term and intensive emission of new particles of D = 4–500 nm took place in anticyclones in February–March and October–November at transportation of cold air masses from Northern Atlantic. In 70% of the spring ENA episodes the new particles concentration reached $N = 10^5$ cm^{-3}, sometimes up to $N = 5.10^5$ cm^{-3}. Typical daily variations of size spectra dN/dlogD (the top diagram) and total concentration N are illustrated by the Figure 1. The complete spectrum dN/dlogD with 19 differential size channels was recorded during 20 min.

At nighttime of 4–6 h (GMT + 3) the inversion depth was minimized by levels Z_i = 70–100 m. Hereinafter Z_i is estimation from sonar and meteorological sensor measurements. Total aerosol concentration decreased to identical minimum $N = 10^3$ cm^{-3}. Aerosol particles by sizes D = 50–150 nm were prevailed.

During sunrises from 8 to 10 h it was observed the nanoparticle concentration growth from $N = (1–3).10^3$ cm^{-3} up to $N = (1–5).10^4$ cm^{-3}.

Averaged growth rate of surface concentration S has appeared rather high dS/dt ≈ 1–2 μm^2 cm^{-3} s^{-1} (for sizes of D = 10–150 nm). The known models of conversion "gas-aerosol" cannot provide such values dS/dt [3].

Acrding the sonar data during 8–10 h the top border of inversion was displaced usually from 50–70 m to 200–250 m. Then, it is more logical to explain ENA effects

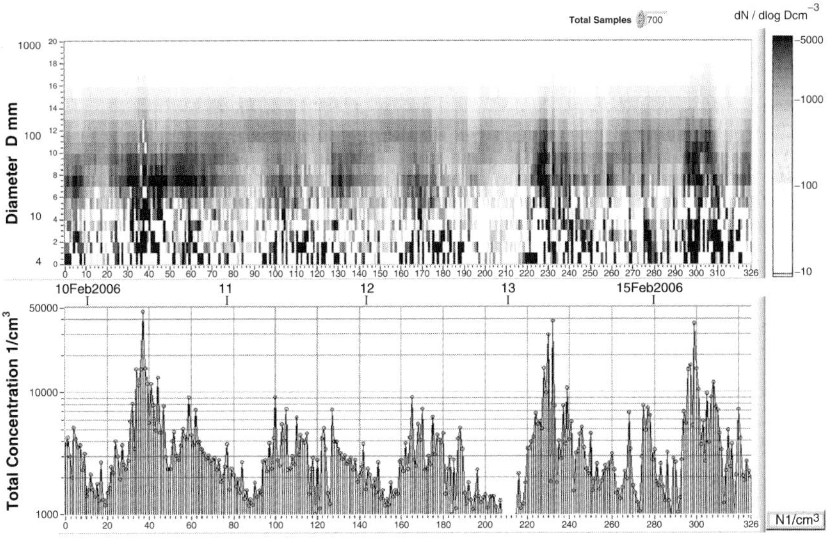

Figure 1 The 3D diagram «aerosol size D – concentration dN/dlogD – current days/sample number» and daily variations of total concentration N. Measurements during 09–14 February 2006 at site 1. Air masses from Northern Atlantic. Clouds and precipitation were absent. Snow cover depth is 30 cm. Air temperature is –(32–10) °C, relative humidity is 65–85%, wind speed is 0.5–2 m/s, and atmospheric pressure 756 ± 6 mm Hg

by improvement of air mixing and inflow of new particles from the overlying layers of troposphere.

It would seem, that the further air warming up and intensification of mixing processes in boundary layer should was stimulate the vertical transport of new particles. In practice, as can see from Figure 1, after the noon there was stabilization or dissipation of nanoparticles (D = 3–12 nm) and Aitken nuclei (D = 10–50 nm). To explain these facts it is possible by the limited concentration of the aerosol and gas precursors, which are grouped near to the top border of inversion.

Regular ENA in Cloudy Atmosphere

As is known, weather of the Central and Eastern Europe in December and January 2007 was warmer on 7–8°C than usually. The sunny day quantity has decreased by factors of 5–6 if to average for last 100 years. It is useful to consider a reaction the aerosol components of continental atmosphere on these meteorological anomalies.

It is noticeable from Figures 1 and 2 the character of daily aerosol variations is similar. However, pays attention the interesting facts. So, aerosol size modes 4–6, 9–12, 30–40 and 50–70 nm are better identified in cloudy atmosphere.

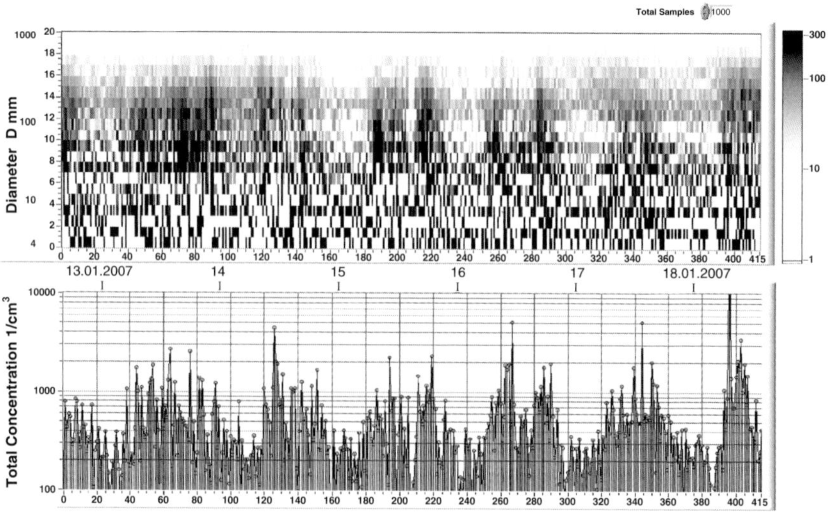

Figure 2 Measurements during 12–19 January 2007 at site 1. Air masses from Northern Atlantic. Cloudiness during 95% of all time. Southern and southwest wind speed is 0–1 m/s. Air temperature from −1°C to +5°C, relative humidity at noon is 80–85%, at midnight is 93–96%. Atmospheric pressure is 726–740 mm Hg. Spectra № 207–210 were measured at the switched off unipolar charger of electric aerosol analyzer

Particles of the accumulative fraction D = 70–150 nm are displaced in greater sizes area D = 150–300 nm that it is reasonable to connect with increasing of relative humidity. During the daytime emissions the size spectrum becomes unimodal in wide interval of D = 30–500 nm. More detailed comment is demanded also the impulse nature of ENA, very low nighttime aerosol concentration N = 100–200 cm^{-3}, S = 3 μm^2 cm^{-3} and other.

References

1. Nilsson, E.D. and Kulmala, M., *Atmos. Chem. Phys. Discuss.*, **6**, 10425–10462 (2006).
2. Smirnov, V.V., Salm, J., Makela, J., and Paatero, J., *Russian Meteorology and Hydrology*, **4**, 30–42 (2005).
3. Smirnov, V.V. Nature and evolution of ultrafine aerosol particles in the atmosphere, in: *Atmospheric and Oceanic Physics*, USA: Pleiades, **6**, pp. 663–687 (2006).

Modeling the Non-ideal Thermodynamics of Mixed Organic/Inorganic Aerosols

A. Zuend, C. Marcolli, B.P. Luo, and Th. Peter

Abstract Tropospheric aerosols contain mixtures of inorganic salts, acids, water, and organic compounds. Interactions between components in such a mixture lead to discrepancies from ideal thermodynamic behavior. Using component activities instead of molar concentrations, one can account for this nonideal behavior. The LIFAC model by Yan et al.,[1] was modified and parameterized to better describe atmospheric relevant conditions and mixture compositions. With these modifications the activities of inorganic salt solutions and acids are well represented up to high ionic strength. The parameterization of direct inorganic/organic interactions of polyol/water/salt solutions strongly improves the agreement between experimental and modeled activity coefficients.

Keywords Aerosol thermodynamics, activity coefficients, organic/inorganic interactions

Introduction

Organic chemical species are ubiquitous in tropospheric aerosols. Field measurements show that organics are not just traces in individual aerosol particles, but rather that about 30% to over 80% of the aerosol mass in the free troposphere are carbonaceous material. Most of this carbonaceous material is probably organic.[2] This is comparable to the other well-known tropospheric aerosol species, sulfuric acid and ammonia, which are mostly internally mixed with organics. The composition of organic aerosols is highly complex, containing hundreds of compounds with a large fraction still unidentified.[3] Such a wide variety of organic compounds in the liquid and solid aerosol phases demands for some classification. This is often done considering different functional groups implying that the semi-volatile organics

Institute for Atmospheric and Climate Science, ETH Zurich, Switzerland

C.D. O'Dowd and P.E. Wagner (eds.), *Nucleation and Atmospheric Aerosols*, 782–786.

tend to contain a high degree of functionality, including hydroxyl, carbonyl, and carboxy groups.[4] Furthermore, the aerosol phase contains also inorganic constituents such as nitrate, sulfate, ammonium, chloride, etc., which all interact with each other.

While the thermodynamics of aqueous inorganic systems at atmospheric temperatures is well established, little is known about the physicochemistry of mixed organic/inorganic particles. Required is a thermodynamic model that predicts the phase partitioning of organic compounds between the gas and the condensed phases at thermodynamic equilibrium.

In this study, we show an activity coefficient model, which explicitly accounts for molecular interaction effects between the constituents, both organic and inorganic. This approach allows the calculation for activity coefficients of all components and hence, the prediction of liquid–liquid phase separations, crystallization in the particle phase and gas/particle partitioning. To reduce the complexity and make use of already measured data, we focus in this report on liquid mixtures containing water, inorganic salts and primary, secondary, or tertiary alcohols or polyols.

Method

Modeling Organic/Inorganic Interactions

Interactions between different chemical components of a mixed solution change the physicochemical properties of these components and thus, those of the solution compared to an ideal solution, where the constituents act as in a pure liquid of that single component. This nonideal mixing leads to distinct deviations from Raoult's law. To correct for nonideal behavior, activity coefficients were introduced. This allows for the computation of the partial derivative for the Gibbs free energy of a system by a molar component, the so-called chemical potential. Pitzer ion-interaction models are well-known for their ability to calculate activity coefficients in aqueous electrolyte solutions, ranging up to a high ionic strength.[5,6] For organic systems, the UNIFAC model – a group-contribution method for predicting liquid-phase activity coefficients of organic mixtures – is widely used.[7] The LIFAC model by Yan et al.[1] merges a Pitzer with a UNIFAC model to calculate the activity coefficients of mixed organic/inorganic systems. This framework enables the computation of chemical potentials and the Gibbs free energy of mixed systems, and therefore possible phase separations. The original LIFAC model was developed for chemical engineering purposes, which differ mainly in the selection of chemical species, and the concentration and temperature range from the needs of aerosol science. In this study, a modified LIFAC model was developed to describe atmospheric relevant aqueous solutions up to high concentrations at room temperature.

Modified LIFAC

The excess Gibbs free energy is an expression for the nonideality of a thermodynamic system. Different interaction types and ranges between solution molecules can be identified and related to the Pitzer model part and the UNIFAC part. Figure 1 shows the three considered interaction ranges in LIFAC. Long range (LR) and middle range (MR) interactions are described within a Pitzer-like part. In which the former are described by an extended Debye-Huckel term. Short range (SR) interactions of noncharged components are calculated in the UNIFAC part. In addition to the original LIFAC we used a binary three-body interaction term in the middle range part, to achieve better results at high electrolyte concentrations. Even supersaturated conditions in ionic solutions are thereby described by the model. We also introduced more realistic values of the ionic group volume and surface area parameters in the UNIFAC part considering dynamic hydration of ions. The sulfuric acid – ammonium sulfate system was introduced, taking the partial dissociation of bisulfate into account. Some other ions were excluded because their abundance in aerosols is marginal.

In the UNIFAC part, changes were made to better meet the specific properties of atmospheric semi-volatile organics, where several strongly polar functional groups can be found in a molecule.[8] Therefore the relative distances of a molecule's functional groups are taken into account.

Binary aqueous electrolyte solutions, as well as ternary organic/inorganic aqueous mixtures, require specific interaction parameters. The semi-empirical expressions of the MR part contain freely adjustable parameters, describing interactions between ions, organic functional groups and water. The Debye-Huckel (LR) part and the UNIFAC (SR) part contain no adjustable parameters for ions. Organic–organic and organic–water interactions are parameterized in the SR part. These UNIFAC parameters were taken from the literature.[8] Due to the many

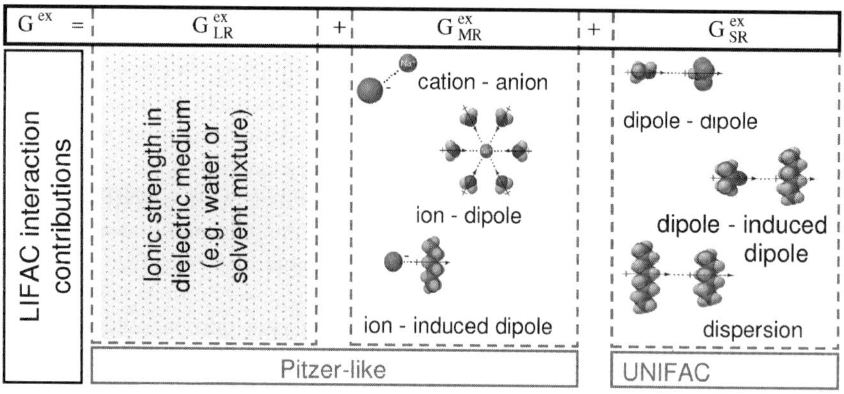

Figure 1 Three major terms, representing different types of molecular interactions, add up to the excess Gibbs free energy (G^{ex}) of a certain system in LIFAC

modifications in the MR part, all pertinent interaction parameters had to be recalculated by fitting the model to experimental data.

Measurements of water activity and activity coefficients comprising several single salt solutions at room temperature were used to parameterize the inorganic part of the model. Experimental data of ternary mixtures containing organics (by now just alcohols and polyols), water and inorganic ions were then used to fit the specific binary interaction coefficients between ions and organic functional groups. Vapor–liquid equilibria (VLE) data, as well as liquid–liquid equilibria (LLE) compositions, salt solubilities and electromotive force (EMF) measurements were used to establish a broad and reliable database.

Results

Water activities, and thus, the relative humidity of the gas phase in equilibrium with aqueous salt solutions are accurately represented by modified LIFAC; e.g., ammonium sulfate solutions are well described up to a salt weight fraction of 85% (water activity: 0.35) and sodium chloride up to at least 45% (water activity: 0.45). Inclusion of organic functional groups – ion interactions strongly improves the agreement between measured and modeled activity coefficients of alcohol/water/salt solutions. Figure 2 shows the comparison between measured and modeled

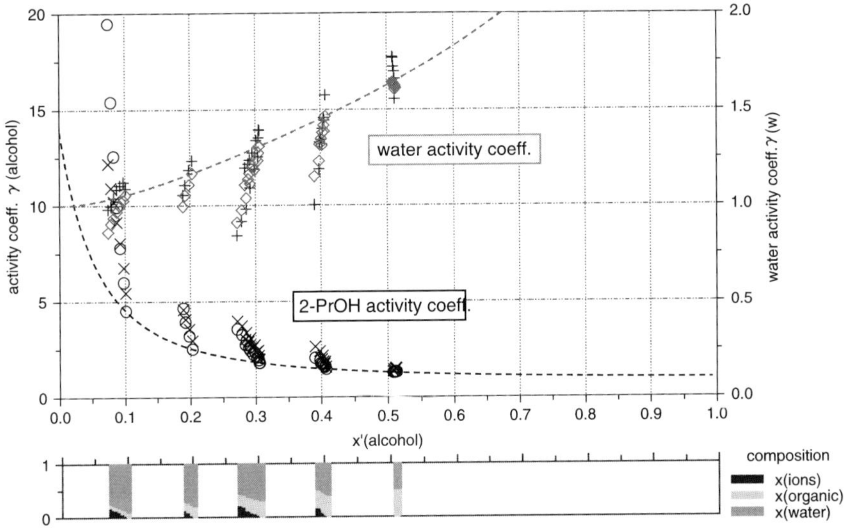

Figure 2 VLE activity coefficients measured by Morrison et al.[9] (**crosses**) and modeled with modified LIFAC (**open symbols**) of NaCl-1-propanol-water solutions. The dashed curves show the model results of the salt-free mixture. The solution composition is plotted below in terms of molar fractions of the constituents with respect to the dissociated salt

activity coefficients for ternary NaCl-1-propanol-water mixtures – one data set of over a hundred used to parameterize the model. Using a broad-based data set and concurrent fitting of parameters reduces the possible influence of inaccurate data and over fitting due to existing inaccuracies in UNIFAC.

Acknowledgment This work was supported by the Swiss National Foundation under No. 200020–103651.

References

1. Yan, W., Topphoff, M., Rose, Ch., and Gmehling, J., *Fluid Phase Equilibria*, **162**, 97–113 (1999).
2. Murphy, D.M., Cziczo, D.J., Froyd, K.D., Hudson, P.K., Matthew, B.M., Middlebrook, A.M., Peltier, R.E., Sullivan, A., Thomson, D.S., and Weber, R.J., *J. Geophys. Res.*, **111**, D23S32, doi:10.1029/2006JD007340 (2006).
3. Rogge, W.F., Mazurek, M.A., Hildemann, L.M., Cass, G.R., Simoneit, B.R.T., *Atmos. Environ.*, **27A**, 1309–1330 (1993).
4. Decesari, S., Faccini, M.C., Fuzzi, S., and Tagliavini, E., *J. Geophys. Res.*, **105**, 1481–1489 (2000).
5. Pitzer, K.S., *Activity Coefficients in Electrolyte Solutions*, 2nd edn., edited by K.S. Pitzer, Boca Raton, FL: CRC Press, pp. 75–153 (1991).
6. Clegg, S.L., Brimblecombe, P., and Wexler, A.S., *J. Phys. Chem. A*, **102**, 2155–2171 (1998).
7. Fredenslund, A., Jones, R.L., and Prausnitz, J.M., *AIChE J.*, **21**, 1086–1099 (1975).
8. Marcolli, C. and Peter, Th., *Atmos. Chem. Phys.*, **5**, 1545–1555 (2005).
9. Morrison, J.F., Baker, J.C., Meredith III, H.C., Nuewman, K.E., Walter, T.D., Massie, J.D., Perry, R.L., and Cummings, P.T., *J. Chem. Eng. Data*, **35**, 395 (1990).

^{7}Be–^{210}Pb Concentration Ratio in Ground Level Air in Málaga (36.7° N, 4.5° W)

C. Dueñas, M.C. Fernández, S. Cañete, and M. Pérez

Abstract Concentration of ^{7}Be and ^{210}Pb in air were continuously monitored, using a high-volume air sampler and a high-resolution gamma-ray spectrometer, at the University of Málaga 36.7° N, 4.5° W from 2000 to 2006. The average monthly concentrations of ^{7}Be ground level air were in the range of 0.36–14.9 mBq/m^{3} and exhibited one or two maxims in sprig and summer and one minimum in autumn. The maximum concentrations for ^{210}Pb were observed in summer, with the average concentrations in the range of 1.4×10^{-4} to 14.4×10^{-4} Bq/m^{3}. The mean monthly concentrations exhibited normal distribution for ^{7}Be and lognormal distribution for ^{210}Pb. The $^{7}Be/^{210}Pb$ ratios were in the range of 4.9–44. The concentration data of ^{7}Be and ^{210}Pb with meteorological variables were used for a comprehensive regression analysis of monthly variation of radioactivity in air.

Keywords ^{7}Be and ^{210}Pb activity concentrations, meteorological factors, tropospheric aerosols

Introduction

Long-term measurements of cosmogenic and atmospheric radionuclides such as ^{7}Be and ^{210}Pb provide important data in studying global atmospheric processes and comparing environmental impact of radioactivity from man-made sources to natural ones. Taken together, ^{7}Be and ^{210}Pb yield information about vertical movements in the atmosphere and the scavenging of aerosols. Several features make ^{7}Be and ^{210}Pb to highly suitable tracers for improving the general circulation model (GCM) for aerosol simulation. ^{7}Be and ^{210}Pb are natural radionuclide tracers of aerosols originating over a range of altitudes in the atmosphere. Following production, they attach to available aerosols and are, therefore, spread evenly over the ambient aerosol

Department of Applied Physics I, Faculty of Sciences, University of Málaga, 29071 Málaga (Spain)

C.D. O'Dowd and P.E. Wagner (eds.), *Nucleation and Atmospheric Aerosols*, 787–791.

size distribution, with respect to area. We report 7 years measurements 2000–2006 of ^{7}Be and ^{210}Pb concentrations in air surface. As ^{7}Be and ^{210}Pb are different origin, their ratio should depend on the altitude from which the air was transported, on continental influences and on removal process.

Material And Methods

Airborne dust samples were collected at a height of 12 m above the ground in Málaga (36° 43′; 40′; N; 4° 28′ 8′; W). Aerosol samples were collected weekly in cellulose membrane filters of 0.8 μm pore size with an air flow rate of approximately 40 L min^{-1}. The air sample was lodged in an all weather sampling station. Measurements by gamma-spectrometry were performed to determine the ^{210}Pb and ^{7}Be activities of the samples using an intrinsic REGe detector. A monthly composite sample containing 4–5 filters was formed (average volume 1,600 m^3) for the ^{210}Pb and ^{7}Be determination. The errors reported are propagated errors arising from the one sigma counting uncertainty due to detector calibration and background correction. The concentrations were corrected for decay to the mid-collection period.

Results

Frequency Distributions of ^{7}Be and ^{210}Pb Specific Activities

The results from individual measurements of ^{7}Be and ^{210}Pb concentrations were analysed to derive the statistical estimates characterizing the distributions. Table 1 provides arithmetic mean (AM) and related statistical information such as geometric mean (GM), standard deviation (SD), dispersion factor of geometric mean (DF), maximum and minimum values and the coefficient of variation (CV). These values are given in mBq m^{-3}.

^{7}Be data and the GM for ^{210}Pb data should be used to characterize average values. Plots of the frequency distribution show histograms for ^{7}Be and ^{210}Pb. Normal distribution for ^{7}Be is significant at the 0,1 level. ^{210}Pb concentration appears log-normal at the 0.1 level Figure 1. Assuming these types of distribution, the AM for $^{7}Be$and the GM for ^{210}Pb should be used to characterized average values.

Table 1 Statistical parameters

	AM	GM	SD	DF	Max	Min	CV (%)
^{7}Be	4.757	4.217	1.809	1.420	14.900	2.470	39.9
^{210}Pb	0.574	0.555	0.026	0.620	1.443	0.140	37.5

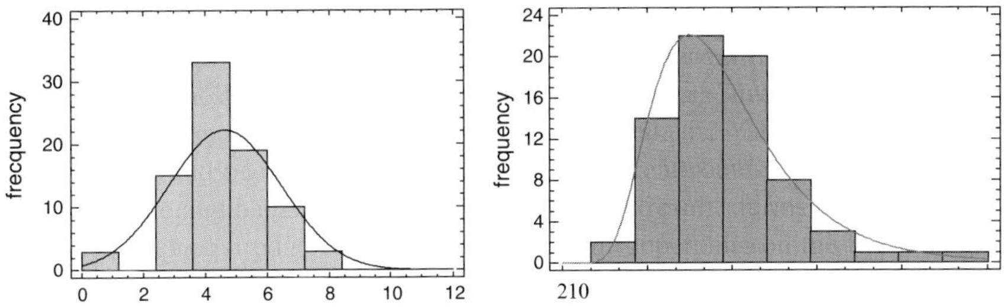

Figure 1 Plots of the frequency distribution of ^{7}Be and ^{210}Pb activities

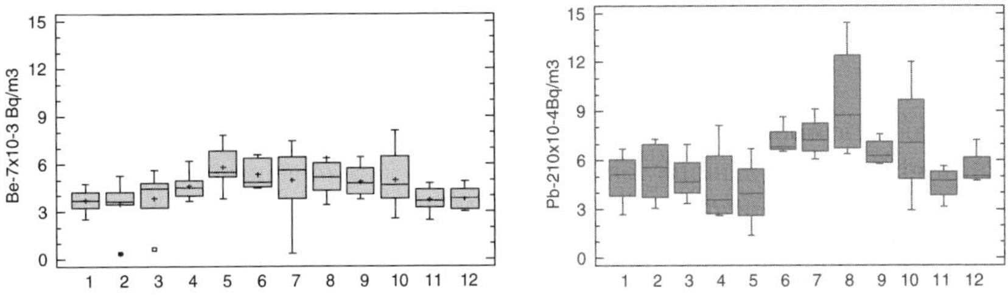

Figure 2 Monthly variations of the activities for the 2000–2006 period

Temporal Variations of ^{7}Be and ^{210}Pb Specific Activities

Figure 2 shows the box and whisker diagrams for monthly variation of ^{7}Be and ^{210}Pb in air activities. The activities of ^{210}Pb shows a marked monthly variation with higher values of activity in summer months and lower values in other months. The results of Figure 2 indicate significant differences in the activity levels of ^{7}Be not only between summer and the rest of the seasons but also between spring and other seasons. The high activities in the summer months are caused by increased vertical transport of ^{7}Be activities from the upper troposphere due to decreased stability of the troposphere during the summer months. The ^{7}Be concentrations in spring are higher than the activities in autumn and winter and are normally assigned to the thinning of the tropause which takes place at midlatitudes, resulting in air exchange between stratosphere–troposphere during late in winter and early spring. In other words, the monthly behaviour of the mentioned tracers shows similar evolution during the summer months but a very different evolution in other months.

From a visual inspection of the data, annual changes seem to be produced and are commonly attributable to different factors such as temperature, atmospheric stability or frequency and amount of precipitation. In order to find the meteorological factors that influence the specific activities of ^{7}Be and ^{210} Pb, we performed a simple regression with meteorological factors; meteorological data were supplied by the Meteorological Institute located 500 m away from the sampling site. In our

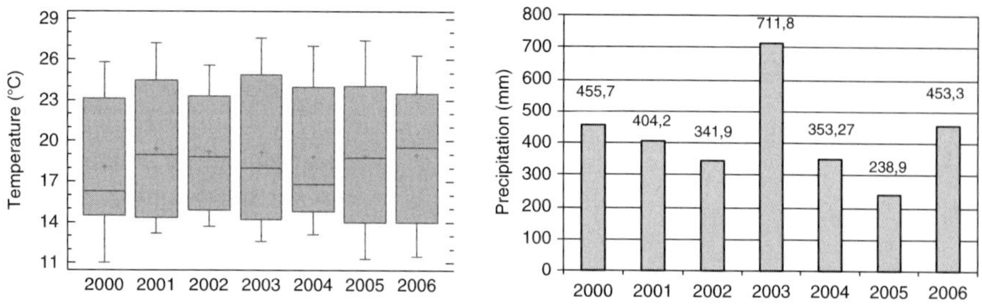

Figure 3 Plots of the air temperature and precipitation for the 2000–2006 period

analyses we used monthly average of the temperature expressed in °C, monthly average of precipitation expressed in mm, monthly average of the relative humidity expressed in % and monthly average wind speed (expressed in m/s).

Judging from the linear coefficient values, it can be said that a high activity of ^{7}Be and ^{210}Pb are present under conditions of high temperature and inverse correlation with precipitation, wind speed and relative humidity. Temperature is the variable most strongly correlated to the activities of mentioned radioisotopes. High temperatures often lead to the setting up of upward convection currents in the atmosphere favouring the radon emanation and dispersion of aerosol particles.

Figure 3 shows the box and whisker diagrams of the air temperature during 6 years and also the amount of precipitation per year. Examining the diagram we conclude that there are differences among the studied years, mainly concerning the precipitation.

Concentrations of ^{7}Be/ ^{210}Pb Activities Ratio as Atmospheric Tracer

Taken together, ^{7}Be and ^{210}Pb yield information about vertical motions in the atmosphere and the scavenging of aerosols. Several features make ^{7}Be and ^{210}Pb highly suitable tracers for improving general circulation models (CGM) aerosol simulation (Koch et al. 1996). Since the production source functions of both tracers are distinctly different, a relatively high correlation between the specific activities of ^{7}Be and ^{210}Pb will provide information on the atmospheric removal behaviour on these nuclides as well as information of whether these two radionuclides can be used as independent atmospheric tracers (Baskaran 1995). The specific activities of ^{210}Pb plotted against ^{7}Be in all aerosol filter samples indicate that there is a positive correlation between the two tracers ($r = 0.611$, $p > 99\%$). The ^{7}Be/^{210}Pb activity ratios varied between a maximum value of 44 and a minimum of 4.9. There is correlation between ^{7}Be/^{210}Pb activities ratio and some air stability classes, such as Pasquill–Gifford method for air stability classification (Kathren 1984; Table 2).

Table 2 Linear correlation coefficients between ^{7}Be and ^{210}Pb specific activities and some meteorological factors

	^{7}Be		^{210}Pb	
	r	P (%)	r	P (%)
Temperature	0.495	>99	0.497	>99
Precipitation	−0.262	>99	0.290	>99
Relative humidity	−0.435	>99	−0.404	>99
Wind speed	−0.100	>99	−0.226	>99

References

Baskaran, M., *J. Geophys. Res.*, **100,** 2833–2840 (1995.).

Kathren, R.L., *Radiactivity in the Environment: Sources, Distribution and Surveillance* (1984).

Koch, D.M., Jacob, D.J., and Graustein, W.C., *J. Geophys. Res.*, **101,** 18151–18666 (1996).

Chemical Composition and Size Distribution of Fine Aerosol Particles Measured with AMS on the East Coast of the Baltic Sea

J. Ovadnevaite[1], **D. Čeburnis**[2,1], **K. Kvietkus**[1], **I. Rimšelyte**[1], **and E. Pesliakaite**[1]

Abstract The size distribution and chemical composition of ambient aerosols were analyzed using an Aerodyne aerosol mass spectrometer (AMS) at Preila station during 3–15 September 2006. The major observed components of the aerosol were sulfates and organics with a smaller amount of nitrates and ammonium. Large contribution of organics was established in all air masses, but it reached 60% of all aerosol mass in North Atlantic marine air masses. The origin of chemical aerosol components was interpreted when size distribution spectra were analyzed. Mainly two modes were registered both for sulfates and organics – one in the accumulation range, the other in the supermicron range. The diameter of sulfates in accumulation mode differed from that of organics in clean marine air mass; they were about 270 nm and 170 nm respectively. This difference showed different sources or transformation mechanisms of sulfates and organics. In the case of polluted air masses the anthropogenic origin of both components was dominant, thus diameters became equal and were about 400 nm, showing dominant secondary production mechanism.

Keywords Aerosols, composition, size distribution, Aerodyne aerosol mass spectrometer

Introduction

The importance of atmospheric particulate matter (PM) is becoming more widely recognized throughout the world because of its impact on global climate, heterogeneous chemistry,[1] air quality,[2] and human health.[3] Understanding of the composition

[1]*Atmospheric Pollution Research Laboratory, Institute of Physics, Savanoriu 231, LT-02300 Vilnius, Lithuania*

[2]*Department of Experimental Physics & Environmental Change Institute, National University of Ireland, Galway, University Road, Galway, Ireland*

C.D. O'Dowd and P.E. Wagner (eds.), *Nucleation and Atmospheric Aerosols*, 792–796.

of atmospheric aerosol particles is necessary for identifying their sources and predicting their effect on various atmospheric processes. Aerosol mass spectrometry offers near real-time measurement capability for the analysis of chemical composition of aerosols. All other techniques require long sampling time and expensive analytical methods to measure aerosol chemical composition. AMS enables to study physical and chemical characteristics of aerosol particles in real time. The aim of this study was to give a short overview of size resolved chemical and physical properties of atmospheric aerosols on the east coast of the Baltic Sea.

Experimental

The Environmental background research station of the Institute of Physics in Preila (55° 55′N and 21°00′E) was chosen for determination of aerosols chemical composition dependence on different air masses. The Preila station has no local pollution sources. Preila background station is located on the coast of the Baltic Sea on the Curonian Spit and is between two water basins: the Curonian Lagoon in the east and the Baltic Sea in the west. It is possible to investigate marine aerosols either originating over North Atlantic after brief passage over Scandinavia or over the Baltic Sea itself.

Instruments

An aerosol mass spectrometer, developed at Aerodyne Research, was used to obtain real-time quantitative information on particle size-resolved mass loadings for volatile and semi-volatile chemical components present in/on ambient aerosol particles.[4]

Aerosols enter the instrument through a sampling inlet that restricts the flow with a 100 μm (or similar diameter) critical orifice and then through a lens, which focuses the aerosols into a tight beam using six apertures while removing most of the atmospheric gas. As the aerosols exit the lens, they are accelerated in supersonic expansion caused by the difference in pressure between the sampling and sizing chambers, which gives different velocities to aerosols of different sizes. After passing through the lens, the aerosols enter the particle sizing chamber. At this point there are two modes of operation, mass spectrometer (MS) mode and time-of-flight (TOF) mode. In MS mode particles fly unimpeded until they impact on a resistively heated surface where the volatile and semi-volatile portions of aerosols are vaporized and then immediately ionized by electron impact. A standard quadrupole mass spectrometer detects the positive ion fragments generated by the electron impact ionization and determines the mass-to-charge (m/z) distribution of the particle beam.

Results And Discussion

The Aerodyne Aerosol Mass Spectrometer has been used to obtain size and chemical composition information about volatile and semivolatile species in ambient aerosol particles. Sampling was performed at the Preila station during 3–15 September 2006. Mass spectrum, mass concentration, as well as size and chemically resolved mass distribution data were obtained.

The major observed components of the aerosol were sulfates and organics with smaller amount of nitrates and ammonium (Figure 1), but their concentration distribution depended on the air mass origin. The concentrations of other chemical species were generally low in the whole observation period.

At the very beginning of this experiment (September 3–4) when the air masses were coming mainly from Western Europe, high concentrations of aerosol and its chemical components were measured (Figure 1), especially organics. But later (September 5–6), when trajectories were coming from Scandinavia, the concentration considerably decreased.

The general behavior of concentration both for total PM concentration and for its constituents was registered: concentrations were low when air mass came from the sea side, and concentrations instantly increased as soon as air mass shifted from the continent. Furthermore, the sensitivity of AMS enabled us to register differences in composition and size in quite similar, at firs sight, air masses. For example, during two days (from 7 to 8 of September) the air mass was generally from the clean oceanic sector, but it passed over England and it considerably enhanced the total PM concentration and concentrations of its components, especially, sulfates and nitrates. But those concentrations were noticeably lower than concentrations registered when air mass originated in Southern Europe (13–15 September, Figure 1). Generally, measurements revealed large variations in concentrations of all chemical species ranging within an order of magnitude depending on air mass origin and history.

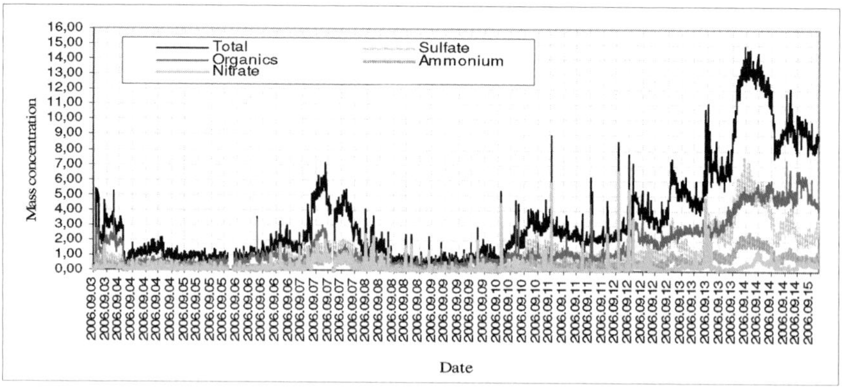

Figure 1 Mass loadings (μg m^{-3}) of nitrates, sulfates, ammonium, organics, chlorides, and total mass

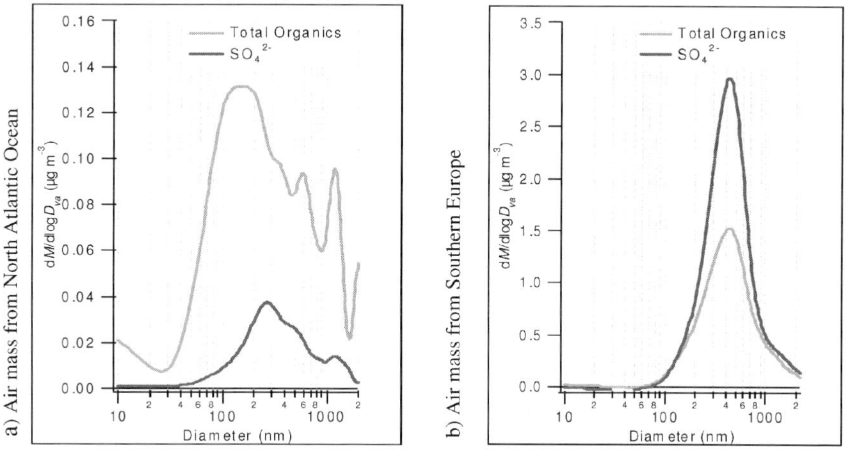

Figure 2 Size distribution of sulfates and organic aerosol parts in very clean air mass from the North Atlantic Ocean (a) and in polluted air mass from Southern Europe (b)

The origin of chemical aerosol components was interpreted when size distribution spectra were analyzed. The difference between distribution of sulfate and organics in clean (Figure 2a) and polluted (Figure 2b) air masses as well as difference between sulfates and organics distributions in the same air mass was demonstrated. Mainly two modes were registered, both for sulfates and organics – one in the accumulation range, the other in the supermicron range. The diameter of sulfates in accumulation mode differed from that of organics; they were about 270 nm and 170 nm respectively (Figure 2a). Difference between diameters of sulfates and organics in clean air mass suggests different sources or transformation mechanisms of these materials. Therefore, sulfates were formed by secondary processes but organic matter likely had different origin probably originating primarily from the sea surface. In the case of polluted air masses, the anthropogenic origin of both components was dominant, thus diameters became equal and were about 400 nm, showing the same secondary production mechanisms including condensation of precursor gases onto existing particles.

Conclusions

The Aerodyne Aerosol Mass Spectrometer has been used to obtain size and chemical composition information about volatile and semi-volatile species in ambient aerosol particles. The major observed components of the aerosol were sulfates and organics with a smaller amount of nitrate and ammonium. Large contribution of organics was established in all air masses, but it reached 60% of all aerosol mass in clear Atlantic masses.

Mainly two modes were registered, both for sulfates and organics – one in the accumulation range other in the supermicron range. The diameter of sulfates in the accumulation mode differed from that of organics in clean air mass; they were about 270 and 170 nm respectively. This difference showed different sources or transformation mechanisms of sulfates and organics. In the case of polluted air masses the anthropogenic origin of both components was dominant, thus diameters became equal and were about 400 nm, showing the same dominant source or transformation mechanisms.

Acknowledgments We gratefully acknowledge financial support by Lithuanian State Science and Studies Foundation. We also thank Jonas Didzbalis for help to test and calibrate AMS.

References

1. Jacob, D.J., *Atmos. Environ.*, **34**, 2131–2159 (2000).
2. Fenger, J., *Atmos. Environ.*, **33**, 4877–4900 (1999).
3. Dockery, D.W. et al., ***N. Engl. J. M.***, **329**, 1753–1759 (1993).
4. Jayne, J.T., Leard, D.C., Zhang, X., Davidovits, P., Smith, K.A., Kolb, C.E., and Worsnop, D.R., *Aerosol Sci. Technol.*, **33**, 49–70 (2000).

Uptake of Nitric Acid on NaCl Single Crystals Measured by Backscattering Spectrometry

Maurus Hess, Ulrich K. Krieger, Claudia Marcolli, and Thomas Peter

Abstract The uptake of nitric acid in sea salt aerosols via the reaction HNO_3 (g) + NaCl (c) $\rightarrow$ $NaNO_3$ (c) + HCl (g) is a main contributor to the chlorine balance in the troposphere, especially in polluted, coastal regions. Backscattering spectrometry was used to measure oxygen concentration profiles on the surface of NaCl single crystals (100) after exposure to nitric acid and water vapor at room temperature. Comparison with the chlorine signal leads to the conclusion that the replacement of chlorine by nitrate is the only significant chemical reaction that occurs. The nitric acid vapor pressure was varied between 10^{-4} to 10^{-2} torr and was found to determine the magnitude of the nitric acid uptake. The relative humidity is revealed to be a crucial parameter, because it determines the kinetics of the uptake. At 65% RH, the nitrate concentration increases almost linearly with time and shows no saturation at all. At 20% RH, after a fast uptake within the first hours, it slows down to a square root dependence of time implying diffusion control.

Keywords Sea salt aerosols, nitric acid uptake, acid-ion exchange reactions, relative humidity, backscattering spectrometry

Introduction

The tropospheric halogen budget is strongly influenced by halogens released from sea salt aerosols.[1] As a matter of fact, sea salt aerosols in polluted regions of the marine troposphere have been often found to display a depletion of chlorine compared to the original composition of seawater. This was generally attributed to the displacement of alkali halides by strong inorganic acids, such as nitric acid.[2] The reaction HNO_3 (g) + NaCl (c) $\rightarrow$ $NaNO_3$ (c) + HCl (g) as a main contributor to chlorine depletion in the condensed phase (c) leads to formation of $NaNO_3$ on the aerosol and the release of the relatively unreactive HCl into the gas phase (g). HCl is

Institute for Atmospheric and Climate Science (IAC), ETH Zurich, Switzerland

C.D. O'Dowd and P.E. Wagner (eds.), *Nucleation and Atmospheric Aerosols*, 797–801.

photochemically inert and highly soluble. The importance of this reaction stems mainly from its potential to compete with other reactions of nitric oxides and NaCl producing ClNO, $ClNO_2$ and Cl_2, which can undergo photolysis and form highly reactive chlorine radicals.[3]

The uptake and the reaction of nitric acid on NaCl has been intensively studied.[4] However, the reported values differ by two orders of magnitude, which has partially been attributed to the adsorption of NO_3^- on the NaCl surface leading to a passivation of the surface.[3] It has been shown that exposure of such passivated surfaces to water vapor causes recrystallization and the formation of $NaNO_3$ crystallites.[3] This allows the exposure of fresh NaCl surface and a steady-state condition of the nitric acid uptake.[3]

Measured chlorine depletions of up to 80% in sea salt aerosols[5] can only be explained in an environment with sufficiently high humidity allowing the nitric acid uptake to proceed into a depth of microns. However, such depth information is still missing up to now, as only indirect information is available from surface monolayer, crystal bulk or gas-phase measurements.

Experimental

Rutherford backscattering spectrometry[6] (RBS) is based on measuring the energy of projectile ions (e.g., He^+) after backscattering from the nuclei of the target. The backscattering energy depends on the mass of the target atom and its depth. Hence, it can be used to measure elemental concentration profiles to a depth of about 1 μm of NaCl crystals.[7] In the study presented here, a modification of this technique based on resonant backscattering of He^+ projectiles on O nuclei is used to measure the oxygen concentration versus depth in NaCl single crystal surfaces (100) after exposure to different nitric acid and water vapor pressures.

Results and Discussion

The reaction HNO_3 (g) + NaCl (c) → $NaNO_3$ (c) + HCl (g) has been studied by measuring nitrate concentration profiles on NaCl surfaces (100), which were exposed for certain time periods to different nitric acid and water vapor atmospheres. By comparing O profiles with Cl profiles of the same sample after exposure, it is evident that replacement of chlorine by nitrate is the only reaction that occurs, in agreement with a previous study.[8] Hence, the measured oxygen concentration is directly converted into nitrate concentration in the following discussion.

Figure 1 shows a selection of such profiles after exposure to a nitric acid vapor pressure of 10^{-3} torr at 20% RH (left panel) and 65% RH (right panel). At low humidity (left panel), the nitrate concentration first increases on the surface within about 100 to 200 nm approaching a value of pure $NaNO_3$. However, significant

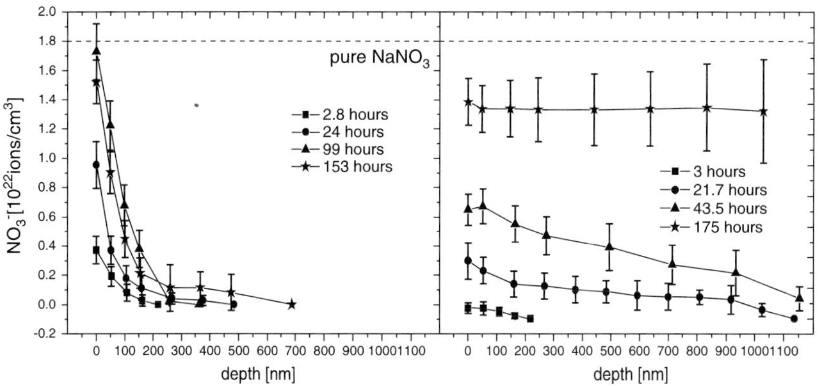

Figure 1 Nitrate concentration profiles after exposure of a NaCl single crystal to a HNO3 vapor pressure of 10^{-3} torr and a relative humidity of 20% **(left panel)** and 65% **(right panel)** for different exposure times. **Abscissa:** probed depth. **Ordinate:** concentration of nitrate with the dashed horizontal line giving the concentration of pure NaNO3. Each profile corresponds to a different crystal, exposed for the indicated period of time

transport to larger depths is not observed. In contrast, at high humidity the uptake depth exceeds 100 nm after 3 h and 1 μm after 22 h. At 175 h, a constant depth distribution of nitrate is observed, corresponding to a replacement of about two thirds of the top micrometer of the original NaCl crystal by $NaNO_3$.

Figure 2 shows the total amount of nitrate ions taken up per square centimeter of exposure area as a function of exposure time. This was obtained by integrating the area under the concentration profiles for samples with nitrate uptake not exceeding the sampling depth of the method (1 μm). Solid symbols represent data at 65% RH and a p[HNO_3] of 10^{-4}, 10^{-3}, and 10^{-2} torr. Note that at 10^{-2} torr and 65% RH, a NaCl bulk solution is stable and this experiment might represent a special case.

The 65% RH data show a slope of about 1 in Figure 2, indicating a linear uptake and no signs of saturation. The uptake rate clearly depends on the vapor pressure of nitric acid. As shown in Figure 1, complete coverage of the surface by nitrate does not occur at such high humidities, but the nitrate is rather transported efficiently into the crystal. The combination of uptake and transport kinetics results in a practically constant uptake rate with time. Within the time scale of our experiments, the nitrate uptake at 65% RH increases constantly without any indications of being diffusion limited.

This can be explained by the strong water adsorption at this humidity leading to the formation of an acidic liquid layer, which causes erosion of the surface. HNO_3 reacts with each chloride ion in the crystal and propagates to larger depth creating a network of pores. This propagation might be faster than the solid-state diffusion of ions and could explain why no decrease in the uptake rate is observed in our experiments.

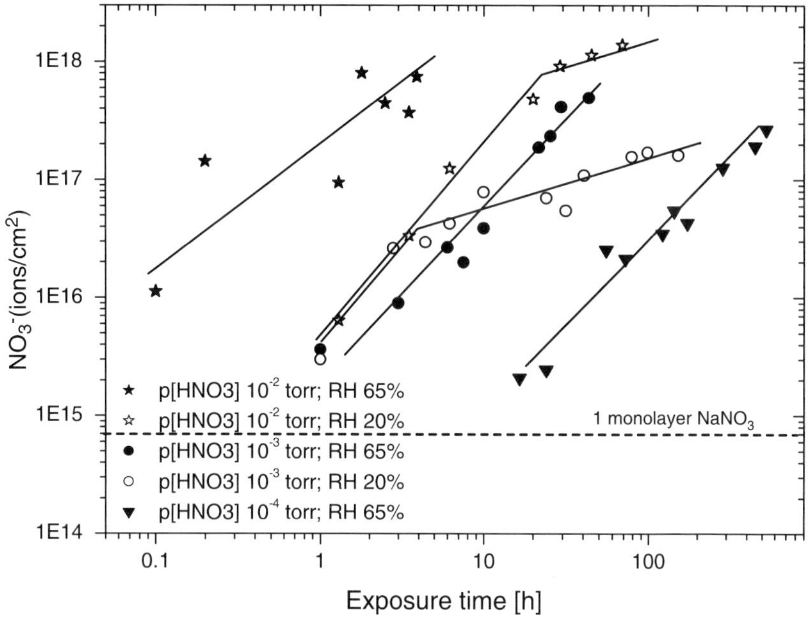

Figure 2 Total nitrate uptake on NaCl as a function of time for different exposure conditions as indicated in the legend. Total uptake is obtained by integration of concentration profiles as shown in Figure 1. Only samples with a nitrate concentration of zero at 1 μm depth are considered. Lines are drawn by hand

At 20% RH, the oxygen concentration at the surface approaches a value corresponding to that of pure $NaNO_3$. The uptake rate is constant within the first two hours at 10^{-3} torr nitric acid vapor pressure and within the first 10 h at 10^{-2} torr. Afterwards, the slope in the log–log plot of Figure 2 ceases to about 0.5, which corresponds to a square root dependence on time implying a diffusion limited process. Therefore, we believe that nitric acid uptake at 20% RH can be explained by adsorption on the NaCl surface to full coverage, along with formation of $NaNO_3$ crystals mediated by the presence of some adsorbed water molecules. This process is limited by the diffusion of nitrate and chlorine through the crystal.

Conclusions

Exposure of an NaCl surface to an atmosphere with nitric acid leads to the reaction HNO_3 (g) + NaCl (c) → $NaNO_3$ (c) + HCl (g). The uptake of nitric acid increases with increasing HNO_3 vapor pressure. At 20% RH, the uptake is restricted to the near surface region. Since the composition of the surface approaches that of pure $NaNO_3$, this reaction becomes limited by solid-state diffusion. At 65% RH, the uptake proceeds into a depth of several micrometers. The uptake rate is linear in time, indicating a process fast enough to provide unlimited ion exchange between HNO_3 and NaCl. This implies that NaCl as the main component in sea salt aerosols can be completely altered into $NaNO_3$ at sufficiently high humidity due to exposure to nitric acid.

Acknowledgments We gratefully thank Art Haberl and Wayne Skala from the Ion Beam Laboratory, SUNY at Albany for the excellent support concerning all questions around the SUNY particle accelerator. This research has been funded by ETH Zurich under grant TH-2202–2.

References

1. O'Dowd, C.D., Smith, M.H., Consterdine, E.I., and Lowe, J.A., *Atmos. Environ.*, **31**, 73 (1997).
2. Junge, C.E., *Tellus*, **8**, 127 (1956).
3. Finlayson-Pitts, B.J. and Hemminger, J.C., *J. Phys. Chem. A*, **104**, 11463 (2000).
4. Rossi, M.J., *Chem. Rev.*, **103**, 4823 (2003).
5. Zhuang, H., Chan, C.K., Fang, M., and Wexler, A.S., *Atmos. Environ.*, **33**, 4223 (1999).
6. Chu, W.K., Mayer, J.W., and Nicolet, M.A., *Backscattering Spectrometry*, New York: Academic Press (1978).
7. Hess, M., Krieger, U.K., Marcolli, C., Huthwelker, T., Ammann, M., Lanford, W. A., and Peter, T., *J. Phys. Chem. A*, in press.
8. Davies, J.A. and Cox, R.A., *J. Phys. Chem. A*, **102**, 7631 (1998).

Investigation of the Heterogeneous Reactions Between Ammonia and Nitric Acid Aerosols

Thomas Townsend and John R. Sodeau

Abstract Acidic troposphere aerosols are mainly composed of inorganic specie such as nitric acid. As the main alkaline species, ammonia (NH_3) plays an importan role in the heterogeneous neutralization of these acidic aerosols. Using a flow tub operated at standard atmospheric pressure and room temperature, the interaction between ammonia and nitric acid (HNO_3) were studied at different humiditie and concentrations. On heating nitric acid, gaseous nitrogen dioxide (NO_2) wa observed whereas at higher humidities and when using a nebulizer, nitrate io (NO_3^-), either in the form of ammonium nitrate (NH_4NO_3) or as a dissociatio ion was produced. Increasing the humidity also leads to an increase of NO/NO production, however, due to competition between HNO_3 and NH_4^+, NO and NO data could not be analyzed accurately. Size distributions obtained correlated wel with the various stages of particulate compositional development.

Keywords Acidic troposphere aerosols, nitric acid, ammonia, heterogeneous neu tralization, flow tube, humidity, concentration, nitrogen dioxide, ammonium nitrate

Introduction

The emission of reactive nitrogen to the troposphere occurs primarily as NC and NO_2 (NO_x). The important sources of NOx are combustion of fossil fuel anc biomass, lightning and biological activity. Nitric acid can be introduced tc the atmosphere from sources of combustion such as industry and car exhausts bu the amounts generated by these means are small. Hence in the troposphere HNO_3 is present mainly due to its formation from free nitrogen species. Two reactions (Eqs (1) and (2)) which may generate it are:

$$NO_2 + OH^{\cdot} + M^1 \rightarrow HNO_3 + M \quad (1$$

Centre for Research into Atmospheric Chemistry, Department of Chemistry, National University of Ireland Cork, and Environmental Research Institute, Cork, Ireland

C.D. O'Dowd and P.E. Wagner (eds.), *Nucleation and Atmospheric Aerosols*, 802–808.

$$N_2O_5 + H_2O(aq)^2 \rightarrow 2HNO_3 \quad (2)$$

The first reaction is based on photochemical reactions for the NO_x-hydrocarbon ystem in daytime. The second reaction occurs only at night time since the reservoir initrogen pentoxide is easily photolyzed during the day. In the troposphere, nitric acid onstitutes the major nitrogen component, and its main sink is thought to be provided y its adsorption to cirrus cloud surfaces. However, uptake of HNO_3 is inefficient and ubsequently the nitric acid is lost from the atmosphere by precipitation [1].

Ammonium nitrate is one of the most important constituents of atmospheric erosols and accounts for 10–20% of the fine aerosol mass [2]. It often coexists with mmonium sulfate and has important environmental impacts with respect to visibility, egradation, radiative forcing, and climate change [3]. In fact, the emission of NO_x, ne of the precursors of atmospheric nitrate, has been predicted to triple while SO_2 mission is expected to decline slightly [4]. Hence, the importance of ammonium itrate in atmospheric aerosols will likely increase substantially in the future.

$$NH3(g)^3 + HNO_3(g) \rightarrow NH_4NO_3(s)^4 \quad (3)$$

NH_4NO_3 is unstable under atmospheric conditions such as low RH (relative umidity), and exists in reversible phase equilibria with its gaseous precursors, nitric cid, and ammonia (Eq. (3)). The dissociation constants of the equilibria depend not nly on the concentrations of acidic gases and ammonia, but also heavily on the hemical composition of the particles, ambient temperatures, and local RH [5].

Of particular note, for less polluted periods, is the fact that most NH_4^+ is neutral-zed by SO_4^{2-} to form ammonium sulfate. Whereas in more polluted periods, when mmonium ion levels and RH are high, then more NH_4^+ is neutralized by NO_3^- to orm ammonium nitrate [6]. The effects of RH on the chemistry of these pollutants nd the efficiency of the heterogeneous conversion under such conditions have been ttle explored and requires detailed study of the aerosol processing involved.

Experimental Methology

Flow-tube Reactor

'he apparatus designed to conduct the experiments is based on an aerosol flow-ıbe instrument, adapted for NH_3 monitoring instrumentation (Figure 1.).The ow-tube reactor is made of a Pyrex glass tube with an inner diameter of 7.5 cm

Iydrocarbon molecule

Aqueous

Gas

Solid

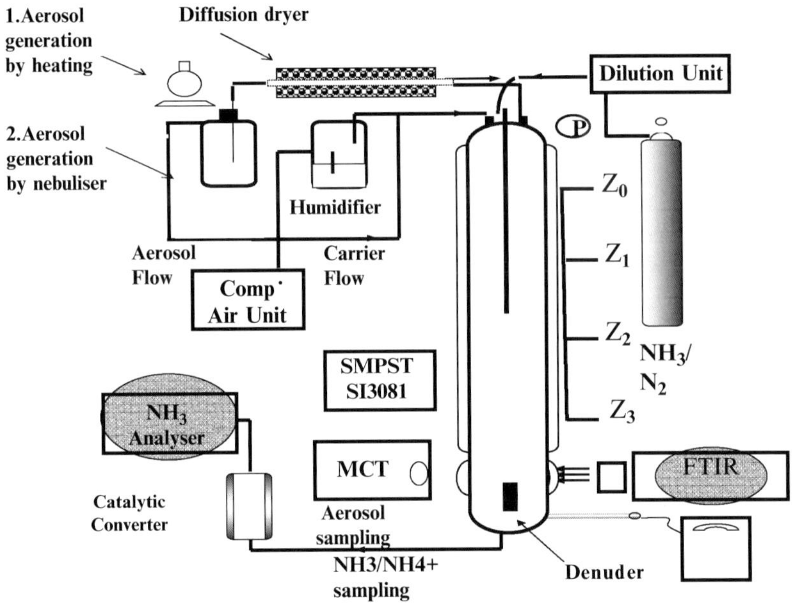

Figure 1 Flow-tube instrument used for experiments

and a 100 cm long reactive length. It is fitted at its base with BaF_2 windows t allow for an in situ FTIR (Fourier Transform Infra Red) monitoring of both ga and condensed phase (BioRad Excalibur coupled with external photoconductiv HgCdTe detector).

Experimental flow conditions used (typically 2 L/min, flow velocity v = 1 cm/s ensured that the necessary laminar flow conditions were always prevailing in th reactor (Reynolds Number Re = 50–100). The pressure probe inlet allows a Baratro pressure gauge to monitor pressure. The relative humidity of the carrier flow throug the tube (IR-grade compressed air) can be tuned between 1 and 99% and is controlle by a digital hydrometer (Rotronic Hygromer). All experiments carried out to da were performed at room temperature (T = 293 K) and atmospheric pressur The apparatus consists of three main regions: aerosol generation, a reaction zone (th flow-tube section with several inlet and outlet ports) and a detection region.

Nitric acid particles were generated in a condensation-type particle generatc where a flow of (~200–500 ccm) dry filtered air was passed over a small reservo of hot liquid 99% concentrated sulfuric acid (~1 g, temperature of 120–140°C usin a stirrer/hotplate and heating oil). Aerosol suspensions were produced using a TS 3076 Atomizer, from 0.1–1% wt aqueous solution in distilled water. The nebulise droplets generated through compressed air expansion are flown through a diff sion-dryer to remove excess humidity and insure complete recrystallization of th dissolved salts before admission into the reactor.

Gas-phase ammonia is admitted into the flow tube via a 6 mm diamete movable, glass injector. From the 100 ppm standard (NH_3/N_2 brand), the ammc ia concentration can be diluted down to the ppb range using a range of mass flowm-ters. A chemiluminescence NO_x technique coupled with catalytic oxidation of mmonia to NO was chosen to monitor the concentration of ammonia and the ormation of products (ammonium-containing species). A denuder coated with xalic acid is fitted in the detection region of the flow tube to quench the reaction y removing gaseous ammonia from the sample allowing for direct detection of roducts via NO conversion. The denuder is dimensioned in order to fit the flow equirements set by the NO_x analyzer (0.6 L/mn) and totally remove gas-phase mmonia without affecting particle concentration.

Aerosol suspensions were characterized using a TSI SMPS (Scanning Mobility Particle Sizer) 3081 instrument, which determines key parameters such as mode the peak diameter in the size distribution, typically 100 nm), number distribution, urface area and particulate mass.

Results and Discussion

FTIR Spectroscopy

On heating nitric acid directly, it was found that at low humidities, gaseous nitrogen lioxide was present. A resultant brown gas was observed in both the conditioner nd the flow tube and the overall process is summarized by Eq. (4).

$$4HNO_3(g)(heated) \rightarrow 4NO_2(g) + 2H_2O(g) + O_2(g) \quad (4)$$

Nitrate ions were observed in the aerosol and became more concentrated by ncreasing the humidity. This was achieved by either passing the main flow through humidifier or reducing the flow concentration. (On addition of ammonia, the well-established deformation peak at 1,400 cm^{-1} of the ammonium ion is unclear lue to the appearance of the nitrate absorption at roughly the same wavenumber, ,451 cm^{-1} [7]). Generating nitric acid aerosol by means of the nebulizer produced itrate absorptions only, regardless of humidity, flow concentration and ammonia ddition. Table 1 shows the wavenumber absorptions and their corresponding ssigned components for the nitric acid generated in the aerosol.

Figure 2 illustrates a typical FTIR spectrum of both nitrogen dioxide and nitrate on, the latter either as a dissociation ion or in the form of ammonium nitrate.

Table 1 Absorption assignments for nitric acid [8]

Wavenumber range (cm^{-1}) predominant vibrations
1,724–1,653 NO_2 asymmetric stretch
1,458–1,382 NOH bend
1,378–1,211 NO_2 symmetric stretch
760–720 NO_2 out-of-plane bend

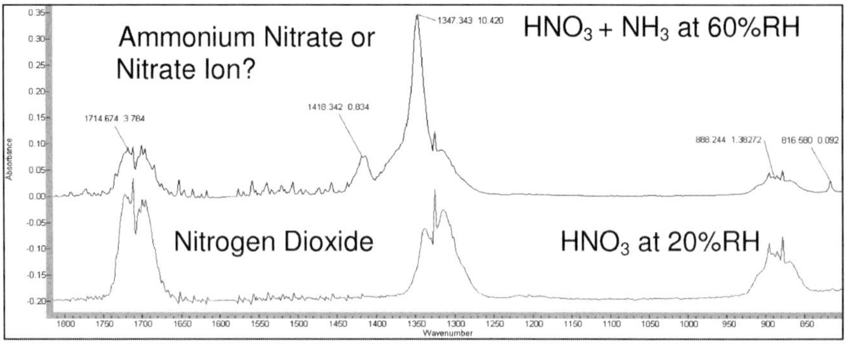

Figure 2 FTIR spectra of nitric acid and ammonia at various conditions

$[NH_4]^+$ Production

$[NH_4]^+$ levels at ppm (parts per million) detection was hampered by the fact that a competition between ammonium ions and NO and NO_2 gas from nitric acid exists. NO_2 levels were particulary high when the nitric acid was heated directly. As the humidity was increased, the $[NO_2]$ dropped and consequentially the [NO] increased. It was also noted that the [NO] increased dramatically on addition of ammonia to the system. Nebulizer generated aerosols did not produce much NO_2, however NO levels were consistently high both before and after ammonia addition. Humidity appeared to increase the amount of NO produced with and without ammonia and also when the flow concentration was lowered. The presence of nitrogen dioxide absorptions on the FTIR spectra correlated well with corresponding NO and NO_2 levels. Figure 3 shows data obtained for one of these experiments.

SMPS Data

It was found that sizes of the particles compared well with the various conditions of the experiments. Not many particles were observed initially when the nitric acid was heated directly, however, on increasing the humidity and introducing ammonia, size distributions were observed in the region of 10^4 to 10^6 numbers of particles, possibly indicating the formation of a hydrated nitrate particle or ammonium nitrate aerosol. Nebulizer generated aerosols produced better data as can be seen from Figure 4. On several occasions bimodal peaks were recorded indicating that large numbers of two different sizes of particles were being produced. An increase in mode (the average diameter of the size distribution) and a decrease in particle number suggest more extensive incorporation of water and ammonia into the aerosols.

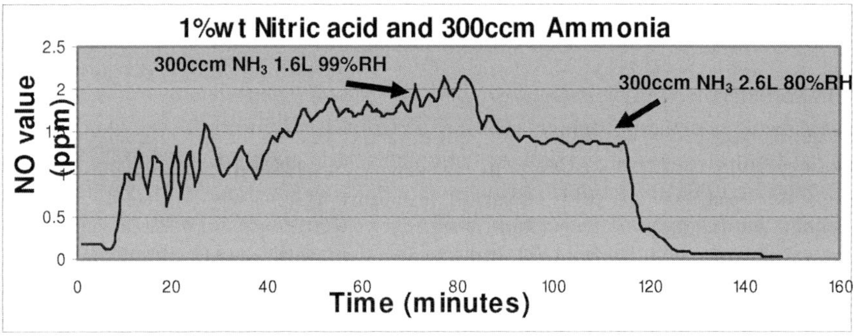

Figure 3 Effect of flow concentration and humidity on NO (ppm) values

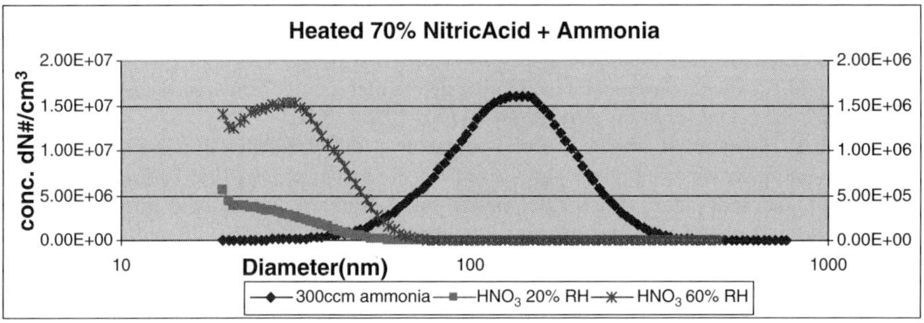

Figure 4 Size distributions of heated nitric acid and ammonia

Summary and Conclusions

From our investigation into the interactions of nitric acid and ammonia, the main products formed were nitrogen dioxide and nitrate ions, the latter possibly either in the form of ammonium nitrate or as a dissociation ion.

FTIR spectroscopy showed that NO_2 was dominant at low humidity whereas nitrate was prevalent under wetter conditions. The NO_x data cannot be assumed to be accurate due to competition between NH_4^+, NO, and NO_2. Size distributions compared well to the various stages of the experiments.

Future Work

The interaction of nitric acid and oxalic acid is currently being studied along with their interaction with ammonia. The ammonia interaction times will be used to calculate the kinetic rate coefficients of these heterogeneous reactions. Future experiments will also deal with the interaction of nitrous acid aerosols and other common dicarboxylic acids in the troposphere such as malonic and succinic acid. These results can then be compared to sulfuric and sulfurous systems which have already been investigated.

Acknowledgments We are indebted to A. Allanic and C. Noonan for their instrumental support and help during experiments. This work was financially supported by the Higher Education Authority Grant PRTLI-3 scheme.

References

1. Shepherd, J.A., Coe, H., Choularton, T.W., McFiggans, G., Comparative study of the uptake of nitric acid to aerosol and cloud particles in the mid and upper troposphere, *Geophys. Res. Abstracts*, **5**, 09355 (2003).
2. Wexler, A.S. and Seinfield, J.H., The distribution of ammonium-salts among a size and composition dispersed aerosol, *Atmos. Environ.*, **24**, 1231–1246 (1990).
3. Jacobson, M.Z., Global direct radiative forcing due to multicomponent anthropogenic and natural aerosols, *J. Geophys. Res. – Atmos.*, **106**, 1551–1568 (2001).
4. Adams et al., General circulation model assessment of direct radiative forcing by the sulfate-nitrate-ammonium-water inorganic aerosol system, *J. Geophys. Res.*, **106**, 1097–1111 (2001).
5. Matsumoto, K. and Tanaka, H., Formation and dissociation of atmospheric particulate nitrate and chloride: an approach based on phase equilibrium, *Atmos. Environ.*, **30**, 639–648 (1996).
6. Zhang et al., Properties of nitrate, sulfate and ammonium in typical polluted atmospheric aerosols (PM10) in Beijing, article in press.
7. Hudson et al., A spectroscopic study of atmospherically relevant concentrated aqueous nitrate solutions, *J. Phys. Chem. A*, 554–548 (2007).
8. Ortega et al., The structure and vibrational frequencies of crystalline nitric acid, *Chem. Phys. Lett.*, **378**, 218–223 (2003).

Carbonaceous Materials in Size-segregated Atmospheric Aerosols in Urban and Rural Environments in North-western Portugal

C.L. Mieiro[1], A. Penetra[1], R.M.B. Duarte[1], C.A. Pio[2], and A.C. Duarte[1]

Abstract The concentration and size distribution of carbonaceous materials in atmospheric aerosols was investigated for two Portuguese areas during the summer of 2004: a rural area (Moitinhos) and an urban area (Oporto). Concentrations of airborne particulate matter (PM), total carbon (TC), organic carbon (OC), elemental carbon (EC), and water-soluble organic carbon (WSOC) were determined for the following particle size range: <0.49, 0.49–0.95, 0.95–3.0, and 3.0–10 µm. For both locations, the highest concentrations of TC, EC, and WSOC were found in the submicrometer size range. On the other hand, the OC exhibited a maximum at 0.49–0.95 µm size range. Samples collected in the urban area showed the highest concentrations of EC, highlighting the contribution of local primary sources (mostly traffic emissions) to the carbonaceous material. From 6% to 24% and from 10% to 46% of the gravimetric PM collected in Moitinhos and Oporto, respectively, was accounted for by the carbonaceous material. Absorption properties of the chromophoric WSOC observed by UV-Vis spectroscopy were different for the various size ranges and also for each sampling location.

Keywords Carbonaceous aerosol, aerosol size distribution, water-soluble organic carbon, urban/rural

Introduction

Over the last 10–20 years there has been an increasing interest concerning the role of atmospheric aerosols on global climate change and public health. All these climatic and human health effects associated with aerosols depend on the concentration, size, and chemical composition of particles themselves.[1] Efforts have been made over the years to elucidate the chemical composition of atmospheric

[1] *Department of Chemistry & CESAM – Campus de Santiago, 3810–193 Aveiro, Portugal*

[2] *Department of Environment and Planning & CESAM – Campus de Santiago, 3810–193 Aveiro, Portugal*

C.D. O'Dowd and P.E. Wagner (eds.), *Nucleation and Atmospheric Aerosols*, 809–813.

aerosols as a function of size, and to achieve a chemical mass closure of aerosols.[2,3] Nevertheless, our knowledge on concentrations, sources and formation mechanisms is far from complete. Detailed investigations of the chemical characteristics of aerosols are therefore important for better understanding their role in several processes occurring in the atmosphere. In the present study, the mass size distributions of the particulate matter (PM), total carbon (TC), organic carbon (OC), elemental carbon (EC), and water-soluble organic carbon (WSOC) were determined for a rural (Moitinhos) and an urban (Oporto) Portuguese areas. An important objective was to achieve an aerosol carbonaceous mass balance as a function of particle size on the basis of the measured concentrations of the various aerosol components. The preliminary results of the absorption properties of the WSOC fractions measured by ultraviolet-visible (UV-Vis) spectroscopy are also discussed.

Experimental Section

Atmospheric aerosols samples were taken at an urban (Boavista roundabout, Oporto, 41° 12′ N, 8° 42′ W) and rural (Moitinhos, 40° 57′ N, 8° 64′ W) areas located in the north-western part of Portugal near the Atlantic Ocean. Particulate matter (PM) was sampled over pre-fired (500°C for 8 h) quartz fiber filters (Whatman QM-A) with an adapted Sierra high-volume cascade impactor (Sierra Instruments Inc.), consisting in three stages and a backup filter, in the following size ranges: <0.49, 0.49–0.95, 0.95–3.0, and 3.0–10 μm. Four and eight samples were obtained in Moitinhos and Boavista, respectively, with an air flow of 1.13 m^3 min^{-1} and collection times of approximately 7 days. Samples acquisitions were carried out from July to mid-September 2004. The concentrations of PM were determined by weighting the filters under controlled moisture conditions (at aproximatelly 50% Relative Humidity) before and after exposure. The concentration of TC and EC were measured for the collected aerosols by means of a thermo-optical method.[4] The WSOC was extracted from each filter with ultra-pure (100 ml) water by mechanical agitation plus ultrasonication, and the final mixture was filtered through a membrane filter (PVDF, Gelman Sciences) of 0.22 μm pore size. The dissolved organic carbon content of each aqueous extract was measured with an automated segmented flow analyser based on a UV-persulphate oxidation method[5].

The UV-Vis spectra of the WSOC fractions were registered on an UV-Vis spectrophotometer Shimadzu (Dusseldorf, Germany) Model UV 2101PC in 1, 2 and 5 cm path length quartz cells. Ultra-pure (Milli-Q) water was used as a blank.

Results and Discussion

The average concentrations of PM, TC, OC, EC, and WSOC, as a function of particle size, are shown in Figure 1. From looking at the entire data set, and for both locations, it could be concluded that PM exhibited a minimum in the

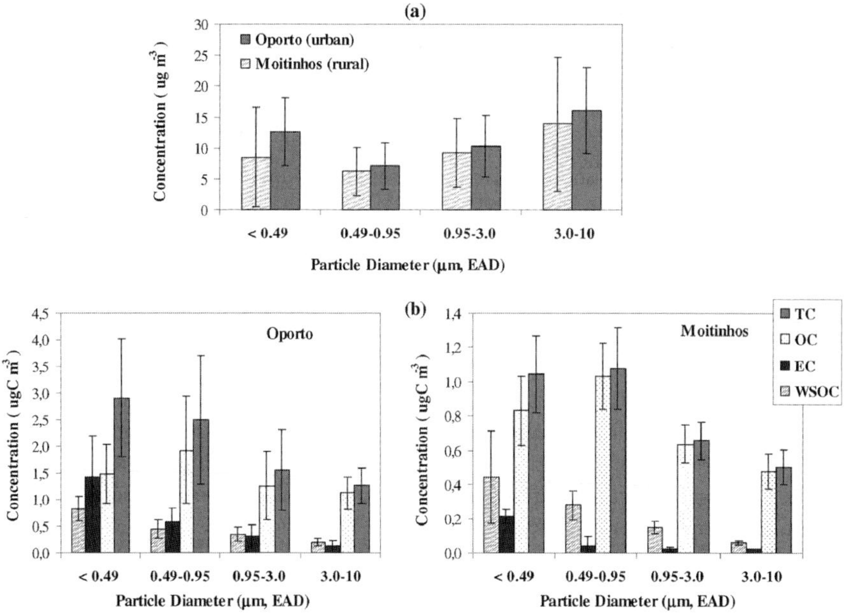

Figure 1 Average concentrations of (a) airborne PM and (b) TC, OC, EC, and WSOC as a function of particle size during Summer 2004 at Oporto (urban) and Moitinhos (rural)

0.49–0.95 μm size range and maxima values in the 3.0–10 and <0.49 μm size ranges. The carbonaceous material was predominantly associated to fine particles, with the highest concentrations typically found in the <0.49 μm size range. However, in Moitinhos, particles in the 0.49–0.95 and <0.49 μm ranges exhibited similar overall content of TC. EC was also predominantly associated to fine particles (<0.49 μm), accounting on average for 48% and 21% of TC at Oporto and Moitinhos, respectively. For the entire size range, the highest concentrations of EC were found in Oporto (11–48% against 3.3–21% in Moitinhos). EC is essentially a primary pollutant, emitted directly during incomplete combustion of fossil and biomass fuels; therefore, EC is expected to have a large contribution from traffic at Oporto. For both sampling sites, the average concentrations of OC exhibited a maximum in the 0.49–0.95 μm size range; the minimum average concentrations were found for the coarse size range. For the entire size range, the OC fraction accounts for 51–89% and 79–97% of TC in Oporto and Moitinhos, respectively. In addition, and also for the entire size range, 16–31% of TC in Oporto and 13–29% of TC in Moitinhos were found to be water-soluble. The WSOC was mostly found in the fine size range.

For the aerosol mass balance calculations, three aerosol carbonaceous components were considered: (1) water-soluble organic matter (WSOM), which was estimated as 1.8*WSOC and 1.6*WSOC for the rural and urban areas, respectively; (2) EC; and (3) water-insoluble organic matter (WINSOM), estimated as 1.2*water-insoluble organic carbon (WINSOC = OC-WSOC). The percentage atribution (average over all samples) of the gravimetric PM to each of the three aerosol components as a function of particle size at Moitinhos and Oporto is shown in Figure 2. In both locations, the organic aerosol (WSOC plus WINSOC) provide a large contribution in the submicrometer size range than they do in the supermicrometer size range. As expected, the percentage contribution of EC to the gravimetric PM is higher in Oporto than in Moitinhos, and increases clearly with deacreasing particle size at both locations. Besides, the percentage contribution of WSOM increases towards small size particles, whereas the WINSOM provides the highest contribution in the 0.49–0.95 μm size range. From 6% to 24% and from 10% to 46% of the gravimetric PM collected in Moitinhos and Oporto, respectively, is accounted for by the aerosol components considered.

The average UV specific absorptivities at 280 nm (ε_{280}) for the WSOC fractions extracted from each particle size range are plotted in Figure 3. Absorbance in this UV region is attributed to π-π^* electron transitions in phenolic arenes, aniline derivatives, polyenes and polycyclic aromatic hydrocarbons with two or more rings.[6] The results show that in Oporto the ε_{280} increases towards submicrometer size range. An opposite trend is observed for samples collected in Moitinhos. Such behaviour suggests that the chromophoric WSOC exhibit a different distribution

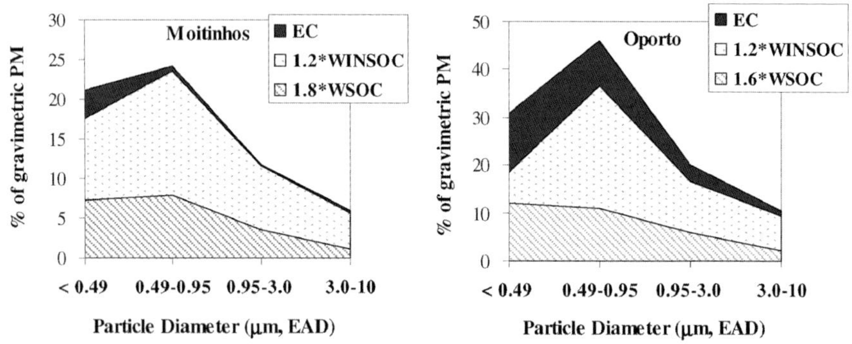

Figure 2 Average percentage attribution of the gravimetric PM to three aerosol carbonaceous components, as a function of particle size during summer 2004 at Moitinhos (rural) and Oporto (urban)

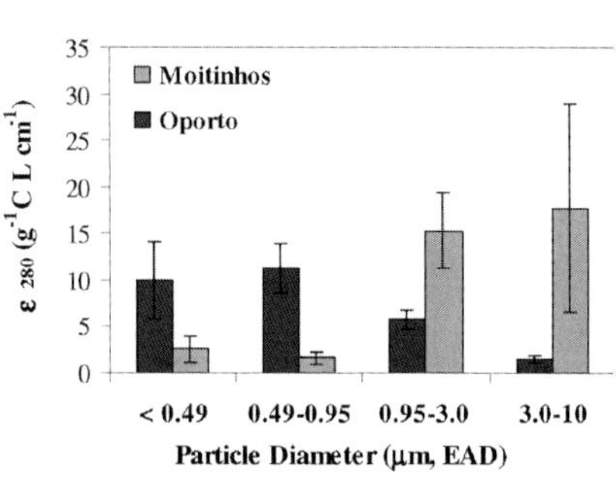

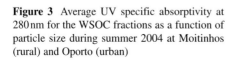

Figure 3 Average UV specific absorptivity at 280 nm for the WSOC fractions as a function of particle size during summer 2004 at Moitinhos (rural) and Oporto (urban)

over the different aerosol sizes. Besides, it highlights also the impact of different levels of pollution on the absorptive properties of aerosol chromophoric WSOC fractions. Additional studies on the chemical features of these chromophoric WSOC will be undertaken.

Acknowledgments "Centre for Environmental and Marine Studies-CESAM" (University of Aveiro) is acknowledged for financial support and for post-graduate scholarships awarded to C.L. Mieiro and A. Penetra.

References

1. IPCC, Aerosols their direct and indirect effects, in *Climate Change 2001: The Scientific Basis*, pp. 289–348 (2001).
2. Krivácsy, Z. and Molnár, Á., *Atmos. Res.*, **46**, 279–291 (1998).
3. Sciare, J., Oikonomou, K., Cachier, H., Mihalopoulos, N., Andreae, M.O., Maenhaut, W., and Sarda-Estève, R., *Atmos. Chem. Phys.*, **56**, 2253–2265 (2005).
4. Pio, C.A., Castro, L.M., and Ramos, M.O., in: *Physico-Chemical Behaviour of Atmospheric Pollutants*, Report EUR 15609/2 EN, Vol. 2, edited by G. Angeletti, and G. Restelli, *Proceedings of the Sixth European Symposium*, , Varese, pp. 706–711 (1993).
5. Lopes, C.B., Abreu, S., Válega, M., Duarte, R.M.B.O., Pereira, M.E., and Duarte, A.C., *Anal. Lett.*, **39**, 1979–1992 (2006).
6. Traina S.J., Novak, J., and Smeck, N.E., *J. Environ. Qual.*, **19**, 151–153 (1990).

Using Föhn Conditions to Characterize Urban and Regional Sources of Particles

D. Mira-Salama, R. Van Dingenen, C. Gruening, J.-P. Putaud, F. Cavalli, P. Cavalli, N. Erdmann, A. Dell'Acqua, S. Dos Santos, J. Hjorth, F. Raes, and N. R. Jensen

Abstract Measurements for the characterization of urban and regional sources of particulate matter (PM) were performed in the Milan area (North of Italy) during Föhn and non-Föhn conditions. The measurements were performed at two different places: in an urban area North of Milan (Bresso) and in a regional area at the EMEP-GAW station in Ispra (about 65 km NW from Milan) during the years 2002–2007. Single particle characterization of aerosols is combined with other physical and chemical aerosol properties obtained during Föhn and non-Föhn conditions, to obtain information about the chemical content of the particles and their mixing state. When the Föhn wind is present, aged background aerosol is removed, and local sources are determining aerosol properties. Aerosol analysis during Föhn and non-Föhn conditions can thus help to elucidate the origin of particulate pollution and to evaluate local urban and regional sources of particles. It was observed that during Föhn events, the accumulation mode in the size range 50 nm < d < 300 nm has practically completely disappeared and that the size fraction below 50 nm is dominating the total number distribution. Single particle measurements indicate that the locally emitted particles appear to have a relatively high content of carbonaceous compounds. Overall it was concluded that the urban emitted particles have mainly been attributed to local emissions from traffic, domestic heating and industrial activities, whereas the regional emitted particle mainly were attributed to domestic heating.

Keywords Urban pollution, Föhn, single particles, aerosols size distributions, aerosol characteristics

Introduction

Atmospheric particles play an important role in chemical as well as physical processes related to air pollution, human health and climate change (Novakov and Penner 1993; Tegen et al. 1997; Lippmann et al. 2000).

European Commission – DG Joint Research Centre, Institute for Environment and Sustainability, Climate Change Unit, TP 290, I - 21020 Ispra (VA), Italy

C.D. O'Dowd and P.E. Wagner (eds.), *Nucleation and Atmospheric Aerosols*, 814–818.

High levels of particulate matter (PM, the sum of the mass of particulate matter with a diameter below 10 μm) have been a serious problem in big cities for decades. The levels of PM at a certain location can be considered as a sum of the regional background and locally emitted particles. Traffic is known to be a very significant local PM10 contributor in the urban environments, be it rather in particle number than in particle mass.

The Po Valley in Northern Italy is known for its high PM levels, especially during winter, when stagnant meteorological conditions prevail. During the cold season, PM10 levels frequently exceed 50 μg m^{-3} (the EU 24 h limit values is not to exceed 50 μg m^{-3} more than 35 days per year, Van Dingenen et al. 2004), invoking traffic ban measures in the urban centers. In this presentation we show an investigation of the contribution of the regional background and the local urban emissions to the total PM load at an urban location (Bresso) in Milan, Italy during winter and also at a regional EMEP-GAW station in Ispra (EMEP-GAW: Cooperative programme for monitoring and evaluation of the long-range transmissions of air pollutants in Europe – Global Atmospheric Watch).

We take advantage of particular meteorological conditions, the "Föhn episodes". Föhn is a katabatic wind flow, typically occurring for some days per month in winter time, which is characterized by warm and extremely clean air flowing over the Alps from north to south. This clean air mass reduces regional PM10 levels from approximately 50 μg m^{-3} down to the order of 1–5 μg m^{-3} during a typical winter day. In such conditions, only local (urban) emissions contribute to the urban background since the regional background aerosol has been removed by the Föhn. The comparison of physical and chemical properties of the urban aerosol during Föhn and non-Föhn conditions allows us to evaluate the contribution of local and long range transported PM on the urban background aerosol in the Milan area.

Experimental

Single particle measurements: The operational set-up of the mobile Single Particle Analysis and Sizing System (SPASS) has been described by Erdmann et al. (2005). In brief, aerosols are introduced into the SPASS using an aerodynamic lens system to obtain a well-collimated particle beam with little divergence. Particles with an aerodynamic diameter in the range of approximately 300 nm to 3 μm are then sized with a two beam laser velocimeter. This information is further used to trigger a frequency quadrupled Nd:YAG laser (wavelength λ = 266 nm, pulse length T = 8 ns, pulse energy E = 20 mJ) which desorbs and ionizes the particle once it enters the ionization region of a bipolar mass spectrometer.

Two time-of-flight (TOF) mass spectrometers, one linear and a reflectron TOF that allows higher mass resolution ($m/\Delta m_{fwhm}$ = 150 and 1,000, respectively), are used in a bipolar mode so that both the positive and negative ions from a single particle are simultaneously registered. Thus the size and chemical composition of single aerosol particles can be characterized simultaneously and online. Particle spectra are analysed using *k*-means clustering, which classifies them in a limited number of groups.

Aerosol *number size distributions:* the size range from 6 nm to 600 nm was measured with a custom-built medium size Vienna-type differential mobility analyser (DMA), operated with a TSI model 3010 condensation particle counter (CPC) as particle detection system. The system is operated in a closed-loop arrangement with dry sheath air (relative humidity < 20%) and hence delivers dry size distributions of the aerosols.

Bulk analysis of the aerosol chemistry: Aerosol particles were collected for chemical analyses at a flow rate of 20 L min^{-1} onto 47 mm filters. Organic and elemental carbon (OC and EC) were measured from an 8 mm diameter of the quartz filters using a multistep thermal method with CO_2 detection, according to the method described by Putaud et al. (2002). Carbon evolved at up to about 340°C in He/O_2 and up to 650°C in pure He was classified as OC. EC was defined as the carbonaceous fraction remaining after this treatment and evolved at 650°C in a He/O_2 carrier gas. The remaining pieces of quartz filters were extracted in ultra pure water (18.2 MΩcm) and the extracts analysed by ion chromatography (IC) for major species (Na^+, NH_4^+, K^+, Mg^{2+}, Ca^{2+}, Cl^-, NO_3^- and SO_4^{2-}). Mineral dust was determined by weighting with a microbalance the residue obtained after ashing in quartz crucible at 600°C the paper filters previously extracted in ultra pure water.

Results, Discussions, and Conclusions

Measurements were performed at two different places during Föhn and non-Föhn conditions: At an urban background site (Urban BG) in Bresso, Milan, and at a regional background site (Regional BG) in Ispra, 65 km to the NW of Milan. Figure 1a–c shows for both sites the median number size distributions measured during winter time for Föhn-conditions (particular situation) and stagnant (i.e., typical non-Föhn conditions), respectively. During the Föhn-conditions, the size distributions are averaged for the duration of each Föhn event which typically lasts between 6 and 12 h. Night-time refers to 0200–0500 and daytime to 0800–1800 local time.

Figure 1 illustrates the dynamics of the regional and local contributions to the size distribution. During the night, local emissions are minimal, and the urban distributions (Urban BG) under normal, stagnant conditions, are similar to the ones at the Regional BG (Figure 1a), indicating that the urban accumulation mode represents the regional background contribution. During daytime, local emissions contribute significantly to the Aitken mode (particle diameter < 100 nm). At both sites, a clear increase in the number of Aitken mode particles is observed (Figure 1b, c, full line). Under the particular Föhn conditions, the accumulation mode is mostly scavenged at both sites, (dashed lines in Figure 1b, c), and the locally produced particles are standing out clearly. Hence, Föhn conditions offer a unique occasion to investigate more in detail the properties of the urban background aerosol, consisting of a mixture of emissions from traffic (in particular at Urban BG), domestic heating and industrial activities.

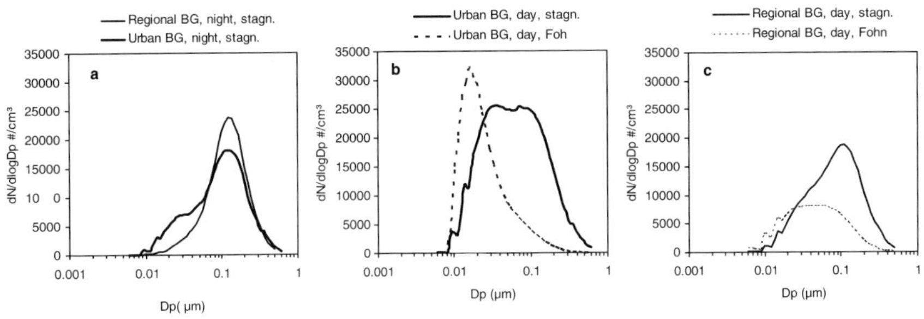

Figure 1 Median number size distributions during Föhn and non-Foehn conditions at the urban (Bresso) and regional (Ispra) background site

The dramatic change in the number size distribution and PM mass during Föhn is also reflected in the chemical composition of the particles. Compared to stagnant conditions, secondary inorganic compounds like ammonium nitrate are strongly reduced, and sometimes even virtually absent, during Föhn episodes, with black carbon becoming the major components.

This observation is confirmed by the single particle instrument. Here, the spectra are classified in terms of specific peak combinations occurring in each particle. The major cluster classes *during Föhn conditions* are carbonaceous classes at both sites. At the urban site, all spectra contain carbon clusters (of which 30% mixed with sulphate/nitrate), and no externally mixed sulphate/nitrate particles are found. At the regional site, 82% of the spectra are predominantly carbon containing, and 16% contain sulphate and/or nitrate. Interestingly, a further distinction can be made between the carbonaceous classes at both sites: at the urban site, the presence during daytime of Ca/CaO peaks together with carbon peaks in many of the spectra indicate that a significant amount of the particles originate from traffic (as observed in several studies, i.e., Vogt et al. 2003), whereas at the regional site, carbon clusters during Föhn are observed as "pure" carbon spectra, or associated with potassium, which indicates a different local source than traffic (i.e., stoves, domestic heating, wood burning) for the particles.

During *stagnant conditions* on the other hand, a mixed carbonaceous class with Ca/CaO peaks appears as a major class at the regional BG site, representing 50% of the spectra, with the other 50% consists of (externally mixed) inorganic compounds (sulphate and/or nitrate). At the urban site, 60% of the spectra appear as nitrate and/or sulphate, and 40% as predominantly carbonaceous.

Overall it was concluded that the urban emitted particles have mainly been attributed to local emissions from traffic, domestic heating and industrial activities, whereas the regional emitted particle mainly were attributed to domestic heating.

More detailed figures, tables, and conclusions will be presented at the conference.

References

Erdmann, N., Dell'Acqua, A., Cavalli, P., Gruening, C., Omenetto, N., Putaud, J.-P., Raes, F. Van Dingenen, R., A.S.H. *Aerosol Sci. Technol.*, **39**, 377–393 (2005).

Lippmann, M., Ito, K., Nadas, A., Burnett, R.T., Association of particulate matter components with daily mortality and morbidity, in: *Urban Pollution, Research Report*, Vol. 95, Cambridge, MA: Health Effects Institute (2000).

Novakov, T. and Penner, J., RE. *Nature*, **365**, 823–826 (1993).

Putaud, J.-P., Van Dingenen, R., and Raes, F., *J. Geophys. Res.*, **107**, 8198, doi:10.1029/2000JD000111 (2002).

Tegen, I., Hollrig, P., Chin, M., Fung, I., Jacob, D., and Penner, J., *J. Geophys. Res.*, **102**, 23895–23915 (1997).

Van Dingenen, R. (28 authors) et al. *Atmos. Environ.*, **38**, 2561–2577 (2004).

Vogt, R., Kirchner, U., Scheer, V., Hinz, K.P., Trimborn, A., and Spengler, B., *J. Aerosol Sci.*, **34**, 319–33 (2003).

Spatial Distribution of Nanoparticles in the Free Troposphere over Siberia

M. Yu. Arshinov[1], B.D. Belan[1], Ph. Nedelec[2], J.-D. Paris[3], and T. Machida[4]

Abstract Two first airborne campaigns of the "YAK-AEROSIB" Russian-French Project were conducted in April and September, 2006, over a vast territory of West and East Siberia. The main goal of the Project is to study spatial distribution of trace impurities, which are responsible for the global worming effect. In the framework of this project French partners provided continuous measurements of CO, CO_2, and O_3 while Russian scientific group measured number concentration of ultrafine and fine aerosols as well as performed aerosol sampling for the chemical analysis. Spatial distribution of aerosol number concentration observed during two different seasons is presented.

Keywords Tropospheric aerosols, airborne measurements, vertical distribution, number concentration of nanoparticles, YAK-AEROSIB

Introduction

At present the confidence level in understanding of greenhouse gases effect on climate change is relatively high (IPCC), which is not the case with tropospheric aerosols, especially their indirect effect. At the same time there is a problem in verification of modeling results due to the lack of observational data over Russia, which occupies a significant area in the northern hemisphere.

Recently this problem has been resolved by establishing Siberian ground-based aerosol monitoring network, which is a part of Global AERONET, and Russian-Japanese Cooperative Institute for the Greenhouse Gases Research in Siberia.

[1]*Institute of Atmospheric Optics, Siberian Branch og the Russian Academy of Sciences, Tomsk, Russia*

[2]*Laboratoire d'Aérologie UPS-CNRS Toulouse, France*

[3]*Laboratoire des Sciences du Climat et de l'Environnement IPSL, CEA-CNRS, Saclay, France*

[4]*National Institute for Environmental Studies, Tsukuba, Japan*

C.D. O'Dowd and P.E. Wagner (eds.), *Nucleation and Atmospheric Aerosols*, 819–823.

The first provides observational data on aerosol optical depth. The second one involves observations of GHG vertical distribution from four aircrafts, aerosol measurements from one aircraft, and eight towers for precise measurements of the surface carbon dioxide and methane.

In spite of the partial progress achieved in solving the observational problem, it did not allow to reconstruct fully a three-dimensional distribution of the above-mentioned species. The main goal of the YAK-AEROSIB project is analysis of seasonal and interannual variations of CO_2, O_3 and aerosol content over Siberia by obtaining information on the cross section of their distribution and calculating vertical gradients between the atmospheric boundary layer and the troposphere at different parts of the path from West to East Siberia.

Experimental

The measurements were carried out from Antonov-30 airborne laboratory.[1] It was used an eight-channel synthetic screen automated diffusion battery (ADB) with a CPC to measure the concentration of nanoparticles.[2] The ADB has been designed and manufactured at ICK&C SB RAS[3]. Thus it were obtained data on particle concentration in two size ranges 3nm $d_p < 70$nm and $d_p > 70$nm as well as the total number concentration of aerosol particles. All data on particle concentration at different altitudes were converted into concentrations at pressure and temperature observed at the ground level. This made it possible the comparisons to be done between the aerosol concentration at different altitude levels.

To achieve the goals of the project the investigations are carried out along the route from Novosibirsk to Yakutsk and back to Novosibirsk. The schematic diagram of the flight is shown in Figure 1.

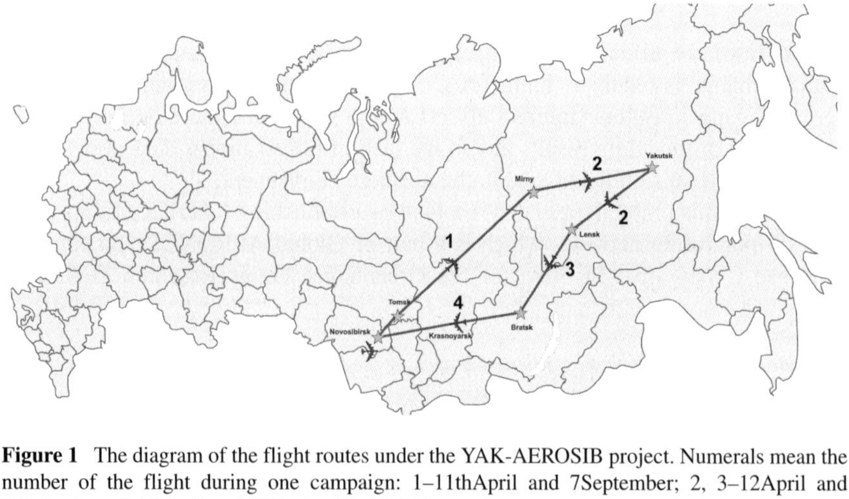

Figure 1 The diagram of the flight routes under the YAK-AEROSIB project. Numerals mean the number of the flight during one campaign: 1–11thApril and 7September; 2, 3–12April and 8September; 4–14April and 10September (2006)

To obtain the cross section of the atmosphere over the entire route, the flight height varied from the minimal possible height of 500 m over the terrain and 100 m in the airport zone and up to 7,000 m along the flight route. Seven such descents are performed along both flight legs (Novosibirsk–Yakutsk and Yakutsk–Novosibirsk). The flights to and back were performed using different routes: the flight to along the north route and back along the south route. As a result, during each round trip 14 profiles are acquired spaced along a horizontal (depending on the flight height) by 50–250 km. Such a flight route allows one to assess the gradients of the measured parameters with a step of 50 to 5,000 km.

Results and Discussion

Both campaigns revealed an inhomogeneity in the spatial distribution of nanoparticles concentration in the troposphere over Siberia (see Figures 2–5). A significant difference in cross sections obtained during cold and warm seasons.

The main sources of enhanced concentrations observed in the lower troposphere in April are the heat and power plants of the big cities (see left panels of Figures 2–5), because of heating season in Siberia lasts until the beginning of May and there were no such a pronounced plumes from cities in the concentration fields observed on September (right panels of Figures 2–5).

High concentration regions of nanoparticles in the boundary layer and middle troposphere measured between 90 and 110 longitudes during September campaign were caused by plumes from intensive forest fires and that is proved by MODIS fire data. Preliminary analysis of the 11 April route between the same longitudes has shown that it was just in the region where the arctic air invaded from the northern regions where typical aerosol number concentration varies with altitude

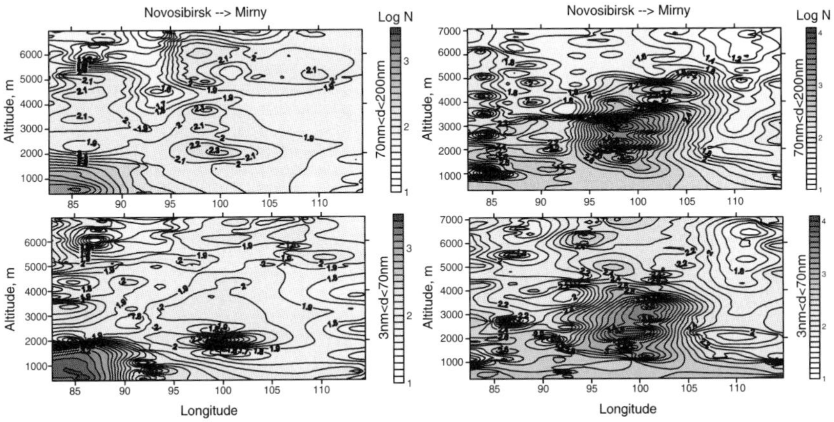

Figure 2 Spatial distribution of nanoparticles concentration along the 1st flight route (here and below: left panel – April 2006, right panel – September 2006)

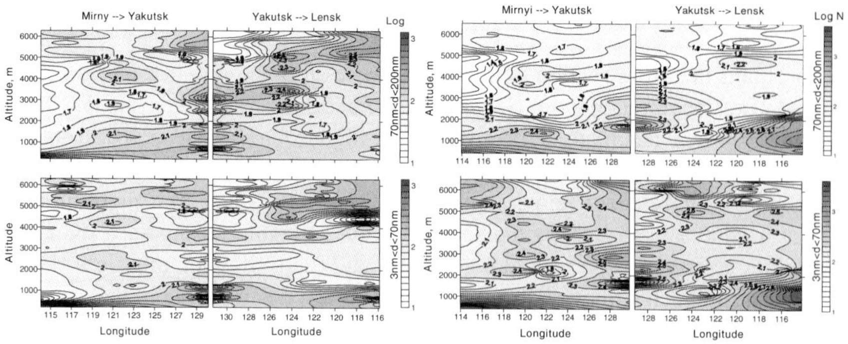

Figure 3 Spatial distribution of nanoparticles concentration along the 2nd flight route

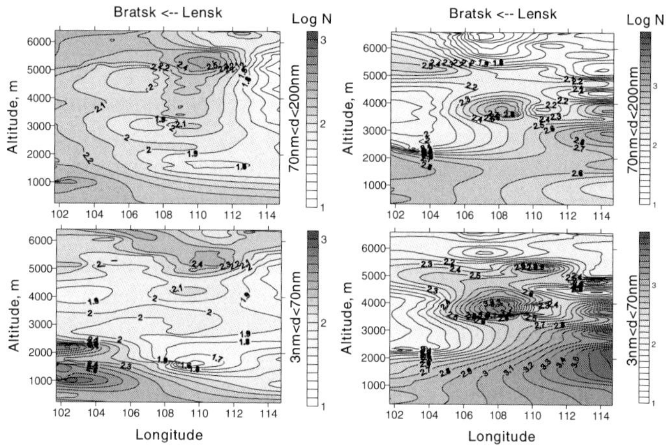

Figure 4 Spatial distribution of nanoparticles concentration along the 3rd flight route

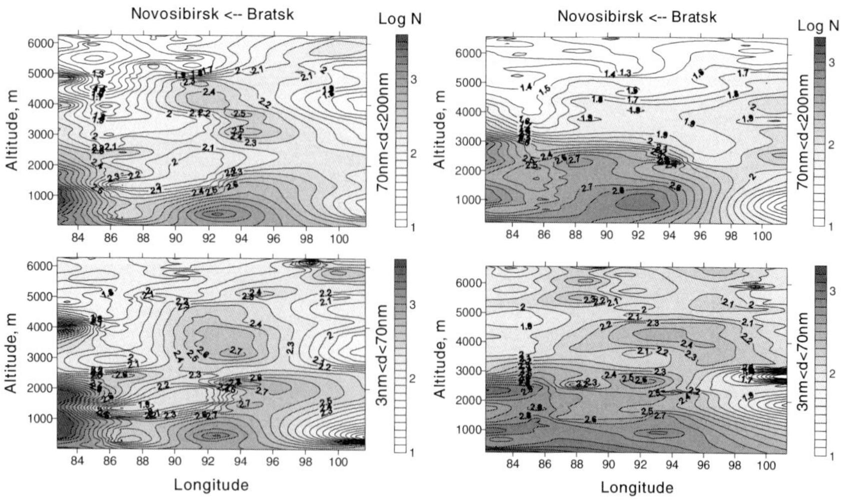

Figure 5 Spatial distribution of nanoparticles concentration along the 4th flight route

insignificantly in the range from 100 to 200 cm^{-3} that reflected in small gradients of concentration observed during this flight.

Aerosol layers of nanoparticles in the upper troposphere can be a result of in situ nucleation from aerosol forming substances transported from the remote sources.

Acknowledgments The study has been done under financial support of CNRS (France), the No. 24.3 Program of the Siberian Branch RAS, with the support from the Presidium of RAS Program No. 16, the programs of the Department of Terrestrial Sciences RAS No. 9 and No. 11, grants of the RFBR 04–05–64559, 06–05–79036, 06–05–08104, and 07–05–00645.

References

1. Zuev, V.E. et al., *Atmos. Ocean.*, **5**(10), 1012–1021 (1992).
2. Arshinov, M.Yu. and Belan, B.D., *J. Aerosol Sci.*, Abstracts of the European Aerosol Conference 2004, **2**, 987–988.
3. Reischl, G.P., Majerowicz, A., Ankilov, A., Baklanov, A.M., Eremenko, S., and Mavliev, R., *J. Aerosol Sci.*, **22**(2), 223–228 (1991).

EC/OC at Two Sites in Prague

Jaroslav Schwarz[1], Xuguang Chi[2], Willy Maenhaut[2], Martin Civiš[3], Jan Hovorka[3], and Jiří Smolík[1]

Abstract Organic and elemental carbon (OC and EC) content in PM_{10}, respectively $PM_{6.5}$ at two urban background sites in Prague was studied. Similar overall average levels were found at the two sites for both species (5.0/5.5 and 0.83/0.74 µg/m^3 for OC and EC, respectively), but substantial differences were found between the sites in some seasons and/or meteorological situations.

Keywords Organic and elemental carbon, urban aerosols

Introduction

Carbonaceous species form typically 10–50% of the total PM_{10} mass.[1] Organic carbon (OC) can be of both primary and secondary origin, while elemental carbon (EC) is only a primary species. Several studies on the composition of the PM_{10}, including measurements for OC and EC in the urban environment, have already been made in the western part of Europe.[2–5] However, for the Prague environment, only one published study[6] is currently available. Sillanpää et al.[6] studied the PM_{10} composition including OC/EC during one winter campaign in Prague (PAMCHAR project). To fill the gap in our knowledge of the PM in Prague, the project "Comprehensive size-resolved characterization of atmospheric particulate matter in Prague" was carried out.

Experimental

Parallel samplings were conducted at two urban background sites in Prague. The first site was located in the northwest suburbs of Prague on the roof of a building in the campus of the Institute of Chemical Process Fundamentals (ICPF), 285 m a.s.l.

[1] *Institute of Chemical Process Fundamentals AS CR, Prague, Czech Republic*

[2] *Ghent University, Department of Analytical Chemistry, Gent, Belgium*

[3] *Charles University, Institute for Environmental Studies, Prague, Czech Republic*

C.D. O'Dowd and P.E. Wagner (eds.), *Nucleation and Atmospheric Aerosols*, 824–828.

The second site was located downtown of Prague city on the roof of the four-storey building of the Institute for Environmental Studies (IES) of the Faculty of Science, Charles University in Prague, in Benátská street, 225 m a.s.l. The distance between both sites is about 7 km. Both the ICPF and IES sites were equipped with the following sampling devices and online instruments: Scanning Mobility Particle Sizers (SMPS), PM_1, $PM_{2.5}$ and PM_{10} samplers, a Gent PM_{10} stacked filter unit (SFU) sampler, and a 10-stage Berner low pressure cascade impactor (BLPI). Online gas analyzers provided data for NO, NO_2, NO_x, O_3, CH_4, and non-methane hydrocarbons. Meteorological data were recorded by meteo-stations. Here, we report the results of the chemical analysis of PM_{10} samples from both sites in Prague using thermal-optical transmission (TOT) analysis for OC/EC determination.

The samples analyzed for OC/EC were obtained using a PM_{10} low-volume sampler (17 l/min) for the ICPF site. The filter material here was 47 mm diameter Whatman QM-A quartz fiber filter. At the IES site, the samples were obtained using a PM_{10} Harvard impactor sampler with a flow rate of 10 L/min employing a pump with automatic regulation, and the filters were 37 mm quartz microfiber discs (QMF grade, Filtrak). At each of the two sites, two quartz fiber filters were used in series to assess the importance of positive artifacts and to correct for them. The face velocities to the filter were similar at both the sites (0.22 and 0.25 m/s). Unfortunatelly, cut-off diameters at the ICPF and IES were not identical. Due to a technical reason, the sampling line upstream of the Harvard impactor was not vertical. The calculated sampling losses showed, that fraction sampled at the IES had a cut point at around 6,5 µm. Thus, these data are interpreted as $PM_{6.5}$. All front and back filters were analyzed at Ghent University, Department of Analytical Chemistry, by the thermal-optical transmission (TOT) method[7] using a Sunset Laboratory OCEC analyzer and the NIOSH temperature program.[8] As OC data we used the difference between the OC on the front filter and the OC on the back filter. All reported data are such corrected data (unless otherwise indicated). Total carbon (TC) was calculated as the sum of (corrected) OC and EC.

Results

The time series for OC and EC at both sites is shown in Figure 1.

The large temporal variability in the data is clearly visible. While the OC data of the two sites are well correlated with each other (correlation coefficient 0.74), the correlation for the exclusively primary EC is worse (correlation coefficient 0.51). Seasonal averages (including the data from 23 to 27 February 2004, which are not shown in Figure 1) are given in Table 1, together with the overall averages over all samples. The overall averages of the two sites are quite similar for both OC and EC. Nevertheless, we have to have to keep in mind, that the size fraction sampled at IES was $PM_{6.5}$ versus PM_{10} at ICPF. This may have an influence, especially at dry conditions and at times when large amounts of primary biogenic particles can be expected. Thus, the summer and spring samples may be more influenced than

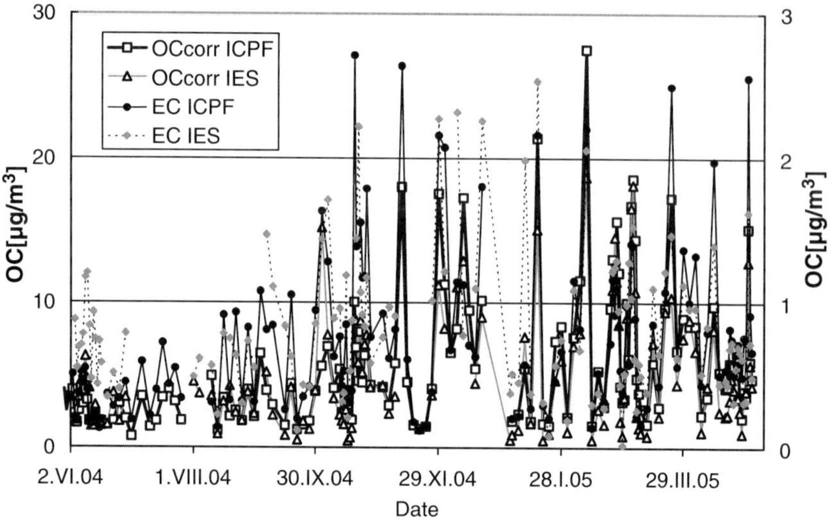

Figure 1 Time series for organic and elemental carbon at the ICPF (PM10) and IES (PM6.5) urban background sites

Table 1 Average concentrations of OC, EC, and TC per site and season

	Average [µg/m3]	Summer [µg/m3]	Autumn [µg/m3]	Winter [µg/m3]	Spring [µg/m3]
OC/ICPF	5.53	2.68	5.81	7.73	6.33
EC/ICPF	0.74	0.52	1.00	0.70	0.94
TC/ICPF	6.27	3.20	6.80	8.43	7.27
OC/IES	4.95	2.76	5.41	6.26	4.12
EC/IES	0.83	0.65	1.07	0.82	0.74
TC/IES	5.75	3.41	6.48	7.04	4.80

the winter or autumn samples. However, from the seasonal average PM mass size distributions obtained with the BLPI at both sites (but not with identical time base) and shown in Figure 2 for the IES site, it appears that particles larger than 6.5 µm were of relatively minor importance, even during spring and summer. This may be because of the sampling height, which was about 20 m above ground. Thus, based on the PM mass size distribution measurements, we may conclude that the differences between the two sites for winter and spring (higher OC at the ICPF suburban site) cannot be explained by the differences in PM size fraction collected. In the case of EC, the impact of the difference in size fractions can be assumed to be negligible, so that we can directly compare the results. Table 2 gives the average meteorological parameters and the ratios for EC/TC and OC2/OC1 (OC found on the back filter over uncorrected OC on the front filter).

For most parameters listed in Table 2, substantial seasonal differences and/or differences between the two sites are noted. The systematically higher EC/TC ratio for the IEC site is in agreement with the much higher traffic intensity in the city

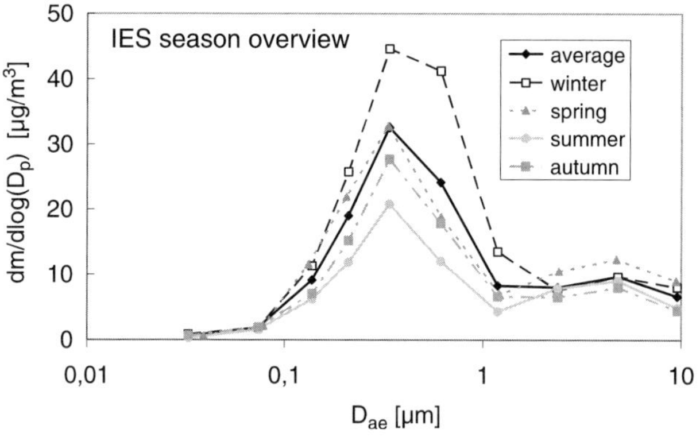

Figure 2 Seasonal average PM mass size distributions for the IEC site

Table 2 Average ratios for EC/TC and OC2/OC1 and average meteorological parameters

	Average	Summer	Autumn	Winter	Spring
ICPF EC/TC	0.13	0.16	0.16	0.10	0.12
ICPF OC2/OC1	0.14	0.22	0.12	0.07	0.18
IES EC/TC	0.18	0.20	0.20	0.16	0.18
IES OC2/OC1	0.30	0.38	0.27	0.28	0.31
Wind velocity	1.7	1.8	1.7	2.0	1.4
Temperature	8.2	18.2	8.3	−1.4	14.1
RH	70.1	57.8	81.0	73.7	63.3

center compared to the ICPF site at the NW suburb. Interesting is the large difference between the OC2/OC1 ratios. At both sites it seems to be temperature dependent, but much less so at the IES site. There are two possible explanations for both facts. First, the VOC concentration downtown might be higher compared to that at the ICPF site and, secondly, at the IEC site the sampling cassettes were placed under a roof inside an air-conditioned space. As a consequence, the negative artifact (loss of semi-volatile OC) could be larger, especially in winter, compared to sampling at ambient conditions. This may also partially explain the lower values of OC found in winter at the IES site.

Finally, the ratios between OC at the IES site and OC at the ICPF site (and similar ratios for EC) for main wind directions and for the heating (HS) and non-heating seasons (NHS) are shown in the Figure 3. While the primary EC data are similar at the two sites for SE winds when the ICPF site is downwind of the city, the EC level is almost three time higher at the IES site when the ICPF site is upwind of the city at NW winds during NHS. The situation is different in HS when a much smaller difference in EC between the two sites is noted for the NW wind direction. This is probably due to the impact from local and regional residential heating using coal and wood. This type of heating is rare in the downtown of Prague, but it is common at suburbs and countryside.

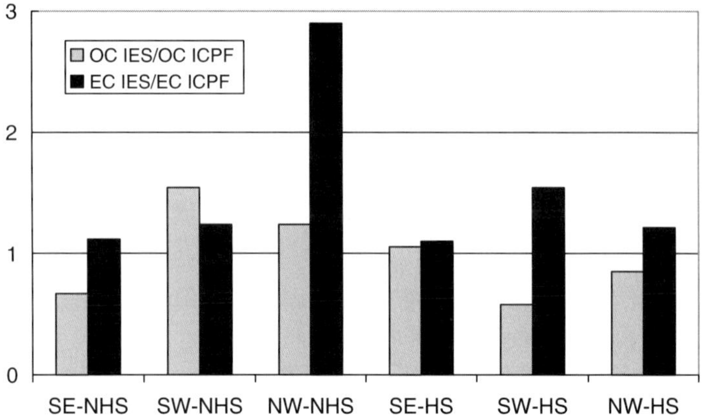

Figure 3 Ratios between OC at the IES site and OC at the ICPF site (and also for EC) for the main wind directions and for the heating and non-heating seasons (HS, NHS)

Acknowledgments The support by grants GA CR No. 205/03/1560 and Ministry of Environment CR No. VaV-SM/9/86/05 is gratefully acknowledged. Support was also provided by the Research Foundation – Flanders and the Belgian Federal Science Policy Office.

References

1. Seinfeld, J.H. and Pandis, S.N., *Atmos. Chem. and Phys.*, New York: Wiley, pp. 491–544 (1998).
2. Puxbaum, H., Gomiscek, B., Kalina, M., Bauer, H., Salam, A., Stopper, S., Preining, O., Hauck, H., *Atmos. Environ.*, **38**, 3949–3958 (2004).
3. Querol, X., Alastuey, A., Ruiz, C.R., Artinano, B., Hansson, H.C., Harrison, R.M., Buringh, E., ten Brink, H.M., Lutz, M., Bruckmann, P., Straehl, P., Schneider, J., *Atmos. Environ.*, **38**, 6547–6555 (2004).
4. Hueglin, C., Gehrig, R., Baltensperger, U., Gysel, M., Monn, C., Vonmont, H., *Atmos. Environ.*, **39**, 637–651 (2005).
5. Viana, M., Chi, X., Maenhaut, W., Cafmeyer, J., Querol, X., Alastuey, A., Mikuška, P., Večeřa, Z., *Aerosol Sci. Technol.*, **40**, 107–117 (2006).
6. Sillanpää, M., Frey, A., Hillamo, R., Pennanen, A.S., Salonen, R.O., *Atmos. Chem. Phys.*, **5**, 2869–2879 (2005).
7. Birch, M.E., Cary, R.A., *Aerosol Sci. Technol.*, **25**, 221–241 (1996).
8. Schauer, J.J., Mader, B.T., Deminter, J.T., Heidemann, G., Bae, M.S., Seinfeld, J.H., Flagan, R.C., Cary, R.A., Smith, D., Huebert, B.J., Bertram, T., Howell, S., Kline, J.T., Quinn, P., Bates, T., Turpin, B., Lim, H.J., Yu, J.Z., Yang, H., Keywood, M.D., *Environ. Sci. Technol.*, **37**, 993–1001 (2003).

Submicrometric Aerosol Size Distributions in Southwestern Spain: Relation with Meteorological Parameters

M. Sorribas[1], V.E. Cachorro[2], J.A. Adame[1], B. Wehner[3], W. Birmili[3], A. Wiedensohler[3], A.M. de Frutos[2], and B.A. de la Morena[1]

Abstract The aim of this work was to analyze statistically the atmospheric particle number size distributions, at the sampling station El Arenosillo located at the southwest of Spain, in terms of their relation to meteorological parameters. With this purpose, surface wind data and submicrometric particle size distributions were collected over one year period. According to the main wind directions the studied zone may be divided in two areas: polluted and nonpolluted areas, with (7,106 ± 281) cm^{-3} and (5,623 ± 400) cm^{-3}, respectively. From the wind database patterns of de sea–land breeze have been identified at El Arenosillo. The relationship between the breeze and the particle number size distributions was analyzed. The average diurnal total concentration showed the greater value with no pure sea–land breeze with synoptic wind forcing from north with (8,683 ± 2,240) cm^{-3}. The results in the case of pure breeze and breeze with synoptic wind forcing from south were an inferior 26% and 12%, respectively. Short time events with high concentrations were identified during local recirculations. They were called Events by Breeze. As regards the wind speed, high concentrations of nucleation and Aitken modes particles were observed during spring time, with low wind velocities. This condition favors the photochemical mechanism leading to the formation of new aerosol particles.

Keywords Tropospheric aerosols, particle size distributions, SMPS, coastal breeze

Introduction

All meteorological phenomenons play a key role in the evolution of number size distributions of atmospheric aerosol. One of the effects of wind speed is to dilute continuously released aerosols at the point of emissions or the transport of particles

[1] *Atmospheric Sounding Station 'El Arenosillo', Earth Observation, Remote Sensing and Atmosphere Department, INTA, Mazagón, 21130, Huelva, Spain*

[2] *Atmospheric Optics Group (GOA-UVA), University of Valladolid, 47071, Valladolid, Spain*

[3] *Leibniz-Institute for Tropospheric Research, 04318, Leipzig, Germany*

C.D. O'Dowd and P.E. Wagner (eds.), *Nucleation and Atmospheric Aerosols*, 829–833.

from their source is determinate by the wind direction at the source. And other examples are the temperature which affect the atmospheric mechanisms either growth or new formation particles increasing the ultrafine mode, and the local recirculations which produce accumulation of particles with influence human health risks.

To investigate the particle number size distribution at El Arenosillo is a difficult task because of geographical location at southwest of Spain. The coastal zone represents a case of mixing: marine aerosols generated at the sea surface are added to continuous continental contribution from dessert intrusions and from Europe, and anthropogenic sources. For example, urban and industrial emissions advected from the northwest (Huelva City and their industrial areas, 35 km) strongly influenced the site occasionally, resulting in higher particle concentration over background level. Moreover at the coast, with frequency are developed mesoscale circulations which are recirculatory processes that produce an accumulation of particle on the see which during the day can be measured over ground.

This study contributes to improve our understanding of the evolution of the particle number size distributions, measured in a forest and coastal place, as a function of local meteorology. These studies are required to identify the concentrations of both locally generated and long-range transported aerosols.

Description Site and Data Set

Measurement Site Localization

The Atmospheric Sounding Station "El Arenosillo" is a platform to investigate various topics about atmospheric sciences. As regards in situ aerosol properties, submicrometric and supermicrometric size distribution, scattering and absorption coefficients and the concentration levels and chemical composition of PM10 and PM2.5 are monitoring continuously (Mogo et al. 2005; Sorribas et al. 2007). The station is located at Huelva, southwest Atlantic coast of Spain (37.1°N, 6.7°W, 40 m a.s.l). The surrounding is a homogeneous *Pinus Sylvestris* forest extending several kilometres to all directions except the southwest where the Atlantic Ocean is at a distance less than 1 km. The nearest small village is Mazagón about 8 km and Huelva City (160,000 inhabitants) is located 35 km to the northwest of the station with an important industrial environment (three important industrial complexes).

The sampling inlet is located 8 m above ground level and 3 m over the top of the trees. The Aerosol Sampling System was designed with a well characterization of the efficiency, which is applied to the entire data set (Sorribas et al. 2006).

Instrumentation and Database

Particle number size distributions from 16.5 to 604 nm were measured with a scanning mobility particle sizer (SMPS, TSI Model 3081 DMA and TSI 3022A CPC).

The sheath air stream was dried. The sampling flow rate of the SMPS was 0.3 lpm with a sheath flow rate of 3 lpm. CPC was operating with low flow. Aerosol Instrument Manager software (version 4.3, TSI INC., St. Paul, Minnesota, USA) was used for data reduction and analysis of the SMPS output. The submicrometric number size distributions have been measured continuously (10 min resolution) from 15 July 2004 so far. For this study, the time period from 15 July 2004 to 25 July 2005 was selected, a total of 375 days. The SMPS was operational on 331 days out of a total, which have 46,135 number size distributions.

Ground-based meteorological parameters have been monitored at approximately the same height and corresponding data were stored and averaged every 10 min.

Result and Discussion

To start the analysis of the atmospheric aerosol during database, the average values are presented. The average total concentration during ten minutes database (ATCM) is (6,607 ± 5,199) cm^{-3}. Table 1 shows the average concentration and 10–90 percentiles. The highest concentrations, shown by 90 percentile, do occur during daytime.

The frequency distribution of the total concentrations shows that 58% of the observations are below 6,000 cm^{-3} and 1% are above 24,000 cm^{-3}. The interval (4,500–6,000) cm^{-3} presents the maximum frequency with value 22%. Daily average of the total concentration during all database is (6,615 ± 2,982) cm^{-3}.

Wind Data

In order to characterize the wind regimen have been used wind roses of 16 directions. The relation between the ATCM and the presented one by the different sectors allowed dividing the studied zone in two areas: polluted and nonpolluted areas.

The nonpolluted hemisphere is that in which the concentration is equal or lower than ATCM. It is formed by the sectors from N to SWS in clockwise. Sector S contains the smaller particle number with a difference with respect to the ATCM of 21%. Average total concentration in the nonpolluted hemisphere was (5,623 ± 400) cm^{-3}.

Table 1 Average total and modal particle concentration during ten minutes database

	Total (cm^{-3})	Nucleation (cm^{-3})	Aitken (cm^{-3})	Accumulation (cm^{-3})
NTotal	6,607	1,502	3,445	1,660
Percentile 10	2,541	151	1,149	455
Percentile 90	11,509	3,659	6,438	3,048

The polluted hemisphere contained bigger concentration than the ATCM. It includes the sectors from SW to NWN in clockwise. The most concentration is in the sector W with a 13% more than the ATCM. All sectors of the polluted hemisphere present variability greater than nonpolluted hemisphere, displaying higher and lower values of total concentration. The average total concentration in the polluted hemisphere is (7,106 ± 281) cm^{-3}. A detailed study of the origin of the particle in terms of the particle number size distributions and the wind flows also will be presented in this work.

As regards wind speed during spring time, high concentration of the nucleation and Aitken modes are observed, when the wind speed is low. During the rest of the time the particle concentration in these modes are lower.

Behavior of Size Distributions Under Breeze Conditions

In Adame et al. (2005) three patterns of coastal breeze were identified. The first pattern identified *Breeze 1*, is considered such as a pure sea breeze since shows perpendicular directions to the coastline, low wind speed and a diurnal-nocturnal regime of twelve hours. The second, *Breeze 2*, is developed under wind synoptic forcing from the northwest which affect strongly the nocturnal regime, blowing from the northwest, not perpendicular to the coastline, and cause higher range of mean wind speed and lower duration of the marine breeze. The third, *Breeze 3* present similar characterises to *Breeze 1*, but with a major duration of sea breeze and higher wind speed, associated to wind synoptic forcing from south.

Daily averages of the total concentration during *Breeze 2* presents the biggest values, with (8,683 ± 2,240) cm^{-3}. The concentrations in the case of *Breeze 1* and *Breeze 3* are 26% and 12% lower, respectively. The lower variability is presented by *Breeze 1* with standard deviation of 1,565 cm^{-3}. In this work a study of the modal concentrations is presented in function of breeze pattern.

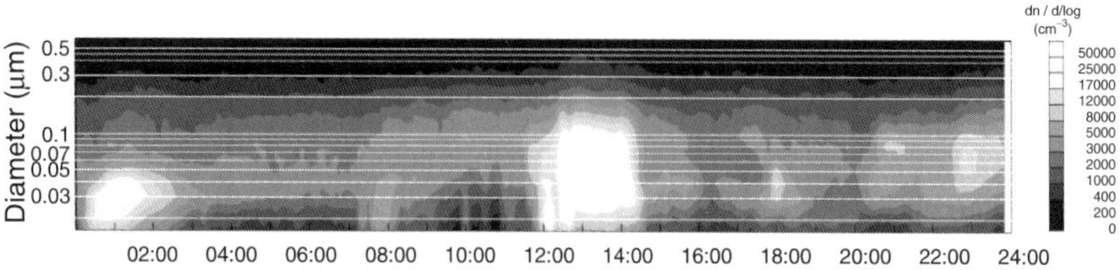

Figure 1 Daily evolution of the particle number size distribution happened on 12October 2004. One hour then, an Event by breeze started and its duration was 0250 h. The average concentrations in the different modes were: nucleation mode (4,911 ± 2,922) cm−3, Aitken mode (10,708 ± 5,416) cm^{-3} and accumulation mode (2,710 ± 1,088) cm^{-3}

vent by Breeze

he carefully manual inspection of the size distributions permitted to distinguish arious types of short time events with high concentration. They were called Event y breeze. The land breeze presents a penetration distance on the sea, and in this orm the particles are accumulated during the night. When the onset of marine reeze takes place, the accumulated particles brought to El Arenosillo. These aero-ols increase the number particle size distributions in all diameters, Figure 1.

cknowledgments This work has been supported by the Spanish "Ministerio de Educación y iencia" by means of project CGL2005–05693-C03–02/CLI and European Union as part of the roject AEROTOOLS.

eferences

dame, J.A., *Caracterización y Comportamiento del Ozono Superficial en la Provincia de Huelva*, Ph.D. dissertation, University of Huelva, Spain (2005).

Iogo, S. et al., Measurements of continuous spectra of atmospheric absorption coefficients from UV to NIR via optical method, *Geophys. Res. Lett.*, **32**, L13811, doi: 10.1029/2005GL022938 (2005).

orribas, M. et al., General characterization of the monitorings station of the in situ atmospheric aerosols properties at El Arenosillo (Huelva), In *Proceedings, 5ª AHLGG*, Seville (2006).

orribas, M. et al., Characterization of sub-micron number particle size distribution measurements at El Arenosillo Station (south-western, Spain) during periods with prevalence of the accumulate mode, *Atmos. Environ.*, submitted for publication.

Characterization of a Propane Soot Generator

E. Barthazy, O. Stetzer, C. Derungs, S. Wahlen, and U. Lohmann

Abstract A diffusion flame propane soot generator is presented. It can operat
under different burning conditions and produce a variable aerosol output. The aeroso
flow can be changed gradually from an unimodal flow of large soot particles to
bimodal flow of soot particles mixed with small droplets of PAHs. The existence o
the small PAH droplets suggests that no sulfuric acid nuclei (as observed for Diese
engines) are necessary to nucleate PAHs.

Keywords Soot, propane burner, PAH, homogeneous nucleation, hydrophobic
water CPC

Introduction

Soot particles are of major importance among the anthropogenic atmospheri
aerosols. Their interaction with radiation as well as their ability to act under certai
circumstances as condensation nuclei mark their importance for cloud microphysics
precipitation physics, radiative forcing and climate change (e.g., Kaufman an
Koren 2006; Highwood and Kinnersley 2006). The production of soot for labora
tory studies, however, is complicated. Either real engines are used which are bulk
and expensive or a soot generator, typically burning some gaseous fuel, is used
Here, a soot generator is presented that can produce pure soot output as well as soo
mixed with small droplets of PAHs.

Instrumentation

Soot particles are generated with a CAST (Combustion Aerosol STandard) soo
generator (http://www.sootgenerator.com), a co-flow (propane/oxidation air) dif
fusion flame propane burner (Figure 1). The flame is quenched with nitrogen t

Institute for Atmospheric and Climate Science, ETH Zurich, 8092 Zürich

C.D. O'Dowd and P.E. Wagner (eds.), *Nucleation and Atmospheric Aerosols*, 834–839.

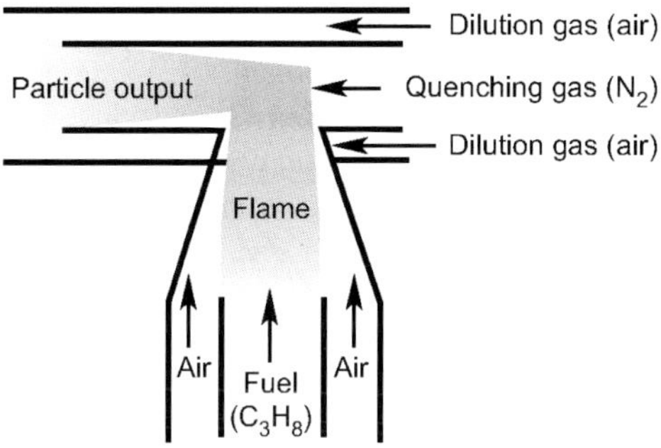

igure 1 Propane and air are burned 'ithin the CAST. The flame is quenched y a flow of nitrogen. The aerosol flow is ırther diluted with air

ıterrupt the combustion process and the particle stream is diluted with com-ressed air.

By varying the relative flow of propane to oxidation air, flames can be generated /ith C/O ratio from 0.21 to 1.4. When the absolute flow of propane/oxidation air ; varied, the size of the flame can be changed. Since quenching happens at a fixed istance above the burner, small flames are quenched at their upper part and large lames close to the base of the flame. By this, the production environment of the erosols can be adjusted twofold: as to the flame chemistry (C/O ratio) and as to ıe place of their origin (relative position within the flame). This leads to aerosols f different number concentration, size distribution and organic/elemental carbon omposition.

:xperiment

'he complete experimental setup consists of four parts: the particle generation, the ilution, a thermodenuder and the particle sampling. Right behind the exhaust of ıe CAST, the particle output is diluted with one to three dilution stages (one Palas 'KL-10 and two VKL-100) by a factor of $10–10^5$.

erosol Size Distributions

igure 2 shows the aerosol size distribution for flames with a fixed C/O-ratio of 0.6 s measured with an SMPS 3080 (DMA 3085 with CPC 3010, TSI Inc.). Flame ze is changed by adjusting the gas flows of both, the propane and the oxidation ir with C/O remaining constant. For practical purposes results are discussed elated to the propane flow only. However, oxidation air flow is always adjusted ıch that C/O remains constant. When the propane flow is high, aerosols are ampled close to the base of the flame and a bimodal size distribution is observed. s the propane flow decreases and sampling moves upward through the flame, the

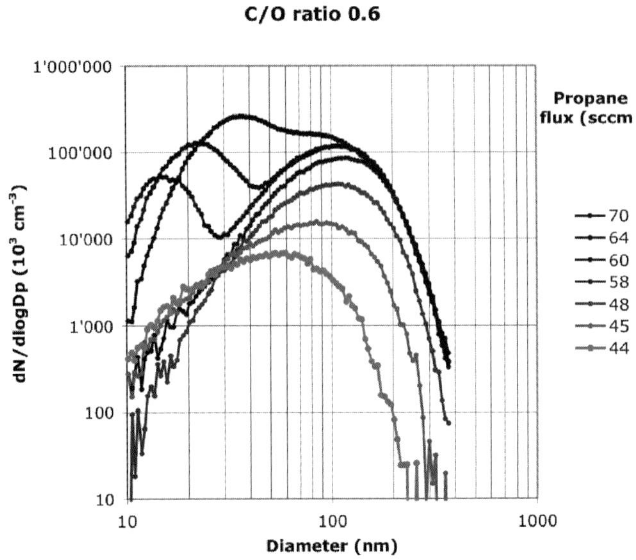

Figure 2 Size distribution of soot particles generated with the CAST with a C/O ratio of 0.6. Th propane flow stands for the total flow of propane/oxidation air

small mode decreases in size and number until it disappears completely. The larg mode remains constant for a while and then decreases slowly in size and numbe Hayashida et al. (2006) describe an experiment where soot, polycyclic aromati hydrocarbons (PAHs) an OH radicals are observed within a propane diffusio flame. They observe close to the base of the flame strong PAH production. Withi the upper part of the flame soot is predominant. Transferred to our experiment w can conclude that the small mode in the size distribution corresponds to condense droplets of PAH and the large mode corresponds to soot particles.

Total Number Concentration

Total aerosol number counts with two CPCs (butanol and water) for flames wit C/O = 0.6 are shown in Figure 3. As the flow of propane increases, the number c particles measured with the two CPCs increases, too. However, at a propane flo rate of about 55 sccm, the count rate of the water CPC plunges suddenly. The di ference in the count rates for a propane flow >60 sccm of the water and the butan CPC is in the order of 4–5 magnitudes.

Obviously, the water CPC is not able to measure the particles adequately whic are generated at a propane flow of 60 sccm or higher. When the flow rate of propan is low, the flame is very small and quenching takes part at or above the tip of th flame. Almost no particles are measured. As the flow rate of propane increases, th

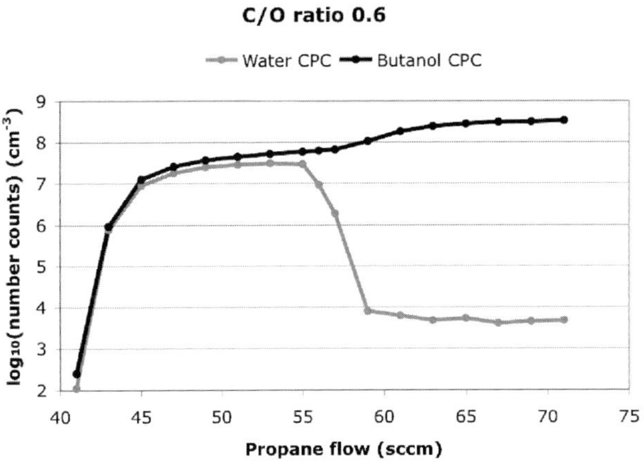

Figure 3 Total aerosol number counts as measured with a butanol and a water CPC

flame height increases and the flame is quenched within its upper part. Now many aerosols are observed, the size distribution shows one mode (Figure 2). These aerosols are soot particles. Moving downward through the flame (by increasing the total flow rate) more and more PAHs are produced and their vapor pressure rises and they start to condense on the large soot particles. This coating renders the soot particles hydrophobic and it is difficult to activate them with the water CPC which results in the plunge of the count rate at a propane flow rate of 55 sccm. However, the size distribution still shows just one mode. Moving further downwards within the flame, vapor pressure of the PAHs increases even more and small droplets of PAHs start to form. This leads to the bimodal size distribution observed for propane flow rates larger than 60 sccm (Figure 2). Now, the aerosol particles are very hydrophobic, only one out of 10^4 to 10^5 can be activated with the water CPC.

Effects of the thermodenuder (inserted between the dilution stages and the aerosol sampling) on the size distribution are shown in Figure 4. As the temperature rises, the small mode decreases and disappears at about 200°C while the large mode is not affected at all. This is a clear indication that the small mode consists of droplets of condensed PAHs.

Conclusions

The bimodal size distributions obtained with the CAST look similar to bimodal distributions observed under certain conditions in diesel exhaust (e.g., Schneider et al. 2005). This gives confidence that CAST soot may be used as a realistic laboratory soot surrogate for different applications where soot particles from engine combustion are needed.

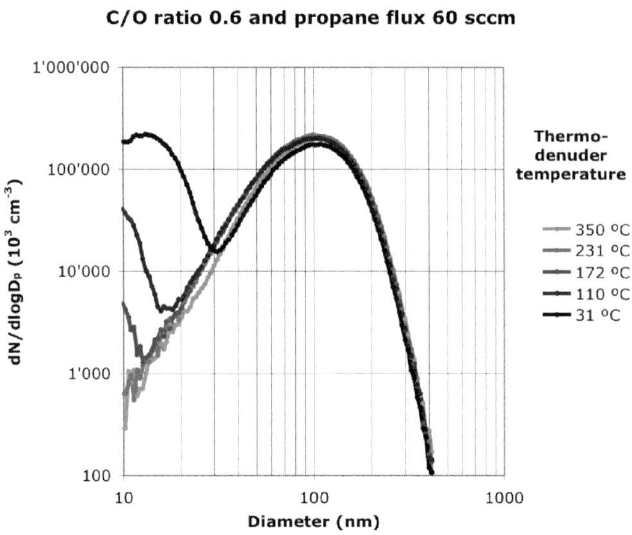

Figure 4 Size distribution of soot particles generated with the CAST with a C/O ratio of 0.6 After dilution, the aerosol flow is directed through a thermodenuder

Schneider et al. (2005) report a bimodal size distribution of the exhaust gas of a diesel passenger car when the engine is operated with a high-sulfur fuel. The small mode disappears completely when they use a thermodenuder. From this they conclude that the small mode is a nucleation mode and does not contain refractory material such as soot. When the engine is operated with low-sulfur fuel, no nucleation mode is observed. Therefore, they assume that the formation of the nucleation mode is triggered by sulfuric acid. The same conclusion is drawn by Tobias et al (2001). They calculate that a small organic particle would have a much higher vapor pressure than a sulfuric acid/water particle of the same size. Therefore, the Kelvin effect would favor nucleation of sulfuric acid/water over organics. Once nucleation has occured, particles would grow by condensation of organic compounds. In our experimental setup the CAST burner is clean, no sulfuric acid nuclei are available. Since we observe a bimodal size distribution with droplets of PAH the conditions must allow nucleation. Either the vapor pressure is high enough for homogeneous nucleation or some very small particles below the detection limit of the SMPS (<10 nm) must act as condensation nuclei. It seems that no impurity within the fuel is needed to produce bimodal size distributions with a small mode of nucleated PAHs.

At last, the water CPC is not a good choice to measure freshly generated soot aerosols. If burning is incomplete, PAHs may form and either coat large soot particles or form pure PAH droplets. In either case the supersaturation within the CPC (S = 1.85, private communication from TSI) is not enough to activate the particles for detection.

References

Hayashida, K., Amagai, K., Satoh, K., and Arai, M., *J. Engineering for Gas Turbines and Power-transations of the ASME*, **128**, 241–246 (2006).
Highwood, E.J., Kinnersley, R.P., *Environ. Int.*, **32**, 560–566 (2006).
Kaufman, Y.J., Koren, I., *Science*, **313**, 655–658, doi: 10.1126/science.1126232 (2006).
Schneider, J., Hock, N., Weimer, S., Borrmann, S., Kirchner, U., Vogt R., and Scheer, V., *Environ. Sci. Technol.*, **39**, 6153–6161 (2005).
Tobias, H.J., Beving, D.E., Ziemann, P.J., Sakurai, H., Zuk, M., McMurray, P.H., Zarling, D., Waytulonis, R., and Kittelson, D.B., *Environ. Sci. Technol.*, **35**, 2233–2243 (2001).

Aerosol Particle Formation Events at Two Siberian Stations

M. Dal Maso[1], L. Sogacheva[1], A. Vlasov[2], A. Staroverova[2], A. Lushnikov[1], M. Anisimov[3], V.A. Zagaynov[4], T.V. Khodzher[5], V.A. Obolkin[5], Yu. S. Lyubotseva[1], I. Riipinen[1], V.-M. Kerminen[6], and M. Kulmala[1]

Abstract We present one-year data of atmospheric aerosol particle size distributions covering the ultrafine size range from two measurement stations situated in Central Siberia: Tomsk and Listvyanka. Analysis of the size distribution time series revealed ca. 30 of new particle formation and growth events at both stations. The events occurred predominantly during spring and summertime, and the formation and growth rates were comparable with observations in the western part of the boreal forest.

Keywords Aerosol particle formation, growth, field measurements

Introduction

Formation of new atmospheric aerosol particles by nucleation events is a significant source of new atmospheric particles.[1] The formed particles have, after growing by condensation, the potential to act as cloud condensation nuclei and thus affect the Earth's radiative balance as well as the hydrological cycle.

The western part of the boreal forest area has been an area where particle formation has been studied intensively, in part because it has been an area where time series of atmospheric submicron size distributions have been measured continuously for years.[2] Studies on these measurements have revealed that the

[1]*Division of Atmospheric Sciences, Dept. of Physical Sciences, 00014 University of Helsinki, Finland*

[2]*Russian State Hydrometeorological University, 195196 St Petersburg, Russia*

[3]*Institute of Catalysis, SB RAS, 630090, Novosibirsk, Russia,*

[4]*Karpov institute of physical Chemistry, Vurontsovo Pole, 105064 Moscow, Russia*

[5]*Limnological institute, Siberian branch of RAS, 3. Ulanbatroskaja, 664043 Irkutsk, Russia*

[6]*Finnish Meteorological Institute, P.O. Box 503–00101 Helsinki, Finland*

C.D. O'Dowd and P.E. Wagner (eds.), *Nucleation and Atmospheric Aerosols*, 840–844.

boreal forest is a frequent source of new particles.[3] However, these studies are concentrated on the western part of the European boreal forest area, while data of the vast forested areas of the Siberian forests is still scarce.

In this work we present two one-year data sets from the Siberian boreal forest, where submicron aerosol size distributions have been measured. The analysis of the time series demonstrates that new particle formation and growth is occurring also in Central Siberia, providing evidence that the studies performed on atmospheric particle formation in the Nordic countries can be extrapolated to the Siberian boreal forest as well.

Materials and Methods

Measurement Stations

Aerosol Size Distribution Measurements

The aerosol size distribution data was measured using Diffusion Aerosol Spectroscopes (DAS).[4] The DAS consists of a set of grid diffusion batteries and a condensation particle counter. The particle concentration at both the inlet and outlet of the diffusion battery is measured to determine the penetration though the diffusion grid. By measuring the penetration with varying numbers of grids, and with knowledge of the size-dependent particle diffusion, one can obtain the size distribution of atmospheric particles.

The diameter size range covered by the DAS measurements is from 3 nm to ca. 250 nm for the Tomsk setup and from 3 nm to 50 nm at Listvyanka. The time resolution was 7 and 3 min per size distribution for the Tomsk and Listvyanka stations, respectively. The size and time resolutions were such that they enabled the analysis of the size distribution time series for the occurrence of particle formation events.

Identification of Particle Formation

To quantify the occurrences of new particle formation, the size distributions resulting from the data inversion were analyzed visually. Because the particle formation events were mostly occurring near noontime, and more than one event a day was rarely observed, the data analysis was performed on a day-to-day basis.

In the Tomsk data set, each day was classified into one of three classes: event, nonevent or undefined. This classification is similar to the one used in analyzing size distribution data measured in Nordic countries.[2,5] To be classified as an event day, a day had to fulfill the following criteria:

- A new mode of particles had to appear in the measured size distribution
- The mean diameter of the new particles must initially be less than 25 nm

- The mode should prevail for several hours
- The mean diameter of the particles in the mode must grow

Nonevent days were days during which clearly no regional new particle formation could be observed. Days during which nucleation mode particles were present but they were present for only a short time (less than 1 h), not a clearly new mode or their diameter was not increasing, were classified as undefined days. This was done to separate the event and nonevent days as clearly as possible for analysis of atmospheric properties leading to new particle formation.

For the Listvyanka data set we concentrated only on identifying the particle formation days. This was done because the coverage of the data was only around 50%, and the full classification would have lead to an excessive number of undefined days (as no knowledge of whether a NPF occurred could be obtained). Additionally, local pollution (inferred from SO2 data, available for about one half of measurement time) often lead to high numbers of ultrafine particles, which made the detection of nonevent days very difficult.

Results

General Overview

The mean particle concentration during the measurements at the Tomsk station was 2,480 cm^{-3}, while the median was 1,950 cm^{-3} and the standard deviation 2,000 cm^{-3}. In Listvyanka, the mean particle concentration measured by the DAS was 4,690 cm^{-3}. The median concentration was 4,090 cm^{-3} and the standard deviation 3150 cm^{-3}. The Tomsk total numbers are comparable with measurements at Finnish background sites, while the Listvynaka measurements, in contrast, show that the site is influenced by another particle source, very probably local pollution.

Particle Formation Events

Using the classification criteria described in the previous section we found 32 days that fulfilled the criteria of a regional-scale particle formation event for the Tomsk station. This is approximately 10% of all days that could be classified. 132 days (40% of the total) were classified as undefined, 34 (10% of the total) of those because of part of the day was not measured. 168 days (50%) were classified as nonevent days. If undefined days are left out of the classification, the event fraction is 16%.

The great majority of the classified formation events occurred in springtime, during April and May. Of the 32 events, 18 (56%) occurred during these two months. Thus, spring is the main period of event occurrence. The number of nonevents is high during winter and summer. Autumn (September–October) is a time with little particle formation but also less non-event days.

Table 1 Overview of the stations and the results

Station	Tomsk	Listvyanka
Coordinates	56.5°N, 85.1°E	51.9°N, 104.7°E
Measurement period	7.3.2005–15.3.2006	22.3.2005–30.3.2006
DAS size range	3–250 nm	3–50 nm
Mean concentration	2,480 cm^{-3}	4,690 cm^{-3}
Median concentration	1,950 cm^{-3}	4,090 cm^{-3}
Number of NPF events	33	31
Mean growth rate	5.5 nm/h	1.8 nm/h
Average particles formed	5,700 cm^{-3}	3,340 cm^{-3}
Mean formation rate	0.4 cm^{-3} s^{-1}	0.4 cm^{-3} s^{-1}

In Listvyanka, the number of events observed was 31, which is 12% of the days that size distributions were recorded (253 in total). 22 events, which is 67% of all the events, were recorded in the period from April to July, while only four were seen in the period from October to February.

Event Properties

Of the 32 event days observed in Tomsk, 18 were classified in class I and analyzed for the properties of the burst. In Listvyanka 18 events were analyzed with respect of growth rate and formation rate of the nucleation mode particles. The results of the analysis are summarized in Table 1.

In Tomsk, the growth rate of the geometric mean diameter of the measured size distributions was on average 5.5 nm/h, the median being 3.9 nm/h. This value is influenced quite much by the very high growth rate observed on April 4, 2005, when the growth rate was 23 nm/h. Without taking this day into consideration, the average growth rate is 4.5 nm/h, and the median 3.5 nm/h. The formation rates of new particles varied between 0.04 and 1 cm^{-3} s^{-1}, with a mean and median of 0.4 cm^{-3} s^{-1}. The number of new particles formed by the formation events was on average 5,700 cm^{-3}, with a median of 4,700 cm^{-3}.

Of the Listvyanka dataset, 18 days were analyzed. The growth rates were found to be lower, with mean and median values of 1.8 and 1.7 nm/h, respectively. The formation rates were on average similar to Tomsk, 0.4 cm^{-3} s^{-1} in mean and median. The number of particles produced during the particle formation was 3,340 cm^{-3}, ranging from 930 to 7,360 cm^{-3}.

Conclusions

We present one-year time series of aerosol size distribution measurements measured at two Central Siberian sites. On of the sites, Tomsk, can be considered a background site, while the other is influenced by anthropogenic sources.

We found occurrences of particle formation events, in which new particles are nucleated and subseqently grow by condensation, at both stations. The occurrences were recorded predominantly during springtime. The formation and growth rates of new particles were of the same order than observed at other stations in the boreal forest.

These findings give tentative evidence that not only the western part, but also the wide areas of Central and eastern Siberia are sources of aerosols produced by processes involving both biogenic and anthropogenic precursors.

Acknowledgments Miikka Dal Maso wishes to thank the Maj and Tor Nessling Foundation for funding this work (Grant No. 20072XX).

References

1. Kulmala, M. et al., *J. Aer. Sci.*, **35**, 143–176 (2004).
2. Dal Maso, M. et al., *Bor. Env. Res.*, **10**, 323–336 (2005).
3. Tunved, P. et al., *Science*, **14**, 261–263 (2006).
4. Julanov, Yu. V. et al., *Atmos. Res.*, **62**, 295–302 (2002).
5. Dal Maso, M. et al., *Tellus B*, in press (2007).

Characterization of Rural Aerosol in Southern Germany (HAZE 2002)

J. Schneider[1], N. Hock[1], S. Borrmann[1,2], G. Moortgat[1], A. Römpp[1,3], T. Franze[4], C. Schauer[4], U. Pöschl[4,1], C. Plass-Dülmer[5], and H. Berresheim[5,6]

Abstract A detailed study of chemical and microphysical aerosol properties was performed in May 2002 at the Meteorological Observatory Hohenpeissenberg in rural southern Germany (Hohenpeissenberg Aerosol Characterization Experiment, HAZE 2002). The measurements included quantitative mass spectrometric (AMS) analysis of submicron particles, number concentration and size distribution between 3 nm and 9 μm diameter, and filter sampling followed by several offline analysis methods including the determination of protein mass, PAHs, and dicarboxylic acids. In addition, ambient levels of OH, H_2SO_4, monoterpenes, ozone, and other gases were measured. Comparison between the submicron mass composition identified by AMS (ammonium: 11%, nitrate: 19%, sulfate: 20%, organics (OM_1): 50%) and the $PM_{2.5}$ derived from high volume filter samples showed that 62% of the $PM_{2.5}$ consisted of non-refractory compounds in the <1 μm diameter size range. The average OM_1:$OC_{2.5}$ ratio was 2.07. OM_1 and PM_1 were highly correlated and showed that constantly about 50% of the submicron mass fraction consisted of organic material. New particle formation occurred during daytime with peak values of up to 14,000 particles/cm^3. These events were most likely triggered by fast photochemical formation of H_2SO_4 followed by ternary H_2SO_4/H_2O/NH_3 nucleation. Ambient temperature ranging between 14°C and 32°C during the campaign was inversely correlated with the amount of ammonium nitrate in the aerosol consistent with corresponding thermodynamic calculations.

Keywords Rural aerosol characterization, organic aerosol constituents, aerosol nucleation

[1] *Max Planck Institute of Chemistry, Mainz, Germany*

[2] *Institute of Atmospheric Physics, Johannes Gutenberg University, Mainz, Germany*

[3] *Now at: Institute of Inorganic and Analytical Chemistry, University Giessen, Germany*

[4] *Technical University of Munich, Germany (U.P. now at: 1)*

[5] *German National Weather Service (DWD), Observatory Hohenpeissenberg, Germany*

[6] *Now at: National University of Ireland Galway, Dept. of Physics, Galway, Ireland*

C.D. O'Dowd and P.E. Wagner (eds.), *Nucleation and Atmospheric Aerosols*, 845–849.

Introduction and Overview of the Haze 2002 Campaign

Aerosol particles influence the radiative balance and water cycle of the atmosphere and can also affect air quality and human health. A significant fraction of atmospheric aerosol particles are formed by gas-to-particle conversion. Previous studies have shown that condensation of sulfuric acid in conjunction with ammonia and water vapor is a major source of new particle formation (NPF) in the lower troposphere (e.g., review by Kulmala 2004). In contrast, the contribution of low volatile hydrocarbons to NPF and particle growth is still very uncertain. However, in recent years sophisticated measurement techniques such as quadrupole aerosol mass spectrometry (Q-AMS) have become available to study the size-resolved composition of submicron (<1 µm diameter) particles in relation to gaseous precursor species.

A three-year study conducted in rural southern Germany (Birmili et al. 2000, 2003) including first long-term measurements of OH and H_2SO_4 (Berresheim et al. 2000) and of monoterpenes showed the dominant importance of photochemical production of H_2SO_4 for NPF and the lack of evidence for participation of hydrocarbons (HC) in this process despite significant HC emissions from surrounding forest. The study was conducted at the Meteorological Observatory Hohenpeissenberg approximately 50 km north of the Alps (47°48' N, 11°01' E, 985 m a.s.l.). The station is operated by Deutscher Wetterdienst (DWD) and is part of the WMO Global Atmosphere Watch program.

Subsequently, in May 2002, the present study (Hohenpeissenberg Aerosol Characterization Experiment, HAZE 2002) was conducted at the same site including Q-AMS, high volume filter sampling (HVS; Digitel DHA 80, 500 L/min, 3–24 h sampling intervals) of $PM_{2.5}$ particles with offline chemical analysis (gravimetric weighing of the filters; analysis of elemental carbon, $EC_{2.5}$, and total carbon, $TC_{2.5}$, the difference yielding organic carbon, $OC_{2.5}$), and a number of aerosol size distribution measurement techniques (optical particle counter OPC PALAS model PCS 2010 for particles with diameters between 270 nm and 9.5 µm; scanning mobility particle sizers SMPS, models 3081 and 3085, TSI Inc., for particle size ranges 7–300 nm, 3–65 nm, respectively; ultrafine particle counters UCPC >3 nm and >14 nm diameter, TSI Inc.).

Results

In Figure 1, panels (a–(c) reveal that the temperature (14–32°C), the concentrations of total and ultrafine particles, and of H_2SO_4 showed a pronounced diurnal cycle, indicating photochemical particle formation. The mass concentrations measured with the Q-AMS data, panel (e), showed a large variation with the diurnal averaged total concentration ranging from 3 µg/m^3 to 13 µg/m^3. Levels of all four species (NO_3, SO_4, NH_4, Organics) increased significantly between May 21, 2002 and May 23, 2002. Backward trajectory calculations confirmed that the advected air masses

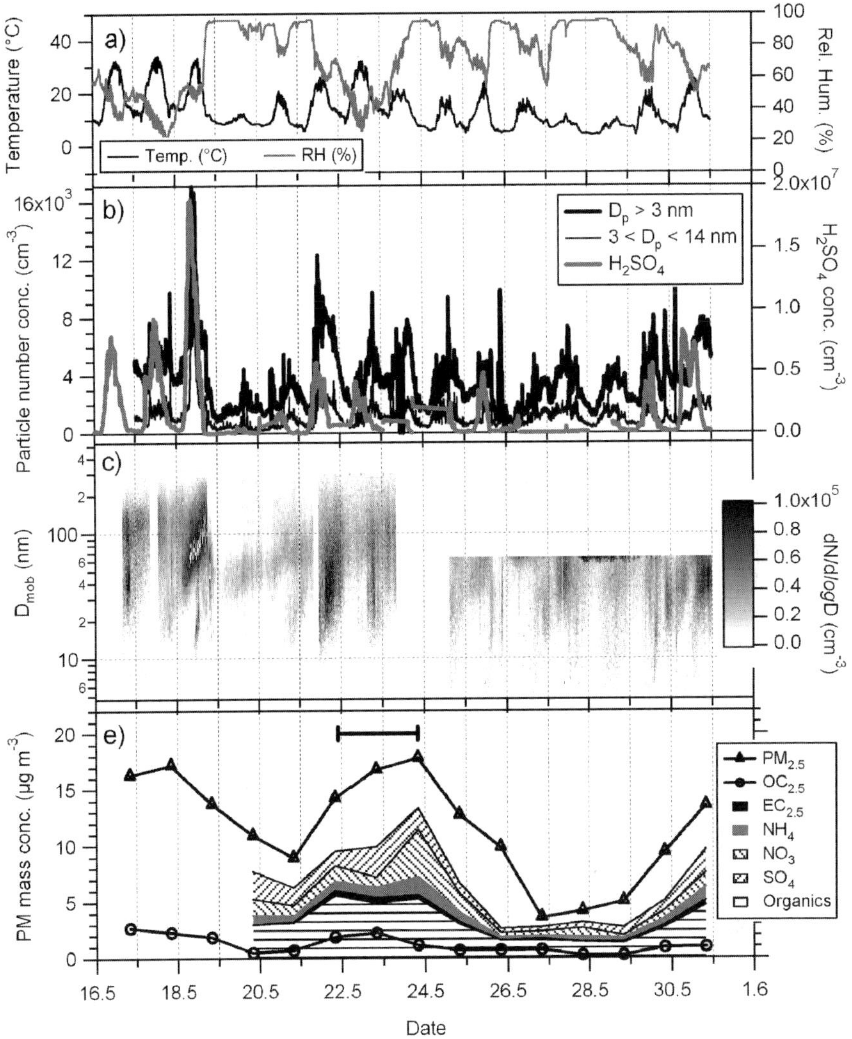

Figure 1 Overview of physical and chemical aerosol parameters measured during HAZE 2002. The horizontal bar in panel (e) denotes the arrival of polluted air from northern Italy (Po Valley)

had encountered pollution from the Po Valley area in northern Italy between 18 and 48 h prior to arrival. The ratio of m/z 44 (marker for oxygenated organic aerosol) to total organics (Figure 2) showed an increased fraction of non-oxygenated "fresh" anthropogenic organic aerosol during this pollution episode.

Chemical resolved mass concentrations were calculated from the HVS sample analyses for organic carbon ($OC_{2.5}$), based on the difference of total carbon ($TC_{2.5}$) and elemental carbon ($EC_{2.5}$). The correlation between the total organic mass concentration measured with the Q-AMS (= total organic matter, OM_1) and $OC_{2.5}$ was quite strong

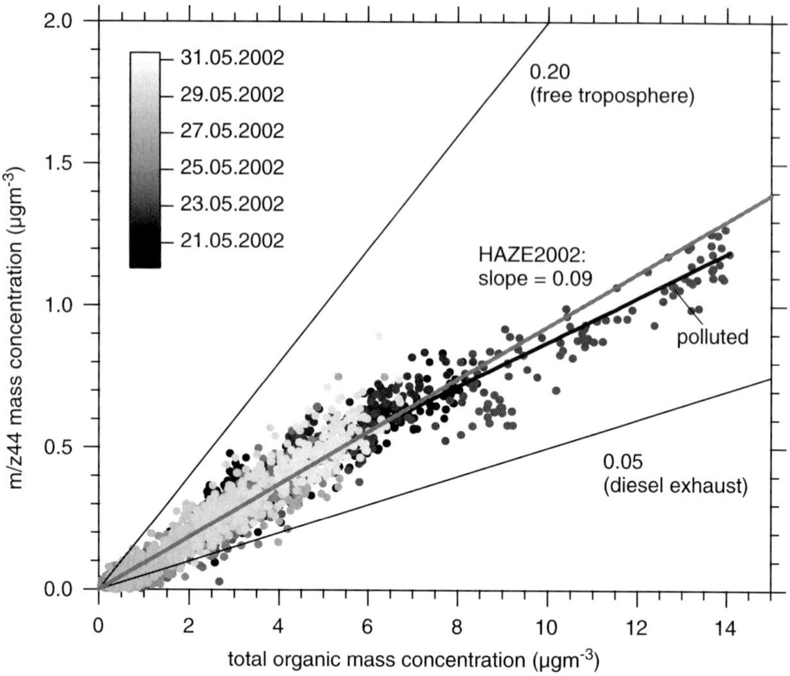

Figure 2 Correlation between m/z 44 (marker for oxygenated organic aerosol, OOA) and total organics inferred from AMS data. The label "polluted"" denotes only the data between 21 May 2002, 2200 h and 23 May 2002, 2000 h, when the air masses originated from the Po Valley (see also Figure 1)

($r^2 = 0.63$) with a linear regression slope between OM_1 and $OC_{2.5}$ of 2.07. This is at the upper end of the approximated OM/OC ratios reported by Turpin and Lim (2001) of 1.6 for urban organic aerosol and of 2.1 for aged organic aerosol, and indicates that the average organics aerosol measured during HAZE was substantially aged and processed. A high correlation between OM_1 and PM_1 ($r^2 = 0.90$) indicated that a constant amount of 50% of the total PM_1 during the HAZE campaign was composed of organic material.

Acknowledgments We would like to thank T. Böttger and the GAW/MOHp staff for technical and data support during the campaign. T. Elste and G. Stange performed the OH and H2SO4 measurements, U. Kaminski provided the GAW MOHp particle data. We also thank J. D. Allan and all contributors for the Q-AMS evaluation software.

References

Berresheim, H., Elste, T., Plass-Dulmer, C., Eisele, F.L., and Tanner, D.J., Chemical ionization mass spectrometer for long-term measurements of atmospheric OH and H_2SO_4, *Int. J. Mass Spec.*, **202**, 91–109 (2000).Birmili, W., Wiedensohler, A., Plass-Dulmer, C., and Berresheim, H.,

Evolution of newly formed aerosol particles in the continental boundary layer: a case study including OH and H_2SO_4 measurements, *Geophys. Res. Lett.*, **27**, 2205–2208 (2000).
Birmili, W., Berresheim, H., Plass-Dulmer, C., Elste, T., Gilge, S., Wiedensohler, A., and Uhrner, U., The Hohenpeissenberg aerosol formation experiment (HAFEX): a long-term study including size-resolved aerosol, H^2SO^4, OH, and monoterpenes measurements, *Atmos. Chem. Phys.*, **3**, 361–376 (2003).
Kulmala, M., Vehkamaki, H., Petajda, T., Dal Maso, M., Lauri, A., Kerminen, V.M., Birmili, W., and McMurry, P.H., Formation and growth rates of ultrafine atmospheric particles: a review of observations, *J. Aerosol Sci.*, **35**, 143–176 (2004).
Turpin, B.J., and Lim, H.J., Species contributions to PM2.5 mass concentrations: Revisiting common assumptions for estimating organic mass, *Aerosol Sci. Technol.*, **35**, 602–610 (2001).

Chemical Composition, Regional Sources, and Seasonal Patterns of TSP Aerosols at Mace Head

D. Ceburnis, C. Mulroy, S.G. Jennings, and C.D. O'Dowd

Abstract Chemical composition of TSP matter has been studied during long-term measurement campaign at Mace Head during 2001–2004. Aerosol particles in clean marine air masses were sampled in parallel with continuous samples to reveal relative contribution of natural sources. Inorganic composition changes revealed strong seasonal patterns, with dominant contribution of sea salt (40–95%) and anthropogenic nitrate. Natural biogenic sources contributed significantly to total aerosol budget. Wind speed and sea salt mass relationship has been examined and used to predict nssSO4 concentration in marine environment. Predicted nssSO4 was overestimated by 34%, but correlation between predicted and measured values was 0.61 at $P << 0.01$.

Keywords Aerosol, north Atlantic, sulfate, sea salt, nitrate, ammonium, regional sources

Introduction

Airborne particles are produced from a variety of sources, including incomplete combustion processes, industry and construction, as well as naturally as a result of resuspension of surface soil material, sea spray, volcanic activity, biomass burning, organic debris, and reactions leading to condensation of volatile precursors. There are major difficulties and uncertainties associated with identifying and quantifying the impact of different source regions and the relative importance of natural and anthropogenic aerosols to climate forcing. No aerosol chemistry data exists at the Mace Head site since the AEROCE measurement program ceased in 1994.

Department of Experimental Physics & Environmental Change Institute, National University of Ireland, Galway, University Road, Galway, Ireland

C.D. O'Dowd and P.E. Wagner (eds.), *Nucleation and Atmospheric Aerosols*, 850–854.

This study presents updated study of TSP (total suspended particulates) inorganic chemical composition and the relative contribution of natural clean versus anthropogenic air masses to aerosol budget at Mace Head.

Results and Discussions

In this study, analysis of TSP (total suspended particulates) filter samples collected between August 2001 and December 2004 is presented. Samples were taken on the 22 m tower at Mace Head using high volume samplers at a flow rate of 40 m^3/h. Sectored samples representing clean marine air masses were collected in parallel with unsectored samples. Samples were changed on a daily basis when possible. Samples were analyzed by ion chromatography for Na^+, Cl^-, NO_3^-, SO_4^{2-}, NH_4^+, and ash content. Samples were thoroughly inspected for their consistency.

Seasonal Patterns

The highest concentrations of non sea salt species were normally observed during September, December, May and April months, which was due to high pressure systems bringing polluted air from continental Europe and UK. However, pollution events were sporadic in their nature and typically lasted couple of days or less. Sea salt concentration pattern exhibited seasonal pattern with higher concentrations during winter, when stronger winds generated more of primary sea salt particles. Clean sector concentrations showed less of variability (except for sea salt) with higher concentrations of non sea salt species during summer months. Concentrations of chemical species presented in this study were somewhat higher than in other studies. However, it must be taken into account that in most other studies, aerosols were sampled using a range of impactors or size-selective devices. These impactors prevent larger particles to be collected. Na^+, Cl^- and SO_4^{2-} concentrations are affected most, but it also applies to some extent even to NH_4^+, NO_3^- and nss SO_4^{2-}, which can be enriched in large particles as well, due to condensation sink of primary gaseous species.

A seasonal cycle was observed in both sectored and unsectored samples, but the cycles were very different. Unsectored samples showed elevated mass concentrations of all major ions during winter, early spring and late autumn seasons due to polluted air masses advecting from UK and continental Europe. Clean sector seasonal cycle of aerosol chemical species was very different from the unsectored one. Ammonium did not show a pronounced pattern, but generally spring and early summer concentrations were slightly higher than during other seasons. NssSO_4 and partly NO_3 exhibited pronounced seasonal cycle with peak values during summer (Figure 1). This is related to phytoplankton life cycle with nssSO_4 coming from oxidation of DMS. Yoon et al. (2006) have shown, that nssSO_4 is highly correlated with MSA, which is a precursor species leading to production of nssSO_4.

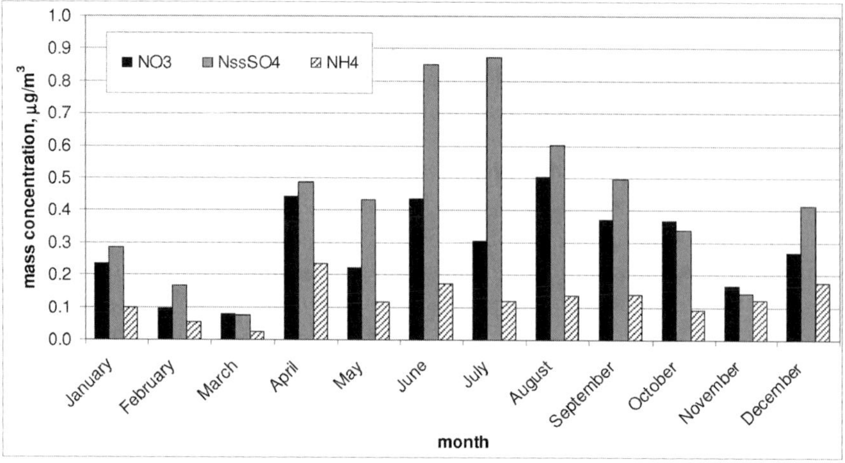

Figure 1 Monthly concentrations (µg m^{-3}) of chemical species for *sectored* samples in 2003 at Mace Head

Contribution of Regional Sources

Unsectored samples were collected regardless of wind direction, therefore, it is difficult to estimate contribution of regional pollution sources based on concentrations levels alone. In contrast, sectored samples clearly represent most of the western sector (190–300°). In order to characterize easterly air masses, backward air mass trajectories were studied and samples characterized by easterly air flow were attributed to the polluted (east) sector. Chloride and sodium concentrations differed by about a factor of two, which was rather obvious, considering Mace Head geographical position on the west coast of Ireland. However, even in easterly air masses, sea salt concentration was rather high, about 7 µg/m^3. Note, that all samples were collected at 22 m height, which is not affected by the surf zone and is representative of a marine boundary layer (Geever et al. 2005).

The difference between concentrations of other ions was much larger: about 20 times for NO_3 and NH_4, and about 8 times for $nssSO_4$. There were even more pronounced differences in the ratio when looking at different seasons. This huge difference is a clear indication that Ireland acts as a sink of pollutants emitted in continental Europe and UK. However, with about 40% of air masses arriving from clean Atlantic sector, Ireland possesses a rather clean air.

Percentage contribution of analyzed major chemical species to the total TSP mass at Mace was totally dependent on air mass sector. However, even in polluted easterly air masses, sea salt (Cl, Na, and sea salt SO_4) contributed to about 40% of the total TSP mass. The rest of the mass was shared between NO_3, NH_4 and $nssSO_4$ with the largest contribution NO_3. This is an indication that emissions of nitrogen compounds are exceeding those of sulfur, which is a result of growing car emissions and scrubbing of sulfur out of fuel. There was very different picture in westerly air masses. Sea salt contributed to over 95% of TSP mass during winter while contribution of remaining species was negligible. However, contribution of sea salt in summer was still high but significantly lower, about 80%. This was due to increased contribution of NO_3 and $nssSO_4$ originating from biogenic activity in oceanic surface waters. Interestingly, concentration of $nssSO_4$ during summer in westerly air masses was about 30% of the $nssSO_4$ level in easterly polluted air masses. This indicates that marine sources contribute substantially to the total burden of $nssSO_4$ over Europe.

Wind Speed and Sea Salt Mass Relationship

Historically, $nssSO_4$ has been monitored by measuring total sulfate. This does not introduce significant error over the continent, where sea salt mass is generally very low. However, sea salt contributes significantly to sulfate mass at marine sites. Wind speed (WS) and sea salt mass relationship has been examined to test a feasibility of recovering $nssSO_4$ concentration from measured total sulfate. Power law relationship has been found between sodium concentration and wind speed in clean marine air masses ($C_{Na} = 0.018^{*}WS^{2.52}$). Then $nssSO_4$ concentration was calculated using predicted Na concentration and standard equation ($nssSO_4 = totSO_4 - 0.25Na$). Figure 2 shows regression between predicted and measured $nssSO_4$. Predicted $nssSO_4$ was overestimated by 34%, but correlation was statistically significant.

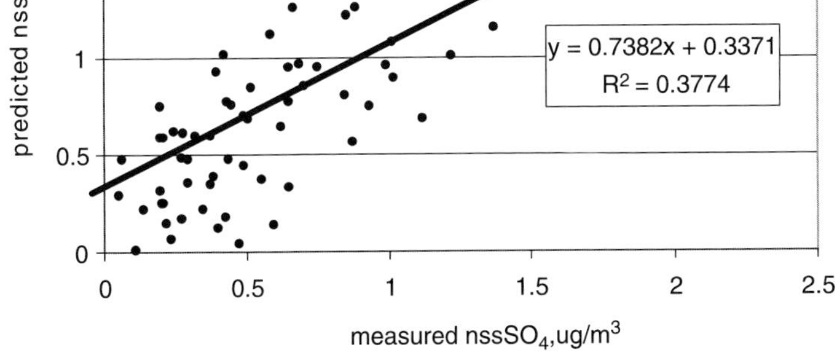

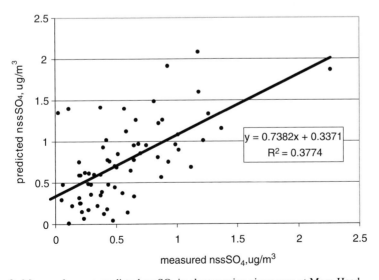

Figure 2 Measured versus predicted $nssSO_4$ in clean marine air masses at Mace Head

Conclusions

A seasonal cycle was observed in both sectored and unsectored samples, but the cycles were very different. Unsectored samples showed elevated mass concentrations of all major ions for winter, early spring and late autumn seasons. Clean sector seasonal cycle of aerosol chemical species showed peak during summer, which was attributed to increased biogenic activity. Even in polluted easterly air masses, sea salt (Cl, Na, and sea salt SO_4) contributed to about 40% of the total TSP mass. Concentration of $nssSO_4$ during summer in westerly air masses was about 30% of the $nssSO_4$ level in easterly polluted air masses. This indicates that marine sources contribute substantially to the total burden of $nssSO_4$ over Europe. Wind speed and sea salt mass relationship allowed to predict $nssSO_4$ concentration in marine environment with reasonable accuracy. $NssSO_4$ was over-predicted by 34%, but correlation between predicted and measured values was 0.61 at $P \ll 0.01$.

Acknowledgment This work was supported by EPA Ireland grant 2003-FS-CD-LS-12-M1.

References

Geever, M., O'Dowd, C.D., van Ekeren, S., Flanagan, R., Nilsson, E.D., de Leeuw, G., and Rannik, U., *Geophys. Res. Lett.*, **32**, L15810, doi:10.1029/2005GL023081 (2005).

Yoon, YJ., Ceburnis, D., O'Dowd, C.D., Jennings, S.G., Jourdan, O., Cavalli, F., Facchini, M.C., Decesari, S., Emblico, L., Fuzzi, S., *J. Geophys. Res.*, doi:10.1029/2006JD007044 (2006).

The Formation of Radiatively Active Aerosol from Coastal Nucleation Events

Regis G. Dupuy, Colin D. O'Dowd, and S. Gerard Jennings

Abstract Two similar aerosol nucleation events sampled at the Mace Head Atmospheric Research station during 2–3 September 2005 resulted in the production of >105 particles cm^{-3}. The contribution of coagulation of these particles to radiatively active sizes (>50 nm) were examined and were found to be radically different. The 3rd September event, with a coagulation factor two times the 2nd September, was found to have a strong impact on the radiative and CCN properties of the observed aerosol population.

Keywords Marine aerosols, coagulation, cloud condensation nuclei, scattering coefficient

Introduction

Coastal zones have been shown to provide a massive source of new, tidal related, aerosol particles in the atmospheric boundary layer with concentrations exceeding 1,000,000 cm^{-3} during nucleation bursts sustained over many hours. While it has been demonstrated that these particles contribute to aerosol-related radiative flux,[1] the relative importance of coagulation and condensation has yet to be assessed.

This study focuses on two events where significant growth was observed – the 2nd and 3rd September 2005. During these two days, a change in the type of air mass can be seen starting from clean air on the 2nd to polluted air on the 3rd September which leads to strong differences on the aerosol CCN behavior and radiative properties.

The measurements were taken at the Mace Head atmospheric research station. Size spectrum measurements are made with two Scanning Mobility Particle Sizer (SMPS) which scan particle sizes from 10 to 500 nm and 3 to 15 nm with a time

---

Department of Experimental Physics & Environmental Change Institute, National University of Ireland, Galway, University Road, Galway, Ireland

C.D. O'Dowd and P.E. Wagner (eds.), *Nucleation and Atmospheric Aerosols*, 855–859.

resolution of 300 and 30 sec respectively. The particle scattering coefficients for 450 nm (blue), 550 nm (green) and 700 nm (red) wavelengths are determined using a TSI 3563 integrating Nephelometer. The total concentration of the particles larger than 10 nm (CN) is also measured by a CPC3010 from TSI. These measurements are made at dry particle sizes for Relative Humidity (RH) lower than 40%. In parallel, an important factor in the estimation of the indirect radiative forcing of the aerosol is measured with a Cloud Condensation Nuclei (CCN) counter from Droplet Measurement. It allows continuous measurement of CCN concentrations for three supersaturations (SS) set at 0.25, 0.5, and 0.75% and the resulting activated particle size spectrum in the size range of 0.75–10 mm.

Growth of Newly Formed Particles

By merging the two SMPS size spectrum, we obtain a particle size range which covers the particle sizes which are relevant for nucleation and growth processes from 3 to 500 nm. We can see in Figure 1 that both nucleation events occurred before noon and provide particle concentration exceeding 100,000 cm^{-3}. However, the 2nd September size spectrum shows much less particles on the accumulation mode (particle sizes over 100 nm) before the nucleation event than that on the 3rd.

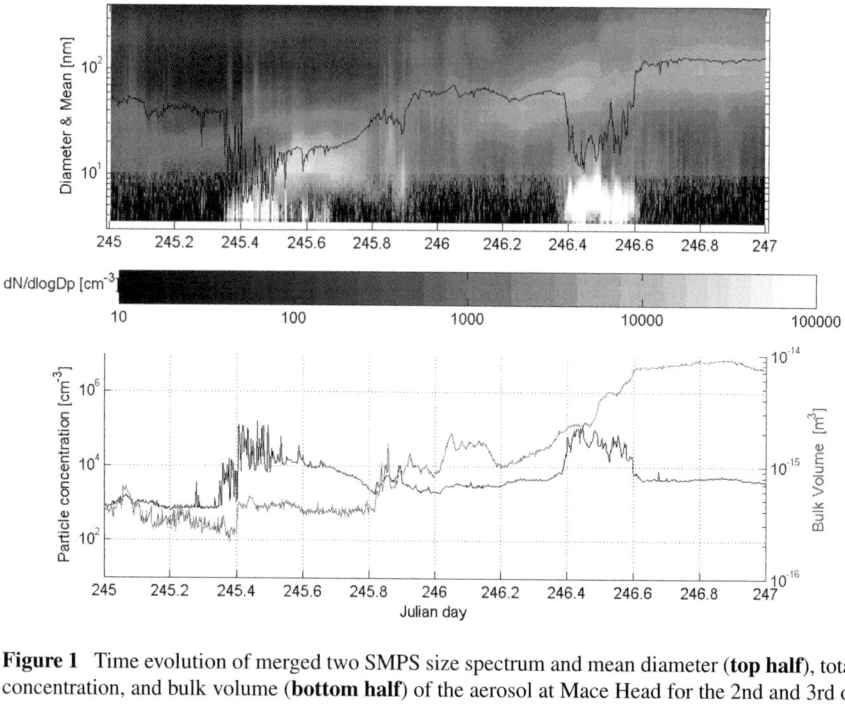

Figure 1 Time evolution of merged two SMPS size spectrum and mean diameter (**top half**), total concentration, and bulk volume (**bottom half**) of the aerosol at Mace Head for the 2nd and 3rd of September 2005

This difference is due to a change of air mass origin clearly seen on the backtrajectories for these days. The event on 2nd September shows clear growth of the nucleation mode, most likely through a combination of coagulation and condensation growth, while on the 3rd September, no significant growth of the nucleation mode is seen, while the accumulation mode undergoes clear growth. In the latter case, it is likely that condensation drives the accumulation mode growth. This assumption is supported by the fact that the aerosol bulk volume increases during the nucleation event. The rapid growth of the nucleation mode on the 2nd is through to be aided by self-coagulation due to the low accumulation mode coagulation sink.

Impact on the Radiative and CCN Properties of the Aerosol

As the CCN capacities[2] and radiative properties[3] of an aerosol arc mainly driven by the accumulation mode particles (at constant chemical properties), the modification of that mode by coagulation of newly produced particles and/or condensation of existing particles should lead to strong changes in radiative properties. This can be seen in the variation of the CCN concentration and aerosol scattering coefficient with time (Figure 2). They are correlated most of the time and an increase of both parameters occurs after a nucleation event. But the behavior of these parameters is completely different during the 3rd September event compared to the 2nd September events.

Table 1 Evolution of the mean diameter before and after a particle nucleation event

Average D_{mean} (nm)	September 2, 2005	September 03, 2005
Before event	40	60
After event	60	120

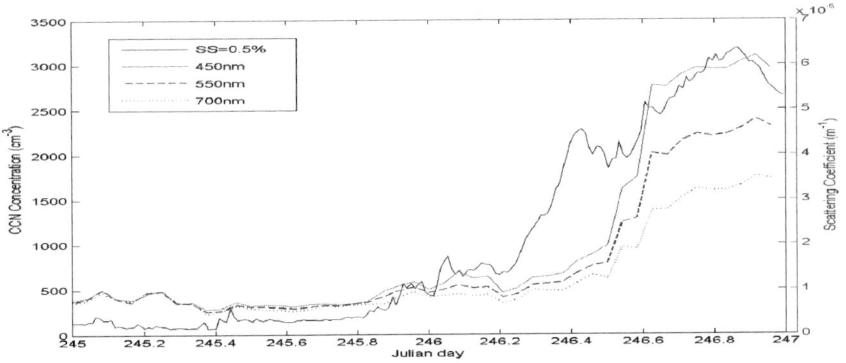

Figure 2 Time evolution of CCN concentration and scattering coefficient at three wavelengths of the aerosol at Mace Head for the 2nd and 3rd of September 2005

Table 2 Ratio of CCN in the aerosol for different supersaturation before and after the 3rd September event

S (%)	F_{CCN} before the event	F_{CCN} after the event
0.25	0.08	0.24
0.5	0.25	0.63
0.75	0.32	0.72
1	0.4	0.75

The increase of CCN concentration and aerosol scattering coefficient occurred at the same time for both parameters on the 2nd, but not on the 3rd September. The CCN concentration increases before the aerosol scattering coefficient. Furthermore, the CCN concentration starts to increase in the middle of the event and not while the particles coagulate. On top of that, for the 2nd, the aerosol scattering coefficients increased after the aerosol nucleation event with the same factor for the three wavelengths used by the Nephelometer. But after the 3rd September event, the aerosol scattering coefficient increased by 3.5, 4.5, and 6 for the 700, 550, and 450 nm wavelengths respectively. As the wavelength of the radiation and the diameter particle are correlated, the difference in the impact on the aerosol scattering coefficient is due to the increase of the smallest particles of the accumulation mode after the nucleation event.

In order to better assess the impact of growth processes on the capacities of aerosol particles to become CCN, we sought to find better representative parameters. For that purpose, we calculated the ratio of CCN (F_{CCN}) in the aerosol for each supersaturation by dividing the CCN concentration to the CN concentration. This ratio is quite stable before and after the 3rd September event and variable during it. Table 2 shows the clear impact of the event on the CCN capacities of the aerosol. The aerosol became two times more CCN after the 3rd September event for most of the supersaturations while it did not change after the 2nd September event.

Conclusions

Measurements of nucleation events in the coastal environment are compared for two cases: one case where the nucleation mode is seen to grow rapidly in size and another where the nucleation mode does not grow but the preexisting accumulation mode is observed to increase. For both cases, an increase of a factor of 2 is observed in both scattering enhancement and CCN enhancement, thus indicating that the coastal zone is an important source of radiatively active aerosol particles. The increase in CCN concentrations suggest that significant water soluble material is condensing in addition to what is assumed to by insoluble iodine oxides driving the actual nucleation process.

Acknowledgment This work was supported by the Marie Curie development host fellowship grant HPMD-CT-2000-00031.

References

1. O'Dowd, Biogenic coastal aerosol production and its influence on aerosol radiative properties, J. Geophys. Res., **106**(D2), 1545–1549 (2001).
2. O'Dowd, C.D., Lowe, J., Smith, M.H., and Kaye, A.D., The relative importance of sea-salt and nss-sulphate aerosol to the marine CCN population: an improved multi-component aerosol-droplet parameterisation, *Q.J. Roy. Met. Soc.*, **125**, 1295–1313 (1999).
3. Kleefeld, Christoph, O'Dowd, C., O'Reilly, S., Jennings, S.G., Alto, P., Becker, E., Kunz, G., Gerrit de Leeuw, The relative contribution of sub and super micron particles to aerosol light scattering in the marine boundary layer (MBL), *J. Geophys. Res.*, **107**(D19), 8103 (2002).

Funneling of Meteoric Material into the Polar Winter Vortex

Linda Megner

Abstract Every day of the order of 100 tonnes of meteoric material enters the atmosphere and ablates at 80–100 km. This material is expected to recondense into nanometer-sized meteoric "smoke" particles, and sediment to the lower atmosphere. The meteoric smoke particles are likely to be the dominant aerosol in the upper stratosphere where the temperatures are high enough for the droplets of sulfuric acid solution to evaporate. Meteoric material has been suggested to be of importance for stratospheric nucleation, heterogeneous chemistry and positive ion chemistry. Studies concerning these processes have so far been based on one-dimensional (1D) models of meteoric material, which cannot properly handle atmospheric transport. The first 2D model, which includes both the coagulation and transport of meteoric material, shows that this material is effectively transported to the winter stratosphere. The majority of the global influx of material is thus funneled into the winter vortex. The number and size distribution of meteoric smoke are therefore, unlike what is implicitly assumed with the use of 1D models, highly dependent on latitude and season. We here present the new estimates of number densities and particle area in the stratosphere, and discuss possible consequences for stratospheric processes.

Keywords Stratosphere, heterogeneous nucleation, meteoric material, NAT, PSC

Introduction

Meteors decelerate as they reach the Earth's atmosphere, whereby they are heated and ablate in the 80–100 km altitude region. The ablated material is then thought to recondense and form nanometre-sized meteoric "smoke" particles. These particles are likely to be the dominant aerosol in the upper stratosphere where the temperatures are high enough for sulfuric acid droplets to evaporate (Hunten et al. 1980). They may be of importance for nucleation processes, heterogeneous chemistry and for the stratospheric sulfur budget. Studies of these processes have generally been

Department of Meteorology Stockholm University

C.D. O'Dowd and P.E. Wagner (eds.), *Nucleation and Atmospheric Aerosols*, 860–864.

based on number densities and size distributions of meteoric particles, obtained from 1D models. Applying such models to a 3D world implicitly results in a smoke layer that is horizontally homogeneous over the entire earth. Results from a 2D model however, show a prominent effect of the atmospheric circulation, which transports the bulk of the meteoric material to the winter stratosphere (Megner et al. 2006). This funneling into the polar vortex results in a globally nonuniform smoke distribution which is consistent with the idea that meteoric material is deposited on the earth surface mainly in the polar regions (Gabrielli et al., 2004). Here we present the expected stratospheric concentration and size distribution of meteoric particles and discuss the possible implications for stratospheric processes.

Model Description

The CARMA-CHEM2D model (CC2D) consists of a coupling between a zonally averaged chemical dynamical model, CHEM2D (Summers et al. 1997), used to transport the material, and a microphysical model, CARMA (the Community Aerosol and Radiation Model for Atmospheres) (Turco et al. 1979), which coagulates the particles. Only CARMA's coagulation algorithm is used, in which the particles undergo coalescence as a result of collisions, mainly due to Brownian motion. The CC2D model has 28 radius size bins ranging from 0.2 nm, which corresponds roughly to molecular sized particles, to 80 nm. The coagulation time step is 30 min and the dynamical time step is 2h.

The coupling of the two sub-models is done by first letting the CHEM2D model transport the particles for a simulation period of 2 h, after which the CARMA model coagulates the particles, i.e., advects them between the size bins, for the same amount of simulation time. Ablated material, corresponding to a global amount of 44 metric tonnes per day is deposited at all latitudes and at altitudes between 70 and 110 km as described by an ablation profile calculated by Kalashnikova et al. (2000). The freshly ablated material is continuously fed into the smallest size bin. As coagulation continues the material is transferred to the bigger size bins, acquires a higher sedimentation speed, and eventually collects in the lowest altitude bins. Above 20 km a steady state is reached where the amount of material entered in the mesosphere, in the smallest size bin, equals the amount accumulated in the lowest altitude bins so that the meteoric smoke profile from year to year is roughly constant. The model is described in detail by Megner et al. (2006).

Results

The highest number densities of meteoric material are found at the altitude at which the ablation takes place, i.e., the upper mesosphere. This material however, is mainly in the form of molecular sized clusters, which have not yet had time to recondense. As these

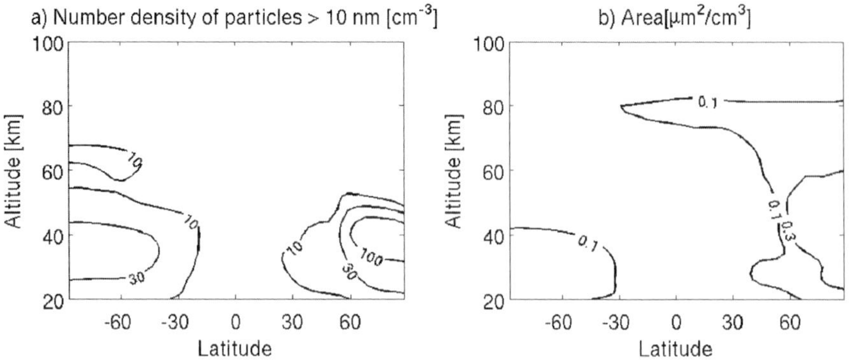

Figure 1 Distribution of meteoric material for December conditions. (a) Number density of particles larger than 10 nm radius. (b) Surface area of meteoric material

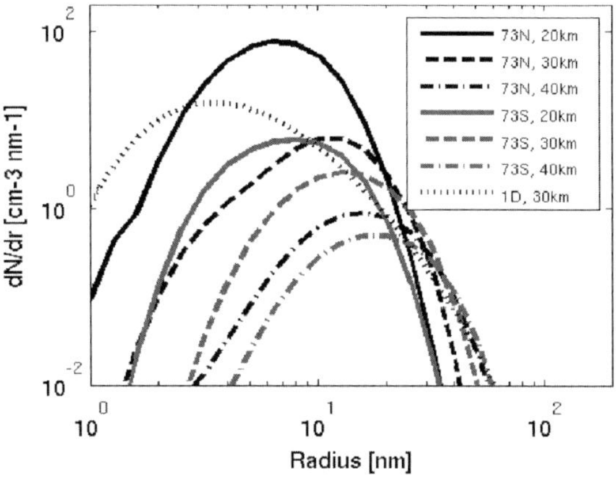

Figure 2 Size distribution (December conditions) of meteoric smoke particles in the north and south polar regions for various altitudes. The dotted line shows the size distribution at 30 km obtained by a 1D model. (From Hunten et al. 1980.)

small particles coagulate they are subject to an efficient transport toward the winter pole, so that the larger particles mainly appear at the poles. Figure 1a shows the number density of particles larger than 10 nm radius. The maximum is reached in the winter stratosphere, whereto the residual circulation is directed. High concentrations are also reached in the summer stratosphere. These particles consist of material, which was deposited here during the previous winter. As opposed to polar latitudes, the mid-latitudes are virtually free from meteoric particles. Figure 1b shows the surface area of meteoric smoke material. We note that, even though the highest total number densities are found in the mesosphere, maximum surface area is reached in the winter stratosphere. As the particles grow larger than ~5 nm radius they are believed to form loosely packed aggregates (Saunders et al. 2006). Moreover, in the lower stratosphere the particles may interact with H_2SO_4 droplets. Both these processes would increase the surface area. Hence, the values in Figure 1b are most likely underestimates.

Figure 2 shows December size distributions at 73°N, 73°S for various altitudes. As coagulation continues, the particles are transported downward, resulting in a decrease of radii with altitude and an increase of number densities. The variation with latitude is relatively insignificant as long as the latitude remains above 60°. Again we note that the number densities at the current winter pole are higher than those of the summer pole, where the material has had 6 months more to coagulate and sediment. This is also the reason why the particles at the current winter pole are smaller. The figure also shows the 30 km size distribution obtained by a 1D model (Hunten et al.1980). We note that it peaks at significantly smaller radii.

Discussion

The number densities of particles larger than 10 nm are about one order of magnitude larger than the ~10 cm^{-3} earlier 1D studies suggested (Hunten et al. 1980) These number densities are potentially high enough to make them important for the formation of the stratospheric condensation nuclei (CN) layer. Based on the 1D studies Zhao et al. (1995) concluded that "micro-meteoric particles themselves are unlikely to change the CN-concentration, or to form a distinct CN layer." This conclusion should therefore be revalidated using the higher number densities. The surface area of meteoric particles in the winter polar vortex may be important for nucleation of NAT (Nitric Acid Trihydrate) particles. Voigt et al. (2005) find that their observations of PSC can be explained neither by homogeneous NAT nucleation nor by NAT nucleation on ice. Thus, they consider meteoric smoke particles to be "favorable candidates for triggering NAT nucleation." Despite the funneling of meteoric material our 2D model shows that the surface area in the polar winter vortex is of similar magnitude to that suggested by 1D models (Hunten et al. 1980). This can be explained by a combination of strong down-welling and more rapid sedimentation due to the faster coagulation accompanying a higher concentration. The surplus in area from the 10 nm sized particles is canceled by a lack of smaller particles, whereas the number of particles larger than 10 nm, because of the more rapid sedimentation, stays roughly constant. Inside the polar winter vortex, the estimate of meteoric particle surface area therefore happens to be similar to what 1D models suggest.

Conclusion

Our 2D estimates for meteoric smoke particles are significantly different from earlier results based on 1D models, especially at mid-latitudes. The distribution is moderately defendant on season and highly dependent on latitude, so that the majority of particles are found in the polar stratosphere and virtually none at mid-latitudes. Hence, it is crucial to take these variations into consideration both in models and when designing experiments to measure meteoric smoke particles.

References

Gabrielli, P. et al., *Nature*, **432**, 1011–1014 (2004).
Hunten, D.M., Turco, R.P., Toon, O.B., *J. Atmos. Sci.*, **37**, 1342–1357(1980).
Kalashnikova, O. et al., *Geophys. Res. Lett.*, **27**, 3293–3296 (2000).
Megner, L., Siskind, D., Rapp, M., Gumbel, J., *J. Geophys. Res.*, submitted (2006).
Saunders, R.W. and Plane, J.M.C., *J. Atmos. Sol.-Terr. Phys.*, **68**, 2182–2202 (2006).
Summers, M. E. et al., *J. Geophys. Res.*, **102**, 3503–3526 (1997).
Turco, R. P. et al., *J. Atmos. Sci.*, **36**, 699–717 (1979).
Voigt, C. et al., *Atmos. Chem. Phys.*, **5**, 1371–1380 (2004).
Zhao, J., Toon, O.B., and Turco, R.P., *J. Geophys. Res.*, **100**, 5215–5227 (1995).

Connection Between Atmospheric Aerosol Optical Depth and Aerosol Particle Number Concentration in the Air in Tbilisi

A.G. Amiranashvili[1], V.A. Amiranashvili[1], D.D. Kirkitadze[1], and K.A. Tavartkiladze[2]

Abstract The preliminary results of study of the connection between the aerosol particle number concentration in surface air (N) and atmospheric aerosol optical depth (AOD) in Tbilisi is carried out. It is obtained that with an increase of aerosols diameter from 0.4 to 0.8 μm the correlation between AOD and N rapidly weakens. The correlation between AOD and N in the range of the aerosols sizes from 0.8 to 10 μm is absent. The correlation between AOD and N for the aerosols with size more than 10 μm is again revealed. Thus, the atmospheric aerosol optical depth can be the sufficiently representative characteristic of the surface air pollution by aerosols with the diameter less than 0.8 μm and more than 10 μm.

Keywords Atmospheric aerosol optical depth, aerosol particle number concentration

Introduction

The atmospheric aerosol optical depth is the integral characteristic of the aerosol pollution of the entire thickness of the atmosphere. At the same time it is important to know, as value of AOD is connected with the surface air pollution. In the work [1] the direct connection AOD with the concentration of surface aerosol is indicated. In the work [2] the inverse correlation between AOD in Tbilisi and surface air electric conductivity in Dusheti (about 45 km northern of Tbilisi) was obtained. Therefore the task of investigation of the connection between AOD and aerosol particle number concentration with different size in surface air was set. The special experiments for the solution of this problem it was not carried out. However, we found 11 cases of the simultaneous measurements of the atmospheric aerosol optical depth and aerosol particle number concentration in surface air in the data archives.

[1] *Mikheil Nodia Institute of Geophysics, 1, M. Aleksidze St., 0193 Tbilisi, Georgia, USA*

[2] *Vakhushti Bagrationi Institute of Geography, 8, M. Aleksidze St., 0193 Tbilisi, Georgia, USA*

C.D. O'Dowd and P.E. Wagner (eds.), *Nucleation and Atmospheric Aerosols*, 865–868.

Below the results of studies of the relationship between AOD and N are represented.

Methods

AOD values were calculated by the method [3,4] using to standard actinometric observations of intensity of clear sky integral direct solar radiation. Values of the AOD are given for the wavelength $\lambda = 1\,\mu m$ and midday hours. The measurements were conducted at the meteorological station of Tbilisi city.

The measurement of the aerosol particle number concentration by the standard photoelectric counter of aerosol particles AZ-5 [5] were carried out. This counter measures the content of aerosols in the following 12 ranges of sizes: 0.4–0.5, 0.5–0.6, 0.6–0.7, 0.7–0.8, 0.8–0.9, 0.9–1.0, 1.0–1.5, 1.5–2.0, 2.0–4.0, 4.0–7.0, 7.0–10.0, and ≥10.0 μm. The range of the measurement of N is from 0.001 to 300 cm^{-3}. An error of measurement does not exceed 20%. The measurements at the territory of the Thermobarochamber of the Institute of Geophysics in 1987–1988 were conducted (the height was 8 m of the earth's surface).

The following designations (besides those pointed out above) will be used below: d – aerosol particles diameter, Cv – the coefficient of variation, R – linear correlation coefficient, α – the level of significance, CONF – confidence interval.

Results

The results of the analysis in Tables 1, 2, and Figure 1 are presented. Table 1 presents the data about the statistical characteristics of N and AOD. In particular from this table follows that the values of N for aerosols with diameter 0.4–0.5 and ≥10 μm respectively comprise: average – 9.95 cm^{-3} and 0.002 cm^{-3}, maximum – 20 cm^{-3} and 0.004 cm^{-3}, minimum – 2 cm^{-3} and 0.001 cm^{-3}. The greatest variation is observed for the particles with sizes 4–7 μm (Cv = 113.3%), smallest – for the particles with sizes 0.5–0.6 μm (Cv = 45.8). The value of the AOD changes from 0.126 to 0.254 with average value 0.188. The coefficient of variation is equal 23.6%.

The relative variation range is equal 68.2%. The correlation between AOD and N with an increase of the particles sizes is rapidly weakens (respectively R = 0.60, 0.45, 0.39, and 0.29 for the particles with the diameter from 0.4 to 0.8 μm). The correlation between AOD and N in the range of the aerosols sizes from 0.8 to 10 μm is insignificant. The correlation between AOD and N for aerosol with diameter more than 10 μm is again revealed (R = 0.46).

For the illustration in Figure 1 the correlation dependences between AOD and N for the particles by sizes 0.4–0.5 μm and 0.6–0.7 μm are represented.

The values of the coefficients of the linear regression equations between N and AOD are represented in Table 2. According to the data of Tables 1 and 2 it is

able 1 The statistical characteristics of aerosol particle number concentration and atmospheric erosol optical depth in Tbilisi

, µm	Max	Min	Range	Average	St Dev	Cv,	R with AOD %	α
.4–0.5	20	2	18	9.95	4.69	47.1	0.60	0.05
.5–0.6	5	1	4	2.96	1.36	45.8	0.45	0.20
.6–0.7	3	0.3	2,7	1.34	0.80	60.0	0.39	0.25
.7–0.8	2	0.1	1,9	0.81	0.64	79.4	0.29	0.35
.8–0.9	1.5	0.2	1,3	0.58	0.48	82.7	0.11	No correlation
.9–1	1.1	0.1	1	0.42	0.38	92.0	-0.02	No correlation
–1.5	1.2	0.05	1.15	0.350	0.395	11.8	0.14	No correlation
.5–2	0.8	0.048	0.752	0.239	0.256	10.3	0.18	No correlation
–4	0.5	0.032	0.468	0.179	0.175	97.6	0.07	No correlation
–7	0.15	0.003	0.147	0.038	0.043	113.3	-0.04	No correlation
–10	0.012	0.001	0.011	0.005	0.004	87.1	0.14	No correlation
10	0.004	0.001	0.003	0.002	0.001	58.4	0.46	0.2
OD	0.254	0.126	0.128	0.188	0.044	23.6		

Table 2 Value of the coefficients of the equations of the linear regression between N and AOD in Tbilisi (AOD = a·N + b, N – cm^{-3})

d, µm	Coefficient	The value of coefficient	95% CONF (+/-)
0.4–0.5	a	0.0057	0.0057
	b	0.1307	0.0620
0.5–0.6	a	0.0148	0.0219
	b	0.1436	0.0709
0.6–0.7	a	0.0218	0.0383
	b	0.1585	0.0590
0.7–0.8	a	0.0197	0.0499
	b	0.1717	0.0506
≥10	a	0.0201	0.0295
	b	0.1529	0.0582

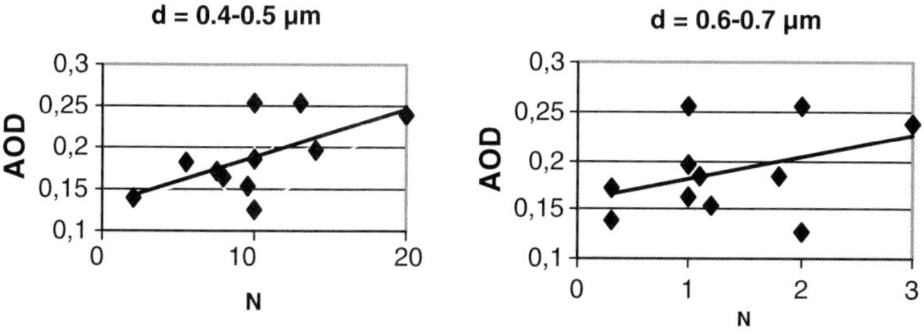

igure 1 Connection between the aerosol optical depth of the atmosphere and aerosol particle umber concentration in air of diameter 0.4–0.5 µm (**left**) and 0.6–0.7 µm (**right**)

possible to estimate the share of the changeability of the aerosol particle numbe concentration in surface air of different sizes to the changeability of the values o the atmospheric aerosol optical depth. In the limits of variation range this shares fo the aerosols with diameter 0.4–0.5 µm, 0.5–0.6 µm, 0.6–0.7 µm, 0.7–0.8 µm an ≥10 µm accordingly comprises: 54.8%, 31.6%, 31.4%, 19.9%, and 3.2%. In th limits of standard deviation this shares respectively comprises: 28.6%, 21.5% 18.6%, 13.5% and 21.6%. Thus, the atmospheric aerosol optical depth is the suffi ciently representative characteristic of the surface air pollution by aerosols with th diameter less than 0.8 µm and more than 10 µm.

Conclusion

The relationship between atmospheric aerosol optical depth and aerosol particl number concentration in surface air in Tbilisi is studied. The correlation betwee AOD and N with an increase of the particles sizes is rapidly weakens (respectivel R = 0.60, 0.45, 0.39, and 0.29 for the particles with the diameter 0.4–0.5, 0.5–0.6 0.6–0.7, and 0.7–0.8 µm). The correlation between AOD and N in the range of th aerosols sizes from 0.8 to 10 µm is insignificant. The correlation between AOD an N for aerosol with diameter more than 10 µm is again revealed (R = 0.46). It i assumed, that the atmospheric aerosol optical depth can be the sufficiently repre sentative characteristic of the surface air pollution by aerosols with the diamete less than 0.8 µm and more than 10 µm.

References

1. Sztyler, A., Relationship between spectral solar radiation extinction and dust pollution in the a of industrial and urban areas in Poland, *J. Aerosol Sci.*, **23**(4), 407–413 (1992).
2. Amiranashvili, A.G., Amiranashvili, V.A, Kalaijeva, L.L., Karauli, N.D., Khunjua, A.T Nodia, A.G., and Vachnadze, J.I., Characteristics of air conductivity in Dusheti, *Proceeding of 12th International Conference on Atmospheric Electricity*, Vol.1, Versailles, France, p 353–356, June 9–13, 2003.
3. Amiranashvili, A.G., Amiranashvili, V.A., Gzirishvili, T.G., Kharchilava, J.F., and Tavartkiladz K.A., Modern climate change in Georgia. Radiatively active small atmospheric admixture Institute of Geophysics, Monograph, *Trans. of M. Nodia Institute of Geophysics of Georgi Academy of Sciences*, **LIX**, pp. 1–128 (2005).
4. Tavartkiladze, K.A., *Modelling of the Aerosol Attenuation of Radiation and Atmospher Pollution Control Methods*, Tbilisi, GA, pp. 1–204 (1989) (in Russian).
5. Beliaev, S.P., Nikiforova, N.K., Smirnov, V.V., and Shelchkov, G.I., Optical-Electron Methods of Studying of Aerosols, Moscow: Energoizdat, pp. 1–230 (1981) (in Russian).

Part VIII
Aerosol Formation, Dynamics, and Growth

Regional Air Quality–Atmospheric Nucleation Interactions

JaeGun Jung[1], Peter J. Adams[2], and Spyros N. Pandis[1,3]

Abstract The creation of new atmospheric particles from in situ nucleation influences climate through cloud-aerosol interactions and may negatively impact human health. Although recent observations show that nucleation is widespread over most continents, the corresponding pathways remain uncertain. A computationally efficient multicomponent aerosol dynamics model (DMAN) that simulates the full aerosol size distribution and composition starting at a diameter of 0.8 nm has been developed. Several proposed nucleation rate expressions for binary (H_2SO_4–H_2O), ternary (NH_3–H_2SO_4–H_2O), and ion-induced nucleation are evaluated using DMAN against ambient measurements from the Pittsburgh Air Quality Study (PAQS). The ternary NH_3–H_2SO_4–H_2O nucleation model is successful in predicting the presence or lack of nucleation on 19 out of 19 days with complete data sets in July 2001 and on 25 out of 29 days in January 2002. DMAN has been added to the three-dimensional (3D) Chemical Transport Model PMCAMx-UF and is tested in the Eastern USA. Reductions of ammonia emissions are predicted to decrease the frequency of nucleation events during both summer and winter, with a more dramatic effect during the summer. The response to changes in emissions of sulfur dioxide during the summer is counterintuitive. Reductions of sulfur dioxide and the resulting sulfate by up to 40% actually increase the frequency of the summer nucleation events. Modeling predicts the opposite effect in winter, with reductions of sulfur dioxide leading to fewer nucleation events.

Keywords Ultrafine PM, tropospheric aerosols, ammonia, Eastern United States, PMCAMx-UF

[1]*Department of Chemical Engineering, Carnegie Mellon University, Pittsburgh, USA*

[2]*Department of Civil and Environmental Engineering, Carnegie Mellon University, Pittsburgh, USA*

[3]*Department of Chemical Engineering, University of Patras, Patra, Greece*

C.D. O'Dowd and P.E. Wagner (eds.), *Nucleation and Atmospheric Aerosols*, 871–877.

Introduction

The two major processes introducing new particles into the atmosphere are in situ nucleation and direct emission from sources such as combustion, sea spray, dust, etc. A variety of field measurements show that in situ nucleation occurs in many places around the globe [1], including urban areas where it was previously thought particularly unlikely due to high concentrations of aerosol surface area, which suppress nucleation. These events alter the number and chemical composition of ultrafine particles and cloud condensation nuclei, with implications for human health [2] and climate [3]. For example, recent studies comparing the health effects of differently sized particles show that ultrafine (<100 nm) atmospheric particles can have negative health effects, possibly due to their high surface area. Climate can be affected as these particles grow to larger than 100 nm in size.

Several mechanisms have been proposed for the production of new atmospheric particles including sulfuric acid–water binary nucleation, sulfuric acid–ammonia–water ternary nucleation, nucleation of organic vapors, ion-induced nucleation, and halogen–oxide nucleation. Another important issue is what species are involved in the growth of these nuclei. Field measurements and model simulations have indicated that the condensation of sulfuric acid alone is often not sufficient to grow these nuclei to detectable sizes. To aid in the growth of these particles, the condensation of organic species, heterogeneous reactions, and ion-enhanced condensation have been suggested.

In this work data taken during July 2001 and January 2002 during the Pittsburgh Air Quality Study (PAQS) [4] is used to evaluate a model using several different nucleation parameterizations for the simulation of these regional nucleation events. The most successful parameterization is then incorporated in a 3D CTM and used for the description of nucleation events in the Eastern USA. The effect of emission control strategies on the frequency of these events is discussed.

Modeling Nucleation and Aerosol Dynamics

Model Development

The Dynamic Model for Aerosol Nucleation (DMAN) [5] simulates nucleation, coagulation, and condensation/evaporation for a multicomponent aerosol population assuming that they are internally mixed. The module is based on the TwO-Moment Aerosol Sectional (TOMAS) algorithm [3] tracking both mass and number concentrations simultaneously. The aerosol size distribution is described with 42 size sections with the lowest boundary at 3.75×10^{-25} kg dry aerosol mass per particle. That corresponds to 0.8 nm dry diameter assuming a density of $1.4\,gcm^{-3}$. Each successive boundary has double the mass of the previous one. The largest bin corresponds to about 10 µm. Both the major inorganic and organic aerosol components are simulated.

DMAN allows its user to select the nucleation parameterization that will be used in the simulations. Choices include the ternary NH_3–H_2SO_4–H_2O nucleation parameterization of Napari et al. [6], the binary H_2SO_4–H_2O parameterization of Vehkamaki et al. [7], the binary parameterization of Jaecker–Voirol and Mirabel [8] multiplied with a 10^7 nucleation "tuner" [9], the semiempirical first order in sulfuric acid concentration expression proposed by Spracklen et al. [10], the ion-induced nucleation parameterization of Modgil et al. [11], and the barierless rate expression of Clement and Ford [12]. Additional details about DMAN can be found in [5].

The accuracy of DMAN and TOMAS algorithms in the atmospheric nanoparticle size range was tested by comparing the model predictions against analytical solutions of the coagulation equation (assuming size-independent coagulation coefficients) and condensation equation (assuming a log-normal initial size distribution and constant gas-phase concentration gradient between the gas and particulate phases). The algorithm conserves mass during coagulation and number during condensation to machine precision. The predicted error for the magnitude of the size distribution at any size was less than 5% for all tests. Although DMAN uses 42 size sections it consumes only 0.23 CPU seconds per simulation hour on a 2.4 GHz AMD CPU.

DMAN is used here as the aerosol module in two frameworks: a box model [13] and inside the 3D-CTM PMCAMx replacing the preexisting aerosol module [14] that described just the aerosol mass distribution above 49 nm. The version of PMCAMx that describes in detail the ultrafine atmospheric PM is called PMCAMx-UF.

Application to Nucleation Events in Western Pennsylvania

The PAQS data set [4] includes a wide range of nucleation events that took place during 2001 and 2002 in Western Pennsylvania [15]. Overall, significant new particle formation was observed on roughly one out of 3 days during the whole year. The frequency of the events was higher during the spring and fall, but was significant during the summer and winter. During the summer most of the nucleation events started just a few hours after sunrise while in the winter they started at noon or in the early afternoon. Nucleation tended to occur on days with below average $PM_{2.5}$ concentrations and clear skies. Most of these events were regional in nature, taking place over several hours with smooth aerosol growth.

The predictions of DMAN using six different nucleation parameterizations for July 15, 2001 are shown in Figure 1. Two of the parameterizations (ternary NH_3–H_2SO_4–H_2O and the Clement and Ford [12] expression for barrierless nucleation) reproduce the major features of the event, while the other four are not successful in this day. All parameterizations were tested for 9 days (Table 1), three with nucleation and six without. The only parameterization that reproduced the observations in all 9 days was the ternary expression of Napari et al. [6]. The barrierless expression of Clement and Ford [12], as it should be expected, predicted nucleation events on most of the days, while the ion-induced nucleation expression and the binary H_2SO_4–H_2O parameterization of Vehkamaki et al. [7] predicted that there

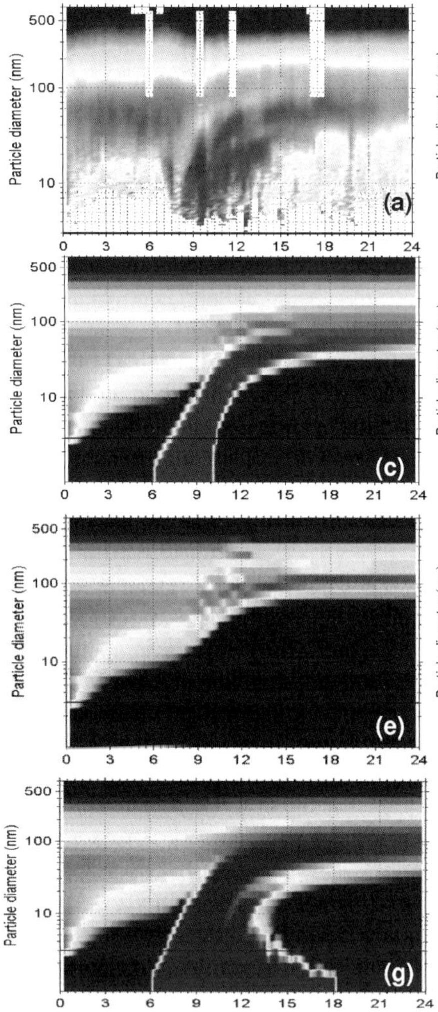

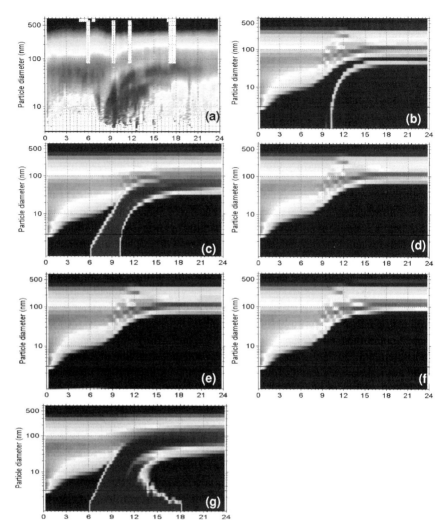

Figure 1 Evolution of the aerosol number distribution for July 15, 2001 in Pittsburgh, Pennsylvania for: (a) observations, (b) binary H_2SO_4–H_2O [8, 9], (c) ternary NH_3–H_2SO_4–H_2O [6], (d) binary H_2SO_4–H_2O [7], (e) first order in sulfuric acid [10], (f) ion-induced [11], and (g) barrierless nucleation [12] parameterizations

should have been no nucleation events. The binary Jaecker–Voirol and Mirabel [8,9] parameterization (after its rate increase by 10^7) reproduces two out of the three events, but also predicts nucleation in two out of the six nonevent days. The semiempirical expression of Spracklen et al. [10], based to a large extent on measurements in the boreal forest, does not reproduce the events in this sulfur-rich relatively polluted environment.

The ternary nucleation parameterization was tested more completely for 19 days in July 2001 and 29 days in January 2002. The model was successful in predicting

Table 1 Summary of DMAN predictions with different nucleation parameterizations

Date	Observed Nucleation	Ternary [6]	Binary [8, 9] ($\times 10^7$)	Binary [7]	1st Order [10]	Barrierless [12]	Ion [11]
2 July	yes	YES[1]	YES	no	No	YES	no
4 July	no	NO	NO	NO	NO	yes	NO
7 July	no	NO	NO	NO	NO	yes	NO
15 July	yes	YES	YES	no	No	YES	no
19 July	no	NO	NO	NO	NO	NO	NO
27 July	yes	YES	no	no	No	YES	no
15 August	no	NO	yes	NO	NO	yes	NO
28 August	no	NO	NO	NO	NO	yes	NO
1 October	no	NO	yes	NO	NO	yes	NO
CORRECT PREDICTIONS		9	6	6	6	4	6

[1]Capitals are used if the model reproduced the observations and lowercase if it failed to capture the occurrence or lack of a nucleation event.

the presence or lack of nucleation in 44 out of the 48 days [13]. The 4 days where it failed were all in January and can probably be explained by the inability of the model to capture the slow gas-phase photochemistry during these periods.

Application in the Eastern USA

The DMAN module, using the ternary NH_3–H_2SO_4–H_2O [6] parameterization, has been incorporated in the PMCAMx 3D CTM to create the version of the code focusing on ultrafine PM, PMCAMx-UF. PMCAMx-UF simulates the emission of size- and composition-resolved PM including ultrafine particles, advection, dispersion, wet and dry deposition, gas-phase chemistry, aerosol processes (nucleation, coagulation, inorganic aerosol condensation and evaporation, secondary organic aerosol growth), and aqueous phase chemistry. The ability of the model to reproduce the fine PM concentration and composition in the Eastern USA has been recently evaluated by [14]. The model has been applied to the Eastern USA modeling domain for July 2001 (Figure 2).

Preliminary results of the models suggest that the nucleation events observed in Pittsburgh during that period often cover an area extending over more than 1,000 km.

Effects of SO_2 and NH_3 Emission Controls

Several scenarios were simulated corresponding to changes in SO_2 and NH_3 emissions. Although SO_2 emissions have been reduced substantially since the 1980s by the Clean Air Act, further reductions are planned to help reduce fine

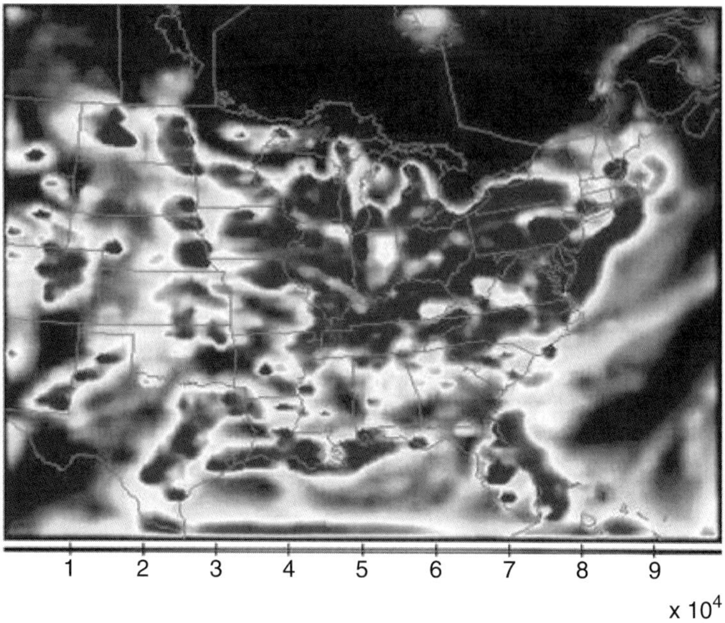

Figure 2 Predicted aerosol number concentration (cm^{-3}) in the Eastern USA, in the morning of July 15, 2001 by PMCAMx-UF

particulate matter mass concentrations in the Eastern USA. Reductions in SO_2 are predicted to increase the frequency and strength of new particle formation, as the lower aerosol sulfate concentrations result in higher gas-phase NH_3 concentrations and decrease the mass and surface area of aged sulfate aerosols, both making nucleation more likely. For the July of 2001 in Pittsburgh reducing SO_2 by 40% will increase the frequency of nucleation events by approximately 50%. Although additional nucleation events are predicted, the decrease in SO_2 results in slower growth of the nucleated particles. With SO_2 concentrations in the Eastern USA having decreased by about a factor of two since 1980, it is also interesting to look at how previous reductions in SO_2 have affected new particle formation. To determine what effect these reductions have had, the model was run with double the SO_2 and increased preexisting aerosol sulfate. The model predicts that the frequency of the nucleation events has doubled in that area during these two decades.

Increasing or decreasing the total NH_3 also has a strong effect on the number of nucleation events and the number concentration of particles, but of an opposite sign than that of SO_2. For example, a 40% decrease in NH_3 during the summer is predicted to eliminate approximately half of the observed new particle formation events. Conversely, if the amount of NH_3 were increased, more nucleation events and higher number concentrations would be seen.

Emissions reductions strategies designed to lower particle mass can increase or decrease the frequency and strength of new particle formation. Given the recent appreciation of the disproportionate negative health impacts of ultrafine particles, a precautionary air quality strategy would seek to decrease fine mass concentrations without a corresponding increase in number concentrations. A decrease of NH_3 together with the planned SO_2 emission reductions may achieve just such a goal.

References

1. Kulmala, M., et al., *J. Aerosol Sci.*, **35**, 143–176 (2004).
2. Oberdorster, G., et al., *J. Toxicol. Environ. Health A*, **65**, 1531–1543 (2002).
3. Adams, P.J. and Seinfeld, J.H., *J. Geophys. Res.*, **107**, 4370 (2002).
4. Wittig, A.E., et al., *Atmos. Environ.*, **38**, 3107–3125 (2004).
5. Jung, J., Adams, P.J., and Pandis, S.N., *Atmos. Environ.*, **40**, 2248–2259 (2006).
6. Napari, I., Noppel, M., Vehkamaki, H., and Kulmala, M., *J. Geophys. Res.*, **107**, 4381 (2002).
7. Vehkamaki, H., et al., *J. Geophys. Res.*, **107**, 4622–4632 (2002).
8. Jaecker-Voirol, A. and Mirabel, P., *Atmos. Environ.*, **23**, 2053–2057 (1989).
9. Russell, L.M., Pandis, S.N., and Seinfeld, J.H., *J. Geophys. Res.*, **99**, 20989–21004 (1994).
10. Spracklen, D.V., Carslaw, K.S., Kulmala, M., Kerminen, V.M., Mann, G.W., and Sihto, S.L., *Atmos. Chem. Phys.*, **6**, 5631–5648 (2006).
11. Modgil, M.S., Kumar, S., Tripathi, S.N., and Lovejoy, E.R., *J. Geophys. Res.*, **110**, 19205 (2005).
12. Clement, C.F. and Ford, I.J., *Atmos. Environ.*, **33**, 489–499 (1999).
13. Gaydos, T.M., Stanier, C.O., and Pandis, S.N., *J. Geophys. Res.*, **110**, D07S12 (2005).
14. Gaydos, T.M., Pinder, R., Koo, B., Fahey, K.M., Yarwood, G., and Pandis S.N., *Atmos. Environ.*, **41**, 2594–2611 (2007).
15. Stanier, C.O., Khlystov, A.Y., and Pandis, S.N., *Aeros. Sci. Tech.*, **38**, 253–264 (2004).

Atmospheric Nucleation

Markku Kulmala

Abstract Aerosol particles influence the quality of our life in many different ways, e.g., via their climatic and health effects, visibility, and acid rain. Aerosol-cloud interactions are the most uncertain process in climate change point of view. Atmospheric aerosol formation is known to occur almost all over the world and the potential importance of these particles to climate and air quality has been widely recognized. Observational data on this phenomenon is based on a large number of measurements that cover the particle sizes larger than about 3 nm diameter. However, almost all the processes driving aerosol formation take place below 3 nm, a size range for which little information is available. Recent instrumental development enables us to study atmospheric concentrations of both neutral and charged nanometer-size clusters. By applying these instruments in the field, we can detect atmospheric nucleation, receive new insight into new particle formation and initial steps of aerosol growth and test different theories and hypothesis.

Keywords Atmospheric aerosols, aerosol formation and growth, nucleation rate

Introduction

An important phenomenon associated with the atmospheric aerosol system is the formation of new atmospheric aerosol particles. Atmospheric aerosol formation consists of a complicated set of processes that include the production of nanometer-size clusters from gaseous vapours, the growth of these clusters to detectable sizes, and their simultaneous removal by coagulation with the pre-existing aerosol particle population.[1,2] Once formed, aerosol particles need to grow further to sizes >50–100 nm in diameter until they are able to influence climate, even though smaller particles may have influences on human health and atmospheric chemistry. While aerosol formation has been observed to take place almost everywhere in the atmosphere,[2,3]

University of Helsinki, Department of Physical Sciences, P.O. Box 64, FI-00014 University of Helsinki, Finland

C.D. O'Dowd and P.E. Wagner (eds.), *Nucleation and Atmospheric Aerosols*, 878–887.

serious gaps in our knowledge regarding this phenomenon still exist. These gaps range from the basic process-level understanding of atmospheric aerosol formation to its various impacts on atmospheric chemistry, climate, human health and environment.

Atmospheric aerosol formation is tied strongly with chemistry, particularly the formation of sulphuric acid and other vapours of very low volatility such as multi-functional organic compounds and iodine vapours. Pre-existing aerosol particles, on the other hand, act as a sink for these vapours and nucleated clusters, thus inhibiting atmospheric aerosol formation. Aerosol formation seems to also be related to several meteorological parameters and phenomena, including the magnitude of solar radiation and atmospheric mixing processes such as the evolution of a continental boundary layer or the mixing of stratospheric and tropospheric air near the tropopause.

Aitken[4] was the first one to report evidence for new particle formation in the atmosphere already in the 19th century. Misaki (1964)[5] observed (the process of) the evolution of mobility distribution of air ions (in a diameter range of 14–100 nm): a gradual shift of a spectral peak towards larger sizes up to about 40 nm after the nucleation burst in the New Mexico semi-desert. However, quantitative measurements of aerosol formation and growth rates have required the development in instrumentation for measuring size distributions down to sizes as small as 3 nm in diameter.[6] On the other hand these sizes are not small enough in order to be able to study nucleation processes, since the phase change between vapour and liquid will occur typically at around 1–2 nm in diameter.[2]

Information on the very first steps of the atmospheric particle formation and growth is required in order to understand the nature of secondary atmospheric particle formation. The minimum particle size detected by the current commercial instruments is approximately 3 nm in diameter, but the atmospheric nucleation and cluster activation processes take place close to 1.5–2 nm[7] (mobility equivalent diameter). Air ions even smaller than 1 nm, on the other hand, can be detected with ions spectrometers such as the Air Ion Spectrometer (AIS)[8] and Balanced Scanning Mobility Analyzer (BSMA),[9] both manufactured by Airel Inc. The concentrations of the small ions, however, are usually so low that they cannot alone explain the observed new particle formation events.

In 2002 during first Finnish–Estonian air ion and aerosol workshop several open scientific questions were asked, namely: How often and in how large areas new particle formation occurs? What takes place below 3 nm diameter? Are there neutral clusters in the atmosphere? What is the composition of nucleation mode aerosol particles? Here I present some answers to those questions.

On Atmospheric Observations

The first question has a clear answer: over the last decade or so, aerosol formation has been observed at a large number of sites around the world.[3] Such observations have been performed on different platforms (ground, ships, aircraft) and over different time periods (campaign or continuous-type measurements).

It appears that several types of atmospheric processes lead to aerosol formation. In the continental boundary layer "regional nucleation events" are common. In such events the growth of nucleated particles continues throughout the day, increasing the particle number concentrations over distances of hundreds of kilometres. Regional nucleation events have been observed in boreal forests in Northern Europe,[10–15] over other remote and rural continental areas,[16,17] in various urban centers,[18–21] and even in heavily polluted regions and megacities.[22–25]

In the marine boundary layer, intense yet highly localized aerosol formation bursts have been reported over coastal environments around Europe and USA.[26–28] Aerosol formation in the remote marine boundary seems to be very rare,[29,30] which suggests that in general, aerosol formation is not a very significant source of particles in the marine boundary layer.

In addition to coastal zones, localized aerosol formation has been observed in different kinds of plumes. Plumes capable of producing new aerosol particles range from large urban and industrial plumes containing SO_2[31,32] to small exhaust plumes associated vehicular traffic.[33–35] Aerosol particles produced by plumes may be important locally or in regional scales.

Aerosol formation has also been observed in the free troposphere, being a frequent phenomenon in the upper troposphere and in cloud-outflow regions.[36,37] Aerosol particles formed in the outflows of convective clouds are likely to be of global significance due to the large volumes of air near the inter-tropical convergence zone where such nucleation events are routinely observed.[38,39] Similar phenomena have also been observed in outflows of mid-latitude convective storms.[36,40]

With few exceptions,[14,41] observed aerosol formation events always occur during daytime, suggesting that photochemistry plays a central role in this process. Other factors that favour atmospheric aerosol formation include a sufficiently low pre-existing aerosol concentration, low relative humidity, high vapour source rate, e.g., via biogenic emissions, and sometimes also a low temperature and weak mixing of air. The most important of these factors seem to be the high vapour source rate, typically via active photochemistry and biogenic source, and the low sink due to pre-existing particles.[3] On the other hand many observed aerosol formation events are related to sulphuric acid.[3,7,16] However, the aerosol production related to the deep convection and cloud outflows is recently claimed to be connected to nucleation of low volatile organic compounds.[40]

Atmospheric aerosol formation events are usually characterized by two quantities, the aerosol formation and growth rate. The former quantity is the rate (particles $cm^{-3}\ s^{-1}$) at which new aerosol particles are being produced in the atmosphere. Due to practical reasons this is not exactly the nucleation rate, but the rate at which particles appear into the smallest measurable sizes, typically a few nanometers (quite often 3 nm). The particle growth rate ($nm\ h^{-1}$) tells us how rapidly the nucleated particles grow to larger sizes. Calculating the aerosol formation rate from observational data is relatively straightforward, whereas determining particle growth rates is more complicated because of potential inhomogeneties in measured air masses, the processes that cause "apparent" particle growth in measured size spectra that is not real, and the fact that the particle growth rate may be both size and time dependent.

Methods that have been applied to determine aerosol formation and growth rates vary from simple visual and fitting-based tools to sophisticated numerical programmes.[15,21,42]

Typical observed formation rates of 3 nm particles associated with regional nucleation events are in the range 0.01–10 cm^{-3} s^{-1}, even though rates up to 100 cm^{-3} s^{-1} have been often observed in urban areas. Locally, aerosol formation rates may be even higher, and rates as high as 10^4–10^5 cm^{-3} s^{-1} have been observed in coastal areas and industrial plumes.[3]

The vast majority of reported particle growth rates[3] lie in the range 1–10 nm h^{-1}, implying that it usually takes 0.5–3 days before the nucleated particles reach sizes at which they may act as cloud condensation nuclei. Growth rates in the range 0.1–1 nm h^{-1} have occasional been observed in remote locations,[14] whereas growth rates larger than 10 nm h^{-1} can be seen in polluted environments.[21,23,43] Intermittently, very high growth rates in excess of 100 nm h^{-1} can be reached in plumes and in coastal zones.[26]

On Atmospheric Clusters

In Figure 1 air ion spectra (naturally charged cluster ions and charged aerosol particles) measured with AIS are presented. The data has been measured at the SMEAR II station (61°51′N, 24°17′E, 181 m a.s.l.)[11] in Hyytiälä, southern Finland, in spring 2006. One of the most important features revealed by the ion instruments is that cluster ions (ions smaller than 1.5 nm) appear to be always present. This seems to be the case both during days with clear aerosol formation (event days) and also on non-event days. Therefore we can confidently say that there are always ion clusters present in the boreal forest environment. Several recent measurements carried out at different locations (marine atmosphere, urban, free troposphere, coastal) confirm this observation: we can state that ion clusters are always present in the atmosphere. In atomic mass units the upper size of these ion clusters is 1,000–1,200 amu. Since the basic result from ion spectrometer is mobility distribution, and there is no unique relationship between mobility and diameter,[44] it is better to use the direct connection between mass and mobility.[45]

Recently significant effort has been put in developing instruments that measure neutral particles below 3 nm. These new instruments such as CPCB (Condensation Particle Counter Battery),[46,47] UF-02proto[48] CPC pair, ion DMPS (Differential Mobility Particle Sizer),[49] and NAIS (Neutral Cluster and Air Ion Spectrometer) can be used to study neutral clusters and calculate estimates for atmospheric particle formation rates close to 2 nm. The dynamics of on cluster and the formation rates of 2–3 nm ions can be investigated using BSMA and AIS. Using this instrumentation the other questions presented in the introduction can be answered.

The composition and therefore the activation properties of the freshly formed nucleation mode can be investigated using CPCB.[46,47] The instrument consists of four CPCs, two with butanol (TSI3010 and TSI3025), and two with water (TSI3785

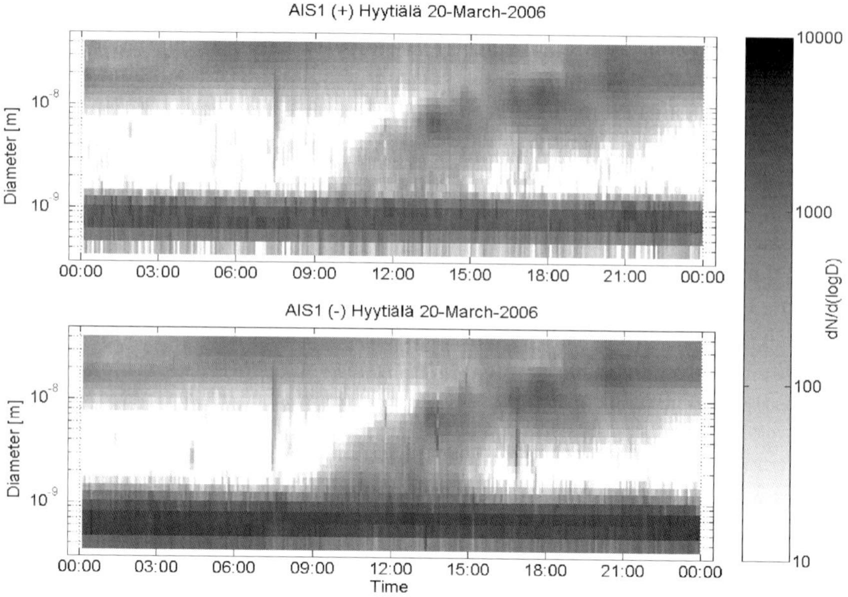

Figure 1 A typical aerosol formation and subsequent growth event measured using an AIS in SMEAR II station, Hyytiälä, Finland. The colour bar indicates concentrations in (ions/cc)

and TSI3786) used as the working fluid. In the case of highly water-soluble particles, the water CPCs detect them at smaller sizes than the instruments based on butanol (see Figure 2). The clear increase of concentration ratio between water and butanol CPCs can be seen when event starts in the morning. At ultrafine range the ratios increases earlier, indicating that small particles (or clusters) in the size range 2–3 nm are water soluble. Somewhat later the same can observed for nucleation mode particles. During the particle formation event the cut-off diameter of the CPCB has been estimated to be close to 2 nm.[46] Therefore the water solubility of 2–3 nm particles is clearly seen during the particle formation event.

Since atmospheric new particle formation seems actually to be a two-step process,[50] information on these steps can be obtained with different approaches. The first step is the nucleation process itself including ion–ion recombination, producing atmospheric (neutral or ion) clusters. The main way to find out the dynamics of the nucleation process is detection of clusters. However, the chemical composition of these clusters might depend on gas concentrations and atmospheric chemistry and ab initio calculations are needed to find out the chemical composition of the clusters. The second step would be the activation of the clusters.[7] The above-mentioned recently developed instruments will give us new insight mainly to the second step.

Recent quantum chemical calculations on cluster composition[51–53] show that ammonia significantly enhances the addition of sulphuric acid molecules to sulphuric acid - water clusters. The NH_3:H_2SO_4 mole ratio of nucleating clusters is predicted[54] to be between 1:3 and 1:1. However, ammonia probably plays no role in

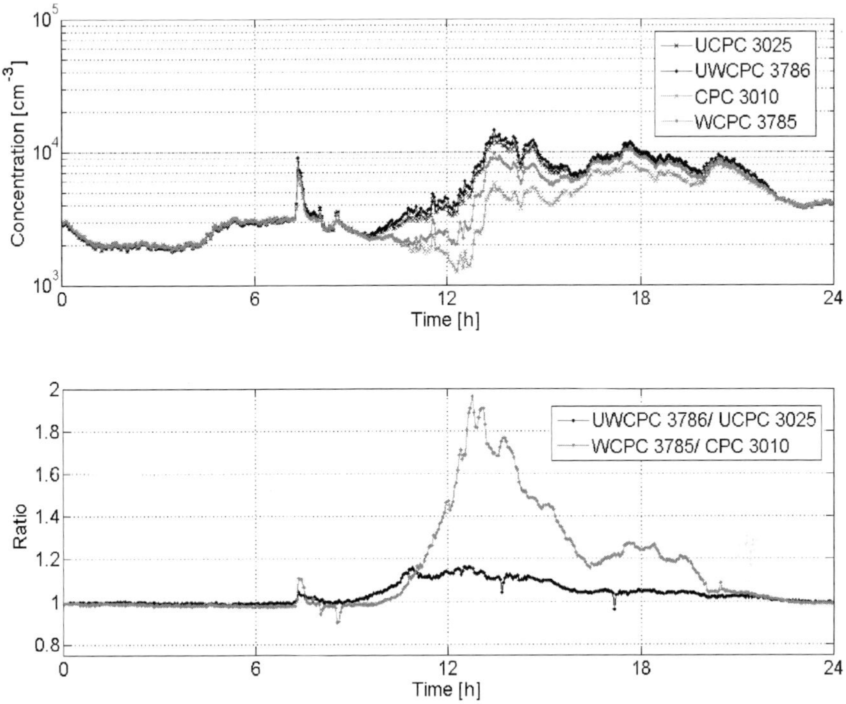

Figure 2 A typical aerosol formation and subsequent growth event (20 March 2006) measured by a CPCB in SMEAR II station, Hyytiälä, Finland. U refers to ultafine CPCs and W to water CPCs. Upper figure shows the concentrations and lower figure the concentration ratios

ion-induced nucleation of the HSO_4^- core ion.[55] The reaction between sulphuric acid and biogenic stabilized Criegee Intermediates, on the other hand, may form[56] significant amounts of nucleation precursors in atmospheric conditions close to sesquiterpene emission sources.

On Formation Rates

The atmospheric nucleation seems to occur around 1.5–2.0 nm in diameter (in Millikan diameter). The total formation rates of the 2 nm particles (J_2) have been calculated from the particle concentrations measured by CPCB and formation rate of negative and positive intermediate ions at 2 nm using BSMA data. In Figure 3 the J_2 values calculated for different instruments continuously during the particle formation event on 20 March 2006 are shown. As seen from the Figure 3, the total formation rate is mainly ion induced during morning hours, and then during the afternoon the neutral fraction starts to dominate being *ca.* 100 times larger.

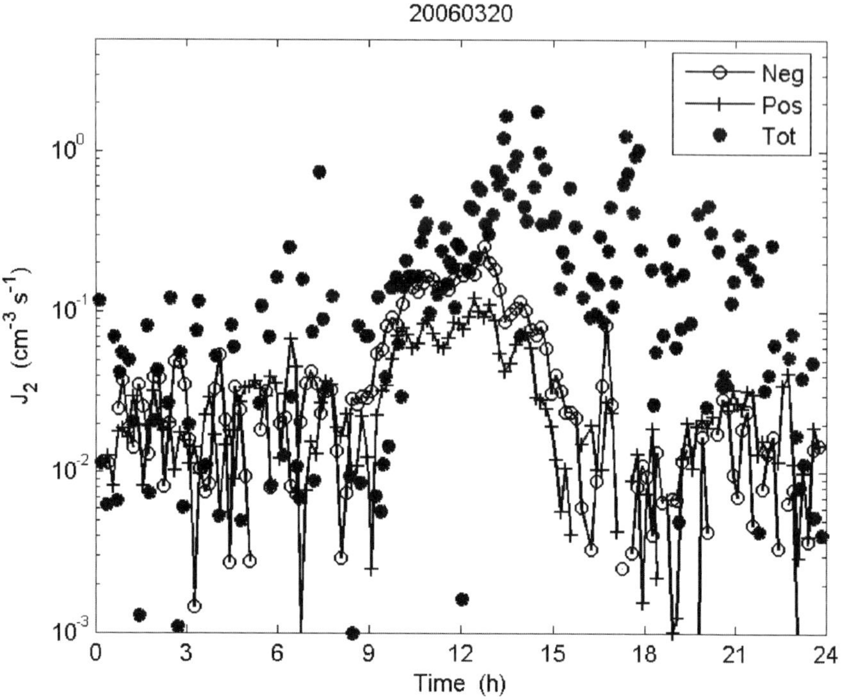

Figure 3 Comparison of 2 nm formation rates during new particle formation on 20 March 2006 in Hyytiälä, Finland. Black dots: CPC-Battery (ions and neutral particles); lines: BSMA (negative and positive ions)

The number concentrations of neutral and charged clusters can be estimated by solving simple balance equations. As a bonus, we get estimations on the relative roles of neutral and ion induced nucleation on observed particle formation events.

The rate at which clusters are activating can be written as

$$\frac{dN_{act}}{dt} = IN_{clus} \cong J_2, \tag{1}$$

where N_{act} is the number concentration of activated clusters, or aerosol particles, N_{clus} is the cluster number concentration, and I is the heterogeneous nucleation rate or activation per one cluster, and J_2

$$J_2 = J_{2,n} + J_{2,i} = I_i N_i^{\pm} + I_n N_n, \tag{2}$$

where $J_{2,n}$ and $J_{2,i}$ refer to the activation of neutral clusters and ions, respectively. Here $N_i^{\pm}$ is number concentration of ion clusters of sign ± and N_n is the number concentration of neutral clusters that have not yet been activated. These neutral clusters may originate from homogeneous nucleation, ion–ion recombination or some chemical reaction(s).

Next, let us make an order of magnitude estimate on the minimum number concentration of neutral clusters, $N_{n,min}$, and the maximum concentration of neutral clusters due to ion–ion recombination, $N_{n,rec,max}$. According to our measurements during the afternoon on 20 March 2006, the sulphuric acid concentration was ca 10^{-6} cm^{-3}, $N_i^{\pm} \approx 500\,\text{cm}^{-3}$ and J_2 and $J_{2,i}$ were about 0.5–1 and 0.01–1 cm^{-3} s^{-1}, respectively. For thermodynamic reasons, $I_n \leq I_i$. The value of $N_{n,min}$ is obtained by setting $I_n = I_i$, which leads to $N_{n,min} \approx 0.5–1 \times 10^4$ cm^{-3}. Using the set of balance equations describing the time evolution of the neutral cluster concentration due to ion–ion recombination we can estimate that recombination makes less than 2,500 neutral clusters per cc. However, since it is not very probable that all ion–ion recombination products stay as stable clusters, the previous number is most probably an overestimation.

On the other hand the formation rate can also be estimated using

$$J_2 = A\times[H_2SO_4] \cong k[H_2SO_4]N_n \tag{3}$$

Here $[H_2SO_4]$ is sulphuric acid concentration, and k is the collision rate (or reaction coefficient). Knowing values for J_2 and $[H_2SO_4]$ and using $N_{n,min}$ we find that k gets values around 10^{-10} cm^3/s, that J_2 is 1 cm^{-3} s^{-1} and values for A (10^{-6} 1/s) are similar as observed recently in the atmosphere.[57,58]

Atmospheric models cannot usually describe efficiently the whole aerosol particle size spectrum. Therefore models are typically not able to start their simulations at 2 nm in diameter. Using recently developed parameterizations any diameter can in principle be modelled. The relation between J_2 and $J(d_p)$ can be estimated based on the theory describing the competition between condensation growth and cluster scavenging[59]:

$$J(d_p) = J_2 \exp\left\{\gamma\left(\frac{1}{d_p} - \frac{1}{2}\right)\frac{CS}{GR}\right\} \tag{4}$$

Here CS (unit m^{-2}) is directly proportional to the condensation sink caused by the pre-existing aerosol particle population and GR (nm h^{-1}) is the growth rate of nucleated particles. The parameter γ depends on many factors but can usually be approximated by assuming it to be equal to 0.23 nm^2 m^2 h^{-1}. The relation given by Eq. (4) assumes implicitly that nucleated clusters grow in size at a constant rate, which may not be the case in the atmosphere. A more general form of Eq. (4), allowing GR to depend on the particle size, has been introduced by Kerminen et al.[60].

Conclusions

Although observed all over the world,[3] and although the theory of two-step formation process[50] has some observational support,[7] the understanding of atmospheric new particle formation is still far from complete. Particularly the nucleation mechanism itself is open.[2,3,7,61]

On the other hand, several steps forwards have been taken in last 5 years as presented here:

1. Atmospheric observations show that ion clusters[62,63] are always around and there are clear indications of existence of neutral clusters.[64]
2. The size at which atmospheric nucleation occurs seems to be 1.5–2 nm in mobility equivalent diameter, which corresponds 1–1.5 nm in mass diameter. This is less than 1,200 amu as cluster mass.
3. Formation of secondary atmospheric particles seems to be a two-step process, where formation of clusters and their activation can be separated.
4. The mechanism itself (see Eqs. 3 and 4) is simple enough to be included in all regional and global models, where health or climate effects of aerosol particles are studied.
5. The effect of ions on particle formation can be estimated using the recently developed instruments and models.

Acknowledgments This work was carried out within the frameworks of the Finnish Centre of Excellence funded by the Academy of Finland, and the NOS-N Nordic Centre of Excellence BACCI. V.-M. Kerminen, I. Riipinen, M. Sipilä, H. Manninen, T. Nieminen, T. Kurtén, and several other people in University of Helsinki are acknowledged for their assistance and clarifying discussions.

References

1. Kerminen, V.-M. et al., *J. Geophys. Res.*, **106**, 24,119–24,126 (2001).
2. Kulmala, M., *Science*, **302**, 1000–1001 (2003).
3. Kulmala, M. et al., *J. Aerosol Sci.*, **35**, 143–176 (2004).
4. Aitken, J.A., *Trans. Roy. Soc.*, **XXX** (1897).
5. Misaki M., *J. Geophys. Res.*, **69**, 3309–3318 (1964).
6. McMurry, P.H., *Atmos. Environ.*, **34**, 1959–1999 (2000).
7. Kulmala, M. et al., *Atmos. Chem. Phys.*, **6**, 787–793 (2006).
8. Mirme, A. et al., *Boreal Environ. Res.*, in press (2007).
9. Tammet, H., *Atmos. Res.*, **82**, 523–546 (2006).
10. Mäkelä, J.M. et al., *Geophys. Res. Lett.*, **24**, 1219–1222 (1997).
11. Kulmala, M. et al., *Tellus*, **53B**, 324–343 (2001).
12. Komppula, M. et al., *J. Geophys. Res.*, **108**(D9), 4295, doi:10.1029/2002JD002939 (2003).
13. Held, A. et al., *J. Geophys. Res.*, **109**, D23204, doi:10.1029/2004 JD005346 (2004).
14. Vehkamäki, H. et al., *Atmos. Phys. Chem.*, **4**, 2015–2023 (2004).
15. Dal Maso, M. et al., *Boreal Env. Res.*, **10,** 323–336 (2005).
16. Weber, R.J. et al., *J. Geophys. Res. D*, **102**, 4375–4385 (1997).
17. Birmili, W. et al., *Atmos. Chem. Phys.*, **3**, 361–376 (2003).
18. Shi, Q. Continuous measurements of 3 nm to 10 μm aerosol size distributions in St. Louis, M.S. thesis, Department of Mechanical Engineering, University of Minnesota, Minneapolis, MN 55455 (2003).
19. Jeong, C-H. et al., *Environ. Sci. Technol.*, **38**, 1933–1940 (2004).
20. Stanier, C.O. et al., *Aerosol Sci. Technol.* **38**(S1), 253–264 (2004).
21. Stolzenburg, M.R. et al., *J. Geophys. Res.*, **110**, D22S05, doi:10.1029/2005JD005935 (2005).
22. Dunn, M.J. et al., *Geophys. Res. Lett.*, **31**, L10102, doi:10.1029/2004GL019483 (2004).

23. Wehner, B. et al., *Geophys. Res. Lett.*, **31**, L22109, doi:10.1029/2004GL021596 (2004).
24. Mönkkönen, P. et al., *Atmos. Chem. Phys.*, **5**, 57–66 (2005).
25. Laaksonen, A. et al., *Geophys. Res. Lett.*, **32**, L06812, doi:10.1029/2004GL022092 (2005).
26. O'Dowd, C.D. et al., *Geophys. Res. Lett.*, **26**, 1707–1710 (1999).
27. O'Dowd, C.D. et al., *J. Geophys. Res.*, **107**(D19), doi: 10.1029/2001JD000555 (2002).
28. Wen, J. et al., *J. Geophys. Res.*, **111,** D08207, doi:10.1029/2005JD006210 (2006).
29. Covert, D.S. et al., *J. Geophys. Res.*, **101**, 6919–6930 (1996).
30. Heintzenberg, J. et al., *Tellus*, **56B**, 357–367 (2004).
31. Woo, K.S. et al., *Aerosol Sci. and Technol.*, **34**, 75–87 (2001).
32. Brock, C.A. et al., *J. Geophys. Res.*, **107**(D12), doi: 10.1029/2001JD001062 (2002).
33. Bukowiecki, N. et al., *Atmos. Environ.*, **36,** 5569–5579 (2002).
34. Kittelson, D.B. et al., *Atmos. Environ.*, **38**, 9–19 (2004).
35. Virtanen, A. et al., *Atmos. Chem. Phys.*, **6**, 2411–2421 (2006).
36. Twohy, C.H. et al., *J. Geophys. Res.*, **107**, doi:10.1029/2001JD000323 (2002).
37. Singh, H.B. et al., *J. Geophys. Res.*, **107**, doi:10.1029/2001JD000486 (2002).
38. Clarke, A.D. *J. Geophys. Res.*, **98**, 20633–20647 (1993).
39. Clarke, A.D. et al., *J. Geophys. Res.*, **103**, 16397–16409 (1998).
40. Kulmala, M. et al. *J. Geophys. Res.*, **111**, D17202, doi:10.1029/2005JD006963 (2006).
41. Wiedensohler, A. et al., *Atmos. Environ.*, **31**, 2545–2559 (1997).
42. Verheggen, B. and Mozurkewich, M., *Atmos. Chem. Phys.*, **6**, 2927–2942 (2006).
43. Kulmala, M. et al., *Atmos. Chem. Phys.*, **5**, 409–416 (2005).
44. Mäkelä, J.M. et al., *J. Chem. Phys.*, **105**, 1562–1571 (1996).
45. Kilpatrick, W.D., *Proc. Annu. Conf. Mass Spectosc.*, **19**, 320 (1971).
46. Kulmala, M. et al., *J. Aerosol Sci.*, **38**, 289–304.
47. Manninen, H.E. et al., these proceedings.
48. Mordas, G., et al., *Boreal Env. Res.*, **10**, 543–552 (2005).
49. Laakso, L., et al., *Atmos. Chem. Phys.*, **7**, 1333–1345 (2007).
50. Kulmala, M. et al., *Nature*, **404**, 66–69 (2000).
51. Kurtén, T. et al., *J. Phys. Chem. A*, **110**, 7178–7188 (2006).
52. Kurtén, T. et al., *J. Geophys. Res.*, in press (2007).
53. Torpo, L. et al., The significant role of ammonia in atmospheric nanoclusters. *Manuscript in preparation* (2007).
54. Kurtén, T. et al., *Atmos. Chem. Phys. Discus.*, **7**, 2937–2960 (2007).
55. Kurtén, T. et al., *Boreal Env. Res.*, in press (2007).
56. Kurtén, T. et al., *J. Phys. Chem. A*, in press (2007).
57. Sihto, S.-L et al., *Atmos. Chem. Phys.*, **6**, 4079–4091 (2006).
58. Riipinen et al., these proceedings.
59. Kerminen, V.-M. and Kulmala, M., *J. Aerosol Sci.*, **33**, 609–622 (2002).
60. Kerminen, V.-M. et al., *Aerosol Sci. Technol.*, **38**, 1001–1008 (2004).
61. Yu, F. and Turco, R.P., *Geophys. Res. Lett.*, **27**, 883–886 (2000).
62. Hirsikko, A. et al., *Boreal Env. Res.*, **10**, 357–369 (2005).
63. Hörrak, U. et al., *J. Geophys. Res. D*, **103**, 13909–13915 (1998).
64. Kulmala, M. et al., *Boreal Env. Res.*, **10**, 79–87 (2005).

Density of Boreal Forest Aerosol Particles as a Function of Mode Diameter

J. Kannosto[1], A. Virtanen[1], T. Rönkkö[1], P.P. Aalto[2], M. Kulmala[2], and J. Keskinen[1]

Abstract Detailed information of particle properties is needed to predict the effects of aerosols in the atmosphere. Chemical composition carries information about the formation mechanisms of the particles. We have developed a method to detect the density of outdoor particles and have applied it to boreal forest aerosols. The average density value of boreal forest nucleation mode particles is 1.1 g/cm^3, and degreases with increasing mode GMD. The density value for Aitken mode is approximately 0.8 g/cm^3. The Aitken and nucleation mode particles seem to have the same origin, while the nucleation mode particles seem to grow into Aitken mode by condensation of some lighter compound. The density obtained for accumulation mode is ca.1.5 g/cm^3 and it seems to increase slightly with increasing particle size.

Keywords Boreal forest aerosol, nucleation, particle density

Introduction

Detailed characterization of atmospheric aerosol particles is needed to understand their effects. Chemical composition of the smallest particles is not easy because of small particle size and low particle concentration. In spite of the recent development of analytical single particle measurement methods, they still do not effectively reach much below 50 nm in particle diameter. Measuring the density of the particles offers an indirect way of obtaining information of the composition of ultrafine particles, but there are not many studies on that either (Saarikoski et al.

[1]*Aerosol Physics Laboratory, Institute of Physics, Tampere University of Technology, P. O. Box 692, FIN-33101 Tampere, Finland*

[2]*Department of Physical Sciences, P.O. Box 64, FIN-00014, University of Helsinki, Finland*

C.D. O'Dowd and P.E. Wagner (eds.), *Nucleation and Atmospheric Aerosols*, 888–891.

2005). We report density values down to lass than 20 nm, obtained from aerosol instrument data with a computational method (Virtanen et al. 2006; Kannosto et al. 2006).

Density Estimation Method

The method is based on simultaneous measurement by ELPI (Electrical Low Pressure Impactor) and SMPS (Scanning mobility particle sizer) and further on the relationship between particle aerodynamic size, mobility size, and effective density. First, multimode distribution is fitted to the measured SMPS data. This is done to get separate modes that are then each allocated an average density value. Using the density values, the size distribution modes, and a mathematical model of the instrument, the current values on each ELPI channel are then simulated. The simulated values are compared to the measured ones. The modal density values providing the best agreement between simulated and measured signals are then sought computationally. The method has been reported by Virtanen et al. (2006).

The method was tested in laboratory for bimodal size distribution using two test oils as particle material: Fomblin (perfluorinated polyether inert fluid, Ausimont Ltd.) and di-octyl sebacate (DOS) (Virtanen et al. 2006). The density of Fomblin is 1.9 g/cm^3 and the density of DOS is 0.91 g/cm^3. Bimodal distributions with one mode consisting of Fomblin and the other of DOS were generated using a tube furnace for Fomblin and a nebulizer with condensation–evaporation cycle for DOS. The geometric mean diameters of DOS distributions were varied between 40–50 nm and of Fomblin distributions between 90–150 nm. The method produced density estimates of 0.8 ± 0.08 g/cm^3 and 1.8 ± 0.26 g/cm^3 for DOS and Fomblin, respectively. These values are within 15% of the true bulk densities.

Results

To our knowledge, we have the first measured values for particle density of boreal forest aerosols in the range of 10–150nm. The density values are not constant but vary as a function of particle size as shown in Figure 1. The particle density values of nucleation mode range from 0.5 to 1.7 g/cm^3 and generally degrease as the mode GMD increases. For Aitken mode the average density value is approximately 0.8 g/cm^3 and for accumulation mode approximately 1.5 g/cm^3. The density value for accumulation mode is similar to the values reported by Saarikoski et al. (2005) but the density value for Aitken mode is lower. For comparison, the density values reported by Saarikoski et al. (2005) are presented in Figure 1 using gray squares.

The average density value of nucleation mode particles is 1.1 g/cm^3. It should be noted that the density of the nucleation mode is not constant but decreases with increasing particle size and finally reaches the value around 0.8 g/cm^3 as modes sift

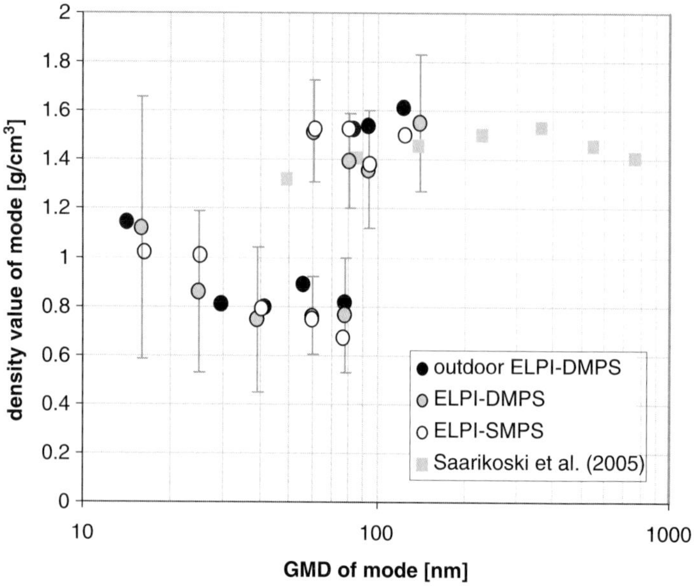

Figure 1 Density values of boreal forest aerosol particles as a function of mode diameter. Three different instrument pairs. The results of Saarikoski et al. (2005) are shown for comparison

Table 1 Average density results of boreal forest aerosols

	ρ(g/cm3)
Nucleation	1.1
Aitken	0.8
Accumulation	1.5

from nucleation range to Aitken range. This indicates that the nucleation mode particles grow into Aitken mode by condensation of some lighter compound.

The density of accumulation mode particles seems to increase slightly with increasing particle size. The density values of accumulation mode particles differ clearly from Aitken mode particles due to the different origin of the particles. It seems that the particle density (or composition) could be a suitable way to divide particles into different modes and could be used instead of particle size range-based mode segregation.

These values are preliminary and further analysis of the data is underway. These density results are average values of 2 weeks measurement campaign in Hyytiälä, Finland (Table 1). Currently, we are looking at the possibility to follow the evolution of particle density of nucleation mode particles during an event.

To test the sensitivity of the method to differences in instrument individuals, a comparison between different instrument pairs was done (Figure 1.) Here the density values are divided into seven size bins and the shown density values are average densities of each size bin. The difference between the calculated density values for three instrument pares were less than 25% (Kannosto et. al 2006).

References

1. Kannosto, J., Ristimäki, J., Virtanen, A., Keskinen, J., Aalto, P.P., and Kulmala, M., Density analysis of boreal forest aerosols. *Chem. Eng. Trans.*, **10**, 95–99 (2006).
2. Saarikoski, S., Mäkelä, T., Hillamo, R., Aalto, P.P., Kerminen, V-M., and Kulmala, M., Physicochemical characterization and mass closure of size-segregated atmospheric aerosol in Hyytiälä, Finland, *Boreal Env. Res.*, **10**, 385–400 (2005).
3. Virtanen, A., Rönkkö, T., Kannosto, J., Ristimäki, J., Mäkelä, J.M., Keskinen, J., Pakkanen, T., Hillamo, R., Pirjola, L., and Hämeri, K., Winter and summer time size distributions and densities of traffic-related aerosol particles at a busy highway in Helsinki, *Atmos. Chem. Phys.*, **6**, 2411–2421 (2006).

Effects of Photochemistry and Convection on the UT/LS Aerosol Nucleation: Observations

David R. Benson[1], Shan-Hu Lee[1], Li-Hao Young[1], William M. Montanaro[1], Heikki Junninen[2], Markku Kulmala[2], Teresa L. Campos[3], David C. Rogers[3], and Jorgen Jensen[3]

Abstract Recent studies show that new particle formation is very active in the upper troposphere and lower stratosphere (UT/LS). And, these results lead to a new question: when does new particle formation *not* occur? Here, we show how photochemistry, surface area and convection affect new particle formation, using the measured aerosol size distributions during the NSF/NCAR GV Progressive Science Missions in December 2005. Three days of sunrise and sunset experiments were made at the latitudes from 18° N to 52° N and altitudes up to 14 km. This is the first time that intensive nighttime aerosol measurements were made in the UT/LS. Aerosol size distributions with diameters from 4 to 2,000 nm were obtained, along with other trace gas species including water vapor, ozone, and carbon monoxide. Surprisingly high concentrations of ultrafine particles were seen continuously during the day and nighttime with high aerosol growth rates, indicating unknown sources of aerosol precursors and new particle formation at night. Also, for air masses that had new particle formation events, it was shown that convection was associated with most of these air masses. On the other hand, for the cases where no new particle formation events were observed, air masses did not experience a vertical motion and there were also high surface area densities. Latitude dependence of new particle formation is also discussed, by comparing with previous studies.

Keywords New particle formation, nighttime nucleation, convection, surface area, UT/LS

[1] *Kent State University, Department of Chemistry, Kent, OH, USA*

[2] *University of Helsinki, Department of Physical Sciences, Helsinki, Finland*

[3] *National Center for Atmospheric Research, Research Aviation Facility, Broomfield, CO*

C.D. O'Dowd and P.E. Wagner (eds.), *Nucleation and Atmospheric Aerosols*, 892–896.

Introduction

Recent aircraft studies showed active new particle formation in the UT/LS with high frequencies (up to 80%) [1] and strong magnitudes (up to 45,000 cm^{-3}) [2]. New particle formation even takes place near the orographic clouds [3,4] and in cirrus clouds [8]. As new particle formation was observed in a wide range of UT/LS conditions [1,2,5–8], it is also important to understand when particle formation *does not* occur. This paper attempts to address this important atmospheric question.

We present results from new particle formation studies during the National Science Foundation (NSF) and National Center for Atmospheric Research (NCAR) NSF/NCAR GV Progressive Science Missions. The objective of these new particle formation studies is to investigate the effects of sun exposure and the latitude and altitude dependence of new particles. Here, we report the measured high ultrafine particles during the nighttime, effects of vertical convection on new particle formation, and the latitude dependence of new particles in the UT/LS. We have also shown enhanced new particle formation in the tropopause region because of the stratosphere and troposphere exchange, using the same data sets [1].

NSF/NCAR GV Progressive Science Missions

Two days of sunrise experiments and 1 day of sunset experiments were made during the GV Progressive Science Missions in December 2005, in Colorado. During each flight, GV flew along the same flight track before and after sunrise (or sunset) over an 8 h period. Aerosol sizes and concentrations were measured with the University of Denver nuclei mode aerosol size spectrometer (NMASS) and forward cavity aerosol spectrometer (FCAS). These instruments are described in detail elsewhere [1,7–8]. NMASS has five condensation nucleus counters that measure cumulative number concentrations of aerosols larger than 4, 8, 15, 30, and 60 nm, respectively. FCAS is a light scattering instrument and sizes aerosols from 90 to 2,000 nm. Using an inversion algorism, size distributions from 4 to 2,000 nm are obtained. The criteria of new particle formation are (i) concentrations of particles from 4 to 9 nm ($N_{4–9}$) > 1 cm^{-3}, (ii) more than 15% of total particles ($N_{4–2000}$) are $N_{4–9}$, and (iii) particles from 4 to 6 nm ($N_{4–6}$) are higher than those from 6 to 9 nm ($N_{6–9}$) [1].

Results and Discussion

Overall Results

Table 1 summaries the measured particle concentrations and meteorological conditions during this mission. A large fraction of the total particles were in the size range from 4 to 9 nm. Surface area densities, in general, were low in this region (4–5 μm^2 cm^{-3}).

Table 1 Summary of the NSF/NCAR GV Progressive Science Mission measurements. N_{4-9}, particle concentrations from 4–9 nm; N_{4-2000}, particle concentrations from 4–2,000 nm. Standard deviation values (1σ) are also included for these measurements

	All days	NPF	Non-NPF
N4–9 (cm−3)	650 ± 1250	670 ± 1270	70 ± 480
N4–2,000 (cm−3)	830 ± 1420	920 ± 1470	170 ± 630
Surface area (μm2cm−3)	4.7 ± 39.1	3.6 ± 18.0	16.1 ± 132.7
Temperature (K)	233.5 ± 19.8	228.7 ± 13.5	248.4 ± 28.0
Relative humidity (%)	22.6 ± 31.1	17.6 ± 21.2	40.1 ± 49.5
Potential temperature (K)	323.9 ± 22.8	326.4 ± 17.4	316.8 ± 34.4
H_2O mixing ratio (ppmv)	580 ± 1120	290 ± 540	1580 ± 1850
Altitude (km)	8.75 ± 3.63	9.52 ± 2.49	6.54 ± 5.28

However, for those nonevent samples, surface area densities were much higher (~16 μm^2 cm^{-3}).

Nighttime Aerosol Formation

High number concentrations of ultrafine particles were observed continuously both during the day and nighttime. Figure 1 shows the measured N_{4-6}, and the calculated coagulation sink and the fraction, in which N_{4-6} disappears via coagulation with large preexisting aerosols, as a function of time during the sunset experiment on December 2, 2005. The time shown here indicates the time differences between the day and night when GV flew the same locations. These coagulation calculations show that higher particle concentrations are present at night than predicted from coagulation calculations. Calculated growth rates were also high, even up to 30 nm h^{-1}, again indicating efficient particle formation during the nighttime. We have also seen similar features during another 2 days of sunrise experiments. Nighttime aerosol formation reported here is consistent with other reports [3–4,9].

When Does New Particle Formation Not Occur?

To answer this unique nucleation question, we have conducted backward trajectory calculations with the NOAA HYSPLIT models [10]. It was common that new particle formation often took place when air masses experienced strong convection, especially in the tropopause region. Vertical convection can bring higher concentrations of aerosol precursors and water vapor to higher altitudes where temperatures and surface area densities are lower. These factors all together create an ideal condition for aerosol nucleation. We also focused on the cases where no new particle formation was observed, and these nonevent air masses did not have a clear vertical

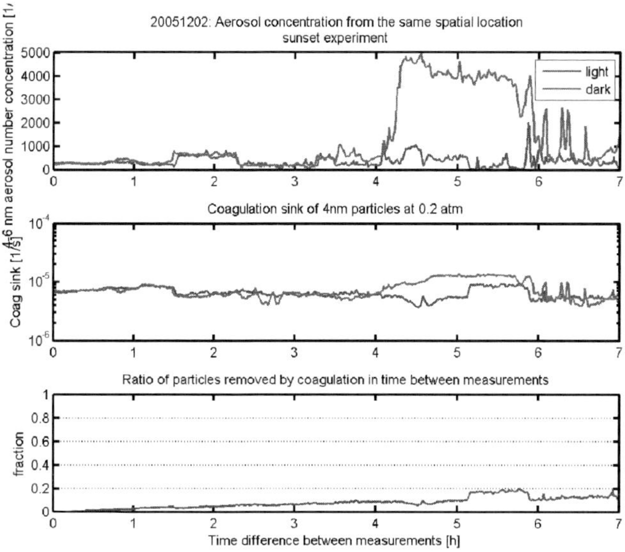

Figure 1 The measured aerosol number concentrations with diameters from 4 to 6 nm, the calculated coagulation sink, and the fraction in which these small particles will disappear via coagulation with large preexisting aerosols, as a function of time. The sunset experiment on December 2, 2005 is shown here. GV flew from Colorado (40° N) south to the 18° N latitude. Sunset occurred in the southernmost spot so the entire southern flight was made before sunset and the northern flight after sunset. The time on the *x*-axis shows the differences between the times of the day and night when flying at the same locations

motion. In addition, surface area densities were also much higher for nonevent cases (Table 1). Our observations are consistent with previous studies [1–2,7].

Because of lower temperatures in the UT/LS, new particle formation is active, if there are sufficient aerosol precursors. And, vertical convection helps bring higher concentrations of precursors to this cold region. At the same time, surface area also becomes very sensitive because of low precursor concentrations in this region. Thus, both convection and low surface area are important to the UT/LS nucleation. And, these two factors play an important role on whether new particle formation occurs or not, in such unique atmospheric conditions.

Latitude Dependence of New Particles

Figure 2 shows the latitude dependence of the measured particles in the upper troposphere near the tropopause. These results show particle concentrations were higher in the subtropics and midlatitudes than in the tropics. This trend is different from the previous report by Lee et al. [7] which showed higher concentrations of ultrafine particles in the lower latitudes. Our results are, however, consistent with

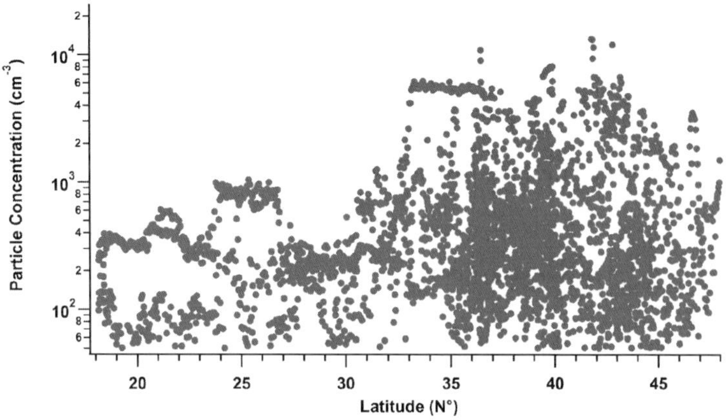

Figure 2 Latitude dependence of particles from 4 to 2,000 nm measured during the GV Progressive Science Missions. Most of these measurements were made in the upper troposphere near the tropopause region. Majority of these particles are ultrafine particles (Table 1.)

the Hermann et al. [5] trend. This is because both the present study and the Hermann et al. [5] measurements were conducted at similar altitude ranges (near the tropopause) and also air mixing induced by convection and the stratosphere and troposphere exchange can be stronger in the midlatitudes than in the tropics. Young et al. (2007) have shown very high frequency (86–100%) and high magnitude (~700–3,960 cm^{-3} N_{4-9} and ~1,000–3,990 cm^{-3} N_{tot}) of new particle formation in the midlatitude tropopause region.

Acknowledgments This study was supported by NSF grants awarded to KSU (ATM-0507709; ATM-0645567) and the KSU startup fund to SHL. NCAR is supported by NSF. We thank James C. Wilson for providing NAMSS and FCAS, and the scientists, engineers, and pilots involved in the NSF/NCAR GV Progressive Science Missions.

References

1. Young L.-H. et al., *J. Geophys. Res.*, in press (2007).
2. Twohy, C.H. et al., *J. Geophys. Res.*, **107**, Doi:10.1029/2001JD000323 (2002).
3. Wiedensohler, A., *Atmos. Environ.*, **31**, 2545–2559 (1997).
4. Mertes, S. et al., *Atmos. Environ.*, **39**, 4233–4245 (2005).
5. Hermann, M. et al., *J. Geophys. Res.*, **108**, doi:10.1029/2001JD001077 (2003).
6. Minikin, A., *Geophys. Res. Lett.*, **30**, Doi:10.1029/2002GL016458 (2003).
7. Lee, S.H. et al., *Science*, **301**, 1886– 1889 (2003).
8. Lee, S.-H. et al., *J. Geophys. Res.*, **109**, Doi:10.1029/2004JD005033 (2004).
9. Mauldin, R.L. III et al., *J. Geophs. Res.*, **108**, Doi:10.1029/2003JD003410 (2003).
10. Draxler, R.R. and Rolph, G.D., *HYSPLIT (HYbrid Single-Particle Lagrangian Integrated Trajectory) Model access via NOAA ARL READY Website* (http://www.arl.noaa.gov/ready/hysplit4.html), Silver Spring, MD: NOAA Air Resources Laboratory (2003).

Estimating Nanoparticle Growth Rates from Size-Dependent Charged Fractions – Analysis of New Particle Formation Events in Mexico City

Kenjiro Iida[1], **Mark R. Stolzenburg** [1], **James N. Smith**[2], **and Peter H. McMurry** [1]

Abstract A method to estimate nanoparticle diameter growth rates during new particle formation (NPF) events from the measured dependence of charged fraction, *f*, on size, D_p, is introduced. The method is especially useful when the mode of the nanoparticle size distribution does not increase monotonically with time, as occurs for regional nucleation events. The growth rate obtained by this method agrees reasonably well with the measured growth rate during regional NPF events for which growth rates were obtained independently. The method was then applied to characterize more complex NPF events observed at Tecamac, Mexico.

Keywords Heterogeneous nucleation, tropospheric aerosols, cluster

Method

When ion concentrations during nucleation and growth events in the atmosphere are not high enough to ensure that stationary state charge distributions are established, changes in charged fractions with size are predominantly affected by rates of both ion–aerosol attachment and condensational growth. Therefore, knowing the ion concentration, the diameter growth rate can be estimated from the measured charged fraction, *f*, versus size, D_p. We accurately measured charged fractions in the 3.7–25 nm range using our Radial-SMPS system [1] and obtained ion concentrations from the ion mobility distributions in the 0.5–6.3nm range measured by Inclined Grid Mobility Analyzer (IGMA) [2]. To facilitate analysis, we make the following simplifying assumptions:

- The charge distributions of freshly nucleated particles are equal for positively and negatively charged particles. This assumption is consistent with the charged fractions measured during the NPF events analyzed.

[1]*Particle Technology Laboratory, University of Minnesota, Minneapolis, MN*

[2]*Atmospheric Chemistry Division, National Center for Atmospheric Research, Boulder, CO*

C.D. O'Dowd and P.E. Wagner (eds.), *Nucleation and Atmospheric Aerosols*, 897–901.

- The mobility and concentration of small ions equals the geometric average values for positive and negative small ions measured with the IGMA. We have confirmed that values of growth rates obtained using this new approach do not change significantly if the mobilities and concentrations of small positive and negative ions measured by the IGMA are used instead. The use of a single value for both polarities significantly simplifies the analysis.
- Growth rates are independent of size and time during the growth from 3.7–25 nm.
- The enhanced growth rate of charged particles due to charge–dipole interaction with condensing vapor is negligible for particles above 3.7 nm [3].
- The enhanced scavenging and coagulation of the charged nanoparticles due to electrostatic interactions are neglected. The analysis of observed NPF events in Boulder and Tecamac showed that the estimated growth rates using this method are about 5–15% higher and 0.1–3.5% lower when scavenging and coagulation are neglected, respectively.

With these assumptions, the rate of change of the charged fraction, f, under steady state or along the growth trajectory is given by

$$GR_f \frac{df}{dD_P} = \beta_0 c_{ion} \left(1 - \frac{f}{f_{stationary}}\right) \tag{1}$$

The attachment coefficients, β, are evaluated by the theory of Fuchs [4] for collisions between ions and neutral particles, β_0, and between ions and charged particles having opposite polarity, β_1. c_{ion} is the representative ion concentration, and GR_f is the diameter growth rate. The stationary state charged fraction, $f_{\text{stationary}}$, is defined as $\beta_0/(\beta_1 + 2\beta_0)$. This expression gives accurate values for sizes below 25 nm based on the first two assumptions mentioned previously.

Equation 1 shows that particle charge distributions approach $f_{\text{stationary}}$ at a given size as they grow by condensation regardless of the initial value of f. Since β_0 is roughly proportional to D_P^2, f asymptotically approaches $f_{\text{stationary}}$ as particles grow. Since the ion mobility and ion mass estimated by the method of Tammet [5] do not vary significantly among different nucleation events, the value of β for a given size also does not vary much. It follows that the difference in the profile of f versus D_P among different events is predominantly caused by the ratio GR/c_{ion}.

The method is verified by applying it to the NPF events where growth rates can be evaluated independently from time-dependent measurements of particle size distribution (PSD). Figure 1 shows measured values of f versus D_P for the NPF events from which the diameter growth rates from PSD, GR_{PSD}, are obtained. $f_{\text{stationary}}$, [6] are shown for comparison. The higher the value of GR_{PSD}/c_{ion}, the greater the deviation of f from $f_{\text{stationary}}$. Initially, the nucleated particles are expected to be almost all electrically neutral for these events [1], and the charged fraction approaches $f_{\text{stationary}}$ as particles acquire charge as they grow. Nucleation events with higher values of GR_{PSD}/c_{ion} would better "remember" their initial charge state, and measured values of f would deviate more significantly from $f_{\text{stationary}}$. These bservations show qualitatively that the measured f versus D_P is consistent with the rend predicted by Eq. 1.

In order to quantitatively support the method, growth rates obtained from PSD luring NPF events, GR_{PSD}, are compared with those obtained by applying Eq. 1 to he measured f versus D_P. There are two major effects that can cause the apparent growth rate of PSD to be greater or smaller than the condensational diameter growth rate. First, intra- and intermodal coagulation of growing mode generally ncrease the apparent particle growth rate [7]. Second, the particle current, , through the upper and lower bounds of the PSD generally decreases the apparent growth rate. In this study, the condensational growth rates were obtained from the pparent growth rates after these effects are taken into account. Table 1 compared GR_{PSD} and GR_f for six regional NPF events measured in Boulder and Tecamac. The correlation coefficient between the two approaches equals 0.963, which supports our argument that charged fractions in 3.5 to 25 nm diameter range can be used to estimate growth rates.

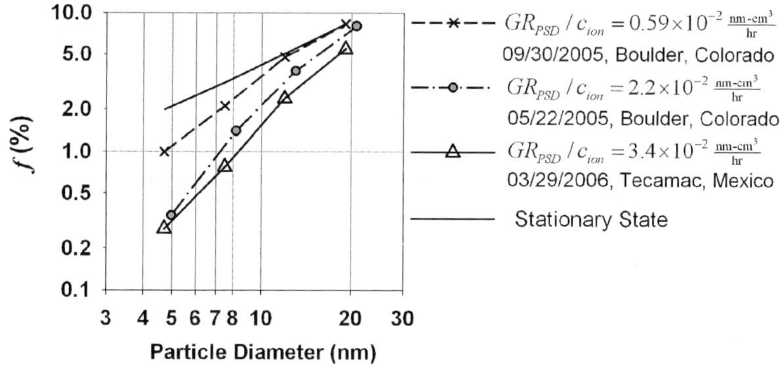

igure 1 Charged fraction, f, versus size D_P during three NPF events which clearly showed the rowing mode of particle size distribution

able 1 Comparison of diameter growth rates calculated from f versus D_P with measured ondensational growth rates obtained from particle size distribution, GR_{PSD}

1M/DD/YYYY	Location	GRPSD, nm/hr †	GRf, nm/hr	Positive/negative ion concentration, cm⁻³	Positive/negative ion electrical mobility, cm²/V-s
/15/2005	Boulder	2.8	4.0	586/401	1.53/1.78
/22/2005	Boulder	9.4	10	486/361	1.55/1.93
/13/2005	Boulder	4.0	3.2	370/248	1.45/1.89
/30/2005	Boulder	5.4	6.2	1,059/878	1.36/1.79
/20/2006	Tecamac	9.1	8.7	409/239	1.44/1.77
/29/2006	Tecamac	8.9	9.0	302/233	1.42/1.88

Application of the Method

Figure 2 shows an example of the type of NPF event that was sometimes observed in Tecamac, Mexico. Concentrations of freshly nucleated particles in the 3–6 nm range $N_{3\text{–}6nm}$, exceeded 10^5 cm^{-3} and remained high for several hours. As a result, it is difficult to follow the peak of the evolving particle size distribution; therefore, it is difficult to obtain the modal diameter growth rate. Note also that during the particle production periods, the time lag between changes in H_2SO_4 and $N_{3\text{–}6nm}$ are short or negligible which qualitatively implies that the growth rates from the size of initial nuclei (~1 nm) to 3–5 nm in Tecamac, Mexico are higher than those typically observed in rural or remote environments [8,9]. The growth rates estimated from the measured *f* versus D_p ranged from 20–38 nm/h. Nucleation rate estimated at 1 nm [8,10] ranged from 1,800–3,300 particles/cm³-s, and are comparable to values estimated from the data taken during a recent study in New Delhi [11,12]. The fraction of sulfur species in the 5–60 nm range measured by the Thermal Desorption Chemical Ionization Mass Spectrometer [13] during a similar NPF event on March 16th was about 10%

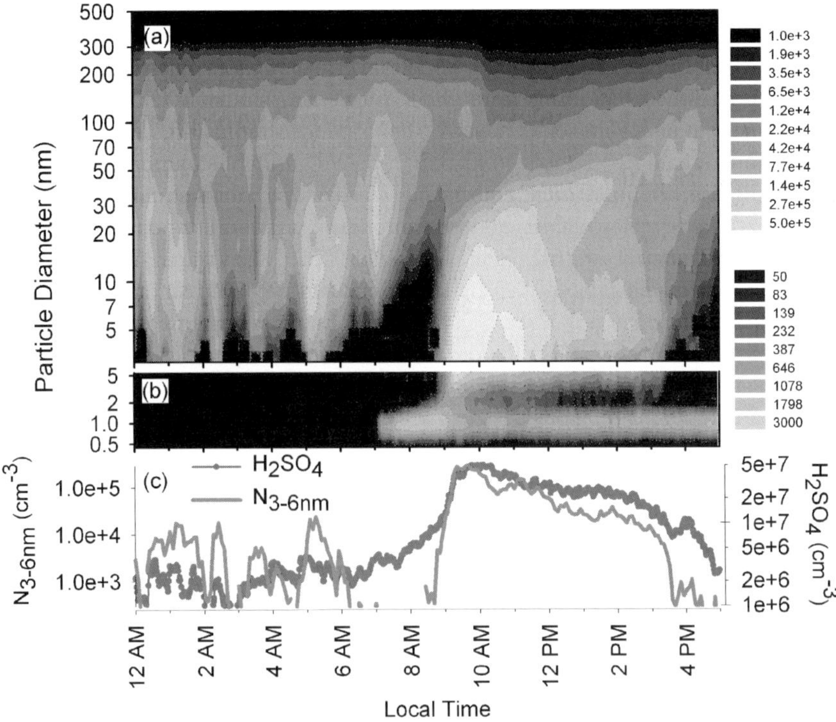

Figure 2 Typical NPF events on March 21, 2006 in Tecamac, Mexico. Contour plots of the particle number distribution ($\Delta N/\Delta \log D_p$, cm^{-3}) measured by the SMPS system and positive small intermediate ions measured by the IGMA are shown respectively in (a) and (b), while the concentration of 3–6 nm particles and H_2SO_4 vapor are shown in (c)

which is consistent with our analysis based on the estimated *GR* and the contribution of *GR* that is due to condensation of H_2SO_4.

Acknowledgment This study article was supported by EPA's NCER STAR Research Program, Agreement No. R82962001, and DOE Grant DE-FG-02-05ER63997.

References

1. Iida, K., Stolzenburg, R.M., McMurry, P.H., Dunn, M.J., Smith, J.N., Eisele, F.L., and Keady, P., *J. Geophys. Res.*, **111**, D23201 (2006).
2. Tammet, H., Inclined grid mobility analyzer: the plain model, Sixth International Aerosol Conference, International Aerosol Research Assembly, Taipei/Taiwan (2002).
3. Tammet, H. and Kulmala, M., *J. Aerosol Sci.*, **36**(2), 173–196 (2005).
4. Reischl, G.P., Mäkelä, J.M., Karch, R., and Necid, J., *J. Aerosol Sci.*, **27**(6), 931–949 (1996).
5. Tammet, H., *J. Aerosol Sci.*, **26**(3), 459–75 (1995).
6. Wiedensohler, A., *J. Aerosol Sci.,* **19**(3), 387–389 (1988).
7. Stolzenburg, M.R., McMurry, P.H., Sakurai, H., Smith, J.N., Mauldin III, R.L., Eisele, F., and Clement, C.F., *J. Geophys. Res.*, **110**(D22), D22S05/1-D22S05/10 (2005).
8. Weber, R.J., Marti, J.J., McMurry, P.H., Eisele, F.L., Tanner D.J., and Jefferson, A., *J. Geophys. Res.*, **102**(D4), 4375–4385 (1997).
9. Sihto, S.-L., Kulmala, M., Kerminen, V.M., Dal Maso, M., Petäjä, T., Rippinen, I., Korhonen, H., Arnold, F., Janson, R., Boy, M., Laaksonen, A., and Lehtinen, K.E.J., *Atmos. Chem. Phys.*, **6**, 4079–4091 (2006).
10. McMurry, P.H., Fink, M., Sakurai, H., Stolzenburg, M.R., Mauldin III, R.L., Smith, J., Eisele, F., Moore, M., Sjostedt, S., Tanner, D., Huey, L.G., Nowak, J.B., Edgerton, E., and Voisin, D., *J. Geophys. Res.*, **110**(D22), D22S02/1-D22S02/10 (2005).
11. Kulmala, M., Petäjä, T., Mönkkönen, P., Dal Maso, M., Aalto, P.P., Lehtinen, K.E.J. and Kerminen, V.M., *Atmos. Chem. Phys.*, **5**, 409–416 (2005).
12. Mönkkönen, P., Koponen, I.K., Lehtinen, K.E.J., Haemeri, K., Uma, R., and Kulmala, M., *Atmos. Chem. Phys.*, **5**, 57–66 (2005).
13. Smith, J.N., Moore, K.F., Eisele, F.L., Voisin, D., Ghimire, A.K., Sakurai, H., and M.P.H., *J. Geophys. Res.*, **110**(D22), D22S03/1-D22S03/13 (2005).

Ions and Charged Aerosol Particles in a Native Australian Eucalypt Forest

Tanja Suni[1], Markku Kulmala[2], Larisa Sogacheva[2], Anne Hirsikko[2], Tommi Bergman[2], Pasi Aalto[2], Marko Vana[2,3], Urmas Horrak[3], Aadu Mirme[3], Sander Mirme[3], Lauri Laakso[2], Miikka Dal Maso[2], Ray Leuning[1], Helen Cleugh[1], Steve Zegelin[1], Dale Hughes[1], Richard Hurley[1], Eva van Gorsel[1], Mark Kitchen[1], Melita Keywood[4], Jason Ward[4], Hannele Hakola[5], Jaana Bäck[6], Carol Tadros[7], John Twining[7], and Jussi Paatero[5]

Abstract We measured atmospheric ion and charged particle concentrations and biogenic aerosol formation with an Air Ion Spectrometer (AIS) in a Eucalypt forest in Tumbarumba, South-East Australia, from July 2005 to October 2006. The measured size range was 0.34–40 nm. Daytime aerosol formation took place on 52% of days with acceptable data. This is approximately twice as often as in a pine forest in Hyytiälä, southern Finland. Possible reasons are the lack of a cold, dark winter which would suppress precursor concentrations and photochemical reactions; weather that is generally drier and sunnier than in Finland; and weaker anthropogenic influence which leads to lower average background concentrations of aerosols. Average growth rates (GR) for 1.3–3 nm and 3–7 nm particles were the same as in Hyytiälä but the 7–20 nm particles grew 50% faster in Tumbarumba than in Hyytiälä. Also, unlike in Hyytiälä, organic vapours are abundant all year round in Tumbarumba so no clear seasonality was evident in GR. Average cluster (0.34–1.8 nm) and intermediate ion (1.8–7.5 nm) concentrations were about 2–4 times those in Hyytiälä which is probably the result of Tumbarumba's high radon concentrations. A new phenomenon, *nocturnal formation* of aerosols (NF), appeared in summer, peaked in autumn, and gradually weakened towards winter. NF lead to intermediate and negative large ion volume concentrations twice as high as during normal daytime events. Therefore it appears that in summer and autumn, nocturnal production was the major aerosol formation mechanism in Tumbarumba.

Keywords Biogenic aerosol formation, atmospheric ions, eucalypt forest, nucleation

[1] *CSIRO Marine and Atmospheric Research, GPO Box 1666, Canberra ACT 2601, Australia*

[2] *Department of Physical Sciences, P.O. Box 64, FIN-00014 University of Helsinki, Finland*

[3] *Institute of Environmental Physics, University of Tartu, Ülikooli St. 18., EE2400 Tartu, Estonia*

[4] *CSIRO Marine and Atmospheric Research, PMB1 Aspendale, VIC 3195, Australia*

[5] *Finnish Meteorological Institute, Erik Palmenin aukio 1, FIN-00560 Helsinki, Finland*

[6] *Department of Forest Ecology, P.O. Box 27, FIN-00014 University of Helsinki, Finland*

[7] *Institute for Environmental Research, ANSTO, PMB 1, Menai NSW 2234, Australia*

C.D. O'Dowd and P.E. Wagner (eds.), *Nucleation and Atmospheric Aerosols*, 902–905.

Introduction

Determining the magnitude and driving factors of biogenic aerosol production in different ecosystems is crucial for future development of climate models. So far, most studies of biogenic aerosol production have taken place at continental and coastal sites in the Northern Hemisphere.[1] Our project is the first to study ion/aerosol characteristics and aerosol production in a temperate forest in the Southern Hemisphere. The study site is an evergreen Eucalypt forest in Tumbarumba, NSW, Australia. Our aim was to determine the characteristics of atmospheric ions and charged particles and the magnitude and driving factors of aerosol production in Tumbarumba and to compare the results to those obtained in a boreal forest in Hyytiälä, southern Finland. Our project is a step towards determining the contribution of forests to the global aerosol load.

Materials and Methods

The Tumbarumba flux station is located in a tall open Eucalypt forest in south eastern New South Wales (35° 39' 20.6" S 148° 09' 07.5" E, 1200 m a.s.l.). The dominant species are *E. delegatensis* (Alpine Ash) and *E. dalrympleana* (Mountain Gum), and the average tree height is 40 m. The instrument tower is 70 m tall and is used to measure fluxes of heat, water vapour and carbon dioxide. Supplementary measurements above the canopy include temperature, humidity, wind speed, wind direction, rainfall, incoming and reflected shortwave radiation and net radiation.[2]

We measured the size distribution of air ions (naturally charged clusters and aerosol particles) with an Air Ion Spectrometer (AIS) from July 2005 to October 2006. The AIS (Airel Ltd., Estonia) measures the concentrations of both negative and positive air ions in the range of 0.34–40 nm, that is, from cluster ions (0.34–1.8 nm) to intermediate (1.8–7.5 nm) and to large (15–40 nm), the larger ones already belonging to the Aitken mode. The AIS consists of two Differential Mobility Analysers (DMA), one for positive and one for negative ions.[3] The AIS was located in a shed on the ground with inlet at 1.5 m height. The total concentration of ultrafine aerosol particles (lower detection limit ~14 nm) was measured with a condensational particle counter (CPC, TSI3010), at the height of 70 m on the tower.

During the EUCAP I campaign in May–July 2006, we measured the size distributions of particles ranging from 15 to 780 nm with a Scanning Mobility Particle Sizer (SMPS) that consisted of a DMA (TSI3071) combined with a CPC (TSI3010). The SMPS was located next to the AIS on ground level. We used the SMPS results to calculate the condensational sink (CS).

New particle formation was classified as normal (ions grew from clusters all the way to large sizes, Figure 1); interrupted (ions only grew from cluster to intermediate sizes, not shown); Aitken (growth started from intermediate sizes and continued to large sizes, Figure1); and nocturnal – sudden nocturnal appearance of large quantities of ions at the same time at least in cluster and intermediate ions (Figure1).

Results and Discussion

On average, negative (positive) cluster ion concentrations were about 2,500 (1,700) cm^{-3} which is 2–4 times the 600–800 cm^{-3} in Hyytiälä.[4] This high concentration is probably due to very high radon concentrations that ranged from 9 to 102 Bqm^{-3} over two days in November (not shown). In Hyytiälä, typical radon concentrations at a few metres height from the ground are only 1–2 Bqm^{-3}.

On average, a daytime new particle formation event took place on 52% of days with acceptable data (Figure 2). This is twice as often as in Hyytiälä. The daytime events exhibited a weak peak from late winter to early summer (Aug–Dec) and a weak minimum from summer to autumn (Jan–May) (Figure 2). The most favourable conditions for new particle formation in summertime were cool, sunny days, but in all other seasons warm days produced more particles than cool days. RH of air masses arriving at the site was

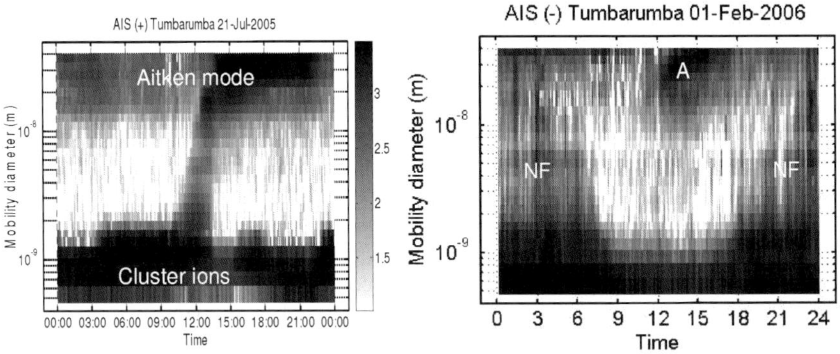

Figure 1 **Left**: A normal new particle formation event extending from cluster ions to the Aitken mode in Tumbarumba in July, 2005. **Right**: a nocturnal formation event (NF; 00–8:00 and 17–24) and an Aitken event (A; 11–18). *x*-axis: time (24 h), *y*-axis: particle diameter. Colour indicates concentration ($dN/dlogD_p$) in logarithmic scale

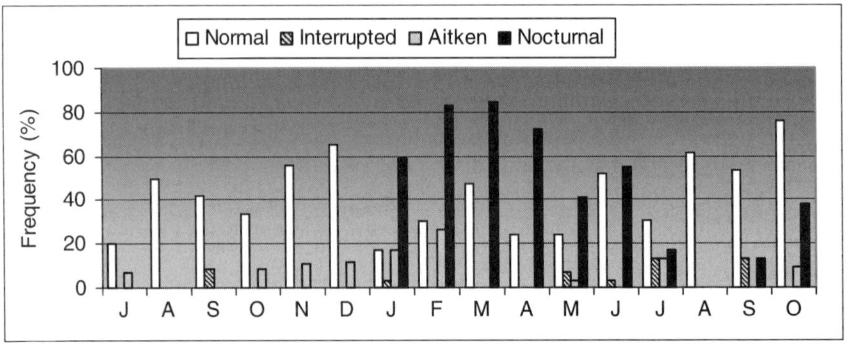

Figure 2 Monthly frequency of daytime (normal, interrupted, and Aitken) and nocturnal new particle formation events as a percentage of available data days during July 2005–October 2006

Table 1 Comparison of daytime and nocturnal aerosol formation strength. Average volume concentrations of intermediate (cv, im) and large (cv, la) ions during November–December 2005 (purely daytime production) and March–April 2006 (mainly nocturnal production)

	$c_{v,im}$ neg/pos (10^{-16} cm^{3}cm^{-3})	$c_{v,la}$ neg/pos (10^{-16} cm^{3}cm^{-3})
Purely daytime (Nov–Dec 2005)	1.1/1.0	78/88
Mainly nocturnal (Mar–Apr 2006)	2.1/1.7	121/90
Ratio nocturnal/daytime	1.9/1.7	1.6/1.0

clearly lower on event days than on non-event days. During EUCAP I, the CS was about 0.0062 s^{-1} on event days (9–15) and about 0.0057 s^{-1} on other days. These numbers are larger than the average 0.0017 (event), 0.0029 (non-event) in Hyytiälä,[5] but the number of samples was only ~10 compared to Hyytiälä's ~800.

Likely reasons behind the frequent aerosol formation in Tumbarumba compared to Hyytiälä are the lack of a long cold winter that would suppress precursor concentration and photochemistry; weather that is generally drier and sunnier than in Finland; and weaker anthropogenic influence that leads to low background concentrations (average 1,200, median 900 cm^{-3} in Tumbarumba; average 2,300, median 1,850 cm^{-3} in Hyytiälä).

The average growth rates (GR) for 1.3–3 nm and 3–7 nm particles were the same as in Hyytiälä (1.5–1.7 and 2.8–3.4 nmh^{-1}, respectively), but the 7–20 nm particles grew 50% faster in Tumbarumba (6.3–7.4 nmh^{-1}) than in Hyytiälä (4–5 nmh^{-1}).[6] In Tumbarumba, organic vapours are abundant all year round so no clear seasonality is evident in GR.

From summer to early winter we observed numerous nocturnal formation events (Figure 1 right) where large numbers of ions appeared in most size classes at the same time. Unlike daytime events, these nocturnal events had a clear seasonal pattern (Figure 2). This seasonality could be related to a rather similar pattern that radon exhalation from the ground exhibits at least in Finland.[5] The volume concentrations of intermediate and negative large ions resulting from nocturnal events were about double compared to daytime events (Table 1). Therefore, it appears that in summer and autumn, nocturnal production was the major aerosol formation mechanism in Tumbarumba.

Acknowledgments This work was supported by the Academy of Finland, Maj and Tor Nessling Foundation, and the Centennial Foundation of Helsingin Sanomat.

References

1. Kulmala, M., Vehkamaki, H., et al., *J. Aerosol Sci.*, **35**(2), 143–176 (2004).
2. Leuning, R., Cleugh, H.A., Zegelin, S.J., and Hughes, D., *Agric.For. Meteorol.*, **129**, 151–173 (2005).
3. Mirme, A., Tamm, E., Mordas, G., Vana, M., Uin, J., Mirme, S., Bernotas, T., Laakso, L., Hirsikko, A., and Kulmala, M., *Bor. Env. Res.* (submitted) (2007).
4. Hirsikko, A., Laakso, L., Horrak, U., Aalto, P., Kerminen, V-M., and Kulmala, M., *Bor. Env. Res.*, **10**(5), 357–369 (2005).
5. Dal Maso, M., et al., *Tellus*, in press (2007).
6. Hirsikko, A., Paatero, J., Hatakka, J. and Kulmala, M., *Bor. Env. Res.*, (in press) (2007).

Factors Controlling Spring and Summer Time Aerosol Size Distributions in the Arctic: A Global Model Study

Hannele Korhonen, Dominick V. Spracklen, Kenneth S. Carslaw, and Graham W. Mann

Abstract A global size-segregated aerosol model is used together with aerosol size distribution measurements to study the observed spring to summer transition of aerosol properties in the Arctic. Previous studies have suggested that the shift from accumulation to Aitken mode dominated distribution is driven by decreasing particle surface area or increasing solar radiation flux that favour nucleation in the atmosphere. We find that binary nucleation mechanism, which forms new particles in free troposphere and is typically considered the main source of marine Aitken mode particles, cannot explain the high summer concentrations of ultrafine particles in the boundary layer. However, when a simple boundary layer nucleation mechanism is included into the model, a transition similar to observations is simulated for the spring months. Our simulations suggest that summer time DMS emissions play a role in the spring to summer transition of the aerosol distribution. The model runs also show that the formation of new particles is highly sensitive to the treatment of aerosol wet scavenging and other aerosol–cloud interaction mechanisms in the model.

Keywords Arctic aerosol, new particle formation, global modelling

Introduction

The Arctic with its sensitive ecosystem is particularly vulnerable to global climate change as current climate scenarios predict for this region a warming rate that is much higher than the global mean. However, many of the processes and feedback mechanisms affecting the Arctic climate are insufficiently understood, among them the factors controlling the state and seasonal changes of the Arctic boundary layer aerosol. Without a thorough understanding of the processes that influence the particle

---

School of Earth and Environment, University of Leeds, Leeds LS2 9JT, UK

C.D. O'Dowd and P.E. Wagner (eds.), *Nucleation and Atmospheric Aerosols*, 906–910.

size distribution, it is impossible to quantify changes in aerosol direct and indirect effects due to climate change and changes from anthropogenic sources.

One of the poorly understood features of the Arctic aerosol is a rapid shift from an accumulation mode dominated aerosol to an Aitken mode dominated distribution in late spring.[1] Earlier work[2] has suggested that this transition is caused by a combination of several effects: change in atmospheric transport patterns, enhanced wet deposition, and increased solar radiation, all of which favour homogeneous new particle formation in the atmosphere. In this paper, we use particle size distribution measurements together with a global aerosol model to study the role of these suggested effects as well as to test our current understanding of the aerosol processes taking place in the Arctic atmosphere in spring.

Methods

GLOMAP[3] is a global model of aerosol processes and an extension to the TOMCAT[4] 3D chemical transport model which is forced by meteorological analysis from European Centre of Medium-Range Weather Forecasts. The aerosol module uses a sectional two-moment scheme (number and mass) to describe the aerosol size distribution. The runs presented here describe the aerosol composition with two compounds: one soluble compound representing sulphate and sea spray aerosol, and one insoluble compound representing primary organic and black carbon aerosol. These two compounds are assumed to mix instantaneously, i.e., they are simulated in one internally mixed distribution.

The aerosol processes simulated in the baseline runs are primary emissions of sulphate, sea spray and OC/BC particles, binary homogeneous nucleation, condensation, hygroscopic growth, coagulation, wet and dry deposition, transport, and cloud processing. Updates to the original GLOMAP code include: Primary sea salt emissions are calculated according to Mårtensson et al.[5] for size range 20 nm–2 μm and according to Monahan et al.[6] for sizes larger than 2 μm. Emissions of SO_2 and carbonaceous aerosol are taken from AEROCOM model intercomparison emission data base (http://nansen.ipsl.jussieu.fr/AEROCOM). It is assumed that 2.5% of SO_2 is emitted as primary sulphate particles at particle sizes proposed by Stier et al.[7] Cloud drop activation is described with a mechanistic scheme of Nenes and Seinfeld.[8] Monthly sea ice cover is taken from a database of British Atmospheric Data Centre.

We compare the model results in the Arctic region to long-term aerosol size distribution measurements from Zeppelin station[1] in Spitsbergen (79N, 12E) as well as to summer time campaign measurements from high Arctic.[9] Our main aim is to understand the processes that control the clear transition from spring time accumulation mode dominated to summer time Aitken mode dominated size distribution as observed at several sites in the Arctic every May/June.

Results

Baseline Model Runs

Throughout the simulated spring of year 2001, the model predicts realistic monthly mean number concentrations of 100–300 cm^{-3} for Spitsbergen. The baseline model runs do not, however, capture the observed features of the size distribution very well. In contrast to observations, the simulated monthly mean particle mode shifts from smaller to larger sizes towards summer (Figure 1a). Furthermore, the simulations show particles smaller than 10 nm in the Arctic boundary layer in early spring but no clear nucleation or Aitken mode in summer, also contrary to observations. The main reasons for the modelled trend are high SO_2 concentrations and cold temperatures in the spring time free troposphere. Furthermore, although the model predicts significant new particle formation in the summer free troposphere, the formed particles are not mixed down to the boundary layer as efficiently as during spring months.

Wet Scavenging of Particles

The main discrepancy between the observations and baseline model results for the polluted months, i.e., March and April, is that the simulated mean size of accumulation mode particles is clearly too small. This could in principle be due to uncertainties in particle injection size of anthropogenic primary emissions, or insufficient growth of particles during transport from continental sources to the Arctic. The most likely reason, however, is too efficient particle wet deposition in the model.

Overall, the monthly modelled precipitation rates are in a fair agreement with the measurements, typically within a factor of 2. The treatment of nucleation scavenging in the model is, however, very simplistic: frontal rain formation is assumed

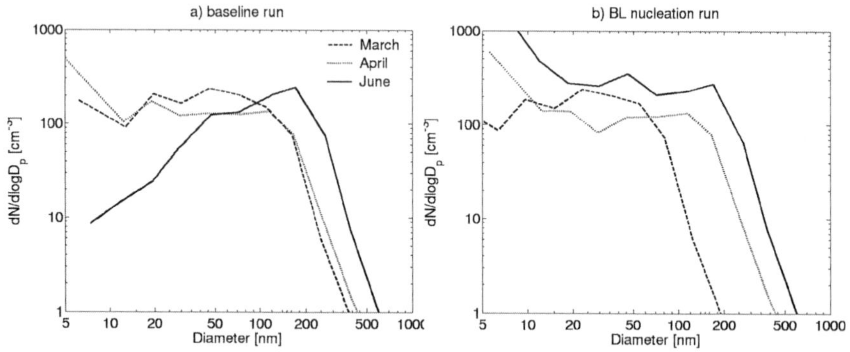

Figure 1 Modelled monthly mean size distributions for Zeppelin station (**a**) in baseline run and (**b**) when new particle formation mechanism for the boundary layer is included

to scavenge 100% of particles whose wet diameter is larger than 206 nm. Reducing this removal efficiency by 20% has negligible effect on the modelled distributions. On the other hand, doubling the cut-off size for nucleation scavenging shifts the modelled accumulation mode towards larger sizes and at the same time suppresses the concentration of particles smaller than 10 nm in March and April, thus clearly improving the agreement with observations for these months (no figure). For June, however, the agreement deteriorates as the accumulation mode particles were already larger than observed in the baseline simulation. Further research is needed to find out whether a seasonal shift in the nucleation scavenging diameter is observed in reality.

New Particle Formation in Boundary Layer

In the baseline simulations particle formation by binary nucleation occurred only in the free troposphere and the downward transport of these new particles could not explain the enhanced Aitken mode in the summer boundary layer. It seems therefore likely that the observed particles either have a primary source that is active only in summer or are formed by nucleation in the boundary layer. According to our simulations a possible summer time primary flux of Aitken mode particles would need to be over 30 times stronger than the flux of ultrafine sea salt particles. We therefore studied the boundary layer nucleation hypothesis by including a simple particle formation mechanism which assumes that new particle formation rate in the boundary layer depends linearly on sulphuric acid vapour concentration, as has been observed at continental sites. [10] These simulations were run using the original nucleation scavenging diameter.

Figure 1b shows that with this mechanism the model reproduces the observed spring to summer transition fairly well. For March and April the modelled sulphuric acid concentration in the Spitsbergen region is of the order of 10^4–10^5 cm^{-3} which is too low to form a significant amount particle in the Aitken mode size. In June, however, high amount of DMS is released into the atmosphere and the model predicts sulphuric acid concentrations of the order of 10^6 cm^{-3}. This concentration is sufficient to produce an Aitken mode that is very similar to the measured one both in size and number concentration. It should be noted that the size distribution measurements at Zeppelin station start from 20 nm and would not be able to capture a separate nucleation mode below this size. Late summer measurements from high Arctic show, however, a clear mode peaking at around 10 nm[9] which is in accordance with our model results.

Outlook

Future work will take a closer look at how the scavenging and ageing processes affect the concentration of accumulation mode particles in the Arctic. We will also extend the simulations to autumn and winter months.

Acknowledgment This work was supported by UK Natural Environment Research Council.

References

1. Ström, J., Umegård, J., Tørseth, K. et al., *Phys. Chem. of the Earth*, **28**, 1181–1190 (2003).
2. Engvall, A.-C., Krejci, R., Ström, J. et al., *Atmos. Chem. Phys. Discuss.*, **7**, 1215–1260 (2007).
3. Spracklen, D., Pringle, K., Carslaw, K. et al., *Atmos. Chem. Phys.*, **5**, 2227–2252 (2005).
4. Stockwell, D. and Chipperfield, M., *Q. J. R. Meteorol. Soc.*, **125**, 1747–1783 (1999).
5. Mårtensson, M., Nilsson, D., de Leeuw, G. et al., *J. Geophys. Res.*, **108**, 4297 (2003).
6. Monahan, E., Spiel, D., and Davidson, K., A model of marine aerosol generation via whitecaps and wave disruption, in: *Oceanic Whitecaps and their role in air-sea exchange processes*, edited by E. C. Monahan and G. MacNiocaill, Dordrecht: D. Reidel Publishing, pp. 167–174 (1986).
7. Stier, P., Feichter, J., Kinne, S. et al., *Atmos. Phys.*, **5**, 1125–1165 (2005).
8. Nenes, A. and Seinfeld, J., *J. Geophys. Res.*, **108**, doi: 2002JD002911 (2003).
9. Heintzenberg, J., Leck, C., Birmili, W. et al., *Tellus*, **58B**, 41–50 (2006).
10. Sihto, S.-L., Kulmala, M., Kerminen, V.-M. et al., *Atmos. Chem. Phys.*, **6**, 4079–4091 (2006).

The Impact of Boundary Layer Nucleation on Global CCN

K.S. Carslaw[1], D.S. Spracklen[1], M. Kulmala[2], V.-M. Kerminen[3], S.L. Sihto[2], and I. Riipinen[2]

Abstract Nucleation of new particles is observed frequently in the atmospheric boundary layer and can substantially increase the total concentration of particles. The importance of nucleation events for cloud condensation nuclei (CCN) concentrations is poorly understood, although some observations suggest that, at least locally, CCN can be greatly enhanced. We have used a global aerosol microphysics model to quantify the impact of boundary layer nucleation events on CCN. Results show that nucleation events can increase mean summertime boundary layer CCN concentrations by 25–60% over remote continental regions and by 10–30% over more polluted regions. Observations show that the nucleation rate varies from region to region by several orders of magnitude. This uncertainty has a substantial impact on model-predicted total particle concentrations but a relatively small impact on predicted CCN: the predicted CCN enhancement over central Europe changes from 12% to 17% for a two order of magnitude increase in the rate. These results demonstrate that boundary layer nucleation is a significant global source of CCN.

Keywords Nucleation, boundary layer, CCN, global model

Introduction

Boundary Layer nucleation events have been observed at many locations around the world ranging from the sub-Arctic through boreal forests to polluted industrial and coastal regions (Kulmala et al. 2004). Localized observations have shown that these particles can grow to CCN sizes (Lihavainen et al. 2003;

[1] *School of Earth and Environment, University of Leeds, UK*

[2] *Dept of Physical Sciences, University of Helsinki, Finland*

[3] *Finnish Meteorological Institute, Helsinki, Finland*

C.D. O'Dowd and P.E. Wagner (eds.), *Nucleation and Atmospheric Aerosols*, 911–915.

Kerminen et al. 2005; Laaksonen et al. 2005). In Spracklen et al. (2006) we used a global aerosol model to quantify the effect of boundary layer nucleation events on global CN (all particles larger than 3 nm diameter). We showed that nucleation can increase continental boundary layer concentrations by up to a factor of 8 above that derived from primary natural and anthropogenic sources alone. Here we extend these simulations to explore the contribution of nucleation to potential CCN.

Both primary and secondary (nucleated) aerosol particles are potential sources of CCN, but the relative contribution of each has not been quantified. Nucleated particles enter the atmosphere at nanometer sizes and must therefore grow considerably before they can act as CCN at ~50 nm dry diameter. The loss by various scavenging processes during growth and transport greatly limits the number of nucleated particles that can grow that large (Pierce and Adams 2006). A global calculation of the impact of nucleation on CCN therefore requires a size-resolved global or regional aerosol microphysics model.

Model Description

We use the GLobal MOdel of Aerosol Processes (GLOMAP; Spracklen et al. 2005a, b, 2006). GLOMAP is a 3D chemical transport model with detailed microphysics. Large-scale atmospheric transport is specified from European Centre for Medium-Range Weather Forecasts (ECMWF) analyses at 2.8° × 2.8° horizontal resolution. Aerosol distributions are described by a two-moment sectional scheme with 20 size bins spanning 3 nm to 25 μm dry diameter. The model includes the processes of particle formation, gas phase and aqueous sulfur chemistry, coagulation, condensation, and dry and wet deposition.

The model includes sulfate, sea salt, black carbon (BC), and organic carbon (OC). To minimize computational expense we treat all particles as acidic sulfate. Size-resolved CCN spectra (Dusek et al. 2006) suggest that this simplified treatment of particle composition will be less important than the changes in the particle size distribution, which the model represents well.

The nucleation scheme is described in detail in Spracklen et al. (2006). The formation rate of 1 nm molecular clusters is given by $j_1 = A[H_2SO_4]$ where $[H_2SO_4]$ is the gas-phase sulfuric acid concentration and A (s^{-1}) is an empirically derived activation coefficient (Kulmala et al. 2006). Evaluation of nucleation events in remote and polluted conditions suggests that A can vary between about 10^{-7} and $10^{-5}\ s^{-1}$ (Sihto et al. 2006; Riipinen et al. 2007). Here we investigate the sensitivity of CCN formation to this range. In the model, we use the analytical formula of Kerminen et al. (2002) to calculate the effective production rate of 3 nm particles, taking into account the loss of nucleated particles through coagulation scavenging. Particles grow by condensation of sulfuric acid and oxidized organics, formed from monoterpenes.

Results

The impact of nucleation on global CCN in June 2002 is shown in Figure 1. CCN is defined to be the number of particles with dry diameter greater than 70 nm, equivalent to 0.2% supersaturation for a soluble particle. The global distribution of potential CCN in the run without nucleation varies between 20 and 250 cm^{-3} in remote marine regions, 500–2,000 cm^{-3} in continental regions and more than 2,000 cm^{-3} in the most polluted regions.

In remote continental locations of the Northern Hemisphere (NH) particle formation enhances CCN concentrations up to a factor of 2, with increases of 30–60% typical over large regions of boreal Asia in June. Increases are typically less than 30% over the most polluted regions, and virtually nil over most oceanic regions.

Figure 2 shows the sensitivity of our predicted CCN concentrations to uncertainty in the nucleation coefficient A. Over central Europe the enhancement increases from 12% with $A = 2 \times 10^{-6}\ s^{-1}$ to 17% with $A = 2 \times 10^{-6}\ s^{-1}$.

Conclusions and Implications

We estimate that nucleation events contribute 20–50% of potential CCN over many remote continental regions and 10–30% over polluted continental regions. The mechanism behind BL particle formation events is not fully understood. We have assumed that sulfuric acid controls the nucleation rate, as observed, and have used an empirical nucleation rate coefficient derived from two sites in Europe. The

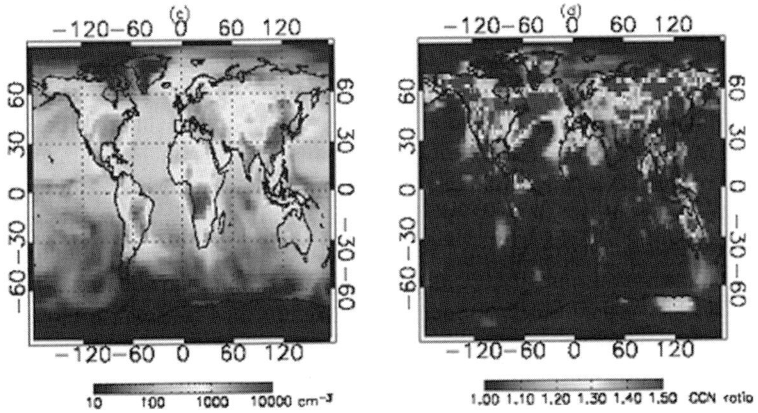

Figure 1 The impact of boundary layer nucleation on global CCN. Left panel shows CCN concentrations and the right panel shows the enhancement that is due to nucleation

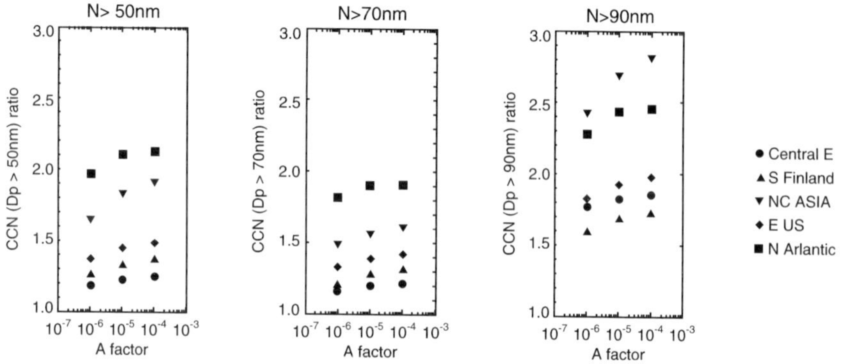

Figure 2 The effect of uncertainty in the nucleation rate coefficient *A*. The mean enhancement in CCN due to nucleation is shown for a number of regions (Central Europe, South Finland, North Central Asia, East USA, North Atlantic) and for three different assumptions of CCN diameter (indicated above each panel)

observed variability in the nucleation rate coefficient *A* suggests that additional factors are important that we do not account for here. Nevertheless, the sensitivity of our results to uncertainty in the nucleation rate (within the limits derived from analyzed observations) appears to be relatively small.

These results show that particle formation events are an important global source of CCN which needs to be included in global and regional atmospheric models. To accurately account for the long-term historical change in aerosol direct and indirect forcing will require a better understanding and treatment of particle formation events.

Acknowledgments We acknowledge funding from the Natural Environment Research Council and the Royal Society in the UK and the Finnish Academy.

References

1. Dusek, U., Frank, G., Hildebrandt, L., Curtius, J., Schneider, J., Walter, S., Chand, D., Drewnick, F., Hings, S., Jung, D., Borrmann, S., and Andreae, M., Size matters more than chemistry for cloud-nucleating ability of aerosol particles, *Science*, **312**, 10.1126/science.1125,261 (2006).
2. Kerminen, V.-M. and Kulmala, M., Analytical formulae connecting the "real" and the apparent nucleation rate and the nuclei number concentration for atmospheric nucleation events, *J. Aerosol Sci.*, **33**, 609–622 (2002).
3. Kulmala, M., Vehkamaki, H., Petajda, T., Dal Maso, M., Lauri, A., Kerminen, V., Birmili, W., and McMurry, P., Formation and growth rates of ultra_ne atmospheric particles: a review of observations, *J. Aerosol Sci.*, **35**(2), 143–176 (2004).
4. Kulmala, M., Lehtinen, K.E.J., and Laaksonen, A., Cluster activation theory as an explanation of the linear dependence between formation rate of 3nm particles and sulphuric acid concentration, *Atmos. Chem. Phys.*, **6**, 787–793 (2006).

5. Laaksonen, A., Hamed, A., Joutsensaari, J., Hiltunen, L., Cavalli, F., Junkermann, W., Asmi, A., Fuzzi, S., and Facchini, M., Cloud condensation nucleus production from nucleation events at a highly polluted region, *Geophys. Res. Lett.*, **32**(L06812), doi:10.1029/2004GL022,092 (2005).
6. Lihavainen, H., Kerminen, V.-M., Komppula, M., Hatakka, J., Aaltonen, V., Kulmala,V, and Viisanen, Y., Production of "potential" cloud condensation nuclei associated with atmospheric new-particle formation in northern Finland, *J. Geophys. Res. Atmos.*, **108**(D24) (4782), doi:10.1029/2003JD003,887 (2003).
7. Pierce, J. and Adams, P., Efficiency of cloud condensation nuclei formation from ultrafine particles, *Atmos. Chem. Phys. Discuss.*, **6**, 10991–11023.
8. Riipinen, I., Sihto, S.-L., Kulmala, M., Arnold, F., Dal Maso, M., Birmili, W., Saarnio, K., Teinila, Kerminen, V.-M., Laaksonen, A., and Lehtinen, K., Connections between atmospheric sulphuric acid and new particle formation during QUEST IV campaigns in Heidelberg and Hyytiala, *Atmos. Chem. Phys. Discuss.*, **6**, 10387–10882.
9. Sihto, S.-L., Kulmala, M., Kerminen, V.-M., Dal Maso, M., Petaja, T., Riipinen, I., Korhonen H., Arnold, F., Janson, R., Boy, M., Laaksonen, A., and Lehtinen, K., Atmospheric sulphuric acid and aerosol formation: implications from atmospheric measurements for nucleation and early growth mechanisms, *Atmos. Chem. Phys.*, **6**, 4079–4091 (2006).
10. Spracklen, D., Pringle, K., Carslaw, K., Chipperfield, M., and Mann, G., A global offline model of size-resolved aerosol microphysics; I. model development and prediction of aerosol properties, *Atmos. Chem. Phys.*, **5**, 2222–2252 (2005a).
11. Spracklen, D., Pringle K., Carslaw, K., Chipperfield, M., and Mann, G., A global offline model of size-resolved aerosol microphysics; II. identification of key uncertainties, *Atmos. Chem. Phys.*, **5**, 3233–3250 (2005b).
12. Spracklen, D., Carslaw, K., Kulmala, M., Kerminen, V.-M., Mann, G., and Sihto, S.-L., The contribution of boundary layer nucleation events to total particle concentrations on regional and global scales, *Atmos. Chem. Phys.*, **6**, 7323–7368 (2006).

Relative Humidity Dependence of Light Extinction by Mixed Organic/Sulfate Particles

Melinda R. Beaver[1,2], Tahllee Baynard[2,3], Rebecca M. Garland[1,2,4], Christa Hasenkopf[2,5], A.R. Ravishankara[1,3], and Margaret A. Tolbert[1,2]

Abstract Light extinction by particles is important locally and regionally for visibility reduction, and globally for climate. Light extinction by particles is strongly dependent on the size, chemical composition, and water content of the aerosol. Here we investigate the humidity dependence of light extinction, fRH_{ext}, by particles of varying organic composition. We have found a linear trend between fRH_{ext} and mass fraction of organic species for a variety of water-soluble organic compounds. We are currently extending the studies to include partially soluble and insoluble organic compounds.

Introduction

Atmospheric aerosols play an important role in the Earth's radiation balance by scattering and absorbing incoming solar radiation. Though progress has been made in quantifying the direct effect of aerosols on radiative forcing, uncertainties still remain. These uncertainties are the effects of relative humidity and particle composition on light extinction by particles.

It is known via ground and air-based atmospheric particulate measurements that tropospheric aerosols are complex internal mixtures of components including inorganic species as well as elemental and organic carbon [1,2]. The hygroscopicity of particles is highly dependent on the particle composition. Traditionally, the water uptake of both inorganic and organic particles has been studied using a Humidified Tandem Differential Mobility Analyzer (HTDMA) to determine the

[1] *Department of Chemistry and Biochemistry, University of Colorado, Boulder, CO, USA*

[2] *Cooperative Institute for Research in the Environmental Sciences, University of Colorado, Boulder, CO, USA*

[3] *NOAA Earth Systems Research Laboratory, Chemical Sciences Division, Boulder, CO, USA*

[4] *Now at Biogeochemistry Department, Max Planck Institute for Chemistry, Mainz, Germany*

[5] *Department of Atmospheric and Oceanic Sciences, University of Colorado, Boulder, CO, USA*

C.D. O'Dowd and P.E. Wagner (eds.), *Nucleation and Atmospheric Aerosols*, 916–919.

aerosol growth factors, Gf [3–5]. These growth factors, which are ratios of humidified particle diameter to the dry particle diameter, then are combined with Mie scattering calculations to determine the effect of water uptake on optical properties. Another approach, taken here, is to directly measure the optical effect of water uptake by particles of varying chemical composition. Here, we have studied the humidity dependence of light extinction of a wide range of organic compounds (dicarboxylic acids and sugars), pure and mixed with ammonium sulfate. We currently are also investigating aromatic and long-chain organic acids, thus spanning a range of compounds with different physical properties.

Experimental Details

The dependence of aerosol extinction, at 532 nm, on relative humidity (RH) has been studied with a tandem cavity ring-down aerosol extinction spectrometer (CRD-AES) and is shown in Figure 1 [6–8]. Extinction within the cavity is related to the ratio of the cavity optical and sample lengths, R_L, the speed of light, c, and the ring-down time constants with, τ, and without, τ_0, sample present, according to Eq. 1.

$$\sigma_e = \frac{R_L}{c}\left(\frac{1}{\tau} - \frac{1}{\tau_0}\right) \tag{1}$$

Particles are generated via atomization, dried (RH < 10%), and size selected to produce a monodisperse aerosol flow.

The dry, size selected, particles enter the first cell of the CRD-AES where the reference aerosol extinction coefficient, σ_{ep}(Dry) is measured. The particle flow then enters a temperature-controlled humidifier to force deliquescence at a set relative humidity, RH (typically 80%). The liquid particles then enter the second cell of the tandem CRD-AES where σ_{ep}(RH) and RH are measured. The two extinction values are then used to calculate fRH_{ext} defined in Eq. 2,

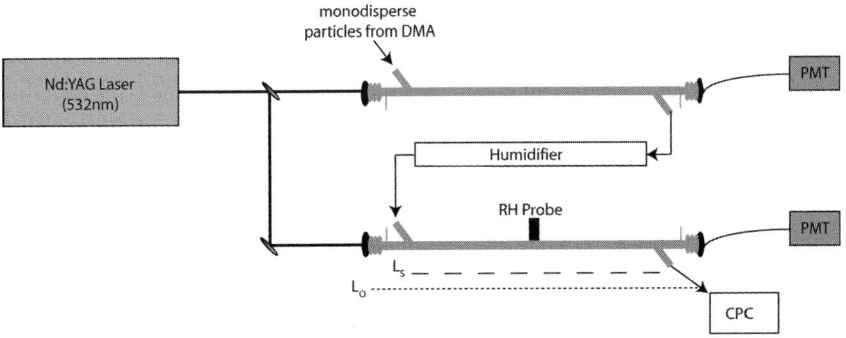

Figure 1 Diagram of experimental tandem CRD-AES used for fRH_{ext} measurements. $R_L = L_O/L_S$

$$fRH_{ext}(80\%RH, Dry) = \frac{\sigma_{ep}(80\%RH)}{\sigma_{ep}(Dry)} \tag{2}$$

to describe the relative humidity dependence of light extinction. Finally, the particle concentration is measured with a condensation particle counter (CPC). With simultaneous measurements of extinction, σ_{ext} (Mm^{-1}) and particle concentration, *N*, we can also determine particle extinction cross sections $C_{ext} = \sigma_{ext}/N$ (cm^2).

Results and Discussion

Typical ring-down signals from the CRD-AES are shown in Figure 2. A typical value for τ_0, the ring-down time constant without sample present, is 100 µs. Then, with aerosol present, τ values are much shorter, due to the extinction of light by the particles.

Extinction cross-section measurements have been made for polystyrene spheres to check the absolute measurements made with this system. Measurements agree within 10% of those calculated using Mie scattering theory [9].

Overall, the organic compounds studied thus far, dicarboxylic acids (adipic, DL-malic, succinic, malonic, and maleic acid) and sugars (glucose, levoglucosan, and mannitol) all exhibited significantly less water uptake at 80%RH than ammonium sulfate. For example, fRH_{ext}(80%RH, Dry) for ammonium sulfate was measured to be 3.29 ± 0.10 for 254 nm dry particles. While, fRH_{ext}(80%RH, Dry) for the ensemble of organics was 1.42 ± 0.36 for 347 nm dry organic particles. Figure 3 summarizes the fRH_{ext}(80%RH, Dry) values 347 nm dry particles for all the internal mixtures of organics/ammonium sulfate studied thus far. A linear fit of fRH_{ext}(80%RH, Dry) = 2.90–0.015*(wt% organic) can be used to describe all the data.

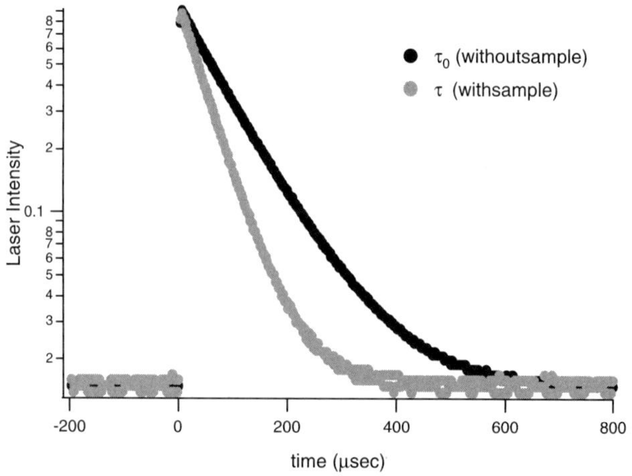

Figure 2 Cavity ring-down signals with and without sample present. τ_0, without sample present is shown in the black trace, and is 100 µs. τ, with particle sample is shown in gray and is 56 µs

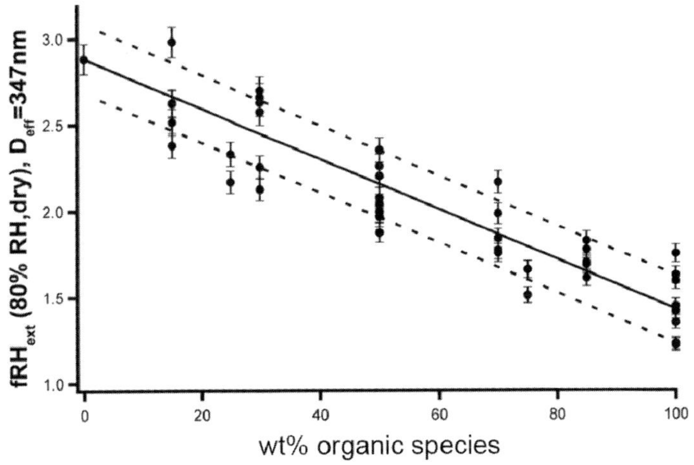

Figure 3 Summary of fRH_{ext}(80%RH, Dry) versus wt% organic compound. The solid line is the best fit to the data, and the dotted lines are ± 1 standard deviation of the data

Planned experiments include testing the extension of this parameterization to less soluble organic species as well as to insoluble organic compounds. Together these data will provide useful information for climate models to include in their prediction of light extinction by organic-containing atmospheric particles.

Acknowledgments This work was supported by NASA grant NNG06GE79G. MRB acknowledges support from the EPA STAR fellowship program, grant FP-91654601. The research described in this paper has been funded in part by the United States Environmental Protection Agency (EPA) under the Science to Achieve Results (STAR) Graduate Fellowship Program. EPA has not officially endorsed this publication and the views expressed herein may not reflect the views of the EPA.

References

1. Murphy, D.M., Czizco, D.J., Froyd, K.D., Hudson, P.K., Matthew, B.M., Middlebrook, A.M., Peltier, R.E., Sullivan, A., Thomson, D.S., and Weber, R.J., *J. Geophys. Res.*, **111**, D23S32, doi:10,1029/2006JD007340 (2006).
2. Saxena, P. and Hildemann, L.M., *J. Atmos. Chem.*, **24**, 57–109 (1996).
3. Svenningsson, B., Rissler, J., Swietlicki, E., Mircea, M., Bilde, M., Facchini, M.C., Decesari, S., Fuzzi, S., Zhou, J., Monster, J., and Rosenorn, T., *Atmos. Chem. Phys.*, **6**, 1937–1952 (2006).
4. Prenni, A.J., DeMott, P.J., and Kreidenweis, S.M., *Atmos. Environ.*, **37**, 4243–4251 (2003).
5. Brooks, S.D., DeMott, P.J., and Kreidenweis, S.M., *Atmos. Environ.*, **38**, 1859–1868 (2004).
6. Pettersson, A., Lovejoy, E.R., Brock, C.A., Brown, S.S., and Ravishankara, A.R., *J. Aerosol Sci.*, **35**, 995–1011 (2004).
7. Baynard, T., Garland, R.M., Ravishankara, A.R., Tolbert, M.A., and Lovejoy, E.R., *Geophys. Res. Lett.*, **33**, L06813, doi:10.1029/2005GL024898 (2006).
8. Garland, R.M., Ravishankara, A.R., Lovejoy, E.R., Tolbert, M.A., and Baynard, T., *J. Geophys. Res.*, accepted (2007).
9. Bohren, C.F. and Huffman, D.R., *Absorption and Scattering of Light by Small Particles*, New York: Wiley (1983).

Evaporation Rates and Saturation Vapour Pressures of C3–C6 Dicarboxylic Acids

Ilona Riipinen[1], Ismo K. Koponen[2], Merete Bilde[2], Anca I. Hienola[1], and Markku Kulmala[1]

Abstract We have measured the evaporation rates of droplets consisting of aqueous solutions of C3–C6 dicarboxylic acids. We have derived expressions for subcooled liquid saturation vapour pressures of the acids by analysing the measurement data with a numerical condensation/evaporation model. At 299 K (close to the measurement temperatures), the saturation vapour pressures are of the order of 10^{-4}–10^{-3} Pa

Keywords Dicarboxylic acids, vapour pressure, condensation/evaporation

Introduction

Malonic, succinic, glutaric, and adipic acids are dicarboxylic acids which are often found in atmospheric aerosol samples.[1] This implies that they might have a role in the formation and growth of new atmospheric aerosol particles. In order to make quantitative analysis on the role of organics in particle formation and growth, however, information on the thermophysical properties of these compounds is needed.

Riipinen et al. (2006)[2] have recently described a method that is suitable for determining thermophysical properties, such as saturation vapour pressures, of organic compounds in aqueous solutions. The method combines the use of well-defined evaporation experiments and accurate binary evaporation modelling.

We have applied the method for determining the saturation vapour pressures of malonic (C3), succinic (C4), glutaric (C5), and adipic (C6) acids in aqueous solutions. All of the pure acids are solid in atmospheric conditions, but once

[1]*University of Helsinki, Department of Physical Sciences, P.O. Box 64, FI-00014 University of Helsinki, Finland*

[2]*University of Copenhagen, Department of Chemistry, Universitetsparken 5, DK-2100, Copenhagen Ø, Denmark*

C.D. O'Dowd and P.E. Wagner (eds.), *Nucleation and Atmospheric Aerosols*, 920–923.

deliquesced, they stay in liquid solutions at relative humidities even below 60%.[3] The subcooled liquid vapour pressures are therefore one of the key properties describing the formation and growth behaviour of droplets consisting of these compounds.

Materials and Methods

Evaporation Rate Measurements

The evaporation rates of binary droplets containing water and the investigated acids were measured with TDMA technique at temperatures near the room temperature (295–301 K), and relative humidities above the deliquescence points of the acids (63–80%, depending on the acid). The droplet sizes considered in this study were typically 80–130 nm. Droplets were generated by atomizing aqueous solutions of adipic and malonic acids, and an almost monodisperse size fraction was selected with a DMA. The selected droplets were allowed to evaporate in a laminar flow tube during well-defined time, and the decrease in particle size was monitored with a second DMA. Temperature and relative humidity were controlled throughout the experimental system. For detailed descriptions of the measurement setup see Bilde et al. (2003)[4] and Koponen et al. (2007).[5] A schematic picture of the experimental setup is presented in Figure 1.

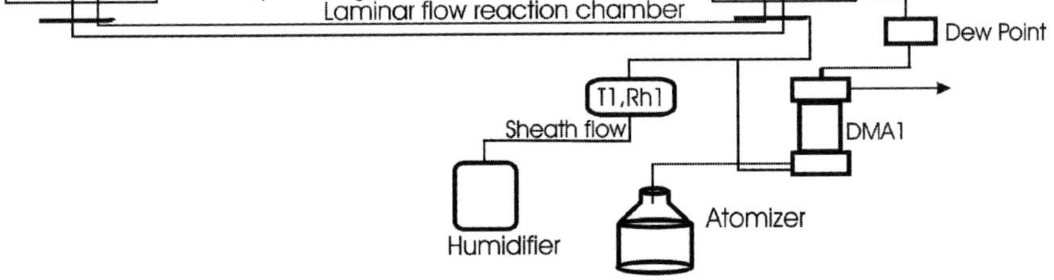

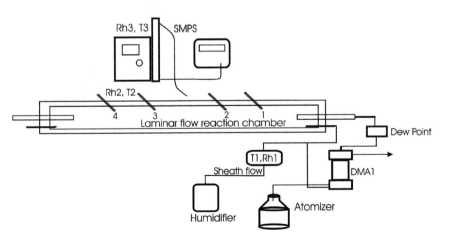

Figure 1 The experimental setup for measuring the evaporation rates of aqueous solution droplets

Evaporation Modelling

The saturation vapour pressures were derived from the evaporation rates by modelling the experimental data with a numerical condensation/evaporation model BCOND.[6] Model input parameters, such as relative humidities, temperatures and droplet concentrations, were obtained from the experiments. The mass and thermal accommodation coefficients were assumed to be unity. The UNIFAC Dortmund model[5] was used for the activity coefficient predictions. The saturation vapour pressure values for the pure subcooled acids were inferred from each data set (corresponding to different temperatures and relative humidities) by matching the modelled evaporation with the experimentally observed reduction in the particle diameter. A temperature dependent expression of form

$$\ln p_{sat}\ [\mathrm{Pa}] = A - \frac{B}{T} - C \ln T \tag{1}$$

was fitted to the epxerimental data, taking into account also the normal boiling points reported for the acids.

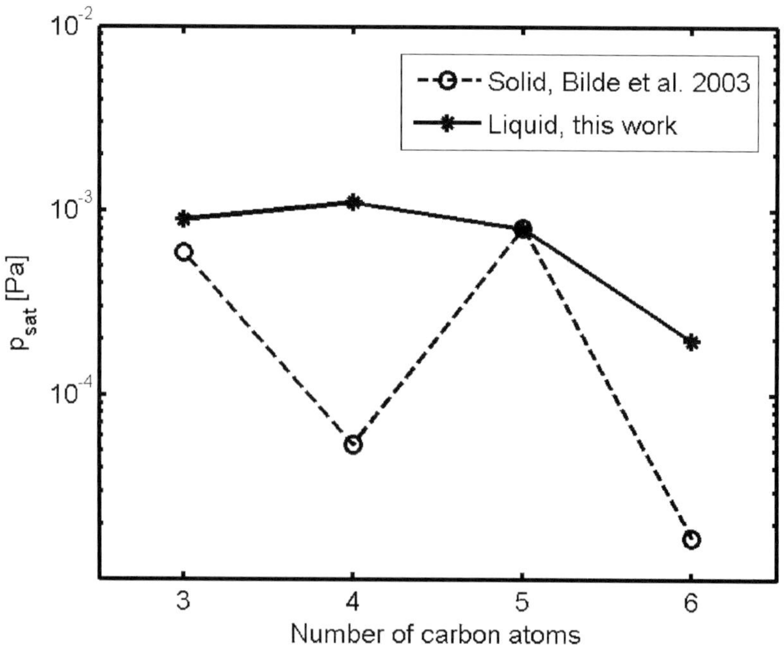

Figure 2 The subcooled liquid and solid state saturation vapour pressures of C3–C6 dicarboxylic acids at 299 K. **Solid curve**: subcooled liquid phase (this work); Dashed curve: solid phase.[4] Number of carbon atoms in the acids: 3 – malonic, 4 – succinic, 5 – glutaric, and 6 – adipic

Table 1 Parameters for the temperature dependent saturation vapour pressure expressions for subcooled liquid state dicarboxylic acids

Acid	*A*	*B*	*C*
Succinic (C4)	119.3281	16,278.443	12.576
Glutaric (C5)	125.7555	16,776.978	13.468
Adipic (C6)	140.3818	18,207.194	15.441

Results and Discussion

The obtained subcooled liquid vapour pressures at $T = 299\,K$ for the investigated C3–C6 dicarboxylic acids are presented in Figure 2 (solid curve). The solid state vapour pressures[4] are also presented for comparison (dashed curve). For all the investigated dicarboxylic acids, the obtained liquid state vapour pressures are higher than (even acids) or close to (odd acids) the solid state values. Also, the even–odd alternation visible for the solid state vapour pressures is not observed for the subcooled liquid phase. The parameters *A*, *B*, and *C* for the temperature dependent expression (Eq. 1) for the studied acid are presented in Table 1. We did not fit the temperature dependence for malonic acid, since currently we have studied the malonic acid droplets only at two temperatures.

Acknowledgment The authors kindly acknowledge BACCI for financial support.

References

1. Kanakidou, M. et al., *Atmos. Chem. Phys.*, **5**, 1053–1123 (2005).
2. Riipinen, I. et al., *Atmos. Res.*, **82**, 579–590 (2006).
3. Peng, C. et al., *Environ. Sci. Technol.*, **35**, 4495–4501 (2001).
4. Bilde, M. et al., *Environ. Sci. Technol.*, **37**, 1371–1378 (2003).
5. Koponen, I.K. et al., *Environ. Sci. Technol.*, submitted.
6. Vesala, T. et al., *J. Aerosol Sci.*, **28**, 565–598 (1997).

Aerosol Formation from Plant Emissions: The Jülich Plant Chamber Experiments

M. Dal Maso[1,2], T. Mentel[1], A. Kiendler-Scharr[1], T. Hohaus[1], E. Kleist[1], M. Miebach[1], R. Tillmann[1], R. Uerlings[1], R. Fisseha[1], P. Griffiths[3], Y. Rudich[4], E. Dinar[4], and J. Wildt[1]

Abstract We have performed measurements of particle formation and growth in a setup consisting of a plant and a reaction chamber, using live plants as well as an α-pinene source. The nucleation rates observed varied between 0.04 and 260 $cm^{-3}s^{-1}$, while the growth rates were 10–30 nm/h. We found that the formation and growth rates of particles increased with increasing amounts of carbon emitted by the plants, but there was significant variation between the plants. We have also modeled the formation of the aerosol using a continuously stirred tank reactor concept, and found that the basic physics and chemistry of the chamber are captured well.

Keywords Aerosol formation, nucleation rates, modeling

Introduction

Biogenic volatile organic compounds (BVOC) are major participants in atmospheric chemistry, both due to their large emissions and their influence on O_3 and OH radical budgets. In addition to affecting the oxidizing capacity of the atmosphere, their chemical degradation leads to the formation of semi- or nonvolatile compounds that considered to be a major source of tropospheric aerosol mass in the submicron size range. They also to play an important part in atmospheric particle formation events, in which new particles are produced from precursor vapors and then grow to larger sizes.[1] Atmospheric aerosol size distribution measurements have shown that such events, often involving biogenic organics, can be a significant aerosol source.[2]

[1] *Institut für Chemie und Dynamik der Geosphäre, Forschungszentrum Jülich, D-52425 Jülich, Germany*

[2] *Division of Atmospheric Sciences, Department of Physical Sciences, 00014 University of Helsinki, Finland*

[3] *Cambridge University, Center of Atmospheric Science, Cambridge, UK*

[4] *Weizmann Institute of Science, Department of Environmental Science, 76100 Rehovot, Israel*

C.D. O'Dowd and P.E. Wagner (eds.), *Nucleation and Atmospheric Aerosols*, 924–927.

Plant emissions, e.g., monoterpenes and sesquiterpenes, are a major source of VOCs in the troposphere.[3] BVOC emissions commonly depend on light and temperature.[3] So far most laboratory investigations on the potential to form secondary organic aerosols (SOA) from plant emissions focused on single VOCs such as α-pinene which is probably the most investigated monoterpene.[4] In this study we investigated the formation of SOA from the mixture of VOCs emitted by spruce, pine, and birch trees by oxidation with ozone and OH.

Materials and Methods

Chamber Setup and Measurements

The experiments were performed in the Jülich plant chamber in order to provide well-defined conditions for plants. The experimental setup consisted of two chambers, both approximately 1.5 m^3 in volume, which are constantly being stirred. The plants were put in one of the chambers and fed with CO_2. They were also irradiated in a 24 h cycle to cause emissions of VOCs.

A fraction of the air carrying plant emissions was transferred from the plant chamber to a second camber, the reaction chamber. The reaction chamber was constantly flushed with ozone, and was equipped with a UV-lamp. When the UV lamp was switched on, photolysis of O_3 produced OH-radicals and particle formation was initiated. This led to a short nucleation peak, producing a rapid increase in particle number. After the nucleation peak, no new particles were produced and the condensable vapors condensed on the existing particle surface.

VOC measurements were conducted both in the plant and reaction chamber with a Proton-Transfer-Reaction Mass Spectrometer (PTR-MS, IONICON),[5] to determine the emission kinetics, and an online-GC-MS system for compound identification. VOC mixing ratios were in the lower ppbv to pptv range. Identification by GC-MS was based on mass spectra and retention times of pure chemicals (Fluka/Aldrich, purity > 93%). The total number of particles formed was measured with an UCPC (TSI3025A). The size distribution of the aerosols in the chamber was measured by a TSI SMPS3936. The resulting SOA was analysed with an Aerodyne aerosol mass spectrometer (Q-AMS).

Measurement Results

The nucleation rates observed during the particle formation bursts varied between 0.04 and 260 $cm^{-3}\,s^{-1}$, derived from the UCPC data. These values are high compared to ambient observations which range 0.01–10 $cm^{-3}\,s^{-1}$.[1] The growth rates were of the order of a few tens of nm h^{-1}, calculated from the SMPS size distributions. They compare reasonably well with growth rates observed in midlatitudes with a range of 1–20 nm h^{-1}.[1]

The maximum SOA volume produced during VOC oxidation was used as the quantity determining the SOA formation potential. We compared these results to those obtained using α-pinene as single VOC. Spruce, pine, and birch were used as model plants representing the Boreal forest. Changing temperature in the plant chamber led to increase of the VOC emissions and furthermore, to an increase of the maximum SOA volumes in the reaction chamber. Plots of maximum SOA volumes versus the total amount of carbon fed into the reaction chamber led to approximately linear relationships.

Also the growth rates of particles showed a linear dependency of the total carbon entering the reaction chamber. The behaviour was quite similar for the pine, spruce, and α-pinene events, while birch produced more effectively growing particles. The nucleation rates also increased with increasing total carbon, but the dependency varied much more between the species. Again, birch was the most effective particle producer with respect to total carbon emitted.

Another interesting feature of the chamber events was their self-regulating nature. The chamber setup lead to a constant dilution of the particle concentration after the particles had been formed. As the particle concentration decreased, also the available condensation surface, or condensation sink,[6] decreased and in several cases, a new burst of nucleation can be observed; the nucleation rates for the later particles, however, were significantly lower than the first time around. The formed number size distributions were also notably wider.

Modeling

The chamber setup, with inflow of VOC-laden air form the plant chamber and outflow from the reactor chamber to the instrumentation, leads to dilution of the aerosol concentration over time, with the aerosol lifetime being of the order of 70 min. Due to this, a model taking this effect into account has to be used in evaluating the chamber data.

The chamber setup can be considered a continuously stirred tank reactor (CSTR[S1]). This means that we assume that there exists perfect mixing inside the tank volume, and the concentration in the reaction chamber are determined by the inflow and outflow rates. Changes in the inflow will result in changes in the chamber concentration after a time lag determined by the in- and outflow. A simple example of modeling an event can be seen Figure 1, where the monoterpenes emitted by a pine (represented by α-pinene in the model) is introduced to the reaction chamber. Some of it is oxidized by O_3, but no particles were seen. When UV is turned on, OH is produced (we assumed a steady-state OH concentration of $3.10^7 cm^{-3}$) and significantly more monterpene is oxidized. This results in a short burst of particle formation, which is then quenched, and the formed particles start to grow. The model predicts the gas-phase concentration of monoterpenes in the chamber from the measured plant chamber data quite well.

The aerosol was simulated with a simple sectional moving center model with only condensation taken in account. Assuming a accommodation coefficient of unity and a negligible vapor pressure for the condensing vapor, we found that 1.7% of the oxidized monoterpene was enough to explain the observed growth rate (14 nm/h in this case) of the particles.

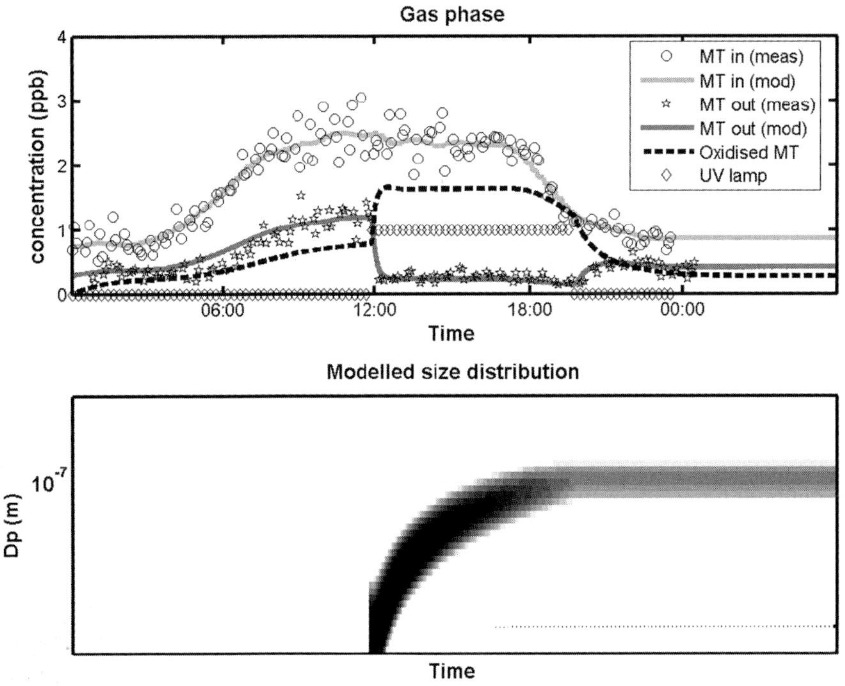

Figure 1 An example of modeling the gas-phase oxidation of VOC in the reaction chamber and the resulting growth in the size distribution. A steady-state OH field of $3 \times 10^7 cm^{-3}$ is assumed. **Top panel**: gas-phase concentrations of monterpenes at the inflow and outflow of the reaction chamber (measured and modeled), modeled oxidation products and the period when the uv lamp was on. **Lower panel**: the evolution of the modeled size distribution. Darker color signifies higher concentrations

The case shows that the basic chemistry and physics in the reaction chamber is captured well. The model serves as a good basis for future studies on more complex, realistic systems with secondary aerosol formation.

Acknowledgments This work was supported by the ACCENT EC-NETWORK OF EXCELLENCY and the ESF INTROP programme. Miikka Dal Maso would like to thank the Maj and Tor Nessling foundation for financial support.

References

1. Kulmala, M. et al., *J. Aerosol Sci.*, **35**, 143–176 (2004).
2. Tunved, P. et al., *Science*, **14**, 261–263 (2006).
3. Guenther, A. et al., *J. Geophys. Res.*, **100**, 8873–8892 (1995).
4. Saathoff, H. et al., *Aerosol Sci. Tech.*, 1297–1321 (2003).
5. Lindinger, W., Hansel, A., and Jordan, A., *Int. J. Mass Spec. and Ion Processes*, **173**, 191–241 (1998).
6. Kulmala, M. et al., *Tellus*, **53B**, 479–490 (2001).
7. Schmidt, Lanny D., *The Engineering of Chemical Reactions*, New York: Oxford University Press (1998).

Upward Fluxes of Particles over Forests: When, Where, Why?

S.C. Pryor[1,2], R.J. Barthelmie[1,2,3*], L.L. Sørensen[2], and S.E. Larsen[2]

Abstract A substantial number of particle fluxes over forest are upwards. Analyses of data from a beech forest in Denmark indicate these "apparent emission" fluxes are frequently statistically different from zero flux and are not solely observed during periods when other micro-meteorological fluxes are ill-defined, which implies that they derive from a/multiple physical cause/s. Upward fluxes are slightly more frequent at night, but do not appear to be dependent on wind direction or speed. The rate of upward fluxes (emission velocity) scales with prevailing geometric mean diameter (GMD) but not with changes in GMD.

Keywords Apparent emission fluxes, forests, atmosphere–surface exchange

Research Objectives

Particle number fluxes over forests are typically downwards (negative), but multiple previous studies have indicated the presence of a substantial number of positive (upward) fluxes [1,2]. Upward particle number fluxes at the Hyytiälä forest have been ascribed to emissions from the research station [1], and/or formation of new particles near to or within the forest canopy [3], and/or random uncertainty of the fluxes [4]. Here we present fluxes from a beech (*Fagus silvatica L.*) forest along with analyses focused on identification of possible causes of upward ("apparent emission") fluxes.

[1]*Atmospheric Science Program, Department Geography, Indiana University, Bloomington, IN 47405, USA*

[2]*Department of Wind Energy and Atmospheric Physics, Risø National Laboratory, Roskilde, Denmark*

[3]*School of Engineering and Electronics, University of Edinburgh, Edinburgh, UK * = presenting author*

C.D. O'Dowd and P.E. Wagner (eds.), *Nucleation and Atmospheric Aerosols*, 928–932.

Table 1 Overview of instrumentation

Instrument	Sampling frequency	Sampling height
Condensation particle counter (TSI CPC 3010)	10 Hz	35 m
Scanning mobility particle sizer (TSI SMPS3936L25)	150 sec	43, 31, and 2 m
Aerodynamic particle sizer (TSI APS 3321)	150 sec	25 m

Site Description and Instrumentation

The measurements described herein were conducted at the CarboEuroFlux experimental forest site at Sorø in Denmark. The measurements presented here were collected during a field experiment conducted in May–June 2004 (see Table 1 and [2]).

Flux Calculation

Data from all half-hour periods when rain was observed or any instrument malfunction occurred were removed from the time series. The 10 Hz time series were then subject to despiking, detrending, and coordinate rotation, and used to compute particle number fluxes using eddy covariance. The resulting fluxes were subject to three corrections: the WPL correction to correct for the variation in particle concentrations due to density variations caused by fluxes of heat/water vapor [5], for the attenuation of CPC response at frequencies above 1 Hz [6], and for the influence of correlation of fluctuations in the saturation ratio with vertical wind speed [7].

Investigating Causes of Upward Fluxes

Over 40% of the resulting half-hour average particle number fluxes from Sorø are positive (upwards) (Figure 1). Many of the upward fluxes exceed the uncertainty bounds computed based on work by Wyngaard [8] using:

$$\sigma^2 = \frac{2\Im_c}{T}\left(\overline{(w'C')^2} - \overline{w'C'}^2\right) \tag{1}$$

Where σ is the inherent uncertainty, T is the integration time interval, and $\Im_c$ is the averaging time to determine the turbulence properties to a given accuracy ($\Im_c \approx$ z/<u>, where z is the effective measurement height, <u> is the mean wind speed).

As shown in Figure 1, 25% of measurement periods exhibited upward fluxes that lie beyond 1–σ of 0.

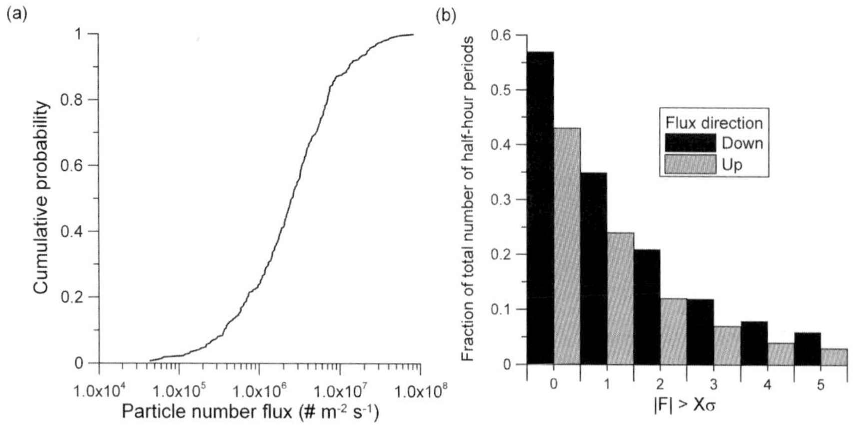

Figure 1 (a) Cumulative probability distribution of all upward half-hour number fluxes and (b) fraction of the total number of flux estimates those are upward or down (x-axis value of 0) and where the F exceeds the specified number of multiples (X) of σ

Periods with upward particle fluxes are also not associated with anomalous fluxes of momentum or heat (Figure 2a), and hence have a/multiple physical cause/s. To investigate this further we analyzed the half-hour average particle number concentrations and fluxes with respect to:

- Prevailing wind direction in 30° sectors. This analysis is focused on determining whether a local point or line ground-based source of particles is the cause of the upward fluxes.
- Hour of the day. This analysis is focused on determining whether the cause of the upward fluxes exhibits a diurnal cycle such as might be the case with stability or nucleation events.
- Wind speed. This analysis is focused on (indirectly) assessing whether there is evidence that frequency with which upward fluxes are observed is dependent on mechanical turbulence, or edge effects.
- Number geometric mean diameter (GMD). Particle diffusivity (and hence mobility) is a nonlinear function of diameter, additionally changes in the particle ensemble due to processes such as nucleation will be manifest by changes in prevailing GMD.

As shown in Figure 2b downward fluxes are more frequently observed than upward fluxes in all wind direction sectors except in the case of the ESE sector in which there are slightly more upward flux cases (180 periods versus 158 periods of downward flux). However, the relative uniformity of the fraction of upward fluxes with wind direction implies these fluxes are not related to a local ground-based particle emission source. Upward fluxes are observed in all hours of the day (Figure 2c) but are more abundant at night. There is little or no variability of the abundance of upward fluxes across a range of wind speed bins from 1 to 8 m s^{-1} (Figure 2d). Analysis of data with respect to the prevailing number GMD computed from the SMPS data is more ambiguous (Figures 2a, 3). The emission velocity (i.e., negative v_d)

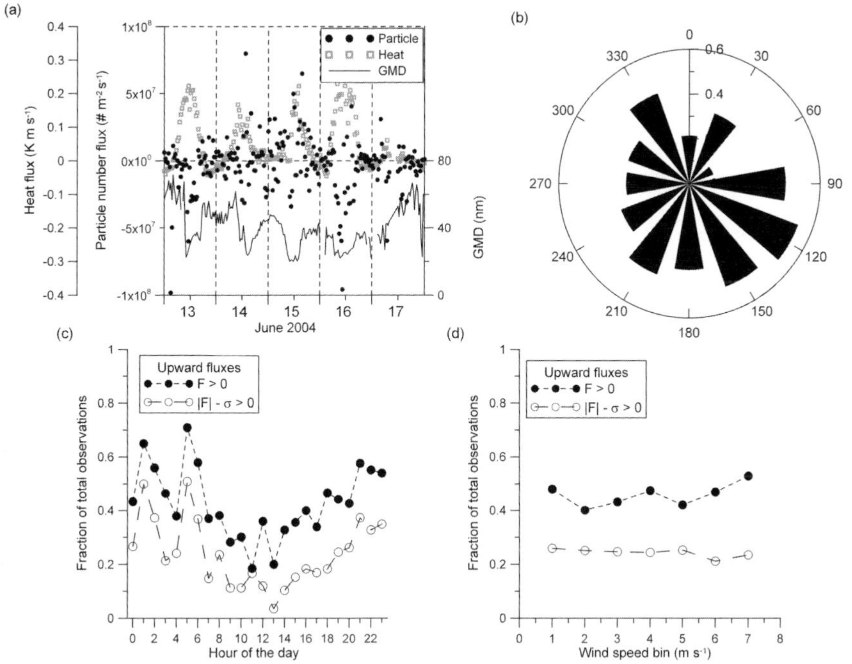

'igure 2 (a) Time series of particle flux, heat flux and prevailing GMD during a sample 5 day eriod. Half-hour average GMD were computed using the SMPS data and pertain to a height of 3 m. (b) The fraction of periods within 30° wind direction sectors when upward fluxes were bserved. (c) Fraction of the total number of observations in each hour of the day when upward luxes were observed (and when fluxes exceeded the associated uncertainty bounds). (d) As c) but for data binned by wind speed

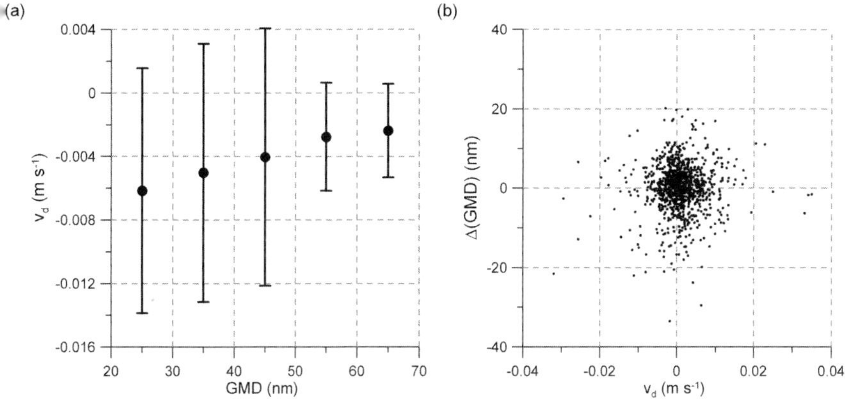

'igure 3 (a) Deposition velocities (v_d) during periods of upward fluxes (negative v_d) bin-averaged y prevailing GMD (the dot shows the mean, the error bars depict the standard deviation). (b) Ialf-hour average v_d plotted against the change in GMD from the previous half-hour period. Note: n (b) data are shown for all periods for which valid observations of Δ(GMD) and v_d exist not olely upward flux periods increases with decreasing GMD, but does not appear to be linked to changes i GMD from the previous half-hour period.

At the current time no definitive explanation for upward particle fluxes can b advanced. It may be that there are two or more mechanisms involved and that th relative importance of those mechanisms varies with (for example) hour of the day Given the prevalence of upward fluxes, further research is clearly warranted.

Acknowledgments This research was funded by grants from NSF (ATM 0334321 and ATM 0544745), the Nordic Centre of Excellence on Biosphere–Aerosol–Cloud–Climate Interaction (BACCI) and the EU funded EUCAARI and ACCENT projects.

References

1. Buzorius, G. et al., *J. Geophys. Res.*, **105**, 19905–19916 (2000).
2. Pryor, S.C. et al., *Environ. Poll.*, in press (2007).
3. Buzorius, G. et al., *J. Aerosol Sci.*, **29**, 157–171 (1998).
4. Gaman, A. et al., *J. Atmos. Oceanic Tech.*, **21**, 933–943 (2004).
5. Webb, E., Pearman, G. and Leuning, R., *Quarterly J. Royal Meteorological Society*, **106**, 85–100 (1980).
6. Horst, T., *Boundary-Layer Meteorology*, **82**, 219–233 (1997).
7. Kowalski, A., *Atmos. Environ.*, **35**, 4843–4851 (2001).
8. Wyngaard, J.C., On surface-layer turbulence, in: *Workshop on Micrometeorology*, edited b D.A. Haugen, Boston, MA: American Meteorological Society, pp. 101–149 (1973).

Observations of Winter-time Nucleation and Particle Growth over/in a Forest

S.C. Pryor[1,2] and R.J. Barthelmie[1,2,3]

Abstract This paper describes results from continuous measurements of particle size distributions at three levels in and over a deciduous forest in the Ohio River Valley. Here we focus on winter-time (leaf-off) periods with high number concentrations of ultrafine particles and show that nucleation events appear to occur above the canopy, and may be linked to the breakdown of the nocturnal inversion. The resulting ultrafine particles have relatively low growth rates of approximately 2 nm h^{-1}, and that the growth rate is apparently independent of incident radiation. Accordingly almost all of the increase in geometric mean diameter (GMD) can be attributed to coagulation.

Keywords Nucleation,dry deposition, nocturnal inversion, forest

Research Objectives

We are conducting continuous measurements of particle size distributions above and within a forest in the Ohio River Valley, and analyzing the data to quantify the frequency and characteristics of nucleation events and high ultrafine particle concentrations, including the chemical composition of the ultrafine particles, the principal mechanisms of nucleation, limitations on nucleation, and growth and the ultimate fate of the resulting ultrafine particles. In this analysis we focus on the following research hypotheses:

- Despite relatively high-average background particle concentration and hence condensational sink, nucleation events are observed on at least 10% of winter days.

[1] *Atmospheric Science Program, Department Geography, Indiana University, Bloomington, IN 47405, USA*

[2] *Department of Wind Energy and Atmospheric Physics, Risø National Laboratory, Roskilde, Denmark*

[3] *School of Engineering and Electronics, University of Edinburgh, Edinburgh, UK*

C.D. O'Dowd and P.E. Wagner (eds.), *Nucleation and Atmospheric Aerosols*, 933–937.

- Winter-time nucleation events are linked to the breakdown of the nocturnal inversion, and vertical transport of precursor gases and/or nucleated particles. Thus, nucleation occurs above the forest canopy and the resulting particles are subsequently turbulently transported to and through the tree corona.
- Growth rates during wintertime nucleation events are low relative to previously reported rates due to the low availability of condensable vapors.

Site Description and Instrumentation

Data presented herein are drawn from instrumentation deployed in the Morgan-Monroe State Forest (MMSF) in southern Indiana [1]. MMSF (39°19′ N, 86°25′ W), is an extensive deciduous forest dominated by sugar maple (*Acer saccharum*), tulip poplar (*Liriodendron tulipifera*), sassafras (*Sassafras albidum*), white oak (*Quercus alba*), and black oak (*Quercus nigra*), which has a total area of 95.3 km^2. This site is part of the AmeriFlux network and hence is equipped with a 45.7 m measurement tower. The tower is fully instrumented with three levels of eddy correlation systems for heat, momentum, water, and carbon dioxide fluxes and a full suite of profiles of radiation, temperature, and relative humidity. A schematic of the particle sampling system as deployed at MMSF is shown in Figure 1.

As shown, air is drawn down three 50 m copper sampling lines to a manifold and from there is supplied to two scanning mobility particle sampler (SMPS) systems:

- SMPS1: Comprising an EC 3080, long-DMA (3081) and CPC 3025A. The range of particle sizes quantified is O(10–400 nm).
- SMPS2: Comprising an EC 3080, nano-DMA (3085), and UCPC 3786 (water-based). The range of particle sizes quantified is O(3–100 nm).

Air is drawn sequentially from each level using a switching system (and the exhaust returned to that manifold to prevent pressure fluctuations) on a rotation of 10 min, such that each height will be sampled by both SMPS systems within a 30 min period. Sampling heights are 46, 28, and 6 m, i.e., above canopy, at the canopy top and below the canopy in the trunk-space. The control system was designed such that, as shown in Figure 1, the sampling system is accessible by remote access via use of Symantec pcAnywhere software and the Internet.

The sampling lines and flow rates were selected to maintain a Re ≤ 2,000 and hence avoid turbulence and reduce particle losses. Nevertheless tubing losses (or incomplete particle transmission through the sampling lines) are a major issue with experimental setups such as ours. In February 2007 an experiment was conducted to estimate sampling losses due to the experimental setup. Sampling losses are estimated using transport efficiencies derived experimentally by switching between sampling air directly into the inlet and through the tubing. All particle concentrations presented herein have been corrected for tubing losses. Note that because of the efficient scavenging of 3–5 nm particles, in practice our system samples only particles with diameters greater than approximately 5 nm.

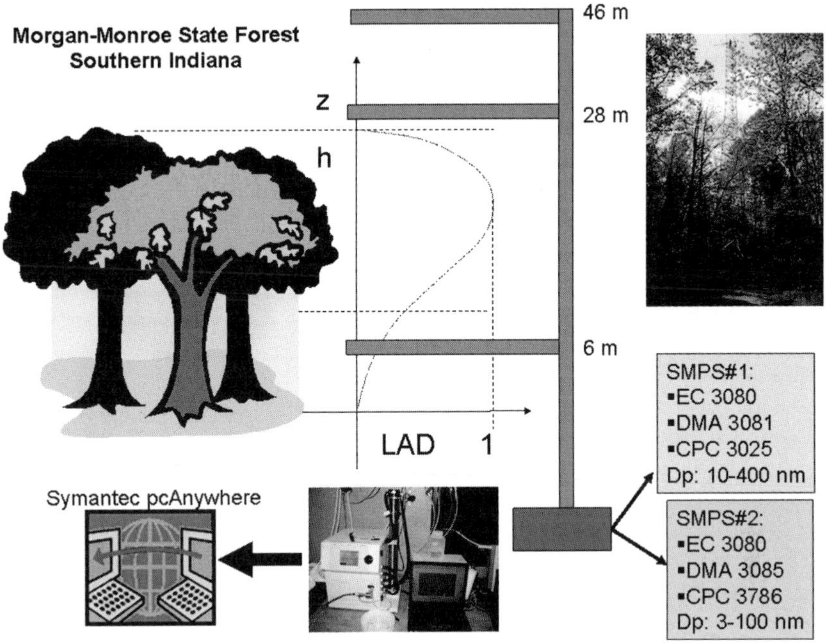

Figure 1 Design of the particle sampling system at MMSF

Data Analysis

We have observed multiple nucleation "event" days (defined as an increase in total particle concentrations by at least a factor of five concomitant with a reduction in geometric mean diameter (GMD)) since the initiation of sampling in October 2006, but for the sake of brevity here present analysis of a single day – 15 December 2006 (Figures 2 and 3) that is representative of winter-time events at this site. As shown in Figure 2, there is relatively good closure of particle size distributions and total particle counts between the two independent SMPS systems. The following further assertions may also be drawn:

- Size distributions and total number concentrations indicate highest ultrafine particle concentrations are observed at the 46 m level. For example, the ratio of total particle concentrations at 46 m relative to 28 m was approximately 1.2 during 12–1 pm on 15 December, and the ratio of 10 nm diameter particles at that time was over 1.6 (Figure 2). Increases in ultrafine particle concentrations are first observed at 46 m. Under the presumption of horizontal homogeneity, these results imply an elevated source of ultrafine particles.
- At the start of an event net radiation (NR) levels are below the threshold of 300 W m^{-2} that has been proposed as characterizing nucleation events, but during the period of highest observed particle concentrations NR > 400 W m^{-2} (Figure 3).

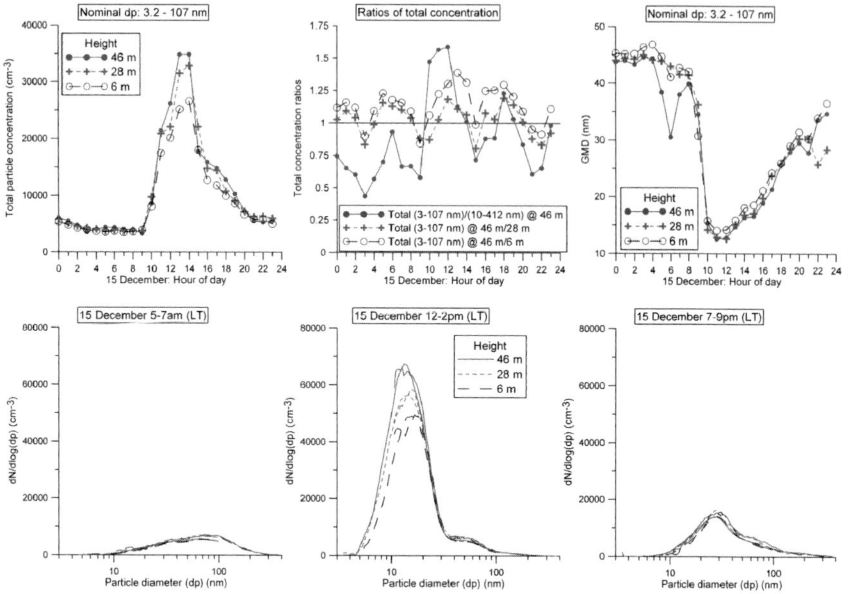

Figure 2 Observed particle concentrations on 15 December 2006. The upper left hand panel shows hourly average total number concentrations at the three-sampling levels from SMPS2. The upper middle panel shows ratios of hourly average total number concentrations from the two SMPS systems, and at the different sampling levels as observed using SMPS2. The upper right hand panel shows the number GMD at the three sampling levels computed using data from SMPS2. The lower panels show the average particle size distributions from the two SMPS systems during 5–7 am, 12–2 pm, and 7–9 pm

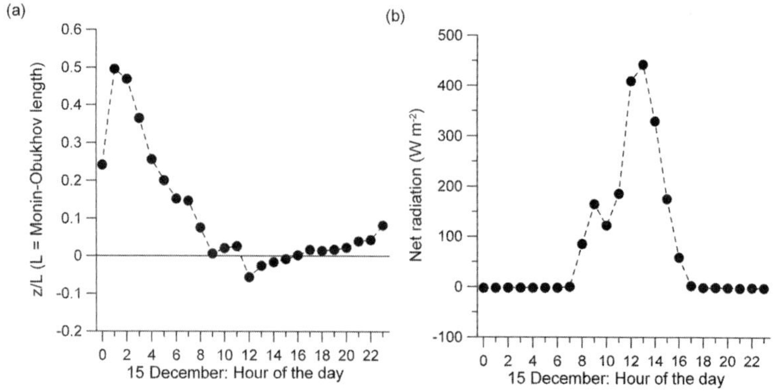

Figure 3 (**a**) Stability conditions during 15 December 2006 as characterized by the ratio of effective measurement height (z) to Monin–Obukhov length (L). (**b**) Hourly average net radiation (W m^{-2})

- Initiation of increased particle concentrations appears to be coincident with a transition from highly stable conditions over night toward unstable conditions. This transition is consistent with erosion of a nocturnal inversion, and vertical transport of precursor gases and/or nucleated particles (Figure 3).
- Growth rates observed at MMSF during the wintertime are at the lower end of previously reported growth rates (which are typically in the range of 1 and 20 nm h^{-1} except in coastal locations (where higher rates have been observed) and in clean polar regions (where lower rates have been observed) [2]). For example, in the case of the 15 December event the GMD from SMPS2 increased from 12.7 nm at noon to 29.5 nm at 8 pm, which implies a growth rate of 2.1 nm h^{-1} (Figure 2). Further, no change in growth rate is observed with decreases in net radiation. Coagulation can account for the majority of the observed increase in GMD which implies the production of condensable vapors was relatively low, in keeping with low biogenic VOC emissions during tree senescence.

Acknowledgment Grants from NSF (ATM 0334321 and ATM 0544745) and the EUCAARI project.

References

1. Pryor, S.C., Barthelmie, R.J., and Jensen, B., Nitrogen dry deposition at an AmeriFlux site in a hardwood forest in the MidWest, *Geophys. Res. Lett.*, **26**(6), 691–694 (1999).
2. Kulmala, M. et al., Formation and growth rates of ultrafine atmospheric particles: a review of observations, *J. Aerosol Sci.*, **35,** 143–176 (2004).

Ion-mediated Nucleation as an Important Source of Global Tropospheric Aerosols

Fangqun Yu[1], Zifa Wang[2], and Richard P. Turco[3]

Abstract Aerosol nucleation events have been observed at a variety of locations worldwide, and may have significant climatic and health implications. While ions have long been suggested as favorable nucleation embryos, their significance as a global source of particles has remained uncertain. Here, an ion-mediated nucleation (IMN) mechanism, which is supported by recent measurements of the overcharging of newly formed particles, has been integrated into a global chemical transport model to study ion mediated particle formation in the global troposphere. We show that predicted annual spatial patterns of particle nucleation agree well with land-, ship-, and aircraft-based observations throughout the troposphere. The ratio of particle number annual source strength due to IMN to those associated with primary particle emission suggests that IMN contribution is significant. This analysis represents the first confirmation of a significant, predictable global-scale source of new tropospheric aerosols.

Keywords Ion-mediated nucleation, tropospheric aerosols, global modeling

Ion-Mediated Nucleation

New particle formation frequently observed at various locations worldwide is an important source of atmospheric aerosols. Measurements indicate that H_2SO_4 and H_2O are clearly involved in many nucleation events observed in the atmosphere. Due to the difficulty of H_2SO_4–H_2O binary homogeneous nucleation (BHN) theory in explaining the observed particle formation rate, it has been

[1]*Atmospheric Sciences Research Center, State University of New York, Albany, New York*

[2]*NZC/LAPC, Institute of Atmospheric Physics, Chinese Academy of Sciences, Beijing, China*

[3]*Department of Atmospheric and Oceanic Sciences, University of California, Los Angeles, California*

C.D. O'Dowd and P.E. Wagner (eds.), *Nucleation and Atmospheric Aerosols*, 938–942.

proposed that others species including ammonia,[1] ions,[2] and certain organic compounds[3] may enhance the H_2SO_4–H_2O nucleation. Analysis of laboratory measurements shows that the contribution of ternary H_2SO_4–H_2O–NH_3 homogeneous nucleation (THN) to the formation of new particles is also likely to be very limited.[4] While certain organics have been shown to be able to enhance H_2SO_4–H_2O nucleation in the laboratory,[3] the level of enhancement is similar to that of ammonia, and the significance of organics in enhancing nucleation rates in the atmosphere is unclear.

Ions, which are generated continuously and ubiquitously in the atmosphere by cosmic radiation and radioactive decay, have long been known to promote nucleation. However, the recent availability of detailed field data on nanometer-sized particles during aerosol nucleation "events" led Yu and Turco[2,5,6] to develop a more comprehensive approach for studying nucleation processes involving ion clusters. They utilized a kinetic model that explicitly treats the complex interactions among small air ions, neutral and charged clusters of various sizes, precursor vapor molecules, and the preexisting aerosol. Compared to homogeneous nucleation, which involves the formation of small, transient neutral molecular clusters, nucleation onto ions is favored because: (1) small charged clusters are typically more stable thermodynamically than their neutral counterparts; (2) the initial growth rates of small ion clusters are enhanced by the dipole–charge interaction between the core ion and the strongly dipolar condensing molecules; and (3) there is a continuous and ubiquitous supply of stable, fast growing ionic embryos. Further, the properties of ions are well determined by extensive laboratory studies, unlike the situation with respect to their neutral counterparts. Yu and Turco[2] refer to the coupled formation and evolution of aerosol size distributions, including both charged and neutral clusters, under the influence of ionization, recombination, neutralization, condensation, evaporation, coagulation, and scavenging as ion-mediated nucleation (IMN). It is clear that the IMN mechanism is different from classical ion-nucleation theory[7,8] which is based on a simple modification of the free energy associated with the formation of a "critical nucleation embryo".

Our understanding of the role of ions in atmospheric nucleation has advanced recently, both from experimental and theoretical points of view. Experimental progress includes measurements of the mobility spectra of atmospheric cluster ions beginning at molecular sizes, as well as the charged fraction of freshly nucleated nanoparticles.[9–12] The theoretical progress includes the quantification of the role of dipole–charge interaction in ion clustering thermodynamics[13–16] and the development of up-to-date IMN models incorporating recently available thermodynamic data and schemes.[17] The long-term (multiple years) measurements taken in Hyytiala, Finland indicate that ions are involved in more than 90% of the particle formation days that can be clearly identified.[11,12] While Laakso et al.[11] concluded that the average contribution of ion nucleation to total nucleation rate is small, we find, based on a conservative analytical analysis and detailed kinetic simulations, that the same measurements presented in Laakso et al.[11] may actually indicate the dominance of IMN in the observed nucleation.

Global Simulation of Ion-Mediated Nucleation

To study particle nucleation in the global atmosphere, we couple our nucleation sub-module to GEOS–Chem model which is a global 3D model of atmospheric composition driven by assimilated meteorological observations from the Goddard Earth Observing System (GEOS) of the NASA Global Modeling Assimilation Office (GMAO). Meteorological fields include surface properties, humidity, temperature, winds, cloud properties, heat flux, and precipitation. For the results presented in this paper, the GEOS-3 grid with 2° × 2.5° horizontal resolution and 30 vertical levels was used. The first 15 levels in the model are centered at approximately 10, 50, 100, 200, 330, 530, 760, 1,100, 1,600, 2,100, 2,800, 3,600, 4,500, 5,500, and 6,500 m above surface. A detailed description of the model (including the treatment of various emission sources, chemistry and aerosol schemes) can be found in the model webpage (http://www.as.harvard.edu/chemistry/trop/geos/index.html).

In order to study aerosol nucleation in the context of 3D models, the nucleation calculations must be simplified or parameterized to reduce computing costs. We have developed an efficient IMN nucleation module based on nucleation rate lookup tables that are derived from detailed nucleation model simulations (8 Yu, 2006c). The IMN nucleation rate (JIMN) depends on the ambient sulfuric acid vapor concentration ($[H_2SO_4]$), temperature (T), relative humidity (RH), ionization rate Q, and the surface area of preexisting particles, S_0 (i.e., JIMN = f($[H_2SO_4]$, T, RH, Q, S_0). For a given set of conditions, JIMN is determined from the look-up table using an efficient multiple-variable interpolation scheme.

We run the GEOS-Chem coupled with nucleation model for 1 year from July 1, 2001 to June 30, 2002. The time step for transport is 15 min and for chemistry (and nucleation) is 60 min. We have compared our simulated annual spatial patterns of particle nucleation with land-, ship-, and aircraft-based observations relevant of particle formation and found that the agreement is quite well. The generally good agreement strongly supports the important role of IMN in generating new particles in global troposphere.

Figure 1 gives the ratio of annual mean IMN rates integrated in the lowest 3 km of atmosphere above surface (e.g., source strength due to IMN SS_{IMN0-3}, #/cm^{2}day) to the annual mean number emission of primary aerosols (e.g., source strength due to primary particle emission $SS_{primary}$, #/cm^{2}day). Figure 1 clearly indicates that IMN is a significant source of particle number in the lower troposphere. At high latitude (30N–90N, 30S-90S), SS_{IMN0-3} to $SS_{primary}$ ratio is above 300 over oceans and between 30 and 300 over land. In the tropic regions (30S–30N), SS_{IMN0-3}/$SS_{primary}$ is above 10 over Pacific Ocean and relatively small over continentals, Atlantic and Indian Oceans. It should be noted that the diameters of nucleated particles are only a few nanometer while those of primary particles are generally above 50 nm. The fraction of nucleated particles survives and is able to grow to CCN size depend on the growth rates (and hence precursor gas concentrations) and the concentration of preexisting particles. Pierce and Adams (2006) found that the probability of a nucleated particles generating a CCN varies from <0.1% to >90%

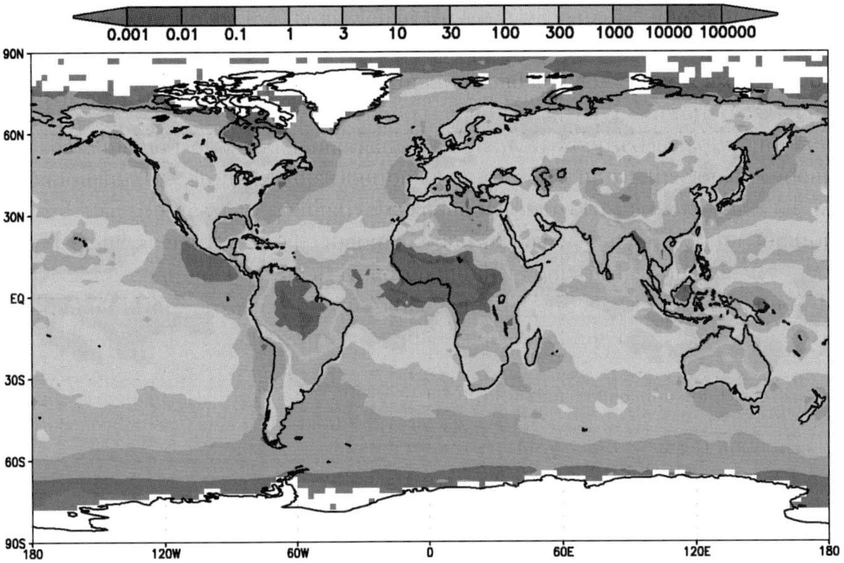

Figure 1 The ratio of annual mean IMN rates integrated in the lowest 3 km of atmosphere above surface (e.g., source strength due to IMN SS_{IMN0-3}, #/cm^{2}day) to the annual mean number emission of primary aerosols (e.g., source strength due to primary particle emission $SS_{primary}$, #/cm^{2}day). The primary aerosols considered in GEOS-Chem and the corresponding assumed (fixed) sizes (radius) are: dust (4 sizes: 0.7, 1.5, 2.5, 4 μm), sea salt (3 sizes: 0.732, 5.67 μm, and ultrafine sea salt mode with radius of 40 nm), black carbons (39 nm), and organic carbons (70 nm)

in different regions of the atmosphere and is between 5% and 40% for a large fraction of nucleated particles in the boundary layer. Clearly IMN is likely to be a significant source of particles that can have important climatic impacts.

Acknowledgments This study is supported by the NOAA/DOC under grant NA05OAR4310103 and NSF under grant 0618124. Z.W. acknowledges the support of National 973 Project (2005CB422205). The GEOS-Chem model is managed by the Atmospheric Chemistry Modeling Group at Harvard University with support from the NASA Atmospheric Chemistry Modeling and Analysis Program.

References

1. Weber, R.J. et al., *Chem. Eng. Comm.*, **151**, 53–62 (1996).
2. Yu, F. and Turco, R.P., *Geophys. Res. Lett.*, **27**, 883–886 (2000).
3. Zhang, R. et al., *Science*, **304**, 1487–1490 (2004).
4. Yu, F., *J. Geophys. Res.*, **111**, D01204, doi:10.1029/2005JD005968 (2006).
5. Yu, F. and Turco, R.P., *Geophys. Res. Lett.*, **24**, 1927–1930 (1997).
6. Yu, F. and Turco, R.P., *J. Geophys. Res.*, **106**, 4797–4814 (2001).
7. Hamill, P., Turco, R.P., Kiang, C.S., Toon, O.B., and Whitten, R.C., *J. Aerosol Sci.*, **13**, 561–585 (1982).
8. Raes, F., Augustin, J., and Vandingenen, R., *J. Aerosol Sci.*, **17**, 466–470 (1986).

9. Vana, M. et al., *Atmos. Res.*, **82**, 536–546 (2006).
10. Iida, K. et al., *J. Geophys. Res.*, **111**, D23201, doi:10.1029/2006JD007167 (2006).
11. Laakso, L. et al., *Atmos. Chem. Phys. Discuss.*, **7**, 1333–1345 (2007).
12. Hirsikko, A. et al., *Atmos. Chem. Phys. Discuss.*, **7**, 201–210 (2007).
13. Nadykto, A., Makela, J., Yu, F., Kulmala, M., Laaksonen, A., *Chem. Phys. Letts.*, **382**/1–2, 6 (2003).
14. Nadykto A. and Yu, F., *J. Geophy. Res.*, **108**, 4717, doi:10.1029/2003JD003664 (2003).
15. Nadykto, A. and Yu, F., *Phys. Rev. Letts.*, **93**, 016101 (2004).
16. Yu, F., *J. Chem. Phys.*, **122**, 084503 (2005).
17. Yu, F., *Atmos. Chem. Phys.*, **6**, 5193–5211 (2006).
18. Pierce, J.R. and Adams, P.J., *Atmos. Chem. Phys.*, **7**, 1367–1379 (2007).

Atmospheric Aerosol and Ion Characteristics during EUCAP (Eucalypt Forest Aerosols and Precursors)

Tanja Suni[1,4], Zoran Ristovski[2], Lidia Morawska[2], Alex Guenther[3], Andrew Turnipseed[3], Larisa Sogacheva[4], Markku Kulmala[4], Hannele Hakola[5], and Jaana Bäck[6]

Abstract We measured the characteristics and dynamics of atmospheric ions, aerosol particles, and their precursors in an intensive field campaign in a Eucalypt forest in Tumbarumba, South-East Australia, in November 2006. The measured size range of ions was 0.34 to 40 nm and that of aerosol particles approximately 10 to 168 nm, and for observing their size distributions we used an Air Ion Spectrometer (AIS) and a Scanning Mobility Particle Sizer (SMPS). We also measured the hygroscopic and chemical properties of the particles with a Volatility-Humidity Tandem Differential Mobility Analyser (VH-TDMA). The total concentration of ultrafine aerosol particles was measured with a Condensational Particle Counter (CPC). Furthermore, we measured ambient concentrations of volatile organic compounds (VOC), SO_x, NO_x/NO_y, and O_3. Finally, we modelled the 96 h back trajectories of air masses arriving at the site and observed that the arrival directions varied greatly and included trajectories that travelled only over land as well as ones that travelled most of the time over the ocean. On the most polluted day, the air masses arrived approximately from the direction of greater Sydney/Newcastle coal mine area. The total concentration of ultrafine aerosol particles was approximately 3,500 cm^{-3}, and daytime aerosol formation took place on 64% of days with acceptable data. The dominant VOCs were isoprene, eucalyptol, *a*- and *b*-pinene, camphene, and limonene. The measured hygroscopic growth factors (G_h) at RH of 90% varied from 1.1 to 1.5. The smallest G_h were observed for aged accumulation mode particles in early mornings, and the largest G_h occurred for the freshly nucleated particles on 10 November, the day with the highest concentration of SO_2.

[1] *CSIRO Marine and Atmospheric Research, GPO Box 1666, Canberra ACT 2601, Australia*

[2] *Queensland University of Technology, 2 George St., Brisbane 4000 QLD, Australia*

[3] *National Center for Atmospheric Research 1850 Table Mesa Drive Boulder, CO 80305, USA*

[4] *Department of Physical Sciences, P.O. Box 64, FIN-00014 University of Helsinki, Finland*

[5] *Finnish Meteorological Institute, Erik Palmenin aukio 1,FIN-00560 Helsinki, Finland*

[6] *Department of Forest Ecology, P.O. Box 27, FIN-00014 University of Helsinki, Finland*

C.D. O'Dowd and P.E. Wagner (eds.), *Nucleation and Atmospheric Aerosols*, 943–947.

Keywords Biogenic aerosol formation, atmospheric ions, VOC; Eucalypt forest, nucleation

Introduction

Determining the magnitude and driving factors of biogenic aerosol production in different ecosystems is crucial for future development of climate models. So far, most studies of biogenic aerosol production have taken place at continental and coastal sites in the Northern Hemisphere.[1] The EUCAP field campaign in November 2006 was a step towards understanding the process of biogenic aerosol formation from organic vapour precursors in a native Eucalypt forest in South East Australia. Our aim was to determine the characteristics and dynamics of atmospheric ions and charged particles from 0.34 to 168 nm and the composition, concentrations, and fluxes of biogenic VOC that react with atmospheric oxidants to form condensable vapours that, in turn, are able to form new aerosol.

Materials and Methods

The Tumbarumba flux station is located in a tall open Eucalypt forest in south eastern New South Wales (35° 39′ 20.6″ S 148° 09′ 07.5″ E, 1,200 m a.s.l). The dominant species are *E. delegatensis* (Alpine Ash) and *E. dalrympleana* (Mountain Gum), and the average tree height is 40 m. The instrument tower is 70 m tall and it is used to measure fluxes of heat, water vapour, and carbon dioxide. Supplementary measurements above the canopy include temperature, humidity, wind speed, wind direction, rainfall, incoming and reflected short-wave radiation, and net radiation.[2]

We measured the size distribution of air ions (naturally charged clusters and aerosol particles) with an Air Ion Spectrometer (AIS; Airel LTD, Estonia)[3] from 0.34 to 40 nm. The sum of neutral and charged aerosol particles from 10 to 168 nm was measured with a Scanning Mobility Particle Sizer (SMPS, TSI 3936). The volatile and hygroscopic properties were measured with a Volatility-Humidity Tandem Differential Mobility Analyser (VH-TDMA).[4] The AIS, SMPS, and VH-TDMA were located in a shed on the ground with inlet at the height of 2 m. The total concentration of ultrafine aerosol particles down to ~14 nm was measured with a condensational particle counter (CPC, TSI3010) at the height of 70 m on the tower. Ambient VOCs at the heights of 10–50 m were sampled in stainless steel adsorbent tubes filled with Tenax-TA and Carbopack-B to catch both monoterpenes and isoprene and analysed using GC-MS techniques.

New particle formation was classified as normal (ions grew from clusters all the way to large sizes, Figure 1); interrupted (ions only grew from cluster to intermediate sizes, Figure 1); Aitken (growth started from intermediate sizes and continued to large sizes, Figure 1)[5]; and nocturnal – sudden nocturnal appearance of large

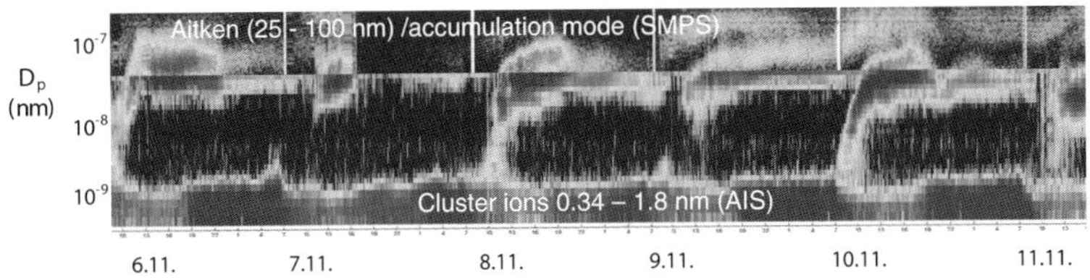

Figure 1 New particle formation measured with an AIS (0.34–40 nm) and an SMPS (figure shows 40–168 nm) during 6–11 November 2006. Three normal particle formation events took place on the 6th, 8th, and the 10th of November. Two interrupted events with subsequent Aitken mode growth occurred on the 7th and the 9th. In addition, growth in only the Aitken/accumulation mode was observable on the 9th and 10th. A less clear formation event took place on the 11th. *x*-axis: time at 3 h intervals, *y*-axis: particle diameter. Dark colour in the growth plumes indicates high and light colour indicates low concentration in logarithmic scale. However, the darkest colour on the background indicates very low concentration of particles

quantities of ions usually in all size classes at the same time but always at least in cluster and intermediate ions (details presented elsewhere[6]).

To determine the source and transport pathways of air masses arriving at Tumbarumba, we analysed back trajectories for the measurement period with the HYSPLIT_4 model, developed by NOAA/ARL.[7] The back trajectories were calculated 96 h backwards in time at a 70 m arrival height above ground level hourly from 0800 to 1600 local time to include the main part of most particle formation events.

Results and Discussion

Between November 2 and 29, the total concentration of ultrafine aerosol particles was approximately 3,500 cm^{-3}. A daytime formation event took place on 18 (64%) of days with acceptable data, and the strongest and clearest event took place on 10 November, the day with the highest SO_2 concentration (Figure 1). Nocturnal formation was observable on four nights. Only 2 days could be classified as entirely aerosol-formation free; the rest of the days were classified as undefined. The air mass arrival directions varied greatly and included trajectories that travelled only over land as well as ones that travelled most of the time over the ocean. On the day with most pollution (Nov 10), the air masses arrived approximately from greater Sydney/Newcastle coal mine area. The dominant VOCs during the campaign were isoprene, eucalyptol, *a*- and *b*-pinene, camphene, and limonene. On colder days, also para-cymene became an important monoterpene. The average midday concentrations at canopy level varied between 11 (*b*-pinene) and 1,011 ng m^{-3} (isoprene).

A detailed discussion on the volatile and hygroscopic behavior of nucleation mode particles is presented elsewhere.[8] In general, the measured hygroscopic growth factors (G_h) at relative humidity of 90% in the VH-TDMA varied from 1.1 to 1.5. The smallest G_h were observed for aged accumulation mode particles measured

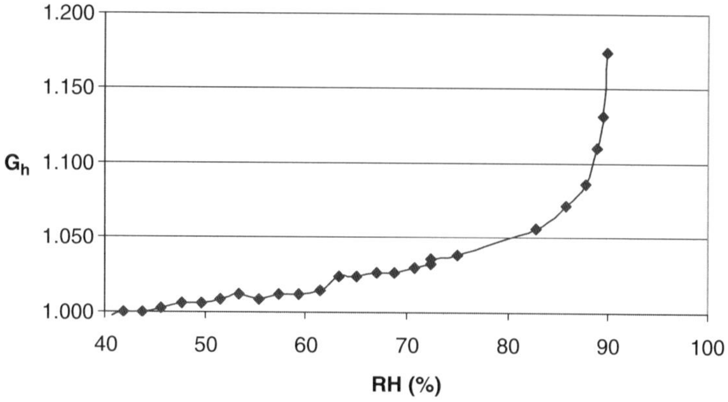

Figure 2 Dependence of the hygroscopic growth factor on relative humidity for a 32 nm particle

in the early mornings. The freshly nucleated particles showed G_h varying from 1.15 to 1.5. The largest G_h occurred on 10 November, the day with the highest concentration of SO_2 (800 pptv as compared to 200 pptv). This is in agreement with previous observations made in a pine forest in Hyytiälä, southern Finland, where a clear correlation between G_h and gaseous H_2SO_4 concentration was observed.[9] Once the nucleation mode has aged and moves into the Aitken mode, the measured hygroscopic growth factor decreases. On 10 November for example, G_h decreased from 1.5 to 1.4 in a period of 2 h. As the particles grow the amount of organic, less hygroscopic material increases, reducing G_h.

The dependence of G_h on the relative humidity in the VH-TDMA (deliquescence curves) of both nucleation mode particles and accumulation mode particles were also measured. All measured particles showed a continuous growth without exhibiting any deliquescence for RH in the range from 40% to 90%. Figure 2 shows such an example for a 32 nm particle measured several hours after a nucleation event on 8 November.

Acknowledgments This work was supported by the Academy of Finland, Maj and Tor Nessling Foundation, BACCI, and the Centennial Foundation of Helsingin Sanomat.

References

1. Kulmala, M. and Vehkamaki, H. et al., *J. Aerosol Sci.*, **35**(2), 143–176 (2004).
2. Leuning, R., Cleugh, H.A., Zegelin, S.J., and Hughes, D., *Agric. For. Meteorol.* **129**, 151–173 (2005).
3. Mirme, A. et al., *Bor. Env. Res.*, (submitted) (2007).
4. Johnson, G.R. et al., *J. Aerosol. Sci.*, **35**(4), 443–455 (2004).
5. Hirsikko, A., Laakso, L., Horrak, U., Aalto, P., Kerminen, V-M., and Kulmala, M., *Bor. Env. Res.*, **10**(5), 357–369 (2005).

6. Suni et al., Ions and Charged Aerosol Particles in a Native Australian Eucalypt Forest, *to be presented at the 17th International Conference on Nucleation & Atmospheric Aerosols* (2007).
7. Draxler, R. and Hess, G.D., *Aus. Met. Mag.*, **47**, 295–308 (1998).
8. Ristovski, Z.D. et al., Hygroscopic and Volatile Properties of Ultrafine Particles in the Eucalypt Forests: Comparison with Chamber Experiments and the Role of Sulphates in New Particle Formation, *to be presented at the 17th International Conference on Nucleation & Atmospheric Aerosols* (2007).
9. Ehn, M., et al., *Atmos. Chem. Phys.*, **7**, 211–222 (2007).

Revising the Fuchs "Boundary Sphere" Method

Vladimir Smorodin

Abstract The Fuchs "boundary sphere" method used in theory of aerosol coagulation and condensation in the transition Knudsen regime ($0.1 \leq Kn \leq 10$) is critically analyzed. Based on the Poisson-type model of Brownian motion of colliding particles and a "probabilistic theorem", we demonstrate that the Knudsen layer near the "sphere of influence" is characterized with a finite probability to find moving particles there, if space is isotropic. Therefore Fuchs' hypothesis about "vacuum" in the Knudsen layer is not physically correct.

Keywords Aerosols, coagulation, condensation, Fuchs' boundary sphere method, Knudsen's layer, nanoparticles, Poisson's distribution, transition flow regime

Introduction

Coagulation and condensation dynamics of nano-sized aerosols presents a great interest in modern nano-science and nano-technology. Here we are going to analyze critically theoretical fundamentals of coagulation theory of nanoparticles. For suspended in air particulates with radii $r \approx 1 \div 100\,\text{nm}$ corresponding Knudsen numbers at normal conditions are: $Kn = l / r \approx 0.1 \div 46.4$ (where l is the effective correlation length of moving particle in medium). This Knudsen range includes the transition flow regime ($Kn \in [0.1,10]$) between the diffusion and the kinetic one. In the transition regime a rigorous study of the particle dynamics requires solving Boltzmann's kinetic equation. Since this equation does not have a general analytical solution, many approximate approaches have been proposed to analyze aerosol transport processes. One of the most conventional is a famous Fuchs' method of a "boundary sphere".

Climate Change Institute, 316 Bryand Global Sciences Center, University of Maine, Orono, ME 04469–5790, USA

C.D. O'Dowd and P.E. Wagner (eds.), *Nucleation and Atmospheric Aerosols*, 948–952.

Concerning the "Boundary Sphere" Method (BSM) in Theory of Coagulation

By Fuchs' model, two coagulating particles may stick after reaching the "absorbing sphere" which equals sum of the particle radii, r. The mean distance, δ', from the surface of the sphere which is reached by the particles after covering a distance l depends on Kn. An imaginary spherical surface of radius $2r + \delta$ concentric with the absorbing sphere presents the "boundary sphere". *In the Knudsen boundary layer between them "the particles move as if in a vacuum, that is in straight lines with a mean thermal velocity,* $\overline{G}$" (Fuchs 1964, p. 291). Based on the continuum condition of the particle flux, after "sewing" in- and out- fluxes at the surface of the "boundary sphere" with the radius $2r + \delta$, a formula for the correction in the transition regime, $C_F = f(Kn)$, has been derived (Fuchs 1934, 1964). If $Kn << 1$, $C_F \approx (1 + 0.75Kn)^{-1}$, that is similar to Cunningham's correction formula. At the other opposite limit, $Kn >> 1$, $C_F \approx 2\sqrt{2}/\pi Kn$, which decreases if r increases, and finally it leads to the coagulation kernel under gas-kinetic conditions. Fuchs' correction describes the transition from diffusive to a gas-kinetic coagulation regime like Cunningham's correction expresses the transition from the hydrodynamic to the gas-kinetic regime of particle motion. The Fuchs approximation gives correct formulas of the flux onto the probe sphere both in the diffusion and kinetic regimes (at $Kn << 1$, and $Kn >> 1$, respectively). In a range of Kn ε {0.1 ÷ 10}, the Fuchs factor bridges both the Stokes-Cunningham slip factor (at $Kn \leq 0.1$) and the Hidy-Brock correction, at Kn ≥ 0.1. At $Kn \sim 1$ different variants of the "boundary" sphere method seems to give a total error of up to 10%.

The BSM was criticized and modified by Hidy and Brock (1965; 1970), Sedunov (1974), and others. In part, Smirnov (1969) found that a major correction contribution at $Kn >>1$ is given by not the "medium" collisions, at $r \sim l$, but by the closest ones. This fact contradicts to the BSM, and shows its non-applicability at large Kn, even the extreme asymptotic flux intensity, if $Kn \to \infty$, is determined correctly. Fuchs and Sutugin (1970) concluded that the simplified flux-matching method is not consistent with near-free molecule theory based on the Boltzmann equation. Analyzing Fucks' method, one may see that it contains two conceptual elements. First, it is a model of the "sphere of influence" which is based on a hypothesis about the "quasi-vacuum" properties of the Knudsen layer. Second, it is the techniques of "sewing" solutions of the limit theories and involving interpolation equations.

Critical Revising the Fuchs "Vacuum" Hypothesis

(A). Regarding Fuchs' model of the Knudsen layer around the "sphere of influence" in frameworks of the coagulation problem, one has to point out that the idea about a "quasi-vacuum" property of the boundary layer is related with an assumption about "inhomogeneity" of space near interfaces. Indeed, near the sphere of influence, space is treated as empty ("vacuum") in a range $R \leq r \leq \delta$, and as continuum if $r > \delta$. Such a "diluted" zone in the Knudsen layer might be explained with any repulsive forces around the "sphere of influence" treated as a "quasi-particle", and only in any condensed structured medium. In fact, in air, any virtual space anisotropy in the particle distribution will be quickly compensated with free diffusion motion. There is also no analogy between the concentration (temperature) jump in the condensation problem, and the space property around the sphere of influence in the coagulation problem; one cannot treat the Knudsen layer around the sphere of influence in coagulation as the "non-steady state zone". There is no necessity to suppose the "quasi-vacuum" layer around the sphere of influence, just because the reflected particle may lose its "memory" after the correlation length distance. The characteristic length can be applied to any space point continuously. In a case of "quasi-vacuum" zone the δ might be paradoxically doubled. Thus an assumption about inhomogeneity ("quasi-vacuum") properties of the Knudsen layer around the sphere of influence cannot be justified physically.

(B).Let us analyze another alternative: homogeneous space, with uniformly distributed ("spread") particles in air when a probability to meet a particle in a given point of space depends only on the particle space concentration, n. Let us imagine two identical spherical particles colliding from a distance $\delta \sim l$. These idealized particles cannot interact until colliding and next sticking or reflecting. In a general case, both particles can have different radii, but for simplicity, we investigate a case of particles with equal radii, $r_1 = r_2 = r$. A moving particle (2) can cover the correlation length, l, and reach a "probe" particle (1), by passing (a) the shortest way or (b) the longest way (touching each other tangentially).

What is the probability for the particle moving to the absorbing sphere to collide with other particles inside of the Knudsen layer, if space is regarded as isotropic? Since the coagulation is a stochastic process, for solving this probabilistic problem we can apply the Poisson distribution: $P(X) = e^{-\mu} \mu^X / X!$ Here $P(X)$ is the probability of X occurrences in an interval of space (or time), and μ is the average number (or expected value) of occurrences in the specified interval, a finite volume, V, at the given mean particle concentration, n. (Regarding a condensation problem, instead of particles, one has to consider vapor and gas molecules in air). Interpreting the quantities in the Poisson formula in a context of our problem, let us treat X as the number of collides of the moving particle with possible particles in the Knudsen layer, and μ as the average number of all particles inside the limited volume covered with the moving particle until its final colliding with the probe particle. At a given free particle path, l, a mean particle concentration in air can be estimated as $n \approx l^{-3}$.

After elementary geometrical calculations, the average number of all particles (or molecules of the Lorenz gas), μ, inside the cylindrical particle "track" of radius r between the moving particle and the probe particle of the same radius, can be presented as:

$$\mu_{\min} \leq \mu \leq \mu_{\max}, \tag{1}$$

$$\mu_{\min} = V_{\min} n \approx \pi r^2 l / l^3 = \pi (r/l)^2 = \pi / Kn^2, \tag{2}$$

$$\mu_{\max} = V_{\max} n \approx 2\pi r^3 \sqrt{(1+0.5l/r)^2 - 1} / l^3 = 2\pi\sqrt{(1+0.5Kn)^2 - 1} / Kn \qquad (3)$$

Averaging *P(X)* over μ one obtains:

$$\overline{P}(X, Kn) = \frac{1}{(\mu_b - \mu_a)} \int_{\mu_a}^{\mu_b} \frac{e^{-\mu}\mu^X}{X!} d\mu \qquad (4)$$

Such an integral can be calculated analytically for different values of *X*. First two integrals for *X* = *0* (no collisions) and *X* = *1* (one collision) can be easily calculated analytically. In part:

$$\overline{P(0, Kn)} = \frac{Exp(-\pi / Kn) - Exp\left[-2\pi\sqrt{(1+0.5Kn)^2 - 1} / Kn^3\right]}{2\pi\sqrt{(1+0.5Kn)^2 - 1} / Kn^3 - \pi / Kn} \qquad (5)$$

Then a complementary probability, $\overline{P(0, Kn)}$, to meet some amount of particles, is:

$$\overline{P(\sum_i X_i, Kn)} = 1 - \overline{P(0, Kn)} \qquad (6)$$

These results demonstrate that the Knudsen layer cannot be treated as "quasi-vacuum", as it is suggested by the Fuchs hypothesis. Such a derivation with a final conclusion, based on an analysis of the Poisson-type model of colliding particles, can be called as *the "probabilistic theorem"* about the "non-vacuum" properties of the Knudsen layer around the "sphere of influence".

Summary

Within a range of the "free path track" of the particle moving to the sphere of influence in Brownian coagulation, there is a finite probability to collide with any "foreign" particle inside the Knudsen layer, if space is homogeneous. Therefore, the Knudsen layer around the sphere of influence cannot be treated as "quasi-vacuum", by Fuchs' hypothesis. Since at any space range, including the free-path scale, a moving particle can meet another one with a finite probability, there is no criterion to divide a space as it is done in the BSM or any its modifications. Efficacy of the BSM is explained by two circumstances: (1) introducing Knudsen's number as a missed factor in both limit theories, and (2) the fact that by properly choosing a boundary layer one can fairly approximate any experimental results near intermediate regime. However nature, sign, and magnitude of corresponding errors arising from the BSM interpolation techniques cannot be established within the limits of such an idealized model.

Table 1 Numerical data of the probabilistic functions, *P(X, Kn)*

KnP(X = i)	P(0)	P(1)	P(2)	P(3)	P(4)	P(5)
0.10	–	–	–	–	–	–
0.25	<0.0001	<0.0001	<0.0001	<0.0001	<0.0001	<0.01
0.50	0.00059	0.00043	0.00160	0.00406	0.00792	0.01277
0.75	0.00154	0.00797	0.02140	0.04020	0.05983	0.07598
1.00	0.01090	0.04425	0.09350	0.13378	0.15970	0.15544
2.00	0.32314	0.33824	0.24766	0.12110	0.00445	0.01310
3.00	0.45876	0.35246	0.14037	0.03855	0.00818	0.00140
4.00	0.59403	0.30305	0.03598	0.01640	0.00254	0.00033
5.00	0.67728	0.25798	0.05500	0.00853	0.00105	–

References

Fuchs, N.A., *Z. Physik*. Bd. 89, H. **6**, 736–742 (1934); *Z. Phys. Chim.*, **171**, 34, 199–208 (1934).

Fuchs, N.A., *The Mechanics of Aerosols*, New York: Pergamon Press (1964).

Fuchs, N.A. and Sutugin, A.G., *Highly Dispersed Aerosols*, Ann Arbor and London: Ann Arbor Science (1970).

Hidy, G.M. and Brock, J.R., *J. Colloid Sci.*, **20**(6), 477–491 (1965).

Hidy, G.M. and Brock, J.R., *The Dynamics of Aerocolloidal Systems*, Oxford: Pergamon Press (1970).

Sedunov, Yu. S., *Physics of Drop Formation in the Atmosphere*, New York, Toronto, and London: Wiley (1974).

Smirnov, V.I., *Trudy TSAO*, **92** (1969).

Atmospheric Charged and Total Particle Formation Rates below 3 nm

Ilona Riipinen, Hanna E. Manninen, Tuomo Nieminen, Mikko Sipilä, Tuukka Petäjä, and Markku Kulmala

Abstract We have calculated total and charged particle formation rates at approximately 2 nm using data from CPCB, UF-02proto CPC pair, NAIS, and ion spectrometer measurements conducted at a boreal forest site in Hyytiälä, Finland. The charged particle formation rates are typically 10–100 times lower than the corresponding total particle formation rates, implying that ion-induced nucleation is not usually dominating in the investigated environment. However, the effect of charge on the activation properties of the particles is observed from the fact that a rise in the total 2 nm formation rate is typically preceded by a rise in the ion formation rate.

Keywords Atmospheric nucleation, aerosol formation and growth, nucleation rate

Introduction

Information on the very first steps of the atmospheric particle formation and growth is required in order to understand the nature of secondary atmospheric particle formation. The minimum particle size detected by the current commercial instruments is approximately 3 nm in diameter, but the atmospheric nucleation and cluster activation processes take place close to 1–2 nm[1] (mobility equivalent diameter). Air ions even smaller than 1 nm, on the other hand, can be detected with ions spectrometers such as the Air Ion Spectrometer (AIS) and Balanced Scanning Mobility Analyzer (BSMA), both manufactured by Airel Inc. The concentrations of the small ions, however, are usually so low that they cannot alone explain the observed new particle formation events.

Recently significant effort has been put in developing instruments that measure neutral particles even below 3 nm. In this study we use such instruments, namely CPCB (Condensation Particle Counter Battery),[2,3] UF-02proto[4] CPC pair and NAIS (Neutral Cluster Air Ion Spectrometer), to calculate estimates for atmospheric

University of Helsinki, Department of Physical Sciences, P.O. Box 64, FI-00014 University of Helsinki, Finland

C.D. O'Dowd and P.E. Wagner (eds.), *Nucleation and Atmospheric Aerosols*, 953–956.

particle formation rates close to 2 nm. The formation rates of 2–3 nm ions are also calculated for comparison, based on BSMA and AIS data. We aim to study the magnitude of the particle formation rates closer to the sizes where the real atmospheric nucleation and activation occurs, as well as investigate the relative contribution of neutral and charged particles to the particle formation.

The composition and therefore the activation properties of the freshly formed nucleation mode can be investigated using CPCB. The instrument consists of four CPCs, two with butanol (TSI3010 and TSI3025) and two with water (TSI3785 and TSI3786) used as the working fluid. In the case of highly water-soluble particles, the water CPCs detect them at smaller sizes than the instruments based on butanol. During new particle formation in boreal forest (Hyytiälä, Finland) the cut-off diameter of the CPCB has been estimated to be close to 2 nm.[2] The UF-02proto is a novel condensation particle counter, which, according to recent calibrations, has the potential to detect particles as small as 1.8 nm. NAIS is an electrical mobility spectrometer similar to AIS, but the aerosol sample is charged with a corona charger before the electrical detection.

Materials and Methods

The data studied in this work has been measured at the SMEAR II station (61°51′ N, 24°17′ E, 181 m a.s.l.)[5] in Hyytiälä, southern Finland, during 10 weeks in spring 2006. We have calculated the total formation rates of the 2 nm particles (J_2) from the particle concentrations measured by CPCB, UF-02proto pair and NAIS in the size range of 2–3 nm (N_{2-3}), using the formula

$$J_2 = \frac{dN_{2-3}}{dt} + CoagS_{2-3} \cdot N_{2-3} + \frac{GR_3}{1\text{ nm}} N_{2-3}, \tag{1}$$

where t is time, $CoagS_{2-3}$ refers to the coagulation sink for the investigated size range, and GR_3 is the growth rate our of the 2–3 nm range. Eq. (1) therefore takes into account the coagulation losses to the pre-existing particles, as well as the growth out of the size range. In the case of charged particles, also the ion–ion recombination and the charging of neutral particles were corrected for. The formation rate of 2–3 nm ions was therefore estimated as

$$J_2^{\pm} = \frac{dN_{2-3}^{\pm}}{dt} + CoagS_{2-3} \cdot N_{2-3}^{\pm} + \frac{GR_3}{1\text{ nm}} N_{2-3}^{\pm} + \alpha \cdot N_{2-3}^{\pm} N_{<3}^{\mp} - \beta \cdot N_{2-3} N_{<3}^{\pm}, \tag{2}$$

where $N^{\pm}_{2-3}$ refers to the 2–3 nm charged particle concentration, and $N^{\pm}_{<3}$ is the ion concentration below 3 nm, consisting mainly of cluster ions. α is the ion–ion recombination coefficient and β refers to the ion-neutral attachment coefficient.[5] The ion concentrations and the particle growth rates were obtained from the ion spectrometer data and the coagulation sink was calculated from the DMPS data. The values of α and β were 1.6–10^{-6} cm^3 s^{-1} and 0.01–10^{-6} cm^3 s^{-1}, respectively.[5]

Results and Discussion

Table 1 shows examples of average formation rates J_2 calculated for the different instruments during four particle formation events (23.4.2006, 24.4.2006, 30.4.2006, and 10.5.2006) in the investigated time series. On two of the days (24.4. and 10.5.) clear overcharging of the freshly formed aerosol was observed with an ion-DMPS, whereas one event (30.4.) was undercharged and one close to neutral (23.4.). In Figure 1, on the other hand, the J_2 values calculated for different instruments continuously during the particle formation event on 23.4.2006 are shown. As can be seen also from Table 1 and Figure 1, both the total formation rates as well as the charged particle formation rates calculated from independent experimental data are typically in quite a good agreement, giving us further confidence on the measured data.

Date	CPCB	UF-02proto pair	NAIS(−)	AIS(−)
23.4.2006	1.58	1.97	1.41	0.02
24.4.2006	0.55	0.72	0.96	0.07
30.4.2006	4.01	–	2.42	0.06
10.5.2006	0.52	–	0.70	0.06

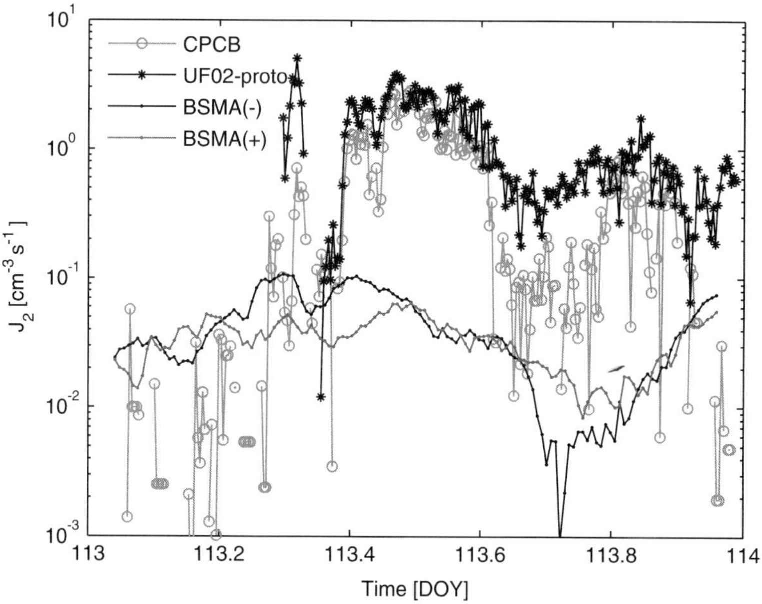

Figure 1 Comparison of 2 nm formation rates during new particle formation on 23.4.2006 in Hyytiälä, Finland. Grey line with circles: CPC-Battery; Black line with asterisks: UF-02proto CPC pair; Black line with dots: BSMA (negative ions); Grey line with dots: BSMA (positive ions)

First, looking at the average formation rates, it can be concluded that the charged particle formation rates are typically 10–100 times lower than the corresponding total formation rates, being relatively higher on days when aerosol overcharging is detected with the ion-DMPS as well. These numbers imply that ion-induced nucleation is usually not dominating in the investigated boreal forest environment.

Second, from the daily data it can be observed that the formation rates of 2 nm charged particles, particularly negative ions (see also Nieminen et al. 2007[7]), typically start to rise before the corresponding rise is seen in the total formation rates (Figure 1). This observation reflects the tendency of the ions to activate for growth easier than neutral particles of the corresponding size, i.e., lower concentrations of the vapour condensing on the particles is needed to growth them to the 2–3 nm size range.

References

1. Kulmala, M. et al., *Atmos. Chem. Phys.*, **6**, 787–793 (2006).
2. Kulmala, M. et al., *J. Aerosol Sci.*, **38**, 289–304.
3. Manninen, H.E. et al., *these proceedings.*
4. Mordas, G. et al., *Bor. Env. Res.*, **10**, 543–552 (2005).
5. Tammet, H. and Kulmala, M., *J. Aerosol Sci.*, **36**, 173–196 (2005).
6. Laakso, L. et al., *Atmos. Chem. Phys.*, **7**, 1333–1345 (2007).
7. Nieminen, T. et al., *these proceedings.*

An Algorithm for Automatic Classification of Two-dimensional Aerosol Data

Heikki Junninen, Ilona Riipinen, Miikka Dal Maso, and Markku Kulmala

Abstract Continuous measurements of aerosol size distribution are coming more and more common, but we are missing an objective way to classify and compare the data measured from different places, by different instruments and groups. In this work we are presenting an algorithm that can be used to classify aerosol size distributions in automatic manner from large data sets. First we emphasized the problem of separating automatically days with the regional new particle formation events (RNPF) and days without particle formation. The performance of the algorithm was verified against manually made classification, and a very good agreement was found. Although, the new algorithm can provide more complex classification here we present the results of classification into three classes, RNPF events, nonevents and the days that do not fit to previous two classes

Keywords Regional new particle formation, automatic classification, self-organizing map

Introduction

Atmospheric aerosols are long studied and well understood phenomenon. But still, the highest uncertainties in understanding the climate change are associated with atmospheric aerosols, and further, the mechanisms of the aerosol formation are not well established. Extensive measurement campaigns are important to understand the mechanisms behind the new particle formation, but only long-term measurements give us information on the event frequency and the means for evaluating its importance to climate change. This motivates research groups around the world to continuously monitor atmospheric aerosol size distributions.

Regional new particle formation (RNPF) appears with great regularity during the day time and very rarely, if ever, during the night. In order to observe particle

Department of Physical Sciences, University of Helsinki, P.O. Box 64, FIN-00014, Helsinki, Finland

C.D. O'Dowd and P.E. Wagner (eds.), *Nucleation and Atmospheric Aerosols*, 957–961.

formation and subsequent growth up to 100 nm the air masses have to remain the same for a long time (up to 12 h). For this reason one can analyze the continuous data a day at a time and classify days as RNPF event days and nonevent days. Dal Maso et al.[1] have developed a classification method for classification of days and the method is applied to data measured in many measurement stations, Hyytiälä, Värriö, Pallas, Aspvreten,[2] and PoVall.[3] The method is based on visual observation of size distributions, in order reduce subjective aspect of the result the analysis are performed in group of 3–4 persons. While being a powerful method and a good first attempt to analyze many data sets, it is very difficult to adapt to different group of people and to get comparable results, say between two different research groups. To address this problem a new automatic classification algorithm was developed.

Algorithm

The algorithm is based on self-organizing maps (SOM) and a decision tree. Self-organizing map is the most popular unsupervised artificial neural network algorithm. It solves difficult high-dimensional and nonlinear problems and makes projection into a low-dimensional space while keeping most of the information.[3] In this work we used SOM toolbox developed by Helsinki University of Technology.

Decision tree is simple combination of decision models with outcome of yes or no answers. The tree consists of branches and leaves, branches are connections between two decision model and leaves are the final stops of decision path. Finally the frequency of cases of specific class (in teaching data) in a leave gives probability for an unknown sample to be a member of this class. Decision tree is simple to understand and interpret. Differently from SOM decision tree is a supervised method, means that it needs some data with ready classification to learn the classification principles. After the teaching, it can be used to classify new data using the learned principles.

For the RNPF classification algorithm we taught five SOMs with different teaching data and one decision tree. The decision tree is taught with classification result from SOMs and traditional event classification.[1] For the final classification of unknown day we search the best matching unit (SOM neuron with smallest Euclidean distance) from each SOM and these Euclidean distances are feed to decision tree. From the final tree leave we get the probability for the day to belong to a class event, nonevent or undefined.

Method

For teaching the algorithm, we used 11-year aerosol size distribution data measured in Hyytiälä, Finland. The sampling site is rural site surrounded by boreal forest and closest city is around 80 km away. Aerosol size distribution was measured from

3–1,000 nm (around 30 size bins) and at 10 min time interval. For the classification the time resolution was decreased to 1 h and size resolution to 15 size bins. Size bins were equally spaced in log-space, with higher resolution bellow 20 nm (10 bins) and lower resolution in sizes above 20 nm (5 bins). Input data to classification algorithm was the whole day of size distribution measurements (15×24 matrix).

Data was divided randomly into three parts, teaching (3/5), validation (3/20), and testing (1/4) data. Over-fitting of the algorithm was avoided by monitoring errors in both, teaching and validation data.

We taught five SOMs in order to capture better the daily variation of aerosol size distributions. Combination of these five SOM gave better performance than using only one SOM. Only difference between the SOMs was the teaching data, each of them is tuned for a specific day type. The teaching data for the five SOMs were: mixture of all type of days, only event days, one nonevent days, only undefined days and last one was only events days but small <20nm sizes were weighted by factor of 10.

As an example of SOM, the one taught with only event days is presented in Figure 1. Each subplot represents a SOM neuron and can be considered as a prototype of an event day.

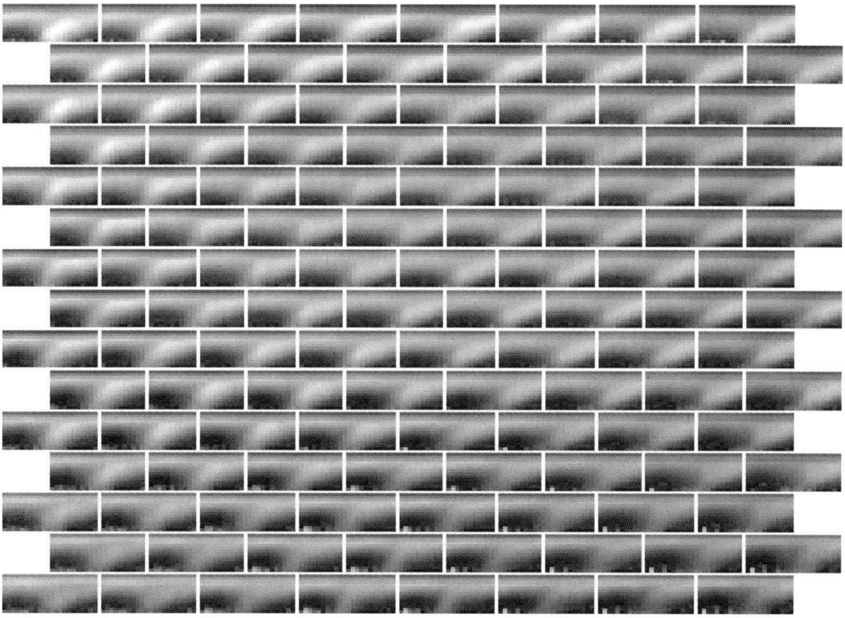

Figure 1 Self-organizing map of RNPF event days. Each small subplot represents a SOM neuron and can be considered as a prototype of the day. Color is equivalent to number concentration of aerosols, white is high and black is low concentration

Results

The new algorithm performed very well compared to traditional classification method (Table 1). The days with RNPF-event are clearly different from all other type of days and the performance of the new algorithm the best for these days. 92% of these days was classified similarly by the two methods. Less than 1% of the event days were classified as nonevent days and <8% as undefined days. When verifying the misclassified days manually we found that these misclassified days were problematic also for human to classify. These days were on the edge of being in one or in other class.

The biggest disagreement between methods was found in the class of undefined days. This class is actually left over class and not a real class, it contains great number of different type of days and there is no clear feature to describe this class.

Not only the number of days, but also the early distribution of the events and non-events was in good agreement (Figure 2).

Conclusions

If considering all the data over all performance of the new algorithm was 79.6% days agreed with the traditional method. If consider only event and nonevent days the two methods agree on 99.1% of the cases. This assures that newly developed

Table 1 The new classification compared to the traditional method. Percentage of days in years 1996–2006

	Traditional method		
New algorithm	Event %	Nonevent %	Undefined %
Event	92	0.8	13
Nonevent	0.7	79	15
Undefined	7.3	20	72
Total	100	100	100

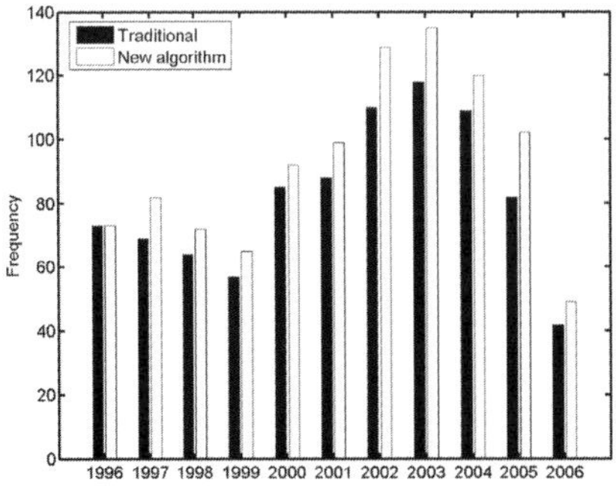

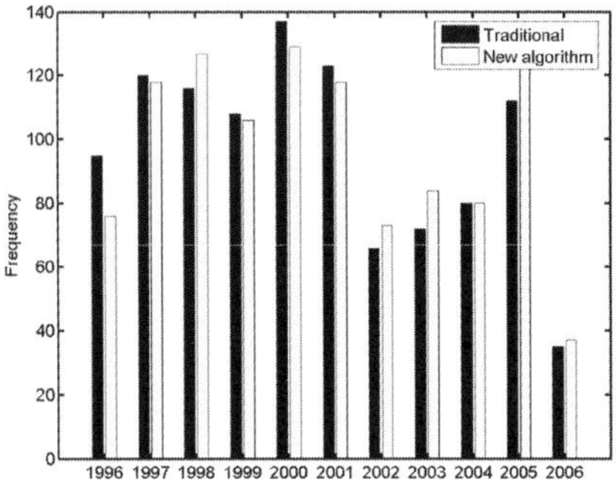

Figure 2 Comparison of frequency of events (**left**) and non-events (**right**) classified by traditional method[1] and by new proposed algorithm. Year 2006 is only from January–June

algorithm gives virtually the same classification result as traditional method, but makes it faster and without subjective opinion of people. The proposed algorithm is also easily adoptable to different sampling sites and even different instruments.

Addition to classification, the SOM can be also used for studying features in the data. From Figure 1 we can see clearly different types of event, e.g., events with high or low nucleation mode number concentration, different starting time of the events, different background number concentration. The utilization if property of SOM will be further investigated in near future.

References

1. Dal Maso, M., Kulmala, M., Riipinen, I., Wagner, R., Hussein, T., Aalto, P.P., Lehtinen, K.E., J. *Boreal Environ. Res.*, **10**, 323–336 (2005).
2. Dal Maso, M., Sogacheva, L., Aalto, P.P., Riipinen, I., Komppula, M., Tunver, P., Korhonen, L., Suur-Uski, V., Hirsikko, A., Kurten, T., Kerminen, V.M., Lihavainen, H., Viisanen, Y., Hansson, H.C., and Kulmala, M., *Tellus*, in Press.
3. A. Hamed, Joutsensaari, J., Mikkonen, S., Sogacheva, L., Dal Maso, M., Kulmala, M., Cavalli, F., Fuzzi, S., Facchini, M.C., Decesari, S., Mircea, M., Lehtinen, K.E.J., and Laaksonen, A., *Atmos. Chem. Phys.*, **7**, 355–376, (2007).
4. Kohonen, T., *Self-Organizing Maps*. Series in Information Sciences, Vol. 30, Heidelberg: Springer, 2nd edn. 1997 (1995).

Capturing the Effect of Sulphur in Diesel Exhaust

M. Lemmetty[1], L. Pirjola[2,3], E. Vouitsis[4], and J. Keskinen[1]

Abstract A set of actual dynamometer measurements of a heavy-duty diesel vehicle with different lubricant oils and fuels is reproduced using Tampere University of Technology Exhaust Aerosol Model (TUTEAM), and the AEROFOR model. TUTEAM, a sectional model considering the evolution of nucleation and soot mode separately, was specifically created for this application and is presented in the article. TUTEAM and AEROFOR are capable of reproducing measurement results with good accuracy. The predicted particle composition is in agreement with the measurements of Schneider et al.[1]

Keywords Heterogeneous exhaust, nucleation mode, aerosol modelling, dilution

Introduction

Diesel emissions are on of the most important sources of particulate matter in urban environment. The diesel aerosol consists of two modes: a solid, fractal-like soot mode formed during combustion, and a liquid, volatile nucleation mode which forms during the dilution of aerosol.

The formation of the nucleation mode takes place rapidly during the dilution, the processes affecting the formation nucleation mode being condensation, nucleation, and coagulation. Although there is rather large amount of experimental data, the

[1]*Institute of Physics, Tampere University of Technology, P.O. Box 629, FIN-33101 Tampere, Finland*

[2]*Department of Technology, Helsinki Polytechnic Stadia, P.O. Box 4020, FIN-00099 City of Helsinki, Finland*

[3]*Department of Physical Sciences, University of Helsinki, P.O. Box 64, FIN-00014 University of Helsinki, Finland*

[4]*Laboratory of Applied Thermodynamics, Aristotle University of Thessaloniki, P.O. Box 458, GR-54124 Thessalonki, Greece*

C.D. O'Dowd and P.E. Wagner (eds.), *Nucleation and Atmospheric Aerosols*, 962–965.

quantitative understanding of the nucleation mode formation is still lacking. The existing models either consider only a single mode[2] or concentrate on studying two monodisperse particle modes.[3] In this work, two models are presented and used to reproduce a set of measurement data.

The Tampere University of Technology Exhaust Aerosol Model (TUTEAM) is a sectional model which considers two externally mixed modes, namely the soot mode and the nucleation mode, tracking their time development and size-dependent chemical compositions separately. The interaction of the two modes via inter-mode coagulation is accounted for. The AEROFOR[4] model is a sectional model developed for atmospheric aerosols which considers a single particle number size distribution. The chemical models used in AEROFOR are updated for this study to reflect the situation in diesel exhaust.

Modelling

The experimental data was selected from the measurements of Vaaraslahti et al. (2005).[5] The data was a set of SMPS measurements using a porous tube dilution system. The measurements were conducted on a EURO II heavy-duty diesel engine and Catalytic Regenerative Trap (CRT) aftertreatment. Five different lubricant sulphur–fuel sulphur concentration combinations were measured at 100% load. In all cases, a nucleation mode attributed to sulphur was observed.

The models used the simple flow model,[6] where the dilution process is described by two time constants: one for dilution and one for cooling. The chemical composition of the exhaust gas was calculated from the stoichiometry of the situation, assuming 100% conversion of sulphur to sulphuric acid. The dilution time constant was held constant at all runs, while the cooling time constant was adjusted to fit the measurements, separately for TUTEAM and AEROFOR. This adjusting parameter was fitted for a single measurement point and used for all measurement points.

Results and Discussion

Both TUTEAM and AEROFOR are capable of capturing the trends exhibited by the nucleation mode and reproducing the measured size distribution (Figure 1). The sulphuric acid–water nucleation mechanism and subsequent condensation and coagulation are sufficient to explain the development of the nucleation and soot mode observed in the experiment. The semivolatile organic component concentrations needed to affect the particle size distribution in the models exceed the organic gas concentration measured in the experiment. The simulation results indicate that the S(IV)-to-S(VI)-conversion efficiency of the CRT is somewhat lower at large fuel sulphur contents than at the lowest ones.

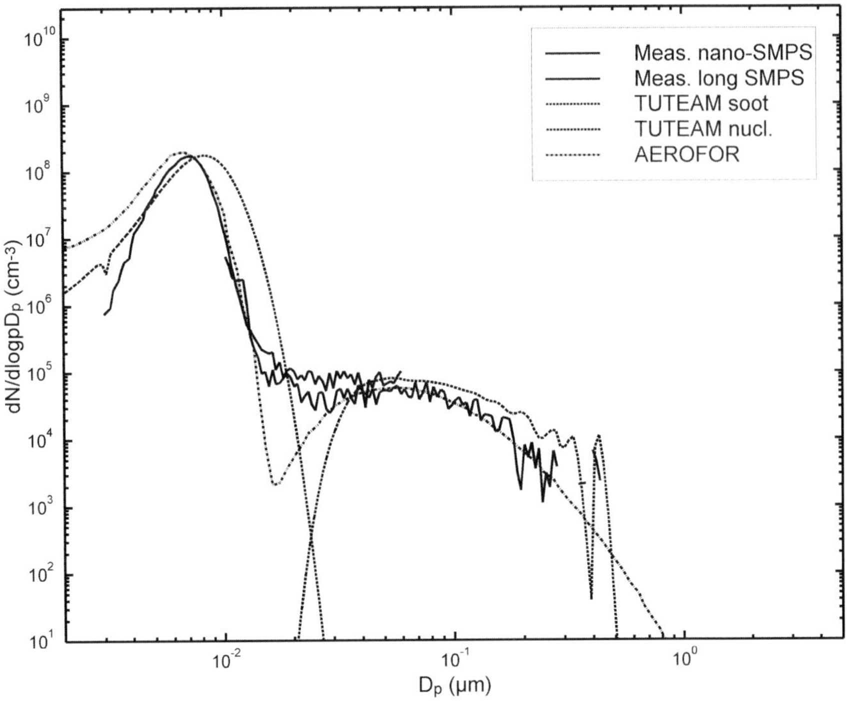

Figure 1 The measured and modelled size distributions using 1 ppm sulphur fuel and lubricant with 5,100 ppm sulphur

The predicted particle composition agrees quite well with the experimental results of Schneider et al.,[1] who measured a nucleation mode consisting mainly of sulphate compounds. The study shows that the particle number size distribution of the diesel exhaust can be quantitatively explained using water–sulphuric acid chemistry in cases where the sulphur-to-sulphuric acid conversion in the after-treatment system is high.

Acknowledgments This work was supported by Finnish Academy of Science and Letters, Vilho, Yrjö, and Kalle Väisälä Fund. We thank Dr. Vaaraslahti of Dekati Ltd (formerly of Tampere University of Technology) for giving the experimental data. The results summarized in this abstract are reported in more detail in a paper by the same authors, submitted to *Aerosol Science and Technology*.

References

1. Schneider, J., Hock. N., Weimer, S., Borrmann, S., Kirchner, U., Vogt, R., and Scheer, V. *Environ. Sci. Technol.*, **39**, 6153–6161(2005).
2. Kim, D.-H., Gautam, M., and Gera, D., *J. Colloid Interface Sci.*, **249**, 96–103 (2002).

3. Vouitsis, E., Ntziachristos, L., and Samaras, Z., *Atmos. Env.*, **39**, 1335–1345 (2005).
4. Pirjola, L., *J. Aerosol Sci.*, **30**(3), 355–367 (1999).
5. Vaaraslahti, K., Keskinen, J., Gieschaskiel, B., Solla, A., Murtonen, T., and Vesala, H., *Environ Sci Technol.*, **39**, 8497–8504.
6. Lemmetty, M., Pirjola, L., Rönkkö, T., Mäkelä, J. M., Keskinen, J., *J. Aerosol Sci.*, **37**(11), 1596–1604 (2006).
7. Tammet, H., *J. Aerosol Sci.*, **26**, 459–475 (1995).
8. Tammet, H. and Kulmala, M., *J. Aerosol Sci.*, **36**, 173–196 (2005).

Conditions Favoring New Particle Formation in A Polluted Environment: Results of the QUEST-Po Valley Experiment 2004

M.C. Facchini[1], L. Emblico[2], F. Cavalli[2], S. Decesari[1], M. Mircea[1], M. Rinaldi[1], S. Fuzzi[1], and A. Laaksoonen [3]

Abstract In the framework of the QUEST-EC project, an intensive measurement campaign was carried out in early spring 2004 at San Pietro Capofiume, a rural site located in the Po Valley. The aim was to investigate the mechanisms of new particle formation in a polluted environment, and to assess the role of aerosol chemical composition in the formation and growth of aerosol particles. The results confirm that new particle formation was preferentially associated with relatively clean air masses, and evidence that new particle formation in this polluted environment is due to a synergy between particular meteorological conditions and the physico-chemical characteristics of aerosol particles.

Keywords New particle formation, nucleation, aerosol chemistry

Introduction

The formation of new aerosol particles in the atmosphere and their growth to CCN size have received considerable attention over recent years. Nucleation appears to be a frequent phenomenon in the continental boundary layer, while evidence of new particle formation has been reported for a variety of environments.[1,2]

As part of the QUEST-EC project, an intensive measurement campaign was carried out at San Pietro Capofiume, a rural site located in the Po Valley (northern Italy, 44° 39′ N, 11° 37′ E), from 13 March to 4 April 2004. The purpose was to investigate the mechanisms of new particle formation in a polluted environment and to assess the role of aerosol chemical composition in the formation and growth of new particles.

[1]*Istituto di Scienze dell'Atmosfera e del Clima – CNR, via P. Gobetti 101, 40129, Bologna, Italy*

[2]*Institute for Environment and Sustainability, Joint Research Center, Ispra, Italy,*

[3]*Department of Applied Physics, University of Kuopio, Kuopio, Finland*

C.D. O'Dowd and P.E. Wagner (eds.), *Nucleation and Atmospheric Aerosols*, 966–968.

Methods

Continuous measurements of particle size distributions were performed using a Twin Differential Mobility Particle Sizer (TDMPS), scanning diameter ranges from 3 nm up to 600 nm. Meteorological parameters and trace gas concentrations (NO_x, NO, NO_2, SO_2, O_3) were also measured. Size-segregated aerosol samples were collected with cascade impactors and, in parallel, PM_{10} were collected with a Sierra Andersen Hi-Vol sampler. Mass concentration, major inorganic ions, water soluble organic matter (WSOM) and total carbon (TC) concentrations were measured in size-resolved samples.

Organic functional group analysis by HNMR was performed on Hi-Vol filters. To study the main features of WSOC, a further characterization of the polar organic fraction was obtained by ion exchange chromatography, separating mono and dicarboxylic acids and polyacidic organic compounds.

On the basis of TDMPS data, aerosol samples, collected on a daily basis, were categorized as event (of different intensity) or nonevent samples.

Results

General Conditions Favoring Nucleation Events

New particle formation was preferentially associated with relatively clean air masses: the aerosol mass concentrations during nucleation events ranged between 12.1 and 21.4 $\mu g\ m^{-3}$, while during nonevent periods mass concentration was higher, between 22.2 and 71.1 $\mu g\ m^{-3}$. On event days, winds had both an easterly and south-westerly component, while they were preferably westerly or north-westerly on nonevent days, when air masses carried high burdens of NO_x and PM_{10}. Overall, during the QUEST-Po Valley campaign, nucleation events occurred in the presence of air masses from the south-westerly and easterly directions, which followed episodes of ventilation of the boundary layer and were characterised by low NO_x levels, low PM_{10} concentrations (< 20 ug/m^3), and relatively high sulphate concentrations The results generally confirm that new particle formation is preferentially associated with relatively clean air masses.

Differences in Size Segregated Chemical Composition Observed Between Nucleation and Non-nucleation Conditions

Figure 1 reports the size-segregated chemical composition for two selected events and nonevent cases. The two cases appear quite different, both in terms of the absolute mass concentration of the individual components and their size distributions.

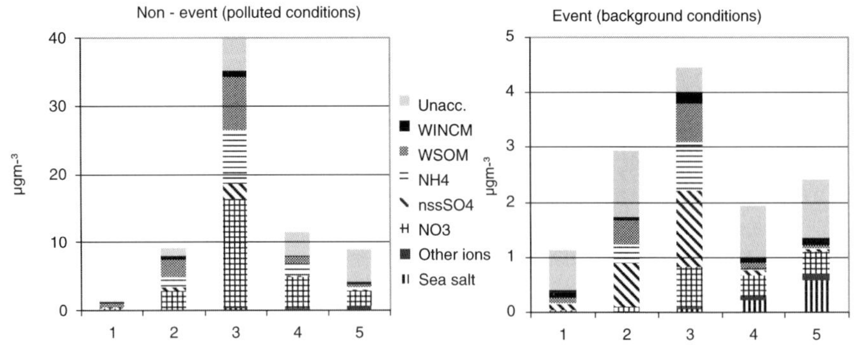

Figure 1 Size-segregated aerosol chemical compositions for the nonevent (**left**) and event (**right**) cases. Water-soluble inorganic species, water-soluble organic fraction (WSOC) and unaccounted mass concentration are reported

The non-event exhibits a monomodal distribution strongly centered on the accumulation mode for NO_3^-, NH_4^+, WSOM, and SO_4^{2-}. It can also be noted that NO_3^- represents the main individual component of the aerosol particles over the whole investigated size range.

The event aerosol shows a bimodal distribution with a dominating accumulation mode and a less-pronounced coarse mode. SO_4^{2-}, NH_4^+, and WSOM are peaked in the accumulation mode, while NO_3^- is also distributed over the larger sizes. The coarse mode exhibits a significant contribution of the “others ions”, mainly consisting of sea salt, tracing a maritime fingerprint of the air mass advected during the nucleation event.

Furthermore, during intense nucleation events, size-segregated samples are characterized by higher 1st stage/2nd stage ratios for several components with respect to nonevent days. On average, the ratios for PM, organic anions (mostly oxalate), minor inorganic ions (Cl, Na, K, Mg, Ca), NH_4^+, WSOM, and WINSOM show a pronounced enrichment in the first stage during nucleation events and less-pronounced enrichment in the case of SO_4 and NO_3.

HNMR analysis of WSOC shows that secondary organic aerosols are rich in aliphatic unsaturated oxygenated structures (H-C-C = O), mostly carbonyls in the early stage, then evolving to more oxidized forms (carboxyls).

Acknowledgment The present study was funded by QUES-EU project.

References

1. Kulmala, M., Vehkamaki, H., Petaja, T., Dal Maso, M., Lauri, A., Kerminen, V.-M., Birmili, W., and McMurry, P.H., *J. Aerosol Sci.*, **35**, 143–176 (2004).
2. Laaksonen, A., Hamed, A., Joutsensaari, J., Hiltunen, L., Cavalli, F., Junkermann, W., Asmi, A., Fuzzi, S., and Facchini M.C., *Geophys. Res. Lett.*, **32**, doi:10.1029/2004GL022092 (2005).

New Method for Simulation of Supersaturated Vapor Condensation

N.M. Kortsensteyn[1] and A.K. Yastrebov[2]

Abstract Method of direct numerical solution of the kinetic equation for the drop size distribution function was proposed for simulation of the bulk condensation of supersaturated vapor. Unlike the method of moments, the approach proposed is unrestricted in the Knudsen number. This approach was tested on a relaxation problem with an instantaneous increase in the supersaturation ratio to a given value. Results for large Knudsen number are in good agreement with the results obtained by the method of moments. It is shown that method of moments can be used for solution of the kinetic equation at small Knudsen numbers only if macro parameters should be found.

Keywords Bulk condensation, distribution function, kinetic equation, direct numerical solution

Introduction

The bulk condensation of supersaturated vapor is modeled by solving the kinetic equation for the drop size distribution function. For homogeneous condensation in an immovable medium, this equation has the form (see, e.g. [1])

$$\frac{\partial f}{\partial t}+\frac{\partial(\dot{r}f)}{\partial r}=\frac{I}{\rho_\Sigma}\delta(r-r_{cr}), \tag{1}$$

where f is the mass distribution function of drop sizes, r is the drop radius, $\dot{r}$ is the drop growth rate, I is the nucleation rate, ρ_Σ is the density of the vapor–gas–drop mixture, δ is the delta function, and r_{cr} is the critical drop radius.

[1] *Krzhizhanovsky Power Engineering Institute, Leninskii pr. 19, Moscow, 119991 Russia*

[2] *Moscow Power Engineering Institute, Krasnokazarmennaya ul. 14, Moscow, 111250 Russia*

C.D. O'Dowd and P.E. Wagner (eds.), *Nucleation and Atmospheric Aerosols*, 969–973.

If value of $\dot{r}$ does not depend drop radius (i.e., at large Knudsen numbers), Eq. (1) can be transformed to a set of equations for the first four moments of the distribution function [2]. The moments of the distribution function are defined as

$$\Omega_n = \int_{r_{cr}}^{\infty} r^n f dr. \tag{2}$$

The moment Ω_0 is equal to the number of drops per mass unit; the moment Ω_3 is equal to their total volume, etc. These moments are obtained as solution, and the distribution function can be calculated from these moments if necessary.

The advantages of the method of moments are relative simplicity and sufficient accuracy. However, at small Knudsen numbers, the moment equations are integer-differential and difficult to solve. In this case we propose to use the experience of the direct numerical solution of the Boltzmann kinetic equation.

Method of Direct Numerical Solution

A modern approach to analysis of evaporation–condensation processes on surfaces uses the Boltzmann kinetic equation. This equation and Eq. (1) have a similar form, so they can be solved by similar methods. To date, a number of methods have been proposed for solving the Boltzmann equation. One of them is the method of direct numerical solution, which was developed in the Dorodnitsyn Computing Center, Russian Academy of Sciences, Moscow, Russia [3,4]. We developed similar method to solve the kinetic equation for the drop size distribution function (1).

The method of direct numerical solution is based on splitting the process into individual stages. For Eq. (1) such stages are nucleation and drop growth, corresponding equations are written as

$$\frac{\partial f}{\partial t} = \frac{I}{\rho_\Sigma}\delta\left(r - r_{cr}\right), \tag{3}$$

$$\frac{\partial f}{\partial t} + \frac{\partial\left(\dot{r}f\right)}{\partial r} = 0. \tag{4}$$

The solution of the Eq. (3) is the initial condition for the Eq. (4) and the solution of the Eq. (4) is the initial condition for the Eq. (3) at the next time step.

For numerical solution of Eqs. (3) and (4), the drop radius calculation region was limited. The natural lower limit is the critical radius, and the upper limit can be taken to be some sufficiently large radius r_{max}. In the calculation region, a computational grid was introduced; the first node of the grid was taken to be the critical radius. Eqs. (3) and (4) were replaced by finite-difference schemes:

$$\frac{f_1^{j+1} - f_1^j}{\Delta t}\Delta r_1 = \frac{I}{\rho_\Sigma}, \tag{5}$$

$$\frac{f_i^{j+1} - f_i^j}{\Delta t} + \frac{\dot{r}_i^j f_i^j - \dot{r}_{i-1}^j f_{i-1}^j}{r_i - r_{i-1}} = 0. \tag{6}$$

Formulation of the Relaxation Problem

We used our method of direct numerical solution to solve the relaxation problem [5]. It was assumed that, at the initial moment of time, the immovable vapor–gas mixture contains no liquid drops and, at $t = 0$, the supersaturation ratio instantaneously takes a value $s = s_0 > 1$. In this case changing of the supersaturation ratio is described by the following equation:

$$\frac{d \ln s}{dt} = -\alpha\pi r_d^2 n_d v_T \left(1 + g_v \frac{L}{C_p T}\left(\frac{L\mu_v}{RT} - 1\right)\right). \tag{7}$$

Here α is condensation coefficient, n_d is the drop number density, L is latent heat, r_d is the average drop size, v_T is the thermal velocity of vapor molecules, μ_v is molar mass of vapor, g_v is the mass concentration of the vapor, and C_p is the specific heat of the vapor–gas–drop mixture. Values of n_d and r_d are calculated using moments of distribution function Ω_0 and Ω_1. Nucleation rate was calculated with help of the Frenkel–Zel'dovich theory, and drop growth rate was calculated by Fuks formula, which does not have any restrictions in the Knudsen number:

$$\dot{r} = \dot{r}_0 \left(1 + \frac{\alpha}{D}\sqrt{\frac{RT}{2\pi\mu_v}}\frac{r^2}{r + \lambda}\right)^{-1} \tag{8}$$

where $\dot{r}_0$ is the drop growth rate for free molecular regime, D is diffusion coefficient, λ is mean free path of vapor molecules.

The set of equations was supplemented by equations of mass and heat balance. We assumed that temperature of drops was equal to one of the vapor–gas mixture.

Results and Discussion

We performed calculations for two mixtures. First of them was mixture of cesium (vapor) and argon (gas). In this case $r << \lambda$, so method of moments can be used to solve the Eq. (1). Figure 1 gives an example of the distribution function versus drop radius at various moments of time. Dependencies of macro parameters (temperature, supersaturation ratio and number density of drops) are presented also. Results in Figure 1 are near to ones obtained with using of method of moments [5].

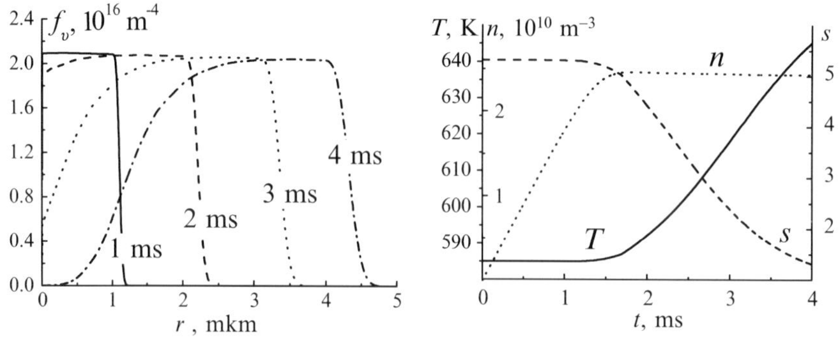

Figure 1 Distribution function and macro parameters for mixture of cesium and argon

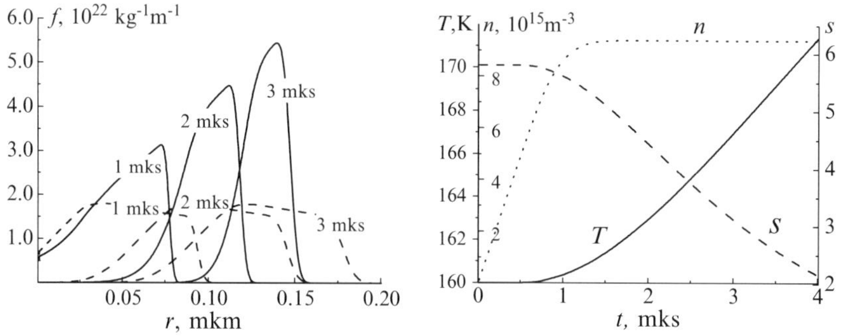

Figure 2 Distribution function (**solid lines** are for r in the Eq. (8), **dashed lines** are for r_d in the Eq. (8)) and macro parameters for mixture of ethane and argon

In the beginning of process (during the induction period [5]), the formation and growth of drops are not accompanied by a noticeable change in temperature and supersaturation ratio. The nucleation rate is almost constant and the number of drops increases linearly (see Figure 1). When the drops first formed become relatively large, the temperature begins to rise and the supersaturation ratio decreases.

Second mixture was one of ethane (vapor) and argon (gas). In this case $r \sim \lambda$, so method of moments was difficult to use. However, method of moments can be made usable if the drop radius r in the Fuks formula (8) is replaced by the average drop size r_d. We used our method to verify accuracy of such approach. Results for mixture of ethane and argon are shown in Figure 2. Macro parameters are presented only for r in the Eq. (8), distribution function is presented for both r and r_d in the Eq. (8).

Dependencies of macro parameters are the same for both r and r_d in the Eq. (8). They have similar form for both mixtures. However, dependencies of distribution function on drop radius differ strongly. The drop growth rate does not depend on drop radius for r_d in the Eq. (8) as well as for free molecular regime; and it decreases with increase of drop radius for r in the Eq. (8). So in the first case profile of distribution function moves without deformation, and in the second case it becomes narrower.

Thus, method of direct numerical solution was proposed to solve the kinetic equation for the drop-size distribution function. Testing of this method demonstrated that it can give qualitatively and quantitatively correct results at large Knudsen numbers. It is shown that method of moments can be used for solution of the kinetic equation (1) at small Knudsen numbers only if macroparameters should be found. Further prospects of using this method are related to the solution of bulk condensation problems unrestricted in the Knudsen number.

Acknowledgment This work was supported by the Russian Foundation for Basic Research, projects no. 05-08-01512 and 07-08-00082.

References

1. Sternin, L.E., *Osnovy gazodinamiki dvukhfaznykh techenii v soplakh (Fundamentals of Gas Dynamics of Two-Phase Flows in Nozzles)*, Moscow: Mashinostroenie (1974).
2. Frenclach, M. and Harris, S.J., *J.Coll.Interface Sci.*, **118**, 252–261 (1987).
3. Aristov, V.V. and Tcheremissine, F.G., *Pryamoe chislennoe reshenie kineticheskogo uravneniya Bol'tsmana (Direct Numerical Solution of the Boltzmann Kinetic Equation)*, Moscow: Vychisl. Tsentr Ross. Akad. Nauk (1992).
4. Kortsenshtein, N.M. and Samuilov, E.V., *Dokl. Phys. Chem.*, **397**, part 2, 169–173 (2004).

Simulations of Iodine Dioxide Nucleation

H. Vuollekoski[1], M. Kulmala[1], V.-M. Kerminen[2], T. Anttila[2], S.-L. Sihto[1], I. Riipinen[1], H. Korhonen[3], G. McFiggans[4], and C.D. O'Dowd[5]

Abstract Aerosol dynamical simulations of coastal new particle formation have been performed in order to investigate the nucleation and growth mechanisms in this environment. In the simulations, it is assumed that the nucleating vapour is iodine dioxide. Both Eulerian and Lagrangian type simulations have been performed and compared to observations. We have found it difficult to achieve the observed, enormous growth rates, and simulations agree with the experiments only when OIO concentration is very high and growth time has been prolonged

Keywords Aerosol modelling, nucleation, particle formation and growth, iodine dioxide

Introduction

Observations of coastal new particle formation have been made, e.g., on the west coast of Ireland.[1] The phenomenon overlaps with low tide, during which algae in the shore line are exposed to ozone in air and emit iodine vapours. These vapours take part in a complex set of chemical reactions forming high atmospheric conditions of iodine dioxide (OIO), which has been suggested to be the key compound behind coastal nucleation events.

[1] *Division of Atmospheric Sciences, Department of Physical Sciences, University of Helsinki, P.O. Box 64, FI-00014, University of Helsinki, Finland*

[2] *Climate and Global Change Unit, Finnish Meteorological Institute, P.O. Box 503, FI-00101, Helsinki, Finland*

[3] *School of Earth and Environment, University of Leeds, LS2 9JT, Leeds, UK*

[4] *School of Earth, Atmospheric and Environmental Sciences, University of Manchester, Williamson Building, Oxford Road, M13 9PL, Manchester, UK*

[5] *Department of Experimental Physics and the Environmental Change Institute, National University of Ireland, University Road, Galway, Ireland*

C.D. O'Dowd and P.E. Wagner (eds.), *Nucleation and Atmospheric Aerosols*, 974–978.

We have simulated coastal new particle formation and growth using an aerosol dynamics box model. The implemented nucleation mechanisms have several unknown parameters and our main goal has been to optimize these parameters by comparing the model results to observations.

Methods

Our simulations have been performed with UHMA (University of Helsinki Multi-component Aerosol model[2]), which is a sectional multi-component aerosol dynamics box-model that simulates tropospheric aerosol particle populations and has all basic aerosol dynamical mechanisms for clear sky conditions implemented: nucleation, condensation, coagulation, and dry deposition. For this study we have modified the model to coastal conditions and to include iodine specific dynamics.

So far three OIO nucleation mechanisms have been tested: cluster activation mechanism,[3] kinetic type nucleation, or sulphuric acid-induced activation, giving for the nucleation rate, respectively:

$$J_{1_{Act}} = A \times [OIO] \tag{1}$$

$$J_{1_{Kin}} = K \times [OIO]^2 \tag{2}$$

$$J_{1_{ActSA}} = A' \times [OIO] \times [H_2SO_4] \tag{3}$$

where A, K, and A' are the activation, kinetic, and sulphuric acid-induced activation nucleation coefficients and [OIO] and $[H_2SO_4]$ are the OIO and sulphuric acid concentrations.

The nucleated particles grow by coagulation and condensation of OIO, sulphuric acid and an organic vapour. OIO is assumed to be formed during low tide and only on the coastline. At this point the OIO production rate is prescribed, i.e., there is no OIO formation chemistry implemented. Sulphuric acid is formed by a chemical reaction between SO_2 and OH, the concentration of which is dependent on the zenith angle of the Sun. Organic vapour concentration is taken as a free parameter, which we intend to study more in detail. Meteorological conditions and available parameters have been chosen to agree with experiments performed in day of the year 163 in 1999 at Mace Head, Ireland.[4]

Coastal areas often have windy conditions, which poses difficulties in simulations, because it decreases the time of influence with the coastline as well as growth time in general. The measurement station is usually stationary, whereas aerosol dynamical reactions occur in the air mass moving with the wind. To account for both aspects, two different models have been developed.

In the Lagrangian model, the simulated box is moving with the wind from the Atlantic Ocean, and OIO is injected into the system during low tide for about 10 min, roughly corresponding to the time of influence of coastal algae.

In the Eulerian model, we simulate air masses individually from coastline to measurement station, i.e., there is no mass transfer between air masses of different ages. Effectively this corresponds to choosing the age of the measured particles, which is equal to the time it takes for the wind to carry them from the coastline to the measurement station.

Results and Discussion

An example of a Lagrangian simulation of a new particle formation event can be seen in Figure 1, in which activation type nucleation was implemented with activation coefficient 0.0001 s^{-1}. Related gas and particle concentrations are illustrated in Figure 2. Unfortunately, to our knowledge there are no experimental observations that could be directly compared with our windy-condition Lagrangian simulations. The current model is unrealistic at high wind speeds since the conditions in the simulated air mass are assumed to be homogeneous, although OIO production should be very local.

Eulerian simulations have given promising results similar to experimental observations. However, the simulations do not show the observed, enormous growth rates. This is probably explained to some degree by limitations of the model (e.g., cloudy conditions not accounted for, yet). A particle size distribution of one of these simulations can be seen in Figure 3, with related gas and particle concentrations in

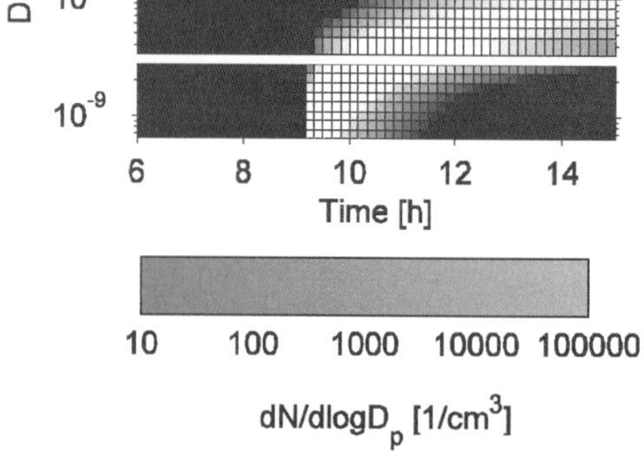

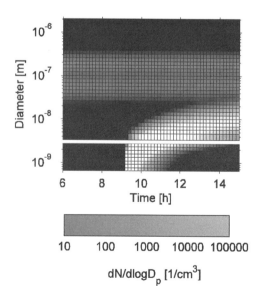

Figure 1 Particle size distribution as a function of time using activation type nucleation mechanism in a Lagrangian simulation

Figure 2 Vapour, 3–6 nm particle and total particle concentrations as functions of time during the simulation of Figure 1

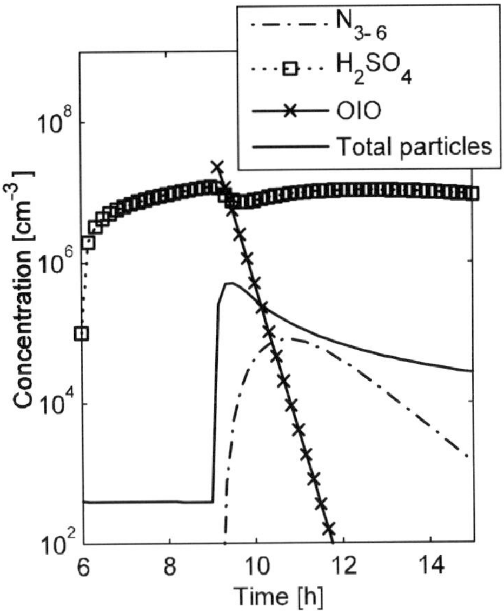

Figure 3 Particle size distribution as a function of time using sulphuric acid induced activation type nucleation mechanism in an Eulerian simulation. Transport of particles from the shoreline to the measurement site was assumed to take 800 s

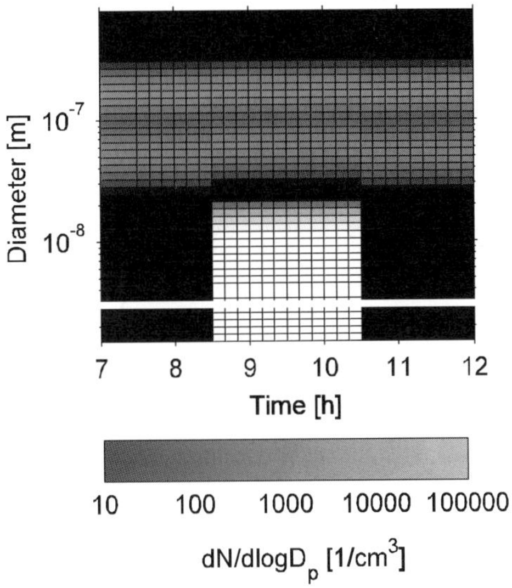

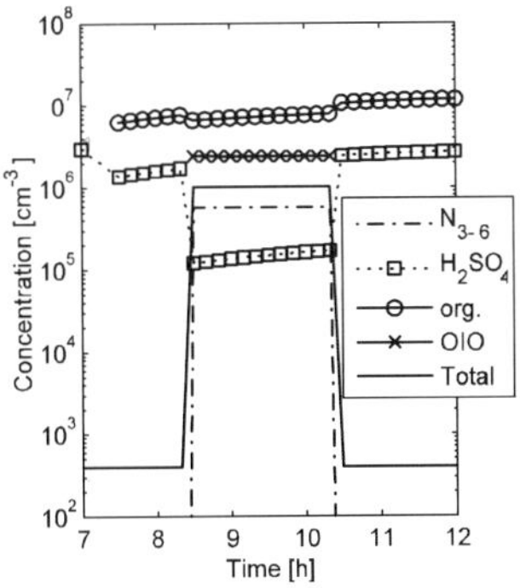

Figure 4 Vapour, 3–6 nm particle and total particle concentrations as functions of time at the "measurement station" location during the simulation of Figure 3

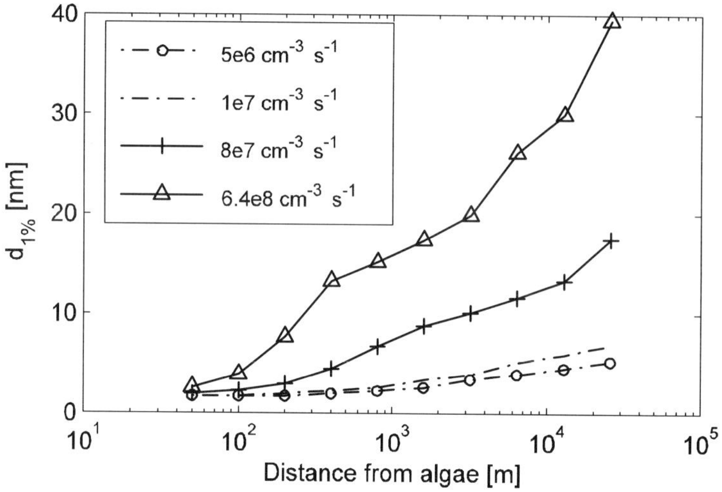

Figure 5 Sizes below which 99% of particles lie as functions of OIO production rate and distance between shoreline and measurement station in an Eulerian simulation using activation type nucleation

Figure 4. A long distance (6.2 km) was set between the shoreline and measurement station to allow for longer particle growth. Although the measurement station is actually situated at about 100 m from the coastline, wind does not always blow directly from the sea, which can lead to much longer growth times. To investigate this relationship further, a series of simulations were performed. Figure 5 shows, as functions of OIO production rate and distance from the shoreline, the size $d_{1\%}$, below which 99% of particles lie (excluding background). One can see that more available OIO paired with longer growth times can result in significantly bigger particles.

We will perform more sensitivity studies by varying model parameters, some of which are currently poorly known. More experimental data is needed for comparison with our results in order to achieve more accurate parameterization for coastal nucleation.

References

1. O'Dowd, C.D. and Hoffmann, T., *Environ. Chem.*, **2**, 245–255 (2005).
2. Korhonen, H., Lehtinen, K.E.J., and Kulmala, M., *Atmos. Chem. Phys.*, **4**, 757–771 (2004).
3. Kulmala, M., Lehtinen, K.E.J., and Laaksonen, A., *Atmos. Chem. Phys.*, **6**, 787–793 (2006).
4. O'Dowd, C.D., Hämeri, K., Mäkelä, J., Väkevä, M., Aalto, P., Leeuw, G., Kunz, G. J., Becker, E., Hansson, H.-C., Allen, A.G., Harrison, R.M., Berresheim, H., Kleefeld, C., Geever, M., Jennings, S.G., and Kulmala, M., *J. Geophys. Res.*, **107**, 8107–8123 (2002).

Development and Estimation of an Expression of Particle Dry Deposition Process Using the Moment Method

S.Y. Bae[1], C.H. Jung[2], and Y.P. Kim[1]

Abstract An expression for the dry deposition process for polydisperse particles that is related to the gravitational settling, impaction, and Brownian diffusion is obtained and tested. The dry deposition velocity expression of Raupach et al. (2001) is approximated and then the moment method is used to derive a set of analytical solutions for polydisperse particles employing the moment method with the assumed lognormal particle size distributions. The applicability of the derived equation is demonstrated for three model cases; hazy, urban, and clear case, respectively.

Keywords Dry deposition velocity, polydisperse particle, lognormal particle size distribution, moment method

Introduction

The dry deposition process is one of the most efficient particle sinks in the air. Due to its simplicity, the inferential method has been widely used to describe the dry deposition process. In the inferential method, dry deposition flux is calculated by a product of dry deposition velocity and local atmospheric concentration at a reference height.

Usually, the dry deposition velocity is calculated based on the resistance model (Wesely 1989) or assumed to be a constant. In air quality models, usually constant dry deposition velocity is used due to computational burden. Thus, it is desirable to apply a detailed expression of dry deposition process which is computationally efficient in an air quality model.

The moment method has been widely used in aerosol dynamics modeling (Whitby and McMurry 1997). Jung et al. (2004, 2006), Jung and Kim (2006), and

[1]*Department of Environmental Science and Engineering, Ewha Womans University, Korea*

[2]*Department of Environmental Health, Kyungin Women's College, Korea*

C.D. O'Dowd and P.E. Wagner (eds.), *Nucleation and Atmospheric Aerosols*, 979–983.

Bae et al. (2006) developed an aerosol dynamics algorithm for condensation, coagulation, and below-cloud scavenging processes based on the modal approach. Since the dry deposition velocity expression based on the resistance model is not in the form of power law, the modal approach cannot be applied.

In this study, an analytical expression for dry deposition process is developed and tested. The dry deposition velocity form proposed by Raupach et al. (2001) is applied for lognormal particle size distributions.

Dry Deposition Velocity for Polydisperse Particles

Approximated Dry Deposition Velocity

The dry deposition velocity proposed by Raupach et al. (2001) is as follows:

$$V_d = V_t + G\left[f_{form} a_f E + \left(1 - f_{form}\right) a_v Sc^{-2/3}\right], \qquad (1)$$

where,

$$V_t = \frac{\rho_p d_p^2 g Cc}{18\mu_a}, G = \frac{u_*^2}{u_r}, E = \left(\frac{St}{St + 0.8}\right)^2,$$

$$St = \frac{\rho_p d_p^2 u_*^2}{18\mu_a v}, Sc = \frac{u_a}{\rho_p D_{diff}}, D_{diff} = \frac{k_b T Cc}{3\pi_a d_p}.$$

Here V_t is the settling velocity in cm s^{-1}, ρ_p and ρ_a the density of particle and air in g cm^{-3}, respectively, d_p the particle diameter in cm, g the gravity acceleration in cm s^{-2}, *Cc* Cunninghum correction factor, μ_a the viscosity of air in g cm^{-1} s^{-1}, $u*$ friction velocity in cm s^{-1}, u_r wind velocity in cm s^{-1}, E the collision efficiency, St the Stokes number, v the kinematic viscosity in cm^2 s^{-1}, Sc the Schmidt number of particle, D_{diff} the vapor diffusivity in cm^2 s^{-1}, k_b the Boltzmann's constant in g cm^2 K^{-1} s^{-2}, T the temperature in K, f_{form} (= 0.75) the fraction of the total canopy drag exerted as from drag, and a_f (= 2) and a_v (= 8) are adjustable parameters to be determined empirically. The collision efficiency should be expressed in the power law form to utilize Eq. (1) in the moment method. For Stoke number < 1, the approximated collision efficiency as a power function of Stokes number is $E \approx 0.33St^{6/5}$. For Stokes number > > 1, the collision efficiency can be approximated as 1. We have applied the harmonic mean approach (Pratsinis 1988; Jung and Kim 2006).

$$E = \left(\frac{St}{St + 0.8}\right)^2 \cong \frac{1}{\dfrac{1}{0.33St^{6/5}} + 1} \tag{2}$$

Mean Number-dry Deposition Velocity for Polydisperse Particles Using the Moment Method

Atmospheric particle size distribution can be assumed as lognormal functions. Moment equation of the *k*th moment of particle size distribution is defined as follows:

$$M_k = \int_0^\infty \upsilon_p^k n(\upsilon_p) d\upsilon_p, \tag{3}$$

where k is an arbitrary real number. The governing equation of dry deposition of particles is

$$\frac{\partial C}{\partial t} = -\frac{V_d C}{H}, \tag{4}$$

where C is the mass concentration of particles in the range of υ_p to $\upsilon_p + d\upsilon_p$ in μg m^{-3} and H the height in cm. Integrating Eq. (4) for particle volume after substituting the mass concentration ($C = \upsilon_p \upsilon_p n(\upsilon_p) d\upsilon_p$) into Eq. (4), the change of total particle number concentration due to dry deposition can be expressed as:

$$\frac{\partial N_p}{\partial t} = -\frac{\int V_d n(\upsilon_p) d\upsilon_p}{HN_p} N_p \equiv -\frac{V_d' N_p}{H}, \tag{5}$$

where V_d' is the mean number-dry deposition velocity. Finally, applying the definition of moment for particles with lognormal function to Eq. (5), the equation of particle dry deposition for polydisperse particle can be expressed as:

$$V_d' = \frac{\left(\gamma_1 M_{2/3} + G\left(\dfrac{M_{4/5} M_0}{\dfrac{1}{\gamma_2} M_0 + \gamma_3 M_{4/5}} + \gamma_4 M_{-2/9}\right)\right)}{M_0} \tag{6}$$

Governing Equations for Aerosol Dynamics for Dry Deposition

The temporal variations of M_0, M_1, and M_2 are calculated based on Eq. (7) and then the parameters of the particle size distributions can be estimated from them.

$$\frac{\partial M_0}{\partial t} = -\frac{1}{H}\left(\gamma_1 M_0^{2/9} M_1^{8/9} M_2^{-1/9} + G\right.$$

$$\frac{\partial M_1}{\partial t} = -\frac{1}{H}\left(\gamma_1 M_0^{-1/9} M_1^{5/9} M_2^{5/9} + G\left(\frac{M_0^{-2/25} M_1^{9/25} M_2^{18/25} M_1}{\frac{1}{\gamma_2} M_1 + \gamma_3 M_0^{-2/25} M_1^{9/25} M_2^{18/25}} + \gamma_4 M_0^{11/81} M_1^{77/81} M_2^{-7/81}\right)\right)$$

$$\frac{\partial M_2}{\partial t} = -\frac{1}{H}\left(\gamma_1 M_0^{5/9} M_1^{-16/9} M_2^{20/9} + G\left(\frac{M_0^{18/25} M_1^{-56/25} M_2^{63/25} M_2}{\frac{1}{\gamma_2} M_2 + \gamma_3 M_0^{18/25} M_1^{-56/25} M_2^{63/25}} + \gamma_4 M_0^{-7/81} M_1^{32/81} M_2^{56/81}\right)\right)$$

$$\left(\frac{M_0^{3/25} M_1^{24/25} M_2^{-2/25} M_0}{\frac{1}{\gamma_2} M_0 + \gamma_3 M_0^{3/25} M_1^{24/25} M_2^{-2/25}} + \gamma_4 M_0^{110/81} M_1^{-40/81} M_2^{11/81}\right)\Bigg)$$

Simulation of the Particle Size Distribution by Dry Deposition Process

The particle size distribution change by dry deposition is simulated by using the expression developed in this study and by the Wesely's expression modified for particles (Seinfeld and Pandis 1998) using the fourth-order Runge-Kutta method. The model cases are hazy, urban, and clear cases shown in Pryor and Binkowski (2004). Time step is 18 s and number of size cut is 1,000 for the Wesely's expression. Figure 1 shows the particle number size distributions between this study and Wesely (1989) are similar. The particle number concentration of Wesely's expression for particles with $d_p < 0.5\,\mu\text{m}$ decreases faster than this study. In this size range, Brownian diffusion mainly affects the dry deposition velocity of this study, and quasi-laminar layer resistance affects that of Wesely. The difference of the deposition velocity between this study and Wesely (1989) is $u^*(1-2u^*/u_r)Sc^{-2/3}$.

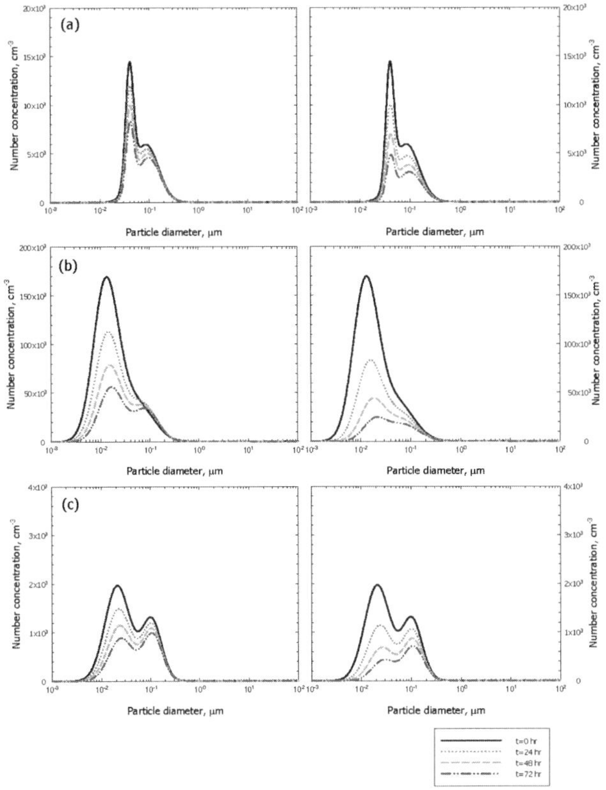

Figure 1 Comparison of the particle number size distribution variations between this study (**left**) and the Wesely's expression (**right**) for three cases: (a) Hazy, (b) Urban, and (c) Clear

Acknowledgments This work was supported by the Korea Science and Engineering Foundation (KOSEF) through the National Research Lab. Program funded by the Ministry of Science and Technology (No. M10600000221-06J0000-22110).

References

1. Bae, S.Y., Jung, C.H., and Kim, Y.P., *J. Aerosol Sci.*, **37**, 1507–1519 (2006).
2. Jung, C.H., Kim, Y.P., and Lee, K.W., *Env. Eng. Sci.*, **21**, 437–450 (2004).
3. Jung, C.H., Park, S.H., and Kim, Y.P., *J. Aerosol Sci.*, **37**, 1400–1406 (2006).
4. Jung, C.H. and Kim, Y.P., *J. Aerosol Sci.*, **37**, 143–161 (2006).
5. Pratsinis, S.E., *J. Colloid Interface Sci.*, **124**, 416–427 (1988).
6. Pryor, S.C. and Binkowski, F.S., *Aerosol Sci. Techonol.*, **38**, 1091–1098 (2004).
7. Raupach, M.R., Briggs, P.R., Ahmad, N., and Edge, V.E., *J. Environ. Qual.*, **30**, 729–740 (2001).
8. Seinfeld, J.H. and Pandis, S.N., *Atmospheric Chemistry and Physics*, New York: Wiley, pp. 958–996 (1998).
9. Wesely, M.L., *Atmos. Environ.*, **23**, 1293–1304 (1989).
10. Whitby, E.R. and McMurry, P.H., *Aerosol Sci. Techonol.*, **27**, 673–688 (1997).

Investigating the Chemical Composition of Growing Nucleation Mode Particles with CPC Battery

H.E. Manninen[1], M. Kulmala[1], I. Riipinen[1], T. Petäjä[1], M. Sipilä[1], T. Grönholm[1], P.P. Aalto[1], and K. Hämeri[1]

Abstract Formation and growth of fresh newly formed atmospheric aerosols were investigated using a condensation particle counter battery (CPCB). From the measurement data we identified days when the two CPC pairs within the CPCB showed differential signals as well as days when no differences in the signals were detected. This categorization was complemented by nucleation event classification. We found out that growing nucleation mode particles were water-soluble both at 3 and 10 nm. During the measurement period water insoluble and water soluble pollutants were detected by the CPCB system. The measurements for this study were performed at SMEAR II station in Hyytiälä, Finland, between March and May 2006.

Keywords Aerosol formation, nanoparticles, chemical composition, atmospheric aerosols

Introduction

New particle formation by nucleation of supersaturated vapours and subsequent particle growth to detectable sizes is one of the key processes controlling the number concentration of aerosols in the atmosphere.[1] A direct determination of the composition of these nucleation mode particles is difficult due to minuscule amounts of matter available for chemical analysis. A Condensation Particle Counter Battery (CPCB) examines the activation properties of atmospheric nanoparticles in different vapours.[2] It provides a method to estimate the chemical composition of growing nucleation mode particles through different activation diameters.

The CPC battery used in this study consisted of two butanol and two water CPCs. The cut-off diameter depends mostly on an activation efficiency inside the CPC.[3] Water soluble and butanol soluble aerosol particles have different heterogeneous activation diameters inside the CPC. In other words, indirect information on the

[1] *Department of Physical Sciences, University of Helsinki, P.O. Box 64, FI-00014 Helsinki, Finland*

C.D. O'Dowd and P.E. Wagner (eds.), *Nucleation and Atmospheric Aerosols*, 984–988.

chemical composition of the particles is obtained as the solubility of the particles to the condensing vapour lead to changes in the cut-off sizes of the CPCs and to the differences in their concentration readings. More details can be found in Kulmala et al. (2006).

Materials and Methods

Our CPCB consists of four CPCs: two butanol CPCs (TSI-3025; TSI-3010) and two water CPCs (TSI-3785; TSI-3786) with cutoff diameters 3 and 10 nm, respectively. The cutoff sizes of the two water CPCs were adjusted in laboratory using insoluble silver particles by varying the temperature difference between the saturator and growth tube so that they matched very closely the cut-off sizes of the corresponding butanol CPCs.[4] During the field measurements TSI-3025 and TSI-3786 measured number concentrations of larger than 3 nm particles and correspondingly TSI-3010 and TSI-3785 counted concentrations of particles larger than 10 nm in diameter.[2]

CPCB field measurements were performed in a rural background station in Southern Finland (SMEAR II station located in Hyytiälä[5]) between March 7 and May 31, 2006. The measuring station is surrounded mainly by a Scots pine dominated forest. DMPS system was used to measure the aerosol size distribution continuously and it was the basis of nucleation event classification.[6] The Hyytiälä station has also a continuous meteorological monitoring.

Results and Discussion

The measurements showed that atmospheric particles that were newly formed during a nucleation burst were more efficiently activated in the water CPCs as compared to the butanol CPCs. The results indicated that growing nucleation mode particles were water-soluble both at 3 nm and at 10 nm.

During the measurement period we observed several new particle formation events. One of those nucleation events occured on May 16, 2006 in Hyytiälä. Figure 1 shows the diurnal cycle of particle size distribution. New particle formation event was observed on that day starting around 9:30 a.m. local time.

Figure 2 shows the variation of the particle number concentrations measured by the CPCB. During the event the ratio between water and butanol CPCs increased in both pairs. The increase in concentrations measured by the TSI 3786 and TSI 3785 had a 35 min time difference. After 1 p.m., the concentration difference between water and butanol CPCs started to decrease again. At night time the ratio was normalized to one.

As this setup was used to measure number concentration of atmospheric aerosol particles, only the composition of the particles and the resulting change in heterogeneous activation diameter can explain the difference in their concentration readings.

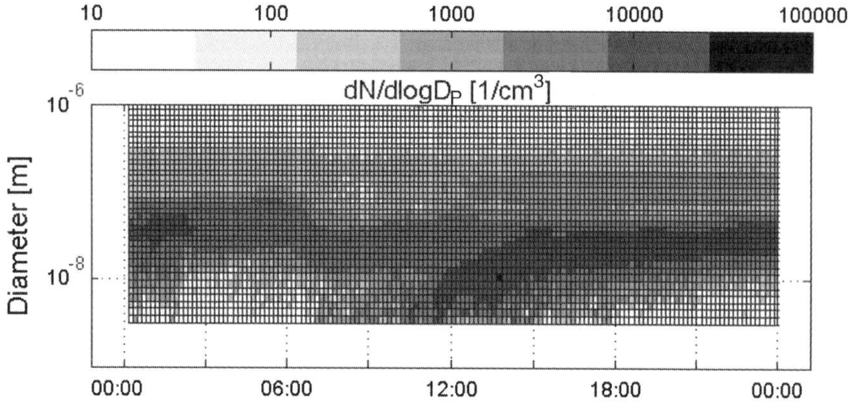

Figure 1 The aerosol particle size distribution during a nucleation event in Hyytiälä on May 16, 2006

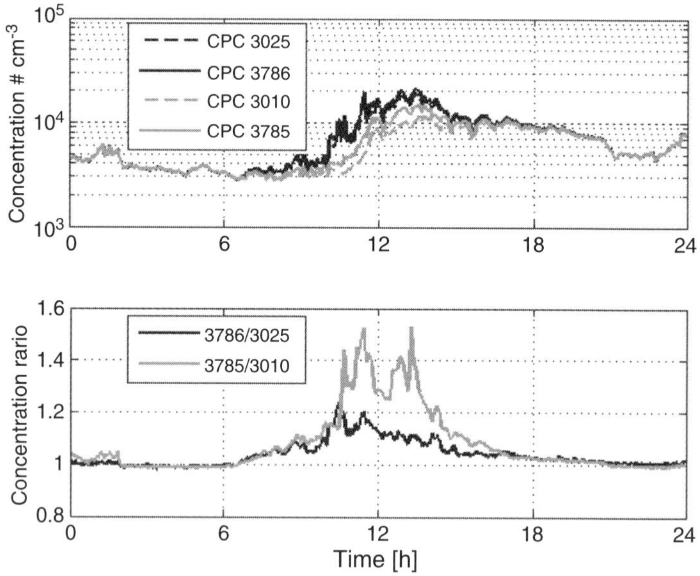

Figure 2 The aerosol particle number concentration with the CPCB during a nucleation event on May 16, 2006 (**top**) and the ratio of the particle number within the pairs of water and butanol CPCs (**bottom**)

Thus water-soluble (hygroscopic) particles will be detected down to lower particle size in the water CPCs due to the increased activation probability in water vapour. Therefore, an increased count rate will be measured in comparison with the butanol CPCs. The detection efficiency of the CPC depends both on the saturation ratio inside the CPC and on chemical composition of sampled aerosol particles.

An increase in the ratio between water and butanol CPCs was detected during every nucleation event (in 32 cases out of 32) within the investigated period. During

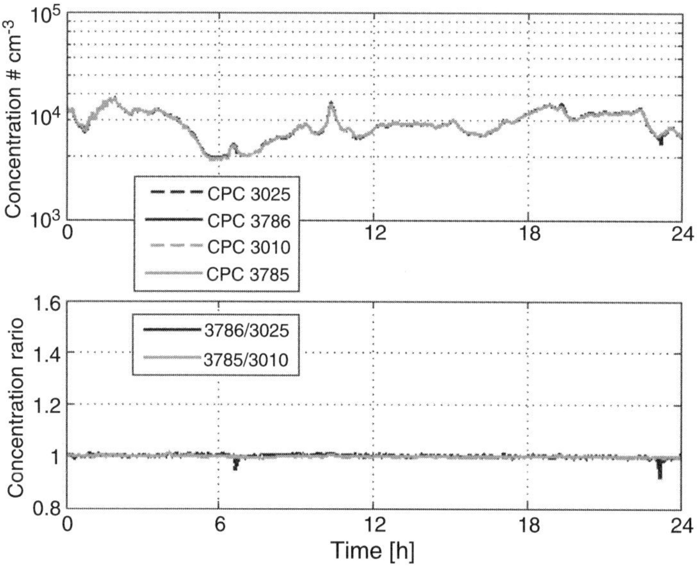

Figure 3 The aerosol particle number concentration with the CPCB on May 23, 2006 (**top**) and the ratio of the particle number within the pairs of water and butanol CPCs (**bottom**). No new particle formation occurred

non-event days the ratio was usually close to unity (see Figure 3). The only exception was pollution episode when water soluble material was present (e.g., SO_2). Water insoluble pollutants (e.g., NO_x, CO) did not affect the CPCB readings.

Conclusions

During a nucleation event the growing nucleation mode particles were detected more efficiently by the water CPCs than by the butanol CPCs. The CPCB measurement data indicates that freshly formed particles at 3 and 10 nm contain water soluble material. The observation of water soluble and water insoluble material during pollution episodes shows that the concentration difference between the water and butanol CPCs within the CPCB is function of particle chemical composition, not a function of total particle concentration.

References

1. Kulmala, M., How particles nucleate and grow, *Science*, **302**, 1000–1001 (2003).
2. Kulmala, M. et al., The condensation particle counter battery (CPCB): a new tool to investigate the activation properties of nanoparticles, *J. Aerosol Science*, **38**, 289–304 (2007).
3. Stoltzenburg, M.R. and McMurry, P.H., An ultrafine aerosol condensation nucleus counter, *Aerosol Sci. Technol.*, **14**, 48–65 (1991).

4. Petäjä, T. et al., *Aerosol Sci. Technol.*, **40**, 1090–1097 (2006).
5. Kulmala, M. et al., Overview of the international project on biogenic aerosol formation in the boreal forest (BIOFOR), *Tellus, B*, **53**, 324–343 (2001).
6. Dal Maso, M. et al., Formation and growth of fresh atmospheric aerosols: eight years of aerosol size distribution data from SMEAR II, Hyytiälä, Finland. *Boreal Env. Res.*, **10**, 322–336 (2005).

Aerosol Formation from Isoprene: Determination of Particle Nucleation and Growth Rates

B. Verheggen[1,2], A. Metzger[2], J. Duplissy[2], J. Dommen[2], E. Weingartner[2], A.S.H. Prévôt[2], and U. Baltensperger[2]

Abstract A particle nucleation and growth event following isoprene photooxidation (initial mixing ratio 350 ppbv) has been analyzed using a novel inverse modeling procedure. Diameter growth rates were found to reach 100 nm h–1, while the nucleation rate was relatively low, around 1 cm–3 s–1. These empirical rates were determined based on the measured aerosol size distributions, independent of classical nucleation theory. Continuous evaporation after the lights were turned off indicates that 10 pptv is a lower limit of the effective saturation vapor pressure of its condensable oxidation products. The aerosol yield was 0.8%, in agreement with other studies.

Keywords Secondary organic aerosol, nucleation rate, growth rate

Introduction

Isoprene has only recently been considered to be potentially important for the formation of Secondary Organic Aerosol (SOA).[1] Isoprene is one of the most abundant non-methane hydrocarbons emitted into the troposphere. Because of its large biogenic source strength of ~500 Tg/year,[2] even a very low aerosol yield has the potential to significantly affect SOA formation. Many smog chamber experiments have since been performed to elucidate the mechanism and degree to which isoprene oxidation products may partition to the aerosol phase.

At atmospherically relevant concentrations of a few tens of ppbv, isoprene oxidation does not lead to the nucleation of new aerosol particles. However, SOA formation by growth of seed aerosol has been observed down to these low concentrations with an aerosol yield of ~1%,[3] while higher yields (up to 5%) were obtained for higher isoprene concentrations.[3,4] The same two studies reported on the observation of oligomers in aerosols from isoprene photooxidation.[3,4] These large molecular

[1] *Institute for Atmospheric and Climate Science, ETH Zürich, 8092 Zürich, Switzerland*

[2] *Laboratory for Atmospheric Chemistry, Paul Scherrer Institut, 5232 Villigen PSI, Switzerland*

C.D. O'Dowd and P.E. Wagner (eds.), *Nucleation and Atmospheric Aerosols*, 989–993.

weight compounds have typically low vapor pressures, which could enable the partitioning of relatively volatile compounds into the aerosol phase, thus increasing the aerosol yield.

Measurements of aerosol formation and growth following isoprene oxidation have been performed in the PSI smog chamber[5] to further investigate its SOA formation potential. Using only the measured aerosol size distributions as input, we determined empirical particle nucleation and growth rates using the recently developed inverse modeling procedure PARGAN.[6]

Measurements

Isoprene oxidation experiments were carried out in a 27 m^3 Teflon chamber at 20°C and 50% RH. A known amount of isoprene (Fluka, 99.5%) was evaporated in a heated glass sampling bulb and continuously flushed with pure air into the chamber and allowed to equilibrate for 30 min. Four xenon arc lamps were used to simulate the solar light spectrum and start the photochemistry.

Aerosol size distributions were measured using a twin-SMPS system, one with a short classifier and a TSI 3786 UWCPC (4–150 nm) and one with a long classifier and a TSI 3022 CPC (16–700 nm). From these we made composite size distributions covering the whole size range from 4 to 700 nm diameter which we used for further analysis.

The experiment was started by producing particles through ozonolysis of isoprene. This aerosol was diluted to reduce the particle concentration to around 500 cm^{-3}. We then injected 350 ppbv of isoprene in the presence of NO and turned on all chamber lights. An hour later, the preexisting particle mode started to significantly grow in size, indicative of condensation. Another 2 h later a modest nucleation event occurred, producing approximately 4,000 new particles cm^{-3}. Figure 1 shows a contourplot of the measured size distributions.

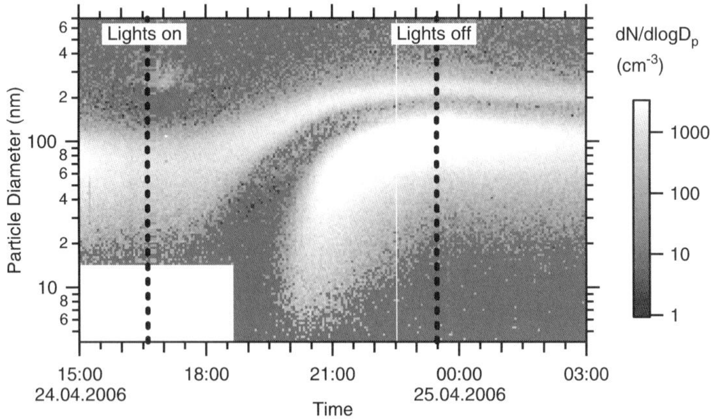

Figure 1 Measured gas-phase species (**upper panel**) and size distributions (**lower panel**) following isoprene ozonolysis in the morning (preexisting mode) and photo-oxidation in the evening

Analysis Procedure

A stable particle is said to be nucleated when its rate of condensation first exceeds its rate of evaporation; the size of this critical cluster is thought to be on the order of 1 nm. Classical nucleation theory, based on bulk liquid properties, does not adequately describe the nucleation rate, as evident from the discrepancies between its predictions and measurements.[7,8,9] These discrepancies illustrate a need to empirically determine the nucleation rate from measurements, independent of classical nucleation theory.

To this end, we used a recently developed inverse modeling procedure called PARGAN (Particle Growth and Nucleation) to determine nucleation and growth rates directly from the sequence of measured aerosol size distributions. First, the size dependent wall loss is determined by fitting the observed decay in particle number, taking into account coagulation. A highly time-resolved (size dependent) particle growth rate is found by regression analysis of the General Dynamic Equation (GDE). These empirical growth rates are then used to estimate the time of nucleation for particles in each measured size bin, defined as the time when their calculated diameter surpassed 1 nm. Their number density at the time and size of nucleation is determined by integrating the particle losses (due to coagulation and diffusion to the walls) that occurred in the time interval between nucleation and measurement. The nucleation rate is then given by the rate at which particles grow past the critical cluster size.

Results

There are some advantages to having preexisting aerosol present during nucleation: It allows the particle growth rate – and thus the time of nucleation – to be determined before the nucleated particles grew to a detectable size. It is also more representative of the atmosphere, where aerosol is always present when nucleation occurs. The diameter growth rate determined using PARGAN is shown in Figure 2. The growth rate is proportional to the concentration of the condensing species above saturation; its equivalent value is given on the right axis.

After the lights had been turned off (23:30), the aerosol evaporated continuously at a rate of 17 nm h^{-1}, presumably due to wall losses of semi-volatile species. This corresponds to a mixing ratio in the chamber of 10 pptv below saturation. Therefore 10 pptv is a lower limit for the effective saturation vapor pressure of the condensable products of isoprene.

The running mean of the growth rate (Figure 2) was used to determine the nucleation rate based on each measured size bin. The results, grouped together in a statistical plot for clarity, are shown in Figure 3. Each symbol is representative of 300 size bins, tracked backwards in time to their estimated time of nucleation.

For an initial mixing ratio of isoprene of 350 ppbv, we found a modest nucleation rate of approximately 1 cm^{-3} s^{-1} (Figure 3), while the condensational growth rate

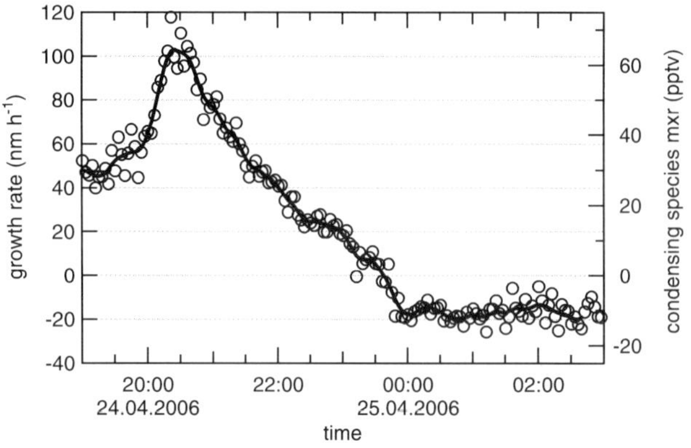

Figure 2 Particle diameter growth rate during photooxidation of isoprene. Solid line is seven-point running mean. The equivalent mixing ratio of the condensing vapor (in excess of saturation) can be read off the right axis

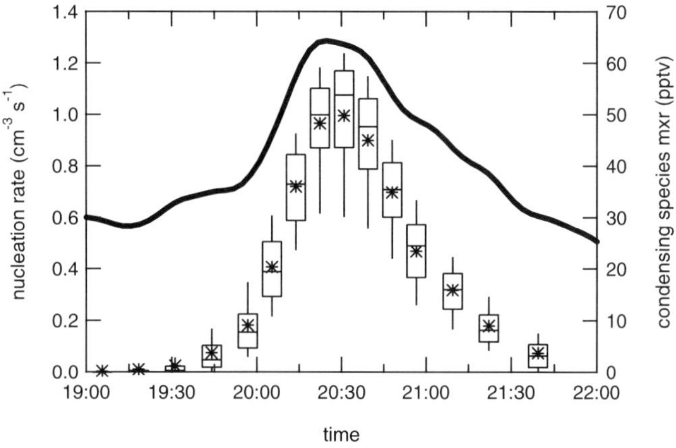

Figure 3 Nucleation rates (average denoted as a double cross; box plot giving 10, 25, 50, 75, and 90 percentile values) determined using PARGAN. Solid line gives the mixing ratio of condensable species (above saturation), proportional to the growth rate

reached a maximum of 100 nm h^{-1} (Figure 2). This suggests that the isoprene oxidation products have a too high saturation vapor pressure to effectively induce nucleation at atmospheric concentrations. However, they probably do contribute to SOA formation in regions where these oxidation products can condense on preexisting aerosol.

The aerosol mass yield (corrected for wall losses; particle density assumed to be 1.4 g cm^{-3}) reached 0.8% at an aerosol volume of close to 4 μm^3 cm^{-3}. This value fits well within the range of measured yields without preexisting aerosol present.[4]

Acknowledgments This work was supported by the Swiss National Science Foundation as well as the EC projects EUROCHAMP and POLYSOA.

References

1. Claeys, M. et al., *Science*, **303**, 1173–1176 (2004).
2. Guenther, A. et al., *J. Geophys. Res.*, **100**, 8873–8892 (1995).
3. Kroll, J.H. et al., *Environ. Sci. Technol.*, **40**, 1869–1877 (2006).
4. Dommen, J. et al., *Geophys. Res. Lett.*, **33**, doi: 10.1029/2006 GL026523 (2006).
5. Paulsen, D. et al., *Environ. Sci. Technol.*, **39**, 2668–2678 (2005).
6. Verheggen, B. and Mozurkewich, M., *Atmos. Chem Phys.*, **6**, 2927–2942 (2006).
7. Wyslouzil, B. et al., *J. Chem. Phys.*, **94**, 6842–6850 (1991).
8. Weber, R.J. et al., *J. Geophys. Res.*, **103**, 16385–16396 (1998).
9. Verheggen, B. and Mozurkewich, M., *J. Geophys. Res.*, **107**, 4123–4134 (2002).

Heterogeneous Reactivity of Sulfate-coated $CaCO_3$ Particles with Gaseous Nitric Acid

A. Morikawa, T. Ishizaka, and S. Tohno

Abstract Laboratory experiments of the reaction between surface sulfate-coated $CaCO_3$ particles and gaseous nitric acid were carried out. Sulfate-coated $CaCO_3$ particles were generated by exposing $CaCO_3$ particles synthesized through liquid reaction to sulfuric acid vapor. Reaction of gaseous HNO_3 with the particles did not show rapid morphological change as with pure $CaCO_3$ particles, but gradually produced deliquescent $Ca(NO_3)_2$. It was revealed that $CaCO_3$ particles whose surfaces were coated with $CaSO_4$ reacted with gaseous nitric acid and formed nitrate although surface $CaSO_4$ hindered the evolution of the reaction between $CaCO_3$ and HNO_3.

Keywords Heterogeneous reaction, modification, calcite, nitric acid, sulfate coating

Introduction

Heterogeneous reactions on the surfaces of aerosol particles such as sea salt or mineral dust are important for their effects on chemical balance of troposphere or on radiative process relevant to the physicochemical properties.[1] Calcite ($CaCO_3$) is one of the most reactive components in mineral dust[2] and the typical reactions with acidic vapors are described as follows:

$$CaCO_3 + H_2SO_4 \rightarrow CaSO_4 + H_2O + CO_2 \tag{1}$$

$$CaCO_3 + 2HNO_3 \rightarrow Ca(NO_3)_2 + H_2O + CO_2 \tag{2}$$

These two reactions have different behavior that sulfate formation of reaction (1) is a diffusion-limited process,[3] while the reaction (2) is not surface-limited

Graduate School of Energy Science, Kyoto University

C.D. O'Dowd and P.E. Wagner (eds.), *Nucleation and Atmospheric Aerosols*, 994–998.

and continues until full consumption of reactants.[2,4] We have applied double thin film method[5] to identify the individual mixing states of sulfate and nitrate of Kosa (yellow sand) particles simultaneously collected at Korea and Japan.[6] Difference of the dominant species in the two countries (sulfate in Korea and nitrate in Japan) suggested the heterogeneous reaction of HNO_3 vapor with Kosa particles whose surfaces are modified to calcium sulfate. $CaSO_4$ shows no reaction to HNO_3, however, internal diffusion of gaseous HNO_3 in mineral particles may lead to the formation of calcium nitrate with the reaction of unreacted calcite in the particle interior. We have examined the heterogeneous reactivity of gaseous nitric acid with laboratory-generated $CaCO_3$ particles whose surfaces were modified to sulfate by sulfuric acid vapor treatment.

Method and Materials

Preparation of Standard Particles

$CaCO_3$ particles were synthesized through liquid phase reaction. Equivalent amount of solutions (2.2 mM, 500 mL) of $CaCl_2$ and $(NH_4)_2CO_3$ were well mixed with magnetic stirrer for 10 h under the constant temperature of 298K and white crystallized rhombohedral $CaCO_3$ particles were generated in the solution. The particles were collected on the filter by filtration and were dispersed in ethyl alcohol. Each particle diameter was approximately 1–5 µm. Drops containing the particles were put onto a silicon wafer of an about 15 mm square and were dried in nitrogen stream. Total weight of $CaCO_3$ particles on the silicon wafer was adjusted to 60–100 µg to keep the particles in dispersed and nonoverlapping condition on the wafer as possible.

Sulfate coating on the surface of $CaCO_3$ particles was carried out according to the method by Ro et al.[7] Sulfuric acid (Nacalai Tesque, purity 97%) was filled in the glass Petri dish, and the silicon wafer where the particles were adhered was fixed downward to face the liquid surface. The distance from the wafer to the liquid surface was about 5 mm. The reaction coating was performed at 303 K for 24 h and surface sulfate-coated $CaCO_3$ particles were produced.

Exposure Reaction Experiment

Reaction of the sulfate-coated calcite particles with nitric acid vapor was carried out in the experimental system as shown in Figure 1. Teflon tubes were used for nitric acid vapor supply line and the inner wall of the reaction chamber was coated with Teflon. Prepared standard particles on the silicon wafer was set in the flow-type reaction chamber (volume: 4.30 L) and nitrogen stream containing gaseous nitric acid was supplied from the bottom of the chamber with the flow rate of 0.5 L/min. Nitric acid vapor was generated by bubbling method using nitrogen as a carrier gas.

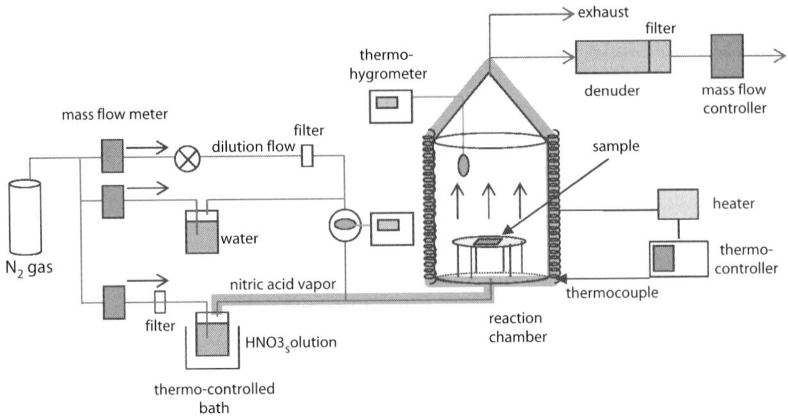

Figure 1 HNO_3 exposure reaction system

HNO_3 supply system and the reaction chamber were all kept at constant temperature of 298 K using thermo-controlled bath or heater with thermo-controller. Concentration of gaseous nitric acid was controlled by changing the flow rate ratio of HNO_3 to dilution flow of nitrogen gas or concentration of nitric acid solution. Relative humidity in the reaction chamber was varied from 20% to 80% by changing the gas flow rate of water bubbling.

After exposure for a definite reaction time, the wafer was set on the sample stage of scanning electron microscope (SEM) (Hitachi, S-3000H) and was subjected to the observation of the particle morphology. Elemental analysis was also performed using energy-dispersive X-ray (EDX) analyzer (EDAX, UTW-CDU-GE). The produced amount of NO_3^- in the sample particles was determined by the analysis of the filtrate after extraction and filtration using Ion Chromatography (IC Shimadzu, HIC-10A). The amount of nonreactive HNO_3 captured at the exit of the chamber by a NaCl coated annular denuder was quantified by IC. Same experiments were also performed for pure $CaCO_3$ particles or pure $CaSO_4$ particles.

Results and Discussion

Figure 2 shows SEM images before exposure to gaseous HNO_3 and after exposure for a given time in case of $CaCO_3$ particles and sulfate-coated $CaCO_3$ particles. HNO_3 vapor pressure was calculated to be about 13 ppm by bubbling 5.0 mol/kg HNO_3 solution and relative humidity was about 80%.[8] In Figure 2(a), almost all $CaCO_3$ particles immediately began to react with HNO_3 to form deliquescent product of $Ca(NO_3)_2$ in two minutes. On the other hand, sulfate-coated $CaCO_3$ particles showed different reactivity as indicated in Figure 2(b). The original particle morphology of rhombohedral crystal was hardly changed after 5 min exposure

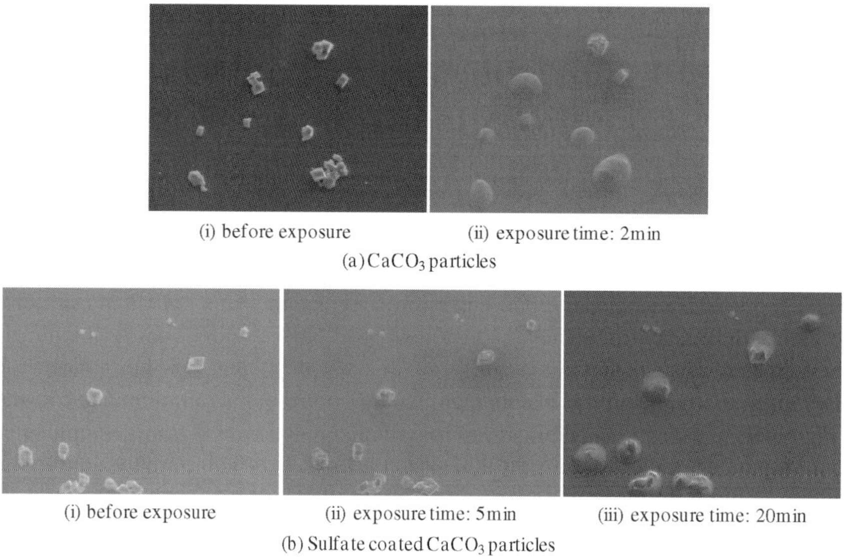

Figure 2 SEM images of $CaCO_3$ particles before/after gaseous HNO_3 exposure: (a) pure $CaCO_3$ particles, (b) sulfate-coated $CaCO_3$ particles. All images were captured at accelerating voltage of 15 kV

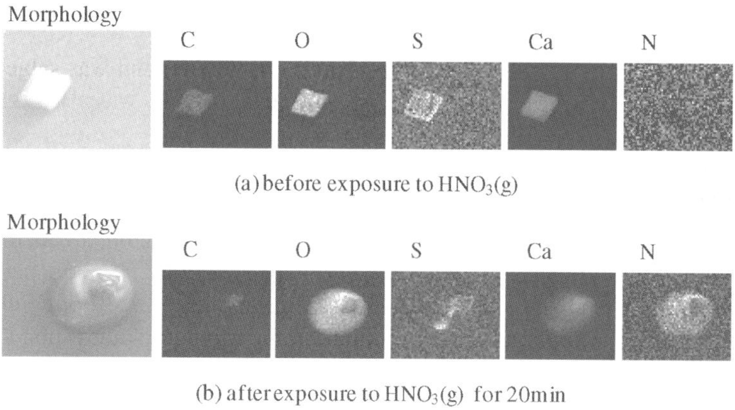

Figure 3 SEM images and elemental mappings of individual sulfate-coated $CaCO_3$ particle: (a) before exposure to HNO_3(g) (b)after exposure to HNO_3(g) for 20 min. All images were captured at accelerating voltage of 15 kV

however after a lapse of 20 min significant percentage of the particles showed the droplet shape implying the formation of deliquescent $Ca(NO_3)_2$. No obvious change of $CaSO_4$ particles was observed after the exposure to HNO_3 vapor for 20 min in terms of morphology and elemental map. Elemental mapping analysis (Figure 3) suggested that HNO_3 reacted with not surface $CaSO_4$ but inner $CaCO_3$ components to form $Ca(NO_3)_2$. $CaCO_3$ particles whose surfaces were coated with $CaSO_4$ reacted with gaseous nitric acid and formed nitrate although surface $CaSO_4$ hindered the evolution of the reaction between $CaCO_3$ and HNO_3.

Acknowledgments This work was partly carried out under the aid of the program "Establishment of COE on Sustainable-Energy System."

References

1. Grassian, V.H., *J. Phys. Chem.* A, **106**, 860–877 (2002).
2. Krueger, B.J., Grassian, V.H., Cowin, J.P., and Laskin, A., *Atmos. Env.*, **38**, 6253–6261(2004).
3. Song, C.H. and Carmichael, G.R., *Atmos. Env.*, **33,** 2203–2218 (1999).
4. Hodzic, A., Bessagnet, B., and Vautard R., *Atmos. Env.*, **40**, 4158–4171 (2006).
5. Tohno, S., Chatani, S., and Kasahara, M., *J. Aerosol Res., Japan*, **13**, 230–236 (1998) (in Japanese).
6. Tohno, S. and Hayakawa, S., Research Report 2001–2005 of Grant-in-Aid for Scientific Research on Priority Areas (A) No.416 *Atmospheric Environmental Impacts of Aerosols in east Asia*, pp. 61–68 (2006).
7. Ro, C.-U., Oh, K.-Y., Osa'n, J., Hoog, J., Worobiec, A., and Van Grieken, R., *Anal. Chem.*, **73**, 4574–4583 (2001).
8. Tang, I.N., Munkelwitz, H.R., and Lee J.H., *Atmos. Env.*, **22**(11), 2579–2585 (1988).

Physical, Chemical and Optical Properties of Fine Aerosol as a Function of Relative Humidity at Gosan, Korea during ABC-EAREX 2005

Kwang-Joo Moon[1,2], Jin-Seok Han[1], and Yoo-Duck Hong[1]

Abstract In order to understand the effect of air pollution in Asia to the formation of the brownish haze which is widely spread over the Asian continents, the water uptake by fine aerosol in the atmosphere has been investigated at Gosan background site, Korea during ABC-EAREX 2005. The concentrations of inorganic ions and carbon components, size distributions, and light scattering coefficients in normal and dry conditions were simultaneously measured by using a parallel integrated monitoring system. The ambient fine aerosols collected at Gosan was dominated by water-soluble ionic species (35%) and carbonaceous materials (18%). Size distribution of particulate surface area showed that the elevation of RH make the ambient fine aerosol grow to be the droplet mode, around 0.6 um, and coarse mode, larger than 2.5 um. The hydroscopic growth was mainly observed in larger size range than 10 um. However, when RH is lower than 80%, the increase of droplet mode area was apparently observed in fine size range ($D_p < 2.5\,\mu m$). Growth factor calculated from scattering coefficients of $PM_{2.5}$ in dry and normal conditions revealed that the water uptake in fine size range began at intermediate RH range, from 40% to 60%, with the average growth factor 1.10 for 40% RH, 1.11 for 50% RH, and 1.17 for 60% RH, respectively. Finally, average chemical composition of $PM_{2.5}$ and the corresponding growth curves were analyzed in order to investigate the relationship between carbonaceous material fraction and hygroscopicity of fine aerosol. As a result, the positive correlation of them was observed suggesting that carbonaceous components also make increase the aerosol water uptake in fine size range as much as water soluble ionic species. In conclusion, the hygroscopic growth of fine aerosol was dominated by the growth of droplet mode aerosol. Especially, the hygroscopic behavior of the droplet mode particles was definitely different with the coarse mode ones. This study suggests that the hygroscopic growth of fine aerosol can be mainly impacted by the organic carbon fraction. In addition, the large fraction of particulate organic materials in the fine size range could accelerate the formation

[1] *Department of Air Quality Research, National Institute of Environmental Research, Incheon, Republic of Korea*

[2] *Department of Environmental Engineering, Inha University, Incheon, Republic of Korea*

C.D. O'Dowd and P.E. Wagner (eds.), *Nucleation and Atmospheric Aerosols*, 999–1003.

of haze which contains large amount of carbonaceous aerosol, especially in the moderately humidified environment (RH < 80%).

Keywords Hygroscopic aerosol, growth factor, scattering coefficient

Introduction

Recently, INDOEX revealed the "brown cloud" phenomenon over the Northern Indian Ocean region with a large impact on the solar radiative heating of the region. The result suggests that urban haze can spread over an entire subcontinent and an ocean basin due to the long-range transport and widely perturb the radiative energy budget of the region and global climate (Ramanathan and Crutzen 2003). In this point of view, Northeast Asia has attracted much attention due to its increasingly high emission of anthropogenic air pollutants. An international program entitled Asian Brown Cloud (ABC) started with Asia in 2002 in order to understand the effects of air pollution and mineral dust in this region to the earth's climate and environment. This study focused on understanding the impact of organic carbon on the hygroscopic properties of ambient fine aerosol in order to find the cause of ABC formation by analyzing the result of ABC-EAREX 2005 field study.

Description of Monitoring

Aerosol monitoring was performed at Gosan, Jeju Island, Korea (33° 17′ N, 126° 10′ E, 70 m a.s.l.) during ABC-EAREX 2005 intensive measurement periods, from 8 March to 6 April 2005. Gosan has served as a super site of ABC project (Ramanathan et al. 2001). The mass and chemical composition of $PM_{2.5}$ was continuously monitored with the interval of 1 h. Anderson FH-62 using β-ray attenuation method was operated to measure $PM_{2.5}$ mass concentration. The real-time monitoring of water-soluble inorganic ions such as SO_4^{2-}, NO_3^-, Cl^-, NH_4^+, Na^+, K^+, Ca^{2+}, Mg^{2+} were conducted by using an Ambient Ion Monitor (AIM) (URG-9000B, URG Co.). The mass concentrations of thermal organic carbon (OC) and black carbon (BC) of $PM_{2.5}$ were thermally measured by semi-continuous carbon analyzer developed by Sunset Laboratory using modified NIOSH 5040 protocol and involved gas TOT method was used for monitoring (Arhami et al. 2006). The hygroscopic and optical properties of ambient aerosols were measured by using a parallel integrated monitoring system which consists of two nephelometers (NGN-3A, Optec Inc.) and two particle sizers (APS 3320, TSI Inc. and PDM 1108, Grimm, Co.). One set is operated in normal condition while the other pre-treats the same aerosols with the diffusion dryer in which particles are exposed to the decreased relative humidity, less than 40%. Two nephelometers measured total

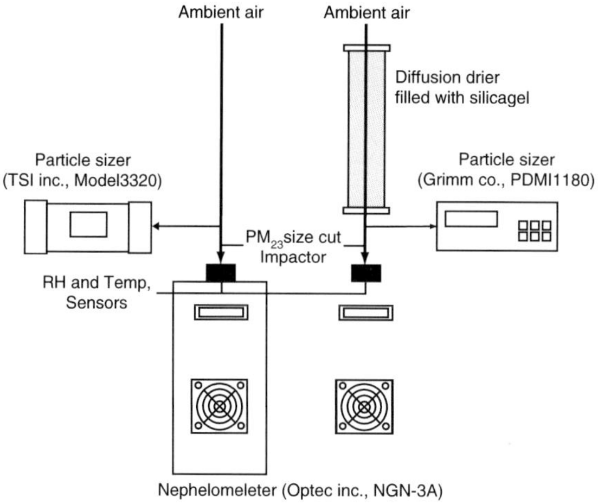

Figure 1 Schematic diagram of parallel integrated monitoring system

scattering of $PM_{2.5}$ at 550 μm, and particle sizers simultaneously counts average number concentration of particles with 10 min integration.

Result and Discussion

Cross-Sectional Area Distributions as a Function of RH

Size distributions of particulate surface area are compared in different RH ranges in order to study the influence of high RH and water uptake on aerosol size distribution. The average size distribution classified by RH in the fine size range (D_p < 2.5 μm) reveals that the particulate surface area increases according to the RH elevation when RH was less than 80%. Especially, the increase of surface area was mainly observed in the droplet mode, 0.3~0.6 μm, as shown in Figure 2.

Variation of Scattering Coefficients According to the Atmospheric RH

Generally, water vapor may be absorbed and condensed on hydrophilic aerosols when the atmospheric relative humidity (RH) is increased. The aerosol water uptake make particle size and the concentration of hydrophilic chemical species including inorganic salts increased. In addition, the optical properties of aerosol

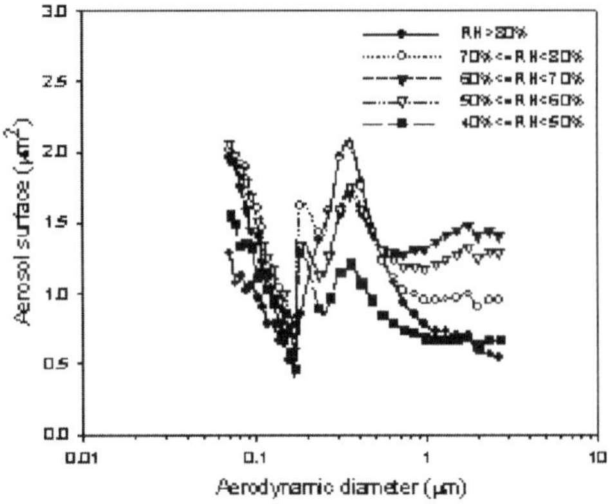

Figure 2 Average size distribution in ambient condition classified by RH

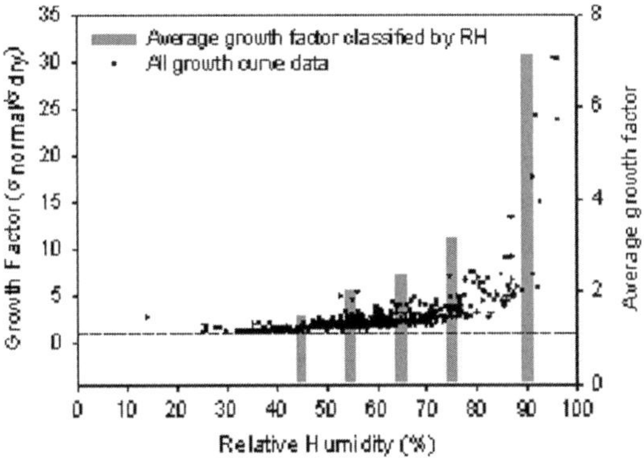

Figure 3 Growth curve data at Gosan during ABC-EAREX 2005

such as scattering and absorbing properties are then modified (Chazette and Liousse 2001). Figure 3 shows the result of the growth factors defined by the ratios of $\sigma_{scat,\,normal}/\sigma_{scat,\,dry}$ as a function of RH at Gosan during ABC-EAREX 2005. Overall, growth factor data of all measurements made a similar appearance with the general shape of the compilation growth curves. Especially, the growth factor curve indicates that the aerosol appeared to start absorbing some water somewhere between 45% and 60% RH with the average growth factor 1.10 for 40% RH, 1.11 for 50% RH, and 1.17 for 60% RH, respectively.

Relationship between Hygroscopicity and Chemical Composition

In order to investigate the influence of chemical components on the hydroscopic growth, the average chemical composition and the corresponding growth curves were studied as shown in Figure 4. As a result, the positive relationship between carbonaceous material fraction and hygroscopicity of fine aerosol was observed

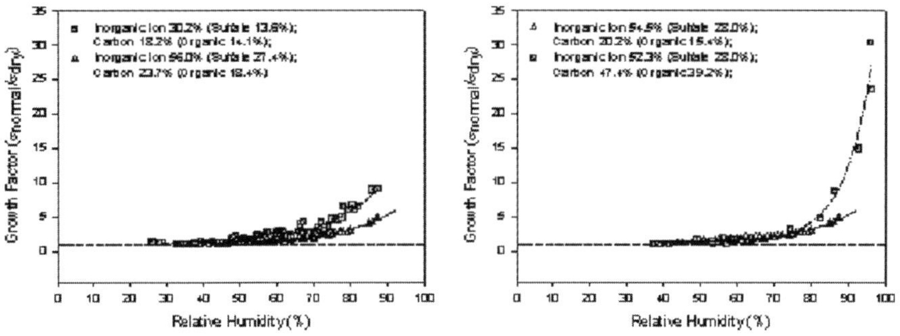

Figure 4 The variability of growth factor curves for different chemical compositions

suggesting that carbonaceous components also contribute as much as water soluble ionic species to the aerosol water uptake at Gosan super site. This result suggests that the hygroscopic growth of fine aerosol can be mainly impacted by the organic carbon fraction, and the large fraction of particulate organic materials could accelerate the formation of haze which consists of fine carbonaceous aerosol, especially in the moderately humidified environment (RH < 80%).

References

Arhami, M., Thomas, K., Fine, P.M., Delfino, R.J., and Sioutas, C. Environ. Sci. Technol., 40, 945–954 (2006).
Chazette, P. and Liousse, C., Atmos. Environ., 35, 2497–2506 (2001).
Ramanathan, V. and Crutzen, P.J., Atmos. Environ., 37, 4033–4035 (2003).
Ramanathan, V. et al., J. Geophs. Res., 106, 28398 (2001).

Austral Summer Particle Formation Events Observed at the King Sejong Station

Y.J. Yoon[1], B.Y. Lee[1], T.J. Choi[1], T.G. Seo[2], and S.S. Yum[3]

Abstract Secondary particle formation events observed at the King Sejong Station (KSJ, 62.22° S, 58.78° W), Antarctica are reported. Climatic effects of these newly formed particles are also proposed from the detected increases in the accumulation mode particle number concentrations.

Keywords Antarctic aerosols, particle formation, nanoparticles, growth

Introduction

Polar regions are located far from the anthropogenic activities and provides favorable environment to study long-term climate change and anthropogenic influence. Atmospheric composition studies under this environment are important to understand a natural long-term variation, and to elucidate the human influence.

Aerosols, which is one of the atmospheric compositions, affect the global radiation budget (Houghton et al. 2001), directly through scattering incoming solar radiation (Charlson et al. 1992), and indirectly through modulation of cloud albedo (Twomey 1974).[1,2,3] Secondary particle formation events are frequently observed over coastal environments (Yoon et al. 2006; O'Dowd et al. 2002),[4,5] the marine boundary layer (Clarke et al. 1998),[6] boreal forests (Makela et al. 1997),[7] Antarctica (Koponen et al. 2003),[8] and Arctic areas (Wiedensohler et al. 1996).[9]

In this paper, we report secondary aerosol formation events observed at the King Sejong Station, Antarctica. The observed contribution of the newly formed particles to the accumulation mode particle population is also reported.

[1] *Korea Polar Research Institute, KORDI, Songdo Techno Park, 7–50, Songdo-dong, Incheon 406–540, Korea*

[2] *Korea Meteorological Administration, 460–18, Sindaebang-dong Seoul 156–720, Korea*

[3] *Department of Atmospheric Sciences, Yonsei University, 134 Sinchon-dong, Seoul 120–749, Korea*

C.D. O'Dowd and P.E. Wagner (eds.), *Nucleation and Atmospheric Aerosols*, 1004–1007.

Experiments

A bank of Condensation Particle Counters (CPC) with different cutoff diameters (TSI CPC model 3025: D > 3 nm and 3010: D > 10 nm) and an Optical Particle Counter (OPC, Grimm model 1.108) were deployed at the King Sejong station (KSJ, 62.22° S, 58.78° W) during austral summer (6 December 2005–7 January 2006). Time resolution of the CPC measurements was set as 1 sec to detect nanometer-size particle formation. Air samples were taken through 30 cm long stainless tubing from outside of the laboratory. The inlet was located about 2 m high from the ground. The OPC measured particle number size distribution for 32 channels ranging from 0.3 µm to 20 µm with a time resolution of 5 min. The OPC was attached to a mast tower, sampling air directly the ambient with 25 cm long unheated inlet.

Meteorological data, irradiance, air temperature, wind speeds, and direction, atmospheric pressure, were obtained from the KSJ meteorological observation tower.

The KSJ station is located in Barton Peninsula, King George Island. The station is facing Marian Cove and the sampling location is about 500 m from the shore. The open ocean around the station has strong biological activities during austral spring and summer seasons. The station has its own pollution source – power station – and only data when are mass was not passing over the pollution source were analyzed in this study.

Results

Particle formation events were observed 18 days out of 33 days (55%) during the measurement period. During the events the total particle number concentration increased from 200 to 300 cm^{-3} up to 50,000 cm^{-3}, and the nucleation lasted for more than 4–5 h depending on the availability of the direct solar radiation. Figures 1 and 2 show an example of particle formation event, on 1 January 2006 and air mass back trajectory for the day, respectively. The air mass originated from the Antarctic continent and traveled over the biologically fertile open ocean before arriving at the sampling site. The particle number concentrations from both CPCs were below 500 cm^{-3} before the formation events. The event started around 4 a.m., as soon as direct solar radiation was available, and lasted for several hours. Under clean marine air mass condition, nucleation events were observed only when the direct solar radiation is available, implying that photochemical reactions of the precursor gases are required to induce the observed events. When air mass traveled over biologically fertile coastal regions, nucleation occurred not only under clear sky days, but also under cloudy conditions, showing higher number concentrations for cloud-free days.

Accumulation mode aerosol number concentration increased dramatically after nucleation events. Figure 3 illustrates an example of increase in the accumulation mode concentration on the 1st January 2006, after a particle formation event. These

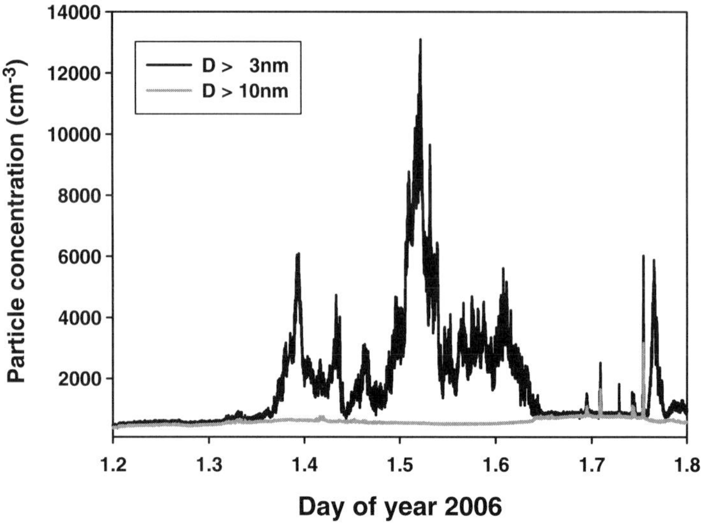

Figure 1 Particle formation event (1 January 2006)

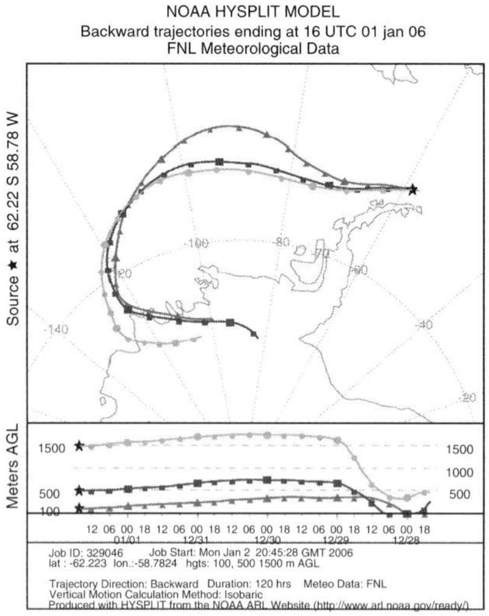

Figure 2 5 day air mass back trajectory on the nucleation event day (1 January 2006)

results suggest effects of particle formation events in the Antarctic regions on climate through direct aerosol effect.

The sources of precursor gases for the secondary particle formation at the Antarctic coastal areas (during austral summer) are thought to be marine biota derived sulfur compounds. To test this hypothesis, long-term measurements of particle formation characteristics should be made.

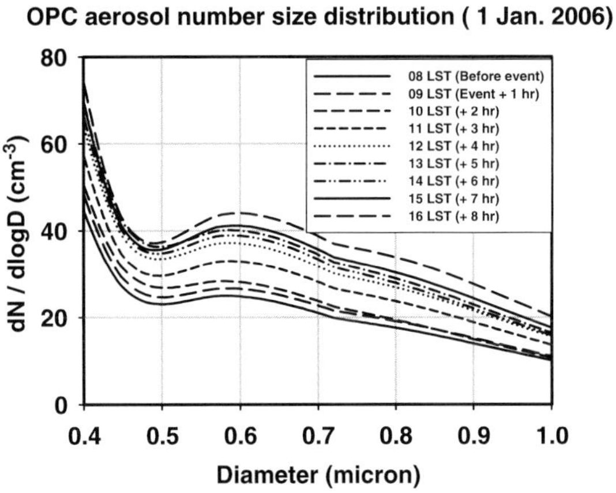

Figure 3 Aerosol number size distribution from KSJ OPC for a nucleation event day (1 January 2006)

Acknowledgments This study was supported by "Integrated research on the Composition of Polar Atmosphere and Climate Change (COMPAC)" (Project PE07030 of Korea Polar Research Institute). S.S. Yum is supported by Grant No. R01–2005–000–11228–0 from the Basic Research Program of the Korea Science and Engineering Foundation. The authors acknowledge the NOAA Air Resources Laboratory (ARL) for the provision of the HYSPLIT transport and dispersion model (http://www.arl.noaa.gov/ready.html).

References

1. Houghton et al., *Climate Change 2001: The Scientific Basis*, **944**, Cambridge: Cambridge University Press (2001).
2. Charlson, R.J., Lovelock, J.E., Andreae, M.O., and Warren, S.G., *Nature*, **326**, 655–661 (1987).
3. Twomey, S., *Atmos. Environ.*, **8**, 1251–1256 (1974).
4. Yoon, Y.J., O'dowd, C.D., Jennings, S.G., and Lee, S.H., *J. Geophys. Res.*, **111**, D13, D13204 (2006).
5. O'Dowd, C.D. et al. *J. Geophys. Res.*, **107**(D19), 8108, doi:10.1029/2001JD000555 (2002).
6. Clarke, A.D., Davis, D., Kapustin, V.N., Eisele, F., Chen, G., Paluch, I., Lenschow, D., Bandy, A.R., Thornton, D., Moore, K., Mauldin, L., Tanner, D., Litchy, M., Carroll, M.A., Collins, J., and Albercook, C., *Science*, **282**, 89–92 (1998).
7. Makela, J.M., Aalto, P., Jokinen, V., Pohja, T., Nissinen, A., Palmroth, S., Markkanen, T., Seitsonen, K., Lihavainen, H., and Kulmala, M., *Geophys. Res. Lett.*, **24**, 1219–1222 (1997).
8. Koponen, I.K., Virkkula, A., Hillamo, R., Kerminen, V.M., and Kulmala, M., *J. Geophys. Res.*, **108**(D18), 4587, doi:10.1029/2003JD003614 (2003).
9. Wiedensohler, A. et al., Occurrence of an ultrafine particle mode less than 20 nm in diameter in the marine boundary layer during Arctic summer and autumn, *Tellus*, **48**, 213–222 (1996).

Analysing the Number Concentration of 50 nm Particles with Multivariate Mixed Effects Model

S. Mikkonen[1], K.E.J. Lehtinen[1,2], A. Hamed[1], J. Joutsensaari[3], and A. Laaksonen[1]

Abstract More than three years of measurements of aerosol size distribution and different gas and meteorological parameters made in Po Valley, Italy were analysed for this study to examine which of the meteorological and trace gas variables effect on the number concentration of 50 nm particles. As the analysis method we used multivariate non-linear mixed effects model. Hourly averages of gas and meteorological parameters measured at SPC were used as predictor variables; the best predictive model was attained with a combination of relative humidity, nucleation event probability, concentrations of SO_2 and ozone, wind speed, and condensation sink. Seasonal variation and the effect of wind direction were also taken cognizance in the mixed model structure.

Keywords Atmospheric aerosols, particle concentration, multivariate analysis, mixed effects model

Introduction

It is well known, that atmospheric aerosols have a great effect on radiation budget, formation of clouds and on climate change. Despite of several years of research, many of the factors affecting the new particle formation and the growth of new formed particles remain unclear. Kulmala et al. (2004) studied several physical and chemical properties affecting particle growth and suggested that the factors affecting new particle formation and growth vary between locations. The purpose of this study is to use statistical data analysis methods to find factors that affect the growth of the particles to detectable sizes in Po Valley area, Italy.

[1] *Department of Physics, University of Kuopio, P.O.B 1627, FIN-70211 Kuopio, Finland*

[2] *Finnish meteorological institute, P.O.B 1627, FIN-70211 Kuopio, Finland*

[3] *Department of Environmental Sciences, University of Kuopio, P.O.B 1627, FIN-70211 Kuopio, Finland*

C.D. O'Dowd and P.E. Wagner (eds.), *Nucleation and Atmospheric Aerosols*, 1008–1012.

Methods

Our data set consists of measurements made between 24 March 2002–30 April 2005 at San Pietro Capofiume (SPC) station in the Po Valley area, Italy. The concentration of particles was measured by using a twin Differential Mobility Particle Sizer (DMPS) system. More details of the measurements can be found in Hamed et al. (2007).

The aim of our study was to predict the number of 50 nm particles with various different factors, i.e., to find the factors affecting the growth of freshly nucleated particles to 50 nm size. Most of our models were combinations from hourly averages of gas and meteorological parameters measured at SPC, including temperature, relative humidity, radiation, O_3, SO_2, NO_2, condensation sink, wind speed and -direction and the probability that the day is a non-event day, i.e., a day when significant new particle formation cannot be seen (Hamed et al. 2007). The probability of a non-event day (PrNE) was calculated with discriminant analysis, details of the method can be found from Mikkonen et al. (2006). PrNE was favoured instead of probability of event day due to better predicting ability. The calculated non-event probability was used instead of observed event classification because otherwise we would have had to exclude the unclassified days, which would have subsidised the data drastically. In addition, the probabilities of a non-event day can be estimated also for those days where the visual event classification has not been made at all.

Since the condensation sink is computed from the size distribution of the particles, there is a risk of circular argumentation when using it in our model. That is why we used only the number of particles bigger than 50 nm in computation of the condensation sink (even if the correlation between the estimates of condensation sinks was 0.99).

Due to complex structure of processes affecting to concentration of small particles it is not reasonable to use general linear effect models in the analysis. We chose to use generalized linear models with logarithmic link function and combine it with mixed model structure (McCulloch and Searle 2001). The main idea of a mixed model is to estimate, not only the mean of the measured response variable y, but also the variance–covariance structure of the data. Modelling the (co-)variances of the variables reduces the bias of the estimates and prevents autocorrelation of the residuals.

Results

We found out that RH, PrNE and the concentration of SO_2 had significant additional variance components for different times of year (Table 1). When the additional variance is taken into account, the model suggests that the effect of RH is negative in January, July, and December and positive for the rest of the year. The decreasing effect of PrNE is on its highest in January, June, and July, and the effect of SO_2 concentration is negative in winter months and positive for the rest of the year.

Table 1 Monthly variation of the intercept term and RH, PrNE, and SO_2 concentration parameters

Month	Intercept	RH	PrNE	SO_2
1	1.4005	−0.0133	−1.5158	0.0094
2	−0.1200	0.0012	−0.3302	−0.0188
3	−0.5569	0.0068	−0.5927	−0.0228
4	−0.4405	0.0039	−0.1994	−0.0464
5	−0.2110	0.0008	−0.3625	0.0802
6	0.1062	0.0020	−1.3562	0.0231
7	0.2364	−0.0022	−1.1124	0.0573
8	−0.5847	0.0068	−0.7054	0.0638
10	−0.1658	−0.0008	−0.5090	0.0545
11	0.0476	0.0013	−0.7333	−0.0399
12	0.2883	−0.0065	−0.7835	−0.0164

Table 2 Variation of fixed effects of wind speed and condensation sink as a function of wind direction

Wind direction	Wind speed	Condensation sink
E	0.0024	19.4326
N	0.0626	15.1709
NE	−0.0221	18.3387
NW	0.0418	16.6365
S	−0.0435	20.8750
SE	0.0181	19.2150
SW	−0.0583	20.7797
W	0.0377	17.6785

Significant weekday effects have been reported for several pollutants (e.g., Marr and Harley 2002). This reflects also to particle concentrations: Tuesday, Wednesday and Thursday seem to have the highest effects on the concentration of 50 nm particles. Effect varies slightly between months.

Wind direction showed out to have a significant effect on the behaviour of some of the predictor variables (Table 2). The effect of local wind speed is mainly positive, i.e., the particle number is higher when the wind speed is higher, except for the winds coming from northeast, south and southwest. The effect of Condensation Sink is on its lowest on north, north-west, and west, respectively. These are the directions to the highly populated areas but also the directions to the mountains.

Effect of Ozone varies within the day (Figure 1); positive effect can be detected on daytime and negative effect on night.

The other significant fixed parameters of the model were (Table 3) relative humidity, which has negative effect on the particle concentration; the growth of the condensation sink, i.e., the difference of two consecutive CS values, which had a positive effect and the yearly intercept terms (2005 is set to reference year).

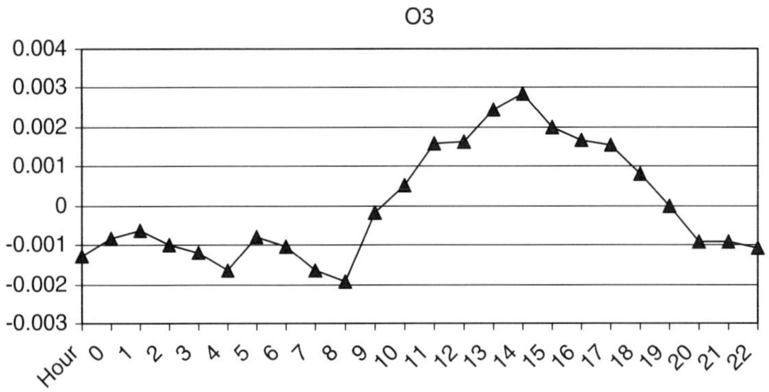

Figure 1 Variation of the random effect of ozone concentration within a day

Table 3 Fixed parameters of the model not dependent on wind direction

Effect	Estimate
Intercept	9.4995
RH	–0.01095
O3	–0.00661
CS50_600diff	12.7232
Year 2002	0.2701
Year 2003	0.2246
Year 2004	0.4118
Year 2005	0

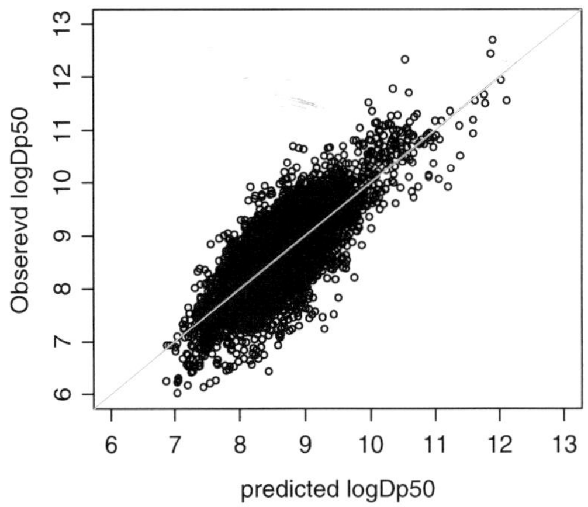

Figure 2 Observed versus predicted number concentrations. Grey diagonal line represents the perfect fit

The coefficient of determination R^2 for the model is 0.61 which indicates that the model explains roughly 61% of the total variation of the particle concentration. The model predicts adequately even the highest peaks of the number concentration (Figure 2) but overestimates slightly the smallest observations.

Acknowledgment This work was supported by Graduate school in Physics, Chemistry, Biology and Meteorology of Atmospheric composition and climate change.

References

Hamed, A. et al., *Atmos. Chem. Phys.*, **7**, 355–376 (2007).
Kulmala, M. et al., *J Aerosol Sci*, **35**,143–176 (2004).
Marr, L.C. and Harley, R.A., *Atmospheric Environment*, **36**, 2327–2335 (2002).
McCulloch, C.E. and Searle, S.R., *Generalized, Linear, and Mixed Models*, New York: Wiley (2001).
Mikkonen, S. et al., *Atmos. Chem. Phys.*, **6**, 5549–5557 (2006).

Aerosol Dynamics Box Model Studies on the Connection of Sulphuric Acid and New Particle Formation

S.-L. Sihto[1], H. Vuollekoski[1], J. Leppä[2], I. Riipinen[1], V.-M. Kerminen[2], H. Korhonen[3], K.E.J. Lehtinen[4], and M. Kulmala[1]

Abstract We have performed a series of simulations with an aerosol dynamics box model to study the connection between new particle formation and sulphuric acid concentration. For nucleation we assumed either activation or kinetic mechanism with linear and square dependence on the sulphuric acid concentration, respectively. We investigated the effect of different simulation parameters on the relationship between sulphuric acid concentration and 3–6 nm particle number concentration or formation rate. The simulations showed that the most important factor affecting the dependence is the growth process, including the amount of condensable vapours, their diurnal profiles, and the saturation vapour pressure of condensable organic vapour.

Keywords Aerosol dynamics, cluster activation, modelling, particle formation, sulphuric acid

Introduction

Aerosol particle formation from gaseous precursors is an important source of new particles in the atmosphere. Measurements at several locations (Idaho Hill, USA[1]; Hyytiälä, Finland[2]; Heidelberg, Germany[3]) have shown a close connection between sulphuric acid and new particle formation: particle formation seems to be a function of gaseous sulphuric acid concentration to the power of 1–2. This dependence is in

[1] *University of Helsinki, Department of Physical Sciences, P.O. Box 64, FI-00014 Univ. of Helsinki, Finland*

[2] *Finnish Meteorological Institute, Climate and Global Change, P.O. Box 503, FI-00101 Helsinki, Finland*

[3] *University of Leeds, School of Earth and Environment, Leeds LS2 9JT, United Kingdom*

[4] *Finnish Meteorological Institute and University of Kuopio, Department of Applied Physics, P.O. Box 1627, FI-70211 Kuopio, Finland*

C.D. O'Dowd and P.E. Wagner (eds.), *Nucleation and Atmospheric Aerosols*, 1013–1017.

contrast with predictions by the theory of ternary sulphuric acid–water–ammonia nucleation which has been considered as the most probable pathway for new particle formation in the planetary boundary layer. To explain the observations, Kulmala et al.[4] proposed the so-called activation and kinetic nucleation mechanisms, with nucleation rate linearly and square dependent on sulphuric acid concentration. The application of these nucleation mechanisms in a global aerosol microphysics model showed good agreement with measurements performed at SMEAR II station in Hyytiälä, Finland.[5]

Particle number concentration measurements start from 3 nm, when actual nucleation, i.e., formation of stable clusters, is assumed to take place near 1 nm. The dependence on sulphuric acid is primarily in the nucleation rate, and due to aerosol dynamical processes such as condensation growth and coagulation the dependence may change as the particles grow to larger sizes. Aerosol dynamical simulations offer a tool to investigate the factors that affect the relationship between sulphuric acid and particle formation during the growth from 1 to 3 nm. We have performed a series of simulations with an aerosol dynamics box model UHMA (University of Helsinki Multicomponent Aerosol model) assuming either activation or kinetic mechanism for nucleation. We studied the effects of different simulation parameters to the dependence on sulphuric acid concentration, by varying, e.g., amount of condensable vapours, saturation concentration of condensable organic vapour and background aerosol distribution.

Methods

The aerosol dynamics model UHMA (University of Helsinki Multicomponent Aerosol model[6]) is a sectional box model that has all the basic aerosol dynamical processes for clear sky conditions implemented: nucleation, condensation, coagulation, and dry deposition. The condensation of a soluble, semi-volatile organic vapour was calculated with the nano-Köhler mechanism.[7]

For nucleation we used the recently proposed activation and kinetic mechanisms with linear and square dependence on the sulphuric acid concentration ($[H_2SO_4]$)[4]:

$$J_{act} = A\left[H_2SO_4\right], \qquad (1)$$
$$J_{kin} = K\left[H_2SO_4\right]^2,$$

where A and K are empirical nucleation coefficients. The values of A and K were taken from measurements made during the QUEST II campaign in Hyytiälä, Finland.[2]

The analysis of measurement data has shown that the number concentration of 3–6 nm particles (N_{3-6}) and the formation rate of 3 nm particles (J_3) correlate with the sulphuric acid concentration after some time delay:

$$N_{3-6}(t) \sim [H_2SO_4]^{n_{N3-6}} (t - \Delta t_{N_{3-6}}),$$
$$J_3(t) \sim [H_2SO_4]^{n_{J3}} (t - \Delta t_{J_3}). \tag{2}$$

Here $n_{N_{3-6}}$ and n_{J_3} are the exponents for N_{3-6} and J_3 correlations, respectively, and $\Delta t_{N_{3-6}}$ and Δt_{J_3} the time delays after which N_{3-6} and J_3 follow the changes in sulphuric acid concentration. Observations from field experiments suggest values $n_{N_{3-6}} = 1$–2 and $n_{J_3} = 1$–3 for the exponents and $\Delta t_{N_{3-6}} = 0.2$–4.1 h and $\Delta t_{J_3} = 0.0$–3.0 h for the time delays.[3] Moreover, the data analysis showed that $n_{N_{3-6}} \leq n_{J_3}$ and that $n_{N3-6} \leq n_{J_1}$, where n_{J_1} is the exponent of nucleation.

Here we studied similar correlations in the simulation data, assuming the activation or kinetic nucleation mechanism, i.e., exponent 1 or 2 in nucleation rate. The formation rate J_3 was calculated from the simulation data as follows:

$$J_3 = \frac{N_3}{\Delta d_p} GR_3, \tag{3}$$

where N_3 is the particle number concentration in a size bin of width Δd_p around particle diameter 3 nm and GR_3 is the growth rate of 3 nm particles. The connections of N_{3-6} and J_3 with sulphuric acid (Eq. (2)) were studied with a numerical fitting procedure by searching through pairs of exponent and time delay, and choosing the combination that gives the greatest correlation coefficient. From the time delay we can estimate the growth rate from 1 to 3 nm:

$$GR_{1-3} = \frac{2\ nm}{\Delta t}, \tag{4}$$

where Δt is assumed as an average time for the clusters to grow from nucleated size of about 1 nm to experimentally observable size of about 3 nm. This method of determining the growth rate is called the time shift analysis. We compare this growth rate to the one calculated from the simulated volume change rate for 1–3 nm particles to validate how good estimate the time shift analysis gives for the real particle growth rate, which varies as a function of both time and particle size.

Results and Discussion

An example of a simulated particle formation event and the fitting of number concentration N_{3-6} with sulphuric acid as a function of time are shown in Figure 1. Typically the exponent for J_3 correlation (Eq. (2)) was greater or equal than the exponent for N_{3-6}, and the time delay for J_3 shorter than for N_{3-6}. These results are in line with the analysis of measurement data reported by Riipinen et al.[3] The time shift analysis has been used to estimate the growth rate of 1–3 nm sized particles from the measurement data (e.g., Sihto et al.[2]). Comparison of the growth rate

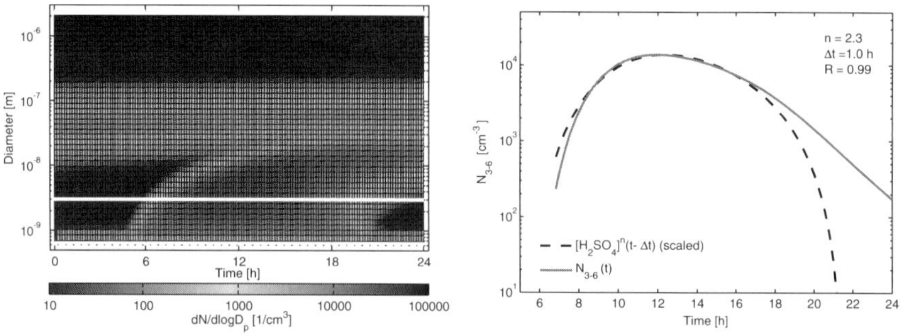

Figure 1 Example of a simulated particle formation event (**left**) and corresponding fit for number concentration in 3–6 nm size range (N_{3-6}) with sulphuric acid concentration (**right**). The exponent, time delay, and correlation coefficient are indicated in the figure

Table 1 Effect of organic vapour saturation concentration (c_{sat}) on the correlation exponents of N_{3-6} and J_3 with the sulphuric acid concentration

	Activation		Kinetic	
Organics c_{sat} (cm^{-3})	$n_{N_{3-6}}$	n_{J_3}	$n_{N_{3-6}}$	n_{J_3}
10	2.1	2.2	3.5	3.6
10^2	2.2	2.3	3.6	3.8
10^3	2.3	2.4	3.8	4.0
10^4	2.5	2.6	4.0	4.1
10^5	2.8	2.9	4.2	4.5
10^6	3.0	3.4	4.4	4.9

calculated from the time delay $\Delta t_{N_{3-6}}$ using Eq. (4) with the growth rates calculated from the simulated volume change rate, indicated that the time delay for N_{3-6} gives, on average, a reasonable estimate for the growth rate of freshly nucleated particles.

The simulations indicate that the most important factor, that affects the relationships of particle concentration and formation rate at 3–6 nm with the sulphuric acid concentration, is the growth process: the amount of condensable vapours (sulphuric acid and an organic vapour), their diurnal profiles, and the saturation vapour concentration of organics. The correlation exponents were observed to decrease as the saturation concentration of the condensable organic vapour was lowered (see Table 1), i.e., when the condensation of organic vapour becomes less limited by the saturation concentration. In order to preserve the exponent in nucleation rate (either 1 or 2), the growth below 3 nm must be fast and not limited by the saturation vapour concentration. In general, in the simulations the exponents for N_{3-6} and J_3 were typically somewhat higher than suggested by the experimental observations. The reasons for this discrepancy need to be studied in more detail. We performed also case studies of some selected days using measured sulphuric acid data as an input.

References

1. Weber, R.J., Marti, J.J., McMurry, P.H., Eisele, F.L., Tanner, D.J., and Jefferson, A., *J. Geophys. Res.*, **102**, 4375–4385 (1997).
2. Sihto, S.-L., Kulmala, M., Kerminen, V.-M., Dal Maso, M., Petäjä, T., Riipinen, I., Korhonen, H., Arnold, F., Janson, R., Boy, M., Laaksonen, A., and Lehtinen, K.E.J., *Atmos. Chem. Phys.*, **6**, 4079–4091 (2006).
3. Riipinen, I., Sihto, S.-L., Kulmala, M., Arnold, F., Dal Maso, M., Birmili, W., Saarnio, K., Teinilä, K., Kerminen, V.-M., Laaksonen, A., Lehtinen, K.E.J., *Atmos. Chem. Phys. Discuss.*, **6**, 10837–10882 (2006).
4. Kulmala, M., Lehtinen, K.E.J., and Laaksonen, A., *Atmos. Chem. Phys.*, **6**, 787–793 (2006).
5. Spracklen, D.V., Carslaw, K.S., Kulmala, M., Kerminen, V.-M., Mann, G.W., and Sihto, S.-L., *Atmos. Chem. Phys.*, **6**, 5631–5648 (2006).
6. Korhonen, H., Lehtinen, K.E.J., and Kulmala, M., *Atmos. Chem. Phys.*, **4**, 757–771 (2004).
7. Kulmala, M., Kerminen, V.-M., Anttila, T., Laaksonen, A. and O'Dowd, C.D., *J. Geophys. Res.*, **109**, D04205 (2004).

Effect of Vegetation on Aerosol Formation in South-East Australia

Tanja Suni[1,4], Hannele Hakola[2], Jaana Bäck[3], Richard Hurley[1], Taina Ruuskanen[4], Markku Kulmala[4], Larisa Sogacheva[4], Ray Leuning[1], Helen Cleugh[1], Eva van Gorsel[1], and Heather Keith[1]

Abstract We compared new particle formation, concentrations of biogenic volatile organic compounds, and air mass arrival routes in a native Australian Eucalypt forest in Tumbarumba, South-east Australia, in 2005–2006. Our aim was to shed light on the effect of local and regional vegetation on biogenic aerosol formation in a remote area largely free of anthropogenic influences. Previous work on aerosol precursors in native Australian forests is all but nonexistent: VOC emissions have been studied only on 15 of the 500–800 Australian Eucalypt species. Formation events were most frequent when wind arrived from the agricultural/pasture lowlands dotted with native Eucalypt woodlands. However, the largest *volumes* of intermediate particles came from the native Eucalypt forest. The high limonene concentration observed at a pasture site suggested that Australian croplands and pastures could be a strong source of secondary aerosol.

Keywords Biogenic aerosol formation, secondary organic aerosol, VOC, eucalyptus

Introduction

The contribution of aerosol formation in forests to the global aerosol load is still uncertain but the phenomenon is clearly widespread and occurs all around the world.[1] Understanding the underlying processes, volatile organic compound (VOC) emission, atmospheric oxidation, and condensation of the resulting less volatile compounds on pre-existing nuclei is important for future development of climate models. So far, most studies of biogenic aerosol production have taken place at

[1]*CSIRO Marine and Atmospheric Research, GPO Box 1666, Canberra ACT 2601, Australia*

[2]*Finnish Meteorological Institute, Erik Palmenin aukio 1,FIN-00560 Helsinki, Finland*

[3]*Department of Forest Ecology, P. O. Box 27, FIN-00014 University of Helsinki, Finland*

[4]*Department of Physical Sciences, P. O. Box 64, FIN-00014 University of Helsinki, Finland*

C.D. O'Dowd and P.E. Wagner (eds.), *Nucleation and Atmospheric Aerosols*, 1018–1022.

continental and coastal sites in the Northern Hemisphere.[1] Our project is the first to study aerosol formation and precursors in a southern hemisphere temperate forest.

Earlier work on biogenic VOCs from Eucalypts is all but nonexistent. There are over 500 native Eucalypt species in Australia, but the most extensive study on their VOC emissions included only 15 of those species.[2] There are no previous measurements of the subalpine Eucalypt species, Alpine Ash and Mountain Gum, widespread in the SE Australian mountains. As it is, making reliable estimates of Australia's continental BVOC emissions and, therefore, of biogenic SOA formation there is difficult.

Materials and Methods

The Tumbarumba flux station[3] is located in the Bago State forest in south eastern New South Wales: 35° 39′ 20.6″ S 148° 09′ 07.5″ E. The forest is classified as a tall open Eucalypt forest. The dominant species are *E. delegatensis* (Alpine Ash) and *E. dalrympleana* (Mountain Gum), and average tree height is 40 m. Elevation of the site is 1,200 m and mean annual precipitation is 1,000 mm. The Bago and Maragle State Forests are adjacent to the south west slopes of southern New South Wales and the 48,400 ha of native forest have been managed for wood production for over 100 years.

We detected new particle formation by measuring size distributions of air ions (naturally charged clusters and aerosol particles) with an Air Ion Spectrometer (AIS).[4,5] The AIS (Airel Ltd., Estonia) measures the size distribution of both negative and positive air ions in the range 0.34–40 nm. Concentrations of VOCs were sampled on stainless steel adsorbent tubes filled with Tenax-TA and Carbopack-B to catch both monoterpenes and isoprene. In front of each tube we used a copper mesh coated with MnO_2 to destroy ozone and prevent further oxidising reactions inside the tube. In Tumbarumba, the tubes were elevated to canopy height (40 m). During field trips up to 1,000 km from Tumbarumba, we used a customised 4WD with a portable meteorological measurement system and a 1.5-m stand and helium balloons for elevating the sample tubes and a Vaisala RH/T sensor to varying canopy heights. We made two field trips to surrounding areas, one in April and one in October 2006, in order to measure VOCs in varying ecosystems around South East Australia. In Tumbarumba, we took VOC samples in connection with two intensive measuring campaigns, EUCAP I and II (EUCalypt forest Aerosols and Precursors) in May–July and November 2006, corresponding to Southern Hemisphere winter and spring.

We analysed back trajectories of air masses arriving at Tumbarumba with the HYSPLIT_4 model, developed by NOAA/ARL.[6] The back trajectories were calculated 96 h backwards in time at a 70-m arrival height above ground level hourly from 8:00 to 16:00 local time to include the main part of most particle formation events. The area within a radius of 1,000 km around Tumbarumba was divided into sectors and characterised according to its dominant vegetation. For this study, the hourly trajectories were manually classified simply into (1) NF: those travelling mainly over native forest (2) all others. We also calculated the distance and the azimuth angle from Tumbarumba for each trajectory at 6, 12, 24, 48, 72, and 96 h from their arrival time.

Results and Discussion

The dominant biogenic VOCs in Tumbarumba were 1,8-cineol (eucalyptol), *a*- and *b*-pinene. Isoprene was hardly present in winter but was abundant in spring (Table 1). Profile measurements showed that the pinenes were produced by the Eucalypts in Tumbarumba, not by the pines a few kilometres further NW. Measurements in varying ecosystems within 1,000 km around Tumbarumba showed great variation in VOC concentrations and compositions. Two extreme examples were a crop/pasture site 400 km south-west from Tumbarumba where limonene concentrations exceeded those of 1,8-cineol ninefold (Table 2) and a pine plantation 60 km east from Tumba where *a*- and *b*-pinene concentrations were 50–60 times those in Tumba. However, these measurements were made inside the canopy and not above it as in Tumbarumba. It is well known that pinenes contribute to secondary aerosol formation[7] but limonene has also recently been shown to have a significant aerosol yield, even greater than a-pinene at least in certain laboratory conditions.[8] If this is the case also in ambient conditions, our results suggest that Australian croplands and pastures could be a strong source of secondary aerosol.

Formation events were most frequent when wind direction was from 260° to 270°, the agricultural/pasture lowland direction dotted with native Eucalypt woodlands (Figure 1). They did not occur when wind was coming from 30° to 60°, the direction largely composed of native grasslands and minimally modified pastures.

Table 1 Dominant biogenic VOC concentrations at 40 m in Tumbarumba in winter and summer

Dominant BVOC concentrations (ngm^{-3})	*a*-pinene	*b*-pinene	1,8-cineol (eucalyptol)	Isoprene
Winter (N = 18)	75	112	146	8
Spring (N = 21)	170	52	280	912

Table 2 Dominant biogenic VOCs at a crop/pasture site and at a pine plantation 400 and 60 km from Tumbarumba, respectively

Dominant BVOC concentrations (ngm^{-3})	Maude (Annual crops and highly modified pastures)	Brindabellas (pine plantation)
2-methyl-3-buten-2-ol	–	355
a-Pinene	–	4695
b-Pinene	11	6294
3-Carene	118	–
Limonene	925	–
1,8-Cineol	110	20

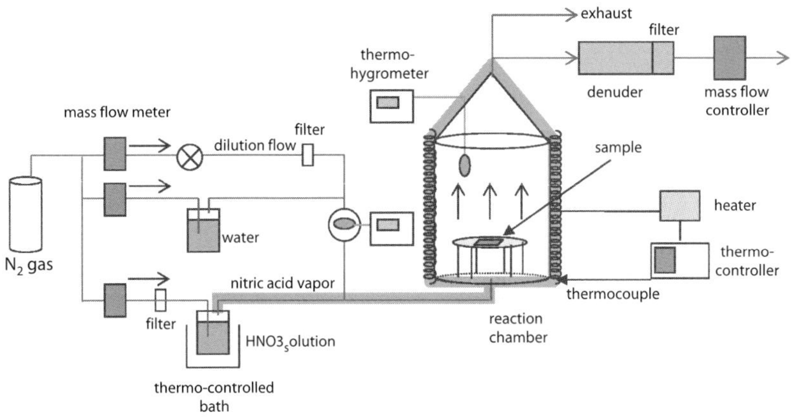

Figure 1 Wind direction during formation events (**left**), all other times (**centre**), and all times during July 2005 to October 2006 (**right**)

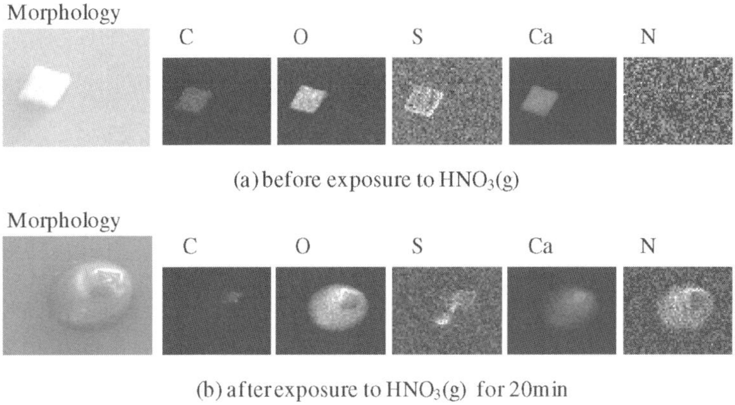

Figure 2 Median volume of 2–14 nm negative ions as a function of trajectory direction at 6 h before arrival to Tumbarumba

However, the largest volumes of intermediate particles came from 200° to 220°, the native Eucalypt forest (Figure 2). The combination of agricultural lowlands and Eucalypt woodlands probably produce the greatest concentrations and variation of combinations of VOCs early in spring when temperatures are higher than on the mountains. Even in winter when the the crops have been harvested but the woodlands are still photosynthesising, these areas are likely to produce significant amounts of VOCs. This could explain the frequency of events from there. However, the tall open Eucalypt forests with their large biomass are likely to be the strongest VOC emitters and therefore the largest volumes of intermediate, nucleation-mode ions are produced there.

Acknowledgment This work was supported by the Academy of Finland, Maj and Tor Nessling Foundation, and the Centennial Foundation of Helsingin Sanomat.

References

1. Kulmala, M., Vehkamaki, H., et al., *J Aerosol Sci*, **35,** 2, 143–176 (2004).
2. He et al. (2000).
3. Leuning, R., Cleugh, H.A., Zegelin, S.J., and Hughes, D., *Agric.For.Meteorol.*, **129**, 151–173 (2005).
4. Mirme, A., Tamm, E., Mordas, G., Vana, M., Uin, J., Mirme, S., Bernotas, T., Laakso, L., Hirsikko, A., and Kulmala, M., *Bor. Env. Res.*, submitted (2007).
5. Hirsikko, A., Laakso, L., Horrak, U., Aalto, P., Kerminen, V-M., and Kulmala, M., *Bor. Env. Res.*, **10,** 5,. 357–369 (2005).
6. Draxler, R. and Hess, G.D., *Aus Met Mag*, **47**, 295–308 (1998).
7. Bonn, B. and Moortgat, G.K., *Atmos. Chem. Phys.*, **2**, 183–196 (2002).
8. Saathoff, H., et al., INTROP/EUROCHAMP/ACCENT Joint Workshop on Organics, 8–11 January 2006, Alpe d'Huez, France, http://imk-aida.fzk.de/abstract/Saathoff_INTROP_2006_poster_final_SOA..pdf

Linear Model of Nucleation Burst in the Atmosphere

A.A. Lushnikov[1], M. Kulmala[1], and Yu. S. Lyubovtseva[1,2]

Abstract A simple linear (with respect to aerosol particle size distribution) model of nucleation bursts in the atmosphere is proposed. The model includes two sources of nonvolatile species, one of which nucleates producing the aerosol particles and the other one condenses onto the particles giving rise to their growth. The most important consequence of the linearity is that the particle size distribution can be presented as a superposition of different regimes. In particular, if the source-enhanced regime is combined with a free one, the latter produces a runaway mode in the particle size distribution. The model serves for estimating the CCN productivity by nucleation bursts.

Keywords Nucleation burst, modelling, condensational growth, tropospheric aerosol

Introduction

Now it becomes more and more evident that the nucleation bursts can contribute substantially to CCN production and can thus exert the climate and the weather conditions on our planet. Commonly accepted opinion connects the nucleation bursts in the atmosphere with an additional production of non-volatile substances that can then nucleate and/or condense on newly born particles, foreign aerosols, or atmospheric ions. The production of nonvolatile substances, in turn, demands some special conditions to be fulfilled imposed on the emission rates of volatile organics from vegetation, current chemical content of the atmosphere, rates of stirring and exchange processes between lower and upper atmospheric layers, presence of foreign aerosols (submicron fraction, first of all) serving as the condensational sinks

[1]*Department of Physical Sciences, University of Helsinki, P.O. Box 64, FI-00014, Helsinki, Finland*

[2]*Geophysical Center of RAS, 3, Molodezhnaya str. 119269 Moscow, Russia*

C.D. O'Dowd and P.E. Wagner (eds.), *Nucleation and Atmospheric Aerosols*, 1023–1027.

for trace gases and the coagulation sinks for the particles of nucleation mode, the interactions with air masses from contaminated or clean regions. Such a plethora of very diverse factors most of which have a stochastic nature prevents direct attacks of this effect. A theoretical modelling of the nucleation bursts is thus of primary importance. It is not surprising therefore a swift growth in the number of efforts directed to modelling the formation of secondary atmospheric aerosols, especially because there is appearing a considerable amount of new information on the mechanisms responsible for the production of secondary aerosols in the atmosphere.

During last two decades numerous attempts of modelling the nucleation processes in the atmosphere have been undertaken. All of them (with no exception) started with commonly accepted opinion that the chemical reactions of trace gases are responsible for the formation of nonvolatile precursors which then give the life to subnano- and nanoparticles in the atmosphere. In their turn, these particles are considered as active participants of the atmospheric chemical cycle leading to the particle formation. Hence, any model of nucleation bursts included (and includes) coupled chemical and aerosol blocks.

Linear Model

The main idea of our model is to decouple the aerosol and chemical parts of the particle formation process and to consider here only the aerosol part of the problem. We thus introduce the concentrations of nonvolatile substances responsible for the particle growth and the rate of embryo production as external parameters whose values can be found either from measurements or calculated independently, once the input concentrations of reactants and the pathways leading to the formation of these nonvolatile substances are known. Next, introducing the embryo production rate allows us to avoid rather slippery problem of the mechanisms responsible for embryos formation. Because neither the pathways nor the mechanisms of production of condensable trace gases and the embryos of condensed phase are well established so far, our semiempirical approach is well approved. Moreover, if we risk to start from the first principles, we need to introduce too many empirical (fitting) parameters.

Aerosol particles throughout entire size range beginning with the smallest ones (with the sizes of order 1 nm) and ending with sufficiently large particles (submicron and micron ones) are formed by some well established mechanisms. These are: condensation and coagulation. Little is known, however, on atmospheric nucleation. This is the reason why this very important process together with intramode coagulation is introduced here as a source of the particles of the smallest sizes. The final productivity of the source is controlled by two these processes simultaneously and is thus always lower than the productivity of the nucleation mechanism alone. Next, coagulation produces the particles distributed over a size interval, rather than the monodisperse ones of a critical size (like in the case of pure nucleation). Respectively, the productivity should be introduced as a function of the particle size and time. In principle, the size dependence of the source can be found theoretically, but it is better to refuse of this idea and to introduce it as the product of a lognormal function and the time dependent total production rate.

The condensational growth depends on the concentrations of condensable vapors, with the condensational efficiencies being known functions of the particle size. The concentrations of condensable trace gases are introduced as known functions. They can also be calculated, once the reaction chains responsible for conversion of volatile trace gases to low volatile ones and respective reaction rates are known (+ stoichiometry of the reactions + initial concentrations of all participants and many other unpleasant things). Of course, neither of this is known and there is not a chance to get this information in the near future.

The losses of particles are caused mainly by preexisting submicron and micron particles. There are also other types of losses: deposition of particles onto leaves of trees, soil losses, scavenging by deposits and mists. Here the loss term is introduced as a sink of small particles on preexisting submicron and micron aerosol particles.

Self-coagulation of particles with sizes exceeding 3 nm is entirely ignored in the model. Many authors estimated the characteristic times of the coagulation process and found them to exceed 10^5 *s*. In what follows we ignore this process. On the contrary, the intermode coagulation (the deposition of newly born particles onto preexisting aerosols) is of great importance and should be taken into account.

The linearity of the model means that the particle size distribution can be schematically presented as follows:

$$n(a;t) = GJ + Fn(a,t=0)\,. \tag{1}$$

Here *G*, *F* are linear evolution operators allowing for restoring the full size distribution by the particle source *J* and the initial conditions $n(t = 0)$. As we will see the first term does not produce a "burst-like" picture, although diurnal increases in the detectable particle concentration are well reproduced. The second term is of special significance. We show that if the source does not work at night time, but a highly disperse (undetectable) aerosol appears from somewhere, then a running-wave type picture typical for the nucleation burst arises. We incline to associate the nucleation events to this very mechanism. Moreover, a fairly recent work [1] already introduced the idea on existence of tiny undetectable aerosol comprising water, sulfuric acid, and ammonium. Our model clearly demonstrates that this invisible mode can produce a clearly expressed hump moving to the right along the size axis.

In the absence of the protoaerosol when the source *J* alone is responsible for the particle formation the picture of the nucleation burst does not remind a run-away mode. The right wing of the particle size distribution crosses the visibility threshold, but then these particle die in interacting with the accumulation mode (the coagulation sink).

Almost all articles on the nucleation burst emphasize the significance of this phenomenon as an important factor of CCN production. Our model allows for rather simple estimation of the productivity of the nucleation burst. CCN particles (with size exceeding $0.1\,^{\mu m}$) are formed by the condensational growth and intramode coagulation. The latter process can be taken into account within the lowest approximation: the rate of the particle production is proportional to $0.5 \int K(r_1,r_2)n(r_1,t)n(r_2,t)dr_1dr_2$, with the size distribution n calculated without accounting for the intramode coagulation process. The final rate is shown to depend critically on the coagulation sinks which are mainly responsible for the losses of newly born particles.

Discussion and Conclusion

Nucleation bursts of the types shown in Figures 1 and 2 were observed in the atmosphere by many authors and in many places (see, e.g. [2]). There appeared a number of explanations of this remarkable phenomenon. In addition to the

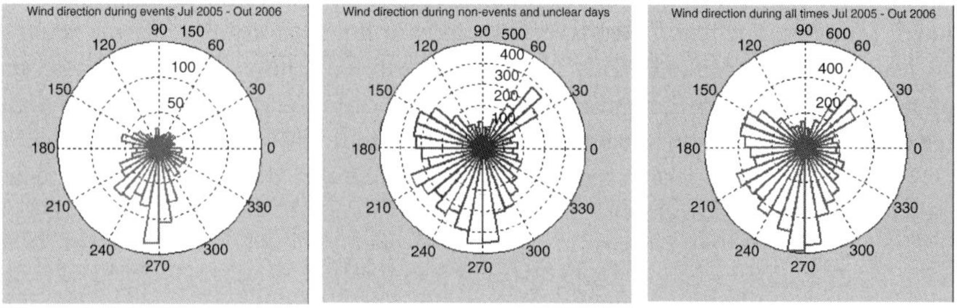

Figure 1 Time evolution of the particle size spectrum. Tiny aerosol particles (protoaerosol) presented in air before the source began to produce fresh aerosol are seen to grow by condensation of low volatile vapors. The resemblance with observed pictures of the nucleation burst is clearly seen

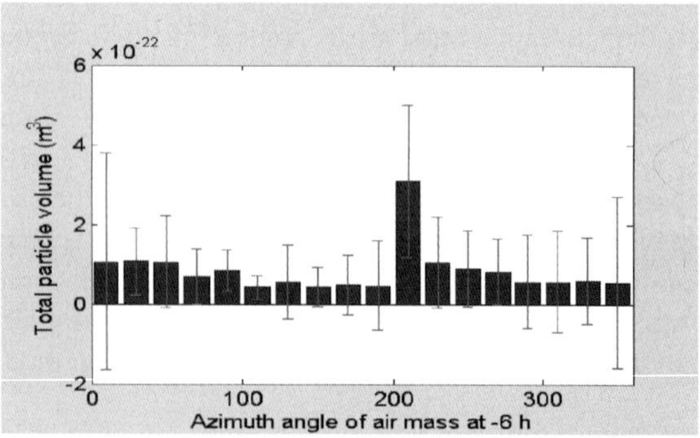

Figure 2 Time evolution of the particle size spectrum. No protoaerosol has presented before the source began to produce fresh particles. Nevertheless, the hump in the spectrum appears in the detectable part of the size spectrum (above 3 nm, vertical dotted line). This is also a nucleation burst, but the picture is qualitatively different from that displayed in Figure 1

photochemical mechanism the "meteorological" mechanisms leading to intense nucleation in the atmosphere were discussed in [3]. The authors of this paper link the nucleation bursts with mixing the layers having different chemical and thermodynamical conditions. It is almost apparent that the meteorological factors play an important role in the formation of the nucleation bursts. But the question comes up how to model all these processes. An attempt to unify existing photochemical models with the thermo– hydrodynamic transport in the atmosphere will hardly give an analyzable results, for both the models are very complicated. We hope that a simplified model of the type described above could help in this situation.

References

1. Kulmala, M., Pirjola, L., and Mäkelä, J.M., Stable sulfate clusters as a source of new atmospheric particles, *Nature*, 66–69 (2000).
2. Kulmala, M., Vehkmäki, H, Petaja, T., Dal Maso, M., Lauri, A., Kerminen, V.M., Birmili, W., and McMurry, P.H., Formation and growth rates of ultrafine atmospheric particles: a review of observations, *J. Aerosol Sci.*, 143–176, (2004).
3. Nilsson, E.D., Rannik, Ü., Kulmala, M., Buzorius, G., and O'Doud, C.D., *Tellus*, B, 441–461 (2001).

Condensational Growth of *n*-Propanol and *n*-Nonane Droplets: Experiments and Model Calculations

Ilona Riipinen[1], Paul M. Winkler[2], Paul E. Wagner[2], Anca I. Hienola[1], Kari E.J. Lehtinen[3], and Markku Kulmala[1]

Abstract We have studied unary and binary condensational growth of propanol-nonane droplets. The condensational growth has been experimentally observed using CAMS-technique. To evaluate the thermophysical properties of the studied compounds, the observations have been compared to model results. The general agreement between the experiments and model is satisfying (maximum differences between modelled and measured radii < 6%), using literature expressions for, e.g., saturation vapour pressures and activity coefficients of the the studied compounds, and assuming mass and thermal accommodation coefficients of unity.

Keywords Condensation, propanol, nonane

Introduction

The condensational growth and evaporation of aerosol particles are one of the theoretically best understood of the aerosol dynamical processes if compared, for instance, with nucleation. Accurate dynamical models describing vapour condensation/evaporation to/from atmospheric aerosol particles are available,[1] given that the thermophysical properties of the investigated compounds are known. Modelled condensational processes are particularly sensitive to saturation vapour pressures and diffusion coefficients, mass and thermal accommodation coefficients, and in the case of mixtures, the activity coefficients. Combined with the accuracy of the condensation theories, this enables one to use condensation models to investigate the preceding thermodynamic properties for different compounds by comparing the

[1] *University of Helsinki, Department of Physical Sciences, P.O. Box 64, FI-00014 University of Helsinki, Finland*

[2] *Institut für Experimentalfysik, Universität Wien, Boltzmanngasse 5, A-1090 Wien, Austria*

[3] *Department of Applied Physics and Finnish Meteorological Institute, University of Kuopio, P. O. Box 1627, FIN-70211 Kuopio, Finland*

C.D. O'Dowd and P.E. Wagner (eds.), *Nucleation and Atmospheric Aerosols*, 1028–1032.

modelled behaviour with well-defined experimental data.[2] Many of the microphysical phenomena related to condensation and nucleation are the same: in both processes the condensing/nucleating substance undergoes a phase transition from gas to condensed phase. The same thermophysical properties are therefore important in both processes, and investigating the condensation behaviour can give valuable information needed also in nucleation research, particularly in applications of classical nucleation theory.[3]

n-propanol and *n*-nonane are organic model compounds that have been subject to several nucleation and condensation studies, both separately[4,5] and as aqueous solutions.[6,7,8,9,10] Recently Gaman et al. (2005)[11] and Viisanen et al. (1998)[12] have presented investigations on the homogeneous nucleation of organic mixtures of *n*-propanol and *n*-nonane. The authors stress the significance of correct thermophysical data in reliable nucleation modelling and address the uncertainties in their calculations partly to uncertainties in the used thermophysics.

In this study we investigate the condensation of *n*-nonane and *n*-propanol both experimentally and theoretically by studying the growth of unary and binary droplets at different ambient conditions. By comparing the modelled condensation with the experimental observations we can obtain valuable information on thermodynamical properties of the studied compounds.

Materials and Methods

Condensational Growth Measurements

The growth of droplets was observed using the constant-angle Mie scattering detection method (CAMS[13]), which provides absolute, time-resolved and non-invasive simultaneous determination of droplet diameter and number density. An accurate experimental approach to the condensational growth of *n*-propanol and *n*-nonane was obtained by studying droplets nucleated on Ag particles. The required vapour supersaturations were achieved by adiabatic expansion chamber resulting in well defined uniform thermodynamical conditions in the measuring volume. Details of the experimental system are presented by, e.g., Winkler et al. (2004).[14] Temperatures of the investigated gas–vapour mixtures after the expansion were 3–7°C (276.15–280.15K), and typical pressures 756–797 hPa. The saturation ratios of *n*-propanol and *n*-nonane inside the expansion chamber ranged from 0.4 to 1.4 and from 0.5 to 1.9, respectively.

Condensation Modelling

The theoretical droplet growth calculations were made using a numerical condensation model (BCOND[1]). The model simulates binary condensation/evaporation by solving coupled differential equations describing mass and heat transfer to/from the droplet. The transition regime corrections and the Kelvin effect were taken into account. The UNIFAC Dortmund[15] model was used for the activity predictions. Four different saturation vapour pressure expressions[16,17,18,19] for *n*-nonane were tested. For propanol, saturation vapour pressure reported by Schmeling and Strey (1983)[20] was used. The mass accommodation coefficients of both of the studied compounds were set to unity. Model input parameters, such as ambient vapour concentrations, temperatures, and droplet concentrations, were obtained from the experiments.

The observed and modelled growth curves were intercompared by setting the first experimentally observed droplet radius of each data set as the initial value for the modelled droplet growth and comparing the subsequent growth processes.

Results and Discussion

Condensational Growth of Pure Compounds

Expressions for the saturation vapour pressures of pure *n*-propanol and *n*-nonane were tested by analysing the growth processes of pure compound droplets with three different condensable vapour concentrations for both vapours. Figures 1a and 1b demonstrate the results of this analysis. In Figure 1a the observed growth of pure propanol droplets is presented along with the modelled growth curve obtained using the saturation vapour pressure expression reported by Schmeling and Strey (1983).[20] A satisfying agreement between the modelled and experimentally observed droplet growth is found. In Figure 1b, on the other hand, similar analysis is presented for *n*-nonane, using all of the four reported literature expressions for the saturation vapour pressure. The best agreement between the experimental and

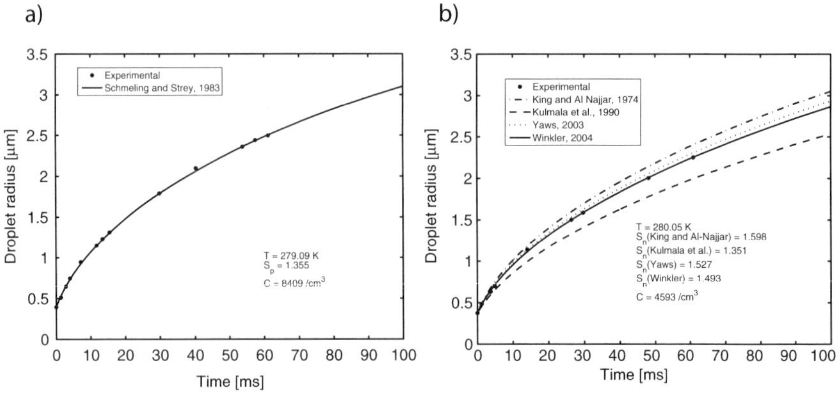

Figure 1 Condensational growth of droplets of pure *n*-propanol (a) and *n*-nonane (b) droplets. The asterisks refer to experimental points and the lines to modelled curves using different saturation vapour pressure expressions

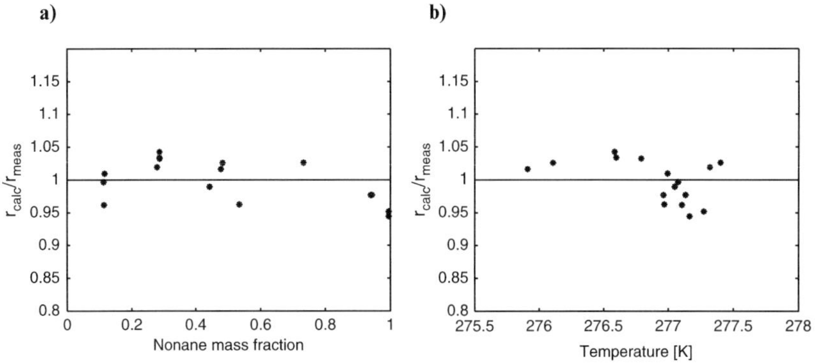

Figure 2 Binary condensation of nonane and propanol: maximum deviations between the modelled and observed droplet radii as a function of droplet composition (a) and temperature (b) corresponding to the whole analysed data set

modelled growth was observed in all studied cases with the saturation vapour pressure reported by Winkler (2004).[19] The expression by Winkler (2004)[19] was thus used for nonane saturation vapour pressure in all binary calculations.

Binary Growth of n-Propanol–n-Nonane Mixture Droplets

The analysed data set of binary propanol–nonane condensation consisted of 16 cases of experimentally observed growth data differing in the vapour supersaturations, temperatures and droplet compositions and concentrations. Comparisons between all experimentally observed and modelled data sets are presented in Figure 2. The maximum deviations between modelled and measured droplet radii are presented in the figure as a function of droplet composition (a) and temperature (b). The maximum deviations have been described by plotting the ratio between calculated and measured droplet radii for the values deviating most in each case. No systematic composition or temperature dependence was found for the maximum deviations between the modelled and experimental data. In general, the agreement between experiments and model calculations is satisfying: even the largest differences are less than 6%. However, the obtained data can be used to optimize, e.g., the activity coefficient predictions further.

References

1. Vesala, T. et al., *J. Aerosol Sci.*, **28**, 565–598 (1997).
2. Riipinen, I. et al., *Atmos. Res.*, **82**, 579–590 (2006).
3. Vehkamäki, H., *Classical Nucleation Theory in Multicomponent Systems*, Springer (2006).
4. Adams, G.W. et al., *J. Chem. Phys.*, **81**, 5074–5078 (1984).

5. Hung, C. et al., *J. Chem. Phys.*, **90**, 1856–1865 (1989).
6. Kulmala, M., *Aerosol Sci.Technol.*, **19**, 381–388 (1993).
7. Strey, R. et al., *J. Chem. Phys.*, **103**, 4333–4345 (1995).
8. Wagner, P.E. and Strey, R., *J. Phys. Chem. B*, **105**, 11656–11661 (2001).
9. Kulmala, M. et al., *J. Phys. Chem. B*, **105**, 11800–11808 (2001).
10. Wagner, P.E. et al., *Phys. Rev. E*, **67**, 021605 (2003).
11. Gaman, A.I. et al., *J. Chem. Phys.*, **123**, 244502 (2005).
12. Viisanen, Y. et al., *J. Chem. Phys.*, **108**, 4257–4266 (1998).
13. Wagner, P.E., *J. Colloid Interf. Sci.*, **105**, 456–467 (1985).
14. Winkler, P.M. et al., *Phys. Rev. Lett.*, **93**, Art. No. 07570 (2004).
15. Gmehling, J. et al., *Fluid Phase Equilibr.*, **54**, 147–165 (1990).
16. Kulmala, M., Vesala, T., and Kalkkinen, J., *Data for Phase Transitions in Aerosol Systems*, Manuscript for laboratory use, University of Helsinki (1991).
17. King, M.B. and Al-Najjar, H., *Chem. Eng. Sci.*, **29**, 1003–1011 (1974).
18. Yaws, C.L., *Yaws' Handbook of Thermodynamic and Physical Properties of Chemical Compounds*, Knovel (2003).
19. Winkler, P.M., Experimental study of condensation processes in systems of water and organic vapors employing an expansion chamber, *Ph.D. thesis*, Universität Wien, Fakultät für Naturwissenschaften und Matematik (2004).
20. Schmeling, T. and Strey, R., *Ber. Bunsen. Phys. Chem.*, **87**, 871–874 (1983).

Connections Between Ambient Sulphuric Acid and New Particle Formation in Hyytiälä and Heidelberg

Ilona Riipinen[1], Sanna-Liisa Sihto[1], Markku Kulmala[1], Frank Arnold[2], Miikka Dal Maso[1,3], Wolfram Birmili[4], Veli-Matti Kerminen[5], Ari Laaksonen[5,6], and Kari E.J. Lehtinen[7]

Abstract The connections between sulphuric acid and particle numbers and formation rates have been investigated during QUEST II-IV campaigns in Hyytiälä and Heidelberg. The particle formation rates typically depend on the sulphuric acid concentration to the power of one or two. The empirical proportionality coefficients of these dependencies range from $3.3 \cdot 10^{-8}$ to $3.5 \cdot 10^{-4}$ (linear dependence) or $2.4 \cdot 10^{-15}$ to $1.3 \cdot 10^{-10}$ (square dependence). The activation of stable clusters and kinetic nucleation mechanisms are capable of explaining the observed particle formation characteristics.

Keywords Sulphuric acid, particle formation and growth, atmospheric nucleation

Introduction

Sulphuric acid has been identified as a key component in aerosol formation and growth.[1,2,3] The exact role of sulphuric acid and the processes limiting the observed new particle formation, however, are still under discussion. Several studies report a

[1] *University of Helsinki, Department of Physical Sciences, P.O. Box 64, FI-00014 University of Helsinki, Finland*

[2] *Max Planck Institute for Nuclear Physics (MPIK), Atmospheric Physics Division, P.O. Box 103980, D-69029 Heidelberg, Germany*

[3] *Forschungszentrum Jülich GmbH, ICG-2: Troposphäre, D-52425 Jülich, Germany*

[4] *Leibniz Institute for Tropospheric Research, Permoserstrasse 15,D-04318 Leipzig, Germany*

[5] *Finnish Meteorological Institute, Erik Palmenin Aukio 1, P.O. Box 503,FI-00101 Helsinki, Finland*

[6] *University of Kuopio, Department of Applied Physics, P.O. Box 1627, FI-70211 Kuopio, Finland*

[7] *Finnish Meteorological Institute and University of Kuopio, Department of Applied Physics, P.O. Box 1627, FI-70211Kuopio, Finland*

C.D. O'Dowd and P.E. Wagner (eds.), *Nucleation and Atmospheric Aerosols*, 1033–1037.

close connection between measured atmospheric sulphuric acid and new particle formation at different locations.[4,5,6]

Kulmala et al. (2006)[7] have proposed the activation of stable clusters[8] to be one of the mechanisms governing the atmospheric particle formation, along with kinetic nucleation.[9,10] The mechanism involves sulphuric acid either as the activating vapour or as a constituent of the activated clusters. Spracklen et al. (2006)[11] have implemented the cluster activation scheme as the particle formation mechanism in a global aerosol microphysics model. The model reproduces the observed secondary aerosol formation with good accuracy. From a modeling point of view, an important advantage of the cluster activation theory is its simplicity: the particle formation rates can be calculated directly from the sulphuric acid concentration, if an estimate for the strength of the coupling is available.

In this work we investigate the role of sulphuric acid in atmospheric particle formation during the QUEST II-IV campaigns conducted in Hyytiälä, Finland, and Heidelberg, Germany. On the one hand, we compare the conditions at the two different sites, Heidelberg representing a polluted environment surrounded by deciduous forest, and Hyytiälä a remote boreal forest site. On the other hand, the data allow for a comparison between two different springs in Hyytiälä: spring 2003 (when the QUEST II took place) has the most particle formation event days so far, whereas the particle formation events in spring 2005 (during QUEST IV) are much fewer in number.

We study the correlations between particle concentrations and formation rates and ambient sulphuric acid concentrations, and investigate the magnitude of empirical nucleation coefficients determining the strength of coupling between sulphuric acid concentrations and particle formation in both locations. For more detailed description of the work, see Sihto et al. (2006)[12] and Riipinen et al. (2006).[13]

Materials and Methods

In this work we used the data sets collected during the QUEST II (15 particle formation events), QUEST III (11 particle formation events) and BACCI/QUEST IV (22 particle formation events) campaigns. The QUEST II has been conducted during March–April 2003 at the SMEAR II[14] station in Hyytiälä (61° 51′ N, 24° 17′ E, 181 m a.s.l.), Finland. The QUEST III campaign has been carried out 28 February–3 April 2004 at the Max Plack Institute for Nuclear Physics in Heidelberg (49° 23′ N, 08° 41′ E, 350 m a.s.l.)[6], Germany, and the BACCI/QUEST IV campaign 5 April–16 May 2005 at the SMEAR II station in Hyytiälä. The SMEAR II station represents a rural site with extensive areas of Scots pine dominated forests surrounding it, whereas the Heidelberg station is situated at a polluted site surrounded by deciduous forest. The utilised data included particle size distributions measured with Twin-DMPS systems, sulphuric acid concentration measured with chemical ionization mass spectrometers (CIMS[15]) and meteorological data, such as temperature and relative humidity.

We studied correlations of form

$$
\begin{aligned}
N_{3-6}\left(t+\Delta t_{N36}\right) &\propto \left[H_2SO_4\right]^{n_{N36}} \\
J_3\left(t+\Delta t_{J3}\right) &\propto \left[H_2SO_4\right]^{n_{J3}} \qquad (1)\\
J_1 &\propto \left[H_2SO_4\right]^{n_{J1}},
\end{aligned}
$$

where N_{3-6} refers to the particle number concentration in the size range 3–6 nm, J_3 and J_1 are the particle formation rates at 3 and 1 nm, t is the time of day and $[H_2SO_4]$ refers to the sulphuric acid concentration. Δt refers to the time lag between the correlated quantities and is related to the time that it takes for the freshly formed particles to grow to detectable sizes, in our case to 3 nm. The exponents n_{N36}, n_{J3}, n_{J1} and the time delays Δt_{N36} and Δt_{J3} were determined from the measured data by a two-parameter fitting procedure, where the correlation coefficients of the relations expressed in Eq. (1) were maximised.

Using J_1 estimated from the particle measurement data we tested the cluster activation mechanism, where the atmospheric nucleation rate depends linearly on sulphuric acid

$$J_1 = A\left[H_2SO_4\right], \qquad (2)$$

and the kinetic nucleation hypothesis, where the particle formation has a square dependence on sulphuric acid concentration

$$J_1 = K[H_2SO_4]^2. \qquad (3)$$

A and K are empirical nucleation coefficients containing the physics and chemistry of the nucleation processes and describing the strength of coupling between particle formation and sulphuric acid concentration. The values of the nucleation coefficients during the QUEST II-IV campaigns were determined by comparing J_1 and J_3 calculated from sulphuric acid data with Eqs. (2) and (3) with those estimated from particle measurements.

Results and Discussion

On all new particle formation days at both locations the number concentrations and formation rates of freshly formed particles correlated with the sulphuric acid concentration. On most (91%) of the particle formation event days the exponents n_{N36}, n_{J3} and n_{J1} were between 1 and 2, implying that the cluster activation scheme or kinetic nucleation are capable of explaining the characteristics of the observed particle formation. Figure 1 shows a plot illustrating the 3 nm particle formation rate and sulphuric acid concentration on 13.4.2005, and the corresponding fitting procedure.

The average nucleation coefficients A and K determined for the QUEST II-IV campaigns are presented in Figure 2. It can be seen that differences exist firstly

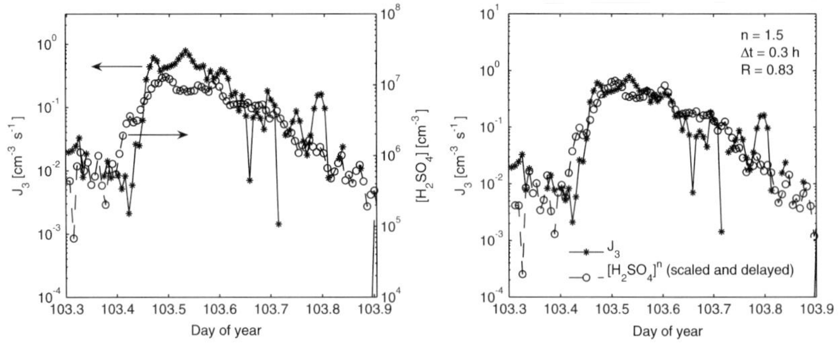

Figure 1 Left: The 3 nm particle formation rate and sulphuric acid concentration on 13.4.2005 in Hyytiälä, Finland. Right: Fitting of sulphuric acid data to the J_3 data, with sulphuric acid raised to the power 1.5 and delayed by 0.3 h. The correlation coefficient between the curves is 0.83

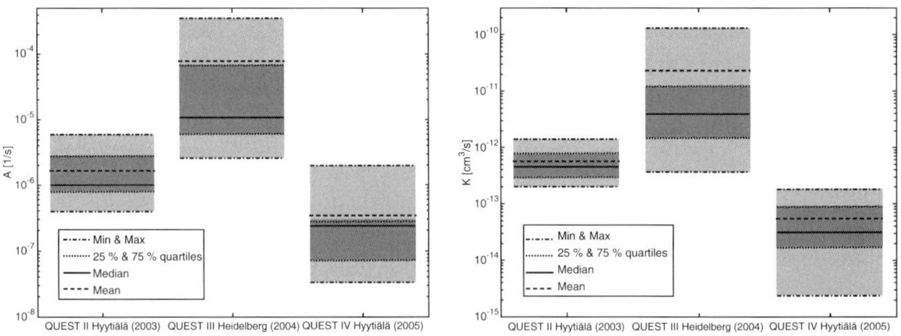

Figure 2 The empirical nucleation coefficients A and K during QUEST II-IV campaigns

between the sites, and secondly between the two different springs in Hyytiälä. Possible reasons for the difference between the campaigns could be, for instance, different concentrations or composition of the activated clusters, different variety or amount of condensable vapours, or other environmental factors related to, e.g., the meteorological conditions or the condensational sink.

Generally it can be concluded that the simple models based on sulphuric acid concentrations and particle formation by cluster activation or kinetic nucleation can predict the occurrence of atmospheric particle formation and growth well. However, future work and longer data sets from different locations are crucially needed to pinpoint the dependencies of the nucleation coefficients.

References

1. Berndt, T. et al., *Science*, **307**, 698–700 (2005).
2. Kulmala, M., *Science*, **302**, 1000–1001 (2003).
3. Kulmala, M. et al., *Atmos. Chem. Phys.*, **4**, 2553–2560 (2004).

4. Weber, R. et al., *J. Atmos. Sci.*, **52**, 2242–2257 (1995).
5. Weber, R. et al., *J. Geophys. Res.*, **102**, 4375–4385 (1997).
6. Fiedler, V. et al., *Atmos. Chem. Phys.*, **5**, 1773–1785 (2005).
7. Kulmala, M. et al., *Atmos. Chem. Phys.*, **6**, 787–793 (2006).
8. Kulmala, M. et al., *Nature*, **404**, 66–69 (2000).
9. McMurry, P.H. and Friedlander, S.K., *Atmos. Environ.*, **13**, 1635–1651 (1979).
10. Lushnikov, A.A. and Kulmala, M., *Phys. Rev. E.*, **58**, 3157–3167 (1998).
11. Spracklen, D.V. et al., *Atmos. Chem. Phys.*, **6**, 5631–5648 (2006).
12. Sihto, S.-L. et al., *Atmos. Chem. Phys.*, **6**, 4079–4091 (2006).
13. Riipinen, I. et al., *Atmos. Chem. Phys. Discuss.*, **6**, 10837–10882 (2006).
14. Hari, P. and Kulmala, M., *Boreal Environ. Res.*, **10**, 315–322 (2005).
15. Hanke, M. et al., *Int. J. Mass Spectr.*, **213**, 91–99 (2002).

Do We Miss Fragile Particles in Particle Size Distribution Measurements?

Boris Bonn[1,2], Michael Boy[1], Hannele Korhonen[3], Tuukka Petäjä[1], and Markku Kulmala[1]

Abstract Currently particle size distribution and number concentration measurements are considered as reference methods for any aerosol particle observations and effects in the ultrafine size range. We simulated a monoterpene–ozone reaction conducted in a smog chamber and the resulting aerosol formation. The measured particle size distribution and number concentration could be obtained only, when we assumed the particles containing a sufficient amount of fragile compounds such as esters, which got lost during the analysis. The study also indicated that peroxy radicals were the key compounds in activating large reactive molecules (secondary ozonides) close to one nm in diameter. Assuming the loss of the "fragile" particle components during detection it was possible to explain the most intense maximum of particle size distribution measurements to occur beyond 10 nm in diameter, not below. This is in contrast for example to any sulphuric acid derived nucleation. Hence, there are indications for an underestimation of secondary particle formation by VOCs.

Keywords Organic nucleation, peroxy radicals, smog chamber, fragile particles

Introduction

Secondary organic aerosol formation from anthropogenic and biogenic volatile organic compounds (VOCs) has been studied for a rather long time (e.g., Seinfeld and Pandis 1998). However, the nucleating compounds and the activation process have not been clarified so far. Homogeneous nucleation of a single low-volatility product fails to explain the onset of new particle formation (e.g., Kamens et al. 1999). It is believed that nucleation starts by the formation of large secondary ozonides (Kamens

[1] *Department of Physical Sciences, Helsinki University, Helsinki, Finland*

[2] *now at: Department of Plant Physiology, Estonian University of Life Sciences, Tartu, Estonia*

[3] *School of Earth and Environment, Leeds University, Leeds, UK*

C.D. O'Dowd and P.E. Wagner (eds.), *Nucleation and Atmospheric Aerosols*, 1038–1042.

et al. 1999; Bonn et al. 2002), which can be suppressed by the addition of water vapour, forming hydroxy-hydroperoxides instead of secondary ozonides. Nevertheless the compound(s) growing the large secondary ozonide molecules ($D_p \approx 0.9$ nm, M = 352 g mol^{-1}) to detectable particle sizes remains unknown. And why is the maximum in particle size distribution measurements seen above 10 nm in diameter instead of at the smallest particle size measured (normally 3 nm in diameter)? One aspect is the significant Kelvin effect at small particle sizes, below about 10 nm.

In this study (Bonn et al. 2007) we link these questions to the reactive growth of secondary ozonide molecules and subsequently aerosol particles by organic peroxy radicals (RO_2), forming esters or ester type complexes, which are easily destroyed by the addition of energy (latent heat release in the particle counter, charging of the particles) as it is known from mass spectroscopy studies.

Methods

Simulation

For theoretical studies the University of Helsinki Multi-component Aerosol box model UHMA was used and extended by an oxidation scheme for tropospheric chemistry and for the investigated monoterpene α-pinene. The UHMA model (Korhonen et al. 2004) contains the simulation of the aerosol processes such as nucleation (formation of secondary ozonides assumed), condensation, coagulation and dry deposition between 0.35 and 500 nm in cluster or particle radius. The chemistry extension (-KAS) treated a reduced tropospheric oxidation scheme as described by Bonn et al. (2005) with added reactions of the stabilized Criegee intermediate and of SO_2 resulting in 215 reactions in total. Therein, the oxidation of SO_2 to sulphuric acid was treated only to check the possibility of sulphuric acid derived clusters, which are activated by organic oxidation products. However, attempts with feasible trace amounts of SO_2 failed in reproducing the onset of nucleation events observed.

Activation of secondary ozonides and the subsequent growth of clusters were considered to occur by heterogeneous reactions of peroxy radicals (HO_2 and RO_2) with the cluster compounds. Therefore, the chemical constituents of the aerosol phase were classified in three groups: (i) acids and hydroperoxides, behaving as HO_2 + peroxy radicals, (ii) nitrates, behaving as NO + peroxy radicals, and (iii) the remainder, behaving as RO_2 + peroxy radicals. The molecular reaction rates have been taken from the master chemical mechanism web page (http://mcm.leeds.ac.uk/MCM). In order to take into account the difference in collision rates (*collision*) of different molecule, cluster or particle sizes, we divided the MCM reaction rate constant by the collision rate of the gas molecules and multiplied with the rate of RO2 with molecules, clusters or particles depending on the size. In order to get the heterogeneous reaction rate constant we assumed a homogeneous mixture of the compounds in the aerosol phase, an assumption that may not strictly hold true.

In order to distinguish between stable or detectable compounds and fragile ones we classified the aerosol products as follows:

Stable compounds: carboxylic acids, carbonyl compounds, nitrates and formats
Fragile compounds: secondary ozonides, hydroperoxides and ester type products
Particle detection was assumed for the stable fraction only.

Experimental

Experiments were performed at the EUPHORE smog chamber facility in Valencia (Spain) within the European project OSOA (Hoffmann (ed) 2002) on the 15th of March 2001. The monoterpene α-pinene was injected twice into a previously cleaned smog chamber of about 200 m^3, which already contained 110 ppbv of ozone. A first injection of 13 μL (corresponding to 9 ppbv) was made at 10:15 and followed by second one (26 μL $\cong$ 18 ppbv) at 14:45. Particle size distribution measurements were performed by a twin DMPS system: A Vienna type DMA and an ultrafine particle counter were used for the lowest size range between 3 and 15 nm and a second DMA was used for aerosol classification between 15 and 400 nm and particles were detected by a CPC3010.

Results

The measured and simulated particle size distributions are displayed in Figure 1. To prevent misinterpretation the simulation output time steps were identical with the ones of the measurement. It appears that the measured particle size distributions can be reproduced fairly well, when assuming the fragile particles to be destroyed by

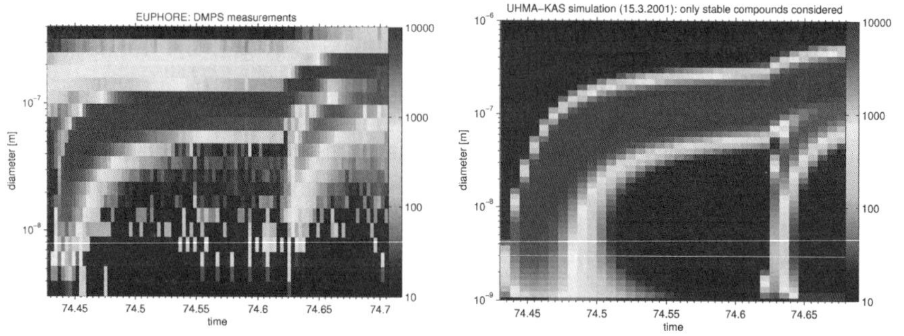

Figure 1 Displayed are the measured particle size distributions of the α-pinene ozonolysis experiment in Valencia (15.3.2001) the left. The UHMA-KAS simulation including the stable aerosol constituents for detection only is shown on the right

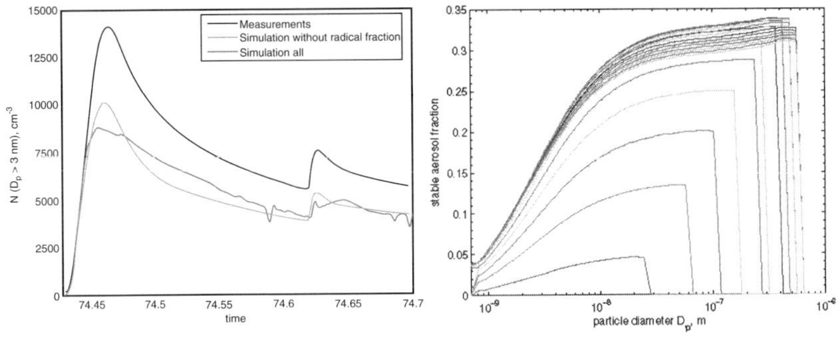

Figure 2 Left: Particle number concentration above 3 nm in diameter: (i) measured, (ii) simulation (all compounds) and Simulation (stable compound only). Displayed on the right is the stable fraction only

the detection method. Otherwise a clear underestimation – especially for ultrafine particles– is obvious.

The very same applies for the particle number concentrations simulated and observed (Figure 2). Especially important is the ratio of HO_2 and RO_2 molecules, which speeds up the initial growth at the second injection. The latter effect is caused by the reaction of HO_2 with the aerosol products transferring them to acid or hydroperoxide type products, for which the reaction with RO_2 is fastest. This emphasizes the role of fragile and detectable particles and therefore particles seem to be detected later during their growth.

Discussion

We were able to simulate secondary organic aerosol formation by nucleation of secondary ozonides and activation by peroxy radicals, if we consider the fragile aerosol constituents to be lost during detection.

But if the assumption of fragile particles to be lost during detection is correct, it would cause serious consequences for the measured and considered particles in VOC environments. This effect is likely to be apparent not only in this smog chamber study of biogenic monoterpene derived SOA but also for anthropogenic VOC studies as well as for ambient conditions, causing an underestimation of organic contributions to new particle formations especially for smaller anthropogenic VOCs in, e.g., urban conditions. Hence we would miss organic particles by the present particle detection method. However, this hypothesis needs to be validated by other methods too in order to extrapolate to atmospheric conditions.

Acknowledgments We kindly acknowledge the support from the staff at Helsinki University the Finnish Met Institute (FMI) and of the staff at the EUPHORE smog chamber facilities in Valencia. Finally the European Union and the Scandinavian states are kindly acknowledged for financial support within the OSOA and the BACCI project.

References

1. Bonn, B., Schuster, G., and Moortgat, G.K., *J. Phys. Chem. A*, **106**, 2869–2881 (2002).
2. Bonn, B., von Kuhlmann, R., and Lawrence, M.G., *J. Atmos.. Chem.*, **51**, 237–270 (2005).
3. Bonn, B., Petäjä, T., Boy, M., Korhonen, H., and Kulmala, M., *Atmos. Chem. Phys. Diss.*, submitted (2007).
4. Finnlayson-Pitts, B and Pitts, *J., Chemistry of the Upper and Lower Atmosphere*. 2nd edn., New York: Academic Press (2000).
5. Hoffmann, T. (ed), *Origin and Formation of Secondary Organic Aerosol (OSOA)*, European project, Bruxelles (2002).
6. Kamens, R.M., Lehtinen, K.E.J., and Laaksonen, A., *Atmos. Chem. Phys.*, **6**, 787–793 (1999).
7. Korhonen, H., Lehtinen, K.E.J. and Kulmala, M., *Atmos. Chem. Phys.*, **4**, 757–771 (2004).
8. Seinfeld and Pandis, S.N., *Atmospheric Chemisty and Physics.*, New York: Wiley Interscience (1998).

Part IX
Marine Aerosol Production

Organic Marine Aerosol: State-of-the-Art and New Findings

Maria Cristina Facchini[1] and C. D. O'Dowd[2]

Abstract The present work reviews the state of knowledge on organic marine aerosol, starting with an historic overview of the topic from the early 1970s to the most recent years. New results from the MAP-EU project are then reported, showing the main features of the organic fraction, and illustrating how primary and secondary components of marine aerosols can be distinguished by means of chemical analysis. The findings also have potential implications for the atmospheric chemistry of aerosol particles and clouds.

Keywords Organic aerosols, primary aerosol, secondary organic aerosol

State of the Art

Marine aerosols contribute significantly to the global aerosol load, thus having an important impact on both the Earth's albedo and climate. Until a few years ago research on marine aerosol was centred on the production of aerosol from sea salt by bubble bursting processes and on non-sea salt sulphate from DMS emitted by phytoplankton, even if the potential role of organics transferred from the ocean surface layer through primary mechanisms and the oxidation of gaseous precursors were hypothesized.[1]

Recently several studies have resulted in a significant improvement of knowledge on organic marine aerosol.[2–4] Such studies have evidenced, on the basis of intensive measurements carried out over the North Atlantic in the period 2002–2005, a strong seasonal variation in the physical and chemical aerosol properties connected to biological oceanic activity. In particular, a

[1]*Istituto di Scienze dell'Atmosfera e del Clima – CNR, via P. Gobetti 101, 40129, Bologna, Italy*

[2]*Department of Experimental Physics & Environmental Change Institute National University of Ireland Galway, Ireland*

C.D. O'Dowd and P.E. Wagner (eds.), *Nucleation and Atmospheric Aerosols*, 1045–1049.

dominant sea salt contribution to sub micron aerosol particle mass was observed during winter, while, moving towards the period of high biological activity, the sea salt signal was progressively replaced by organic matter (mainly insoluble but with a still relevant polar and hydrophilic character) and nss-sulphate as main constituents of the fine aerosol fraction. The organic component was mainly attributed to bubble bursting processes, due to the predominantly insoluble and surface active character of organic carbon in marine aerosol particles.[2,3]

In a recent paper Meskhidze and Nenes[5] observed an appreciable increase in cloud droplet number concentration over the Southern Ocean during phytoplankton bloom, and proposed that oxidation products of isoprene are the potential secondary organic aerosol species responsible for the changes in chemical composition of cloud condensation nuclei, thus influencing cloud droplet number concentration.

The present paper reports some very recent results obtained in the three MAP experiments carried out during 2006. The new findings support the hypothesis that both primary and secondary components are present in organic marine aerosol. The primary component is characterized by low solubility and by surface active properties, while the secondary component appears as a complex mixture of oxidation products of biogenic VOCs.

New Findings on Marine Organic Aerosol in the Framework of the Project Map

Primary Marine Organic Aerosol from Bubble-mediated Aerosol Production Field Experiment

A bubble-mediated aerosol production field experiment was carried out within the Project MAP (Marine Aerosol Production Project) during a 4 week ship-borne campaign over the North Atlantic, from 11 June to 5 July 2006 in the laboratory on board the Celtic Explorer.

The main objective of the experiment was to reproduce in laboratory the primary process of bubble bursting and, by applying the same analytical methodology used for environmental samples, to obtain the fingerprint of the chemical composition of primary marine aerosol during the period of high biological activity. Aerosol samples were produced in a stainless steel bubble tank specially designed for the MAP experiment: The tank was continuously supplied with fresh oceanic water rich in organic and biological species. Two Berner low-pressure impactors were used for aerosol sampling from bubbles tanks, allowing a full characterization of organic and inorganic soluble and insoluble components. In Figure 1 an example of sea salt and total organic carbon chemical composition is shown for a typical bubble bursting sample; nss-sulphate and nitrate concentration are not reported in the figure, since their concentrations were close to the detection limit,

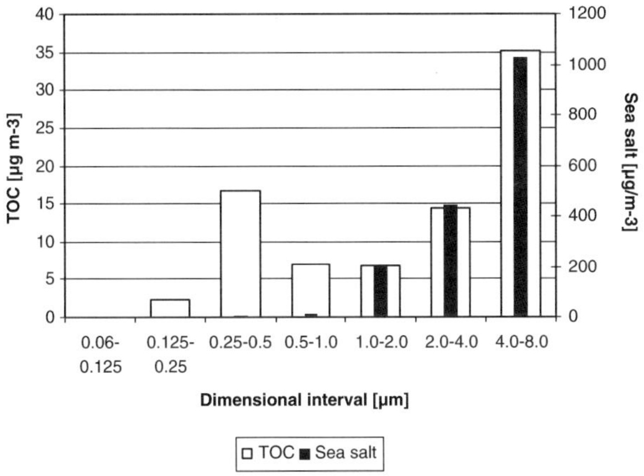

Figure 1 Total organic carbon and sea salt concentration as a function of size in an aerosol sample produced in the MAP bubble bursting experiment

proving that the bubble bursting experiment carried out during the MAP cruise was free of contaminations.

Figure 1 shows that sea salt dominates the coarse fraction, as expected and, conversely, total organic carbon is significantly enriched in the submicrometer size range, representing from 60% to 80% of the total aerosol mass in the 0.05–0.5 µm size interval, and up to 20–30% in the 0.5–0. 1 µm size range, where the contribution of sea salt starts to dominate. Further chemical analysis of organic fraction showed that organic carbon transferred into the aerosol particles by the bubble bursting experiment is mainly water insoluble and strongly surface active.

A similar pattern for water insoluble organic carbon and sea salt size segregated relative composition was observed in the case of clean marine samples collected in previous experiments as well as during the MAP experiment. The observed increasing enrichment of the aerosol organic fraction with decreasing size is also consistent with thermodynamic predictions of the bubble bursting process under conditions in which the ocean surface layer is rich in surfactant material of biogenic origin that can be incorporated into sea spray drops.[6]

These results are in agreement with the hypothesis that primary biogenic sources contribute significantly to the marine aerosol organic carbon concentration observed during periods of phytoplankton bloom, when the WIOC concentration represents a relavant fraction of organic marine aerosol.[4]

The results indicate that organic insoluble carbon in clean marine aerosol can be mainly associated to primary mechanisms and, as a direct consequence of this observation, the water soluble fraction will be mainly accounted for by secondary organic aerosol formation processes.

Secondary Organic Aerosol in Marine Clean Conditions

One main simplification derived from the findings of the bubble bursting experiment is the straightforward association of the two main organic carbon categories, water insoluble and water soluble, to primary and secondary marine aerosol sources, respectively.

To investigate WSOC chemical composition, Proton Nuclear Magnetic Resonance (1 H NMR) was used as the main tool of analysis: the spectra show that purely alkylic groups, CHCH, account for the main part of the WSOC hydrogen content, 55% (percentage contribution to the sum of the identified functional groups). The aromatic/alkene content of WSOC in North Atlantic samples is particularly low, accounting for only 1–4% of the total. Interestingly, the clean marine aerosol samples share the aforementioned HNMR features with samples of biogenic SOA obtained from oxidation of monoterpenes in chamber experiments and during the QUEST-2 field experiment in the boreal forest.[7] One difference is that marine samples also contain aliphatic C-H groups bound to a heteroatom, either nitrogen, sulphur, or a halogen atom, which were not found in the continental biogenic SOA. This functional group accounts for up to 7% of the total organic hydrogen content in aerosol samples collected during the MAP campaign. Larger amounts (up to 20%) were observed during past experiments.[2] Although the source of these peculiar organic components of marine aerosol is yet to be understood, oxidation of aliphatic amines following the reaction proposed by Murphy et al.[8] provides a possible mechanism of secondary formation of aerosol N-containing aliphatic compounds.

Atmospheric Implications

The new results have some potential atmospheric implications. It is easy to imagine that the fate of primary and secondary marine organic aerosol will be quite different depending on aerosol–water interactions: aerosol hygroscopicity and aerosol activation (CCN and ICN). The role of surface active species, mainly emitted from primary sources, should be reevaluated in conjunction to solubility as discussed by.[9] In addition, a general consistency of WS OC composition with that of SOA from biogenic VOCs was found here. However, gaseous precursors comprise N-containing species as well as terpenes and VOCs other than isoprene, thus complicating the picture with respect to the modeling approach proposed by Meskhidze and Nenes.[5]

Acknowledgments The entire MAP team is acknowledge for contributing to the success of the MAP experiments. The current study if funded by the MAP EU project (GOCE-018332).

References

1. Blanchard, D.C, *Science*, **146**, 396–397 (1964).
2. Cavalli, F., Facchini, M.C., Decesari, S., Mircea, M., Emblico, L., Fuzzi, S., Ceburnis, D., Yoon, Y.J., O'Dowd, C.D., Putaud, J.P., and Dell'Acqua, A.J., *Geophys. Res.*, **109**, doi: 10.1029/2004JD005137 (2004).
3. O'Dowd, C.D., Facchini, M.C., Cavalli, F., Ceburnis, D., Mircea, M., Decesari, S., Fuzzi, S., Yoon, Y.J., and Putaud, J.P., *Nature*, **431**, 676–680 (2004).
4. Yoon, Y.J., Ceburnis, D., Cavalli, F., Jourdan, O., Putaud, J.P., Facchini, M.C., Decesari, S., Fuzzi, S., Jennings, S.G., O'Dowd, C.D., *J. Geophys. Res.*, in press, (2006).
5. Meskhidze, N. and Nenes, A., *Science*, **314**, 1419–1423 (2006).
6. Oppo, C. et al., *Marine Chem.*, **63**, 235–253 (1999).
7. Cavalli, F., Facchini, M.C., Decesari, S., Emblico, L., Mircea, M., Jensen, N.R., and Fuzzi, S., Size-segregated aerosol chemical composition at a boreal site in southern Finland, during the QUEST project, *Atmos. Chem. Phys.*, **6**, 993–1002 (2006).
8. Murphy, S.M., Soorooshian, A., Kroll, J.H., Ng, N.L., Chhabra, P., Tong, C., Surrat, J.D., Knipping, E., Flagan, R.C., and Seinfeld, J.H., *Atmos. Chem. Phys. Discuss.*, **7**, 289–349 (2007).
9. Sorjamaa, R., Svenningsson, B., Raatikainen, T., Henning, S., Bilde, M., and Laaksonen, A., *Atmos. Chem. Phys.*, **4**, 2107–2117 (2004).

Physicochemical Characterisation of Marine Boundary Layer Aerosol Particles during the Sea Spray, Gas Fluxes, and Whitecaps (SEASAW) Experiment

Justin J.N. Lingard, Barbara J. Brooks, Sarah J. Norris, Ian M. Brooks, and M.H. Smith

Abstract Ship-borne, Aerosol Time-of-flight Mass Spectrometry (ATOFMS) and volatility (thermal) analysis measurements of the physicochemical composition of marine boundary layer aerosol particles were conducted as part of SEASAW: observations of Sea Spray, Gas Fluxes, and Whitecaps in the NE Atlantic Ocean. Of particular interest were those within the size range of $d_p = 0.2–1.5\,\mu m$, which are associated with cloud formation. Preliminary results showed the sampled particles to be mixture of Na (sea salt), Ca and alumino-silicate rich particles, the latter representing wind-blown dusts and soil. K-rich particles were also detected and were believed to be representative of aerosols derived from biogenic sources. Minor components included organic carbon (OC) and oxygenated carbon compounds that may also have been originated from biogenic sources. The open-ocean sampled aerosol size distribution exhibited a tri-modal size distribution. The results suggest that the MBL aerosols were dominated by two key source types: sea-spray and wind-blown dusts and soil, with a possible minor contribution from biogenic particles.

Keywords ATOFMS, volatility analysis, marine aerosols, physicochemical analysis, SEASAW

Introduction

Marine boundary layer (MBL) aerosol particles represent a chemical and mass-transfer pathway between the ocean surface and lower atmosphere. Reactions in or on sea salt aerosol particles are believed to exert a strong influence on oxidation processes in the MBL through the production of halogen radicals, possibly via Cl atoms produced in the reaction between HCl and OH after the volatilisation of HCl from acidified (H_2SO_4/HNO_3) sea salt particles. Mineral aerosols, e.g., wind-blown dusts

Institute for Atmospheric Science, School of Earth and Environment, University of Leeds, LEEDS, LS2 9JT, UK

C.D. O'Dowd and P.E. Wagner (eds.), *Nucleation and Atmospheric Aerosols*, 1050–1054.

and soils, may significantly affect the N–, S– and atmospheric oxidant cycles, and tend to be less efficient absorbing moisture. Their alkalinity also favours the uptake of SO_2 and NO_2.[1]

Modification of the hydrophilic properties of aerosols can be due to adsorption of organic surfactants. These act to reduce surface tension, enhancing small particle growth at lower RHs in accordance with the Kelvin effect, and thereby promoting cloud condensation nuclei (CCN) activation.[2] Previous research suggests that the key source of organic material in the marine environment is phytoplankton.[3] The impact of biogenically derived organic coatings on the formation and properties of MBL aerosol remains uncertain, but may be substantial and is likely to be largest for the smaller particles. Particles in the size range of $d_p = 0.2–1.0\,\mu m$ account for the majority of CCN number concentrations, whereas particles $d_p = 0.5–2.0\,\mu m$ tend to be the most optically efficient, dominating the direct effect of aerosol-light scattering.[4] SO_4^{2-} particles are the main cooling agents among aerosols, though their effect can be counteracted by the presence of soot.[5] Anthropogenic activity not only furthers atmospheric soot loading, but enhances atmospheric NO_2 concentrations, increasing heterogeneous chemical reactions between sea salt particles and gas-phase HNO_3, leading to particle-phase $NaNO_3$ production and the liberation of HCl (g), contributing further to the production of atmospheric halogen atoms as noted earlier.[6]

Recent advances in the field of single-particle mass spectrometry (SPMS) allow near real-time measurements of the physicochemical properties of MBL aerosol, thus providing improved insights into the sources and role of such particles in atmospheric chemistry.

Method

Field Observations

Measurements were taken of MBL aerosol as part of the field observations of the Sea Spray, Gas fluxes And Whitecaps (SEASAW) campaign during November 2006 onboard RRS Discovery. Sampling was conducted in both open-ocean (NE Atlantic Ocean: 55–59° N, 8–14° W) and the coastal waters off NW Scotland. Detailed qualitative physicochemical analysis of individual airborne MBL aerosol particles was obtained using a TSI Model 3800 Aerosol Time-Of-Flight Mass Spectrometer (ATOFMS)[7,8] fitted with an aerodynamic focusing lens (AFL-100).[9] Parallel analysis of both d_p and composition of accumulation mode particles ($d_{va} = 0.2–1.5\,\mu m$) was provided by a light scattering technique coupled with dual-ion (bipolar) time-of-flight mass spectrometry. These measurements were supplemented with compositional and particle mixing state observations inferred from a high-temperature volatility system (VACC: Volatile Aerosol Chemical Composition), composed of a PMS ASASP-HS and tube heater.[10]

Further aerosol spectrometers: TSI and Grimm CPCs, coupled with PMS PCASP/ASASP-X, FSSP, OAP, and a Grimm 1.108 Portable Dust Monitor, were employed to provided simultaneous measurement of the background aerosol number concentration and size distribution within the size range of d_p = 0.003–150 µm. Atmospheric soot (carbon) mass loadings were measured using a Magee Scientific Aethalometer. Standard meteorological measurements of temperature, pressure, RH, wind speed and direction were also made.

Data Analysis

Mass spectra of 195,485 particles were acquired during the field observations (out of a total of 2,285,819 sampled). The d_{va} and the bipolar mass spectra were recorded for each ablated or "hit" particle. MS-Analyze* was used to query the generated data to produce "peak-lists" of each hit particle. The resulting peak lists were interrogated using MS-Analyze and/or YAADA v.2 (www.yaada.org)†, allowing a range of searches to be carried out, e.g., based on particles of a specific d_{va}, composition, date/time, etc. Data analysis was limited to sampling times when the air flow was within 90° of the bow to prevent the inclusion of sampled aerosol from on-board sources located aft of the aerosol sample inlet. By choosing suitable ion markers in the individual particle mass spectra, the ATOFMS was capable of describing changes in particle composition for a range of chemical species in real time.[11]

Results and Discussion

Figure 1 clearly demonstrates the existence of a tri-modal aerosol population for particles sampled over the open-ocean, with a minor peak at d_{va} = 0.26, accompanied by major peaks at d_{va} = 0.37 and 1.24 µm.

Preliminary analysis of the results shows the sampled aerosol to be dominated by Na-based particles (sea salt particles). Ca and alumino-silicate rich particles were also found to be present, representing wind-blown dusts and soil. K-rich particles were also detected and are believed to be representative of aerosols derived from biogenic sources.[1] Minor components included organic carbon (OC) and oxygenated carbon compounds which may also have been derived from biogenic sources. It is noteworthy

---

*MS-Analyze is a Microsoft Windows®-based, C + + developed by Markus Gälli of TSI Inc., Shoreview, MN., USA. It provides data analysis and display tools, statistical lists, file management capabilities, and export functions for ATOFMS datasets. This allows both the + ve and −ve mass spectrum, as well the d_p, for each particle to be displayed. Dataset interrogation is carried out via Microsoft Access® databases.

†YAADA is a package of data management and analysis functions written for Matlab by Jonathan Allen of Arizona State University, Tempe, AZ, USA, which are designed to process ATOFMS datasets. YAADA includes functions to import, query, plot, and quantitatively analyse ATOFMS data.

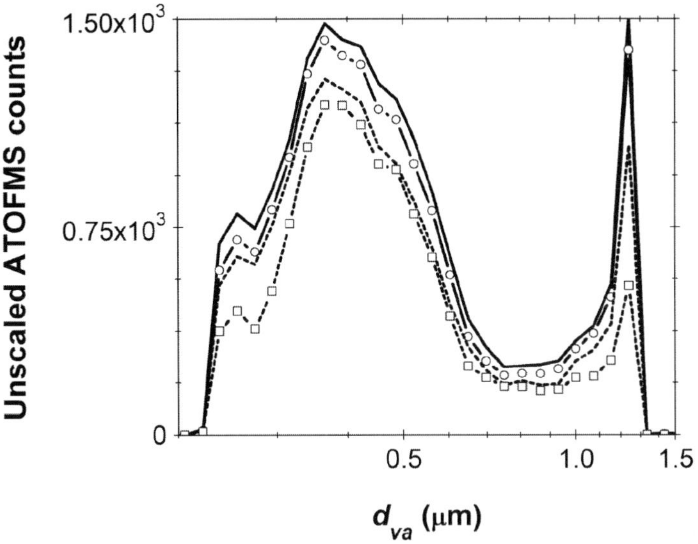

Figure 1 Measured aerosol size distribution of: –: Total ablated; ⊄–⊄: Na-fresh; --: K-fresh; and, ϒ–ϒ: Alumino-silicate particles, sampled over the open ocean (23–25 November 2006)

that the latter particle type was measured during the initial period of the field measurements when air masses originated from upper latitudinal sources (North Canada and Greenland); however, they were absent towards the latter portion of the campaign with air masses originating from South Greenland and Iceland. These subsequently skirted across North-west Europe/mainland UK, mixing with polluted or aged air masses, before arriving at the sampling location – indicated by the ratio of $n(NO_3^-)/n(Cl^-)$ detected.[6] High ATOFMS particle counts were noted corresponding to those measured by the other aerosol spectrometers, especially under high wind speed conditions. The results suggest that MBL aerosol is dominated by two key sources: sea-spray and wind-blown dusts and soil, with a possible minor contribution from biogenic sources.

Presentation of detailed results from the ATOFMS alongside VACC and background measurements will be made. In particular comparisons will be presented between these results and that of a forthcoming cruise in March–April 2007. The role of long-range transport, local sources and the influence of meteorology will be investigated.

Acknowledgments This work was supported by funding from the Natural Environment Research Council (NERC) under grant NE/C001842/1. The assistance of the Captain and crew of RRS Discovery and the technical support of Dan Comben and Chris Barnard from National Marine Facilities in making these measurements possible is gratefully acknowledged. Additional thanks are extended to Markus Gälli of TSI Inc., Shoreview, Minnesota, USA, for instrumental and analytical assistance. Jonathan Allen of Arizona State University, Tempe, Arizona, USA, provided assistance with YAADA. Thanks to Gourihar Kulkarni of IAS, Leeds, for helpful discussions regarding MATLAB and YAADA.

References

1. Andreae, M.O. and Crutzen, P.J., *Science*, **276**, doi: 10.1126/science.276.5315.1052 (1997).
2. Ellison, G.B., Tuck, A.F., and Vaida, V., *J. Geophys. Res.*, **104**(D9), doi: 10.1029/1999JD900073 (1999).
3. O'Dowd, C.D., Facchini, M.C., Cavalli, F., Ceburnis, D., Mircea, M., Decesari, S. Fuzzi, S., Yoon, Y.J., and Putaud, J.P., *Nature*, **431**, doi:10.1038/nature02959 (2004).
4. Russell, L.M., Maria, S.F., and Myneni, S.C.B., *Geophys. Res. Lett.*, **29**, doi: 10.1029/2002GL014874 (2002).
5. Buseck, P.R. and Pósfai, M., *P. Natl. Acad. Sci. USA*, **96**, doi:10.1073/pnas.96.7.3372 (1999).
6. Gard, E.E., Kleeman, M.J., Gross, D.S., Hughes, L.S., Allen, J.O., Morrical, B.D., Fergenson, D.P., Dienes, T., Gälli, M.E., Johnson, R.J., Cass, G.R., and Prather, K.A., *Science*, **279**, doi: 10.1126/science.279.5354.1184 (1998).
7. Gard, E.E., Mayer, J.E., Morrical, B.D, Dienes, T., Ferguson, D.P., and Prather, K.A., *Anal. Chem.*, **69**, 4083–4091 (1997).
8. Noble, C.A. and Prather, K.A., *Environ. Sci. Tech.*, **30**, 2667–2680 (1996).
9. Liu, P., Ziemann, P.J., Kittelson, D.B., and McMurry, P.H., *Aerosol Sci. Technol.*, **22**, 293–324 (1995).
10. Smith, M.H. and O'Dowd, C.D., *J. Geophys. Res.*, **101**(D14), 19583–19591 (1996).
11. Dall'Osto, M., Beddows, D.C.S., Kinnersley, R.P., Harrison, R.M., Donovan, R.J., and Heal, M.R., *J. Geophys. Res.*, **109**(D21302), doi: 10.1029/2004JD004747 (2004).

Iodine Speciation in Marine Boundary Layer

Senchao Lai, Nicola Springer, Julia Münz, and Thorsten Hoffmann

Abstract Iodine chemistry is important in marine boundary layer especially for new particle formation. Here we present the iodine speciation results including the inorganic and organic iodine in $PM_{2.5}$ and size fractionated particles and the gaseous I_2 in oceanic air. Aerosol samples are from the campaign at Mace Head, Ireland, and the research cruise in North Atlantic during June–July 2006. Denuder samples for I_2 measurement is from the North Atlantic. I_2 concentration was extremely low in the oceanic atmosphere. Water soluble organic iodine (SOI) and non-water soluble iodine (NSI) were shown to be the major parts in iodine content in marine aerosol. The speciation of SOI and NSI is still open and the high percentages of SOI and NSI hint that there are still unknown processes exist in marine iodine formation.

Keywords Iodine speciation, iodide, iodate, soluble organic iodine, size distribution

Introduction

Over the past few years, there has been increasing evidence that iodine does have an important influence on marine atmospheric chemistry. The current understanding is that volatile iodine precursors, such as diiodomethane or even molecular iodine are released by marine algae into the atmosphere, where they are rapidly photolyzed during daytime and form low volatile iodine oxides which finally self-nucleate. However, many open questions remain although numerous studies have been done to investigate the release mechanism of the volatile iodine compound from phytoplankton, the preliminary speciation of iodine in aerosol, sea-air iodine cycling and so on. Therefore, sampling, identification and quantification of those iodine species have become necessary for the understanding of atmospheric iodine chemistry in MBL.

Institute of Inorganic and Analytical Chemistry, Johannes Gutenberg University of Mainz Duesbergweg 10–14, D-55128 Mainz, Germany

C.D. O'Dowd and P.E. Wagner (eds.), *Nucleation and Atmospheric Aerosols*, 1055–1059.

Methodology

Sample Collection

In order to obtain spot aerosol samples, onshore and offshore campaigns have been performed during June–July 2006. At Mace Head Atmospheric Research Station, Ireland, a virtual impactor $PM_{2.5}$ (aerosol with aerodynamic diameter lower than 2.5 µm) sampler and a five-stage cascade sampler, namely Berner Impactor with fractionated sizes of 0.085–0.25, 0.25–0.71, 0.71–2.0, 2.0–5.9, 5.9–10 µm, were used to take $PM_{2.5}$ marine aerosol samples and size fractionated particles simultaneously. Offshore sampling with one virtual impactor $PM_{2.5}$ sampler and a self-developed denuder technique was conducted at Celtic Explorer scientific vessel over the North Atlantic Ocean to collect $PM_{2.5}$ samples and gaseous I_2 in the oceanic atmosphere (Figure 1).

Sample Analysis

After sampling, ultrasonic assisted water extraction was applied to collect water soluble iodine species from the filter samples. Inductively Coupled Plasma Mass Spectrometry (ICP-MS) was used to measurement water soluble iodine, while a new online analytical technique of Gel Electrophoresis (GE) coupled with ICP-MS was developed for iodide and iodate separation and accurate quantification in marine aerosol (Brüchert et al. 2007). The water extracted filter was followed by the Tetra-methyl-ammonium Hydroxide (TMAH) extraction to extract non-water soluble iodine (NSI). The extract was analyzed by ICP-MS. ^{129}I isotope dilution was used in all iodine quantification with ICP-MS. Detection limits are 0.1 ng/ml for

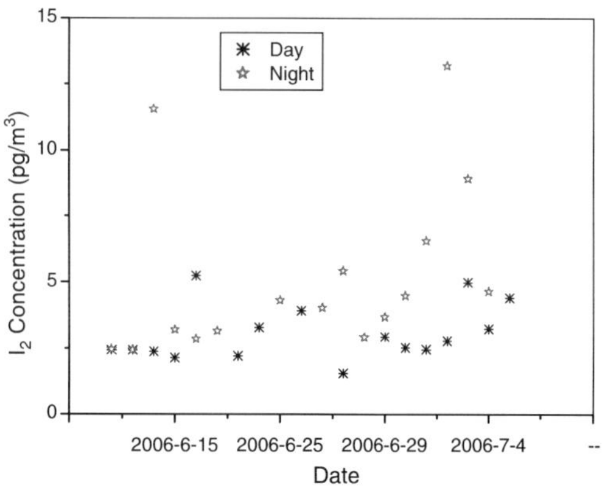

Figure 1 I_2 in Celtic explorer campaign

ICP-MS and GE-ICP-MS (expressed as iodine) iodine measurements, respectively. Soluble organic iodine (SOI) is calculated by the difference between water soluble iodine and the sum of iodide and iodate. Total iodine in particle phase is the sum of water soluble iodine and non-water soluble iodine. Gaseous I_2 was collected by a self-made starch denuder and followed by a TMAH extraction and ICP-MS detection.

Results

The observed total iodine concentration in $PM_{2.5}$ varied from 10.58 to 100.20 ng/m^3 during the offshore campaign on North Atlantic. At Mace Head station, total iodine was found with a range of 7.77–97.47 ng/m^3 in $PM_{2.5}$ samples (Table 1). However, very low concentrations of iodate and iodide were found in $PM_{2.5}$ samples in both campaigns. Iodide concentration was in the range of 0.26–5.57 ng/m^3 in $PM_{2.5}$ in open Atlantic Ocean atmosphere and 0.93–6.00 ng/m^3 at Mace Head station. Iodate concentrations were lower than iodide concentrations in all $PM_{2.5}$ samples with the range of N.A.–0.44 ng/m^3 during Celtic Explorer Campaign and N.A.–7.15 ng/m^3 during onshore campaign (Table 2). Molar ratio of iodate/iodide was in the range of 0–0.33 with the median of 0.10 in the North Atlantic oceanic air while it was in the range of 0–0.67 with the median of 0.24 at Mace Head. The similar trends have been found in other researches.

Table 1 Concentrations of iodine species in PM2.5 at Mace head (ng/m^3)

	Morning		Afternoon		Night	
Iodine species	Range	Median	Range	Median	Range	Median
Iodide	0.93–4.88	2.09	2.21–6.00	3.44	1.03–2.23	1.55
Iodate	N.A.-1.23	0.74	0.21–7.15	1.25	N.A.-3.31	0.72
SOI	15.95–97.04	36.08	11.59–44.60	16.09	1.42–17.56	7.19
NSI	N.A.-35.32	17.27	6.82–37.37	18.38	0.99–7.84	5.86
Total iodine	40.37–97.96	56.96	30.55–88.13	35.72	7.77–25.92	15.93

*N.A. means not available. **Morning, afternoon, and night sample numbers are all 5.

Table 2 Concentrations of iodine species in PM2.5 during Celtic explorer campaign (ng/m^3)

	Day		Night	
Iodine species	Range	Median	Range	Median
Iodide	0.47–5.57	0.86	0.26–1.95	1.12
Iodate	N.A–0.44	0.13	N.A.–0.42	0.20
SOI	1.82–48.17	8.82	8.97–50.66	13.39
Total iodine	10.58–66.73	20.74	15.71–100.20	33.88

*N.A. means not available. **Day sample number is 7 and night sample number is 6.

In this work, SOI and NSI account for the predominate parts in $PM_{2.5}$ during the onshore and offshore campaigns. Variable soluble organic iodine was obtained with the percentage from 17.2% to 80.4% in offshore samples and from 14.5% to 87.9% in onshore samples. Although SOI and NSI have not been specified but the high percentages in $PM_{2.5}$ indicate that more atmospheric reactions about iodine especially those relating to the organic iodine formation and so on should take into account.

The I_2 concentrations obtained in the Atlantic atmosphere were extremely low. The day samples had lower level of I_2 than the night samples. Some of the night samples had several times higher concentrations than other samples.

Interesting diurnal variation of iodine species were shown at Mace Head. Lowest concentrations of iodine species were observed at night. For iodide and iodate, the samples in afternoon have higher concentrations, whereas the highest concentration of SOI was in the morning samples. However, all iodine species in night samples were higher than those in day samples in the ship campaign.

Size distribution information was also investigated in this study by the five-stage cascade sampler (Berner Impactor). Our results show that more iodide resided in fine particles in day samples but reverse trend were found at night, while iodate was clearly presented in the coarse ones. SOI and NSI both distributed mainly in fine mode particles. The size distribution may reveal the different uptake processes of the different iodine species (Figure 2).

Conclusion

Our results show that the soluble organic iodine and the non-soluble iodine are the major iodine species in $PM_{2.5}$. The presence of SOI in coastal and open oceanic atmosphere suggests that it is important to clarify the possible pathway to form

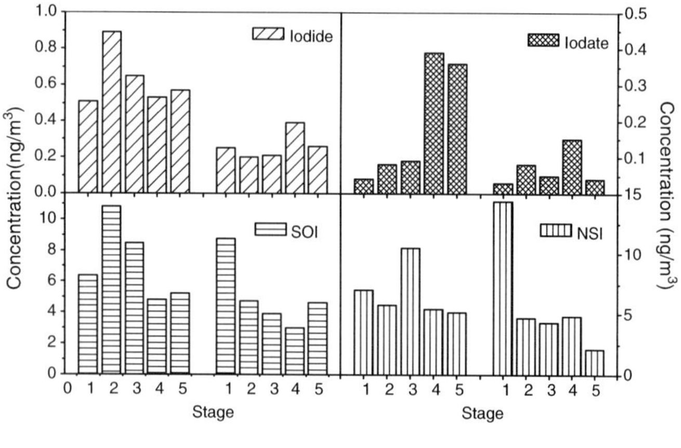

Figure 2 Size distribution of iodine species

SOI. This finding still brings more peradventure concerning to the complexity of the atmospheric iodine reactions because atmospheric iodine chemistry models currently do not include any aerosol-phase reactions with organic substances. Moreover, since NSI is still unknown for the researchers until now, its existence implies that more attentions should be paid to it. However, the Inorganic iodine accounts for considerable low part with relatively high iodide compared with iodate, which is believed to be the opposite due to iodide recycling reactions and accumulation of iodate in aerosol particles as it ages. In coastal atmosphere, sunlight intensity may be favorable for iodide and iodate formation and uptake to the aerosol since higher levels iodide and iodate were observed in the afternoon samples. It is different from SOI with its highest concentration in the morning sample, which indicates its uptake process to aerosol may relate to an unknown mechanism. The data from the Celtic Explorer present different trend with high concentrations of the iodine species including inorganic, organic iodine in particles phase and I_2 in gas phase at night. The complicated influence from the different factors such as meteorological conditions and so on would make it difficult for understanding. For further understanding of the whole mechanism for iodine-containing emission and their role in particle formation process, more works are still needed in the future.

Acknowledgments This work was supported by the European Union integrated project "Marine Aerosol Production from Natural Source" and the Deutsche Forschungsgemeinschaft (DFG) Interdisciplinary Research Training Group program ("Graduiertenkolleg") on "Trace analysis of elemental species: Development of methods and applications". The authors would also like to thank Dr. Jörg Bettmer, Mr. Jens Heilmann, Mr. Wolfram Brüchert, Mr. Andereas Helfrich and their groopmates for their assistance in ICP-MS measurements.

References

Baker, A.R., *Geophys. Res. Lett.*, **31**, L23S02 (2004).
Baker, A.R., *Environ. Chem.*, **2**, 295–298 (2005).
Bruchert, W., Helfrich, A., Zinn, N., Klimach, T., Breckheimer, M. Chen, H., Lai, S., Hoffmann, T., and Bettmer, J., *Anal. Chem.*, in press (2007).
Chen, H., *Ph.D. Dissertation*, University of Mainz (2005).
O'Dowd, C. and Hoffmann, T., *Environ. Chem.*, **2**, 245–255 (2005).

DOAS Measurements of Iodine Oxides in the Framework of the MAP (Marine Aerosol Production) Project

Katja Seitz

Abstract Recent field and laboratory studies indicate a great relevance of reactive iodine in new particle formation processes. Since particle in the marine atmosphere affect the microphysical properties of clouds, they have a potential impact on climate. Objective of the MAP (Marine Aerosol Production) project was to quantify the key processes associated with primary (PMA) and secondary marine aerosol (SMA) production from natural sources. Iodine oxides were detected with the Differential Optical Absorption Spectroscopy (DOAS). Results from an intensive campaign at Mace Head and onboard the Celtic Explorer and their correlation with particle bursts will be presented.

Keywords Iodine oxides, particle formation

Doas Measurements of Iodine Oxides and the Correlation with Particle Bursts

Introduction

The measurement principle is based on the Long Path Differential Optical Absorption Spectroscopy (LP-DOAS) determining the absorption of different trace gases along an extended light path in the open atmosphere (e.g., Platt 1994). For the wavelength range from 510 to 590 nm the trace gases OIO and I_2 feature strong absorptions whereas the wavelength range between 390 and 470 nm is best for the detection of IO. There are different ways to apply DOAS. One can either use an artificial light source (active DOAS) or use (scattered) sunlight (passive DOAS). While the advantage using LP-DOAS is the well-defined lightpath that allows the easy derivation of mixing ratios, the technique cannot be applied, e.g., onboard a ship. Here passive DOAS measurements yield optical densities of the investigated trace gases which can then be converted to mixing ratios with the aid of radiative transfer models.

C.D. O'Dowd and P.E. Wagner (eds.), *Nucleation and Atmospheric Aerosols*, 1060–1063.

Methodology

Typical DOAS instruments either emit the light of an artificial light source (typically Xe-Arc lamp) as a parallel light beam to a cube retro reflector and simultaneously receive the reflected light (LP-DOAS) or receive scattered or direct sunlight. In both application the received light contains information of the trace gas absorptions, typically in the UV and visible wavelength range. Quantitatively the absorption of radiation is expressed by Lambert-Beers Law:

$$I(\lambda) = I_0(\lambda)\exp\left(-\int_0^L \sum_i \sigma_i(\lambda)\cdot c_i \cdot dl\right) \tag{1}$$

where $\sigma_i(\lambda)$ denotes the absorption cross section at the wavelength λ, while $I_0(\lambda)$ is the initial intensity emitted from the source and $I(\lambda)$ is the radiation intensity after passing through an air volume with a length L. Since atmospheric scattering and broadband absorption can hardly be separated, the absorption cross section σ is divided in two parts:

$$\sigma(\lambda) = \sigma_{broad}(\lambda) + \sigma'(\lambda) \tag{2}$$

$\sigma(\lambda)$ describes only the characteristic narrow band absorption structures of different trace gases. For a DOAS setup Lambert-Beers Law can be expressed as:

$$I(\lambda) = I_0(\lambda)\exp\left(-\int_0^L \sum_i \sigma_{i,broad}(\lambda)\cdot c_i \cdot dl\right) \times \exp\left(-\int_0^L \sum_i \sigma_i'(\lambda)\cdot c_i \cdot dl\right) \tag{3}$$

$$I(\lambda) = I_0'(\lambda)\exp\left(-\int_0^L \sum_i \sigma_i'(\lambda)\cdot c_i \cdot dl\right) \tag{4}$$

$I_0'(\lambda)$ contains the initial intensity and the broad band absorption structures and is computed from the measured spectrum $I(\lambda)$ by applying a low pass filter. The in laboratory measured absorption cross sections $\sigma_i(\lambda)$ have to be treated with the same high pass filter algorithm to achieve the differential absorption cross sections $\sigma_i'(\lambda)$.

With this technique it is possible to determine average concentrations along the investigated light path of many different trace gases simultaneously like O_3, SO_2, NO_2, HONO, HCHO, BrO, NO_3, ClO, IO (e.g., Platt 1994; Platt et al. 1979).

LP-DOAS

Figure 1 shows a typical LP-DOAS set-up. The light of the Xe-lamp is sent to the reflector as a parallel beam and then reflected back to the main mirror from where it is focused on a quartz fibre through which the light is then sent to the spectrometer. After passing an Analog-Digital-Converter the obtained spectra can be analysed

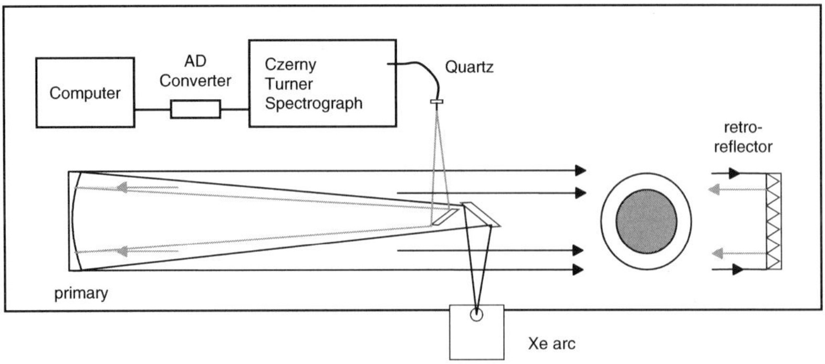

Figure 1 Set-up of a LP-DOAS

in the computer. The lamp reference is received by taking a short-cut spectra of the light without passing the atmosphere.

MAX-DOAS

Passive DOAS can either be done by looking directly into the sun or by using scattered sunlight. The latter one was used onboard the Celtic Explorer research vessel as well as at Mace Head during the MAP intensive summer campaign. The instrument's telescope receives scattered sunlight under different elevation angles (Figure 2).

The reference spectra are taken looking into the zenith. Using MAX-DOAS and radiative transport models one cannot only determine mixing ratios but also create vertical profiles of the trace gases of interest.

Measurements of Iodine Oxides at Mace Head, June 2006

During the intensive campaign in June 2006 LP measurements as well as MAX-DOAS measurements have been performed at Mace Head. Meanwhile MAX-DOAS measurements have been performed onboard the Celtic Explorer. While no iodine oxides could be detected in first evaluations for the Celtic Explorer data, IO could be detected with both, the MAX-DOAS and the longpath DOAS at Mace Head. High concentrations of IO are correlated with particle bursts what is in good agreement with recent studies. Figure 3 shows a time series for IO taken with the LP-DOAS. Results of both measurement sites will be presented.

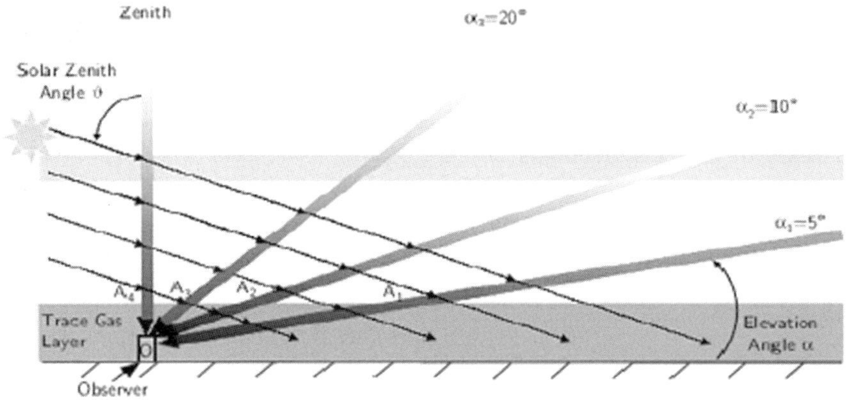

Figure 2 Max-DOAS set up: Spectra are taken under different elevation angles. The zenith scan is used as reference spectrum

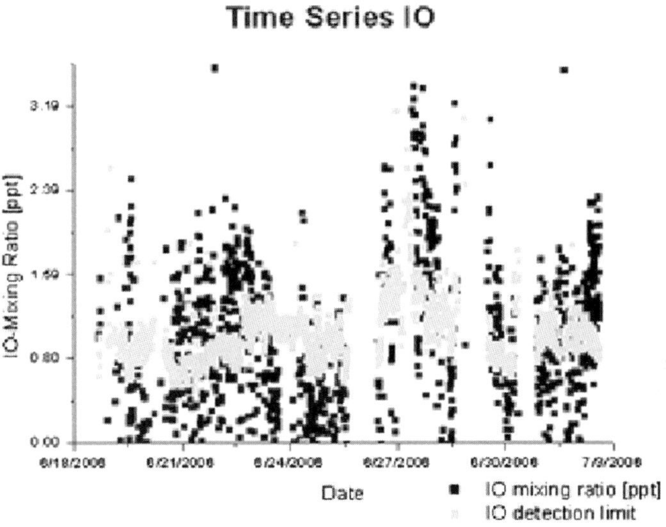

Figure 3 Time series of IO measured with the LP-DOAS during the intensive campaign in June 2006

References

Platt, U., Differential Optical Absorption Spectroscopy (DOAS), *Monitoring by Spectroscopic Techniques, M.W. Sigrist J.* (1994).Wiley, New York.

Platt, U., Perner, D., and Pätz, H., Simultaneous measurements of atmospheric CH2O, O3 and NO2 by differential optical absorptions, *J. Geophys. Res.* (84) 6329–6335 (1979).

Chemical Fluxes in North-east Atlantic Air

D. Ceburnis[1], C.D. O'Dowd[1], M.C. Facchini[2], L. Emblico[2], S. Decesari[2], J. Sakalys[3], and S.G. Jennings[1]

Abstract An alternative method for the estimation of chemical aerosol fluxes, based on a vertical concentration gradient measurement, has been developed and executed at the coastal research station in Mace Head, Ireland. Total gravimetric PM1.0 mass, sea salt and water insoluble organic carbon (WINSOC) concentration profiles showed a net production, while $nssSO_4$ and water soluble organic carbon (WSOC) concentration profiles showed a net removal at the surface. It was concluded that WSOC was predominantly of secondary origin, while WINSOC was predominantly of primary origin. Potential influence of the surf zone emissions was limited to 5–20% and the measured fluxes were characteristic of the coastal zone of up to 5 km. Sea salt concentration and mass flux relationships with the wind speed were established and fitted with power function (C = a*U b). A comparison with the mass flux, derived from eddy correlation measurements yielded good agreement.

Keywords Aerosol, North Atlantic, concentration gradient, flux, sulphate, sea salt, water soluble organic carbon, water insoluble organic carbon

Introduction

Direct observations of aerosol chemical composition do not allow to unambiguously define production sources and mechanisms. Reliable marine aerosol emission inventories require direct experimental evidence, accounting for both production

[1]*Department of Experimental Physics & Environmental Change Institute, National University of Ireland, Galway, University Road, Galway, Ireland*

[2]*Istituto di Scienze dell'Atmosfera e del Clima - CNR, Via Gobetti 101, 40129 Bologna, Italy*

[3]*Institute of Physics, Savanoriu 231, LT-02300 Vilnius, Lithuania*

C.D. O'Dowd and P.E. Wagner (eds.), *Nucleation and Atmospheric Aerosols*, 1064–1069.

and removal of aerosol chemical species. Therefore, novel methods are needed to address these problems.

Aerosol chemical composition is normally obtained using off-line chemical analysis and only recently by online mass spectrometry techniques. Relatively long sampling times of several days are required to obtain sufficient particulate matter, especially for the analysis of organic carbon species in clean marine air masses. A novel method for the estimation of aerosol chemical fluxes, based on a vertical concentration gradient measurement, has been developed. In this study we present the first vertical concentration measurements of aerosol chemical species and their corresponding fluxes at the coastal research station in Mace Head which is representative marine atmosphere of North Atlantic.

Experimental Methods

According to theoretical approach, stable pollutant emissions within a limited area must form a vertical concentration profile, because natural air turbulence, caused by wind or temperature difference, unable to evenly distribute pollutants in the boundary layer within limited space and time. The emission of particulate chemical species from the surface can be estimated using an approach described by Valiulis et al. (2002). Therefore, having measured concentration gradient it is relatively easy to calculate flux of any chemical species. The approach, however, does not allow distinguishing between upward and downward fluxes, but rather enables estimation of the net fluxes.

Vertical concentration profiles are normally non-linear in nature and, therefore, at least 3 sampling points are required to represent the profile. Aerosols were sampled at 3, 10, and 30 m levels. The tower was located about 80 m from the coast line at high tide. PM1.0 size selective inlets (PMX inlet, Sven Leckel Ingenieurburo GmbH, Germany) were used to prevent collection of large amounts of sea salt, which may interfere with other species and because main focus was on submicron aerosol particles. An automated clean sector sampling system ensured sampling only of air mass of marine origin, which spent 4–6 days over the ocean without contact with land. Each set of samples was deployed for 7 days.

Samples were collected during the period April–October 2005, which covered the period of high biological activity over the North-east Atlantic. All together, 18 vertical profile samples were collected. Samples were analysed for gravimetric mass, Na^+, SO_4^{2-}, water soluble (WSOC) and insoluble carbon (WINSOC). Analytical method details can be found in Cavalli et al. (2004). Due to short sampling time or inconsistent chemical analysis about half of the profiles were discarded.

Results and Discussion

Vertical Concentration Profiles

Aerosol mass and chemical species concentrations spanned over a wide range of values due to apparent seasonality in physicochemical properties of marine aerosols (Yoon et al. 2006) and an apparent influence of wind speed. Therefore, it was necessary to normalize the data in order to accommodate all species in all samples in one graph for comparison. Normalization by the sum of concentrations at three levels was chosen as the most effective one. Figure 1 shows the averaged normalized concentration profiles of different species for the whole sampling period. Differences between concentrations at all three heights were generally statistically significant. PM1 mass, sea salt, and water insoluble carbon exhibited a decreasing with height concentration profile, while $nssSO_4^{2-}$ and water soluble carbon showed an increasing profile. The decreasing profile represents an upward mass flux and, therefore, respective chemical species should be produced at the sea surface.

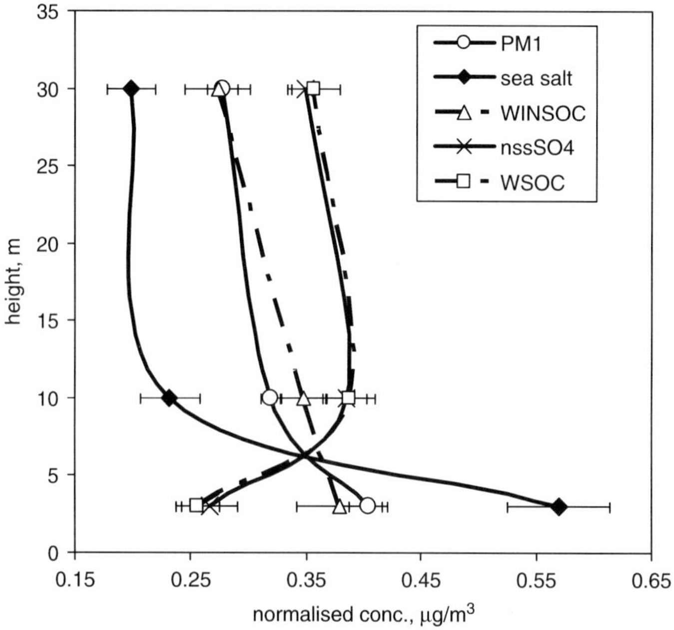

Figure 1 Normalized vertical concentration profiles of PM1 mass, sea salt, $nssSO_4^{2-}$, water soluble (WSOC) and insoluble carbon (WINSOC) at Mace head. Error bars represent standard deviation

Therefore, concentration patterns clearly show whether sea surface is a net source or a net sink of chemical species. The increasing profile means that $nssSO_4^{2-}$ and water soluble carbon species are removed at the surface and, hence, must be produced high in the atmosphere, e.g., clouds. In general, the presented concentration profiles are consistent with the current knowledge that sea salt particles are produced by sea spray at the surface, while $nssSO_4^{2-}$ is mainly produced by aqueous reactions in clouds and therefore removed at the surface. It should be noted that the removal of certain species at the surface is caused by dry deposition only, as wet deposition below cloud evenly removes aerosol particles. A new result was that water soluble and insoluble organic carbon species exhibited the opposite profiles. Water soluble organic carbon showed consistently decreasing concentration profile, while individual water insoluble carbon profiles were more complicated.

The potential influence of the surf zone and the representative area were estimated using a methodological approach described by Valiulis et al. (2002). It was estimated, that the difference in concentration (or gradient) between 3 and 10 m reaches 90% of its value at 1,170 m, while between 10 and 30 m – at 4,840 m. Emissions from greater distances have not affected the gradient, but mainly enhanced the absolute concentration. Surf zone emissions (80–180 m) could have contributed up to 20% of the difference in concentration between 3 and 10 m level and <5% between 10 and 30 m level.

Mass Flux Relationship with Wind Speed

It is a wind stress at the surface which is driving the primary flux of aerosol particles in marine environment through the entrainment of air below wave crest and bubbles rising and bursting at the surface. Thus the produced particles not only contain sea salt, but also organic material, which may be present in the sea water. If the wind stress is a driving force for the primary sea salt flux, a relationship between wind speed and sea salt mass should be established. Indeed, the sea salt mass concentration relationship with the wind speed was following a power law (C = 8.4 × 10 – 5 U 4.19). Having obtained a sea salt concentration relationship with wind speed, it was attempted to calculate mass flux from individual vertical concentration profiles, which in turn were obtained at different wind speeds. Figure 2 shows PM1 flux relationship with the wind speed, again fitted with power function

$$(C = 1.62 \times 10^{-4}\ U^{3.98}).$$

In order to compare the mass flux with the number flux of Geever et al. (2005), particle diameter and density had to be considered. The number flux obtained using EC technique gives only the total number of particles, e.g., >10 nm or >100 nm (Geever et al. 2005). The number flux can be converted into mass flux assuming modal diameter and density of submicron particles. Density of the particles was set at 1.56 g/cm^3 based on long term observation of aerosol chemical composition at

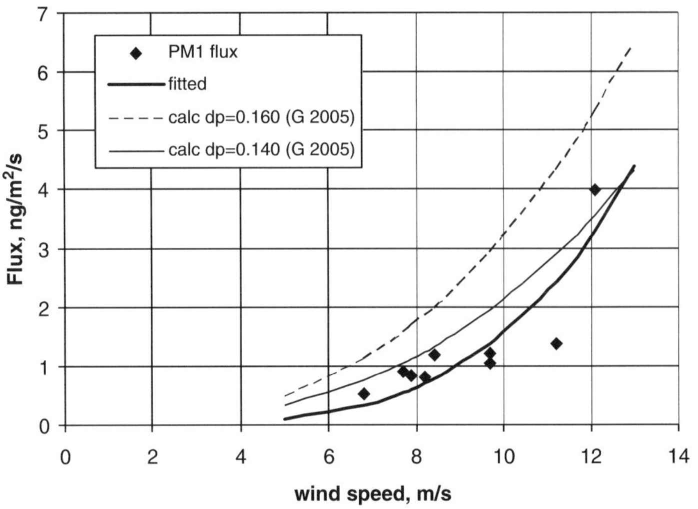

Figure 2 PM1 flux at 22 m relationship with wind speed. Calculated fluxes for the same height obtained from Geever et al. (2005) for submicron mode diameters of 0.160 and 0.140 μm

Mace Head (Cavalli et al. 2004; Yoon et al. 2006). The average diameter of accumulation mode particles during biologically active season at Mace Head is around 160 nm, varying between 100 and 200 nm over the course of the year (Yoon et al. 2006). However, considering primary and secondary processes, responsible for particle diameter changes, it is fair to assume, that primary particles produced at the sea surface are smaller in diameter, perhaps 140 nm in diameter, while secondary processes make up the rest. Therefore, Figure 2 contains two curves using two different diameters. It is clear that using the modal diameter of 140 nm the agreement is better.

It is possible to get the sea salt mass flux relationship with the wind speed in exactly same way. However, it was not possible to obtain the relationship of WINSOC with the wind speed, due the fact that only in limited number of cases WINSOC concentration profile was resembling production mechanism, thus extremely limiting range of different wind speed values.

Conclusions

Gradient fluxes of aerosol mass and inorganic components in marine air over North-east Atlantic were calculated using vertical concentration profiles measured at Mace Head. Total gravimetric mass, sea salt and WINSOC exhibited an upward flux, pointing to the primary source at the sea surface. $NssSO_4^{2-}$ and WSOC showed an opposite, downward flux, pointing to the removal of the species, which also indicates that it is produced higher up in the atmosphere. The WINSOC profile was considered highly dependent on biological activity in the gradient formation zone. It has been demonstrated that the potential influence of the surf zone is limited to less than 5% or 20% (varying with height) and the measurements are representative of the open water in the coastal zone of 5 km. Calculated fluxes are the first experimental estimates of the production (removal) rate of chemical aerosol components in the marine atmosphere at Mace Head. The flux relationship with wind speed fitted by power law ($C = a^*U^b$) agreed well with EC measurements performed at the same site.

Acknowledgment This work was supported by EPA Ireland grant 2003-FS-CD-LS-12-M1 and Lithuanian State Science & Study Foundation grant C-23/2005.

References

Cavalli, F., Facchini, M.C., Decesari, S., Mircea, M., Emblico, L., Fuzzi, S., Ceburnis, D., Yoon, Y.J., O'Dowd, C.D., Putaud, J.-P., and Dell'Acqua, A., *J. Geophys. Res.*, **109**(D24), doi:10.1029/2004JD005137 (2004).

Geever, M., O'Dowd, C.D., van Ekeren, S., Flanagan, R., Nilsson, E.D., de Leeuw, G., and Rannik, U., *Geophys. Res. Lett.*, **32**, L15810, doi:10.1029/2005GL023081 (2005).

Valiulis, D., Ceburnis, D., Sakalys, J., and Kvietkus, K., *Atmos. Environ.*, **36** (39–40), 6001–6015 (2002) .

Yoon, Y.J., Ceburnis, D., O'Dowd, C.D., Jennings, S.G., Jourdan, O., Cavalli, F., Facchini, M.C., Decesari, S., Emblico, L., and Fuzzi, S., *J. Geophys. Res.*, doi:10.1029/2006JD007044 (2006).

On the Contribution of Isoprene Oxidation to Marine Aerosol over the Northeast Atlantic

Tatu Anttila[1], Baerbel Langmann[2], Saji Varghese[2], Claire Scannell[2], and Colin D. O'Dowd[2]

Abstract The secondary organic aerosol (SOA) formation potential of isoprene has been investigated with a regional scale climate model REMOTE. The focus was on the contribution of isoprene oxidation products to marine aerosol over the Northeast Atlantic. As a first stage of the work, the aerosol dynamic module of REMOTE, M7, was extended to treat the SOA formation. In addition, an isoprene source function over the oceans was included in REMOTE. The preliminary results suggest that isoprene oxidation is a potentially important source of marine aerosol over the Northeast Atlantic.

Keywords Marine aerosol production, secondary organic aerosol

Introduction

Marine aerosols exert considerable influence on the global radiative budget by acting as nuclei on which cloud droplets are formed. In order to quantify the effect, the sources of marine aerosols need to be characterized comprehensively. O'Dowd et al.[1] have found significant amounts of both water soluble and water insoluble organic aerosol mass in clean marine air; however, the relative fraction produced by primary and secondary processes remains to be evaluated. One potentially significant but poorly characterized source of marine aerosol is secondary organic aerosol (SOA) formation resulting from isoprene oxidation.[2,3] Until recently, the SOA formation potential of isoprene has been thought to be marginal, and the latest experimental information has been used only in the global modeling study of Henze and Seinfeld[4] who concluded that isoprene oxidation is indeed a potentially significant

[1] *Finnish Meteorological Institute, Research and Development, Climate and Global Change, P. O. Box 503, FI-00101 Helsinki, Finland*

[2] *Department of Physics & Environmental Change Institute, National University of Ireland, Galway, University Road, Galway, Ireland*

C.D. O'Dowd and P.E. Wagner (eds.), *Nucleation and Atmospheric Aerosols*, 1070–1074.

source of SOA. However, they utilized a global climate model and therefore modeling studies focusing on marine areas and employing a smaller modeling domain with enhanced resolution are warranted. To this end, we have investigated the net and relative contributions of isoprene oxidation products to the marine aerosol mass using the regional scale online climate model REMOTE.[5] The model domain covers Europe and northern Atlantic in particular.

The aerosol dynamics model M7[6,7] has been implemented in REMOTE recently. The standard version of M7 does not include a scheme for SOA formation, and therefore developing such a scheme was the first stage of the work. Also an isoprene source function over the oceans, which was lacking prior to this study, was incorporated into the model.

Model Development

Extension of M7

Since isoprene oxidation products consist of a large number of individual chemical compounds, we lumped them into two model compounds. It is also assumed that the two model compounds are produced in the gas phase with constant yields, the values of which are chosen according to recent smog chamber studies.[8,9] Since the available evidence suggests that reaction of isoprene with the OH radical is the only pathway that leads to notable SOA formation,[4] no other reaction channels are considered. Finally, it is assumed that aerosols present in any mode containing water, excluding the nucleation mode, can absorb isoprene oxidation products. In total, six aerosol components were thus added to M7.

The gas/particle partitioning of semi-volatile isoprene oxidation products is calculated using the Henry's Law equilibrium[10] because it is suitable for treating water-soluble compounds and because available information on semi-volatile oxidation products of isoprene suggest that these compounds are highly water soluble.[2,11,12] According to Henry's Law, the equilibrium partitioning of an ideally behaving water-soluble compound X between the gas and aqueous phase is given by the following equation:

$$C_{X,aq} = H_X(T) \times C_{X,gas}, \tag{1}$$

where $C_{X,aq}$ and $C_{X,gas}$ are the aqueous- and gas-phase concentrations of X, respectively, H_X is the Henry's Law constant for X, and T is the temperature.

The Henry's Law constants of the model compounds are chosen according to the available data on isoprene oxidation products and on polyols in particular.[11] The temperature dependency of the Henry's Law constant is accounted for using the Clausius-Clapeyron equation. Furthermore, it is assumed that the model compounds behave ideally. Other required parameters are the molecular weights and densities of the isoprene oxidation products, and their values are chosen according to the available data for polyols.[13,14]

It is expected that isoprene oxidation products, being water soluble, contribute to the particle water content. This is accounted for by applying the ZSR relation according to which solutes take up water independently of each other.[10] The needed water activity data is taken from the literature.[14]

Implementation of the Source Function for Isoprene

The marine isoprene flux has been estimated from the ocean chlorophyll-a concentration (derived from satellite imageries) which is used as a proxy for the marine biological activity. Based on the study of Meskhidze and Nenes,[3] we assume that the average isoprene seawater concentration in nM is proportional to chlorophyll-a concentration in mg/m^3 as measured from SeaWiFs. The air-sea flux of isoprene is determined as the product of the isoprene seawater concentration and a gas exchange coefficient according to Palmer and Shaw[15] where the gas exchange coefficient depends on the wind speed at a height of 10 m and Schmidt number which in turn depends on surface temperature.

Results

The first preliminary results presented here are based on simulations for June 2002 (Figure 1). The REMOTE model has been applied in 0.5° horizontal resolution with 19 vertical levels and driven by ECMWF analysis data at its lateral

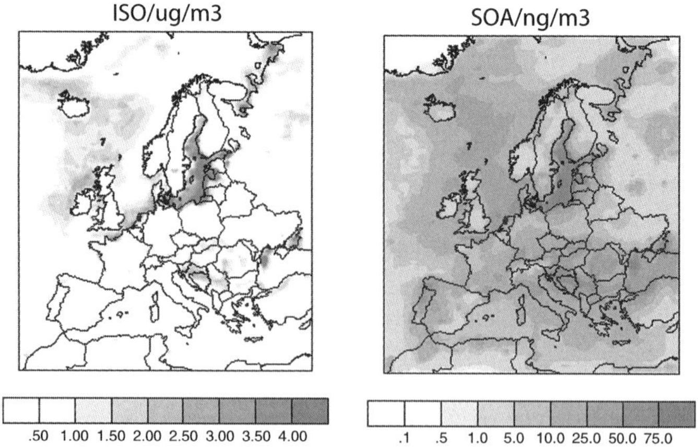

Figure 1 Predicted gas-phase concentrations of isoprene (**left**) and aerosol-phase concentrations of isoprene oxidation products (**right**) over the modeling domain

boundaries. Anthropogenic emissions are taken from the EMEP data sets and biogenic isoprene emissions from deciduous forests as described in Langmann.[5] To summarize, the near surface isoprene concentrations are predicted to be below 4 $\mu g\ m^{-3}$ and the maximum concentrations were found over the Baltic Sea due to the high chlorophyll-a concentration. The total near surface SOA concentrations varied typically in the range 1–50 $ng\ m^{-3}$, which is comparable with the results presented by Henze and Seinfeld.[4] Over the Northeast Atlantic, the SOA concentrations resulting from isoprene oxidation were around 50 $ng\ m^{-3}$ at maximum. The largest concentrations were found in the coarse mode which is explained by the abundance of hygroscopic sea salt aerosol in the mode. Location wise, highest SOA concentrations were predicted over the Baltic Sea due to high isoprene concentrations.

Conclusions

The results suggest that isoprene oxidation is potentially important source of marine aerosol over Europe. The future work includes performing a set of simulations aiming to quantify the sensitivity of the results to various assumptions regarding the properties of isoprene oxidation products.

Acknowledgments The authors would like to acknowledge Environmental Protection Agency of Ireland, Enterprise Ireland, and the European Commission (MAP project, http://macehead.nuigalway.ie/map). One of the authors (T.A.) acknowledges the financial support from the ACCENT network (Atmospheric Composition Change - European Network of Excellence, http://www.accent-network.org).

References

1. O'Dowd, C.D., Facchini, M.C., Cavalli, F., Ceburnis, D., Mircea, M., Decesari, S., Fuzzi, S., Yoon, Y.J., and Putaud, J.P., *Nature*, **431**, 676–680, doi:10.1038/nature02959 (2004).
2. Claeys, M., Graham, B., Vas, G., Wang, W., Vermeylen, R., Pashynska, V., Cafmeyer, J., Guyon, P., Andreae, M.O., Artaxo, P., and Maenhaut, W. *Science*, **303**, 1173– 1176 (2004).
3. Meskhidze, N. and Nenes, A., *Science*, **314**, 1419–1423 (2006).
4. Henze, D.K. and Seinfeld, J.F., *Geophys. Res. Lett.*, **33**, L09812, doi:2006GL025976 (2006).
5. Langmann, B., *Atmos. Environ.*, **34**, 3585–3598 (2000).
6. Vignati, E., Wilson, J., and Stier, P., *J. Geophys. Res.*, **109**, D22202, doi:10.1029/2003JD004485 (2004).
7. Stier, P., J. Feichter, J., Kinne, S., Kloster, S., Vignati, E., Wilson, J., Ganzeveld, L., Tegen, I., Werner, M., Balkanski, Y., Schulz, M., Boucher, O., Minikin, A., and Petzold, A., *Atmos. Chem. Phys.*, **5**, 1125–1156 (2005).
8. Dommen, J., Metzger, A., Duplissy, J., Kalberer, M., Alfarra, M.R., Gascho, A., Weingartner, E., Prevot, A.S.H., Verheggen, B., and Baltensperger, U., *Geophys. Res. Lett.*, **33**, L13805, doi:2006GL026523 (2006).
9. Kroll, J.H., Ng, N.L., Murphy, S.M., Flagan, R.C., and Seinfeld, J.H., *Environ. Sci. Technol.*, **40**, 1869–1877 (2006).

10. Seinfeld, J.H. and Pandis, S.N. *Atmospheric Chemistry and Physics: From Air Pollution to Climate Change*, New York: Wiley (1998).
11. Sander, R., http://www.mpch-mainz.mpg.de/~sander/res/henry.html (1999).
12. Edney, E.O., Kleindienst, T.E., Jaoui, M., Lewandowski, M., Offenberg, J.H., Wang, W., and Claeys, M., *Atmos. Environ.*, **39**, 5281–5289 (2005).
13. Dean, J.A., *Lange's Handbook on Chemistry*, 13th edn., New York: McGraw-Hill (1985).
14. Marcolli, C. and Peter, T., *Atmos. Chem. Phys.*, **5**, 1545–1555 (2005).
15. Palmer P.I. and Shaw, S.L., *Geophys. Res. Lett.*, **32**, doi: 10.1029/2005GL022592 (2005).

Organic Fraction in Recently Formed Nucleation Event Particles in Mace Head Coastal Atmosphere during Map 2006 Summer Campaign

P. Vaattovaara[1], **A. Kortelainen**[1], **and A. Laaksonen**[1,2]

Abstract Marine secondary organic compounds have a huge potential to effect on ultrafine particles properties and thus climatologically important issues. However, the role of organic compounds in recently formed nucleation event particles is widely undetermined. In this work, organic fraction in recently formed nucleation event particles has been experimentally studied in North Atlantic coast (Mace Head, Ireland) during MAP 2006 summer campaign. The results indicate the importance of organic compounds in the nucleation and the lower end of Aitken mode particles formation and define quantitative amount of secondary organic compounds in those newly formed clean marine air particles at the coastal site.

Keywords Particle formation, marine coast, biogenic secondary organics fraction

Introduction

Newly formed nanometer-sized particles have been observed at coastal and marine environments worldwide.[1] Such nanoparticles can grow into larger sizes,[2] being able to scatter incoming radiation and contribute a direct and an indirect (via clouds) cooling effect to the Earth's radiation budget. However, because of difficulties related to the determination of the chemical composition of ultrafine aerosol particles, the organic contribution to the nucleation mode ($d < 20\,nm$) and the lower end of Aitken mode particles ($d < 50\,nm$) in different coastal and marine environments has been widely undetermined. From the viewpoint of coastal and marine climate processes, it is important to know those particles better. Recently, we indicated[3] a significant marine secondary organic contribution to those sized marine

[1] *Department of Physics, University of Kuopio P.O. Box 1627, FI-70211, Kuopio, Finland*

[2] *Finnish Meteorological Institute*

C.D. O'Dowd and P.E. Wagner (eds.), *Nucleation and Atmospheric Aerosols*, 1075–1078.

particles, in addition to iodine compounds, in West Ireland coast (Mace Head). Additionally, we suggested using satellite figures and earlier nucleation event observations around the world that a marine biota originated secondary organic contribution would be worldwide phenomena at biologically active waters. That suggestion has got support at Southern Ocean by work of Meskhidze and Nenes.[4] In this study, we applied the UFO-TDMA (ultrafine organic tandem differential mobility analyzer) method [5] again at Mace Head coastal station to get additional data of an organic fraction in recently formed coastal nucleation mode and the lower end of Aitken mode particles during clean marine new particle formation event days.

Methods

In this EU-funded MAP (marine aerosol production) summer campaign, the UFO-TDMA (ultrafine organic tandem differential mobility analyzer) method[5] was applied to study an organic fraction in recently formed coastal nucleation mode and the lower end of Aitken mode particles at the Mace Head (Ireland) research station. The coastal events were typical for the Mace Head region and they occurred at low tide conditions during efficient solar radiation and a biological active period (i.e., an enhanced mass concentration of chlorophyll a of the North Atlantic Ocean) in summer 2006.

Also many types of additional information were available and used for aiding the UFO-TDMA data analysis. Thus, the presence of nucleation mode, Aitken mode, and bigger particles both particle concentration were detected by Mace Head atmospheric research station DMPSs (differential mobility particle sizer) and CPCs (condensation particle counter); air mass origin was solved based on the HYSPLIT (HYbrid Single-Particle Lagrangian Integrated Trajectory) model[6, 7] back trajectory analysis, and cleanness (black carbon, $BC < 50\,ng/m^3$) of the air was examined from the research station BC data. Ocean chlorophyll a concentration and thus, ocean biological activity was checked by Sea-WiFS (Sea-viewing Wide Field-of-view Sensor) satellite data (NASA/Goddard Space Flight Center and ORBIMAGE); the station data for low tide times, wind direction (180°–300° clean sector) and strength, solar radiation intensity, and relative humidity were used to follow the local conditions. Consequently, all presented UFO-TDMA data have been measured during clean marine air.

Results and Conclusions

Figure 1 shows 10 nm growth factors during two separate clean marine particle formation events. The events happen around same low tide but the first comes from more open ocean direction and is not so intense than the second one. Consequently,

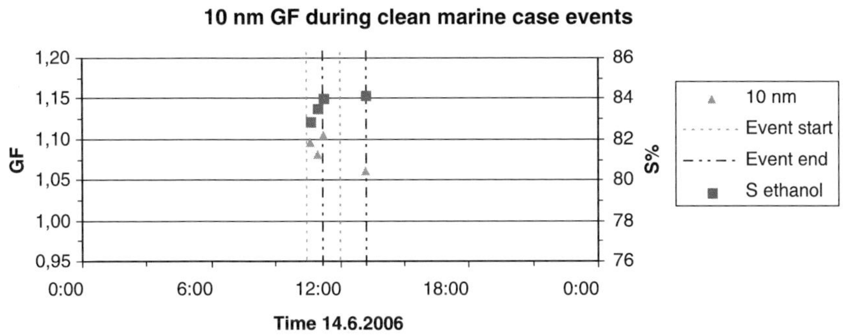

Figure 1 Growth factors (GF) of 10 nm particles measured during noon new particle formation events 14 June 2006. The start and the end times of these clean marine air mass events have been marked by **vertical dashed lines**

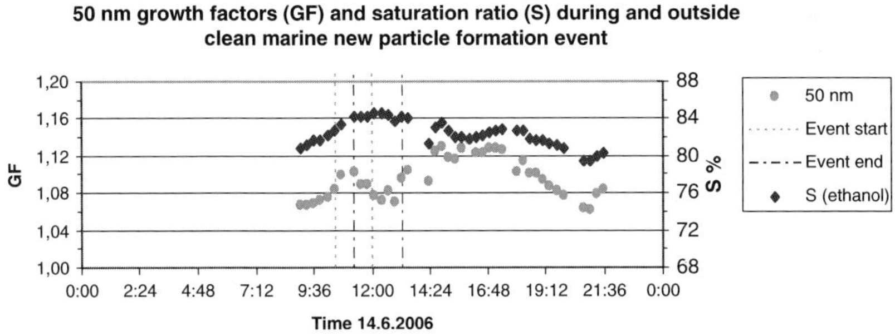

Figure 2 Growth factors (GF) of 50 nm particles measured during and outside noon new particle formation event 14 June 2006. The start and the end time of these clean marine air mass event have been marked by **vertical dashed lines**

nucleation mode particles have different composition in those two events. A probable reason for difference in growth factors is higher iodine oxide content in the second event because more coastal region is expected to include more iodine gas producing brown seaweeds. The second event has growth factor about 1.06 which correspond clean marine air nucleation event growth factors which we measured during earlier campaign at Mace Head 2002.[3] Similar values (1.06–1.08) were also measured during 15–19 June 2006 clean marine air mass events. Thus, it means that if organic material is very soluble in ethanol, like for example marine biota produced precursor gas isoprene, organic fraction is typically near 50% in 10 nm particles. That interpretation is in agreement with our earlier conclusion where role of iodine oxides is taken into account.[3]

Figure 2 shows 50 nm growth factors during same two clean marine new particle formation events. In order to follow particles ethanol solubility properties, ethanol saturation ratios have been described. It is seen that growth factors change during events but subsaturation stay quite stabile at about 84%. It is also seen that growth factors decrease when the first event ends and increases again when the second event starts. Thus, it is expected that more coastally produced iodine oxides have not so big effect on 50 nm than 10 nm particles composition.

Figure 2 also shows how 50 nm growth factors keep growing after the second event ends. Air mass is still clean marine and passes through coastal area with low tide (minimum at noon) and intense solar radiation. Thus, secondary organic vapors would be able to condensate to preexisting particles and change their properties. Sulfate compounds are not expected to increase growth factors because ammonium sulfate has growth factor 1 and ammonium bisulfate about 1.05 at 84% subsaturation (and 1 at 82% subsaturation). On the other, clean marine air include typically much more ammonium than sulfate at Mace Head region, thus a high sulfuric acid contribution is not possible in those particles. When ethanol subsaturation is decreased from 82% toward 76%, it is seen a growth factors decrease. Thus, because 50 nm particles are so sensitive to ethanol subsaturation change, those 50 nm particles have to include some fraction very ethanol soluble organic compounds. However, the smaller particles the higher content of ethanol soluble organic material.

Acknowledgments This work was supported by EU (European Union) 6th Framework Programme through the MAP (marine aerosol production) project and Vilho, Yrjö, and Kalle Väisälä foundation. We wish to thank the staff of National University of Ireland (Galway, Ireland) about helpfulness and for delivering Mace Head research station data during this MAP campaign. The authors also gratefully acknowledge Sea-WiFS project (NASA/Goddard Space Flight Center and ORBIMAGE) for satellite data and the NOAA Air Resources Laboratory (ARL) for the provision of the HYSPLIT transport and dispersion model and READY web site (http://www.arl.noaa.gov/ready. html) used in this work.

References

1. Kulmala, M., Vehkamäki, H., Petäjä, T., Dal Maso, M., Lauri, A., Kerminen, V.-M., Birmili, W., and McMurry, P.H., *J. Aerosol. Sci.*, **35**, 143–176 (2004).
2. O'Dowd, C.D, *J. Geophys. Res.*, **106**, 1545–1550 (2001).
3. Vaattovaara, P., Huttunen, P., Yoon, Y.J., Joutsensaari, J., Lehtinen, K.E.J., O'Dowd, C.D., and Laaksonen, A., *Atm. Phys. Chem.*, **6**, 4601–1616 (2006).
4. Meskhidze, N. and Nenes, A., *Science*, **314**, 1419–1423 (2006).
5. Vaattovaara, P., Räsänen, M., Kühn, T., Joutsensaari, J., and Laaksonen, A., *Atmos. Chem. Phys.*, **5**, 3277–3287 (2005).
6. Draxler, R.R. and Rolph, G.D., HYSPLIT (HYbrid Single-Particle Lagrangian Integrated Trajectory). Model access via NOAA ARL READY Website (http://www.arl.noaa.gov/ready/hysplit4.html), Silver Spring, MD: NOAA Air Resources Laboratory (2003).
7. Rolph, G.D., Real-time Environmental Applications and Display sYstem (READY) Website (http://www.arl.noaa.gov/ready/hysplit4.html), Silver Spring, MD: NOAA Air Resources Laboratory (2003).

A Global Emission Inventory of Submicron Sea-spray Aerosols

Claire Scannell and Colin D. O'Dowd

Abstract A global emission inventory of accumulation mode sea-spray aerosol is presented. This unique inventory also includes the organic fraction associated with sea-spray aerosol as well as the inorganic sea salt component. The organic fraction is significant in regions of high biological activity over the oceans and in many regions can comprise up to 60–70% of the accumulation mode fractional mass. The emission inventory is suitable for inclusion in global chemical transport models lacking an integrated sea-spray source function.

Keywords Emission inventory, sea spray, source function

Introduction

Marine aerosols are produced over large areas as approximately 70% of the earth's surface is covered by ocean, hence they contribute significantly to the global aerosol load and to climate forcing via the direct aerosol affect, by scattering incoming solar radiation and the indirect aerosol affect, by acting as cloud condensation nuclei (CCN). Sea spray is a type of marine aerosol which comprises of inorganic sea salt and organic matter to different degrees. Submicron sea spray is produced from whitecaps resulting from bubble bursting processes during which film and jet drops are produced. Although early research has documented the organic enrichment in sea spray,[1] this process has been difficult to quantify.

It is also believed that organic matter plays a role in secondary aerosol formation. Studies of coastal nucleation events at Mace Head have shown some evidence of organic matter visible in coastal nucleation mode particles. It was found that in certain types of nucleation events a significant percentage of the mass fraction could be accredited to the organic fraction.[2,3] Many recent studies have suggested isoprene from plankton emissions as a precursor to such secondary organic aerosols.[4]

Department of Experimental Physics & Environmental Change Institute, National University of Ireland, Galway University Road, Galway, Ireland

C.D. O'Dowd and P.E. Wagner (eds.), *Nucleation and Atmospheric Aerosols*, 1079–1082.

Traditionally, interest in marine aerosols has been on large particles, (>1 μm) and on the production of sea salt and non sea salt sulphates. Research has shown that in remote oceanic regions marine aerosols dominate the particulate mass concentration with a significant contribution to this concentration from the submicron size range.[5] Further research has shown that organic matter amounts to a significant fraction of the submicron aerosol mass.[6] Studies have also shown that modal diameters and the concentration of organic matter follow a seasonal variation.[7] The modal diameter increases from winter to summer for both the aitken and accumulation modes. During periods of high plankton blooms, in the summer months, organic matter is the dominant component of the submicron marine aerosol, contributing approximately 63% of the total aerosol mass in the submicron size range. Outside this blooming period, in winter, the organic fraction decreases to about 15% and sea salt becomes the dominant component to the accumulation mode.[6] Studies have shown that the majority of the organic fraction within the sea spray was water insoluble organic carbon (WIOC) which is driven by biological activity in the ocean.[8] There is therefore a need when building a marine aerosol emission inventory that both components, sea salt and organic matter are considered and quantified. Such an emission inventory has to date been lacking.

Source Functions

This study uses the combined inorganic–organic sea-spray source function in the accumulation mode size range recently developed from eddy-correlation flux measurements,[9] aerosol chemical measurements,[6] and satellite-derived oceanic chlorophyll-a measurements.[10] The physical source function is however, reanalysed in terms of a power law as presented in Figure 1. This combined organic–inorganic sea spray is used to develop a global sea-spray emission inventory for sizes between 100 and 500 nm.

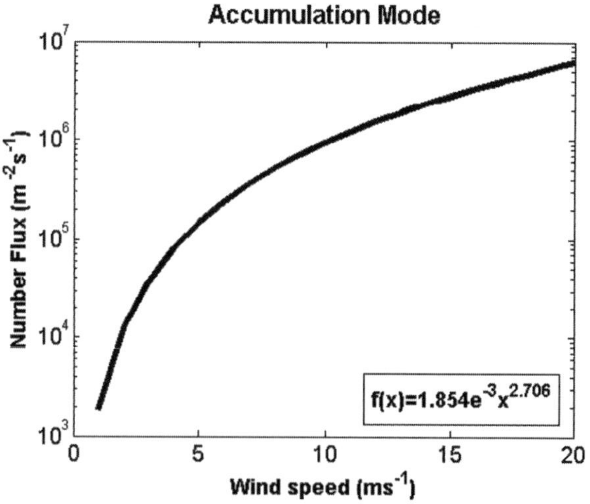

Figure 1 Accumulation mode Geever source function modified according to a power fit 10

Results

The global accumulation mode number flux, mass flux and organic fraction mass flux for the winter and summer of 2006 is illustrated in Figure 2.

During winter, the highest fluxes, (3×10^{6} m^{-2} s^{-1}), occurs in the North Atlantic, although there is a significant flux over the southern oceans too. The accumulation mode mass flux peaks at 7×10^{-11} kg m^{-2} s^{-1} in the North Atlantic and in parts of the southern oceans. The organic carbon fraction peaks at 0.78 mg m^{-3} in the southern oceans during winter and over northern latitudes in summer. It should be noted that the North Atlantic is not totally visible in the winter months, nor parts of the southern oceans in the summer, due to the line of sight of the MODIS (Aqua) satellite. These

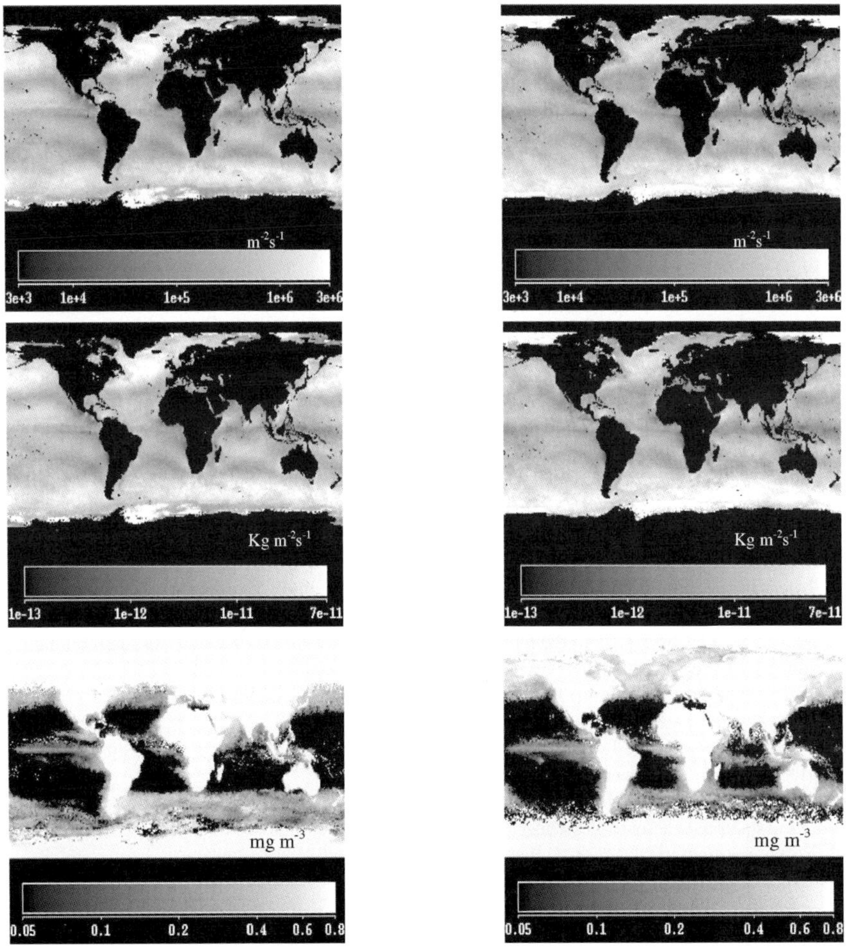

Figure 2 **Left panel**: winter; **right panel**: summer; **top panel**: accumulation mode number flux; **middle panel**: accumulation mode mass flux; **bottom panel**: organic fraction mass flux

results point to significant global sources of primary marine aerosol which are an important source of cloud condensation nuclei[12]; however, they also point to significant changes in the primary marine aerosol chemical composition which can also impact on cloud nucleating properties.[6] It is important to include this combined inorganic–organic sea-spray source function in global climate and chemical transport models. This combined organic–inorganic sea-spray emission inventory is suitable for chemical transport models lacking an integrated or dynamic source function.

Acknowledgment The Environmental Protection Agency and the European Commission (MAP project) are acknowledged for funding this project.

References

1. Blanchard, D.C., Sea to air transport of surface active material, *Science*, **146**, 396–397 (1964).
2. Vaattovaara, P., Hutten, P.E., Yoon, Y.J., Joutsensaari, J., Lehtinen, K.E.J., O'Dowd, C.D., and Laaksonen, A., The composition of nucleation and Aitken modes particles during coastal nucleation events: evidence for marine secondary organic contribution, *Atmos. Chem. Phys.*, **6**, 4601–4616 (2006).
3 O'Dowd, C.D. and de Leeuw, G., Marine aerosol production: a review of the current knowledge, *Phil. Tran. of the Roy. Soc. A* (2007).
4. Meskhidez, N. and Nenes, A., Phytoplankton and cloudiness in the southern ocean, 10.1126/science.1131779, *Science* (2006).
5. Intergovernmental Panel on Climate Change (IPCC), *Climate Change 2001: The Scientific Basis*, edited by Houghton et al., New York: Cambridge University Press (2001).
6. O'Dowd, C.D., Fachini, M.C., Cavalli, F., Ceburnis, D., Mircea, M., Decesari, S., Fuzzi, S., Yoon, Y.J., and Putaud, J.P., Biogenically driven organic contribution to marine aerosol, *Nature*, doi: 10.1038/nature02959 (2004).
7. Yoon, Y.J., Ceburnis, D., Cavalli, F., Jourdan, O., Putaud, J.P., Facchini, M.C., Descari, S., Fuzzi, S., Sellegri, K., Jennings, S.G., and O'Dowd, C., Seasonal characteristics of the physico-chemical properties of the North Atlantic marine atmospheric aerosols, *J. Geophys. Res.*, **112**, D04206, doi:10.1029/2005JD007044 (2007).
8. Monahan, E.C., Spiel, D.E., and Davidson, K.L., A model of marine aerosol generation via whitecaps and wave disruption, in: *Oceanic Whitecaps and Their Role in Air-sea Exchange Processes*, edited by E.C. Monahan and G. MacNiochaill, Dordrecht, The Netherlands: Reidel, pp. 167–174 (1986).
9. Geever, M., O'Dowd, C., van Ekeren, S., Flanagan, R., Nilsson, D., de Leeuw, G., and Rannik, U., Sub-micron sea-spray fluxes. GRL, 32, 2005, L15810.2005.
10. O'Dowd, C.D., Langmann, B., Varghese, S., Scannell, C., Ceburnis, D., and Facchini, C., A combined organic–inorganic sea spray source function, *Geophys. Res. Lett.*, submitted (2007).
11. O'Dowd, C.D. and de Leeuw, G., Marine aerosol production: a review of the current knowledge, *Phil. Tran. of the Roy. Soc. A* (2007).
12. O'Dowd, C.D., Lowe, J., Smith, M.H., and Kaye, A.D., The relative importance of sea-salt and nss-sulphate aerosol to the marine CCN population: an improved multi-component aerosol-droplet parameterization. *Q. J. Roy. Met. Soc.*, **125**, 1295–1313 (1999).

A Combined Organic–Inorganic Sea-spray Source Function

Colin D. O'Dowd[1], Baerbel Langmann[1], Saji Varghese[1], Claire Scannell[1], Darius Ceburnis[1], and Maria Cristina Facchini[2]

Abstract This study presents a novel approach to determine sea-spray generation by a combined organic–inorganic submicron source function. It requires wind speed and surface ocean chlorophyll-a concentration as input parameters. The combined organic–inorganic source function is implemented in the REMOTE regional climate model and sea-spray fields are predicted with particular focus on the Northeast Atlantic. The model predictions, using the new source functions, compare well with observations of total sea-spray mass and organic carbon fraction in sea-spray aerosol.

Keywords Marine aerosol, sea-spray source function, organic aerosol

Introduction

Primary marine aerosol, or sea spray, represents one of the most important aerosols on the global scale due to the ocean covering more than 70% of the Earth's surface. Sea-spray aerosols contribute significantly to the radiative budget in terms of scattering of incoming solar radiation (the direct aerosol effect) and the modification of cloud microphysics and radiative properties (the indirect aerosol effect). Estimates for top-of-atmosphere, global-annual radiative forcing due to sea salt are −1.51 and −5.03 W m^{-2} for low and high emission values, respectively[1] (IPCC 2001).

Although submicron sea salt occurs in significant number concentrations,[2] many attempts to include a submicron component rely on arbitrary extensions of supermicron source functions,[3] although recent improved parameterizations for submicron sea salt are now becoming available (for a review[4]). Moreover, there has been a serious omission of the organic aerosol component of sea spray. It has been recognized since the 1950s that organic matter concentrated at the ocean surface can be entrained

[1]*Department of Experimental Physics & Environmental Change Institute, National University of Ireland, Galway, University Road, Galway, Ireland*

[2]*Istituto di Scienze dell'Atmosfera e del Clima – CNR, Italy*

C.D. O'Dowd and P.E. Wagner (eds.), *Nucleation and Atmospheric Aerosols*, 1083–1087.

into sea spray through the bubble bursting process,[5] however, this component of sea spray has been particularly difficult to incorporate into a source function since there is a scarcity of understanding on the factors which influence this process.

Recent studies into marine aerosol physicochemical properties have, however, resulted in a significant improvement of our understanding of its properties which are elucidating the various production mechanisms. Extensive studies on Northeast Atlantic marine aerosol over a three year period found that the submicron mode contained a dominant sea salt contribution in winter and a dominant organic contribution in summer.[6–8] This contrast was linked to seasonality in biological activity as shown from comparison with seasonal concentrations of surface-water chlorophyll-a derived from satellite.[6,8] The predominant organic fraction was water-insoluble organic carbon (WIOC) with properties similar to that expected for organic matter derived from a primary bubble-bursting source.[7] The remaining water soluble organic carbon (WSOC) component of the aerosol mass is thought to be more likely produced via secondary aerosol formation mechanisms for the and can comprise up to 20–25% of the mass. This study aims to develop a combined organic–inorganic sea-spray source function which can account for the seasonality in the WIOC fraction.

Methodology

Although the processes-level understanding is not fully developed, oceanic biological activity can be quantified using chlorophyll-a as a surrogate. For the Mace Head dataset,[7,8] the correlation between chlorophyll-a and WIOC was explored as a method of estimating the impact of biological activity on the fractional contribution of WIOC to the sea-spray aerosol. Chlorophyll-a concentrations (taken from MODIS) in oceanic waters upwind of Mace Head were correlated to the observed WIOC fraction in aerosol. This was achieved through selecting a grid area 1,000 × 1,000 km to the west of Mace Head and in which air parcels would have advected over for a period of approximately 48 h. The average chlorophyll-a concentration in this grid was then calculated and correlated with the fractional concentration of WIOC to sea-spray mass measured over the same period, assuming that sea spray comprised predominantly of sea salt and WIOC although with a minor WSOC component (arbitrarily set to 5% of the WIOC fractional contribution). The relationship between fractional WIOC and chlorophyll concentration is shown in Figure 1. The parameterization of WIOC fraction as a function of chlorophyll-a leads to 0% at very low chlorophyll-a concentrations and saturates at 70% for a concentrations above 5 mg m^{-3}.

The derived fractional composition of sea-spray particles produced as a function of chlorophyll-a fields is then integrated into a physical sea-spray source function[9] derived from eddy-correlation measurements.[9] This source function is regenerated as a power-law in the form of $N_{Flux}(10^6\ m^{-2}\ s^{-1}) = 1.854\times10^{-3}\ U_{22}^{2.706}$ ($r^2 = 0.8$) to provide more realistic concentrations at low wind speeds. The seasonal variation in the sea-spray diameter has been prescribed based on Mace Head data.[8] With the combination of the modal diameter and number flux, we effectively get a volume

Figure 1 Correlation between the fractional WIOC component of sea spray as a function of grid-average chlorophyll-a concentration.

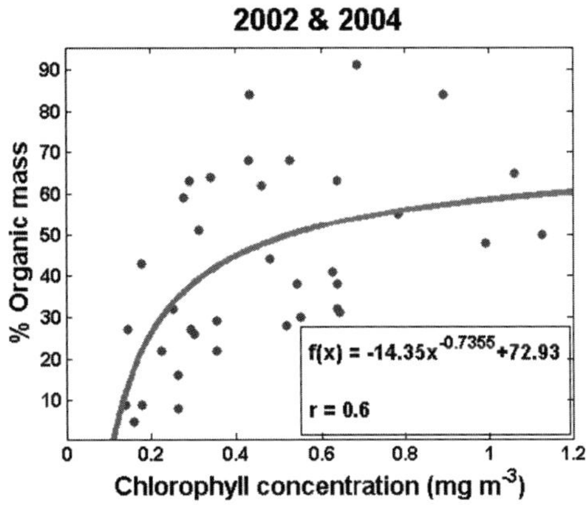

flux, which is transformed into a sea-spray mass flux based on density estimates of the WIOC/sea salt ratio: density of WIOC = 1g cm^{-3} and density of sea salt = 2.165 g cm^{-3}. Assuming that the sea spray is internally mixed, we can constrain the sea-spray flux to a log-normal flux mode with known diameter, mass, number and chemical composition and this log-normal source function is directly interfaced with the M7 modal aerosol module[10] in the REMOTE (REgional MOdel with Tracer Extension) regional climate model.[11]

Regional Scale Simulations of Sea Spray

Simulations were conducted over an area covering the Northeast Atlantic and Europe for three 1 month periods in late spring (May 2002), winter (January 2003) and early summer (June 2003) using a horizontal resolution of 0.5° and boundary conditions provided by ECMWF analysis data. The monthly average submicron mass distribution and the WIOC fractional contribution, resulting from the combined inorganic and organic sea spray is displayed in Figure 2 for the three selected months. In winter (January 2003), maximum accumulation mode sea-spray mass concentrations of ~3 μg m^{-3} are seen over the Northeast Atlantic and to the west of Ireland (and Mace Head). For this period, biological activity is at its lowest level and the WIOC fraction is more or less zero. By contrast, for the May and June periods of high biological activity (e.g., late spring and early summer), total sea-spray mass is lower, due to generally lower wind speeds, and lies in the range of 1–2 μg m^{-3}. For these periods, the WIOC fraction has increased to between 50–60%. It should be noted that the lower sea-spray mass in May 2002 compared to June 2003 is due to 30% higher precipitation during the former period rather than to differences in wind speed (wind speeds were comparable during both periods). Similar seasonal contrasts are seen in over the North Sea and the Baltic Sea. The seasonal variation and magnitudes of sea salt mass and WIOC are in good agreement with those concentrations reported for Mace Head over the same period.[8]

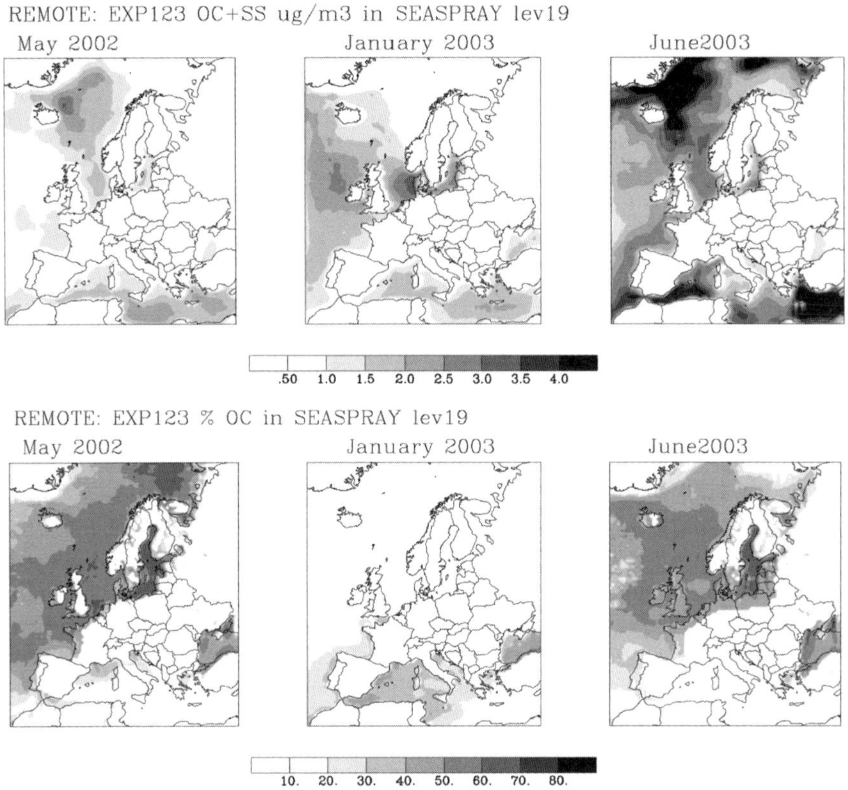

Figure 2 (**Top panel**) Near surface sea-spray mass concentrations around the European regions. (**Bottom Panel**) Percentage primary organic contribution to sea-spray mass.

Conclusions

A combined inorganic–organic submicron source function has been developed and successfully implemented in a regional climate model. The source function leads to good replication of observations of inorganic and organic aerosol components over the Northeast Atlantic. The results illustrate a significant contribution of natural primary organic aerosol to marine aerosol distributions which are driven by marine biota.

References

1. Intergovernmental Panel on Climate Change (IPCC), *Climate Change 2001: The Scientific Basis*, edited by Houghton et al., New York: Cambridge University Press (2001).
2. O'Dowd, C.D. and Smith, M.H., Physico-chemical properties of aerosol over the North East Atlantic: evidence for wind speed related sub-micron sea-salt aerosol production. *J. Geophys. Res.*, **98**, D1, 1137–1149 (1993).

3. Monahan, E.C., Spiel, D.E., and Davidson, K.L., A model of marine aerosol generation via whitecaps and wave disruption, in: *Oceanic Whitecaps and Their Role in Air-sea Exchange Processes*, edited by E.C. Monahan and G. MacNiocaill, Dordrecht, The Netherlands: Reidel, 167–174 (1986).
4. O'Dowd, C.D. and de Leeuw, G., Marine aerosol production: a review of the current knowledge, *Phil. Tran. of the Roy. Soc. A* (2007).
5. Blanchard, D.C., Sea to air transport of surface active material, *Science*, **146**, 396–397 (1964).
6. O'Dowd, C.D., Facchini, M.C., Cavalli, F., Ceburnis, D., Mircea, M., Decesari, S., Fuzzi, S., Yoon, Y.J., and Putaud, J.P., Biogenically-driven organic contribution to marine aerosol, *Nature*, doi:10.1038/nature02959 (2004).
7. Cavalli, F., Facchini, M.C., Decesari, S., Mircea, M., Emblico, L., Fuzzi, S., Ceburnis, D., Yoon, Y.J., O'Dowd, C.D., Putaud, J.-P., and Dell'Acqua, A., Advances in characterization of size resolved organic matter in marine aerosol over the North Atlantic, *J. Geophys. Res.*, doi:10.1029/2004JD0051377 (2004).
8. Yoon, Y.J., Ceburnis, D., Cavalli, F., Jourdan, O., Putaud, J.P., Facchini, M.C., Descari, S., Fuzzi, S., Jennings, S.G., and O'Dowd, C.D., Seasonal characteristics of the physico-chemical properties of North Atlantic marine atmospheric aerosols *J. Geophys. Res.*, doi:10.1029/2005JD007044 (2007).
9. Geever, M., O'Dowd, C., van Ekeren, S., Flanagan, R., Nilsson, D., de Leeuw, G., and Rannik, U., Sub-micron sea-spray fluxes. *Geophys. Res. Letts.*, doi:10.1029/2005GL023081 (2005).
10. Vignati, E., Wilson, J., and Stier, P., M7: an efficient size-resolved aerosol microphysics module for large-scale aerosol transport models, *J. Geophys. Res.*, doi:10.1029/2003JD004485 (2004).
11. Langmann, B., Numerical modelling of regional scale transport and photochemistry directly together with meteorological processes, *Atmos. Environ.*, **34**, 3585–3598 (2000).

Observations of Oceanic Whitecap Coverage in the North Atlantic during Gale Force Winds

Adrian Callaghan[1], Gerrit de Leeuw[2], and Leo Cohen[3]

Abstract Preliminary percentage whitecap coverage results calculated from digital images obtained during the 2006 MAP field campaign in the North Atlantic are presented. Observations of whitecap coverage in the literature are scarce for wind speeds exceeding 20 m s^{-1} and here results are given for a period of gale force winds from June 20 to 21. Sustained wind speeds of up to 27 m s^{-1} were recorded. Whitecap coverage increased with increasing wind speed with an approximately cubic power law relationship. It was found that segregation of wind speed data into increasing and decreasing wind speed resulted in two distinct whitecap coverage wind speed relationships. Percentage whitecap coverage calculated during periods of increasing wind speed was proportional to the wind speed to the power of 3.41. Percentage whitecap coverage calculated during periods of decreasing wind speed was proportional to the wind speed to the power of 2.74. Breaking waves and resulting whitecap coverage provide an important mechanism for generation of marine aerosols and these results have implications for a whitecap coverage based sea spray source function (S3F).

Keywords Whitecap coverage, wind speed, marine aerosols

Introduction

The interaction of the wind and the sea surface produce waves. As the wind speed increases the wave field develops and at a certain critical wind speed threshold the waves begin to break forming whitecaps. This threshold typically occurs between 4

[1]*Department of Earth and Ocean Sciences, National University of Ireland, Galway*

[2]*Finnish Meteorological Institute, R&D, ClimateChange Unit, Helsinki, Finland & University of Helsinki, Dept. of Physical Sciences, Helsinki, Finland & TNO Build Environment, Apeldoorn, The Netherlands*

[3]*TNO Defence, Security and Safety, The Hague, The Netherlands*

C.D. O'Dowd and P.E. Wagner (eds.), *Nucleation and Atmospheric Aerosols*, 1088–1092.

and 7 m s^{-1}. Whitecaps are the surface manifestations of the air entrained into the water column during breaking. Many datasets of open ocean percentage whitecap coverage W, can be found in the literature.[1–3] Each data-set typically results in a W to wind speed relationship. These relationships are typically characterised by a large data scatter but it is generally accepted that whitecap coverage scales with the cube of the wind speed. However, much of the published data are for low to moderate wind speed conditions and little data exist for wind speeds in excess of 20 m s^{-1}.

Breaking waves and resulting whitecap coverage are responsible for the injection of seawater droplets into the air which is the primary mechanism controlling the rate of sea salt aerosol generation over the ocean.[2] Air entrained by breaking waves forms a sub-surface bubble plume. The bubble plume undergoes various stages from formation, injection, rise and senescence with associated changes in the bubble size distribution.[4] The rising bubbles reach the sea surface and burst, thereby producing marine aerosols. The largest bubbles produce film drops due to breaking of the bubble surface film after protruding the surface.[5] Bubbles smaller than 1.7 mm also produce jet drops when the cavity left after rupture of the bubble film shoots up.[6] When wind speeds exceeds about 9 m s^{-1} the drag becomes strong enough to directly tear off sea spray droplets from the wave crests which are called spume drops.

One way to parameterise primary marine aerosol production is to use the sea spray source function (S3F). This describes the sea spray flux per unit area and per unit of time, parameterised as function of environmental parameters. One such environmental parameter is W. W to wind speed relationships are poorly constrained at high wind speeds due to the inherent difficulty of making whitecap observations in these conditions.

One of the goals the EU FP6 project MAP (Marine Aerosol Production) is "to quantify the number and size flux of primary inorganic and organic marine sea-spray aerosol (PMA)". Data to achieve this objective were collected during the MAP cruise aboard the Celtic Explorer, from June 11 to July 5, 2006. Various types of experiments were conducted to provide data to derive the S3F. Here we present the first preliminary results for whitecap measurements made during a period of gale force winds.

Data Collection and Image Analysis

Whitecap observations were made with a video camera mounted on the monkey deck of the R.V. Celtic Explorer. The video signal was split into individual images which were stored on a p.c. in the ship's dry lab. The recording was automatically started after day break and stopped before dusk. Frames were continuously recorded, with a frequency of 50 Hz, during 55 min each hour. Images collected over the course of an hour were used to form a single value of whitecap coverage. Due to the severity of the sea conditions and presence of raindrops on the lens of the camera, certain images were omitted from the analysis. Images that contained the horizon were not analysed due to highly uneven illumination between foreground and background. Wind Speed was recorded every 5 min and hourly averages were calculated to compare with hourly estimates of W.

Each individual image was analysed for the presence of whitecaps. Whitecap identification was based on an automated thresholding technique developed at NUIG. The technique provided a very efficient method of analysis and removed the subjectivity of a human analyst. All pixels in an image above a certain threshold were classified as whitecap pixels and the remaining pixels were classified as background water.

Results

The MAP field campaign on board the R.V. Celtic Explorer took place in the North Atlantic Ocean, northwest of Ireland during the month of June 2006. A variety of wind speeds were encountered. The results presented here focus on a period of gale force winds on June 20 and 21. Sustained wind speeds of up to 27 m s^{-1} were encountered. Data recorded with the video camera throughout this gale were used for the initial analysis of the whitecap coverage because a wide range of wind speeds was encountered, including periods with increasing and decreasing wind speed.

Figure 1 shows W as a function of wind speed recorded during this period. Data were fitted in a least squares sense and the power law relationship and R squared value are both displayed on the figure. It can be seen that the power law approaches a cubic relationship. Included for comparison is the Monahan and O'Muircheartaigh 1980, power law relationship ($W = 3.84\times10^{-4} \times U^{3.41}$) which is referred to as MOM from hereon.

In Figures 2 and 3 the data have been partitioned into two categories. Figure 2 shows W for the times when wind speed was increasing and Figure 3 shows W for when the wind was decreasing. As in Figure 1, the resulting power law and

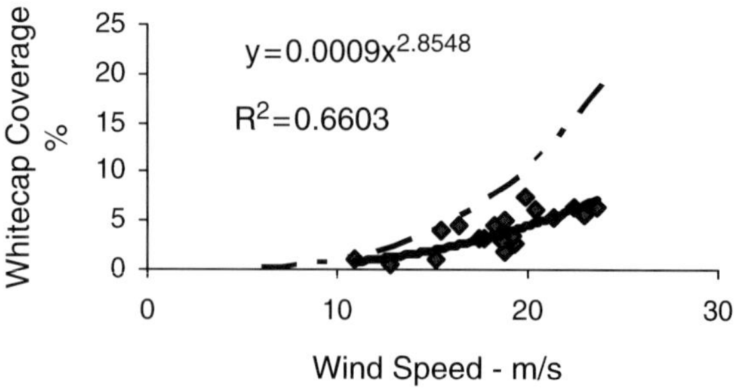

Figure 1 W as a function of wind speed. The solid line is the least squares fit for this study and the dashed line is the power law of MOM

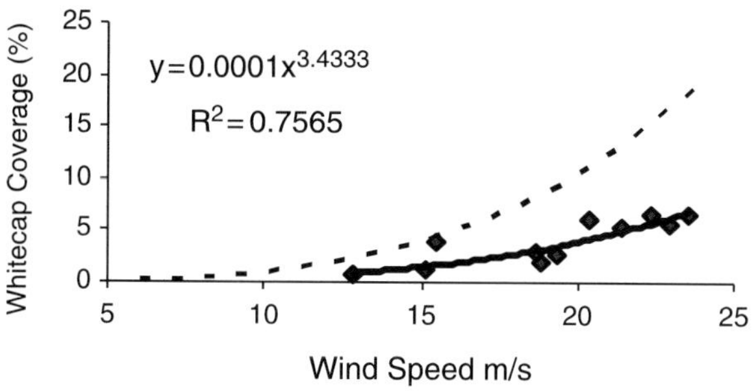

Figure 2 W as a function of increasing wind speed only. The solid line is the least squares fit for this study. The dashed line represents MOM

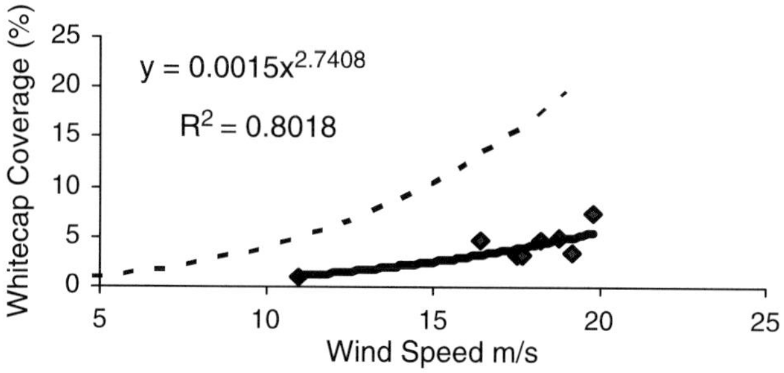

Figure 3 W as a function of decreasing wind speed only. The solid line is the least squares fit for this study. The dashed line represents MOM

correlation coefficient from the least squares analysis are displayed in each figure. It is seen that partitioning the wind data into increasing and decreasing periods has a clear impact on the resulting power law relationships and R squared values. In both cases a higher R-squared value is obtained than in the case where the wind speed data are not partitioned. There are fewer data points in the decreasing wind speed curve in Figure 3 with a gap between 10 and 16 m s^{-1}. This gives the 10 m s^{-1} data point a relatively high weight. Therefore the data will be further processed to obtain more data points over the whole range. The current data clearly show the difference in the wind speed dependence of the whitecap coverage for increasing wind speed.

Conclusion and Further Work

Preliminary results have been presented from whitecap-made measurements during gale force winds on the MAP summer cruise, 2006. The results clearly show the difference between the effects of increasing and decreasing wind speed on

W. Image processing and analysis is ongoing and the results will be related to direct measurements of sea spray and sea salt aerosol fluxes using the eddy covariance method to derive a parameterisation for the sea spray aerosol flux based on detailed analysis of processes described by a variety of environmental parameters.

Acknowledgments The MAP project is financially supported by the European Commission (FP6, project number 018332) and participating institutes. The Celtic Explorer cruise was supported by the Marine Institute. The skill of the Celtic Explorer crew is greatly appreciated.

References

1. Lewis, E.R. and Schwartz, S.E., *Sea Salt Aerosol Production: Mechanisms, Methods, Measurements and Models – A Critical Review*, Vol. Geophysical monograph 152, Washington, DC: American Geophysical Union (2004).
2. Stramska, M. and Petelski, T., Observations of oceanic whitecaps in the north polar waters of the Atlantic, *J. Geophys. Res.*, **108**(C3), 3086, doi:10.1029/2002JC001321 (2003).
3. Monahan, E.C. and O'Muircheartaigh, I.G., Optimal power-law description of oceanic whitecap coverage dependence on wind speed, *J. Phys. Ocean*, **10**, 2094–2099 (1980).
4. Leifer, I. and de Leeuw, G., Bubbles generated from wind-steepened breaking waves: 1. Bubble plume bubbles, *J. Geophys. Res.*, **111**, C06020, doi:10.1029/2004JC002673 (2006).
5. Spiel, D.E., On the birth of film drops from bubbles bursting on seawater surfaces, *J. Geophys. Res.*, **103**, 24907–24918 (1998).
6. Spiel, D.E., More on the births of jet drops from bubbles bursting on seawater surfaces, *J. Geophys. Res.*, **102**, 5815–5821 (1997).

Primary Marine Aerosol Produced during Bubble Bursting Experiments Using Baltic Sea, North Sea, and Atlantic Waters

Kim Hultin[1], E. Douglas Nilsson[1], Radovan Krejci[2], Gerrit de Leeuw[3], and Monica Mårtensson[1]

Abstract In this work, an experimental approach has been used to investigate the aerosol production from the bubble-bursting mechanism using water from four different seas. The influence of water temperature, salinity, and dissolved oxygen on primary aerosol particles emitted was investigated. The first results point in the direction of biology as the key variable affecting the particle production, while the effects of both temperature and oxygen seem to be disguised whenever the biological activity in the surface water is enhanced.

Keywords Marine aerosols, sea salt, bubble bursting

Aerosols from Bubble Bursting

The major influence by aerosols on our climate is in complex manner through direct and indirect effects. Marine aerosols in particular are of high interest, since they are produced over large areas and hence are representative of a significant fraction of aerosols found in the atmosphere.[1] Moreover, marine aerosols control cloud microphysical properties over remote parts of the ocean where the sensitivity of cloud properties to slight changes in cloud nucleus properties through albedo enhancement is high. Bubble bursting from whitecaps is considered to be the most effective mechanism for particulate matter to be ejected into the atmosphere from

[1]*Department of Applied Environmental Science (ITM), Stockholm University, 106 91 Stockholm, Sweden*

[2]*Department of Meteorology, Stockholm University, 106 91 Stockholm, Sweden*

[3]*TNO Physics and Electronics Laboratory, P.O. Box 96864, 2509 JG The Hague, The Netherlands*

Corresponding author: kim.hultin@itm.su.se

C.D. O'Dowd and P.E. Wagner (eds.), *Nucleation and Atmospheric Aerosols*, 1093–1097.

ocean surface. The marine aerosol can consist of sea salt, organic matter as well as bacteria. This leads to the fact that aerosol particles can have an impact on such separate areas as the climate or the health of living organisms.[2]

Recent laboratory work performed by Mårtensson et al.[4] suggests seasonality and climate zone effects in the production of primary marine aerosols, where the number and size distribution of particles produced depends on the water temperature. Field work from the North Atlantic has shown that both the aerosol's physical and chemical properties exhibited clear seasonal patterns following water biological activity.[5] The production of aerosol droplets by laboratory generated whitecaps has also been found to increase when the water was supersaturated with oxygen.[6]

The overall embracing aim of this study is to quantify, and improve the parameterization of the primary marine aerosol emissions with respect to environmental variables including temperature, salinity and oxygen saturation.

Experimental Methods

The Skagerrak and North Sea surface waters were collected in September/October 2003. The two Baltic experiments were conducted in 2005, the first one in May/June (Askö experiment, 100 km southeast of Stockholm) and the second in September (Garpen, Kalmar strait experiment, southeast Sweden). The North Atlantic data is part of the EU project MAP, and sampling was made west of Ireland during a cruise in June and July 2006 with the Irish research vessel Celtic Explorer. Sampling methods varied among different waters; while the water collected in 2003 were frozen prior to investigation, Askö water was collected using a small boat and investigated right after. Garpen and North Atlantic waters on the other hand, were sampled continuously, the latter during a ship cruise.

The experiments were performed using a carefully sealed and slightly over pressurized tank filled with approximately 20 L of water, in which a waterfall jet simulated the action of a breaking wave. The sheath air supplied to the tank was particle free to ensure measurements of marine aerosol from bubble bursting only. The aerosol size distribution was measured with a DMPS (Differential Mobility Particle Sizer) and an OPC (Optical Particle Counter), together covering the range between 0.20 μm up to 2.2 μm D_p (up to 5 μm D_p for Skagerrak- and North Sea waters). In case of the Baltic- and the North Atlantic experiments, the DMA simultaneously scanned through the aerosol spectra with sampled air either at ambient temperature or heated to 300°C. The purpose of this was to measure both the total and the nonvolatile (sea salt) aerosol size distribution. The water temperature, salinity and dissolved oxygen were measured in or right after the bubble tank continuously. Part of the data was obtained during periods with high water biological activity and thus most probably making the influence of organic material more significant compared to other periods.

Results

As verified by Table 1, differences between experimental setups yields differences in mean number concentration. The differences are not as large in the modes measured, which implies that despite the differences in experimental setups, the waters can be compared to each other.

Looking at an example of the volume size distributions, the waters show a similar slope toward 2 μm (Figure 1a, b), even though they differ much in salinity. Despite that, we will focus on the differences between the waters used. One must first underline that there seems to be a surprisingly strong similarity between the waters.

Table 1 Main features of the different waters used

Water	Aerosol number concentration (pt/cm^3)	Aerosol mean number concentration (pt/cm^3)	Mode 1 (nm)	Mode 2 (nm)	Mode 3 (nm)
Askö, the Baltic	46–785	344	80	200	–
Garpen, the Baltic	30–257	162	80	200	645
Skagerrak	26–181	61	–	300	–
North Sea	17–168	72	–	300	740
North Atlantic	109–7,179	2,224	80	200	–

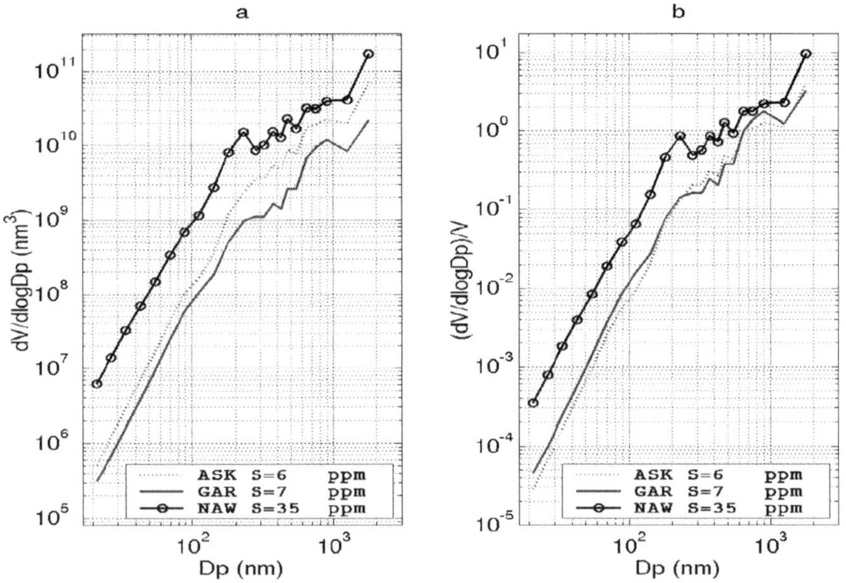

Figure 1 Volume size distributions of three of the waters used at Tw = 287 K. ASK = Askö, GAR = Garpen, and NAW = North Atlantic water. Left-hand figure (a) is not normalized; right-hand figure (b) is (to the total volume). S = salinity of the water

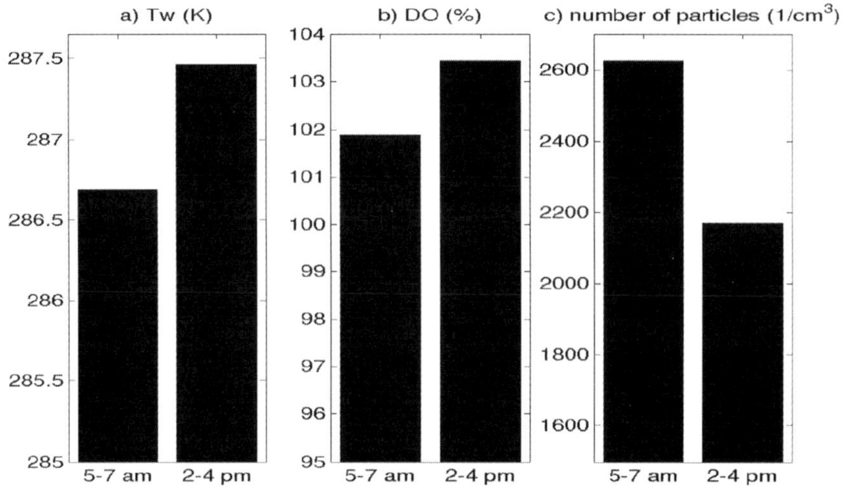

Figure 2 Water temperature (a), dissolved oxygen (b) and aerosol number concentration (c) during high and low biological productivity hours during the North Atlantic cruise in 2006

Bubble bursting appears to be a stable and predictable "aerosol generator", which gives hope for a successful parameterization of this aerosol source.

Further results point in the direction of biology as the number one variable affecting the particle production. Biological activity results in altered chemistry at the oceans' surface, which may change important properties such as surface tension and result in a changed particle production. Biological activity also results in changes in the dissolved oxygen concentration, which in turn can cause changes in the particle production. Temperature variations can also affect the particle production, but the effects of both oxygen and temperature seem to be disguised whenever the biology/chemistry of the surface water is enhanced. An example of the biological activity importance in surface sea waters is shown in Figure 2. During the cruise in the North Atlantic the water temperature changed during the day due to solar radiation, and the level of dissolved oxygen followed its pattern as a result of the increased biological activity in the water during the light hours of the day. The number concentration of aerosol particles produced by the bubble bursting process reveals close to an anti-correlated pattern, which could be due to changes in surface tension from the increased biological activity.

Acknowledgments This work is supported by the Swedish Research Council for Environment, Agricultural Science and Spatial Planning (FORMAS), The Swedish Research Council, and the European Union through the project MAP.

References

1. Matthias-Maser, S., Brinkmann, J., and Schneider, W., *Atmos. Environ.*, **33**, 3569–3575 (1999).
2. Blanchard, D.C., The production, distribution, and bacterial enrichment of the sea-salt aerosol, in: *Air-sea Exchange of Gases and Particles*, edited by P.S. Liss and W.G.N. Slinn, pp. 407–454 (1983).
3. Nilsson, E.D., Rannik, Ü, Swietlicki, E., Leck, C., Aalto, P.P., Zhou, J., and Norman, M., *J. Geophys. Res.*, **106**(D23), 32139–32154, (2001).
4. Mårtensson, E.M., Nilsson, E.D., de Leeuw, G., Cohen, L.H., and Hansson, H.-C, *J. Geophys. Res.* **108**(D9) (4297) (2003).
5. O'Dowd, C.D., Facchini, M.C., Cavalli, F., Ceburnis, D., Mircea, M., Decesari, S., Fuzzi, S., Yoon, Y.J., and Putaud, J.-P., *Nature*, **431**, 676–680 (2004).
6. Stramska, M., Marks, R., and Monahan, E.C., *J. Geophys. Res.* **95**(C10), 18281–18288 (1990).

Similarity Between Aerosol Physicochemical Properties at a Coastal Station and Open Ocean over the North Atlantic

M. Rinaldi[1], M.C. Facchini[1], C. Carbone[1], E. Finessi[1], S. Decesari[1], M. Mircea[1], S. Fuzzi[1], D. Ceburnis[2], and C.D. O'Dowd[2]

Abstract In order to assess the potential influence of the costal environment on marine aerosol chemical composition, two parallel aerosol data sets, collected during two campaigns (coastal site campaign at Mace Head station and open ocean campaign in North Atlantic Ocean), are compared. Aerosol samples were analyzed for Water Soluble Organic Carbon (WSOC), Total Carbon (TC), and main inorganic ions. In addition, a detailed chemical characterization of WSOC was performed.

Results show a strong similarity between size-segregated chemical composition in the submicron size interval, at the coastal station and open ocean site, thus excluding potential effects of the shore line environment. Only the coarse fraction seems in some way influenced by the proximity of the shoreline, the sea salt concentration being higher in this size interval for the coastal site.

Keywords Aerosol chemistry, biogenic particles, coastal particles, marine aerosols

Introduction

Organic marine aerosol (of both primary and secondary origin) represents an important component of the climate feedback system involving aerosols and clouds.[1,2] Given the availability of large data sets of marine aerosol physicochemical properties at coastal sites, the evaluation of the potential influence of coastal environment on aerosol population is essential to extrapolate the results to open ocean conditions, thus understanding how much the sampling site characteristics could bias the results.[2,3]

[1]*Istituto di Scienze dell'Atmosfera e del Clima – CNR, via P. Gobetti 101, 40129, Bologna, Italy*

[2]*Department of Experimental Physics & Environmental Change Institute, National University of Ireland Galway, Ireland*

C.D. O'Dowd and P.E. Wagner (eds.), *Nucleation and Atmospheric Aerosols*, 1098–1101.

In order to assess the potential influence of the costal environment on marine aerosol organic and inorganic chemical composition, two parallel aerosol data sets, collected during two campaigns in the framework of the EC project MAP, are compared.

Methods

The field experiments were carried out from 5 June to 5 July 2006, during the period of high oceanic biological activity, at Mace Head Atmospheric Research Station on the west coast of Ireland (costal site) and on board the oceanographic vessel "Celtic Explorer" (open ocean site) sailing in the North Atlantic between 100 and 300 Km off the west coast or Ireland. At the coastal site the sampling was performed on the field mast at 10 m height from the ground, on the ship the sampling height was fixed approximately at thc same height (10 m) from the sea level.

Aerosol samples were collected, under clean marine sector conditions[1] by means of eight-stages Berner impactors equipped with Tedlar foils, collecting particles in eight size fractions between 0.060 and 16 µm diameter.

In order to obtain a detailed chemical characterization of the organic fraction, aerosol samples were also collected by high volume virtual impactors, segregating fine (a.d. less than 1 µm diameter) and coarse particles (a.d. between 1 and 10 µm diameter) on quartz filters.

Three parallel aerosol samples were collected during the campaigns as a result of sampling time of the order of 50 h each. WSOC (Water Soluble Organic Carbon) and main inorganic ions analyses were performed on Tedlar foils, while high-volume samples were used for Total Carbon (TC) analyses and organics chemical characterization: HPLC speciation according to acidic properties,[4] HNMR functional group analyses[5] and surface tension measurements.[6]

Results

As shown in Figure 1, non-sea salt (nss) sulfate is the dominant species in the submicron fraction in both samples sets, showing a similar pattern both in coastal and open ocean samples: close to detection limit concentrations in the smallest particles and maximum concentration for accumulation mode particles (0.25–0.5 µm). For each stage of the fine fraction the coastal site samples show a higher average $nssSO_4^{-2}$ concentration than the open ocean samples, but in the range of the observed variation in the concentration observed during high biological activity period.[7]

Conversely, the average WSOC concentration in the fine fraction is very similar for samples collected at Mace Head and on the open ocean: 0.02 µg m^{-3} and 0.02 µg m^{-3} for the smallest particles, 0.07 µg m^{-3} and 0.05 µg m^{-3} for small accumulation

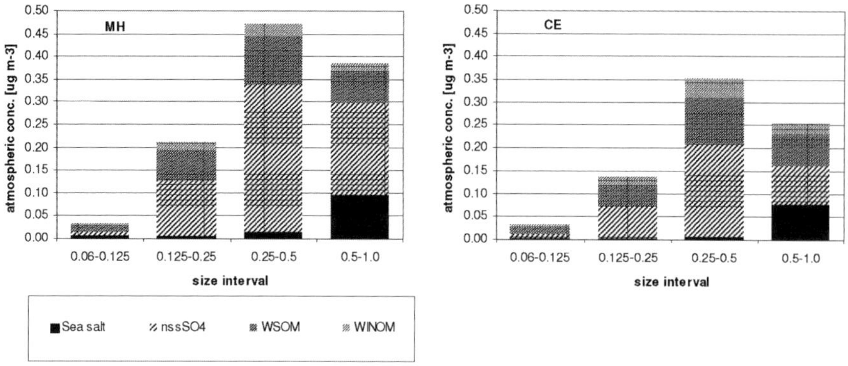

Figure 1 Average composition of submicron aerosols collected at the coastal site (labeled MH on the left) and in open ocean conditions (labeled CE on the right)

mode particles, 0.11 μg m^{-3} and 0.10 μg m^{-3} for particles in the range 0.25–0.5 μm and 0.07 μg m^{-3} and 0.07 μg m^{-3} for large accumulation mode particles. The ratio between WSOC and WIOC (Water Insoluble Organic Carbon) is slightly lower in the samples collected at the coastal site, WSOC being the dominant fraction of organics in both sample sets.

The coarse fraction is dominated by sea salt in all size intervals. However, in terms of absolute mass, the coastal samples show higher sea salt concentration compared to the open ocean ones. In both sample sets nitrate is present in the coarse fraction, as a result of the reaction with gaseous nitric acid (chlorine displacement).

Concerning WSOC chemical characterization, HNMR analyses show that the functional group distribution in samples collected during the cruise is consistent with the one of samples collected at Mace Head, suggesting the same chemical composition and the same origin.

Moreover, the measurements of surface tension in fine fraction isolated WSOC highlight the same pattern for samples collected in mid-ocean and at the coastal site (Figure 2).

Conclusion

In conclusion, our results show a strong similarity between sub micron aerosol physicochemical properties at the coastal station and open ocean site, particularly with regard to the main organic components, thus excluding potential effects of the shoreline environment on marine aerosol. Only the coarse fraction seems in some way influenced by the proximity of the shoreline, the sea salt concentration higher being for the coastal site.

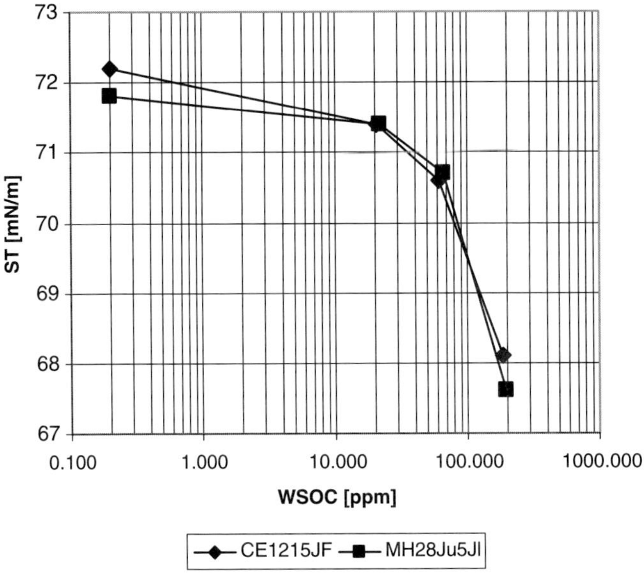

Figure 2 Surface tension of fine aerosol water extracts as a function of WSOC concentration: coastal (MH) and open ocean (CE) samples

Acknowledgment The present study is funded by the MAP-EU project (GOCE-018332).

References

1. Cavalli, F., Facchini, M.C., Decesari, S., Mircea, M., Emblico, L., Fuzzi, S., Ceburnis, D., Yoon, Y.J., O'Dowd, C.D., Putaud, J.P., and Dell'Acqua, A., *J. Geophys. Res.*, **109**, doi: 10.1029/2004JD005137 (2004).
2. O'Dowd, C.D., Facchini, M.C., Cavalli, F., Ceburnis, D., Mircea, M., Decesari, S., Fuzzi, S., Yoon, Y.J., and Putaud, J.P., *Nature*, **431**, 676–680 (2004) and on-line supporting material included.
3. Coe, H., Allan, J.D., Alfarra, M.R., Bower, K.N., Flynn, M.J., McFiggans, G.B., Topping, D.O., Williams, P.I., O'Dowd, C.D., Dall'Osto, M., Beddows, D.C.S., and Harrison, R.M., *Atmos. Chem. and Phys.*, **6**, 3289–3301 (2006).
4. Mancinelli, V., Rinaldi, M., Finessi, E., Emblico, L., Mircea, M., Fuzzi, S., Facchini, M.C., and Decesari, S., *J. Chromatogr. A*, in press.
5. Decesari, S., Facchini, M.C., Fuzzi, S., and Tagliavini, E., *J. Geophys. Res.*, **105**, 1481–1489 (2000).
6. Facchini, M.C., Decesari, S., Mircea, M., Fuzzi, S., and Loglio, G., *Atmos. Environ.*, **34**, 4853–4857 (2000).
7. Yoon, Y.J., Ceburnis, D., Cavalli, F., Jourdan, O., Putaud, J.P., Facchini, M.C., Decesari, S., Fuzzi, S., Jennings, S.G., and O'Dowd, C.D., *J. Geophys. Res.*, in press.

Marine Aerosol and Secondary Particle Formation over the North Atlantic

M. Ehn[1], T. Petäjä[1], P. Aalto[1], G. de Leeuw[1,2], C.D. O'Dowd[3], and M. Kulmala[1]

Abstract During the summer of 2006 a measurement campaign was conducted on a ship in the North Atlantic. We measured the concentration, size distribution and hygroscopicity of the atmospheric aerosol particles. The size distribution consisted of two separate modes, one peaking at 50 nm and the other at 200 nm. Based on the hygroscopicity measurements, the first mode was internally mixed and had a hygroscopic growth factor of 1.6–1.7. In the afternoon of 13 June we observed a mode of small particles, which grew during the evening to 30–40 nm. These particles are believed to be secondary particles, which would make this a very rare observation. Further analysis will be made once more supporting data becomes available.

Keywords Marine aerosols, nucleation

Introduction

Oceans are known to produce aerosol particles[1] and aerosol particles are known to influence the global radiation balance.[2] As oceans cover over 70% of the earth's surface, the total impact on climate from this natural source is very important. Particles are produced by two separate processes: primary particles are emitted directly into the atmosphere from breaking waves, secondary particles are formed through gas-to-particle conversion in the atmosphere. An accurate source function for marine aerosols, both concerning number and composition, is worthwhile.

[1] *Division of Atmospheric Sciences, Dept. of Physical Sciences, University of Helsinki, P.O. Box 64, FI-00014 University of Helsinki, Finland*

[2] *Finnish Meteorological Institute, Research and Development, Climate and Global Change Unit, P. O. Box 503, FI-00101 Helsinki, Finland*

[3] *Department of Experimental Physics and Environmental Change Institute, National University of Ireland, Galway, Ireland*

C.D. O'Dowd and P.E. Wagner (eds.), *Nucleation and Atmospheric Aerosols*, 1102–1105.
Springer 2007

The Marine Aerosol Production (MAP) EU-project set out to improve the quantitative understanding of both primary and secondary production processes. As part of the project, a field campaign was conducted during June–July 2006. Measurements were made on the west coast of Ireland at the atmospheric research station in Mace Head, and simultaneously on a ship west and north of Ireland. We measured the total number concentration, number size distribution and hygroscopicity of particles over the ocean on the R/V Celtic Explorer between 11 June and 5 July.

Measurements

Our instruments were located in a sea container placed on the foredeck of the ship. A TSI 3022 Condensation Particle Counter (CPC) was used to measure the number concentration, a twin-Differential Mobility Particle Sizer (DMPS) to measure the size distribution of 3 nm–1 μm particles and a Humidity Tandem Differential Mobility Analyzer (HTDMA) to measure the hygroscopicity of particles with dry sizes of 25, 50, 75, and 100 nm.

Based on weather forecasts, the ship tried to stay in clean marine air masses throughout the cruise, to reduce continental influence. Furthermore, to avoid sampling ship exhaust, the vessel traveled into the wind as often as possible.

Results and Discussion

The clean marine background size distribution consisted of two separate modes, one larger mode peaking at around 50 nm and a second, slightly smaller at about 200 nm. An example of this can be seen in Figure 1, and this specific day also contains one of the most interesting results of our measurements, namely the mode of small particles detected in the evening.

This mode is believed to be due to recently formed secondary particles. The spectra are very different from those we see when sampling ship exhaust, and based on backward trajectories (Figure 2), there should be no continental influence as the air mass has traveled above water for at least 36 h.

Our HTDMA could only measure particles up to 100 nm, and thus we only sampled one of the two modes in the background size distributions. For the sampled mode, the growth was monomodal with a hygroscopic growth factor (GF) of 1.6–1.7, which was very close to the GF for ammonium sulfate used for daily calibrations during the cruise.

Particles that are this soluble will fairly easily activate as cloud condensation nuclei (CCN). At a supersaturation of 0.25%, the activation diameter of ammonium sulfate is around 75 nm.[3] This may partly explain why the particle number starts decreasing after 50–60 nm; more and more particles start forming cloud droplets as their size increases. Even though the water from these droplets might later

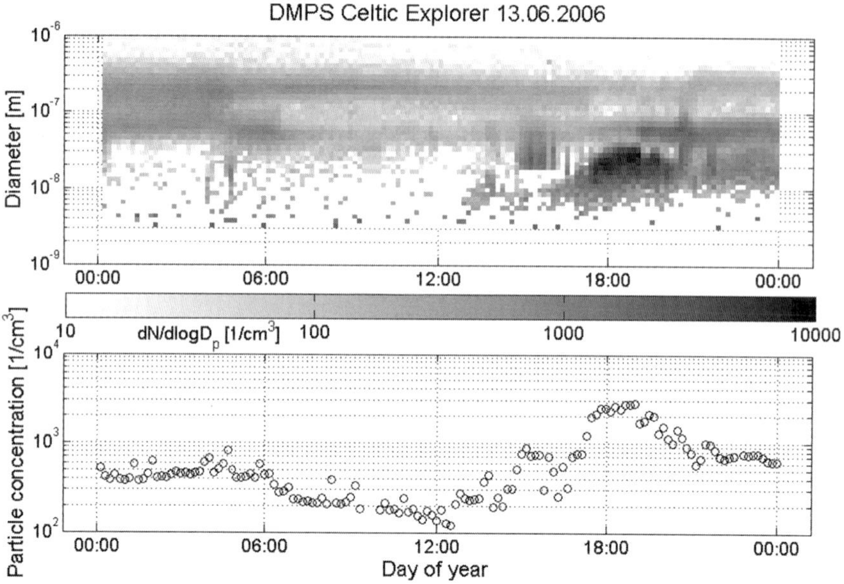

Figure 1 Size distribution (**upper plot**) and number concentration (**lower plot**) measured by the twin-DMPS on June 13, 2006. In the upper plot we can discern the two modes commonly seen in the clean marine air. At around 1500 a third mode develops at around 10 nm, and by 1800 the mode has grown to 30–40 nm. The lower plot indicates that particle concentrations were below 150 cm^{-3} at noon. Later, when the new particles appeared, concentrations went up to 3,000 cm^{-3}

evaporate, they will have acquired more mass during this process and thus be part of the mode of larger particles.

Unfortunately, the HTDMA was not running on the day when we observed the possible new particle formation event, and therefore we do not have supporting data for further analysis. However, other groups conducting measurements on board will certainly have interesting data to compare with, once data has been properly evaluated. Furthermore, the air mass passing over the Celtic Explorer during the event passed over Mace Head about 10 h later. This gives us the possibility to study the evolution of these new particles over a longer period of time.

Conclusions

The typical clean marine size distribution consists of two modes at 50 nm and 200 nm, respectively. The existence of two separate modes may be due to CCN activation, which fits fairly well with the results of the hygroscopicity of the particles (GF = 1.6–1.7).

In the afternoon of 13 June a mode of small particles was detected, and these particles are believed to originate from secondary particle formation above the ocean. Observations of secondary particle formation in the marine boundary layer

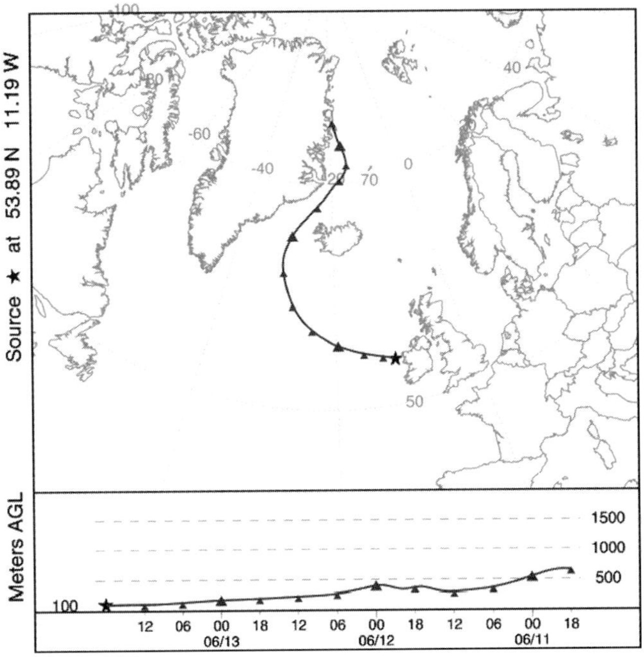

Figure 2 72 h backward trajectory at 1800 on 13.6. The air mass has traveled above the open ocean for at least 36 h, perhaps up to 72 h, before reaching the ship

over the North Atlantic are very rare. Further analysis will be made once more supporting data becomes available.

Acknowledgments We gratefully acknowledge the assistance of the captain and crew of the R/V Celtic Explorer both before and during the cruise. This work was funded by the EU (FP6, project number 018332).

References

1. O'Dowd, C., Smith, M., Consterdine, I., and Lowe, J., Marine Aerosol, sea-salt, and the marine sulphur cycle: a short review, *Atmos. Environ.*, **31**, 73–80 (1997).
2. Intergovernmental Panel on Climate Change, *Climate Change 2007: The Physical Science Basis – Summary for Policymakers* (2007).
3. Giebl et al., Giebl, H., Berner, A., Reischl, G., Puxbaum, H., Kasper-Giebl, A., and Hitzenberger, R., Ccn activation of oxalic and malonic acid test aerosols with the university of Vienna cloud condensation nuclei counter, *J. Aerosol Sci.*, **33**, 1623–1634 (2002).

Role of the Volatile Fraction of Marine Aerosol on its Hygroscopic Properties

Karine Sellegri[1], Paolo Villani[1], David Picard[1], Regis Dupuy[2], Colin D. O'Dowd[2], and Paolo Laj[1]

Abstract The hygroscopic growth factor (HGF) of 85 and 20 nm particles marine aerosols was measured during January 2006 for a three-week period within the frame of the MAP (Marine Aerosol Production) winter campaign, using the TDMA technique. The results were compared to aerosol produced in a simulation tank by bubbling air through sea water sampled at the coastal site of Mace Head during the campaign, and through synthetic sea water (exempt of organic substances). This simulation was assimilated to primary production. The 85 nm HGF observed in the atmosphere during clean marine sectors were lower than the ones measured from the bubbling processes: the sea salt HGF mode is slightly lower and there is an additional 1.5 HGF mode on the atmospheric aerosol. This would indicate that either the sea water sampled near Mace Head was not as rich in hydrophobic matter as further up wind or that secondary processes have occurred during transport. The role of the volatile fraction of the aerosols was then studied by gently heating the particles at 90°C (without particle size change) and measuring the subsequent HGF change, with a combination of Volatility and Hygroscopicity TDMA (i.e., the VH-TDMA). We observed that the volatilization of less than 10% by diameter of the particles lead to an inhibition of the 1.5 GF mode for 85 nm particles but not for 20 nm particles. This result would indicate that secondary condensing processes, implying volatile substances, would have influenced the 85 nm particles. These results only apply to low biological activity periods.

Keywords TDMA, growth factor, primary production, secondary production, volatile compounds

[1] *Laboratoire de Météorologie Physique, Observatoire de Physique du Globe de Clermont-Ferrand, Université Blaise Pascal, France*

[2] *Physics Department & Environmental Change Institute, National University of Galway, Ireland*

C.D. O'Dowd and P.E. Wagner (eds.), *Nucleation and Atmospheric Aerosols*, 1106–1109.

Introduction

The chemical composition and hygroscopicity of marine aerosols have been widely studied during ship measurement campaigns over a large variety of oceans and seas. However, we have only a limited knowledge of the variability over time of these properties. The chemical composition and concentration of marine aerosols is depending on parameters such as wind speed and white cap coverage, but also depending on the sea water composition. Recent findings have shown that submicron marine aerosols not only contain sea salt and sulfate, but also contain substantial levels of organic carbon. This organic content has a seasonal variability with maximum concentrations during the spring and autumn seasons, corresponding to the blooming periods of phytoplankton [1].

Yet we do not know how these organic species are integrated into the particulate phase. Two possible processes are:

1. Primarily, during the mechanical process of bubble bursting: the sea surface layer is known to be enriched with organic matter and film/jet drops are likely to be enriched as well.
2. Secondarily, through condensation of low volatile organic gases released in the marine atmosphere by microorganisms.

It is possible to partly investigate whether the organic particulate matter is primarily or secondarily produced by observing the mixing state of particles. The MAP campaign's main goal is to understand what the origin of the large organic fraction of submicron aerosol. The hygroscopicity of the marine aerosols can give new insight into this question. Hygroscopicity of aerosol particles is classically studied using aerosol classifiers in tandem, separated by a humidifier (HTDMA, Humidity Tandem Differential Mobility Analyzer). Recently, it has been shown by several workers that the surface of the particles [2] as well as mixture effect [3]) influence the particles growth rate when they are exposed to a humid environment. Theoretical calculations of sub-saturation growth mainly involves the solubility of the particle bulk, the role of the particle surface being taken into account through the surface tension of the solute, and through the accommodation coefficient of water on the hydrated surface. However, surface tensions and accommodation coefficients are poorly documented and it is still an open question whether this theoretical approach is accurate or not. In order to study mixture effects and surface effects on particles hygroscopicity, we used a VHTDMA, which aim is to modify the surface of particles by gently heating them, and then measuring their hygroscopic properties before and after surface treatment.

Site and Methods

The VHTDMA has been operated for the winter campaign (January 2006) of the MAP (Marine Aerosol Production) project at the Mace Head Research Station, Ireland. The instrument newly developed at the LaMP is composed of two DMAs and CPCs, separated by an oven and a hydration device in series. A complete study of an aerosol size consists in measuring the size change obtained after slight heating (up to 110°C) (Volatility-Scan), then the size change due to exposure to a humid flow (90%) (Humidity-Scan), and finally the size change due to heating followed by humidifying (Volatility-Humidity-Scan). The winter MAP campaign offered long periods of steady clean marine air masses which enabled us to study primary marine aerosol production. In clean marine air masses 85 and 20 nm particles were selected for studying their hygroscopic properties. The results were compared to aerosol produced in a simulation tank by bubbling air through sea water sampled at the coastal site of Mace Head during the campaign, and through synthetic sea water (exempt of organic substances).

Results

Slight HGF differences were observed between the aerosol generated from bubbling in natural sea water and bubbling in synthetic sea water, both for 85 and 20 nm particles. The aerosol resulting from natural sea water was less hygroscopic than synthetic sea water (Figure 1a, b). This indicates that the non-sea salt compounds present in natural sea water decreases the HGF compared to pure sea salt, already when the primary aerosol is generated.

The HGF of atmospheric aerosols sampled within clean marine air masses was lower than the HGF of particles generated by bubbling in sea water (natural or synthetic) (Figure 1a): the sea salt mode is slightly decreased and an additional HGF mode appears at around 1.5. This is significant for 85 nm particles but not so obvious for 20 nm particles (Figure 1b). Because the primary production of 20 and 85 nm particles by bubble bursting would lead to the same chemical composition of both sizes, this result indicates that secondary processes are likely to have influenced the 85 nm particle in a larger extent than the 20 nm particles.

Gentle volatilization had no effect on the HGF of synthetic sea salt bubble-generated aerosols. It is then likely that non-sea salt compound in sea water responsible for lower hygroscopicity than sea salt is not volatile at 90°C. On the contrary, the hygroscopic behavior of atmospheric marine aerosols shows a significant change when gentle volatilization is applied: the 1.5 HGF mode is inhibited for the 85 nm particles. Hence we can conclude that the volatile fraction evaporating from the marine aerosol, most probably originating from secondary organic process, was responsible for a lower hygroscopic behavior as compared to the remaining salts. This result is representative for the low biological activity season, and would need to be compared with the spring period.

Acknowledgment This work was supported by ACCENT.

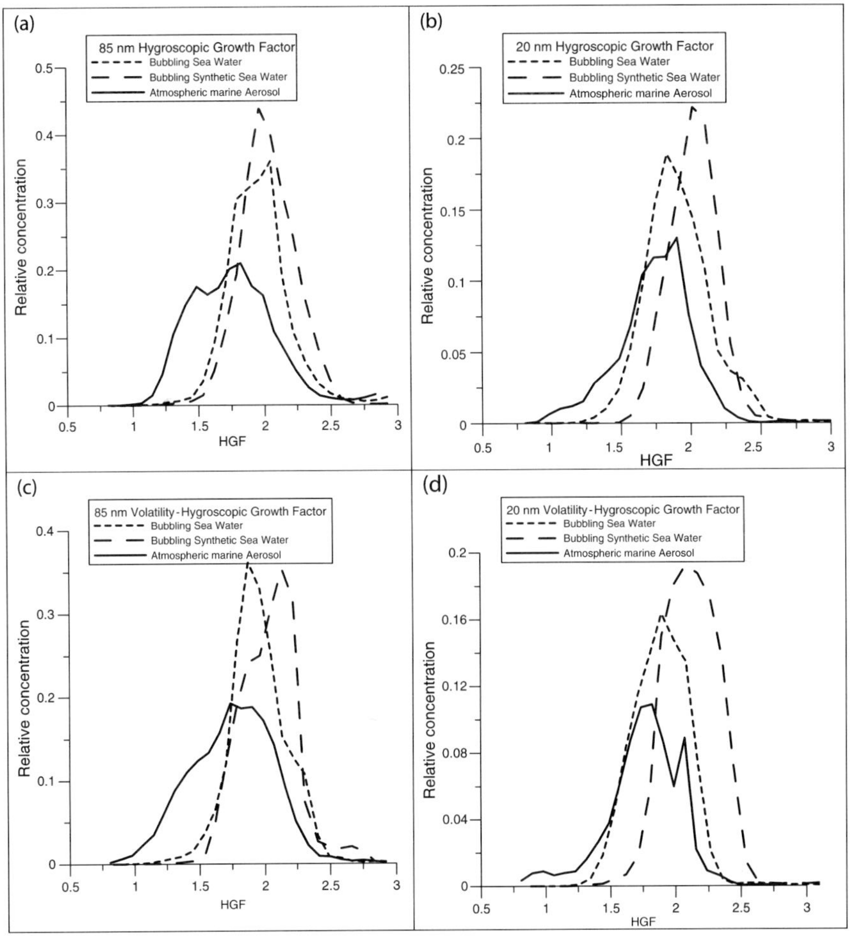

Figure 1 Hygroscopic Growth Factor (HGF) measured at 85% RH for (a) 85 nm particles, (b) 20 nm particles, and HGF measured at 87% RH after a gentle volatilization at 90°C for (c) 85 nm particles and (d) 20 nm particles. Atmospheric marine aerosols have been averaged over a stable marine air mass period of two days (from January 17, 2006 1200 h to January 19, 2006 1800 h).

References

1. O'Dowd et al., Biogenically driven organic contribution to marine aerosol, *Nature*, **431** (7009), 676–680 (2004).
2. Sorjamaa, R. and Laaksonen, A., The influence of surfactant properties on critical supersaturations of cloud condensation nuclei, *J. Aerosol Sci.*, **37**, 1730–1736 (2006).
3. Sjogren et al., Hygroscopic growth and water uptake kinetics of two-phase aerosol particles consisting of ammonium sulfate, adipic and humic acid mixtures, *J. Aerosol Sci.*, **38**, 157–171 (2007).

Sea Salt Production and Distribution over the North-east Atlantic

Saji Varghese, Baerbel Langmann, and Colin D. O'Dowd

Abstract This paper presents the results from the implementation of a new submicron sea salt parameterization in a regional climate model (REMOTE). Evaluation of sea salt production, its regional distribution, and transport are carried out in this study for the periods May 2002, January 2003, and June 2003 over the domain of the North-east Atlantic and Western Europe. The monthly mean mass concentrations simulated for coarse mode are up to 70 times more than the accumulation mode with maximum mode values of approximately 56 and 5.5 $\mu g\ m^{-3}$, respectively.

Keywords Sea salt, marine aerosols, regional climate model

Introduction

Sea salt is one of the main components of primary marine aerosols. Sea salt aerosols are mainly formed from sea spray generated by bubble bursting due to winds over the oceans. They affect the radiative balance of the atmosphere directly by reflection, scattering and absorption (direct effect). The cloud microphysics and radiative properties are also altered which is known as the indirect effect. The magnitude of indirect effect and its uncertainty are several fold compared to the direct radiative effect of sea salt aerosols. In general, the level of scientific understanding on the contribution of aerosols to the different radiative forcing components remains poor.[1]

The development of a robust submicron sea salt source function has remained a challenge for a long period. Recently one[2] has been developed for sea spray based on measurements carried out at Mace Head. Most of the earlier attempts by other researchers are based on laboratory measurements or experiments carried out under

Department of Experimental Physics & Environmental Change Institute, National University of Ireland, Galway, University Road, Galway, Ireland

C.D. O'Dowd and P.E. Wagner (eds.), *Nucleation and Atmospheric Aerosols*, 1110–1114.

highly idealized conditions (for example[7]). Some of them even extrapolate measurements made for super-micron particles to submicron sea salt particles. These inconsistencies are overcome by[2] to a great extend, which explicitly provide the number flux of accumulation and Aitken mode sea salt particles. These modes are important in terms of understanding cloud microphysics and its development.

In this study the new sea-spray source function[2,3] is incorporated and the production mechanism, regional distribution and transport are evaluated for both submicron[2,9] and super-micron[9] sea salt particles.

Model Description and Set-up

The three-dimensional high resolution regional climate model REMOTE (Regional Model with Tracer Extension[4]) is used in the study presented here. It includes coupled climate-chemistry-aerosol dynamic modules. The dynamics is based on the regional weather forecast model system of the German Weather Service. The physical parameterizations are based on the ECHAM-4 model.[8] An Arakawa-C grid is used for the prognostic equations to represent the general meteorological variables and trace species. The model uses 19 vertical layers from the surface up to 10 hPa with a horizontal resolution of 0.5° represented by 91 × 81 grid points. Geographically, the simulation model domain covers North-east Atlantic and Western Europe. At the lateral boundaries, the REMOTE is driven by ECMWF re-analysis data every 6 h. The aerosol dynamics is based on the modal aerosol model M7.[5,6] The aerosol model treats five aerosol species including sea salt. Four different modes (nucleation, Aitken, accumulation, and coarse) are available to deal with size distribution and number concentration. In this study, the sea-spray production mechanism after[2] is implemented in REMOTE.

Results

The regional climate model is first evaluated to determine its capability to simulate wind speed and total sea salt. Figures 1 and 2 show a comparison of observed data with model results for June 2003. The model simulated wind speed agree quite closely with measurements for about 50% of the comparison period. For the rest of the period, it is underestimated by 10–25%. Generally, there is good agreement between measured and simulated total sea salt mass. However, there is an underestimation for certain periods due to underprediction of wind speed and the mismatch in the sampling periods. It can be concluded that the sea salt mass is predicted well for the evaluated periods.

Simulations are carried out for the periods of May 2003, January 2003, and June 2003.

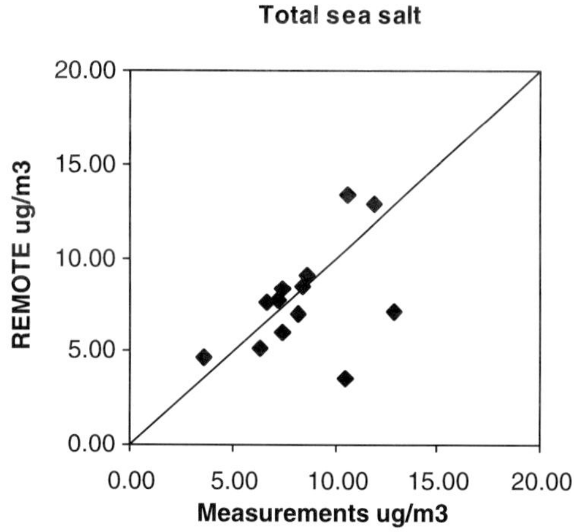

Figure 1 Comparison of measured and modelled sea salt mass at Mace Head for June 2003

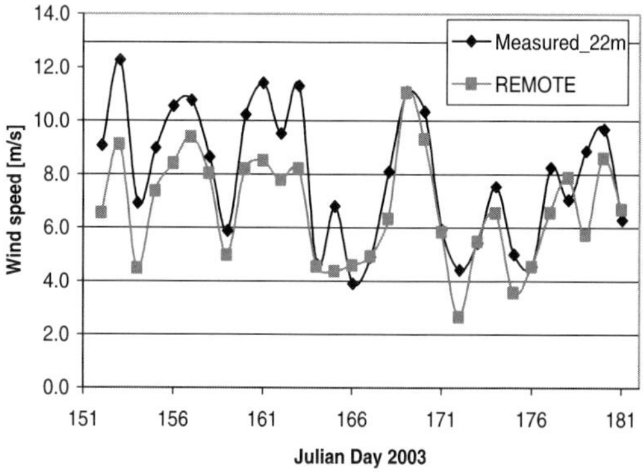

Figure 2 Comparison of measured and modelled wind speed at Mace Head for June 2003

The monthly mean values of sea salt concentration in the accumulation and coarse modes are shown in Figure 3. For accumulation mode, in May 2002, the concentration is quite low over the domain with a maximum of 2.5 $\mu g\ m^{-3}$ occurring over the Mediterranean and north-east of Iceland. The low concentrations are mainly due to the low wind speeds in this month and efficient precipitation over the ocean. For January 2003, the highest concentrations are found in the North Sea, Baltic Sea, and in the Atlantic Ocean west of Ireland. In late spring and early summer, an opposite pattern of low concentration is seen over the west of Ireland with values less than 1 $\mu g\ m^{-3}$ owing to low wind speeds. However areas in the Meditteranean and north-east of Iceland show remarkable increases with a maximum of more than 5 $\mu g\ m^{-3}$ which can again be attributed to high wind speeds.

In the case of coarse mode distribution, the concentrations are at times 70-fold more compared to accumulation mode. For June and January 2003, the distribution patterns

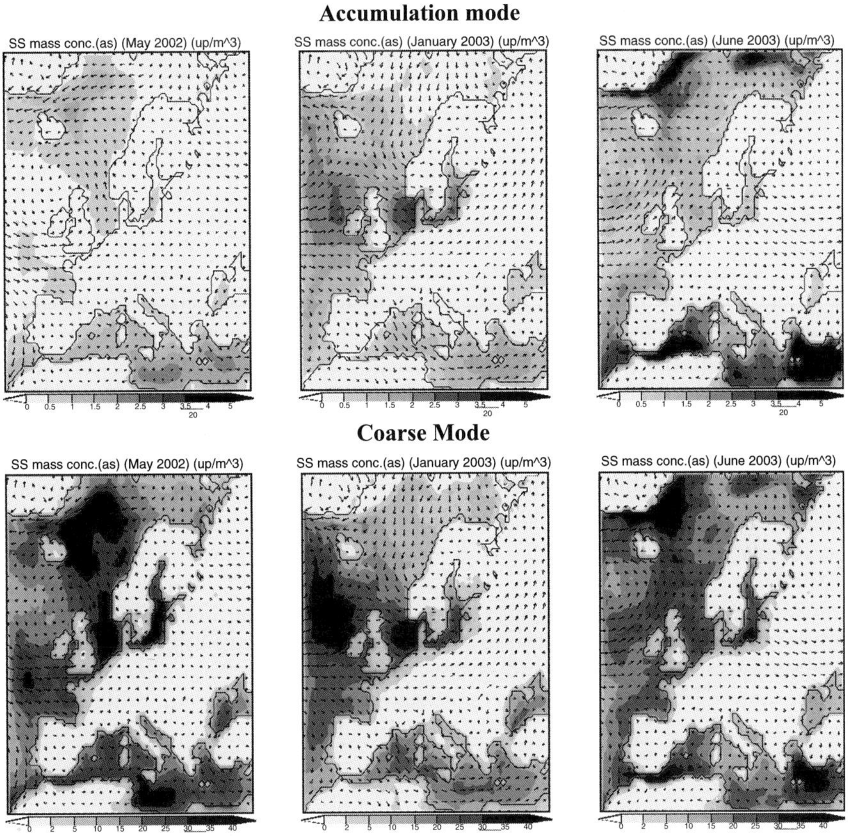

Figure 3 Monthly mean distribution of sea salt particles in the accumulation and coarse modes for the months May 2002, June 2003, and January 2003 in μg m^{-3}. Plots on the upper panel indicate accumulation mode and that below indicate coarse mode

Accumulation mode
Coarse Mode

remain similar to that of the accumulation mode but on an increased scale. During May 2002, there is a pronounced increase in concentration particularly in the areas north-east of Iceland, North Sea, Baltic Sea, and the Mediterranean region. This peculiar behaviour can be explained only after further investigation into the deposition and sedimentation processes. Highest concentration of 56 μg m^{-3} occur over east of Iceland during this period. In January, coarse mode distribution is 5–20% more than accumulation mode with the highest difference in the west Mediterranean whereas in June, maximum differences of up to 40% is seen in the Baltic Sea and 30% in the west of Ireland. In general the differences ranged from 5–40% in this month. Of the three months studied, May showed the highest variability between the two modes with more than 70% increase over Baltic, Irish, and North seas. Over Atlantic, 20 to >70% variation is seen while over the Mediterranean it is 10–25% in this period. In all the simulations, inland transportation of sea salt is confined to the coastal regions.

Conclusion

An attempt is made to account for the contribution of sea salt aerosols in the accumulation and coarse modes. The new sea-spray source function for submicron particles implemented in REMOTE gives agreeable results against observations available for comparison. However, further investigation and comparisons against observations as well as an intercomparison of previous sea salt source functions are required to have robust conclusions. Aerosol–cloud interactions and the impact of sea salt aerosols on the number and concentration of prevailing aerosols will be future topics for study. Furthermore, its contribution to radiative forcing components will also be investigated.

Acknowledgments The authors would like to acknowledge the support of Enterprise Ireland, EPA (Ireland) and the European Commission under the ACCENT Network-of-Excellence and the specifically targeted research project Marine Aerosol Production project.

References

1. IPCC report, *Climate Change 2007: The Physical Science Basis* (Summary for Policymakers) (2007).
2. Geever, M., O'Dowd, C., van Ekeren, S., Flanagan, R., Nilsson, E. D., de Leeuw, G., and Rannik, Ü., *Geophys. Res. Lett.*, **32**, doi: 10.1029/2005GL023081 (2005).
3. O'Dowd, C., Langmann, B., Varghese, S., Scannell, C., Ceburnis, D., and Facchini, M.C., submitted to GRL (2007).
4. Langmann, B., *Atmos. Environ.*, **34**, 3585–3598 (2000).
5. Vignati, E., Wilson, J., and Stier, P., *J. Geophys. Res.*, **109**, doi: 10.1029/2003JD004485 (2004).
6. Stier, P., Feichter, J., Kinne, S., Kloster, S., Vignati, E., Wilson, J., Ganzeveld, L., Tegen, I., Werner, M., Balkanski, Y., Schulz, M., Boucher, O., Minikin, A., and Petzold, A., *Atmos. Chem. Phys.*, **5**, 1125–1156 (2005).
7. Clarke, A.D., Owens, S.R., and Zhou, J., *J. Geophys. Res.*, **111** (2006).
8. Roeckner, E., Arpe, K., Bengtsson, L., Christoph, M., Claussen, M., Dümenil, L., Esch, M., Giorgetta, M., Schlese, U., and Schulzweida, U., Report No. 218, Max Planck Institute for Meteorology, Hamburg, Germany (1996).
9. Gong, S.L., *Global Biogeochem. Cycles*, **17**(4), 1097, doi:10.1029/2003GB002079 (2003).

Evaluation of Measured and Predicted Cloud Condensation Nuclei in Mace Head

John Byrne, Colin D. O'Dowd, S. Gerrard Jennings, and Regis Dupuy

Abstract Measurements of cloud condensation nuclei (CCN) were made in Mace Head during the winter and summer periods of Marine Aerosol Production (MAP) project in 2006. CCN measurements were performed using a DMT Continuous-flow CCN Instrument. The aerosol size distribution and aerosol microphysics measurements were measured with a TSI SMPS. Using the size and by implying inferred composition information, detailed CCN closure analyses were performed for the data sets from the winter and summer MAP campaigns. For the winter campaign, idealized compositions of pure sea salt and non-sea salt sulfates (nss-S) were assumed. The analysis indicates that in this case, there was good general agreement between the predicted and observed CCN concentrations and that the marine aerosol over this period comprised predominantly sea salt and nss-S. This approach was repeated for the data sets for the summer campaign. To achieve closure between predicted CCN and observed CCN for the summer period, an insoluble organic mass fraction of 0.6 had to be assumed, thereby confirming a significant organic fraction to marine CCN during periods of high biological activity.

Keywords Cloud condensation nuclei, critical diameter, insoluble core, nss

Introduction

The importance of clouds in the climate system is well established; clouds play a vital role in the global radiation budget and hydrological cycle. Clouds form when a parcel of air becomes supersaturated with respect to water vapor and the excess water condenses rapidly on ambient particles to form cloud droplets. For this rapid condensation (termed activation) to occur at a given supersaturation the particle must have sufficient soluble mass; this subset of the aerosol population is called cloud condensation nuclei, denoted CCN.

Department of Experimental Physics and the Environmental Change Institute, National University of Ireland, Galway, University Road, Galway, Ireland

C.D. O'Dowd and P.E. Wagner (eds.), *Nucleation and Atmospheric Aerosols*, 1115–1119.

Marine aerosol contributes significantly to the global radiative budget due to the extensive coverage of ocean surface to the Earth's total surface area, and consequently, changes in marine aerosol abundance and/or chemical composition will impact on both the Earth's albedo and climate. To date much of the research on marine aerosol has centered on the production of aerosol from sea salt and non-sea salt sulphates (nss-S).[1–4] Recent field studies and papers, however, have shown that known aerosol production processes for inorganic species cannot account for the entire aerosol mass that occurs in sub micrometer sizes. Several experimental studies have pointed to the presence of significant concentrations of organic matter in marine aerosol.[5] There is some knowledge available about the composition of organic matter but in many areas, there is a lack of information about the contribution of organic matter to marine aerosol, as a function of aerosol size, as well as its characterization as hydrophilic or hydrophobic.

The Marine Aerosol Production (MAP) project has as one objective to elucidate the cloud drop forming potential of marine aerosols over the North Atlantic and to examine the influence of organic matter on this process. Extensive measurements of aerosol physical and chemical properties were undertaken in clean marine air at Mace Head during the year 2006. Over the year, the period of low biological activity (LBA) can be regarded as winter (winter MAP campaign) and the period of high biological activity (HBA) can be regarded as spring through to autumn (summer MAP campaign). In this study, we make assumptions on the aerosol chemical composition and look for closure with measured CCN concentrations, thereby elucidating the chemical composition of the aerosol. In general, aerosol-CCN closures have been difficult to achieve,[6] most likely because of incomplete knowledge of the aerosol composition and its effect on activation, although some studies have achieved a reasonable degree of closure for certain marine environments.[7,8]

Results

Aerosol-CCN closure studies involve exposing an atmospheric aerosol population to a particular super saturation or set of super saturations (in this case 0.25%, 0.50% and 0.75%) under controlled conditions and measuring the CCN concentration. The CCN concentration is compared to Kohler theory predictions based on concurrent measurements of dry particle size distributions and inferred chemical compositions. A successful closure study will accurately predict the concentration of CCN, i.e., the CCN predicted/CCN observed ratio will be 1.0.

Air masses arriving at Mace Head under clean-sector selection criteria (clean sector, 180°–300°, black carbon mass concentration $<50\,\mathrm{ng\ m^{-3}}$, and total aerosol number concentration below $700\,\mathrm{cm^{-3}}$) typically spent four days traveling over the North Atlantic Ocean, more often emerging from the Canadian, Greenland, and Arctic regions and less frequently from northern USA and subtropical regions. Analysis of available data indicates that coastal influences are negligible.

Valid size distributions and concurrent CCN measurements were performed on 20 days between January 9 and 29, 2006, for the winter campaign and for 35 days (June 5–July 10) for the summer campaign. In the analysis, the critical diameter (D_{crit}) corresponding to sea salt and nss-S was calculated for each super saturation and then the number of particles larger than D_{crit} was calculate from the SMPS measurements and compared to the observed CCN. The scatter plot of derived CCN versus measured CCN for the winter period is shown in Figures 1 and 2 for sea salt and nss-S, respectively at a supersaturation of 0.5%. While the agreement between calculated and observed CCN is very good, the calculated CCN is slightly greater (~10–15%) than that observed for the sea salt assumption, while it is slightly smaller (~10–15%) for the nss-S assumption. Comparison of the two results suggests that the winter marine aerosol comprised a mix of sea salt and nss-S.

For the summer period, the calculated sea salt CCN concentration was significantly higher than that for the observed. The same pattern is seen, to a lesser degree for the nss-S assumption (Figure 3) where the calculated CCN is of the order of 20–25% greater than that observed. Chemical composition analysis of summer time, or high biologically active periods suggest an insoluble organic component of the order of 60–70%. Therefore, a third assumption of a 0.6 insoluble mass fraction core and nss-S was made and the analysis repeated. The comparison of the calculated CCN based on an insoluble organic fraction leads to a significant improvement of the correlation and the slope of the regression more or less equal to 1.

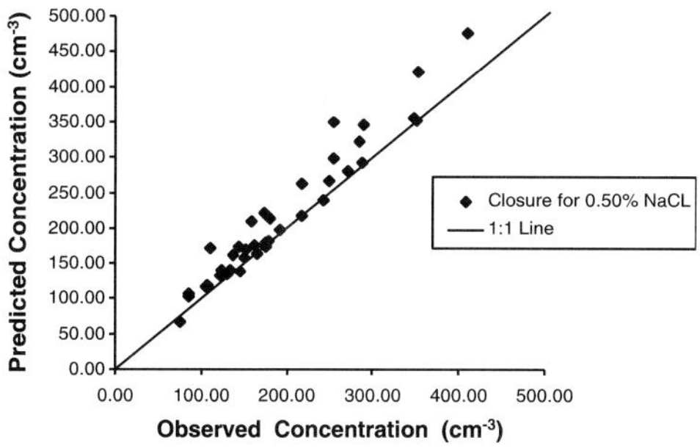

Figure 1 Scatter plot of the simplified closure analysis for clean sector data in the MAP winter campaign at S = 0.50%, idealized composition of pure sea salt assumed (R2 = 0.93)

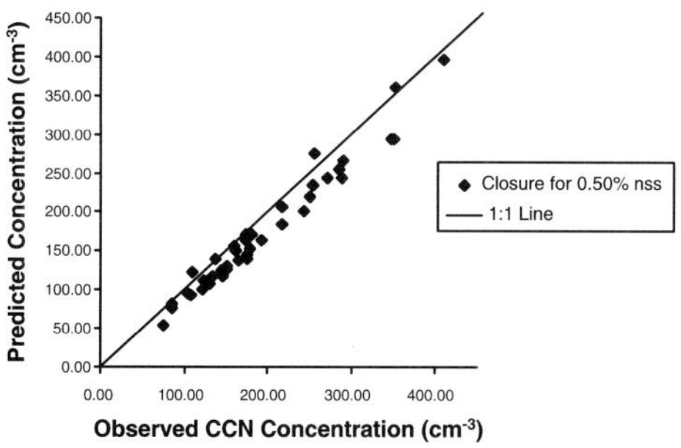

Figure 2 Scatter plot of the simplified closure analysis for clean sector data in the MAP winter campaign at S = 0.50%, idealized composition of nss assumed (R2 = 0.97)

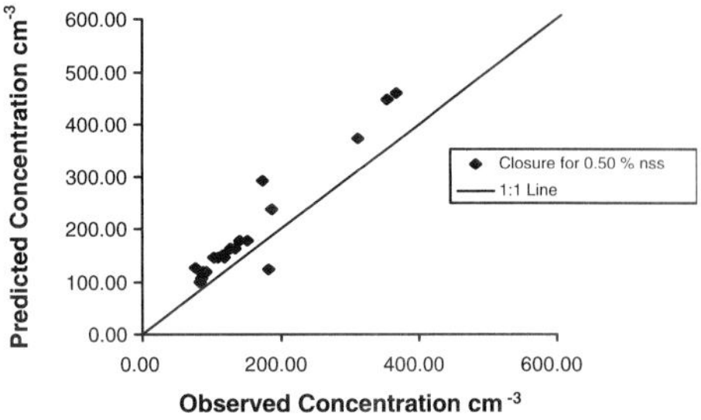

Figure 3 Scatter plot of the simplified closure analysis for clean sector data in the MAP summer campaign at S = 0.50%, idealized composition of nss assumed (R2 = 0.92)

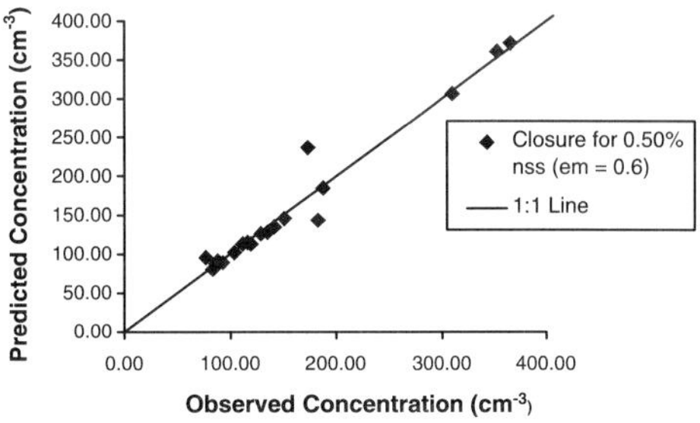

Figure 4 Scatter plot of the improved closure analysis for clean sector data in the MAP summer campaign at S = 0.50%, assumed composition of nss and an organic mass fraction of 0.6 assumed (R2 = 0.96)

Conclusions

A preliminary closure study of measured and calculated CCN concentrations is presented for clean North Atlantic marine aerosol for winter and summer periods. Winter is regarded as a period of low biological activity while summer is regarded as a period of high biological activity and aerosols present in the latter period are expected to contain a significant amount of organic mater. The closure study illustrates that in winter, assumptions of the aerosol chemical composition comprising sea salt and nss-S will lead to good closure, however, for summer, this assumption will lead to an over prediction of calculated CCN to the tune of 30%. Repeating the analysis using a 0.6 insoluble mass fraction leads to almost perfect closure, corroborating previous suggestions that during periods of high biological activity, the marine aerosol contains significant amounts of primary organic insoluble matter.

Acknowledgments We would like to acknowledge the European Commission Marine Aerosol Production project and the EPA, Ireland, for their support.

References

1. Shaw, G., Bio-controlled thermostasis involving the sulfur cycle. *Clim. Change*, **5**, 297–303 (1983).
2. Charlson, R.J., Lovelock, J.E., Andreae, M.O., and Warren, S.G., Oceanic phytoplankton, atmospheric sulfur, cloud albedo and climate, *Nature*, **326**, 655–661 (1987).

3. O'Dowd, C.D., Lowe, J., Smith, M.H., and Kaye, A.D., The relative importance of sea-salt and nss-sulphate aerosol to the marine CCN population: an improved multi-component aerosol-droplet parameterisation. *Q.J. Roy. Met. Soc.*, **125**, 1295–1313 (1999).
4. O'Dowd, C.D., Lowe, J.A., and Smith, M.H., Coupling sea-salt and sulphate interactions and its impact on predicting cloud droplet concentrations, *Geophys. Res. Lett.*, **26**, 1311–1314 (1999).
5. O'Dowd, C.D., Facchini, M.C., Cavalli, F., Ceburnis, D., Mircea, M., Decesari, S., Fuzzi, S., Yoon, Y.J., and Putaud, J-P, Biogenically driven organic contribution to marine aerosols, *Nature*, **431**, 676–680 (2004).
6. Wood, R., Johnson, D.W., Osborne, S.R., Bancy, B.J., Andreae, M.A., O'Dowd, C.D., Glantz, P., and Noone, K.J., Boundary Layer, aerosol and chemical evolution during the thrid Lagrangian experiment of ACE-2, *Tellus*, **52B**, 401–422 (2000).
7. VanReken, T.M., Rissman, T.A., Roberts, G.C., Varutbangkul, V., Jonsson, H.H., Flagan, R.C., and Seinfeld, J.H., Towards aerosol/cloud condensation nuclei (CCN) closure during CRYSTAL-FACE, *J. Geophys. Res.*, **108**(D20), 4633 (2006).
8. Chuang, P.Y., Collins, D.R., Pawlowska, H., Snider, J.R., Jonsson, H.H., Brenguier, J.-L., Flagan, R.C., and Seinfeld, J.H., CCN measurements during ACE-2 and their relationship to cloud microphysical properties, *Tellus*, **52B**, 843–867 (2000).

Part X
Remote Sensing of Aerosols and Clouds

Aerosol Impact on Remote Sensing in Coastal Environment

G.A. Kaloshin

Abstract Last version of aerosol model of the marine and coastal atmosphere surface layer is considered. The model distinctive feature is parameterization of amplitude and width of the modes as functions of fetch and wind speed. Last version of developed code MaexPro 5.0 for spectral profiles of aerosol extinction coefficients calculations is presented. The comparison of the results calculated with help MaexPro 5.0 with available experimental data and with the forecast of aerosol extinction on the code NAM and ANAM is given.

Keywords Coastal aerosol extinction, mie scattering, fetch

The Code MaexPro 5.0

Extinction of IR radiation in the marine boundary layer is dominated by scattering and absorption due to atmospheric aerosol. This is important to optical retrievals from satellite, remote sensing, backscatter of light to space (including climate forcing), cloud properties etc. In unpolluted regions the greatest effects on near shore scattering extinction will be a result of sea salt from breaking waves and variations in relative humidity. The role of breaking waves appears to be modulated by wind, tide, swell, wave spectra, and coastal conditions. These influences will be superimposed upon aerosol generated by open ocean sea salt aerosol that varies with wind speed.

The focus of our study is the extinction and optical effects due to aerosol in a specific coastal region. This involves linking coastal physical properties to oceanic and meteorological parameters in order to develop predictive algorithms that describe 3D aerosol structure and variability.

V.E. Zuev Institute of Atmospheric Optics SB RAS, Tomsk, 634055, Russia

C.D. O'Dowd and P.E. Wagner (eds.), *Nucleation and Atmospheric Aerosols*, 1123–1128.

The Basic Features of the Code MaexPro 5.0

The key element of MaexPro 5.0 is empirical microphysical model MaexPro [1].

The particle size distribution of the code MaexPro 5.0 is similar to models MEDEX, NAM and ANAM [2] is submitted as the sum of four modified lognormal functions. In against available models NAM and ANAM amplitude and width of various modes is parameterized as functions of the fetch (distance that an air mass travels over water).

Calculation of Aerosol Scattering and Extinction Coefficient

Using known Mie programs for spheres, microphysical model MaexPro allows one to make calculations of the aerosol scattering and extinction coefficients spectral profiles $\alpha(\lambda)$, $\sigma(\lambda)$. The refractive index for sea salt and the waters as proposed by F. Volz, 1973 is used in the code MaexP ro 5.0. It is necessary to note, that in the overwhelming majority of cases in the marine atmospheric surface layer distinction between and $\alpha(\lambda)$ $\sigma(\lambda)$ in the visible band are not beyond tool mistakes.

Extrapolation of the Humidity Grown Factor f_H

Besides calculation of $\alpha(\lambda)$ is carried out with use of the following extrapolation connected to the profile of the humidity growth factor f_H:

$$\left(\frac{\alpha_{\mathrm{H}}}{\alpha_{0\mathrm{m}}}\right)=\left(\frac{0.037}{1.017-f_{\mathrm{H}}/100}\right)^{0.84}$$

where α_{0m} coefficients of aerosol extinction at $\mathrm{H} = \mathrm{H}_0$.

Calculation of structures f_H is carried out under following conditions:

- if $20\,\mathrm{m} \le \mathrm{H} \le$ of $25\,\mathrm{m}$ then $\mathrm{f_H} = \mathrm{f}_{25\,\mathrm{m}}$
- if $\mathrm{H} \le 20\,\mathrm{m}$ and $\mathrm{f_H} \le \mathrm{f}_{25\,\mathrm{m}}$, then $\mathrm{f_H} = \mathrm{f}_{25\,\mathrm{M}}$
- otherwise, if $\mathrm{H} \le 20\,\mathrm{m}$, $\mathrm{f_H} = (\mathrm{f}_{25\,\mathrm{m}} + 7) \times \mathrm{H}^{-0.03}$

Extrapolation suits at $40\% < \mathrm{RH} < 98\%$

Area of Applicability of the Code MaexPro 5.0

The code suitable for the particles sizes spectrum 0.01–100 microns on radius and advanced by present time for the range of heights 0–25 m, up to heights where in our opinion, there are the most essential changes of microphysical structure.

Ranges of change of a wind speed make 3–18 m s^{-1}, sizes of fetch up to 120 km, RH – 40–98%. The resolution at spectral interval 0.2–12 μk is 0.001 μk.

Interface of the Code MaexPro 5.0

The MaexPro code develops for calculation spectral and vertical profiles of aerosol extinction coefficient $\alpha(\lambda)$, aerosol sizes distribution, area distribution, volumes distribution, modes aerosol extinction spectra is submitted. The program carries out calculation $\alpha(\lambda)$, as functions of atmospheric effects using standard meteorological parameters, aerosol microphysical structure, a spectral band and a height of the sensor location place.

In Figure 1 the composite window "extinction spectra" of the code MaexPro 5.0 is submitted.

The top part of the window is intended for the task of input parameters, the bottom – for results of calculation. Here with use of a service command "the switch OverPlot" aerosol extinction spectra $\alpha(\lambda)$ calculated on a basis *dN/dr* are resulted which can be received with the help of aerosol counter AZ-5 with a range of aerosol particles measurement on radius $\Delta\lambda$ = 0.4–10 microns and dust counter OMPN-10.0 (OPTEK Spb), carrying out the control of fractions in standard PM 10, PM 2.5, and PM 1. Figure 1 shows that for three spectra of the aerosol sizes in wavelength band $\Delta\lambda$ = 2–12 μm $\alpha(\lambda)$ differ approximately twice and in wavelength band $\Delta\lambda$ = 0.2–2 μm of distinction in values $\alpha(\lambda)$ reach 10 and more.

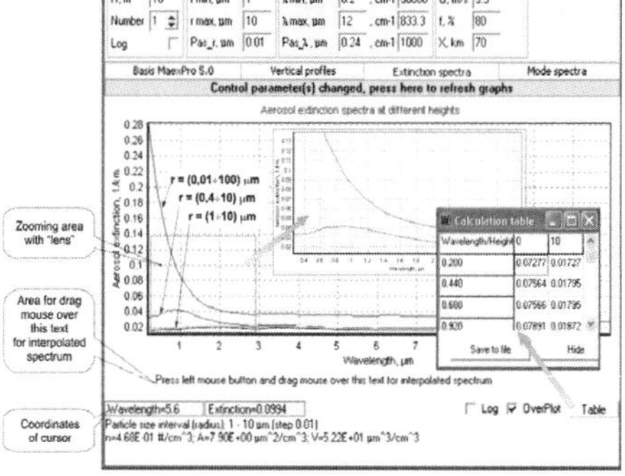

Figure 1 Composite window "Extinction spectra" including service commands of the code MaexPro 5.0 interface

Results of Comparison with the Experimental Data Received in IAO

On Figure 2 are show results of comparison between the aerosol extinction calculated by the code MaexPro 5.0 with experimental data received in IAO SB RAS at transmission measurements of a spectral transparency in coastal environments [3]. Results of calculation are received for entrance meteorological parameters and heights of a sea level close to conditions of the experiment.

Figure 2 illustrate an agreement between the $\alpha(\lambda)$ spectral behavior and the experimental results received in IAO in the coastal environment.

Results of Comparison with the Code ANAM

On Figure 3 are show results of comparison between the aerosol extinction calculated by the code MaexPro 5.0 and by the code ANAM.

Despite of different key entrance parameters of models the fetch for MaexPro and the air mass parameter (AMP) for ANAM Figure 3 illustrates a good agreement between the $\alpha(\lambda)$ spectral behavior for visible and near-infrared spectrum band as prediction by the code MaexPro 5.0 at fetch, equal 3, and the results received on code ANAM at AMP equal 5, 7, and 8.

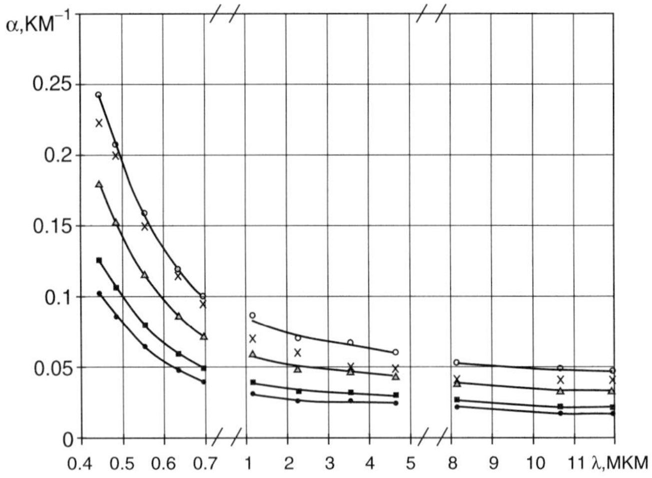

Figure 2 Aerosol extinction spectrum $\sigma(\lambda)$ at height H = 4 m for fetch X = 30 km, wind speed $U = 3.3\,\text{m s}^{-1}$ and for various relative humidity: ♦ – RH = 66%, ■ – RH = 75%, ▲ – RH = 85%, o – RH = 90%

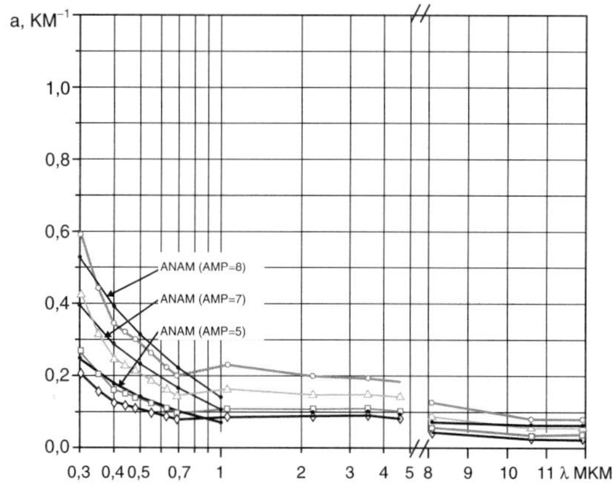

Figure 3 Aerosol extinction $\sigma(\lambda)$ spectrum at height H = 20 m for fetch X = 3 km, wind speed $U = 3.3\,\mathrm{m\,s^{-1}}$ and for various relative humidity: ♦ – *RH* = 66%, ■ – *RH* = 75%, ▲ – *RH* = 85%, o – *RH* = 90%

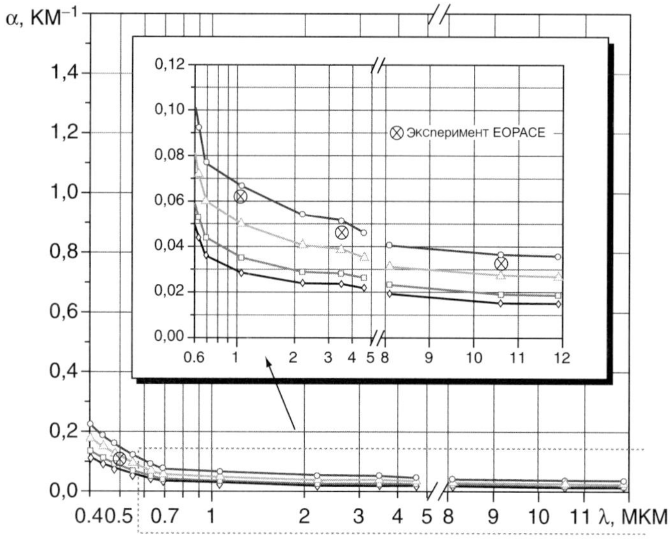

Figure 4 Aerosol extinction spectrum $\sigma(\lambda)$ at height H = 20 m for fetch X = 30 km, wind speed $U = 3.3\,\mathrm{m\,s^{-1}}$ and for various relative humidity: ♦ – *RH* = 66%, ■ – *RH* = 75%, ▲ – *RH* = 85%, o – *RH* = 90%

Results of Comparison with Experiment Eopace

Figure 4 also show the agreement between the $\alpha(\lambda)$ spectral behavior as prediction by the code MaexPro 5.0 and the available results in near- and mid-infrared spectrum band received during the 5 year program EOPACE [4].

Summary

The paper presents the modeling data of the aerosol extinction coefficient predicted by the code MaexPro depending of fetch, height of sea level, and meteorological parameters.

The forecast of $\sigma(\lambda)$ by the code MaexPro 5.0 realistic reflects the influence on aerosol extinction of known effects of a coastal zone.

The $\sigma(\lambda)$ prediction by the code MaexPro 5.0 correspond to the results received on the code ANAM.

The forecast of $\sigma(\lambda)$ profiles by the code MaexPro 5.0 is in an agreement with available experimental data.

The code MaexPro may assess the propagation conditions at using satellite remote sensing data to monitor coastline variation and to analysis the eroding, depositing features, and evolution process will be of great significance for the river mouth regulation, river course planning, and coastal protective project program and trend prediction of coastal evolution.

References

1. Kaloshin, G. and Piazzola, J., *Proc. of the 23rd International Laser Radar Conference*, 423–426 (2006).
2. Gathman, S., van Eijk, A.M.J., and Leo, H., *Cohen, Proc. SPIE*, **3433**, 41 (1998).
3. Kabanov, M.V., Panchenko, M.V., Pkhalagov, Yu. A. et al., *Optical Properties of the Coastal Atmospheric Hazes*, Novosibirsk: Nauka (1988).
4. Jensen, D.R., Gathman, S.G., Zeisse, C.R., et al., *Opt. Eng.*, **40**(8), 1486–1498 (2001).

Full Column Aerosol Optical Depth Observations in the Vicinity of Clouds

Jens Redemann[1], Qin Zhang[1], Philip B. Russell[2], and John M. Livingston[3]

Abstract Studying the spatial variability of aerosol properties in the vicinity of clouds is essential for our ability to separate aerosol and cloudy pixels in remotely sensed data and hence to determine aerosol effects on climate at different spatial scales. In this paper, we describe aerosol observations collected near cloud edges by the NASA Ames Airborne Tracking Sunphotmeter, AATS-14 in the EVE (Extended-MODIS-λ Validation Experiment) experiment off the Northern CA coast in April 2004. We find that in about 75% of the cases there was an increase of 5–25% in mid-visible aerosol optical depth (AOD) in the closest 2–3 km near the clouds. Interestingly, the majority of AOD increases at the cloud edges all occurred without increases in aerosol particle size (as inferred from AOD wavelength dependence). These observations can be used for assessing the ability of passive satellite aerosol sensors to detect aerosol variability near clouds.

Keywords Aerosol remote sensing, aerosol–cloud interaction

Introduction

Our ability to assess aerosol radiative effects using remote sensing data depends on the discrimination between cloudy and cloud-free viewing elements. In the case of the direct aerosol radiative forcing of climate it is essential to avoid sub-pixel cloud contamination of pixels used for aerosol retrievals, while including areas of increased aerosol concentration in the vicinity of clouds. In the case of the indirect and semi-direct aerosol effects on climate it is also mandatory to distinguish cloudy from cloud-free pixels, while ensuring spatial correspondence and hence interaction of aerosol and cloud air masses studied. Therefore, studying the spatial variability

[1] *BAER Institute, Sonoma, CA, USA*

[2] *NASA Ames, Moffett Field, CA, USA*

[3] *SRI International, Menlo Park, CA, USA*

C.D. O'Dowd and P.E. Wagner (eds.), *Nucleation and Atmospheric Aerosols*, 1129–1133.

of aerosol properties in the vicinity of clouds is essential for our ability to determine aerosol effects on climate at different spatial scales.

The current generation of satellite aerosol sensors, such as MODIS (Kaufman et al. 1997) and MISR (Kahn et al. 2001) aboard the Terra satellite, MODIS aboard Aqua, and OMI aboard Aura (Levelt et al. 2006) are much more capable of detailed global aerosol observations than the previous generation. Spatial variability in radiance fields observed by passive satellite sensors is frequently used in the separation of cloud and aerosol signals, under the assumption that most clouds exhibit larger spatial variability than aerosols (Martins et al. 2002). However, Kaufman et al. (2006) pointed out that aerosol and cloud fields are often spatially correlated and that therefore, "rigorous cloud screening can systematically bias toward less cloudy and drier conditions, underestimating the average aerosol optical thickness". Also, several studies (e.g., Kaufman et al. 2005; Redemann et al. 2006), found that differences between MODIS and suborbital AOD measurements were correlated with MODIS-derived cloud fraction, indicating the need for a renewed focus on cloud screening. There are many pitfalls to passive satellite remote sensing of aerosols in the vicinity of clouds. For example, Wen et al. (2006) found that 3D radiative transfer effects in a broken cumulus cloud field embedded in background biomass burning aerosol with an AOD of 0.1 could reach as far as 3 km away from clouds and cause an overestimate of the apparent clear sky reflectance that would lead to an overestimate in AOD by 40%.

Only a small number of studies (e.g., Redemann et al. 2006) have been directed at assessing the variability of aerosol optical properties in cloud halos (hereafter defined as areas of increased aerosol particle size or concentration in the immediate vicinity of clouds). Among the methods for studying the spatial variation of aerosol optical properties are in situ measurements at cloud level, satellite remote sensing, and airborne remote sensing of either the full atmospheric column or extended layers. In situ aircraft measurements of cloud halos are often available at much finer spatial resolution and provide more detail on aerosol optical and microphysical properties than the remotely sensed data. However, there is rarely information available on the full column variation of aerosol properties above or below aircraft level, making the in situ measurements valuable for studying the aerosol–cloud interaction itself, but less relevant for the testing of satellite remote sensing in the vicinity of clouds. Provided that a given airborne remote sensing method is not subject to the same 3D effects that the satellite sensor is subject to, the airborne measurements have the advantage of providing a quasi-full-column view of the atmosphere near clouds, albeit with much less detail on spatial variability and optical properties of aerosols by comparison to the in situ observations.

In this paper we describe our recent efforts in studying the aerosol-cloud boundary using airborne observations of full-column AOD collected by the NASA Ames Airborne Tracking Sunphotmeter, AATS-14 in the EVE (Extended-MODIS-λ Validation Experiment) experiment aboard the CIRPAS Twin-Otter aircraft off the Northern CA coast in April 2004.

Measurements

AATS-14 measures direct solar beam transmission in narrow channels by using detectors in a tracking head that is mounted externally to the aircraft. From the measured slant-path transmissions we derive the aerosol optical depth (AOD), in 13 wavelength bands at (354, 380, 453, 499, 519, 604, 675, 778, 864, 1,019, 1,240, 1,558, and 2,139 nm) and the columnar amounts of H_2O and O_3. AATS-14 data are corrected for Rayleigh scattering and absorption by O_3, NO_2, H_2O, and O_2–O_2. Measurements in previous deployments and methods for data reduction and error analysis have been described previously (e.g., Russell et al. 1999). Radiometric calibration of AATS-14 is determined using the Langley plot technique. For EVE, AATS-14 was calibrated at the Mauna Loa Observatory, Hawaii, in March and June of 2004, bracketing the EVE campaign. Redemann et al. (2006) provide a detailed description of the AATS-14 measurements in EVE. After consideration of all possible sources of error, the AATS-14 derived AOD had the highest uncertainties for those channels with the largest difference between pre- and post-mission calibration. For example, the uncertainties in AOD at 380 nm on March 26, 2004, at a mean aerosol air mass factor of 1.2, yielded a mean value of 0.008, while the average uncertainty in the 1,240 nm AOD was 0.005.

Results and summary

The results presented in this section are based on data collected by AATS-14 in EVE at aircraft altitudes of 30–50 m. As such, they represent the full atmospheric column. Figure 1 shows the time traces of AATS-14 raw signals, AOD, screened AOD, and modified Ångstrom exponent for a segment of flight CIR06 on April 30, 2004 off the Northern California coast. The modified Ångstrom exponent, α, is here defined as the coefficient of the linear term in the quadratic fit, $\ln(AOD) = \beta - \alpha * \ln(\lambda) - \gamma * \ln^2(\lambda)$, where λ is wavelength.

Figure 1 shows four areas that are affected by clouds, most clearly identifiable by Ångstrom exponents that deviate noticeably from the background aerosol values of ~0.25. In between these clouds, the AOD decreases with increasing distance from the cloud edge. Figure 2 combines the four near-cloud AOD observations shown in Figure 1 (now labeled clouds 4–7) with observations near six additional clouds in EVE. In Figure 2, only the absolute and relative AOD at 499 nm are shown (first and second panel), again along with the modified Ångstrom exponent, α. The left panels show the data as a function of distance before the cloud edge, while the right panels show the data as a function of distance after cloud edge. In all, a total of 17 cloud edges are shown in Figure 2 (not all clouds had useable data before AND after cloud passage).

Figure 2 shows that the 17 cloud edges presented have an increase in mid-visible AOD between 5 and 25% in the closest 2–3 km near the clouds. These 17 cloud edges represent about 75% of the available cases in EVE. The remaining 25%

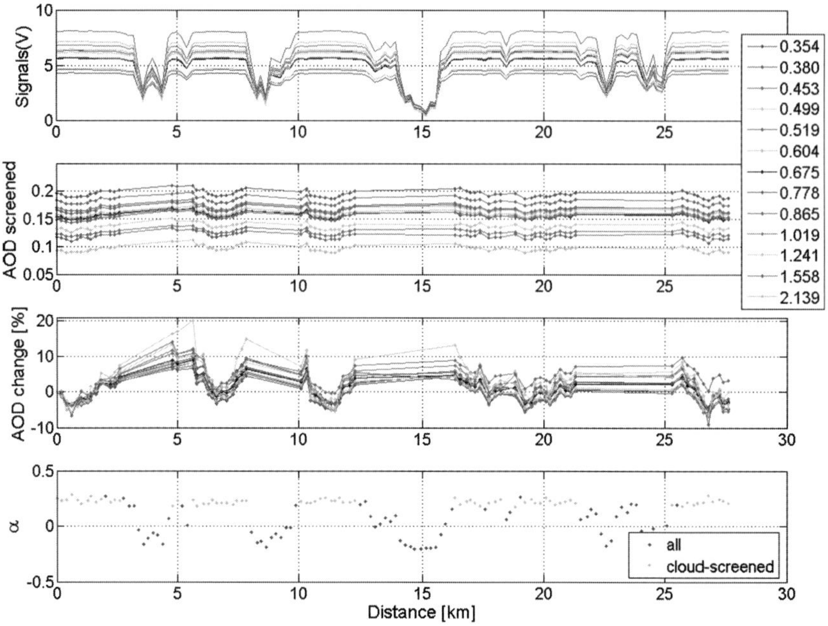

Figure 1 Time traces of AATS-14 signals, screened AOD, normalized AOD and modified Ångstrom exponent, α, for April 30, 2004

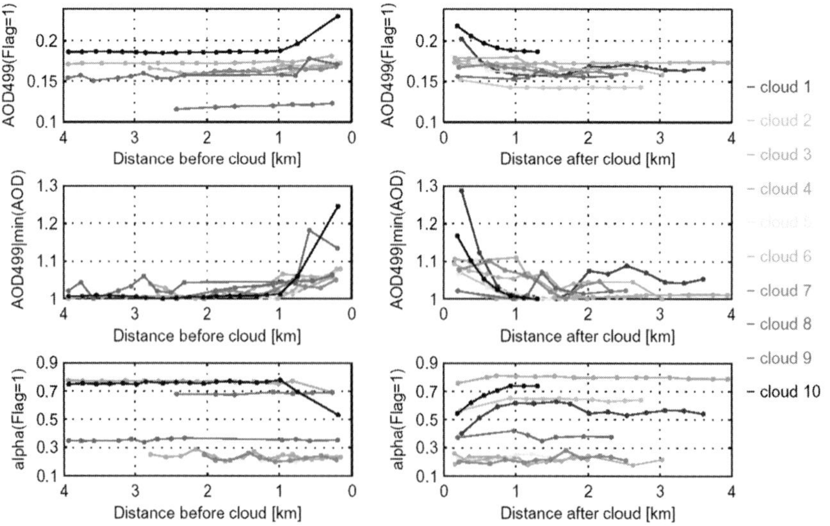

Figure 2 **First row** – Mid-visible (499 nm) AOD as a function of distance before (**left panels**) and after (**right panels**) cloud edge for ten different clouds observed during EVE. **Second row** – same as first row, but normalized to the lowest AOD for each transect. **Third row** – Modified Ångstrom exponent as a function of distance away from the cloud edge

(not shown here) indicated no clear increases of AOD near clouds. Interestingly, the majority (i.e., 12 out of 17 cases) of AOD increases at the cloud edges occurred without increases in aerosol particle size (as inferred from decreases in α). These observations are especially useful for assessing the ability of passive satellite aerosol sensors to detect aerosol variability near clouds, because unlike ground-based sensors, the airborne observations provide a better definition of the distance away from clouds.

References

Kahn, R., Banerjee, P., and McDonald, D., *J. Geophys. Res.*, **106**, 18219–18238 (2001).

Kaufman, Y.J., Gobbi, G.P., and Koren, I., *Geophy. Res. Lett.*, **33**, L07817 (2006).

Kaufman, Y.J., Remer, L.A., Tanre, D., Li, R.-R., Kleidman, R., Mattoo, S., Levy, R., Eck, T., Holben, B.N., Ichoku, C., Martins, J., and Koren, I., *IEEE TGRS*, **43**(12), 2886–2897 (2005).

Kaufman, Y.J., Tanré, D., Remer, L.A., Vermote, E., Chu, A. and Holben, B.N., *J. Geophys. Res.*, **102**, 17051–17067 (1997).

Redemann, J., Zhang, Q., Schmid, B., Russell, P.B., Livingston, J.M., Jonsson, H., and Remer, L.A., *Geophys. Res. Lett.*, **33** (2006).

Russell, P.B., Livingston, J.M., Hignett, P., Kinne, S., Wong, J., and Hobbs, P.V., *J. Geophys. Res.*, **104**, 2289–2307 (1999).

Wen, G., Marshak, A., and Cahalan, R.F., *IEEE Geo. Rem. Sens. Lett.*, **3**, 169–172 (2006).

Dust Intrusion Influence on Atmospheric Boundary Layer Using Lidar Data

A. Nemuc[1], S. Stefan[2], D. Nicolae[1], C. Talianu[1], V. Filip[2], and J. Ciuciu[1]

Abstract The PBL (Planetary Boundary Layer) is the lowest part of the troposphere and plays an important role in our everyday live. Diverse applications of boundary-layer meteorology require better understandings of the PBL. The depth of the PBL varies greatly in space and time, but knowledge of the depth of PBL aid in the validation of parameterizations of the PBL in general circulation models and improve our abilities to model the coupling between the atmosphere and the Earth's surface. Lidar (Light Detection and Ranging) measurements provide useful information on the structure of PBL. The aim of this paper is to study the influence of dust intrusion on mixed layer. To detect the mixed layer depth we used the gradient method on two cases: dust intrusion and background aerosol. We discuss the results using complementary information as synoptic maps, backtrajectories and radio soundings. Similar results for the PBL height obtained from lidar data and radio sounding showed in this paper demonstrates the importance of the remote sensing measurements in atmosphere.

Keywords Lidar, dust intrusion, PBL

Introduction

PBL has an important role for the whole atmosphere–earth system because it acts as the interface where the coupling between the atmosphere and the Earth's surface occurs, and the depth of the PBL controls the transfer of momentum, heat and moisture between them. The planetary boundary layer (PBL) contains most of the aerosol and water vapor in the atmosphere and thus has a major influence on radiative fluxes. Studies of climate sensitivity, therefore, require careful consideration of the role of the PBL.

[1]*National Institute for R&D of Optoelectronics, Magurele, P.O. Box MG-5, Magurele, R077125, Romania*

[2]*University of Bucharest, Faculty of Physics, Atmospheric Physics Department, Magurele, Romania*

C.D. O'Dowd and P.E. Wagner (eds.), *Nucleation and Atmospheric Aerosols*, 1134–1138.

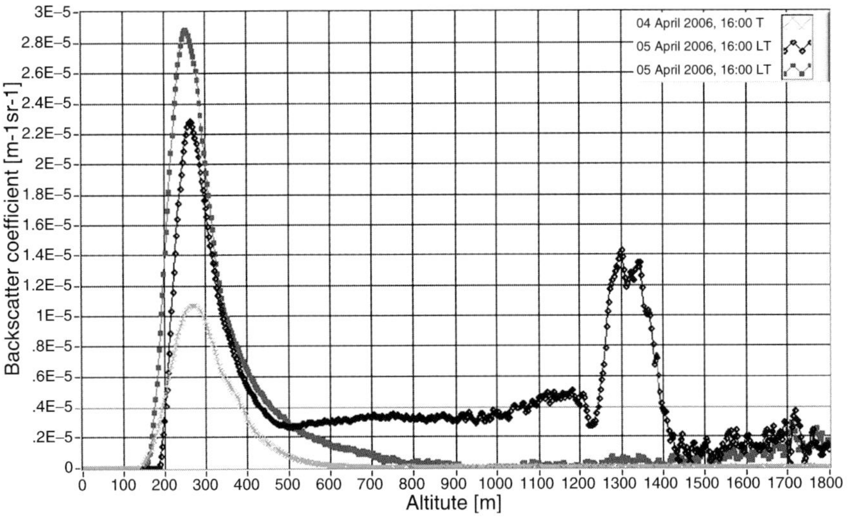

Figure 1 Backscattering coefficient profiles on April 4, April 5, and April 6 – 1,064 nm sounding wavelength – dust intrusion at 1,200–1,400 m penetrating from the free troposphere in the PBL on April 5

Strawbridge, K. B. and B. J. Snyden shown that lidar has provided for many years accurate measurements of the top of the boundary layer [1]. Aside from being clearly visible on lidar plots, determining the PBL height from lidar data does not require the more difficult task of obtaining an absolute calibration of the system. Gradient method [2] has been employed to extract the PBL height from lidar data (Figure 1), obtained with the system described in Section 2.

But interpreting data from lidar must take account on the advection of air mass which can change the vertical mixing. In Section 3 we have compared the results from lidar profiles with the ones from radio soundings. The analysis of air mass backtrajectories and, synoptic maps confirmed the advection of Saharan dust over Romania in the day under observation with lidar system. The results are discussed in Section 4.

Remote Sensing of the PBL

Elastic backscattering lidar system used for our field experiment at ROMEXPO site in northwestern part of Bucharest (44.25 N, 26.05 E), is based on a Nd:YAG laser working at the 1,064 nm fundamental wavelength and at 532 nm second harmonic delivering pulse with short pulses (nanoseconds) at 20 Hz repetition rate. The system can detect micron size aerosols from far distances, with a very good spatial resolution, therefore a very good tool for PBL monitoring (3).We are using backscattering coefficient dependence of altitude to determine the height of atmospheric boundary layer (ABL) by using the aerosols like tracers [4].The fundamental premise takes

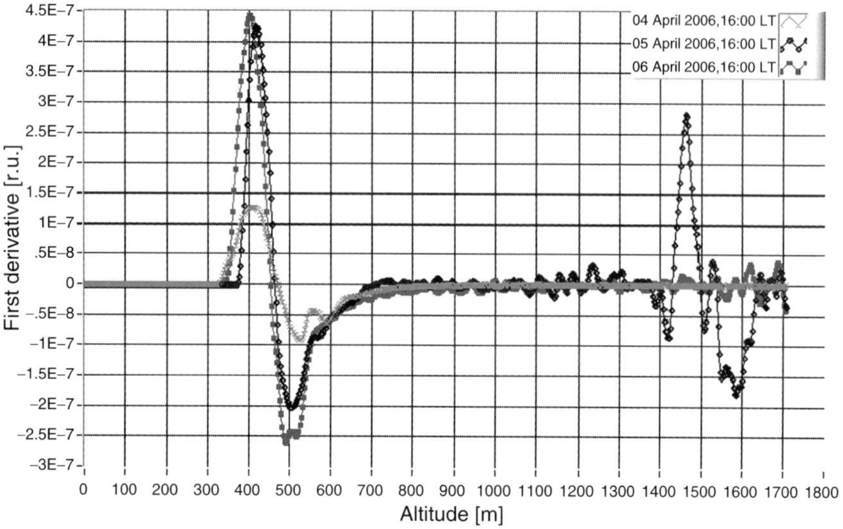

Figure 2 First derivative of the backscattering coefficient profile on April 4, April 5, and April 6 – 1,064 nm sounding wavelength

advantage of the large gradient in aerosol concentration generally evident between the boundary layer aerosols and those found in the free troposphere.

Data Analysis

Lidar Data

Figure 2 clearly shows the different heights for the PBL for the three consecutive days: before, during and after a dust intrusion on April 5th.The values have been determined as the first maximum of the minima of the first derivative for the backscattering coefficient profile.

To interpret data from LIDAR we have to take into account that meteorology and consequently origin of air mass influenced the state of the atmosphere and the height of local ABL.

Radio Sounding

Change in air mass and the temperature inversion (height of mixed layer) can be observed on the available radio soundings performed by National Meteorological Service at 00 h (Figure 3) on April 6, 2006.The 500 m level shows the residual layer

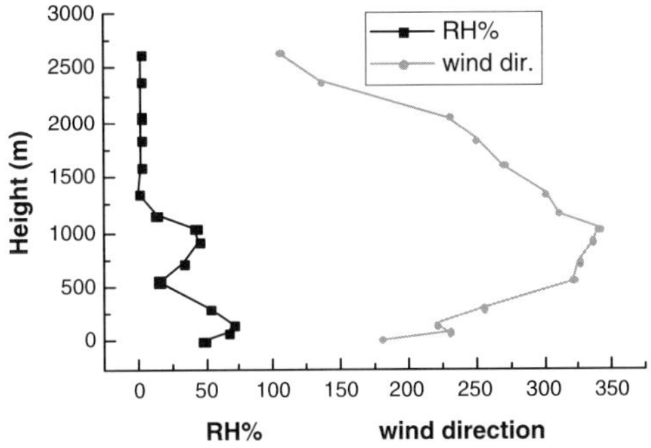

Figure 3 Relative humidity and wind direction vertical profiles on April 5 from radio sounding data

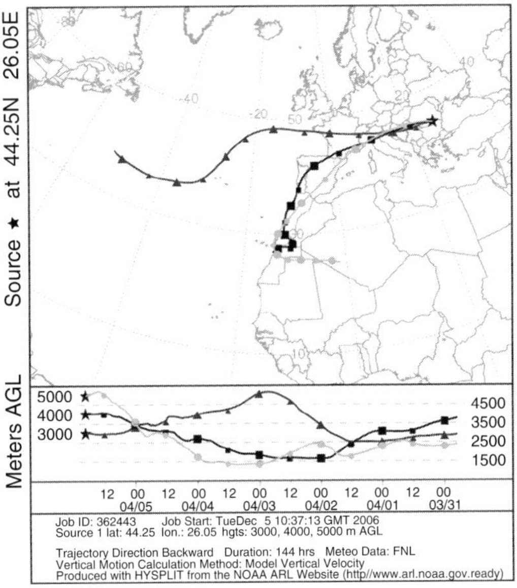

Figure 4 Five days backtrajectories arriving over Bucharest on April 5 showing the origin of air masses

height of the atmospheric boundary layer. Vertical distribution of wind direction confirms the transport of the Saharan dust at the same level as is determined from the lidar data (~1,200 m).Also humidity drops abruptly when the dust is transported into the continental atmosphere.

In addition, to emphasize the change in air mass and consequently the intrusion of Saharan dust, kinematics backward trajectories have been invoked. The version 4 of the Hybrid Single –Particle Lagrangian Integrated Trajectory model (HYSPLIT) [5,6], developed by the National Oceanic and Atmospheric Administration (NOAA)'s Air Resources Laboratory (ARL) was used. The backward trajectories are shown in Figure 4 and can be observed the change in trajectory at high altitudes on April 05, 2006. The map clearly identifies the Saharan origin of the upper air masses. Saharan dust layers reach the southern part of Romania predominantly by cyclonic circulation due to the strong trough observed at all the levels from a cyclonic system located in northwestern part of Africa. The type of air circulation and the change of air mass can be observed in synoptic and absolute and relative geopotential maps (not shown).

Summary and Conclusions

Similar values of the PBL height were determined from lidar data (gradient method) and radio soundings (temperature inversion method) for April 5 when a dust intrusion modified the vertical structure of PBL. For April 5 and 6, 2006 the height of the detected layer is the results of the advection of air mass and not to vertical mixing. The layer was formed when aerosol was injected by air mass from northern part of Africa (Saharan dust).

References

1. Strawbridge, K.B. and Snyden, B.J., Planetary boundary layer height determination during Pacific 2001 using the advantage of scanning lidar instrument, *Atmos. Environ.*, **38**, 5861–5871 (2004).
2. Menut L., Flamant C., Pelon J., and Flamant P.H., Urban boundary layer height determination from lidar measurements over Paris, *Appl. Opt.*, **38**, 945–954 (1999).
3. Talianu C., Nicolae, D., Ciuciu, J., Ciobanu, M., Babin, V., Planetary boundary layer height detection from LIDAR measurements, *J. Optoelectron. Adv. Mater.*, **8**(1), 243–246 (2006).
4. Nicolae D. and Cristescu, C.P., Laser remote sensing of tropospheric aerosol, *J. Optoelectron. Adv. Mater.*, **8**(5), 1781–1795 (2006).
5. Draxler, R.R. and Hess, G.D., An overview of the Hysplit_4 modeling system for trajectories, dispersion, and deposition, *Aust. Met. Mag.*, **47**, 295–308 (1998).
6. Draxler, R.R., Boundary layer isentropic and kinematic trajectories duing the August 1993 North Atlantic regional experiment intensive, *J. Geophys. Res.*, **101**(D22), 29255–2926 (1996).

Alignment of Atmospheric Dust Observed by High-Sensitivity Optical Polarimetry

Z. Ulanowski[1], J. Bailey[2], P.W. Lucas[1], J.H. Hough[1], and E. Hirst[1]

Abstract Optical polarimetry observations on La Palma, Canary Islands, during a Saharan dust episode showed dichroic extinction, indicating the presence of vertically aligned particles in the atmosphere. It is postulated that the alignment was due to an electric field of the order of 1 kV/m or greater. Two alternative mechanisms for the origin of the field are examined, including the effect of reduced atmospheric conductivity and the presence of charged dust.

Keywords Tropospheric aerosol, mineral dust, alignment, electric field, polarimetry

Measurements

Mineral dust in the atmosphere exerts significant influence on radiation, both directly and indirectly. Yet large gaps in the understanding of its role exist. It is now strongly argued that the nonsphericity of mineral dusts should be taken into account, particularly in remote sensing retrievals.[1] Polarimetry has a special role to play here because it is sensitive to particle shape. However, polarimetry is also sensitive to particle orientation – for example, the scattering of orthogonal polarization components from aligned nonspherical particles is generally unequal (dichroic). Here we present the results of polarimetric observations during a dust episode over the Canary Islands.

---

[1]*Science and Technology Research Institute, University of Hertfordshire, Hatfield AL10 9AB, United Kingdom. e-mail z.ulanowski@herts.ac.uk*

[2]*Australian Centre for Astrobiology, Macquarie University, Sydney NSW 2109, Australia*

C.D. O'Dowd and P.E. Wagner (eds.), *Nucleation and Atmospheric Aerosols*, 1139–1144.

Polarimetry

The observations were carried out from 27 April 2005 to 8 May 2005 with a new high-sensitivity astronomical polarimeter Planetpol,[2] mounted on the William Herschel Telescope located on the island of La Palma at an altitude of 2,340 m. The measurements extended over 590–1,000 nm wavelengths and were of polarized flux from four nearby stars which normally show little polarization. It was noted that observations from 3 to 7 May were characterized by increased linear dichroism, manifested by excess horizontal polarization component of transmitted light, which rose with the zenith angle, with a value of almost 5×10^{-5} on the 4th of May. In contrast, observations immediately before and after that period showed polarization at or below the detection level, $\sim 10^{-6}$ – Figure 1. The total flux was also reduced. The nearby Carlsberg Meridian Telescope (CMT) showed that the zenith extinction increased from a typical value of 0.1 at 625 nm by between 0.08 and 0.23. The excess polarization can be interpreted as being due to the interaction of the starlight with nonspherical particles having their long axes preferentially oriented in the vertical direction (hence causing the vertical component of polarization to be scattered and/or absorbed more strongly).

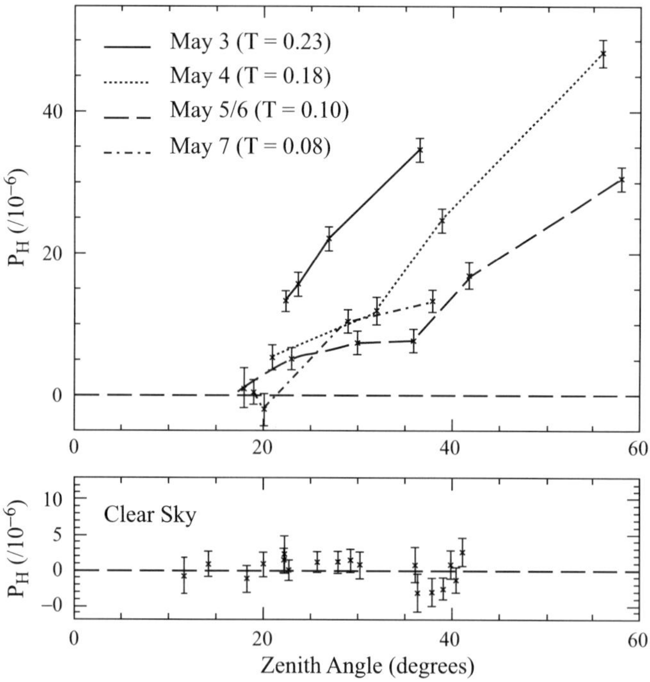

Figure 1 Clear sky (**lower panel**) and dust affected (**upper panel**) excess linear polarization in parts per million as a function of zenith angle. Optical depth τ is shown in brackets

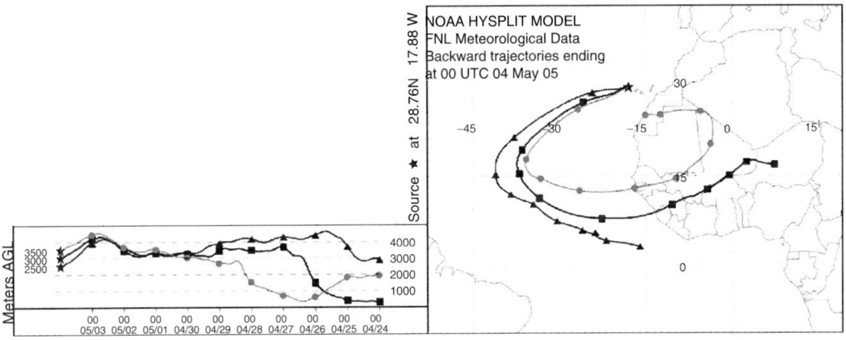

Figure 2 Backtrajectories ending at 00 UTC on 4 May 2005, computed for final altitudes of 2,500, 3,000, and 3,500 m. The left panel shows the altitude and the right one the geographical location

Dust Properties

Ten day backtrajectories were computed using the HYSPLIT model.[3] They indicate that the air mass during the dust episode originated in the arid western Sahel and Sahara – Figure 2. The plume reached La Palma via an indirect route over the Atlantic and was 6–10 days old. For brevity, we will refer to the dust as Saharan dust (SD).

Optical properties of the SD layer were obtained from the AERONET sun photometer in Santa Cruz on the nearby island of Tenerife (52 m altitude). Size distribution, optical depth and effective radius were retrieved.[1] The optical depths were similar to those from the CMT, and the distributions showed a coarse mode centered on ~4 μm size, with up to 0.7% of particles by surface area in the 10–20 μm range.

Analysis

The presence of vertically aligned particles is unexpected, because aligned atmospheric particles are typically oriented horizontally due to aerodynamic forces. We will therefore examine whether atmospheric electric fields can produce such alignment.

Atmospheric Electric Field Due to Aerosol

The fair weather electric field strength is on average ~120 V/m in clear atmosphere near ground level[4] – too low to cause particle alignment. However, the presence of aerosol, fog or haze lowers atmospheric conductivity because large particles "scavenge" small ions. The drop in conductivity leads to increased potential difference across the aerosol layer in accordance with Ohm's law, and hence in increased field gradient.[4]

First we need to establish a numerical relationship between the optical properties of an uncharged aerosol layer and the electric field gradient within it. The derivation is too lengthy to be shown here, however, by assuming that aerosol attachment dominates over direct ion recombination, the aerosol layer has large horizontal extent and that the conductivity within the layer is much lower than outside it, we arrive at the expression:

$$E = \frac{U}{\Delta z\left(1 + \frac{R_{c0}\, e^{-}\, \mu_{\pm}\, q\, \bar{s}_{ext}}{\beta_{eff}\, \tau}\right)} \tag{1}$$

where U is the ionospheric potential (280 kV), Δz the layer thickness, R_c the columnar resistance of the atmosphere, e^- is the elementary charge, $\mu_{\pm}$ the mean positive and negative ion mobility, q the total ion (+ and –) production rate, $\bar{s}_{ext}$ the effective extinction cross section, β_{eff} is the effective ion–aerosol attachment coefficient, and τ the optical depth. We also consider the effect of topography. Electrostatics predicts threefold field enhancement at the top of a hemispherical mountain, if homogeneous air conductivity is assumed.[5] The enhancement factor can be written as $1 + 2(z_0/z)^3$ where z is the altitude above base level and z_0 mountain height. By integrating this expression from the mountain top to an altitude z_1 we obtain an average factor $1 + z_0/z_1 + (z_0/z_1)$.^{2} The enhancement factors calculated from this formula are 2.5 and 2.2 for a 500 m and 1,000 m column, respectively, above a 2.3 km mountain. In the absence of data specific to the site we apply an enhancement factor of 2 to Eq. (1). The electric field calculated for given optical depth τ and geometric layer thickness is shown in Figure 3.

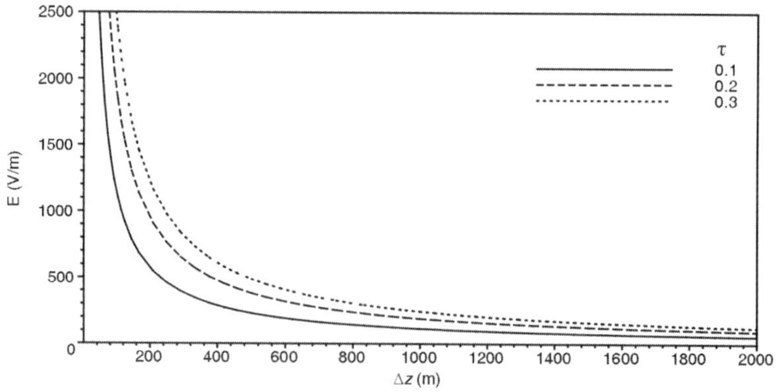

Figure 3 Electric field from Eq. (1) as a function of aerosol layer thickness Δz for optical depth τ, with the topographic enhancement factor of 2. For calculating the effective attachment and extinction coefficients, trimodal lognormal size distributions were fitted to the sun photometer retrieved ones

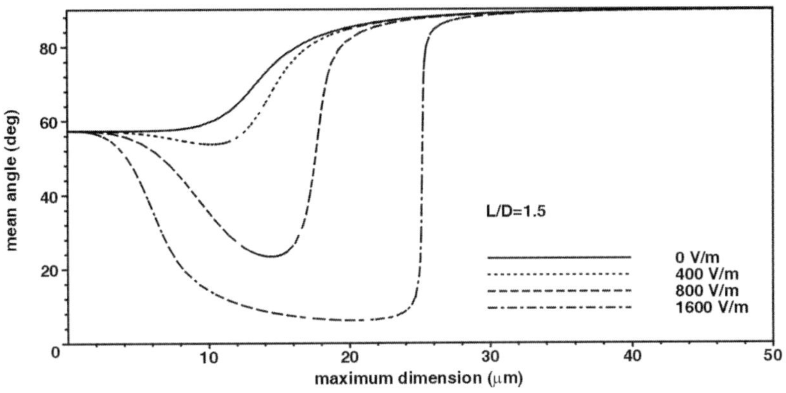

Figure 4 Mean orientation angle as a function of maximum dimension of conducting, prolate ellipsoids with aspect ratio 1.5 and density 2.6 g/cm^3, falling under gravity in a vertical electric field

Aerosols can also produce electric fields if the particles are charged. The fields can be enhanced if charge separation takes place due to size-dependent sedimentation rates, and field strengths as high as –20 kV/m have been observed during dust storms.[6,7]

Particle Alignment in Electric Fields

The alignment was modeled by considering Boltzmann statistics for ellipsoidal particles subjected to torques due to electric field and air flow (sedimentation), both in the vertical direction, and to rotational diffusion. Example results are shown in Figure 4.

Conclusions

The observed dust alignment could have been caused by an electric field within the dust layer. From Figure 4, the minimum field required to align slightly nonspherical ~10–20 μm particles is of the order of 1 kV/m. The field may have been present because of the depletion of atmospheric conductivity due to the dust. However, Figure 3 shows that the layer needs to be rather thin to produce sufficient field strength – which is unlikely for the aged dust cloud in question. In contrast, fields accompanying charged dust clouds can be stronger, and could in principle produce the required alignment.

Acknowledgments We thank the NOAA Air Resources Laboratory for the provision of the HYSPLIT transport model, and the AERONET network for the sun photometer data.

References

1. Dubovik, O., Sinyuk, A., Lapyonok, T., Holben, B.N. et al., *J. Geophys. Res.*, **111**, D11208 (2006).
2. Hough, J.H., Lucas, P.W., Bailey, J.A., Tamura, M. et al., *Publ. Astron. Soc. Pac.* **118**, 1302–1318 (2006).
3. Draxler, R.R. et al., http://www.arl.noaa.gov/ready/hysplit4.html NOAA Air Resources Laboratory (2003).
4. Gringel, W., Rosen, J.M., and Hofmann, D.J., Electrical structure from 0 to 30 kilometers, in: *The Earth's Electrical Environment*, Washington: National Academy Press, pp. 166–182 (1986).
5. Schottky, W., *Z. Physik A*, **14**, 63–106 (1923).
6. Kamra, A.K., *J. Geophys. Res.*, **77**, 5856–5869 (1972).
7. Smirnov, V.V., *Izvestiya Akad. Nauk Fiz. Atmosf. Okeana*, **35**, 616–623 (1999).

Retrieval of Aerosol Distributions by Multi-Axis Differential Absorption Spectroscopy (MAX-DOAS)

R. Sinreich, U. Frieß, T. Wagner*, S. Yilmaz, and U. Platt

Abstract We present and demonstrate a method which uses a set of slant column density (SCD) measurements of oxygen dimers (O_4) at several different elevation angles to determine the atmospheric aerosol extinction within the atmospheric boundary layer. In addition, the height of the atmospheric boundary layer can usually be derived. Furthermore, the technique can be readily extended to determine concentrations of several other trace gases including NO_2, SO_2, or CH_2O. The method is based on the comparison of measurements with O_4 SCDs from an accurate radiative transfer model.

Keywords New measurement techniques, tropospheric aerosols, spectroscopy, MAX-DOAS

Introduction

We introduce a novel method for the determination of the aerosol optical density and vertical distribution using scattered sunlight. The method is based on the Differential Optical Absorption Spectroscopy (DOAS), which has been applied for three decades to measure a wide variety of gaseous species relevant to atmospheric chemistry (e.g., NO_2, O_3, SO_2, aldehydes, OH, NO_3, ClO, BrO[1]) by measuring scattered sunlight. DOAS is inherently calibrated, contact free, identifies the particular molecules unequivocally, and allows real time measurements.[2] Moreover, it derives its outstanding sensitivity from the approach to measure "differential" intensities, i.e., minute differences between the intensity of neighboring spectral intervals.

DOAS also can be operated in an active (i.e., using artificial light sources) mode which has the advantage of a well defined absorption path length, but suffers from relatively complex optics and high-power requirements (typically several 100 W).[1]

Institute of Environmental Physics, University of Heidelberg, Heidelberg/Germany

* *also at Max Planck Institute for Chemistry, Mainz/Germany*

C.D. O'Dowd and P.E. Wagner (eds.), *Nucleation and Atmospheric Aerosols*, 1145–1149.

Passive instruments, on the other hand, are inherently simple and require little power (below 1 W in some cases). However, the absorption path is not readily defined for measurements of scattered sunlight and has to be calculated by atmospheric radiation transport models.

DOAS and MAX-DOAS Principle

DOAS relies on the law of Lambert-Beer which describes the attenuation of light traversing matter or, like in our case, the atmosphere:

$$I(\lambda) = I_0(\lambda)\exp\left(-\int_0^L \sum_i \sigma_i(\lambda).c_i.dl\right) \tag{1}$$

with $\sigma_i(\lambda)$ as the absorption cross section and c_i as the concentration of the absorbing trace gas at the wavelength λ. $I_0(\lambda)$ denotes the initial intensity emitted from the sun as light source and $I(\lambda)$ is the attenuated intensity after passing through the atmosphere along the light path L. However, atmospheric scattering is overlaid to the absorption structures of the trace gases. This restriction can be overcome by separating the absorption cross section σ in a narrow and a broad band part:

$$\sigma(\lambda) = \sigma_{broad}(\lambda) + \sigma'(\lambda) \tag{2}$$

$\sigma'(\lambda)$ describes only the characteristic narrow band absorption structures of different trace gases while $\sigma_{broad}(\lambda)$ includes the broad band part. The law of Lambert-Beer can now be expressed as:

$$I(\lambda) = I_0(\lambda)\exp\left(-\int_0^L \sum_i \sigma_{i,broad}(\lambda)\cdot c_i \cdot dl\right) \times \exp\left(-\int_0^L \sum_i \sigma_i'(\lambda)\cdot c_i \cdot dl\right) \tag{3}$$

The DOAS method concentrates on the narrow band absorption features

$$I(\lambda) = I_0'(\lambda)\exp\left(-\int_0^L \sum_i \sigma_i'(\lambda)\cdot c_i \cdot dl\right) \tag{4}$$

where $I_0'(\lambda)$ contains the initial intensity and the broad band structures, which is computed from the measured spectrum $I(\lambda)$ by applying a low pass filter. The differential absorption cross sections $\sigma_i'(\lambda)$ can be removed by applying a high pass filter to the absorption cross sections $\sigma_i(\lambda)$.

The primary quantity of a DOAS measurement is the "slant column density" (SCD) which is the integrated concentration along the light path, or more realistically the average integrated concentration over many possible light paths. In contrast, the "vertical column density" (VCD) is the integrated concentration over a vertical path through the atmosphere. Thus, the VCD is independent of the particular measurement geometry and the sunlight conditions.

Some years ago the Multi-Axis-DOAS (MAX-DOAS) technique was developed, which observes scattered sunlight from a variety of viewing elevations. So, in particular at observation directions pointing slightly above the horizon, high sensitivity for gases close to the ground is obtained.[3,4] Moreover, some degree of vertical resolution can be derived. Assuming a well-mixed layer of gas (e.g., within the atmospheric boundary layer), its vertical extent can be determined with good accuracy.[3,5]

Retrieval Method

The advantages of the MAX-DOAS principle can be extended to the measurement of aerosols. Recent work has demonstrated that properties of the atmospheric aerosol can be determined from differential absorptions of atmospheric constituents with known distributions (like the oxygen dimer O_4) by the MAX-DOAS technique.[4,6] Since the concentration of O_4 is proportional to the square of the oxygen (O_2) concentration and thus to the air density, its absorption features are most sensitive to changes in the radiation transport in the lower troposphere.

The retrieval method is based on the comparison of measurements with O_4 SCDs from an accurate radiative transfer model (see Figure 1). Thereby, a large number

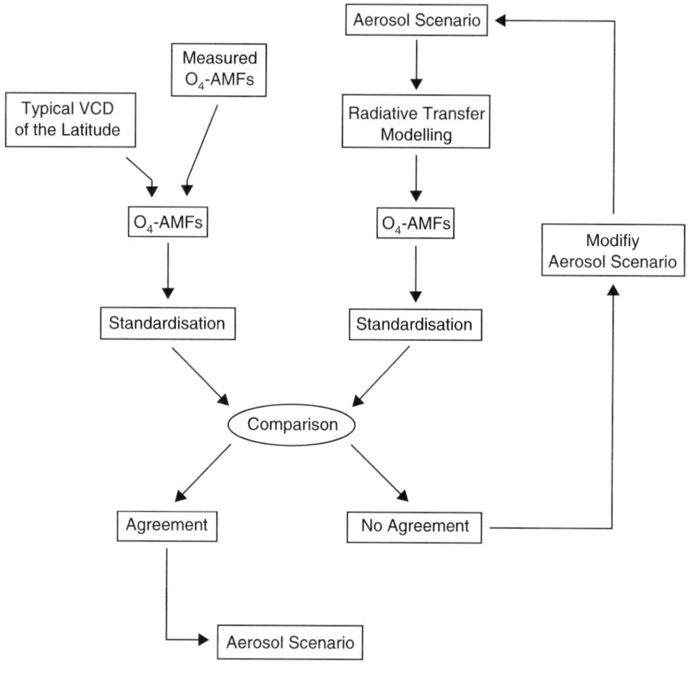

Figure 1 Retrieval algorithm of the properties of aerosol using O_4. The calculated values are modified until they match the measured ones

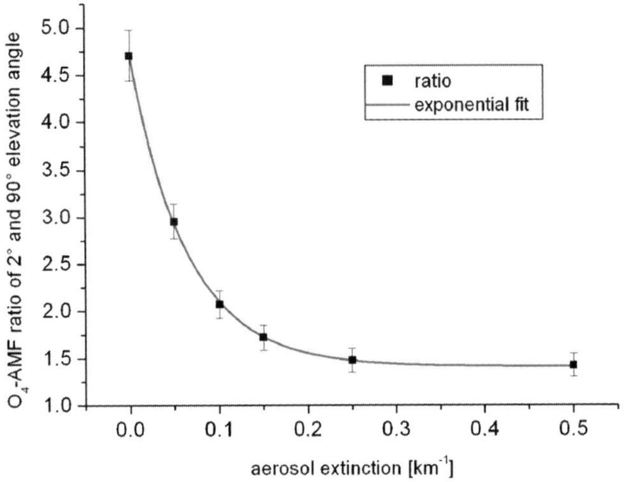

Figure 2 Ratio of O_4 air mass factors calculated for 2° and 90° elevation angles versus the aerosol extinction coefficient assumed for a box profile with 2 km height. The sensitivity is remarkable up to 0.25 km^{-1} aerosol extinction (corresponding to a visibility range of about 16 km)

of photon trajectories through the atmosphere are simulated. Height profiles of pressure, temperature and ozone as well as ground albedo are the most important parameters which have to be determined in advance for a realistic modeling of the atmosphere. Additionally, aerosols have a large impact on the light path. Thus, the aerosol profile and properties can be altered until the modeled values agree with the measurements. The output of the modeling is the Air Mass Factor (AMF) which is defined as the ratio between the SCD and the VCD.

To demonstrate the potential of the method, we show an example of an application of this algorithm in Figure 2.[5] The modeled O_4 ratio of 2° and 90° elevation angles (angle between horizon and viewing direction) against the aerosol extinction coefficient is plotted for an assumed aerosol box profile with 2 km height. Now, a comparison with a measured ratio leads to the actual aerosol extinction (here we assume horizontal homogeneity of the aerosol parameters, however an additional azimuth scan would yield information on the aerosol phase function and size distribution). The plot in Figure 2 demonstrates the sensitivity of O_4 AMFs to the aerosol extinction, which is particularly high at lower aerosol optical densities.

Summary

We will present examples for the retrieval of aerosol vertical distributions based on the above described method. The knowledge of the aerosol distribution enables the determination of concentrations of several other trace gases including NO_2, SO_2, or CH_2O by MAX-DOAS. Future steps will aim on the application of optimal estimation on MAX-DOAS measurements which will allow for a better quantitative assessment of the retrieved quantities.

References

1. Platt U., Differential optical absorption spectroscopy (DOAS), air monitoring by spectroscopic techniques, in: *Chemical Analysis Series*, Chemical Analysis Series, Vol. 127, edited by M.W. Sigrist, , New York: Wiley (1994).
2. Platt U., Modern methods of the measurement of atmospheric trace gases, *J. Phys. Chem. Chem. Phys*. "PCCP" ,**1**, 5409–5415 (1999).
3. Hönninger G. and Platt U., The role of BrO and its vertical distribution during surface ozone depletion at alert, *Atmos. Environ*., **36**, 2481–2489 (2002).
4. Wagner, T., Dix, B., Friedeburg, C. v., Frieß, U., Sanghavi, S., Sinreich, R., Platt, U., MAX-DOAS O_4 measurements: a new technique to derive information on atmospheric aerosols – Principles and information content, *J. Geophys. Res*., **109**, D22205, doi:10.1029/2004JD004904 (2004).
5. Sinreich, R., Frieß, U., Wagner, T., and Platt, U., Multi axis differential optical absorption spectroscopy (MAX-DOAS) of gas and aerosol distributions, *Faraday Discuss*., **130**(08), doi:10.1039/b419274 (2005).
6. Frieß, U., Monks, P.S., Remedios, J.J., Rozanov, A., Sinreich, R., Wagner, T., and Platt, U., MAX-DOAS O_4 measurements: a new technique to derive information on atmospheric aerosols: 2. Modeling studies, *J. Geophys. Res*., **111**, D14203, doi:10.1029/2005JD006618 (2006).

Variations of Desert Dust Optical Properties over Solar Village, KSA

F.M. Hasan[1] and I. Sabbah[2]

Abstract Aerosol optical properties over solar village, KSA have been studied using ground-based observations through Aerosol Robotic Network (AERONET). Our analysis covers 8 years record started from February 1999 through January 2007. These data are categorized into two groups in terms of the values of aerosol optical thickness (AOT) observed at 500 nm channel (τ_{500}). One group consists of 248 dusty days with high values of AOT ($\tau_{500} \geq 0.55$). The modal value of the Ångstrom wavelength exponent (α), for this group is 0.1. The other group is relatively dust free, it contains the rest of the data days (2,206 days) with modal value ~0.4 for the exponent α. Diurnal, monthly seasonally, and annual variations are studied. The time series plot of the daily average values of both τ_{500} and α exhibit seasonal variations. A notable correlation ($r = 0.51$) is obtained between the values of AOT and water vapor content (WVC) during relatively dust free days. In dusty days, a significant correlation ($r = 0.56$) is obtained between values of α and water vapor. A strong negative correlation ($r = -0.68$) is found between the mean monthly values of τ_{500} and α. Significant change in the size distribution of coarse particles is notable.

Keywords Aerosols, Ångstrom exponent, aerosol time series, volume size distribution

Introduction

Dust refers to airborne mineral particles that enter the atmosphere directly through action of wind, assisted in many cases by human activities. Dust source regions are desert, semiarid desert, dry lake beds and desertificated areas. The atmospheric life time of dust depends on particle size; large particles are quickly removed from

[1] *Environmental Sciences Program, Faculty of Graduate Studies, Kuwait University, Kuwait,*

[2] *Physics Department, Faculty of Science, Kuwait University, Kuwait*

C.D. O'Dowd and P.E. Wagner (eds.), *Nucleation and Atmospheric Aerosols*, 1150–1154.

atmosphere by gravitational setting, while submicron size particles can have atmospheric lifetime of several weeks. Dust particles can be transported long distances from their source. Dust composition plays an important role in determining dust aerosol optical properties and prior knowledge of the real variability of aerosol optical parameters over different areas is of interest for many studies.[1]

Aerosol optical thickness which may be derived from measurements of attenuated direct solar radiation as well as the aerosol size distribution and single scattering albedo (SSA), which may be derived from combined aerosol optical thickness and solar almucantar sky radiance data, are pivotal parameters defining the aerosol optical state of the atmosphere.[2]

The solar village is located 50 km North West of Riyadh area, capital of Saudi Arabia. The site coordinates are 24.91°N 46.41°E and elevation of 650 m a.s.l.

In this work, we study aerosol optical properties over solar village in Saudi Arabia. We analyze diurnal, monthly, seasonally and annual variations. And we study the volume size distribution dynamics.

Data Analysis

Aerosol Robotic Network data has been used to study the variations of aerosol optical thickness, Ångstrom wave length exponent and perceptible water column in solar village for duration of 88 months starting from February 1999. The Ångstrom wavelength exponent α is derived using the spectral optical measurements of channel 500–870 nm. The AOT (τ_{500}) measurements were relatively high through the years with τ_{500} exceeds 0.2. The 6 month period of March to August is characterized by slightly higher values of τ_{500} due to significant increase in dusty days. A notable correlation ($r = 0.51$) is obtained between the daily values of AOT at 500 nm wave length and water vapor content (WVC) during relatively free dust days. In dusty days the correlation between AOT & water vapor vanish. A significant positive correlation ($r = 0.56$) is obtained between the daily values of α and WWC in dusty days. The monthly mean values of aerosols optical properties and some weather parameters are plotted in Figure 1. For 2,206 days with $\tau_{500} < 0.5$ relatively dust-free days during the interval February 1999 to January 2007. The numbers of days used in each month are shown in Figure 1g. We see that the values of AOT in Figure 1a and those of Angstrom wave length exponent in Figure 1b exhibit 8 year cycle variation. The values of α were very low around the months of May, indicated by vertical large dashed lines, when the values of AOT were maxima. They reached the highest values around January, indicated by vertical small dashed line, when the values of AOT were minima. The dots in Figure 1f represent the values of the correlation coefficient between AOT and the exponent α for the 71 month with more than 20 days is ($r = -0.68$). The mean values of the monthly wind speed in m s^{-1} (WS) are shown in Figure 1c (dots connected with lines). We see that the values of (WS) increase during months with low values of the exponent α.

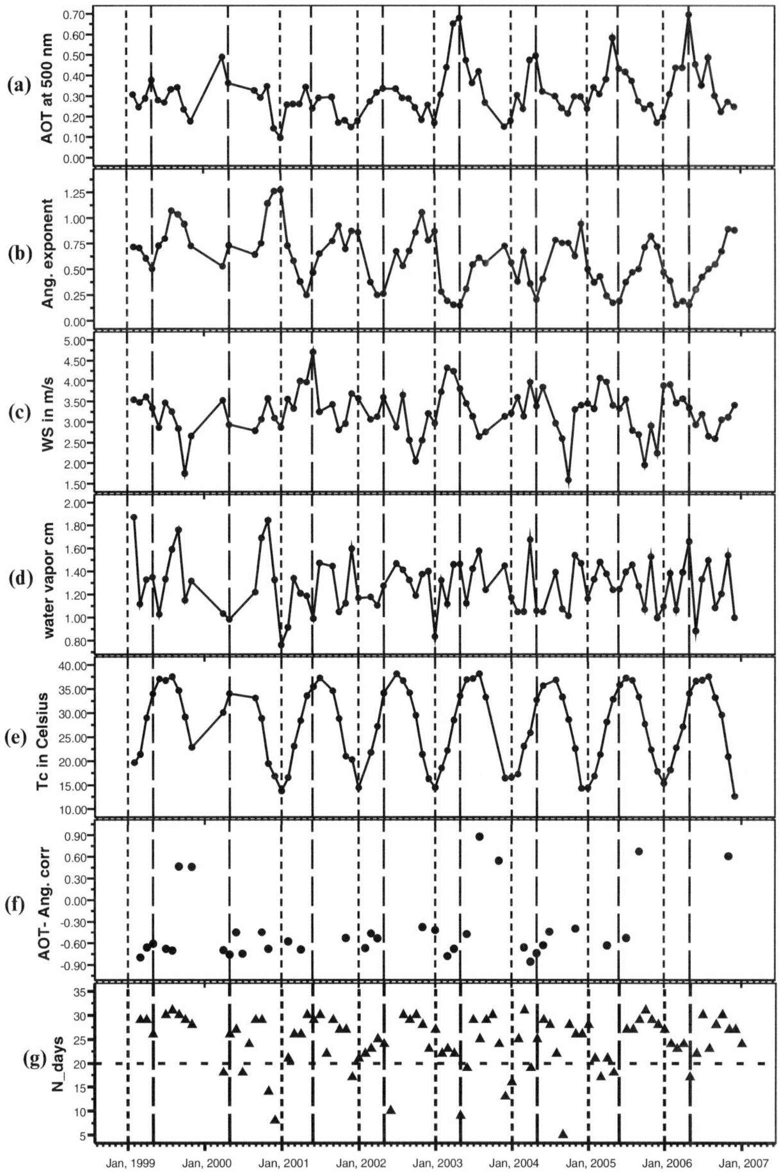

Figure 1 Time series of the mean monthly values of: (a) AOT at 500 nm, (b) Ångstrom exponent, (c) wind speed (m s^{-1}), (d) water vapor content (cm), and (e) average temperature (°C). (f) Correlation coefficients between the daily values of AOT and α (relatively free dust days), (g) The numbers of days used in each month. Dashed line in (g) represents 20 days. The months of May with lowest values of α are represented with vertical large dashed lines, while months of January with the highest values of α are represented with vertical small dashed lines

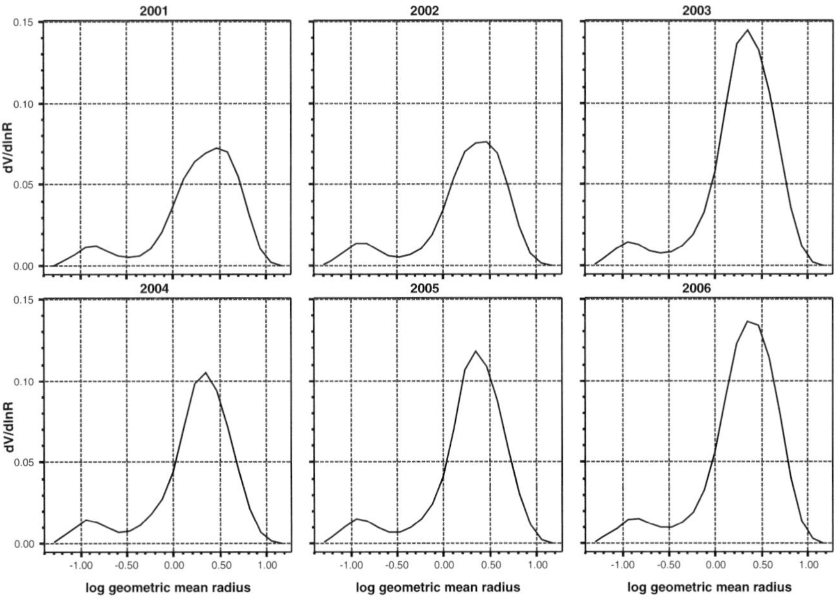

Figure 2 Variations of the daily averaged size distributions (dV/dlnR) for years 2001–2006

Figure 2 illustrates the yearly variation of size distributions (dV/dlnR). The values of aerosol volume size distribution are affected with variations in the concentration of coarse aerosol fraction. The fine aerosol fraction change also, but there is no significant change in the amplitude of the size distribution. The geometric average values for the fine and coarse aerosol effective radii are 0.14 nm ± 0.022 and 1.59 nm ± 0.23, respectively. The geometric standard deviations (σ) for fine and coarse aerosol effective radii are 0.494 and 0.62 respectively. The size distribution is maximum during year 2001 at coarse radius of 3.2 μm. And at ~2 μm during years 2002–2006. The maximum size distributions occur in years 2003 and 2006.

Conclusions

We conclude that the monthly variation of the Ångstrom wavelength exponent which is associated with the changes in aerosol optical thickness is mainly due to substantial load of dust. The direction of the wind blowing over Riyadh area during dusty days was mostly from North (Shamal winds). These winds blow over Arabian Peninsula coming from Iraqi deserts. In relatively free dust days South East winds (Al Kaus) are usually dominant. We found a notable correlation between AOT and WVC in relatively free dust days. The changes in the amplitude of the volume size distributions are mainly associated with the variations in the concentration of coarse particles. Coarse aerosols radii are shifted toward lower values starting from year 2002 to year 2006. The highest volume size distribution amplitudes are found in year 2003 and 2006.

References

1. King, Y., Kaufman, J., Tanre, D., and Nakajima, T., Remote sensing of tropospheric aerosols from space: past, present, and future. *Bull. Amer. Meteor. Soc.*, **80**, 2229–2259 (1999).
2. Smirnov, A., Holben, B., Dubovik, O., Neil, N., and Eck, T., Atmospheric aerosol optical properties in the Persian Gulf. *J. Atmos. Sci.*, **59**, 620–634 (2002).

Atmospheric Aerosols Optical Properties and Climate over Solar Village, KSA

I. Sabbah[1] and F.M. Hasan[2]

Abstract Interaction between aerosols optical properties and climate over solar village, KSA has been studied during the 8 year interval 1999–2007. We use ground-based observations of aerosol optical thickness (AOT) through Aerosol Robotic Network (AERONET) in conjunction with weather parameters measured in Riyadh. Superposed epoch analysis has been applied for days with high aerosol optical thickness (AOT) ($\tau_{500} \geq 0.55$). The mean daily value of the wind speed was very high during dusty days, while that of the exponent$^{\alpha}$ was the lowest. The mean daily value of the difference between the maximum and minimum daily temperature values ΔT (°C) dropped significantly during dusty days.

Keywords Aerosol optical thickness, Ångstrom wavelength exponent, wind speed, air temperature

Introduction

Dust refers to airborne mineral particles that enter the atmosphere directly through action of wind, assisted in many cases by human activity. Desert dust is considered to be one of the major sources of tropospheric aerosol loading, and play an important role in climate forcing studies. Widely prevalent in the tropics, dust aerosols are effective in reflecting solar energy back to space thereby "cooling" the earth's surface. Besides their radiative impact in the shortwave portion of the electromagnetic spectrum, dust aerosols also have an important radiative effect in the long wave (LW). Having mean particle sizes on the order of several micrometers, dust aerosols can effectively reduce the earth's LW emission by re-emitting at a colder temperature when compared to the surface and thereby "warming" the earth.[1] The solar village is located 50 km North West of Riyadh area, capital of Saudi Arabia.

[1] *Physics Department, Faculty of Science, Kuwait University, Kuwait*

[2] *Environmental Sciences Program, Faculty of Graduate Studies, Kuwait University, Kuwait*

C.D. O'Dowd and P.E. Wagner (eds.), *Nucleation and Atmospheric Aerosols*, 1155–1158.

The site coordinates are 24.91 N, 46.41 E and elevation of 650 m a.s.l. Riyadh climate is affected by latitude, elevation above the sea level, wind patterns, and land forms. There is no rule for ocean currents on Riyadh climate. Riyadh is located in highly arid area that lies in subtropical zone that has extremely low yearly precipitation, receiving much less rain annually than what needed to satisfy the climatological demand for evaporation and transpiration.[2]

Data Analysis

Values of AOT observed over solar village have been divided into two groups. The dividing line is taking to be the mean value of AOT measured at 500 nm plus the standard deviation during the 9 year time interval (1999–2007). A total of 248 days are classified as dusty days with a modal value of 0.1. The values of AOT and the exponent α exhibit an 8 year cycle. The exponent α reaches the lowest value during the months of May when that of AOT is the highest. The reverse is true around January, AOT reaches the lowest value when that of α is the highest. The values of AOT are anti correlated with those of the exponent α for the relatively dust-free days. Superposed epoch analysis is applied for the first time to the optical and weather parameters measured over solar village in Saudi Arabia. The values of AOT and those of the wind speed were the highest during the 248 dusty days with $\tau_{500} \geq 0.55$. In contract the values of the Ångstrom exponent α and the temperature values were the lowest. This shows clearly that dust aerosols particles blown over Saudi Arabia had scattering effect that reduces the temperature difference.

Figure 1 displays time series plots of diurnal variations of the values of: (a) AOT, (b) the Ångstrom exponent α, (c) wind speed (m s^{-1}), (d) the difference between the daily maximum and minimum temperature values ΔT (°C). Daily weather parameters were considered only for days with AOT measurements. Dashed horizontal line in Figure 1a represents the mean value of τ_{500} plus its associated standard deviation over the whole period. We see a significant drop in temperature difference ΔT and an increase in the wind speed for the dusty days. Mean and modal values of several optical and weather parameters are listed in Table 1 for each group. We see a significant drop in temperature difference ΔT and an increase in the wind speed for the 248 dusty days.

Conclusions

The values of AOT and the exponent α exhibit a one year cycle. The exponent α reaches the lowest value during the months of May when that of AOT is the highest. The reverse is true around January, AOT reaches the lowest value when that of α is the highest. The values of AOT are anticorrelated with those of the exponent α for the relatively dust-free days. The values of AOT and those of the wind speed were

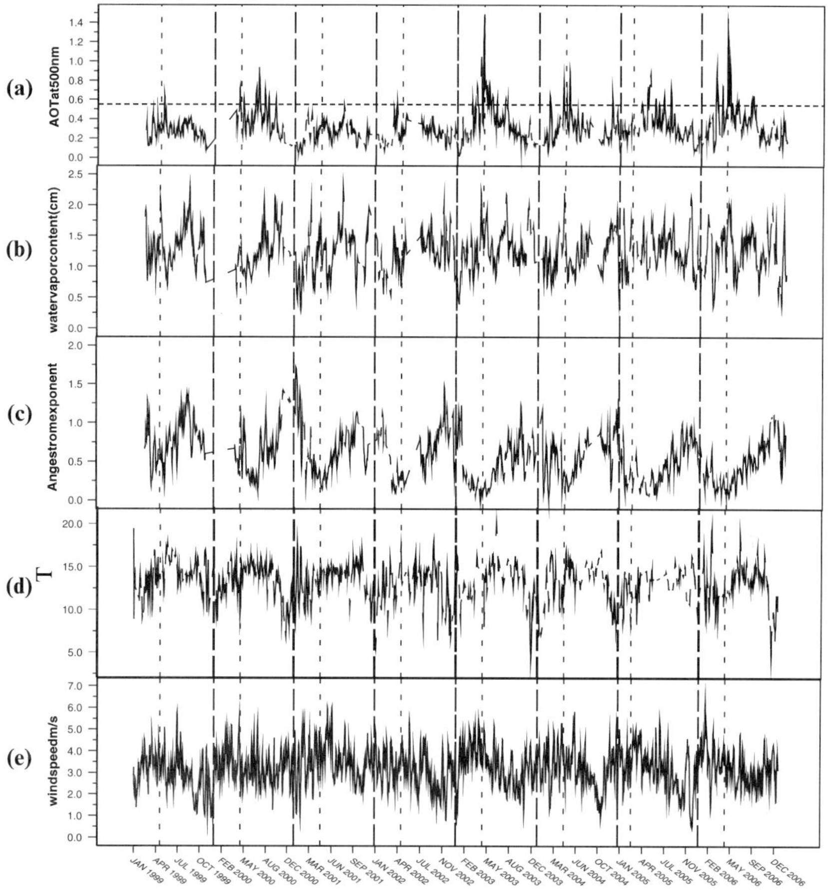

Figure 1 Time series plots of: (a) aerosol optical thickness, (b) Ångström exponent, (c) temperature difference ΔT (°C) and (d) wind speed (m s^{-1}). The reference line shown in (a) is a threshold (τ_{500} = 0.55) above which the values of AOT are considered to be high. The months of May with lowest values of α are represented with vertical dashed lines, while January with the highest values are represented with vertical solid lines

Table 1 Comparison between mean and modal values of the optical and weather parameters for the two aerosols groups

	Number of days	Descriptive statistics	τ_{500}	$\alpha_{500/870}$	ΔT (°c)	Wind speed (m s^{-1})
$\tau_{550} \geq 0.55$	248	Mean ± St error	0.82 ± 0.03	0.22 ± 0.02	12 ± 0.22°C	3.8 ± 0.08 m s^{-1}
		Mode	0.22	0.1	13.3	4
$\tau_{550} < 0.55$	2,206	Mean ± St error	0.27 ± 0.03	0.62 ± 0.01	13.6 ± 0.07°C	3 ± 0.02 m s^{-1}
		Mode	0.55	0.4	16.1	2.5

the highest during dusty days. In contract the values of the Ångstrom exponent α and the temperature values were the lowest. This shows clearly that dust aerosols particles blown over the solar village had a scattering effect that reduces the temperature difference.

References

1. Zhang, J. and Christopher, S.A., Longwave radiative forcing of dust aerosols over the saharan desert estimated from MODIS, MISR, and CERES observations from Terra, *Geophys. Res. Letts.*, 30(23).
2. Steward Edgell, H., *Arabian Desert: Nature Origin and Evolution*, New York: Springer (2006).

Aerosol Optical Depth Determination by Combination of Lidar and Sun Photometer

T. Evgenieva[1,2], N. Kolev[1,2], I. Iliev[1,2], and I. Kolev[1,2]

Abstract Two campaigns of investigation of the atmosphere over the urban area of the Sofia City were carried out. Active (lidar) and passive (sun photometer) remote sensing were used to study various optical characteristics (extinction coefficient, aerosol optical depth and Angstrom parameters) during the convective boundary layer formation. The aerosol optical depth values obtained from the two devices during simultaneous measurements performed in clear sunny days are juxtaposed. The results show that the values of the aerosol optical depth taken by the sun photometer exceed those retrieved from the lidar data and depend on the aerosol structure in the stage of the planetary boundary layer development.

Keywords Lidar, sun photometer, aerosol optical depth

Introduction

Atmospheric aerosols play an important role in many atmospheric processes. Most aerosols of anthropogenic origin are found in the lower troposphere and contribute significantly to the haze often visible during early morning and after sunset near to the ground surface.[1] This study aimed at: (i) determining the atmospheric boundary layer (ABL) height over an urban area situated in a mountain valley using the aerosol as a tracer; (ii) determining the variation in the aerosol optical characteristics over the area in question during the ABL development (convective and stable boundary layer formation) using a lidar and a sun photometer.

[1]*Institute of Electronics, Bulgarian Academy of Sciences, 72, Tsarigradsko shosse Blvd., 1784 Sofia, Bulgaria*

[2]*Central Laboratory of Solar–Terrestrial Influences, Bulgarian Academy of Sciences*

C.D. O'Dowd and P.E. Wagner (eds.), *Nucleation and Atmospheric Aerosols*, 1159–1163.

Methods and Apparatus

The main parameters of the two devices are: Specifications of the lidar: transmitter – a standard Nd-YAG laser (operational wavelength 532 nm, pulse duration and energy 15–20 ns and 10–15 mJ, repetition rate 12.5 Hz); receiving antenna – a Cassegrainian telescope (main mirror diameter 150 mm, equivalent focal length 2,250 mm); photodetector – a PMT with an interference filter (1 nm FWHM); data acquisition and processing set – 10 bit 20 MHz ADC and a PC.[2] Specification of the sun photometer: optical channels: $\lambda = 380$ nm, $\lambda = 500$ nm, $\lambda = 675$ nm, $\lambda = 936$ nm, and $\lambda = 1,020$ nm, viewing angle – 2.5°, dynamic range $>3 \times 10^5$, computer interface – RS 232.[3]

The extinction coefficient of atmospheric aerosol is determined from the lidar data using common inversion Fernald method[4] for solving of the lidar equation. One of the most basic parameters of the aerosols is the optical depth obtained from solar extinction measurement with sun photometers. The calculations are based on the Beer–Lambert–Bouguer law.[5]

Results and Discussion

The experiments were carried out in the south-eastern part of the city of Sofia, where the Central Laboratory of Solar-Terrestrial Influences and the Institute of Electronics (aerosol lidar and sun photometer) are located. In the present work, as we have already mentioned, we present experimental data from two experimental campaigns – autumn (2006) and winter (2006/2007).

Lidar Data

The synoptic situation during the whole period of the present autumn campaign was almost constant, without significant changes, and we had the possibility to make observations in the second half of October 2006. Taking into account the mixing layer height (MLH) the data from this campaign shows two different types of convective boundary layer (CBL): CBL with maximum height $H_{CBL} = 600$ m and CBL with maximum height $H_{CBL} = 1,000$ m.

The experimental lidar data for one day obtained in the second half of October, namely 18 October 2006, are shown on the Figure 1. The experiment started at 0845 LST and finished at 1330 LST. On the first HTI, the SBL with $H_{SBL} = 150–200$ m remaining from the previous night is seen, as well as the RL with height $H_{RL} = 1,050$ m. The new ML begins to form at 0910 LST and its height is about $H_{ML} = 100$ m; its height is about $H_{ML} = 200$ m at 1030 LST and reaches $H_{ML} = 500–550$ m at 1100 LST. The next few images show that the MLH reaches $H_{ML} = 750$ m at 1200 LST, decreasing to $H_{ML} = 600$ m at 1300 LST. In the next half hour MLH reaches $H_{ML} = 1,000$ m.

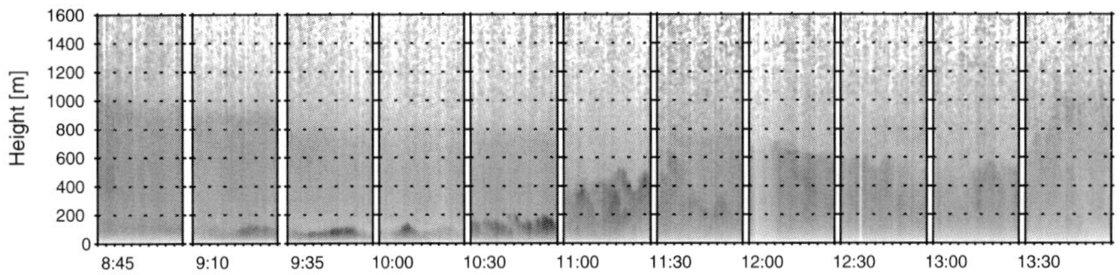

Figure 1 Height–time images obtained from the lidar data on 18 October 2006

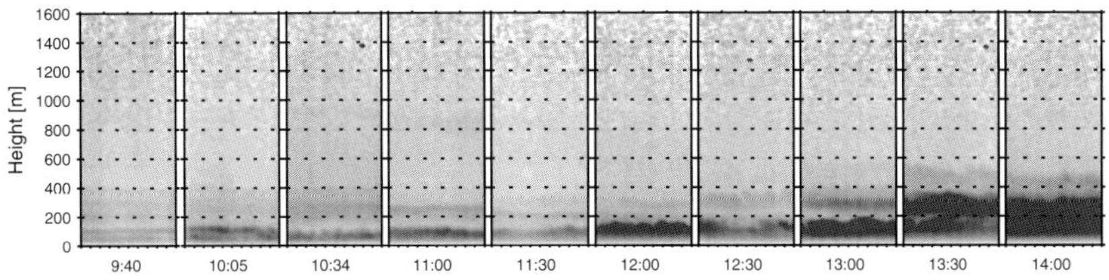

Figure 2 Height–time images obtained from the lidar data on 10 January 2007

In the next part of the study we present the lidar data obtained on 1 day of the winter campaign 2007. Figure 2 shows 10 HTIs obtained on 10 January 2007.

The experiment started at 0940 LST and was completed at 1400 LST. On the first two HTIs one can see the SBL remaining from the previous night with height $H_{SBL} = 400$ m, with four layers being situated in it. Two of them are located in the first 100–150 m and the others are at heights ranging from H = 200 m to H = 350 m. The RLH is $H_{RL} = 1{,}000$ m and stays constant during the experiment. The beginning of the formation of the new ML occurs at 1034 LST and the first two aerosol layers are destroyed. The MLH remain constant around $H_{ML} = 100$ m until 1130 LST, and reaches $H_{ML} = 200$ m at 1230 LST. The SBL is destroyed completely at 1330 LST when the MLH reaches H = 550 m and remains about $H_{ML} = 500$ m at 1400 LST.

Sun Photometer Data

We present here the experimental data for AOD obtained by using a Microtops II sun photometer. Generally, the data from the autumn campaign show constant decrease of the AOD at the various wavelengths used, especially at wavelength $\lambda = 500$ nm. The comparison with the lidar data given below is precisely for that wavelength $\lambda = 500$ nm which is the closest to the laser operational wavelength $\lambda = 532$ nm.

On Figure 3 (a) variations in the AOD on 18 October 2006 are shown. The experiment started at 0920 LST and finished at 1620 LST. The mean value at

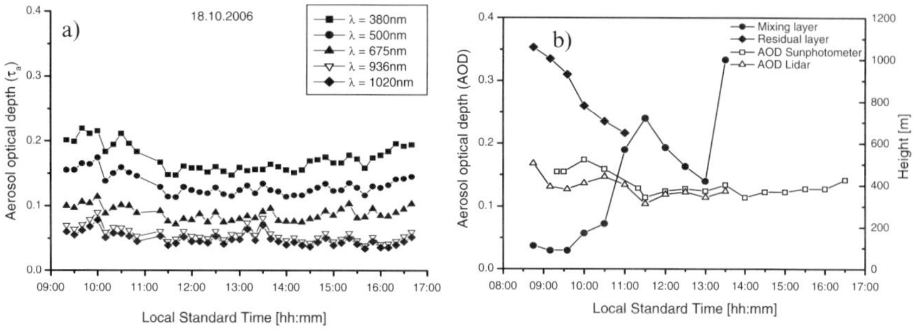

Figure 3 (a) Variations in the AOD at λ = 380 nm, λ = 500 nm, λ = 675 nm, λ = 936 nm, and λ = 1,020 nm obtained from sun photometer on 18 October 2006; (b) Variations in the height of the different layers and in the aerosol optical depth determined from the lidar, and sun photometer data on 18 October 2006

λ = 500 nm is $\tau_a^{mean} = 0.13$. Fluctuations in AOD between 0920 LST and 1130 LST are observed. In this period AOD at λ = 500 nm varies between $\tau_a = 0.18$ and $\tau_a = 0.13$. The mean AOD for the rest of the days in October varies from $\tau_a^{mean} = 0.17$ to $\tau_a^{mean} = 0.33$. In November the mean AOD values are $\tau_a = 0.22$ and $\tau_a = 0.30$. In January the mean AOD varies from $\tau_a = 0.15$ to $\tau_a = 0.27$. Actually, the weather in January was dry and warm which resulted in the formation of an ABL that is not typical for this season characterizing by aerosol pollution in the whole layer. Figure 3 (b) present a comparison between the AOD values obtained from the lidar and the sun photometer data for 1 day during the ABL formation, namely 18 October 2006. In the beginning of the experiment the RLH is $H_{RL} = 1{,}050$ m. The RL is fully destroyed at 1100 LST at height $H_{RL} = 650$ m. The formation of the new ML starts around 0830 LST at height around $H_{ML} = 100$ m and reaches maximum height $H_{ML} = 1{,}000$ m at 1330 LST. The AOD values obtained from the lidar data are lower but close to sun photometer one. The maximum values obtained from the two devices are around $\tau_a = 0.17$. The minimum values are $\tau_{al} = 0.10$ and $\tau_{as} = 0.11$ obtained from the lidar and the sun photometer, respectively. The behaviour of the AOD obtained from the two devices is similar but depends on the aerosol structure and relative humidity in the stage of the PBL development.

Conclusion

Despite of the preliminary character of the analysis, the data obtained by simultaneous measurements using a lidar and a sun photometer reveal a relation between the ML development, the atmospheric aerosol optical characteristics and the concrete meteorological parameters of the atmosphere over the observed region.

Acknowledgments Part of the presented results obtained from the lidar and sun photometer observations were obtained within the framework of the ACCENT subproject WP12 (Aerosols). The investigations concerning the aerosol optical depth measurements were partially supported by the National Science Fund at the Ministry of Education and Science of Republic of Bulgaria under contract DSI 12/05 "Infrastructure".

References

1. Balis, D., Papayannis, A., Galani, E., Marenco, F., Santaceraria, V., Hamonov, E., Chazette, P., Ziomas, I., and Zarefos, C., *Atmos. Environ.*, **34**, 925–932 (2000).
2. Kolev, I., Savov, P., Kaprielov, B., Parvanov, O., and Simeonov, V., *Atmos. Environ.*, **34**, 3223–3235 (2000).
3. Solar Light Company, Inc., "User's guide MICROTOPS II Sunphotometer Version 5.5", USA (2003).
4. Fernald, F.G., *Appl. Opt.*, **23**, 652–653 (1984).
5. Devara, P.C.S., Pandithurai, G., Raj, R.E., and Sharma, S., *J. of Aerosol Sci.*, **27**(4), 621–632 (1996).

Wind Speed Influences on Aerosol Optical Depth in Clean Marine Air

J.P. Mulcahy, C.D. O'Dowd, S.G. Jennings, and D. Ceburnis

Abstract The relationship between the aerosol optical depth (AOD) and wind speed at Mace Head, Ireland is determined for clean marine conditions in a steady state wind speed environment. Carrying out the analysis in such specific conditions reduced the number of external factors (e.g., changing air masses, relativity humidity, advection processes) contributing to the total measured AOD. A high correlation, in terms of a power law, was determined for both 862 nm and 500 nm (r^2 = 0.97) and a strong anti-correlation was observed between the Ångstrom exponent and the wind speed (r^2 = 0.93).

Keywords Aerosol optical depth, marine aerosol

Introduction

Marine aerosol, one of the largest sources of natural aerosol, plays an important role in the Earth's climate system.[1] Many studies have highlighted the relatively significant contribution of primary produced sea salt particles to the chemical composition of both the fine and coarse marine aerosol modes.[2,3] Consequently, sea salt particles influence the radiative balance of the atmosphere both *directly* by acting as efficient scatterers of the incoming solar radiation and *indirectly* through their role as cloud condensation nuclei, thereby affecting the microphysical properties of marine boundary layer (MBL) clouds, including importantly the cloud albedo.[4–6]

Sea salt concentrations in the marine atmosphere are primarily controlled by the primary production mechanisms of bubble bursting and spume drop production, which in turn are largely dependent on the surface wind speed. The total columnar aerosol optical depth (AOD) is an important optical measurement of the aerosol loading in the atmosphere and the corresponding scattering and absorption properties

Department of Experimental Physics & Environmental Change Institute, National University of Ireland, Galway, University Road., Galway, Ireland

C.D. O'Dowd and P.E. Wagner (eds.), *Nucleation and Atmospheric Aerosols*, 1164–1168.

of the dominating aerosols. Previous studies in the literature suggest only a weak correlation between the AOD and wind speed.[7–9] However, in many instances the dominating background aerosol is influenced by a wide range of factors, including air mass history, variable meteorological parameters, relative humidity effects in the atmospheric column, advection processes, and coupled with the relatively short residency time of the larger coarse mode sea salt aerosol particles makes the determination of a wind–AOD relationship difficult. This study attempts to reduce the number of external contributing factors by reporting the relationship between the wind speed and the total columnar AOD for clean marine and steady-state wind speed conditions alone.

Methodology

The study was strictly limited to periods of low oceanic biological activity (October–March) and cleans marine air masses, in order to maximize the impact of sea salt scattering. Sea salt dominates all aerosol size ranges during the low biological period and the contributing fractions of biogenically produced inorganic and organic matter to the marine aerosol are relatively small.[10] This is not the case during periods of high biological activity due to the seasonal variation in the chemical composition of marine aerosol at Mace Head.[10,11] In addition, clean marine air masses contain by definition the largest fraction of sea salt aerosol.

The measurements were carried out at the Mace Head atmospheric research station on the west coast of Ireland (53.3°N, 9.9°W). The station is ideal for making measurements in a clean marine environment as it is situated directly in the path of the prevailing westerly–southwesterly airflow approaching from the Atlantic Ocean. Over 52% of air masses sampled at Mace Head are in this clean marine wind sector.[8] In order to select time periods corresponding to clean marine conditions only, various criteria were applied. These criteria are based on a number of previous studies, which investigated the cleanest marine air conditions at the station and the concomitant microphysical and physico-chemical aerosol characteristics.[12–14]

Periods of stable, steady-state wind speed conditions were selected in the described conditions between January 2002 and December 2004. In this study, a "steady state" wind speed event was defined by a low variability in the daily wind speeds (daily standard deviation $<2\,m\,s^{-1}$) with a maximum standard deviation of $1\,m\,s^{-1}$ allowed for the period AOD measurements were made. Due to the larger variability observed in high wind speed events ($>10\,m\,s^{-1}$) the above maximum value was increased to $2\,m\,s^{-1}$ in such cases.

AOD measurements at Mace Head were made using a Precision Filter Radiometer (PFR).[15] The AOD is calculated from direct solar radiation measurements every 2 min at four wavelengths (862, 500, 412, and 368 nm). The Ångstrom exponent (α) is derived by linear regression through the measured AOD values at all four wavelengths using the Ångstrom Power Law.

Results

The AOD measurements and the corresponding wind speed measurements from a total of 14 selected steady state events were merged together and the resultant dataset was sorted into bins based on the wind speed. Each bin was ± 0.5 m s^{-1} wide and spaced in 1 m s^{-1} increments. Figure 1 shows the correlation of the AOD at 862 and 500 nm with wind speeds ranging from 4 to 18 m s^{-1}. A power law curve was fitted to the data and a strong (r^2 = 0.97) dependence of the AOD on the square of the wind speed (U^2) was found for both wavelengths. This significantly high correlation is in contrast to previous studies at Mace Head,[8] which report only a weak increasing trend in the calculated MBL AOD with increasing wind speeds above 5 m s^{-1} (r^2 = 0.24). However, these measurements of the MBL AOD were determined from scattering and absorption measurements, where the scattering data suffer from significant particle losses in the sampling system at high wind speeds, with a 50% cut-off diameter calculated to be approximately 6 μm at 6.8 m s^{-1} decreasing further with increasing wind speeds.[16] As a result, any increase in the concentration of the larger sea salt particles in high wind speed conditions may not be detected by the sampling system used. Smirnov et al. (2003) report a correlation coefficient, r, of 0.37 at 500 nm. A higher correlation of $r = 0.52$ is reported for 1,020 nm, which is expected considering that sea salt particles have larger diameters (from ~0.1–100's μm) and therefore have a greater contribution to the scattering in the infrared wavelength region. These correlations were also based on linear fits and not a power law fit. Fitting a linear regression line to the data in Figure 1 yielded marginally smaller r^2 values of 0.95 and 0.94 for 862 nm and 500 nm respectively. It is also worth noting, that this study reports measurements for a much larger wind speed range (4–18 m s^{-1}) than other already highlighted studies (maximum 14 m s^{-1}).[7–9]

The similar correlations observed for the 500 nm and 862 nm can be explained by the neutral spectral dependence invariably exhibited by sea salt particles.[17] This is evident in Figure 2, where α levels off close to 0. The Ångstrom exponent is a qualitative indicator of particle size. Therefore, larger particles such as sea salt typically have values of $\alpha < 1$, decreasing further with an increasing dominance of

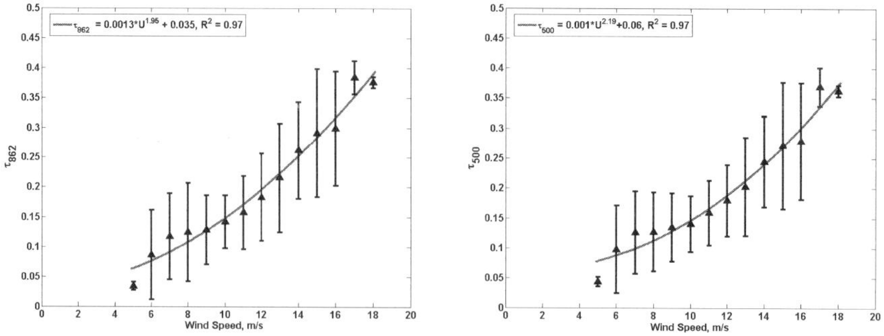

Figure 1 Correlation of AOD (τ) with wind speed (U) for 862 nm (**left**) and 500 nm (**right**)

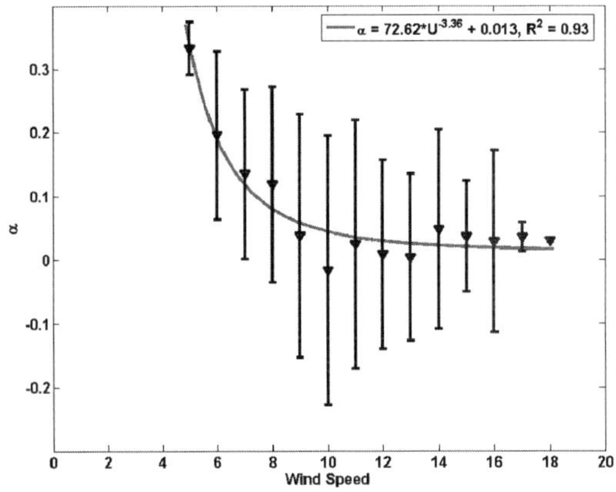

Figure 2 Correlation of α with wind speed

the coarse mode. This is again reflected in the strong anti-correlation observed in Figure 2 ($r^2 = 0.93$). This further supports the theory that the increase in AOD at larger wind speeds is primarily as a result of enhanced scattering by super-micron particles.

Conclusions

For clean marine, steady state wind speed (U) conditions, the AOD was found to be dependent on U^2, with a significantly high correlation ($r^2 = 0.97$). This result importantly expands upon previous studies by limiting the amount of scatter in the correlation and by increasing the wind speed range over which the relationship is determined and thus highlights the significant contribution of sea salt particles to the total AOD at high wind speeds and it's important influence on the Earth's radiative climate.

References

1. Fitzgerald, J.M, *Atmos. Environ. Part A*, **25**, 533–545 (1991).
2. O'Dowd, C.D., Smith, M.H., Consterdine, I.E., and Lowe, J.A., *Atmos. Environ.*, **31**, 73–80 (1997).
3. O'Dowd, C.D. and Smith, M.H., *J. Geophys. Res.*, **98**, 1137–1149 (1993).
4. Murphy, D.M., Anderson, J.R., Quinn, P.K., McInnes, L.M., Brechtel, F.J., Kreidenweis, S.M., Middlebrook, A.M., Posfai, M., Thomson, D.S., and Buseck, P.R., *Nature*, **392**, 62–65 (1998).
5. Quinn, P.K. and Coffman, D.J, *J. Geophys. Res.*, **104**, 4241–4248 (1999).
6. Latham, J. and Smith, M.H., *Nature*, **347**, 372–373 (1990).
7. Villevalde, Y.V., Smirnov, A.V., O'Neill, N.T., Smyshlyaev, S.P., and Yakovlev, V.V., *J. Geophys. Res.*, **99**, 20983–20988 (1994).

8. Jennings, S.G., Kleefeld, C., O'Dowd, C.D., Junker, C., Spain, T.G., O'Brien, P., Roddy, A.F., and O'Connor, T.C., *Boreal Env. Res.*, **8**, 303–314 (2003).
9. Smirnov, A., Holben, B.N., Eck, T.F., Dubovik, O., and Slutsker, I., *J. Geophys. Res.*, **108**, 4802, doi:10.1029/2003JD003879 (2003).
10. O'Dowd, C.D., Facchini, M.C., Cavalli, F., Ceburnis, D., Mircea, M., Decesari, S., Fuzzi, S., Yoon, Y.J., and Putaud, J.-P., *Nature*, **431**, 676–680, doi:10.1038/nature02959 (2004).
11. Yoon, Y.J., Ceburnis, D., Cavalli, F., Jourdan, O., Putaud, J.-P., Facchini, M.C., Decesari, S., Fuzzi, S., Jennings, S.G., O'Dowd, C.D., *J. Geophys. Res.*, doi:10.1029/2005JD007044 (2007).
12. Cooke, W.F., Jennings, S.G, and Spain, T.G., *J. Geophys. Res.*, **102**, 25339–25346 (1997).
13. Jennings, S.G., Geever, M., McGovern, F.M., Francis, J., Spain, T.G., and Donaghy, T., *Atmos. Environ.*, **31**(17), 2795–2808 (1997).
14. O'Dowd, C.D., Smith, M.H., and Jennings, S.G., *J. Geophys. Res.*, **98**, 1123–1135 (1993).
15. Wehrli, C., GAW Report No. 162, pp. 36–38 (2004).
16. Kleefeld, C., O'Dowd, C.D., O'Reilly, S., Jennings, S.G., Aalto, P., Becker, E., Kunz, G., and de Leeuw, G., *J. Geophys. Res.*, **107**, 8103, doi:10.1029/2000JD000262 (2002).
17. Smirnov, A., Holben, B.N., Kaufman, Y.J., Dubovik, O., Eck, T.F., Slutsker, I., Pietras, C., and Halthore, R.N., *J. Atmos. Sci.*, **59**, 501–523 (2002).

Validation of Satellite Retrieved Aerosol Optical Depth with Ground-based Measurements at Mace Head, Ireland

J.P. Mulcahy, S.G. Jennings, C.D. O'Dowd, W. von Hoyningen-Huene*, and J.P. Burrows*

Abstract An extensive validation study was undertaken to assess the feasibility of operationally retrieving aerosol optical depth (AOD) over the north-eastern Atlantic region using the latest aerosol satellite remote sensing techniques. The Bremen Aerosol Retrieval (BAER) algorithm was used to derive AOD from the Sea-viewing Wide Field-of-view Sensor (SeaWiFS) over the coastal location of Mace Head, Ireland and its performance over land, ocean and coastal surfaces was independently assessed. In addition, Level-2 aerosol products from the Moderate Resolution Imaging Spectroradiometer (MODIS) were also evaluated. The MODIS AOD showed an overall better agreement with ground-based measurements at Mace Head giving $r^2 = 0.72$ at 550 nm. BAER AOD revealed an underestimating trend over ocean surfaces, with the best agreement observed at 412 nm over the coastal and land surfaces ($r^2 = 0.69$).

Keywords Aerosol optical depth, satellite remote sensing, SeaWiFS, MODIS, PFR

Introduction

Satellite observations provide a unique perspective from which to observe the highly variable aerosol fields over large regions of the Earth's surface and have been increasingly investigated as the most viable method of accurately determining the global impact of aerosols on climate change.[1,2] Numerous aerosol retrieval algorithms have been developed for both land and ocean surface retrievals and for a variety of different satellite systems.[3–8] However, the relative impact of satellite remote sensing technologies and techniques on the further advancement

Department of Experimental Physics & Environmental Change Institute, National University of Ireland, Galway, University Road, Galway, Ireland

* *Institute of Environmental Physics & Remote Sensing, University of Bremen, Germany*

C.D. O'Dowd and P.E. Wagner (eds.), *Nucleation and Atmospheric Aerosols*, 1169–1173.

of aerosol characterization and expertise is critically dependent on the continual improvement and development of the aerosol retrieval algorithms. This requires extensive validation of the satellite-derived aerosol properties with in situ measurements from ground-based networks across the globe. This work presents the results of one such validation study carried out over the Mace Head Global Atmospheric Watch (GAW) research station on the westernmost periphery of Europe.[9]

The Bremen Aerosol Retrieval (BAER) algorithm[8] was used to derive the aerosol optical depth (AOD) from the Sea-viewing Wide Field-of-view Sensor (SeaWiFS) over both land and ocean surfaces. In addition, Level-2 aerosol products from the MODerate resolution Imaging Spectroradiometer (MODIS) were also evaluated.[10] The combined land and ocean AOD dataset at 550 nm is discussed below. The two MODIS instruments on the Terra and Aqua satellites were treated separately in this study. Ground-based optical depth measurements were provided by a Precision Filter Radiometer (PFR) at the Mace Head station.[11] The goal of the study was to evaluate the performance of all three satellite sensors and their respective aerosol retrieval algorithms, with the ultimate purpose of operationally applying these satellite remote sensing techniques over the north-east Atlantic area. Unfortunately, the climate at Mace Head is regularly cloudy and windy, making both in situ and remote AOD measurements difficult. Such high levels of cloud cover severely limited the number of collocated clear satellite scenes and ground-based measurements available for comparison.

Methodology

The validation procedure[12] was carried out for the months of April 2003 and May 2004. In order to investigate any possible biases introduced when retrieving over coastal areas, separate independent validations were carried out over coastal, ocean and land surfaces.

An area (or *window*) corresponding to 50 × 50 km centered on the Mace Head (*coast*) geographical coordinates (53.3° N, 9.9° W) was selected from the original satellite scene. Due to the larger pixel resolution of the MODIS aerosol product (10 × 10 km) compared to the BAER-derived AOD (1 × 1 km), an array of this size ensured a sufficiently large statistical sample. An equal-sized window was selected centred 30 km offshore from the station to represent an open *ocean* surface. Over *land*, a smaller window size of 12 × 12 km was chosen due to the presence of numerous small inland lakes in the area surrounding Mace Head. Accordingly, the SeaWiFS data was only analyzed in the case of the latter. A minimum of $N/5$ AOD values were required before the average AOD and standard deviation for the window could be calculated, where N is the number of pixels in the selected window. This spatial average was then compared with a 1 h temporal average of the PFR measurements (30 min before and after the satellite overpass).[12]

Results

The results of the linear regression analysis between the satellite-derived AOD (τ_{SAT}) and the PFR measurements (τ_{PFR}) at 555 nm (550 nm for MODIS) are shown in Table 1. Overall, MODIS generally showed a good agreement with the ground-based data with r^2 values of up 0.72 and 0.83 determined over the coastal and ocean surfaces respectively. Over coastal regions, a slight overestimation by MODIS, indicated by the high intercept values of both Terra and Aqua in Table 1, is more than likely due to water pixels directly adjacent to the coast being retrieved using the MODIS land algorithm rather than the ocean algorithm.[11] This is further supported by Figure 1b where both datasets show a distinct peak at +0.08. In the case of Terra-MODIS, a comparison of the combined land/ocean data-set with the ocean only data-set for the ocean surface, revealed a larger number of valid retrievals for the former suggesting further uncertainties introduced by the incorrect use of the

Table 1 Results of the linear regression analysis at 555/550 nm, where $\tau_{sat} = a^*\tau_{PFR} + b$

Satellite	A	b	r^2	N
	Mace Head			
SeaWiFS-BAER	0.70	0.07	0.45	16
(412 nm)	(0.72)	(0.03)	(0.69)	(16)
MODIS-Terra	0.95	0.12	0.72	21
MODIS-Aqua	0.82	0.16	0.53	20
	Ocean			
SeaWiFS-BAER	0.76	0.01	0.49	16
MODIS-Terra	0.99	0.01	0.66	18
MODIS-Aqua	1.06	0.004	0.82	14
	Land			
SeaWiFS-BAER	0.74	0.1	0.53	16
(412 nm)	(0.73)	(0.1)	(0.64)	(16)

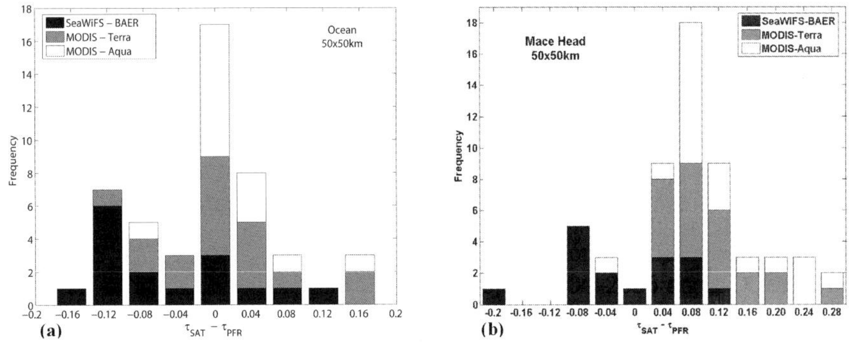

Figure 1 Histogram of the AOD differences between the three satellite sensors and PFR for (a) the ocean surface and (b) the Mace Head coastal surface

land algorithm. This is possibly due to the occurrence of some small islands within the 50 × 50 km window. This leads to a smaller r^2 value of 0.66 for the Terra MODIS instrument in this case.

BAER derived SeaWiFS AOD exhibits consistently lower correlations than MODIS (Table 1). The best agreement was observed at 412 nm (shown in parentheses in Table 1) over the coastal and land surfaces with r^2 values of 0.69 and 0.64 respectively. It is noted however that the significance of these correlations is affected by the smaller number of collocated measurements (N in Table 1) for the SeaWiFS comparison than for MODIS. This is due to the larger number of daily overpasses of the Terra and Aqua satellites in this area. Over the ocean, the BAER AOD tended to underestimate the PFR values as observed in Figure 1a, where a clear peak of the SeaWiFS-BAER histogram is found at –0.12. This is more than likely caused by the overestimation of the underlying surface reflectance and subsequent overestimation of the retrieved AOD. Over Mace Head, the distribution of τ_{BAER}-τ_{PFR} in Figure 1b reveals a more inconsistent behaviour of both over- and underestimations. This could lead one to believe that the r^2 value of 0.69 in this case is misleadingly high.

It is also worth noting that the performance of the BAER algorithm in this region of the north-east Atlantic could have been affected by the presence of cloud contaminated pixels. An additional cloud screening test (not reported in detail here) was carried out on the BAER AOD and strongly suggested that the cloud screening currently implemented in the BAER algorithm is not sufficient to deal with the high levels of cloud, in particular thin cirrus, encountered in this region. Certain dubious data values as highlighted by this test were not included in the regression analysis reported here.

Conclusions

The MODIS Level 2 AOD product out-performed the BAER-derived SeaWiFS AOD in this validation study over the Mace Head coastal location in the north-east Atlantic. Although potential uncertainties in the MODIS retrieval over direct coastal regions were observed, it showed excellent agreement over ocean surfaces. BAER AOD gave best agreement in the UV spectral region over the coastal and land surfaces but showed a tendency to underestimate ground-based AOD over the ocean.

References

1. King, M.D, Kaufman, Y.J., Tanré, D., and Nakajima, T., *Bull. Am. Meteorol. Soc.*, **80**(11), 2229–2259 (1999).
2. Kaufman, Y.J., Tanré, D., and Boucher, O., *Nature*, **419**, 215–223 (2002).
3. Husar, R.B., Prospero, J.M., and Stowe, L.L., *J. Geophys. Res.*, **102**, 16889–16909 (1997).

4. Kaufman, Y.J., Tanré, D., Remer, L.A., Vermote, E.F., Chu, A., and Holben, B.N., *J. Geophys. Res.*, **102**, 17051–17067 (1997).
5. Stowe, L.L., Ignatov, A.M., and Singh, R.R., *J. Geophys. Res.*, **102**, 16923–16934 (1997).
6. Tanré, D., Kaufman, Y.J., Hermann, M., and Mattoo, S., *J. Geophys. Res.*, **102**, 16971–16988 (1997).
7. Veefkind, J.P. and de Leeuw, G., *J. Aerosol. Sci.*, **29**, 1237–1248 (1998).
8. von Hoyningen-Huene, W., Freitag, M., and Burrows, J.P., *J. Geophys. Res.*, **108**, 4260, doi:10.1029/2001JD002018 (2003).
9. Jennings, S.G., Kleefeld, C., O'Dowd, C.D., Junker, C., Spain, T.G., O'Brien, P., Roddy, A.F., and O'Connor, T.C., *Boreal Env. Res.*, **8**, 303–314 (2003).
10. Remer, L.A., Kaufman, Y.J., Tanre, D., Mattoo, S., Chu, D.A, Martins, J.V., Li, R.R., Ichoku, C., Levy, R.C., Kleidman, R.G., Eck, T.F., Vermote, E., and Holben, B.N., *J. Atmos. Sci.*, **62**(4), 947–973, doi:10.1175/JAS3385 (2005).
11. Wehrli, C., GAW Report No. 162, pp. 36–38 (2004).
12. Ichoku, C., Chu, D.A., Mattoo, S., Kaufman, Y.J., Remer, L.A., Tanre, D., Slutsker, I., and Holben, B.N., *Geophys. Res. Lett.*, **29**(12), doi:10.1029/2001GL013206 (2002).

Part XI
Chemical Composition and Cloud Drop Activation

Laboratory Studies on the Properties and Processes of Complex Organic Aerosols

Yinon Rudich[1], Elad Dinar[1], Ilya Taraniuk[1], Ali Abo Riziq[1], Ellen R. Graber[2], Caryn Erlick[3], Tatu Anttila[4,‡], and Thomas Mentel[4]

Abstract Aerosol particles affect climate directly and indirectly. The direct climatic effect is via scattering and absorption of radiation and the indirect effect is mostly by affecting the properties of clouds and precipitation. These climatic effects impose one of the largest uncertainties in our understanding of the climate system. Recent studies have shown that organic matter, natural and anthropogenic, is dominating the tropospheric aerosol mass and affects its properties. We will discuss laboratory studies that focus on studying the properties and processes of organic aerosols with emphasis on their interaction with vapor pressure, physical properties, optical properties, and density. We will emphasize the experimental approach and highlight the increased complexity of the studied systems.

Introduction

Atmospheric aerosols influence climate and human health on regional and global scales.[1–3] The climatic effect of aerosols is among the largest uncertainties in understanding earth's climate and in predictions of future climatic changes.[4] Aerosols are also implicated in serious health consequences; however, the role of aerosol chemical composition in those health effects remains uncertain. Some aerosols are directly emitted from various stationary and mobile sources (*primary*

---

[1]*Department of Environmental Sciences, Weizmann Institute of Science, Rehovot 76100, Israel,e-mail: yinon.rudich@weizmann.ac.il*

[2]*Institute of Soil, Water and Environmental Sciences, Volcani Center, A.R.O., Bet Dagan 50250, Israel*

[3]*Department of Atmospheric Sciences, The Hebrew University of Jerusalem, Jerusalem, 91904, Israel*

[4]*Institute for Tropospheric Chemistry, Research Center Jülich, Jülich Germany*

‡*current address: Research and Development, Finnish Meteorological, Institute, 00101 Helsinki, Finland*

C.D. O'Dowd and P.E. Wagner (eds.), *Nucleation and Atmospheric Aerosols*, 1177–1181.

aerosols) such as combustion and biomass burning,[5] vegetation,[6] ocean surfaces[7] and deserts.[8] Aerosol mass also forms in the atmosphere (*secondary aerosols, SOA*) through reactions that transform volatile organic and inorganic species into low vapor pressure compounds that can condense on existing particles. Organic particulate matter can also be directly emitted to the atmosphere, such as from the ocean surface or in biogenic aerosols.[7,11] In many environments, a major fraction of the particulate mass is composed of organic compounds.[12,13] These organics significantly influence aerosol properties; therefore they directly modulate the role aerosols play in the environment.

More then 4 decades ago, Went[14] suggested that photochemical oxidation of primary volatile and semi-volatile organic compounds can produce secondary polymers of high carbon content that may be involved in cloud droplet and haze particle formation. Recent studies of rural and urban particles,[15,16] fogwater,[17] marine particulate samples[18], and biomass burning aerosols[19] concluded that 20–70 wt % of the WSOC fraction consists of high molecular weight polycarboxylic acids.[20] These compounds, consisting of a heterogeneous mixture of structures containing aromatic, phenolic and acidic functional groups,[16,21–23] have certain similarities to humic substances (HS) from terrestrial and aquatic sources. Therefore, these aerosol-associated compounds are referred to in the atmospheric chemistry literature as HUmic-LIke Substances (HULIS). Accumulating evidence shows that HULIS may form in atmospheric particulate matter via photooxidation of primary biogenic and anthropogenic precursors, or may be directly emitted from soils, vegetation, biomass burning and soot automotive exhaust.[24–26]

In this talk we will describe laboratory experiments with aerosols that contain significant amounts of complex organic matter. The goal of these experiments is to better understand the physical and chemical properties of these aerosols and their processes, aiming at getting a better model of the atmospheric system. Specifically we will describe experiments which contain actual atmospheric HULIS as well ad biogenic aerosols.

Experimental Methods

We will describe several different studies using humidity tandem DMA (HTDMS), with cloud simulation chamber to measure hygroscopic properties, experiments with a dual wavelength cavity ring down aerosol spectrometer aiming at measuring the optical properties of organic aerosols.[27–29] We will discuss the surface tension behavior of aqueous solutions containing HULIS and the possible influence of surface tension in activation. We have used several different aerosol types in these experiments:

- Aerosols containing HULIS extracts from collected aerosol particles. The samples are composed of HULIS from fresh and aged smoke particles as well as those extracted from pollution aerosols. The properties of these aerosols were

compared to the behavior of aerosols composed of molecular weight fractionated Suwanee River Fulvic Acid (SRFA).
- Single component organic aerosol generated by nebulizing aqueous solutions of the studied compounds.
- Mixtures of organic and inorganic aerosols.
- Organic aerosols that formed by photochemical oxidation of biogenic organic compounds emitted from plants in a simulation chamber.

Results and Discussion

We studied the activation to cloud droplets and hygroscopic growth of aerosols containing atmospheric HULIS extracted from fresh, aged and pollution particles and compared their activation to size fractionated Suwannee River fulvic acid, and correlated it to the estimated molecular weight and measured surface tension. A correlation was found between CCN-activation diameter of SRFA fractions and number average molecular weight. The lower molecular weight fractions activated at lower critical diameters, which is explained by the greater number of solute species in the droplet with decreasing molecular weight. The three aerosol-extracted HULIS samples activated at lower diameters than any of the size-fractionated or bulk SRFA. By considering number average molecular weight (MN), measured surface tension (ST) and activation diameters, the Köhler model was found to account for activation diameters, provided that accurate physicochemical parameters are known. Sampled that aged in the atmosphere had lower molecular weight, higher density and lower activation diameters. This is attributed to chemical reactions with OH and ozone that lead to smaller, more oxygenated products.

Cavity ring down (CRD) spectrometry was used for measuring the optical properties of pure and mixed laboratory-generated aerosols. The extinction coefficient (α_{ext}), extinction cross section (σ_{ext}) and extinction efficiency (Q_{ext}) were measured for polystyrene spheres (PSS), ammonium sulphate ($((NH4)_2(SO_4))$), sodium chloride (NaCl), glutaric acid (GA), and Rhodamine-590 aerosols. The refractive indices of the different aerosols were retrieved by comparing the measured extinction efficiency of each aerosol type to the extinction predicted by Mie theory. Aerosols composed of sodium chloride and glutaric acid in different mixing ratios were used as model for mixed aerosols of two nonabsorbing materials, and their extinction and complex refractive index were derived. Aerosols composed of Rhodamine-590 and ammonium sulphate in different mixing ratios were used as model for mixing of absorbing and nonabsorbing species, and their optical properties were derived. The refractive indices of the mixed aerosols were also calculated by various optical mixing rules and a core plus shell Mie model. We found that for nonabsorbing mixtures, the mixing rules calculations give comparable results, with the linear mixing rule giving a slightly better fit than the others. For absorbing mixtures, the differences between the refractive indices calculated using the mixing rules and those retrieved by CRD are generally higher.

The interaction between aerosols and radiation depends on their complex index of refraction, which is related to the particles' chemical composition. The contribution of light absorbing organic compounds, such as humic like substances (HULIS) to aerosol scattering and absorption is among the largest uncertainties in assessing the direct effect of aerosols on climate. Using a cavity ring down aerosol spectrometer (CRD-AS), the complex index of refraction of aerosols containing HULIS extracted from pollution, smoke, and rural continental aerosols, and molecular-weight fractionated fulvic acid was measured at 390 nm and 532 nm. The imaginary part of the refractive index (absorption) substantially increases towards the UV range with increasing molecular weight and aromaticity. At both wavelengths, HULIS extracted from pollution and smoke particles absorb more than HULIS from the rural aerosol. Sensitivity calculations for a pollution type aerosol containing ammonium sulfate, organic carbon (HULIS), and soot suggests that accounting for absorption by HULIS leads in most cases to a significant decrease in the single scattering albedo and to a significant increase in aerosol radiative forcing efficiency, towards more atmospheric absorption and heating, This indicates that HULIS in addition to black carbon in biomass smoke and pollution aerosols can contribute significantly to light absorption in the ultraviolet and visible spectral regions.

References

1. Ramanathan, V., Crutzen, P.J., Kiehl, J.T., and Rosenfeld, D., *Science*, **294**, 2119 (2001).
2. Kaufman, Y.J., Tanre, D., and Boucher, O., *Nature*, **419**, 215 (2002).
3. Lelieveld, J., Crutzen, P.J., Ramanathan, V., Andreae, M.O., Brenninkmeijer, C.A.M., Campos, T., Cass, G.R., Dickerson, R.R., Fischer, H., de Gouw, J.A., Hansel, A., Jefferson, A., Kley, D., de Laat, A.T.J., Lal, S., Lawrence, M.G.; Lobert, J.M., Mayol-Bracero, O.L., Mitra, A.P., Novakov, T., Oltmans, S.J., Prather, K.A., Reiner, T., Rodhe, H., Scheeren, H.A., Sikka, D., and Williams, J., *Science*, **291**, 1031 (2001).
4. Houghton, J.T. and Ding, Y., *Climate Change 2001: The Scientific Basis* (2001).
5. Andreae, M.O., Rosenfeld, D., Artaxo, P., Costa, A.A., Frank, G.P., Longo, K.M., and Silva-Dias, M.A.F., *Science*, **303**, 1337 (2004).
6. Andreae, M.O. and Crutzen, P.J., *Science*, **276**, 1052 (1997).
7. Tervahattu, H., Hartonen, K., Kerminen, V.M., Kupiainen, K., Aarnio, P., Koskentalo, T., Tuck, A.F., and Vaida, V., *J. Geophys. Res.*, **107**, art. no (2002).
8. Prospero, J.M., *Proceedings of Natl. Acad. Sci. USA*, **96**, 3396 (1999).
9. Kulmala, M., *Science*, **302**, 1000 (2003).
10. Zhang, Q., Worsnop, D.R., Canagaratna, M.R., and Jimenez, J.L., *Atmos. Chem. Phys.*, **5**, 3289 (2005).
11. Ellison, G.B., Tuck, A.F., and Vaida, V., *J. Geophys. Res.*, **104**, 11633 (1999).
12. Jacobson, M.C., Hansson, H.C., Noone, K.J., and Charlson, R.J., *Rev. Geophys.*, **38**, 267 (2000).
13. Kanakidou, M., Seinfeld, J.H., Pandis, S.N., Barnes, I., Dentener, F.J., Facchini, M.C., Van Dingenen, R., Ervens, B., Nenes, A., Nielsen, C.J., Swietlicki, E., Putaud, J.P., Balkanski, Y., Fuzzi, S., Horth, J., Moortgat, G.K., Winterhalter, R., Myhre, C.E.L., Tsigaridis, K., Vignati, E., Stephanou, E.G., and Wilson, J., *Atmos. Chem. Phys.*, **5**, 1053 (2005).
14. Went, F.W., *Nature*, **187**, 641(1960).
15. Samburova, V., Szidat, S., Hueglin, C., Fisseha, R., Baltensperger, U., Zenobi, R., and Kalberer, M., *J. Geophys. Res.*, **110** (2005).

16. Decesari, S., Facchini, M.C., Matta, E., Lettini, F., Mircea, M., Fuzzi, S., Tagliavini, E., and Putaud, J.P., *Atmos. Environ.*, **35**, 3691 (2001).
17. Krivacsy, Z., Kiss, G., Varga, B., Galambos, I., Sarvari, Z., Gelencser, A., Molnar, A., Fuzzi, S., Facchini, M.C., Zappoli, S., Andracchio, A., Alsberg, T., Hansson, H.C., and Persson, L., *Atmos. Environ.*, **34**, 4273 (2000).
18. Tervahattu, H., Juhanoja, J., and Kupiainen, K., *J. Geophys. Res.*, **107**, art. no (2002)
19. Hoffer, A., Gelencser, A., Guyon, P., Kiss, G., Schmid, O., Frank, G., Artaxo, P., and Andreae., M.O., *Atmos. Chem. Phys. Discuss.*, **5**, 7341–7360 (2005).
20. Graber, E.R. and Rudich, Y., *Atmos. Chem. Phys.*, **6**, 729 (2006).
21. Varga, B., Kiss, G., Ganszky, I., Gelencser, A., and Krivacsy, Z., *Talanta*, **55**, 561 (2001).
22. Kiss, G., Varga, B., Galambos, I., and Ganszky, I., *J. Geophys. Res.*, **107**, 8339 (2002).
23. Gysel, M., Weingartner, E., Nyeki, S., Paulsen, D., Baltensperger, U., Galambos, I., and Kiss, G., *Atmos. Chem. Phys.*, **4**, 35 (2004).
24. Kalberer, M., Paulsen, D., Sax, M., Steinbacher, M., Dommen, J., Prevot, A.S.H., Fisseha, R., Weingartner, E., Frankevich, V., Zenobi, R., and Baltensperger, U., *Science*, **303**, 1659 (2004).
25. Hoffer, A., Kiss, G., Blazso, M., and Gelencser, A., *Geophys. Res. Lett.*, **31**, L06115 (2004).
26. Gao, S., Ng, N.L., Keywood, M., Varutbangkul, V., Bahreini, R., Nenes, A., He, J.W., Yoo, K.Y., Beauchamp, J.L., Hodyss, R.P., Flagan, R.C., and Seinfeld, J.H., *Environ. Sci. Tech.*, **38**, 6582 (2004).
27. Dinar, E., Mentel, T.F., and Rudich, Y., *Atmos. Chem. Phys.*, *in press* (2006).
28. Dinar, E., Taraniuk, I., Graber, E.R., Anttila, T., Mentel, T.F., and Rudich, Y., *J. Geophys. Res.*, *in press* (2006).
29. Dinar, E., Taraniuk, I., Graber, E.R., Katsman, S., Moise, T., Anttila, T., Mentel, T.F., and Rudich, Y., *Atmos. Chem. Phys.*, **6**, 2465 (2006).

Internal Mixing of Organic and Elemental Components in DYCOMS-II Cloud Drop Activation

Lelia N. Hawkins, Lynn M. Russell[1], and Cynthia H. Twohy

Abstract Microphysical and chemical aerosol measurements were collected during the 2001 Dynamics and Chemistry of Marine Stratocumulus II (DYCOMS-II) research flights conducted 300 km off the coast of San Diego, California. Research flights were conducted during day and night in marine stratocumulus clouds lying within a stable boundary layer. For most flights, the mass concentration of compounds in bulk aerosol measurements and the residual mass of those compounds from cloud droplets were highly correlated ($R2 > 0.8$). This relationship is expected for particle populations of internally mixed aerosol particles. The experimental mean mass scavenging coefficient was found to be 0.33±0.08 in cloud by linear regression. Chemical composition of the submicron dry particles shows that particles collected in and below marine stratocumulus clouds during DYCOMS-II research flights consist mainly of sea salt with ammonium, sulfate, and organic compounds. Droplet composition data indicate little distinction between chemical components during activation and scavenging. In particular, organics compose a large fraction of both particle and droplet residual mass. Transmission and scanning electron microscopy (TEM and SEM) single particle analysis show that the composition of particles of diameter between 0.2 and 1.3 μm are internally mixed with sea salt. Most particles larger than 0.2 μm contained both reacted and unreacted sea salt. A comparison of Forward Scattering Spectrometer Probe (FSSP) data to Counter flow Virtual Impactor (CVI) data provided insight into the efficiency of the CVI to capture DYCOMS-II cloud droplets. In one flight, up to 47% of droplets were smaller than 9 μm and were therefore not seen by the inlet. The lack of dependence on fractional chemical composition provides evidence that for the particle sizes that activated in the clouds studied here, particles contained similar, internally mixed components such that size and composition, and hence CCN activity, did not vary independently.

[1]*Scripps Institution of Oceanography, University of California, La Jolla, California, USA*

College of Oceanic and Atmospheric Sciences, Oregon State University, Corvallis, USA

C.D. O'Dowd and P.E. Wagner (eds.), *Nucleation and Atmospheric Aerosols*, 1182–1184.

Keywords Droplet activation, chemical composition, cloud scavenging

Introduction

Here we present the results of the Dynamics and Chemistry of Marine Stratocumulus-II(DYCOMS-II) experiment from 2001. Physical properties such as entrainment and drizzle are detailed in Stevens et al. (2003). Those physical properties are not independent of the particle chemical composition. Twohy et al. (2005) explored the aerosol indirect effect of DYCOMS-II particles (0.1–3.0 μm) and determined that below-cloud aerosol concentrations were correlated with droplet number concentrations and anti-correlated with droplet size and drizzle concentrations. This study focuses on the physical, chemical, and CCN properties of a subset of those particles between 0.2 and 1.3 μm in diameter.

Results

The 0.2–1.3 μm particles sampled in and below cloud during DYCOMS-II activate to droplets as if they are internally mixed. The functional groups, mineral components, and trace metals sampled in the these particles in eastern Pacific stratocumulus-topped boundary layers provides an estimated 33% particulate mass scavenging ratio in droplets larger than 9 μm in diameter. The droplet residual mass is consistently around 33% of the particle mass, even for compounds of varying hygroscopicity and in flight 4 is sufficiently similar to the below-cloud particle composition to indicate internal mixtures. TEM and SEM data show that greater than 90% of particles above 0.26 μm are internal mixtures of sea salt, sulfate and nitrate. The lack of dependence of cutoff diameter on composition suggests that DYCOMS-II particles in this size range were internally mixed. The C-130 CVI missed particles below 9 μm in diameter while previous DYCOMS-II studies use measurements from droplets ranging from 2 to 47 μm, resulting in significant differences between those particles and the particles studied here. The size and chemical composition of aerosols dramatically impact their ability to act as CCN. For the particle compositions sampled during DYCOMS-II, there is sufficiently small variation in composition that the fraction of particle mass activated appears to depend exclusively on supersaturation and particle size.

Acknowledgments This work was supported by National Science Foundation under grant ATM01–04707and National Oceanic and Atmospheric Administration under grant NA17RJ1231. SEM data were graciously provided by Jim Anderson. We thank Monica Rivera and SteveMaria for assistance in sample collection and analysis and Jean-Louis Brenguier and Frederic Burnet for the flight 5 Fast-FSSP data. We also acknowledge NCAR and RAF, including Krista Laursen and Chris Webster for project and field assistance, respectively.

References

Stevens, B., Lenschow, D.H., Vali, G., Gerber, H., Bandy, A., Blomquist, B., Brenguier, J.-L., Bretherton, C.S., Burnet, F., Campos, T., Chai, S., Faloona, I., Friesen, D., Haimov, S., Laursen, K.,Lilly, D.K., Loehrer, S.M., Malinowski, S.P., Morley, B., Petters, M.D., Rogers, D.C., Russell, L.,Savic-Jovcic, V., Snider, J.R., Straub, D., Szumowski, J., Takagi, H., Thornton, D.C., Tschudi, M.,Twohy, C., Wetzel, M., and van Zanten, M.C., Dynamics and Chemistry of Marine Stratocumulus –DYCOMS-II, *Bull. Am. Met. Soc.*, **84**, 579–593 (2003).

Twohy, C.H., Petters, M.D., Snider, J.R., Stevens, B., Tahnk, W., Wetzel, M., Russell, L.M., and Burnet, F., Evaluation of the Aerosol Indirect Effect in Marine Stratocumulus Clouds: Droplet Number, Size, Liquid Water Path, and Albedo, *J. Geophys. Res.*, **110**, DOI 10.1029/2004JD005116 (2005).

Optical Particle Counter Measurement of Marine Aerosol Hygroscopic Growth

Jefferson R. Snider[1] **and Markus D. Petters**[2]

Abstract A technique is developed for the determination of the hygroscopic growth factor of dry particles with diameter between 0.3 and 0.6 μm and is applied to measurements made during the second Dynamics and Chemistry of Marine Stratocumulus (DYCOMS-II). Two optical particle counters (OPC) are utilized; one measures the aerosol size spectrum at ambient relative humidity and the other dries the particles prior to light scattering detection. Growth factors are based on measurements made in the region of the Mie scattering curve where scattered light intensity increases monotonically with particle diameter, i.e., D < 0.9 μm. Growth factors at approximately 90% ambient relative humidity in a marine air mass sampled over the eastern Pacific Ocean range between 1.4 and 1.7 and suggest that upwards of 30% of the particle mass is non-hygroscopic.

Introduction

The equilibrium size of an aerosol particle at a specified relative humidity, compared to its dry size, defines what is commonly called the hygroscopic growth factor (GF). Accurate knowledge of the growth factor is important in studies of visibility and direct aerosol forcing of climate. Growth factors are also linked to cloud condensation nuclei activity and as such help to constrain the first indirect effect of aerosol on climate. Most measurements of the GF have focused on particles smaller than 0.3 μm diameter and in this situation the hygroscopic tandem differential mobility analyzer (HTDMA) is the tool of choice. We propose a technique based on concurrent measurements of aerosol size spectra obtained from a forward scattering spectrometer probe (Baumgardner et al. 1992, hereafter the F300) and a passive cavity aerosol spectrometer probe (Liu et al. 1992, hereafter

[1] *University of Wyoming*

[2] *Colorado State University*

C.D. O'Dowd and P.E. Wagner (eds.), *Nucleation and Atmospheric Aerosols*, 1185–1189.

the PCASP). While the former detects particles at ambient RH, the latter dries particles to RH < 40% (Strapp et al. 1992). Results are presented for 31 constant-altitude circles flown within and above a summertime marine boundary layer.

Measurements

Field studies were conducted during seven research flights (RF1, RF2, RF3, RF4, RF5, RF7, and RF8) flown over the eastern Pacific during DYCOMS-II (Stevens et al. 2003). Measurements were made on the NCAR C-130 during July 2001 and the approximate location of the target area was 200 km west of San Diego, California. The time interval used to average the F300 and PCASP data was approximately 30 min. This corresponds to sampling conducted along one of many 60 km (diameter) constant-altitude circles flown by the C-130 above and below stratocumulus cloud layers during DYCOMS-II. Hereafter, we refer the 60 km circles as segments.

Three to five below-cloud segments, and one above-cloud segment, are available for each flight. Averaged values of RH ranged from 6% to 36% and from 70% to 96% during the above-cloud and below-cloud segments. To avoid particles produced by the impact and shattering of drizzle we selected a drizzle-free subset of the below-cloud segments.

The PCASP and F300 were mounted below the right wing of the C-130 and both used the signal-processing hardware and firmware developed by Droplet Measurement Technologies (Boulder, CO). The OPCs detect light scattered from particles crossing a helium–neon laser beam ($\lambda = 0.633\,\mu m$); we refer to this process as a transit event.

The F300 samples the particles unobtrusively and hence can be used in conjunction with a dry particle size measurement to derive the hygroscopic growth factor. Because particle warming that occurs within the PCASP, prior to a transit event, particles sampled from a humid boundary layer are detected at an RH estimated to be no larger than 40% (Strapp et al. 1992).

Bias in the particle sizing performed by the PCASP results in GF error limits that range between 13% overestimate and 15% underestimate. This error includes the effects of both residual water and particle shape on PCASP sizing accuracy. In addition, amplifier time response leads to particle undersizing in the F300; this error causes an 11% underestimate of the GF.

Hygrosopic Growth Factor

Spectral densities derived for the two OPCs were compared over a restricted diameter range; this was selected by identifying a section of the size spectrum with constant slope in a double-logarithmic presentation of spectral density versus particle diameter. Channels lying within the constant-slope sections are indicated by the small black and grey rectangles in the left panel of Figure 1; for the dry particles measured by the PCASP the diameter range is 0.3–0.6 µm. The averaged PCASP spectral densities ($f(D)$), restricted to the constant-slope section, were fitted to a power function of the form

$$f(D)_P = A_P \cdot D^C$$

where A_P and C are fit parameters. For the F300 data, a single parameter, A_F, was fitted with the "slope parameter" (C) prescribed by the PCASP fit

$$f(D)_F = A_F \cdot D^C$$

The hygroscopic growth factor was derived as

$$GF = (A_F / A_P)^{(1/C)}$$

Values of GF for the seven above-cloud segments are not expected to be substantially different from unity. A check of this expectation is reported here since it corroborates our assertion that the relative absolute bias in the GF measurements is less than 20%. With the exception of RF8, the GFs derived from the above-cloud segments plot within ± 10% of GF=1. Averaged above-cloud GF values are 1.0 ± 0.1 ($n = 7$).

Values of GF derived for the twenty four below-cloud segments are presented in the right panel of Figure 1. Apparent for most flights is GF increasing ambient RH, and that this increase is nearly consistent with one of three model predictions. For

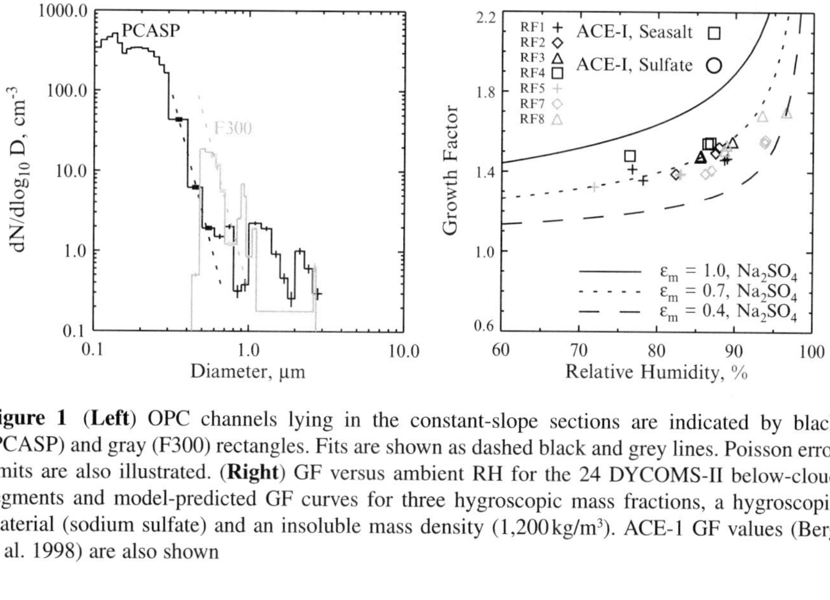

Figure 1 (**Left**) OPC channels lying in the constant-slope sections are indicated by black (PCASP) and gray (F300) rectangles. Fits are shown as dashed black and grey lines. Poisson error limits are also illustrated. (**Right**) GF versus ambient RH for the 24 DYCOMS-II below-cloud segments and model-predicted GF curves for three hygroscopic mass fractions, a hygroscopic material (sodium sulfate) and an insoluble mass density (1,200 kg/m³). ACE-1 GF values (Berg et al. 1998) are also shown

the model-predicted growth factors we assume the following: (1) the hygroscopic material is sodium sulfate, (2) hygroscopic mass fraction (ε_m) = 0.4, 0.7, or 1.0, and (3) non-hygroscopic mass density = 1,200 kg/m^3.

Also illustrated are HTDMA GF averages reported by Berg et al. (1998) for ship-based measurements made within an unpolluted marine air mass sampled south and southwest of Tasmania during the first Aerosol Characterization Experiment (ACE-1). The ACE-1 data is for particles of dry diameter 0.15 μm and is shown for GF modes corresponding to sulfate-dominant and sea salt-dominant particles. At RH = 90%, a relative humidity commonly used for comparing HTDMA GF measurements, the GF values we report plot 25% smaller than the value for the ACE-1 sulfate-dominant particles.

Discussion

The GF values we report for the ambient RH interval 87 to 93% range from 1.4 to 1.7. The right panel of Figure 1 demonstrates that these growth factors are significantly smaller than measurements made in a marine southern hemisphere air mass thought well-removed from the influence of continental aerosols (Berg et al. 1998). The comparison GF values (Berg et al. 1998) are 1.8 ± 0.1 and 2.1 ± 0.1 for growth modes attributed to sulfate- and sea salt-dominant particles, respectively (RH =90%). These results come from HTDMA measurements conducted on board a ship.

The ACE-1, and other ship-based HTDMA studies, show that the sea salt- to sulfate-dominant particle ratio is less than 0.1 for dry particle diameters less than 0.15 μm when sampling at wind speeds less than 10 m/s. These ship-based data correspond to measurements made at 10 m above sea level (a.s.l.). For our data set, restricted to measurements made at or above 95 m a.s.l., the segment-averaged wind speeds ranged between 5 and 12 m/s and were usually 7 m/s. Since the DYCOMS-II wind speeds at 95 m a.s.l. were typically smaller than 10 m/s we assert that sea salt-dominant particles contributed minimally to the count of particles sizing between the dry diameters 0.3 and 0.6 μm.

Since sea salt-dominant particles are thought to have made a minor contribution to the GF values we document, we conclude that the most appropriate comparator for our result is the GF reported for sulfate-dominant particles during ACE-1. Comparing our result to ACE-1, we arrive at two limiting interpretations of the growth factor data for DYCOMS-II : (1) After accounting for the amplifier time response of the F300 (11% GF underestimate, Section 2), and considering only data points with ambient RH values between 87% and 93%, our GF is 14% smaller than the ACE-1 result. This −14% bias is nearly consistent with the GF underestimate attributed to PCASP sizing inaccuracy (Section 2). We conclude, based on our GF measurements and model calculations that the sampled particles had a hygroscopicity consistent with that of pure sodium sulfate. (2) After accounting for the amplifier time response of the F300, and again considering only flight segments with RH values between 87% and 93%, we consider the 13% GF overestimation due to PCASP sizing inaccuracy (Section 2). For this interpretation the effects of correcting for F300 response time bias (+11%) and the PCASP sizing bias (−13%) nearly cancel. It follows from our GF data, and the model, that the hygroscopic mass fraction is $\varepsilon_m \sim 0.7$. The latter conclusion is consistent with submicron residual mass fractions (here presumed non-hygroscopic) reported for the Pacific Ocean latitude interval 40–60 °N and for marine air mass sampling conducted at Cheeka Peak, Washington (48 °N and 125 °W) (Quinn et al. 2000).

Summary

Hygroscopic growth factors derived from data acquired by two concurrently operated airborne OPCs are reported. The advantages of our approach are that it is based on instruments that are common to aerosol physics aircraft operated in North America and Europe and that it is now available for evaluation of the hygroscopic response of dry particles sizing between 0.3 and 0.6 μm. Results from the DYCOMS-II campaign suggest that a significant fraction of the particulate mass is non-hygroscopic.

References

Baumgardner, D. et al., Interpretation of measurements made by the forward scattering spectrometer probe (FSSP-300) during the airborne arctic stratospheric expedition, *J. Geophys. Res.*, **97**, 8035–8046 (1992).

Berg, O.H. et al., Hygroscopic growth of aerosol particles in the marine boundary layer over the Pacific and Southern Oceans during the First Aerosol Characterization Experiment (ACE 1), *J. Geophys. Res.*, **103**, 16,535–16,545 (1998).

Liu, P.S.K. et al., Response of particle measurement systems airborne ASASP and PCASP to NaCl and latex particles, *Aerosol Sci. and Technol.*, **16**, 83–95 (1992).

Quinn, P.K. et al., Surface submicron aerosol chemical composition: what fraction is not sulfate? *J. Geophys. Res.*, 105, 6785–6805 (2000).

Stevens, B. et al., Dynamics and chemistry of marine stratocumulus – DYCOMS-II, *Bull. Amer. Meteor. Soc.*, **83**, 579–593 (2003).

Strapp, J.W. et al., Hydrated and dried aerosol-size-distribution measurements from Particle Measuring Systems FSSP-300 probe and deiced PCASP-100X probe, *J. Atmos. Oceanic Technol.*, **9**, 548–555 (1992).

Aerosol–CCN Closure at a Semi-rural Site

R.Y.-W. Chang[1], P.S.K. Liu[2], W.R. Leaitch[2], and J.P.D. Abbatt[1]

Abstract Given the importance of cloud formation to climate, it is crucial that the aerosol-to-droplet activation process be well understood. In this study, aerosol–cloud condensation nuclei (CCN) closure was evaluated by comparing CCN concentrations measured using a CCN counter with concentrations predicted from aerosol size distributions and size-resolved aerosol compositions measured by an Aerodyne Quadrupole Aerosol Mass Spectrometer. The study took place in November 2005 at the Centre for Atmospheric Research Experiments, a semi-rural site located outside of Egbert, which is 80 km north of Toronto, Canada, and the objective was to characterize the sensitivity of our Köhler model to assumptions made about the organic component, such as water solubility and surface tension. Comparable CCN concentrations were obtained in the base case, which assumes that the organic fraction of the aerosol is insoluble and that the surface tension of water applies to the growing droplet. However, for specific periods during which the organic fraction dominated the aerosol mass, CCN concentrations were consistently underpredicted. Sensitivity analyses indicate that including a water-soluble organic component and/or reducing the surface tension is necessary to obtain the best agreement between modelled and measured CCN concentrations over all conditions in this study. A general conclusion is that significant uncertainties arise in predicting CCN levels using the base case only when the fraction of soluble inorganic species is below approximately 25% by mass.

Keywords Cloud condensation nuclei, closure study, organic aerosols, Köhler theory

[1]*Department of Chemistry, University of Toronto, 80 St. George St., Toronto, ON, M5S 3H6, Canada*

[2]*Science & Technology Branch, Environment Canada, 4905 Dufferin St., Toronto, ON, M3H 5T4, Canada*

C.D. O'Dowd and P.E. Wagner (eds.), *Nucleation and Atmospheric Aerosols*, 1190–1194.

Introduction

Aerosols can indirectly affect climate by acting as cloud condensation nuclei (CCN). The droplet activation of the inorganic aerosol mass has been well-characterized using Köhler theory, however, the role of the organic fraction is more difficult to assess. Aerosol–CCN closure studies, which compare measured and predicted CCN concentrations in the field, test our understanding of the parameters that affect droplet activation. Closure is achieved if the predicted and measured CCN concentrations are comparable. A series of successful studies conducted under varied aerosol types and loading but using the same Köhler model assesses the model's robustness and the appropriateness for inclusion in a climate model. The objective of this study was to conduct a closure study at a semi-rural site that is influenced by aged aerosols and not impacted by local sources, with a focus on determining how the organic fraction needs to be represented in the model in order to achieve the best degree of closure.

Closure Study

The sampling campaign took place in November 2005 at Environment Canada's Centre for Atmospheric Research Experiments (CARE) near Egbert, Ontario (44.23 N, 79.78 W), a semi-rural site approximately 80 km north of Toronto. The site was chosen because it is often the recipient of polluted aerosols from the south, as well as cleaner air masses from the north. In both cases, the ambient aerosols are relatively aged and are not dominated by local emissions.[1]

The instruments used in this study were an Aerodyne Quadrupole Aerosol Mass Spectrometer in particle time-of-flight mode, which measures the size-resolved aerosol mass concentrations of the non-refractory components; a TSI 3081 and 3010 Scanning Mobility Particle Sizer, which measures the aerosol size distribution; and a CCN counter, which measures the CCN concentration.

Droplet activation was modelled using Köhler theory, which was modified to generalize the inorganic component by assuming that each mole of sulphate and nitrate contributes 3 mol of ions and 2 mol of ions, respectively. A soluble organic component was also added for the sensitivity analysis. CCN concentrations were predicted by integrating the size distribution upward from the critical diameter, which was calculated from the size-resolved chemical composition data using the Köhler equation.

Results and Discussion

In the base case, the organics are completely insoluble and the surface tension of the growing droplet is that of water. Despite CCN concentrations spanning over two orders of magnitude, the correlation is high. However, data points with a high organic mass fraction were consistently underpredicted. The sensitivity of our Köhler model was tested to the major assumptions made about the properties of the aerosols by varying the input parameters. A summary of the corresponding changes in the predicted CCN concentration and the variance in the closure plots is given in Table 1, where the uncertainty in the slopes is approximately ± 20%. A change in the slope indicates the model's sensitivity to that parameter, while an increase in the linear correlation coefficient demonstrates that the changed parameter describes the aerosol population better during this study. The factors that were tested were the presence of a water-soluble organic carbon (WSOC) component, a reduction in the surface tension due to a surface-active species, and the use of a time-averaged, size-resolved chemical composition.

A WSOC component was included by modifying the Köhler equation. As in the study by Broekhuizen et al.,[2] this component was represented by a sparingly soluble diacid and the fraction of the organic mass that is water soluble was assumed to be constant at all sizes. Including a WSOC component selectively increases the data points with a higher organic fraction, which had been underpredicted in the base case. Overall CCN concentrations only increase by 9% for a 60% WSOC component, which is relatively low compared to an urban environment.[2] This is consistent with Köhler theory, which is generalized in Figure 1. This figure shows that CCN activity is relatively insensitive once the soluble inorganic mass fraction is greater than approximately 25%, as was the case at Egbert, whereas the CCN activity is more sensitive in an environment dominated by insoluble organics, such as an urban site.

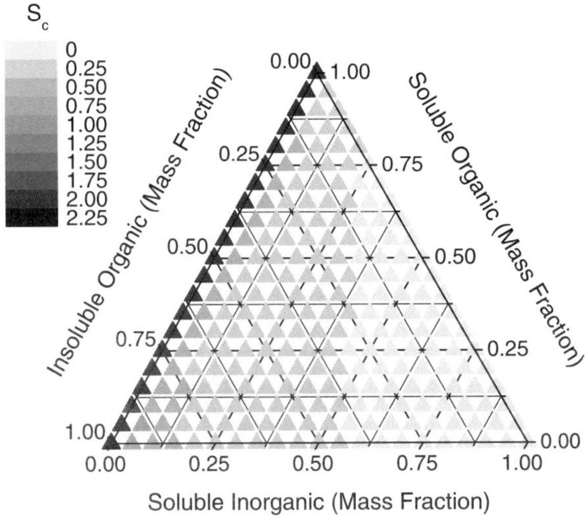

Figure 1 Critical Supersaturation (S_c) of a 90 nm Particle in a Three-Component Model. The three components are: a soluble inorganic (M = 0.132 kg mol^{-1}, ρ = 1770 kg m^{-3}, solubility = 0.43 kg kg^{-1}, ν = 3), an insoluble organic (M = 0.200 kg mol^{-1}, ρ = 1,000 kg m^{-3}), and a soluble organic (M = 0.146 kg mol^{-1}, ρ = 1360 kg m^{-3}, solubility = 0.02 kg kg^{-1}, ν = 1), where M is the molecular weight, ρ the density and ν the van't Hoff factor of the species

Table 1 Summary of sensitivity analysis

Condition	Slope	Intercept (cm^{-3})	r
Base case	0.86	−51	0.962
60% WSOC	0.94	−18	0.975
−20% Surface tension	1.15	−71	0.970
Average composition	0.71	73	0.968

The presence of a surface-active organic compound can lower the surface tension and decrease the critical supersaturation. As seen in Table 1, this increases the predicted CCN concentrations dramatically, as compared to the WSOC case. This is primarily because the surface tension of all the particles is assumed to decrease uniformly, regardless of composition. The surface tension could be reduced from 10% to 25% in polluted air,[3] however, it is difficult to know if this suppression is generally applicable. Our model is most sensitive to changes in the surface tension, which is in agreement with other closure studies.[4]

In light of a recent study which suggests that the average chemical composition is less important than the measured size distribution when predicting CCN concentrations,[5] the activation diameter corresponding to the average particle composition in the size range of activation diameters was used to predict CCN concentrations. This is shown in Table 1 as "Average Composition" and has the lowest slope. Although predicted CCN concentrations are more sensitive to changes in the size distribution and particle number density, the chemical composition cannot be discounted.[2,6]

Conclusions

The overall degree of closure is found to be within experimental uncertainties assuming that the organics are insoluble and that the droplet has the surface tension of water. However, during periods when the organic content of the aerosol is high, the predicted numbers of CCN using the base model are significantly lower than the measured values. The overall variance in the data set is reduced if the water-soluble fraction of the organic component of the aerosol is increased.

The general behaviour of the Köhler model is that the critical supersaturation will strongly depend on the degree of water solubility only at high insoluble organic fractions. Such situations will arise with traffic- and biogenic-dominated emissions. However, when a sizable fraction of soluble inorganic material is present, this sensitivity is diminished.

A surface tension reduction of 20% can have a strong effect on the number of predicted CCN. However, there is currently no easy way to predict the surface tension that is applicable. That being said, excellent closure is obtained by assuming the surface tension of water applies, with some degree of organic solubility, which could form the basis of a simple Köhler treatment that can be included in climate models.

Acknowledgments We acknowledge support from NSERC and CFCAS. We also thank Frank Froude and the staff at the CARE of Environment Canada for their help during the field study.

References

1. Rupakheti, M., Leaitch, W.R., Lohmann, U., Hayden, K., Brickell, P., Lu, G., Li, S.-M., Toom-Sauntry, D., Bottenheim, J.W., Brook, J.R., Vet, R., Jayne, J.T., and Worsnop, D.R., *Aerosol Sci. & Tech.*, **39**, 722–736 (2005).
2. Broekhuizen, K., Chang, R. Y.-W., Leaitch, W.R., Li, S.-M., and Abbatt, J.P.D., *Atmos. Chem. Phys.*, **6**, 2513–2524 (2006).
3. Facchini, M.C., Mircea, M., Fuzzi, S., and Charlson, R.J., *Nature*, **401**, 257–260 (1999).
4. Roberts, G.C., Artaxo, P., Zhou, J., Swietlicki, E., and Andreae, M.O., *J. Geophys. Res.*, **107**, doi:10.1029/2001JD000583 (2002).
5. Dusek, U., Frank, G.P., Hildebrandt, L., Curtius, J., Schneider, J., Walter, S., Chand, D., Drewnick, F., Hings, S., Jung, D., Borrmann, S., and Andreae, M.O., *Science*, **312**, 1375–1378 (2006).
6. Medina, J., Nenes, A., Sotiropoulou, R.-E.P., Cottrell, L.D., Ziemba, L.D., Beckman, P.J., and Griffin, R.J., *J. Geophys. Res.*, **112**, doi:10.1029/2006JD007588 (2007).

Characterization of Sesquiterpene Secondary Organic Aerosol: Thermodynamic Properties, Aging Characteristics, CCN Activity, and Droplet Growth Kinetic Analysis

A. Asa-Awuku[1], G. Engelhart[2], B.H. Lee[2], S. Pandis[2,3], and A. Nenes[1,4]

Abstract Secondary Organic Aerosols (SOA) can be a considerable source of ambient organic aerosol. The generation of SOA from different hydrocarbon precursors yields different low volatile particulates that could affect aerosol hygroscopicity, CCN activity, and cloud droplet growth kinetics. In this work we investigate SOA from monoterpene and sesquiterpene ozonolysis experiments and characterize the ability of the organic component to depress surface tension, act as CCN, and affect growth kinetics. In addition, we provide experimental evidence that the aging of SOA yields hygroscopic materials.

Keywords Tropospheric secondary organic aerosols, droplet growth kinetics, chemical composition, cloud droplet activation, sesquiterpenes, CCN, aging aerosols

Introduction

Aerosols have the ability to impact the earth's climate and hydrological cycle because of their potential to activate and become cloud droplets. The presence of these types of aerosols, cloud condensation nuclei (CCN), may enhance cloud reflectivity and life time. Carbonaceous particulate matter has been observed to comprise 70–90% of total aerosol mass[1,6,23]; the water soluble organic compounds (WSOC) may attribute 10–70% of the total organic fraction. WSOC can significantly affect the ability of CCN to interact with water vapor and form clouds, influence CCN hygroscopicity, surface tension, and possibly, droplet growth kinetics,[7,19,22] properties all central to assessing the aerosol-indirect effect.

[1]*School of Chemical and Bimolecular Engineering, Georgia Institute of Technology, USA*

[2]*Department of Chemical Engineering, Carnegie Mellon University, USA*

[3]*Department of Chemical Engineering, University of Patras, Greece*

[4]*School of Earth and Atmospheric Science, Georgia Institute of Technology, USA*

C.D. O'Dowd and P.E. Wagner (eds.), *Nucleation and Atmospheric Aerosols*, 1195–1199.

The WSOC component of CCN may be emitted as primary organic aerosol or formed in the atmosphere through oxidation processes with volatile organic carbon (VOC) to yield secondary organic aerosol (SOA). Natural VOC emissions (e.g., monoterpernes and sesquiterpenes), the major source of biogenic SOA, are estimated to be on the order of 1150 Tg per year,[12] and are believed to rival that of anthropogenic emissions.[10,11,15] As a consequence studies suggest that the SOA fraction can contribute significantly to the total organic particulate mass, at least 50% of the primary organic aerosol emissions,[15,21] and hence could potentially be a significant source of WSOC.

Additionally, it has been suggested that some secondary organics are oligomers[5,14] with characteristics similar to HULIS.[5] If this is the case, humic-like SOA may act as strong surfactants, depressing droplet surface tension and enhancing growth kinetics[4,9,16]; both of which can increase droplet number.[4] Yet little is known about the chemical composition of SOA, only roughly 10%[13,18,20] of the chemical identity of all organic mass is known. As a consequence the thermodynamic properties, solubility, surfactant characteristics, and droplet growth kinetics of organic aerosol properties necessary to constrain cloud droplet formation have remained elusive.[15] A robust description of important average thermophysical characteristics for SOA would provide better constraints for GCM parameterizations of the aerosol indirect effect.

In this study we experimentally investigate the in situ CCN activity of aging SOA generated in dark chamber ozonolysis chamber experiments. From filter samples obtained during these experiments we also investigate surfactant characteristics and droplet growth kinetics of the water-soluble component. Employing chemical composition and CCN activity data from our filter experiments we estimate thermodynamic properties (e.g., molar mass) using Kohler Theory Analysis (KTA).[17] The characterization of the aggregate organic component from SOA will aid in the improvement of our understanding and assessment of climate change.

Experimental and Analysis Methods

The SOA generated in this study was from unseeded dark ozonolysis reactions with a sesquiterpene biogenic hydrocarbon, beta-caryophyllene. Oxidation was carried out at 22°C up for up to 12 h. During the in situ experiments, the CCN activity of the SOA is measured throughout the whole aging experiment. Oxidation was carried out with and without the presence of a hydroxyl radical scavenger (2-butanol). To differentiate the CCN activity of volatile and nonvolatile components, the SOA was at times passed through a thermal denuder before introduction to aerosol sizing and CCN instruments (Figure 1). As shown in Figure 1, the sample aerosol is introduced into a TSI Scanning Mobility Particle Sizer (SMPS 3080) using a TSI 3081 Differential Mobility Analyzer; the concentration (TSI 3010 CPC) and CCN activity of the monodisperse aerosol is subsequently measured by a DH1 Static Diffusion CCN Counter and a Continuous-Flow Streamwise Thermal Gradient CCN Counter developed by

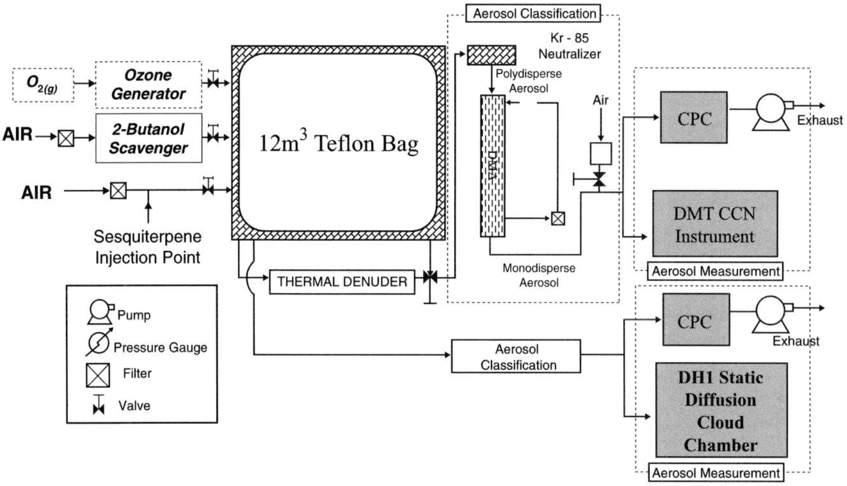

Figure 1 In situ experimental setup for SOA formation from sesquiterpenes

Droplet Measurement Technologies. For given supersaturations the ratio of CCN over CN is calculated, and cut off diameters (d_{50}) are identified, which correspond to the smallest dry particle size that activates at the instrument supersaturation.

We also characterized offline the chemical make-up of the water-soluble component of SOA using Kohler Theory Analysis.[4,17] SOA filter samples were collected for experiments without 2-butanol scavenger. The filters are extracted in water, atomized, dried, size selected, and introduced into the CCN counter. This procedure is repeated for varying mass fractions of SOA and $(NH_4)_2SO_4$. From the CCN activation of the filter sample experiments we infer the molecular weight of the water-soluble components,[4,17] probe for strong surfactants,[4] and infer characteristics of SOA droplet growth kinetics and compare them to $(NH_4)_2SO_4$.[2] We compare our experimental results with SOA from ozonolysis of terpinolene, 1-methlycycloheptene, and cylcoheptene.[3] In addition, we also compare beta-caryophyllene SOA with the behavior of SOA obtained from monoterpene sources that were similarly generated.

Initial Results and Discussion

Figure 2 presents the cutoff diameter of CCN as a function of aging, supersaturation, and the presence of OH. Clearly, the SOA age over time significantly and the CCN activity increases dramatically during the process. The presence of OH facilities activation, as CCN first appears a half hour into the experiment (see "1% no scavenger" data). In the absence of OH, CCN do not appear for 5 h; however beyond the sixth hour, the difference in CCN activity between the "no scavenger" and "with scavenger" SOA is very small. If the "no scavenger" SOA is introduced

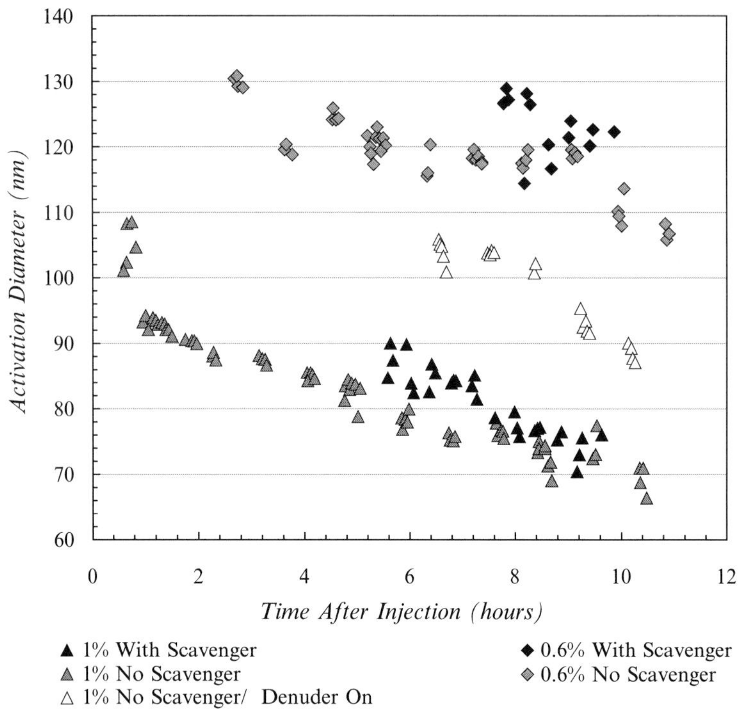

Figure 2 CCN Activity of beta-caryophyllene SOA

into the thermal denuder at 35°C (slightly higher than the operating temperature of the DMT CCN instrument), no CCN are measured for 6 h (Figure 2). This suggests that the oxidation products are volatile and that the ten degree increase in CCN instrument may partially evaporate the SOA, inhibiting their activation into cloud droplets. The DH1 CCN measurements support this as no time delay is seen and the instrument is operated close to room temperature. After sufficient aging, volatility decreases to the point where the CCN activity is consistent between the instruments.

The water soluble fraction of SOA extracted from SOA chamber samples may be hydrophilic and can behave similarly to the significantly aged aerosol during the in situ smog chamber experiments at CMU. In addition the WSOC may contain low molecular weight compounds observed in chemical speciations[8] and estimated from KTA.[17] In future work, investigation of SOA from other sesquiterpenes and aromatic hydrocarbons may indicate the presence of surface active oligomers and thus must be investigated. Applying a rigorous experimental approach to evaluate the CCN activity of SOA from various parent hydrocarbons will provide valuable chemical insight that will constrain the source of global WSOC from SOA.

Acknowledgments This work was supported by a National Science Foundation CAREER award and a NASA Earth System Science Fellowship.

References

1. Andreae, M.O. and Crutzen, P.J., *Science*, **276**, 1052–1058 (1997).
2. Asa-Awuku, A. and Nenes, A., in preparation.
3. Asa-Awuku, A., Nenes, A., Gao, S., Flagan, R.C., and Seinfeld, J.H., in preparation.
4. Asa-Awuku, A., Nenes, A., Sullivan, A., Hennigan, C.J., and Weber, R.J., *Atmospheric Chemistry and Physics and Discussions*, in review.
5. Baltensperger, U., Kalberer, M., Dommen, J., Paulsen, D., Alfarra, M.R., Coe, H., Fisseha, R., Gascho, A., Gysel, M., Nyeki, S., Sax, M., Steinbacher, M., Prevot, A.S.H., Sjoren, S., Weingartner, E., and Zenobi, R., *Faraday Discussions*, **130**, 265–278 (2005).
6. Cachier, H., Liousse, C., Buatmenard, P., and Gaudichet, A., *J. Atmos. Chem.*, **22**, 123–148 (1995).
7. Decesari, S., Facchini, M.C., Mircea, M., Cavalli, F., and Fuzzi, S., *J. Geophys. Res. Atmos.*, **108**, 4685 (2003).
8. Gao, S., Keywood, M., Ng, N.L., Surratt, J., Varutbangkul, V., Bahreini, R., Flagan, R. C., and Seinfeld, J.H., *Journal of Physical Chemistry A*, 10147–10164 (2004).
9. Graber, E.R. and Rudich, Y., *Atmos. Chem. Phys.*, **6**, 729–753 (2006).
10. Guenther, A., Archer, S., Greenberg, J., Harley, P., Helmig, D., Klinger, L., Vierling, L., Wildermuth, M., Zimmerman, P., and Zitzer, S., *Physics and Chemistry of the Earth Part B-Hydrology Oceans and Atmosphere*, **24**, 659–667 (1999).
11. Guenther, A., Geron, C., Pierce, T., Lamb, B., Harley, P., and Fall, R., *Atmos Environ*, **34**, 2205–2230 (2000).
12. Guenther, A., Hewitt, C. N., Erickson, D., Fall, R., Geron, C., Graedel, T., Harley, P., Klinger, L., Lerdau, M., McKay, W. A., Pierce, T., Scholes, B., Steinbrecher, R., Tallamraju, R., Taylor, J., and Zimmerman, P., *J. Geophys. Res. Atmos.*, **100**, 8873–8892 (1995).
13. Kalberer, M., *Analytical and Bioanalytical Chemistry*, **385**, 22–25 (2006).
14. Kalberer, M., Paulsen, D., Sax, M., Steinbacher, M., Dommen, J., Prevot, A. S. H., Fisseha, R., Weingartner, E., Frankevich, V., Zenobi, R., and Baltensperger, U., *Science*, **303**, 1659–1662 (2004).
15. Kanakidou, M., Seinfeld, J. H., Pandis, S. N., Barnes, I., Dentener, F. J., Facchini, M. C., Van Dingenen, R., Ervens, B., Nenes, A., Nielsen, C. J., Swietlicki, E., Putaud, J. P., Balkanski, Y., Fuzzi, S., Horth, J., Moortgat, G. K., Winterhalter, R., Myhre, C. E. L., Tsigaridis, K., Vignati, E., Stephanou, E. G., and Wilson, J., *Atmos. Chem. Phys.*, **5**, 1053–1123 (2005).
16. Kiss, G., Tombacz, E., and Hansson, H. C., *J. Atmos. Chem.*, **50**, 279–294 (2005).
17. Padro, L. T., Asa-Awuku, A., Morrison, R., and Nenes, A., *Atmospheric Chemistry and Physics Discussions in review*.
18. Rogge, W. F., Mazurek, M. A., Hildemann, L. M., Cass, G. R., and Simoneit, B. R. T., *Atmospheric Environment Part a-General Topics*, **27**, 1309–1330 (1993).
19. Saxena, P., and Hildemann, L. M., *J. Atmos. Chem.*, **24**, 57–109 (1996).
20. Seinfeld, J. H., and Pandis, S. N., *Atmospheric Chemistry & Physics: From Air Pollution to Climate Change*: John Wiley & Sons. (1998).
21. Seinfeld, J. H., and Pankow, J. F., *Annual Review of Physical Chemistry*, **54**, 121–140 (2003).
22. Shulman, M. L., Jacobson, M. C., Carlson, R. J., Synovec, R. E., and Young, T. E., *Geophys. Res. Lett.*, **23**, 277–280 (1996).
23. Yamasoe, M. A., Artaxo, P., Miguel, A. H., and Allen, A. G., *Atmos Environ*, **34**, 1641–1653 (2000).

Cloud Condensation Nucleus Activity of Secondary Organic Aerosol Particles Mixed with Sulfate

Stephanie M. King, Thomas Rosenoern, John E. Shilling, Qi Chen, and Scot T. Martin

Abstract Cloud condensation nuclei (CCN) activity of mixed organic-sulfate particles was investigated using a steady-state environmental chamber. The organic component consisted of secondary organic aerosol (SOA) generated in the dark from 22 ppb α-pinene at conditions of 300 ppb ozone and 40% relative humidity at 20°C. CCN analysis was performed for 80 to 150 nm particles having variable SOA–sulfate volume fractions. AMS measurements also determined an effective SOA density of $1.4 \pm 0.1\, g\, cm^{-3}$. Critical supersaturation, which increased for greater SOA volume fraction and smaller particle diameter, was well predicted by a two-component Köhler model that used ammonium sulfate and SOA as the two components and an effective molecular weight of $230\, g\, mol^{-1}$ for the SOA component. Results from this study further imply that, for the range of conditions studied, the particles may not reach a non-liquid state even at very low water activities, which suggests that the effect of limited solubility may not be a necessary consideration in the parameterization of cloud droplet formation in global climate models.

Keywords Cloud condensation nuclei, secondary organic aerosol, α -pinene

Introduction

In ambient particulate matter, organic compounds are often internally mixed with inorganic species, typically sulfates, and consideration of the inorganic fraction alone in atmospheric particles does not sufficiently explain activation (Novakov and Penner 1993; Facchini et al. 1999). Consequently, several laboratory studies have investigated the CCN activity of internally mixed particles that included both inorganic and organic species (Cruz and Pandis 1998; Raymond and Pandis 2003; Shantz et al. 2003; Bilde and Svenningsson 2004; Broekhuizen et al. 2004; Lohmann et al. 2004; Henning et al. 2005). However, the organic fraction in these studies consisted of at most a few organic species, whereas SOA in the atmosphere

Harvard University

C.D. O'Dowd and P.E. Wagner (eds.), *Nucleation and Atmospheric Aerosols*, 1200–1204.

may contain thousands. To address this phenomenon, CCN properties of SOA have been investigated using environmental chambers, with the focus thus far on pure SOA (Hartz et al. 2005; VanReken et al. 2005; Prenni et al. 2007). In the study described in this paper, we provide additional atmospheric relevance through a chamber investigation of internally mixed SOA–sulfate particles and Köhler model parameters that accurately describe the observed CCN behavior.

Experimental

Experiments were carried out in an environmental chamber that consists of a constant temperature room housing a flexible Teflon bag of approximately 5 m^3. The chamber is operated in a feedback-controlled dynamic mode, for which the flow in equals the flow out (i.e., a steady-state volume). Throughout the duration of these experiments, temperature, RH, and ozone concentration were maintained at 20°C, 40% RH, and 300 ppb, respectively. Dry $(NH_4)_2SO_4$ seed particles were introduced into the bag using an injection system based on a TSI model 3076 atomizer. Commercially obtained pure α-pinene (Aldrich) was introduced using a syringe pump that injected a mixture of 22 ppb α-pinene and 23 ppm 1-butanol (1:600 by volume) into a bulb that was flushed with clean air. Upon exit from the environmental chamber, the flow was split for simultaneous sampling by an Aerodyne high-resolution time-of-flight aerosol mass spectrometer (HR-ToF-AMS) and a Droplet Measurement Technologies continuous-flow streamwise thermal gradient cloud condensation nucleus counter (CCNC).

Multiple-Component Köhler Model

Modeling of the CCN activity of internally mixed sulfate and SOA particles was based on Köhler theory (Köhler 1936). The vapor pressure of water over an aqueous droplet can be expressed as:

$$s = a_w \exp\left(\frac{4\sigma M_w}{RT\rho_w d_{aq}}\right), \tag{1}$$

where s is the saturation ratio relative to a flat surface of liquid water, a_w is the water activity of the solution, σ is the solution-vapor surface tension, M_w is the molecular weight of pure water, R is the universal gas constant, T is the solution temperature, ρ_w is the density of pure water, and d_{aq} is the aqueous particle diameter. The water activity for a multicomponent solution is given by:

$$a_w = \frac{n_w}{n_w + \sum_i \nu_i n_i}, \tag{2}$$

where n_w is the number of moles of water, n_i is the number of moles of species i in solution, and ν_i is the corresponding van't Hoff factor. The possibility of limited solubility can be taken into account when determining the number of moles of species i by using an expression analogous to that of Henning et al. (2005):

$$n_i = \min\left[\frac{\left(d_{aq}^3 - d_0^3\right)\rho_w C_{sat,i}}{M_i}, \frac{\varepsilon_i d_0^3 \rho_i}{M_i}\right]\frac{\pi}{6}, \quad (3)$$

where $C_{sat,i}$ is the saturation concentration of species i (g g^{-1} H_2O), ε_i is the volume fraction in the dry particle, ρ_i is the density, d_0 is the diameter of the dry particle, and M_i is the molecular weight. The maximum supersaturation (s_{max}) in the Köhler curve resulting from Eqs. (1–3) defines the critical supersaturation as $S_c = s_{max} - 1$.

Results

CCN activity was investigated for particle mobility diameters of 80, 100, 120, and 150 nm, each of which varied in organic volume fraction according to the corresponding sulfate seed particle diameter. S_c values for each organic volume fraction (ε^d_{SOA}) are shown in Figure 1. The measured S_c values were used to constrain a least-squares fit of a two-component model that uses Eqs. (1–3). The model, consisting of ammonium sulfate as one chemical component and SOA approximated as one effective chemical component, was employed in a global optimization for the entire data set (solid lines in Figure 1).

Since SOA is a mixture that consists of many organic species, effective values for the organic component parameters were used (Table 1). Effective density (ρ_{SOA}) was determined as 1.4 g cm^{-3} by comparing the mode of the vacuum-aerodynamic size distribution to the mode of the mobility-size distribution and assuming a spherical particle shape, including a correction of the density of ammonium sulfate. The effective van't Hoff factor (ν_{SOA}) and surface tension (σ) were assumed to be 1 and 0.0725 N m^{-1}, respectively (Hartz et al. 2005). The parameters that were numerically optimized were the effective molecular weight of the organic component (M_{SOA}) and the effective saturation concentration of the organic component ($C_{sat,SOA}$).

A comparison of the observed S_c values to the two-component model S_c values including a 1:1 line is shown in the inset of Figure 1. Results of the fitting procedure also revealed that the chamber particles behaved as though they were fully solvated at activation, up to the largest measured organic fraction, which implies that consideration of the state of internal mixing within the particles, e.g., coated or fully mixed, is unnecessary for this study. No inflection point is apparent in the observed S_c curve in Figure 1, which would be expected if solubility is a limiting factor (Henning et al. 2005). Therefore, the calculated $C_{sat,SOA}$ value is the lower limit of the effective SOA solubility (Hartz et al. 2005). However, based on our current knowledge of the composition of α-pinene SOA, the effective solubility should be below the lower limit. To explain this apparent contradiction, we suggest that this

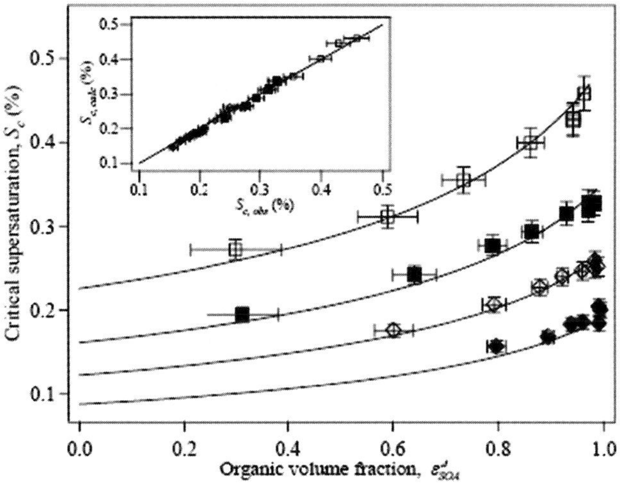

Figure 1 Critical supersaturation (S_c) for 50% CCN activation. Data are shown for four-particle mobility diameters as a function of organic volume fraction (ε^d_{SOA}). Symbols represent measured values. Lines represent modeled values using one set of globally optimized parameters (cf. Table 1). Lines for each diameter are drawn from $\varepsilon^d_{SOA} = 0$ to the largest measured ε^d_{SOA}. Error bars are 95% confidence intervals. **Inset**: Comparison of modeled S_c to observed S_c. Fitting these data with a 1:1 line (**solid line**) yields an r-squared value of 0.99.

Table 1 Parameters used in Kohler model to calculate S_c

Parameter	Value
Surface tension, σ	0.0725 N m^{-1}
van't Hoff factor of $(NH_4)_2SO_4$, ν_{AS}	2.2
Effective SOA density, ρ_{SOA}	1.4 g cm^{-3}
Effective SOA van't Hoff factor, ν_{SOA}	1
Effective SOA molecular weight, M_{SOA}	230 g mol^{-1}
Saturation concentration (lower limit) of SOA, $C_{sat,SOA}$	0.085 g g^{-1} H_2O

behavior may be explained either by the hypothesis that the particles are supersaturated solutions of organic material (Bilde and Svenningsson 2004), or by the hypothesis that the particles are liquids (or amorphous solids) even at very low water activities. The latter hypothesis is further supported by a previous study of internally mixed particles, which found that the liquid state is the thermodynamically stable phase of particles containing a sufficiently high number of miscible components (Marcolli et al. 2004).

Acknowledgments This material is based upon work supported by the National Science Foundation under Grant No. ATM-0513463. SMK acknowledges support from the EPA STAR fellowship program. TR acknowledges support from the Danish Agency for Science Technology and Innovation under Grant No. 272–06–0318.

References

Bilde, M. and Svenningsson, B., CCN activation of slightly soluble organics: the importance of small amounts of inorganic salt and particle phase, *Tellus Ser. B-Chem. Phys. Meteorol.*, **56**(2), 128–134 (2004).

Broekhuizen, K., Kumar, P.P., and Abbatt, J.P.D., Partially soluble organics as cloud condensation nuclei: role of trace soluble and surface active species, *Geophys. Res. Lett.*, **31**(1), L01107, doi:10.1029/2003GL018203 (2004).

Cruz, C.N. and Pandis, S.N., The effect of organic coatings on the cloud condensation nuclei activation of inorganic atmospheric aerosol, *J. Geophys. Res.-Atmos.*, **103**(D11), 13111–13123 (1998).

Facchini, M.C., Mircea, M., Fuzzi, S., and Charlson, R.J., Cloud albedo enhancement by surface-active organic solutes in growing droplets, *Nature*, **401**(6750), 257–259 (1999).

Hartz, K.E.H., Rosenorn, T., Ferchak, S.R., Raymond, T.M., Bilde, M., Donahue, N.M., and Pandis, S.N., Cloud condensation nuclei activation of monoterpene and sesquiterpene secondary organic aerosol, *J. Geophys. Res.-Atmos.*, **110**, D14208, doi:10.1029/2004JD005754 (2005).

Henning, S., Rosenorn, T., D'Anna, B., Gola, A.A., Svenningsson, B., and Bilde, M., Cloud droplet activation and surface tension of mixtures of slightly soluble organics and inorganic salt, *Atmos. Chem. Phys.*, **5**, 575–582 (2005).

Köhler, H., The nucleus in and the growth of hygroscopic droplets, *Trans. Faraday Soc.*, **32**(2), 1152–1161 (1936).

Lohmann, U., Broekhuizen, K., Leaitch, R., Shantz, N., and Abbatt, J., How efficient is cloud droplet formation of organic aerosols? *Geophys. Res. Lett.*, **31**, L05108, doi:10.1029/2003GL018999 (2004).

Marcolli, C., Luo, B.P., and Peter, T., Mixing of the organic aerosol fractions: liquids as the thermodynamically stable phases, *Journal of Physical Chemistry A*, **108**(12), 2216–2224 (2004).

Novakov, T. and Penner, J.E., Large Contribution of Organic Aerosols to Cloud-Condensation-Nuclei Concentrations, *Nature*, **365**(6449), 823–826 (1993).

Prenni, A.J., Petters, M.D., Kreidenweis, S.M., and DeMott, P.J., Cloud droplet activation of secondary organic aerosol, *J. Geophys. Res.*, in press (2007).

Raymond, T.M. and Pandis, S.N., Formation of cloud droplets by multicomponent organic particles, *J. Geophys. Res.-Atmos.*, **108**(D15), 4469, doi:10.1029/2003JD003503 (2003).

Shantz, N.C., Leaitch, W.R., and Caffrey, P.F., Effect of organics of low solubility on the growth rate of cloud droplets, *J. Geophys. Res.-Atmos.*, **108**(D5), 4168, doi:10.1029/2002JD002540 (2003).

VanReken, T.M., Ng, N.L., Flagan, R.C., and Seinfeld, J.H., Cloud condensation nucleus activation properties of biogenic secondary organic aerosol, *J. Geophys. Res.-Atmos.*, **110**, D07206, doi:10.1029/2004JD005465 (2005).

Particle Size Critical Supersaturation Relationships

James G. Hudson and C.F. Rogers

Abstract A small degree of variability of size-Sc measurements has been cited as evidence that CCN might be deduced from particle size measurements alone. However, we present size-critical supersaturation (Sc) measurements with a greater range of variability that casts doubt on this concept.

Keywords aerosol, CCN

Introduction

Cloud condensation nuclei (CCN) size is determined by passing an aerosol through a differential mobility analyzer (DMA) and then to a CCN spectrometer.[1] This separates particle composition (chemistry) from particle size. A small range of variability of particle size-S_c measurements suggested that CCN might be adequately deduced from particle size measurements.[2] Here we present a more extensive set of size-S_c measurements that shows significantly more variability. The measurement methods have been described previously.[3]

Results

Table 1 shows the airborne CCN measurements[1] that were obtained in three different field projects. The first six rows are from the AIRS2 project in continental/polluted air masses, the next four rows are from the RICO project in clean maritime air, and the last three rows are from the MASE project in polluted air over the ocean below and above stratus clouds. Figure 1 shows a composite of these data.

Desert Research Institute, Nevada System of Higher Education, Reno, Nevada 89512-1095

C.D. O'Dowd and P.E. Wagner (eds.), *Nucleation and Atmospheric Aerosols*, 1205–1208.

Figure 1 Measured mean critical supersaturations (S_c) for various mean dry particle sizes. **Solid line** is the theoretical size-S_c relationship for NaCl, **dotted line** is that for ammonium sulfate. **Triangles** with error bars are average and standard deviations of all data in Table 1 at four diameters. **Squares** are minimal and dots are maximal measurements.

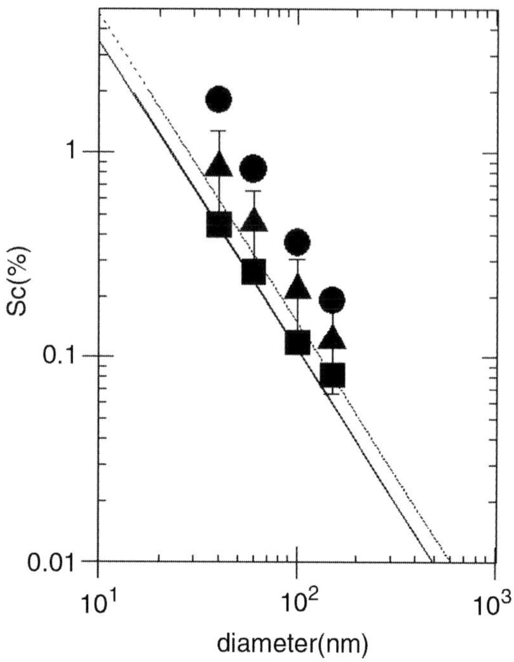

In order to relate dry particle size, r_d, and S_c of any particle, Fitzgerald et al.[4] put a dimensionless solubility parameter, B, into the Köhler equation

$$S = A/r - Br_d^3/r^3 \tag{1}$$

When r_d is expressed in nanometers

$$B \sim 1.94 x 10^3 \, r_d^{-3} \, S_c^{-2} \tag{2}$$

Lines of constant B are parallel; B(NaCl) = 1.23 and B(ammonium sulfate) = 0.70 are shown in Figure 1. Lower B represents less soluble substances.

Figure 2 suggests a possible relationship between B and CN concentration that is similar to Figure 12 of Hudson and Da .[1] Figure 2 generally supports Hudson and Da[1] that the r_d-S_c relationship usually more closely resembles that of inorganic soluble species in cleaner air masses (i.e., RICO, rows 7–10, Table 1). In more polluted air masses, dry particles with the same S_c are larger.

Discussion

Particles with 60 nm r_d had an S_c range of 0.5–0.7%,[2] but we found more than twice this range. Our B range for these particles was six times higher.[2] They showed a 17 nm r_d range for 0.4% S_c particles but we found a 49 nm range. Our B range for these

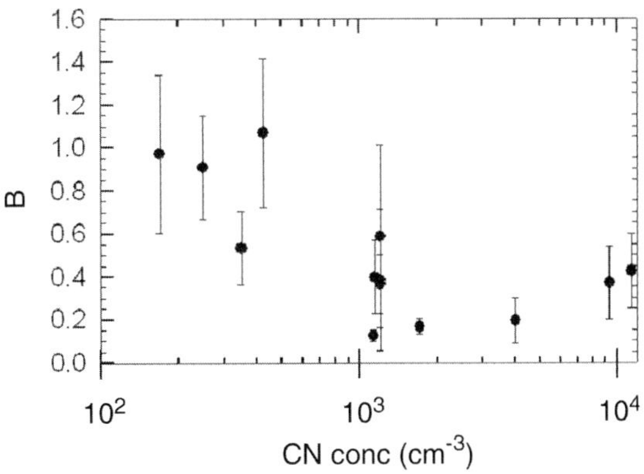

Figure 2 Average and standard deviation of the solubility parameter, B, plotted against the simultaneous average CN concentration during each set of size-S_c measurements

Table 1 Solubility parameter, B, for the 13 sets of size-S_c measurements. Later rows show the average, standard deviations, and ranges of all the measurements. The bottom rows show the B values for sodium chloride and ammonium sulfate

Date	Alt (m)	Location	B (60 nm)	B (0.4%)	Ave and SD of B	B range
Nov. 24,03	600	Syracuse, NY	0.17	0.16	0.19±0.10	0.37
Nov. 24,03	1,550	Syracuse, NY	0.94	0.37	0.58±0.43	1.43
Nov. 25,03	600	Lake Huron	0.13	0.19	0.38±0.33	0.79
Dec. 4,03	1,000	Lake Huron	0.17	0.15	0.16±0.04	0.05
Dec. 4,03	350	Lake Huron	0.34	0.26	0.37±0.17	0.51
Dec. 4,03	350	Lake Huron			0.42±0.17	0.32
Jan. 19,05	400	Caribbean	0.98	1.11	0.97±0.37	1.10
Jan. 24,05	850	Caribbean	0.93	1.00	0.91±0.24	1.05
Jan. 24,05	400	Caribbean	0.97	0.88	1.07±0.35	1.19
Jan. 24,05	100	Caribbean	0.57	0.59	0.53±0.17	0.52
July 25,05	100	California	0.32	0.31	0.40±0.17	0.69
July 25,05	400	California	0.11	0.12	0.12±0.03	0.08
July 25,05	100	California	0.34	0.34	0.36±0.14	0.47
Average			0.50	0.46	0.50	0.66
SD			0.36	0.35	0.31	0.43
Range			0.86	0.99	0.95	1.43
NaCl			1.23	1.23	1.23	
Ammonium Sulfate			0.70	0.70	0.70	

particles is nearly six times higher.[2] The range of average B for the 13 measurement sets (0.95) is five times more than all of Dusek et al.[2] Probably the main reason for the difference was the lack of cleaner air masses.[2] Most of the greater range was for higher B in cleaner air masses that approach that of soluble salts. Nevertheless, there are frequent continental outflows (e.g., MASE, rows 11–13 of Table 1) that result in a large variability over the ocean; compare with RICO (rows 7–10 of Table 1).

Conclusions

The conclusion that size measurements alone could be used to accurately estimate CCN depended on the narrow range of r_d-S_c summertime surface measurements in Germany.[2] Their range is characteristic of polluted air masses and not at all like cleaner air masses.

On the other hand they also said it might be necessary to classify air masses before deducing CCN from size measurements.[2] This admits greater r_d-S_c variability such as shown here. If air mass classification is needed it is an admission of the importance of particle chemistry vis-à-vis particle size. They also seemed to say that although there may be different ranges of r_d-S_c in different air masses, there might be small enough variability in r_d-S_c within each air mass so that particle size measurements might be sufficient to estimate CCN.[2] The apparent systematic variability in particle solubility, B, with air mass (Figure 2) suggests that the large overall r_d-S_c range might give way to smaller ranges within given air masses as was found in Germany.[2] This suggests hope that size measurements might produce sufficiently accurate CCN estimates within individual air masses as long as consistent limited r_d-S_c relationships can be verified.[2] It appears that this probably has a greater likelihood in polluted air masses, but is less likely in mixed air masses such as may be found over the ocean. Since these are the very situations that are of most importance for understanding the Indirect Aerosol Effect this does not bode well for the practicality of deducing CCN from particle size measurements.

Acknowledgments Support for the AIRS2 and RICO measurements was from the US National Science Foundation grants ATM-0313899 and ATM-0342618. Support for the MASE measurements was from the Atmospheric Sciences Program of the US Department of Energy grant DE-FG02-05ER63999. The aircraft in AIRS2 and RICO was the NCAR RAF C-130 and the MASE aircraft was the Battelle G1.

References

1. Hudson, J.G., *J. Atmos. & Ocean. Techn.*, **6**, 1055–1065 (1989).
2. Dusek, U., Frank, G.P., Hildebrandt, L., Curtius, J., Schneider, J., Walter, S., Chand, D., Drewnick, F., Hings, S., Jung, D., Borrmann, S., and Andreae, M.O., *Science*, **312**, 1375–1378 (2006).
3. Hudson, J.G. and Da, X., *J. Geophys. Res.*, **101**, 4435–4442 (1996).
4. Fitzgerald, J.W., Hoppel, W.A., and Vietti, M.A., *J. Atmos. Sci.*, **39**, 1838–1852 (1982).

Aerosol Microphysics in the GISS Climate Model

S.E. Bauer[1,2], D. Wright[3], D. Koch [1,2], S. Menon[4], and R. McGraw[3]

Abstract We are developing aerosol and cloud microphysical schemes within the Goddard Institute for Space Studies (GISS) coupled aerosol-climate model. The quadrature method of moments scheme allows a computationally efficient description of aerosol size distribution and mixing state. New cloud droplet nucleation and hydrometeor spectral schemes will depend upon the aerosol microphysical size and solubility information. Our ultimate objective is to improve our simulation of aerosol impacts on climate. Aerosol mixtures affect aerosol solubility, lifetime, and radiative forcing. Aerosol number and solubility in turn affect cloud properties. This paper presents first preliminary results regarding simulated aerosol mass and number concentrations, in dependence of the chosen aerosol model modes and mixing processes.

Keywords Aerosol microphysics, global climate model, auadrature method of moments

Introduction

The treatment of aerosol properties and processes in present global climate models is highly simplified. In the next generation of global aerosol models, aerosol properties must accurately represent mass and number concentration, particle size and size-dependent composition, optical properties, solubility, and ability to serve as nuclei for cloud particles. To accommodate a better representation of aerosols in the

[1]*The Earth Institute at Columbia University, 2880 Broadway, New York, NY 10025, USA*

[2]*NASA-Goddard Institute for Space Studies, 2880 Broadway, New York, NY 10025, USA*

[3]*Brookhaven National Laboratory, Building 815E, 75 Rutherford Drive, Upton, NY 11973–5000, USA*

[4]*Lawrence Berkeley National Laboratory, 1 Cyclotron Road, MS 51R208, Berkeley, CA 94720, USA*

C.D. O'Dowd and P.E. Wagner (eds.), *Nucleation and Atmospheric Aerosols*, 1209–1213.

GISS GCM, we develop a new aerosol microphysical scheme and use the quadrature method of moments (QMOM) to describe aerosol distributions.

The Aerosol Microphysics Scheme

The QMOM provides a statistically based alternative to modal and sectional methods for aerosol simulation. Key moments of the aerosol population, including number, mass, and mixed moments entering the covariance matrix of a principal components analysis, are tracked in place of the distribution itself. This makes the new approach highly efficient, yet provides the comprehensive representation of natural and anthropogenic aerosols, and of their mixing states and direct and indirect effects, that the GISS GCM requires. We are gradually developing the QMOM, beginning with the tracking of only two moments, particle number and mass. Extension to modes represented with 3–6 moments is planned [1].

The Microphysical aerosol scheme that is used in our global climate model model is called MATRIX (Multiconfiguration Aerosol TRacker of mIXing state), a new aerosol microphysical module. It is designed to support model calculations of the indirect effect and permit detailed treatment of aerosol activation and cloud formation. MATRIX comprises a family of several alternative models, or aerosol "mechanisms", each representing a choice of 8–16 internally mixed particle populations or modes. For each mode the tracked species are number concentration and mass concentrations of sulfate, nitrate, ammonium, water, black carbon, organic carbon, mineral dust, and sea salt. Each mechanism includes modes representing sulfate, insoluble mineral dust, soluble mineral dust, sea salt, insoluble black carbon,

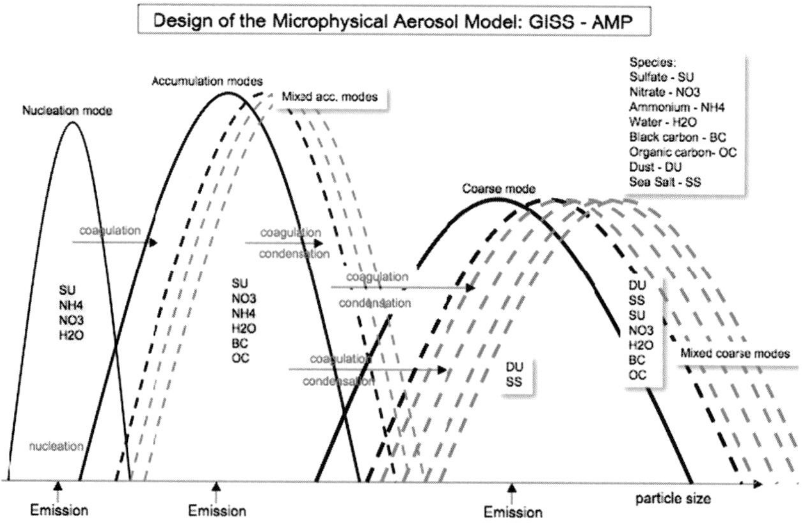

Figure 1 Design of the microphysical aerosol model MATRIX

soluble black carbon, organic carbon, and a mixed mode containing all aerosol chemical species. Most mechanisms also include additional modes including those formed by coagulation among these populations to better characterize the mixing state of the aerosol. MATRIX dynamics represent new particle formation [2,3,4,5,6,7], particle emissions, gas-particle mass transfer, aerosol-phase chemistry, condensational growth, and coagulation within and between modes.

The thermodynamical model EQSAM [8], to calculate the gas/liquid/ solid partitioning of the sulfate-nitrate-ammonium system (H2SO4/HSO4-/SO2–4 - HNO3 /NO3- – NH3/NH4 + –HCL/CL- – Na + Ca2 + – Mg2 + – K + – H2O) is implemented into MATRIX. MATRIX includes as a further option the much more complex ISOROPIA [9] aerosol chemistry scheme for nitrate formation; however it is also computationally expensive.

The new cloud microphysics in the GISS GCM are an overall improvement to the cloud microphysical scheme, due to better account for aerosol–cloud interactions via the implementation of prognostic equations for cloud droplet number and ice crystal number and a two-moment bulk cloud scheme. The 2Mom scheme [10], predicts the number concentration and mixing ratio of four hydrometeors (cloud drop, cloud ice, rain, snow) and the interactions between them (e.g., melting, freezing, collection, sedimentation, autoconversion, accretion, evaporation, condensation, deposition, sublimation).

Preliminary Model Results

A preliminary simulation is carried out with the most complex mechanism (16 modes and 51 species, e.g., mixtures). Results are shown in Figure 2 for a winter season.

Table 1 Shows the total mass concentration per species for five different mechanisms. The noMic case excludes microphysical processes. NM1 is the most complex mechanism and NM2–NM4 more simplified mechanisms.

The burden of condensate species, sulfate and nitrate are very similar among the simulations, only NM4, the mechanism excluding the aitken mode shows slightly lower sulfate concentrations. BC and OC concentration vary significantly since their removal depends on solubility (from mixture with soluble species) and size. The coarse aerosols, sea salt and dust are less effected by aerosol particle size changes, but the solubility of mineral dust depends on its condensates.

Summary and Conclusions

A multi-configurational aerosol microphysical model, based on the quadrature method of moments has been implemented into the GISS climate model. The model includes a series of different nucleation, condensation, coagulation, and aerosol

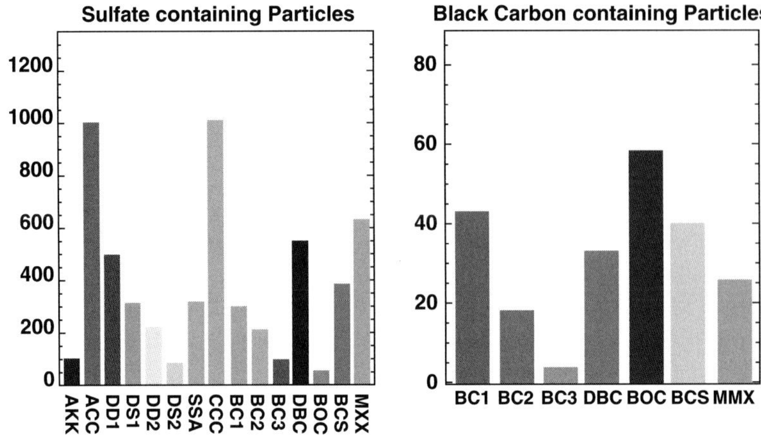

Figure 2 Global column mean mass concentrations for sulfate (**left panel**) and black carbon (**right panel**) species, as present in the following modes: aitken (AKK), accumulation (ACC), accumulation and coarse dust mixtures (DD1, DD2), accumulation and coarse soluble dust mixtures (DS1, DS2), sea salt mixture (SSA), black carbon mixtures (BC1, BC2, BC3), dust-black carbon mixture (DBC), black and organic carbon mixture (BOC), black carbon sea salt mixture (BCS), higher mixtures (MXX). The two plots show the mass of sulfate and black carbon in each of these modes, respectively

Table 1 Mass concentration per species, winter season

	Sulfate Tg/yr	Nitrate Tg/yr	BC Tg/yr	OC Tg/yr	Dust Tg/yr	Sea Salt Tg/yr
No Mic	0.71	0.19	0.097	0.61	28.8	4.1
NM1	0.71	0.19	0.067	0.44	20.0	4.0
NM2	0.71	0.19	0.067	0.44	17.2	4.7
NM3	0.71	0.19	0.081	0.61	23.5	4.4
NM4	0.70	0.19	0.059	0.61	22.5	4.8

phase chemistry modules, than can be all tested in this modeling framework. We presented some first, very preliminary results concerning aerosol mass and aerosol mixing. As a next step, the MATRIX-GCM aerosol scheme will be refined, to optimize selection of mechanisms to use in production simulations. The aerosols will be compared with observation from field campaigns and satellite retrievals under testing of several microphysical modules. The cloud microphysics will be coupled to the aerosol microphysics and the resulting cloud distributions will be validated. Modifications to aspects of the cloud scheme (autoconversion) and radiation will be required.

Acknowledgments This work has been supported by the NASA MAP program Modeling, Analysis and Prediction Climate Variability and Change (NN-H-04-Z-YS-008-N), managed by Don Anderson.

References

1. Wright, D.L., Numerical advection of moments of the particle size distribution in Eulerian models, *J. Aerosol Sci.*, in Press, Corrected Proof, Available online 19 January 2007.
2. Jaecker-Voirol, A., Mirabel, P., Heteromolecular nucleation in the sulfuric acid-water system, *Atmos. Environ.*, **23**, 2053–2057 (1989).
3. Vehkamaki, H., Kulmala, M., Napari, I., Lehtinen, K.E.J., Timmreck, C., Noppel, M., and Laaksonen, A., An improved parameterization for sulfuric acid-water nucleation rates for tropospheric and stratospheric conditions, *J. Geophys. Res.*, **107**, doi:10.1029/2002JD002184 (2002).
4. McMurry, P.H. and S.K. Friedlander, New Particle Formation in the Presence of an Aerosol, *Atmos. Environ.*, **13**, 1635–1651 (1979).
5. McMurry, P.H., New Particle Formation in the Presence of an Aerosol: Steady-State Rates, Time Scales, and Sub-0.01 μm Size Distributions, *J. Colloid Interface Sci.*, **93**, 72–80 (1983).
6. Turco, R.P., Zhao, J.X., and Yu, F., A new source of tropospheric aerosols: ion-ion recombination, *Geophys. Res. Lett.*, **25**, 635–638 (1998).
7. Kerminen, V.-M. and Kulmala, M., Analytical formulae connecting the "real" and the "apparent" nucleation rate and the nuclei number concentration for atmospheric nucleation events, *J. Aerosol Sci.*, **33**, 609–622 (2002).
8. Metzger S.M., Dentener, F.J., Lelieveld, J., and Pandis, S.N., Gas/aerosol partitioning 1: a computationally efficient model, *J. Geophys. Res.*, **107**, D16, 10.1029/2001JD001102 (2002).
9. Nenes, A., Pilinis, C., and Pandis, S.N., Isorropia: a new thermodynamic model for multiphase multicomponent inorganic aerosols, *Aquat. Geochem.*, **4**, 123–152 (1998).
10. Morrison, H., Curry, J.A., and Khvorostyanov, V.I., A new double-moment microphysics parameterization for application in cloud and climate models. Part I: Description, *J. Atmos. Sci.*, **62**, 1665–1677 (2005).

Secondary Organic Aerosol Formation and Online Chemical Composition Analysis by Thermal Desorption Chemical Ionisation Aerosol Mass Spectrometer (TDCIAMS)

G. Eyglunent[1,*], A. Leperson[2], G. Solignac, N. Marchand[1], and A. Monod[1]

Abstract Atmospheric aerosols may have an important impact on climate and also human health. Aerosol Organic fraction, of which a major part is potentially of secondary origin, could play a significant role in these two effects. Formation mechanisms of Secondary Organic Aerosols (SOA) are complicated phenomena which are not completely understood, and the chemical composition of SOA is difficult to determine because of its complexity and temporal variability. Moreover, classical analytical methods, which are off-line, cannot provide real time information about aerosol formation and chemical composition. Recently, aerosol mass spectrometers have been developed to obtain such information.[1] This work presents the application of an aerosol mass spectrometer using thermal desorption and chemical ionisation: the TDCIAMS. This instrument can perform online analysis of SOA chemical composition thus allowing the study of SOA formation and aging.

Keywords Aerosol Mass Spectrometry, SOA, chemical composition, aerosol Formation

Atmospheric aerosols may have an important impact on climate and also human health. Aerosol Organic fraction, of which a major part is potentially of secondary origin, could play a significant role in these two effects. Formation mechanisms of Secondary Organic Aerosols (SOA) are complicated phenomena which are not completely understood, and the chemical composition of SOA is difficult to determine because of its complexity and temporal variability. Moreover, classical

[1]*Laboratoire de Chimie et Environnement, Université de Provence, 3, place Victor Hugo, 13331, Marseille, France*

[2]*Institut de Combustion, Aérothermique, Réactivité et Environnement, CNRS, 1C, Avenue de la Recherche Scientifique, 45071 Orléans Cedex 2, France*

* *Now at Institut de Combustion, Aérothermique, Réactivité et Environnement, CNRS, 1C, Avenue de la Recherche Scientifique, 45071 Orléans Cedex 2, France*

C.D. O'Dowd and P.E. Wagner (eds.), *Nucleation and Atmospheric Aerosols*, 1214–1218.

analytical methods, which are off-line, cannot provide real time information about aerosol formation and chemical composition. Recently, aerosol mass spectrometers have been developed to obtain such information.[1] This work presents the application of an aerosol mass spectrometer using thermal desorption and chemical ionisation: the TDCIAMS. This instrument can perform online analysis of SOA chemical composition thus allowing the study of SOA formation and aging.

The instrument was developed from the modification of a commercial APCI/MS/MS. The advantage to use thermal desorption and chemical ionisation is that it leads to less degradation and fragmentation of organic molecules than other techniques such as LASER ionisation or electronic impact. The inlet of the instrument has been modified in order to introduce aerosol sample (Figure 1).

Volatile organic compounds are trapped in a parallel plate charcoal denuder. The thermo-desorption unit of the TDCIAMS allows volatilisation of organic aerosol. A corona discharge followed by a quadrupole mass spectrometer permit ionisation and analysis of the composition of the aerosol.

Characterisation of the performances of the instrument has shown that the particle transmission efficiency is higher than 85% for a wide range of particle size, and the gas trapping efficiency is higher than 90% in the inlet. The particles volatilisation efficiencies are higher than 90% for different SOA mass and composition, and the optimisation of the detection/ionisation step has allowed us to obtain calibration curves with repeatability higher than 90% for standard products of α-pinene ozonolysis. It has been shown that the TDCIAMS is well adapted to online analysis of the chemical composition of organic particles for diameter of several hundred nanometers, and for a concentration range of 5–1,000 $\mu g\ m^{-3}$.

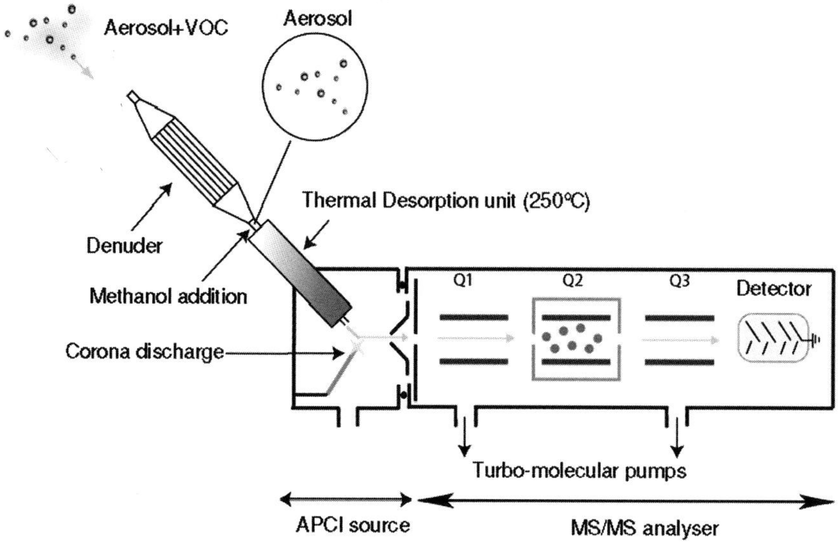

Figure 1 Scheme of the TDCIAMS (Thermal Desorption Chemical Ionisation Aerosol Mass Spectrometer)

SOA have been produced from the ozonolysis of α-pinene, in a Teflon Atmospheric Simulation Chamber, equipped with in situ FTIR for the gas phase analysis and a SMPS for granulometric distribution of the particles. α-pinene/O_3 system has been chosen because it is one of the major sources of SOA on a global scale. The TDCIAMS was directly connected to the chamber, in order to analyse online the chemical composition of particles. Two experiments were performed, with the same initial reactant concentrations, with and without the addition of formic acid. This compound was used in the 2nd experiment to inhibit nucleation,[2,3] and the comparison between both experiments allowed us to study the early stage of the reaction.

The addition of formic acid induced a particle nucleation inhibition. This addition resulted in a strong decrease of the particle number (divided by 7), a conservation of the produced SOA mass, and an increase of the diameter of the particles, in good agreement with previous studies.[2,3] The new results consist in the online analysis of SOA chemical composition. The 3D mass spectra of the SOA obtained is shown in Figure 2.

In Figure 2, one can see the time profiles of three acids formed in the particle phase: pinonic acid (m/z = 181 amu), pinic acid (m/z = 185 amu), and norpinic acid (m/z = 171 amu).[4]

Qualitatively, no modification of the chemical composition of SOA has been observed between the two experiments.

However, quantitative modification of the SOA chemical composition has been highlighted (Table 1). The contribution of the main products (pinic, pinonic, norpinic acids, and pinonaldehyde) to the SOA total mass, their molar and mass yields decreased when formic acid was added.

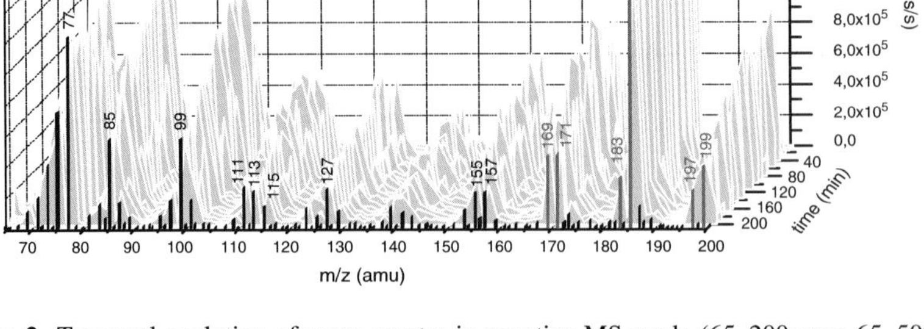

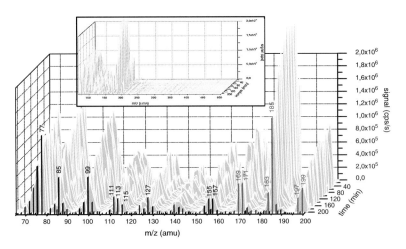

Figure 2 Temporal evolution of mass spectra in negative MS mode (65–200 amu; 65–500 amu in box) during α-pinene ozonolysis without formic acid addition

Table 1 Mass and molar yields, and contributions to total aerosol mass, of major particulate products of α-pinene ozonolysis in the particle phase without and with formic acid addition

Experiment: α-pinene ezonolysis		"without HCOOH"		"with HCOOH"
Initial concentrations	α-pinene (ppb)	482		482
	Ozone (ppb)	680		680
	HCOOH (ppm)	0		7.6
Mass Yield (%)	Pinic acid	7.7		5.3
	Pinonic acid	3.8		1.4
	Norpinic acid (+pinolic acid)	0.6		0.5
	Pinonaldehyde	1.92		0.7
Molar Yield (%)	Pinic acid	5.6		3.9
	Pinonic acid	2.8		1.0
	Norpinic acid (+pinolic acid)	0.5		0.4
	Pinonaldehyde	1.4		0.6
Contribution to Total aerosol Mass (%)	Pinic acid	17.9	27	13.5
	Pinonic acid	7.8	12	4.1
	Norpinic acid (+pinolic acid)	1.4	2	1.1
	Pinonaldehyde	3.8	4.8	1.6
Size range (%)	*nm*	*20.5–835*		*20.9–461*

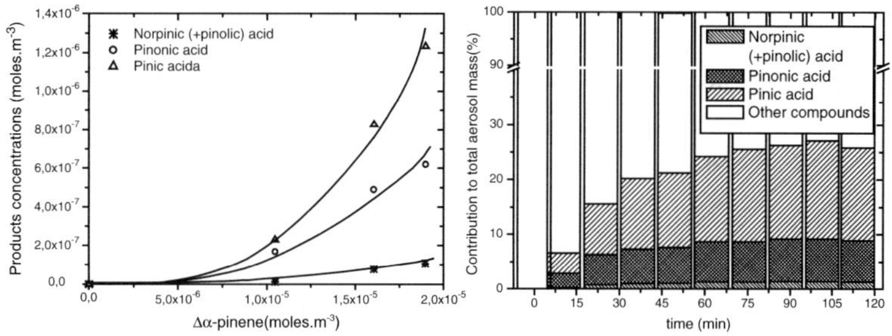

Figure 3a and 3b Secondary apparition and contribution to total SOA mass of pinic, pinonic, and norpinic acids during α-pinene ozonolysis

These decreases are explained by the reaction of formic acid with Criegee biradicals formed during the 1st step of ozonolysis of α-pinene. Furthermore, the concentration ratios between the products are significantly modified when formic acid was added, showing a modification of the SOA chemical composition and a modification of the chemical behaviour of the Criegee biradicals.

For a long time, pinonic, pinic, and norpinic acids have been considered as nucleating precursors of SOA. Here, we show that they are of secondary origin (Figure 3a) and represent only a small fraction of aerosol mass on the early stage of the reaction (Figure 3b).

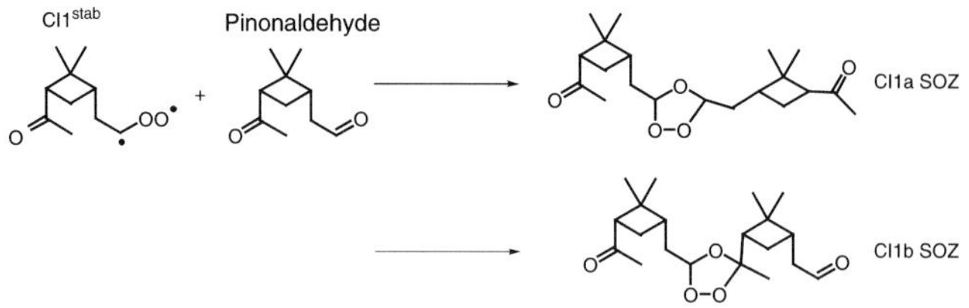

Figure 4 Secondary ozonide formed from reaction of pinonaldehyde with stabilised Criegee intermediate

However, the particles are of primary origin, thus they must involve primary reaction products. The implication of pinonaldehyde (which is the major gas phase reaction product) has been studied. The molecule proposed from previous studies to initiate nucleation is a secondary ozonide formed from pinonaldehyde reaction with stabilised Criegee intermediate (Figure 4) which has a sufficient low vapour pressure (below those of pinic acid) to initiate nucleation.

The potentiality of the TDCIAMS will be presented, as well as the results obtained from the study of α-pinene ozonolysis in terms of SOA chemical composition and aerosol formation.

Acknowledgment This work was supported by the PRIMEQUAL programme and the French ministry of research.

References

1. Sullivan, R.C. and Prather, K.A., *Anal. Chem.*, **77**, 3861–3886 (2005).
2. Lee, S. and Kamens, R.M., *Atmos. Environ.*, **39**, 6822–6832 (2005).
3. Bonn, B., Schuster, G., and Moortgat, G.K., *J Phys Chem A*, **106**, 2869–2881 (2002).
4. Winterhalter, R., Van Dingenen, R., Larsen, B.R., Jensen, N.R., and Hjorth, J., *Atmos. Chem. Phys. Discuss.*, **3**, 1–39 (2003).

Cloud Condensation Nuclei Activity at Jeju Island (Korea) in Spring

Mikinori Kuwata[1], Yuzo Miyazaki[1], Yuichi Komazaki[1*], Yutaka Kondo[1], Jong Hwan Kim[2], and Seong Soo Yum[2]

Abstract The number concentrations of cloud condensation nuclei (CCN) and size distributions of CCN/CN (condensation nuclei) ratios were measured at Jeju Island (Korea) in spring 2005. The CCN/CN ratio increased with the increase in particle diameter. The temporal variation of 50% activation diameter (particle diameter at CCN/CN = 0.5; D50) was correlated with that of the water-soluble fraction of aerosol components at PM2.5. Calculation of the critical dry diameter was performed using simultaneously observed aerosol chemical component data. The calculated critical dry diameter was well correlated with the observed D50. However, the calculated data overestimated the observed D50. This discrepancy could be caused by the size dependence of chemical composition and decrease of surface tension due to a presence of water-soluble organic carbon.

Keywords Cloud condensation nuclei, tropospheric aerosol, aerosol chemical components

Introduction

Some aerosol particles act as cloud condensation nuclei (CCN) and influence the formation processes of cloud droplets (indirect effect). CCN number concentrations depend on number size distributions and chemical composition of aerosol particles. Therefore, the number fraction of CCN active particle is a key parameter relating the concentrations of aerosol and CCN. In many cases, this fraction is defined as CCN/CN (condensation nuclei) ratio because CN counters are used to measure aerosol number concentrations.[1] The CCN/CN ratio of atmospheric particles has been measured in various regions of the world. In many of these studies "bulk" CCN/CN ratios

[1]*Research Center for Advanced Science and Technology, the University of Tokyo, Japan*

[2]*Department of Atmospheric Sciences, Yonsei University, Korea*

[*]*Now at Japan Agency of Marine-earth Science and Technology, Yokosuka, Japan*

C.D. O'Dowd and P.E. Wagner (eds.), *Nucleation and Atmospheric Aerosols*, 1219–1223.

were measured. However, it is necessary to measure size distributions of CCN/CN ratio to characterize the effects of size distribution and chemical composition on CCN activities individually.[2] Thus far, very limited measurements of the CCN/CN ratio size spectra have been reported. In this study, we observed the number concentration of CCN and the size distributions of CCN/CN ratio at the supersaturations (S) of 0.097, 0.27, 0.58, and 0.97% at Gosan (33.17° N, 126.10° E) on Jeju Island, Korea during the Atmospheric Brown Cloud – East Asian Regional Experiment 2005 (ABC – EAREX2005) campaign (March 18–April 5, 2005).

Observations

Figure 1 shows the CCN observation system used in this study. Ambient particles were dried (RH < 5%) by a diffusion dryer, and were charged using a 241 Am neutralizer. The size of charged particles was selected by a differential mobility analyzer (DMA). The voltage applied to the DMA was scanned in a stepwise fashion. The number concentration of selected particles was measured by a condensation particle counter (CPC), and the number concentration of size-selected CCN was measured by a CCN counter. The S in the CCN counter was calibrated using $(NH_4)_2SO_4$ particles. The calibration results are summarized in Table 1. In this study, four different values of S (0.097, 0.27, 0.58, and 0.97%) were used. Inverse analysis was performed for the CCN and CN size distribution data to eliminate the artifact of multiple charged particles on CCN/CN spectra. A three-way valve was placed upstream of the CCN counter so that the CCN counter can also measure the CCN number concentrations of the atmosphere.

The measurements of aerosol chemical compositions were simultaneously conducted. Aerosol inorganic components (NH_4^+, SO_4^{2-}, and NO_3^-) were measured by a particle-into-liquid sampler combined with ion chromatography (PILS-IC), and the concentrations of water-soluble organic carbon (WSOC) were obtained using a PILS-WSOC. The measurements of organic carbon (OC) and elemental carbon (EC)

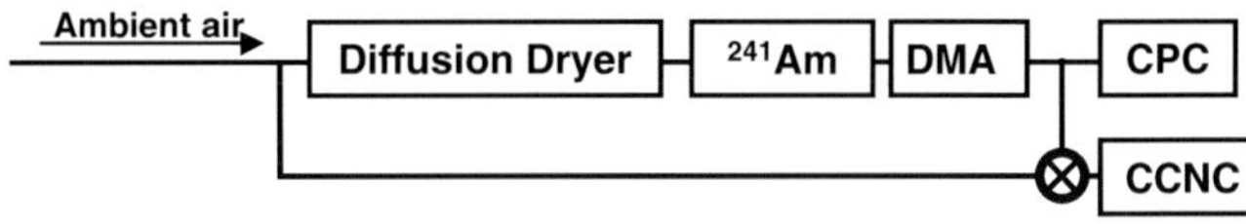

Figure 1 Schematic of CCN observation system used in this study

Table 1 CCN number concentrations and D50 diameters observed in this study

	S = 0.097%	0.27%	0.58%	0.97%
CCN number (p/cc)	1194 ± 746	2543 ± 1277	3496 ± 1510	3996 ± 1686
D50 diameter (nm)	136 ± 17	71 ± 6	44 ± 3	31 ± 3
(cf.) D50 diameter of $(NH_4)_2SO_4$ (nm)	125 ± 3	63 ± 2	38 ± 1	27 ± 1

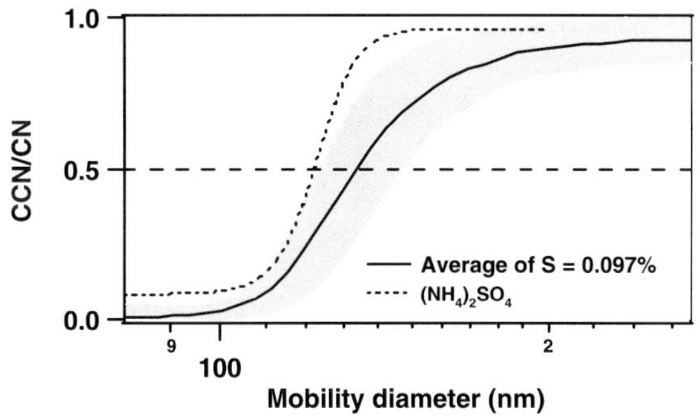

Figure 2 The size distributions of CCN/CN ratio

were conducted using a semi-continuous ECOC analyzer (Sunset Laboratory, Inc.). $PM_{2.5}$ cyclones were used for inlet systems for all instruments described above.

Figure 2 shows the average and standard deviation of CCN/CN ratio at S = 0.097%, and CCN/CN ratio of $(NH_4)_2SO_4$ is also shown for comparison. The CCN/CN ratio strongly depends on particle diameter. It is almost zero at 100 nm and close to unity for 200 nm and higher. We have denoted 50% activation diameter (the diameter at CCN/CN = 0.5) as D50 and regard this diameter as the critical dry diameter (threshold diameter for CCN activation) of atmospheric particle. The D50 diameter and CCN number concentrations obtained in this study are summarized in Table 1.

Results

The observed CCN number concentrations were about a few thousands particles/cm^3. These concentrations are significantly higher than those observed in other regions of the world.[1] Yum et al.,[3] reported the similar high concentration at Anmyeon (west coast of Korean Peninsula) during springtime. Therefore, one can assume that these concentrations are typical for this region. The D50 diameters during the observation period were generally larger than those of $(NH_4)_2SO_4$ by 4–11 nm.

Discussion

Figure 3 shows the temporal variations of aerosol chemical composition and D50 diameter at S = 0.097%. Aerosol chemical compositions were classified into four groups: EC, water-insoluble organic carbon (WIOC), WSOC, and inorganic. WIOC concentration was calculated by subtracting WSOC concentration from OC concentration. Inorganic components are the sum of NH_4^+, SO_4^{2-}, and NO_3^-. It is to be noted that WIOC and WSOC fractions contain only the mass of carbon, and do not

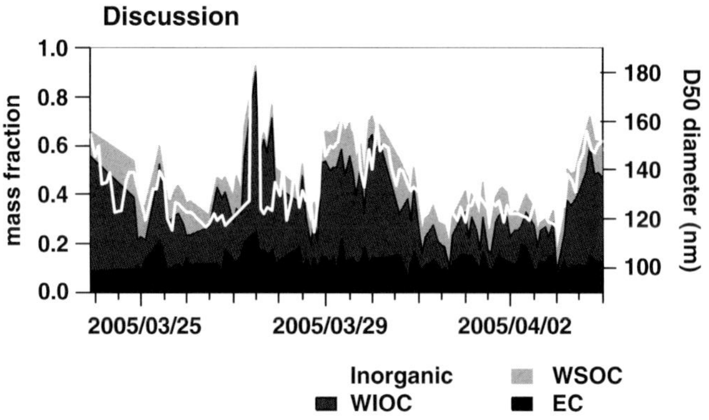

Figure 3 Temporal variations of aerosol chemical composition and D50 diameter at S = 0.097%

contain masses of other elements such as oxygen and hydrogen. It can be seen that the observed D50 diameter decreases with the increase in the mass fraction of water-soluble components (Inorganic components + WSOC). It can be seen that the observed D50 diameter decreases with the increase in the mass fraction of water-soluble components (Inorganic + WSOC). The number of water-soluble molecules and ions included in a particles with a given diameter increase with the increase in fraction of water-soluble components. The decrease of D50 with the fraction of the water-soluble components is qualitatively consistent with the Köhler theory.

The critical dry diameter (Dcrit: the calculated threshold diameter for CCN activation) was calculated using the observed $PM_{2.5}$ chemical composition based on the Köhler theory. For these calculations, we assumed that (1) aerosol chemical composition is size-independent. (2) The chemical characteristics (elemental ratio, molecular weight, and density) of WSOC were equal to that of oxalic acid. Oxalic acid is one of the smallest (in molecular weight) molecules in atmospheric water-soluble organic compounds. Therefore, this estimate gives the upper limit of the Raoult's effect. (3) All carbon atoms of WIOC are assumed to be in methylene group ($-CH_2-$) and density of WIOC was assumed as 0.8 g/cm^3. (4) Surface tension is equal to that of water, and (5) the ideal solution approximation was used for the calculation.

The calculated D_{crit} and the observed D50 are compared in Figure 4. The D_{crit} is moderately correlated ($r^2 = 0.38$) with the observed D50.

However, the D_{crit} is systematically larger (20%) than the observed D50. If the molecular weight of the WSOC is larger than oxalic acid (assumption 2), the D_{crit} becomes lager. Therefore, the assumed chemical composition of WSOC does not explain the discrepancy. The assumed density of WIOC (assumption 3) may be an underestimate. However, increasing the density to 1.2 g/cm^3 decreased the D_{crit} only by 5%. Thus, the assumptions on chemical composition cannot explain the discrepancy.

As discussed above, we assumed size-independent chemical composition (assumption 1) and the surface tension of water (assumption 4). These assumptions

Figure 4 Comparison of D50 and D_{crit}

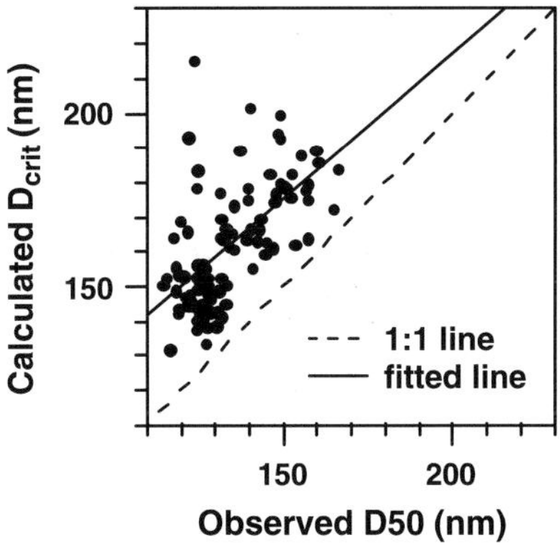

are significant and can explain the discrepancy (assumption 1) and the surface tension of water (assumption 4). These assumptions are significant and can explain the discrepancy.

Acknowledgment This study was supported by the Ministry of Education, Culture, Sports, Science, and Technology (MEXT), and the Japanese Science and Technology Agency (JST). M.Kuwata thanks the Japan Society for the Promotion of the Science (JSPS) for research fellowship as young scientists.

References

1. Seinfeld, J.H. and Pandis, S.N., *Atmospheric Chemistry and Physics* New York: Wiley, 761–827 (2006).
2. Dusek, U., Frank, G.P., Hildebrandt, L., Curtius, J., Schneider, J., Walter, S., Chand, D., Drewnick, F., Hings, S., Jung, D., Borrmann, S., and Andreae, M.O., *Nature*, **312**, 1375–1378 (2006).
3. Yum, S.S., Hudson, J.G., Song, K.Y., and Choi, B-C., *J. Geophys. Res.*, **32**, L09814 (2005).

CCN Properties of Water-soluble Organic Compounds Produced by Common Bioaerosols

Sanna Ekström[1], Barbara Noziére[1], and Hans-Christen Hansson[2]

Abstract Polyols are considered mainly to be formed by bioorganisms (fungi) and have been found in large concentrations in atmospheric aerosols in many environments, both in the fine and coarse mode particles. Osmolality and tensiometry measurements were used to calculate the Köhler curves for C_3–C_6 polyols and their analogue di-acids. The results show that the large affinity of polyols with water lowers the Raoult effect of water droplets and leads to similar CCN properties than the corresponding dicarboxylic acids. But because polyols are considerably more soluble than dicarboxylic acids they would activate droplet growth at a much earlier phase. This indicates for the first time that organic compounds produced almost exclusively by bioorganisms could play a role in cloud formation.

Keywords Bioaerosols, water-soluble organic compounds, polyols, Köhler curve (CCN)

Introduction

Fungi exist as a bioaerosols (Carvalho et al. 2003) in abundance that varies widely but can reach 4×10^5 m^3. Global average emissions over land from the fungi Asco- and Basidiomycota was recently estimated to 17 Tg per year, which can be compared to 12–70 Tg per year of secondary organic aerosol (Elbert et al. 2006) from an estimated total of 1150 Tg global emissions per year of volatile organic compounds from natural sources (Guenther et al. 1995).

Polyols are produced and excreted at different growth stages of fungi (Cor and Jaap 1995) and are abundant in atmospheric aerosols over all land types. They are mainly present in coarse aerosols (diameter > 2.5 μm) in the Amazon forest (Graham et al. 2003) and at a rural (meadow) site (Carvalho et al. 2003). But arabitol and mannitol were

[1]*Department of Meteorology, Stockholm University*

[2]*Department of Applied Environmental Research, Stockholm University Stockholm, Sweden*

C.D. O'Dowd and P.E. Wagner (eds.), *Nucleation and Atmospheric Aerosols*, 1224–1229.

primarily found in the fine mode of aerosols in a boreal forest (Carvalho et al. 2003), showing that these compounds can also be present in cloud condensation nuclei.

Organic acids are considered as the organic compounds the most efficient in activating cloud droplet growth because of their surfactant effect, which has been subject of numerous studies (e.g., Bilde et al. 2004; Hori et al. 2003; Giebl et al. 2002; Prenni et al. 2001; Corrigan et al. 1998; Cruz and Pandis 1997; Shulman et al. 1996). But the very large affinity of polyols with water, increasing with carbon chain length, could affect the other factor controlling droplet growth, the Raoult effect, and thereby also result in effective cloud droplet activation. To investigate this hypothesis, the CCN properties of C_3–C_6 polyols was studied for the first time and compared with those of the corresponding diacids.

Theory

Cloud droplet activation was described with traditional Köhler theory (Köhler 1936), expressing the water vapor pressure in equilibrium with a droplet of given radius. The Köhler equation is:

$$S = \frac{p}{p_0} - 1 = a_w \, e^{\frac{2\sigma_{Solution} M_W}{r\rho_{water} RT}} - 1, \tag{1}$$

where p is the water vapor pressure over the droplet, p_0 the vapor pressure over a flat water surface, a_w water activity in the droplet solution, $\sigma_{Solution}$ the surface tension of the droplet solution, M_w the molecular weight of water, ρ_{water} the density of water, R the universal gas constant, and T temperature.

Kiss and Hansson (2004) suggested a new approach to determine Köhler curves from the measurements of the osmolality and surface tension of solutions. Experimental values of $\sigma_{Solution}$ are replaced directly in Eq. (1) while osmolality, C_{osm} (in Osmol kg^{-1}), is used to calculate water activity;

$$a_w = \frac{1}{1 + \frac{C_{osm} \cdot MW_{water}}{1000}}, \tag{2}$$

where MW_{water} is the molecular weight of water.

Experimental

The osmolality and surface tension of different solutions of C_3–C_6 polyols, glycerol, erythritol, arabitol, and mannitol, were measured as well as those of solutions of the corresponding C_3–C_6 α,ω-dicarboxylic acids, malonic, succinic, and adipic acid. Osmolality was measured with a KNAUER K – 7,000 vapor pressure osmometer and an FTÅ 125 tensiometer was used to determine surface tension.

Results

The Köhler curves for the polyols and corresponding organic diacids are presented in Figures 1–3. Köhler curves for polyols are reported here for the first time and could thus not be compared with previous measurements. But the good agreement between our measurements and earlier studies for the organic acids (Bilde et al. 2004;

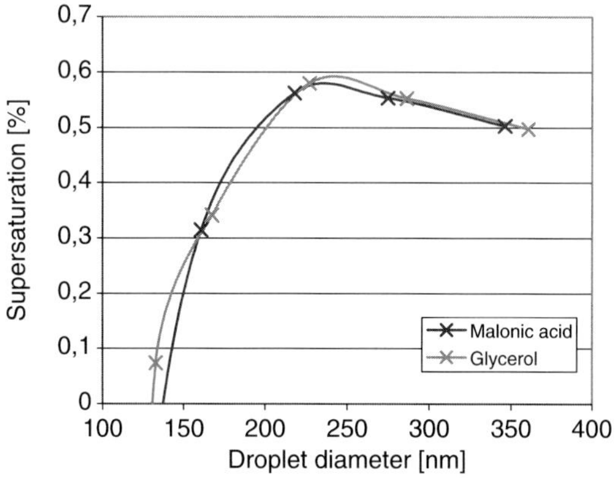

Figure 1 Köhler curves for glycerol and malonic acid (C_3) for particles with d_{dry} = 60 nm using the original Köhler equation

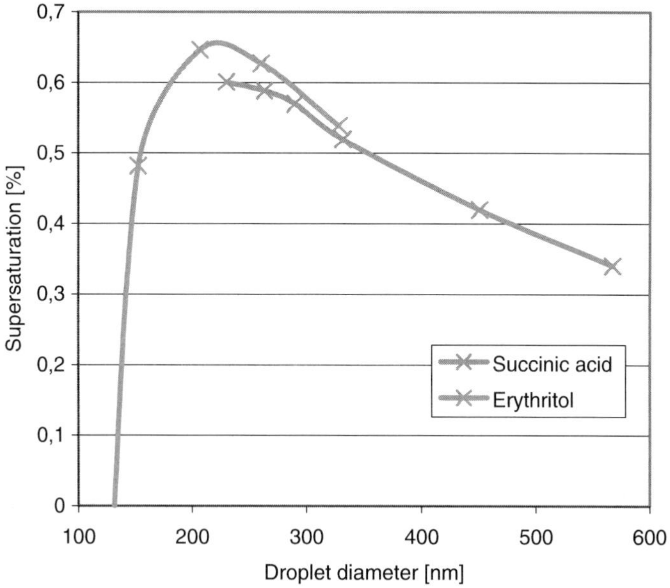

Figure 2 Köhler curves for erythritol and succinic acid (C_4) for particles with d_{dry} = 60 nm using the original Köhler equation

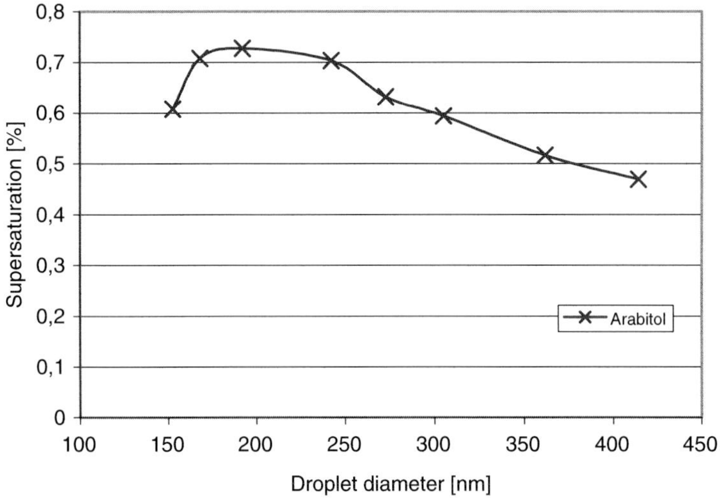

Figure 3 Köhler curves for arabitol (C_5) for particles with d_{dry} = 60 nm using the original Köhler equation

Hori et al. 2003; Giebl et al. 2002; Prenni et al. 2001; Corrigan et al. 1998; Cruz and Pandis 1997) confirmed the validity of our measurements. In fact, for adipic and malonic acid our results were closer to theoretical values than previous experimental results (Giebl et al. 2002; Corrigan et al. 1998). The limited solubility of succinic and adipic acid in water precluded the determination of their Köhler curves for small droplet radius. This indicates the inability of organic acids to activate the growth of small particles. By contrast, the polyols are soluble over the entire range of concentration studied, and thus able to activate droplet growth at a very early stage.

As it can be seen in Figures 1–4 the critical supersaturation for cloud droplet activation in the presence of polyols increases with increasing carbon chain length. Glycerol has similar CCN properties than malonic acid (Figure 1), but erythritol and mannitol have a slightly lower CCN efficiency than succinic (Figure 2) and adipic acid (Figure 4), respectively. The CCN efficiency of D-arabitol (Figure 3) lies between those of erythritol and mannitol.

Conclusions

Polyols were found to have similar CCN properties to their corresponding dicarboxylic acids, the critical supersaturation slightly decreasing with carbon chain length from C_6 to C_3. This CCN efficiency was interpreted to result from the large water affinity of polyols, enforcing the Raoult effect, while organic acids lower the Kelvin effect. The much larger solubility of polyols in water compared to organic acids also means that these compounds can activate the initial phase of droplet growth while organic acids can not.

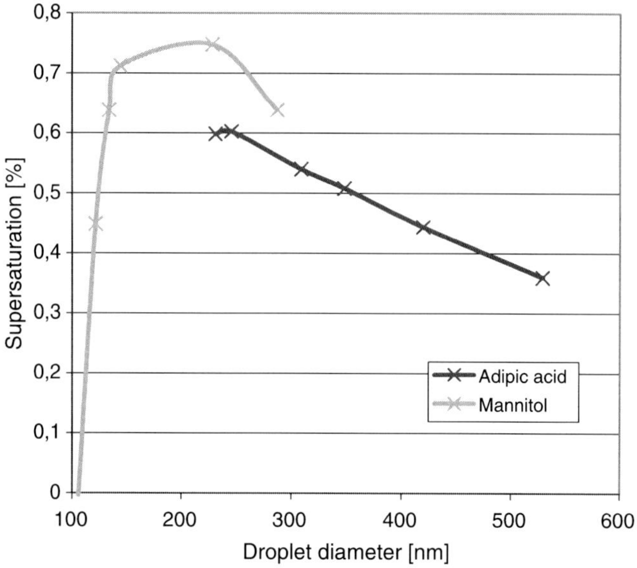

Figure 4 Köhler curves for mannitol and adipic acid (C_6) for particles with d_{dry} = 60 nm using the original Köhler equation

The emissions of polyhydroxylated organic compounds by living organisms could thus participate in cloud formation in remote forest areas, where bioorganisms are abundant and inorganic CCN less available, and possibly also in marine environments during extreme biologic activity such as algae blooms, as confirmed by the recent observations of enhanced cloud droplet numbers by Meskhidze and Nenes (2006) in the Southern Ocean.

Acknowledgments We would like to thank Ulla Wideqvist from the Institute of Applied Environmental Research, Stockholm University, Stockholm, Sweden, for supplying expertise and support during the measurements. BN acknowledges the support of a Marie Curie Chair and an International Reintegration Grant from the European Commission.

References

Bilde M. and Svenningson B., *Tellus*, **56B**, 128–134 (2004).
Carvalho A. et al., *Atmos. Environ.*, **37**, 1775–1783 (2003).
Chen M. et al., *Progress in Natural Sci.*, ***11***, No 9. 681–687 (2001).
Cor F.B. and Jaap V., *Microbiology lett.*, **134**, 57–62 (1995).
Corrigan C.E. and Novakov T., *Atmos. Environ.*, **33**, 2661–2668 (1998).
Cruz, C. and Pandis, S.N., *Atmos Environ.*, **31**, 2205–2214 (1997).
Elbert W. et al., *Atmo. Chem. and Phys. Discuss.*, **6**, 11317–11355 (2006).
Giebl H. et al., *J. Aerosol Sci.*, **33**, 1623–1634 (2002).
Guenther A. et al., *J. Geophys. Res.*, **100**, 8873 (1995).

Graham B. et al., *J. Geophys. Res.*, **108**, AAC 6: 1–13 (2003).
Hori M. et al., *J.Aerosol Sci.*, **34**, 419–448 (2003).
Jones A. M. et al., *Sci. Total Environ.*, **326**, 151–180 (2003).
Kiss, G. and Hansson, H.-C., *Atmos. Chem. Psys. Discuss.*, **4**, 1–23 (2004).
Köhler, H., *Trans. Faraday. Soc.*, **32**, 1152–1161 (1936).
Meskhidze N. and Nenes A., *Science,* **314**, 1419–1423 (2006).
Prenni, A.J., *J. Phys. Chem. A*, **105**, 11240–11248 (2001).
Saxena, P. and Hildemann, L., *J. Atmos. Chem.*, **24**, 57–109 (1996).
Shulman M.L. et al., *Geophys. Res. Lett.*, **23**, 277–280 (1996).

An Annual Study of Organic Atmospheric Aerosol from a Rural Site of Madrid (Spain)

Oscar Pindado, Rosa M. Pérez, Susana García, and Ana I. Barrado

Abstract The organic fraction of aerosols was characterized in two particulate matter fractions, $PM_{2.5}$ and PM_{10} from a rural site of Madrid, since April 2004 to March 2005. Aerosol-associated *n*-alkanes, polycyclic aromatic hydrocarbon (PAH), alcohols and acid compounds were measured in order to evaluate its seasonal variability and sources. An analytical protocol has been developed; samples were Soxhlet extracted, cleaned-up by silica gel column chromatography and subsequently analyzed by gas chromatography mass spectrometry (GC-MS). Polar compounds were previously derivatized with BSTFA. Samples studied contain *n*-alkanes ranged from C_{14} to C_{40}, 13 PAHs from acenaftene to Benz(ghi)perilene have been quantified, and several polar compounds, as alcohols, fatty acids, and some secondary organic aerosol components, mainly some products of degradation of α-pinene. Moreover, distinct seasonal profiles were detected in both particulate fractions studied.

Keywords Atmospheric aerosol, chemical composition, organic compounds, $PM_{10}/PM_{2.5}$

Introduction

The European Directive 1999/30/CE settles down maximum values of different atmospheric pollutants, between which total particulate matter is. The purpose of this Directive was to reduce their concentrations and therefore preventing and/or reducing its effects over environment and human health. In addition, the IPCC (Intergovernmental Panel on Climate Change, 1995) emphasized over the adverse impact on human health and environment produced by atmospheric aerosols. Nevertheless, within the particulate matter of atmospheric aerosol, the organic fraction is less studied and understood than the inorganic one, so it is going to be more difficult diminish its concentration.

Chemistry Division, Department of Technology, CIEMAT, Avda. Complutense 22, 28040 Madrid, Spain

C.D. O'Dowd and P.E. Wagner (eds.), *Nucleation and Atmospheric Aerosols*, 1230–1234.

On the other hand, within atmospheric aerosols, rural aerosols are the less studied. Nowadays, data sets of concentrations reached by organic compounds in urban areas are going to be available. Chemical composition of rural aerosol is mainly conducted by biogenic origin, with a lower anthropogenic contribution. In fact, a study about composition in a rural area is of particular interest for the understanding of background of atmospheric pollution.

Objective

This work is aimed at identify several compounds of organic fraction of atmospheric aerosol collected in a rural area of Madrid (Spain) for a period of one year. There have been simultaneous studied two different fractions, PM_{10} and $PM_{2.5}$.

Experimental Part

Sampling and Organic Analysis

An annual campaign took place in Chapineria, a little town situated in the southwest of the city of Madrid, collecting simultaneously two different fractions, $PM_{2.5}$ and PM_{10}. Sampling was performed using two high-volume samplers and was collected over quartz filters, previously calcinated.

Samples were Soxhlet extracted with a mixture of dichloromethane/acetone (3:1). Later, extracts were concentrated and cleaned-up by silica gel column chromatography and four fractions were obtained, alkanes, PAHs, alcohols, and acids respectively. Polar compound were derivatized with BSTFA and analyzed by GC-MS, as same as alkanes, whereas PAHs were submitted to HPLC with florescence detection.

Target Compounds

It has been analyzed more than 50 organic compounds. In order to facilitate the study, these compounds were divided in four different families:

1. aliphatic hydrocarbons (C_{14}–C_{40} *n*-alkanes, phytane, and pristine)
2. polycyclic aromatic hydrocarbons (Acenaphthene, Fluorene, Phenanthrene, Fluoranthene,, Benz(a)anthracene, Chrysene, Benzo(b)Fluoranthene, Pyrene, Anthracene, Benzo(k)Fluoranthene, Benzo(a)Pyrene, Dibenz(a,h)anthracene, Benzo(g,h,i)perilene)
3. alcoholic compounds (C_{12}–C_{26} alcohols)

4. acids compounds (C_9–C_{28} fatty acids, oleic, linoleic, pinic, pinonic, norpinonic, azelaic acids)

Results And Discussion

Table 1 shows minimum, maximum, and mean values of total concentration of particulate matter in both fractions together its seasonal variability.

In all seasons the mean mass concentrations of $PM_{2.5}$ and PM_{10} was slightly lower than that measured in a urban area, for example, the extent work of Ling-Yan in Beijing during 2002 (Ling-Yan et al. 2006), where mean values raise 66–78 μg/m^3 for $PM_{2.5}$ and the work of Chan (Chan et al. 2005) for $PM_{2.5}$ and PM_{10} fractions. Therefore, an explication of this fact was the lower anthropogenic contribution in the rural area. In addition, values measured for $PM_{2.5}$ and PM_{10} were similar to that collected in a rural area of Portugal (Alves et al. 2001).

The $PM_{2.5}$ fraction showed more seasonal variability than PM_{10} fraction. The highest values for two fractions were reached during last days in September; corresponding with intrusion episode from Sahara.

Aliphatic and Aromatic Hydrocarbons

The first fraction analyzed contains *n*-alkanes ranged from C_{14} to C_{40}. *n*-alkane C_{29} usually showed the highest values in both fractions. It was remarkable the odd to even predominance mainly in PM_{10} fraction. Moreover in PM_{10} fraction, CPI and %WNA values were higher during summer than in winter, due to plants emits less amount of matter in winter than in summer and spring.

In relation with PAHs, 13 compounds were evaluated in PM_{10} and $PM_{2.5}$ samples. The annual average concentrations of each individual PAH in the rural area were within a range of 1–1500 pg/m^3 in PM_{10} fraction and 1–600 pg/m^3 in $PM_{2.5}$ fraction. Benzo(ghi)perilene, Benzo(a)chrisene and Pirene were the most abundant PAH compounds identified in both fractions. The seasonal variability was 5–6 times higher during winter compared to summer in PM_{10} and $PM_{2.5}$ fractions.

Table 1 Minimum, maximum, and mean values for PM_{10} and $PM_{2.5}$ during the sampling period

	$PM_{2.5}$ (μg/m^3)			PM_{10} (μg/m^3)		
	Min	Max	Mean	Min	Max	Mean
Spring	4,7	22,4	13,3	8,2	57,1	26,0
Summer	3,1	28,0	27,1	12,8	94,9	34,1
Autumn	4,2	37,3	18,3	5,9	109,5	34,1
Winter	6,9	64,1	21,8	11,4	87,3	37,0

Polar Compounds

Eighteen alcoholic compounds, within the range of C_{12}–C_{30}, were identified and quantified in both fractions studied with a maximum concentration of 200 ng/m^3. CPI and %WNA values confirm the biogenic contribution to aerosol.

Last fraction contains acidic compounds. There have been identified fatty acids comprised between nonanoic and octacosanoic acid with mean values in $PM_{2.5}$ fraction of 50.8 , 170.3 , 208.7 , and 139.9 ng/m^3 for spring, summer, autumn, and winter respectively. Moreover, some unsaturated fatty acids and three degradation products of α-pinene were quantified in both fractions. In the same way as alcohols, CPI and %WNA values confirm the biogenic contribution to atmospheric aerosol.

Seasonal Variability

For organic compounds, unlike which it happens for total concentration of particulate matter, the seasonal variability was more pronounced in PM_{10} samples (Figure 1). It is remarkable the fact in PM_{10} fraction; acids average concentration low 40% since summer to winter, however *n*-alkanes and PAHs increased a lot. This variability appears in $PM_{2.5}$ fraction slightly, only *n*-alkanes and PAHs increase its average concentrations and acids diminish it.

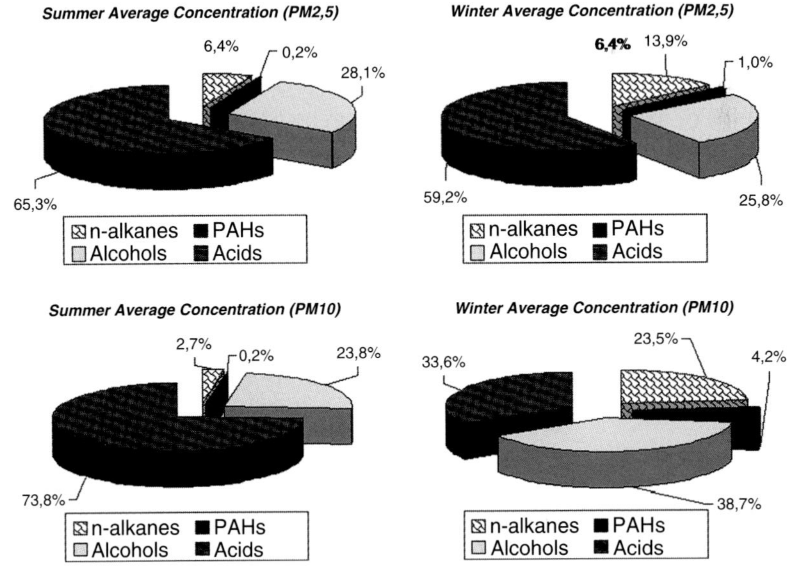

Figure 1 Average concentration of the four families of compounds studied in both fractions during winter and summer

Variability of PAH in samples studied may be due to increase of fossil fuel combustion for domestic heating during winter, meanwhile acids variability was due to lower biogenic emissions by plants during winter.

Acknowledgments This work was developed by the grant provided by "Ministerio de Educación y Ciencia" of Spain (REN2003-08603-C04-02). Authors would like too thank to the "Consejeria de Medioambiente de la Comunidad de Madrid" for their sampling and infrastructure support.

References

1. Alves C., Pio C., and Duarte A., *Atmos. Environ.*, **35**, 5485–5496 (2001).
2. Chan C.Y., Xu X.D., Li Y.S., Wong K.H., Ding G.A., Chan L.Y., and Cheng X.H., *Atmos. Environ.*, **39**, 5113–5124 (2005).
3. Intergovernmental Panel on Climate Change (IPCC), *Climate Change*, New York: Cambridge University Press (1995).
4. Ling-Yang H., Min H., Xiao-Feng H., Yuan-Hang Z., and Xiao-Yan T., *Sci. Total Environ.*, **359**, 167–176 (2006).

Fine Particulate Matter in Apulia (South Italy): Chemical Characterization

M. Amodio, P. Bruno, M. Caselli, G. de Gennaro, P. Ielpo, B.E. Daresta, P.R. Dambruoso, C.M. Placentino, and M. Tutino

Abstract Samples of $PM_{2,5}$ and PM_{10} have been collected in the course of two Italian national projects: SITECOS (PRIN 2004–2006) and PCOST (PRIN 2005), in several sampling sites of Apulia Region (South Italy). Chemical characterization (inorganic components, polyciclic aromatic hydrocarbons, and carbonaceous fraction) has been performed on the PM samples collected. The results obtained do not show a seasonal trend for PM_{10} concentrations. The organic carbon is the principal constituent of $PM_{2.5}$ and PM_{10}. Sulphate, nitrate in certain cases, and ammonia show a similar distribution in both Bari (coastal town of Apulia region) and Taranto (coastal town with an important industrial area) sampling sites. By observing the PAHs concentrations in the PM samples collected in Bari and Taranto and applying the APCS receptor model to the samples collected, it has been possible to note that Taranto industrial area is the predominant PAHs source, while predominant PAHs source in Bari is automotive traffic.

Keywords $PM_{2,5}$ PM_{10}, chemical characterization, receptor models

Introduction

The air we breathe every day can be contaminated by polluting substances emitted by industries, vehicles, or other sources. These polluting substances can have bad effects both on human health and environment. Air pollution control is necessary to prevent the situation from becoming worse in the long run. The air pollution control involves the knowledge of the chemical–physical characteristics of the pollutants and the meteorologic and climatic conditions of the region considered.

The results shown in this contribution arise from two Italian national projects: SITECOS (Integrated Study on national TErritory for the characterization and COntroll of atmospheric pollutants) PRIN project 2004–2006 and PCOST (Fine Particulate in COasTal areas: development and implementation of predictive instruments) PRIN project 2005.

Department of Chemistry, University of Bari, Via E. Orabona, 4, 70126, Bari, Italy

C.D. O'Dowd and P.E. Wagner (eds.), *Nucleation and Atmospheric Aerosols*, 1235–1238.

The SITECOS project deals with a research on national territory to characterize and to control atmospheric pollutants in 10 Italian cities.

The aim is the valuation of different effects of climatic and emission features on pollutants dynamics of North, Centre, and South Italy.

In these sites, in each year, two monitoring campaigns with $PM_{2,5}$ and PM_{10} samplings have been carried out.

The principal goal of this project is the attainment of an organic vision of the air quality on national level, evaluating the effects of the meteoclimatic and emissive differences of pollutants among North, Middle, and South of Italy.

The PCOST project deals with the execution of measurement (one during the winter and another during the summer) in coastal areas in order to characterize the fine coastal particulate.

In the course of SITECOS project, monitoring campaigns in Bari, Taranto, and Mount Pollino (remote site) during the months of October 2005, February, and July 2006, have carried out. During these campaigns $PM_{2.5}$ and PM_{10} daily samples have been collected. In the course of PCOST project the first monitoring campaign has been performed in March 2007 and the second one in June 2007 in three sampling sites of Bari, a coastal town of South Italy. $PM_{2.5}$ and PM_{10} daily samples have been collected. Chemical characterization has been performed on the PM samples. The chemical investigation has dealt with analysis of inorganic component (anions, cations, metals), polyciclic aromatic hydrocarbons (PAHs) and carbonaceous fraction (TC/OC/EC).

Analytical Results

The results obtained in several monitoring campaigns during 2005 and 2006 do not show a seasonal trend for PM_{10} concentrations, moreover in the several sampling sites is evident a common contribution to the PM_{10} amount, apparently independent from local sources.

The organic carbon is the principal constituent of $PM_{2.5}$ and PM_{10}, and is characterized by OC/EC diagnostic ratio higher than 2.5 in all sites, suggesting the secondary organic aerosol formation.[1]

The ionic fraction is a substantial part of PM.

Sulphate, nitrate in certain cases, and ammonia, prevalent components in ionic fraction, show a similar distribution in both Bari and Taranto sampling sites: this suggests a uniform distribution on wide area.

Experimental results, highlighted by characteristic diagnostic ratio magnesium/sodium, show the presence of marine aerosol in particulate matter. Marine salts do not contribute in significant way to PM_{10}, while Saharan dust events cause considerable increments of particulate matter.[2–3]

During such events, confirmed by diagnostic ratio sulphate/calcium, the ionic coarse fraction percentage exceeds the 50%, compared to a mean contribution of 30%.

By observing the PAHs concentrations in the PM samples collected in Bari and Taranto, it has been possible to point out the prominent influence of Taranto industrial area: for these samples is evident a direct relation between wind direction coming from industrial area (North) and high PAHs concentrations.

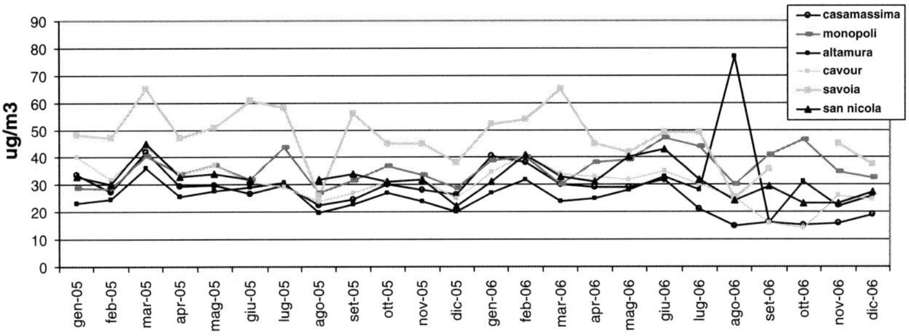

Figure 1 Monthly mean of PM_{10} in several sampling sites in Bari town (Cavour, Savoia, S.Nicola) and Bari Province (Casamassima, Monopoli, Altamura)

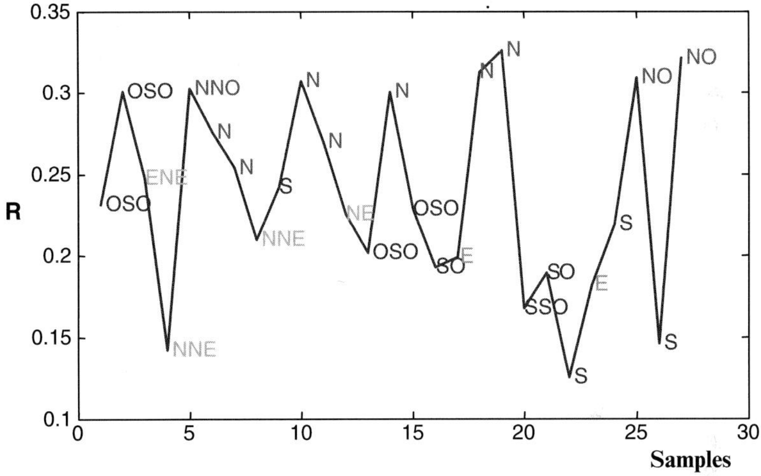

Figure 2 Sulphate and Ammonia concentration in $PM_{2.5}$ samples collected in four sampling sites: two in Bari (Cavour and Japigia) and two in Taranto (Orsini and Dante)

Application of Receptor Models to Data Collected

Multivariate statistical analysis as receptor models (APCS, Absolute Principal component Analysis) has been applied in order to identify the profile sources and their contributions.[4–5] The results for Bari and Taranto showed the presence of recurrent sources in all sites, such as vehicular traffic, marine aerosol, secondary, and crustal particulate matter. The secondary particulate matter gives the most important contribution.

Applying the APCS receptor model to data collected in Taranto it has been possible to identify an additional source that contributes to PAHs concentration.

In Figure 3 the ratio between the contribution of the source named "Industrial" and the total contribution of the sources found applying the APCS model to data collected in Taranto (via Orsini) is shown for all samples considered.

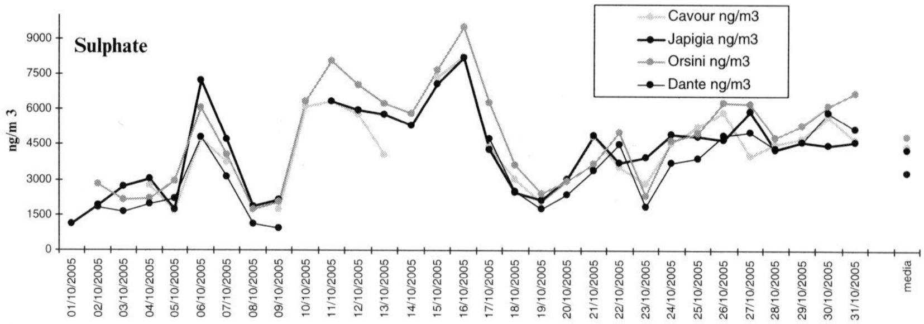

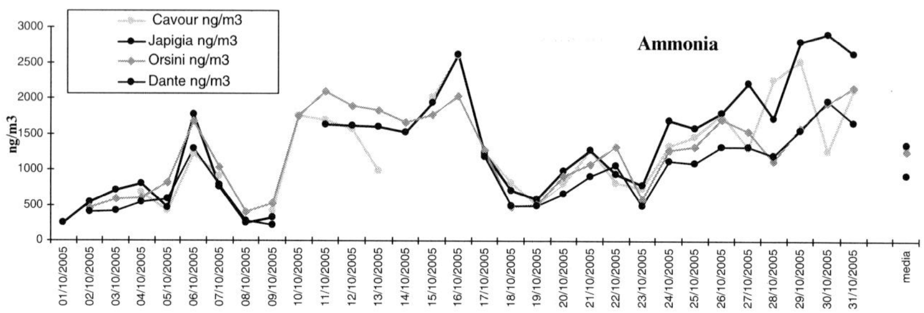

Figure 3 R is the ratio between the contribution of the Industrial source and the total contribution of the sources found applying the APCS model to data collected in Taranto (via Orsini). The daily mean wind directions are shown

As one can see the maximum value of R is obtained when the daily mean wind direction comes from North, that is when the air masses reach the sampling site coming from the industrial area of Taranto city.

References

1. Turpin, B.J., Huntzicker, J.J, *Atmos. Environ.*, **29**, 3527–3544 (1995).
2. Seinfeld J.H. and Pandis S.N., *Atmospheric Chemistry and Physics, from Air Pollution to Climate Change*, New York: Wiley, 444 (1998).
3. Wilson, T.R.S., *Salinity and the Major Elements of Sea Water*. Riley, J.P., and Skirrow, G. (eds), Orlando, FL: Chemical Oceanography Academic, 365–413 (1975).
4. Thurston G.D. and Spengler J.D., *AtmosphericEnvironment*, **19**, 9–25 (1985).
5. Caselli, M., de Gennaro, G., and Ielpo, P., *Environmetrics*, **17**(5), 507–516 (2006).

Author Index

I

J

K

L

Subject Index

B

C

D

E